NEUROANATOMICAL BASIS *of* CLINICAL NEUROLOGY

Second Edition

NEUROANATOMICAL BASIS *of* CLINICAL NEUROLOGY

Second Edition

Orhan E. Arslan
University of South Florida
Tampa, USA

CRC Press is an imprint of the
Taylor & Francis Group, an **informa** business

CRC Press
Taylor & Francis Group
6000 Broken Sound Parkway NW, Suite 300
Boca Raton, FL 33487-2742

Printed on acid-free paper
Version Date: 20140505

Printed and bound in India by Replika Press Pvt. Ltd.

International Standard Book Number-13: 978-1-4398-4833-3 (Hardback)

Library of Congress Cataloging-in-Publication Data

Arslan, Orhan, author.
Neuroanatomical basis of clinical neurology / author, Orhan E. Arslan . -- Second edition.
p. ; cm.
Includes bibliographical references and index.
ISBN 978-1-4398-4833-3 (hardcover : alk. paper)
I. Title.
[DNLM: 1. Nervous System--anatomy & histology. 2. Nervous System Physiological Phenomena. WL 101]

QP361
612.8--dc23 2014002748

Visit the Taylor & Francis Web site at
http://www.taylorandfrancis.com

and the CRC Press Web site at
http://www.crcpress.com

This book is gratefully dedicated to my parents, Zübeyde and Inayet, and to my sister Gülşen and brother Midhat, whose love and unwavering belief in science continue to guide me through the complex journey of life and the perplexing mysteries of the nervous system.

Contents

Foreword xvii
Preface xix
Acknowledgments xxi

SECTION I Basic Neuroanatomy

Chapter 1 Developmental Aspects of the Nervous System 3

Formation of the Neural Tube 3
Neural Tube Defects 4
Development of the Brain Vesicles 10
Differentiation of the Neural Tube 10
Genetic and Molecular Aspects 12
Medulla Spinalis (Spinal Cord) 13
Myelencephalon 14
Metencephalon 15
 Pons 15
 Cerebellum 15
Mesencephalon (Midbrain) 16
Diencephalon (Thalamus, Hypothalamus, Epithalamus and Subthalamus) 16
Telencephalon (Cerebral Hemispheres, Basal Nuclei and Ventricular System) 17
Suggested Reading 19

Chapter 2 Basic Elements of the Nervous System 21

Neuroglia 21
 Macroglia 21
  Astrocytes 21
  Oligodendrocytes 22
  Schwann Cells 22
  Ependymal Cells 23
 Microglia 23
Neurons 24
 Soma (Perikaryon) 24
 Neuronal Processes 25
  Dendrites 25
  Axons 25
Myelin 26
Neuronal Degeneration 28
 Antergrade Degeneration 28
 Retrograde Degeneration 28
 Transynaptic Degeneration 28
Functional and Clinical Consideration 29
Regeneration 31
Demyelinating Metabolic Disorders 32
Classification of Neurons 36
Synaptic Connectivity 37
 Synaptic Disorders 39
Suggested Reading 41

SECTION II Morphologic and Sectional Neuroanatomy

Chapter 3 Spinal Cord ... 45

Blood Supply ... 48
Venous Drainage ... 51
Internal Organization ... 51
Gray Matter ... 51
White Matter ... 52
Spinal Cord Segments ... 53
Spinal Pathways ... 54
Ascending Tracts ... 54
Ascending Tracts in the Posterior Funiculus ... 54
Ascending Tracts in the Lateral Funiculus ... 54
Ascending Tracts in the Ventral Funiculus ... 54
Descending Tracts ... 55
Descending Tracts in the Lateral Funiculus ... 55
Descending Tracts in the Ventral Funiculus ... 55
Suggested Reading ... 55

Chapter 4 Brainstem ... 57

Medulla ... 57
Fourth Ventricle ... 57
Caudal Medulla ... 59
Motor Decussation Level ... 59
Level of Sensory Decussation ... 59
Midolivary Level ... 60
Rostral Medulla ... 62
Pontomedullary Junction ... 62
Pons ... 62
Caudal Pons ... 64
Midpons ... 65
Rostral Pons ... 66
Midbrain ... 66
Caudal Midbrain ... 68
Rostral Midbrain ... 70
Suggested Reading ... 71

Chapter 5 Reticular Formation ... 73

Raphe Nuclei ... 73
Medial Reticular Zone ... 74
Ascending Reticular Activating System ... 74
Lateral Reticular Zone ... 75
Sleep and Associated Disorders ... 77
Suggested Reading ... 79

Chapter 6 Cerebellum ... 81

Morphologic Characteristics ... 81
Blood Supply and Venous Drainage ... 83
Cerebellar Classification ... 84
Cerebellar Cortex ... 84
Cerebellar Nuclei ... 87
Cerebellar Afferents ... 87
Cerebellar Efferents ... 91

Cerebellar Circuits ... 93
Cerebellovestibular Circuit ... 93
Reticulocerebellar Circuit ... 93
Rubrocerebellar Circuit ... 93
Cortico-cerebro-cerebellar Circuit ... 94
Intracerebellar Circuit ... 95
Functional and Clinical Consideration ... 95
Cerebellar Lesions and Associated Diseases ... 98
Suggested Reading ... 99

Chapter 7 Diencephalon ... 101

Thalamus ... 101
Thalamic Nuclear Group ... 102
Anterior Nucleus ... 103
Ventral Nuclear Group ... 103
Lateral Geniculate Body (LGB) ... 103
Medial Geniculate Body (MGB) ... 104
Ventral Posterior Nucleus (VPN) ... 104
Ventral Anterior Nucleus (VA) ... 105
Ventral Lateral Nucleus (VL) ... 106
Lateral Nuclear Group ... 107
Lateral Dorsal (LD) ... 107
Lateral Posterior ... 107
Pulvinar ... 107
Medial Nuclear Group ... 108
Dorsomedial Nucleus ... 108
Intralaminar Nuclear Group ... 109
Midline Nuclear Group ... 110
Reticular Nuclear Group ... 110
Functional and Clinical Consideration ... 110
Hypothalamus ... 111
Hypothalamic Areas and Nuclei ... 112
Functional and Clinical Consideration ... 115
Hypothalamic Afferents ... 119
Hypothalamic Efferents ... 120
Pituitary Gland (Hypophysis Cerebri) ... 122
Functional and Clinical Consideration ... 123
Epithalamus ... 123
Functional and Clinical Consideration ... 123
Pineal Gland ... 124
Stria Medullaris ... 125
Habenula ... 125
Posterior Commissure ... 126
Subthalamus ... 126
Suggested Reading ... 127

Chapter 8 Telencephalon ... 129

Cerebral Hemispheres ... 129
General Characteristics ... 129
Frontal Lobe ... 131
Prefrontal Cortex ... 132
Parietal Lobe ... 134
Temporal Lobe ... 136
Functional and Clinical Consideration ... 136
Occipital Lobe ... 137

Central Lobe (Insular Cortex) ... 138
Limbic Lobe ... 138
Corpus Callosum ... 139
Cerebral Cortex (Gray Matter) ... 140
Sensory Cortex ... 143
Primary Sensory Cortex ... 143
Secondary Sensory Cortex ... 144
Motor Cortex ... 144
Primary Motor Cortex ... 145
Supplemental Motor Cortex ... 145
Premotor Cortex ... 146
Association Cortex ... 147
Cortical Afferents ... 147
Cortical Efferents ... 147
Cerebral White Matter ... 149
Commissural Fibers ... 150
Association Fibers ... 152
Projection Fibers ... 153
Cerebral Dysfunctions ... 153
Aphasia ... 153
Apraxia ... 157
Agnosia ... 158
Dementia ... 159
Seizures ... 161
Coma ... 163
Cerebral Dominance ... 165
Autism ... 166
Basal (Ganglia) Nuclei ... 167
Blood Supply of the Cerebral Hemispheres ... 167
Venous Drainage of the Cerebral Hemispheres ... 172
Meninges ... 175
Dura Mater ... 175
Dural Sinuses ... 178
Posterosuperior Group of Dural Sinuses ... 178
Anteroinferior Group of Dural Sinuses ... 179
Arachnoid Mater ... 182
Pia Mater ... 183
Brain Barrier ... 184
Functional and Clinical Significance ... 186
Circumventricular Organs ... 188
Ventricular System and Cerebrospinal Fluid ... 188
Hydrocephalus ... 191
Suggested Reading ... 193

SECTION III Peripheral Neuroanatomy

Chapter 9 Autonomic Nervous System (ANS) ... 197

Autonomic Neurons and Synaptic Connections ... 197
Sympathetic Nervous (Thoracolumbar) System ... 200
Paravertebral Ganglia ... 200
Prevertebral Ganglia ... 204
Pattern of Distribution of the Sympathetic Fibers ... 204
Parasympathetic Nervous (Craniosacral) System ... 205
Cranial Part ... 208
Sacral Part ... 208

Higher Autonomic Centers....208
Autonomic Reflexes....208
Autonomic Plexuses....210
Enteric Nervous System....213
Afferent Components of the Autonomic Nervous System (ANS)....214
Disorders of the Autonomic Nervous System....214
Suggested Reading....220

Chapter 10 Spinal Nerves....223

Formation, Distribution, and Components of the Spinal Nerves....223
Cervical Spinal Nerves....226
Cervical Plexus....227
Terminal Branches....227
Brachial Plexus....229
Trunk Injuries....230
Terminal Branches of the Roots....232
Terminal Branches of Superior Trunk....233
Terminal Branches of Lateral Cord....234
Terminal Branches of Medial Cord....234
Terminal Branches of Posterior Cord....244
Thoracic Spinal Nerves....247
Lumbar Spinal Nerves....249
Lumbar Plexus....249
Terminal Branches of the Lumbar Plexus....249
Sacral Spinal Nerves....256
Sacral Plexus....256
Terminal Branches of the Sacral Plexus....257
Spinal Reflexes....264
Superficial Reflexes....264
Deep Reflexes....265
Suggested Reading....266

Chapter 11 Cranial Nerves....267

Olfactory Nerve....267
Optic Nerve....269
Oculomotor Nerve....270
Trochlear Nerve....274
Trigeminal Nerve....276
Abducens Nerve....282
Facial Nerve....285
Vestibulocochlear Nerve....290
Glossopharyngeal Nerve....292
Vagus Nerve....294
Accessory Nerve....299
Hypoglossal Nerve....301
Suggested Reading....303

SECTION IV Functional Neuroanatomy

Chapter 12 Neurotransmitters....307

Amino Acid Neurotransmitters....307
Gamma-Amino Butyric Acid....307
Glutamic Acid....308
Glycine....309

Acetylcholine....310
Monoamines....311
Catecholamines....311
Norepinephrine....311
Epinephrine....312
Dopamine....312
Indolamines....313
Serotonin....313
Histamine....315
Neuropeptides....315
Enkephalins....316
Endorphin....316
Substance P....316
Cholecystokinin (CCK)....316
Hypothalamic Peptides....316
Suggested Reading....316

SECTION V Special Somatic Sensations

Chapter 13 Visual System....321

Peripheral Visual Apparatus....321
Eyeball....321
Tunica Fibrosa....321
Cornea....321
Sclera....321
Tunica Vasculosa....321
Choroid Layer....321
Ciliary Body....321
Lens....322
Iris and Pupil....323
Anterior Chamber of the Eye....324
Refractive Disorders....324
Tunica Nervosa....325
Retina....325
Optic Nerve....329
Optic Chiasma....331
Optic Tract....332
Lateral Geniculate Nucleus (LGN)....334
Optic Radiation....334
Visual Cortex....336
Primary Visual Cortex....336
Secondary Visual Cortex....338
Tertiary Visual Cortex....338
Ocular Movements....339
Disconjugate (vergence) Movement....340
Conjugate (version) Movement....340
Saccadic Eye Movement....340
Vestibulo-Ocular Eye Movement....341
Smooth Pursuit Eye Movement....341
Disorders of Ocular Movements....341
Ocular Reflexes....342
Gaze Centers....344
Suggested Reading....344

Chapter 14 Auditory System ... 347

Peripheral Auditory Apparatus ... 347
External Ear ... 347
Middle Ear ... 348
Inner Ear ... 352
Spiral Ganglion ... 354
Cochlear Nerve ... 354
Central Auditory Pathways ... 355
Cochlear Nuclei ... 355
Acoustic Striae ... 355
Lateral Lemniscus ... 356
Inferior Colliculus ... 357
Medial Geniculate Nucleus ... 357
Auditory Radiation ... 357
Auditory Cortices ... 357
Auditory Dysfunctions ... 359
Conductive Deafness ... 359
Sensorineuronal Deafness ... 359
Tinnitus ... 360
Auditory Tests ... 361
Audiometry ... 361
Brainstem Auditory Evoked Response ... 361
Weber Test ... 361
Rinne Test ... 362
Suggested Reading ... 362

Chapter 15 Vestibular System ... 363

Peripheral Vestibular Apparatus ... 363
Semicircular Canals ... 363
Vestibule ... 364
Saccule ... 364
Ultricle ... 364
Vestibular Ganglion ... 364
Vestibular Nerve ... 364
Central Vestibular Pathways ... 364
Vestibular Nuclei ... 364
Secondary Vestibulocerebellar Fibers ... 366
Vestibulospinal Tracts ... 367
Vestibulo-Ocular Fibers ... 367
Vestibular Dysfunctions ... 368
Vestibular Tests ... 371
Suggested Reading ... 373

SECTION VI Special Visceral Sensations

Chapter 16 Olfactory System ... 377

Peripheral Olfactory Apparatus ... 377
Olfactory Receptors ... 377
Olfactory Nerve ... 379
Olfactory Bulb ... 379
Anterior Commissure ... 382
Olfactory Pathways ... 382
Olfactory Cortices ... 383
Olfactory Cortical Dysfunctions ... 384
Suggested Reading ... 386

Chapter 17 Limbic System 387

Hippocampal Formation 387
Hippocampal Gyrus 387
  Schizoprenia 389
  Dentate Gyrus 390
Subicular Complex 390
Connections of the Hippocampal Formation 391
  Afferents of the Hippocampal Formation 391
  Efferents of Hippocampal Formation 392
Functional and Clinical Consideration 393
  Memory and Amnesia 393
Septal Area 396
Induseum Griseum 399
Amygdala 399
Limbic Lobe 403
Cingulate Gyrus 403
Prefrontal Cortex 404
Hypothalamus 404
Suggested Reading 406

Chapter 18 Gustatory System 407

Peripheral Gustatory Receptors 407
Gustatory Neurons and Associated Ganglia 409
Gustatory Pathways 410
Gustatory Dysfunctions 410
Suggested Reading 411

SECTION VII General Somatic Sensations

Chapter 19 Cortical and Subcortical Sensory Systems 415

Sensory Fibers 415
Sensory Receptors 416
Cortical Sensations from the Body 419
Cortical Sensations from the Head 432
Subcortical Sensations 432
Suggested Reading 433

SECTION VIII Motor Systems

Chapter 20 Upper and Lower Motor Neuron Systems 437

Upper Motor Neuron (UMN) Palsies 439
Lower Motor Neurons (LMNs) 445
Lower Motor Neuron (LMN) Palsies 446
Suggested Reading 461

Chapter 21 Extrapyramidal Motor System 463

Basal (Ganglia) Nuclei 463
Corpus Striatum 463
  Dorsal Striatum 463
  Ventral Striatum 465
  Dorsal Pallidum 465
  Ventral Pallidum 466

Claustrum ... 466
Substantia Nigra ... 466
Red Nucleus ... 467
Subthalamic Nucleus ... 468
Reticular Formation ... 468
Connections of the Corpus Striatum ... 470
Functional Loops and Pathways ... 471
Dysfunctions of the Extrapyramidal System ... 472
Suggested Reading ... 480

Index ... 481

Foreword

This new and greatly enhanced second edition of the *Neuroanatomical Basis of Clinical Neurology* builds on the strengths of the first edition published in 2001. The book is an invaluable resource for understanding the central and peripheral neuroanatomy underlying normal and abnormal function. Recent advances uncovering the anatomic underpinnings of complex behaviors, including memory, autism, and dementia, are integrated in this new edition. Expanded clinical correlations, with the addition of many clinical vignettes, and the inclusion of multiple radiology images further enhance the value of this book to medical students and postgraduate students. While many neuroanatomy books give scant attention to the autonomic and peripheral nervous systems, this book highlights the many clinical disorders resulting from their dysfunction. Two decades of discovery in neuroscience since the publication of the first edition have yielded profoundly deepened insights into the neuronal circuitry, and mechanisms in both health and disease and are skillfully integrated in this new edition.

Warren J. Strittmatter, MD
Professor of Neurology
Duke University Medical Center
Durham, North Carolina, USA

Preface

The continued interest in the publication of a new edition of this textbook by neuroscience educators and students in several medical schools and allied health institutions in the United States and abroad has given the impetus to embark on this project. The support of the editorial staff of Taylor & Francis undoubtedly facilitated this work and provided the means for a successful conclusion. The constructive input of a great many colleagues and outstanding individuals in the field of neuroscience and neurology helped to establish the foundation for the direction of this new edition. Consideration has been given to the many helpful suggestions of students with regard to scope and contents of particular chapters.

Toward that end, a great effort was directed at the revision of a significant number of schematic drawings and the modification of a number of brain images. The introduction of numerous brain sections and applied radiographic images including angiograms was aimed at enhancing the reader's ability to interpret both structural and functional features of the nervous system within the context of neurologic disorders. The inclusion of multiple colors in the illustrations has facilitated the interpretation of the anatomy of the nervous system and rendered the illustrations an integral part of the documented text. The distinguishing characteristic of this textbook has remained in the presentation of the peripheral and central nervous systems as a continuum within the context of neurological disorders.

The massive and ever-changing information in neuroscience required a review of a large number of articles, research papers, and authoritative resources in neuroscience and neurology and incorporation of the most significant facts into the 21 chapters of this work. This is reflected in the expanded clinical correlations introduced into many of the chapters and the number of references listed as suggested reading. We aimed at adding clarity, depth, and relevance to the text and diagrams. We remained cognizant of the immense value of the key clinical correlations in demonstrating the importance of understanding the neuronal circuitry and structural organization of the nervous system. Throughout this revision, bridging the gap between the peripheral and central systems and the integration of structural organization with clinical disorders remained a primary focus of this work.

The reader will notice that the basic outline of the contents persists; an introductory coverage of the developmental and cellular aspects of the nervous system is followed by the morphology and the internal organization of the central nervous system. The peripheral nervous system with its somatic and autonomic components has been extensively discussed, with emphasis on nerve entrapments and neuropathies. Detailed discussion of a number of autonomic dysfunctions strengthens the overall understanding of fundamentals of neuronal interconnectivity of the autonomic nervous system in view of their manifestations in systemic diseases. To enrich the discussed topics, a concerted effort is directed at incorporating several common and relevant clinical conditions into most chapters with a careful attempt to limit related yet uncommon neurologic diseases. Since the localization of the central nervous system provides key insights to understanding the sensory and motor pathways, the functional organization has, for the most part, been preserved.

New topics such as autism have been introduced into the discussion of the role of prefrontal cortex in behavior and attention. Discussion of various forms of cortical dysfunctions, such as seizures, disconnection syndrome, coma, and dementia, has been significantly expanded. The reader will observe the introduction of substantial additional information into the auditory, vestibular, gustatory, and limbic systems. To reflect the significance of memory, associated disorders, and the structural and neuronal chemistry of the hippocampal gyrus, a balanced coverage of the limbic system including associated neurotransmitters have been maintained. Similarly, structural components of the extrapyramidal system and the neurochemical basis of movement disorders have been given particular attention as mastery of neuroanatomical facts and the associated pathways helps frame the therapeutic approaches to these diseases.

The importance of neurochemical characteristics and the structural organization of the central and peripheral nervous systems and localization of lesions in therapeutic and surgical approaches underscores the importance of understanding the internal organization, neuronal circuitry, and structural relationships of these systems. This work kept these fundamental facts in view when it was first written and when the revision was planned.

The interest shown, through the years, by students in medical schools and the allied health programs as well as by residents of neurology, neurosurgery, and physical therapy and rehabilitation medicine encouraged the revisions introduced into this project. We are confident that this edition of the textbook and atlas will continue to address the educational needs of and remain an effective source of information for students of neuroanatomy and neuroscience.

Orhan E. Arslan

Acknowledgments

I have had the privilege of working with and benefiting from the experience of many distinguished administrators, academicians, and scientists whose contributions were crucial for the successful completion of this project. With appreciation I distinctly recognize Bryan A. Bognar, MD, Vice Dean and Chief Academic Officer, University of South Florida Morsani College of Medicine, for his guidance, discerning ideas and thoughtful observations. I gratefully acknowledge Paul M. Wallach, MD, Vice Dean for Academic Affairs at Georgia Regents University for his inspirational leadership, unwavering support, and indispensable contributions to this academic endeavor. I would like to recognize Kevin Sneed, Pharm D, Senior Associate Vice President and Dean, University of South Florida College of Pharmacy for his encouragement and continued support of this project. I extend a note of gratitude to Steven Specter, PhD, Associate Dean for Student Affairs at the University of South Florida Morsani College of Medicine for his helpful gestures and perceptive suggestions. I am particularly thankful to William S. Quillen, DPT, PhD, Associate Dean, University of South Florida Morsani College of Medicine, and Director of the School of Physical Therapy and Rehabilitation Sciences, for his support and meaningful ideas. I express my appreciation to David Birk, PhD, Professor of Molecular Pharmacology and Physiology, for his useful critiques. My thanks certainly goes to Peter Dunne, MD, Professor and Chair Emeritus of Neurology, University of South Florida Morsani College of Medicine for the thorough review of the book and for many gracious comments and ideas.

I had the distinct privilege of working with Warren Strittmatter, MD, Professor and Chief of Neurology at Duke University Medical School, since the publication of the first edition of this book. He has generously reviewed this current edition, wrote the forward, and provided numerous critical suggestions to which I remain grateful.

Several of my eminent colleagues and scientists that I had the good fortune to work with at Rosalind Franklin University/Chicago Medical School deserve special recognition. I am profoundly grateful to Charles E. McCormack, PhD, Professor of Physiology and Biophysics, for his exceptional and generous gestures, undivided support, and firm belief in the success of this undertaking. I wish to pay special thanks to Richard A. Hawkins, PhD, Professor of Physiology and Biophysics and Former President and Chief Academic Officer, for his mentorship, support, and meaningful suggestions. I convey my thanks to Vel Nair, PhD, Distinguished Professor of Pharmacology and former Dean of the Graduate and Postdoctoral Studies, for the positive contribution to this revision that he has offered. I would like to also convey my appreciation to William N. Frost, PhD, Professor and Chair, Department of Cell Biology and Anatomy, for his support and constructive input.

This work would not have been possible without the splendid artistic talents of Frederick R. Weller and Deborah Rubenstein, and the excellent photography of Joseph NadaKapadam. I extend my thanks to the Editorial staff at Taylor & Francis Publishing Group/CRC Press and, in particular, Lance Wobus, Senior Editor; Amy Blalock, Supervisor of Editorial Project Development; Robert Sims, Project Editor for Production; Amor Nanas, Project Manager with Manila Typesetting Company, and many others in the production team for the numerous courtesies and patience that brought this work into a successful conclusion. Last, but not least, I thank my brother, Talat Arslan, MD, for accepting the responsibility of refining this work and patiently providing key clinical facts that highlight the neuroanatomical circuitry and disorders.

Section I

Basic Neuroanatomy

1 Developmental Aspects of the Nervous System

Despite the ongoing and exciting research, our understanding of the intricate connections of neurons, their functions, and associated supporting cells remains limited. Since almost all neurons are connected in one way or another to contiguous or distant neurons, investigation of specific neuronal collections or specific parts of the nervous system pose limitations on our understanding of the nervous system. However, our ability to correlate structural and molecular characteristics with diseases entities has been broadened by the technological advancement in research methodologies and the immense documented data. Understanding the development of the nervous system from genetic, molecular, and morphologic aspects will likely to further enhance the ability to understand the basis of congenital anomalies of the nervous system and the myriad of neurodegenerative disorders through advancements in stem cell genetic research.

Neural development is the process by which the nervous system with all its components comes into existence. It entails the cellular basis and the underlying mechanism that guide the developmental process, including neural induction, cellular differentiation, migration, axonal guidance, and synapse formation. This process starts during the third week of gestation, when the neural ectoderm forms the neural plate, which eventually becomes the neural tube. The neural tube is the origin of most of the neurons and glial cells. Dorsolateral to the neural tube, the neuroectodermal cells constitute a distinct group of cells, the neural crest cells, which give rise to the autonomic and sensory ganglia as well as other structures. Closure of the neuropores initiates the development of the brain vesicles, leading to the formation of various compartments of the central nervous system (CNS). Failure of closure of the neuropores results in variety of malformations ranging from anencephaly to spina bifida. Further differentiation of the primitive neural tube leads to the formation of the ependymal, mantle, and marginal layers. Division of the mantle layer into alar and basal plates accounts for the selective pattern of localization of the sensory and motor nuclei in the developed CNS. The neural canal converts into various parts of the ventricular system. Developmental defects of the nervous system can lead to motor, sensory, and cognitive dysfunctions.

The nervous system is derived from the neural ectoderm, which derives from the undifferentiated ectoderm by the mesodermal signals. At the onset of gastrulation, presumptive mesodermal cells migrate along the dorsal midline to give rise to the notochord.

FORMATION OF THE NEURAL TUBE

During the third week of gestation (day 16), the notochord (chordamesoderm) induces (neural induction) the overlying ectodermal cells in the rostral part of the embryonic disc to differentiate into the neural plate (Figures 1.1, 1.2, and 1.3). During neural induction, notochord-derived noggin and chordin diffuse into the overlying ectoderm, inhibiting the activity of bone morphogenetic protein 4 (BMP4) and initiating the differentiation into neural cells. Differentiation of the

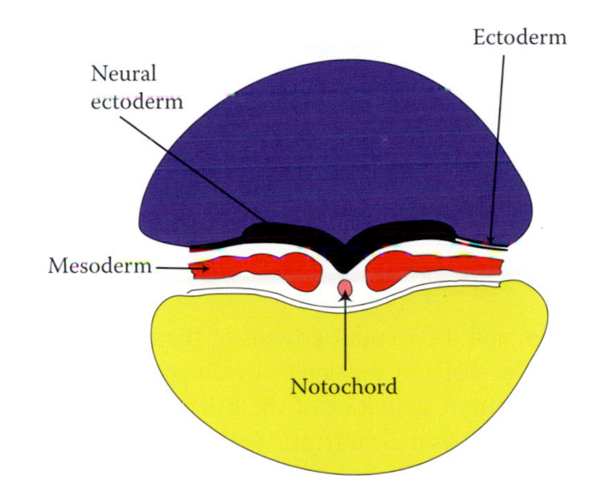

FIGURE 1.1 Transverse section of the embryo showing the location of the neural ectoderm relative to the notochord and mesoderm.

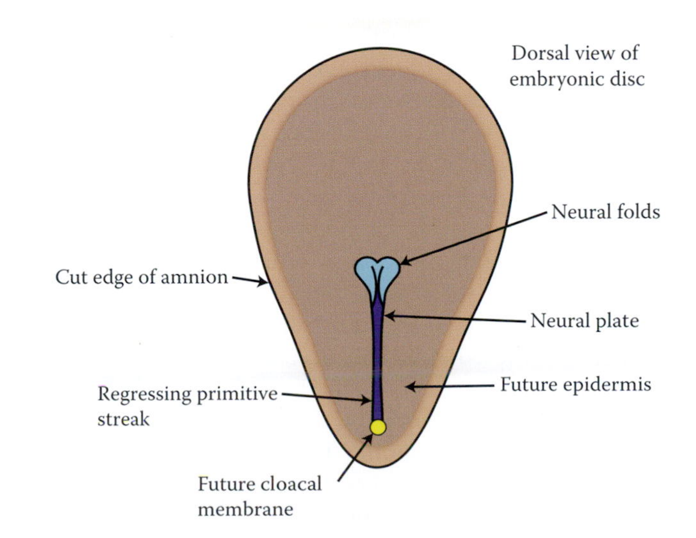

FIGURE 1.2 Schematic drawing (dorsal view) of the neural plate, neural folds, and their relationships to the future epidermis and primitive streak.

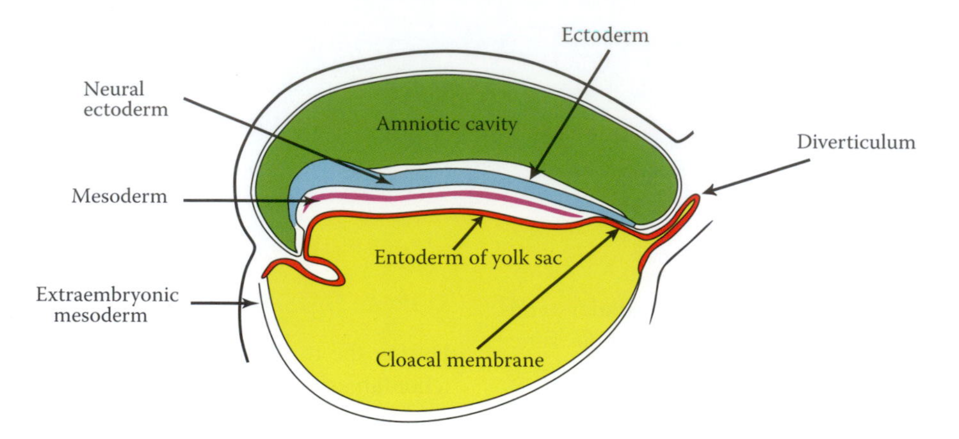

FIGURE 1.3 Diagram of the neural ectoderm (cephalocaudal) in relationship to the notochord and intraembryonic coelom. Note the location of the mesodermal tissue.

pluripotent stem cells into neural tissue occurs upon inhibition of both the transforming growth factor-β (TGF-β) and BMP activities. The ventral and dorsal parts of the neural plate are controlled by the notochord and ectodermal plate, respectively. It has been documented that the ventral part of the neural plate, which is considered as the "organizer" produces follistatin, noggin, and chordin, which block BMPs and thus enable neural differentiation of the ectoderm.

NEURAL TUBE DEFECTS

Proliferation and differential growth of the neural plate cells, changes in the shape of the cells, stretch posed by the rapidly developing embryo, activities of the microtubules and microfilaments, as well as their intrinsic movements ultimately lead to the formation of the neural groove (day 18). The neural groove acts as a median hinge point around which the neural folds expand (Figures 1.2 and 1.5). Initially, fusion of the neural folds occurs in the cervical region, gradually expanding in both rostral and caudal directions, leading to the formation of the neural tube (Figure 1.5). Failure of the neural folds to fuse, differentiate, and detach from the surface ectoderm may lead to *rachischisis*, which is, as described later in this chapter, a group of malformations that assume a variety of forms depending upon the involved part of the neural tube. The last portions of the neural tube to close are the rostral and caudal neuropores, which maintain connections with the amniotic cavity.

Detachment of the neural tube from the surface ectoderm (future epidermis) and its assumption of a more ventral position are believed to be enhanced by nerve cell adhesion molecule (N-CAM) and N-cadherin molecules synthesized by the neural tube itself. As the neural folds fuse, a specialized population of cells on the dorsolateral part of the neural tube forms the *neural crest cells* (Figures 1.4 and 1.5). Neural crest cells lose their epithelial-specific adhesion molecules and express a new group of cell adhesion molecules such as integrin and laminin. With the help of pseudopodia that develop from the basal aspects, the neural crest cells are pulled through the basal membrane of the neural tube, enabling the surface ectoderm and the basal membrane of the neural tube to guide the migration of the neural crest cells. Migration of these cells occurs in a craniocaudal direction, with the more cephalic cells departing before the closure of the cranial neuropore.

The process that encompasses the formation of the neural plate, floor plate, and neural sulcus, as well as closure of the neuropores with the final configuration of the neural tube, is known as primary neurulation. Secondary neurulation refers to the development of the caudal neural tube, a process that is initiated by the formation of a solid tube or mass caudal to somite 31 (first or second lumbar somite) and the appearance of ectodermal-lined vacuoles. These vacuoles eventually coalesce and open into the end of the neural tube. Later, canalization and final union of the caudal mass with the rest

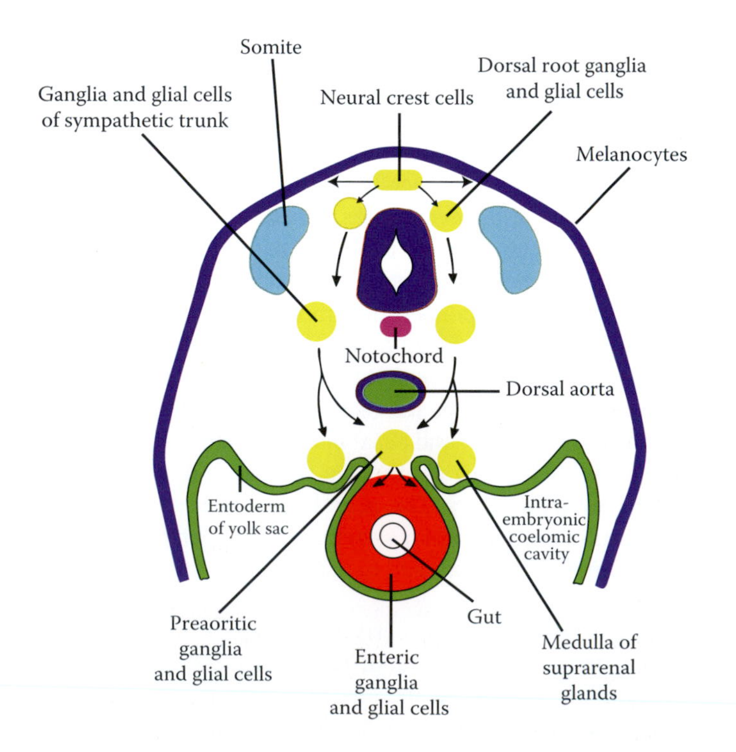

FIGURE 1.4 Path of migration of the neural crest cells. Note the locations of the somites, notochord, and gut.

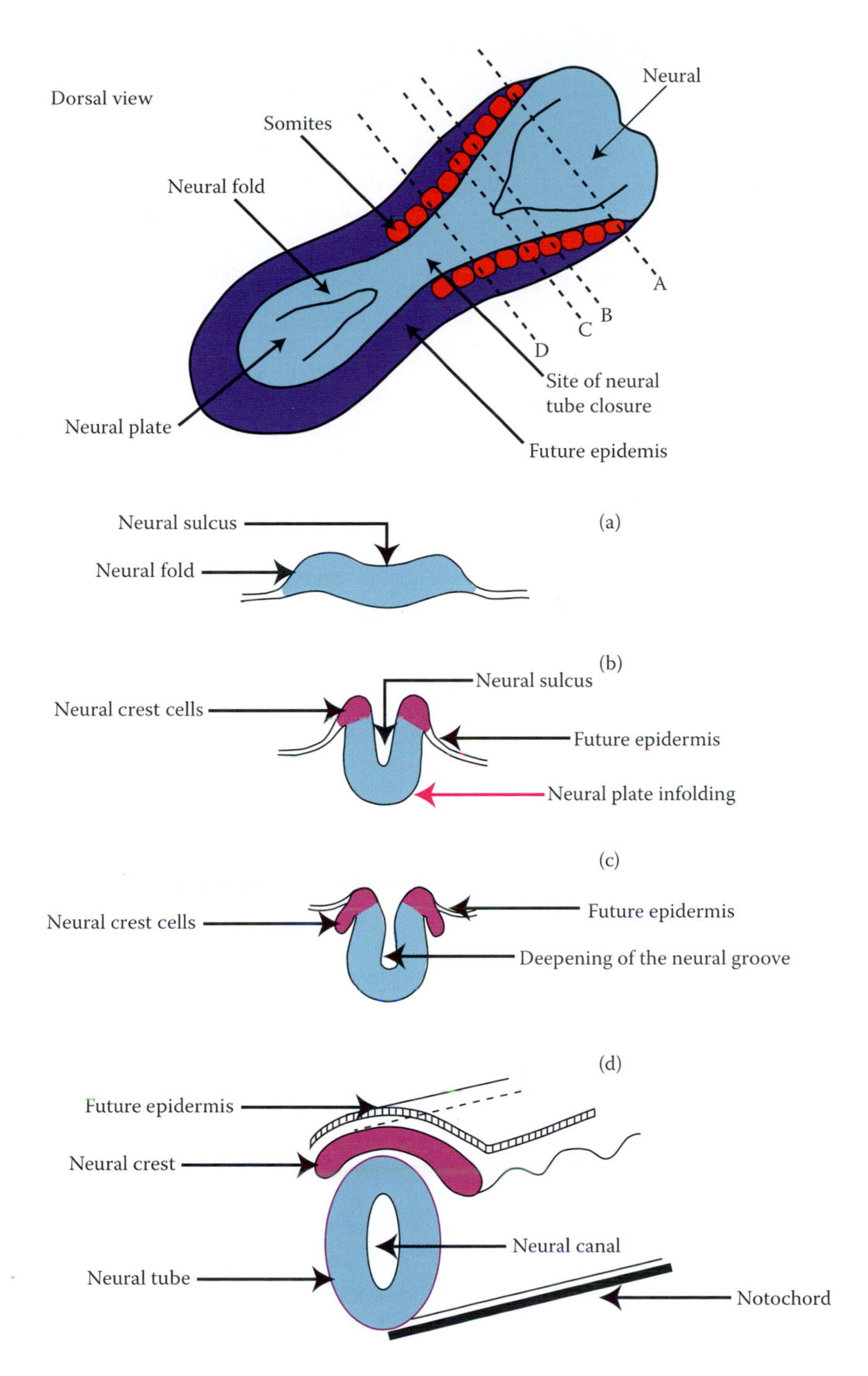

FIGURE 1.5 Neural crest cells and neural canal. Neural crest cells eventually separate from the newly formed neural tube. (a) Formation of the neural plate and the beginning of the neural folds and neural sulcus. (b) Appearance of the neural crest cells and prominence of the neural folds. (c) Deepening of the neural sulcus, movements of the upper end of the neural folds medially and continued formation of the neural crest cells. (d) Development of the neural tube and neural canal and separation from the neural crest cells.

of the neural tube occurs. In 33–35% percent of embryos, the caudal neural tube remains forked.

Neural crest cells contribute to the formation of the sensory ganglia of the dorsal roots; trigeminal, facial, glossopharyngeal, and vagus nerves; autonomic ganglia; and satellite cells of these ganglia. They also give rise to the Schwann cells, adrenal medulla, melanocytes, auditory nerve, neurolemma of the peripheral nerves, cells of the pia and arachnoid mater of the occipital region and spinal cord, intraocular muscles, ciliary body, and the carotid bodies. The target tissue that receives the migrating cells may determine the development of these structures (Table 1.1).

The target tissue may also influence secretion of catecholamines and acetylcholine by structures of neural crest origin. Eventually, neural crest cells spread segmentally along the entire length of the neural tube, contributing to the formation of the structures associated with future peripheral nervous system. Failure of neural crest cells to migrate along the wall

TABLE 1.1
Derivatives of the Neural Crest Cells

Neuroectodermal Cells	Derivatives
Neural crest cells	1. Dorsal root ganglia and sensory ganglia of cranial nerves
	2. Sympathetic and parasympathetic ganglia
	3. Schwann cells
	4. Adrenal medulla
	5. Melanocytes
	6. Neurolemma of the peripheral nerves
	7. Cells of the pia and arachnoid mater of the occipital lobe and spinal cord
	8. Intraocular muscles
	9. Ciliary bodies
	10. Carotid bodies

of the developing intestinal tract produces congenital loss of parasympathetic ganglia in Meissner's and Auerbach's plexuses and signs of Hirschsprung's disease (congenital aganglionic megacolon).

Hirschsprung's disease (congenital megacolon), a congenital anomaly with a male-to-female ratio of 3:1, is associated with failure of neural crest cells to migrate. This condition commonly affects the rectum and the sigmoid colon (three-fourths of cases), but rarely involves the entire colon. It may also occur in more proximal locations, depending upon the migratory defect of neural crest cells. It is characterized by impaired peristaltic movement at and beyond the affected part of the colon, followed by bowel stasis, chronic constipation, abdominal distention, hypertrophy, constriction of the aganglionic segment, and dilatation of the colon proximal to the affected area. It usually manifests itself in the newborn by an inability to pass meconium, followed by intestinal obstruction. The etiology of this disorder may include mutation of the rearranged (RET) receptor tyrosine kinase during transfection. RET is a tyrosine kinase that undergoes enzymatic activation and initiates intracellular signaling upon binding the glial-derived neurotrophic factor and the endothelium B receptor.

The posterior neuropore closes at the level of the first or second lumbar somite, which marks the future first or second lumbar spinal segment. PAX-3, sonic hedgehog, and open brain genes, as well as folic acid (vitamin B_{12}) and cholesterol, play an important role in the closure of this tube. As mentioned earlier, failure of closure of the neuropores produces various forms of dyspharic defects collectively known as *rachischisis*. The site of closure of the anterior neuropore is represented in the newborn by the lamina terminalis, a vestigial structure located rostral to the hypothalamus.

Failure of closure of the anterior neuropore results in *cranioschisis*, an anomaly that is associated with a lack of development of the brain (anencephaly) and the skull. Anencephaly may be seen in Meckel (Meckel–Gruber) syndrome, a rare autosomal recessive and fatal condition associated with MSK1 and MSK3 genes that is characterized by a sloping forehead, occipital meningoencephalocele (80%), a polycystic dysplastic kidney (90%), hepatic defects, pulmonary hypoplasia subsequent to oligohydraminos, and polydactyly. It may also be seen in brachydactyly syndrome, an autosomal dominant disorder that exhibits abnormal shortening of the fingers and toes.

The neural tube maintains a connection with the amniotic cavity via the anterior and posterior neuropores, which close by the 25th and 27th day, respectively.

Anencephaly (exencephaly) is one of the most severe forms of congenital anomalies that occur between days (18–24) of embryonic development. It is relatively common, with an incidence of 1 per 1000 births in the United States. It is most commonly seen in female infants and is usually fatal. The exposed neural tissue undergoes degeneration and becomes converted into a mass of vascular connective tissue, intermixed with masses of degenerated brain and choroid plexus. Due to these defects and the exposure to infectious agents, death usually occurs shortly after delivery. The vault of the skull of an anencephalic fetus fails to form, and its base is usually covered with a vascular membrane. Orbits appear shallow, and the eyes tend to bulge externally. The fetus exhibits wide shoulders, a short trunk, and a neck that is commonly absent, giving the impression that the head is stemming directly from the body. Anencephaly may also be associated with the absence of vertebral arches, amyelia (absence of the spinal cord), and hydramnios (excess of amniotic fluid) due to the lack of a swallowing reflex. Fifty percent of anencephalic fetuses are aborted spontaneously. Pregnancy with an anencephalic fetus is usually complicated by delayed onset of labor (most deliveries after 40 weeks gestational age).

Neural tube development induces the formation of somites (a paired segmented division of paraxial mesoderm that develops along the length of the early embryo). The underlying paraxial mesoderm also starts segmentation on day 20 (Figure 1.6). Failure of development of the neural tube may impair development of the surrounding somites (e.g., bone, muscles, and skin) and vice versa. Therefore, malformation of the brain and its coverings may also be accompanied by abnormal ossification of the bony skull. Herniation of brain

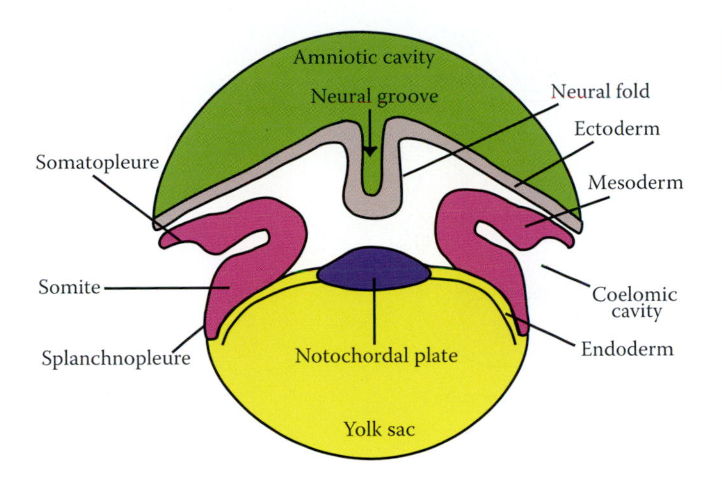

FIGURE 1.6 Schematic drawing illustrating the neural tube with the dividing sulcus limitans, neural crest cells, and the future epidermis.

tissue through a gap in the posterior midline of the skull (cranium bifidum) is known as *encephalocele*.

Meningoencephalocele occurs when the herniated brain tissue is associated with meningeal coverings. Cranial meningocele refers to the protrusion of the meninges into the sac. The bony defect, which results from incomplete closure of the calvaria, may be confined to the occiput or extend to the arch of the atlas.

Encephalocele is seen in Klippel–Feil deformity, a malformation that is commonly associated with fusion of vertebrae, a decrease or sometimes increase in the number of vertebrae, and hydrocephalus (enlargement of the brain and skull subsequent to an excess of cerebrospinal fluid [CSF]). The prognosis of this condition depends upon the degree of involvement of the brain tissue and associated meninges. Encephalocele should be corrected surgically unless death is eminent or if it concurrently occurs with dextrocardia, laryngomalacia, renal agenesis, or pulmonary hypoplasia. Encephaloceles, which make up 10%–20% of all craniospinal malformations, are predominant in female fetuses. The anatomic location of an encephalocele may vary with the geographic region; in Southeast Asia, they are more commonly located in the anterior cranial vault, whereas in Europe and the United States, they are commonly located in the occiput. The amount of brain tissue within the herniated sac varies and may involve parts of cerebral hemispheres, the cerebellum, and even the brainstem. Protrusion of a large amount of brain tissue into the encephalocele may reduce the size of the brain, resulting in microcephaly.

Microcephaly may also be due to failure of brain growth, producing a smaller brain with less prominent gyri and a considerably smaller skull than usual. Due to the marked difference between the anterior and posterior ends of an encephalocele, blood vessels may be occluded or ruptured, leading to infarction and hemorrhage. Sclerosis in the herniated sac, as well as impairment of CSF circulation and resultant hydrocephalus, may also occur. The microcephalic brain may be the result of rubella infection or exposure to radiation. At birth, the infant with encephalocele is usually neurologically normal or may exhibit increased flexor motor tone. Spells of apnea and bradycardia may occur in severe cases of occipital encephaloceles involving the brainstem. Microcephaly may be associated with *holotelencephaly* (holoprosencephaly), in which the telencephalon fails to cleave into two cerebral hemispheres and ventricles but instead forms a single structure with a large single ventricular cavity.

An abnormally large brain and skull (*macrocephaly*) is usually a fatal condition associated with syringomyelia, a disease which exhibits cavitation around the central canal of the spinal cord or lower brainstem (syringobulbia), and may occur in one or both cerebral hemispheres, without any detectable change in intracranial pressure. Macrocephaly may also be seen in individuals with Arnold–Chiari malformation or trauma.

Spinal dysraphism (defective fusion) refers to a developmental malformation that occurs when the neural tube fails to completely close and the vertebral arches fail to fully form. Nonclosure of the posterior neuropore at day 27, a less frequent anomaly, leads to the formation of *spina bifida* (myeloschisis) or split spine. It may be ascribed to a variety of factors including excess of vitamin A, high concentration of plasma glucose, deficiency of folate or to the effects of teratogens such as aminopterin, valproic acid, trypan blue and retinoic acid.

This anomaly is classified into spina bifida occulta and spina bifida cystica (Figure 1.7). Amniocentesis, sampling the amniotic fluid by the insertion of a needle through the abdominal wall, is useful in detecting myeloschisis. This procedure is based upon the fact that the levels of alpha-fetoprotein (AFP) at 16–18 weeks of pregnancy in the amniotic fluid are far greater in cases of myeloschisis than in healthy pregnancies.

Spina bifida occulta, the least clinically significant form of spina bifida, is a mesodermal (rather than an ectodermal) abnormality and is rarely associated with neurological dysfunction. It is a closed neural tube *defect, which* is characterized by defective vertebral laminae at certain levels, exposing the meninges through a bony gap, without the actual involvement of the spinal cord or meninges. This malformation is usually asymptomatic, unless a developed lipoma or bony process compresses the spinal cord at the affected area, leading to nocturnal enuresis (neurogenic bladder) and retarded development of the lower limb (asymmetrical or sometimes unilateral shortening of one leg and foot). Compression of L5 motor roots may produce

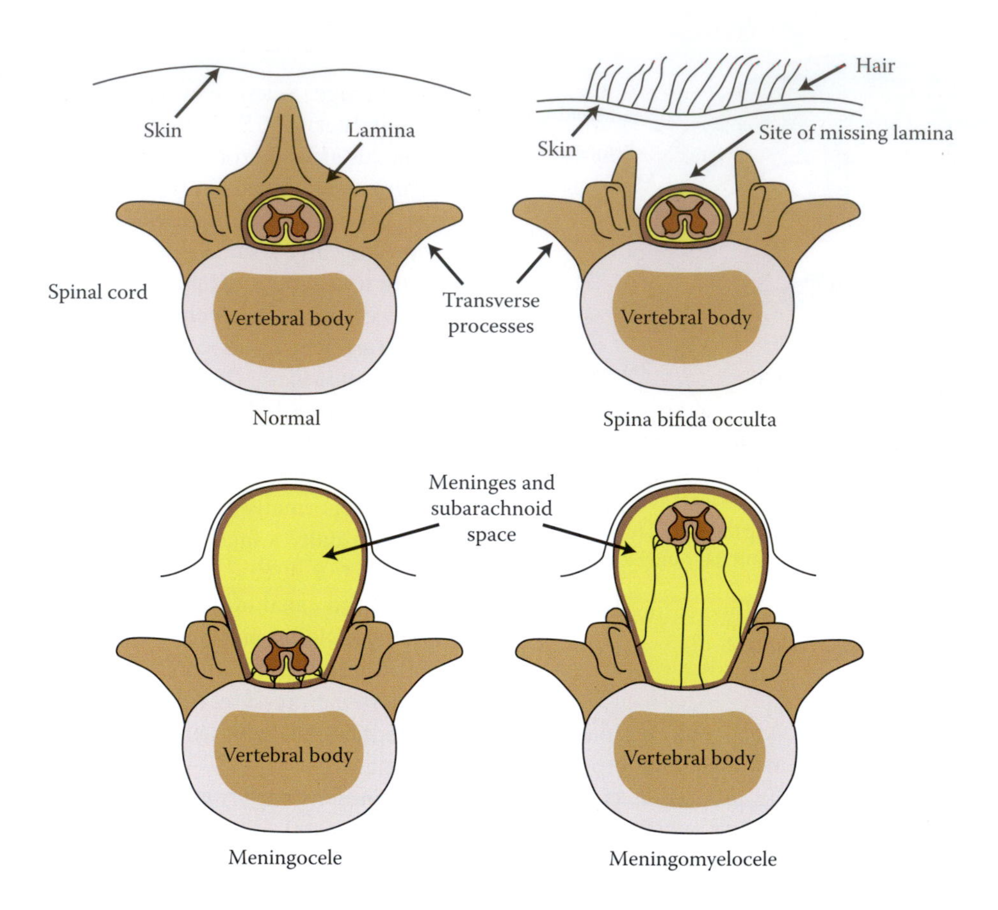

FIGURE 1.7 Various forms of spina bifida. Note the absence of vertebral lamina and the protrusion of the meninges or meninges and spinal cord tissue.

calcaneovalgus, dorsiflexed, everted, and abducted feet. Equinovarus, plantar-flexed, inverted, and adducted feet, may occur as a result of compression of the S1 motor roots. Back pain, impairment of sensations, and sciatica (pain radiating from the back to the leg) may also accompany this condition. Intrauterine ultrasound had not proven useful in the prenatal diagnosis of this condition. Radiographic plain films of the spine in neonates are also of no use in detecting spina bifida occulta since the vertebral laminae at that time are cartilaginous and radiolucent, making it very difficult to reveal defective spinous processes. The most common form of spina bifida occulta is the dermoid sinuses and the lipoma-covered defect in spina bifida.

Dermoid sinuses are deep, epithelial-lined tracts, sometimes containing hair that ascends from an external opening over the spine or scalp. They terminate in the intracranial structures and occasionally communicate with the subarachnoid space, predisposing the patient to meningitis. Dermoid sinuses are most common in the lumbar or sacral region (65%) and less so in the occipital region (30%). These sinuses may go unnoticed and are only detected after repeated bouts of meningitis. Occasionally, brain abscesses occur as a consequence of repeated infections, and the patient may present with hypertension, seizures, and fever. Lipoma-covered defect, a form of spina bifida occulta, occurs in the sacral or lower lumbar region and is very rarely associated with neurological abnormalities.

Spina bifida cystica (spina bifida manifesta, or spina bifida aperta), having an incidence of 1 per 1000 births, is characterized by a saclike protrusion of either meninges (meningocele) or a combination of meninges and spinal cord/nerve roots (meningomyelocele).

A *meningocele* (10% of cases of spina bifida cystica) is covered by thin easily ruptured meninges and occasionally by skin, and it commonly occurs in the lumbosacral portion of the vertebral column. It is frequently associated with myelodysplasia (spinal cord defects), occurring in the majority of cases of spina bifida cystica. Motor, sensory, bowel and bladder functions remain unaffected in this condition, although a tethered spinal cord has been reported. It may be associated with teratoma and Currarino syndrome (triad of a presacral mass, failure of the sacrum to form properly, and anomalous anal canal). It may result from cranial base dehiscence involving the occipital, ethmoidal, or nasal bones.

Meningomyelocele is much more common than *meningocele*, constituting 90% of cases of spina bifida cystica. It may be asymptomatic or manifests severe neurological symptoms if the defect is large. It may contain the cauda equina if it occurs in the lumbosacral region or the spinal cord if the anomaly involves the thoracic or cervical regions. Because of the continuation of the open neural tube with the external surface, infants with this condition are prone to bacterial meningitis. Patients may exhibit pain, hydrocephalus, bowel and bladder dysfunctions, and motor and sensory deficits in the lower extremity. A lipomyelomeningocele appears when fatty tissue and skin cover the herniated sac. In this case, neurogenic bladder is almost always a universal presentation. Meningomyelocele may be accompanied by hydrocephalus, talipes equinovarus (clubfoot, which is plantar flexed, inverted and adducted), sensory and motor deficits, paraplegia, as well as bowel dysfunction. A high percentage (90%) of infants with *meningomyelocele* develop hydrocephalus as a part of Arnold–Chiari syndrome.

Arnold–Chiari syndrome is thought to result from fixation of the developing spinal cord in the sac of a meningomyelocele, creating an undue downward strain on the spinal cord and brainstem. This eventually causes displacement of the cerebellum and hindbrain through the foramen magnum into the cervical part of the vertebral canal. Hydrocephalus may be seen in this syndrome as a result of closure of the foramen of Magendie and foramina of Luschka of the fourth ventricle. Blockage of absorption of the cerebrospinal fluid into the dural sinuses and impairment of the circulation of the fluid around the base and lateral surface of the brain may also contribute to hydrocephalus. Herniation of the cerebellar vermis into the cervical part of the vertebral column may occur as a result of early fusion of its hemispheres, followed by fusion of the neural folds higher in the midbrain. The latter may be associated stenosis of the cerebral aqueduct (a canal that runs through the mesencephalon). Additionally, an anomalous small posterior cranial fossa may not be of sufficient size to accommodate the growing cerebellum, leading to displacement and herniation.

Arnold–Chiari syndrome is divided into four types (I through IV). *Type I* is seen in young adults often as a result of downward displacement of the cerebellar tonsils through the foramen magnum into the cervical part of the vertebral column. Brainstem elongation and medullary kink may be seen. It manifests as hydrocephalus, dysphagia, dysphonia, syringomyelia, and bilateral cerebellar dysfunctions. Patients with Ehlers–Danlos syndrome or Marfan syndrome may acquire Chiari malformation due to connective tissue disorders and the accompanying hypermobility of the atlanto-occipital and atlantoaxial joints.

Type II has its onset in neonates and is associated with meningomyelocele and more displacement of the fourth ventricle and cerebellar vermis into the vertebral column. Due to the low level of the spinal cord, the cervical spinal nerve roots assume an ascending course in order to reach their corresponding intervertebral foramina. Hypoplasia of the clivus, tectal beaking, and low-positioned confluence sinus are structural changes seen in this condition. The lower location of the confluence sinus is a characteristic that distinguishes this condition from Dandy–Walker syndrome, which shows an upward turn. Disproportionate enlargement of the occipital horn of the lateral ventricle (colpocephaly) is also observed. Patients with this type of anomaly exhibit hydrocephalus, syringomyelia, dysphagia, stridor, and dysphonia. *Type III* presents a cystlike fourth ventricle in the posterior cranial fossa and is associated with occipital encephalocele, and consequently, hydrocephalus, syringomyelia, and tethered spinal cord are seen, whereas *type IV, which is not compatible with life*, is characterized by hypoplastic cerebellum. Patients with this kind of malformation may exhibit hydrocephalus, dysphagia, dysphonia, syringomyelia, and cerebellar dysfunctions. Telencephalic developmental anomaly may lead to the absence of gyri, referred to as lissencephaly, or abnormally thick and wide gyri, known as pachygyria.

Approximately half of the individuals with spina bifida cystica may also exhibit a partial or complete division of the spinal cord into two symmetrical parts (*diastematomyelia*). A bony projection, cartilage, or fibrous tissue often separates these two parts. Each part may have its own dural, arachnoidal, and pial coverings, and arterial supply. The part of the spinal cord, cranial or caudal to the site of this anomaly, remains united. Although the transverse diameter of the vertebral bodies at the affected site may show perceptible widening, the anteroposterior dimension is often narrowed. This anomaly is compatible with life and may not pose any serious neurological complications. Radiographic imaging techniques, myelography, and visible cutaneous manifestations may make diagnosis of this condition possible. Patients with scoliosis may be examined for this type of malformation via myelography. Incidence of congenital malformations is approximately 3% in neonates, and CNS anomalies make up approximately one-third of all congenital malformations. Thus, the incidence of CNS congenital malformation is around 1%. Manifestations of these congenital defects may vary widely from those occurring incidentally without apparent symptoms to those that are incompatible with life. Malformations may develop as a result of radiation; metabolic disorders; chromosomal abnormalities (e.g., trisomies); chemical exposure (e.g., illicit drugs and alcohol); and infection by viruses (e.g., cytomegalovirus, herpes, rubella), bacteria (e.g., toxoplasma gondii, treponema pallidum), and so forth.

DEVELOPMENT OF THE BRAIN VESICLES

By the end of the *fourth week*, and following complete separation from the ectodermal surface, the neural tube is composed of a caudal part, which becomes the spinal cord and the expanded rostral brain vesicles (Figure 1.8) that form the brain hemispheres and the brainstem. While the most cephalic portion of the neural tube undergoes drastic differentiation, the caudal portion continues to form. The cephalic portion undergoes flexion at the level of the future mesencephalon. At first, the rostral part of the neural tube consists of three primary brain vesicles: prosencephalon, mesencephalon, and rhombencephalon (Figure 1.9). Rapid growth of the primary brain vesicles during the fifth week results in the formation of the telencephalon, diencephalon, mesencephalon, metencephalon, and myelencephalon as well as associated flexures. The cephalic flexure lies between mesencephalon and rhombencephalon. The pontine flexure, which develops into the transverse rhombencephalic sulcus, separates the metencephalon and the myelencephalon. Furthermore the cervical flexure lies between the rhombencephalon and the spinal cord.

Persistence of the neural canal within the center of the primitive brain vesicles gives rise to a group of interconnected, fluid-filled cavities that compose the ventricular system. The rhombencephalic vesicle develops into the fourth ventricle; the mesencephalic vesicle becomes the cerebral aqueduct; the diencephalic vesicle converts into the third ventricle; and the telencephalic vesicle develops into the lateral ventricle (Table 1.2).

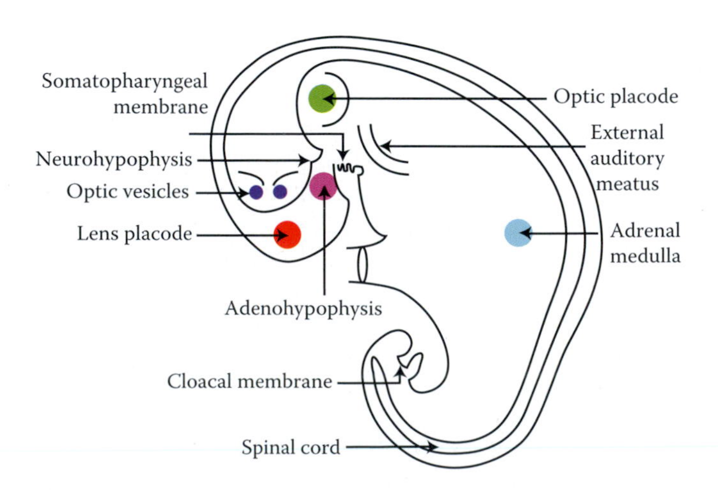

FIGURE 1.8 Neural tube showing its expanded rostral end with optic vesicles. Observe the location of the adrenal medulla, a derivative of the neural crest cells.

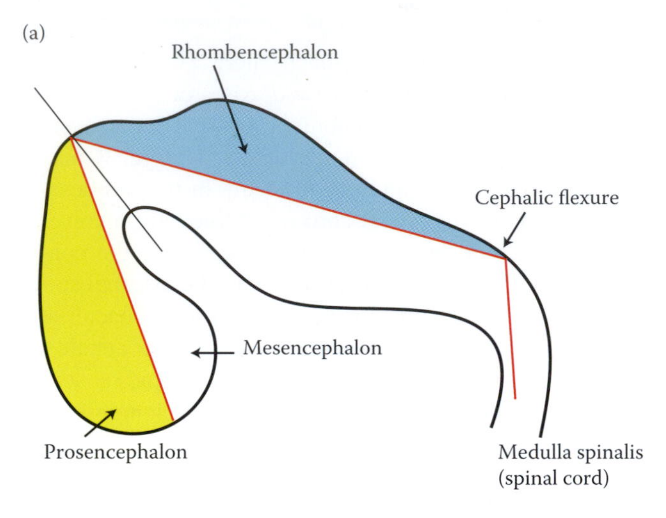

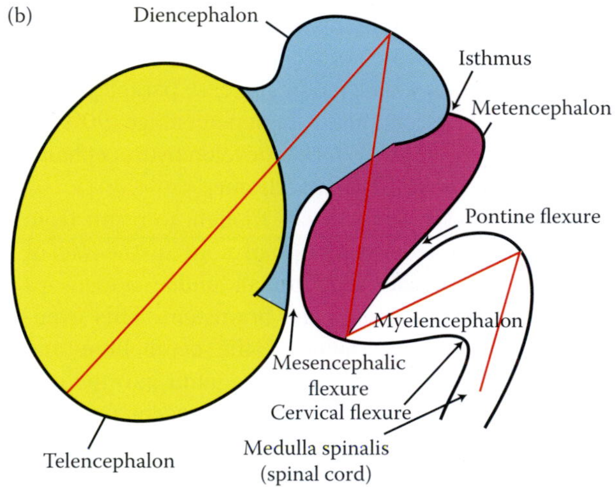

FIGURE 1.9 Initial differentiation of the rostral neural tube into (a) primary and (b) secondary brain vesicles. Flexures that separate certain brain vesicles are also illustrated.

DIFFERENTIATION OF THE NEURAL TUBE

By the middle of the fourth week, the neural tube develops three layers, consisting of the ventricular zone and ependyma (innermost), mantle (intermediate), and marginal (outermost) layers (Figures 1.10 and 1.11). This differentiation commences in the rhombencephalic region and then extends in a craniocaudal direction. The dendritic processes of the mantle neurons (neuroblasts) form the marginal layer (future white matter). In the spinal cord and the rhombencephalon, the ventral part of the mantle layer represents the sites of the motor neurons that appear earlier than the sensory neurons. The mantle layer (future gray matter) consists of a narrow dorsal alar plate and a thick basal plate, separated by the sulcus limitans (Figures 1.10 and 1.11). As it forms the lining of the central canal and ventricular system, the ventricular zone (ependyma) separates the cerebrospinal fluid from the blood vessels of the choroid plexus and neurons of the brain and spinal cord. Mitotic division of the neuroepithelial cells in the ependymal layer forms neuroblasts, which migrate laterally to the mantle layer.

TABLE 1.2
Derivatives of the Primary Brain Vesicles

Primary Brain Vesicles	Secondary Brain Vesicles	Neural Tube (Wall)	Neural Canal (Ventricular System)
Prosencephalon	Telencephalon	Cerebral hemispheres Rostral part of hypothalamus Rostral part of third ventricle Archicortex, paleocortex, and neocortex	Lateral ventricles
	Diencephalon	Thalamus, metathalamus subthalamus, epithalamus, and part of hypothalamus	Caudal part of the third ventricle
Mesencephalon	Mesencephalon (midbrain)	Colliculi, tegmentum, cerebral peduncles, cerebral aqueduct	Cerebral aqueduct
Rhombencephalon	Isthmus rhombencephali	Superior medullary velum and superior cerebellar peduncles	Rostral part of fourth ventricle
	Metencephalon	Cerebellum, pons, and middle cerebellar peduncles	Middle part of the fourth ventricle
	Myelencephalon	Medulla oblongata	Rostral part of the central canal, caudal part of the fourth ventricle, and inferior cerebellar peduncle
Medulla spinalis	Spinal cord	Spinal cord	Central canal

Migration of neuroblasts is guided by the cytoplasmic extensions of the developing neurons, which are anchored, to the pia mater on the outer surface of the CNS. Shortening of these processes also enables some neurons to migrate. This migration is dependent upon the radial glial cells (90% of migration), which is clearly evident in the development of the cerebral cortex. Here, the neuronal stem cells proliferate in the ventricular zone, followed by the formation of the preplate, which develops into the subplate neurons and the Cajal–Retzius cells. This type of migratory pattern involves translocation of the cell bodies to the pial surface via elongation and contraction of the microtubule around the nucleus and its movement (nucleokinesis). Since radial glial cells form a platform for migrating cells, they also undergo displacement and differentiation and allow the migrating neurons to form the cortical plate by splitting the preplate. As each wave of migrating cells move ahead of their predecessors, the most recently migrated neurons occupy positions in close proximity to the surface of the cortex. In the cerebellum, this migration is evident in the course of the developing granule cells parallel to the surface of the neural tube and along the long processes of the Bergmann glial cells. An adhesion protein molecule known as astrotactin may mediate neuroglial interaction during the migration process. Some studies propose the possibility of existence of genetically predetermined entities that may guide the migration process. As the migration from the neuroepithelium or from one of the secondary proliferative zones continues, the differentiated neuronal and glial cell precursors either are ellipsoidal in shape (apolar) or have a single process (unipolar). Few apolar or unipolar neurons are retained in the CNS of the vertebrate, but most continue through a bipolar phase. But a great majority of neurons develop even more processes and are known as multipolar neurons.

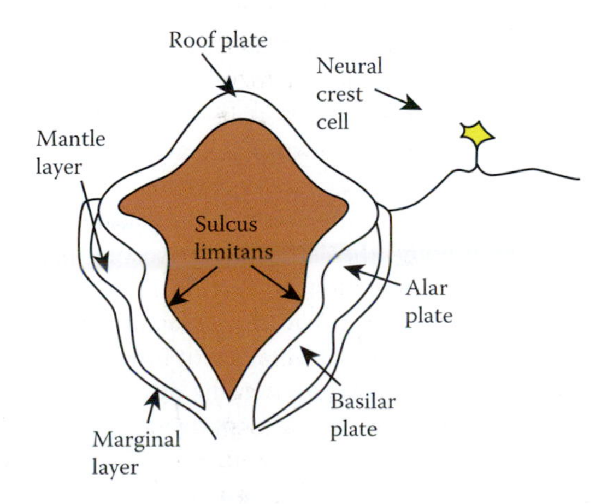

FIGURE 1.10 A section of the neural tube showing the sulcus limitans, alar and basal plates, and the roof and floor plates.

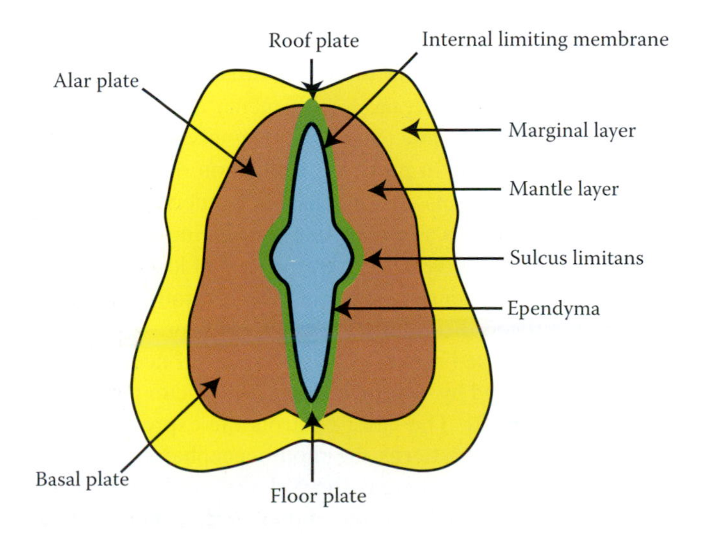

FIGURE 1.11 Cross section of the caudal neural tube (future spinal cord). The basal plate forms the ventral horn and the associated motor neurons, whereas the alar plate forms the dorsal horn and sensory neurons of the adult spinal cord.

In order to attain the appropriate position in the cerebral cortex, majority of interneurons migrate in a tangential manner. This is evident in the migratory wave between the subventricular zone and the olfactory tract. In case of the multipolar neurons, the migration does not follow the schemes described (locomotion of somal translocation); instead, they express neuronal markers and send several thin extensions in different directions to enable their migration.

GENETIC AND MOLECULAR ASPECTS

Expression of certain homeobox genes, at the initial stages of development, may play an important role in the segmentation of the rhombencephalon in which each segment (rhombomeres), a series of eight bilateral protrusions from the rhombencephalic wall, may specify the origin of certain cranial nerve. For example, neurons of rhombomere 2 (r2) form the trigeminal ganglion, while r4 forms the geniculate ganglion. Domains of gene expression, for example, those for the homeobox (HOX)-B genes and the transcription factor Krox20, adjoin rhombomere boundaries. Genes that contain homeoboxes in humans are classified into HOX-A, HOX-B, HOX-C, and HOX-D and are identified by numbers 1–13. These genes may regulate the expression of a number of genes that collectively determine the structure of one body region. HOX genes are expressed in the developing rhombomeres and neural crest. A complement of HOX genes is thought to form axial and branchial codes that specify the locations of somites and neurons along the length of the embryo. An interesting feature of specific homeotic (HOX) genes is their linear order (collinearity) and their transcription in the cephalocaudal and the 3' end to 5' end directions. Since HOX gene expression is affected by teratogenes such as retinoic acid, transformation of rhombomeres may occur. Transformation of a trigeminal nerve to a facial nerve may occur as a result of induced changes of rhombomere 2/3 to a 4/5 disposition. Similar changes may be observed upon the application of retinoic acid to other HOX codes.

Ventrodorsal patterning of the neural tube may be regulated by the *sonic hedgehog* (Shh) gene, a product of the notochord and floor plate. Shh sends signals to cells in the neural tube to maintain proper specification of ventral neuron progenitor domains, and its absence renders this process unattainable. Shh acts as a morphogen inducing the selective cell differentiation to ventral interneurones at low concentration and to motor neurons at higher concentrations. Failure of Shh-modulated differentiation causes holoprosencephaly, which was discussed earlier. The rostrocaudal neural development is governed by the fibroblast growth factor (FGF) and by the retinoic acid. The latter controls the expression of 3' HOX genes, which patterns the rhombencephalon along the anteroposterior axis, whereas the 5′ HOX genes, which are expressed more caudally in the spinal cord, remain outside the sphere of influence of the retinoic acid. It is worth noting that the FGF signaling pathway is implicated in tumor progression and growth via the dysregulation of cell proliferation, differentiation, survival, and angiogenesis in multiple tumor types. The Shh gene acts by inhibiting the suppression of the expression of PAX-3 and dorsalin genes. Non-HOX homeobox genes, a separate group of genes, are also involved in developmental patterning of the embryo, but lack the cephalocaudal expression seen in the HOX homeobox genes and their role in organogenesis.

PAX genes (PAX-1 to PAX-9), a segmentation group of genes, regulate the morphology of the developing embryo. They are transcription factors and are implicated in the specialization of different regions of the CNS. PAX-3 and PAX-7 are expressed in the alar plate and roof plates as well as the neural crest cells. On the other hand, PAX-5 and PAX-8 are expressed in the intermediate gray columns. Ventricular and basal regions of the neural tube are sites where PAX-6 is expressed. Expression of PAX-1 in the ventromedial region of the somite is induced by several factors such as Shh, notochord, and floor plate, whereas expression of PAX-3 and PAX-7 within the dorsolateral region of the somites is induced by the dorsal ectoderm. PAX genes may play important roles in certain genetic diseases, such as Waardenburg's syndrome and aniridia. Mutated PAX-3 and PAX-6 may be involved in Waardenburg's syndrome and aniridia, respectively.

> *Waardenburg's syndrome* is an autosomal dominant condition associated with Mutated PAX-3. It presents with congenital deafness; wide-bridged nose; disorders of pigmentation (white eyelashes, white forelock, and leukoderma); and dystopic canthorum (lateral displacement of the canthi). In *aniridia*, a rare bilateral hereditary anomaly associated with mutated PAX-6, absence or abnormal development of the retina and iris remains a prominent feature.

Extensions of the developing neuronal processes may be governed by many factors, which include the development of growth cones and presence of N-CAMs, neuroglial adhesion molecules (NgCAMs), transiently exposed axonal glycoprotein (TAG-1), actin, extracellular matrix adhesion molecules (E-CAMs), and guidepost cells. Growth cones are able to descry the chemical signals (actin and actin-binding proteins) and test the new environment in all directions via filopodia and lamellipodia. Since polymerization of actin control, to an extent, the movement of the growth cones, any substances that limit this process, such as fungal toxin cytochalasin B, may also hinder their further growth. Additionally, calcium, interaction with other intracellular second messenger systems, and phosphorylation by protein kinase may indirectly affect the direction of neuritic growth by acting upon the actin-binding proteins. Movements of growth cones may also be shaped by N-CAM, a molecule that enhances neuritic fasciculation, and by laminin, fibronectin, and tenascin (cytotactin), which are members of the E-CAMs that act via receptor integrins. NgCAM, integrin, and N-cadherin (calcium-dependent molecule) also share an important role in the process of axonal development.

Synaptic connections among developing neurons occur early in development and undergo constant correction and refinement. Synaptogenesis is an essential facet of neuronal circuitry in the developing CNS and plays an important role in remodeling adult neural networks. Studies have shown that immunoglobulin synaptic adhesion molecule SynCAM 1 dynamically alters synaptic sites and neuronal plasticity. SynCAM is a cell adhesion molecule that is present in both presynaptic and postsynaptic membranes. Enhancement of the excitatory synapse numbers is closely correlated with the overexpression of SynCAM 1; similarly, a reduction in the number of excitatory synapses is related to the loss of SynCAM 1. In view of this reciprocal correlation, heightened or lessened SynCAM 1 expression can affect the mechanism that regulates neuronal plasticity and synapse number as well as activity-dependent neuronal changes. Additionally, SynCAM 1 also alters the neuronal plasticity at mature synapses and regulates spatial learning by impacting long-term depression.

Studies have shown that CNS synaptogenesis, particularly glutamatergic synapses, appears to be mediated by agrin (proteoglycan that aggregates the acteylcholine receptors) signals. This was evident in the unusual and highly dynamic dendritic filopodia that establish contact with axons followed by recruitment of postsynaptic proteins to the site of contact. There is additional evidence that indicates the close relationship between synaptogenesis and astrocytic differentiation. Studies have also shown that neuroligins (NLGNs, a cell adhesion molecule on the postsynaptic membrane), which function via neurexins and SynCAM, can play a role in inducing presynaptic differentiation.

The initial overproduction of neurons may be controlled by subsequent cell death, which is determined by the available trophic substances in the immediate location. Selectivity of the neurotrophic substances in supporting a certain neuronal population may also determine the fate of certain neurons. Extracellular, intracellular, and transmembranous tyrosine kinase domain remain essential in mediating the effects of neurotrophins. Hormones such as testosterone may also govern the extent of development of certain areas of the CNS.

MEDULLA SPINALIS (SPINAL CORD)

The spinal cord develops from the caudal portion of the developing neural tube, extending the entire length of the vertebral column (Figures 1.8 and 1.9). During this stage of development, the spinal nerves also appear. After the third fetal month, the vertebral column outgrows the spinal cord, and as the vertebrae grow caudally, the dorsal and ventral roots, anchored within their appropriate foramina, pursue a longer course within the vertebral canal. This accounts for the formation of the cauda equina and the final position of the dorsal root ganglia and spinal nerves. By birth, the spinal cord terminates at the level of the third lumbar vertebra.

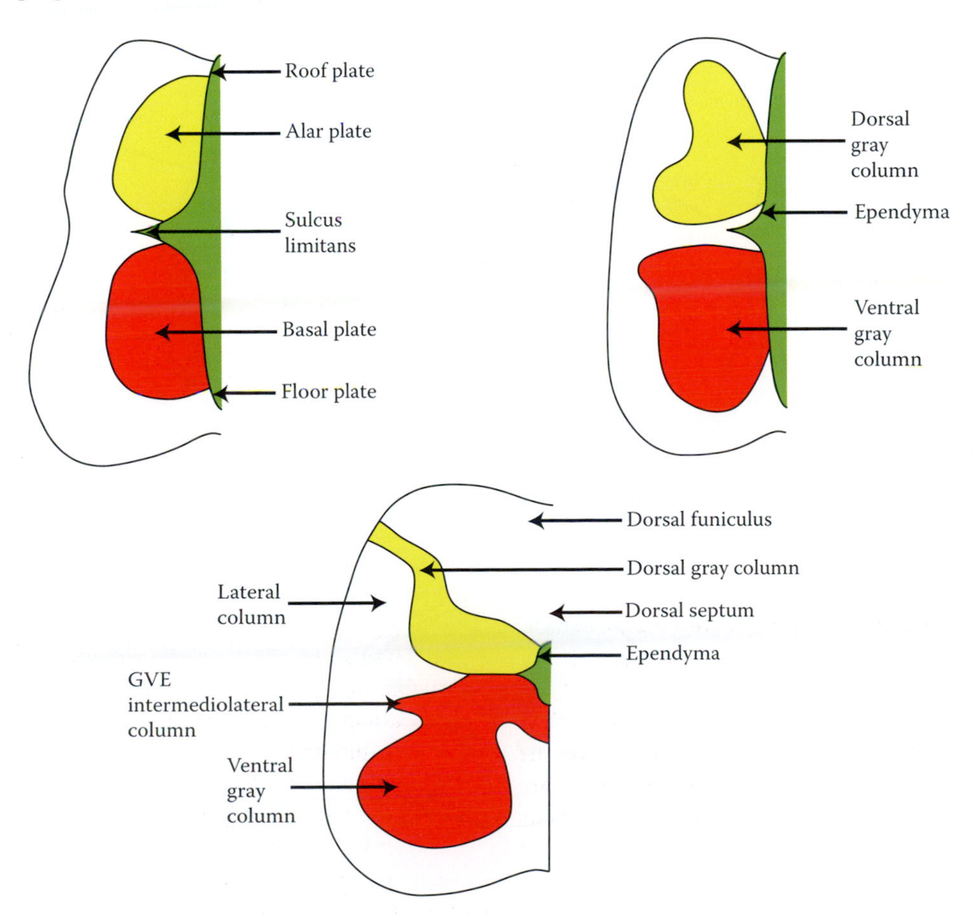

FIGURE 1.12 Development of the spinal cord. Note the formation of the somatic and autonomic neuronal columns

Much of the neural canal is obliterated, and the remaining part forms the central canal. The alar plate forms the dorsal horn, which represents the sensory area, while the basal plate, representing the motor area, becomes the future ventral horn of the adult spinal cord. Quite early, some of the neuroblasts of the basal plates begin to differentiate into the alpha motor neurons that extend to form the ventral roots of the spinal nerves. As the basal plates enlarge, they protrude ventrolaterally on each side of the floor plate. They do not fuse in the midline, thus producing the midline ventral median fissure. The roof plate is obliterates, and the neural canal is reduced in size. The lower end of the spinal cord and the intermediate part of the neural (primitive central) canal, between the level of the second coccygeal and third lumbar vertebra, undergoes necrobiosis (selective death), degenerates, and becomes adherent to the covering pia as the filum terminale. The site of degeneration may, occasionally, give rise to congenital cysts.

Transformation of the dorsal portion of the neural canal, the ependymal cells, and their long basal processes gives rise to the dorsomedian septum. Obliteration of the floor plate and extension of axons of some neuroblasts across the midline result in the formation of the ventral white commissure. Neuroblasts of the neural tube that develop toward the marginal layer of the cord do so primarily by following a ventrolateral direction. The fasciculus proprius develops from the marginal layer into all three funiculi of the spinal white matter. While the transformation of the neural tube into the spinal cord is progressing, a large neural crest–derived mass of axons is added to the cord.

The cells of the dorsal root ganglia are derived from the neural crest and the neural tube. Their central processes grow into the spinal cord and contribute in great quantity to the white matter of the cord, forming the dorsal columns and the dorsolateral fasciculi. General somatic afferents (generated at or near the body surface) and general visceral afferents (generated in or on mucus membranes of visceral structures) lie dorsal to the sulcus limitans and are represented by the dorsal gray and white columns. Conversely, general visceral efferents (GVEs, autonomic parasympathetic) and general somatic efferent cell columns (innervate skeletal muscles) lie ventral to the sulcus limitans (Figures 1.11 and 1.12).

Despite the fact that fibers of the corticospinal tract begin to develop in the 9th week and complete by the 29th week of fetal life, myelination and restoration of their motor function are only achieved by the end of the 2nd year of postnatal life. Those fibers that are destined to the cervical and upper first thoracic segment are in advance of the fibers that reach the lumbosacral segments, which, in turn, are in advance of the fibers that project to the face.

MYELENCEPHALON

The embryonic myelencephalon forms the caudal part of the rhombencephalon, consisting of alar and basal plates (Figure 1.13). Due to widening of the roof plate, the alar and basal

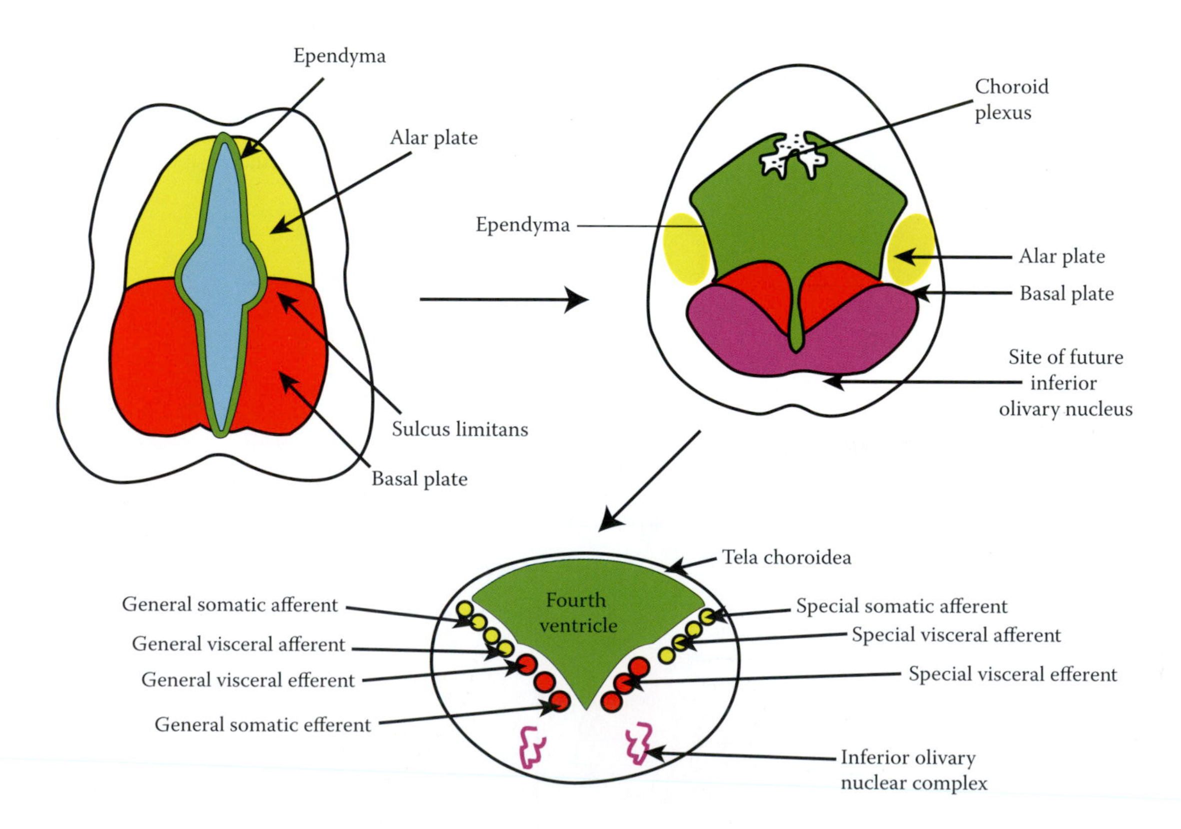

FIGURE 1.13 Cross section of the developing medulla. The alar and basal plates are arranged in a mediolateral direction, signifying the locations of the motor and sensory neurons.

plates eventually assume a more lateral/medial, rather than dorsal/ventral, position. Both the alar and basal plates contribute to the reticular formation, whereas the medullary pyramids remain of telencephalic origin. The alar plate constitutes the sensory part that contributes to the formation of several nuclei (Figure 1.14). The vestibular and auditory nuclei contain neurons that receive *special somatic afferents* transmitting the vestibular and auditory impulses. The spinal trigeminal, gracilis, and cuneatus nuclei receive *general somatic afferent*s that are generated at or near the body surface. The solitary nucleus, which contains *special and general visceral* neurons, receives taste and visceral sensations, respectively, whereas the inferior olivary nuclear complex, another alar plate–derived nucleus, functions as a cerebellar relay nucleus.

Similarly, the *basal plate* (Figure 1.14) gives rise to the hypoglossal nucleus, the source of general somatic efferent fibers to the lingual muscles; the nucleus ambiguus, which supplies special visceral efferent fibers to the laryngeal, pharyngeal, and palatal muscles (derivatives of branchial arches); and to the dorsal motor nucleus of the vagus and the inferior salivatory nucleus that provide general visceral efferent (parasympathetic) fibers.

The *roof plate* persists to form the ependyma of the tela choroidea, inferior medullary velum, and caudal part of the roof of the fourth ventricle. Axons of the marginal layer are derived from neuronal extensions of the medial lemniscus and spinothalamic tract. Attachment of the choroid plexus to the roof of the fourth ventricle is secured by the tela choroidea, which is formed by the ependymal layer of the myelencephalon covered by pia mater; they form the tela choroidea. During the fourth or fifth month of development, the paired foramina of Luschka, at lateral recesses of the fourth ventricle, and the single median foramen of Magendie make their appearance.

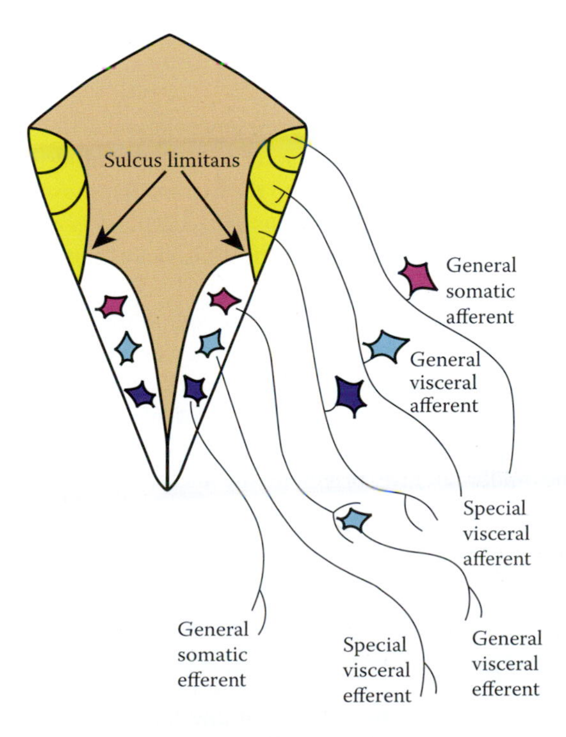

FIGURE 1.14 Schematic diagram of the somatic and visceral neurons associated with the developing medulla.

METENCEPHALON

The metencephalon (Figures 1.15 and 1.16) differentiates into the pons and cerebellum.

Pons

In the pons the basal plate gives rise to primarily efferent nuclei including the abducens, facial, trigeminal, and superior salivatory nuclei (Figure 1.16). The abducens nucleus provides *general somatic efferent fibers* supplying the lateral rectus; the facial and trigeminal nuclei motor nuclei give rise to the *special visceral efferent fibers* that innervate the facial and masticatory (branchial) muscles, respectively. The superior salivatory nucleus provides *general visceral efferent (parasympathetic presynaptic)* fibers to regulate the secretion of the lacrimal, sublingual, and submandibular glands.

Derivatives of the *alar plate* include somatic and visceral sensory nuclei. The vestibular and auditory nuclei transmit *special somatic afferent fibers* from the corresponding receptors; the principal sensory nucleus conveys *general somatic afferent fibers from the head*, whereas the solitary nucleus receives *general visceral* (visceral sensations) and *special visceral* (taste) *afferent* fibers. The pontine nuclei are cerebellar relay nuclei that enable the cortical efferent to affect cerebellar function.

Cerebellum

The cerebellum is derived from the *rhombic lip* of the *alar plate* (Figure 1.17). During the fourth month of development, the posterolateral fissure is the first to appear. The cerebellar cortex develops from the migrating neuroblasts of the external granular layer, which is formed by the germinal cells of the rhombic lip that migrate over the surface of the cortical lip. Both the alar and basal plates contribute to the reticular formation. At about the fifth week of embryonic development, the lateral parts of the alar plates on both sides of the roof

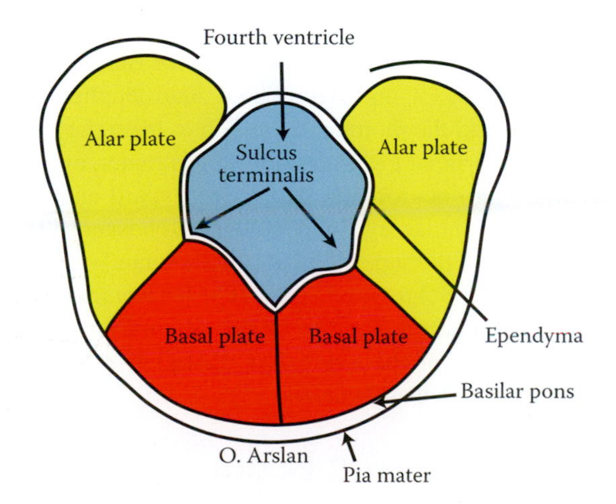

FIGURE 1.15 Developing pons and associated fourth ventricle.

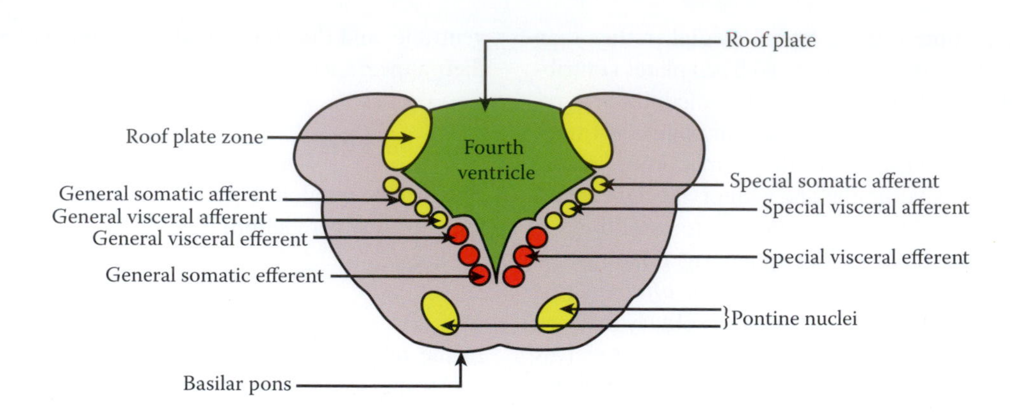

FIGURE 1.16 A more elaborate diagram of the structures in Figure 1.15. Note the afferent and efferent neurons associated with the basal and alar plates.

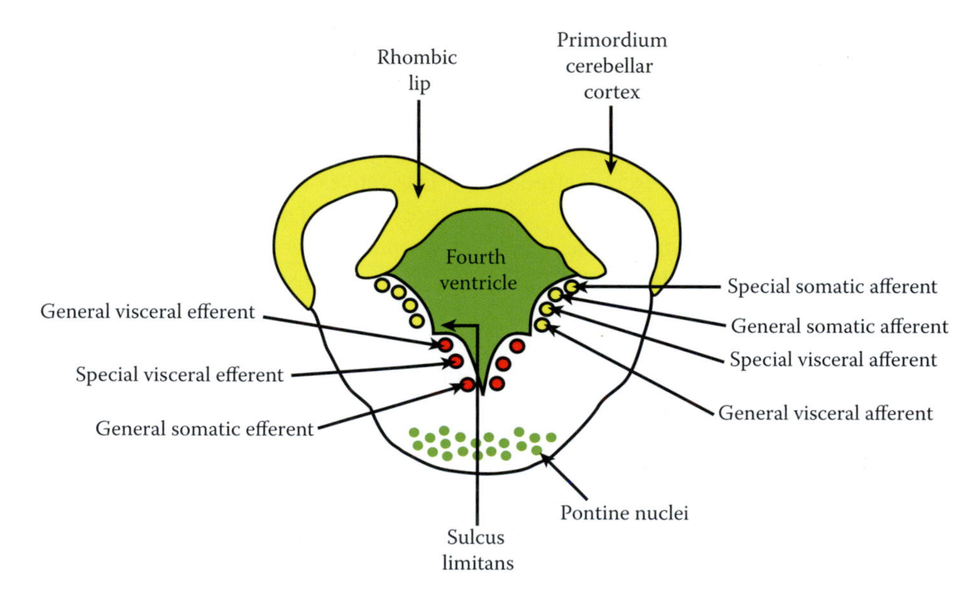

FIGURE 1.17 Transverse section of the metencephalon. Position of the rhombic lips is also illustrated.

of the metencephalon join to form the rhombic lips, which eventually become the cerebellar vermis and hemispheres. The remaining part of the alar plate forms the superior and inferior medullary veli. Some neuroblasts of the mantle layer migrate outward into the marginal layer (towards the surface) to mature and become cerebellar cortical neurons. This migration is guided by the processes of Bergmann glial cells. The groups of undifferentiated neuroepithelial cells that move around the rhombic lip region to form the external granular, a superficial layer beneath the pia mater, eventually differentiate into neuroblasts that move inward and mature into the adult granular layer and stellate and basket cells. The young neurons of the superficial part of the mantle layer form the Purkinje and Golgi type II cells. The periventricular neuroblasts that remain at the site of the original mantle layer become the cells of the cerebellar (fastigial, globose, and emboliform, and dentate) nuclei.

MESENCEPHALON (MIDBRAIN)

The mesencephalon, morphologically the most primitive of the brain vesicles, contains both basal and alar plates (Figure 1.18). The basal plate of the mesencephalon differentiates into the trochlear and oculomotor nuclei, providing *general somatic efferent fibers* (GSE) to the extraocular muscles, and to the Edinger-Westphal nucleus *that supplies general visceral efferent fibers* (GVE) to the constrictor muscle of the pupil and the ciliary muscle.

Differentiation of the alar plate results in the formation of the superior and inferior colliculi, whereas the corticofugal fibers form the crus cerebri. The substantia nigra, red nucleus, and reticular formation are probably of mixed origin from neuroblasts of both basal and alar plates. Neural crest cells of the midbrain give origin to the mesencephalic nucleus. Development of the isthmus (mesencephalo-metencephalic junction) is controlled by FGF 8.

DIENCEPHALON (THALAMUS, HYPOTHALAMUS, EPITHALAMUS AND SUBTHALAMUS)

The diencephalon consists of roof and alar plates but lacks the basal and floor plates. Differentiation of the alar plate here results in the formation of the thalamus, hypothalamus,

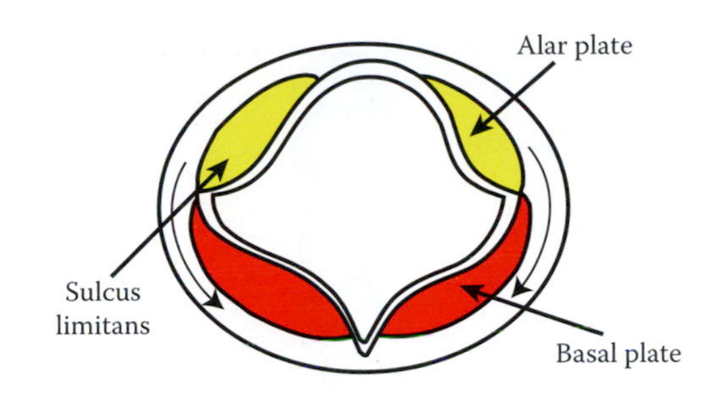

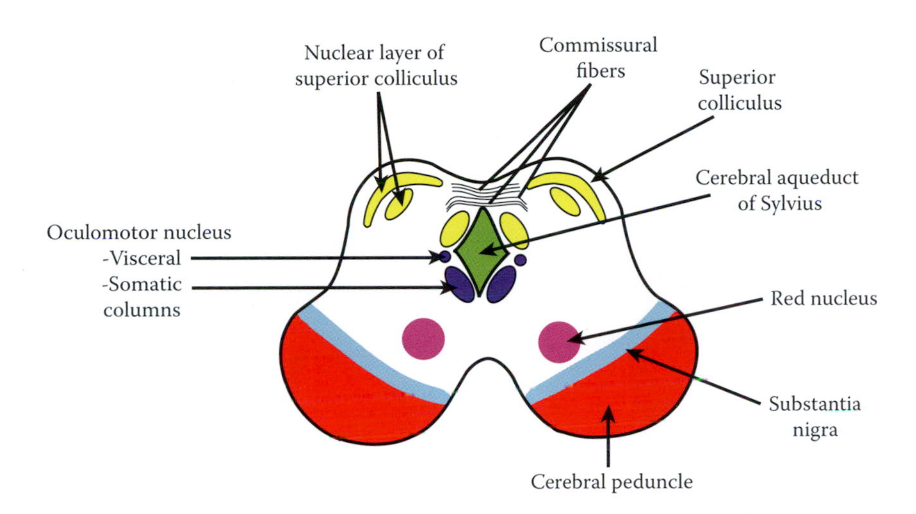

FIGURE 1.18 The developmental stages of the mesencephalon and derivatives of the alar, basal, and roof plates.

neurohypophysis, and infundibulum. A diverticulum of the stomodeum is derived from Rathke's pouch. Derivatives of the roof plate include the epiphysis cerebri, habenular nuclei, and posterior commissure. The ependyma and vascular mesenchyme of the roof plate give origin to the choroid plexus of the third ventricle.

TELENCEPHALON (CEREBRAL HEMISPHERES, BASAL NUCLEI AND VENTRICULAR SYSTEM)

The two lateral outpocketings, which arise from the cephalic end of the prosencephalon, form the telencephalon, constituting the cerebral hemispheres (Figures 1.9 and 1.19). These lateral diverticula evaginate from the most rostral end of the neural tube near the primitive interventricular foramen of Monro and are connected via the midline region known as the telencephalon impar. These diverticula are rostrally in continuity around the foramen of Monro, but caudally remain continuous with the lateral walls of the diencephalon. Enormous positive pressure exerted by the accumulated fluid within the neural canal results in the rapid expansion of the brain volume in the early embryo (3–5 days of development). This is aided by the constriction of the neural tube at the base of the brain via the surrounding tissues. The foramen of Monro forms the rostral part of the third ventricle. At the end of the third month, the superolateral surface of the cerebral hemisphere shows a slight depression anterior and superior to the temporal lobe. This occurs due to the more modest expansion of this site relative to the adjoining cortical surface. This depression, the lateral cerebral fossa, gradually overlapped by the expanding cortical area, converts into the lateral cerebral sulcus (fissure). The floor of this sulcus becomes the insular cortex. Apart from the lateral cerebral and hippocampal, sulci the cerebral hemispheres remain smooth until early in the fourth month, when the parieto-occipital and calcarine sulci appear. During later stages of development (fifth month of prenatal life), the cingulate sulcus and, later (sixth month), the remaining sulci appear on the superolateral and inferior surfaces of the brain. Virtually all sulci become recognizable by the end of the eighth month of development. The ventricular and subventricular parts of the telencephalic lateral diverticula form the ependyma, the cortical neurons, and the glial cells. The intermediate cell layer of the telencephalic diverticula differentiates into the white matter, while the cortical zone differentiates into the various layers of the isocortex.

At the beginning, the wall of the cerebral hemisphere consists of three basic layers that include the inner neuroepithelial, mantle, and marginal layers. Each neuroepithelial cell has a single nucleus and double cytoplasmic extensions. The deep extension extends to the internal limiting laminae, whereas the superficial extension stretches to the external limiting membrane, which itself is covered by the pia mater. Attachment of the superficial and deep extensions is maintained via end feet that contribute also to these membranes or laminae. As the nuclei undergo division, the cytoplasmic processes remain solid. One of the nuclei remains near the ventricular surface, and the other migrates within the cytoplasmic extensions to the

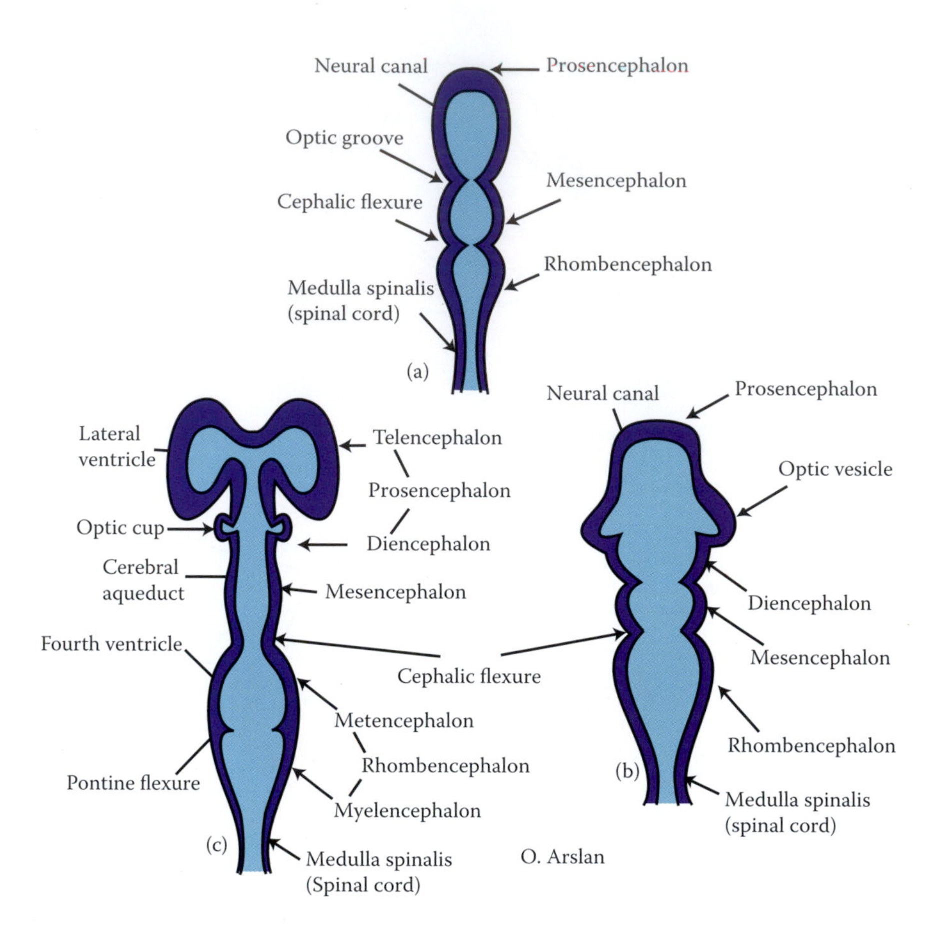

FIGURE 1.19 Formation of the primary and secondary brain vesicles and associated components of the ventricular system. (a), (b), and (c) mark the stages of differentiation of the neural tube and development of the flexures.

pial surface. As it reaches the pial matter, the cytoplasmic process separates from the original cell and begins to surround the newly formed nucleus. Neuroblasts that maintain position near the pia matter are unipolar, with one neuronal extension, which eventually divides into finer processes or dendrites. As the thickness of the cortex increases subsequent to an increase in the number of neuroblasts, the unipolar neuroblasts become deeply located; at the same time, the neuroblasts begin to form axons that stretch to the ventricular surface and dendrites, extending to the subpial layer. Glioblasts, which differentiate into the astrocytes and oligodendrocytes, are derived from the neuroepithelial cells that line the neural canal when the production of the neuroblasts ceases.

Most cortical neurons follow an "inside-out" pattern of migration from the ventricular and subventricular zones through the intermediate zones to the cortical plate, allowing the neurons that form at a later stage of development to migrate and maintain an outward position to the neurons that develop earlier. Thus, the recently formed neurons occupy the basal layers of the cortex, while the older neurons maintain locations in the superficial layers. In the initial stage of migration, the neuroblasts are allowed to proceed to a site between the marginal layer and the white matter. The nuclei of the neuroepithelial cells lie near the ventricle, while the cytoplasm elongates to form deep and superficial processes. Some neuroblasts traverse the initial group of migratory neuroblasts to assume a position in the middle third of the mature cortex, whereas others may pursue different courses among the previous group of neuroblasts to reach more superficial positions. This pattern of migration is in line with the radial columnar organization of the cerebral cortex.

The invagination formed by the attachment of the cerebral diverticula to the roof of the diencephalon leads to the formation of the choroid fissure. The latter fissure allows a narrow strip of thin ependymal roof plate, with the accompanying pial covering (tela choroidea) to invaginate into the lateral ventricle. As the temporal lobe develops, the choroidal fissure, with the invaginating tela choroidea and the choroid plexus, continues to increase in length along the medial wall of the developing temporal lobe. Hence, in the adult, the choroid plexus is a continuous structure found in the third ventricle; interventricular foramen; and the body, trigone, and inferior horns of the lateral ventricles. In fetal life, the choroid plexus occupies most of the lateral ventricle and then gradually decreases in size.

Formation of the anterior commissure (the first commissure to appear during development) is followed by the development of the hippocampal commissure and the corpus callosum (around the fourth month). These commissures are formed by axons that extend from one hemisphere to the other using the embryonic lamina terminalis as a bridge. Initially, the hippocampus appears as a ridge derived from the medial cortical wall.

The basal nuclei are derived from the mantle layer of the telencephalon. Specifically, the caudate and putamen derive from the ganglionic eminences in the ventrolateral part of the telencephalic ventricle, while the pallidum originates from lateral and medial hypothalamic analogs. The corpus striatum assumes a striated appearance following the crossing of fibers that connect the telencephalic vesicles to the diencephalon and brainstem.

SUGGESTED READING

Caldarelli M, Di Rocco C. Diagnosis of Chiari I malformation and related syringomyelia: Radiological and neurophysiological studies. *Childs Nerv Syst* 2004;20(5):332–5.

Emery JL, Lendon RG. Lipomas of the cauda equina and other fatty tumors related to neurospinal dysraphism. *Dev Med Child Neurol* 1969;11:62–70.

Galliot B, Dollé P, Vigneron M et al. The mouse Hox 1.4 gene: Primary structure, evidence for promoter activity and expression during development. *Development* 1989;107:343–59.

Goldstein S, Reynolds CR. *Handbook of Neurodevelopmental and Genetic Disorders in Children*. New York: Guilford Press, 1999.

Graham A, Heyman I, Lumsden A. Even-numbered rhombomeres control the apoptotic elimination of neural crest from odd-numbered rhombomeres in the chick hindbrain. *Development* 1993;119:233–45.

Guo F, Wang M, Long J, Wang H, Sun H, Yang B, Song L. Surgical management of Chiari malformation: Analysis of 128 cases. *Pediatr Neurosurg* 2007;43(5):375–81.

Hankinson TC, Grunstein E, Gardner P, Spinks TJ, Anderson RC. Transnasal odontoid resection followed by posterior decompression and occipitocervical fusion in children with Chiari malformation Type I and ventral brainstem compression. *J Neurosurg Pediatr* 2010;5:549–53.

Hensinger RN, Lang JR, MacEwen GD. Klippel–Feil syndrome: A constellation of associated anomalies. *J Bone Joint Surg* 1974;56A:124.

Hoffman HJ. Comment on Pang D, Dias MS, Ahab-Barmada M. Spinal-cord malformation: Part 1: A unified theory of embryogenesis for double spinal-cord malformations. *Neurosurgery* 1992;31:451–80.

Kim HJ, Kim AY, Lee C et al. Hirschsprung disease and Hypoganglionosis in adults: Radiologic findings and differentiation. *Radiology* 2008;247(2):428–34.

Loukas M, Shayota BJ, Oelhafen K, Miller JH, Chern JJ, Tubbs RS, Oakes WJ. Associated disorders of Chiari Type I malformations: A review. *Neurosurg Focus* 2011;31(3):E3.

Milhorat TH, Bolognese PA, Nishikawa M, McDonnell NB, Francomano CA. Syndrome of occipitoatlantoaxial hypermobility, cranial settling, and Chiari malformation type I in patients with hereditary disorders of connective tissue. *J Neurosurg Spine* 2007;7(6):601–9.

Nadarajah B, Alifragis P, Wong R, Parnavelas J. Neuronal migration in the developing cerebral cortex: Observations based on real-time imaging. *Cereb Cortex* 2003;13(6):607–11.

Rakic P. Mode of cell migration to the superficial layers of fetal monkey neocortex. *J Comp Neurol* 1972;145(1):61–83.

Schoenwolf GC, Smith IL. Mechanisms of neurulation: Traditional viewpoint and recent advances. *Development* 1990;109:243–70.

Serbedzija GN, Fraser SE, Bronner-Fraser M. Pathways of trunk neural crest cell migration in the mouse embryo as revealed by vital dye labelling. *Development* 1990;108:605–12.

Tabata H, Nakajima K. Multipolar migration: The third mode of radial neuronal migration in the developing cerebral cortex. *J Neurosci* 2003;23(31):9996–10001.

Tassabehji M, Read AP, Newton VE, Harris R, Balling R, Gruss P, Strachan T. Waardenburg's Syndrome patients have mutations in the human homologue of the Pax-3 paired box gene. *Nature* 1992;355:635–6.

Willis B, Guthikonda B. Congenital Chiari malformations. *Neurol India* 2010;58(1):6–14.

Yüceer N, Mertol T, Arda N. Surgical treatment of 13 pediatric patients with Dandy–Walker syndrome. *Pediatr Neurosurg* 2007;43(5):358–63.

2 Basic Elements of the Nervous System

Understanding the cellular characteristics of the nervous system enables us to appreciate the extensive synaptic connections and the functional organization of this system. Connectivity between neuronal populations necessitates generation of impulses along the cell membrane and actions of neurotransmitters as well as the support of the nonexcitable glial cells. The shape of the neurons, characteristics of the axons and dendritic arborizations, as well as the receptive zones of the neurons can also affect synaptic connectivity and determine their actions.

The nervous system receives, encodes, integrates and transmits sensory stimuli. It generates motor activity and coordinates movements. In addition, the nervous system also regulates our emotion and consciousness. In summary, it controls all the activities that preserve the individual and species. These functions are accomplished at cellular levels by the neurons and are enhanced via supportive glial cells. Neurons maintain certain common structural and morphological characteristics that enhance their activities and correlate closely with their functions. Following an injury, these structures may undergo changes or assume different positions in the neurons. Myelin, the covering of certain axons, plays an important role in nerve conduction and exhibits degenerative changes in particular diseases. Glial cells form the skeleton of the nervous system, display several forms, and participate in a variety of supportive functions that collectively maintain the optimal environment for neuronal activity.

NEUROGLIA

Neuroglia are nonexcitable supporting cells that outnumber neurons at a ratio of 2:1, forming the skeleton of the central nervous system (CNS). They are of both ectodermal and mesodermal origin and are commonly associated with tumors of the CNS. They have only one type of cell process, do not form synapses, and retain the ability to undergo mitosis. Glial cells provide the optimal milieu for neuronal function by balancing the ionic concentration within the extracellular space. They provide nutrients, discard metabolites and cellular debris, and sharpen neuronal signals, preventing ephapsis or cross talk by forming a protective myelin sheath. The glial cells mediate the extent of impulse flow, activity of neurons, and frequency of excitation. Thus, changes in the glial cell membrane potentials may occur as a result of the fluctuation in the potassium ion concentration, which, in turn, is affected by level of the generated impulses. They also secrete growth-promoting molecules such as the nerve growth factor (NGF); glial-derived neurite-promoting factor (GNPF), which is a protease inhibitor (trypsin, urokinase, and thrombin); and tenascin. Neuronal sprouting may be facilitated by glial-derived nexin (GDN), which prevents the digestion of the extracellular matrix molecules by inhibiting proteases secreted by growth cones. Neuroglia also contributes to the formation of the blood–brain barrier, which selectively permits substances and molecules to enter the CNS. They also allow developing neuroblasts to move to their final destinations. Presence of molecules such as fibronectin, laminin, and cellular adhesion molecules may account for the mechanism by which neurons migrate along processes of certain glial cells and not others. Neuroglial cells are classified as macroglial and microglial cells. Macroglial cells are further classified into astrocytes, oligodendrocytes of the CNS, and Schwann cells of the peripheral nervous system (PNS).

MACROGLIA

Macroglia are classified into astrocytes, oligodendrocytes, and ependymal cells.

Astrocytes

Astrocytes (star cells) are the largest and the most numerous and show the most branching among all the glial cells (Figures 2.1, 2.2, and 2.5). They are present in both gray and white matter and possess processes that branch repeatedly in an irregular fashion and assume starlike configurations. Astrocytes establish contacts with the nonsynaptic parts of the neurons and form perivascular end feet that extend to the blood capillaries. They retain the capacity to multiply. Astrocytes retrieve glutamate and γ-aminobutyric acid (GABA) after their release from the nerve endings. They

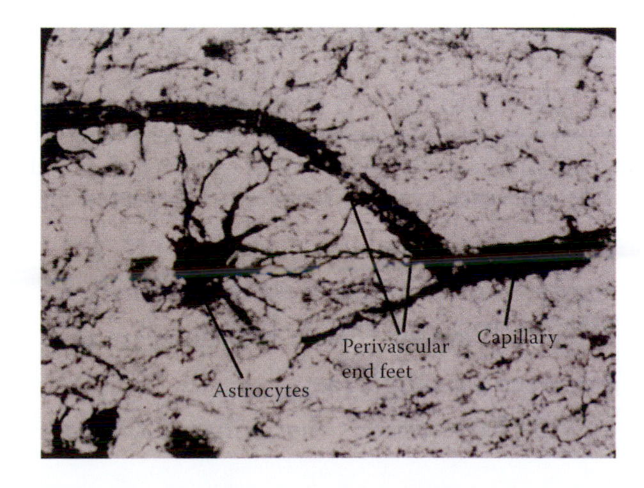

FIGURE 2.1 Photomicrograph of the astrocytes showing their branches and perivascular end feet.

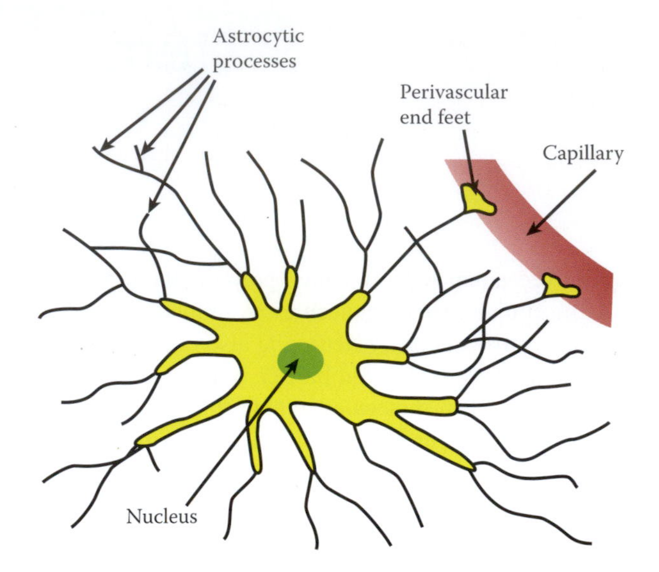

FIGURE 2.2 Cellular characteristics of the astrocytes and perivascular end feet.

invest most of the synaptic neurons and assume a phagocytic function. They also maintain the normal concentration of potassium, which is essential for neuronal activity, by removing and then facilitating its return to the blood. Astrocytes are also considered as the principal glycogen storage site in the CNS. Glycogen breakdown and release of glucose is accomplished by the action of norepinephrine upon β receptor molecules in the astrocytes. During embryonic development, precursors of astrocytes (radial glial cells) guide the migration of the developing neurons. The superficial processes of astrocytes extend to the surfaces of the brain and spinal cord to form the external glial limiting membrane and expansions that attach to the pia mater (pia–glial layer). They play a major role in providing a form of scaffolding, or structural support, on which the neurons and their processes are assembled. Astrocytic processes ensheath the initial segments of axons and the bare segments at the nodes of Ranvier.

Astrocytes are classified into protoplasmic and fibrous astrocytes. Protoplasmic astrocytes are located primarily in the gray matter and have shorter processes, while fibrous astrocytes have longer processes and are located primarily in the white matter. Fibrous astrocytes are the scar-forming cells that bridge the gap between severed ends of axons in pathological conditions involving the CNS. Astrocytes also play a role in the formation of the blood–brain barrier.

> Astrocytoma is the most common form of brain tumor (glioma). This tumor may cause an increase in intracranial pressure, headache, nausea, and vomiting. Supratentorial glioma may produce a shift of the pineal gland, third ventricle, and anterior cerebral arteries, whereas infratentorial tumors are most likely to produce hydrocephalus.

Modified astrocytes are classified into Müller cells and Bergmann glial cells. Müller cells of the retina are elongated cell columns that exhibit expanded foot processes forming the inner limiting membrane of the retina. Bergmann glial cells of the cerebellum lie at the Purkinje layer, sending several processes with short side branches that ascend and envelop the Purkinje cell dendrites.

Oligodendrocytes

Oligodendrocytes, the myelin-forming glial cells in the white and gray matters, are characterized by relatively few branched processes bearing close resemblance to stellate neurons (Figures 2.3, 2.4, and 2.5). Oligodendrocytes that lie within the white matter, especially of fetal brain and in myelin sheath bundles, and are aligned in rows between nerve fibers are known as interfascicular oligodendrocytes (Figure 2.3). They are numerous in the fetus and newborn but rapidly decrease in number (absolutely or relatively) as myelination progresses. Oligodendrocytes that lie closely opposed to neuronal cell bodies in the gray matter are called perineuronal (satellite) oligodendrocytes (Figure 2.4). A few oligodendrocytes that occupy locations near blood capillaries are known as perivascular oligodendrocytes.

Schwann Cells

Schwann cells are the supporting cells in the PNS, which are derived from the neural crest cells, forming the capsular (satellite) cells of the dorsal root and autonomic ganglia.

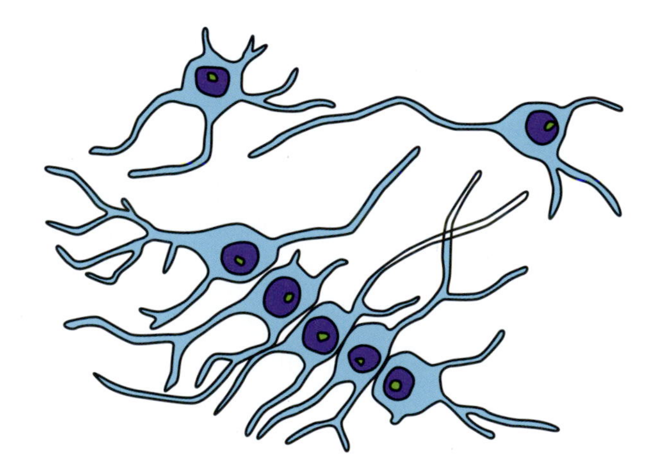

FIGURE 2.3 This is a simplified diagram of the interfascicular oligodendrocyte with its scanty branches.

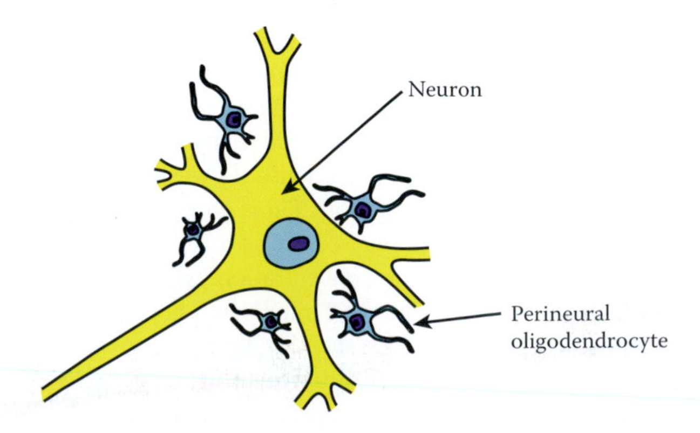

FIGURE 2.4 This drawing shows a perineuronal oligodendrocyte adjacent to a neuron.

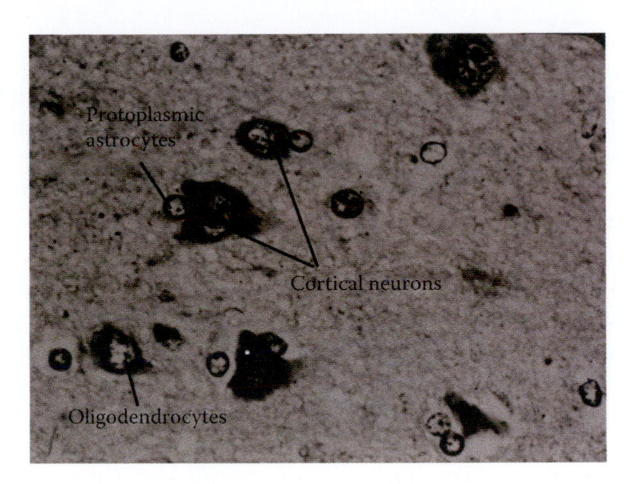

FIGURE 2.5 Photograph of the various types of astrocytes and oligodendrocytes.

Schwannomas and neurofibromas are the most common tumors of the PNS, which are represented in *von Recklinghausen's neurofibromatosis*. This disease exhibits peripheral and central forms. It has an autosomal dominant pattern of inheritance and spontaneous mutations. The gene defect encodes NGF receptor, located on the long arm of chromosome 17. In the peripheral form (type 1), patients exhibit *café-au-lait* flat cutaneous pigmented spots and Lisch nodules in the iris in childhood. Adults with this type of neurofibromatosis show increase in the number and size of the cutaneous spots and also develop neurofibromas. The latter are benign tumors usually located on the skin or in the subcutaneous tissue. Mental retardation, epilepsy, spinal deformities, and tumors such as gliomas and pheochromocytoma may complicate this form of the disease. In the rare central form (type 2), multiple meningiomas and Schwannomas (bilateral acoustic neuromas of the vestibular nerves) occur. It is associated with the loss of specific alleles from chromosome 22. The possibility of involvement of chromosome 17 does exist.

Ependymal Cells

The *ependymal cells* are arranged as a single layer of epithelial-like cells with variable heights in different regions that form the lining of the ventricular system and central canal of the spinal cord. These cells are the remnants of the embryonic neuroepithelium and maintain their original position after the neuroblasts and glioblasts have migrated into the mantle layer. They vary from columnar to cuboidal to squamous, depending upon their location. The variability of their shape often makes their identification difficult. They have processes that penetrate the brain and extend into the pia mater. Embryonic ependymal cells are ciliated, and some adult cells may retain cilia permanently, producing movements of the cerebrospinal fluid. Modified ependymal cells cover the choroid plexus and play a significant role in the secretion of cerebrospinal fluid. Ependymal cells also have numerous microvilli, exhibiting high oxidative activity. Both cellular structure and chemical reactions reflect the secretory and absorptive functions of these cells. The apices of these ependymal cells are joined by junctional complexes, which are not occluding junctions. Since substances can readily pass between these cells, the brain/cerebrospinal fluid (CSF) interface is not a barrier. The deep basal surface of some adult ependymal cells retains processes that extend for a variable distance from the cell body. Many of the shorter processes are intertwined with a heavy concentration of astrocytic processes forming a subependymal (internal limiting) glial membrane. Specialized ependymal cells may be attenuated and form the lining of the circumventricular organs (medial eminence, subfornical organ, subcommissural organ, and the organ vasculosum of the lamina terminalis).

Ependymal cells with long basal processes that project into the perivascular space that surrounds the underlying capillaries are known as tanycytes. Since these capillaries are fenestrated, they do not form a blood–brain barrier, allowing substances to pass from the blood and nervous tissue to the CSF via these specialized ependymal cells (tanycytes). These cells are found around the floor of the third ventricle, and in the lining of the median eminence of the hypothalamus, which suggests a possible role of these cells in the secretions of the adenohypophysis.

Microglia

The microglia are small cells whose primary function is phagocytosis of cellular debris associated with pathological processes in the CNS (Figure 2.6). They probably possess ion channel–linked P_2 purinoceptors, which may be activated by 5′-adeno-sine triphosphate in response to injury. Microglia are considered to be a derivative of the angioblastic mesenchyme. Although the consensus dictates that only mature monocytes enter the postnatal brain, the view that monocytes may differentiate into microglial cells has received a large body of support. Microglial cell migration to the nervous tissue occurs through the walls of the parenchymal and meningeal vessels, accounting for the dramatic increase in their numbers at sites of CNS infections. They undergo transformations in and around the site of infection, which include shortening and thickening of cell processes, increase in the volume of the cell body, and later,

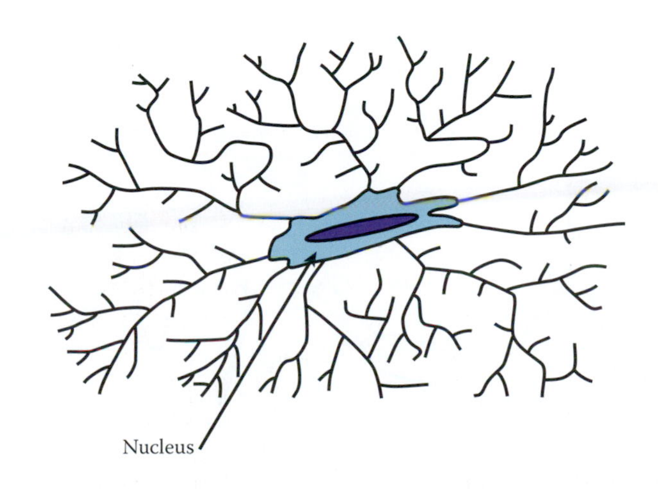

FIGURE 2.6 Schematic drawing of a microglial cell.

retraction of the processes. These changes are followed by complete disappearance of the cellular processes and the conversion of the cell into spherical corpuscular form, known as compound granular corpuscles or gutter cells. Promotion of immune response by the microglia and activated T lymphocytes, which enter the brain by crossing the blood–brain barrier, is evidenced in experimentally induced encephalitis. The small microglial cells are more abundant in the gray matter of the CNS and are also found in the retina.

NEURONS

Neurons form the trophic, genetic, and excitable components of the nervous system, which receive, conduct, and transmit nerve impulses. They detect stimuli and make the appropriate responses. Neurons can be excitatory, inhibitory, sensory, motor, or secretory in function. Constant reduction in the number of neurons after birth and the inability of mature neurons to divide represent some of the main characteristics of these cellular entities. A collection of neurons that serve the same overall function and generally share the same efferent and afferent connections is known as a nucleus within the CNS (e.g., trochlear nucleus) and as a ganglion in the PNS (e.g., dorsal root ganglion). Although many unique features exist that distinguish neurons from each other, there are common characteristics shared by neurons (Figure 2.7).

Soma (Perikaryon)

The soma (perikaryon) represents the expanded receptive zone of the neuron, consisting of an unnmyelinated protoplasmic mass, which surrounds the nucleus. It represents the central machinery for protein synthesis and harbors the organelles needed for the metabolic functions of the neuron. Somata are unmyelinated in humans, have smooth surfaces with the exception of gemmules (postsynaptic projections), and may establish somato-somatic, axosomatic, or dendrosomatic connections. Areas of the perkaryon that do not form synaptic connections are covered by glial processes. Through these diverse modalities of contact, somata produce excitatory or inhibitory action. One (rarely more than one) prominent nucleolus is positioned in the centrally located nucleus. The cytoplasm of the soma contains granular and agranular endoplasmic reticulum as well as free polyribosomes. Polyribosomes coalesce to form visible basophilic RNA-rich masses in the cytoplasm known as chromatin bodies (Nissl bodies). Chromatin bodies are more distinct in highly active spinal α motor neurons and in the large neurons of the dorsal root ganglia. Nissl bodies, which extend along dendrites but not axons, are involved in high cellular activity and protein synthesis. These granules begin to disperse or undergo chromatolysis in response to nerve injury or in degenerative conditions. The anatomic location of the cell body has no

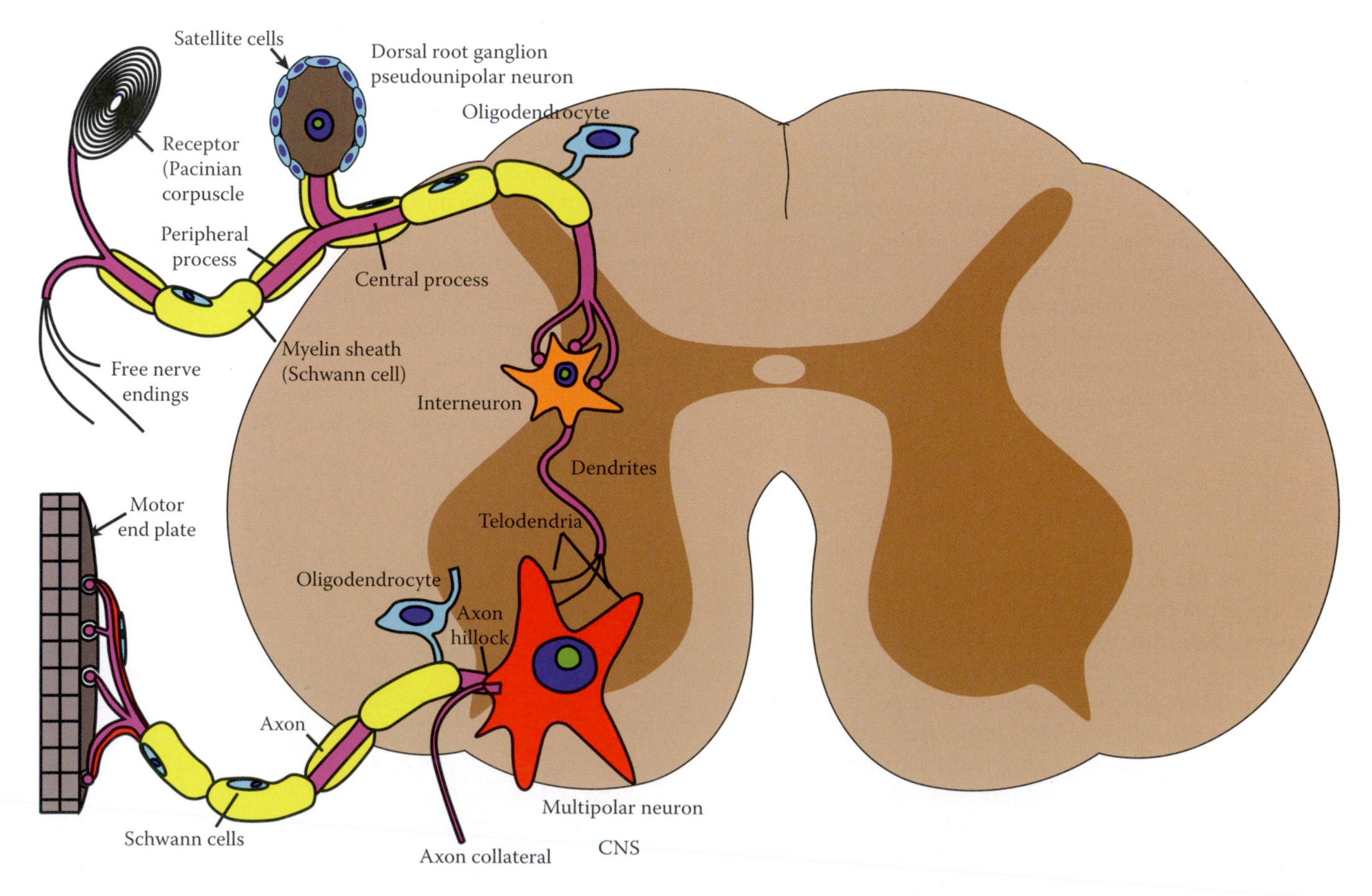

FIGURE 2.7 Various features of neurons and neuroglia cells. Myelin-forming cells within the CNS and PNS, as well as synaptic connections between these neurons, are also shown.

functional significance. The plasma membrane of the soma, although generally smooth, may possess gemmules, which are conspicuous spinous postsynaptic projections. There are numerous enzymes on the surface of the perikaryon that mediate ionic transport, such as adenosine triphosphatase (ATPase) and adenylate cyclase. The soma may engage in axosomatic, dendrosomatic, and soma-somatic synapses. Within the cytoplasm of neuronal soma several cellular structures exist including the neurotubules, neurofilaments, mitochondria, ribosomes, as well as aging pigment, lipofuscin granules (corporal amylacea). The latter consists of lipoprotein and lysosomes and is present in the dorsal root ganglia. Neurotubules are randomly arranged in the perikaryon. At the surface of neuronal soma, various enzymes exist, such as ATPase, which is activated by sodium and potassium. Within the cytoplasm, a relatively large, round nucleus exists with one or more nucleoli. Centrioles that generate and possibly maintain microtubules are practically present in all neurons. Some neurons are rich in pigment, such the substantia nigra and locus ceruleus; others contain zinc (hippocampal gyrus), copper (locus ceruleus), and iron (oculomotor nucleus).

Neuronal Processes

Dendrites

The *dendrites* are branched processes that originate from the soma and form the afferent or receptive zone of neurons. They show a similar pattern of branching in neurons with similar functions guided by the extent of interaction with afferent fibers and the activity of synapses. Functional demand may restrict, enhance, or modify the branching of the dendrites that starts during development. Neuronal plasticity can be essential in determining the overall branching pattern and length of the dendrites which in turn regulate the range and scope of synaptic connectivity. Dendrites have spines that maximize contact with other neurons, mediating excitatory and inhibitory axodendritic as well as dendrodendritic synapses. They contain microfilaments and microtubules, smooth endoplasmic reticulum, ribosomes, and Golgi membrane. In more peripheral dendrites, free ribosomes and rough endoplasmic reticulum become progressively sparse and may be entirely lacking. Microtubules and microfilaments are much more conspicuous in the dendrites than in the soma and more regularly aligned along the axis of the dendrite, forming the most striking feature of the dendrites. The microtubules are believed to be involved in the dendritic transport of proteins and mitochondria from the perikaryon to the distal portions of the dendrites. The dendritic transport, which occurs at a rate of 3 mm/h is comparable to some forms of axoplasmic transport. Destruction of the microtubules by drugs, such as colchicine and vinblastine, inhibits this transport. Dendritic transport may also involve viral glycoprotein that is basolaterally targeted. Dendrites contain exclusively the microtubule-associated protein 2 (MAP-2) but do not contain growth-associated protein 43 (GAP-43). For this very reason, MAP-2 antibodies are utilized in the identification of dendrites via immunocytochemical methods.

Axons

Axons form the efferent portion of the neurons and in general are thinner than dendrites, assuming considerable length. Compared to dendrites, axons are more uniform and contain fewer microtubules and more microfilaments but no ribosomes. Axons are longer than dendrites and may measure up to 6 feet in length, beginning from the axon hillock and giving rise to collaterals that terminate as the telodendria. They provide an avenue for transport of substances to and from the soma. Axons originate from the soma or, less frequently, from the proximal part of dendrites. The axon is divisible into axon hillock, initial segment, axon proper, and the telodendria (axonal terminal). A clearly recognizable elevation, the axon hillock, continues with the soma. The relative absence of free ribosomes and rough endoplasmic reticulum is the most obvious feature of the axon hillock. In myelinated axons, the initial segment extends from the axon hillock to the beginning of the myelin sheath. This segment is unmyelinated, maintains inhibitory axo-axonal synapses, and contains some microtubules, neurofilaments, and mitochondria but lacks rough endoplasmic reticulum. The neurotubules and neurofilaments are gathered into small parallel bundles, connected by electron-dense crossbridges. Here at the initial segment, the axolemma (plasma membrane bounding the axon) is lined by a dense core consisting of spectrin and F-actin, allowing voltage-sensitive channels to attach to the plasmalemma. Each myelin segment is separated from the neighboring node along the length of the axon by nodes of Ranvier. These nodes, where axonal branches arise, contain sodium and possibly potassium channels. Axonal terminals are initially myelinated, but as they repeatedly branch, the myelin sheath will disappear. This will enable terminals to establish synaptic contacts with axons, dendrites, neurons in the CNS, or muscle fibers and glands in the PNS. The endings are characterized by tiny swellings known as terminal boutons. Microtubule-associated proteins (MAPs) such as tau interconnect axonal microtubules. Within the axons, microtubules, neurofilaments, lysosomes, and mitochondria are located. Microtubules have polar ends (+ and −); the (+) ends are directed away from the perikaryon, containing kinesin-coated organelles essential for axonal growth. On the (−) ends dynein-coated organelles are located. Kinesin and dynein bind to membrane receptors.

Neurofilaments are usually found in association with microtubules, as constant components of axons. In the growth cones of the developing axons, filamentous structures finer than neurofilaments exist and are known as microfilaments. These actin filamentous structures facilitate growth and movement and can be inhibited by chemical agents that depolymerize actin. Neurofilaments within the regenerating axons contain a calmodulin-binding membrane-associated phosphoprotein and GAP-43 protein, which may be used as markers to identify these axons.

Proteins, neurotransmitters, mitochondria, and other cellular structures synthesized in the soma or proximal portion of the dendrites are transported to the axon and axon terminals via a process known as axoplasmic transport. This transport may occur in a distal (anterograde) direction toward the axon terminals, while allowing other substances to be transported in the reverse (retrograde) direction from the axon toward the cell body. Axoplasmic transport within the microtubules may be maintained utilizing the proteins dynein and kinesin. This process may involve fast, intermediate, and slow phases.

The fast phase of axoplasmic transport includes transport of selected proteins (e.g., molecules carried by the hypothalamo-hypophyseal tract), vesicles, membrane lipids, or enzymes that act on transmitters. This phase of the transport occurs at a speed of 100–400 mm/day, both in anterograde and retrograde directions, utilizing smooth endoplasmic reticulum and microtubules. The retrograde component of this phase is formed by the degraded structures within the lysosomes and may contain neurotropic viruses such as rabies and herpes simplex. The fast phase is energy dependent and can be inhibited by colchicine, hypoxia, and the inhibitors of oxidative phosphorylation, glycolysis, and the citric acid cycle. It has been suggested that proteins that follow the fast axonal transport must either pass through the Golgi complex or join proteins that do so utilizing clathrin-coated vesicular protein. Activation of kinesin or dynesin can determine the direction of the fast phase of transport. In the *intermediate phase*, mitochondrial proteins are transmitted at a rate of 15–50 mm/day. The *slow phase* of the transport utilizes microtubules, microfilaments, neurofilament proteins, mitochondria, lysosomes, and vesicles, proceeding in the anterograde direction only, at a speed of 0.1–3 mm/day. This phase carries 80% of the substances carried by axoplasmic transport, providing nutrients to the regenerating and mature neurons. The slowest phase deals with the transportation of triplet proteins of tubulin and neurofilaments.

An axon may be myelinated or unmyelinated and ends in the synaptic terminals. Myelinated axons have faster conduction velocity of the impulses generated.

MYELIN

Myelin is formed by the Schwann cells in the PNS and by the oligodendrocytes in the CNS. This is an insulating complex cover of cell membrane with a unique ultrastructural form that

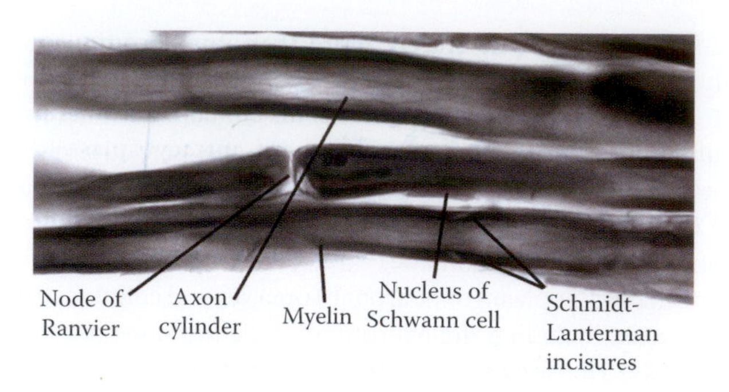

FIGURE 2.9 This is a longitudinal section of a single myelinated nerve fiber.

encircles axons and is composed of two-thirds lipid and one-third protein (Figures 2.8 and 2.9). Categorization of the membrane proteins found in the myelin is based on their locations and hydrophobic characteristics. Myelin basic proteins consist of P_1 and P_2 and are located on the cytoplasmic surface of the membrane. P_1 contributes to the PNS but more to CNS myelin sheath (30% of myelin protein) and is believed to function together with proteolipid protein (PLP) in the CNS, whereas P_2 is a predominant protein in the PNS and is believed to be associated with multiple sclerosis (MS). P_1 exhibits various forms due to changes induced by posttranslational phosphorylation and methylation as well as by messenger RNA (mRNA) splicing. P_1 has been utilized as an antigen in certain CNS demyelinating diseases. P_2 is utilized as an antigen in the experimental development of neuritis seen in Guillain–Barré syndrome (GBS). P_2, as discussed earlier, is present in the PNS in high concentration relative to low concentration in the CNS. It is a fatty acid–binding protein possibly also linked to GBS.

The myelin protein P_0 is a membrane protein that belongs to the immunoglobulin supergene group and is closely associated with myelin-associated protein (MAG). P_0, which forms nearly half of the total protein in the PNS, is glycosylated, in contrast to PLP in the CNS myelin, which is not glycosylated. It functions as a structural equivalent of PLP and its variant Dalton myelin proteolipid protein (DM20) in the CNS. P_0 shows mutation in Charcot–Marie–Tooth neuropathy.

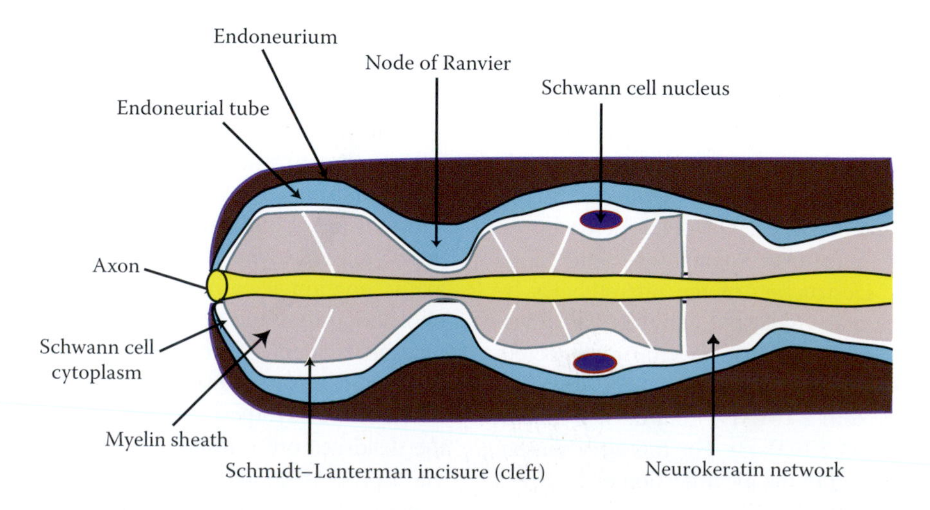

FIGURE 2.8 Myelin sheath of a peripheral nerve, nodes of Ranvier, and associated coverings.

PNS contains, though in small percentage, peripheral myelin protein (PMP22), which is also glycosylated and may act to limit the growth in Schwann cells. MAG, a component of the myelin protein in both CNS and PNS, is found at the mesoaxon, paranodal incisures, and loops and the inner periaxonal membrane where compaction is initiated. MAG's close proximity to F-actin and spectrin appear to support its role in membrane adhesion. Similarly, 2′,3′cyclic nucleotide 3′phosphodiesterase (CNP), which shows more presence in the CNS than the PNS, also co-localizes with MAG at the paranodal regions and incisures and may play a role in the incorporation of P_1. Myelin proteins are extractable with various organic solvents. CNS myelin exhibits tight junction complexes through the intermodal myelin segment.

The lipid portion is primarily cholesterol (40%) in a free form and phospholipids (mostly sphingolipid), which are the phosphate esters of glycerol or sphingosine or their derivatives (glycosphingolipids). Interestingly, the macroscopic difference between gray and white matter is attributed to the lipid content of the myelin. Sphingomyelin consists mainly of C22 and C24 fatty acid residues. The major glycolipid is galactocerebroside, and it represents the main component of the myelin and is characteristically high. Minor lipid species also exist, such as galactosyl glycerides, phosphoinositides, and gangliosides. Gangliosides that contain sialic acid (*N*-acetylneuraminic acid) form 1% of myelin lipid. PNS myelin is LM1 (sialosylparagloboside), whereas GM-4 (sialosylgalactosylceramide-4) is the main ganglioside of CNS myelin. Both PNS and CNS myelin contain, although in low concentration, acidic glycolipids that serve as antigens in myelin. Myelin lipid contains different fatty acid moieties, predominantly those with even numbers. Longer chains of fatty acids and unsaturated residues primarily exist in the myelin of older populations, and their proportion increases with maturation. Changes in the fatty acid composition in Resum's disease and disorders of the oxidation of long-chain fatty acid in adrenoleukodystrophy can cause severe demyelination. Because of the significant presence of lipid in the myelin, metabolic disorders of lipid seen in inherited conditions, such as Neimann–Pick disease, metachromatic leukodystrophy, and Krabbe's disease, are commonly associated with demyelination.

Myelin allows for substances to be transported between the axon and the myelin-forming cells (Schwann cells or oligodendrocytes). It maintains high-velocity saltatory nerve conduction, a mode of conduction that proceeds from one node of Ranvier to another in a faster and more energy-efficient way. Myelin is not a continuous covering but, rather, a series of segments interrupted by nodes of Ranvier. In the PNS, each internodal segment represents the territory of one Schwann cell. These nodes are sites of axonal collaterals and bare areas for ion transfer to and from the extracellular space. Extensions of the myelin on both sides of a node of Ranvier are known as paranodal bulbs. These myelin bulbs may lose contact with the axon and undergo degeneration as a result of crush injury. Interruptions within successive layers of myelin are known as Schmidt–Lanterman incisures. Myelin is formed by the oligodendrocytes or Schwann cells during the fourth month of fetal life, and continues into postnatal life.

Myelination is initiated near the soma of neurons and continues toward the axon terminals. It does not cover the axon hillock, dendrites, or axonal terminals. For the myelination process to begin and, later, be maintained, adhesion/recognition molecules, such as MAG and P_0, must come to action. It is partly determined by the diameter of the axon and occurs in axons that range between 1 μm in the CNS and 1.5 μm in the PNS.

The first step of this process involves surrounding the axon with cytoplasmic membranes of Schwann cells or oligodendrocytes that are detached initially but later fuse together. The double layer of the Schwann cell plasma membrane that wraps the axons forms the meson, which elongates and differentiates into inner and outer parts. It has been suggested that myelination involves deposition of P_0, MAG, and PLP into the lamellae after their transportation as vesicles to the membrane protein at the mesoaxon and also the incorporation of soluble proteins such as myelin basic protein (MBP) at the paranodal regions.

In the process of myelination, several layers of cell membranes surround a given axon in a tight spiral manner, separated by the cytoplasm. The presence of actin and tubulin at the paranodal region and the Schmidt–Lanterman incisures and their contractile effect may play a role in the coiled arrangement of the myelin. Since myelin formation occurs at a particular site, elongation of the axon requires successive layers of myelin to stretch and cover a larger area of the axon. This results in more layers being concentrated near the center of the internode. When the cytoplasmic and external surfaces of cell membranes come into apposition upon receding of the cytoplasm, they form continuous major and minor dense lines, respectively. The minor dense line, also known as the intraperiod line, contains a gap that allows extracellular space to continue with the periaxonal space. This intraperiod gap allows metabolic exchange and serves to accommodate the increasing thickness of the axon by allowing lamellae to slip on one another and thus reduce their numbers. The thickness of myelin is paralleled to an extent by an increase in the diameter of the axon. Myelination is a sporadic process that does not follow a uniform pattern in early postnatal and late fetal life.

In contrast, oligodendrocytes, the myelin-forming cells in the CNS, are associated with more than one axon and with more than one internodal segment (roughly 15–50 internodes). Unlike that in the PNS, the elongation of an axon to an intended site precedes the movement of the originators of the oligondendrocytes. Thus, axonal contact, upregulation of transcription of myelin protein genes, elevation of cyclic adenosine monophosphate (cAMP), and downregulation of suppressor genes appear to be interrelated essential work in cohort in the myelination process of PNS axons. This issue of axonal contact is of no value in the myelination process of the CNS axons, as axonal activity seems to play a more significant role in the proliferation and survival of the oligodendrocytes.

Thus, the multiple associations are maintained by extension of the oligodendrocytes around each axon. Myelination in the CNS begins initially with the vestibular and spinocerebellar tracts. The corticospinal tract and dorsal white column pathways may not be completely myelinated at birth. It should also be remembered that axonal growth and elongation to a destination generally occur before the migration of oligodendrocytes and formation of myelin.

Unmyelinated axons in the CNS lack any form of ensheathment, whereas unmyelinated axons of the PNS are enveloped by Schwann cell cytoplasm. Peripheral axons are lodged in sulci along the surface of Schwann cells. Some Schwann cells in the PNS may encase more than 20 axons through the multiple grooves on their surfaces.

Demyelination may be a primary or secondary process. Primary demyelination affects the thickly myelinated motor fibers and is associated with intact axons, as in MS and myelinopathy that affects the thickly myelinated motor fibers of the lower extremity and spares the small sensory fibers. Demyelination secondary to destruction of the axon may be seen in storage diseases and Wallerian degeneration. Incomplete myelination (hypomyelination) occurs in maple syrup urine disease and in phenylketonuria (PKU).

NEURONAL DEGENERATION

Antergrade Degeneration

Antergrade (Wallerian) degeneration occurs as early as 12–48 hours after the insult and includes axonal fragmentation, retraction and disintegration of the myelin distal to the site of injury, as well as dispersion of the mitochondria and neurofilaments (Figure 2.12). Myelin breakdown passes through a beading stage followed by fragmentation into droplets of lipid and lamellar debris. At the same time, hematogenously derived cells begin to invade the endoneurium. Cellular debris is later absorbed by the macrophages in the PNS and by the microglia, macrophages, and astrocytes in the CNS. Macrophages that enter the endoneurium may induce secretion of NGF from the Schwann cells by releasing interleukin. Macrophages usually disappear from the endoneurium, though foamy macrophages may persist for a long period of time. Absorption of cellular debris and axonal disintegration is aided by the hydrolytic enzymes of the lysosomes and axonal protease such as phospholipases, which are Ca^{2+} dependent. Possible absence of these enzymes can impair axonal degradation and eventual absorption. Axonal propagation and overgrowth at the proximal and distal stumps occur after gliosis by the Schwannn cells. After few weeks, the cytoplasm of Schwann cells forms guidance tunnels, which are tubes within the basal lamina that continue along the course of the damaged axons and guide the regenerating axons. These tubes or bands of Bűgner can persist long enough to guide the regrowth of the axon or may be replaced by the endoneurium. Degenerative changes occur in the terminals earlier if the site of the lesion is adjacent to the synapse.

Retrograde Degeneration

Retrograde (indirect Wallerian) degeneration is seen in the PNS and CNS, although it does not occur in all neurons. In this type of degeneration, proximal disintegration of the axonal cytoskeleton and breakup of the myelin sheath occur. These changes are accompanied by sealing off of the severed ends by the axolemma, preventing leakage of axoplasm. As the dendrites retract from their synaptic contacts, retraction bulbs are formed at the swollen severed ends of the axons. The soma undergoes chromatolysis, where the Nissl bodies break up near the axon hillock, followed by dissolution of the cytoplasm within 3 days. Swelling of the soma is accompanied by deviation of the nucleolus into a peripheral eccentric position. Dispersion of the Golgi apparatus is accompanied by an increase in number of lysosomes, mitochondria, granular endoplasmic reticulum, and free ribosomes. Fully developed neurons may resist the process, allowing slow atrophic changes to occur, whereas poorly developed or immature neurons may die quickly. If regeneration fails, death of the affected neuron can eventually occur, a most probable outcome in CNS injuries.

Transynaptic Degeneration

The *transneuronal (transynaptic) degeneration*, on the other hand, occurs in neurons that provide the sole afferents to a neuron that receives axonal injury. It may also be seen in neurons that originally received input from the injured axon. These reactions, which are manifestations of disuse atrophy, extend slowly beyond the synaptic cleft to the adjacent neurons. Thus, the neuronal changes across the synaptic cleft are the result of lack of trophic substances provided to the adjacent neurons by the damaged neurons.

Contrary to the past established belief that the CNS is a passive entity that lacks the capacity to rejuvenate following axonal injury, it is now evident that in order to restore neuronal function, new synapses, though irregular and less efficient, come to existence following a traumatic experience. It has been shown that these new synaptic connections are formed by other afferents to the denervated site. In order to expand CNS capability to recover from traumatic impact, embryonic neuronal transplant has been introduced. Studies conducted on neuronal transplantation have demonstrated that embryonic tissue with similar characteristics to the affected neurons or genetically modified tissue can establish synaptic linkage with the damaged host neurons. Replacement of the degenerated dopamine-secreting nigral neurons with similar tissue of embryonic origin has had variable success. Axonal interruption in the CNS causes the loss of capacity to regenerate except when a peripheral nerve segment is transplanted to guide and bridge the gap between severed segments of the axon for a limited regenerative process. In this process, the transplanted astrocytes and Schwann glial cells enter and follow the fiber tract of the affected axons. The role of the perivascular microglia is expressing major histocompatibility complex (MHC) I and II antigen and secreting cytokines that, in turn, induce luminal adhesion molecules of the lymphocytes and graft cells to express MHC I antigens. Further stimulation of the perivascular microglia is provided by diapedetic lymphocytes through lymphocyte factors such as gamma interferon (IFNγ). This feedback activation perpetuates the activation process to involve the entire transplanted tissue and is eventually followed by the introduction of MHC II, leukocyte function–associated antigen-1 (LFA-I), and cluster designation-4 (CD4), which express dendritic cells in the perivascular space. Despite all the accumulated data from experiments, there remains a much greater challenge in the ability to pattern the embryonic axonal development after that of the adult, which appears to be significantly different with regard to patterning, direction, and presence of certain contiguous glial cells. Further, CNS immune response to transplanted embryonic tissue is not fully understood, and there is the apprehension that major degenerative disease may ensue as a result of introduction of antigen through transplanted tissue.

FUNCTIONAL AND CLINICAL CONSIDERATION

Guillain–Barré syndrome (GBS) is the most common acute polyneuropathy that affects the peripheral nervous system, producing typically ascending paresis with loss of reflexes. Sensory and autonomic disorders may also occur. In most patients, recovery is achieved with no complications. However, relapses and mortality can occur as a result of pulmonary embolism and pneumonia. Duration and severity of the condition correlate with the protracted nature of the manifestations seen in some patients. In majority of patients, this condition is preceded by a recent infection usually involving the respiratory tract such as mononucleosis. This disease can develop after surgical operations of the abdomen and thorax. Patients most often exhibit symmetric paresis in the lower extremities but rarely in the foot. Paresis may take days to weeks to develop, followed by worsening of symptoms before it stabilizes at. Patients also exhibit symmetric facial weakness that usually accompanies a severe form of quadriparesis but with preserved masticatory muscle function. Unusually severe facial paresis relative to mild extremity paresis distinguishes GBS from myasthenia gravis. Urinary retention may also occur with sacral paresthesia.

Variations of this disease are not uncommon and can be in the form polyneuropathy with ataxia and ocular disorders (gaze palsies and loss of light-blink reflexes), or it takes a descending form of paresis starting with ocular, facial, and pharyngeal paresis (Miller Fisher variant of GBS). In the latter, variant the triad of ophthalmoplegia, ataxia, and areflexia is seen. Mild sensory signs that are not persistent with glove-and-stocking neuropathy can be seen as severe initial manifestations of this disease. Pain, felt as aching and soreness in the muscles, may occur without other sensory deficits, is invariably experienced at night, and precedes muscle paresis. Ataxia may develop subsequent to loss of position sense. Glomerulonephritis with proteinuria and pseudotumor cerebri with elevated CSF protein and pressure as well as elevated venous pressure have also been seen in some patients with GBS. Autonomic manifestations such as consistent nonfluctuating tachycardia, excessive bronchial secretion, lack of tear secretion, nausea, phtophobia, chronic constipation, and hypohidrosis are seen in GBS due to possible histamine release, a fact that must be considered in the administration of succinylcholine. However, ileus and paroxysmal orthostatic hypotension are seen more frequently in patients with breathing disorders and pronounced paralysis. Bradycardia and loss of accommodation are seen in botulism, distinguishing it from GBS. An autoimmune variant of GBS, an acute motor axonal neuropathy (AMAN), prevalent in Mexico and China, is associated with the presence of anti-ganglioside3 (GD3) antibodies. Remitting–relapsing signs of upper motor neuron palsy (Babinski sign, hyperreflexia, spasticity), changes in consciousness, ataxia, and ophthalmoplegia are seen in a rare variant of GBS (Bickerstaff's brainstem encephalitis). This variant is associated with lesions in the brainstem.

Diabetic neuropathy is associated with axonal damage and segmental demyelination. This predominantly sensory polyneuropathy is usually seen in the form of a stocking–glove distribution. The onset may be insidious or sudden, and patients exhibit numbness and tingling in the lower extremities particularly the toes, but severe disabling pain is less frequently encountered. Dysesthesia (paresthesia induced by contact) and spontaneously felt paresthesia in the lower extremities are common deficits in this condition. Decreased sense of proprioception, ankle hyporeflexia, and muscular paresis are also seen depending on the extent of axonal damage. However, diabetics are resistant to sensory and motor deficits induced by ischemia. Decreased vibratory sense due to degeneration of the thickly myelinated nerve fibers can lead to sensory (pseudotabetic) ataxia.

Sensory deficits may be extreme and range from tearing or burning pain and unusual sensitivity to tactile sensation to reduction or insensitivity to nociceptive stimuli. Autonomic dysfunctions also occur but are not common, producing postural hypotension, bradycardia on standing, dumping sign, bladder dysfunction, impotence, and night diarrhea. Mononeuropathy of the oculomotor, trochlear, or abducens as well as femoral or sciatic nerves is seen commonly in older populations with diabetes. It has been demonstrated that segmental demyelination occurs as an independent process not directly related to axonal disintegration and is accompanied by loss of unmyelinated fibers. Microvascular disorders of the vasa nervorum including endoneurial metabolic derangement that affects the permeability at the blood–nerve–nerve barrier can occur subsequent to thickening of the endoneurial capillary wall and may account for focal demyelination in diabetics. Diabetic neuropathy produces impaired nerve conduction by altering the neural metabolism of sorbitol, *myo*-inositol, and sodium–potassium ATPase.

Multiple sclerosis (MS) is the most common demyelinating disease of the CNS, which leaves the axons relatively preserved. Despite the unknown etiology of this disease, epidemiological, genetic, immunological, and viral factors have been implicated. Decreased suppressor T lymphocytes and increased frequency of a particular kind of human leukocyte antigen (HLA) are considered some of the immunological abnormalities associated with this disease. Some investigators claim that contraction of measles at an older age and the presence of high measles antibody titer in the CSF may account for the development of this dreadful disease. Europeans and inhabitants of higher altitudes who have lived at least through the age of 15 in cool northern latitudes (above the 37th parallel) may show greater incidence of this disease. It is very rare in Latin America, Japan, and central Africa. Whites are twice as likely to get the disease than individuals with darker complexion. This disease is more common in females and shows higher incidence in first-degree relatives and monozygotic twins. The mean age of onset is 33 years, with virtually all cases developing between 15 and 50 years. Although rarely advancing from the onset, this disease produces slowly progressive neurological disorders characterized by relapses and remissions.

The severity of MS increases with time, although improvement accompanied by remission is common. In general, 2–3 years pass before remission. Because of the variable course, manifestations range from benign to potentially fatal episodes. Initial symptoms and signs of this disease are intensified by fever and emotional stress and following infection, trauma, or childbirth. Demyelination affects the structures within the CNS (Figure 2.10), with a predilection for the optic nerves, spinal cord, periventricular area, brainstem, and cerebellum. Demeylination, which shows distribution in the periventricular and paraventricular areas, is accompanied by pathologic processes of the glia and vasculature. Demyelination of the optic nerve causes acute optic neuritis and subsequent blurred vision and central scotoma progressing to partial or total blindness. This may be associated with myelopathy in Devic syndrome. Demyelination of the medial longitudinal fasciculus produces internuclear ophthalmoplegia, a common manifestation characterized by ipsilateral paresis of the ocular adduction and contralateral nystagmus. It also affects the dorsal (posterior) columns and the motor pathways of the spinal cord, producing ataxia and spastic paralysis. As a result, patients walk with broad-based gait and sway back and forth and to the sides. Corticospinal involvement often causes weakness and spasticity along with other signs of upper motor neuron syndromes. Lesions of the anterolateral systems are often symmetrical and cause paresthesia (pins and needles and tingling sensation) sequentially in the digits, limbs, and adjacent parts of the trunk. This type symmetric paresthesia is often confused with polyneuropathy. Selective destruction of the lateral spinothalamic tract may account for the loss of pain and temperature sensations in acute cases. An electric-shock–like sensation that radiates down the back and upper extremity upon flexion of the neck (Lhermitte sign) is seen in the early stage of the disease. This sign can also be seen in spinal stenosis, subacute combined system degeneration, and mass lesions. Spinal cord lesions can also result, though rarely, in impotence and bladder dysfunction. Men may experience premature or retrograde ejaculation. Locus minoris resistantia, the tendency for relapses in an area of previous activity, may also be seen in this condition. Cerebellar involvement produces disturbances in the finger-to-nose test (inability to point the finger accurately with the eye closed) and heel-to-shin test as well as intention tremor (unsteady grip and tremor during activity).

The combination of proprioceptive sensory loss, signs of upper motor neuron palsy and cerebellar dysfunction, disorders of eye movement (nystagmus), and history of visual deficits are all considered diagnostic for this disease. Clinicians consider dysarthria, nystagmus, and tremor (Charcot triad) as the cardinal signs of this disease. Depression is common in the initial stage and during remission of this disease. Some exhibit euphoria as a sign of relief when the attack subsides; others may live in psychological denial. Overt intellectual impairment is a late sign of this disease due to the fact that the gray matter of the cortex is spared in this disease and demyelination has to be extensive enough to impede the normal cerebral intellectual process. This fact distinguishes MS from Alzheimer's disease and manifestations of cerebrovascular accidents.

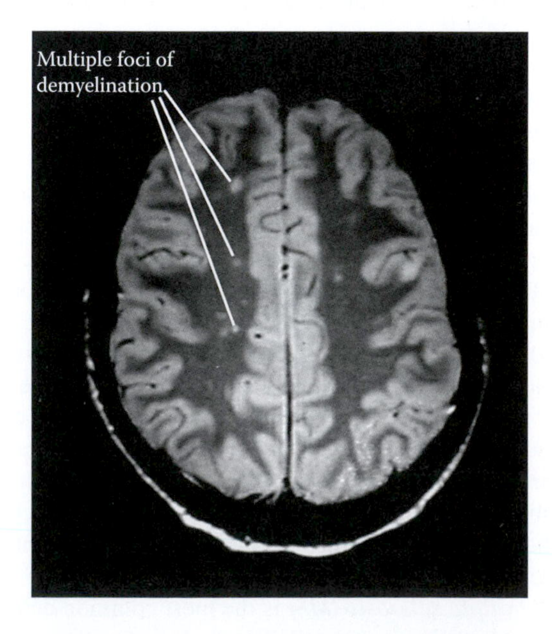

FIGURE 2.10 This is a magnetic resonance image of a patient with multiple sclerosis showing multiple foci of demyelination.

Neuroimaging is the most useful supportive tool in diagnosis, together with clinical history, physical findings, and CSF values. CSF cell count and glucose level remain within normal limits; however immunoglobulinG (IgG) shows an increase in MS patients. Myelin basic protein, which is seen in the CSF of MS patients, correlates with the extent of demyelination. Magnetic Resonance Imaging (MRI) shows lesions in the periventricular area, pons, medulla, and the corpus callosum. T2-weighted and fluid-attenuated inversion recovery (FLAIR) images with contrast T1 images are very useful in demonstrating sites of demyelination. Poor vision can be ascertained by visual evoked responses that show prolonged latency. Brainstem auditory evoked potentials and somatosensory evoked potentials also exhibit abnormality in this disease. A minimum of two episodes of neurologic disorder or abnormal exam results are consistent findings with MS.

IFNβ-1b, adrenocorticotropic hormone (ACTH), cyclosporine, azathioprine, mitaxantrone, cyclophosphamide, methylprednisone or other steroid medications combined with physical therapy have proved to be beneficial. This disease should be differentiated from GBS, which produces demyelination in the PNS, affecting young and middle-aged individuals. MS may mimic signs of brainstem astrocytoma, neurologic abnormalities of acquired immune deficiency syndrome (AIDS); systemic lupus (which exhibits seizures, stroke, and psychosis); as well as combined system disease (vitamin B_{12} deficiency).

REGENERATION

Regeneration refers to the ability of a neuron to restore function following a traumatic injury. In the PNS, regeneration does occur but is influenced by factors such as the site (the degree of regeneration is inversely proportional to the length of the axon), the size of the gap formed between parts of the severed axon, the presence of infection and/or foreign body contamination, temporal proximity of the injury, the type of injury, and whether or not tissue is lost. Since growth cones (growth projections from the severed axon terminal) fail to properly align with the path of the axon that has undergone a transection injury, regeneration is more difficult than in crushing injuries in the endoneurium, where Schwann cells remain intact. It is noteworthy to add that regeneration is more likely if the site of injury is closer to the target site. The rate of growth of the regenerating axon varies, generally ranging between 3 and 4 mm/day in primates. Signs of regeneration start with the formation of growth cones in the distal end of the proximal segment of the severed axon. These growth extensions, which develop during the first week after the nerve injury, reach the distal segment through guidance tunnels formed by the Schwann cells. This bridging may require 3–4 weeks to occur after a week of latency. These changes are later followed by reconnection with the appropriate target; maturation, which requires recognition; establishment of a functional synapse; and myelination as well as increased thickness of the axons. However, neuromas and associated agonizing pain may develop at the ends of the sprouting axons if the distance is long enough not to allow complete approximation of the distal and proximal segments.

Continuity between the severed axonal ends must be maintained with minimal gap in order for regenerating axons to enter the bands of Bünger. Presence of a large intra-axonal gap prevents proper axonal regeneration from the proximal toward the distal stumps, consequently rendering the bands of Bünger functionless, leading to the arrest of Schwan cell proliferation. Axonal budding in peripheral nerves occurs when some fibers within the nerve trunk are damaged while the remaining fibers are intact. These buds will extend into areas originally innervated by the injured fibers and will restore their function. It should be noted that during this regenerative process and few days after the traumatic injury, a dramatic increase in the Schwann cells due to proliferation of endogenous Schwann cells and exogenous myelomonocytic cells occurs. This is possibly in response to mitogens generated by cellular debris, transforming growth factor $beta_1$ ($TGF\beta_1$), fibronectin, and glial growth factor (GGF). The latter acts through tyrosine kinase receptor encoded by proto-oncogene *neu*ERBB2, located on the long arm of chromosome 17. The increase in ERBB2 and GGF after injury can be correlated with their role in the regenerative process. The increase in the expression of neurotrophic substances such as NGF, brain-derived neurotrophic factor (BDNF), and ciliary neurotrophic factor (CNTF) provided by Schwann cells in injured axons is significant. A notable change in the Schwan cells after injury is the molecular transformation and acquiring of the characteristics of the nonmyleinating cells including the upregulation of NGF, neural cell adhesion molecule (N-CAM) genes, and glial acidic fibrillary protein (GFAP) and downregulation of the expression of P_0, P_1, and P_2. Neurotrophic factors released from Schwann cells of the distal stump may enable the regenerating axons from the proximal stump to cross the narrow gap as they grow. Following axonal injury, upregulation of expression of NGF and BDNF, particularly the latter, becomes significant. This change is not confined to these factors; in fact, nerve growth factor receptor (NGF-R), GAP-43, GFAP, and N-CAM genes also show remarkable upregulation in the Schwann cells. This in contrast to CNTF, also secreted by Schwann cells, which follows a reverse pattern, showing a noticeable decrease. This genetic and molecular upregulation combined with the neurotrophic factors shows the important role Schwann cells play in guiding and maintaining axonal growth. In the CNS, regeneration will not become possible, due to the lack of the environment that facilitates this process.

Regeneration is very limited in the CNS, and true growth is almost impossible due to the fact that bands of Bünger are not formed by myelin-forming oligodendrocytes. Additionally, scar and necrotic tissue from trauma or infection may impede the repair process. Growth of axons does not follow a particular pattern to reestablish the connection; thus, functional restitution becomes unattainable. The most significant goal of modern rehabilitative medicine is to prevent atrophy of the muscles in individuals with motor neuron diseases. One of the means to achieve this end is to apply electrical stimulation to the affected muscles, preventing denervation hypersensitivity and reducing atrophy.

Although complete ideal regeneration is not possible, the return of sensations in the PNS is affected by different factors. There is a considerable difference between the rate of regeneration at the proximal and distal stumps of the severed axons. In the proximal stump, the regeneration is around 60 μm per hour and 160 μm daily. Complete recovery of functional properties of a severed axon is proportional to the length of the interstump gap and type of injury. The longer the interstump distance, the longer the recovery time for reestablishing functional recovery. This gap between the severed ends is minimal in incisional injury but becomes extensive in laceration, and deep burns, ischemic wounds, and injuries occurred as a result of stretch, although stretch of up to 6% of the original length can be tolerated. Location of the injury and type of nerve affected are additional factors that affect the outcome of regeneration. It has been reported that recovery of sensation and motor function may vary according to the type of modality and Tinel sign, which is characterized by paresthesia and electrical sensation that radiate in the area of distribution of the tapped nerve, and can serve as an indicator of sensory fiber regeneration but not motor. Paresthesia can be elicited at the site of initially injury and further distally as the axonal sprouting continues. The most distal point where paresthesia is elicited can be considered as the site of most active regenerative process, indicating that the unmyelinated regenerating axons are particularly sensitive to mechanical pressure. However, combination of sensory loss and negative Tinel sign months after axonal interruption may indicate that the regenerative process is ceased or did not occur at all. False-positive results are seen where regeneration could not have occurred or progressed to produce the sign. Recovery of all sensory modalities may not always be possible, and variations do occur when a large nerve trunk is transected. Perception of a single stimulation as dual sensations at different cutaneous sites, disruption of stereognosis (tactile gnosia), and two-point discrimination are observed. The younger the patient is, the better the chance of restoration of sensory modalities. Anatomic variations, overlap in the cutaneous zones of innervation between adjacent nerves, changes in the cortically sensory neurons that are related to the denervated areas, and delegation of sensory conduction to intact fibers within the same nerve trunk may account for these variable regenerative outcomes. Motor functional recovery also shows variations relative to the type of injury. Postrestoration dominance of a muscle's or a group of muscles' activity and involuntary movements occurring in conjunction with purposeful motor activity have been seen in injuries of the trunks or cords of the brachial plexus. Restoration of functions after brachial plexus upper trunk injury may result in the extension of the hand at the wrist, flexion of the forearm at the elbow, and abduction of the arm at the shoulder, resembling the upper extremity of a trumpet player. This is explained on the basis of a defective regenerative process, multiple segmental contributions within the affected trunk or cord that innervate the agonist and antagonist muscles, and the extent of damage to the fibers that perform a particular function.

Neurons are classified according to the chemical nature of the neurotransmitter that they release into cholinergic, adrenergic, noradrenergic, dopaminergic, serotoninergic, GABAergic neurons, and so forth. Cholinergic neurons release acetylcholine (Ach) and are commonly found at neuromuscular junctions. Noradrenergic neurons are abundant in the sympathetic ganglia and the reticular formation, whereas adrenergic neurons are found in the adrenal medulla and within the synaptic dense-cored vesicles. Dopaminergic neurons are present mainly in the substantia nigra, corpus striatum, and cerebral cortex, while serotoninergic neurons occur in the raphe nuclei and in the rounded synaptic vesicles. GABAergic neurons are present in the cerebellar cortex and spinal cord. Neurons may also be classified into pseudounipolar, bipolar, and multipolar neurons.

DEMYELINATING METABOLIC DISORDERS

In certain diseases such as Refsum's disease and metachromatic leukodystrophy (MLD), impairment of α-oxidation and accumulation of phytanic acid lead to demyelination and production of easily degradable abnormal myelin. Demyelination also occurs in acquired neurometabolic disease (Korsakoff–Wernicke syndrome), due to thiamine deficiency, and in lipid storage (lysosomal) diseases including Gaucher's disease, globoid cell leukodystrophy, Fabry's disease, Neimann–Pick disease, and Tay–Sachs disease. These genetic disorders are autosomal recessive conditions with the exception of Fabry's disease, a sex-linked abnormality with no ethnic or gender predilection. They are the result of a deficiency of intracellular lysosomal enzymes that regulate the catabolism of sphingolipids. Patients with these disorders carry enzymatic structures in their tissues that are similar to the normal enzymes but are not capable of degrading lipids (Table 2.1).

TABLE 2.1

Metabolic Diseases, Enzymatic Deficiencies, and Associated Metabolites

Disease	Deficient Enzyme	Accumulated Metabolite
Gaucher's disease	Glucocerebrosidase	Glucocerebrosides
Globoid leukodystrophy	Galactocerebrosidase	Galactocerebrosides
Fabry's disease	Galactosidase A	Ceramide trihexoside
Neimann–Pick disease	Sphingomyelinase	Sphingomyelin
Metachromatic leukodystrophy	Cerebroside sulfatase A	Sulfatide

PKU, a hereditary condition caused by a defect in the phenylalanine decarboxylase, is transmitted as an autosomal recessive trait. It is one the most common aminoacidurias, which occurs in one per 20,000 births. This enzymatic defect results in the accumulation of phenylalanine in the blood that may be further metabolized to phenylacetic acid, which is eventually excreted in the urine. Exposure to excessive blood levels of phenylalanine may affect neuronal maturation and myelin formation by desegregation of brain polysomes. It has also been put forward that high concentrations of phenylalanine may inhibit transport of other neutral amino acids across the blood–brain barrier. Others have stated that the inhibitory role of high intracerebral levels of phenylalanine on synaptosomal Na^+–K^+ ATPase activity that regulates the synthesis of neurotransmitters may be responsible for this condition. Newborns with this disease generally do not exhibit clinical manifestations, and for this very reason, prenatal screening tests of the amniotic cells and chorionic villi samples are essential for detection of this condition.

Affected infants have lighter skin and eye color and are not retarded at birth. Eventually, however, patients show signs of mental retardation, seizures, psychoses, extreme hyperactivity, "musty" body odor, and cutaneous rash (eczema). In rare cases, PKU may be caused by defects in the metabolism of tetrahydrobiopterin (BH_4), which is the electron donor for the phenylalanine hydroxylase that contributes to the formation of tyrosine and dihydrobiopterin (QH_2). BH_4, which is synthesized from guanosine-5′-triphosphate (GTP), hydroxylates tyrosine and tryptophan. Since patients are unable to hydroxylate tyrosine or tryptophan, which is mediated by BH_4, restriction of phenylalanine may not prevent the neurological complications from occurring. Patients exhibit convulsions and other severe neurological manifestations.

Gaucher's disease is an autosomal recessive disease in which a deficiency in the enzyme glucocerebroside results from accumulation of abnormal glucocerebrosides in the reticuloendothelial system. It manifests signs of oculomotor nerve palsy, hepatosplenomegaly, hypertonicity, opisthotonos (a prolonged severe muscular spasm that produces acute arched back), hyperextension of the head and neck, hyperflexion of the arm and hand, tetany, spasticity, and seizures. Individuals with Gaucher's disease may also exhibit expressionless face, and combined with musculoskeletal deficits, giving rise to the description of "wooden figure."

Tangier disease (hereditary high-density lipoprotein deficiency) is a rare autosomal recessive condition in which high-density α-lipoproteins are reduced in the plasma but with normal or high levels of triglycerides. This condition is associated with degeneration of the thinly myelinated and unmyelinated fibers coupled with accumulation of cholesteryl esters and neutral lipids in the Schwann cells. Cholesteryl esters are also deposited in the bone marrow, skin, intestine, spleen, and tonsils. Tonsillar deposition causes enlargement and yellowish discoloration of this lymphoid structure. Patients exhibit paresis of the hand and facial muscles, areflexia or hyporeflexia of the deep tendon reflexes, analgesia, and dissociated sensory loss (bilateral loss of pain and temperature sensations with preservation of tactile and vibratory sensations). Dissociation of sensory loss seen in this condition should be differentiated from syringomyelia.

Globoid cell leukodystrophy (Krabbe's disease), a rare fatal infantile disease that affects children with an onset between the age of 2 and 6 years, is caused by a deficiency in galactosylceramide β-galactosidase and accumulation of galactocerebrosides. In this autosomal recessive disorder, macrophages containing inclusion bodies filled with cerebroside are found in the white matter of the CNS, while similar inclusions are seen in Schwann cells and endoneurial macrophages, but not commonly in the peripheral nerves. Accumulation of an excessive amount of galactocerebrosides leads to disintegration of the myelin in the cerebrum, cerebellum, brainstem, and possibly the spinal cord. This disease is characterized by progressive mental retardation, blindness, convulsion, deafness, signs of pseudobulbar palsy (loss of cranial nerve motor functions due to disruption of cerebral input), and quadriplegia (complete loss of motor functions in the extremities). Rapid cerebral demyelination and presence of globoid cells in the white matter will be seen. Due to generalized rigidity and tonic spasm, the body stiffens, and the hand forms a fist, which would be particularly evident when the affected infant is held. Patients usually die within 2 years.

Maple syrup urine disease (branched-chain ketoaciduria) is another rare condition that results from anomalies of leucine, valine, and isoleucine metabolism as a consequence of a defect of branched-chain keto acid decarboxylase. Patients exhibit convulsions, hypertonicity, characteristic odor of urine and perspiration, changes in reflexes, coma, and possible death. Prenatal diagnosis may be possible through enzyme assay of the anomalous metabolites. Acute cases of this disease may be treated by peritoneal and/or hemodialysis.

The diagnosis of these diseases has been made easier following the recognition that antibodies that bind to the affected enzymes may artificially be prepared and that traces of the above-mentioned nonfunctional enzymes in the skin fibroblasts, leukocytes, an amniotic fluid may be detected. Fragmentation of the myelin sheath, loss of ability to conduct sensory impulses, impairment of motor function, and trophic changes also occur as the nerve undergoes degeneration (Figure 2.11) following axonal injury. The microscopic

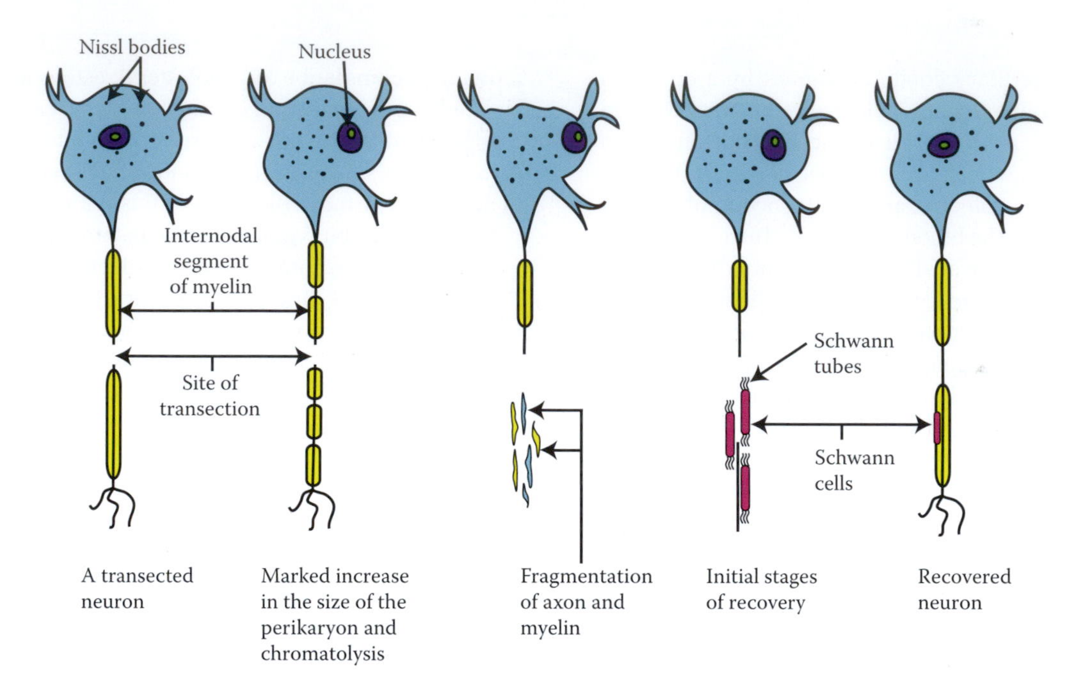

FIGURE 2.11 Cellular changes and stages of recovery of an injured neuron. Note the changes distal and proximal to axonal injury.

alterations (nerve degeneration) in a neuron following damage to its axon may include changes distal to the site of trauma (anterograde degeneration), proximal to the site of damage (retrograde degeneration), or across the axonal terminal into the adjacent neuron (transneuronal degeneration) (Figure 2.12).

Tay–Sachs disease is an autosomal recessive disorder that results from the absence of hexosaminidase A, subsequent to accumulation of ganglioside monosialic2 (GM2) in the perikarya of the neurons. Gangliosides are glycolipids normally present in the plasma membrane of neuronal cell bodies. This disease is common in Jewish infants (3–6 months of age) of eastern European origin and French Canadian parents. Patients may exhibit seizures, blindness, laughing spells, and abnormal acousticomotor reaction. The latter reaction is characterized by brisk extension of legs and arms followed by clonic jerks of all limbs, neck extension, and startled facial expression in response to sudden sharp noise. Cherry-red spots on the macula and progressive intellectual, physical, and neurologic deterioration are additional features of this serious disease. Infants that live longer than 6 months may develop macrocephaly.

Fabry's disease (angiokeratoma corporis diffusum) is an X-linked recessive disorder with linkage to the Xg blood group locus. It results from the lack of α-galactosidase (ceramide trihexosidase), a lysosomal hydrolase, and the accumulation of glycolipid, dihexosidase, and trihexosidase in the autonomic and dorsal root ganglia and in the myelinated fibers of the brainstem and myocardium. Accumulation of glycolipid in the superior cervical ganglion of the sympathetic chain is associated with anhydrosis (lack of sweating). Glycolipid accumulation may also occur in the renal tubules and glomeruli, serving as a diagnostic tool. Antenatal diagnosis of this disease is possible. Deposition of glycolipid also occurs in the hypophysis, eye, smooth muscles of blood vessels, and skin. As the name indicates, scaly hyperkeratotic telangiectactic red to blue skin lesions, known as angiokeratoma corporis diffusum, are seen on the trunk, particularly in the "bathing trunks" area. Patients exhibit corneal dystrophy, engorged conjunctival blood vessels, cerebral ischemia, fever, burning pain in the extremities, and skin lesions in males. Death occurs as a result of renal failure or associated disorders subsequent to hypertension. Ataxia, signs of upper motor neuron palsy, and urinary incontinence are also seen. Demyelination and subsequent loss of thinly myelinated fibers, as a result of the deposited glycolipid in the dorsal root ganglia, may account for the burning pain felt by individuals with this disease. Involvement of the blood vessels may explain the frequency with which cerebrovascular accidents occur in patients with this affliction.

Neimann–Pick disease is a fatal disease of infants, resulting from a lack of sphingomyelinase and accumulation of excessive amounts of sphingomyelin in various tissues. However, it is not yet clear whether abnormal breakdown, stereochemical anomaly, or excessive production of sphingomyelin is responsible for the manifestations of this disease. It is thought that the presence of this substance is responsible for disintegration of the myelin in the white matter of the brain and brainstem. This disease exhibits in pancytopenia

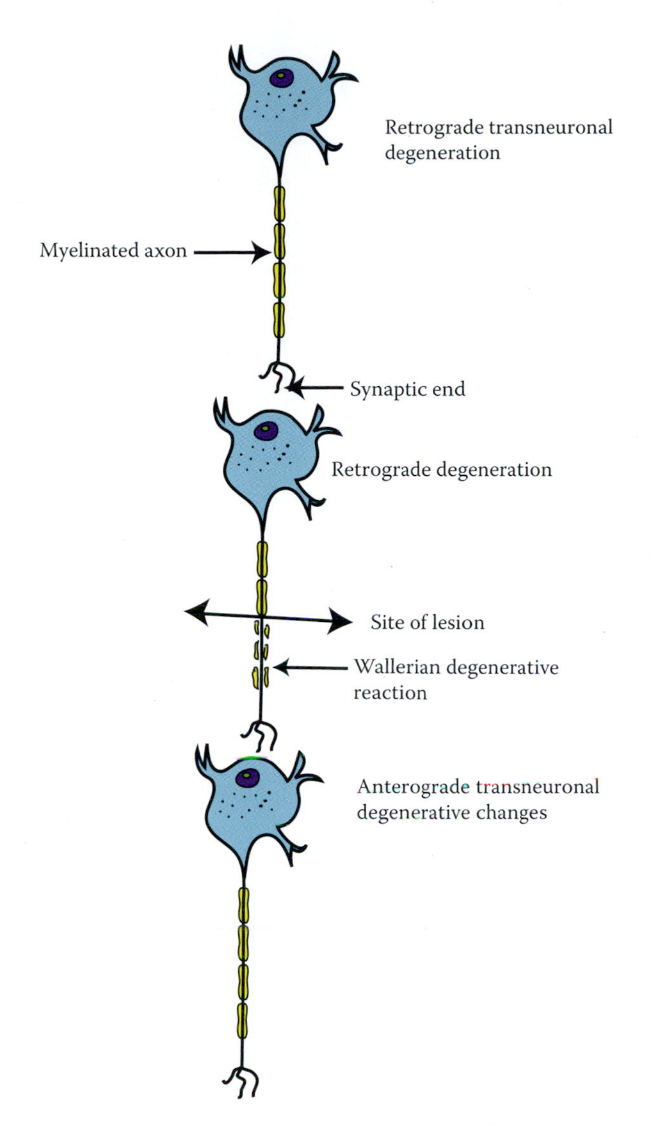

FIGURE 2.12 Summary of major changes shown in Figure 2.11.

(a marked reduction in the number of erythrocytes, leukocytes, and platelets); xanthoma (a benign, fatty, fibrous, and yellowish plaque that develops in the subcutaneous tissue, often around tendons); and feeding problems. It may also manifest in growth and mental retardation, seizures, deafness, and macular irregularity (cherry-red macular spots) that occur in about one-fourth of patients, leading to blindness. Children are cachectic and commonly die between the age of 6 months and 3 years.

Metachromatic leukodystrophy (MLD) is genetically heterogeneous and includes a series of autosomal recessive disorders. This fatal disease results from a deficiency in cerebroside sulfatase, followed by accumulation of sulfatide in excessive amounts in the myelin, Schwann cells, and macrophages. It is seen between the age of 1 and the third decade of life, and affected infants usually die by the age of 6 years. Myelin, with its abnormal sulfatide content, may stain metachromatically brown upon treatment with acidified cresyl violet. In the childhood and late onset of this disease, there is a deficiency of aryl sulfatase A, while multiple sulfatase deficiencies exist in rare variants. Normally, sulfatide is degraded into cerebroside and inorganic sulfate. There is no concrete evidence that correlates between the amount of stored sulfatide and the extent of demyelination. It is believed that accumulated sulfatide could potentially interfere with Schwan cell function and, thus, adversely affect the velocity of motor and sensory nerve conduction. Microscopically, zebra and "tuffstone" inclusion bodies are seen in the peripheral nerves. Chemically formed abnormal myelin in the brain, cerebellum, brainstem, spinal cord, and peripheral nerves does not survive and undergoes disintegration. The most common form of MLD affects infants between the ages of 1 and 2 years, but may also be seen up to the age of 4 years. Poor feeding and irritability followed by gate disturbances with areflexic or hyporeflexic tendons are seen in patients with this condition. As the disease progresses, muscular spasticity palsy become clear. Opisthotonus; myoclonus; mental retardation; difficulty speaking; and then quadriplegia, deafness, and blindness are seen as the disease advances. The juvenile form is much rarer and affects patients between the ages of 3 and 20 years with nearly similar manifestations. The adult form, which occurs in the late 20s, shows cerebellar disorders, peripheral neuropathy, optic atrophy, and cortical dysfunction.

Refsum's disease (heredopathia atactica polyneuritiformis) or (phytanic acid storage) is a rare slowly progressive autosomal recessive neurologic condition. There is an increase of parental consanguinity. It results from accumulation of phytanic acid (tetramethylated 16-carbon-chain fatty acid) in the serum, CNS, and PNS subsequent to a deficiency in the catabolism of fatty acids and failure of alpha oxidation of phytanic acid to α-hydroxyphytanic acid. Phytanic acid, a product of phytol, is present in the diet as a component of animal and plant fat. This condition is characterized by demyelination and hypertrophy of the spinal nerve. Dorsal columns and the motor neurons of the ventral horn are also affected. Manifestations, which usually develop between the first and third decades of life, include polyneuropathy, polyneuritis, sensory deafness, ichthyosis (scaly skin), musculoskeletal disorders, cardiac disorders, cerebellar deficits, and locomotor ataxia. It also includes retinal degeneration, nocturnal blindness, and retinitis pigmentosa, an inflammatory condition of the retina associated with progressive loss of retinal response. Ocular signs may also include pupillary abnormalities, clumping of pigment, and shrinkage of visual field. Death can occur as a sequel of cardiomyopathy. Reduction of phytanic acid and amelioration of certain clinical signs can be achieved by dietary restriction of fruit, vegetables, and butter.

CLASSIFICATION OF NEURONS

Unipolar neurons are the simplest class of neurons, which exhibit a single extension that gives rise to branches, some of which are receptive (dendrites); others function as axons. True unipolar neurons, which are relatively rare in vertebrates, form the dorsal root ganglia, the granule cells of the olfactory system, and the mesencephalic trigeminal nucleus (Figures 2.7, 2.13, and 2.14). Pseudounipolar neurons give off a single process that divides into a peripheral receptive branch (dendrite) and a central extension serving as an axon. Both of these branches maintain a structural resemblance to axons.

Bipolar neurons are also a relatively uncommon class of neurons. They are symmetrical cells with an ovoid or elongated body and with a single dendritic process and an axon arising from opposite poles. These processes are approximately equal in length. They form the vestibular (Scarpa's) ganglia, spiral (auditory) ganglia, and retinal bipolar cells (Figure 2.15).

Multipolar neurons (Figures 2.7, 2.16, and 2.17) are the most common types of neurons in the CNS; they form the autonomic ganglia. They possess a single axon with several symmetrically radiating dendrites. Some neurons have multiple axons or lack axons altogether. Multipolar neurons can be classified on the basis of dendritic branching pattern and shape of the soma into stellate, pyramidal, fusiform, Purkinje, and glomerular cells.

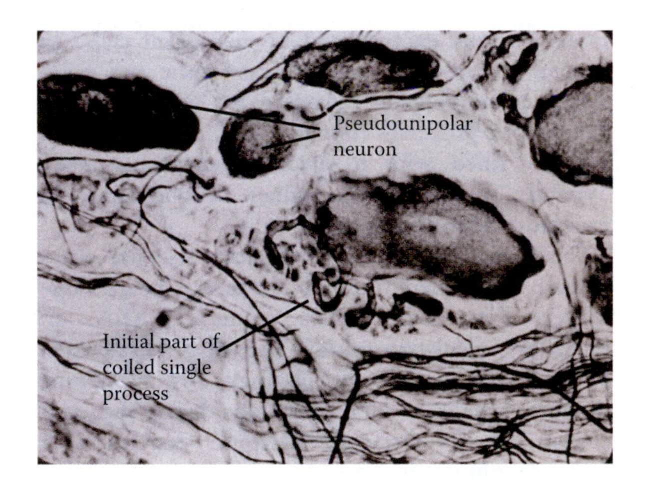

FIGURE 2.13 Photomicrograph of the pseudounipolar neuron and its initial process.

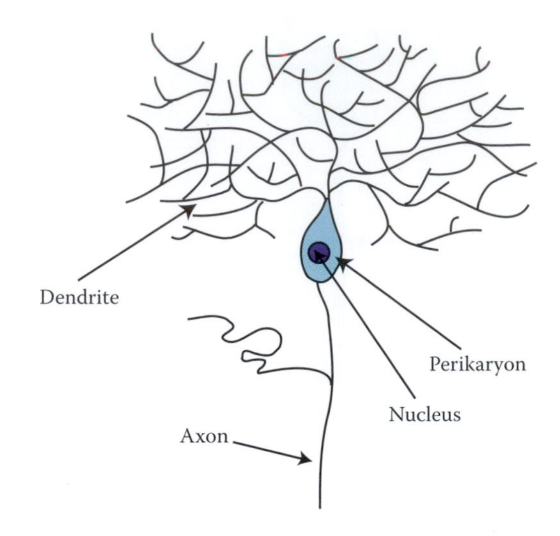

FIGURE 2.15 Bipolar neuron with branching dendrite and axon.

Stellate (star) neurons are found in the spinal cord, reticular formation, and cerebral cortex. They have dendrites of equal lengths (isodendritic) that radiate uniformly in all directions.

Pyramidal neurons are multipolar, exhibiting pyramidal-shape soma with basal dendrites and a single apical dendrite that ascend toward the surface of the cerebellar cortex. They are most abundant in the cerebral cortex and hippocampal gyrus.

Fusiform neurons are distinguished by their spindle-shaped and flattened soma with dendrites at both ends.

Purkinje neurons that form the intermediate layer of the cerebellar cortex have flask-shaped soma with apical treelike dendritic branches, ascending toward the surface of the cerebellum, maximizing synaptic contacts. Purkinje cells are motor neurons that project long axons beyond the area of the soma.

Glomerular neurons have a few convoluted dendritic branches and form the mitral and tufted cells of the olfactory bulb. Mitral cells have an inverted cone-shaped dendritic field and a soma that resembles a bishop's miter.

Anaxonic cells are abundant in the retina (amacrine cells) and the olfactory bulb, where they are known as granule cells.

On the basis of axonal length, multipolar neurons can also be categorized into Golgi type I, with long axons projecting to distant parts of the CNS, and Golgi type II, possessing

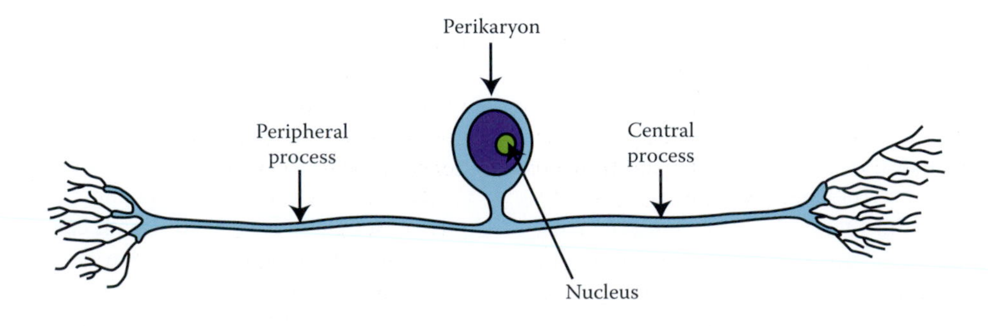

FIGURE 2.14 Schematic drawing of the pseudounipolar neuron. Cellular elements and associated extensions are also shown.

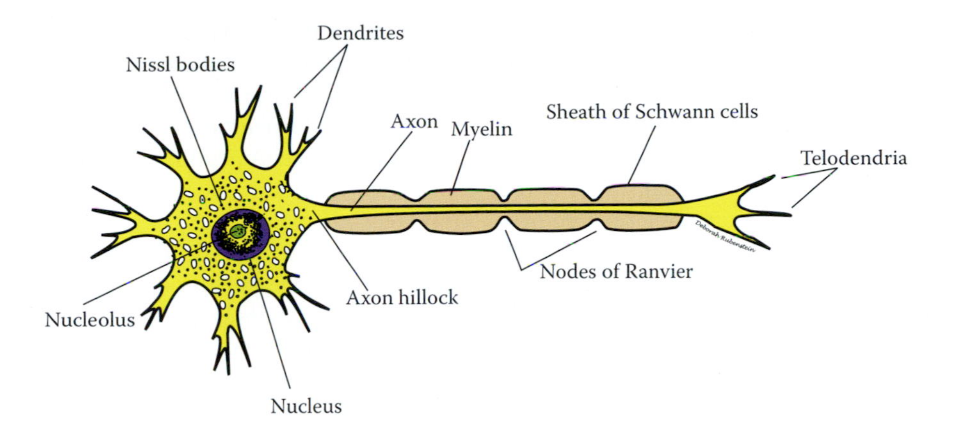

FIGURE 2.16 Multipolar neuron. Observe the numerous branches of dendrites, and a single uniform axon is also shown.

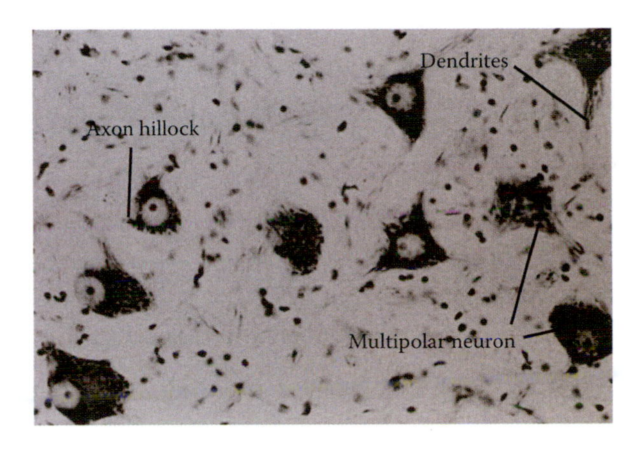

FIGURE 2.17 These multipolar neurons of the cerebral cortex exhibit an axon with an apical dendrite.

short axons that establish contacts with local neighboring neurons. The Golgi type II represents the inhibitory interneurons (such as the periglomerular olfactory neurons), which are activated by the ascending sensory pathways and play an important role in lateral inhibition. Neurons without axons, as mentioned earlier, are known as anaxonic, such as the amacrine cells of the retina and granule cells of the olfactory bulb, which establish synapses with parallel neurons.

Neurons can also be classified based on their functional role into somatic motor, somatic sensory, visceral motor, and visceral sensory neurons.

SYNAPTIC CONNECTIVITY

Synapses are specialized junctional complexes formed by the axonal terminal of one neuron opposing the dendrites, soma, or the axon of another neuron (Figure 2.7), where neuronal communication is maintained. They represent sites of impulse generation (action potentials) and transmission across a population of neurons within the CNS. The most common synapse is between an axon and dendrite, although the axosomatic synapse is also common. Synapse interfaces between neurons provide trophic substances and act as a "gate" for controlling impulses. A single axon may establish a synapse with one neuron (e.g., connections of the olivocerebellar fiber with dendrites of the Purkinje neurons). Multisynapses are seen between the parallel fibers of granule cells and the neurons of the molecular layer of the cerebellum. Synaptic glomeruli in the olfactory bulb and the granular layer of the cerebellum consist of an axon that synapses with dendrites of one or more neurons encapsulated by neuroglial cells. In general, synapses consist of presynaptic and postsynaptic components separated by synaptic clefts.

Presynaptic processes contain round, granular, or flat vesicles filled with a specific neurotransmitter. Typically, round vesicles contain Ach, an excitatory neurotransmitter. Small granular vesicles with electron-dense cores contain norepinephrine, an excitatory neurotransmitter. Flattened vesicles contain GABA, an inhibitory neurotransmitter. The close relationship between the vesicle morphology and functional synaptic type is evident when considering the association of the flattened synaptic vesicles with symmetrical membrane specializations and spherical vesicles with asymmetrical membrane thickenings. Synaptic vesicles are fixed by the cytoskeleton proteins F-actin and spectrin and mobilized when neurotransmission is initiated. This action is aided by the microtubules.

A synaptic cleft is a small gap that separates presynaptic and postsynaptic neurons and is crossed by fine fibrils. This cleft creates a physical barrier for the electrical signal transmitted from one neuron to another.

The postsynaptic membrane may be part of a muscle cell or neuron, upon which neurotransmitter molecules bind after crossing the synaptic cleft. The part of the postsynaptic membrane that lies adjacent to the presynaptic membrane is known as the subsynaptic membrane. Synaptolemma is a term that denotes the combined presynaptic and subsynaptic membranes. An increase in the postsynaptic receptor sites may be responsible for the exaggerated response following denervation (denervation hypersensitivity).

Synapses in the CNS differ morphologically and functionally from their counterparts in the PNS. They are not always cholinergic (as in the PNS), and they utilize several excitatory neurotransmitters such as catecholamines (epinephrine, norepinephrine, and dopamine); amino acid neurotransmitters (glutamine, aspartate, cysteine, etc.); serotonin; histamine;

enkephalin; and so forth. Transmission through the central synapses is governed by factors such as diffusion and reabsorption and may be excitatory or inhibitory (activation drives the membrane potential of the postsynaptic neuron toward or away from its threshold level for firing nerve impulses). Transmission in the peripheral synapses, as in the neuromuscular junction, is generally excitatory, secured by a single presynaptic activation, and is dependent upon the degradation of the neurotransmitters by cholinesterase. Synapses in the CNS occur between one presynaptic ending and several postsynaptic neurons, contrary to the 1:1 synapse ratios in peripheral transmission. The variability and efficiency of transmission and neurotransmitter discharge in the central synapses are dependent upon the number of activated presynaptic endings.

Electrical synapses exhibit close contact between presynaptic and postsynaptic membranes and act through direct ionic coupling. Gap junctions enable the nerve impulses to cross directly from one cell to another and act on the postsynaptic membrane via connexins, a group of gap-junction forming proteins. Gap junctions can function as a timing device as in olivocerebellar connections or exhibit characteristics that enhance transfer of ions or regulate response to trophic factors. These synapses, which are common in lower vertebrate motor pathways, are similar to the electrical junctions (intercalated discs) of the cardiac muscle cells. Electrical synapses act much more rapidly than chemical synapses. Vestibular and inferior olivary nuclei, cerebellar and cerebral cortices, the olfactory bulb, and the retina contain electrical synapses.

Classification of synapses is also, based on the morphological characteristics and the type of action they eventually produce, into Gray's type I and II synapses. Gray's type I is an excitatory synapse with round vesicles in which the synaptic cleft is wide and the presynaptic and postsynaptic membrane densities are asymmetrical, with the subsynaptic zone being thicker than the presynaptic zone. This type of synapse contains a wide variety of neurotransmitters, including Ach, glutamate, and hydroxytryptamine. Gray's type II synapse is an inhibitory synapse with flat vesicles in which the synaptic cleft is narrower, and the presynaptic and postsynaptic membrane densities are symmetrical.

Synapses may also be classified as chemical or electrical. Chemical synapses are unidirectional and slow and involve the release of a neurotransmitter by synaptic vesicles into the synaptic cleft, producing changes in the permeability of the postsynaptic membrane. The effect of the neurotransmitter is controlled by local enzymes and/or by reabsorption. Chemical synapses are further categorized on the basis of the utilized neurotransmitter. Cholinergic synapses use Ach, adrenergic synapses utilize epinephrine or norepinephrine, and dopaminergic synapses utilize dopamine. Asymmetric synapses contain several neurotransmitters, including Ach, glutamate, 5-hydroxytryptamine, dopamine, and adrenaline and noradrenaline, while the symmetric synapses are associated with glycine or GABA. Neurosecretory endings with dense-cored vesicles in the CNS and neurohypophysis are identical to the presynaptic endings of chemical synapses. Synaptic terminals also contain one or more modulators that are stored in dense synaptic vesicles that accompany those that contain the neurotransmitters. These modulators, which are primarily neuropeptides, enhance or inhibit the response of receptors by the neurotransmitters or act directly on the postsynaptic membrane.

Synapses may also be axodendritic, the most common of which may be symmetrical or asymmetrical. Symmetrical axodendritic synapses predominate near the soma on the larger dendritic trunks. Axosomatic synapses occur on the perikaryon, exhibiting both symmetrical and asymmetrical forms. This type of synapse that involves the initial segment of the axon may be inhibitory to cellular discharge. They are commonly symmetrical and may release inhibitory neurotransmitter GABA. Axo-axonic synapses, in general, reduce the amount of neurotransmitter released by the axon and therefore are regarded to mediate presynaptic inhibition. Dendrosomatic and somato-somatic synapses are described in the sympathetic ganglia. Dendrodendritic synapses are, for the most part, of the symmetrical type; however, in the olfactory bulb, the dendrites of the mitral cells form asymmetrical synapses with the dendrites of the granule cells.

The *neuromuscular junction* (motor end plate) is the site of synaptic contacts between the terminal branches of the α motor axon and the skeletal intrafusal muscle fibers (Figure 2.18). The α motor axon terminals give off several short branches over an elliptical area known as the motor end plate. Within the subneural plate, the sarcolemma is thrown into synaptic folds, forming a unique type of neuromuscular junction known as *en plaque* or subneural apparatus, which is abundant in muscle fibers that propagate action potential. Another type of neuromuscular junction where propagation of action potentials does not occur but excitation is carried over branches of long nerve terminals that further divide into small neuromuscular junctions, *en grappe* endings, is characteristically seen in the stapedius and extraocular muscles. A similar arrangement occurs with regard to γ efferent terminals in the intrafusal muscle fibers.

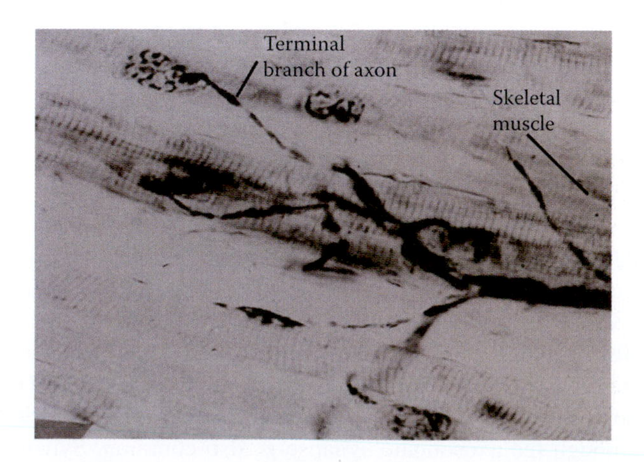

FIGURE 2.18 Photomicrograph of the motor end plate. The synaptic connection between the terminal axon and skeletal muscle fibers is clearly illustrated.

The *neuromuscular junction* is the site where depolarization of muscle fiber membrane and muscular contraction are initiated. Structurally, each motor end plate consists of presynaptic and postsynaptic membranes. The presynaptic membrane is formed by the platelike unmyelinated end of a single motor axon, with numerous membrane-bound Ach-filled vesicles, but with no convergence of synaptic input. The postsynaptic membrane, which is formed by muscle cell invagination that corresponds to the presynaptic vesicles, is separated from the extracellular space by the Schwann cells. The synaptic membranes of the motor end plates are separated by synaptic clefts that are larger than the synaptic membranes in the CNS, and that postsynaptic potential at the neuromuscular junction is much greater than its counterpart in the CNS. The release of Ach is dependent upon the frequency of the action potential and the influx of calcium ions. Once released, the Ach diffuses across the synaptic cleft and increases the permeability of the postsynaptic membrane to the sodium and potassium ions, thus producing depolarization. The endplate potential (EPP) is local, and its amplitude varies with the distance and the amount of Ach. Ach is excitatory at the neuromuscular junction but assumes an inhibitory role at certain sites when the receptor molecules are coupled to a potassium channel. The impact of Ach is rapid and brief due to degradative action of hydrolytic enzymes and diffusion out of the synaptic cleft. Cholinergic receptors show continuous turnover, and a large number of them are replaced.

Synaptic Disorders

Disease processes, drugs, and exposure to toxins may disrupt chemical transmission. Local anesthetics such as procaine, tetrodotoxin, and saxitoxin block the generation of action potentials. Hemicholinium blocks the synthesis of Ach by preventing the reuptake of choline into the cell. A high concentration of magnesium may block the release of Ach and cause paralysis by competing with calcium receptors. Transmission at the presynaptic level of the neuromuscular junction may be blocked by exotoxin produced by strains of the bacterium *Clostridium botulinum*.

Botulism is caused by a neurotoxin released by *C. botulinum*, an anaerobic gram-positive bacillus, which is resistant to heat up to 100°C (212°F) as well as digestive enzymes. There are a number of types of this neurotoxin that are effective in humans: types A (most common), B, E, and F (most rare). The presence of these types of toxins follows, to an extent, a geographic pattern in the United States. This toxin blocks the release of Ach by either binding calcium receptors or preventing entry of calcium ions during the action potential. Botulinum toxin is synthesized in an inactive form, which must be cleaved into heavy and light chains joined by a disulfide bridge to become active. Toxicity is initiated by binding of these chains to specific presynaptic receptors. Ultimately, the toxic component is discharged from the lysosomes into the cytoplasm of the presynaptic terminals. Ingestion of clostridial toxin produces signs and symptoms of botulism. The neurotoxin enters the body either through ingestion of contaminated and improperly preserved home-canned food, vegetables, fruits, and fish (mainly type E) or foil-wrapped food or through a contaminated deep penetrating wound with this particular toxin. It also occurs through the large intestine, particularly in infants. Symptoms may appear within 1–2 days after ingestion. Symptoms include nausea, vomiting and diarrhea, dysphagia (difficulty swallowing that can cause aspiration pneumonia), and dysarthria (difficulty with speech). Gastrointestinal symptoms are not seen in wound-induced toxin. Blurred vision is accompanied by dilated and unresponsive pupils and paralysis of the extraocular muscles, which produces bilateral diplopia. This is followed by descending bilateral and symmetric paresis (weakness and incomplete paralysis) of muscles of the neck, trunk, diaphragm, and extremities. CSF, sensory modalities, and body temperature remain normal. In infants younger than 1 year of age, botulism can occur as a result of colonization of *C. botulinum* following ingestion of, for example, honey and the release of toxin in the large intestine. Patients exhibit constipation followed by neurological and muscular disorders.

Black widow spider (*Latrodectus mactans*) venom produces toxicity through α-latrotoxin, a protein molecule that combines with the presynaptic membrane and allows both sodium and calcium ions to enter the terminal, causing an initial massive release of Ach. This is followed by decline and fast depletion of the transmitter that produces initial contraction and painful spasm, rigidity, and subsequent paralysis of the associated muscles. Patients usually experience a sharp penetrating pain, followed by a dull, sometimes numbing pain in the affected area. Cramping pain with muscle spasm in the trunk and shoulder also occurs. Chest pain may mimic appendicitis, and abdominal pain can be mistaken for appendicitis. Other generalized manifestations observed in this condition include headache, vertigo, hyperhydrosis, cutaneous rash, nausea, and vomiting. This condition is rarely fatal, and affected children and the elderly endure more severe symptoms.

Bungarotoxin, a venom of *Bungaro multicinctus*, is a protein that inhibits the release of synaptic vesicles from the cholinergic and motor nerve terminals by exhibiting phospholipase A_2 activity. Initially, this toxin produces a slight reduction in EPP amplitude followed by intensification and then progressive decrease and final complete blockage of the transmitter.

Tetany, another condition that results from abnormalities at the neuromuscular junction, is associated with hypocalcemia induced by deficiency of parathormone and vitamin D and alkalosis due to hyperventilation, as high pH increases calcium binding by serum proteins and the ionized fraction of plasma calcium decreases without hypocalcemia. It is characterized by generalized muscle spasm; tingling sensation in the lips, tongue, and digits; and facial muscle spasm. Spasm of the forearm muscles and flexion of the hand at the wrist (Trousseau's sign) can be produced by applying a tourniquet and reducing the blood supply to the forearm and hand. The fingers are pressed together, and the thumbs are adducted (obstetrician hand). The lower extremity exhibits carpopedal spasm, in which the thigh and knee are extended, whereas the feet are plantar flexed and inverted. However, Trousseau's sign is also seen in alkalosis, hypokalemia and hyperkalemia, and hypomagnesemia. Administration of a diuretic such as furosemide exacerbates manifestations of latent tetany. In latent tetany, involuntary contractions, twitching, or spasm of the facial muscles can be elicited by light tapping of the facial nerve trunk as it runs in the parotid gland anterior to the pinna (Chvostek's sign). This sign may be elicited in some healthy individuals and can be absent in patients with hypocalcemia. It is treated by administration of calcium, which restores normal transmitter release.

Due to development of multiple and highly sensitive ectopic sites, the sensitivity of the muscle membrane increases dramatically upon denervation. Denervation hypersensitivity is accompanied by random contraction of individual muscle fibers, known as fibrillation. Since these contractions develop individually and are asynchronous, they are not visible through the skin. The postsynaptic membrane contains acetylcholinesterase, an enzyme that limits the duration of action of Ach and curtails the depolarization process by hydrolysis of Ach into choline and acetate.

Antibodies to the (nicotinic) cholinergic receptor may also block transmission at the postsynaptic level of the neuromuscular junction, as in myasthenia gravis (Erb–Goldflam disease). These antibodies bind to the main immunogenic region of the μ-2 subunit receptors, produce cross linkage between cholinergic (nicotinic) receptors, eventually increase the rate of lysosomal degradation and endocytosis. The density of the postsynaptic receptors in this disease may be as low as one-third of normal, and the synaptic cleft is widened. Some speculate that mycocytes and thymic muscle–like cells may express acetylcholine receptor (AChR) on their surface, triggering inflammatory response and subsequent production of cross-reacting antibodies from the thymus to the muscular cholinergic receptors.

Myasthenia gravis (Figure 2.19) is an acquired autoimmune disease of neuromuscular transmission, in which developed antibodies attack the cholinergic receptors at the postsynaptic level in the neuromuscular junction. These autoantibodies may block the binding to the α subunit of the cholinergic receptors or initiate immune mediated reaction via T lymphocytes, for example, complement-mediated receptor lysis that causes degradation the cholinergic receptors and their associated postsynaptic membrane. Depletion of a considerable number of cholinergic receptors reduces the number of muscle fibers that can be depolarized, and thus, a decrease in the generated muscle action potential and muscle contraction will ensue. However, the synthesis of the receptor is normal in myasthenia gravis, but the rate of degradation increases dramatically as the antibodies mark the receptor for accelerated degradation.

This disease, which is most common in women, may accompany other systemic autoimmune disorders, such as systemic lupus erythematosus and pernicious anemia. It is classified as ocular, generalized, neonatal, congenital, or drug induced. It usually develops between the second and fourth decades of life. Within 2 years of appearance of ocular signs, generalized weakness develops. If generalized weakness does not develop within this period of time, the condition may not progress any further. This condition is characterized by bilateral fluctuating weakness of the ocular (50%–60% of patients), masticatory, and pharyngeal muscles and muscles of facial expression, as well as muscles of the neck and upper extremity. Clinical signs show varying intensity and occasionally follow an episodic course. Although these signs may be seen occasionally only on one side, weakness of ocular and palpebral muscles is the first presenting sign and most often remains asymmetric and intermittent. Ptosis (drooping of the upper eyelid) with lid retraction that

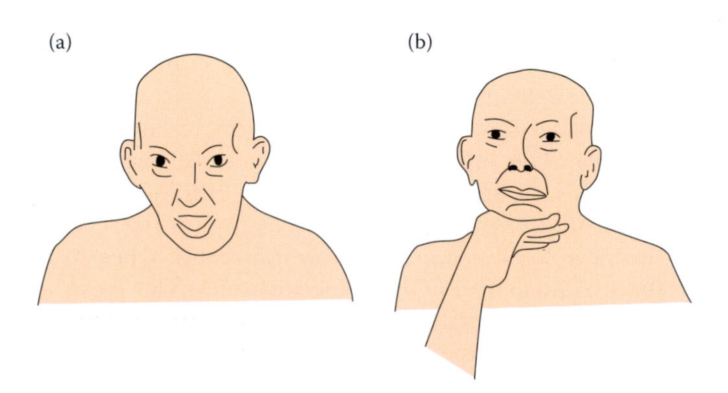

FIGURE 2.19 These schematic drawings show the manifestations of myasthenia gravis. (a) Observe drooping of the head and jaw. (b) The patient attempts to hold the head and jaw up with a hand during conversation.

cannot be masked by contraction of the frontalis is the most prominent and usually the initial manifestation of this disease. Repeated forceful opening and closure of the eyes may exacerbate ptosis (Simpson's test). Although pupillary light reflex (characterized by constriction of the pupil of both eyes when light is applied to one eye) remains intact, deficits such as diplopia upon upward gaze and convergence (medial deviation of both eyes) and mydriasis (dilatation of the pupil) are also seen.

Nasal speech, dysphagia, and inability to hold the head up may form additional early signs of this condition. Voluntary muscle weakness is heightened by repetitive maneuvers, inducing dramatic fatigue. Increasing fatigue of individual muscles is more pronounced toward evening or after physical activity. Rest may provide a temporary relief from the symptoms. Chewing is normal initially but becomes progressively worse as time passes. Expressionless facies develops as the patient's facial muscles become weak, and the patient's smile is characteristic (myasthenia snarl). Respiratory muscle involvement in some patients may lead to death (myasthenic crisis) due to hypovolemia-induced respiratory failure. Sensory transmission and tendon reflexes are preserved in this disease. Diagnosis of this disease is done by intravenous injection of a short-acting anticholinesterase agent known as edrophonium chloride. Edrophonium chloride can be used to distinguish between myasthenia gravis and cholinergic crisis. Edrophonium causes temporary relief in patients with myasthenia gravis but not in patients with cholinergic crisis, which is a diffuse overactivation of the muscarinic receptors and depolarization of nicotinic cholinergic receptors of the skeletal muscles. Overtreatment with anticholinesterase or exacerbation of the disorder may lead to respiratory arrest and possible death.

Myasthenia gravis should be distinguished from Lambert–Eaton syndrome, which is a presynaptic disorder, seen in oat cell carcinoma of the lung, and is associated with pernicious anemia. Antibodies that block the calcium channels essential for the release of Ach cause this syndrome. It manifests itself as weakness in the muscles of the upper extremity, sensory loss, ataxia, and deep tendon areflexia. Sparing of the extraocular muscles is one of the pathognomic features of this disease. It should be noted, however, that the affected muscles show a maximum increase in strength following voluntary exercise (warm-up), which is a unique characteristic of this neurological disorder. The "warm-up" phenomenon in Lambert–Eaton syndrome is the result of concomitant actions of Ach release and then depletion followed by facilitation of transmitter release by repetitive activities. Autonomic deficits such as sexual dysfunction and dry mouth may also be seen.

Curare drugs (D-tubocurarine) reversibly attach to the postsynaptic membrane, thus preventing any reaction to Ach. D-tubocurarine is a short-acting drug that may be used with local anesthetics to promote muscle relaxation during anesthesia. Its action is terminated by the administration of anticholinesterase. Depolarizing blocking agents such as decamethonium bromide and succinylcholine may mimic Ach at the postsynaptic membrane level. Since these agents are not affected by cholinesterase, they induce prolonged depolarization. Anticholinesterase drugs such as physostigmine and neostigmine prolong the action of the Ach by reversibly inactivating the enzyme. Nerve gas (di-isopropyl fluorophosphate) and organic phosphates irreversibly bind to acetylcholinesterase, producing prolonged depolarization, paralysis, and death due to asphyxiation.

α-neurotoxin, a curare mimetic, is a nondepolarizing blocking agent at the postsynaptic cholinergic receptors. This toxin is produced by snakes of the families Elapidae (e.g., cobras, coral snakes, etc.) and Hydrophidae (sea snake). α-neurotoxin consists of the long toxin with 71–74 amino acids and five internal sulfide bonds and a short group with 60–62 amino acids and four internal disulfide bonds. The toxin with the short amino acids exhibits faster binding to the α subunits of the Ach receptor than the long toxin (irreversible binding) and a reversible dissociation capacity.

SUGGESTED READING

Aguayo AJ, Rasminsky M, Bray GM, Carbonetto S, McKerracher L, Villegas-Pérez MP, Vidal-Sanz M, Carter DA. Degenerative and regenerative responses of injured neurons in the central nervous system of adult mammals. *Phil Trans R Soc London* 1991;331:337–43.

Atwood HL, Lnenicka GA. Structure and function in synapses: Emerging correlations. *Trends Neurosci* 1986;9:248–50.

Bhär M, Bonhoeffer F. Perspectives on axonal regeneration in the mammalian CNS. *Trends Neurosci* 14. 1994;17:473–9.

Colman DR. Functional properties of adhesion molecules in myelin formation. *Curr Opin Neurobiol* 1991;1:377–81.

Fawcett JW, Keynes RJ. Peripheral nerve regeneration. *Annu Rev Neurosci* 1990;13:43–60.

Hayashi M, Chernov M, Tamura N et al. Gamma Knife surgery for abducent nerve schwannoma. Report of 4 cases. *J Neurosurg* 2010;113:136–43.

Hirokawao N. Axonal transport and the cytoskeleton. *Curr Opin Neurobiol* 1993;3:724–31.

Johnston J, So TY. First-line disease-modifying therapies in paediatric multiple sclerosis: A comprehensive overview. *Drugs* 2012;72(9):1195–211.

Lee MK, Cleveland DW. Neurofilament function and dysfunction: Involvement in axonal growth and neuronal disease. *Curr Opin Cell Biol* 1994;6:34–40.

Linden DJ. Long-term synaptic depression in the mammalian brain. *Neuron* 1994;12:457–72.

Link H, Huang YM. Oligoclonal bands in multiple sclerosis cerebrospinal fluid: An update on methodology and clinical usefulness. *J Neuroimmunol* 2007;180(1–2):17–28.

Nakajima K, Kohsaka S. Functional roles of microglia in the brain. *Neurosci Res* 1993;17:187–203.

Pittock SJ, Lucchinetti CF. The pathology of MS: New insights and potential clinical applications. *Neurologist* 2007;13(2):45–56.

Rashid W, Miller DH. Recent advances in neuroimaging of multiple sclerosis. *Semin Neurol* 2008;28(1):46–55.

Remahl S, Hildebrand C. Relation between axons and oligodendroglial cells during initial myelination II. The individual axon. *J Neurocytol* 1990;19:883–98.

Torpy JM, Burke AE, Glass RM. JAMA patient page: Neurofibromatosis. *JAMA* 2008;300(3):352.

Vallee RB, Bloom GS. Mechanisms of fast and slow axonal transport. *Annu Rev Neurosci* 1991;14:59–92.

Vincent A, Newsom-Davis J. Disorders of neuromuscular transmission, Chapter 448. In Goldman L, Ausiello D, eds. Cecil *Medicine*, 23rd ed. Philadelphia, PA: Saunders Elsevier, 2007.

Wu L, Saggau P. Presynaptic inhibition of elicited neurotransmitter release. *Trends Neurosci* 1997;20:204–12.

Zinman L, Ng E, Bril V. IV immunoglobulin in patients with myasthenia gravis: A randomized controlled trial. *Neurology* 2007;68(11):837–41.

Section II

Morphologic and Sectional Neuroanatomy

3 Spinal Cord

The spinal cord is derived from the caudal part of the neural tube, and in adults, it occupies the upper two thirds of the vertebral column. It stretches between the upper border of the foramen magnum to the intervertebral disc between the first and second lumbar vertebrae. In newborns, it extends to the level of the third lumbar vertebra. The sites of attachment of the 31 pairs of spinal nerves mark the individual segments of the spinal cord. Each spinal segment consists of a central gray mater and a peripheral white associated with the dorsal and ventral roots. The gray mater is formed by the neuronal cell bodies and divided into Rexed laminae. The white matter, divided into funiculi, is primarily formed by the asecending and descending pathways. Due to selective location of the pathways within the spinal cord, a lesion of one funiculus can disrupt one type of sensory modality and spare another. There are 8 cervical, 12 thoracic, 5 lumbar, 5 sacral, and 1 coccygeal spinal segments. The spinal cord is not uniform, exhibiting a cervical enlargement that corresponds to the fifth cervical through the first thoracic spinal segments (roots of the brachial plexus) and a lumbar enlargement that corresponds to the first lumbar through the third sacral spinal segments (roots of the lumbosacralplexus).

The spinal cord is the cylindrical part of the central nervous system (CNS), occupying the upper two-thirds of the vertebral column. The lower end of the spinal cord (conus medullaris) shows variation relative to the height of the individual, particularly in females. Flexion and extension of the trunk may also produce relative variation of the lower end of the spinal cord. In some individuals, the spinal cord may terminate as high as the twelfth thoracic vertebra or may extend as far down to the level of the intervertebral disc between the second and third lumbar vertebrae.

Due to the difference in the rate of development of the vertebral column relative to the spinal cord, the spinal cord segments do not always correspond to the vertebral levels. In general, the rule of 2 applies to the vertebral levels T1–T10. In other words, the injured spinal segments are determined by adding 2 to the level of the affected vertebrae. Spinous processes of the T11–T12 vertebrae correspond to the lumbar spinal segments. Accordingly, the cervical spinal nerves exit above their corresponding vertebrae, while the remaining spinal nerves emerge from the vertebral column below the corresponding vertebrae. When the dorsal and ventral roots of the lower lumbar and sacral segments assume a longer course around the conus medullaris to reach the corresponding intervertebral foramina, the cauda equina is formed (Figures 3.1 and 3.2).

Lesions of the cauda equina produce cauda equina syndrome, which are caused most commonly by lumbar central disc herniation or subluxation. It can also result from ependymomas and neurinomas, lumbar puncture (LP), spinal anesthesia, Paget's disease, ankylosing spondylitis, epidural abscess or hematoma, bleeding, gunshot wounds, arteriovenous malformation (AVMA), spondylolisthesis with spinal stenosis subsequent to osteoarthritis, burst fractures of the vertebral body, and stab wounds. Patients with this rare disorder manifest saddle anesthesia (external genitalia, urethra, anal canal, and medial thighs) with a well-defined upper border, bilateral or unilateral sciatic pain, and severe low back pain with bilateral progressive weakness of the triceps surae (gastrocnemius and soleus) and intrinsic plantar muscles of the foot. The sympathetic innervation of the sweat glands of the plantar foot is spared. These manifestations are accompanied

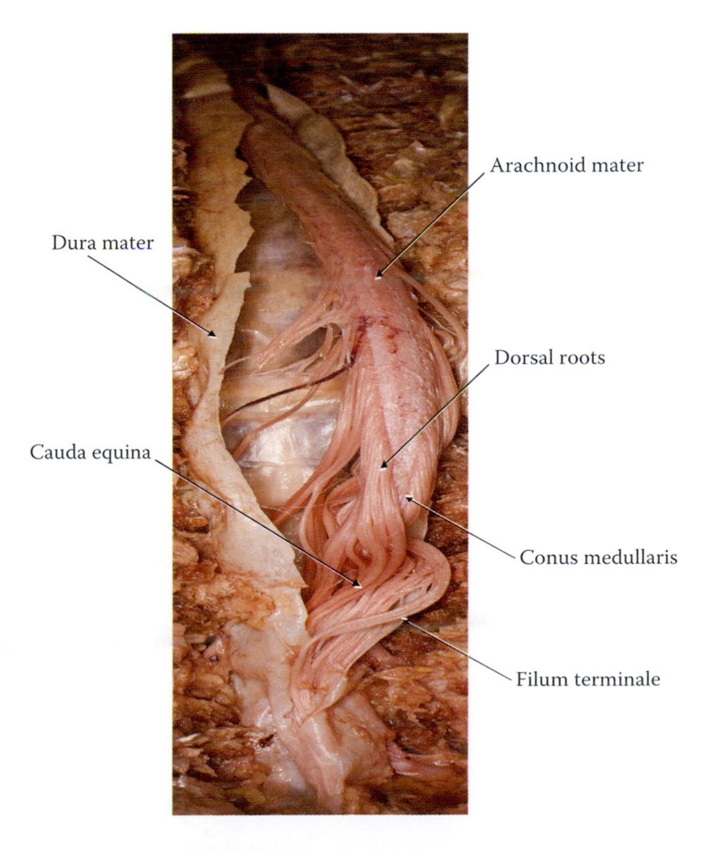

FIGURE 3.1 The caudal end of the spinal cord, filum terminale, and spinal meninges are shown in this picture.

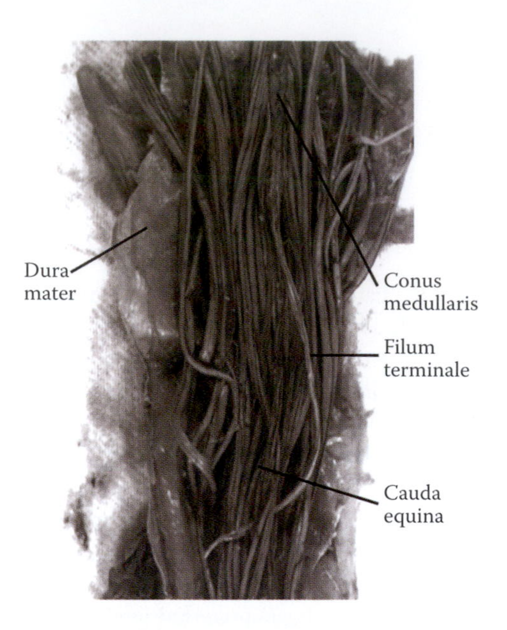

FIGURE 3.2 A more elaborate picture of the caudal spinal cord. The cauda equina and filum terminale are clearly visible.

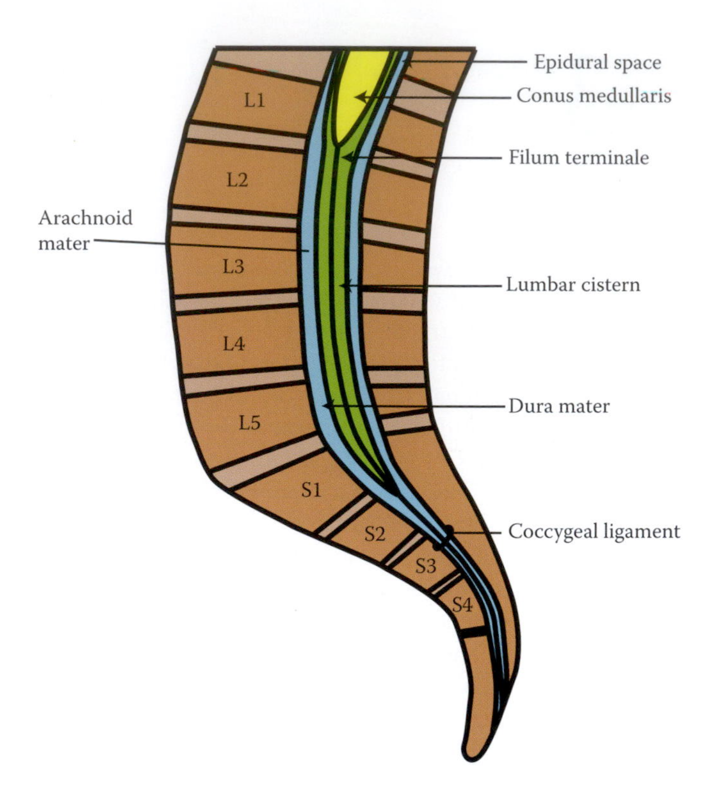

FIGURE 3.4 Caudal part of the spinal cord showing the meninges, filum terminale, and coccygeal ligament. Corresponding vertebral levels are also shown.

by detrusor muscle paresis leading to urinary retention and postvoid residual incontinence, sudden impotence, and decreased anal sphincteric tone and subsequent bowel incontinence. Ependymoma produces distinctly more progressive manifestations with calcaneal, patellar, and adductor tendon reflexes gradually disappearing, while bowel and urinary incontinence occurs late in the course of the disease.

The spinal cord is invested by the dura, arachnoid, and the pia mater. The dura mater (pachymeninx), a thick, collagenous layer, forms the atlantooccipital membranes and consists of an inner meningeal and an outer endosteal layer. The outer endosteal layer forms the periosteum of the vertebral canal and fuses with the epineurium (the outermost covering) of the spinal nerves at or slightly beyond the intervertebral foramina. This fusion is more pronounced in the cervical vertebrae with greater movements and less in the lumbar vertebrae. The endosteal layer is separated from the meningeal layer by the epidural space (Figures 3.3 and 3.4), which contains fat and the internal vertebral plexus, a venous network that maintains connection with the systemic veins. This space, which extends from the cranial base to the sacral hiatus, is bounded anteriorly by the anterior longitudinal ligament, posteriorly by the ligamentum flavum, and laterally by the vertebral pedicles. Dural continuation around the filum terminale and the lumbar cistern forms a tubelike dural sac.

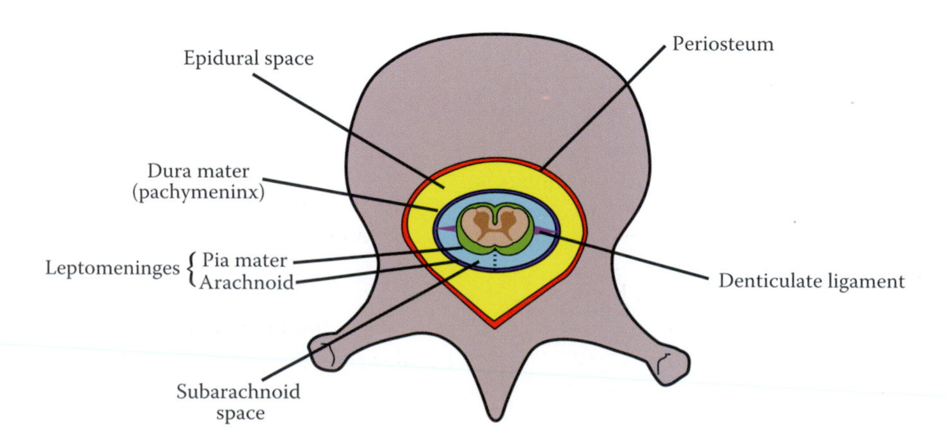

FIGURE 3.3 Spinal meninges, epidural and subdural spaces, as well as the attachments of the dentate ligaments are shown.

In its upper part, the dura adheres to the posterior longitudinal ligament. At the level of the second sacral vertebra, the spinal dura joins the filum terminale to attach to the coccyx as the coccygeal ligament.

The internal vertebral (epidural) plexus affects cerebrospinal fluid (CSF) pressure by forming continuous tamponade of the spinal dural sac. Increased intrathoracic or intra-abdominal pressure (coughing, sneezing, and straining during defecation, or abdominal compression) can thereby increase CSF pressure. Pain associated with parturition, where surgical intervention is not contemplated, can be alleviated by injection of local anesthetics into the epidural (extradural) space.

Epidural anesthesia is a versatile procedure that involves the administration of local anesthetics via a needle into the epidural space. It is utilized to supplement general anesthesia when a deep level of anesthesia is not needed. It provides stable hemodynamics during the surgical procedure and provides better postoperative pain control and rapid recovery. It can reduce pulmonary complications, gastrointestinal disorders and immune system disorders, as well as autonomic dysfunction associated with surgery. As a result, mortality and morbidity are equally reduced. It has been utilized in surgical procedures of prostate and bladder operations, management of chronic pain, cesarean section and labor pain, inguinal hernia, knee and hip surgery, and even coronary revascularization and aortic valve replacement. It is less likely to be associated with major hemodynamic complications when the blockade is done at or below T10. The varying dose of anesthetics used in epidural blockade in the elderly and obese patients may be related to the amount of epidural fat, which decreases with age and increases proportionally with obesity. In order to successfully administer anesthetic into the epidural space, appreciation of the surface structures is important. The spinous processes, which are useful landmarks in determining the level of epidural needle entry, show regional variations in direction. In the cervical and lumbar regions, the spinous processes are horizontal, while the thoracic vertebrae exhibit caudally directed processes. This explains the basis for horizontal needle entry into the cervical and lumbar epidural space and paramedian approach into the thoracic region. Needle entry must be distal to the conus medullaris, which corresponds to the lower border of L1 in adults and lower border of L3 vertebra in children. Epidural anesthetic in adults commonly introduced in the interspinous space between the L3 and L4 vertebrae. The line that connects the upper tubercles of the iliac crests corresponds to the spinous process of L4 or the intervertebral disc between L4 and L5. Diffusion of the anesthetic solution through the dural coverings of the emerging nerve roots acts on the nerve roots, which carry sensory, motor, and sympathetic fibers. The effect of the anesthesia depends on the blocked spinal segments and is first seen in the autonomic fibers and then in the nociceptive and thermal fibers, followed by the proprioceptive and, finally, the motor fibers. This type of anesthesia can be associated with certain complications such as meningitis, arachnoiditis, spinal root injury, epidural hematoma, epidural abscess, and cauda equina syndrome.

In saddle block or caudal anesthesia, the epidural space is reached via the sacral hiatus, utilizing a catheter. This commonly employed regional block in children is a useful adjunct in general anesthesia, but with short postoperative analgesia. It is indicated for infradiaphragmatic surgical procedures. The quality and level of the caudal blockade is dependent on the dose, volume, and concentration of the injected drug. Prior to the administration of the anesthetic, the posterior superior iliac spines and the sacral ligament between the sacral cornua are identified and palpated. Then the needle is passed at a 45° angle through the skin and sacral ligament proximal to the intergluteal line.

Spinal epidural abscesses may occur as a result of the posterior spread of infection directly from tuberculous vertebral bodies or secondary to epidural anesthesia or intervertebral disc surgery. They can also occur as a result of systemic disease and hematogenous spread of *Staphylococcus aureus* infection. Intravenous drug users and immunocompromised individuals have higher incidence of epidural abscess. Low back pain, which may progress gradually to involve motor, sensory, and/or bowel and urinary incontinence, may also be seen in this condition. Infection or abscesses associated with epidural space may be diagnosed via blood and CSF culture as well as computed tomography (CT) myelography of the spinal cord.

Spinal epidural hematoma can occur as a result of trauma or spinal diathesis, which partially or completely blocks the epidural space, producing sudden pain followed by sensory and motor deficits. The dura mater may also be the site of AVMA that is commonly seen in the thoracolumbar segments. Dural AVMA may eventually develop into subarachnoid hemorrhage, producing combined upper and lower motor neuron deficits.

The arachnoid mater is a loose, delicate, irregular, and trabecular layer that is continuous with cranial arachnoid mater. It is generally avascular and surrounds the spinal cord without following the sulci. As the spinal vessels pierce this layer and enter the subarachnoid space to supply the pia mater and spinal cord, it is invested by the arachnoid and pial cells. Like the dura mater, the arachnoid and pia mater fuse with the epineurium of the spinal nerves, thus rendering the spinal subarachnoid a closed space.

Arachnoiditis (inflammation of the arachnoid matter) may be caused by trauma or invasive imaging procedures (e.g., myelography), or it may remain idiopathic. This condition may produce adhesion of the leptomeninges, obliteration of the subarachnoid space, formation of arachnoid cysts, and possible vascular occlusion. Arachnoid cysts are congenital outgrowths that assume positions outside (extradural) or inside the dura (intradural). Extradural cysts, common in the thoracic region, may remain asymptomatic or produce compression of the spinal cord and/or roots. Intradural cysts may or may not communicate with the subarachnoid space.

The subdural space between the arachnoid mater and the dura mater represents a potential interval, which may accidentally be penetrated during induction of epidural anesthesia, producing toxic effects and eventual spinal cord damage.

Subdural abscess may occur as a result of underlying remote or contiguous infections such as dental, retroperitoneal or tuberculous abscesses. It may also be a spontaneous condition, producing fever, and back pain in the thoracic or lumbar region that radiates to areas of the spinal nerve distributions. Compression of the spinal cord as a result of an abscess may produce paraplegia or quadriplegia. Sensory and/or sphincteric deficits may also occur depending on the site and extent of the abscess.

Subdural hematoma may arise as a result of trauma or anticoagulant therapy or following LP. It produces signs and symptoms similar to subdural abscesses. However, paraplegia and quadriplegia usually occur within minutes to hours in individuals with subdural hematoma, compared to subdural abscess in which the deterioration may take days.

The pia mater, the innermost layer of meninges, is highly vascular and follows the anterior median fissure and posterior medial sulcus of the spinal cord. It consists of an epipial layer pierced by arteries that supply the spinal cord and an intima–pial layer that intimately adheres to the walls of the spinal cord, giving rise to the dentate ligaments (Figure 3.3). These ligaments, which are flat, triangular pial extensions that extend to the dura mater, course between the dorsal and ventral roots and act as suspensory ligaments for the spinal cord, resisting trauma-induced spinal displacement. These ligaments are associated with cervical, thoracic spinal nerves, extending from the level of the foramen magnum to the level of the first lumbar spinal nerve. Condensation of the pia mater between the conus medullaris and the second sacral vertebra is known as the filum terminale (Figures 3.1 and 3.2). Both pia and arachnoid mater form the leptomeninges, and they continue around the spinal nerves as perineurium. The subarachnoid space, between the arachnoid and pia mater, contains the cerebrospinal fluid and spinal arteries and veins. This space shows enlargement around the filum terminale and cauda equina and forms the lumbar cistern. Deep to the arachnoid mater is a perforated middle layer that becomes prominent on the dorsal and ventral surfaces of the spinal cord. Condensation of this layer forms the dorsal, dorsolateral, and ventral ligaments that connect the inner and outer parts of the middle layer.

LP is a procedure is performed to aspirate CSF from the subarachnoid space for the evaluation of signs of meningitis or subarachnoid hemorrhage. It may also be utilized to administer medications to the lumbar cistern. LP or spinal tap is contraindicated in individuals with increased intracranial pressure. This is based upon the fact that a LP may precipitate transtentorial herniation by suddenly reducing the pressure in the vertebral canal. The site of puncture in adults is usually in the L3–L4 or L4–L5 vertebral interspace, while in infants, a much lower level is indicated (L5–S1). In this procedure, the skin and interspinous ligaments are anesthetized while the patient lies on his/her side. The dura and arachnoid mater must be pierced to gain access to the subarachnoid space.

Spinal anesthesia is conducted by the injection of anesthetic solution into the lumbar cistern to block the lower thoracic, lumbar, and sacral spinal nerve roots. This procedure is performed when general anesthesia is not desired or to be avoided and yet relaxation of the muscle is required, in cesarean section, circumcision, herniorrhaphy. The anesthetic is introduced in L3–L4 or L4–L5 with the patient in lateral decubitus position, or sitting in a bent-over position.

Myelography, another procedure that utilizes the lumbar cistern, is used to visualize the vertebral column, spinal cord, and the posterior cranial fossa. A myelographic contrast medium is injected percutaneously via a needle into the lumbar cistern distal to the termination of the spinal cord. The spinal cord and spinal roots become discernible through a series of radiographic images. Since the contrast medium is radiopaque, the spinal cord and the nerve roots may appear radiolucent. For a detailed visualization, CT images may be obtained after contrast injection.

BLOOD SUPPLY

The vertebral artery (VA), the principal arterial source to the cervical segments of the spinal cord, is a branch of the subclavian artery, although occasionally, it may arise directly from the aorta, brachiocephalic trunk, or thyrocervical trunk (Figure 3.5). This artery has extracranial and intracranial portions. It runs posteriorly initially anterior to the stellate ganglion and then ascends in the transverse foramina of the upper six cervical vertebrae. The first segment of the VA (V1), which extends between its origin and the transverse foramen of the C6 vertebra, is the most likely site of arterial dissection. The V2 segment encompasses the portion of the VA within the transverse foramina of C6–C6. In order to gain access to the cranial cavity, it leaves the C1 transverse foramen (V3 segment) and courses on the upper border of the atlas in the sulcus for the VA before it makes a sharp posterior bend, piercing the posterior atlantooccipital membrane and dura mater. It enters the cranial cavity through the foramen magnum.

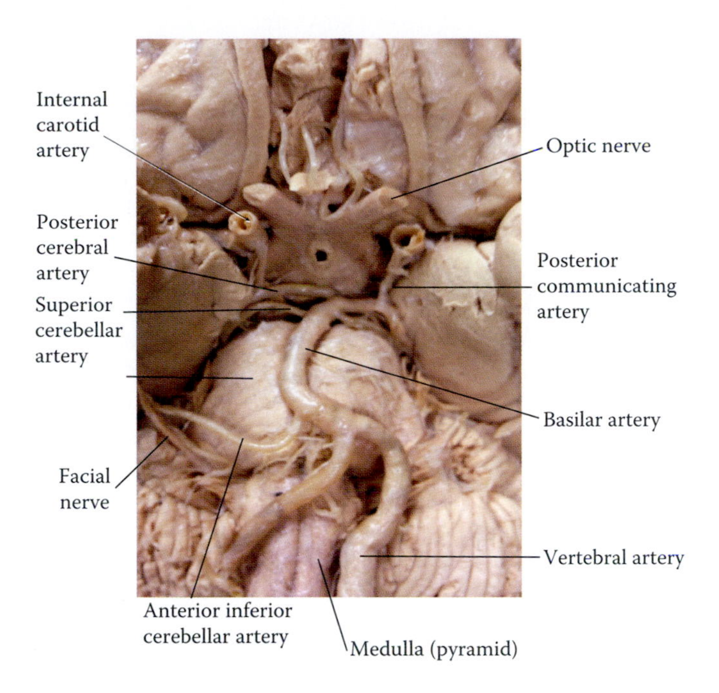

FIGURE 3.5 Vertebral and anterior spinal arteries are shown on the ventral surface of the medulla.

The V3 segment is prone to be affected in temporal arteritis. After it passes through the foramen magnum, it courses lateral to the medulla and anterior to the hypoglossal (V4 segment). It joins the corresponding artery of the opposite at the pontomedullary sulcus to form the basilar artery. It gives rise to the meningeal, anterior, and posterior spinal (Figures 3.6, 3.7, and 3.8) and medullary branches. It joins the VA of the opposite side to form the basilar artery (Figures 3.5 and 3.7). The vertebral arteries establish anastomosis with the multiple radicular arteries through the spinal branches, external carotid artery through the occipital branch, and subclavian artery through branches of the thyrocervical trunk and occipital artery.

The spinal cord is supplied by the anterior and posterior spinal arteries, which arise from the vertebral arteries, and by the multiple radicular arteries. The anterior spinal artery is often a single vessel lodged in the anterior median fissure that arises from the V4 segment of the VA. It is usually formed by the spinal branches of the VA and continues along the length of the spinal cord. It supplies primarily the anterior two-thirds of the spinal cords via central branches and, to a lesser extent, the medial medulla. The central branches of the anterior spinal artery are more abundant in the lumbosacral and cervical segments and less so in the thoracic segments. They supply the ventral horns, intermediate zone of the gray matter, corticospinal tracts, anterolateral system, and the area around the central canal.

Occlusion of the anterior spinal artery produces degeneration of the anterior two-thirds of the spinal cord and disruption of the pain and descending motor pathways bilaterally. This occlusion produces manifestations of Beck syndrome, which will be discussed later with the combined lesions of the sensory and motor systems. The anterior spinal artery in the thoracic segments may become too small, with an extremely narrow lumen

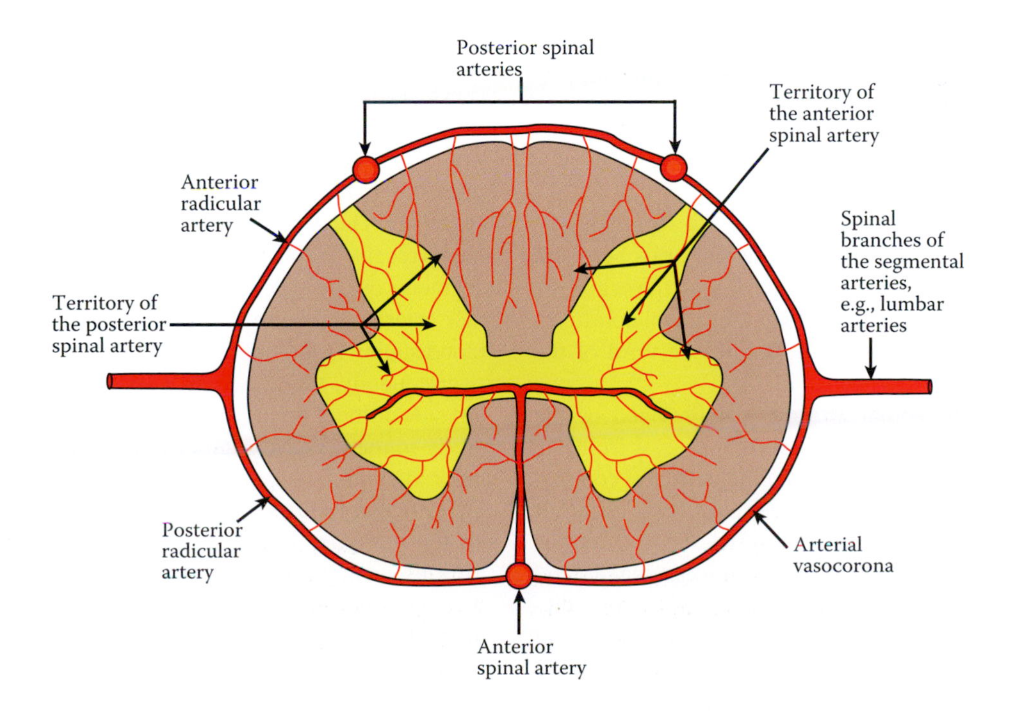

FIGURE 3.6 Distribution of the anterior and posterior spinal arteries and the formation of the arterial vasocorona.

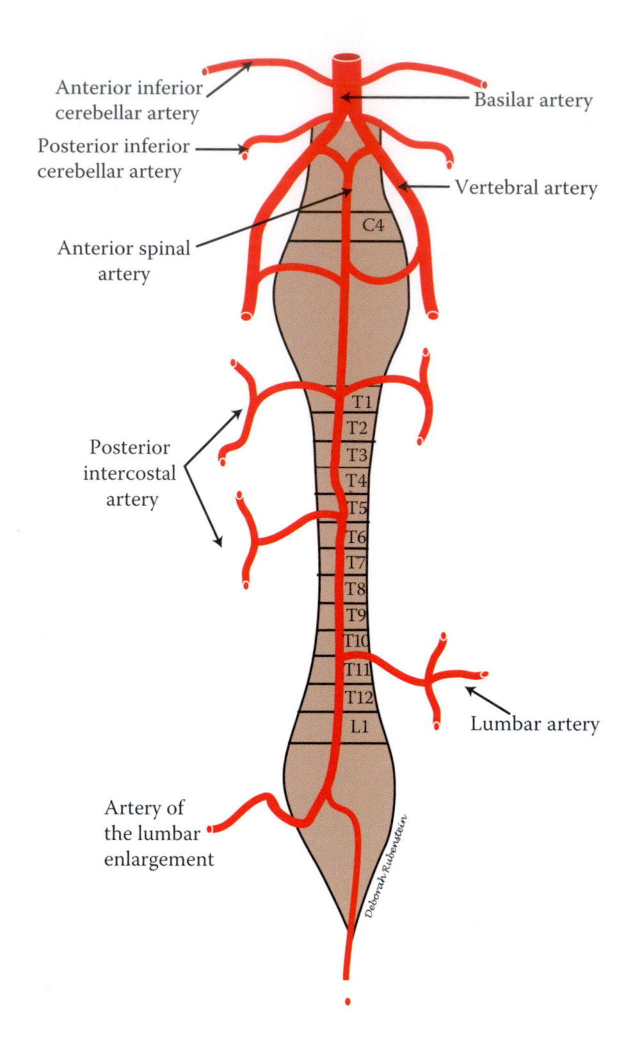

FIGURE 3.7 Schematic drawing of the anterior spinal artery and its connections to the radicular arteries.

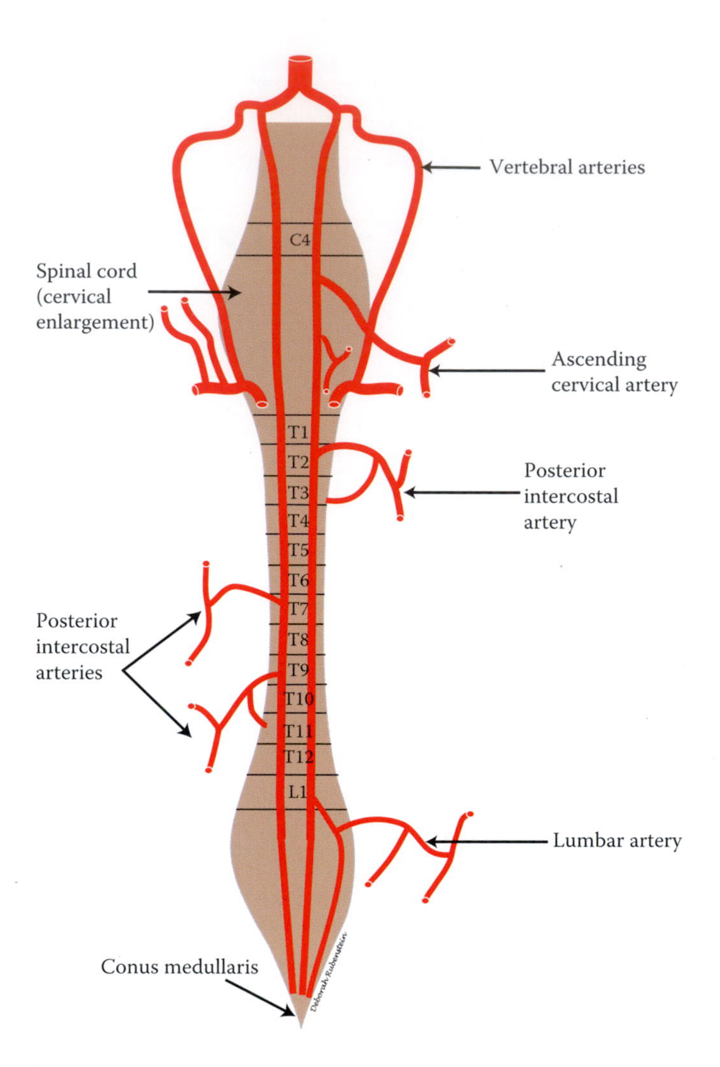

FIGURE 3.8 Posterior spinal arteries and territory of distribution. Note their connections to the multiple radicular arteries.

incapable of providing blood to the spinal cord. This is particularly true of segments T3–T7, which may receive one radicular artery only that accompanies the T4 or T5 roots.

The posterior spinal arteries arise from V4 segments and are reinforced by a number of small radicular branches from the segmental arteries, which vary according to the segments of the spinal cord. They descend along the dorsolateral part of the spinal cord, supplying the posterior one-third of the spinal cord.

Continuation of the spinal arteries is maintained by the radicular arteries that arise at each vertebral level from the neighboring segmental arteries outside the vertebral column, including the ascending cervical, deep cervical, posterior intercostal, lumbar, and lateral sacral arteries. These branches reach the spinal cord via the intervertebral foramina, ascend and descend, and then divide into anterior and posterior radicular branches that follow and supply the ventral and dorsal roots, respectively. The posterior radicular arteries are more numerous and maintain connections with the posterior spinal arteries. Six to ten anterior radicular arteries enter the vertebral canal via the intervertebral foramina of the lower cervical and lower thoracic as well as the upper lumbar part of the vertebral column; reach the spinal cord, joining the anterior spinal artery; and become the principal source of arterial blood to the thoracic, lumbar, sacral, and coccygeal spinal segments. The cervical segments and upper two thoracic segments receive blood supply from the subclavian artery. T3–T7 segments are provided with blood supply from thoracic radicular arteries, while the lower spinal cord (T8–CC1) receives blood supply from the lumbar radicular arteries.

Frequently, the radicular arteries are only present on the left side of the thoracic and lumbar spinal segments and bilaterally in the cervical segments. In 60%–65% of individuals, one radicular artery (artery of Adamkiewicz, arteria radicularis magna, or artery of the lumbar enlargement), generally on the left side, arises from the lower posterior intercostal arteries or upper lumbar arteries and frequently accompanies the ninth thoracic and second lumbar spinal roots. This artery bifurcates into a small ascending and a larger descending branch that join the anterior spinal artery. The descending branch eventually encircles the conus medullaris to join the posterior spinal artery. The artery of Adamkiewicz may supply approximately the lower two-thirds of the spinal cord, which extends from T8 to the coccygeal segment (Figures 3.6, 3.7, and 3.8). Roots that form the cauda equina are supplied by branches derived from the lumbar, iliolumbar, and lateral sacral arteries.

Spinal segments T1–T4 and L1 are predisposed to infarctions due to the lack of sufficient arterial anastomotic channels and the great distance between the radicular arteries. These watershed infarctions may be seen as a sequel to cardiac arrest, clamping of the aorta, or acute local ischemia. Occlusion of the artery of lumbar enlargement (artery of Adamkiewicz) may produce paraplegia (paralysis of the lower extremities and lower parts of the body), urinary incontinence, and loss of sensation from the lower extremities. Occlusive diseases of the anterior spinal artery (Beck syndrome), as a result of aortic dissecting aneurysm or atheroma, produce combined sensory and motor deficits.

VENOUS DRAINAGE

The spinal veins follow the pattern of distribution of the spinal arteries, forming a plexus in the spinal pia mater. The venous blood of the spinal cord first drains into small veins that open into central veins and then into the median, ventrolateral, and dorsolateral longitudinal veins. The ventrolateral and dorsolateral longitudinal veins accompany the corresponding roots of the spinal nerves. Cranially, spinal veins establish communication with the veins of the brainstem and cerebellum through the foramen magnum. Eventually, these venous channels open into the radicular veins and join tributaries of the internal vertebral (epidural) plexus. The epidural venous plexus lies in the vertebral canal and drains the red bone marrow contained in the vertebral bodies by joining the basivertebral veins and the external vertebral plexus. The basivertebral veins occupy the vertebral bodies and emerge as a single vein that drains into the internal vertebral (epidural) plexus. Eventually, the spinal veins drain through the epidural and external vertebral plexus into the intervertebral veins that connect with the vertebral, intercostal, lumbar, and lateral sacral veins.

Numerous connections exist at each intervertebral space between the epidural (Batson's internal vertebral) venous plexus and systemic veins including the pelvic veins, and the superior and inferior vena cavae via the azygos and hemiazygos veins. These valveless venous channels and connections may serve as a potential route of spread of cancer cells from the thyroid gland, breast, and prostate to the vertebral bodies. In a similar manner, Batson's venous plexus may also serves as a venous route for the spread of urinary tract infection to the vertebral bodies and the ensuing osteomyelitis. The internal vertebral (epidural) plexus affects CSF pressure by forming continuous tamponade of the spinal dural sac. Increased intrathoracic or intra-abdominal pressure (coughing, sneezing, and straining during defecation, or abdominal compression) can thereby increase CSF pressure.

INTERNAL ORGANIZATION

Each spinal segment consists of central gray and peripheral white matter that are connected by the corresponding gray and white commissures. The central canal is a tube that pierces the gray commissure of the spinal cord, ascends into the caudal medulla, and continues with the fourth ventricle. This canal does not stretch the entire length of the spinal cord and is frequently obliterated.

Gray Matter

The gray matter is a butterfly-shaped area with anterior and posterior horns that are present at all spinal levels. An additional lateral horn that lodges the intermediolateral columns (preganglionic sympathetic neurons) exists in the thoracic and upper two or three lumbar spinal segments. Gray commissures surround the central canal and separate it from the white matter. Most of the spinal cord neurons are small and propriospinal (90%), linking the ventral and dorsal horns within one segment or interconnecting several segments (intersegmental). The intermediate zone between the dorsal and ventral horns is generally formed by medium-sized neurons, while the largest neurons occupy the ventral horn.

Based upon the cytoarchitecture of the neuronal cell bodies, the gray matter is classified by Rexed into nine laminae and area or lamina X (Figure 3.9). True lamination is evident in the dorsal horn, and considerable overlap exists among certain laminae.

- Lamina I contains the posteromarginal nucleus, consisting of neurons that display horizontal dendrites in order to maximize their contact with the incoming fibers of the dorsal root. The dorsolateral tract of Lissauer separates this lamina from the surface of the spinal cord.
- Lamina II (substantia gelatinosa) has dark appearance on Nissl-stained sections due to the high neuronal population. It is thin in the thoracic but is thicker in cervical and lumbar spinal segments. It consists of Golgi type II neurons, receiving fibers that carry pain and temperature sensations. This lamina is the main processing center for nociceptive (noxious) stimuli in the spinal cord. Axons of these neurons contribute to the formation of the

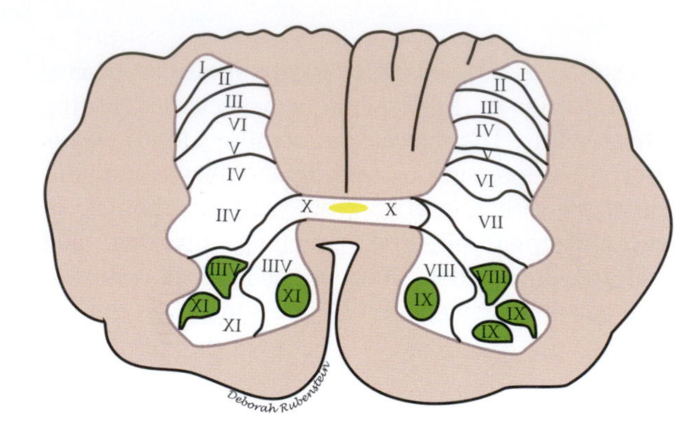

FIGURE 3.9 Rexed laminae and the cytoarchitecture of the gray columns of the spinal cord.

Lissauer zone (dorsolateral fasciculus). The latter, a well-developed bundle in the upper cervical segments, consists of myelinated and unmyelinated fibers that surround the dorsal root fibers, occupying the area between the apex of the dorsal horn and the surface of the spinal cord. This bundle, which also contains propriospinal fibers, ascends one or two segments within the spinal cord, allowing collaterals to be distributed to the posterior gray column.

- Laminae III and IV contain the proper sensory nucleus and occupy a large region of the dorsal horn. This nucleus contributes axons to the lateral spinothalamic tract and receives virtually all sensory modalities carried by the dorsal root.
- Lamina V occupies the neck of the posterior horn and establishes synapses with the corticospinal and rubrospinal tracts. The lateral part of this nucleus is known as the reticular nucleus.
- Lamina VI is present in the spinal cord enlargements, but absent particularly in the fourth thoracic through the second lumbar segments.
- Lamina VII forms the intermediate zone; receives fibers from the corticospinal and rubrospinal tracts; and contains the Clarke's, intermediolateral, and intermediomedial nuclei. Clarke's nucleus extends from the eighth cervical or first thoracic to the second or third lumbar spinal segments, giving rise to the dorsal spinocerebellar tract. The intermediolateral nucleus occupies the lateral horn between the first thoracic and the second or third lumbar spinal segments, providing preganglionic sympathetic axons. At the second, third, and fourth sacral spinal segments, this nucleus provides preganglionic parasympathetic fibers. The intermediomedial nucleus extends the entire length of the spinal cord and receives visceral afferents.
- Lamina VIII occupies the anterior horn in the spinal cord enlargements and contains commissural neurons, which receive axons of the vestibulospinal, pontine reticulospinal, and tectospinal tracts.
- Lamina IX contains α and γ motor neurons that innervate the extrafusal and intrafusal muscle fibers, respectively. The α motor neurons receive excitatory input from the descending pathways and the reflex arcs and inhibitory input from the propriospinal neurons. Excitatory input far exceeds the inhibitory projections by a ratio of 2:1. They give inhibitory recurrent branches to the interneurons (Renshaw cells), thus facilitating their action. In general, α motor neurons are arranged somatotopically, in which the abductor neurons are located anteriorly, the flexor neurons are positioned posteriorly, and the extensors as well as the adductor neurons maintain intermediate positions. In the lumbosacral segments, the neurons for the trunk are medial; the neurons that innervate the foot occupy a lateral position, while neurons for the leg and thigh have intermediate position. In the thoracic segments, lamina IX exhibits a similar somatotopic arrangement whereby the neurons associated with innervation of the abdomen lie medial to the intercostal neurons, and the neurons for the innervation of the back muscles and skin assume an intermediate position. In the cervical segments, the neurons that provide innervation to the hand lie to the lateral side of the neurons that innervate the forearm, whereas trunk neurons are the most medially positioned. Neurons for the arm and shoulder occupy a position medial to the forearm and lateral to the trunk neurons. These neurons are classified into tonic and phasic neurons. The tonic α motor neurons innervate the slow, oxidative–glycolytic muscle fibers, exhibiting slow conduction and the ability to readily depolarize. They are inhibited during rapid movement by the Renshaw cells. Phasic neurons display higher threshold and ability to maintain fast conduction, innervating the fast and oxidative–glycolytic muscles. Phasic neurons also send more recurrent branches to the Renshaw cells than the tonic neurons. The γ neurons are located among the α motor neurons, innervating the contractile parts of the muscle spindles. Both α and γ neurons are involved in voluntary movement via the α–γ coactivation and γ loop.
- Lamina X or area X, according to some unsupported claims, consists of small neurons that form the gray commissures around the central canal. It receives some afferents from the dorsal root fibers and contains neuroglial cells in its ventral part that send cytoplasmic extensions to the adjacent pia mater.

White Matter

The white matter occupies the peripheral part of the spinal cord and consists only of neuronal processes. The anterior white commissure connects the white matter on both sides, representing the site of decussation of the lateral and ventral spinothalamic tracts, as well as the ventral spinocerebellar and the anterior corticospinal tracts. The part of the white matter located between the entering fibers of the dorsal roots is known as the dorsal funiculus, containing the dorsal white columns. The part of the white matter that lies between the dorsal and ventral roots on each side is known as the lateral funiculus, containing the lateral corticospinal, rubrospinal, and lateral spinothalamic tracts. The area of the white matter between the emerging ventral roots is referred to as the ventral funiculus and contains the ventral spinothalamic, tectospinal, and reticulospinal tracts as well as the medial longitudinal fasciculus. A tract refers to a group of nerve fibers that have the same origin, destination, course, and function. A fasciculus shares common features of the tract, but the constituent fibers maintain diverse origins.

SPINAL CORD SEGMENTS

The cervical spinal segments are eight in number, are generally large, and have a greater mass of white matter. The dorsal funiculus is divided into a medial gracilis and a lateral cuneatus fasciculus (Figure 3.10). The thoracic segments (Figure 3.11), which are characterized by a small and distinct lateral horn, contain the intermediolateral cell column, which gives rise to the preganglionic sympathetic fibers. The dorsal funiculi of the upper six spinal segments contain the gracilis and cuneatus fasciculi, while the lower six thoracic segments contain only the gracilis fasciculus. Another important feature of the thoracic segments is the presence of Clarke's nuclear column, which is particularly well developed, in the lower two thoracic spinal segments. The axons of this nuclear column form the ipsilateral dorsal spinocerebellar tract that conveys unconscious proprioceptive information from the muscle spindles and Golgi tendons of the lower extremities. Additionally, the gray matters of the thoracic segments are tapered in an "H" shape. The axons of this nuclear column form the ipsilateral dorsal spinocerebellar tract that conveys unconscious proprioceptive information from the muscle spindles and Golgi tendons of the lower extremities. The lumbar segments (Figure 3.12) contain massive amounts of gray matter and relatively less white matter. The upper two lumbar segments contain the continuation of Clarke's nucleus and the intermediolateral columns. The sacral segments (Figure 3.13) are small compared to other segments and contain large amounts of gray matter. The intermediolateral cell column in the sacral spinal segments provides preganglionic parasympathetic fibers.

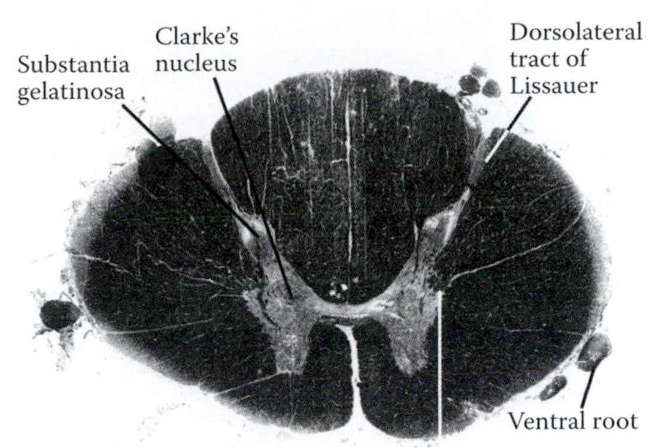

FIGURE 3.11 Transverse section through the fifth thoracic spinal segment, showing Clarke's nucleus and the intermediolateral column, substantia gelatinosa, and dorsal tract of Lissauer.

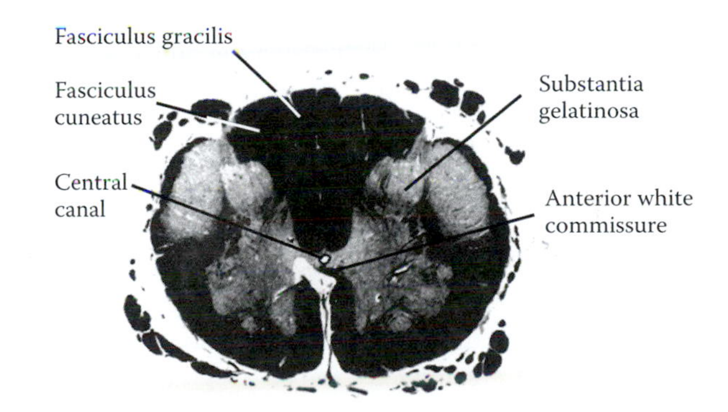

FIGURE 3.12 Section of the fourth lumbar segment showing the main characteristics of this level.

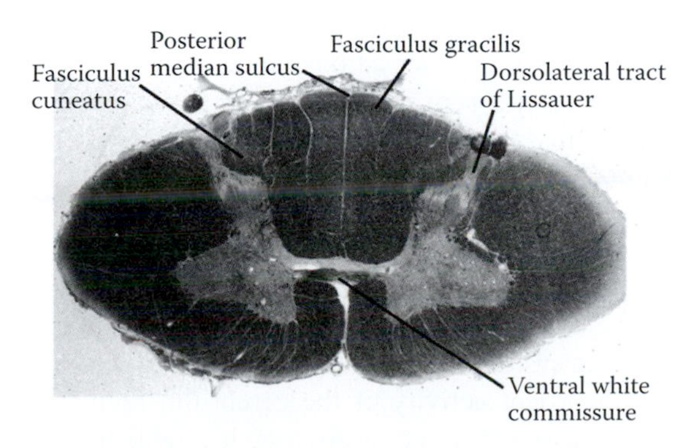

FIGURE 3.10 Fourth cervical segment. Note the wide transverse diameter and large amount of white and gray matter.

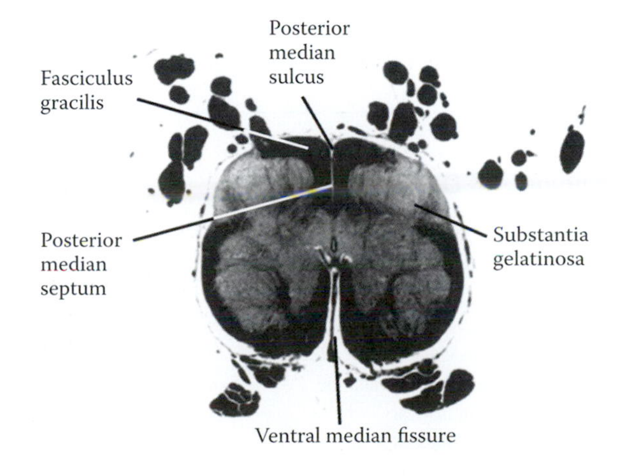

FIGURE 3.13 Section through the fifth sacral spinal segment.

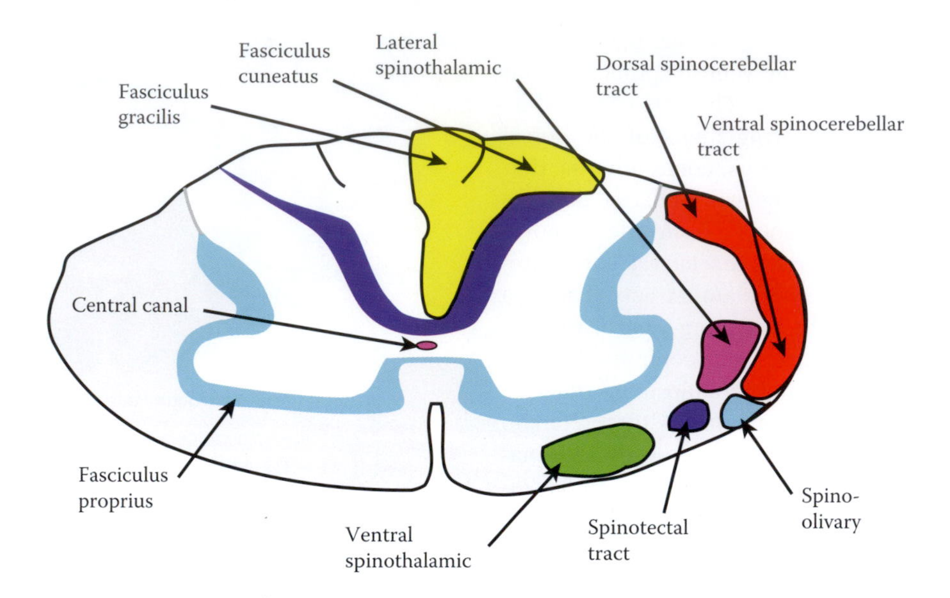

FIGURE 3.14 Principal ascending pathways within the spinal cord.

SPINAL PATHWAYS

Ascending Tracts

These ascending pathways (Figure 3.14) convey conscious and unconscious sensory information to the higher levels of the CNS. The first-order neurons for all ascending tracts from the body are located in the dorsal root ganglion (DRG) of the spinal nerves. The second-order neurons are located either in the gray matter of the spinal cord or in the brainstem. The ventral posterolateral (VPL) nucleus of the thalamus constitutes the third-order neurons for these pathways. The signals for the information conveyed by the ascending pathways are concerned with the regulation of muscle tone, joint sensation (position sense), vibration, pain and temperature sensations, discriminative tactile sensations, and intersegmental reflexes. These pathways may establish monosynaptic connections or utilize an extensive network of neurons and are contained in the funiculi of the spinal cord (Table 3.1).

Ascending Tracts in the Posterior Funiculus

The dorsal white columns transmit fine tactile and vibratory sense via the Pacinian corpuscles and position and movement sense (kinesthesia) from the muscle spindle. They also convey two-point discrimination of simultaneously applied blunt pressure points from the Ruffini corpuscles and stereognosis (ability to recognize form, size, texture, and weight of objects) via a variety of receptors.

TABLE 3.1

Ascending and Descending Tracts

Funiculus	Ascending Tracts	Descending Tracts
Posterior	Dorsal white columns	Interfascicular fasciculus and fasciculus septomarginalis
Lateral	Lateral spinothalamic dorsal and spinocerebellar	Corticospinal and rubrospinal
Ventral	Ventral spinothalamic, spino-olivary, spinoreticular, and spinotectal	Anterior corticospinal, vestibulospinal, reticulospinal, and medial longitudinal fasciculus

Ascending Tracts in the Lateral Funiculus

The lateral spinothalamic tract (neospinothalamic), also known as the lateral system, is a contralateral pathway that conveys thermal and painful sensations from somatic and visceral structures. Pain and temperature, received by the free nerve endings, enter the spinal cord via the lateral bundle of the dorsal root.

In the peripheral parts of the lateral funiculus, the dorsal and ventral spinocerebellar tracts are located, carrying unconscious proprioception from the lower extremity to the cerebellum.

Ascending Tracts in the Ventral Funiculus

The ventral spinothalamic tract (paleospinothalamic or anterior system) runs in the ventral funiculus and transmits signals associated with light touch, and possibly tickling, itching, and libidinous sensations. Since fine touch and discriminative tactile sensation are primarily carried in the dorsal columns, the clinical significance of this pathway is not clear.

The spino-olivary tract is a contralateral tract, conveying cutaneous information and afferents from Golgi tendon organs to the dorsal and medial accessory olivary nuclei. The spinoreticular tract is an integral part of the ascending reticular activating system that plays an important role in changing the electrocortical activity of the cerebrum, regulating the state of consciousness and awareness. It establishes a direct link between the spinal cord and the brainstem reticular formation, extending the entire length of the spinal cord. It also

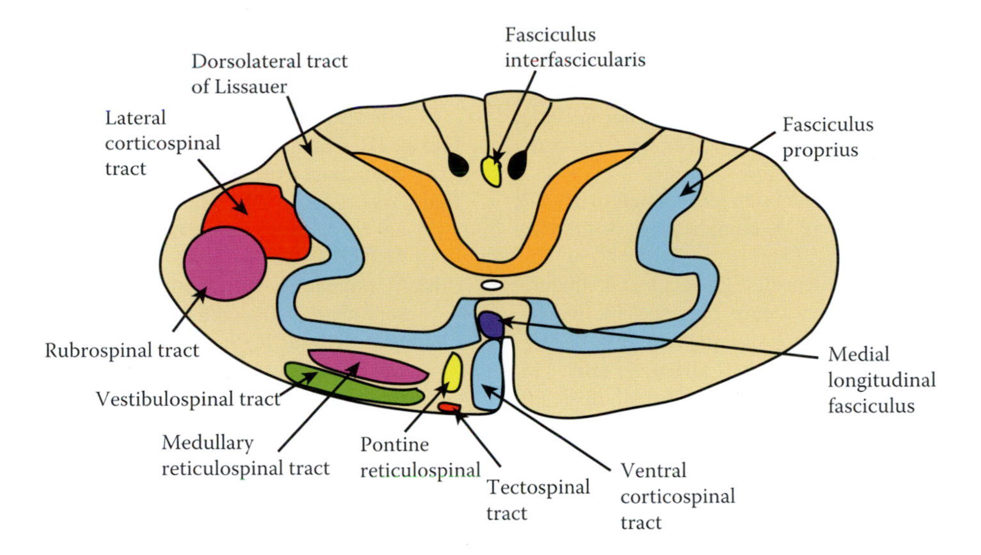

FIGURE 3.15 Section of the spinal cord showing the descending pathways.

contributes to the formation of the spinoreticulothalamic tract. The spinotectal tract is composed of axons of neurons that are derived from lamina VII of the spinal gray matter. Although the functional significance of this pathway is not clear, its role in modulating the transmission of pain, thermal, and tactile sensation awaits further study.

Descending Tracts

The descending pathways (Figure 3.15) deal with maintenance of posture and balance; control of visceral and somatic reflex activity, muscle tone, and motor activity in general; and modification of the sensory signals. The descending pathways include the corticospinal, rubrospinal, tectospinal, and interstitiospinal tracts. They also include the vestibulospinal and reticulospinal pathways, as well as descending autonomic pathways that are derived from the hypothalamus and the brainstem reticular formation.

Descending Tracts in the Lateral Funiculus

The lateral corticospinal tract is a phylogenetically new pathway that exists in man and other mammals. It continues to develop throughout the first 2 years of life. This pathway forms the largest crossed component of the corticospinal tract, controlling voluntary motor functions, especially movement associated with the digits.

The rubrospinal tract is a contralateral tract that may be traced from the superior collicular level of the midbrain to the thoracic or lumbosacral segments. This excitatory pathway, which regulates the neurons of the flexor muscles, originates from the magnocellular part of the red nucleus.

Descending Tracts in the Ventral Funiculus

The anterior corticospinal tract represents approximately 10%–15% of the corticospinal fibers and travels ipsilaterally in the spinal cord near the anterior median fissure. This pathway influences muscles of the upper extremity and the neck via synapses in the corresponding segments.

The vestibulospinal tracts include the lateral and medial vestibulospinal tracts. The lateral vestibulospinal tract is excitatory and runs the entire length of the spinal cord. The medial vestibulospinal tract is monosynaptically inhibitory and extends to the upper cervical segments.

The reticulospinal tracts comprise the ipsilateral pontine reticulospinal and predominantly ipsilateral medullary reticulospinal tracts. These tracts convey information received by the reticular formation from the cerebral cortex, cerebellum, cranial nerves, and hypothalamus to the spinal cord.

The medial longitudinal fasciculus is a composite bundle of ascending and descending fibers that originate from vestibular and reticular nuclei. This tract occupies the dorsal portion of the ventral funiculus.

The spinospinal tract (fasciculus proprius) is an intersegmental tract of ascending and descending fibers (crossed and uncrossed), which mediates the intrinsic reflex mechanisms of the spinal cord. It exists in all spinal funiculi, conveying information to higher segments prior to establishing contact with interneurons.

SUGGESTED READING

Abdel-Maguid TE, Bowsher D. Alpha and gamma-motoneurons in the adult human spinal cord and somatic cranial nerve nuclei. The significance of dendroarchitectonics studied by the Golgi method. *J Comp Neurol* 1979;186:259–70.

Appel NM, Lide RP. The intermediolateral cell column of the thoracic spinal cord is comprised of target-specific subnuclei: Evidence from retrograde transport studies and immunohistochemistry. *J Neurosci* 1988;8:1767–75.

Batson AJ, Sands J. Regional and segmental characteristics of the human adult spinal cord. *J Anat* 1977;123:797–803.

Biglioli P, Spirito R, Roberto M, Grillo F, Cannata A, Parolari A, Maggioni M, Coggi G. The anterior spinal artery: The main arterial supply of the human spinal corda preliminary anatomic study. *J Thorac Cardiovasc Surg* 2000;119:376–9.

Clark RG. Anatomy of the mammalian spinal cord. In Davidoff RA, ed. *Handbook of the Spinal Cord,* Vols. 2/3. New York: Dekker, 1984, 1–45.

Curry WT Jr, Hoh BL, Amin-Hanjani S, Eskandar EN. Spinal epidural abscess: Clinical presentation, management, and outcome. *Surg Neurol* 2005;63:364–71.

Doita M, Marui T, Nishida K. Anterior spinal artery syndrome after total spondylectomy of T10, T11, and T12. *Clin Orthop Rel Res* 2002;405:175–81.

Lubenow T, Keh-Wong E, Kristof K, Ivankovich O, Ivankovich AD. Inadvertent subdural injection: A complication of epidural block. *Anesth Analg* 1988;67:175.

McMenemin IM, Sissons GR, Brownridge P. Accidental subdural catheterization: Radiological evidence of a possible mechanism for spinal cord damage. *Br J Anaesth* 1992;69:417–9.

Rodriguez-Baeza A, Muset-Lara A, Rodriguez-Pazos M, Domenech-Mateu JM. The arterial supply of the human spinal cord: A new approach to the arteria radicularis magna of adamkiewicz. *Acta Neurochir* 1991;109:57–62.

Ross JS, Masaryk TJ, Modic MT, Delamater R, Bohlman H, Wilbur G, Kaufman B. MR imaging of lumbar arachnoiditis. *Am J Roentgenol* 1987;149:1025.

Schneider RC, Crosby EC, Russo RH, Gosch HH. Traumatic spinal cord syndromes and their management. *Clin Neurosurg* 1972;20:424–92.

Steffek M, Owczuk R, Szlyk-Augustyn M, Lasinska-Kowara M, Wujtewicz M. Total spinal anesthesia as a complication of local anaesthetic test-dose administration through an epidural catheter. *Acta Anaesthesiol Scand* 2004;48:1211–3.

Stein B. Intramedullary spinal cord tumors. *Clin Neurosurg* 1983;30:717–41.

Tekkok IH, Cataltepe K, Tahta K et al. Extradural hematoma after continuous extradural anesthesia. *Br J Anaesth* 1991;67:112–5.

Tobin WD, Layton DD. The diagnosis and natural history of spinal cord arteriovenous malformations. *Mayo Clin Proc* 1976; 51:637–46.

Ullah M, Salman SS. Localization of the spinal nucleus of the accessory nerve in the rabbit. *J Anat* 1986;145:97–107.

Vilming S, Kloster R, Sandvik L. When should an epidural blood patch be performed in postlumbar puncture headache? A theoretical approach based on a cohort of 79 patients. *Cephalagia* 2005;25:523–7.

Wakabayashi K, Takahashi H. The intermediolateral nucleus and Clarke's column in Parkinson's disease. *Acta Neuropathol* 1997;94:287–9.

Wiltse LL, Fonseca AS, Amster J, Dimartino P, Ravessoud FA. Relationship of the dura, Hofmann's ligaments, Batson's plexus, and a fibrovascular membrane lying on the posterior surface of the vertebral bodies and attaching to the deep layer of the posterior longitudinal ligament. An anatomical, radiologic, and clinical study. *Spine* 1993;18(8):1030–43.

4 Brainstem

The brainstem represents the infratentorial portion of the central nervous system (CNS), consisting of the mesencephalon (midbrain), pons, and medulla. It lies anterior to the cerebellum and is connected to it via the superior, middle, and inferior cerebellar peduncles, containing the fourth ventricle, cerebral aqueduct, and the central canal. Through the fourth ventricle, the cerebrospinal fluid (CSF) drains into the subarachnoid space and eventually into the systemic venous circulation. This part of the CNS is connected to the 3rd through the 12th cranial nerves and contains the associated motor and sensory nuclei. The central core of the brainstem contains deeply seated reticular nuclei that regulate sleep and control respiratory and cardiovascular centers.

MEDULLA

The medulla (Figure 4.1) represents the caudal part of the brainstem, continuing caudally with the spinal cord through the foramen magnum. The rostral end of the medulla is demarcated ventrally by the pontobulbar sulcus, giving passage to the abducens, facial, and vestibulocochlear nerves. Dorsally, it is bounded by a line joining the lateral recesses of the fourth ventricle.

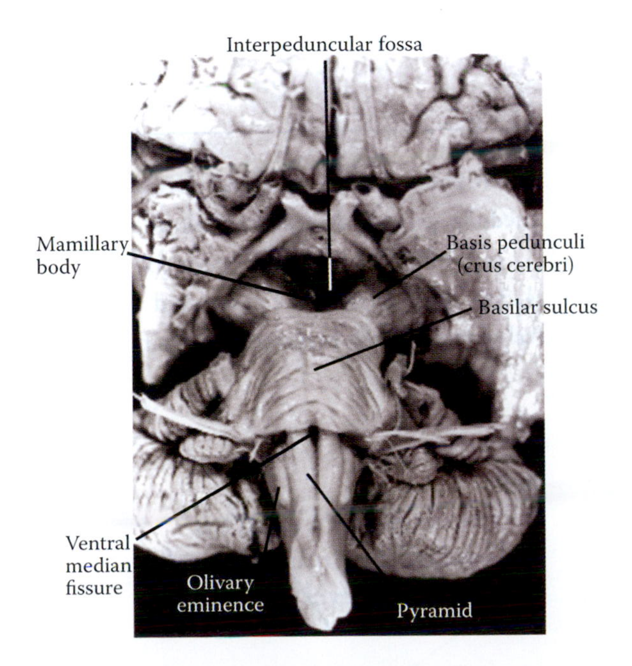

FIGURE 4.1 Ventral surface of the brainstem and its connections to the cerebellum. Some of the important features of the medulla, pons, and midbrain are shown in this picture.

FOURTH VENTRICLE

The fourth ventricle is derived partly from three parts: isthmus rhombencephali, metencephalon, and the myelencephalon. It is lined with ependymal cellsand has a diamondshaped floor (rhomboid fossa) and a roof that assumes the shape of a tent with the apex (fastigium) protruding into the cerebellum. Both dorsal surfaces of the medulla and pons contribute to the rhomboid fossa. The diamond-shaped rhomboid fossa is divided into upper and lower triangular areas, separated by an intermediate zone. The apex of the upper triangular area is defined by the cerebral aqueduct, the lateral walls by the superior cerebellar peduncles, whereas the base is limited by a line that interconnects the lateral recesses. The narrow zone between the upper and the lower triangular areas that extends through the lateral recesses is demarcated as the intermediate zone and is crossed by the stria medullaris. The lower triangular area is flanked by the inferior cerebellar peduncles, with its apex pointing toward the central canal and crossed by the ependymal layer of the obex. The rhomboid fossa is split in the middle by a median sulcus surrounded on both sides by ependymal protrusions, which are limited by additional sulci limitans. Sulci limitans expand to form the superior and inferior fovea in the upper and lower triangular areas, respectively. The vestibular area is located lateral to the inferior fovea, which continues more laterally with the cochlear area. Caudal to the inferior fovea and medial to the area postrema, the vagal trigone becomes visible, which overlies the dorsal motor nucleus of vagus. The ependymal protrusions in the superior triangle are known as the facial colliculi, formed by the abducens nuclei and facial nerve fibers, while the surrounding sulci limitans expand to form a pigmented area known as the locus ceruleus. Similarly and in the lower triangular area, these protrusions form the hypoglossal trigones that contain the hypoglossal nuclei.

The fourth ventricle is connected to the third ventricle via the cerebral aqueduct and to the cerebellomedullary cistern via the foramina of Luschka (lateral apertures) and Magendie (median aperture). It continues caudally with the central canal of the spinal cord and is bounded inferolaterally by the inferior cerebellar peduncles that merge into the medulla and by the gracilis and cuneatus (dorsal column) tubercles and superolaterally by the superior cerebellar peduncles. The roof has two components, a cranial and a caudal part. The cranial part is formed by the superior cerebellar peduncles and the superior medullar velum, while the caudal part is composed of the inferior medullary velum, an incomplete layer, and ependyma, with the pia mater. Within the caudal roof, caudal to the nodule of the cerebellar vermis, a median aperture (foramen of Magendie) connects the fourth ventricle to the cerebellomedullary cistern. The caudal roof also contains a midline choroid plexus, which extends into the lateral recesses of the ventricle into the foramina of Luschka. Foramina of Luschka extend if complete to the subarachnoid space around the pontomedullary junction. The choroid plexus is suspended by the tela choroidea that continues with the tenia. The site of converging tenia at the inferior angle of the rhomboid fossa is known as the obex, which is inferomedial to the area postrema.

The dorsal surface of the rostral half of the medulla (open part) forms the lower part of the rhomboid fossa, while the caudal half (closed part) contains the central canal. The junction of the caudal and rostral medulla is demarcated on the dorsal surface by the obex. The ventral median fissure of the medulla is continuous with the corresponding fissure in the spinal cord, bounded on both sides by the pyramids, and it is obliterated at its caudal part by the decussating fibers of the corticospinal tracts. The olivary eminence lies on the lateral side of the pyramid and is formed by the underlying inferior olivary nuclear complex. The anterolateral (preolivary) sulcus, which contains the filaments of the hypoglossal nerve, separates the pyramids from the olivary eminence. The postolivary sulcus is the site of attachment for the glossopharyngeal, vagus, and accessory nerves.

Caudally, the posterior surface of the medulla contains the posterior median sulcus and is flanked on both sides by the fasciculi gracilis. The medially located fasciculi gracilis are separated from the laterally located fasciculi cuneatus by the cranial extension of the posterior intermediate septum. These fasciculi terminate in the corresponding tubercles that overlie their respective nuclei. The gracilis and cuneate nuclei receive information concerning conscious proprioception, two-point discrimination, vibratory sense, and discriminative tactile sensation from the upper and lower parts of the body. Axons of gracilis and cuneate neurons project to the contralateral ventral posterolateral nucleus of the thalamus via the medial lemniscus. The gracilis fasciculus also acts as a conduit for impulses that originate from stretch receptors and Golgi tendon organs in the lower extremity (unconscious proprioception) and terminates in Clarke's nucleus. The cuneate fasciculus also contains sensory impulses generated from the stretch receptors and Golgi tendon organs of the upper extremity (unconscious proprioception) en route to the accessory (external) cuneate nucleus. The caudal part of the medulla, between the fasciculus cuneatus and the accessory nerve, contains a prominence produced by the underlying spinal trigeminal tract and nucleus. The central canal runs close to the posterior surface and is continuous with the corresponding canal in the spinal cord caudally and the fourth ventricle rostrally.

The *rostral* (open) *part of the medulla* is bounded laterally by the inferior cerebellar peduncles that meet at the obex. It contains the fourth ventricle and forms the lower half of the rhomboid fossa. In this region, the rhomboid fossa contains the hypoglossal trigone medially and the vagal trigone laterally. These trogons overlie the hypoglossal and dorsal motor nucleus of the vagus, respectively. The stria medullaris emanates from the arcuate nuclei, crossing the inferior cerebellar peduncles and the vestibular area of the rhomboid fossa. The vestibular area overlies the medial and inferior vestibular nuclei corresponding to the site of foramen of Luschka.

The medial part of the medulla containing the pyramids, medial lemniscus, hypoglossal nerve and nucleus, and inferior olivary nucleus is supplied by the anterior spinal artery and the bulbar branches of the vertebral artery, while the dorsal and lateral parts of the caudal medulla, which contain the associated cuneatus, gracilis, dorsal vagal, and solitary nuclei and fasciculi, receive blood supply from the posterior spinal and posterior inferior cerebellar branches of the vertebral artery (Figure 4.2). The latter arterial branch provides

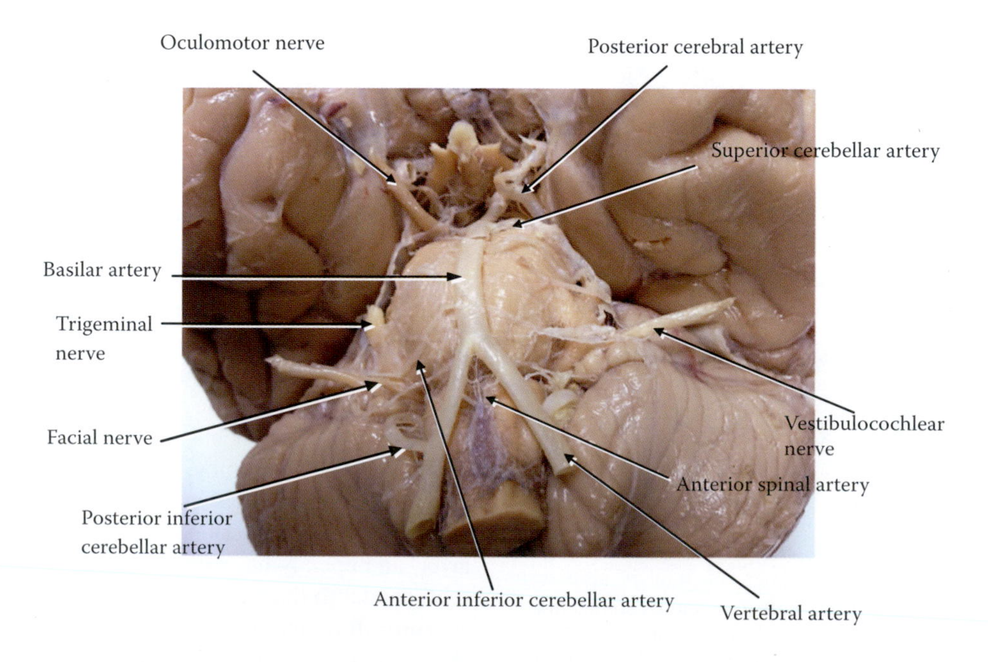

FIGURE 4.2 Relationship of the ventral surface of the brainstem to the vertebrobasilar arterial system is shown. Most of the associated cranial nerves are also seen.

blood supply to the lateral medulla and inferior cerebellar peduncle.

Venous drainage of the caudal medulla is maintained by the anterior and posterior spinal veins. The rostral medulla drains into the sigmoid, superior, or inferior petrosal sinuses. Medullary structures are clearly prominent at three levels, and each level may present unique features and contain nuclei or pathways that extend to a more caudal or rostral levels. These levels are comprised of the *spinomedullary junction, midolivary level, rostral medulla, and pontomedullary junction*.

Caudal Medulla

Motor Decussation Level

This level represents the spinomedullary junction (Figure 4.3) which marks the gradual transition between the spinal cord and the medulla. It contains, for the most part, structures that extend from the spinal cord with same additional unique neurons, specific to this level of the medulla such as the gracile and cuneate nuclei. A unique feature of this level is the *pyramidal decussation*, which marks the site of crossing of the corticospinal fibers.

> A lesion at the site of pyramidal decussation produces cruciate hemiplegia, which is characterized by paralysis of the ipsilateral upper extremity and contralateral lower extremity.

At a more rostral level, the spinal trigeminal nucleus and tract replace the substantia gelatinosa and the dorsolateral tract of Lissauer, respectively. The *spinal trigeminal nucleus* extends from the midpons to the upper segments of the spinal cord, representing the rostral extension of the substantia gelatinosa. It receives thermal, painful, and tactile sensations from the head region through branches of the trigeminal, facial, glossopharyngeal, and vagus nerves. The fibers that terminate in this nucleus form the spinal trigeminal tract lateral to the corresponding nucleus. Within the spinal trigeminal nucleus and tract, the ophthalmic nerve fibers are caudal to the more rostral mandibular nerve fibers, and the maxillary nerve fibers maintain an intermediate position.

> Since the pain and thermal fibers terminate in the most caudal portion of this nucleus and tract, excision of the spinal trigeminal tract (tracteotomy) may be performed, although rarely, at the level of the medulla to alleviate the intractable pain associated with trigeminal neuralgia.

The *gracilis and cuneate fasciculi* are prominent at this level and occupy the corresponding positions to the spinal cord. The most significant characteristics of this level are the decussation of the corticospinal tracts and the appearance of the dorsal column nuclei. The crossed fibers represent nearly 85% of the corticospinal tracts. The uncrossed fibers form the anterior corticospinal tract, descend in the anterior funiculus, and later cross at the anterior white commissure. Some uncrossed fibers pass into the lateral funiculus, forming the anterolateral corticospinal tract. The *medial longitudinal fasciculus* (MLF) is displaced laterally by the pyramidal fibers. The gray matter around the central canal is markedly expanded into the reticular formation. The spinocerebellar and spinothalamic tracts occupy the same position as in the spinal cord.

Level of Sensory Decussation

The *gracilis nucleus* (Figures 4.4, 4.5, and 4.6) occupies a larger area at this level, stretches rostrally to the level of the obex, and is covered by a thin strip of the gracilis

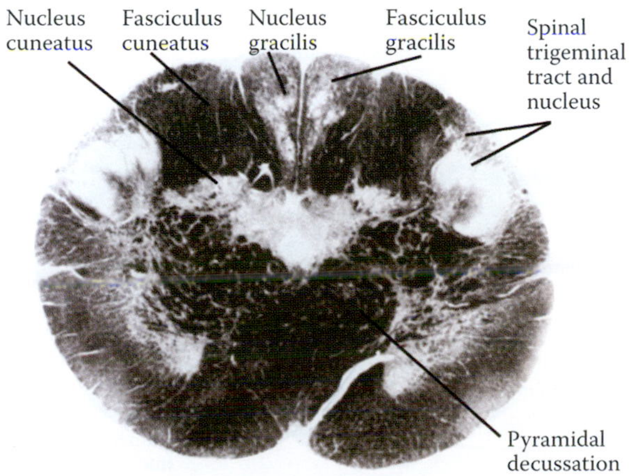

FIGURE 4.3 Level of the decussation of the corticospinal tracts. The spinal trigeminal tract and nucleus and the dorsal column nuclei are clearly demonstrated.

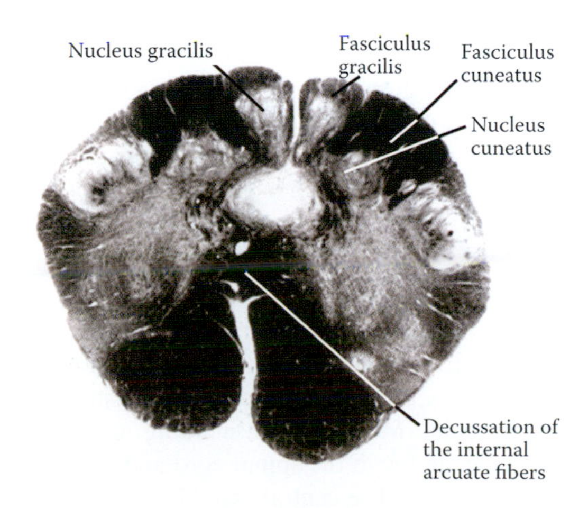

FIGURE 4.4 Section through the caudal medulla at the level of the sensory decussation. At this level, the medial lemnisci are formed. The cuneate and gracilis fasciculi maintain similar positions to Figure 4.3.

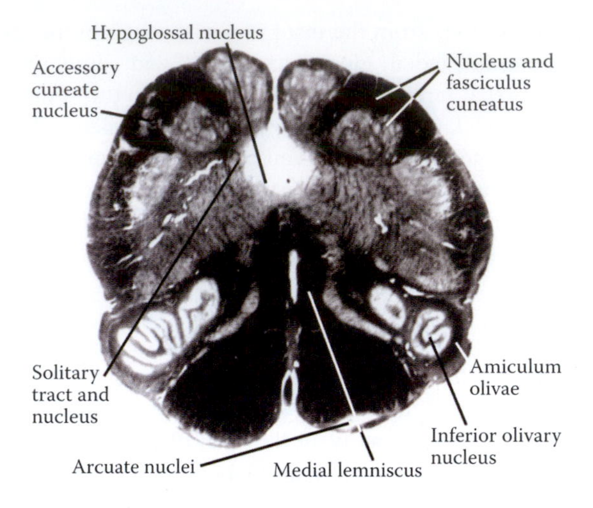

FIGURE 4.5 Caudal medulla rostral to the level shown in Figure 4.4. The inferior olivary, hypoglossal, solitary, gracilis, and cuneatus are clearly evident.

The area postrema, located in the ependyma of the rhomboid fossa immediately rostral to the obex, is a chemoreceptor trigger zone that regulates blood pressure and emesis, and responds to apomorphine and certain glycosides. Stimulation of this area may account for the emetic action of certain medications. Stimulation of this area, which is rich with dopaminergic receptors, by the administration of high doses of levodopa can induce nausea. Dopamine antagonists can suppress emesis by acting on this area. Additionally, the area postrema, an integral part of the circumventricular organs, lacks the blood–brain barrier system and thus allows substances to freely enter the brain. Brain barriers are structures that selectively allow certain substances to enter the CNS and exclude others.

fasciculus. The laterally positioned *cuneate nucleus* also increases in size and extends more rostrally than the gracilis nucleus, reaching the midolivary level. Also, at this level, the fasciculus cuneatus is greatly reduced. The internal arcuate fibers, axons of the dorsal column neurons, run ventromedially through the reticular formation, decussating in the midline and continuing as the medial lemniscus on the opposite side. This sensory decussation (decussation of the internal arcuate fibers) is a landmark for this level of the medulla. The *accessory (lateral) cuneate nucleus* initially appears at this level as a lateral appendage to the cuneate nucleus, receiving information from Golgi tendon organs, muscle spindles, and tactile receptors of the upper extremity. It conveys this information to the ipsilateral cerebellum, via the cuneocerebellar tract, and also to the caudal part of the ventral posterolateral (VPLc) nucleus of the thalamus. The *central gray* expands into the reticular formation, containing the lateral, dorsal, and ventral reticular nuclei. The spinal trigeminal tract and nucleus occupy a dorsolateral position. The hypoglossal nucleus, the dorsal motor nucleus of vagus, the solitary nucleus, the ambiguus nucleus, the inferior olivary nuclear complex, and the arcuate nuclei are first seen at this level. These nuclei will be discussed at the midolivary level.

Midolivary Level

The midolivary level (Figures 4.6 and 4.7) contains important medullary structures such as the medial and inferior vestibular, inferior olivary, hypoglossal, solitary, arcuate, and dorsal motor vagal nuclei. Most of the ascending and descending pathways associated with the spinal cord and lower medulla continue at this level. The central canal is converted into the fourth ventricle at this level of the medulla, while the fourth ventricle expands to maintain a close relationship with certain medullary nuclei.

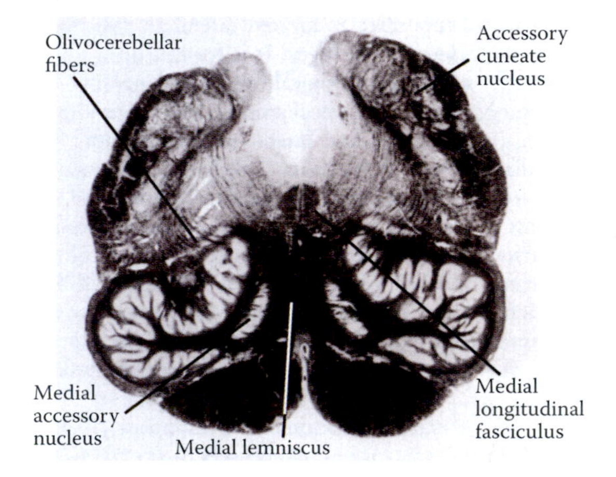

FIGURE 4.6 The prominent inferior olivary nucleus and expansion of the fourth ventricle are characteristics of this level. The medial longitudinal fasciculus and medial lemniscus maintain similar positions to Figure 4.5.

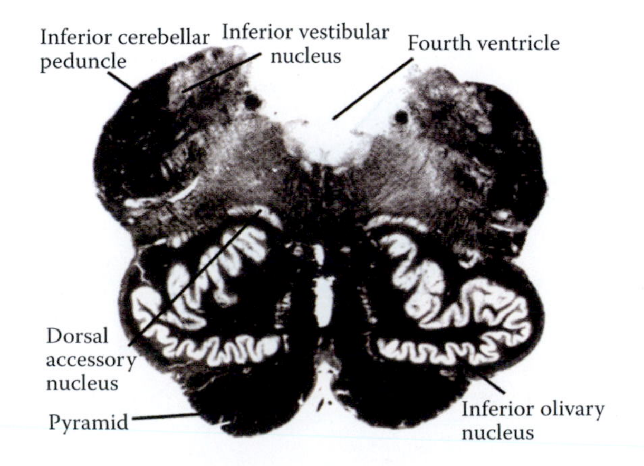

FIGURE 4.7 Medulla at a more rostral level than in Figure 4.6. Note the great expansion of the inferior olivary and vestibular nuclei, as well as the fourth ventricle.

The *medial and inferior vestibular nuclei* replace the dorsal column nuclei at this level of the medulla. The medial vestibular nucleus (MVN) lies medial to the inferior vestibular nucleus, continuing rostrally with the superior vestibular nucleus. Both the medial and inferior vestibular nuclei receive primary vestibular fibers and convey the information through the juxtarestiform body to the cerebellum. Information received by the MVN is transmitted to the cervical segments of the spinal cord through the predominantly ipsilateral medial vestibulospinal tract, a component of the MLF. Ascending fibers from the medial and inferior vestibular nuclei also project to the extraocular motor nuclei via the MLF. The MLF contains axons of the vestibular nuclei that project to the extraocular motor nuclei and to the spinal cord.

The *inferior vestibular nucleus*, the smallest of the vestibular nuclei, extends from the upper end of the nucleus gracilis to the pontomedullary junction. This nucleus is located between the MVN and the inferior cerebellar peduncle. It is characteristically speckled due to crossing of longitudinally oriented primary vestibular fibers. Secondary vestibulocerebellar fibers originate almost exclusively from this nucleus and enter the cerebellum through the juxtarestiform body, terminating in the flocculonodular lobe of the cerebellum.

The *arcuate nuclei* are cerebellar relay nuclei that occupy a position ventral to the pyramids. These nuclei are derived from the rhombic lip and project to the cerebellum via the inferior cerebellar peduncle. Axons of the arcuate nuclei pursue either a midline course, decussating in the midline and form the stria medullaris of the fourth ventricle, or a more lateral course superficial to the inferior olivary nucleus, forming the posterior external arcuate fibers. The arcuate nuclei, like the pontine nuclei, receive cortical input and convey this information to the opposite cerebellum.

The *inferior olivary nuclear complex*, the most characteristic feature of this level, occupies the area above the pyramid. It is the largest nucleus in the medulla, which projects to the cerebellum. It consists of the principal and accessory olivary nuclei. The principal olivary nucleus receives the ipsilateral rubro-olivary component of the central tegmental tract. It also receives bilateral cortico-olivary fibers, projections from the periaqueductal gray matter, crossed fibers from the cerebellar cortex, and projections from the accessory oculomotor nuclei. These fibers enclose the inferior olivary nuclear complex as *amiculum olivae*. The axons of the principal *inferior olivary nucleus* form the largest component of the inferior cerebellar peduncle. This crossed olivocerebellar tract projects to the lateral parts of the cerebellar cortex. The medial and dorsal accessory olivary nuclei receive impulses from the vestibular, gracilis, and cuneate nuclei and from the spinal cord. They convey this information to the medial portions of the cerebellar cortex.

The *hypoglossal nucleus* occupies the hypoglossal trigone in the floor of the caudal fourth ventricle. It consists of multipolar neurons that supply general somatic efferents (GSEs) to the lingual muscles. This nucleus extends from the level of the inferior olivary nuclear complex to the level of the stria medullaris of the fourth ventricle. Caudally, it is ventral and paramedial to the central gray matter. Its axons sweep ventrally, emerging through the anterolateral sulcus between the olivary eminence and the pyramid. The hypoglossal nucleus is divided into ventral or lateral and medial parts subdivided into smaller subnuclei. The motor neurons of the hypoglossal nucleus that innervate the retrusor muscles are located in the dorsal part, whereas neurons that innervate the protrusor muscle are located ventrally in the nucleus. Further study of the organization of this nucleus has revealed that the medial part innervates the transverse and vertical lingua as well as the genioglossus muscles, whereas the lateral part supplies the styloglossus, hyoglossus, as well as superior and inferior longitudinal lingua muscles. The hypoglossal nucleus receives bilateral corticobulbar fibers mainly to the medial part, with contralateral predominance to the rest of the nucleus.

The *perihypoglossal nuclei* are groups of neurons located adjacent to the hypoglossal nucleus. This group includes the nucleus intercalatus, nucleus of Roller, and nucleus prepositus hypoglossi. The latter nucleus occupies the same position as the hypoglossal nucleus, but at a more rostral level. Through their connection to the extraocular motor nuclei, cerebellum, and vestibular nuclei, the perihypoglossal nuclei play an important role in controlling eye movements.

The *dorsal motor nucleus of the vagus* occupies the area dorsolateral to the hypoglossal nucleus and forms the vagal trigone, a triangular eminence in the fossa rhomboidea. This nuclear column extends from the level of the hypoglossal nucleus both caudally and rostrally and gives rise to parasympathetic preganglionic (general visceral efferent [GVE]) fibers destined to the thoracic and abdominal viscera.

The *nucleus ambiguus* lies medial to the lateral reticular nucleus and above the inferior olivary nuclear complex. Neurons of this nucleus gives rise to special visceral efferent fibers (SVEs), which distribute via the glossopharyngeal and vagus nerves, and general visceral efferents (GVEs) via the vagus nerve. The fibers of the ambiguus nucleus join the cranial part of the accessory nerve and travel within branches of the vagus nerve to be distributed to the laryngeal, pharyngeal, and palatal muscles.

The *solitary tract* is located lateral to the dorsal motor nucleus of the vagus. It is formed by fibers of the vagus, glossopharyngeal, and facial nerves. The *solitary nuclear complex* surrounds the solitary tract, comprising the medial and lateral nuclear groups. The medial portions of the solitary nucleus unite caudal to the obex, forming the commissural nucleus of the vagus nerve. The medial subdivision and the caudal portion of the lateral subdivision of the solitary nucleus receive general visceral afferents (GVAs), which convey information from baroreceptors and chemoreceptors, while the dorsal portion of the lateral nucleus receives taste (special visceral afferent) sensation. The solitary nuclear complex conveys general visceral information to the respiratory, cardiovascular, and gastrointestinal centers of the brainstem and also to the autonomic neurons in the spinal cord. The part of the solitary nucleus (gustatory subnucleus) that deals with taste sensation projects to the ventral posteromedial (VPM) nucleus of the thalamus via the solitariothalamic tract.

At this level of the medulla, the *reticular formation* contains neurons for the cardiovascular and respiratory centers. The reticular neurons that remain distinct at this level are the lateral reticular, paramedian reticular, nucleus reticularis gigantocellularis, nucleus reticularis parvocellularis, nucleus raphe obscurus, and raphe pallidus. The *lateral reticular nucleus* lies dorsal to the inferior olivary complex and is primarily a relay nucleus that projects to the cerebellum through the inferior cerebellar peduncle. It receives afferents from the red nucleus, cerebral cortex, and spinal cord. The *paramedian nuclei* are cerebellar relay nuclei that lie parallel to the median raphe. The *nucleus reticularis gigantocellularis* occupies the medial zone of the medulla, modulates somatic and visceral motor activity, as well as controls muscle tone through the medullary reticulospinal tract. The *nucleus reticularis parvocellularis* lies in the lateral (sensory) part of the reticular formation. The *nucleus raphe obscurus and raphe pallidus* consist of serotonergic neurons.

The less visible *inferior salivatory nucleus* can be seen within the reticular formation of the medulla, which provides parasympathetic fibers to the glossopharyngeal nerve. These fibers synapse in the otic ganglion and later supply the parotid gland. At this level, the *inferior cerebellar peduncle* becomes discernible and continues throughout the upper medulla, occupying the area lateral to the spinal trigeminal tract and nucleus. It is formed primarily by the cerebellar afferents such as the spinocerebellar tract, cuneocerebellar tract, and so forth. Lateral to the raphe nuclei and dorsal to the pyramids, the *MLF* occupies a vertical position, containing only descending fibers, which project primarily to the spinal cord.

Rostral Medulla

In the rostral medullary level (Figure 4.8), enlargement of the inferior cerebellar peduncle and the appearance of the dorsal and ventral cochlear nuclei dorsolateral to this peduncle will be noticed. Expansion of the medial and inferior vestibular nuclei will be noted at this level. Additionally, the hypoglossal nucleus is replaced at this level by the nucleus prepositus hypoglossi. The reticular formation is expanded, and the dorsal motor nucleus of the vagus has disappeared. A similar position to the midolivary level is maintained by the spinal trigeminal nucleus, which is crossed by fibers of the glossopharyngeal nerve.

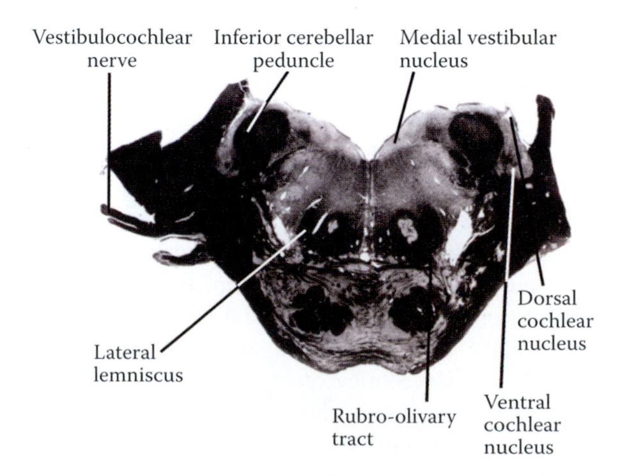

FIGURE 4.8 Rostral medulla. At this level, the inferior olivary nuclei are reduced, the middle cerebellar peduncle is visible, and the cochlear nuclei assume a dorsolateral position to the inferior cerebellar peduncle.

The *nucleus raphe magnus* contains the main serotonergic neurons, projecting bilaterally to the spinal cord. This projection courses in the lateral funiculus, inhibiting the spinal neurons that facilitate pain transmission.

Pontomedullary Junction

At the pontomedullary junction, the cerebellum and the lateral vestibular nucleus appear for the first time; other structures may maintain similar locations and dimensions or may undergo reduction relative to lower levels. The *cerebellum* forms the roof of the fourth ventricle, and the flocculus can be observed. The *inferior cerebellar peduncle* maintains its large size, and is bounded dorsally by the cochlear nuclei. In addition to the *medial and inferior vestibular nuclei*, the *lateral vestibular (Dieter's) nucleus* becomes apparent. The *medial lemniscus* begins to assume a horizontal position and move between the pyramid and the diminishing olivary nuclear complex. You will notice a specific reduction in the size of the *inferior olivary nuclear complex* and, specifically, the accessory olivary nuclei. Furthermore, the *nucleus ambiguus*, the *spinal trigeminal tract and nucleus*, and the solitary nucleus maintain their presence at this level. The *pyramids* begin to disperse, forming fasciculi, as the transition to the basilar pons begins. Additionally, the *reticular formation* shows expansion. The site of junction of the medulla, cerebellum, and pons form the *cerebellopontine angle*, a common site for acoustic neuromas.

PONS

The pons (Figures 4.9 and 4.10) forms the mid portion of the brainstem and the rostral part of the rhombencephalon. It is connected to the cerebellum via the middle cerebellar peduncle. It is bounded rostrally and caudally by the pontocrural (between the pons and midbrain) and the pontobulbar (between the medulla and pons) sulci, respectively. Within the pontocerebellar angle, the pontobulbar sulcus gives passage to the abducens nerve in the midline and the facial and vestibulocochlear nerves more laterally. The dorsal surface of the pons forms the rostral half of the rhomboid fossa, whereas the ventral surface lies adjacent to the basilar part of the occipital bone (clivus) and the dorsum sella of the sphenoid bone. On its ventral surface, the pons is demarcated centrally by the basilar sulcus, which contains the basilar artery. On both sides of the basilar sulcus, the descending cortical motor fibers (corticospinal and corticobulbar tracts) form the pontine protruberances, which are supplied by the pontine branches of the basilar artery.

Occlusion of the pontine arteries may produce signs and symptoms of locked-in syndrome, a motor disorder that affects the limbs and trunk, with the exception of eye movements. For additional information on this condition, see Chapter 20, "Motor Systems."

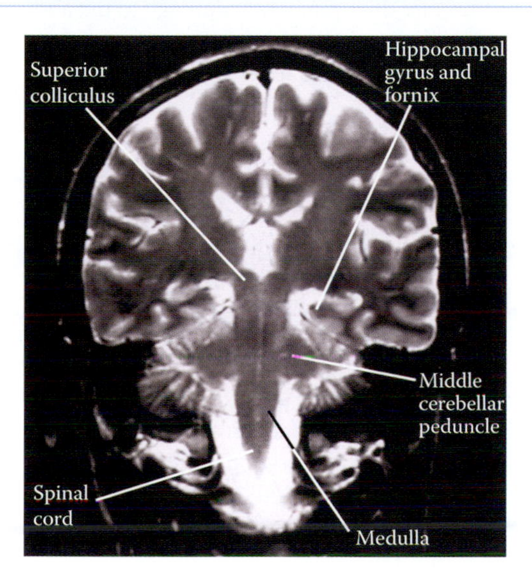

FIGURE 4.9 Magnetic resonance imaging (MRI) scan illustrating the pons and the prominent middle cerebellar peduncle.

Ventral to the middle cerebellar peduncle, the trigeminal nerve emerges from the midpons. The transverse section of the pons reveals a dorsal tegmentum and a ventral basilar portion. The tegmentum is the rostral continuation of the medullary reticular formation, while the basilar pons consists of the longitudinal corticospinal and corticobulbar fibers, as well as the transverse pontocerebellar fibers.

Paramedian branches of the basilar artery supply the medial pons, and the short circumferential branches nourish the pontine nuclei, corticospinal and corticobulbar tracts, and some of the trigeminal nuclei. The long circumferential branches of the basilar artery and some branches from the anterior inferior cerebellar artery supply the caudal pontine tegmentum. The superior cerebellar artery supplies the tegmentum of the rostral pons. At the pontobulbar sulcus, the *basilar artery* is formed by the union of the vertebral arteries, ending in the rostral pons by dividing into the posterior cerebral arteries (Figures 4.11 and 4.12). It supplies the pons, cerebellum, midbrain, and temporal and occipital lobes of the brain. This vessel gives rise to the anterior inferior cerebellar, labyrinthine, pontine, superior cerebellar, and posterior cerebral arteries. Pontine veins open primarily into the sigmoid sinus, or the petrosal sinuses.

The pons exhibits certain unique structures and features, some of which may continue caudally in the medulla and spinal cord and rostrally with the midbrain. All these neuronal structures are best identified at the *caudal pons*, *midpons*, and *rostral pons*.

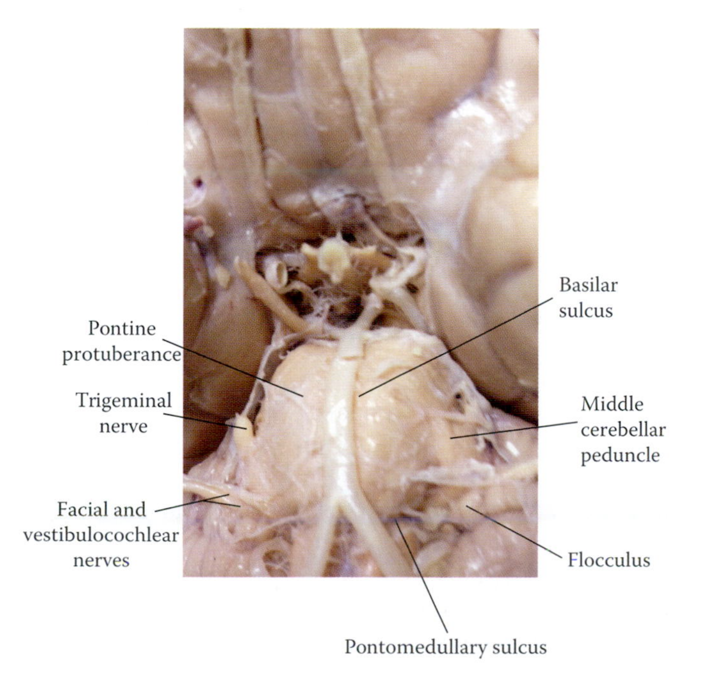

FIGURE 4.10 Ventral surface of the brainstem. The middle cerebellar peduncle is separated from the pons by the trigeminal nerve, and the pontomedullary sulcus marks the exit of the abducens, facial, and vestibulocochlear nerves.

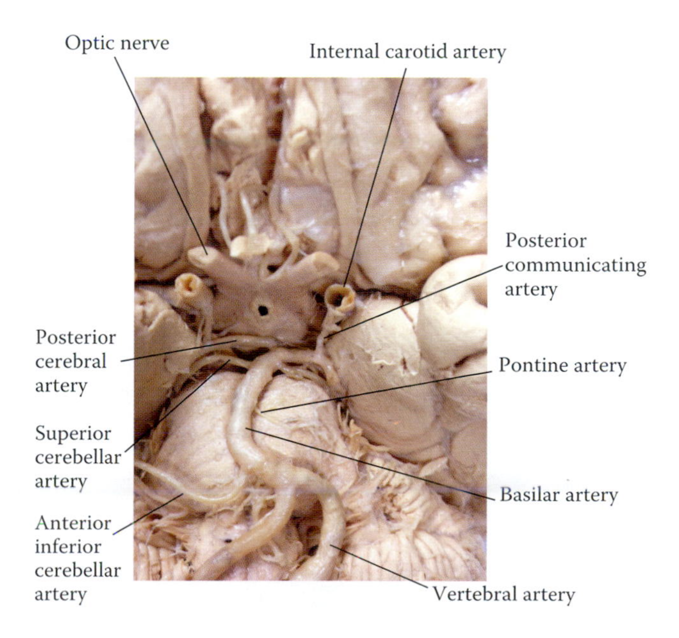

FIGURE 4.11 Basilar artery on the ventral surface of the pons. The anterior inferior cerebellar, superior cerebellar, and posterior cerebral arteries are shown. Notice the connection of the vertebrobasilar system to the internal carotid artery via the posterior communicating artery.

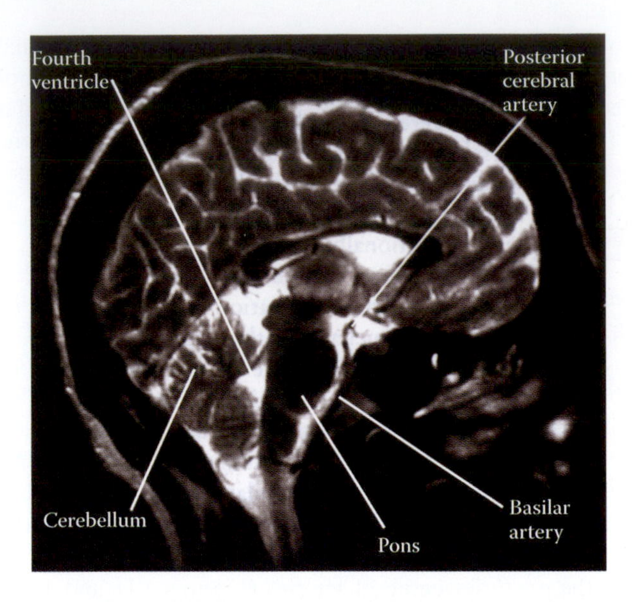

FIGURE 4.12 MRI scan illustrates the pons ventral to the cerebellum separated by the fourth ventricle. The basilar artery is also identified.

Caudal Pons

There are characteristic neuronal elements at the caudal pons, which include the abducens, facial motor, superior vestibular, and superior olivary nuclei (Figure 4.13). Appearance of the lateral lemniscus and continuation of the basilar pons are additional features of this level.

The *abducens nucleus* (GSE) lies deep to the ependyma of the rhomboid fossa and as a component of the facial colliculus. It is surrounded by the motor fibers of the facial nerve. This nucleus is unique among cranial nerve motor nuclei in that it contains two populations of neurons. The larger population consists of the alpha motor neurons, which form the abducens nerve, whereas the smaller population consists of neurons that send axons through the contralateral MLF to the motor neurons of the oculomotor nucleus, innervating the medial rectus muscle during lateral gaze.

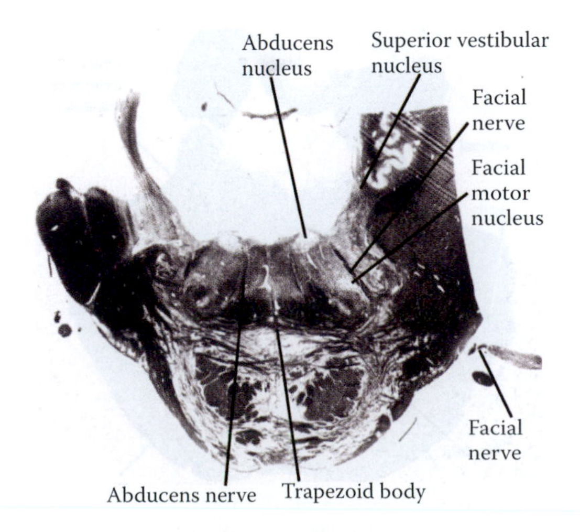

FIGURE 4.13 Caudal pons and the level of the abducens and facial nuclei. The medial lemniscus assumes a horizontal position, and the middle cerebellar peduncle is enlarged. Trapezoid body and superior vestibular nucleus are illustrated.

> The presence of these two neuronal populations within the abducens nucleus may account for the distinct difference in the deficits produced by lesions of the abducens nerve versus the abducens nucleus. A lesion that only damages the abducens nerve results in medial strabismus and diplopia, while damage to the abducens nucleus produces signs of lateral gaze palsy (inability to look to the side of the lesion).

The *facial motor nucleus* (SVE) occupies the lateral tegmentum, adjacent to the superior olivary nucleus, and gives rise to the motor fibers (SVE) that supply the muscles of facial expression, stapedius, stylohyoid, and posterior belly of the digastric muscle. These fibers encircle the abducens nucleus (internal genu), form the facial colliculus, and later emerge through the pontocerebellar angle. The *superior salivatory nucleus* (GVE) lies adjacent to the caudal end of the facial motor nucleus. It provides preganglionic parasympathetic fibers to the pterygopalatine and submandibular ganglia via the greater petrosal and chorda tympani, respectively. Eventually, the generated impulses enhance lacrimation, salivary secretion, and secretion of mucus glands of the palate, nose, and pharynx.

The *lateral vestibular* (Deiters) *nucleus* special somatic afferent (SSA) consists of multipolar giant neurons, which are the source of the principal vestibulospinal tract. This nucleus is located in the lateral portion of the ventricular floor and extends from the rostral end of the inferior vestibular nucleus to the level of the abducens nucleus. It receives primary vestibular fibers and sends both crossed and uncrossed ascending fibers via the MLF, in a symmetric manner, to the abducens, trochlear, and oculomotor nuclei. The *superior vestibular nucleus* (SSA), the most rostral vestibular nucleus, lies inferior to the superior cerebellar peduncle. It receives primary vestibular fibers and sends ipsilateral secondary vestibular fibers through the MLF, predominantly to the trochlear nucleus, and to the intermediate and dorsal cell columns of the oculomotor nuclear complex that innervate the inferior oblique and inferior rectus, respectively. The *ventral and dorsal cochlear nuclei* (SSA) are located dorsolateral to the inferior cerebellar peduncle, with the ventral cochlear nucleus, the largest, appearing to be continuous with the dorsal cochlear nucleus. The dorsal cochlear nucleus forms the acoustic tubercle in the floor of the fourth ventricle. These cochlear nuclei receive auditory impulses via the central processes of the spiral ganglia and convey the processed auditory information through the acoustic striae to the superior olivary nuclei and trapezoid nuclei and eventually to the inferior colliculi via the lateral lemniscus. The *trapezoid body* is formed mainly by the decussating ventral acoustic striae from the ventral cochlear, superior olivary, and trapezoid nuclei. This bundle of fibers occupies a midline position ventral to the medial lemniscus.

The *superior olivary nucleus* is an auditory relay nucleus, which extends from the level of the facial motor nucleus to the level of the trigeminal motor nucleus. It receives collaterals from the acoustic striae of both sides. This nucleus contributes fibers to the trapezoid body and the lateral lemniscus. The *lateral lemniscus* represents the main ascending auditory pathway and is located lateral to the superior olivary nucleus, containing the corresponding nucleus that receives secondary and tertiary auditory fibers.

The *spinal trigeminal nucleus* lies lateral to the facial motor nucleus and ventral to the lateral vestibular nucleus. It continues rostrally with the chief sensory nucleus of the trigeminal nerve. This nucleus receives pain and temperature sensations (GSA) from the head region through branches of the trigeminal, facial, glossopharyngeal, and vagus nerves. It provides axons to the ventral trigeminal nucleus. Descending fibers from the trigeminal ganglion form the spinal trigeminal tract lateral to the corresponding nucleus. The *inferior cerebellar peduncle* reaches its maximum size at this level and connects the medulla to the cerebellum, transmitting most of the cerebellar afferent fibers from the spinal cord. It consists of a medial portion (juxtarestiform body) and a lateral portion (restiform body). The *MLF* contains ascending vestibulo-ocular fibers and axons of the internuclear neurons of the abducens nucleus en route to the ventral cell column of the oculomotor nuclear complex. It also contains fibers of the interstitiospinal, medial vestibulospinal, and pontine reticulospinal tracts.

The *reticular formation* at this level of the pons contains the nucleus reticularis pontis caudalis, which contributes to the pontine reticulospinal tract. The area of the reticular formation adjacent to the abducens nuclei has been given special consideration due to its involvement in lateral gaze palsy. This region, the paramedian pontine reticular formation (PPRF), is thought to contain the center for lateral gaze, receives contralateral cortical input from the frontal eye field directly or via the superior colliculus, and projects to the ipsilateral abducens nucleus. Ischemic degeneration of the paramedian pontine reticular formation may occur as a result of occlusion of the basilar artery. The basilar part of the pons consists primarily of the descending corticofugal fibers such as the corticospinal, corticobulbar, and the pontocerebellar fibers. Fibers of the pontocerebellar arise from the pontine nuclei, constituting an indirect pathway from the cerebral cortex to the cerebellum through the middle cerebellar peduncle. As it descends through the pons toward the medulla, the central tegmental tract deviates laterally, occupying the center of the pontine tegmentum. The *middle cerebellar peduncle* (Figure 4.9), the largest cerebellar peduncle, may remain visible at this level, consisting exclusively of the pontocerebellar fibers. The *medial lemniscus* occupies a horizontal position in the ventral part of the tegmentum, dorsal to the trapezoid body.

Midpons

The midpontine level (Figures 4.14 and 4.15) is characterized by the presence of the trigeminal nerve fibers and principal sensory, mesencephalic trigeminal, and motor trigeminal nuclei. The distinct appearance of the superior cerebellar peduncle and visibility of the middle cerebellar peduncle are additional features of this pontine level.

The *principal* (chief) *trigeminal sensory* (Figures 4.14 and 4.15) nucleus is located lateral to the fibers of the trigeminal nerve and is concerned with pressure and tactile sensations. In this nucleus, the mandibular fibers occupy a dorsal position to the maxillary fibers, which assume an intermediate position, whereas the ophthalmic fibers occupy a ventral position. The ascending fibers from the ventral part of this nucleus contribute to the ventral trigeminal lemniscus, while the dorsal part comprises the dorsal trigeminal lemniscus.

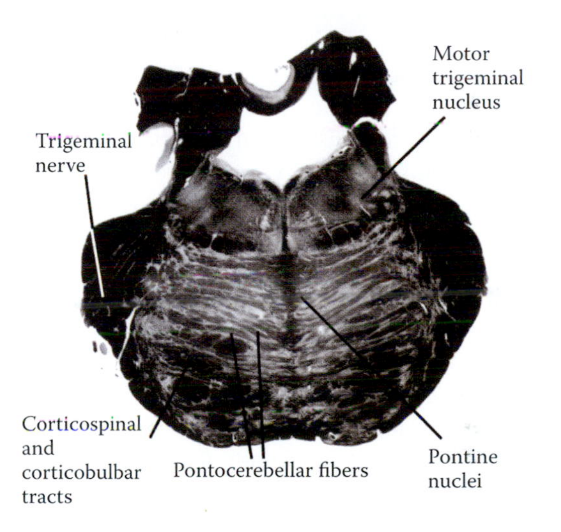

FIGURE 4.14 Section of the pons at the level of the trigeminal nerve. Corticospinal and corticobulbar tracts and the pontine nuclei occupy the entire basilar pons.

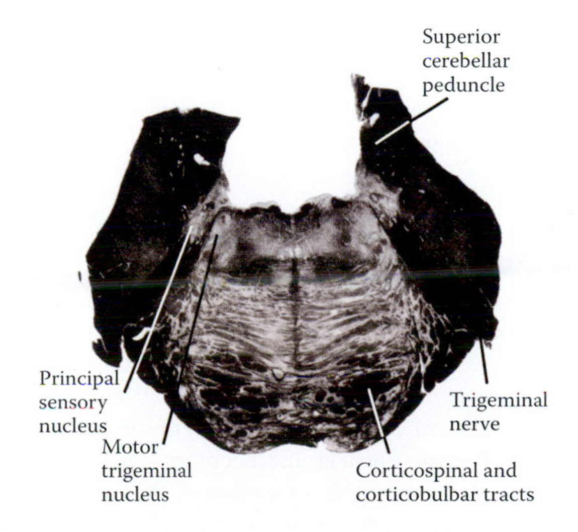

FIGURE 4.15 Section of the pons at the level of the trigeminal nerve showing the motor and principal sensory nuclei.

The most rostral part of the spinal trigeminal nucleus (pars oralis) merges with the principal sensory nucleus.

The *motor trigeminal nucleus* lies medial to entering fibers of the trigeminal nerve and provides innervation through its mandibular division (V3) to the muscles of mastication, tensor tympani, tensor palatini, and anterior belly of the digastric muscle. The motor trigeminal nucleus mediates the jaw jerk reflex and receives bilateral corticobulbar fibers (via interneurons). The *mesencephalic nucleus* is the only sensory nucleus that contains unipolar neurons that are retained within the CNS. This nucleus extends from the level of the trigeminal nerve to the caudal midbrain. It primarily conveys proprioceptive impulses from the muscles of mastication to the motor trigeminal nucleus, mediating the monosynaptic jaw jerk reflex. It also receives input from the facial and ocular muscles, temporomandibular joint, and peridontium. In that regard, it may be considered homologous to some degree to Clarke's nucleus of the spinal cord. The mesencephalic tract is formed by processes of the corresponding nucleus and provides collaterals to the motor root of the trigeminal nerve. The mesencephalic tract is formed by processes of the corresponding nucleus, providing collaterals to the motor root of the trigeminal nerve.

The *superior cerebellar peduncle* lies dorsolateral to the fourth ventricle and contains cerebellar efferents, which have not yet undergone decussation. The *middle cerebellar peduncle* is pierced by the trigeminal nerve, losing its connection to the cerebellum at this level. The *nucleus reticularis pontis* oralis is part of the medial reticular zone that contributes to the pontine reticulospinal tract. The *superior central nucleus* is also seen at this level of the pons. The *basilar pons* expands at this level, containing essentially the same structures seen in the caudal pons. The *fourth ventricle* is narrower and is bounded dorsally by the superior medullary velum.

Rostral Pons

At the level of the isthmus (Figure 4.16), the fourth ventricle terminates, and the cerebral aqueduct begins. It is the site of decussation of the trochlear nerve fibers and the appearance of the pigmented locus ceruleus. The basilar pons assumes its greatest size, and the lemniscal triad occupy a dorsal position in the pontine tegmentum.

> Since the superior medullary velum contains the decussating fibers of the trochlear nerves, a lesion or a mass at this site can produce bilateral trochlear nerve palsy (superior medullary velum syndrome).

At more rostral levels, the central gray matter expands around the fourth ventricle and becomes the periaqueductal gray. At the same time, expansion of the basilar part of the pons and narrowing of the tegmental portion occurs. Additionally, the medially directed fibers of the superior cerebellar peduncles begin entering the tegmentum. The locus ceruleus and the trigeminal mesencephalic nucleus occupy basically the same position.

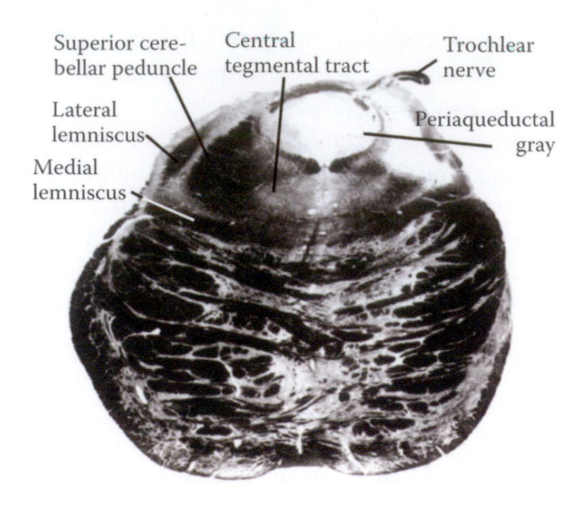

FIGURE 4.16 Rostral pons (isthmus). Note the decussation of the trochlear nerves in the superior medullary velum and crossing of the superior cerebellar peduncles within the tegmentum.

Dorsolateral to the superior cerebellar peduncle, the *lemniscal trigone* occupies the lateral part of the tegmentum. This trigone is formed by the lateral lemniscus dorsally, the medial lemniscus ventrally, and the spinal lemniscus, assuming an intermediate position. The trigeminal lemniscus appears dorsal to the medial lemniscus, and the central tegmental tract becomes a very prominent tegmental structure at this level.

The *reticular formation* at this level of the pons consists of a number of nuclei, which include the reticulotegmental, superior central (median), and dorsal raphe nuclei. The *reticulotegmental nucleus* is a pontine paramedian reticular nucleus that receives input from the dentate nuclei of the cerebellum and the cerebral cortex. This nucleus projects back to the contralateral cerebellar cortex and the ipsilateral vermis. The *superior central* (median) *nucleus* lies dorsal to the reticulotegmental nucleus. It sends impulses to the superior colliculus, pretectum, hippocampal formation, and mammillary bodies. The *dorsal raphe nucleus* projects to the lateral geniculate nucleus, the neostriatum, the substantia nigra, the pyriform lobe, and the olfactory bulb. Both the superior central and dorsal raphe nuclei consist of serotonergic neurons that project via the medial forebrain bundle to the hypothalamus, which, in turn, projects to the substantia nigra, intralaminar thalamic nuclei, and septal area. Since the medulla and pons contribute to the formation of the fourth ventricle and share important relationships with this cavity, a brief account of this ventricle will serve a useful purpose.

MIDBRAIN

The midbrain (Figures 4.17, 4.18, 4.19, and 4.20) represents the shortest part of the brainstem and is derived from the unmodified third brain vesicle. It is bounded rostrally by an imaginary line that passes behind the posterior commissure and the mammillary bodies and caudally by another line that connects the pontocrural sulcus and the posterior borders of the inferior colliculi. The midbrain consists of the tectum, tegmentum, and basis pedunculi. Each half of the midbrain, excluding the tectum, is known as the cerebral peduncle. Four rounded eminences connected by commissures constitute the tectum (quadrigeminal plate). The larger and more grayish upper pair of eminences is known as the superior colliculi, while the smaller lower pair is termed the inferior colliculi.

For descriptive purposes, the midbrain (Figures 4.18, 4.19, and 4.20) may be classified into superior and inferior collicular levels. These two levels contain common features and structures that include the *cerebral aqueduct; substantia nigra; crus cerebri;* and the *trigeminal, spinal* (anterolateral system), and *medial lemnisci*. The *cerebral aqueduct* interconnects the third and fourth ventricles, separating the tectum from the tegmentum of the midbrain at both curricular levels. The periaqueductal gray matter surrounds this duct, containing the accessory oculomotor nuclei and the dorsal tegmental nucleus. This area gives rise to an inhibitory descending pathway for painful stimuli.

> Obstruction of the cerebral aqueduct may occur congenitally, resulting in a noncommunicating hydrocephalus.

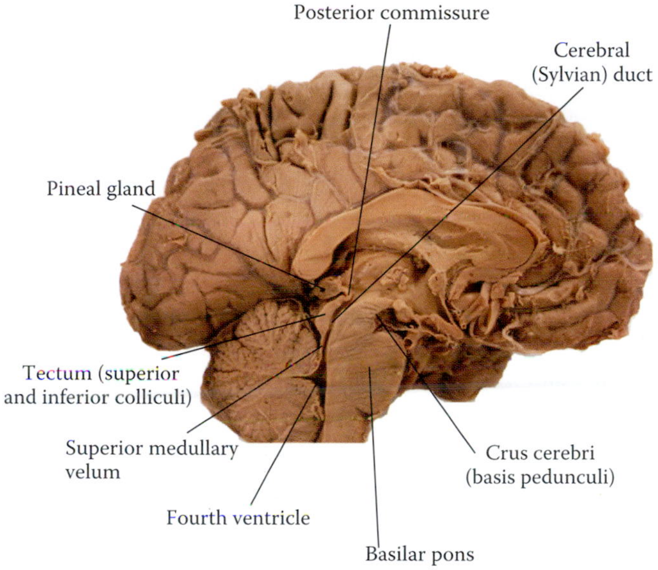

FIGURE 4.17 Midsagittal section of the brainstem. The midbrain with the associated cerebral aqueduct and tectum are clearly visible

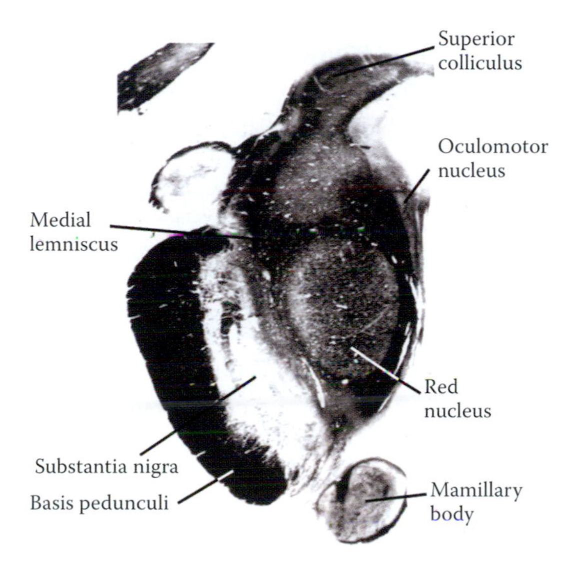

FIGURE 4.19 Section of the left half of the midbrain at the level of the superior colliculus. This photograph illustrates the tectum, tegmentum, substantia nigra, and crus cerebri.

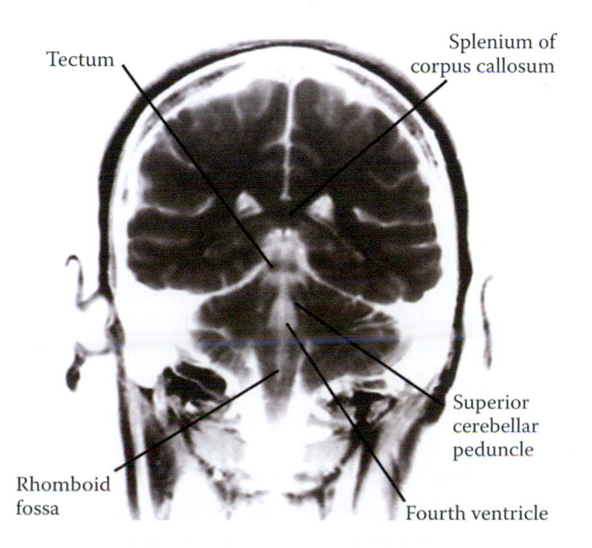

FIGURE 4.18 This MRI scan illustrates the tectum and connection of the midbrain to the cerebellum via the superior cerebellar peduncle.

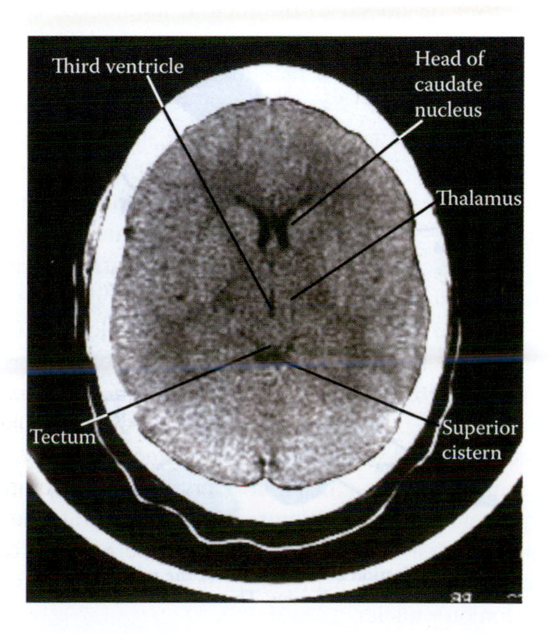

FIGURE 4.20 CT scan of the brain showing the tectum of the midbrain in relationship to the third ventricle and thalamus.

The *substantia nigra* is the largest pigmented nucleus of the midbrain, lying dorsal to the crus cerebri. It consists of the pars reticularis and pars compacta. The pars compacta is the main source of the inhibitory neurotransmitter dopamine, forming with the pars lateralis the group A9 of Dahlstrom and Füxe. Group A9 together with the retrorubral nucleus (group A8) constitutes the principal dopaminergic neurons of the midbrain. Group A10 (paranigral nucleus) interconnects the pars compacta of the substantia nigra on both sides. Acetylcholine is also contained in these neurons. The substantia nigra receives input from the striatum, globus pallidus, and subthalamic nucleus. It projects to the striatum, reticular formation, tectum, pedunculopontine nucleus, and thalamus. Nigral projection to the tectum arises from the pars reticulata, influencing coordination of head and eye movements. In addition to dopamine, the substantia nigra contains a high concentration of substance P, serotonin, and glutamic acid decarboxylase (GAD).

A lesion of the subdstantia nigra is responsible for signs and symptoms of Parkinson's disease, which will be discussed with the extrapyramidal system.

The *crus cerebri* (basis pedunculi) is the most ventral portion of the cerebral peduncles that contains in its medial two-thirds the corticospinal and corticobulbar fibers. Within the crus cerebri, the frontopontine fibers are medial, while the parietopontine, temporopontine, and occipitopontine fibers assume a more lateral position. These peduncles form the boundaries of the interpeduncular fossa that gives passage to the oculomotor nerve. The floor of this fossa forms the posterior perforated substance, allowing passage of the central branches of the posterior cerebral arteries. The *trigeminal lemnisci* lie dorsal to the medial lemniscus, carrying pain, temperature, and tactile sensations to the VPM nucleus of the thalamus from the face and head region. The ventral trigeminal lemniscus, a crossed pathway derived from the spinal trigeminal nucleus and the ventral portion of the principal sensory nucleus, terminates in the VPM nucleus of thalamus. The neuronal axons of the dorsal portion of the principal sensory nucleus form the ipsilateral dorsal trigeminal lemniscus. The *spinal lemniscus* (anterolateral system) is located dorsal to the lateral part of the substantia nigra, bounded dorsolaterally by the lateral lemniscus and ventromedially by the medial lemniscus. It consists primarily of fibers of the lateral spinothalamic tract. The *medial lemniscus* is the principal pathway that conveys conscious proprioception, fine tactile sensation, and vibratory sense to the thalamus. It assumes a horizontal position ventral to the trigeminal and spinal lemnisci. The *MLF* at this level consists of ascending vestibulo-ocular fibers and axons of abducens internuclear neurons, projecting to the trochlear and oculomotor nuclei.

The *central tegmental tract*, a composite bundle of fibers, lies dorsolateral to the red nucleus and lateral to the MLF. The descending components of this pathway originate from the motor cortex, red nucleus, periaqueductal gray, thalamus, and tegmentum of the midbrain en route to the inferior olivary nucleus (where it forms the amiculum or the capsule of the inferior olivary complex). This pathway regulates intrareticular conduction through its short ascending fibers. It also conveys cortical motor input to the contralateral cerebellum through the inferior olivary nucleus, and eventually to the red nucleus and basal nuclei (a collection of subcortical nuclei embedded in the white matter of the cerebral hemispheres) via cerebellar projections. The central tegmental tract is also considered to be the main ascending pathway for the reticular formation, conveying impulses to the subthalamus and the intralaminar thalamic nuclei.

The *dorsal longitudinal fasciculus*, a predominantly ipsilateral tract ventrolateral to the cerebral aqueduct, contains ascending and descending pathways. It connects the hypothalamus to the Edinger–Westphal, solitary, salivatory, facial, and hypoglossal and ambiguus nuclei, as well as to the tectum.

The midbrain is supplied by branches of the posterior cerebral, posterior communicating, anterior choroidal, and superior cerebellar arteries. The paramedian branches of the posterior cerebral and posterior communicating arteries supply midline structures including the oculomotor nuclei, MLFs, and medial parts of the substantia nigra. The lateral tegmentum, medial lemnisci, spinal lemnisci, substantia nigra, and crus cerebri are supplied by the circumferential branches of the posterior cerebral and superior cerebellar arteries. The long circumferential branches of the posterior cerebral artery and small branches from the superior cerebellar artery supply the tectum. Venous blood of the midbrain terminates in the internal cerebral vein or the great cerebral vein of Galen.

Caudal Midbrain

The inferior collicular level (Figures 4.21 and 4.22) contains the inferior colliculi, trochlear nucleus, tegmental nuclei, locus ceruleus, lateral lemniscus, and pedunculopontine nucleus.

The *inferior colliculus* is connected to the medial geniculate body via the inferior brachium, representing the auditory reflex center. It consists of a large central nucleus, responding to binaural impulses, and a pericentral nucleus concerned with the ipsilateral auditory impulses. The principal afferent to the inferior colliculus is the lateral lemniscus that conveys auditory impulses to the medial geniculate body via the brachium of the inferior colliculus. See also the auditory system, Chapter 14.

The *trochlear nucleus* is a round nucleus embedded within the MLF, lying ventral to the periaqueductal gray matter. Its axons, which constitute the trochlear nerve, travel dorsally and decussate completely in the superior medullary velum. These fibers emerge from the dorsal surface of the pons, immediately below the inferior colliculi, innervating the superior oblique muscle.

A lesion involving the trochlear nerve results in extorsion, impairment of downward movement of the affected eye, and vertical diplopia, which increases on attempted downward gaze. Patients compensate for this deficit and ameliorate diplopia by tilting the head to the contralateral side (*Bielschowsky sign*). Infants with this deficit may develop torticollis (spasmodic contracture of the sternocleidomastoid muscle) as a result of continuous tilting of the head. This condition is also discussed in detail in Chapter 11.

The *dorsal tegmental* (supratrochlear) *nucleus* is located between the two trochlear nuclei, receiving input from the mammillary bodies and the interpeduncular nucleus through the mammillotegmental tract. Projections from this nucleus ascend within the dorsal longitudinal fasciculus to be distributed to the limbic system nuclei of the diencephalon and telencephalon.

The *ventral tegmental nucleus* is considered to be a continuation of the superior central nucleus of the pons. Both the dorsal and ventral tegmental nuclei convey impulses to the lateral hypothalamus, the preoptic area, and the mammillary bodies via the dorsal longitudinal fasciculus, mammillary peduncle, and medial forebrain bundle.

The *dorsal raphe nucleus*, which lies adjacent to the dorsal tegmental nucleus, synthesizes and transports serotonin.

The *interpeduncular nucleus* is located dorsal to the interpeduncular fossa receiving the habenulopeduncular tract.

The *pedunculopontine nucleus*, which lies in the lateral tegmentum ventral to the inferior colliculus, modulates the activities of the nigral and pallidal neurons through its connections to the cerebral cortex, globus pallidus, and substantia nigra. Its compact part is mainly cholinergic and projects to the thalamus, regulating locomotion. This nucleus is crossed by fibers of the superior cerebellar peduncle.

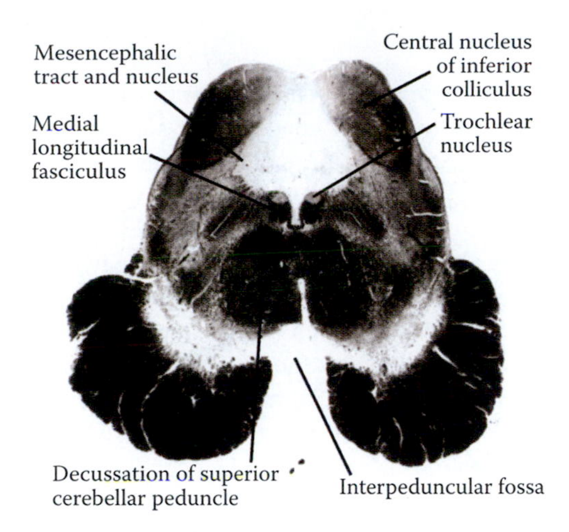

FIGURE 4.22 Section of the midbrain at the level of the inferior colliculus illustrating the trochlear nucleus and the site of decussation of the superior cerebellar peduncles.

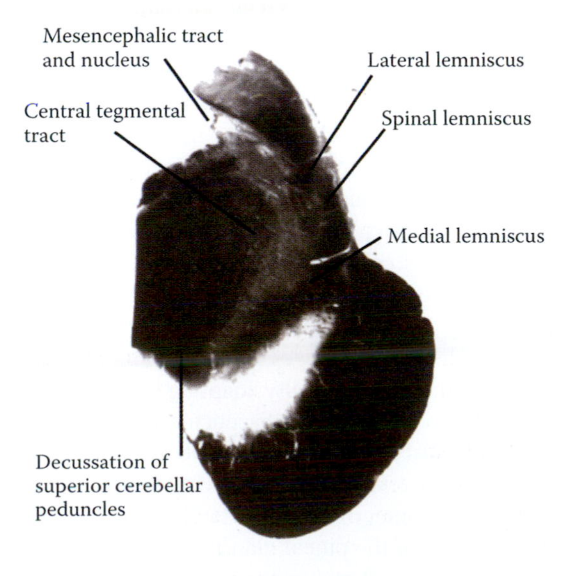

FIGURE 4.21 Image of the right half of the midbrain. Mesencephalic trigeminal nucleus and lemniscal triad (medial, lateral, and spinal lemnisci).

The *parabigeminal nucleus* contains a collection of cholinergic neurons that are located lateral to the lateral lemniscus and ventrolateral to the inferior colliculus. Neurons of these nuclei regulate both moving and stationary visual stimuli via their bilateral connections to the superior colliculus.

The *locus ceruleus nucleus* appears at the rostral pons medial to the trigeminal mesencephalic nucleus. It is a pigmented bluish nucleus that may readily be identified on a gross brain. This nucleus synthesizes and transports norepinephrine to the midbrain, cerebellum, medulla, spinal cord, diencephalon, and telencephalon. The spinal projection of the locus ceruleus descends in the lateral funiculus and exerts direct inhibitory influences upon the neurons that form the lateral spinothalamic tract. These projections use α receptors and not the opiate receptors, as is the case with the raphespinal tract. Noradrenergic neurons of the locus ceruleus project via the medial forebrain bundle, stria medullaris, and mammillary peduncle to the cerebral cortex, hypothalamus, midbrain tegmentum, and telencephalon. Intralaminar thalamic nuclei receive profuse projections from the locus ceruleus. Fibers of the locus ceruleus also terminate on small cerebral vessels and capillaries, accounting for the possible role of this nucleus in the regulation of the cerebral blood flow. The locus ceruleus also regulates the functions of the preganglionic sympathetic neurons of the spinal cord and modulates cerebellar activities. Additionally, this nucleus is involved in the reinforcement mechanism essential for learning, and REM sleep (see also the reticular formation, Chapter 5).

The *superior cerebellar peduncles* complete their decussation within the tegmentum. These peduncles represent the main cerebellar output to the ventral lateral nucleus of the thalamus and the red nucleus.

The *lateral lemniscus* is the principal ascending auditory pathway, which occupies the dorsolateral part of the midbrain, terminating in the inferior colliculus.

Rostral Midbrain

In addition to the common characteristics discussed earlier, the midbrain at this level contains the superior colliculus, red nucleus, and oculomotor nucleus (Figures 4.19, 4.20 and 4.24).

The *superior colliculus* (Figures 4.19, 4.23, and 4.24) consists of seven alternate gray and white laminae, which include the zonal, superficial gray, optic, intermediate gray, deep gray, deep white, and periventricular layers. The deep gray and deep white layers form the parabigeminal nucleus. The first zonal fibers consist of the myelinated and unmyelinated fibers of the external corticotectal tract that emanate from Brodmann areas 17, 18, and 19 of the occipital cortex. Small multipolar interneurons with which cortical fibers synapse constitute the superficial gray layer, whereas the optic layer consists of the retinotectal fibers that give collaterals to the other superficial layers of the superior colliculus representing the contralateral visual field. Retinotectal fibers terminate as clusters, while retinogeniculate fibers end as collaterals in the optic layer, with foveal fibers occupying an anterolateral position in this layer. Most of the axons in the optic emanate from slow W cells and some from fast Y cells, but there are also multipolar neurons that provide axons to the retina. The main receptive zone is formed by the intermediate gray and white layers and receives the medial corticotectal tract emanating from layer V of the ipsilateral secondary visual cortex (area 18), but also receives fibers from other neocortical areas that mediate eye movements. Contralateral spinotectal and spinothalamic fibers as well as input from the inferior colliculus, serotonergic fibers from the raphe nuclei, and noradrenergic fibers from the locus ceruleus also project to the intermediate gray and white layers. The parabigeminal nucleus that is formed by the deep gray and deep white layers near the periaqueductal gray matter consists of neurons with dendrites extending into the optic layer and axons that form the main output from the superior colliculus. Afferent fibers to the superior colliculus convey visual, tactile, and possibly thermal, pain, and auditory impulses through axonal projections from the retina, spinal cord, and occipital and temporal cortices, as well as the inferior colliculus. Efferent fibers from the superior colliculus project to the retina, brainstem, spinal neurons, pulvinar, lateral geniculate nucleus, parabigeminal nucleus, and pretectum. Efferents to the pulvinar are eventually conveyed to the primary and secondary visual cortices as part of the *extrageniculate visual pathway* from the retina to the cortex that mediates visual attention and orientation. This visual pathway plays an important role in the ability to point at an object and identify different levels of luminosity with conscious visualization (blindsight) in patients with a disrupted primary visual cortex (area 17). It is mediated by the tecto-pulvino-cortical pathway.

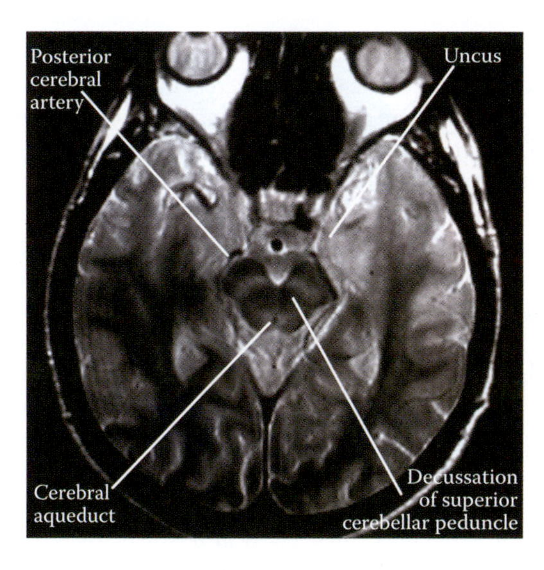

FIGURE 4.23 MRI scan through the midbrain at the level of the inferior colliculus. This image illustrates the relationship of the midbrain to the uncus and posterior cerebral arteries.

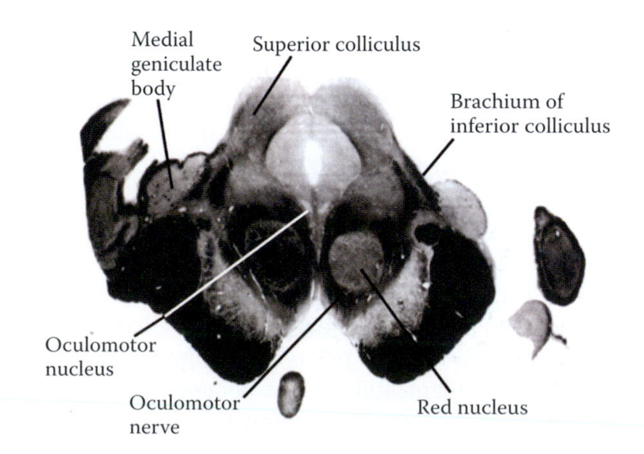

FIGURE 4.24 Section of the midbrain at its junction with the diencephalon. The superior colliculus, oculomotor nucleus and nerve, and red nucleus are principal features of this level.

The superior colliculus also projects to the oculomotor nuclear complex and to the periaqueductal gray matter and, through this, to the abducens nuclei. Tectotegmental fibers project to the substantia nigra, red nucleus, and tegmental reticular nuclei of the brainstem. Tectopontine fibers influence the activities of the cerebellum through the pontocerebellar fibers. Tecto-olivary fibers project to the contralateral medial accessory olivary nucleus, which relays to the posterior vermis to mediate ocular movements. Auditory descending fibers to the superficial layer of the superior colliculus enable integration of the visual and auditory behavior. Stimulation of the superior colliculus produces contralateral head movement toward the source of visual stimuli. Due to the bilaterality of the retinotectal input, experimental division of the tegmentum has shown major visual behavioral changes, including loss of response to threatening stimuli and difficulty adapting to the environment.

The superior colliculus lies adjacent to the rostral interstitial nucleus of the MLF, ventral to the posterior commissure and the pineal gland, a relationship that bears clinical significance in Parinaud syndrome, in which vertical gaze palsy is associated with a tumor of the pineal gland. Movements of the head and neck toward visual and auditory stimuli are mediated by projections of the superior colliculi to the brainstem and spinal cord via the tectobulbar and tectospinal tracts, respectively.

The *red nucleus is* a highly vascularized nucleus in the center of the midbrain tegmentum, which is encircled by fibers of the superior cerebellar peduncle (Figures 4.19 and 4.20). The pinkish color of the nucleus is attributed to a ferric iron pigment in the multipolar neurons. It is crossed by fibers of the oculomotor nerve en route to the interpeduncular fossa. It has a caudal magnocellular and a rostral parvocellular part. The *magnocellular part* gives rise to the contralateral rubrospinal tract, which controls flexor muscle tone. Medial to the red nucleus, the habenulopeduncular tract (fasciculus retroflexus) descends, projecting to the interpeduncular nucleus. It is delivering cerebellar input to the upper three cervical spinal cord segments. Fibers of the rubrospinal tract are incorporated within the lateral corticospinal tract, maintaining identical termination sites. A lesion that disrupts the corticospinal tract in monkeys produces initially complete upper motor neuron palsy, which disappears, and complete recovery occurs as a result of the compensatory effect of the rubrospinal tract. Cerebral cortical input to the red nucleus is conveyed via the corticorubral fibers, while cerebellar input emanating from the contralateral dentate, globose, and emboliform nuclei are carried via the superior cerebellar peduncle. Information received by the red nucleus is delivered to laminae V through VII of the spinal cord. The *parvocellular part* of the red nucleus forms the rubro-olivary tract that projects to the ipsilateral inferior olivary nucleus within the central tegmental tract. Due to the massive connections of the inferior olivary nucleus to the cerebellum through the olivocerebellar tract and to the cerebral cortex through the central tegmental tract and a relay in the red nucleus, the rubro-olivary tract forms part of a feedback loop conveying motor cortical input to the cerebellum and back to the motor cerebral cortex. This loop is bidirectional and able to shift the control of motor activity from the corticospinal to the rubrospinal tract for programmed automation and at the same enables the corticospinal tract to intervene during automated motor activity conducted by the rubrospinal tract in response to environmental effects.

The *oculomotor nuclear complex* has a "V"-shaped configuration and is located medial to the MLF. It consists of somatic and visceral columns. The somatic columns (GSE) provide innervation to the extraocular muscles (with the exception of the lateral rectus and superior oblique) and levator palpebrae muscles, while the visceral columns (Edinger–Westphal nucleus) provide presynaptic parasympathetic fibers (GVE) to the ciliary ganglion, which eventually innervate the constrictor pupillae and ciliary muscles.

SUGGESTED READING

Abbott S, Kanbar R, Bochorishvili G, Coates MB, Stornetta RL, Guyenet PG. C1 neurons excite locus coeruleus and A5 noradrenergic neurons along with sympathetic outflow in rats. *J Physiol* 2012;590(Pt 12):2897–915.

Berry DJ, Ohara PT, Jeffery G, Lieberman AR. Are there connections between the thalamic reticular nucleus and the brainstem reticular formation? *J Comp Neurol* 1986;243(3):347–62.

Brochier T, Ceccaldi M, Milandre L, Brouchon M. Dorsolateral infarction of the lower medulla: Clinical-MRI study. *Neurology* 1999;52(1):190–3.

Buhler AV, Proudfit HK, Gebhart GF. Separate populations of neurons in the rostral ventromedial medulla project to the spinal cord and to the dorsolateral pons in the rat. *Brain Res* 2004;1016(1):12–9.

Dauvergne C, Ndiaye A, Buisseret-Delmas C, Buisseret P, Vanderwerf F, Pinganaud G. Projections from the superior colliculus to the trigeminal system and facial nucleus in the rat. *J Comp Neurol* 2004;478(3):233–47.

Farkas E, Jansen AS, Loewy AD. Periaqueductal gray matter input to cardiac-related sympathetic premotor neurons. *Brain Res* 1998;792(2):179–92.

Field TS, Benavente OR. Penetrating artery territory pontine infarction. *Rev Neurol Dis* 2011;8(1–2):30–8.

Galambos R, Schwartzkopff J, Rupert A. Microelectrode study of superior olivary nuclei. *Am J Physiol* 1959;197:527–36.

Holstein GR, Friedrich VL Jr, Kang T, Kukielka E, Martinelli GP. Direct projections from the caudal vestibular nuclei to the ventrolateral medulla in the rat. *Neuroscience* 2011;175:104–17.

Hsieh JH, Chen RF, Wu JJ, Yen CT, Chai CY. Vagal innervation of the gastrointestinal tract arises from dorsal motor nucleus while that of the heart largely from nucleus ambiguus in the cat. *J Auton Nerv Syst* 1998;70(1–2):38–50.

Imbe H, Murakami S, Okamoto K, Iwai-Liao Y, Senba E. The effects of acute and chronic restraint stress on activation of ERK in the rostral ventromedial medulla and locus coeruleus. *Pain* 2004;112(3):361–71.

Ito H, Seki M. Ascending projections from the area postrema and the nucleus of the solitary tract of Suncus murinus: Anterograde tracing study using Phaseolus vulgaris leucoagglutinin. *Okajimas Folia Anat Jpn* 1998;75(1):9–31.

Lavezzi HN, Parsley KP, Zahm DS. Mesopontine rostromedial tegmental nucleus neurons projecting to the dorsal raphe and pedunculopontine tegmental nucleus: Pychostimulant-elicited Fos expression and collateralization. *Brain Struct Funct* 2012;217(3):719–34.

Lovick TA. Projections from brainstem nuclei to the nucleus paragigantocellularis lateralis in the cat. *J Auton Nerv Syst* 1986;16(1):1–11.

McLaughlin N, Ma Q, Emerson J, Malkasian DR, Martin NA. The extended subtemporal transtentorial approach: The impact of trochlear nerve dissection and tentorial incision. *J Clin Neurosci* 2013;20(8):1139–43.

Park SH, Becker-Catania S, Gatti RA, Crandall BF, Emelin JK, Vinters HV. Congenital olivopontocerebellar atrophy: Report of two siblings with paleo- and neocerebellar atrophy. *Acta Neuropathol* 1998;96(4):315–21.

Romanowski CA, Hutton M, Rowe J, Yianni J, Warren D, Bigley J, Wilkinson ID. The Anatomy of the Medial Lemniscus within the Brainstem Demonstrated at 3 Tesla with High Resolution Fat Suppressed T1-Weighted Images and Diffusion Tensor Imaging. *Neuroradiol J* 2011;24(2):171–6.

Seo JP, Jang SH. Characteristics of corticospinal tract area according to pontine level. *Yonsei Med J* 2013;54(3):785–7.

Sharma Y, Xu T, Graf WM, Fobbs A, Sherwood CC, Hof PR, Allman JM, Manaye KFJ. Comparative anatomy of the locus coeruleus in humans and nonhuman primates. *Comp Neurol* 2010;518(7):963–71.

Zhou J, Shore S. Convergence of spinal trigeminal and cochlear nucleus projections in the inferior colliculus of the guinea pig. *J Comp Neurol* 2006;495(1):100–12.

5 Reticular Formation

On a developmental basis, the reticular formation is considered to be one of the oldest functional units in the central nervous system. It occupies the central core of the brainstem, extending rostrally to include the midline, intralaminar, and reticular thalamic nuclei and the zona incerta of the subthalamus. Reticular neurons mediate local reflexes, receiving collaterals from the ascending and descending pathways with the exception of the medial lemniscus. The reticular formation is bounded ventromedially by the pyramidal tracts and the medial lemnisci and dorsolaterally by the secondary sensory pathways. Regulation of both somatic and visceral (autonomic) motor functions and modulation of the electrocortical activities are maintained by massive reticular connections to the autonomic centers in the brain and the spinal cord. Additional functions of the reticular formation include control of emotional expression, pain transmission, and regulation of reflex activities associated with the cranial nerves.

The *reticular formation* consists of deeply localized and poorly identified nuclear groups, which are scattered predominately in the brainstem. It contains centers that generate motor activities (e.g., walking and running) and regulate conjugate eye movements and cardiovascular activities including regulation of blood pressure. It also contains centers for expiration, inspiration, vomiting, and deglutition. Reticular neurons are classified into median raphe, paramedian, and medial and lateral nuclear columns.

RAPHE NUCLEI

The *raphe nuclei* (Figure 5.1) synthesize and transport serotonin to various areas of the central nervous system through ascending and descending projections. Dahlstrom and Füxe group these serotonergic neurons into nine clusters. Serotonin, the neurotransmitter for raphe nuclei, is involved in the slow (non–rapid eye movement [NREM]) phase of sleep, behavioral regulation, and inhibition of pain transmission. Raphe neurons also contain substance P. Raphe neurons within the pons project inhibitory impulses to the paramedian reticular formation producing rapid eye movement (REM) during sleep. They also send inhibitory fibers to the nuclei reticularis pontine oralis and caudalis, which, upon release from this inhibition, elicit involuntary movements of the trunk and limbs. Inhibitory input to the pontine neurons that project to the lateral geniculate body may be responsible for the pontine–geniculate–occipital spikes and associated visual-natured REM sleep. It has been suggested that the latter phenomenon and dreams may be suppressed by certain medications or upon consumption of dairy product that contain tryptophan. Lesions of the raphe nuclei produce prolonged insomnia. This group of nuclei includes the raphe pallidus and raphe obscurus, raphe magnus, dorsal raphe nucleus, and superior central nucleus

- The *nucleus raphe pallidus* is located in the pons, projecting to laminae I, II, and V of the dorsal horns of all spinal segments, as well as to the intermediolateral columns. The spinal projections of the raphe pallidus convey pain-controlling input from the periaqueductal gray matter to the spinal cord.
- The *nucleus raphe obscurus* lies in the pons and forms the inhibitory intermediate raphe spinal projection to the spinal cord, modulating the sympathetic neurons of the intermediolateral column and thus regulating the cardiovascular function.
- The *nucleus raphe magnus* is located in the rostral medulla and caudal pons and contains B3 neurons, maintaining bilateral projections to the spinal cord within the Lissauer tract and to the posterior part of the spinal trigeminal nucleus within the spinal trigeminal tract. These spinal projections descend in the lateral funiculus and may inhibit the neurons that transmit painful stimuli, acting as an additional endogenous analgesic pathway. The serotonergic component of the spinal projection, which represents 20% of all these fibers, establishes excitatory synaptic contacts with the enkephalinergic neurons and inhibitory contacts with the lateral spinothalamic tract neurons of the substantia gelatinosa. These synaptic contacts at spinal levels produce stimulus-bound profound analgesia, a common phenomenon seen upon stimulation of the periaqueductal and periventricular gray matter. They are also thought to enhance motor response to nociceptive stimuli via fight and flight response.
- The *dorsal nucleus of the raphe* is located in the midbrain, at the pontomesencephalic junction. This nucleus corresponds to group B7, conveying impulses to the hypothalamus, putamen, caudate nucleus, amygdala, intralaminar thalamic nuclei, and septal region via the medial forebrain bundle (MFB). It also projects to the superior colliculus, mammillary bodies, hippocampal formation, substantia nigra, locus ceruleus, olfactory cortex, and lateral geniculate body. Reciprocal connections between the dorsal nucleus of raphe and limbic system are documented. Ascending fibers from this nucleus distribute via the dorsal longitudinal fasciculus and the MFB to the mammillary body, ventomedial hypothalamus, habenula, preoptic area, and suprachiasmatic nucleus.

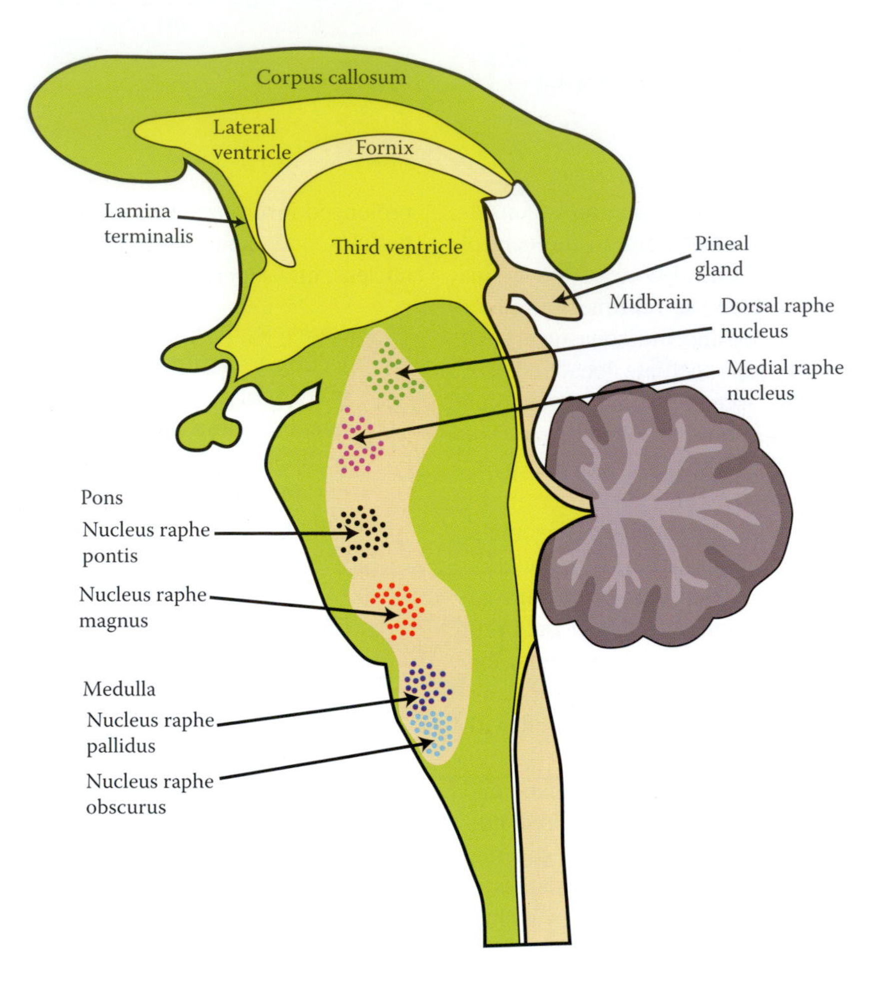

FIGURE 5.1 Schematic drawing of the brainstem illustrating the midline raphe nuclei of the reticular formation.

The *superior central nucleus* (Figure 5.5) is located at the pontomedullary junction, corresponding to groups B6 and B8 of Dahlstrom and Füxe. It projects diffusely to all areas of the cerebral cortex and specific regions of the cerebellar cortex. It also maintains reciprocal connections with the limbic system, projecting also via the MFB and dorsal longitudinal fasciculus to areas that overlap with the dorsal raphe nuclear projections.

The *paramedian nuclei* (Figure 5.3) lie parallel to the MLF and the medial lemniscus, and consist of nuclei, which maintain reciprocal connections with the cerebellum. The reticulotegmental, a paramedian nucleus at the rostral pons, maintains connections with the cerebellum and the globus pallidus.

MEDIAL RETICULAR ZONE

The *medial reticular nuclei* form the efferent zone of the reticular formation, which receives afferents from the spinal cord and collaterals from the spinoreticulothalamic tract, as well as the cochlear, vestibular, and trigeminal nerves. They consist primarily of the *nucleus reticularis gigantocellularis* (Figure 5.3) in the medulla and the *nucleus reticularis pontine caudalis* and *oralis* (Figures 5.4 and 5.5) in the pons. These reticular nuclei convey information received from the cerebral cortex, vestibular nuclei, cerebellum, spinal cord, and lateral zone of the reticular formation to the spinal cord through the inhibitory medullary reticulospinal tract and excitatory pontine reticulospinal tract. Both reticulospinal tracts control posture (sitting and standing) and automatic movements (walking) and exert powerful influences upon the axial and proximal appendicular muscles. Both tracts function in conjunction with the vestibulospinal tract. The *pontine reticulospinal* tract is a massive uncrossed tract that descends through the medial longitudinal fasciculus and anterior funiculus, terminating in laminae VII, VIII, and IX. It activates extensor α motor neurons via the gamma loop. The input of the basal nuclei to the pontine reticular nuclei is evident in the postural disorders exhibited by patients with Parkinson's disease. On the other hand, the medullary reticulospinal tract descends in the lateral funiculi, terminating in laminae VII, VIII, and IX of both sides and laminae IV, V, and VI on the ipsilateral side. This tract synapses upon the flexor α motor neurons.

ASCENDING RETICULAR ACTIVATING SYSTEM

The medial reticular nuclei regulate eye movements through projections via shorter fibers to the extraocular motor nuclei and sensory nuclei of the brainstem. These neurons in the rostral pons and midbrain, which project via ascending fibers of the central tegmental tract (CTT) to the intralaminar thalamic nuclei and hypothalamus, receive multiple sensory inputs. This reticular input to the intralaminar thalamic nuclei and diffuse areas of cerebral cortex via the CTT and re-excitation of the thalamus in a reverbating circuit via a positive feedback loop are utilized in the alterations of the electrocortical activity and regulation of the sleep–wake cycle. The CTT is a tegmental polysnaptic network known as the *ascending reticular activating system.* This system is associated not only with consciousness, but also with memory, emotion, drive, and motivation. It has extrinsic and intrinsic elements; the extrinsic element consists of neurons in the medulla and pons that respond to stimulation generated by the cranial and spinal nerves, without being involved in the sleep–wake cycle. By contrast, the intrinsic element is represented in the mesencephalic neurons, which exhibit cyclic (e.g., diurnal) activity related to the projection of the anterior hypothalamus (suprachiasmatic area) to the midbrain via the MFB.

Damage to the reticular formation at the level of the rostral pons and caudal medulla may lead to coma or akinetic mutism (coma vigil). An electroencephalogtram (EEG) similar to the slow phase of sleep characterizes this condition, with no appreciable change in the autonomic and somatomotor reflexes or eye movements.

LATERAL RETICULAR ZONE

The *lateral reticular zone* represents the sensory reticular area that receives input from the ascending sensory pathways as well as the cerebral cortex. In the medulla, this zone contains the nucleus reticularis parvocellularis and nucleus reticularis lateralis. In the rostral pons and caudal midbrain, it contains the reticulotegmental, *Kölliker-Fuse*, medial parabrachial, and pedunculopontine nuclei. In the caudal medulla, it lies medial to the solitary nucleus, nucleus ambiguous, and dorsal motor nucleus of the vagus, containing at this level the adrenergic C2 and noradrenergic A2 groups. In general, these nuclei are precerebellar relay nuclei, which receive input from a variety of sources such as the cerebral cortex, spinal cord, and vestibular nuclei and maintain reciprocal connections with the cerebellum. The lateral reticular zone also contains L and M micturition centers and noradrenergic A, adrenergic C, and cholinergic Ch neurons, which are scattered in the medulla and upper pons. The medial (M)-or (pontine micturition) center excites the bladder muscle through projections to alpha motoneurons of the S2, S3 and S4 spinal segments and inhibits the urethral sphincter through excitatory projections to the gamma-amino butyric acid (GABA)-immunoreactive interneurons within that inhibit urethral sphincter motoneurons. The lateral (L)-or (pontine storage) center through direct projections, excites urethral sphincter motoneurons. The L-region, through direct projections, excites urethral sphincter motoneurons. The noradrenergic neurons are classified into A1, A2, A4, A5, A6, and A7 groups (A3 is absent in humans). Adrenergic neurons are categorized into C1 and C2, while cholinergic neurons are divided into Ch5–Ch6. C2 and A2 groups that are located in the medulla in close proximity to the nucleus ambiguus. In the lateral tegmental area of the pons, noradrenergic A1, A2, A4, A5, A6, and A7 cell groups; adrenergic C1–C2 groups; and cholinergic Ch5–Ch6 groups are visible. A2, A4, A5, and C1 cellular groups are located ventrolaterally in the pons near the facial motor nucleus. A4 extends along the medial surface of the superior cerebellar peduncle. A5, in conjunction with C1, may act to regulate vasomotor (blood pressure, caliber of the vessels, heart rate, etc.) activities. A1 and A7 are located in the lateral pontine tegmentum. A1, A2, A5, and A7 influence the activities of the amygdala, septal nucleus, hypothalamus, bed nucleus of stria terminalis, and diagonal band of Broca via the CTT and the MFB.

The *nucleus reticularis parvocellularis* is located between the nucleus reticularis gigantocellularis and the spinal trigeminal tract and nucleus. It contains noradrenergic A2 and adrenergic C2 cellular groups. The connections of the hypoglossal to the trigeminal and facial motor nuclei are maintained by the parvocellular reticular nucleus. Afferents to this nucleus arise from the contralateral red nucleus and cerebral cortex.

The *lateral reticular nucleus* (Figures 5.2 and 5.3) occupies the lateral medulla, projecting to the cerebellum via the inferior cerebellar peduncle.

The *reticulotegmental nucleus* (Figure 5.5) is located in the center of the ventral pontine tegmentum at the level of the trigeminal nerve and isthmus. It maintains reciprocal connections to the cerebellum, receiving afferents via the superior cerebellar peduncle and projecting back to the cerebellum through the middle cerebellar peduncle.

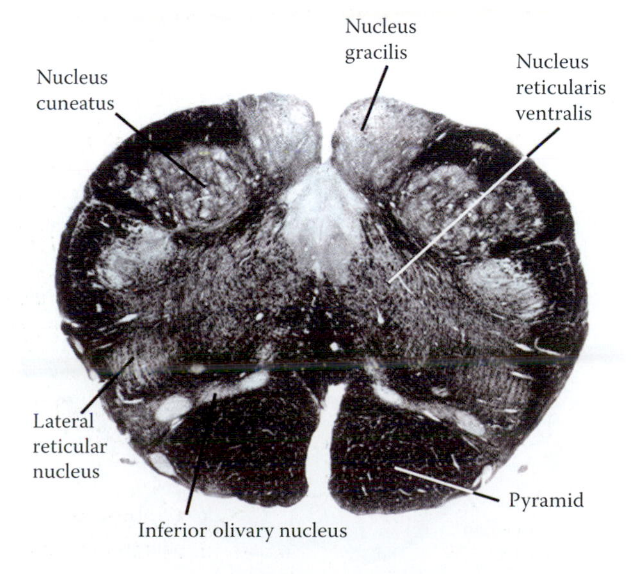

FIGURE 5.2 Section of the caudal medulla at the level of decussation of the internal arcuate fibers. The lateral reticular nucleus is dorsolateral to the inferior olivary nucleus, and the nucleus reticularis ventralis is centrally located.

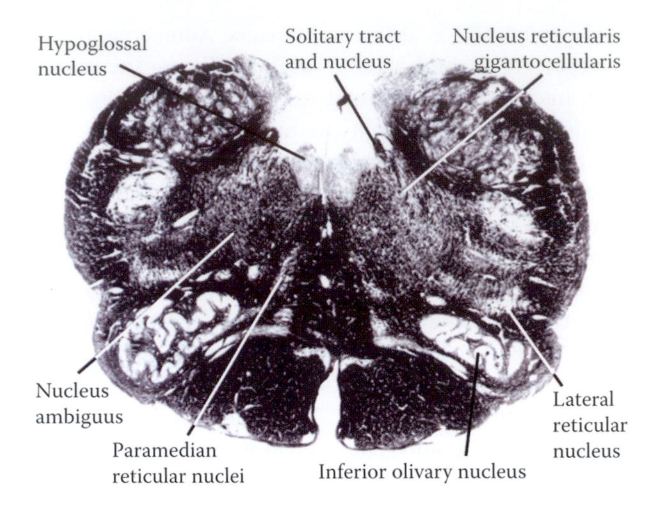

FIGURE 5.3 Section of the medulla at the midolivary level. The paramedian reticular nuclei lie parallel to the midline and lateral to the raphe nuclei. The nucleus reticularis gigantocellularis represents the medial reticular zone and the source of the medullary reticulospinal tract.

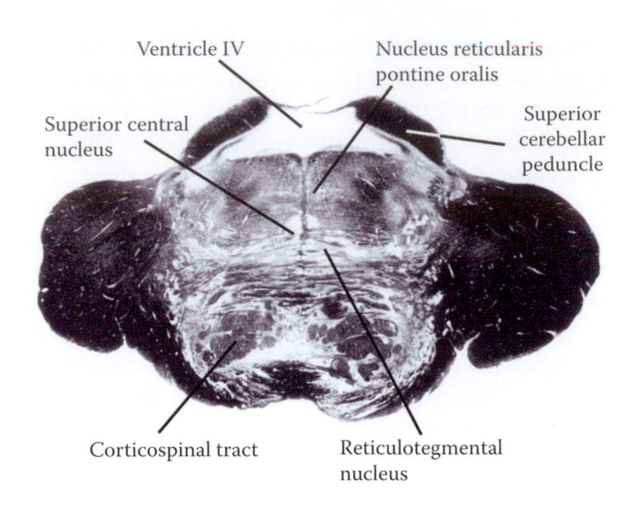

FIGURE 5.5 Section of the pons at the level of the trigeminal nerve. The locations of the superior central and reticulotegmental nuclei, as well as the nucleus reticularis pontine oralis, are illustrated.

The *Kölliker-Fuse nucleus* is located medial and ventral to the superior cerebellar peduncle. It acts as a pneumotaxic center by projecting to the inspiratory and expiratory centers in the medullary and pontine reticular formation. These centers convey the input to the preganglionic sympathetic neurons of the upper three or four thoracic spinal segments, phrenic nucleus, and intercostal neurons.

The *medial parabrachial nucleus* is located medial and ventral to the superior cerebellar peduncle, establishing reciprocal connections with the insular cortex, amygdala, and hypothalamus. This nucleus projects to the pontine micturition M center.

The *pedunculopontine nucleus* (Figure 5.6), which is classified as group Ch5, is comprised of the excitatory cholinergic neurons that lie dorsolateral to the superior cerebellar peduncle at the caudal mesencephalon. It continues with the more caudal cholinergic cell group (Ch6) in the pontine central gray matter. This nucleus is crossed by fibers of the superior cerebellar peduncle and receives input from the globus pallidus, substantia nigra, and primary motor cortex. It maintains a strong connection with the basal nuclei and projects primarily to substantia nigra pars reticulata and compacta, subthalamic nucleus, medial pallidum, and striatum. The firing of the neurons of the pedunculopontine precedes

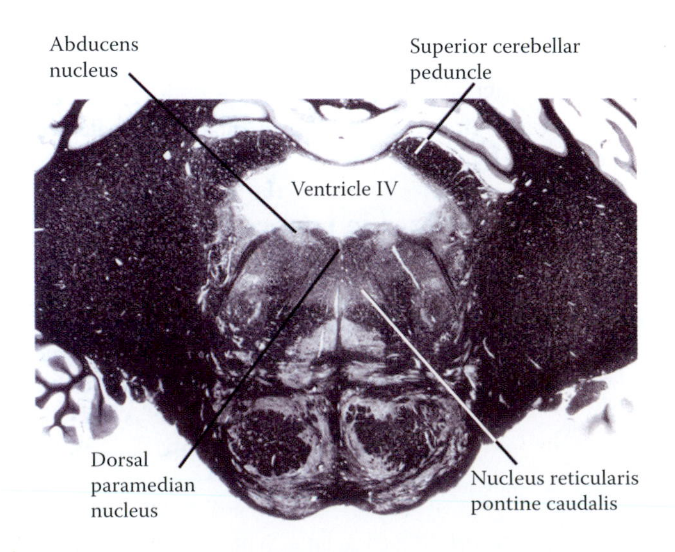

FIGURE 5.4 In this section of the pons at the level of the abducens nucleus, the nucleus reticularis pontine caudalis is shown superior to the medial lemnisci.

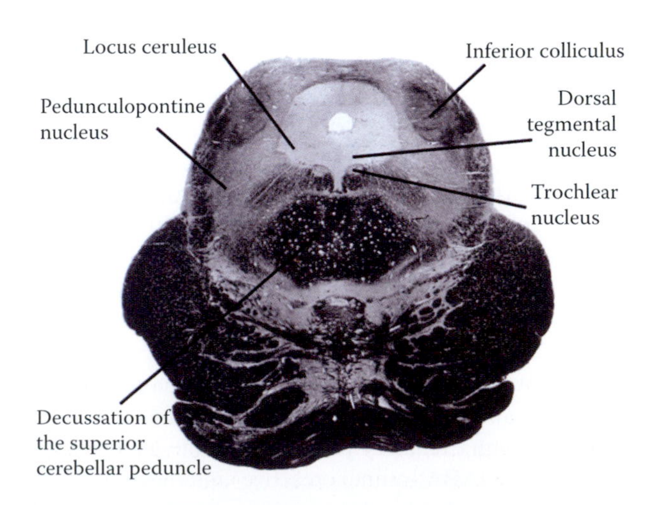

FIGURE 5.6 Section of the pons at the level of the trochlear nucleus. The pigmented locus ceruleus is located ventrolateral to the cerebral aqueduct within the periaqueductal gray matter. The pedunculopontine nucleus is dorsolateral to the superior cerebellar peduncle.

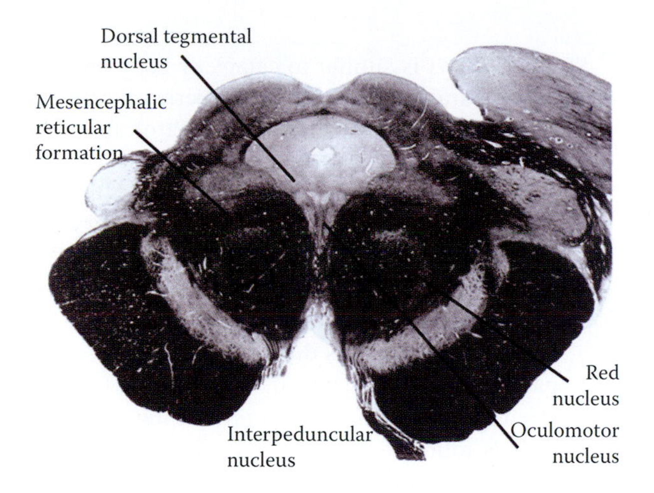

FIGURE 5.7 Section of the midbrain at the level of the superior colliculus. The mesencephalic reticular formation occupies the center of the tegmentum lateral to the red nucleus. The dorsal tegmental nucleus is located dorsal to the oculomotor nucleus, whereas the interpeduncular nucleus is dorsal to the interpeduncular fossa.

the intended movement and is modified by it. It is believed that this nucleus also plays a role in the sleep–wake cycle.

The *locus ceruleus* (noradrenergic group A6) and nucleus subceruleus are pigmented noradrenergic neurons (Figure 5.6) contained in the rostral pons that merge caudally with group A4. These nuclei play an important role in the regulation of the paradoxical (REM) phase of sleep and control of sensory neurons. The locus ceruleus is known to project monosynaptically to the cerebral cortex, spinal cord (intermediolateral lateral column), hippocampal formation, septal area, and diencephalon. It also projects to structures in the telencephalon and the entorhinal area (secondary olfactory cortex) via the stria medullaris thalami, stria terminalis, fornix, and MFB.

The mesencephalic reticular formation occupies the center of the tegmentum lateral to the red nucleus. The dorsal tegmental nucleus is located dorsal to the oculomotor nucleus, whereas the interpeduncular nucleus is dorsal to the interpeduncular fossa (Figure 5.7).

SLEEP AND ASSOCIATED DISORDERS

The locus ceruleus is believed to have a role in the rapid eye movement phase (REM or paradoxical sleep) of sleep and in the control of cortical activity through the intralaminar thalamic nuclei. Rapid conjugate eye movements and increase in intracranial pressure, as well as heart and respiratory rates, characterize REM sleep, which begins 90 min after sleep onset in the early evening. It also involves increase in temperature, nocturnal tumescence, hypotonia, and very brief episodes of facial and limb movement. In this stage, the muscles remain paretic and flaccid, and the deep tendon reflexes cannot be elicited. It represents 20%–25% of total sleep time, consisting of brief episodes (3–4 each night) of rapid eye movements (each episode may last 5–30 min) that follow the predominant cycles of NREM phase, which may last up to an hour and half. In the REM phase, muscle movement and tone in the head, trunk, and extremities are absent (flaccid), and their tendons remain areflexic. Angina and cluster headache may also occur during this phase. Since autonomic activities occur in a motionless individual, the REM phase is also known as paradoxical sleep. These autonomic activities include increase in pulse and blood pressure, metabolism, and blood flow in the brain.

NREM sleep represents 75%–80% of nocturnal sleep in adults and 50% of that in the newborn. In this phase, the eye movements are slow, and muscles maintain tone with normal deep tendon reflexes. Autonomic activities are minimal, although secretion of growth hormone and prolactin increases during this phase. Sleepwalking (*somnambulism*), bed-wetting (enuresis), night terrors, and seizures may occur in NREM sleep. NREM sleep is divided into four stages that are distinguished by the slower and higher-voltage EEG patterns and depth of unconsciousness.

Stage 1 lasts only for 1 min or up to 7 min. Sleep is easily discontinued during this stage (maintains low arousal threshold), for example, by softly calling one's name, closing door, and so forth. The length and percentage of this stage may increase when the sleep is disrupted.

Stage 2 manifests sleep spindles or K complexes on EEG. Arousal in stage 2 may be produced by a more intense stimulus. As this progresses, a gradual appearance of high-voltage slow activity on EEG is seen, which eventually meets the criteria for stage 3.

Stage 3 presents with high-voltage (75 μV) and slow-wave (2 cycles per minute) activity accounting for more than 20% but less than 50% of the EEG activity. This stage usually lasts only a few minutes in the first cycle and is transitional to stage 4.

Stage 4 is identified when the high-voltage slow-wave activity is more than 50% of the EEG activity, and it generally lasts 20–40 min in the first cycle.

Stages 3 and *4* are often referred to as slow-wave sleep, delta sleep, or deep sleep. It is the most restorative sleep. Rising to lighter NREM sleep stages is usually associated with a series of body movements. NREM and REM sleep continue to cycle throughout the night; REM sleep episodes generally become longer, and stage 3 and 4 sleep occupy less time in the second cycle and may disappear altogether from later cycles as stage 2 sleep expands to occupy the NREM portion of the cycle. On average, across the night, the NREM–REM cycle is approximately 90–110 min. In the young adult, slow-wave sleep dominates the NREM portion of the sleep cycle toward the beginning of the night, while REM sleep tends to be greatest in the last one-third of the night.

In addition to serotonin (NREM sleep) and norepinephrine (for REM), numerous other substances are also involved in the sleep mechanism, such as the hypothalamic delta sleep-inducing peptide, corticotropin-releasing hormone (suppresses the slow-wave sleep), and D2 and E2 prostaglandins (act upon the sleep and wake centers in the anterior and posterior hypothalamus, respectively). Lesions of the locus ceruleus suppress the deep phase of sleep. In contrast, long-term arousal may result from increased secretion of the catecholamines from this nucleus. The balanced opposing effects of serotonergic and noradrenergic neurons regulate the sleep states.

Narcolepsy is a condition in which patients enter into brief, often unpredictable, and irresistible episodes of sleep during waking hours (when conducting activities that require constant attention). Sleep attacks during a single day may range between 1 to as many as 20, and each attack may last from a few minutes to as much as a few hours. It has been estimated that up to 50% of these individuals suffer from memory loss. Some narcoleptics may experience automatic behaviors that may include rapid blinking, repetitive motor activities, and irrelevant speech.

This condition starts in the second decade of life between adolescence and age 25 years and may follow traumatic emotional situations. Excessive daytime sleepiness, preceded by a feeling of overwhelming fatigue, may be an initial sign of narcolepsy. Presence of excessive daytime somnolence, cataplexy, sleep (hypnagogic and hypnopomic) hallucinations, and sleep paralysis represents the *narcolepsy tetrad* that is essential for the diagnosis of narcolepsy. Most narcoleptics present with cataplexy, which may be brought on by excitement, anger, or fear.

Cataplexy usually occurs months to years following the initial onset of sleep attacks. Visual and/or auditory hallucinations may be detected in individuals with prolonged cataplectic attacks. It is characterized by sudden loss of muscle tone, which may lead to sudden collapse of the patient to the ground.

Hypnagogic hallucination refers to the hallucinations that occur prior to falling asleep, whereas *hypnopompic* hallucination occurs upon awakening.

Sleep paralysis is similar to cataplexy but refers to the paralysis that may be experienced by the patient slightly before the onset of sleep or shortly after awakening. The individual is aware of the paralysis and often describes the struggle to get up. This paralysis is temporary and may be terminated spontaneously or after mild sensory stimulation.

The gold standard for diagnosis of narcolepsy is sleep study. In this diagnostic method, the patient is allowed an adequate amount of sleep approximately 10 days prior to the study. This is followed by a *multiple sleep latency test* (MSLT) in which the patient is asked to nap every 2 h throughout the day while the time to fall asleep (sleep latency) is measured. Narcoleptics usually require less than 4 min to fall asleep, a considerably smaller amount of time compared to normal individuals. They also usually enter REM sleep within minutes of the onset of sleep. Occurrence of more than two episodes of sleep onset REM period during the MSLT is virtually diagnostic of this disorder. This condition is commonly known to be associated with human leukocyte antigen (HLA) DR2 on the short arm of chromosome 6. Although the inheritance pattern is believed to be autosomal dominant, environmental factors should also be considered in the assessment of this condition. First-degree relatives of narcoleptics may exhibit excessive daytime somnolence and have much greater incidence of narcolepsy. Treatment of this disorder may be accomplished by stimulants such as methylphenidate, which is thought to increase the release of norepinephrine and thus decrease REM sleep. Prolonged administration of this medication may cause insomnia, irritability psychosis, and possible tolerance.

Centrally acting α-1 adrenergic agonists such as modafinil and selegiline, a monoamine oxidase type B inhibitor, although less effective than methylphenidate, may be used as second-line agents. Tricyclic antidepressants (TCAs) and serotonin reuptake inhibitors such as femoxetine and fluxetine may be used for treatment of cataplexy. Amphetamine and methylphenidate may be used as treatment of narcolepsy. Since cataplexy, hypnogogic (sleep) hallucination, and sleep paralysis occur in REM sleep, TCAs and MAO inhibitors that suppress REM are used as a therapeutic measures for these disorders. γ-hydroxybutyrate and behavioral modification may be useful in the treatment of cataplexic episodes.

SUGGESTED READING

Baghdoyan HA, Lydic R. M2 muscarinic receptor subtype in the feline medial pontine reticular formation modulates the amount of rapid eye movement sleep. *Sleep* 1999;22:835–47.

Chamberlin NL, Saper CB. A brainstem network mediating apneic reflexes in the rat. *J Neurosci* 1998;18(15):6048–56.

Dahlström A, Fuxe K. Evidence for the existence of monamine-containing neurons in the central nervous system. *Acta Physiol Scand* 1964;232:1–55.

Dahlström A, Fuxe K. Evidence for the existence of monoamine neurons in the central nervous system. II. Experimentally induced changes in the intraneuronal amine levels of bulbospinal neuron systems. *Acta Physiol Scand* 1965;247:1–36.

Dzirasa K, Ribeiro S, Costa R et al. Dopaminergic control of sleep-wake states. *J Neurosci* 2006;26(41):10577–89.

Flegontova VV. The topography and cellular organization of the reticular formation of the thoracic part of the spinal cord in the cat. *Neurosci Behav Physiol* 2000;30:111–3.

Gerrits N, Voogd J. The projection of the nucleus reticularis tegmenti pontis and adjacent regions of the pontine nuclei to the central cerebellar nuclei in the cat. *J Comp Neurol* 1987;258:52–69.

Guilleminault C, Pelayo R. Narcolepsy in children: A practical guide to its diagnosis, treatment and follow-up. *Paediatr Drugs* 2000;2:1–9.

Hobson JA. Sleep and dreaming: Induction and mediation of REM sleep by cholinergic mechanisms. *Curr Opin Neurobiol* 1992;2:759–63.

Kishi E, Ootsuka Y, Terui N. Different cardiovascular neuron groups in the ventral reticular formation of the rostral medulla in rabbits: Single neurone studies. *J Autonom Nerv Syst* 2000;79:74–83.

Mancia M. One possible function of sleep: To produce dreams. *Behav Brain Res* 1995;69:203–6.

Martin GF, Hostege G, Mettler WR. Reticular formation of the pons and medulla. In Paxinos G, ed. *The Human Nervous System*. San Diego, CA: Academic Press, 1990, 203–20.

McGinty D, Szymusiak R. Keeping cool: A hypothesis about the mechanisms and functions of slow-wave sleep. *Trends Neurosci* 1990;13:480–7.

Sasaki S, Isa T, Naito K. Effects of lesion of pontomedullary reticular formation on visually triggered vertical and oblique head orienting movements in alert cats. *Neurosci Lett* 1999;265:13–6.

Shibata M, Goto N, Goto J, Nonaka N. Nuclei of the human raphe. *Okajimas Folia Anat Jpn* 2012;89(1):15–22.

Siegel JM, Rogawski MA. A function of REM sleep: Regulation of noradrenergic receptor sensitivity. *Brain Res Rev* 1999;13:213–33.

Sugaya K, Nishijima S, Miyazato M, Ogawa Y. Central nervous control of micturition and urine storage. *J Smooth Muscle Res* 2005;41(3):117–32.

Swadling C. Narcolepsy is a rare yet serious disorder. *Nurs Times* 1999;95:41.

Webster HH, Jones BE. Neurotoxic lesions of the dorsolateral pontomesencephalic tegmentum-cholinergic cell area in the cat. II. Effects upon sleep-waking states. *Brain Res* 1988;458:285–302.

Wilson VJ. Vestibulospinal reflexes and the reticular formation. *Prog Brain Res* 1993;97:211–17.

6 Cerebellum

The cerebellum is derived from the rhombic lip of the alar plate and constitutes an important part of the rhombencephalon. It is located in the posterior cranial fossa, covered by the tentorium cerebelli, and separated by the fourth ventricle from the dorsal surfaces of the pons and medulla. The tentorium cerebelli is lodged between the cerebellum and the occipital lobes of the brain. It is connected to circuits that establish a link between the motor ans sensory cortices of the brain. It receives sensory input from the vestibular, reticular, and trigeminal nuclei as well as the spinal neurons and cerebral input via the pontine nuclei. It influences motor activity through projection to the motor cortices via specific thalamic nuclei. The cerebellum, also termed the motor autopilot, coordinates (synchronizes) fine voluntary motor activity, adjustment during motion, and eye movements; regulates phonation; maintains equilibrium; and processes exteroceptive impulses. It is also a critical structure in reflex modification, motor learning, and cognitive and emotional control. Lesions of the cerebellum or its connections produce muscular incoordination in the muscles of the eye (nystagmus), speech (dysarthria), and extremities and trunk (ataxia). Despite the remarkable motor deficits associated with cerebellar lesions, a great deal of functional recovery over time is attainable.

MORPHOLOGIC CHARACTERISTICS

The cerebellum lies caudal and inferior to the occipital lobes and dorsal to the pons and midbrain (Figures 6.1 and 6.2). Gross examination of the cerebellum reveals a corpus cerebelli with inputs from the spinal cord and pontine and trigeminal nuclei, and a flocculonodular lobe that maintains primary connections to the vestibular nuclei. The corpus cerebelli is subdivided into regions that receive spinal and pontine inputs and is connected to the midbrain, pons, and medulla via the superior, middle, and inferior cerebellar peduncles, respectively. The inferior cerebellar peduncle is medial to the middle cerebellar peduncle and consists of a dorsolateral part, the restiform body that contains afferents derived from the spinal cord and medulla (olivocerebellar, spinocerebellar, cuneocerebellar, and trigeminocerebellar tracts), and a medial part, which consists of mainly of efferent fibers (fastigiovestibular, Purkinje, and cerebellovestibular fibers) with some vestibular afferent (secondary and primary vestibulocerebellar fibers) fibers.

The middle cerebellar peduncle is the largest that passes laterally from the basilar pons and contains the fibers of the pontocerebellar tract. The superior cerebellar peduncle consists of efferent fibers that emanate from dentate, emboliform, and globose nuclei with the ventral spinocerebellar tract and some fibers from the fastigial nucleus. It decussates in the caudal mesencephalon and enters the cerebellum to join afferent fibers of the spinocerebellar tract.

The corpus cerebelli comprises two lateral hemispheres, which are connected by the midline vermis. The vermis and cerebellar hemispheres are separated by the paravermal sulcus. These hemispheres are separated posteriorly by the posterior cerebellar notch that contains the falx cerebelli. This deep notch continues inferiorly with the cerebellar vallecula, a median fossa between the two hemispheres. Each cerebellar hemisphere consists of a peripheral cortical gray matter, which is thrown into lamina or cerebellar folia, and a central white medullary substance, containing the cerebellar nuclei. Such an arrangement is also present in the cerebral cortex. In contrast, the spinal gray matter and white matter pursue a reverse arrangement.

The fourth ventricle extends into the white matter as a transverse gap, the fastigium, marking the junction of the superior and inferior medullary velum (white matter bands that extend rostrally and caudally in the roof of the fourth ventricle). The fastigium in the roof of the fourth ventricle contains the fastigial, nuclei whereas the floor of the fourth ventricle contains the vestibular area that overlies the vestibular nuclei. The cerebellar hemispheres are divided into lobes and lobules by the primary, horizontal, prepyramidal and postpyramidal, postlingual and postcentral, as well as the posterolateral fissures (Figures 6.3, 6.4, and 6.5).

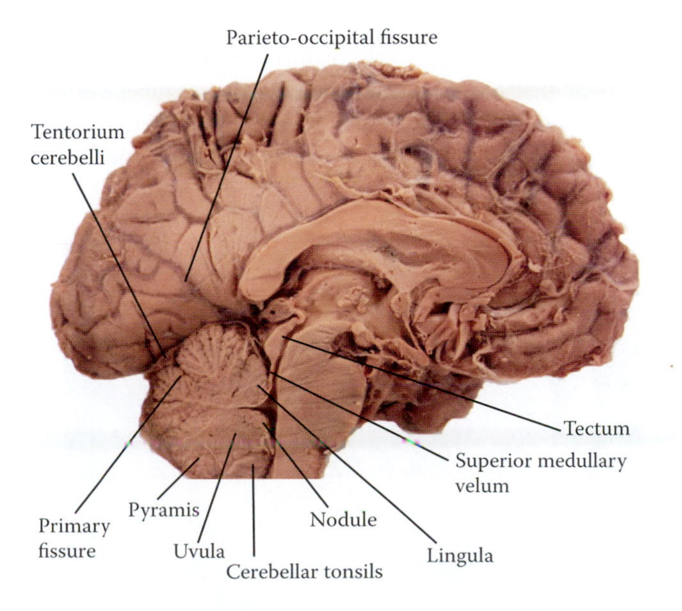

FIGURE 6.1 Midsagittal section of the brainstem and cerebellum. The fourth ventricle separates the cerebellum from the pons and medulla. The components of the cerebellar vermis are illustrated.

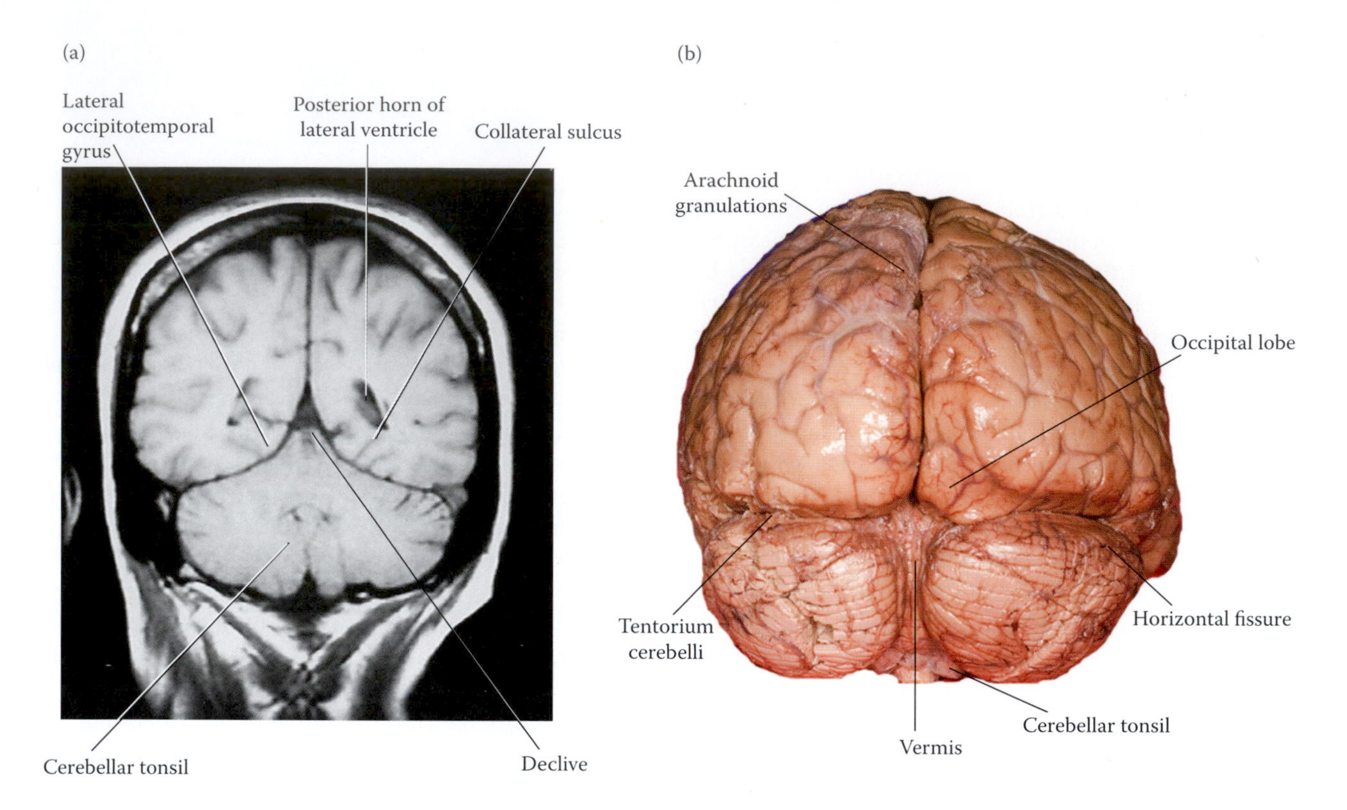

FIGURE 6.2 (a) MRI scan of the brain through the occipitotemporal gyri. The cerebellum lies inferior to the occipital lobe and is separated by the tentorium cerebelli. (b) In this photograph, the relationship of the cerebellum to the occipital is illustrated.

Developmentally, the posterolateral fissure is the first fissure to appear, marking the caudal boundary of the flocculonodular lobe. The primary fissure serves as a landmark that separates the rostrally located anterior lobe from the more caudally positioned posterior lobe. The cerebellum is further divided into superior and inferior surfaces by the horizontal fissure, which also divides the ansiform lobule (cerebellar cortical areas that correspond to the folium and tuber vermis) into superior and inferior semilunar lobules. The midline vermis, which interconnects the cerebellar hemispheres, is divided into segments. Each vermal segment expands laterally into the cerebellar hemispheres, with the exception of the lingula (*Bolk's* nomenclature). In view of this arrangement, the central lobule of the vermis corresponds to the ala of the cerebellar cortex, the culmen to the anterior quadrangular lobule, and the declive to the posterior quadrangular (simple) lobule, whereas both the folium and tuber vermis correspond to the ansiform (superior and inferior semilunar lobules) and biventral lobules of the cerebellar cortex (Figure 6.7).

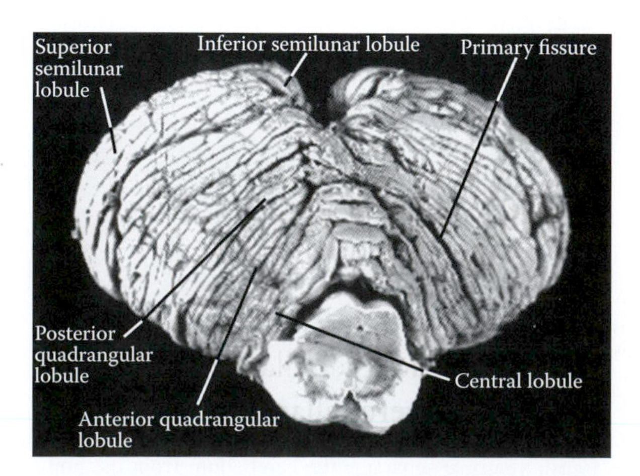

FIGURE 6.3 Superior surface of the cerebellum. The superior vermis and the corresponding parts of the cortical hemispheric areas are shown.

The central lobule is separated from the lingula by a postlingual fissure, whereas the central lobule is demarcated from the anterior quadrangular lobule via the postcentral fissure. Similarly, the pyramis corresponds to the biventral lobules, and the uvula is associated with the cerebellar tonsils and medial belly of biventral lobule (Figures 6.4, 6.5, and 6.6). The flocculi remain attached to the midline nodule and form the flocculonodular lobe. The prepyramidal fissure bounds the tuber vermis caudally, and the postpyramidal fissure separates the uvula from the pyramid.

Larsell classified the vermis in a simplified pattern in which each hemispheric lobule can be related to one of the vermian parts. Lobules I–V of Larsell constitute the anterior lobe, lobule VI extends to the cortex as the simple lobule, and VII corresponds to the ansiform lobule. In the same manner, lobule VIII has a hemispheric extension represented in the paramedian lobule (a combination of the gracile lobule and lateral belly of the biventral lobule). The paraflocculus is shared by the vermian lobules IX and X. The extension of lobule X is the flocculus.

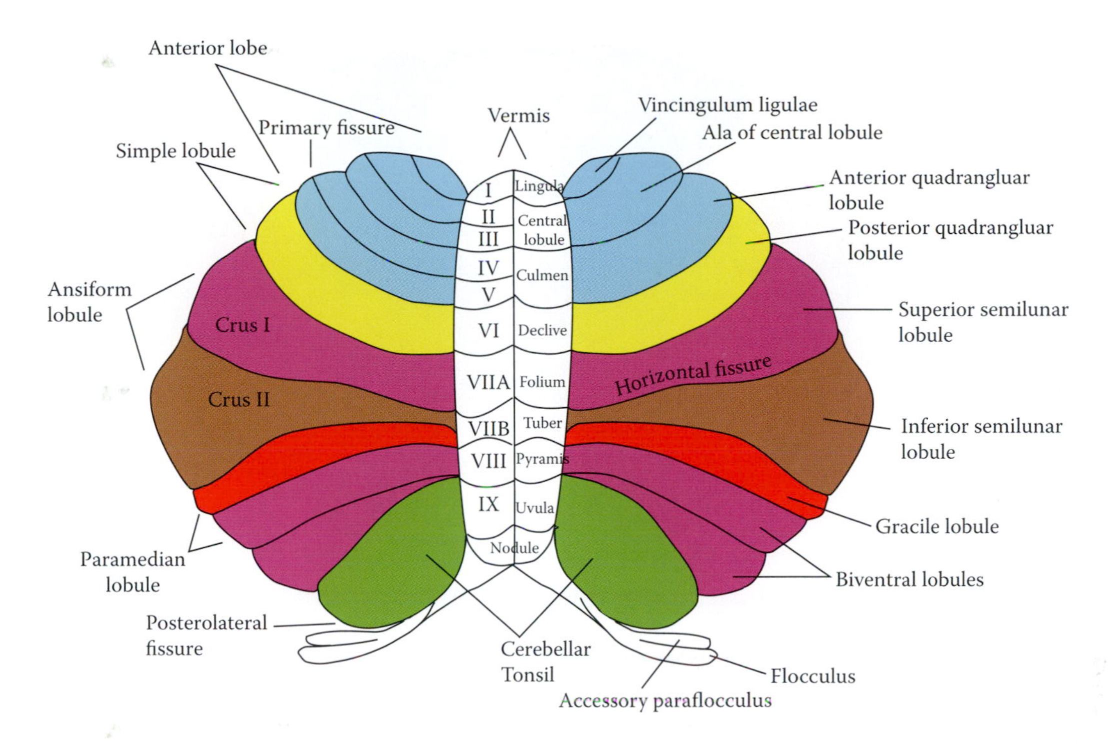

FIGURE 6.4 Various parts of the vermis and associated cortical parts are shown in this diagram.

Voogd divided the cerebellar cortex into longitudinal zone that maintains discrete connections to the cerebellar nuclei. On the basis of this mapping, the efferent fibers from the cerebellar cortex to the cerebellar nuclei appear to arise in cortical zones that belong to the same compartment as the nuclear region to which they project.

The vermis and its cortical expansions are sites of somatotopic representation of the body. Case in point, the head and face are represented in the simple lobule and the leg in the central lobule, whereas the arm occupies a distinct area in the culmen.

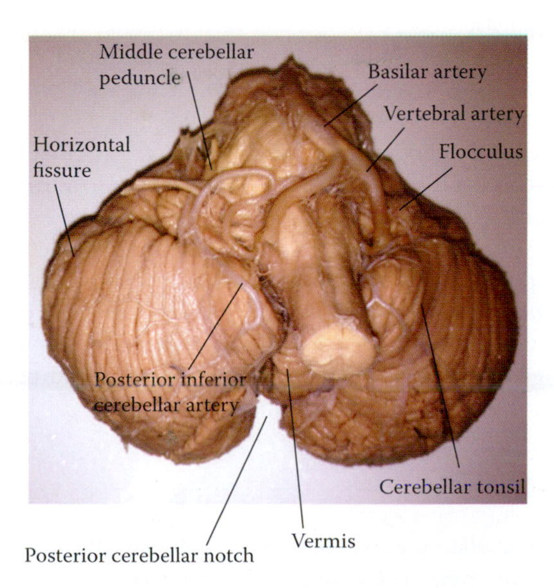

FIGURE 6.5 Inferior surface of the cerebellum. Note the branches of the anterior and posterior inferior cerebellar arteries on this surface. The posterior cerebellar notch is occupied by the falx cerebelli.

BLOOD SUPPLY AND VENOUS DRAINAGE

Blood flow to the cerebellum is sustained by the superior cerebellar, anterior, and posterior inferior cerebellar arteries. The *superior cerebellar artery supplies* the superior portion of the cerebellum, middle and inferior cerebellar peduncles, superior medullary velum, and choroid plexus of the fourth ventricle. This vessel originates from the basilar artery (an artery formed at the pontobulbar sulcus by union of the two vertebral arteries). The *anterior inferior cerebellar artery* (AICA) supplies

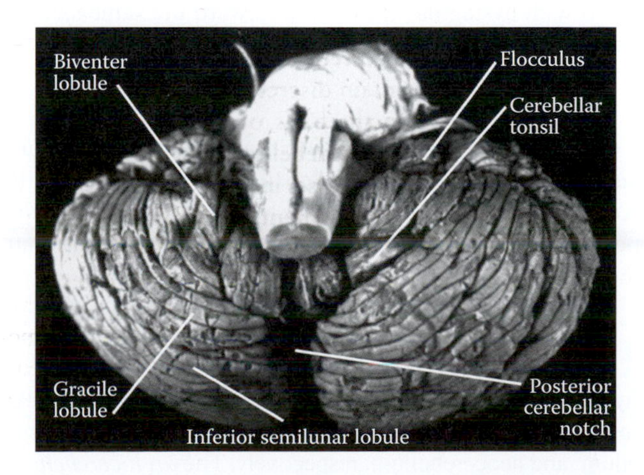

FIGURE 6.6 Inferior view of the cerebellum. Cerebellar tonsils and flocculus are clearly visible. Biventeral, gracile, and inferior semilunar lobules are illustrated.

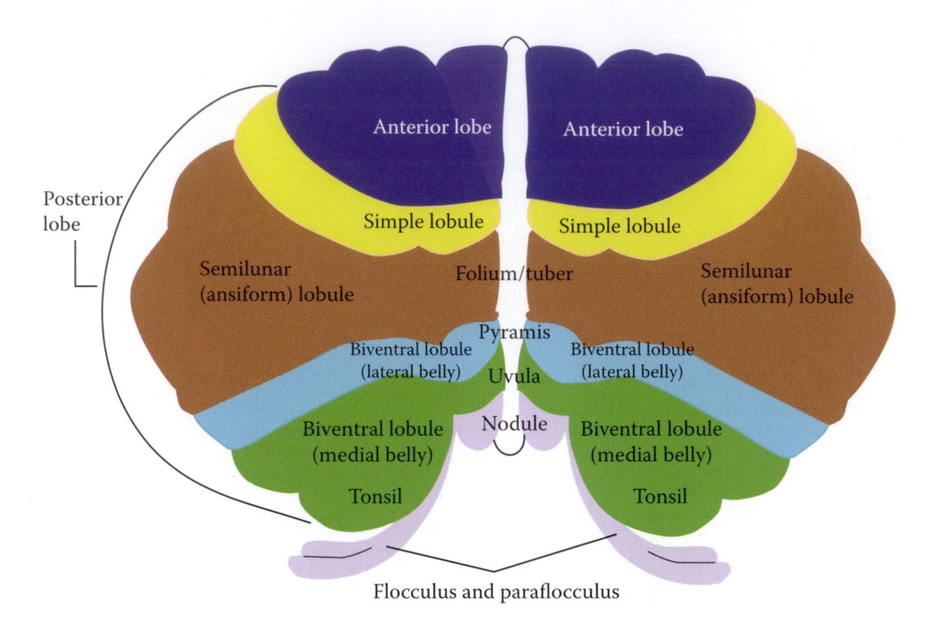

FIGURE 6.7 Various lobes of the cerebellum and associated fissures. Note that the morphologic divisions may correspond to a great extent to the functional classification.

the anterolateral part of the inferior surface of the cerebellum, pyramids, tuber vermis, dentate nucleus, white matter of the cerebellum, and most of the tegmentum of the caudal pons. The AICA arises from the lower part of the basilar artery and forms a loop into the internal acoustic meatus, where it frequently gives rise to the labyrinthine artery. The *posterior inferior cerebellar artery* supplies the cerebellar hemispheres and the inferior vermis, uvula, nodule, cerebellar tonsils, choroid plexus of the fourth ventricle, and dentate nuclei.

Occlusion of the AICA produces lesions of the abducens, facial, and vestibulocochlear nerves; vestibular nuclei; middle and superior cerebellar peduncles; as well as the trigeminal lemniscus, producing symptoms and signs of lateral inferior pontine syndrome. This syndrome manifests lateral gaze and facial palsies, vertigo with nystagmus (fast phase) toward the same side, deafness, ataxia, and contralateral hemianesthesia.

Cerebellar veins are classified into superior, inferior, and lateral veins. The superior cerebellar veins drain into the great cerebral vein of Galen, the inferior cerebellar veins open into the straight sinus, and the lateral veins drain into the transverse, superior, and inferior petrosal sinuses.

CEREBELLAR CLASSIFICATION

Phylogenetically, the cerebellum ranges from the oldest to the most recent parts, consisting of the archicerebellum, paleocerebellum, and neocerebellum, respectively. The *archicerebellum*, which corresponds to the flocculonodular lobe, represents the oldest and most primitive part of the cerebellum. The *paleocerebellum*, the second-oldest cerebellar component, consists of the anterior lobe and part of the anterior vermis. Most of the cerebellar hemispheres (posterior lobe) and part of the posterior vermis are the latest to appear phylogenetically, forming the *neocerebellum*. Functionally, the classification of the cerebellum comprises the vestibulocerebellum, spinocerebellum, and pontocerebellum. The phylogenetic and functional divisions show close similarities. The *vestibulocerebellum* is comprised of the flocculonodular lobe and caudal part of the uvula. It has reciprocal connections with the vestibular nuclei, maintaining equilibrium and mediating vestibulo-ocular reflex. The interrelationship between the vestibular nuclei, oculomotor system, and cerebellum plays an important role in the coordination of voluntary eye movements. The *spinocerebellum* includes most of the anterior lobe, pyramis, and corresponding cortical parts (biventral lobules). It receives input from the spinal cord and mesencephalic trigeminal nucleus and deals with propulsive movements (e.g., walking and swimming).

The flocculonodular lobe deals with activities that govern posture and movement through its connections to the brainstem and spinal cord. The posterior lobe, with the exception of the simple lobule, the gracile and biventral, lobule and the corresponding vermal parts (declive and pyramis), comprises the *pontocerebellum*. This lobe receives cerebral cortical input via the pontine nuclei. It coordinates fine movements through its connections to the cerebral cortex and possibly planning for anticipated movement.

CEREBELLAR CORTEX

The cerebellar cortex consists of granular, Purkinje, and molecular layers. These layers contain interneurons (granule, Golgi, basket, and stellate cells) that are functionally inhibitory (Figures 6.8, 6.9, and 6.10). The *granular layer* (innermost layer) consists of small, densely packed granule cells, Golgi neurons, and mossy fiber rosettes. Granule cells give

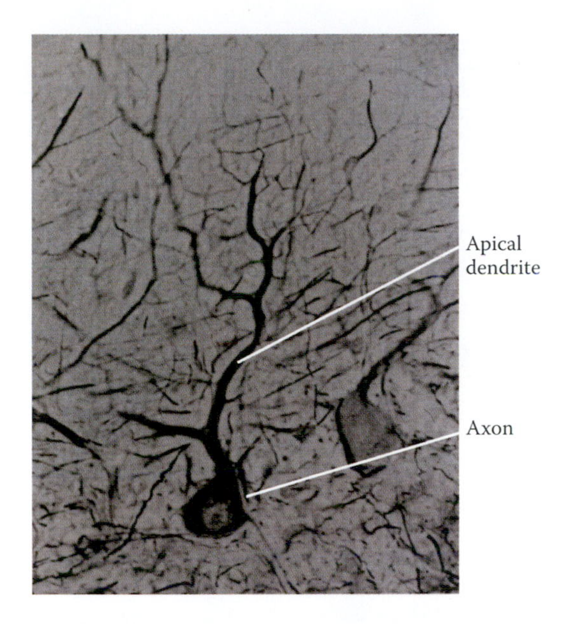

FIGURE 6.8 Photograph of the Golgi neuron with its prominent apical dendrite.

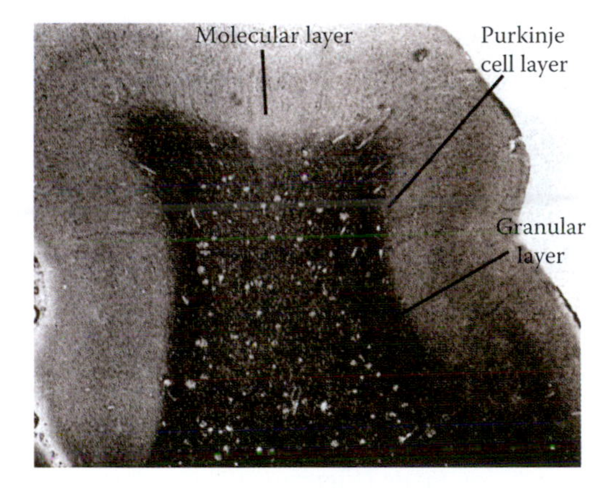

FIGURE 6.9 The cerebellar cortical layers. Note the single Purkinje layer between the outer molecular and inner granular layer.

rise to axons that ascend into the molecular layer through the Purkinje layer, bifurcating into T-shaped fibers that run parallel to the direction of the cerebellar folia (parallel fibers). They have a number of dendrites that form clawlike terminal expansions that establish synapses with the terminals of the mossy fibers and Golgi neuronal axons in the cerebellar glomeruli. N-methyl-D-aspartatereceptors (NMDA), which are voltage dependent and produce slow depolarization and opening of the calcium channels in the postsynaptic neurons, are the predominant ionotropic glutamate receptors in the immature granule cells. However, α-amino-3-hydroxy-5-methyl-4-isolaxone propionic acid (*AMP*) and high-affinity kainate types of non-NMDA glutamate receptors mediate fast excitatory transmission. It must be realized that receptors mediating cholinergic (nicotinic and muscuranic) receptors also exist in the granule cells. During their course, axons of the granule cells establish extensive contacts with the successive Purkinje cells like telephone poles. This transmission between the parallel fibers of the granule cells and the Purkinje dendrites are mediated by metabotropic glutamate receptors (mGluR2 and mGluR7 types), which are coupled to the phosphoinositide hydrolysis second messenger system and by the ionotropic glutamate receptors of the *AMP* type. Transmission between Purkinje and parallel fibers may be blocked by adenosine that binds to A_1-adenosine receptors of the parallel fibers. Desensitization of Purkinje cell AMP and reduction of the synaptic transmission may occur as a result of simultaneous activation of the parallel and climbing fibers, a process known as long-term depression (LTD).

LTD is dependent upon an increase in intracellular calcium and on cyclic guanosine 3′,5′-monophosphate (cGMP)–dependent protein kinase in Purkinje cells. Since synthesis of cyclic AMP is catalyzed by soluble guanylate cyclase (present in all cerebellar cells), which is activated by NO synthetase, inhibition of production of cGMP by NO may abolish LTD. cGMP is located in Bergmann glial cells and astrocytes. Changes in the circuitry of the cerebellum during learning and adaptation processes may be attributed to the LTD. Thus, LTD and adenosine are the two mechanisms by which transmission between granules and Purkinje cells is blocked. Climbing fiber activity determines the extent of release of adenosine. The excitatory input mediated by the parallel fibers and incoming mossy fibers results in repetitive firing of the Purkinje neurons with the conventional type of action potentials.

Golgi neurons (Figure 6.8), which are also located in this layer, have somas that are lodged in the outer portions of the granular layer and dendrites that extend into the molecular layer. Their dendrites and axonal branches extend in a radially symmetrical pattern, in exact contrast to the pattern of branching of other inhibitory neurons and Purkinje cells. In the mossy synaptic glomeruli, the Golgi cells also establish inhibitory axosomatic contacts with the mossy and parallel fibers of granule cells. Golgi cells contain *gamma*-Aminobutyric acid (GABA), especially α_6 subunit and glycine, a feature that distinguishes these cells from the basket and stellate cells. The α_6 subunit is predominant in the Golgi neurons when compared to the granule cells, accounting for the scarcity of benzodiazepine-sensitive $GABA_A$ receptors in the granular layer. Some Golgi cells may also contain enkephalin and somatostatin, as well as choline acetyl transferase, which catalyze the synthesis of acetylcholine. However, calcium-binding proteins are strikingly lacking in these cells. Since most Golgi cells contain the $mGluR_2$ type of metatropic glutamate receptor, which is dependent upon cyclic adenosine monophosphate, inhibition of AMP and, subsequently, Golgi cells may result from stimulation of the glutamatergic parallel or mossy fibers. Mossy fibers constitute the bulk of the cerebellar afferents (nonolivary fibers) and terminate in the granular layer as *mossy rosettes* (Figure 6.10). Sites of linkages between mossy rosettes, axons of the Golgi neurons, and dendrites of the granule cells occur in the synaptic glomeruli.

The *Purkinje cell layer* (Figures 6.9 and 6.10) consists of a single row of flask-shaped neurons with massive and flattened dendritic trees between the molecular and granular

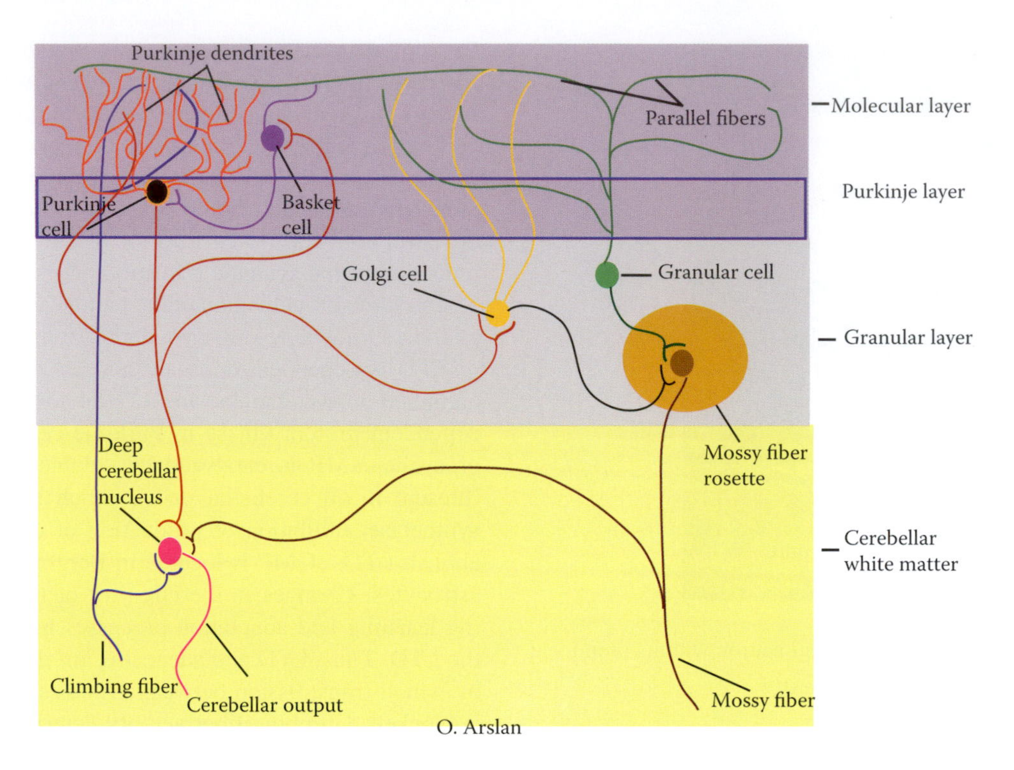

FIGURE 6.10 Cellular organization of the cerebellar cortex. The mossy fibers, cerebellar glomeruli, and climbing fibers are illustrated.

layers that establish contacts with the climbing fibers and the axons of the granule cells. Branches of each Purkinje neuron are confined to a plane horizontal to the long axis of the folium. The proximal first- and second-order dendrites receive the excitatory climbing fibers, whereas the distal dendrites exhibit spines that establish synapses with terminals of the parallel fibers. Purkinje axons pass through the granular layer into the white matter and are initially unmyelinated and surrounded by the distant branches of the basket cells. The axons continue into the white matter, giving rise to recurrent collaterals that end on Purkinje, Golgi, and basket neurons. Then the axons terminate in the vestibular or cerebellar nuclei in the form of plexuses. Subunits of $GABA_A$, which are present in the cerebellar neurons, include α_1, α_3, β_2, and μ_2. The combination of α_1, β_2, and μ_2 subunits displays high-affinity binding for benzodiazepine ligands. α_1 subunits are present in the soma of Purkinje cells opposite the terminals of the basket cells, whereas α_3 subunits occupy the proximal dendritic tree where the stellate cell terminates. Elements of the second messenger system that mobilize calcium from subsurface cisterns (e.g., receptor for $InsP_3$ or inositol 1,4,5-triphosphate) and protein kinase C are contained in the Purkinje neurons. Calcium-binding proteins such as calmodulin, calbindin, and parvalbumin are also located in these cells. In fact, these elements are functionally related to the metabotropic receptors that mediate synaptic transmission between parallel fibers and Purkinje cells. All these factors may play an important role in the increase of calcium concentration in Purkinje cells. Purkinje axons project to the cerebellar nuclei in a mediolateral direction (axons of vermal neurons project to the fastigial nucleus, paravermal neurons to the globose and emboliform nuclei, and lateral hemispheric neurons to the dentate nucleus and the lateral vestibular nuclei). Thus, the cerebellar excitatory input must overcome the tonic inhibitory impulses generated by the Purkinje cells upon the cerebellar nuclei. These projections, which represent the inhibitory corticonucleocerebellar tract, are arranged in symmetrical longitudinal bands that reflect the zebrin-positive cells. The latter is a group of proteins that are contained in certain populations of Purkinje neurons.

The *molecular layer* (Figure 6.10) is a largely acellular outer layer, containing the dendrites of the Purkinje neurons, climbing fibers, axons of the granule cells, stellate cells, and Basket cells. *Stellate* and *basket cells* have similar structures, their dendrites extend toward the surface, and their axons run transversely to the folia and parallel to the dendrites of the Purkinje cells. They receive excitatory input from the parallel fibers, and their somata receive collaterals from the Purkinje neurons, as well as parallel and mossy fibers. These GABAergic cells, which contain NMDA receptors, mediate the parallel fiber–Purkinje inhibition. Basket cell somata establish linkage with Purkinje axon recurrent collaterals and climbing, parallel, and mossy fibers. Basket cell axons run in the molecular layer and give collaterals that ascend with Purkinje cell dendrites toward their somata, forming pericellular basket networks, and then continue wrapping around Purkinje axons in a brush configuration. Large epithelial (Bergmann) glial cells and their radiating branches and processes that surround all cerebellar cortical neurons are also present in this layer.

Bergmann cells are associated with signaling *processes*, and are the primary source of cGMP in the cerebellar cortex. These cells, among all cerebellar neurons, uniquely contain the α_2 subunit of $GABA_A$. The enzyme, which is involved in the hydrolytic cleavage of 5′-nucleotide monophosphates and formation of the adenosine, also resides in these cells. Among other features of Bergmann cells are their possession of kainate receptors with specific ionic arrangement. The release of homocysteic acid, a putative amino acid neurotransmitter, by these cells is dependent upon the climbing fibers. They are capable of glutamate uptake and its conversion into glutamine, a process that enable the synthesis of glutamate by glutamatergic terminals. These terminals form the external limiting membrane.

CEREBELLAR NUCLEI

The cerebellar nuclei are a cluster of gray matter masses embedded in the white matter of the cerebellum, consisting of multipolar neurons of different sizes with long branching dendrites and an axon that travels through the superior cerebellar peduncle, uncinate fasciculus, and juxtarestiform body. This nuclear group includes the fastigial, globose and emboliform, and dentate nuclei. They receive collateral of excitatory projections from mossy and climbing fibers, as well as inhibitory input from the Purkinje cell axons. The fastigial and interpositus nuclei connect to the spinocerebellum; the dentate nucleus communicates exclusively with the lateral parts of the pontocerebellum. The fastigial nucleus is connected bilaterally with the vestibular nuclei and the brainstem reticular formation. There are GABAergic neurons that form the nucleo-olivary tract that connects the fastigial nucleus to the medial accessory olivary nucleus. The main output of the fastigial nucleus is bilateral projection to the vestibular nuclei with contralateral predominance via the uncinate fasciculus. Uncrossed fastigiovestibular fibers occupy the juxtarestiform body. Some fastigial fibers terminate in the deep layers of the superior colliculus. Neurons of these nuclei contain ionotropic glutamate receptors of the NMDA and non-NMDA types, as well as GABAergic ($GABA_A$ α_1-β_2 or 3-μ_2) and glycine receptors.

The *fastigial nucleus* is the most medial and phylogenetically the oldest deep cerebellar nucleus. It is located near the apex of the fourth ventricle where the superior and inferior medullary velum join. It receives input from the vermis; and collaterals of cerebellar cortical afferents; and afferents from the medial and inferior vestibular nuclei, locus ceruleus, and reticulotegmental nucleus. It projects to the vestibular nuclei and the reticular formation.

The *globose and emboliform nuclei* are intermediate in phylogeny and location, receiving input from the paravermal cortex, reticulotegmental nucleus, medial and dorsal accessory olivary nuclei, and collaterals from other cerebellar afferents.

The *dentate nucleus* is the most lateral nucleus and, developmentally, is the most recent. It is an irregularly folded layer of neurons that encloses a mass of white matter, resembling a leather purse. Most cerebellar afferents, cortical fibers from the lateral hemispheric zone, afferents from the pontine nuclei, trigeminal sensory nuclei, reticulotegmental nucleus, inferior olivary nucleus, locus ceruleus, and the raphe nuclei terminate in this nucleus. It forms the main cerebellar output, which projects through the hilum of the nucleus and then continues within the superior cerebellar peduncle to the red nucleus and the ventral lateral nucleus.

CEREBELLAR AFFERENTS

The climbing fibers (Figure 6.10), which utilize L-glutamate as a neurotransmitter, primarily represent the olivocerebellar fibers that establish synaptic contacts with the Purkinje dendrite. In addition to the L-glutamate and L-aspartate, peptides are also contained in certain subpopulation of climbing fibers. Corticotropin-releasing factor is also contained in these fibers. A single axon of the inferior olivary nucleus establishes excitatory connections with about a dozen Purkinje neurons, and each Purkinje neuron, in turn, receives only one climbing fiber. The dendrites of the Purkinje neurons are entirely entwined by the climbing fiber, making roughly over 200 synaptic contacts. A single climbing fiber may evoke an excitatory postsynaptic potential that maintains amplitude greater than 25 mV (complex spikes), which greatly exceeds the Purkinje cell threshold. Due to this very reason, a single impulse in the climbing fibers always elicits an action potential in more than 10 Purkinje cells. Thus, the action potential of a single climbing fiber is not a graded response but an all-or-none action. Thus, the Purkinje cells respond to the generated action potential with complex spikes, in contrast to the simple spikes evoked by the T-parallel fibers of the granule cells. According to some investigators, the climbing fiber is primarily concerned with fast, ballistic movements. Others claim that the climbing fiber system reflects the summation of the inhibitory and excitatory synaptic activity at any instant time. Some also theorize that the signals in the climbing fibers are meant to ascertain error in executing a motor activity (e.g., during learning stages), which suggests that the frequency of firing of the climbing fibers is not in any way related to the direction or the speed of a motor activity but, rather, is related to the disturbances of that activity. Glutamate receptors at the synaptic connections between the Purkinje neurons and the climbing fibers are non-NMDA type, which is responsible for the influx of calcium ions into the dendritic tree of the Purkinje neurons.

The *olivocerebellar tract* (Figures 6.10, 6.11, and 6.14), a contralateral pathway that arises from the inferior and the accessory olivary nuclei, represents the major climbing fiber system. It crosses the medullary reticular formation to be distributed to the opposite vermal and cerebellar hemispheres via the inferior cerebellar peduncle. The accessory olivary nucleus conveys joint, tactile, visual, and vestibular impulses to the cerebellum from certain nuclei of the medulla. The inferior olivary nucleus sends information to the cerebellum, which is derived from the spinal cord via the spino-olivary pathway, motor cortex, and periaqueductal gray and accessory oculomotor nuclei.

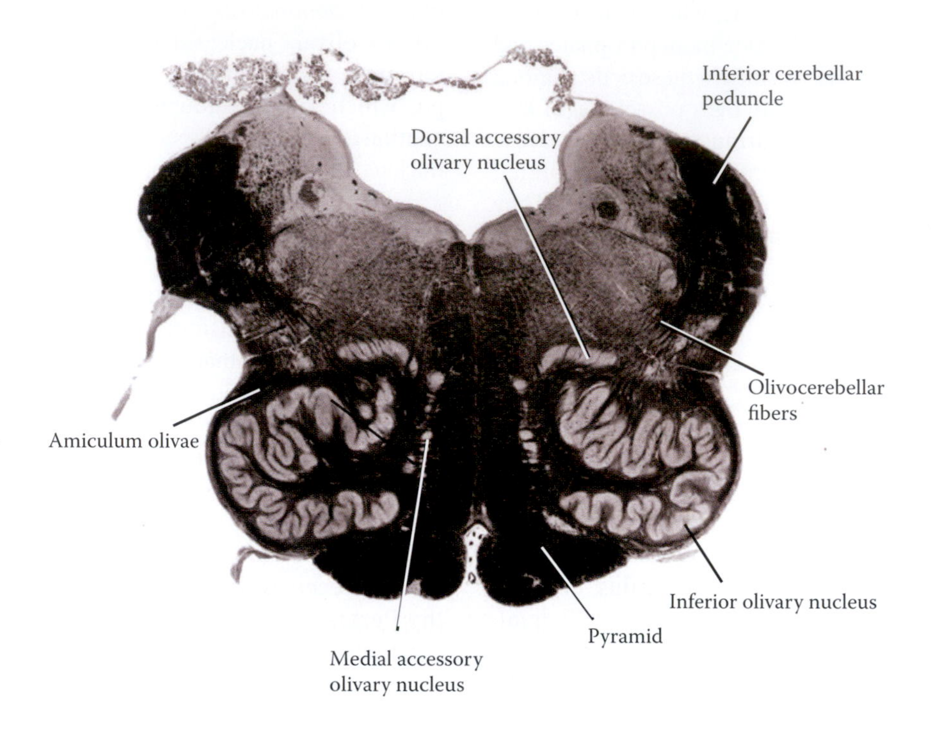

FIGURE 6.11 Section through the medulla at the midolivary level. The massive fibers of the olivocerebellar tract represent the main component of the inferior cerebellar peduncle.

> Secondary transsynaptic degeneration and atrophy of the inferior olivary nucleus may occur as a result of cerebellar cortical degeneration involving the superior vermis. Olivopontocerebellar atrophy, as the name indicates, is associated with degeneration of the inferior olivary nucleus, pontine nuclei, and cerebellar cortex. It is categorized into inherited and sporadic forms. Patients with this condition exhibit progressive cerebellar ataxia, signs of upper motor neuron palsy, involuntary movements, eye movements, bowel and bladder dysfunction, abnormal autonomic disorders, and peripheral neuropathy. A lesion, which disrupts the circuitry between the dentate nucleus and the inferior olivary nucleus, may result in palatal myoclonus, a continuous rhythmic contraction of the posterior pharyngeal muscles that resembles tremor.

The mossy fibers refer to all afferents of the cerebellar cortex with the exception of the olivocerebellar tract. They include the primary and secondary vestibulocerebellar, spinocerebellar, reticulocerebellar, pontocerebellar, tectocerebellar, and trigeminocerebellar tracts. Some of the fibers may establish excitatory synaptic contacts with the cerebellar nuclei, basket, Golgi, and stellate neurons. These fibers may be concerned with slow and tonic motor activity. L-glutamate is the primary neurotransmitter in mossy fibers with the exception of the secondary vestibulocerebellar fibers that utilize acetylcholine. The mossy fibers traverse the white matter and give branches to adjacent folia, and with each folium, mossy fibers provide branches to the granular layer, expanding into grapelike endings (mossy fiber rosettes) that join the cerebellar glomeruli. The latter are spherical or ovoid asymmetric excitatory synaptic sites that consist of mossy fiber rosettes, terminals of Golgi axons (asymmetrically inhibitory), and dendrites of the multiple granule cells. There is one cerebellar glomerulus for every granule cell. The primary vestibulocerebellar fibers (Figures 6.12 and 6.14) are the central processes of the neurons of the vestibular ganglia. These fibers run within the juxtarestiform part of the inferior cerebellar peduncle and terminate in the vestibulocerebellum of the same side.

Secondary vestibulocerebellar fibers (Figures 6.12, 6.14, and 6.19) originate from the superior, medial, and inferior vestibular nuclei; run within the juxtarestiform part of the inferior cerebellar peduncle, and terminate in the paraflocculus and the vestibulocerebellum (flocculus, nodule, and uvula) on both sides.

The spinocerebellar tracts (Figures 6.13 and 6.14), which are represented by two principal pathways, carry proprioceptive, stretch, and tactile sensations, terminating in the spinocerebellum (tactile input mainly terminates ipsilaterally in the simple lobule). The *dorsal* spinocerebellar tract represents the axons of Clarke's column, which extends in the thoracic and upper lumbar segments of the spinal cord. It is an ipsilateral tract that enters the spinocerebellum through the inferior cerebellar peduncle, carrying proprioceptive, tactile, and pressure impulses from individual muscles and joints of the lower extremity and lower half of the trunk. The upper limb equivalent of this tract is the *cuneocerebellar tract* that is derived from the accessory cuneate nucleus and terminates ipsilaterally in the pontocerebellum and spinocerebellum via the inferior cerebellar peduncle (Figure 6.15).

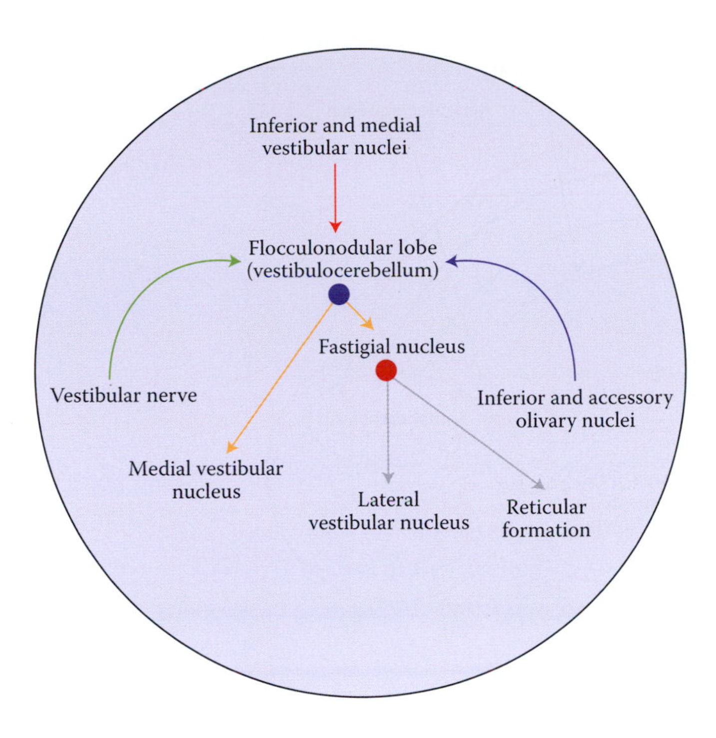

FIGURE 6.12 Schematic drawing of the projections of the vestibular nerve (primary vestibulocerebellar tract) and the vestibular nuclei (secondary vestibulocerebellar tract) to the cerebellum.

The *ventral spinocerebellar tract* (Figure 6.17) is derived from the intermediate gray columns and the border cells of the anterior horn cells of the thoracolumbar and sacral segments. Information conveyed by this crossed tract originates from the whole lower extremity, reaching the cerebellum through the superior cerebellar peduncle. The fibers of this tract cross again within this peduncle and terminate in the ipsilateral spinocerebellum. Lamina VII of the cervical enlargement gives rise to the rostral spinocerebellar, an ipsilateral tract, which represents the upper limb equivalent of the ventral

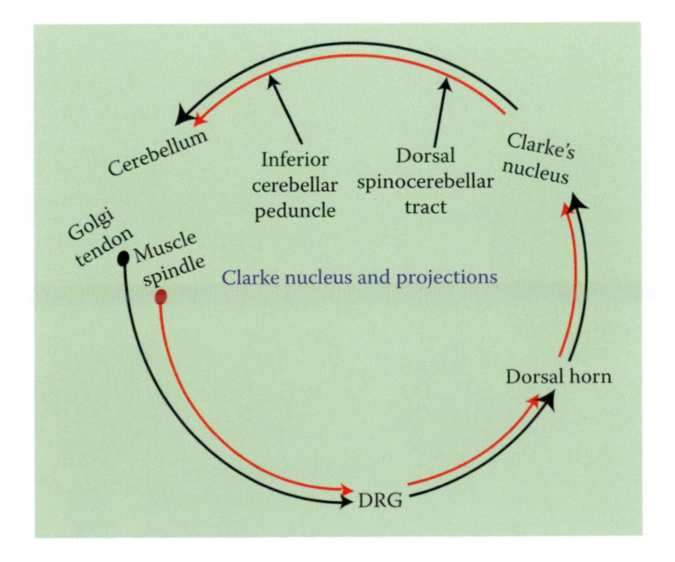

FIGURE 6.13 Clarke's nucleus and its projection to the cerebellum through the dorsal spinocerebellar tract. DRG, dorsal root ganglion.

spinocerebellar tract. It enters the cerebellum through the inferior and superior cerebellar peduncles, to be distributed to the anterior lobe of the cerebellum. The ventral and rostral spinocerebellar tracts act jointly as a relay center reflecting the neuronal activities in the descending motor pathways.

> Severe loss of balance is associated with lesions of the spinocerebellar tracts as well as the median vermis.

The *reticulocerebellar tract* (Figures 6.14, 6.17, and 6.20) originates from the pontine reticulotegmental, and the medullary lateral and paramedian reticular nuclei. The lateral reticular nucleus projects bilaterally via the inferior cerebellar peduncle to the vermis of the spinocerebellum, fastigial, and emboliform nuclei. This projection conveys information from all levels of the spinal cord (spinoreticular) that initially establishes synaptic contacts with the neurons of the lateral cervical nucleus and later projects to the medullary reticular formation, and eventually to the cerebellum. The paramedian nuclei, which receive fibers from the interstitial nucleus, tectum, spinal cord, and cerebral cortex, send efferents to the entire cerebellum with the exception of the paraflocculus. The reticulotegmental nucleus projects to the anterior lobe, simple lobule, folium, and tuber vermis via the middle cerebellar peduncle. Some fibers also terminate in the fastigial, globose, and dentate nuclei.

The *pontocerebellar tract* (Figures 6.15, 6.16 and 6.22) delivers the cortical inputs from the ipsilateral motor, visual, and auditory cortices to the contralateral pontocerebellum (some to the ipsilateral pontocerebellum) by way of the middle cerebellar peduncle. It comprises, by far, the most massive afferent system, which passes through the anterior and posterior limbs of the internal capsule (a massive bundle of fibers, which consists of afferent and efferent fibers, connecting the cerebral cortex to subcortical centers, as well as the spinal cord). It then runs through the basis pedunculi of the midbrain and enters the basilar pons, establishing synapses with the pontine nuclei.

The *arcuatocerebellar fibers* (Figure 6.14) are derived from the arcuate nuclei located ventral to the pyramids of the medulla. These fibers pursue two distinct courses: a superficial ventral and a deeper dorsal course. The superficial (ventral external arcuate) fibers run along the lateral and ventral surfaces of the medulla and then enter the cerebellum via the inferior cerebellar peduncle. Fibers that follow a dorsal course are known as the posterior external arcuate fibers. These fibers run near the midline of the floor of the fourth ventricle, then extend laterally as the stria medullaris of the fourth ventricle, and finally enter the cerebellum via the inferior cerebellar peduncle. The sites of termination of the arcuatocerebellar fibers include the vestibulocerebellum and the pontocerebellum.

The *tectocerebellar* fibers convey auditory and visual information to the pontine nuclei and then to the spinocerebellum, through the superior cerebellar peduncle.

The *trigeminocerebellar tract* (Figure 6.14) originates from the mesencephalic nucleus of the trigeminal nerve, conveying proprioceptive impulses from the muscles of

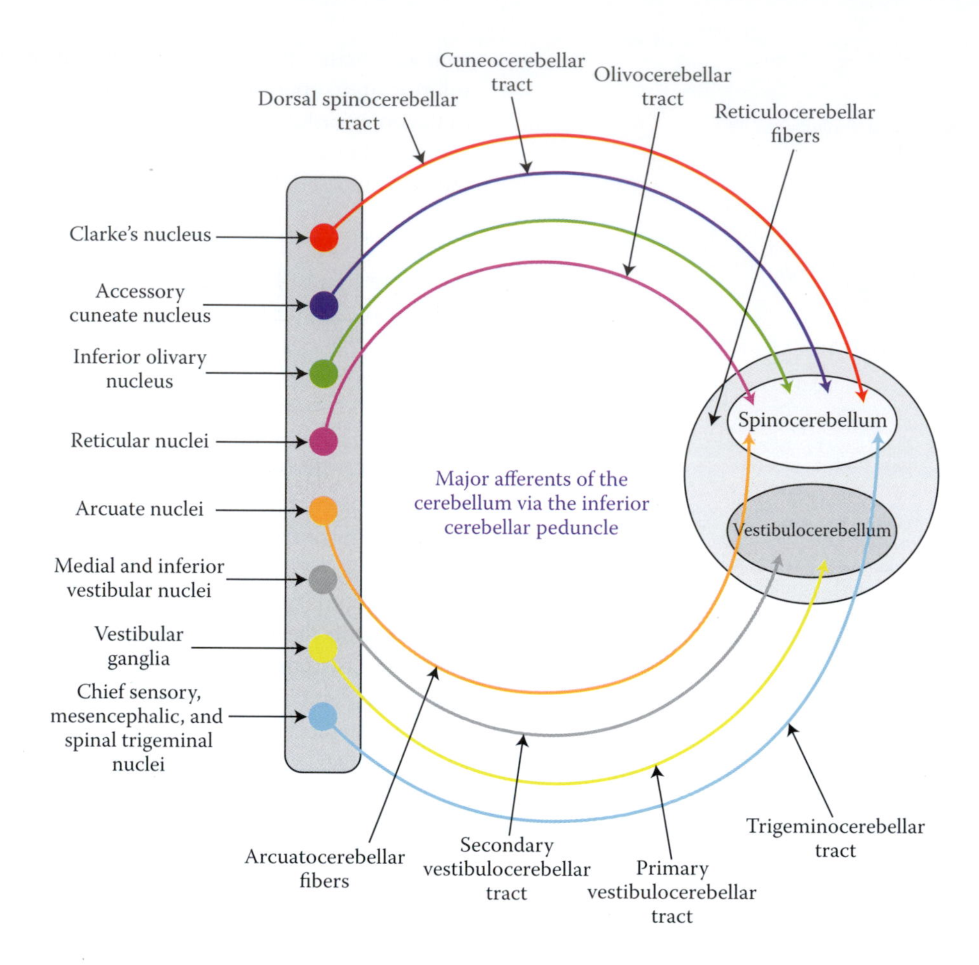

FIGURE 6.14 The major afferents to the cerebellum via the inferior cerebellar peduncle. Note that the bulk of afferents come from the inferior olivary nucleus.

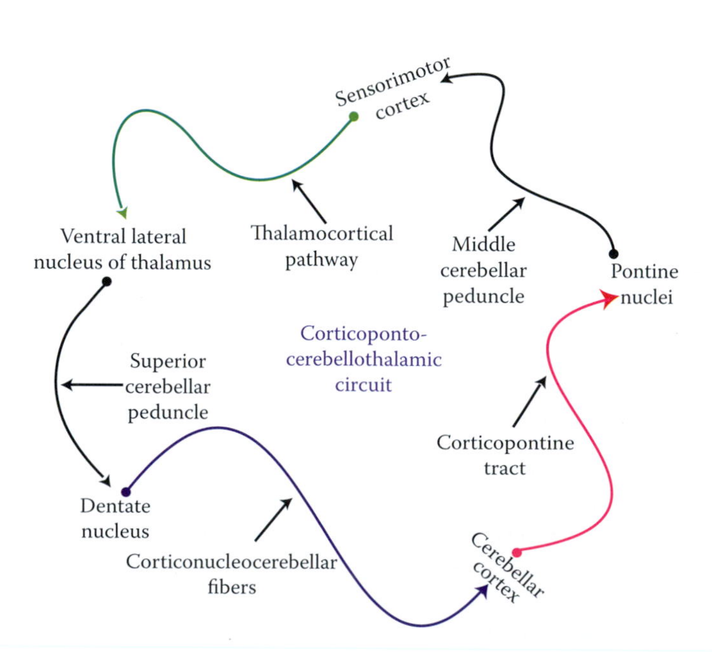

FIGURE 6.15 This circuit is mediated via the connections of the pontine nuclei, which project to the cerebellar cortex. A series of neurons are involved, as a closed-circuit loop, in the transmission of impulses between the cerebellum, thalamus, and motor cerebral cortex.

mastication and muscles of facial expression to the contralateral simple lobule and rostral vermis (spinocerebellum) via the superior cerebellar peduncle. Some trigeminal fibers also originate from the principal sensory and spinal trigeminal nuclei, which receive tactile sensation, and convey this sensation to the anterior lobe via the inferior cerebellar peduncle.

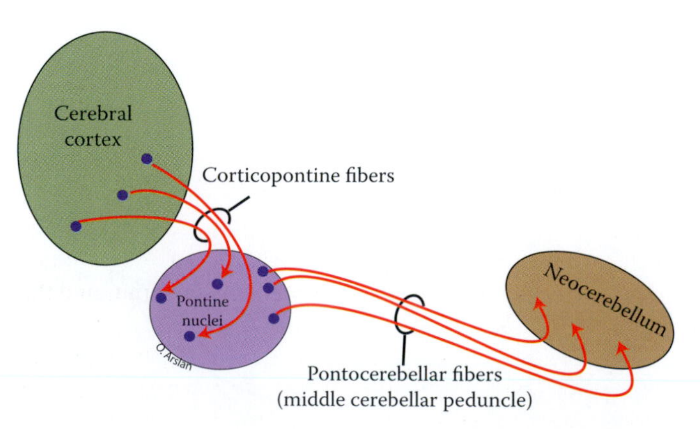

FIGURE 6.16 The neuronal components of the middle cerebellar peduncle and corticopontine fibers. Contralateral axons of the pontine nuclei form this peduncle.

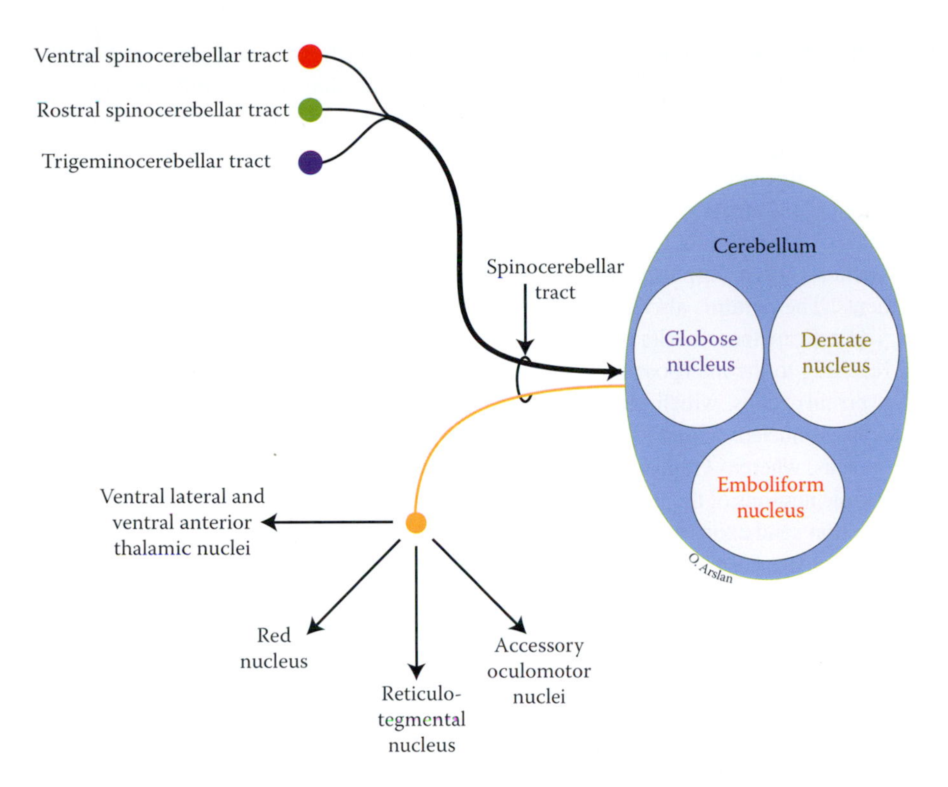

FIGURE 6.17 Afferent and efferent pathways within the superior cerebellar peduncle. The cerebellar projection to the thalamus constitutes the principal tract within this peduncle. Note that some fibers also terminate in the red nucleus and reticulotegmental and accessory oculomotor nuclei.

Aminergic cerebellar afferents include fibers from the locus ceruleus and raphe nuclei that project to the cerebellar cortex via the superior and inferior cerebellar peduncles. Axons of the locus ceruleus form a network in the molecular layer, where they increase the GABA-mediated inhibition of Purkinje cells, which is produced by the cerebellar interneurons (stellate and basket cells). They also sharpen the signals and reduce the background activity by enhancing the release of glutamate from the parallel fibers onto the Purkinje cells. Serotonergic fibers emanate mainly from the medullary reticular formation. Noradrenergic projections to the cerebellum inhibit Purkinje cells by β-adrenergic receptor–mediated inhibition of adenylate cyclase in the Purkinje cells. Dopaminergic neurons to the cerebellum gain origin from the ventral tegmentum and may act upon the D_2 and D_3 receptors of the molecular layer. Some cholinergic fibers may also be found in the Purkinje cell layer.

CEREBELLAR EFFERENTS

Some cerebellar efferents arise from the cerebellar cortex and project to the vestibular nuclei and the deep cerebellar nuclei as the corticonucleo-cerebellar fibers. Others originate from the cerebellar nuclei and project to the thalamus and the brainstem reticular formation.

The *corticonucleo-cerebellar* (Figure 6.18) fibers represent the axons of the Purkinje neurons of the cerebellar cortex that project to the vestibular and cerebellar nuclei. Neuronal axons of the vermal cortex extend to the vestibular and fastigial nuclei with a certain degree of specificity. In general, Purkinje neurons of the nodule and uvula project to the cerebellar nuclei and

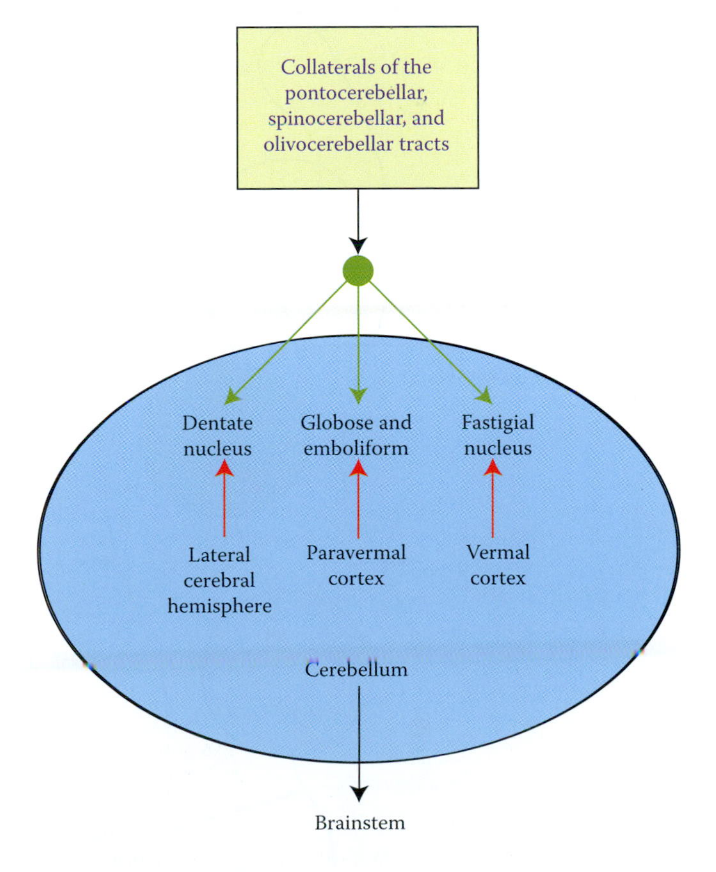

FIGURE 6.18 Somatotopic projections of the Purkinje cell axons to the cerebellar nuclei. These nuclei also process information received via collaterals of some mossy and climbing fibers. Purkinje axons also convey information to the brainstem.

to all of the vestibular nuclei with the exception of the lateral vestibular nucleus, while those of the paleocerebellum project to the lateral and inferior vestibular nuclei. There are A and B parallel zones of Purkinje cells in the vermis of the anterior lobe and simple lobule that project to the rostral part of the fastigial nucleus and the lateral vestibular nucleus, respectively. The caudal part of the fastigial nucleus integrates input from the folium and tuber vermis, which includes visual impulses used to calibrate saccadic eye movements. The pyramis also projects to the same areas that receive input from the anterior lobe. Uvular projections are more far reaching to the interposed and dentate nuclei. Intermediate (paravermal) zones, which include C_1, C_2, and C_3, project to the interposed nucleus. C_1 and C_3 zones send fibers to the emboliform nucleus, whereas C_2 projects primarily to the globose nucleus. Purkinje neurons of the lateral cerebellar cortex form D_1 and D_2 zones that send axons to the caudolateral and rostromedial parts of the dentate nucleus, respectively.

The *dentatorubrothalamic tract* is formed by axons of the dentate nucleus, which runs within the superior cerebellar peduncle, and terminates in the ventral lateral and ventral posterolateral thalamic nuclei. These terminations are specific and do not show any overlap with the pallidal termination. The thalamic nuclei that receive cerebellar output project to the motor cortex via the thalamocortical radiation. This connection enables the cerebellum to exert influence over motor activity. Collaterals of this projection are also given to the red nucleus, oculomotor nucleus, and associated accessory oculomotor nuclei (interstitial nucleus of and Cajal and nucleus of Darkschewitch). Efferents from the globose and emboliform nuclei project to the contralateral ventral posterolateral, intralaminar, and ventral lateral thalamic nuclei via the superior cerebellar peduncle. Collaterals of these fibers terminate in the red nucleus.

The *fastigiovestibular pathway* (Figure 6.19) is an excitatory pathway derived from the fastigial nucleus that projects to the ipsilateral vestibular nuclei via the juxtarestiform body. It also projects to the lateral and inferior vestibular nuclei and to the nucleus reticularis gigantocellularis of the contralateral medulla via the *uncinate fasciculus* (hook bundle of Russel). There are also fastigial nuclear projections to the ventral lateral and ventral posterolateral thalamic nuclei. Cerebellar influences upon motor activity may be mediated via the fastigial projection to the ventral lateral nucleus of thalamus, the primary motor cortex, and the corticospinal tract. The lateral vestibular nucleus conveys this information to the spinal cord, regulating the motor activity of the antigravity muscles. Fastigial projection to the nucleus reticularis

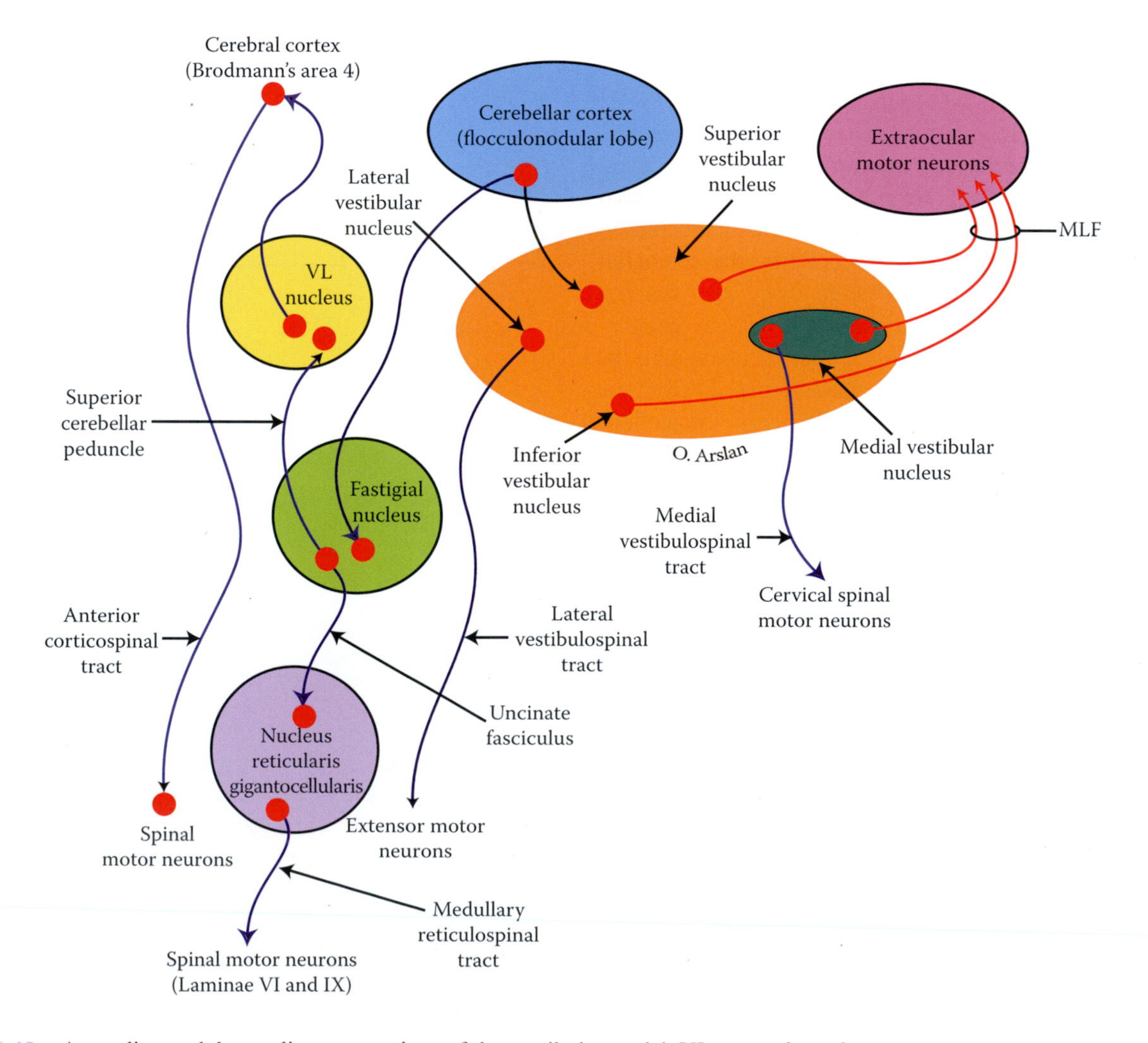

FIGURE 6.19 Ascending and descending connections of the vestibular nuclei. VL, ventrolateral.

gigantocellularis allows the fastigial nucleus to influence motor activity via the medullary reticulospinal tract. This connection serves as an additional route by which the fastigial nucleus regulates motor activity.

Lesions of the efferent fibers of the cerebellum contained in the superior cerebellar peduncles produce severe incapacitating proximal tremor that leads to rhythmic oscillation of the head or trunk that makes standing or sitting without support an impossible task.

CEREBELLAR CIRCUITS

The cerebellar functions are performed by a series of closed-circuit pathways, which indicate the complexities of cerebellar connections and the role each circuit and associated centers play in programming, sequencing, grading, and ultimately coordinating motor activities.

Cerebellovestibular Circuit

The cerebellovestibular circuit includes afferent fibers from the vestibular nerve and nuclei, descending vestibular projections to the spinal neurons, and ascending vestibulo-ocular fibers to the motor nuclei of the eye muscles. Through this loop, impulses generated by the flocculonodular lobe, uvula, and associated folia project to the vestibular nuclei. Also, information received from the vestibular receptors is conveyed to the vestibulocerebellum via the primary (vestibular nerve) and secondary vestibulocerebellar fibers. Vestibular projections to the flocculonodular lobe eventually influence the activities of the ventral lateral nucleus of the thalamus and the motor cerebral cortex via their connections to the fastigial nucleus. Purkinje neurons of the nodule and flocculus project to the medial and inferior vestibular nuclei that receive input from the vestibular nerve. Medial and inferior vestibular nuclei project bilaterally via the medial longitudinal fasciculus (MLF) to the motor nuclei that govern eye muscles. Thus, the flocculus plays an important role in adaptation of compensatory eye movements, fixation of gaze, and generation of smooth pursuit movement of the eye by suppressing the vestibulo-ocular reflex. The lateral vestibular nucleus, which receives input from the anterior vermis, provides ipsilateral excitatory projection through the lateral vestibulospinal tract to the extensor neurons of the spinal cord. The medial and inferior vestibular nuclei also provide a bilateral inhibitory medial vestibulospinal tract through the MLF to the cervical spinal neurons (Figure 6.19). Thus, the vermis exerts an influence on the antigravity postural muscles and vestibular reflexes.

Reticulocerebellar Circuit

Information received by the reticular formation, which originates from diverse sources such as the fastigial nucleus, cerebellar cortex, and spinal cord, is processed and sent back to the cerebellum directly via the inferior cerebellar peduncle

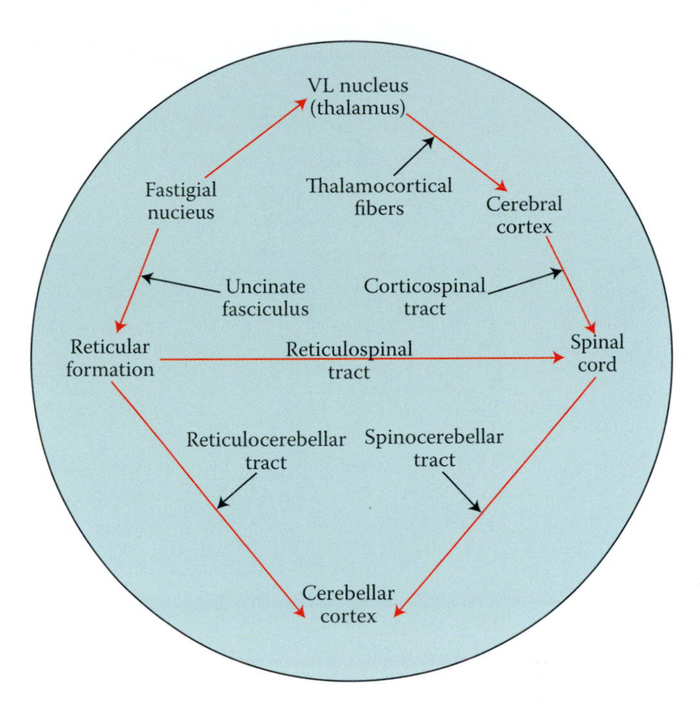

FIGURE 6.20 Efferent and afferent fibers of the reticular formation within the cerebellar feedback loop. VL, ventrolateral.

as a component of the reticulocerebellar tract (Figure 6.20). It is also sent indirectly via the reticulospinal tract to the spinal cord, which projects back to the cerebellum by the spinocerebellar tracts. The spinal cord also conveys impulses to the cerebellum, which are generated in the cerebral cortex and delivered via the corticospinal tract to spinal cord segments.

Rubrocerebellar Circuit

The red nucleus receives input from the ipsilateral cerebral cortex and the contralateral cerebellar nuclei. Through its connection to the inferior olivary nucleus (via the rubro-olivary tract), the red nucleus influences activities of the cerebellar cortex through the massive olivocerebellar fibers (Figure 6.21). Thus, the combination of corticorubral fibers, spinal projections of the red nucleus (rubrospinal tract), spino-olivary fibers (conveying multisensory input to the inferior olivary nucleus from the spinal cord), inferior olivary nucleus itself, and the olivocerebellar fibers; cerebellar nuclei; and finally, the red nucleus serves to complete this feedback circuit. Due to the massive connections of the inferior olivary nucleus to the cerebellum through the olivocerebellar tract and to the cerebral cortex through the central tegmental tract and a relay in the red nucleus, the rubro-olivary tract forms part of a feedback loop conveying motor cortical input to the cerebellum and back to the motor cerebral cortex. This loop is bidirectional and able to shift the control of motor activity from the corticospinal to the rubrospinal tract for programmed automation and, at the same time, enables the corticospinal tract to intervene during automated motor activity conducted by the rubrospinal tract in response to environmental effects.

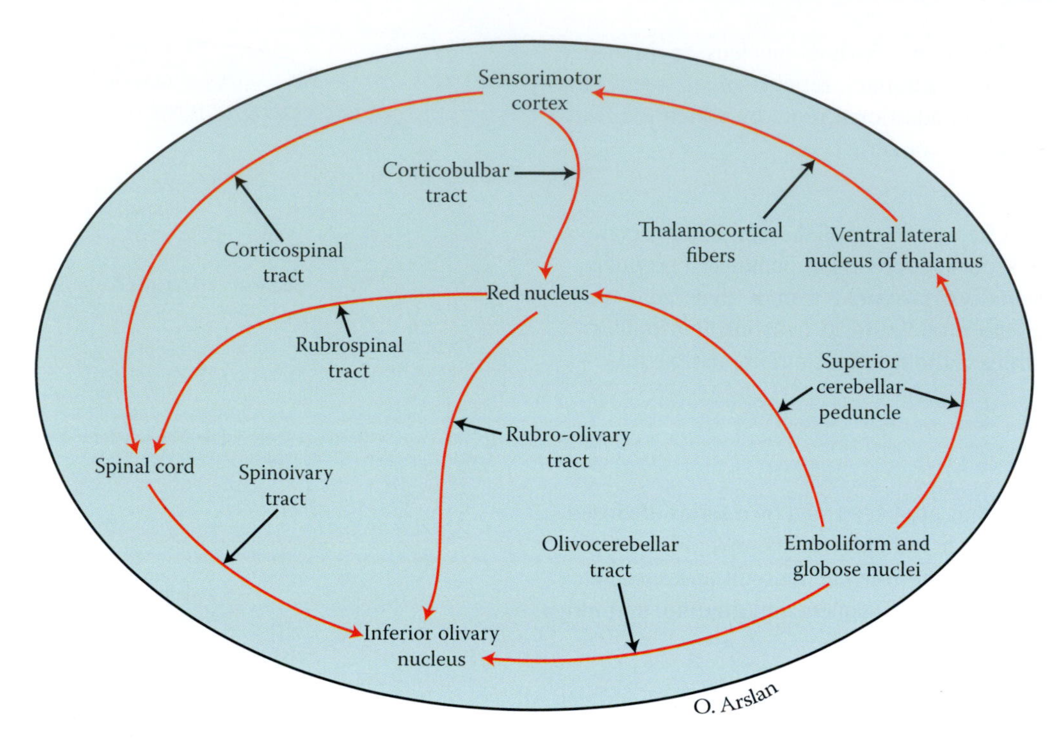

FIGURE 6.21 Schematic drawing of the connections of the red nucleus within a cerebellar feedback circuit.

Cortico-cerebro-cerebellar Circuit

The cortico-cerebro-cerebellar circuit is mediated by the diffuse projections of the cerebral cortex to the pontine nuclei, which sequentially project to the cerebellar cortex via the middle cerebellar peduncle, completing the cerebrocerebellar tract (Figure 6.22). The cerebellar cortex, via its projection to the dentate nucleus (corticonucleo-cerebellar tract), acts upon neurons that influence the ventral lateral thalamic nucleus (dentatorubrothalamic pathway), which eventually affect the cerebral motor cortex. Thus, voluntary movements initiated by the motor cerebral cortex are modulated by this feedback loop.

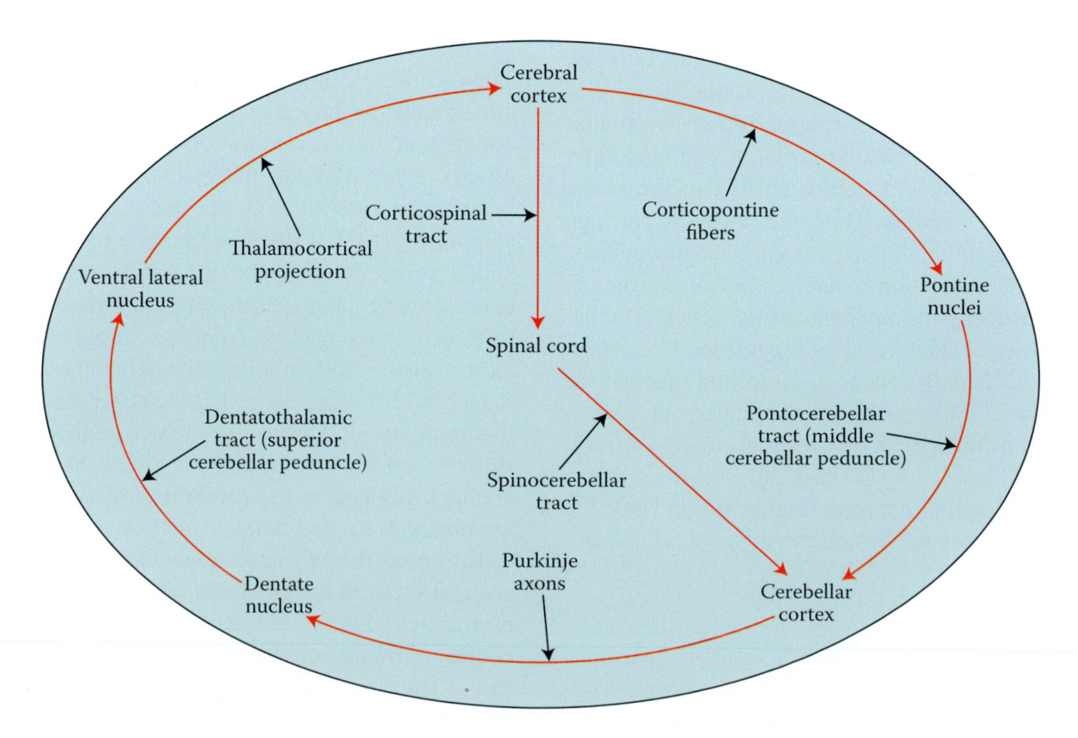

FIGURE 6.22 The cortico-cerebro-cerebellar circuit and role of the pontine nuclei in mediating coordinated motor activity initiated by the cerebral cortex.

Intracerebellar Circuit

The intracerebellar feedback loop utilizes the inhibitory connections of granule cells with the Golgi neurons and the reciprocal inhibition exerted by the Golgi neurons (negative feedback), which is exemplified by the cerebellar glomeruli (Figure 6.10). Granule cells also exert feedforward inhibition upon the Purkinje cells through its connections to other interneurons such as the basket and stellate cells. Within this feedback circuit, collaterals of Purkinje neurons project to the inhibitory cerebellar interneurons, which sequentially send fibers back to the same Purkinje neurons or other interneurons.

FUNCTIONAL AND CLINICAL CONSIDERATION

The cerebellum maintains balance and posture by gradual modulation of muscle tension and by maintaining the orderly sequence of muscular contractions. It also plays an important role in the timing of movements. These complex activities are accomplished by integrating information received from the certain areas of the cerebral cortex that are involved in the planning and command aspects of movements with the sensory feedback arising in the periphery during the course of movement. Cerebellar function is generally based on several signal-processing mechanisms. One of these mechanisms is the feedforward processing in which the signals are processed unidirectionally from the stage of input to the stage of output without reverberatory chains of neuronal excitatory circuits. Cerebral association cortices that subserve higher-order functions are linked with the lateral cerebellar hemispheres of the posterior lobe in feedforward loops via the pontine nuclei and in feedback loops from deep cerebellar nuclei via the thalamus. Thus, the cerebellum is not able to sustain neural activity by itself but provides a quick and sufficient response to any specific input. The second mechanism through which the cerebellum operates is the divergence and convergence in which the signals obtained from a relatively small number of mossy fibers (millions) are conveyed by diverging to billions of granule cells. Then, the parallel fibers of the granule cells converge of on a smaller number of Purkinje cells. This allows the cerebellum to operate through a small number of incoming signals, subject them to extensive internal processing, and then deliver the desired output through a limited number of out channels.

The cerebellum also functions on the basis of modularity in which multiple independent operating modules exist with similar internal structures but with different input and output and connections. This modularity is structured in such a manner that a cluster of neurons in the brainstem, a narrow strip of Purkinje cells, and a group of neurons in the deep cerebellar nuclei work together without being influenced or affected by other neighboring modules. Neuronal plasticity is another significant aspect of cerebellar function that adds a great degree of flexibility to adjust and fine-tune the afferent and efferent components of the cerebellum. Thus, the amplitude of synaptic connectivity between the mossy fibers and the cerebellar nuclei and between the parallel fibers of the granule cells and the Purkinje cells can be constantly modified. The modifiability is evident in the vestibulo-ocular reflex as when the head turns to the left, the eyes, in a reflexive, manner turn to the right to stabilize the image on the retina. This is accomplished by stimulation of the horizontal semicircular duct, which sends impulses via the vestibular nerve to the vestibular nuclei, and eventually, the motor nuclei of the extraocular muscles, leading to an equal movement of the eyes toward the opposite side. Disruption of the vestibulo-ocular connection produces instability of the image on the retina when the head is rotating. The information about the mismatch between vestibular input and eye movement is conveyed to the flocculus by the climbing fibers, which activates the Purkinje neurons to modify the situation.

As mentioned earlier, the cerebellum plays a significant role in motor learning directly or indirectly, which is evident in the fine adjustments that it exerts upon the action that is performed. Motor learning may also include other central nervous system (CNS) structures to which the cerebellum provides signals. It has been proposed that the cerebellum enables the integration of elemental motor activity encoded by climbing fibers with the related sensory context encoded by the mossy fibers and that a cerebellar Purkinje cell functions as a perceptron, an abstract learning center, and the climbing fibers provide a signal that induces synaptic modification in parallel fiber–Purkinje cell synapses, promoting motor learning. According to this notion, the climbing fiber input would strengthen the synchronously stimulated parallel fibers. However, others have claimed that the climbing fiber activity is simply an *error* signal and would cause weakening rather than strengthening the synchronously activated parallel fibers. The latter theory was formulated in the Cerebellar Model Articulation Controller.

In addition to its role in motor coordination, the cerebellum is believed to be involved in cognitive processing and emotional control. Studies have shown that the anterior lobe of the cerebellum contains a primary sensorimotor area, whereas the posterior lobe contains a secondary sensorimotor area. The cortico-cerebrocerebellar loop enables the cerebral hemispheres to influence the neuronal activity of the posterior cerebellar lobe in a feedforward loop via the pontine nuclei and in a feedback loops to the cerebral cortex from dentate nuclei via the thalamus. An additional reciprocal connection that reinforces the role of the cerebellum in sensorimotor control, cognition, intellect, emotion, and autonomic functions is established between the cerebellum and hypothalamus. The role of the cerebellum in these functions seems evident in cerebellar cognitive affective syndrome in patients with lesions of the posterior lobes of the cerebellum due to degenerative processes, stroke, tumor, superficial hemosiderosis (iron deposition associated with demyelination), hypoplasia, and agenesis. Patients exhibit deficits involving executive functions (planning, abstract reasoning, multitasking, and verbal working memory), visual–spatial cognition, and linguistic performance and changes in emotion and personality but without perceptible motor deficits. When the vermis is also involved, additional affective deficit is seen. Patients may also be distractible, easily irritable, and impulsive; suffer from anxiety disorders; and show stereotypical behaviors, disinhibition with proclivity to assign ulterior motives to the behavior of others. These manifestations may attenuate with time. Studies have not reliably confirmed the role of the cerebellum in cognitive functions, and further investigations need to be conducted to document consistent and conclusive data that bear clinical relevance.

The variations in severity of the signs and symptoms of cerebellar dysfunction depend upon the extent of the lesion and duration of the insult. These manifestations are usually seen ipsilaterally and as a constellation of deficits. They emerge as signs of release from the inhibition exerted on intact structures by the cerebellum. Cerebellar deficits may occur as a result of direct compression or invasion of cerebellar tissue by a developing mass, ischemia, tumors, or hemorrhage of the posterior cranial fossa and subsequent obstruction of the cerebrospinal fluid pathway. A developing mass may also produce secondary effects upon other areas of the cerebellum by pressure or compression of vessels. It should be understood, however, that pure cerebellar deficits produced in experimental animals are seldom encountered in man. Most patients exhibit a combination of gait and postural disturbances (ataxia), asynergy, hypotonia, visual disturbances, vertigo (sense of rotation of the environment or self), dementia, headache, nausea, and vomiting.

Neocerebellar lesions are the most common and involve major parts of one cerebellar hemisphere, its efferent fibers, and the corresponding parts of the posterior vermis. In unilateral lesions, hypotonia and muscular incoordination will be seen ipsilaterally. Intention tremor and ataxia are seen when the dentate nucleus or the superior cerebellar peduncles are affected. The dysfunction may be transient, accompanied by rapid improvement even if the cortical lesion is extensive. Signs of dysfunction associated with the appendicular muscles include spooning of the hand (hyperextension of the fingers) and intention tremor, which may be unilateral or bilateral and is noted during movement. The patient may exhibit mild ataxic (broad-based and unsteady) gait, hypotonia, tendency to fall toward the affected side, and asynergy (which includes dysmetria, adiadochokinesis, and rebound phenomenon). Rebound phenomenon is characterized by uncontrollable oscillation of the outstretched arm up and down upon sudden release of pressure by the examiner. Scanning (telegraphic) speech, a form of dysarthria, is characterized by slurred, labored, garbled, hesitating, and monotonous speech with inappropriate pauses. Handwriting may be affected in the same manner (macrographia), showing, characteristically, letters larger than normal. Nystagmus, a late common sign, occurs as a result of destruction of the cerebellar connections to the vestibular nuclei. Horizontal nystagmus, which is seen in neocerebellar lesions, is commonly associated with impairment of tracking movements and becomes markedly visible upon gazing to the side of the lesion.

Paleocerebellar lesions are rare and affect the vermis of the anterior lobe. In this type of lesion, increased extensor muscle tone and postural reflexes accompanied by truncal ataxia (nodding movements of the head and trunk) and signs of decerebrate rigidity (due to involvement of the brainstem) can also be seen. Impairment of gait with relative preservation of the upper extremity is an additional sign of this condition.

Archicerebellar lesions target the flocculonodular lobe and uvula, producing deficits identical to midline (vermal) lesions. These lesions may occur as a result of medulloblastoma, a childhood malignant tumor arising in the roof of the fourth ventricle that occurs between 5 and 10 years of age, multiple sclerosis (MS), chronic alcoholism, tumor, or vascular disease. Deficits usually include truncal ataxia (staggering gait and unsteady posture while standing and tendency to fall backward) and positional nystagmus without appendicular ataxia (ataxia of limb movement). Truncal ataxia necessitates constant support due the inability of the patient to maintain standing position. Midline lesions of the cerebellum restricted to the lingula, superior medullary velum, and superior cerebellar peduncle also produce bilateral trochlear nerve palsy, nystagmus, and ipsilateral tremor of the corresponding limb.

Posture and gait abnormalities (ataxia) resemble drunken gait, which is broad-based, irregular, and staggering. These deficits include tendency to fall (patients become apprehensive and frightened to stand), limb ataxia (past pointing of the extremities), and difficulty in walking in a straight line. Inability to stand may require the patient to seek support and attempt to alleviate the situation by constant adjustment of the extremities and head (*titubation*). Patients keep the feet too widely apart or too closely together, with the head and body deviated toward the side of the lesion during walking.

Asynergy refers to the lack of coordinated action between muscle groups or movements that normally maintain the proper degree of harmony and smooth and accurate sequencing. Asynergistic muscles lack synchronous activity, skill, and speed. Lack of proper sequence and grouping of muscles that are associated with successive components of a motor activity may produce movement that is decomposed (decomposition of movements) and broken down into puppetlike acts. Asynergy of the muscles of the mouth, pharynx, and larynx may lead to disturbance of the mechanism that regulates breathing and phonation (dysarthria). This disturbance produces a peculiar form of speech, scanning (telegraphic) or staccato speech, which is a slow, slurred, explosive, and ataxic speech with prolonged intervals between syllables and wrong pauses. Asynergy may also be manifested in the form of dysmetria, adiadochokinesis, hyperkinesia, and rebound phenomenon of Holmes.

Dysmetria is the inability to gauge the distance, range, rate of speed, or power of movement. Overshooting (hypermetria) or undershooting (hypometria) of the intended target may occur as a result of a lack of appreciation of distance or range. Individuals may perform the act slowly or very rapidly with minimal or maximal power.

Adiadochokinesis is the inability to perform alternate successive pronation and supination of the forearm, opening and closing of the fists, or tapping of the finger, or to properly executing the finger-to-nose or heel-to-shin tests. In the finger-to-nose test, as the examiner asks the patient to put his/her finger on his/her nose, the finger begins to oscillate gradually and then violently as it approaches the nose. In heel-to-shin test, the patient, while in supine position, is asked to touch the knee of one leg with heel of the other extremity and then move the heel downward in front of the shin to the ankle joint.

Hyperkinesia is seen in the form of nonrhythmic, jerky, irregular, uncontrollable, coarse, to-and-fro movements of the limbs (*kinetic* or *intention tremor*) during the course of a movement, or upon command, as in the finger-to-nose and heel-to-shin tests. The amplitude of the tremor increases as the intended target is approached.

Rebound phenomenon of Holmes refers to the lack of normal checks of agonist and antagonist muscles and a tendency to overshoot the target rather than stopping smoothly. For example, sudden release of the a flexed arm against resistance by the examiner may cause the released arm to strike the patient's face due to the delay in contraction of the triceps brachii, a muscle that is ordinarily responsible for arrest of overflexion.

Hypotonia (reduced muscle tone) results in weak, flabby, and fatigued muscles (asthenia). The affected muscles may not resist passive movements of the joints into extreme degrees of flexion or extension. The contraction and relaxation phases of movements become slow, delaying the initiation of voluntary movements. Tendon reflexes are diminished. When the patient is asked to outstretch his/her forearms, the outstretched forearm on the affected side assumes a pronated position and generally maintains a higher position than the limb on the contralateral side.

Other cerebellar deficits include nystagmus, visual and ocular disorders, hyporeflexia, macrographia, dementia, headache, nausea, and vomiting as well as vertigo (sense of rotation of self or environment). Cerebellar disease usually produces anteroposterior or side-to-side movements of the environment and a sense of instability during walking. Occlusion of the subclavian artery, medial to the origin of the vertebral artery, by arteriosclerotic plaques may also produce vertigo, a weaker pulse, and lower pressure in the upper extremity of the affected side relative to the lower extremity. This results from diversion of blood flow from the vertebral artery on the (intact) opposite side, which maintains a higher blood pressure, to the vertebral artery on the occluded side, with a lower blood pressure, and subsequently to the subclavian artery distal to the site of occlusion. Diversion of blood to the subclavian artery (stealing) from the vertebral artery is generally exacerbated during physical effort (increased metabolic demand), leading to manifestations of *cerebellar ischemia*, which include vertigo and dizziness. On this basis, the described condition is known as *subclavian steal syndrome*.

Occlusion or stenosis of the proximal part of the vertebral artery, however, may produce, although rarely, transient ischemic attack (TIA) and vertigo at rest (to be distinguished from vertigo associated with exertion, which is seen in subclavian steal syndrome). Other deficits also seen in brainstem and cerebellar ischemia include diplopia (double vision), oscillopsia (sensation of oscillation of the viewed object), numbness, and hemiparesis. Occlusion or stenosis of the vertebral artery on one side rarely slows the blood flow to the brainstem or cerebellum if efficient collateral circulation in the neck as well as symmetry of vertebral arteries are maintained. Stenosis of the basilar artery may cause symptoms, which vary from vertigo, nausea, dysarthria (difficulty in speech), hemianesthesia, and paresis of conjugate eye movements, to dysphagia, diplopia, occipital headache, and vertical and horizontal nystagmus.

Ocular and visual disorders are the result of disruption of the cerebello-vestibulo-ocular reflexes, including nystagmus, ocular dysmetria, disturbances of conjugate gaze, and diplopia. *Nystagmus* is a rhythmic oscillation of one or both eyes at rest or with ocular movements. Patients with cerebellar dysfunction cannot maintain gaze away from the midline (rest) position, and attempts to do so may result in slow movements of the eyes toward the center (slow component of nystagmus). The rapid corrective movement in the direction of the gaze is considered the fast component of cerebellar nystagmus. The direction as well as amplitude of its fast phase decreases with sustained deviation of the eyes toward the target. Following return of the eyes toward the midline, the nystagmus resumes with the fast phase away from the midline. This transient oscillation of the eyes upon gazing toward a target that is associated with blurred vision may result from ocular dysmetria. Blurred vision that improves by closing one eye is a sign of dysconjugate gaze subsequent to cerebellar dysfunction. Inflammatory and degenerative diseases of the cerebellum may reduce visual acuity or result in transient or permanent blindness. Permanent blindness is seen in certain familial diseases associated with degeneration of the spinocerebellar tracts. *Diplopia*, the most common ocular manifestation, may be constant or transient deficit. Smooth pursuit movements are also impaired, forcing the patient to track moving objects by compensatory saccades.

Dementia may result from obstructive hydrocephalus, which compresses the enlarged brain against the bony rigid wall of the skull. Obstructive hydrocephalus develops from compression of the fourth ventricle by a cerebellar mass or hemorrhage within the posterior cranial fossa. Impairment of memory and transient confusion are also common features of cerebellar degenerative diseases.

Headache, the most common manifestation of cerebellar dysfunction, is typically a severe, persistent, dull pain of the occipital or frontal region that shows no lateralization and remains unresponsive to conventional analgesics. Postural changes may exacerbate headache. Frontal headache is a manifestation of deviation of the tentorium cerebelli, which is innervated by the recurrent meningeal branch of the ophthalmic nerve, a branch that also supplies the skin of the forehead.

Nausea and vomiting are most severe in the morning and may last for months. Postural changes may attenuate or reduce the severity of nausea. Nausea and vomiting may be abrupt, dependent upon the extent and degree of progression of cerebellar deficits. With associated weight loss, these deficits may obscure the true etiology and may lead the physician to undertake extensive abdominal evaluation. Nausea and vomiting may result from compression or irritation of the emetic center in the brainstem.

CEREBELLAR LESIONS AND ASSOCIATED DISEASES

Cerebellar lesions may occur in MS, acoustic neuroma, Benedikt syndrome, cerebellar herniation, *Dandy–Walker syndrome*, Foix's syndrome, Nothnagel's syndrome, cerebellar aplasia, Friedreich's ataxia, frontal lobe tumors, and cerebellar hemorrhage.

MS, as described earlier, is a multifocal demyelinating disease of the CNS. It produces lesions in brain, brainstem, and *cerebellum*. Patients exhibit an unsteady gait, truncal ataxia, intention tremor, and slurred speech. The most severe form of cerebellar ataxia occurs in MS, where the slightest attempt to move the trunk or limbs results in a violent and uncontrollable ataxic tremor. This is usually due to involvement of the dentatorubrothalamic tract and the adjacent structures in the tegmentum of the midbrain. See also Chapter 2.

Acoustic neuroma, a cerebellopontine angle tumor, may also produce cerebellar dysfunctions by compressing the flocculus and the middle cerebellar peduncle. This tumor, although commonly benign, is associated with widening of the internal acoustic meatus. See also the vestibular system.

Benedikt syndrome is a condition that results from destruction of the oculomotor nerve, red nucleus, medial lemniscus, and possibly the spinothalamic tracts. It exhibits signs of ipsilateral oculomotor palsy, contralateral cerebellar dysfunctions, loss of pain, thermal, and position sensations on the opposite side (also described with the motor system and cranial nerves, Chapters 20 and 21). Occlusion of the superior cerebellar artery may produce degeneration of the superior cerebellar peduncle, and the spinal and trigeminal lemnisci, producing some of the signs of Benedikt syndrome. The vascular occlusion causes ipsilateral atonia, ataxia, asthenia, bradyteleokinesis (hesitancy and slowness in completion of a movement), and loss of pain and temperature sensation on the contralateral sides of the face and body.

Cerebellar herniation refers to the bilateral or unilateral downward displacement of the cerebellar tonsils that occurs as a result of a tumor of the posterior cranial fossa or frontal lobe, or from the pressure exerted by an edematous brain. The paraflocculus and possibly part of the temporal lobe may herniate through the foramen magnum into the cervical part of the vertebral column. Cerebellar herniation is characterized by occipital headache, tonic spasmodic contracture of the neck (nuchal rigidity) and back muscles (cerebellar fits), turning of the head toward the lesion side, as well as extension and medial rotation of the extremities. These symptoms are followed later by progressive loss of consciousness as a result of destruction of the ascending reticular activating system. Vasomotor changes such as reduced cardiac contraction and feeble pulse, numbness in the upper extremity, and difficulty of swallowing may all be attributed to compression of the structures and nuclei in the brainstem. Mechanical displacement of the fourth ventricle in this herniation may eventually cause compression of the medullary respiratory center, and death ensues following respiratory arrest. A tumor, which causes upward deviation of the cerebellum, may lead to compression of the fourth ventricle and obstruction of the foramina of Magendie and Luschka, followed by hydrocephalus. These deficits may occur in Dandy–Walker syndrome.

Dandy–Walker syndrome is a congenital sporadic neural disorder caused by failure of development of the cerebellar vermis, midline cerebellar cortex posteriorly, thinning of the posterior cranial fossa, and the formation of a cyst-like midline structure that replaces the fourth ventricle. It may develop slowly or quickly and can be associated with administration of warfarin during pregnancy. Infants may exhibit retardation of motor development and progressive hydrocephalus, whereas adult patients endure increased intracranial pressure, which leads to development of seizures, ataxia, nystagmus, and breathing pattern disorders. The cervical spinal nerve roots assume an ascending course in order to reach their corresponding foramina. This syndrome may also be associated with agenesis (failure of development) of the corpus callosum. Prenatal ultrasound and amniocentesis due to increased risk for fetal karyotype abnormalities help to diagnose this condition prenatally. Shunting and endoscopic ventriculostomy may help control the intracranial pressure.

Foix's syndrome is a manifestation of a lesion involving the red nucleus and the superior cerebellar peduncle, sparing the third cranial nerve. Patients with this condition may exhibit cerebellar ataxia and hemichorea on the contralateral side. However, if the lesion involves the superior cerebellar peduncles at the point of their decussation, cerebellar ataxia will be seen bilaterally.

Nothnagel's syndrome results from an expanding lesion of the tectum that impinges upon and exerts downward pressure

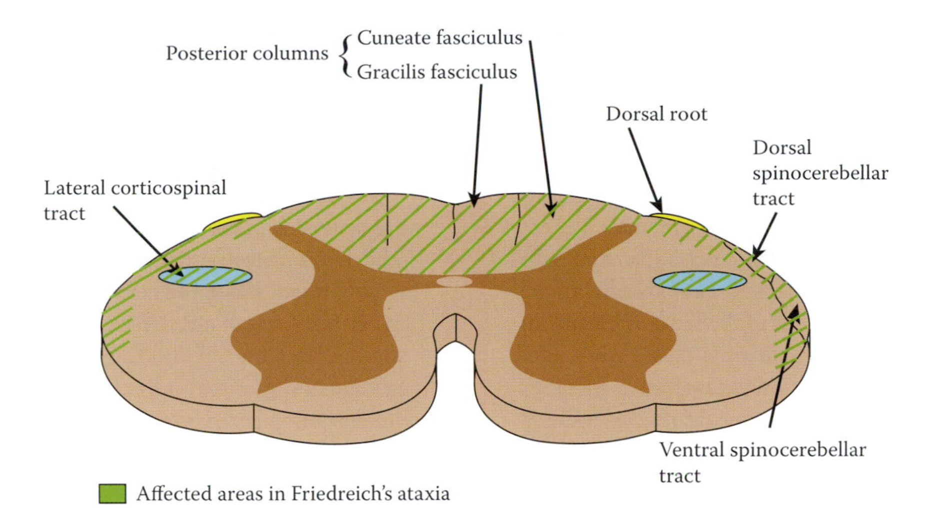

FIGURE 6.23 Section of the spinal cord showing the lesions associated with Friedreich's ataxia. The dorsal columns and lateral corticospinal and spinocerebellar tracts are affected.

upon the superior vermis of the cerebellum. Cerebellar symptoms and paralysis of the extraocular muscles ipsilaterally are the distinguishing features of this condition.

Cerebellar aplasia is a congenital malformation in which part of the cerebellar hemisphere or most of the cerebellum does not develop and is usually associated with anomalies of the contralateral inferior olive. Neonates with this condition exhibit intention tremor and other motor dysfunctions associated with standing and walking. Cerebellar aplasia is also seen in individuals with encephalocele and in Klippel–Feil syndrome.

Friedreich's ataxia (Figure 6.23) is an autosomal recessive disorder that results from degeneration of the spinocerebellar tracts, dorsal white and Clarke's columns, and the thickly myelinated fibers of the dorsal roots as well as the dorsal root ganglia. Atrophy of the glutamatergics neurons of the dentate nucleus and the GABAergic Purkinje neurons also occurs. It is the most common condition among hereditary ataxias. The responsible mutant gene frataxin is located on chromosome 9 and contains expanded guanine-adenine-adenine triplet repeats in the first intron. The onset of this disease is usually before the end of puberty, exhibiting muscular weakness primarily in the lower extremities, gait ataxia, areflexia, and loss of joint sensation from the lower extremity. Patients may also manifest pes cavus (high arched foot and clawing of the toes), scoliosis (exaggerated lateral curvature of the spine), cardiac murmurs, and cardiomyopathy. Mental capacities often decline. Nystagmus, dysarthria, vertigo, and hearing loss may also be seen in this disease. Essential tremor, if present, may be a minor manifestation. Refsum's disease and abetalipoproteinemia (Bassen–Kornzweig syndrome) share some features with Friedreich's ataxia. One-fifth of patients with this condition also exhibit manifestations of diabetes mellitus.

Frontal lobe tumors may produce cerebellar deficits contralateral to the side of the tumor as a result of increased intracranial pressure and possible compression of the corticopontine fibers. Manifestations of mental disorders appear long before any cerebellar signs.

Cerebellar hemorrhage produces headache, vertigo, and ataxia and may also lead to ipsilateral conjugate gaze disturbances.

SUGGESTED READING

Apps R, Garwicz M. Anatomical and physiological foundations of cerebellar information processing. *Nat Rev Neurosci* 2005;6(4):297–311.

Audinat E, Gähwiler BH, Knopfel T. Excitatory synaptic potentials in neurons of the deep nuclei in olivo-cerebellar slice cultures. *Neuroscience* 1992;49:903–11.

Bear MF, Malenka RC. Synaptic plasticity: LTP and LTD. *Curr Opin Neurobiol* 1994;4:389–99.

Chen S, Hillman DE. Colocalization of neurotransmitters in the deep cerebellar nuclei. *J Neurocytol* 1993;22:81–91.

Chumas PD, Armstrong DC, Drake JM et al. Tonsillar herniation: The rule rather than the exception after lumboperitoneal shunting in the pediatric population. *J Neurosurg* 1993;78:568–73.

Dietrichs E, Walberg F. Cerebellar nuclear afferents–Where do they originate? A reevaluation of the projections from some lower brainstem nuclei. *Anat Embryol* 1987;177:165–72.

Dow RS. Cerebellar syndromes including vermis and hemispheric syndromes. In Vinken PJ, Bruyn GW, eds. *Handbook of Clinical Neurology*, Vol. 2. Amsterdam: North Holland Publishing, 1969.

Garthwaite J, Brodbelt AR. Synaptic activation of N-methyl-D-aspartate and non-N-methyl-D-aspartate receptors in the mossy fiber pathway in adult and immature rat cerebellar slices. *Neuroscience* 1989;29:401–12.

Guibaud L, Larroque A, Ville D, Sanlaville D, Till M, Gaucherand P et al. Prenatal diagnosis of 'isolated' Dandy–Walker malformation: Imaging findings and prenatal counselling. *Prenat Diagn* 2012;32(2):185–93.

Kawamura K, Hashikawa T. Projections from the pontine nuclei proper and reticular tegmental nucleus on to the cerebellar cortex in the cat: An autoradiographic study. *J Comp Neurol* 1981;201:395–413.

Kayakabe M, Kakizaki T, Kaneko R, Sasaki A, Nakazato Y, Shibasaki K, Ishizaki Y, Saito H, Suzuki N, Furuya N, Yanagawa Y. Motor dysfunction in cerebellar Purkinje cell-specific vesicular GABA transporter knockout mice. *Front Cell Neurosci* 2014;7:286.

Llinas RR, Walton KD, Lang EJ. Cerebellum, Chapter 7. In Shepherd GM, ed. *The Synaptic Organization of the Brain*. New York: Oxford University Press, 2004.

Lodi R, Tonon C, Calabrese V, Schapira AH. Friedreich's ataxia: From disease mechanisms to therapeutic interventions. *Antioxid Redox Signal* 2006;8(3–4):438–43.

Metry DW, Dowd CF, Barkovich AJ, Frieden IJ. The many faces of PHACE syndrome. *J Pediatr* 2001;139(1):117–23.

Montarolo PG, Palestim M, Strata P. The inhibitory effect of olivocerebellar input on the cerebellar Purkinje cells in the rat. *J Physiol* 1982;332:187–202.

Oscarsson O. Functional units of the cerebellum-sagittal zones and microzones. *Trends Neurosci* 1979;2:143–5.

Reisser C, Schukriecht HF. The anterior inferior cerebellar artery in the internal auditory canal. *Laryngoscope* 1991;101: 761–6.

Simpson JI, Wylie DR, De Zeeuw CI. On climbing fiber signals and their consequence(s). *Behav Brain Sci* 1996;19(3):384–98.

Timmann D, Daum I. Cerebellar contributions to cognitive functions: A progress report after two decades of research. *Cerebellum* 2007;6(3):159–62.

Zhang N, Ottersen OP. In search of the identity of the cerebellar climbing fiber transmitter: Immunocytochemical studies in rats. *Can J Neurol Sci* 1993;20(Suppl 3):S36–42.

7 Diencephalon

The diencephalon is derived from the prosencephalon together with the telencephalon. It is bounded by the lamina terminalis rostrally and the posterior border of the mammillary bodies caudally, lodged between the brainstem and the cerebral hemispheres. It is divided by the hypothalamic sulcus, which extends between the interventricular foramen of Monro and the cerebral aqueduct, into dorsal and ventral portions. The dorsal portion consists of the thalamus and epithalamus, while the ventral portion is comprised of the hypothalamus and subthalamus. The third ventricle separates the diencephalon from the corresponding part of the opposite side. Through various components, the diencephalon exerts a regulatory effect on vital activities, including functions of the pituitary gland; autonomic nervous system; circadian rhythm; feeding centers; sleep–wake cycle; general sensation; visual, auditory, and motor integration; as well as affect and cognition.

The diencephalon, an important part of the central nervous system (CNS) rostral to the brainstem, consists of the thalamus, epithalamus, hypothalamus, and subthalamus. The thalamus and epithalamus are separated from the hypothalamus and subthalamus by the hypothalamic sulcus. The latter sulcus extends between the interventricular foramen of Monro and the cerebral aqueduct. The diencephalon lies between the cerebral hemispheres, caudal to the lamina terminalis. It contains the *third ventricle* (Figure 7.1), which is bounded superiorly by the ependyma and the pia mater that join together to form the tela choroidea. This ventricle stretches from the lamina terminalis rostrally to the cerebral aqueduct caudally and is connected to the lateral ventricle via the interventricular foramen of Monro and to the fourth ventricle through the cerebral (Sylvian) aqueduct. It extends into the pineal stalk as the pineal recess and to the area superior the optic chiasma as the optic recess. It is frequently crossed by fibers of the interthalamic adhesion (massa intermedia) and also extends into the infundibulum as a funnel-shaped infundibular recess. The cerebrospinal fluid within this ventricle is secreted by the choroid plexus, which is attached to the tela choroidea and is supplied by the posterior choroidal artery.

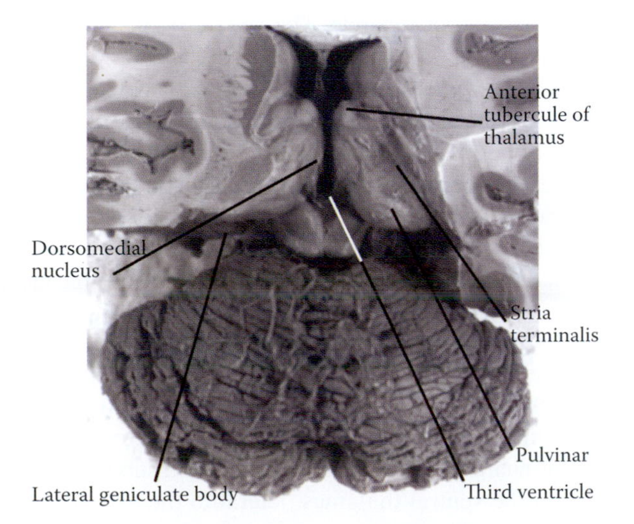

FIGURE 7.1 Dorsal surface of the diencephalon. Thalami are located on both sides of the third ventricle.

THALAMUS

The thalamus is the largest component of the diencephalon, which lies superior to the hypothalamic sulcus. It is separated from the caudate nucleus by the genu of the internal capsule. It forms the floor of the central part of the lateral ventricle, the posterior boundary of the interventricular foramen of Monro, and part of the lateral wall of the third ventricle (Figures 7.1, 7.2, and 7.3). Both thalami may be interconnected in the midline by the massa intermedia (interthalamic adhesion), which extends behind the interventricular foramen of Monro.

It receives direct sensory impulses of all modalities, with the exception of the olfactory sensation. Olfactory information, combined with some other sensations, is conveyed to the olfactory cortex, amygdala, and mammillary body. Impulses that arise from the cerebellum and the basal nuclei are also integrated in the thalamus. These impulses are then projected to the motor and premotor cortices. To the same extent, visceral

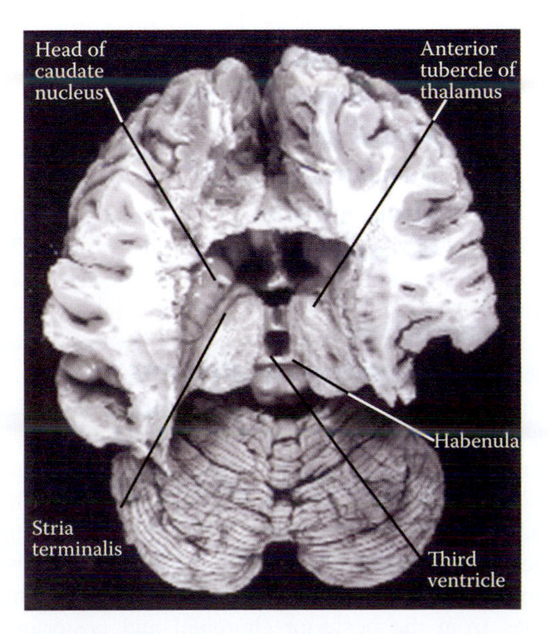

FIGURE 7.2 Dorsal surface of the diencephalon in relation to the basal nuclei. Note the habenular commissure connecting the habenular nuclei of the epithalamus.

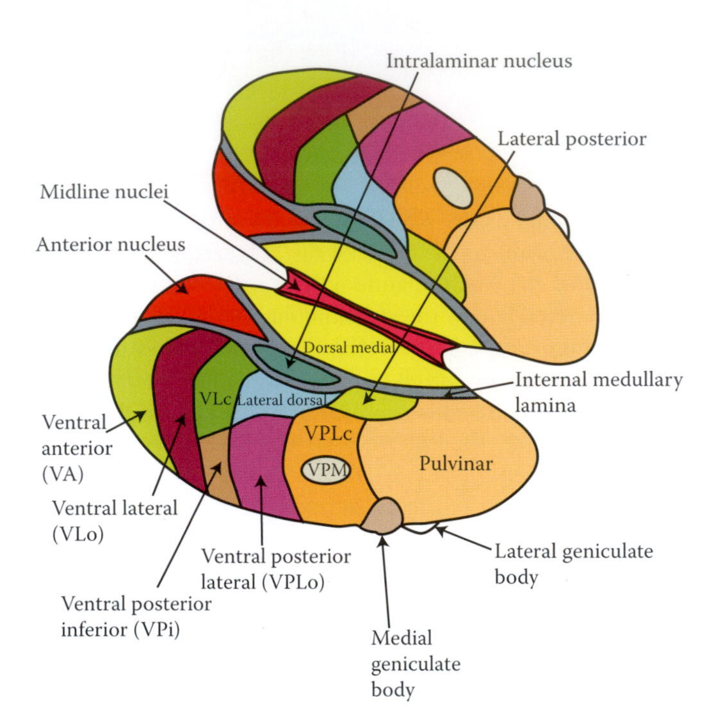

FIGURE 7.3 Three-dimensional diagram of the thalamus. Major nuclear groups are demarcated by the internal medullary lamina. Midline connection of thalami (interthalamic adhesion) is also illustrated. VLc, ventral lateral caudal part; VPM, ventral posteromedial; VPLc, ventral posterolateral caudal part.

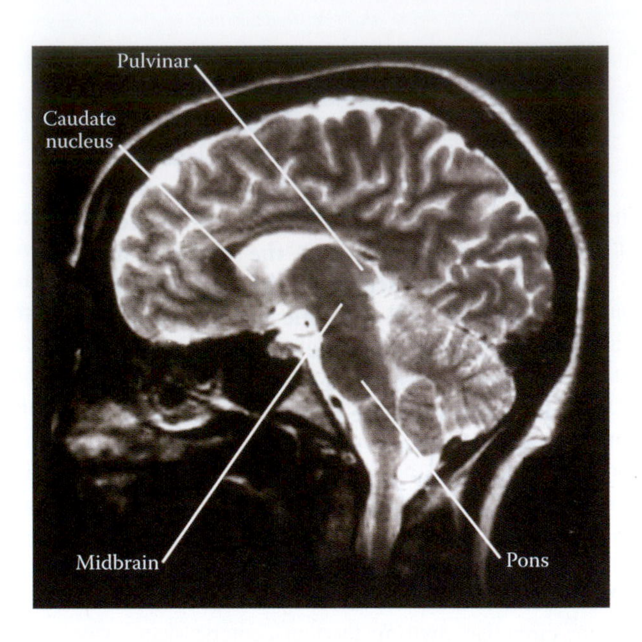

FIGURE 7.4 Magnetic resonance imaging (MRI) scan of the brain showing the thalamus in relation to the caudate nucleus, midbrain, and corpus callosum. The posterior cerebral artery (main source of blood supply to the thalamus) is indicated.

activities are also influenced by the thalamic connections to the hypothalamus and cingulate gyrus. In general, the thalamus integrates and modifies the sensory and motor inputs and selectively tunes the output signals in a manner most efficient for stimulation of the cerebral cortex. Conscious awareness of pain, crude touch, temperature, and pressure sensations may occur at the thalamic level. The thalamus also processes information that influences the electrocortical activity in the sleep–wake cycle through the ascending reticular activating system (ARAS). Additionally, the thalamus is also involved in the modification of the affective component of behavior via its connections to the limbic system.

With respect to the blood supply of the thalamus, the posterior cerebral artery plays an important role (Figures 7.4 and 7.28). As the terminal branch of the basilar artery, it supplies the midbrain and portions of the occipital and temporal lobes, giving rise to the posterior choroidal artery and central branches. These central branches include the posteromedial (thalamoperforating) and posterolateral (thalamogeniculate) arteries. The latter branch may also arise from the posterior communicating artery, which emanates from the internal carotid artery. Anteromedially, the thalamus is supplied by the thalamoperforating arteries, while the caudal parts of the thalamus, including the pulvinar and geniculate bodies, are supplied by the thalamogeniculate arteries. Numerous branches from the posterior choroidal and posterior communicating arteries supply the superior and inferior parts of the thalamus, respectively. Occlusion of the terminal part of the basilar artery, supplying the diencephalon and midbrain, may result in pupillary dilatation or constriction, loss of light reflex, vertical gaze and short-term memory, hallucination, agitation, and coma or hypersomnolence. In one-third of population, the thalamogeniculate arteries are derived from a single branch, artery of Percheron, that supplies the paramedian thalamus and the midbrain including the oculomotor and mesencephalic reticular nuclei.

THALAMIC NUCLEAR GROUP

The thalamus has dorsal and medial surfaces that are separated by the stria medullaris thalami. The dorsal thalamus is divided into medial and lateral nuclear groups by the internal medullary lamina (Figure 7.3). The diverging limbs of the internal medullary lamina surround the anterior nucleus of the thalamus, forming the anterior tubercle. The ventral portion consists of the caudally located medial and lateral geniculate bodies (metathalamus), and ventral anterior (VA), ventral lateral (VLo), and ventral posterior nuclear groups. The ventral posterior nuclear complex comprises the ventral posterolateral (VPLo) and ventral posteromedial (VPM, arcuate) nuclei. The dorsal portion incorporates the lateral dorsal (LD) and lateral posterior (LP) nuclear groups and the pulvinar. Along the periventricular gray matter of the third ventricle, a group of midline thalamic nuclei exist, which are more prominent in the interthalamic adhesion (massa intermedia).

Another group of thalamic nuclei, the intralaminar nuclei, are located in the substance of the internal medullary lamina. The reticular nucleus of the thalamus, which is considered to be a continuation of the zona incerta of the subthalamus, is located in the ventral thalamus, lateral to the external medullary lamina. The latter separates the thalamus from the internal capsule. All thalamic nuclei maintain reciprocal connections with the cerebral cortex. The corticothalamic fibers are part of a feedback circuit that exerts effects upon the thalamus by selectively inhibiting or facilitating thalamic input to the cerebral cortex. On the basis of their connections to the cerebral cortex and ascending pathways, the thalamic nuclei have customarily been classified as specific and nonspecific nuclei. The latter is further subdivided into relay and association nuclei. However, the assumption upon which this classification rests remains tenable. It is clear that both specific and dense projections exist; all cortical areas receive more than one such type of thalamic input. It is equally clear that diffuse, nonspecific projections do not originate from a single, discrete group of thalamic nuclei and that many "specific" nuclei may also convey "nonspecific" projections to widespread cortical areas.

Anterior Nucleus

The *anterior nucleus* (Figures 7.1, 7.2, 7.3, and 7.5) is bounded by the bifurcating limbs of the internal medullary lamina, forming a prominent swelling at the anterior pole of the thalamus (anterior tubercle). It also constitutes the posterior boundary of the foramen of Monro. Profuse reciprocal connections exist between the anterior nucleus and the mammillary body (via the mammillothalamic tract), as well as the hippocampal formation (via the fornix). This nucleus consists of anteroventral, anteromedial, and anterodorsal subnuclei. The anteroventral and anteromedial subnuclei receive input from the ipsilateral medial mammillary nucleus, whereas the anterodorsal subnucleus receives bilateral afferents from the lateral mammillary nuclei and the midbrain reticular formation. Projections from the anterodorsal and anteromedial subnuclei are destined to the anterior and middle portions of the cingulate gyrus. The latter gyrus also maintains reciprocal connections to the various divisions of the anterior thalamic nucleus. Additionally, the anterior nucleus interconnects with other thalamic nuclei of the ipsilateral and contralateral sides. It is an integral component of the limbic system, providing a linkage between the thalamus, hippocampus, hypothalamus, and cingulate gyrus. Constructing a balance between instinctive and volitional behavior is thought to be an important function of this nucleus.

> Due to its close association with the limbic system circuitry, the anterior nucleus is considered to be an essential element in the short-term memory. The significance of this fact is illustrated in the anterograde amnesia observed in lesions of the mammillothalamic tract, a pathway that terminates in the anterior nucleus. This type of amnesia is seen in *Korsakoff's syndrome*, which results from a thiamine deficiency (Chapter 17).

Ventral Nuclear Group

The ventral nuclear group comprises the metathalamus, ventral posterior nuclear complex, the VA nucleus, and the VL nucleus. The metathalamus consists of the lateral and medial geniculate bodies.

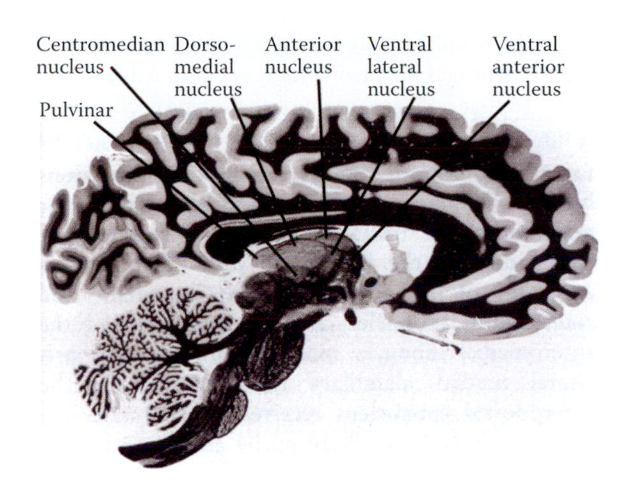

FIGURE 7.5 Midsagittal photograph of the brain. Anterior, ventral anterior, and dorsomedial nuclei as well as the pulvinar are illustrated.

Lateral Geniculate Body (LGB)

The *lateral geniculate* body is a visual relay structure that lies lateral and rostral to the medial geniculate, separated from the crus cerebri by the fibers of the optic tract (Figures 7.1, 7.3, 7.7, 7.9, 7.21, and 7.22). Posteromedially, it receives the brachium of the superior colliculus. It consists of the dorsal lateral geniculate nucleus; its ventral part, which is known as the pregeniculate nucleus belongs to the thalamus. This nucleus, which resembles Napoleon's hat, consists primarily of six laminae, starting with lamina 1 in the innermost ventral and ending with lamina 6 at the outer dorsal part, separated by interlaminar zones. Laminae I and II form the magnocellular part, while laminae III, IV, V, and VI comprise the parvocellular part, with lamina S, the superficial lamina, occupying the most ventral part. The neurons of the magnocellular part are only sensitive to the black and white colors, respond quickly, and possess a high-resolution capacity. They receive projections from the fast-conducting Y-type retinal ganglion cells that provide collaterals to the superior colliculus. The parvocellular neurons, which receive input from the X retinal ganglion cells, are responsive to colors, less sensitive to low-contrast stimuli, and slow reacting to visual stimuli, and project primarily to Brodmann area 17. The S lamina and interlaminar zones of the lateral geniculate nucleus receive input from the W retinal ganglion cells with slow conduction and large receptive filed. They also receive input from the primary visual cortex (Brodmann area 17); the superficial layers of the superior colliculus and laminae 1, 4, and 6 receive visual impulses from the crossed fibers of the optic tract, whereas laminae 2, 3, and 5 only receive fibers from the ipsilateral optic tract. The inferior visual quadrant is represented in the medial portion, whereas the superior quadrant is represented in the lateral part of the lateral geniculate body. Fibers from the macula project to the caudal part of the lateral geniculate body. The lateral geniculate body has reciprocal connections with the primary visual cortex (Brodmann area 17). There is also input from the noradrenergic neurons of the locus ceruleus, serotonergic neurons of the mesencephalic raphe nuclei, and cholinergic neurons of the reticular formation. Due to the fact that retinal fibers terminate in the lateral geniculate body in a nonoverlapping manner from both eyes, a neuronal interaction must occur to achieve binocular vision. To that end, activation of the nonprojecting eye will inhibit the fast-conducting magnocellular Y neurons, possibly through interneurons or dendrites of neurons within the interlaminar zone.

The efferent fibers from this nucleus run within the retrolenticular part of the internal capsule, forming the geniculocalcarine tract (optic radiation). The lower fibers of the optic radiation pursue a course within the temporal lobe and loop backward to join the more dorsal fibers as Meyer's loop. These fibers terminate primarily within lamina 4 of the striate (Brodmann area 17), parastriate (Brodmann area 18), and peristriate (Brodmann area 19) cortices.

Medial Geniculate Body (MGB)

The *medial geniculate body* (MGB) is an auditory relay nucleus between the inferior colliculus and the primary auditory cortex located ventrolateral to the pulvinar separated by the brachium of the superior colliculus (Figures 7.3, 7.6, 7.7, 7.18, 7.22, and 7.23). It consists of medial, ventral, and dorsal nuclei, which are distinguished by their morphologic characteristics, connections, and density. The medial (magnocellular) nucleus receives fibers from the inferior colliculus (detection of intensity and duration of sound) and the deep part of the superior colliculus, indicating the possible role of this nucleus in the mediation of modalities other than sound. Neurons of this part of the MGB are preferentially tuned to certain frequencies and show reduction in response proportional to the sound intensity. It projects diffusely to lamina 6 of the auditory, insular, and opercular cortices. The dorsal (posterior) nucleus overlies the ventral nucleus and consists of principal cells and interneurons but without frequency-based layering. This nucleus receives afferents from the pericentral nucleus of the inferior colliculus and from other auditory relay nuclei. A broad range of frequencies is regulated in the dorsal nucleus, which accounts for the lack of tonotopic organization. Projection of the dorsal nucleus is limited to the secondary auditory cortex (Brodmann area 22). Neurons of the ventral nucleus receive afferents from the ipsilateral inferior colliculus via the brachium of the inferior brachium, responsible for relaying intensity, binaural information, and frequency to the auditory cortex. It exhibits a tonotopic arrangement in which low frequencies project laterally, whereas high-pitched sounds are conveyed medially. This ventral nucleus projects primarily to layer IV of the primary auditory cortex. Commissural neurons do not exist between the medial geniculate bodies.

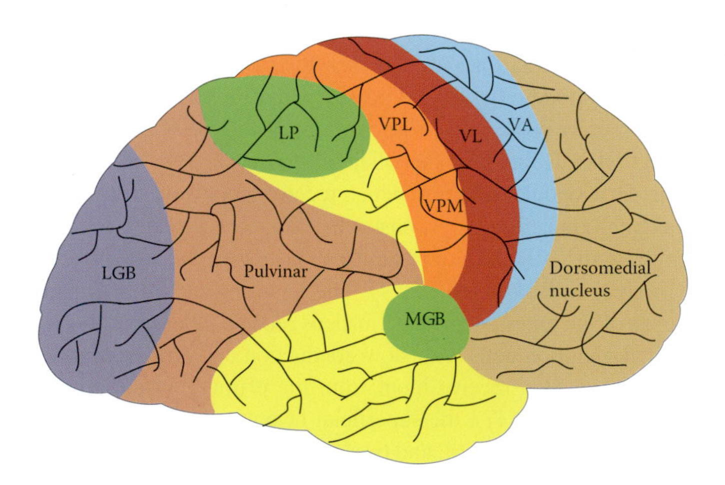

VA Ventral anterior
VL Ventro lateral
VPL Ventral posterolateral
VPM Ventral posteromedial
LP Lateral posterior
LGB Lateral geniculate body (nucleus)
MGB Medial geniculate body (nucleus)

FIGURE 7.6 Cortical areas of the thalamic projections on the lateral surface of the cerebral hemisphere. These projections are specific sites where certain sensory and/or motor impulses are integrated.

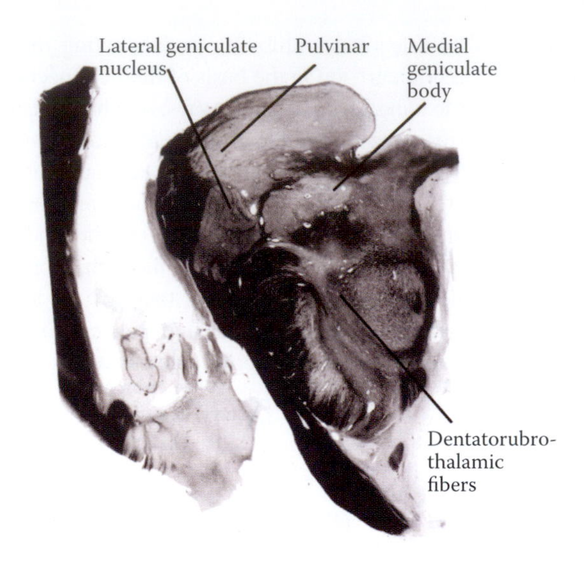

FIGURE 7.7 Section through the thalamomesencephalic junction. Metathalamus (lateral and medial geniculate bodies) is shown. The pulvinar overlies the geniculate bodies, forming the main component of the posterior thalamus.

Ventral Posterior Nucleus (VPN)

The *ventral posterior nucleus* (*VPN*) (Figures 7.3, 7.6, 7.11, and 7.12) consists of the VPL, VPM, and ventral posterior inferior (VPI) nuclei.

The *VPL nucleus* (Figures 7.3, 7.6, 7.11, 7.12, and 7.15) conveys impulses received via the spinothalamic tract, solitariothalamic tract, spino-cervico-thalamic tract, and medial lemniscus to the dorsal and intermediate regions of the postcentral gyrus and the secondary somatosensory cortex. Due to the considerable difference in the density of the peripheral innervation of different body regions, many more neurons tend to respond to stimulation of the hand than the trunk. Similarly, the distorted mapping of the body in this nucleus also reflects the difference in innervation density. Within this nucleus, the cervical fibers terminate medially, the thoracic and lumbar fibers terminate dorsally, and the sacral fibers are positioned laterally. Neurons of the VPL are modality specific and are unaffected by anesthesia.

The VPM nucleus is also known as the arcuate nucleus, because of its crescent shape (Figures 7.3, 7.6, 7.8, 7.10, and 7.12). It lies on the lateral side of the centromedian nucleus and consists of a medial parvocellular part, which receives gustatory impulses through the ipsilateral solitariothalamic tract, and a lateral principal part, which receives general sensation (tactile, thermal) from the head region. The general sensations from the head region ascend via the crossed ventral trigeminal tract and the uncrossed dorsal trigeminal tract.

The ventral trigeminal tract originates from the spinal trigeminal nucleus and the ventral part of the principal sensory nucleus. The uncrossed dorsal trigeminal tract is derived from the dorsal part of the principal sensory nucleus. Information received by the VPM is conveyed to the lower part of the postcentral gyrus and to the secondary somatosensory cortex via the thalamocortical fibers. VPL and VPM projections to the sensory cortex are contained in the posterior limb of the internal capsule.

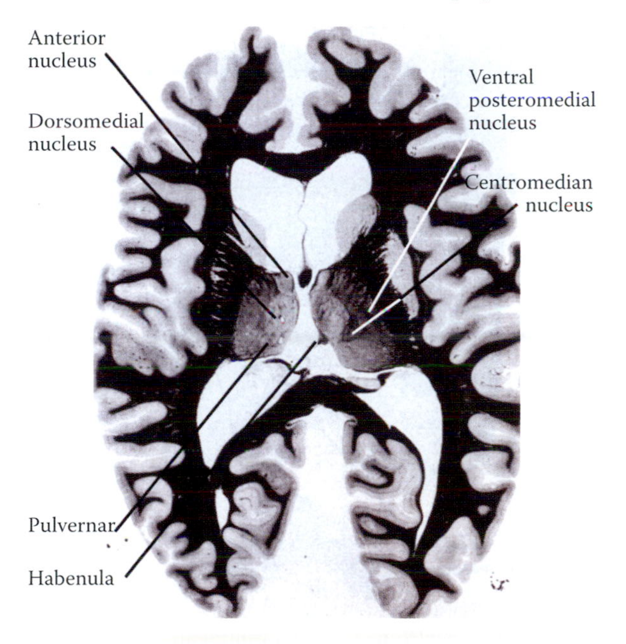

FIGURE 7.8 Section through the mid-level of the diencephalon. The centromedian (intralaminar) and ventral posteromedial nuclei as well as the pulvinar are clearly shown.

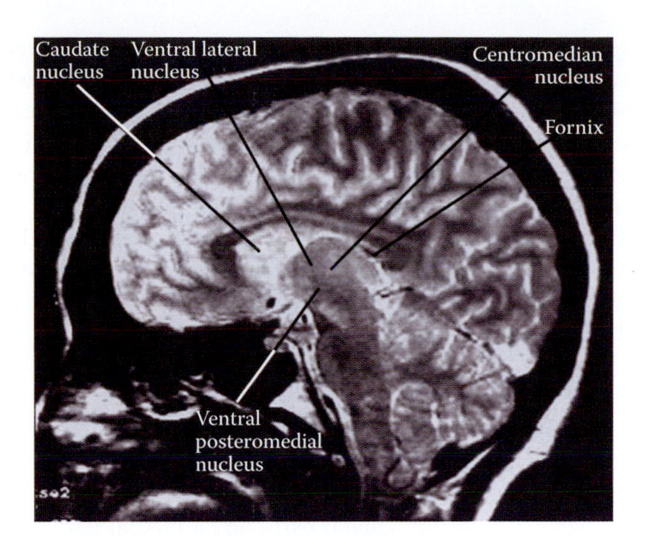

FIGURE 7.10 MRI scan of the brain showing the ventral lateral, centromedian, and ventral posteromedial nuclei.

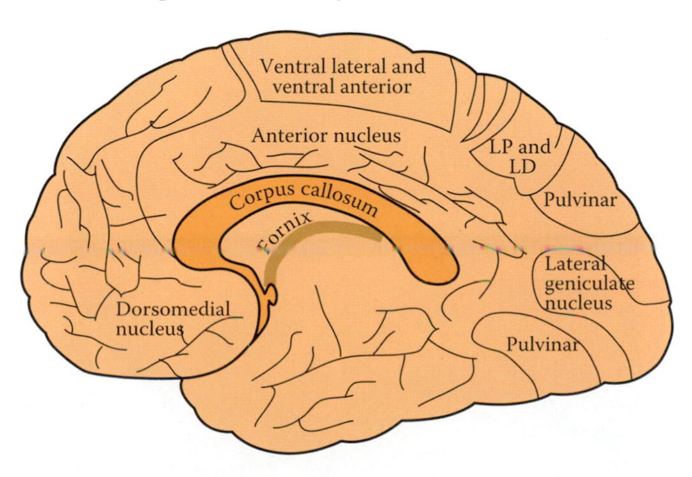

FIGURE 7.9 Schematic diagram of the medial surface of the brain illustrating cortical projections of thalamic nuclei. LD, lateral dorsal; LP, lateral posterior.

The *VPI* nucleus lies in close proximity to the thalamic fasciculus and the reticular nucleus of the thalamus. It receives terminals of the ascending vestibular fibers, which bypass the medial longitudinal fasciculus, delivering this information bilaterally to the vestibular cortical center, which lies adjacent to the facial region in the primary sensory cortex. Neurons of this nucleus respond to deep stimuli, particularly tapping.

The *posterior thalamic zone* (*PTZ*) is a nuclear complex, which is located dorsal to the MGB and medial to the pulvinar. It comprises the posterior nucleus, which is continuous with the VPI nucleus; the suprageniculate nucleus; and the nucleus limitans (which lies between the pulvinar and the pretectum). This nuclear complex has a broad range of connections, and it is neither place nor modality specific. In particular, the posterior nucleus of the PTZ receives pain and nociceptive stimuli via the lateral spinothalamic tract. It also receives tactile, vibratory, and auditory impulses. The posterior thalamic nuclei project principally to the secondary somatosensory cortex of both cerebral hemispheres, particularly area IV.

Ventral Anterior Nucleus (VA)

The *VA nucleus* is bounded anteriorly and laterally by the reticular nucleus (Figures 7.3, 7.5, 7.6, and 7.13). It forms the anterior pole of the ventral nuclear group. It has rich connections with midline and intralaminar nuclei as well as the reticular nucleus. It is interesting to note that no fibers have been traced from the VA nucleus to the contralateral thalamus or to the striatum. The mammillothalamic tract crosses the VA nucleus. The magnocellular part of this nucleus receives input from the midbrain reticular formation, midline, and intralaminar nuclei and projects to the orbitofrontal cortex and Brodmann area 8. Due to its linkage to the intralaminar nuclei, wide areas of the cerebral cortex can be activated, and desynchronization of the electrocortical activity can be achieved by stimulation of this nucleus. The VA nucleus as well as the orbitofrontal cortex is involved in the physiological phenomenon known as the *recruiting response*. The substantia nigra also projects to the magnocellular part of the nucleus, running parallel to the mammillothalamic tract. On the other hand, the principal part of this nucleus conveys input generated by the globus pallidus (via the thalamic fasciculus) and contralateral cerebellar nuclei (via the superior cerebellar peduncle) to the premotor cortex (Brodmann area 6). The connections of the VA nucleus to the premotor cortex, globus pallidus, and substantia nigra (areas thought to be involved in dyskinetic movements) may explain the reduction or abolition of tremor seen in Parkinsonism.

Ventral Lateral Nucleus (VL)

The *VL nucleus* (Figures 7.3, 7.6, 7.10, 7.11, 7.13, and 7.14), also known as the ventral intermediate nucleus, occupies the area posterior to the VA nucleus. It is divided into a rostral (oral part), a posterior (caudal part), and a medial part. The afferents to the VL nucleus are derived from the contralateral dentate nucleus (via the superior cerebellar peduncle), ipsilateral red nucleus, contralateral globus pallidus (through the thalamic fasciculus), and pars reticulata of the substantia nigra, and also from the precentral gyrus and premotor cortex. The afferents from the substantia nigra terminate in the medial part, while the remaining afferents terminate in the oral and caudal parts of the VL nucleus. Projections of the VL nucleus to the precentral gyrus are somatotopically arranged and establish monosynaptic connections with the neuron of this gyrus. The VL nucleus is part of the link between the cerebellum and the motor cortex and between the basal nuclei and the cerebral cortex. Therefore, the cerebellum and the basal nuclei influence motor activity through their projections to the VL nucleus. Surgical ablation of the VL nucleus may be performed in order to relieve the tremor and rigidity associated with Parkinsonism.

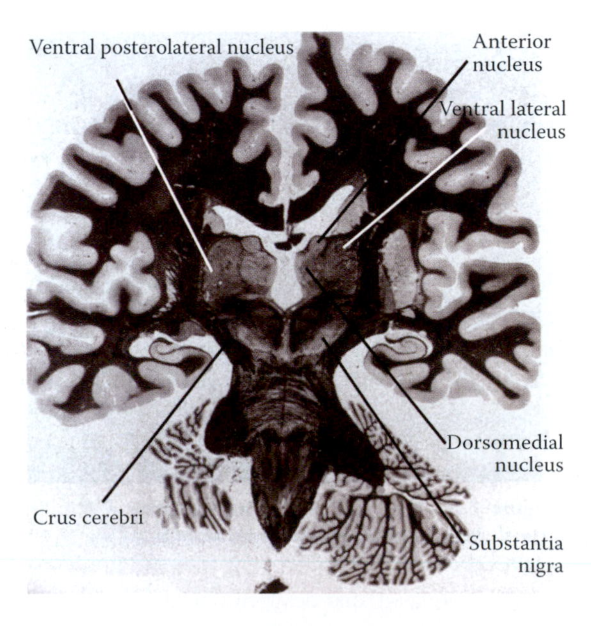

FIGURE 7.11 Coronal section of thalamus through the ventral posterolateral thalamic nucleus. In this view, the anterior, dorsomedial, and ventral lateral nuclei are seen.

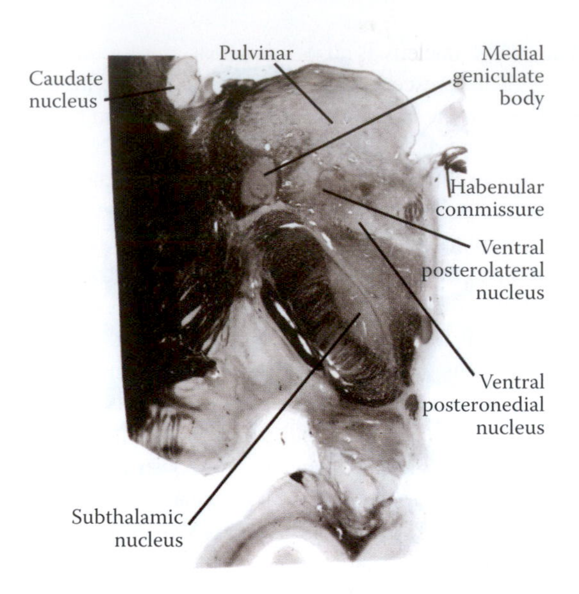

FIGURE 7.12 Section of the thalamus through the habenular commissure. Note the distinct ventral posterolateral and ventral posteromedial nuclei as well as the pulvinar.

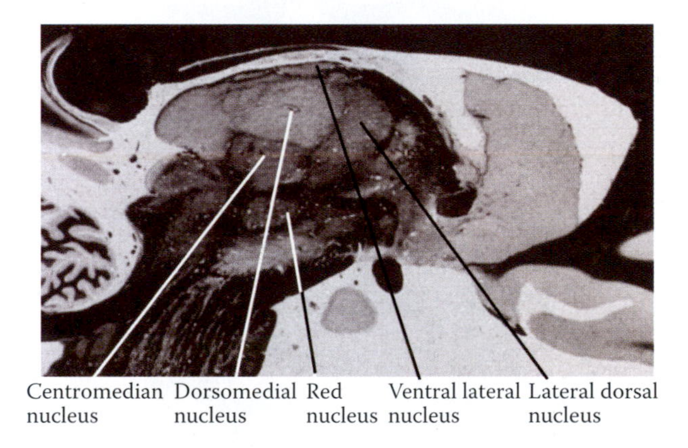

FIGURE 7.13 Sagittal view of the thalamus showing ventral lateral, dorsomedial, centromedian, and ventral anterior nuclei.

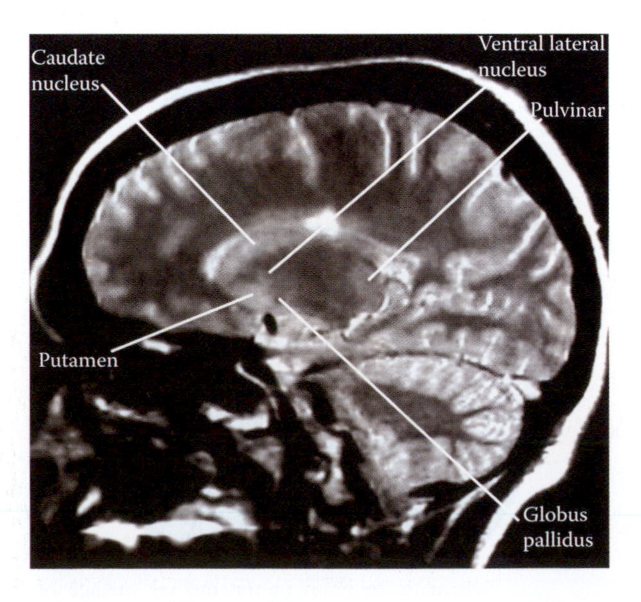

FIGURE 7.14 MRI scan illustrating the ventral lateral nucleus and pulvinar.

LATERAL NUCLEAR GROUP

Lateral Dorsal (LD)

The *lateral group of thalamic nuclei* includes the lateral dorsal (LD) nucleus, lateral posterior (LP) nucleus, and pulvinar. The *LD nucleus* (Figures 7.3, 7.9, 7.13, 7.16, and 7.18) represents the rostral extension of the dorsal group of thalamic nuclei. It lies posterior to the anterior nucleus and receives input from the superior colliculus and pretectal area. Reciprocal connections exist between this nucleus and the cingulate gyrus as well as the precuneus gyri. The LD nucleus also projects to the posterior part of the parahippocampal gyrus.

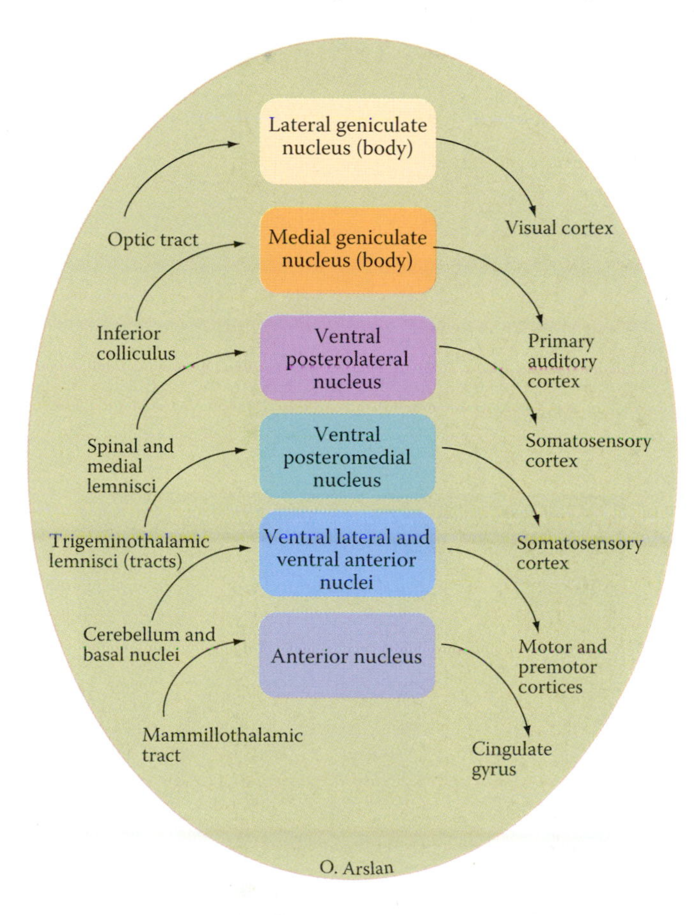

FIGURE 7.15 Summary of the main afferent and efferent connections of the thalamic relay nuclei.

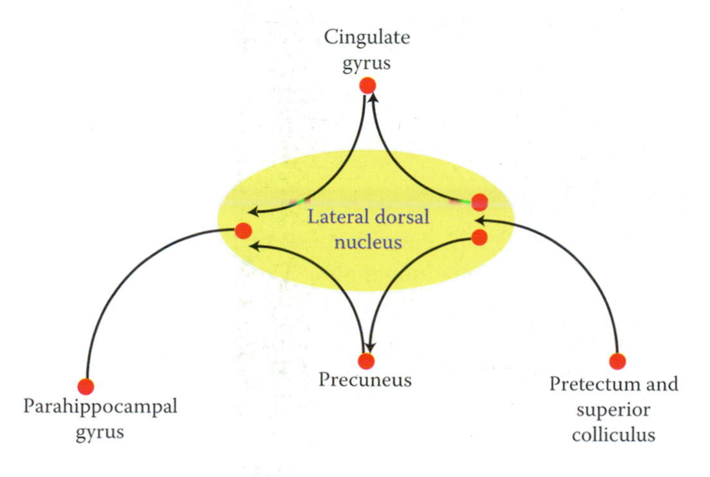

FIGURE 7.16 Schematic drawing of the afferent and efferent connections of the lateral dorsal nucleus.

Lateral Posterior

The *LP nucleus* lies caudal to the LD nucleus (Figures 7.3, 7.9, 7.17, and 7.18). The geniculate bodies, as well as neurons of the ventral group thalamic nuclei, convey information to this nucleus. The parietal and occipital lobes (Brodmann areas 5 and 7) have reciprocal connections with this nucleus. However, no known connections exist between the LD nucleus and the primary sensory, visual, or auditory cortices.

Pulvinar

The *pulvinar* is the largest and phylogenetically the most recent thalamic nucleus (Figures 7.1, 7.2, 7.3. 7.7, 7.8, 7.12, 7.13, and 7.19). It overlies the geniculate bodies, separated

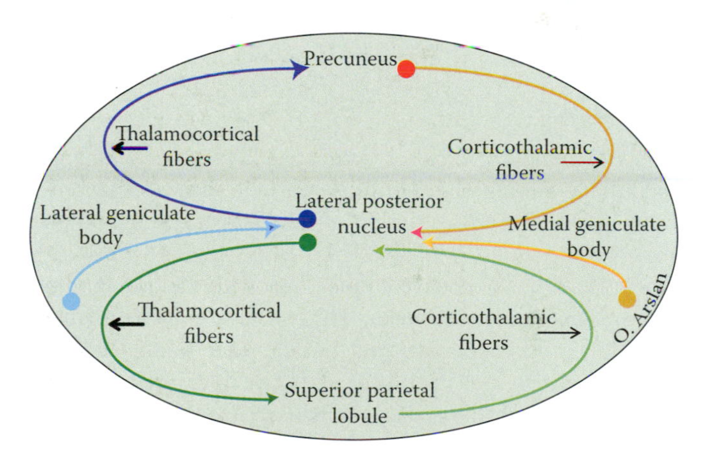

FIGURE 7.17 Simplified diagram of the main projections and afferent of the lateral posterior thalamic nucleus.

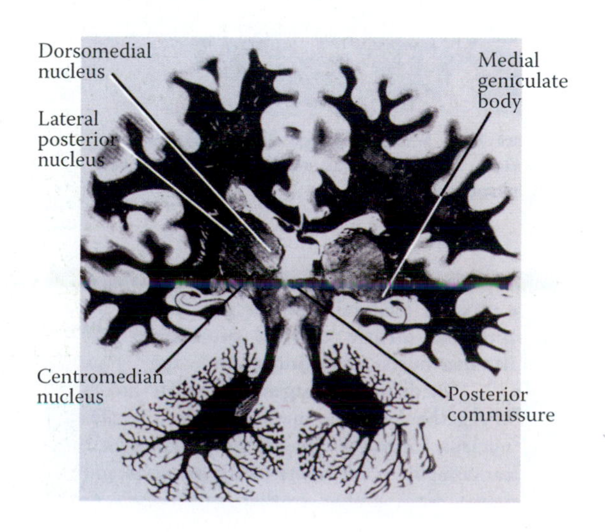

FIGURE 7.18 This section through the posterior commissure illustrates the prominent dorsal medial, lateral posterior, and centromedian nuclei and the medial geniculate body.

from them by the brachium of the superior colliculus. It has multisensory functions, receiving input from the intralaminar nuclei, geniculate bodies, and superficial layers of the superior colliculus, as well as reciprocal connections with the temporal, parietal, and occipital lobes. Wernicke's sensory speech center (Brodmann area 22) maintains a rich connection with this nucleus. Visual input from the retina also reaches the pulvinar via the lateral geniculate body and the superior colliculus. Visual impulses from the inferior and lateral parts of the pulvinar, which projects to the supragranular layers of the primary visual cortex and to layers I, III, and IV of the secondary visual cortex (areas 18 and 19), constitute the *extrageniculate visual pathway*, linking the retina, tectum, and visual cortex. The medial part has primary reciprocal connections with the posterior parietal cortex. Based on these connections, the pulvinar is implicated in visual and oculomotor control, as well as pain and speech modulation.

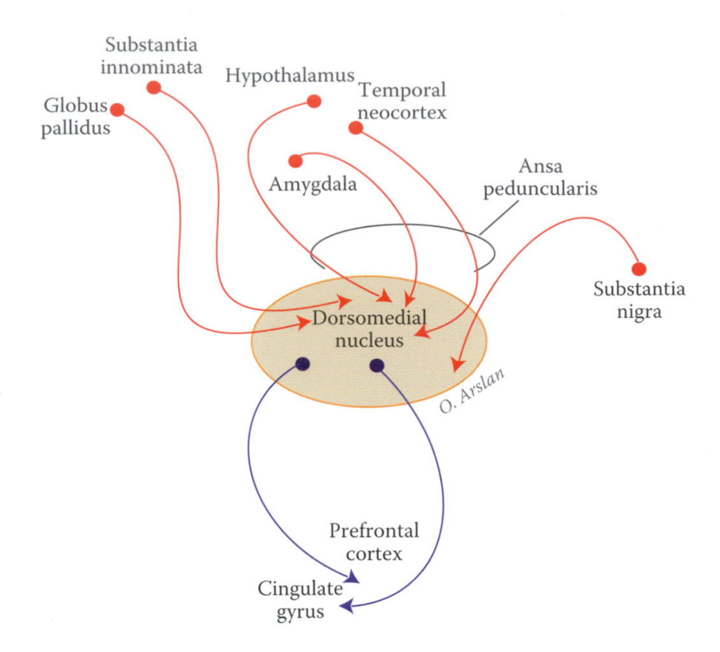

FIGURE 7.20 Schematic diagram of the afferents of the dorsomedial nucleus of the thalamus that contribute to the formation of the ansa peduncularis.

MEDIAL NUCLEAR GROUP

The medial group of thalamic nuclei consists of a single large dorsomedial nucleus.

Dorsomedial Nucleus

The *dorsomedial nucleus* is located between the internal medullary lamina and the third ventricle. It is considered a relay nucleus for transmission of impulses to the hypothalamus (Figures 7.1, 7.3, 7.5, 7.8, 7.11, 7.13, 7.18, 7.19, 7.20, 7.23, and 7.24). It has extensive connections with the intralaminar and midline thalamic nuclei. This nucleus consists of magnocellular, parvocellular, and paralaminar parts. Through its connections, the magnocellular part integrates impulses received through the ansa peduncularis from the amygdala, lateral hypothalamus, diagonal band of Broca, orbitofrontal cortex, and substantia innominata. The ansa peduncularis is comprised of the inferior thalamic peduncle and the interconnecting fibers between the amygdala and the preoptic area. The parvocellular part establishes reciprocal connections with the prefrontal cortex (Brodmann areas 9, 10, 12, and 13) and receives input from the globus pallidus. The paralaminar part maintains reciprocal connections to the premotor cortex (Brodmann areas 6 and 8) and also receives input from the pars reticulata of the substantia nigra.

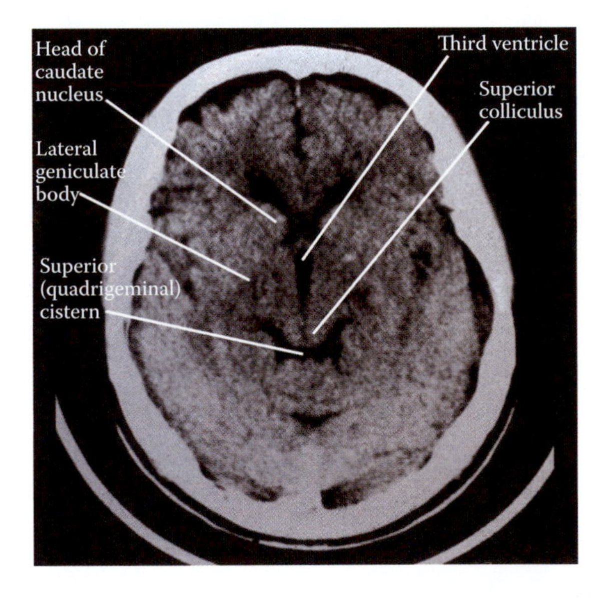

FIGURE 7.21 This computed tomography scan shows the lateral geniculate body and its location rostral and lateral to the superior colliculus.

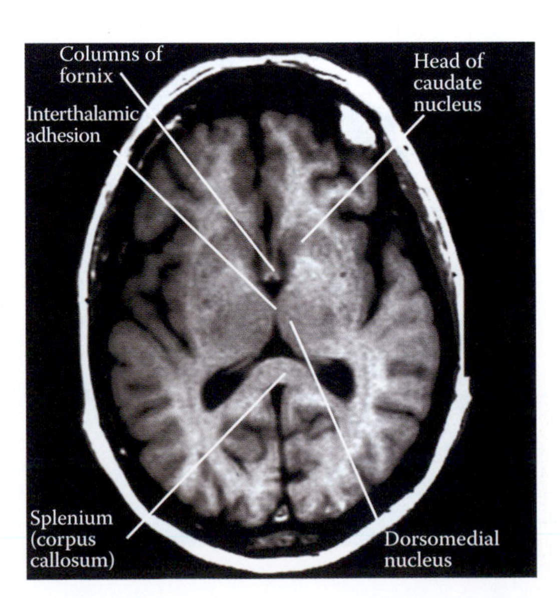

FIGURE 7.19 An inverted MRI scan of the brain showing the dorsomedial nucleus and the interthalamic adhesion where midline nuclei are located.

The dorsomedial nucleus is associated with the affective qualities of behavior and somatic sensation through its connections with the hypothalamus and the prefrontal cortex. It has no connection with any particular sensory nucleus. Removal of this nucleus and/or a prefrontal lobotomy may be used to modify or reduce the emotional stress associated with chronic pain. Patients who have undergone this surgical procedure report feeling pain without being distressed. This is identical to the perception of pain that patients report when given narcotics. Reduction in anxiety, aggressive behavior, or obsessive thinking may result from destruction of this nucleus. Some degree of amnesia and confusion may develop later.

INTRALAMINAR NUCLEAR GROUP

The *intralaminar thalamic nuclei* represent a group of thalamic nuclei embedded in the internal medullary lamina. These include the centromedian (Figures 7.3, 7.8, 7.10, 7.13, 7.18, and 7.23), parafascicular, central lateral, and central medial nuclei. They receive afferents from the globus pallidus, striatum, and dentate nuclei of the cerebellum; pedunculopontine nucleus; and terminals and collaterals from the spinal, medial, and trigeminal lemnisci. Ascending reticular fibers, which are derived from the nucleus reticularis gigantocellularis, nucleus reticularis ventralis, and dorsalis of the medulla and nucleus reticularis pontis, form the principal afferents to the intralaminar nuclei. They project ipsilaterally to the centromedian–parafascicular nuclear (CM–PF) complex and to the paracentral and central nuclei.

The spino-reticulo-thalamic (paleo-spino-thalamic) tract projects bilaterally to the intralaminar nuclei. Despite the diffuse cortical projections of the intralaminar nuclei, this connection is not reciprocal. The centromedian is the largest nucleus in this group, which, together with the medially located parafascicular nucleus, forms the CM–PF nuclear complex. This nuclear complex sends fibers to the putamen and the substantia nigra. Other smaller intralaminar nuclei project to the caudate nucleus. The projection of the precentral gyrus to the centromedian nucleus is predominantly ipsilateral. Some of these fibers run within the internal capsule and crus cerebri. Due to its connection to the striatum (caudate and putamen) and motor cortex and, indirectly, to the globus pallidus, the centromedian nucleus serves as an important element of a feedback circuit between these areas. The habenulopeduncular tract (fasciculus retroflexus) divides the parafascicular nucleus into medial and lateral divisions. The PF nucleus conveys the input received from the premotor cortex (Brodmann areas 6, 8, and 9) to the putamen. Desynchronization of electrocortical activity is attributed to

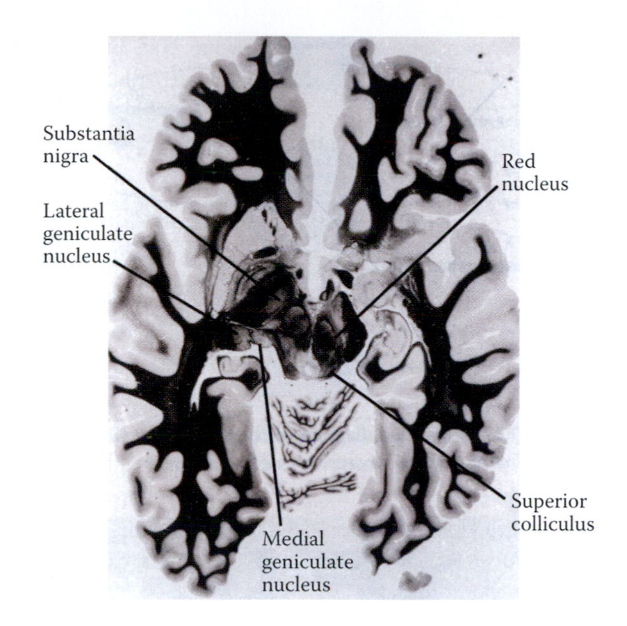

FIGURE 7.22 Section through diencephalomesencephalic junction. The lateral and medial geniculate bodies are prominently illustrated.

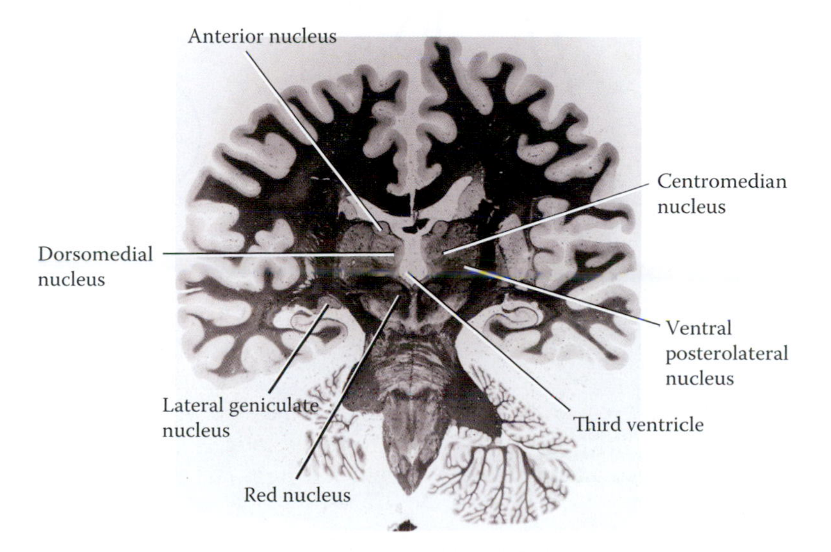

FIGURE 7.23 Section of the thalamus through the anterior nucleus. Dorsomedial, centromedian, and ventral posterolateral nuclei are illustrated. The geniculate bodies are also visible.

the diffuse connections of the intralaminar nuclei to all areas of the cerebral cortex and to their extensive input from the reticular formation. Both the intralaminar thalamic nuclei and the specific thalamic nuclei project to the cerebral cortex, giving off collaterals to the reticular nucleus.

MIDLINE NUCLEAR GROUP

The interthalamic adhesion (massa intermedia) and in the periventricular area of the third ventricle. These nuclei comprise the paleothalamus (relatively new on a phylogenetic basis), consisting of anterior and posterior paraventricular nuclei, as well as the rhomboidal, reuniens, central, and paratenial nuclei. These nuclei receive afferents from the reticular formation, corpus striatum, cerebellum, hypothalamus, and spinothalamic tracts. The main projections of these nuclei are to the cingulate gyrus, amygdala, entorhinal cortex, and prepyriform cortex.

RETICULAR NUCLEAR GROUP

The principal nuclear group of the ventral thalamus is the reticular nucleus.

This nuclear group consists primarily of the *reticular nucleus* which resembles structurally and is continuous with the zona incerta. It lies between the external medullary lamina and the posterior limb of the internal capsule. All areas of the cerebral cortex as well as the midbrain reticular formation and the globus pallidus convey impulses to the reticular nucleus. This nucleus *does not project to the cerebral cortex*, but it does influence the activity of other thalamic neurons that project to the cerebral cortex.

FUNCTIONAL AND CLINICAL CONSIDERATION

Thalamic syndrome (*Dejerine–Roussy syndrome or thalamic apoplexy*) may result from occlusion of the posterior choroidal, the posterior cerebral, and the thalamogeniculate arteries. Hemiparesis on the contralateral side may be seen initially as a result of edema that compresses the corticospinal tracts as they descend through the posterior limb of the internal capsule. Ataxia and clumsiness in the affected limb may occur and is often associated with episodes of severe pain following a transient anesthesia (loss of all sensations) contralaterally. In general, it is attributed to a loss of joint sensation and a loss of the ability to appreciate movement. At the onset of the syndrome, a complete contralateral hemianesthesia and hemianalgesia occurs. Pain, temperature, and crude touch gradually return; however, positional sense and stereognosis are lost permanently. Sensation in the face may or may not be affected, depending upon the involvement of the VPM nucleus of the thalamus. Visual deficits, such as homonymous hemianopsia, may also be detected if the lateral geniculate body is damaged. Hypersensitivity to slight superficial stimuli, which results in severe pain, is common in this condition. Thalamic pain is a diffuse, burning, and lingering type of sensation that may be elicited by mild stimuli of touch, pressure, and vibration and aggravated by emotional conditions. A pinprick may cause agonizing and intolerable pain.

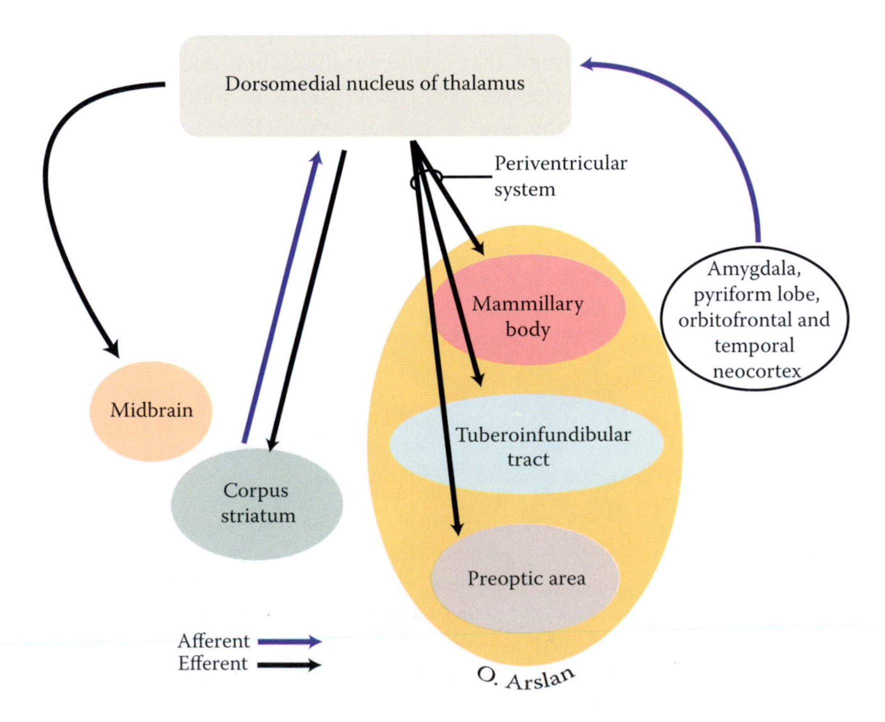

FIGURE 7.24 Schematic diagram of the afferent and efferent fibers of the dorsomedial nucleus of the thalamus. Note that the connection to the corpus striatum is bilateral. The projections of the dorsomedial nucleus to the mammillary body, tuberoinfundibular region, and preoptic area comprise the periventricular system.

Pressure from one's clothing or sound from a musical instrument may produce tremendous discomfort. These sensations occur in response to any stimulus; however, pain threshold is increased, and pain sensation is often prolonged and may be elicited by a stronger stimulus. Occasionally, a condition known as *thalamic hand* may be observed on the contralateral side. It is characterized by flexion of the digits at the metacarpophalangeal joints and extension at the distal phalangeal joints, along with flexion and pronation of the wrist. General hypotonia results in a constant need to readjust posture. Thalamic hematomas may cause prominent sensory deficits contralaterally. Due to the caudal location of the corticospinal fibers, relative to the hematoma, muscular weakness is less likely to occur. However, choreiform movements and ataxia may be seen in the contralateral limb. Ocular deviation, in which one eye exhibits a lower position than the other eye at rest and maintains an exaggerated convergence (pseudoabducens nerve palsy), and an inability to look upward also occur. These ocular changes, which result from compression of the adjacent tectal and pretectal areas by a thalamic hematoma, give the appearance of an individual staring at the bridge of his nose. Hematoma of the left thalamus is generally associated with fluent aphasia. A right-sided thalamic hematoma is accompanied by anosognosia and left-sided visual neglect. Due to involvement of the intralaminar thalamic nuclei and the central tegmental tract, thalamic hemorrhage may initially be associated with decreased consciousness (Table 7.1).

HYPOTHALAMUS

The hypothalamus (Figures 7.25, 7.26, 7.27, and 7.28) lies posterior to the lamina terminalis and optic chiasma, ventral to the dorsal thalamus, medial to the subthalamus, and lateral to the ventral part of the third ventricle. The lamina terminalis separates the paraterminal gyri from the preoptic area. The hypothalamus lies dorsal to the optic tract, midbrain tegmentum, and internal capsule. Its caudal boundary

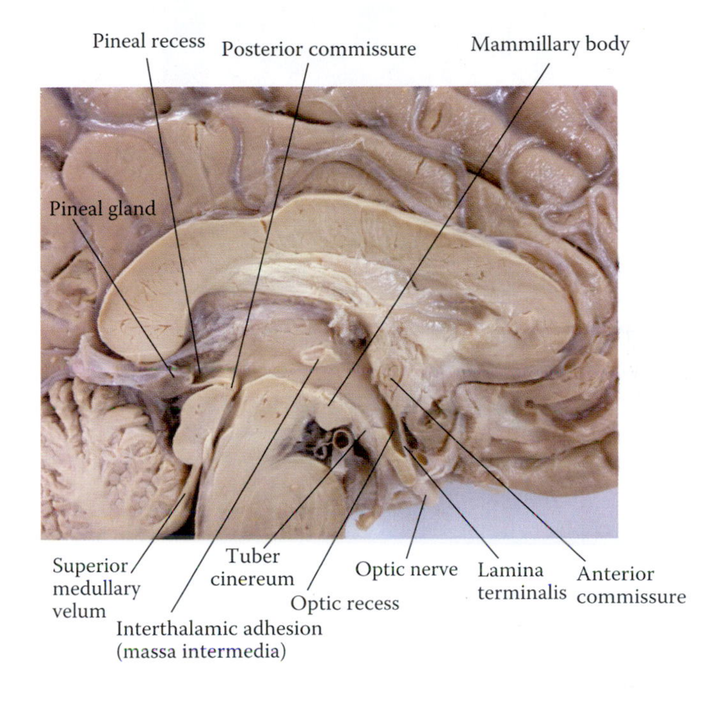

FIGURE 7.25 Midsagittal section of the brain illustrating the components and boundaries of the hypothalamus.

TABLE 7.1
Summary of Connections of Thalamic Nuclei

Afferents	Thalamic Nuclei	Efferents
Medial lemniscus	Ventral posterolateral (VPL)	Somesthetic cortex (postcentral gyrus)
Spinothalamic tract	VPL	Somesthetic cortex (postcentral gyrus)
Trigeminal lemnisci	Ventral posteromedial (VPM)	Somesthetic cortex (postcentral gyrus)
Inferior colliculus	Medial geniculate body	Auditory cortex
Optic tract and visual cortex	Lateral geniculate body	Visual cortex (Brodmann area 17)
Hypothalamus, prefrontal cortex	Dorsomedial nucleus	Prefrontal cortex
Mammillothalamic tract and cingulate gyrus	Anterior	Cingulate cortex
Globus pallidus, motor cortex, and cerebellum	Ventral lateral	Primary motor cortex (precentral gyrus)
Substantia nigra, dentate nucleus, and premotor cortex	Ventral anterior	Premotor cortex
Hypothalamus	Lateral dorsal	Cingulate and precuneate gyri
Parietal and occipital cortices	Lateral posterior	Parietal and occipital cortices
Intralaminar nuclei and tectum	Pulvinar	Temporal, parietal, and geniculate bodies
Ascending reticular activating system	Intralaminar	Putamen, substantia nigra, globus pallidus, and lateral spinothalamic tracts
Reticular formation, globus pallidus, and spinothalamic tracts	Midline	Cingulate gyrus, amygdala, cerebellum, and hypothalamus
Cerebral cortex, globus pallidus, and reticular formation	Reticular	Thalamic nuclei projecting to midbrain

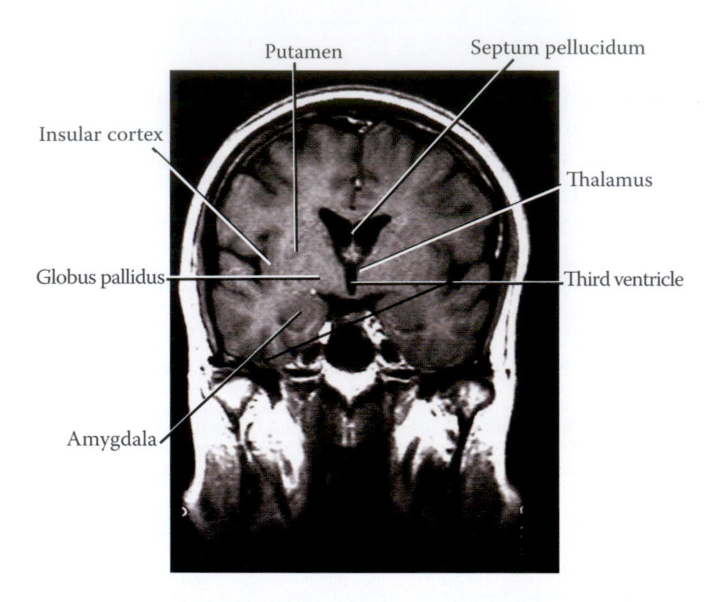

FIGURE 7.26 An inverted MRI scan of the brain. The hypothalamus, as indicated, forms the floor of the third ventricle.

is demarcated by a vertical plane that passes posterior to the mammillary body. It is connected ventrally to the pituitary gland and is divided into lateral and medial parts by the fornix. Morphologically, the hypothalamus is comprised of supraoptic, infundibulotuberal, and mammillary regions, with each region containing specific nuclei.

Within the floor of the interpeduncular fossa, the mammillary bodies appear as a two rounded eminences and contain the corresponding nuclei encircled by the posterior columns of the fornix. Posterior to the optic chiasma and anterior to the mammillary bodies, a shallow gray matter mass, the tuber cinereum, can be seen. A midline tube, the infundibulum, emerges from the tuber cinereum and extends downward to the posterior pituitary. Near the origin of the infundibulum, the median eminence, which receives the tuberoinfundibular tract, makes its appearance.

Hypothalamic Areas and Nuclei

The hypothalamus is categorized into distinct areas that contain various nuclear groups. These areas are arranged in the anteroposterior direction into preoptic, supraoptic, tuberal, and mammillary areas and in the mediolateral direction into medial, intermediate, and lateral zones. Lateral and intermediate areas are separated by the fornix, mammillothalamic tract, and fasciculus retroflexus.

The medial zone lies immediately deep to the ependymal layer of the third ventricle and the periaqueductal gray matter. It contains rostrally the vascular organ of the lamina terminalis, an area devoid of the blood–brain barrier. Within the chiasmatic region of this zone, the preoptic, suprachiasmaic, and periventricular nuclei are distinctly seen, whereas the tuberal region contains the arcuate (tuberal) nucleus. Within the intermediate zone, the paraventricular, supraoptic, ventromedial, dorsomedial, mammillary, intermediate, and tuberomammillary nuclei are well differentiated. The latter nucleus contains γ-aminobutyric acid (GABA), histamine, and galanin, which project diffusely to the neocortex, hypothalamus, and brainstem. The lateral zone contains nuclei

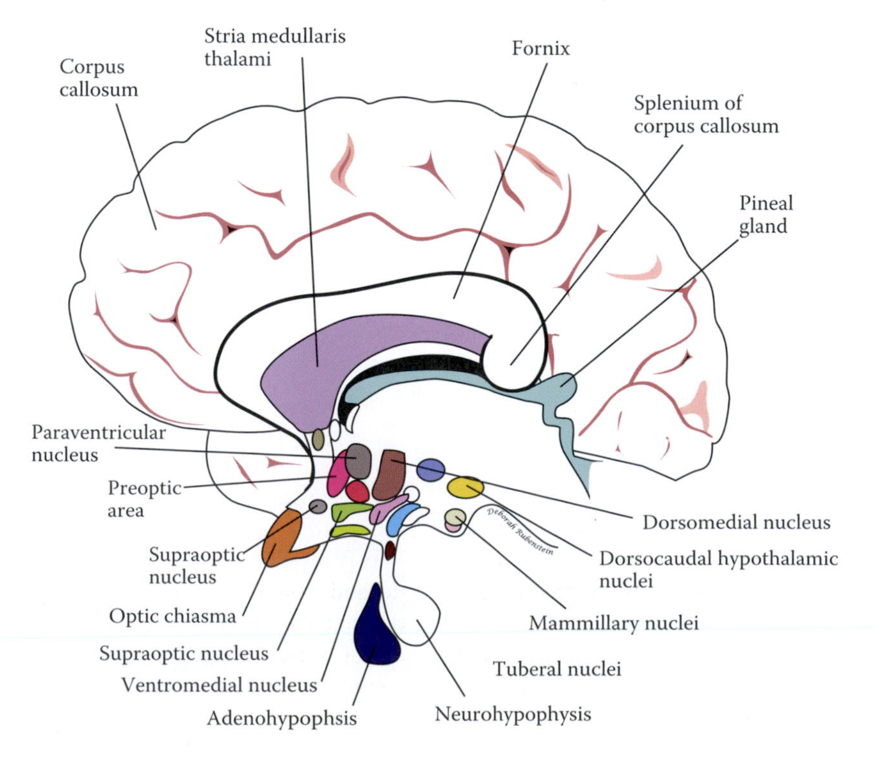

FIGURE 7.27 Hypothalamic nuclei and associated areas.

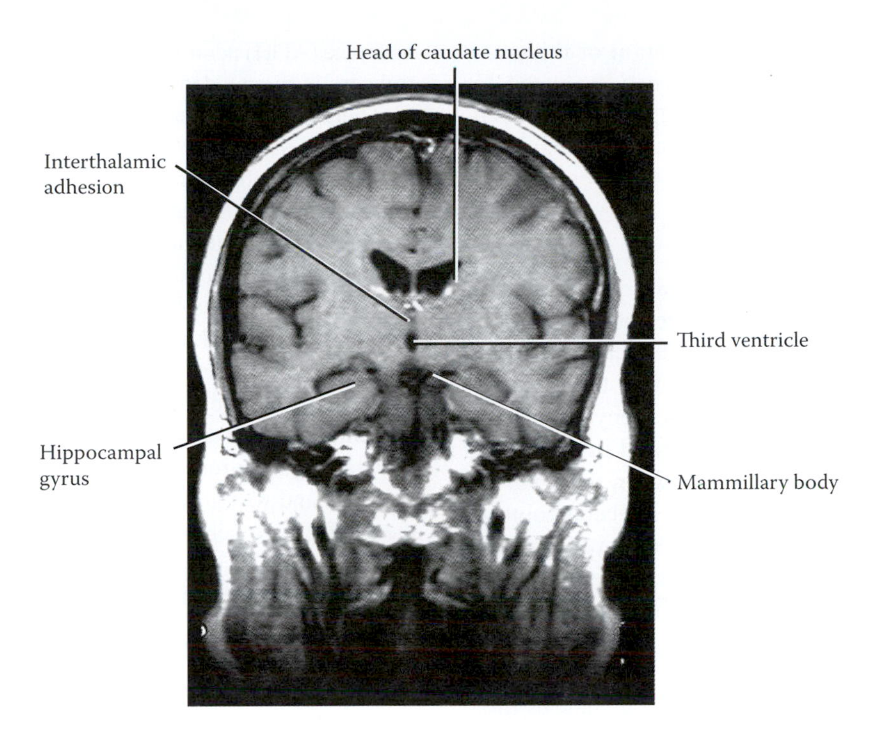

FIGURE 7.28 In this inverted MRI scan, the prominent mammillary region of the hypothalamus is illustrated.

associated with the preoptic, lateral, and posterior areas of the hypothalamus. The lateral tuberal nuclei, which undergo atrophy in Huntington's chorea, are also located in the lateral zone. The preoptic region of this zone contains several neuromodulators including enkephalin, cholecystokinin, vasoactive intestinal peptide (VIP), substance P, angiotensin II, acetylcholine, adrenaline, noradrenaline, endorphin, delta sleep–inducing peptide, dynorphin, α-melanocyte–stimulating hormone (α-MSH), galanin, vasopressin, oxytocin, somatostatin, corticotropin-releasing hormone (CRH), gonadotropin-releasing hormone, neurotensin, glycine, glutamate, and aspartate. Due to the large number of nuclei located within the hypothalamus, our discussion will be devoted to major hypothalamic nuclei that bear known functional significance.

The *suprachiasmatic nucleus* (SCN) is located posterosuperior to the optic chiasma and in close proximity to the third ventricle. This nucleus, which has ventrolateral and dorsomedial subdivisions, is involved in the circadian rhythm that regulates the sleep–wake cycle, body temperature, hormonal plasma levels, and renal secretion. This endogenous clock is associated with photoreception and requires that the rhythmic activities are self-sustaining and are synchronized to external time, maintaining appropriate temporal relationships in a changing environment. The ventrolateral part of the nucleus consists of neurons that are immunoreactive for VIP, receiving serotonergic projections from the mesencephalic raphe nuclei and visual input from neurons of the lateral geniculate nucleus. Mediation of the light–dark cycle rhythm is accomplished by the glutamatergic retinal afferents that project to neurons of the ventrolateral subdivision of the nucleus, which respond to the onset, offset, and intensity of light by changing the firing rate. They are unresponsive to color, movement, and pattern. One of the striking characteristics of the photic neurons of SCN is their ability to express a specific complement of subunits for the *N*-methyl-D-aspartate (NMDA) receptors. It has also been suggested that the expression of c-fos genes within the SCN may be induced by the application of light during the night, which resets the biological clock.

On the other hand, the *dorsomedial subdivision* contains parvocellular neurons that are immunoreactive for arginine vasopressin. Neuronal axons of the SCN project to the paraventricular, tuberal, and ventromedial hypothalamic nuclei. The projections of the SCN to the reticular formation that eventually affect the activities of the sympathetic neurons and secretion of melatonin from the pineal gland may be mediated via the paraventricular nucleus. This nucleus contains vasopressin, VIP, and neurotensin. Vasopressin neurons in the SCN show marked reduction in Alzheimer's disease.

The SCN receives bilateral input from the photoreceptors of the retina, mediating the phase and period of the biological clock. It also receives serotonergic projections from the midbrain raphe nuclei, the geniculohypothalamic (GHT) from the lateral geniculate nucleus, as well as neuropeptide Y (NPY)–containing neurons from the thalamic intergeniculate thalamic leaflet (IGL), which mediates photic and nonphotic signals. With the SCN projecting to 15 regions and receiving direct afferents from about 35 regions and if multisynaptic relay afferents are included, the number expands to approximately 85 brain areas. The fact that IGL, which is involved in the regulation of circadian rhythm, has diffuse bilateral and reciprocal connections to approximately 100 regions magnifies the already expansive involvement of CNS structures in regulation of the circadian rhythm. Few of these sites have been evaluated for their contributions to circadian rhythm regulation. The IGL connections may suggest its possible role in the regulation of eye movements during sleep.

The role of nonphotic cues in circadian rhythm, particularly during development, has been the subject of extensive research. The available data indicate that the biological clock begins to function before the retinal fibers project to the SCN and that synchrony is established through the placenta and expressed in preterm infants. The latter may serve as a guide in the establishment of neonatal care facilities. In a related point, change of work shift and transcontinental traveling can disrupt the synchrony and thus produce depression and cognition and mood disorders. A corollary to this is the fact that patients with clinical depression usually show disorders in sleep–wake and reproductive cycles, which may possibly strengthen the argument that changes in the social environment can form a basis for changes in the biological clock and eventually lead to depression. An important fact needs to be remembered: that the timing mechanism in the biological clock operates independently from the activities that require establishment of synaptic linkages and the generation of action potential within the SCN. It has been suggested that coupled interactions of high-frequency oscillators within this nucleus are most likely responsible for the sustained circadian oscillation. The ability of the transcriptional inhibitors to reset the biological clock during phase shifting may strengthen the thought that preserving the time-keeping process is a self-maintaining transcriptional cycle. Further, the biological clock rhythm remains active even in the tissue grafts containing fetal suprachiasmatic nuclei.

Studies have shown that damage of the SCN leads to the disruption of the circadian rhythms and disturbances in the sleep–wake cycle. Disturbance during sleep may predispose affected individuals to the development of sundowner syndrome, a condition that is characterized by increased confusion, vocalization, sleep apnea, restlessness, agitation, pacing in the early evening, and eventual dementia.

The paraventricular and supraoptic nuclei consist of a distinct population of neurons, magnocellular and parvocellular, which are highly vascularized. The paraventricular nucleus consists of a large population of neurons medial to the fornix that stretch dorsally to the hypothalamic sulcus. These neurons show somatotopic organization in which neurons that project to the neurohypophysis are located laterally, the parvocellular neurons that project to the median eminence and infundibulum occupy a more medial position, and the midsized neurons that project caudally assume a posterior location. The antidiuretic hormone (ADH) neurons of the paraventricular nucleus lie ventrolaterally surrounded by the oxytocin-producing neurons.

There are three subgroups of the magnocellular neurons of the supraoptic nucleus, which, for the most part, are located dorsolateral to the optic tract, with the exception of the retrochiasmatic neurons of this nucleus, which extend medially toward the tuberal nuclei. The magnocellular secretory neurons of the *supraoptic* and *paraventricular* nuclei of the hypothalamus are large and possess extensive rough endoplasmic reticulum and Golgi zones with large secretory granules. These neurons primarily secrete ADH and oxytocin. ADH neurons appear to be sensitive to osmotic fibers that emanate from the subfornical organ, preoptic nucleus, and noradrenergic neurons of the cardiovascular centers in the caudal pons and rostral medulla. ADH neurons are inhibited by the GABAergic neurons and stimulated by stress, as well as by angiotensin II, glutamate, acetylcholine, and α_2-adrenergic neurons. ADH neurons also secrete, though in small quantities, peptides that include dynorphin, cholecystokinin, peptide histidine isoleucine (PHI), and thyrotropin-releasing hormone.

Oxytocin neurons, though lacking glycopeptides, are similar to ADH neurons in that they are bound to neurophysin; inhibited by stress, GABA, and opioid peptides; and excited by glutamate. Oxytocin-secreting neurons contain enkephalin, galanin, cholecystokinin, and dynorphin, receiving afferents from the uterine cervix, vagina, and nipple to mediate the Ferguson reflex (initiated by stretching of the internal ostium of the cervix and subsequent secretion of oxytocin that produces self-sustaining uterine contractions). ADH and oxytocin are synthesized on the ribosomes of cell bodies and packaged in Golgi complexes. Both hormones are peptides linked to the precursor protein, neurophysin. Exocytosis of the secretory granules into the pericapillary spaces occurs following depolarization of the neuroendocrine cells. *ADH* changes the permeability of the distal convoluted and collecting tubules, thus increasing the reabsorption of water into the bloodstream and counteracting dehydration. ADH neurons also secrete other peptides such as dynorphin, galanin, and PHI. In mammals, ADH neurons receive input from the median preoptic area, and from the noradrenergic neurons of the brainstem that carry cardiovascular input. They are inhibited by GABA and excited by angiotensin II, α_2-adrenergic, glutamate, and acetylcholine. *Oxytocin* is secreted in conjunction with small amounts of enkephalins, galanin, and dynorphin. Oxytocin-secreting neurons receive afferents from the uterine cervix, vagina (*Ferguson reflex*), and nipple for milk ejection. Thus, oxytocin causes an increase in the contraction of the smooth muscles of the mammary gland and the uterus. Both ADH and oxytocin have overlapping functions and are delivered to the posterior lobe of the pituitary gland through the supraoptico-paraventriculo-hypophyseal tract.

At the tuberal region, the ventromedial nucleus appears larger and more distinct than the dorsomedial nucleus. The parvocellular neurons of the tuberal nuclei secrete the hormone regulating factors (HRFs), which are delivered to the median eminence (upper portion of the infundibulum) via the tuberoinfundibular tract. The hypophyseal portal system then carries these substances from the median eminence to the anterior lobe of the pituitary, where they may enhance or inhibit the release of the hormones from the anterior lobe of the pituitary gland. The tuberal nuclei contain histamine, galanin, GABA, and cholinesterase, but not choline acetyl transferase and so forth. The mammillary area consists of the medial and lateral mammillary nuclei. Histamine and GABA and galanin are the main content of the medial mammillary nucleus, which maintains a diffuse connection to the cerebral cortex and brainstem.

Certain hypothalamic nuclei exhibit sexual dimorphism due the influence of circulating hormones. However, the differences between males and females are subtle regarding connectivity and chemical sensitivity. A particular nucleus that demonstrates this feature is the intermediate nucleus, which is located between the supraoptic and paraventricular nuclei in the suprachiasmatic area. The number of neurons in this nucleus is estimated to be twice as numerous in males as it is in females. This differential increase starts at the age of 4 years and continues until the fifth decade of life in the male, with a concomitant decrease in the number of neurons in the female. Differences have also been demonstrated in the bed nucleus of the stria terminalis, interstitial hypothalamic nucleus, and SCN. The data suggest that the interstitial nucleus of the hypothalamus is reduced in size in homosexual men and women in general compared to heterosexual men and that the overall size of the SCN in homosexual males is twice as large as that of heterosexual males. It is larger in heterosexual men than women. However, the number of cells in the SCN remains constant, but the shape of the cells shows variations, being spherical in the male and elongated in the female. Odor and appearance are preferred stimulants by males. This preference diminishes if the dimorphic nucleus is affected by a lesion. In general, the hypothalamus of homosexual men and heterosexual women respond to testosterone; in the same manner, the hypothalamus of heterosexual men and homosexual women respond to estrogen. These variations in gross or microscopic features may not be sufficient as more work needs to be done to link these differences with biochemical and hormonal changes.

In the hypothalamus, neuron/glia organization, synaptic linkages, dendritic arborization, and associated receptors are affected by functional changes that play a significant role in modulating the control mechanism. A clear example is seen in the oxytocin magnocellular neurons, during delivery and lactation, which undergo hypertrophy, and their dendrites establish synaptic linkage without the presence of astrocytes. An increase in the number of synapses and the number of the GABAergicsynaptic boutons of dendrodendritic and somatodendritic connections is observed. These rapid changes occur as the functional demand necessitates increased secretion of oxytocin but not with adjacent vasopressin neurons. The changes in the axosomatic synapses are also seen in the magnocellular neurons with increased age. Similarly, the nerve terminals in the posterior pituitary (neurohypophysis) will be surrounded by a very limited number of processes of pituicytes when hormonal secretion is increased, enabling extensive contact with the perivascular basal lamina. These changes also occur in the neurons of the gonadotropin-releasing hormone and their terminals in the median eminence (hormone-releasing or -inhibiting factors). During the reproductive cycle and as a result of an increase in estrogen secretion, the estrogen-receptive neurons of the arcuate (tuberal) nucleus will be covered by astrocytic lamella, and GABAergic axosomatic synaptic connections show marked decrease. Redistribution of oxytocin neurons, increase in the axodendritic synaptic connections, and density of dendritic spines in the ventromedial nucleus of hypothalamus occur in response to estrogen.

The blood supply to the hypothalamus is provided primarily by the posteromedial (thalamoperforating) branches of the posterior cerebral artery, including the mammillary bodies, infundibulum, tuber cinereum, and pituitary gland. The mammillary bodies, tuber cinereum, and optic chiasma are surrounded by the arterial circle of Willis (Figure 7.29), an arterial chamber formed by branches of the internal carotid and vertebral arteries, providing shunting between the vertebrobasilar system and carotid systems. The specific arterial branches involved in this circle are the anterior cerebral, internal carotid, anterior and posterior communicating, and posterior cerebral arteries. However, it does not always show symmetry, as one or more branches may undergo hypoplasia.

Unilateral obstruction of the internal carotid artery is usually asymptomatic, unless it is acute, due to the fact that blood will flow from the opposite side via the arterial circle of Willis. Branches of the arterial circle of Willis are classified into anteromedial, anterolateral, posteromedial, and posterolateral arterial branches.

FUNCTIONAL AND CLINICAL CONSIDERATION

The hypothalamus controls fluid and electrolyte balance, food intake, and reproductive cycle, and integrates visceral and endocrine functions. It is an important component of the limbic system through which emotions gain expression. It contains sympathetic (posterolateral) and parasympathetic (anteromedial) centers. Stimulation of the anteromedial part of the hypothalamus increases gastrointestinal motility and bladder contractions and decreases the heart rate, leading to peripheral vasodilatation. It contains centers for thermoregulation (the anterior center regulates heat dissipation, whereas the posterior center mediates activities that produce and conserve heat). On the other hand, stimulation of the posteromedial part produces mydriasis (dilatation of the pupil), increased heart rate, and peripheral vasoconstriction. The hypothalamus contains receptors that control feeding (feeding centers), which include the hunger center (lateral hypothalamic nucleus) and the satiety center (ventromedial nucleus), as well as centers for sleep–wake cycle (SCN and anterior hypothalamus), memory, and behavioral regulation (ventromedial nucleus). Behavioral regulation may encompass fear, rage, pleasure, sexual attitude, and reproduction.

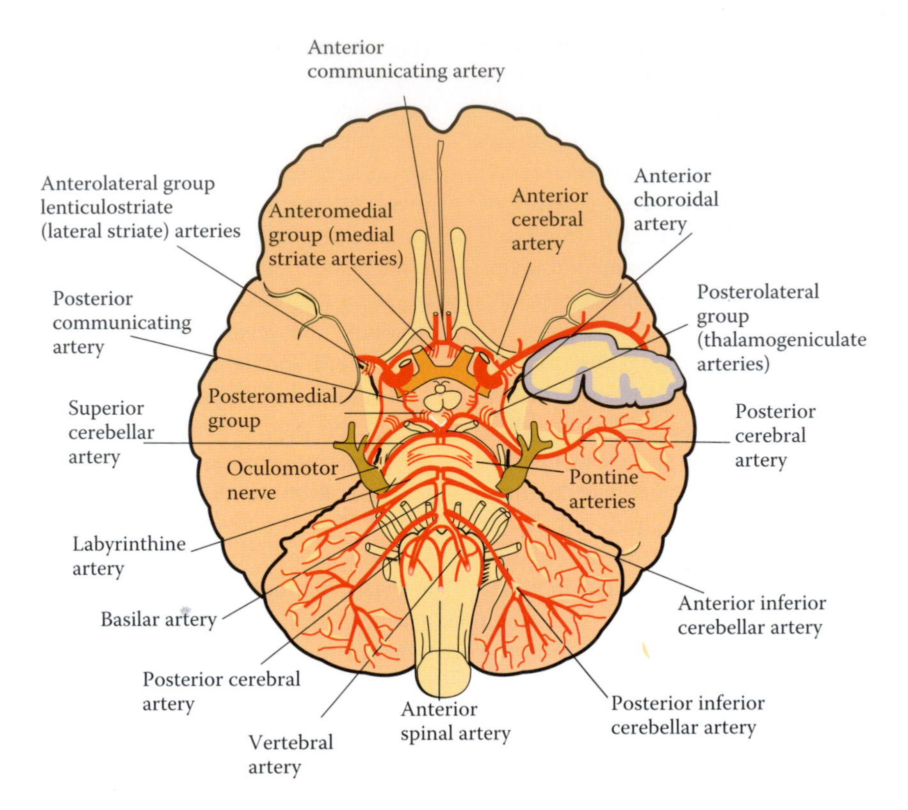

FIGURE 7.29 Diagram of the arterial circle of Willis, which encircles the mammillary bodies, tuber cinereum, and optic chiasma. Observe the main branches that emanate from the contributing arteries of this circle.

The anterior hypothalamus, including the magnocellular vasopressin neurons, is very sensitive to even the slightest changes in osmotic pressure of the blood and activates the neurons in the posterior pituitary to release vasopressin. This is facilitated by the vascular organ of the lamina terminalis and the subfornical organ, which lack brain barriers and allow angiotensin II to enter the anterior hypothalamus and influence the median preoptic nucleus. Additionally, the vagus nerve can exert influence over osmotic regulation through osmoreceptors in the portal vein. Release of vasopressin also occurs in response to emotional stress, pain, and nausea. Traumatic head injury that disrupts the supraopticohypophyseal tract, defect in vasopressin production, or disorders of osmoreceptor set point produces polyuria and polydipsia (diabetes insipidus). Restoration of function occurs following an injury because of the fact that axonal terminals establish new connections with the vasculature of the median eminence. Hypotension can stimulate drinking and the secretion of vasopressin via activation of receptors located in the left atrium (inhibitory) and aortic arch as well as the baroreceptor (excitatory) located in the initial part of the internal carotid artery. Afferents from these receptors converge onto the solitary nucleus via fibers of the vagus and glossopharyngeal nerves, respectively, and via the brainstem ascending monoaminergic pathways to the hypothalamic vasopressin neurons. These stimuli will be heightened if hypovolemia occurs. Vagus nerve input to the solitary nucleus conveys signals associated with gastric distension, which eventually suppresses feeding. Carotid body activation as a result of oxygen and carbon dioxide fluctuation stimulates ADH secretion.

Vasopressin neurons receive afferent fibers that convey osmotic sensation from the noradrenergic neurons of the brainstem that regulate the cardiovascular system, the subfornical organ, and the median preoptic area. GABA is inhibitory, whereas glutamate, acetylcholine, angiotensin, II and α_2-adrenergic neurons are excitatory to the vasopressin neurons. Despite the main secretions of the vasopressin and oxytocin neurons, other peptides are also secreted by the same neurons, such as galanin, dynorphin, thyrotropin-releasing hormone, and cholecystokinin.

Stimulation of the anterior and lateral hypothalamus as well as the paraventricular nucleus can decrease blood pressure and cardiac rate. In contrast, activation of the posterior hypothalamus and the tuberomammillary nuclei produces increase in blood pressure and cardiac rate. As discussed earlier, hypothalamic regulating factors control the secretion of the prolactin and gonadotropin. The latter is initially released at birth and then ceases to start again at puberty in a pulsatile and cyclical manner. Reproductive function can adversely be affected as a result of disruption of gonadotropin-releasing factors subsequent to stress of starvation or body mass. Because of the established link between reproductive function and gonadotropin-releasing factor, damage to the hypothalamus that disrupts these neurons can cause or inhibit sexual development or produce signs of precocious puberty depending on the extent and location of the pathologic process.

Generally speaking, the hypothalamus plays an important role in the activities that preserve the individual and species, such as searching out for a mate, food, and drink; home-building; and the rearing of young. However, the greater role of the neocortex and limbic system over the hypothalamus must be considered. Uterine contraction and contraction of the myoepithelial cells of the mammary gland are induced by the release of oxytocin at the nerve terminals in the posterior pituitary. Uterine contraction, induced through Ferguson reflex, which ceases by child delivery, is mediated by afferents from the cervix that travel through the inferior hypogastric (pelvic) plexus and through the anterolateral system via a relay in the brainstem to the hypothalamic oxytocin neurons. Milk ejection reflex is mediated by the intercostal nerves, conveyed to the anterolateral system, and delivered to the hypothalamic oxytocin neurons via a relay in the reticular formation.

The hypothalamus also influences feeding behavior, endocrine function, and autonomic activity of the gastrointestinal tract and thus balances the energy need and intake. The hypothalamus contains feeding and satiety centers that define the "set point" theory of weight control and respond to the changes in the level of glucose, free fatty acid, and insulin levels. Glucose-sensitive neurons are located in the arcuate and ventromedial nuclei that mediate appetite. The lateral hypothalamic area and ventromedial, arcuate, and paraventricular nuclei appear to be involved in food intake behavior. The lateral hypothalamic area consists of loosely organized nuclei that constitute the feeding center and receives olfactory input, an essential element of gustatory function and initiation of food intake that promotes appetite. A lesion in this area produces anorexia. The ventromedial nucleus, which acts as a "satiety center," receives visceral input from the solitary nucleus of the medulla.

Hyperphagia (excessive eating) associated with obesity is seen in Frölich syndrome (adiposogenital dystrophy) subsequent to bilateral lesion of the ventromedial nuclei. This rare syndrome is a childhood disorder characterized by obesity, retardation of growth, hypogonadism, and poor development of the genital organs associated with depressed secretion of gonadotropin-releasing factors. Since this condition is commonly associated with hypothalamic tumor, compression of the optic tract or chiasma produces visual deficits. When vasopressin-secreting neurons are affected, polyuria and polydipsia (excessive thirst) also develop. Stimulation of the ventromedial nucleus enhances glycogenolysis, gluconeogenesis, and lipolysis, whereas stimulation of the lateral hypothalamus produces the reverse effects. A lesion of the lateral hypothalamus causes aphagia or hypophagia, while stimulation of the same area prolongs food intake.

Leptin, ghrelin, angiotensin, insulin, cytokines, plasma concentrations of glucose, and osmolarity influence the function of the hypothalamus. Ghrelin is a hunger-stimulating peptide and hormone in the circulation and a counterpart of leptin. It is secreted mainly by the lining of the gastric fundus and the epsilon cells of the pancreas. Ghrelin levels increase and decrease before and after meals, respectively. This peptide/hormone stimulates growth hormone secretion by binding to specific receptors in the hypothalamus, anterior pituitary, as well as the nodose ganglia and the vagal terminals in the gastrointestinal tract. Ghrelin provides neurotrophic substances to the hippocampus and plays an important role in learning and memory by altering nerve cell connections and cognitive adaptation to the environment. The fact that ghrelin concentration is higher during the day and before meals may form a basis for the idea that learning can best be achieved during these times.

It increases food intake and, consequently, adipose tissue by activating the orexigenic NPY neurons in the arcuate nucleus. Ghrelin also activates the mesolimbic cholinergic–dopaminergic circuit that enhances the reward system associated with food, alcohol, and addictive drugs. In addition to the central signaling, ghrelin also exerts a peripheral modulatory effect on satiety by modifying the threshold of gastric vagal afferents, rendering them less sensitive to distension, leading to overeating. It has been suggested that ghrelin inhibits inflammatory processes, apoptosis, and oxidative stress, which brings to light its possible role in the treatment of inflammatory gastrointestinal disorders. Despite its therapeutic importance, ghrelin may play a role in enhancing gastrointestinal and pancreatic malignancy. It has been shown experimentally that ghrelin levels are significantly elevated under stressful conditions for prolonged period of time and that this change is associated with exacerbation of depression-like symptoms, such as lack of appetite and social withdrawal. Similarly, sleep deprivation is associated with obesity and high levels of plasma ghrelin, and as the sleep hours increased, the concentration of ghrelin. Short sleep duration is associated with high levels of ghrelin and obesity. An inverse relationship between the hours of sleep and blood plasma concentrations of ghrelin exists; as the sleep hours increase, ghrelin concentrations decrease, accompanied by a reduction in appetite.

Plasma levels of ghrelin appear to be lower in obese compared to nonobese individuals, with the exception of Prader–Willi syndrome–induced obesity. Cachectic cancer patients and those with anorexia nervosa show high levels of ghrelin compared to thin individuals who exhibit no pathologic manifestations. Since ghrelin concentration has been linked to an increase in the concentration of dopamine in the substantia nigra, its possible role in countering the onset of Parkinson's disease has been suggested.

Leptin is a protein hormone derived from adipose tissue that produces long-term appetite suppression. It acts on hypothalamic receptors by decreasing the activity of NPY in the arcuate nucleus and of anandamide, both feeding stimulants, and also promotes the synthesis of α-MSH, an appetite suppressant. Thus, it gives the feeling of satiety and strengthens resistance toward high-calorie foods. Destruction of NPY neurons induces anorexia, whereas disruption of α-MSH induces obesity by promoting food intake. The plasma level of leptin correlates proportionally with the extent of starvation (decrease of food intake) and not with overeating as well as with body fat. It acts on receptors in the mediobasal hypothalamus that balance energy. Leptin works in conjunction with amylin, produced by beta cells in weight control. Homozygous mutations of leptin genes heighten the desire for food, which leads to obesity and can be corrected by the administration of large doses of recombinant human leptin. Poor solubility and low circulating half-life and potency limited its use for therapeutic purposes.

Exogenous leptin increases vascular endothelial growth factor levels and thus enhances angiogenesis. Studies have shown that leptin modulates the immune response to atherosclerosis in obese patients with obesity. Leptin also promotes the surfactant expression of type II pneumocytes. Leptin levels increase and decrease during gestation and postpartum, respectively. Leptin may be responsible for hyperemesis gravidarum (severe morning sickness of pregnancy) and initiation of early menarche.

In summary, leptin shows reduction in response to short-term fasting and physical activity and increase in obese patients with obstructive sleep apnea or individuals under stress or who are sleep deprived. Restful sleep can restore normal levels of leptin. Changes in the level of estrogen and testosterone can also affect the plasma levels of leptin. Resemblance to interleukin-6 and the link between plasma level elevation and elevation of white blood cell count added strength to the idea that leptin has role as an inflammatory marker responding to adipose tissue–derived inflammatory cytokines. Prolonged elevation in leptin levels is associated with several metabolic and cardiovascular diseases as well as with the rise in insulin and cortisol levels. This may relate to the role of leptin in counteracting the overeating-induced cellular stress through inflammatory process. Several studies are being conducted on the possible therapeutic use of metreleptin, an analog of human leptin, in the treatment of diabetes and lipodystrophy.

Changes in body temperature initiate autonomic, endocrine, and metabolic responses through the thermoregulatory centers of the hypothalamus that enable conservation or dissipation of energy. Increased temperature induces heat loss through cutaneous vasodilation, sweating, and reduction in heat production. Reduced body temperature initiates changes such as vasoconstriction, piloerection, shivering, increased thyroid activity, and heat production. The thermosensitive receptors of the preoptic area of the hypothalamus respond to temperature changes of the blood that supplies the hypothalamus contained in the hypophyseal branches of the internal carotid artery. These receptors respond to invading bacteria though developing a fever as part of the cardinal signs of inflammation and also to cytokines including prostaglandins, which is evident in the antipyretic properties of aspirin. Temperature regulation is accomplished via activation of the anteromedial and posterolateral hypothalamic areas. The anteromedial area provides the mechanism for heat dissipation (sweating and dilation of the cutaneous vessels) through hypothalamic projections to the autonomic centers in the brainstem via the medial forebrain bundle (MFB) and, eventually, the preganglionic autonomic neurons at spinal cord level. There are also projections to the food intake center. Therefore, a lesion of the anteromedial hypothalamus subsequent to an expanding pituitary tumor produces hyperthermia. The posterolateral hypothalamic area governs the mechanism of heat production through shivering, piloerection, and vasoconstriction. A lesion in this area produces hypothermia.

The sleep–wake cycle is an integral part of the circadian rhythm and is homeostatically regulated. The sleep cycle is divided into several stages and types. In the first stage, the electroencephalogram (EEG) becomes synchronized with slow EEG activity as a transition from desynchronized in the waking state. In this phase, the person cannot dream but can easily be awakened. This phase is interrupted by three to four episodes of "paradoxical sleep," in which rapid eye movements (REMs) become prominent. In REM sleep, the cervical muscle tone is reduced, EEG waves are fast and have low amplitude, and the ability to attain a wakeful state becomes difficult. REM sleep is mediated by the adrenergic neurons of the locus ceruleus. Sleep is a balance between the ARAS and the hypnogenic centers scattered in the hypothalamus and brainstem, such as the anterior hypothalamus that inhibits the activities in the ARAS and induces EEG synchronization. Accordingly, a lesion that disrupts the anterior hypothalamus produces insomnia, in contrast to a lesion of the posterior hypothalamus that causes hypersomnia (excessive daytime sleepiness). Serotoninergic neurons of the mesencephalic raphe nuclei, which maintain prominent projection to the SCN, are thought to mediate the slow phase of sleep. The latter phase of the sleep is thought to be suppressed by CRH. Stress that produces sleep deprivation activates the hypothalamic–pituitary–adrenal axis and increases the CRH. Prostaglandin D2 acts on the sleep centers of the anterior hypothalamus, and E2 acts on the wake centers of the posterior hypothalamus. Similarly, administration of benzodiazepine, an anxiolytic, hypnotic, and sedative medication, into the anterior hypothalamus induces sleep, whereas administration of the same medication into the dorsal raphe nucleus produces wakefulness. Benzodiazepine acts on the GABAA/BZ-Cl− complex, and possibly on the cholinergic neurons of the basal forebrain, laterodorsal tegmental/pedunculopontine nuclei, caudal raphe and pontine reticular formation, as well as noradrenergic and glutamatergic neurons.

Distinct areas of the brainstem, thalamus, hypothalamus, and cortex mediate consciousness. Studies have shown that the ventrolateral preoptic area (VLPO) and the lateral hypothalamus, respectively, contain GABAergic/galaninergic (Gal) and hypocretin/orexin neurons (mediate global arousal state). Destruction of the VLPO induces insomnia, whereas disruption of the hypocretin/orexin neurons produces narcolepsy. The latter is seen in patients with encephalitis lethargica who exhibit sleep inversion in which their activity becomes restricted to night and they sleep during the day. VLPO inhibits while hypocretin/orexin neurons stimulate the adrenergic neurons of the locus ceruleus, serotoninergic neurons of the dorsal raphe, and histaminergic neurons of the tuberomammillary nucleus that constitute the ascending cortical activating system. There is evidence that alertness is also controlled by dopaminergic neurons of the ventral tegmental area (A10). The descending dopaminergic neuronal projections of A11 may mediate muscle weakness and possibly complete paralysis in the facial, masticatory, and ocular muscles seen in cataplexy (sudden and episodic loss of muscle tone associated with sleepiness and hallucination triggered by emotional conditions) and periodic limb movement. It has been suggested that serotonin (5HT) plays a significant role in the control of the hypoglossal nucleus, and disruption of this delicate relationship may be associated with obstructive sleep apnea.

These functions are mediated through afferent and efferent fibers.

HYPOTHALAMIC AFFERENTS

Hypothalamic afferents emerge from various areas of the CNS, including the hippocampal formation, septal area, amygdala, zona incerta, subthalamic nucleus, tegmental nuclei, periaqueductal gray matter, spinal cord, and retina. Other input originates from the tegmental nuclei, periaqueductal gray matter, nucleus ambiguus, solitary nucleus, superior colliculus, Edinger–Westphal and hypoglossal nuclei, retina, locus ceruleus, and raphe nuclei. Collaterals of the lemniscal systems and gustatory and visceral pathways also project to the hypothalamus. The hypothalamus also receives noradrenergic input from the locus ceruleus, serotonergic input from the raphe nuclei, and cholinergic afferents from the ventral tegmental nucleus. Some of the projections to the hypothalamus are direct and emanate from the ipsilateral and contralateral subthalamic nucleus, zona incerta, and dorsomedial nucleus of thalamus. The origin, course, and termination of the major afferents are discussed and illustrated in Figures 7.30 and 7.31.

The *hippocampal formation* maintains reciprocal connection to the hypothalamus via the fornix, a robust bundle of fibers that emanate from the cornu ammonis I and the subiculum. As it approaches the anterior commissure, the fornix receives fibers from the indusium griseum, cingulate gyrus, and septal area and eventually divides into precommissural and postcommissural columns. The precommissural column terminates in the lateral and preoptic hypothalamic nuclei, and septal area, while the postcommissural column projects to the medial mammillary nucleus with collaterals to the medial and lateral hypothalamic nuclei.

The *septal area*, *orbitofrontal cortex*, and *midbrain reticular formation* project to the hypothalamus via the MFB. The MFB is a loosely aligned group of fibers, which connects the lateral hypothalamus to structures that form the limbic system, such as the septal area and orbitofrontal cortex. It contains acetylcholine, epinephrine, norepinephrine, dopamine, histamine, serotonin, substance P, somatostatin, neurotensin, adrenocorticotropic hormone, and VIP.

The *amygdala* consists of corticomedial, basolateral, and central nuclei and sends fibers to the hypothalamus via the stria terminalis and the ventral amygdalofugal fibers. The stria terminalis originates from the corticomedial nucleus, runs parallel but lateral to the fornix and third ventricle, and projects to the anterior hypothalamic region. Fibers from the corticomedial nucleus also project to the preoptic nucleus

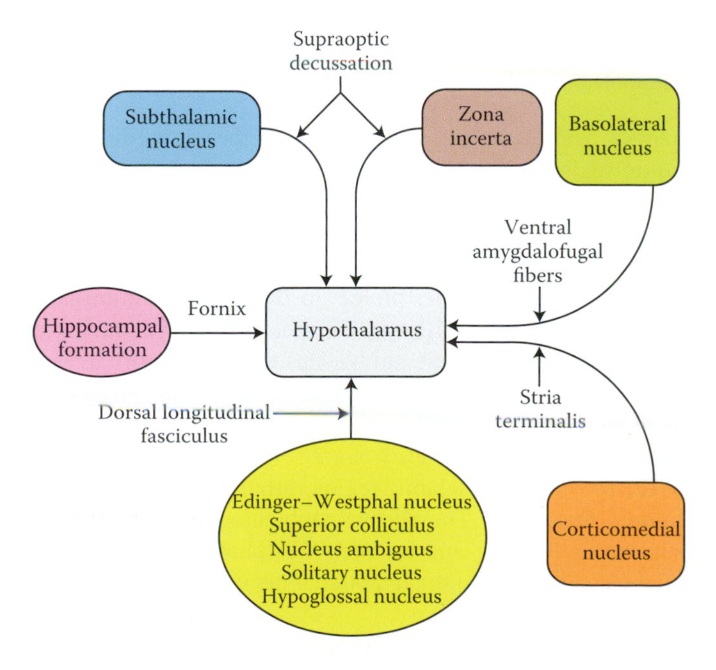

FIGURE 7.30 Schematic drawing of the afferent and efferent fibers of the hypothalamus. Some of these connections, for example, from the hippocampal formation, are bilateral.

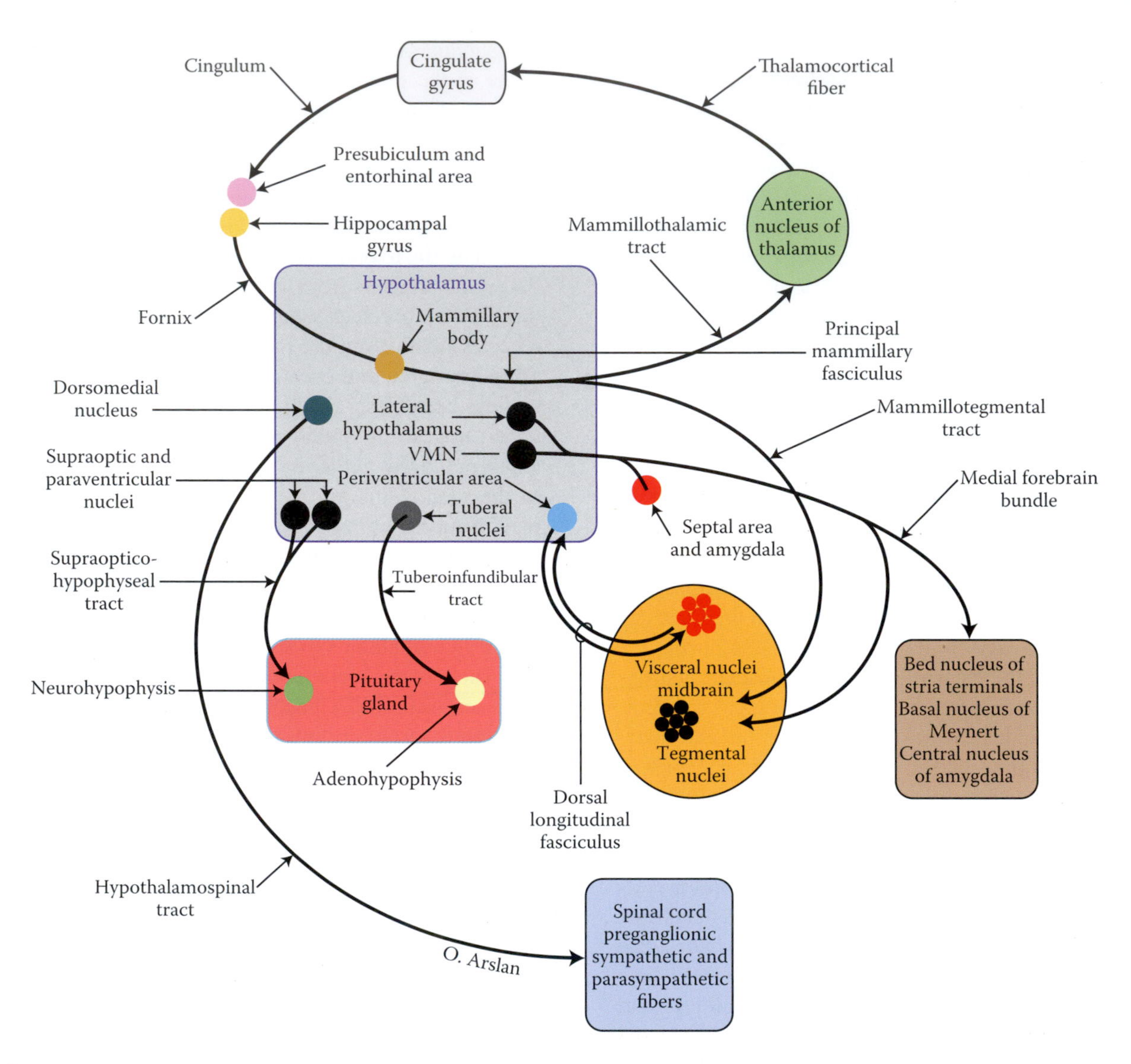

FIGURE 7.31 The various connections of the hypothalamus, some of which form feedback loops such as Papez circuit of emotion. VMN, ventromedial nucleus.

and the ventromedial nucleus. Input to the lateral hypothalamic region emanates from the basolateral nucleus and is carried by the ventral amygdalofugal fibers that pass ventral to the lentiform nucleus and dorsal to the optic tract before it reaches the hypothalamus.

The *tegmental nuclei* project to the lateral mammillary nucleus via the mammillary peduncle, conveying cholinergic input. The ventral tegmental nucleus also sends noradrenergic fibers to the paraventricular and supraoptic nuclei as well as to the lateral hypothalamus.

The *periaqueductal gray matter* of the midbrain projects to the medial hypothalamic region through the *dorsal longitudinal fasciculus*. The nucleus ambiguus, solitary nucleus, superior colliculus, and Edinger–Westphal and hypoglossal nuclei also project to the hypothalamus via the dorsal longitudinal fasciculus.

The *retina* sends visual projections to the SCN of the hypothalamus via the retinohypothalamic pathway. This connection is utilized in the control of activities of the pineal gland (diurnal regulation).

The *locus ceruleus* (adrenergic), *raphe nuclei* (serotonergic), and *mesolimbic dopaminergic neurons* also project to the hypothalamus. Adrenergic group A11 projects to the medial hypothalamic nuclei, whereas groups A13 and A14 convey impulses to the dorsal and rostral hypothalamus.

Additional afferents emanate from the septal nucleus and nucleus accumbens septi, which are derived from the medial olfactory stria. Some cortical input from the insular cortex also projects to the hypothalamus either directly or through the dorsomedial nucleus of thalamus. Direct corticohypothalamic fibers terminate in the mammillary nuclei and posterior hypothalamus.

HYPOTHALAMIC EFFERENTS

The diverse functions of the hypothalamus are maintained through multiple projections to the anterior nucleus of the thalamus, midbrain reticular formation, pituitary gland, and spinal cord (Figure 7.31). These connections are established via the following pathways:

The *principal mammillary fasciculus* emanates from the mammillary nuclei and divides into the mammillothalamic and mammillotegmental tracts. The mammillothalamic tract, as the name indicates, projects to the anterior nucleus of the thalamus and partly to the subthalamus through the lateral part of the hypothalamus. Mammillary projections to the anterior nucleus of the thalamus eventually spread to the cingulate gyrus. Through the cingulum, a cortical association bundle, information that reaches the cingulate gyrus will be delivered to the hippocampal gyrus. Since the hippocampal gyrus projects back to the mammillary nuclei via the fornix, the pathway that starts from the mammillary nuclei constitutes a feedback loop that involves the anterior nucleus of the thalamus, cingulate gyrus, cingulum, hippocampal gyrus, and fornix constitutes the *Papez circuit of emotion*, an element thought to be essential in regulating emotion. During its course to the tegmental nuclei of the midbrain, the mammillotegmental tract runs inferior to the midbrain and ventral to the MLF.

The hypothalamo-hypophyseal tracts

Hypothalamic influence on the pituitary gland is mediated by secretory neurons that produce neurosecretory peptides. Neurosecretory neurons have simple dendritic organization, give rise to axonal collaterals, and form synaptic boutons within the hypothalamus. They secrete several active peptides that are scattered on the dendrites, axons, axonal terminals and neuronal soma, exerting their influence through paracrine or autocrine mechanisms, and are affected by synaptic input and adjacent glial activity. Neurosecretory peptides are synthesized on the rough endoplasmic reticulum of the neuronal cell bodies, converted into neurosecretory granules via the Golgi complexes, and then carried along the axons via anterograde axoplasmic transport. The eventual release of secretory products by exocytosis and absorption into adjacent capillaries occurs as a sequel of a cascade of events that include neuronal electrical activity and generation of calcium influx at terminals. The regulatory mechanism of hypothalamus over the hypophysis is mediated via the supraoptic–hypophyseal and tuberoinfundibular tracts. The former pathway delivers neurosecretory neuronal axons from the magnocellular neurons of the supraoptic and paraventricular nuclei to the neurohypophysis that secrete ADH and oxytocin to the inferior hypophyseal capillary plexus, whereas the tuberoinfundibular tract transmits HRFs (peptides and amines) from parvocellular neurons of the tuberal nuclei to the median eminence and infundibulum and through the superior hypophyseal capillary plexus, and to the portal hypophyseal system to the capillary sinusoids within the adenohypophysis. In addition to their role in the neurohumoral mechanism, the tuberal nuclei are the major source of the dopaminergic HRFs to the adenohypophysis, preoptic, periventricular, dorsomedial, and ventromedial hypothalamic nuclei. Diffuse projections also originate from these nuclei that terminate in areas of the cerebral cortex that contain cholinesterase and typically undergo neurofibrillary degeneration in Alzheimer's disease.

Lesions of the hypothalamus produce a multitude of symptoms depending upon the location and the structures involved. To produce hypothalamic dysfunction, a lesion must be bilateral. One of the most common sources of hypothalamic dysfunction is craniopharyngioma, a tumor of Rathke's pouch. Posterior hypothalamic lesions are more likely to injure the major pathways associated with the hypothalamus. Bilateral damage to the medial forebrain bundle may result in deep coma, whereas lesions of the dorsal longitudinal fasciculus may disrupt the heat production and heat-dissipating mechanisms. Destruction of the supraoptic nuclei or supraoptico-hypophyseal tract may produce diabetes insipidus, a condition that is characterized by polydipsia (excessive drinking of water) and polyuria (copious urination). Involvement of the tuberal nuclei may lead to cessation of the hormone regulating factors and resultant sexual dystrophy. Destruction of the satiety center (ventromedial nucleus) produces obesity as a result of hyperphagia (excessive eating), whereas damage to the feeding center results in anorexia. Hypothermia and hyperthermia may result from destruction of the anterior and posterior nuclei of the hypothalamus, respectively. Selective destruction of the posterior hypothalamic region causes poikilothermia (a condition in which body temperature varies with the environmental temperature) and hypersomnia (excessive sleeping). Bilateral destruction of the medial region of the hypothalamus produces violent behavior in previously docile individuals.

Hypothalamic tumors are often slow growing, achieving a large size prior to the appearance of symptoms. Signs such as hydrocephalus, focal cerebral dysfunction, and hypopituitarism are often seen. Slow-growing tumors produce dementia, disturbances of food intake, and endocrine dysfunctions. Acute destructive processes of the hypothalamus may lead to coma or to autonomic disturbances. Diseases that affect the hypothalamus and pituitary gland may have both endocrine and nonendocrine manifestations.

Lesions of the intermediate hypothalamic area that destroy the mammillary bodies, fornix, and stria terminalis may produce signs of Korsakoff's syndrome. Marked anterograde amnesia (short-term memory loss) and preservation of intermediate and long-term memories characterize this syndrome. Consciousness usually is not altered, but affected individuals have a tendency to fabricate when responding to questions (compensatory confabulation). This syndrome is seen in chronic alcohol abuse associated with thiamine deficiency.

The *dorsal longitudinal fasciculus* consists of fibers, largely contralateral, that project from medial and periventricular areas of the hypothalamus to the central gray of the midbrain; accessory oculomotor, salivatory, and solitary nuclei; nucleus ambiguous; dorsal motor nucleus of the vagus; and facial nuclei transmitting impulses between neurons of the reticular formation and the hypothalamus.

Paraventricular and dorsomedial nuclei provide a *direct hypothalamospinal tract*, which terminates in the intermediolateral columns of the spinal cord. Other hypothalamic neurons that send impulses to the autonomic neurons are the atrial natriuretic peptide neurons of the dorsomedial nucleus, the α-MSH of the lateral hypothalamus, and dopaminergic neurons of the zona incerta.

Some cholinergic fibers are described as descending from the hypothalamus to the cerebellum via the superior cerebellar peduncle. Within this pathway, serotonergic fibers from the raphe nuclei also project to the hypothalamus.

The *ventromedial nucleus* projects via the MFB to the midbrain reticular formation, central nucleus of amygdala, bed nucleus of stria terminalis, and basal nucleus of Meynert. The *Septal area* and *amygdala* maintain reciprocal connections with the hypothalamus.

There also exists a diffuse and direct projection to large areas of the isocortex from the tuberal nuclei and the posterolateral hypothalamus. The neurons associated with the hypothalamocortical connection contain cholinesterase, mediate maintenance consciousness, and undergo neurofibrillary degeneration in Alzheimer's disease.

The hypothalamus shares embryological, morphological, and functional characteristics with the pituitary gland. The hormonal secretions of the pituitary gland are dependent upon the functional integrity of the hypothalamus. This is accomplished by the hypothalamic input to the medial eminence via the tuberoinfundibular tract, which eventually affects the secretion of the adenohypophysis, and by the hypothalamic neurons that secrete the ADH and oxytocin and are stored in the neurohypophysis.

PITUITARY GLAND (HYPOPHYSIS CEREBRI)

The pituitary gland (Figure 7.32) develops from Rathke's pouch (an ectodermal outpocketing of the stomodeum) and the infundibulum (a downward extension of the diencephalon). Rathke's pouch forms the adenohypophysis, the intermediate lobe, and the pars tuberalis.

> Remnants of the Rathke's pouch may give rise to craniopharyngioma, a common tumor in children that extends dorsally and involves the third ventricle, producing dwarfism, visual disturbances, and erosion of the sella turcica.

The diencephalic (neural ectodermal) part of the pituitary gland contains neuroglia and receives fibers from the hypothalamus. It consists of the adenohypophysis and the neurohypophysis. The cavernous sinuses flank this gland, which lies in the hypophyseal fossa of the sella turcica. It is covered partly by the diaphragma sella and is connected to the hypothalamus by the infundibulum.

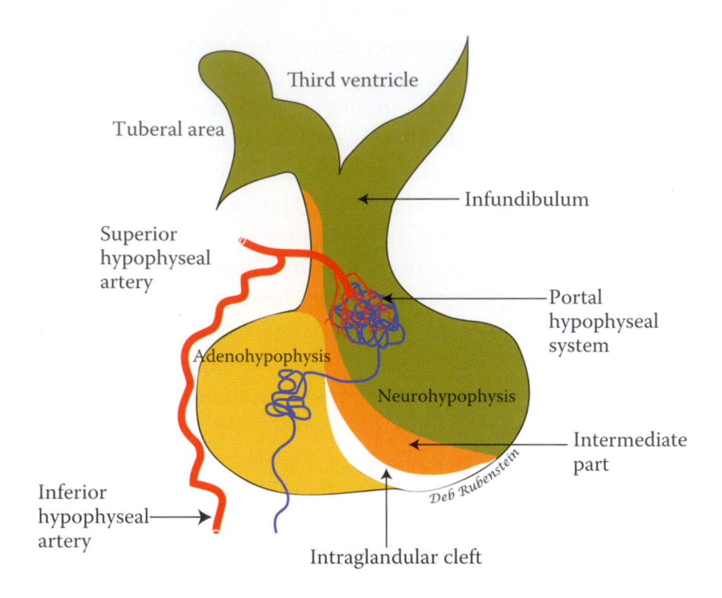

FIGURE 7.32 The adenohypophysis, neurohypophysis, infundibulum, and intermediate part of the pituitary gland. The formation of the portal hypophyseal system, an important link between the hypothalamus and adenohypophysis, is illustrated.

> Based on the above relationships, enlargement of the sella turcica, subsequent to development of a pituitary tumor, may be detected radiographically. The position of the pituitary gland above the sphenoidal sinus may also be utilized in surgical removal of pituitary tumors via transsphenoidal approach. Additionally, the location of the adenohypophysis posterior and inferior to the optic chiasma may account for disruption of the nasal retinal fibers as a sequel to an adenoma, and the development of bitemporal heteronymous hemianopsia (tunnel vision).

The pituitary gland is supplied by the superior and inferior hypophyseal branches of the internal carotid artery. The superior hypophyseal artery supplies the tuberal region, infundibular stalk, optic chiasma, and medial eminence. The inferior hypophyseal arteries form an arterial ring around the infundibular stem and provide blood supply to the lower infundibulum and the posterior lobe. These arteries establish numerous anastomoses and branch repeatedly, terminating into capillaries and capillary sinusoids in the medial eminence and the infundibular stem. These capillary sinusoids are the beginning of the hypophyseal portal system, which conveys blood to the epithelial tissue of the anterior lobe via primary and secondary capillary venous plexuses. The portal hypophyseal system forms two capillary beds, which drain the venous blood of the pituitary gland, carrying the HRFs from the medial eminence of the hypothalamus to the anterior lobe of the pituitary gland.

The *adenohypophysis* comprises the anterior lobe, intermediate part, and tuberal region. It secretes somatotropin (growth hormone), thyrotropin, prolactin or luteotropin, adrenocorticotropin, luteinizing hormone (interstitial cell–stimulating hormone), follicle-stimulating hormone, and melanocyte-stimulating hormone. The hormone regulating factors of the hypothalamic tuberal nuclei regulate the adenohypophysis. The tuberal nuclei project to the medial eminence through the tuberoinfundibular tract and regulate the glandular function of the adenohypophysis via the portal hypophyseal vessels.

Pituitary gland dysfunctions are associated with a variety of pathological conditions and syndromes. Some of these conditions primarily affect the adenohypophysis, while others may be confined to the neurohypophysis. These conditions are described below.

Sheehan's syndrome is a condition that exhibits persistent amenorrhea, asthenia (muscle weakness), visual disorders, episodes of hypotension, and altered consciousness. It is associated with hypopituitarism of the anterior lobe and is observed in individuals with postpartum hemorrhage and spasm of the infundibular arteries.

Adenoma of the anterior pituitary produces a variety of symptoms, depending upon the type of tumor, which may include acromegaly and/or gigantism (enlargement of the face, hand, and feet); impotence; amenorrhea (cessation of menses); galactorrhea; or Cushing disease. The latter is characterized by moon facies, obesity with prominent fat pad in the neck and shoulder, hypertension, renal calculi, and irregular menses.

Pituitary tumors most often arise from the anterior lobe and are classified, according to the degree of their endocrine activity, into endocrine-active and endocrine-inactive tumors.

Endocrine-active tumors are, for the most part, microadenomas and may be detected by radiographic imaging of the sella turcica. They may *increase secretion* of many hormones such as the growth hormone, producing acromegaly or gigantism. On the other hand, some tumors may result in *undersecretion* of hormones, producing, for example, dwarfisim (abnormally short body stature) in childhood, among other deficits.

Endocrine-inactive tumors, by contrast, become clinically significant only upon enlargement, thus compressing the adjacent nerves or brain tissue. These tumors may produce headache, visual disturbances, and extraocular motor palsies. Surgical removal of these tumors may be accomplished by transsphenoidal approach; however, extensive subfrontal or parasellar expansion of the tumor may require a transfrontal approach. Tumors that grow laterally or wedge under the optic nerve may necessitate craniotomy.

FUNCTIONAL AND CLINICAL CONSIDERATION

The *neurohypophysis* is comprised of the posterior lobe and the infundibular stem, and acts as a storage depot for vasopressin and oxytocin. Approximately half the volume of the neurohypophysis consists of axonal swellings; the largest are the Herring bodies, which may be as large as erythrocytes. *Herring bodies* provide a source of secretory granules and have a longer life span (more than 2 weeks). Vasopressin, also known as ADH, enhances water reabsorption by the distal convoluted tubules of the kidneys, whereas oxytocin causes milk ejection from lactating mammary glands and contraction of the uterine muscles. These hormones are secreted by the paraventricular and the supraoptic nuclei of the hypothalamus. During suckling, stimulation of the sensory nerve endings in the female nipple and areola activates the spinoreticular fibers, which subsequently relay the generated impulses to the supraoptic and paraventricular nuclei via the dorsal longitudinal fasciculus. The infundibulum primarily consists of fibers of the hypothalamo-hypophyseal tracts that project to the neurohypophysis.

Lesions of the posterior lobe may occur in skull fractures, suprasellar and infrasellar tumors, tuberculosis, vascular disease, or aneurysms. These lesions produce temporary signs of diabetes insipidus (due to lack of vasopressin), a condition that is characterized by copious and dilute urination and excessive thirst. The onset is generally sudden, and nocturia (excessive urination at night) is a presenting symptom in most patients.

EPITHALAMUS

The epithalamus occupies the dorsolateral part of the diencephalon and consists of the *pineal body* (epiphysis cerebri), *stria medullaris thalami*, *habenula*, and *posterior commissure* (Figures 7.33, 7.34, and 7.35).

The *pineal gland* (Figures 7.33, 7.34, and 7.35) is an endocrine gland, which is located rostral to the superior colliculi and posterior commissure and inferior to the splenium of the corpus callosum. It is connected to the habenula and the posterior commissure via the laminae of the pineal stalk. The latter contains the pineal recess of the third ventricle.

Functional and Clinical Consideration

This gland secretes melanocyte-stimulating hormones (lipotropic hormones), which play an analgesic role due to its endorphin content. It also secretes indolamines such as melatonin and associated enzymes (*N*-acetyltransferase and hydroxyindole-*o*-methyltransferase, which synthesizes melatonin from serotonin) that show sensitivity to the variations in diurnal light and circadian rhythms, and it is believed to influence the secretion of the gonadotropic hormones, maintaining a regulatory role in reproductive development.

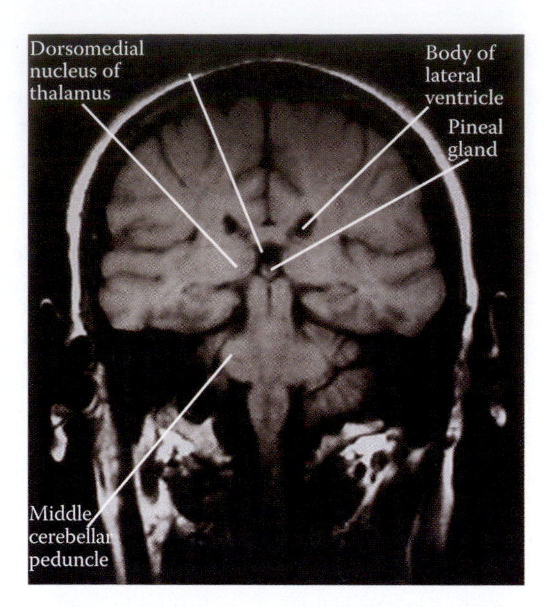

FIGURE 7.33 An inverted MRI scan showing the epithalamus with its main components, the pineal gland and habenula. Observe their relationship to the third ventricle.

Through these secretions, the pineal gland may exert a regulatory influence, modifying the activity of the pituitary, adrenal, and parathyroid glands, as well as gonads. Darkness activates the secretion of melatonin, inhibiting sexual development through a series of neuronal chains in the retina, hypothalamus, reticular formation, and spinal cord. Retinal input to the SCN of the hypothalamus is eventually conveyed to the reticular formation. Activation of the reticular formation produces excitatory effects on the sympathetic neurons

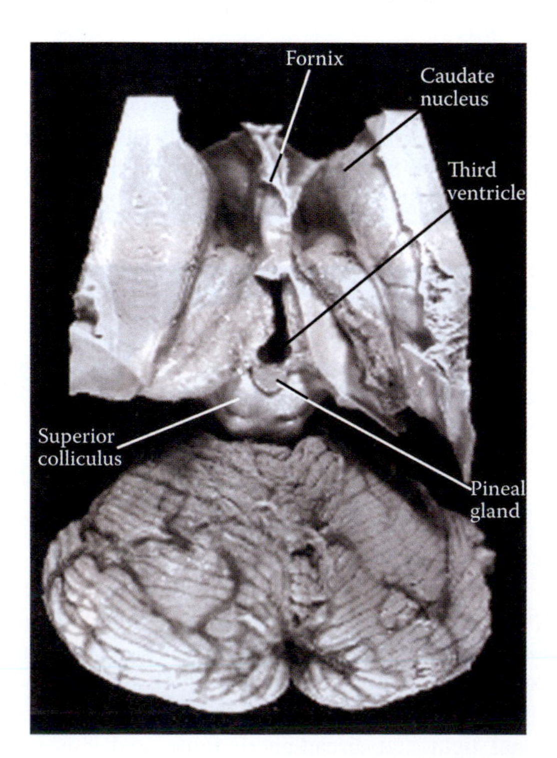

FIGURE 7.34 Dorsal surface of the diencephalon illustrates the epithalamus and its relationship to the third ventricle and tectum.

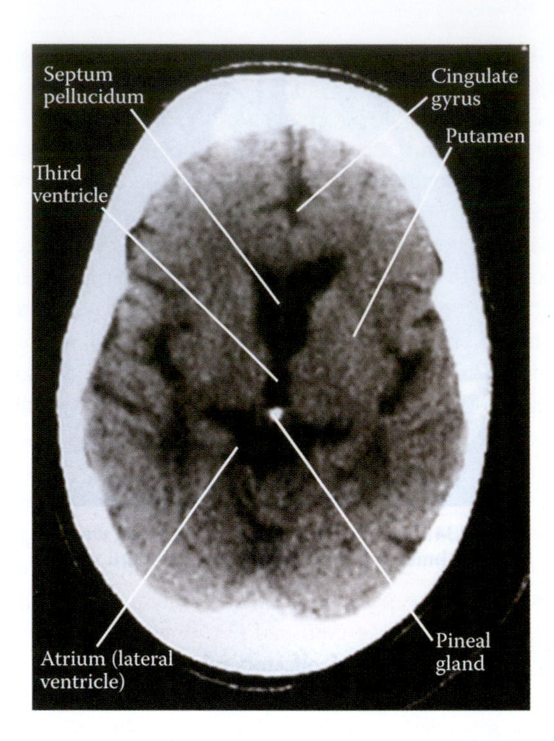

FIGURE 7.35 This computed tomography scan shows the pineal gland with calcareous concretion (brain sand) and the third ventricle.

of the intermediolateral column of the upper two thoracic spinal segments, via the reticulospinal tracts. Sympathetic neurons convey the secretomotor impulses from the reticular formation to the pineal gland through the *nervi conarii* (sympathetic postganglionic fibers), resulting in the release of catecholamines from the pinealocytes and subsequent receptor-mediated (β-adrenergic) increase of cyclic adenosine monophosphate (cAMP).

Bilateral lesions of the suprachiasmatic nuclei may abolish the rhythmic activities of the *N*-acetyltransferase and produce low levels of hydroxyindole-*o*-methyltransferase activity, resulting in disruption of the circadian rhythms associated with the sleep–wakefulness cycle and spontaneous motor activities that pertain to drinking and feeding.

PINEAL GLAND

The pineal gland also secretes norepinephrine and contains significant concentrations of hypothalamic peptides such as luteinizing hormone–releasing factor, thyrotropin-releasing factor, and somatostatin. Pinealocytes also contain tryptophan hydroxylase and aromatic amino acid decarboxylase, which are involved in the synthesis of serotonin. Increase of cAMP evokes augmentation in serotonin *N*-acetyltransferase activity that is followed by increased pineal glandular activity and production of melatonin and eventual inhibition of reproductive development.

The most common malignant tumor of this gland is germinoma (atypical teratoma), which is frequently seen in young males. Pinealoma (pineal gland tumor) may compress the tectum and the posterior commissure, producing signs of *Parinaud syndrome*, which is characterized by bilateral vertical gaze palsy, hydrocephalus, and loss of pupillary light reflex. Cysts or tumors associated with the pineal gland may compress the hypothalamus, resulting in obesity and hypogonadism.

Pineal (sand) concretions or corpora arenacea in the astrocytes were considered as important radiographic sites for detection of a brain shift–associated space-occupying intracranial mass on the contralateral side. Deviation of the gland from a midline position may be considered significant. However, it must also be remembered that due to the relatively large size of the right cerebral hemisphere, a normal pineal gland may slightly deviate to the left side.

STRIA MEDULLARIS

The *stria medullaris thalami* consists of axons that originate from the septal area, runs dorsomedial to the thalamus, and terminates in the habenular nuclei of both sides. It contains afferents to the habenular nuclei from the septal area that receives input from the amygdala, hippocampal formation, and anterior perforated substance.

HABENULA

The *habenulae* (Figures 7.33, 7.34, 7.36, and 7.37) are two triangular eminences comprised of the medial and lateral habenular nuclei, representing the sites of convergence of major limbic system pathways. The lateral habenular nucleus receives afferents from the globus pallidus, substantia innominata, tectum, lateral hypothalamus, prepyriform cortex, lateral preoptic area, basal nucleus of Meynert, midbrain raphe nuclei, olfactory tubercle, and pars compacta substantia nigra. Input from the prepyriform cortex, tectum, nucleus

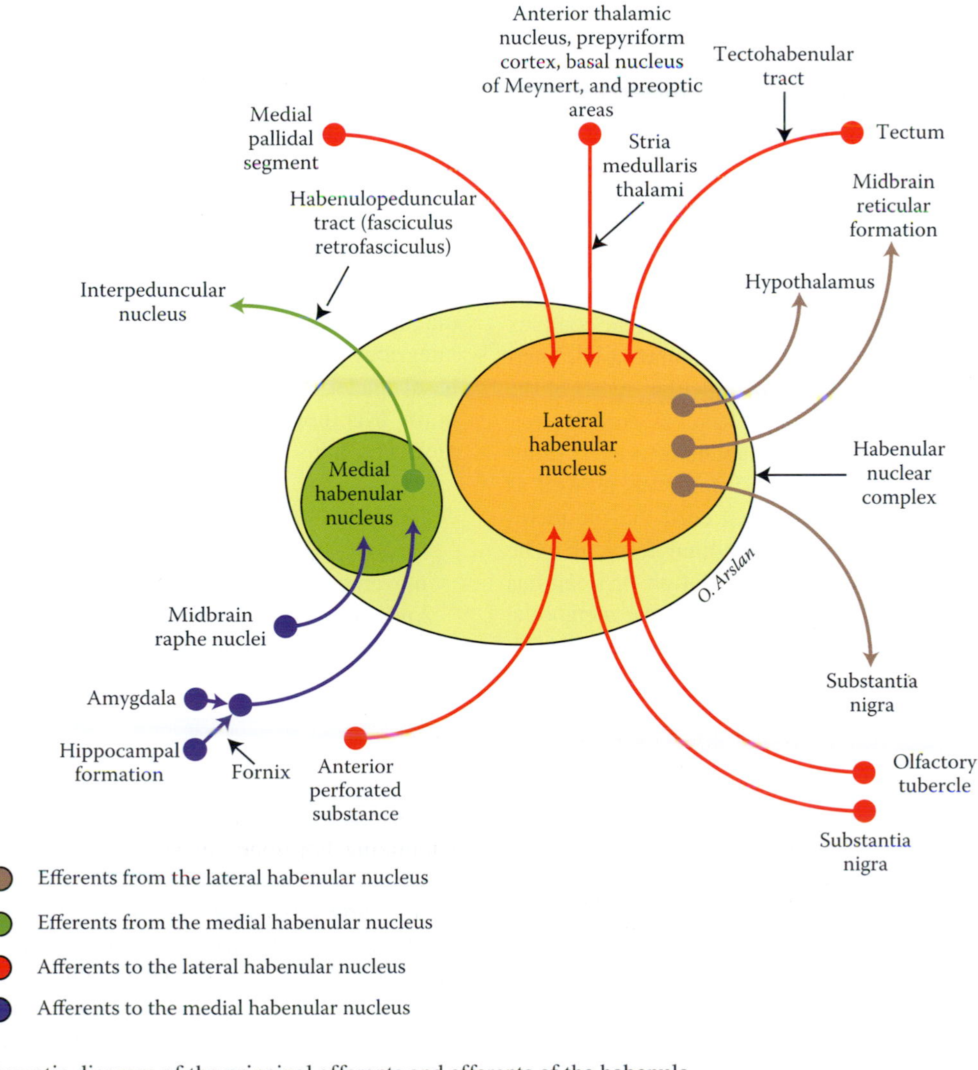

FIGURE 7.36 Schematic diagram of the principal afferents and efferents of the habenula.

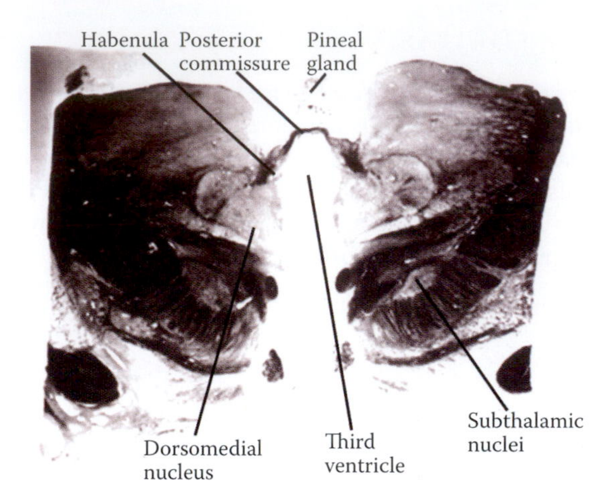

FIGURE 7.37 Section through the posterior commissure showing the pineal gland, habenula, and certain parts of the subthalamus.

basalis of Meynert, and septal and lateral preoptic area travels within the stria medullaris thalami. The *stria medullaris thalami* also contains fibers that transport neuromediators such as acetylcholine, norepinephrine, serotonin, GABA, luteinizing hormone–releasing factor (LHRF), somatostatin, vasopressin, and oxytocin. The lateral habenular nucleus projects back to the substantia nigra and also to the midbrain reticular formation and the hypothalamus.

The medial habenular nucleus, the smallest component of the habenular nuclear complex, receives fibers from the serotonergic neurons of the midbrain reticular formation and from the septofimbrial nucleus. The latter nucleus receives input from the amygdala and the hippocampal formation. Some adrenergic fibers also project to the medial habenular nucleus from the superior cervical ganglion of the sympathetic trunk. The output from the habenular nuclear complex primarily emanates from the medial habenular, a cholinergic nucleus that projects to the interpeduncular nuclei of the midbrain via the habenulopeduncular tract (fasciculus retroflexus). This pathway enables the habenula to exert influences upon the gastric and salivatory secretions as well as the preganglionic neurons of the spinal cord via the tecto-tegmento-spinal and the dorsal longitudinal fasciculi. The possible role of the habenula in the regulation of sleep has been suggested by some investigators. Extensive metabolic, thermal, and endocrine disturbances may accompany damages to the habenular nuclei.

POSTERIOR COMMISSURE

The *posterior commissure* consists of myelinated fibers that cross the midline caudal and within the posterior lamina of the pineal gland. It lies dorsal to the superior colliculi and the cerebral aqueduct. It contains the interstitial nuclei of the posterior commissure, nucleus of Darkschwitsch, and interstitial nucleus of Cajal, which also provide the axons that cross in the posterior commissure. Accessory oculomotor and pretectal nuclei project to the corresponding structures on the contralateral side through this commissure. The subcommissural organ is located ventral to the posterior commissure, which consists of ciliated columnar ependymal cells, transporting substances and secreting them into the cerebrospinal fluid from pinealocytes, capillaries, and neurons.

The posterior commissure may be damaged by pineal tumors. However, no known symptoms are associated with lesions of this commissure, though impairment of visual tracking movement has been reported.

SUBTHALAMUS

The subthalamus is a small area of the diencephalon that lies ventral to the hypothalamic sulcus and lateral and caudal to the hypothalamus. It contains the subthalamic nucleus, zona incerta, cranial portions of the red nucleus, and substantia nigra. It also contains the spinal and medial lemnisci and efferent fibers from the globus pallidus (ansa lenticularis, lenticular, thalamic, and subthalamic fasciculi) and cerebellum, which are destined to the thalamus. In mammals, it also contains the entopeduncular and prerubral nuclei. It continues caudally with the tegmentum of the midbrain, separated from the globus pallidus by the internal capsule (Figure 7.36).

The *subthalamic nucleus* (Figures 7.7, 7.12, and 7.37) is a biconvex nucleus that lies medial to the internal capsule, caudally overlying the substantia nigra. It receives input from the lateral segment of the globus pallidus, reticular formation, and the motor and prefrontal cortices. It projects to both segments of the globus pallidus via the subthalamic fasciculus and to the pars reticulata of the substantia nigra. It is also connected to the ipsilateral red nucleus, mesencephalic reticular formation, and zona incerta. The subthalamic nucleus integrates motor activities through its connections with the basal nuclei, substantia nigra, and tegmentum of the midbrain. It is thought to have an inhibitory influence upon the globus pallidus.

Lesions of this nucleus produce violent, uncontrollable movements of the contralateral extremities, a condition known as hemiballism, which is described with the extrapyramidal motor system.

The *zona incerta* is a thin layer of gray matter ventral to the thalamic fasciculus that extends with the reticular nucleus of the thalamus and the midbrain reticular formation. It contains dopaminergic neurons and receives cholinergic afferents from the midbrain tegmentum.

The *prerubral and entopeduncular nuclei* are located ventral to the zona incerta and adjacent to the posterior limb of the internal capsule. It receives fibers from the globus pallidus, which are destined to the midbrain reticular formation. These nuclei project through the central tegmental nuclei to the inferior olivary nucleus.

SUGGESTED READING

Andrews ZB, Erion D, Beiler R, Liu ZW, Abizaid A, Zigman J, Elsworth JD, Savitt JM, DiMarchi R, Tschoep M, Roth RH, Gao XB, Horvath TL. Ghrelin promotes and protects nigrostriatal dopamine function via an UCP2-dependent mitochondrial mechanism. *J Neurosci* 2009;29(45):14057–65.

Braak H, Braak E. Anatomy of the human hypothalamus (chiasmatic and tuberal region). *Prog Brain Res* 1992;93:3–16.

Cardoso ER, Peterson EW. Pituitary apoplexy: A review. *Neurosurgery* 1994;14:363–73.

Cezário AF. Hypothalamic sites responding to predator threats—The role of the dorsal premammillary nucleus in unconditioned and conditioned antipredatory defensive behavior. *Eur J Neurosci* 2008;28(5):1003–15.

Dermon CR, Barbas H. Contralateral thalamic projections predominantly reach transitional cortices in the rhesus monkey. *J Comp Neurol* 1994;344:508–31.

Fliers E, Unmehopa A. Functional neuroanatomy of thyroid hormone feedback in the human hypothalamus and pituitary gland. *Mol Cell Endocrinol* 2006;251(1–2):1–8.

Ghika-Schmid F, Bogousslavsky J. The acute behavioral syndrome of anterior thalamic infarction: A prospective study of 12 cases. *Ann Neurol* 2000;48:220–7.

Groenewegen HJ, Berendse HW. The specificity of the "nonspecific" midline and intralaminar thalamic nuclei. *Trends Neurosci* 1994;17:52–7.

Guridi J, Obeso JA. The role of the subthalamic nucleus in the origin of hemiballism and parkinsonism: New surgical perspectives. *Adv Neurol* 1997;74:235–47.

Kao YF, Shih PY, Chen WH. An unusual concomitant tremor and myoclonus after a contralateral infarct at thalamus and subthalamic nucleus. *Kaohsiung J Med Sci* 1999;15:562–6.

Krout KE, Loewy AD. Periaqueductal gray matter projections to midline and intralaminar thalamic nuclei of the rat. *J Comp Neurol* 2000;424:111–41.

Mai JK, Kedziora O, Teckhaus L, Sofroniew MV. Evidence for subdivisions in the human suprachiasmatic nucleus. *J Comp Neurol* 1991;305:508–25.

Power BD, Mitrofanis J. Specificity of projection among cells of the zona incerta. *J Neurocytol* 1999;28:481–93.

Romeo RD, Bellani R, Karatsoreos IN, Chhua N, Vernov M, Conrad CD, McEwen BS. Stress history and pubertal development interact to shape hypothalamic-pituitary-adrenal axis plasticity. *Endocrinology* (The Endocrine Society) 2005;147(4):1664–74.

Sato F, Parent M, Levesque M, Parent A. Axonal branching pattern of neurons of the subthalamic nucleus in primates. *J Comp Neurol* 2000;424:142–52.

Sumova A, Travnickova Z, Peters R, Schwartz WJ, Illnerova H. The rat suprachiasmatic nucleus is a clock for all seasons. *Proc Natl Acad Sci USA* 1995;92:7754–8.

Swaab DF. Sexual orientation and its basis in brain structure and function. *PNAS* 2008;105(30):10273–4.

Wicklund MR, Knopman DS. Brain MRI findings in Wernicke encephalopathy. *Neurol Clin Pract* 2013;3(4):363–4.

Wittkowski WH, Schulze-Bonhage AH, Bockers TM. The pars tuberalis of the hypophysis: A modulator of the pars distalis? *Acta Endocrinol* 1992;126:285–90.

Yañez J, Anadón R. Afferent and efferent connections of the habenula in the rainbow trout (Oncorhynchus mykiss): An indocarbocyanine dye (DiI) study. *J Comp Neurol* 1996;372:529–43.

8 Telencephalon

The telencephalon is a derivative of the lateral diverticula, which are interconnected by the median telencephalic impar. It is comprised of the cerebral hemispheres and contains the lateral ventricles. Examination of the cerebral hemispheres reveals an outer cellular gray cortical shell and an inner axonal white matter that contains the basal nuclei. The cerebral hemispheres are connected by the corpus callosum, consisting of the frontal, parietal, temporal, occipital, as the well the limbic and the central (insular cortex) lobes. Fibers that cross the white matter and connect areas within the same and opposite hemisphere form commissural and association fibers. Blood supply of the cerebral hemisphere is provided by the carotid and veretebrobasilar systems.

CEREBRAL HEMISPHERES

General Characteristics

The cerebral hemispheres are mirror-image duplicates that occupy the cranial cavity and are interconnected by the corpus callosum. The corpus callosum, which forms the base of the sagittal (interhemispheric) sulcus, lies ventral and partly caudal to the anterior cerebral vessels and the falx cerebri, consisting of the rostrum, genu, trunk, and splenium. Both cerebral hemispheres are composed of an outer gray matter thrown into folds (gyri) and an inner white matter, containing the basal nuclei. Each cerebral hemisphere is divided into the frontal, parietal, temporal, occipital, central (insular cortex), and limbic lobes via sulci and fissures (Figures 8.1, 8.2, 8.6, 8.7, 8.8, and 8.9). The frontal, parietal, temporal, and occipital lobes are interconnected by the genu, trunk, and splenium of the corpus callosum, respectively. The central sulcus separates the frontal and parietal lobes and contains the Rolandic branch of the middle cerebral artery (MCA). The lateral cerebral (Sylvian) fissure (sulcus) begins in the Sylvian fossa, contains the MCA, and demarcates the temporal lobe from the frontal and parietal lobes.

The lateral cerebral fissure gives rise to anterior, ascending, and posterior rami that divide the inferior frontal gyrus into orbital, angular, and opercular parts. The angular and opercular parts form the Broca's motor speech center. The floor of this fissure is formed by the insular cortex.

Each hemisphere contains a C-shaped cavity, called the lateral ventricle, filled with the cerebrospinal fluid (CSF).

Expansion of the frontal, parietal and occipital lobes in a rostrocaudal direction and downward displacement of the temporal lobe during development are responsible for the curved morphology of this ventricle, which parallels that of the caudate nucleus and fornix. Thus, it extends from the frontal lobe to the occipital lobe and then inferiorly to the temporal lobe. On each side, it consists of an anterior horn that continues with the body (central part), a posterior horn that extends into the occipital lobe, and an inferior horn that curves and continues within the temporal lobe. They communicate with the third ventricle via the interventricular foramen of Monro, which is bounded rostrally by the fornix and caudally by the anterior end of the thalamus. Within the frontal lobe, the anterior horn extends inferior and caudal to

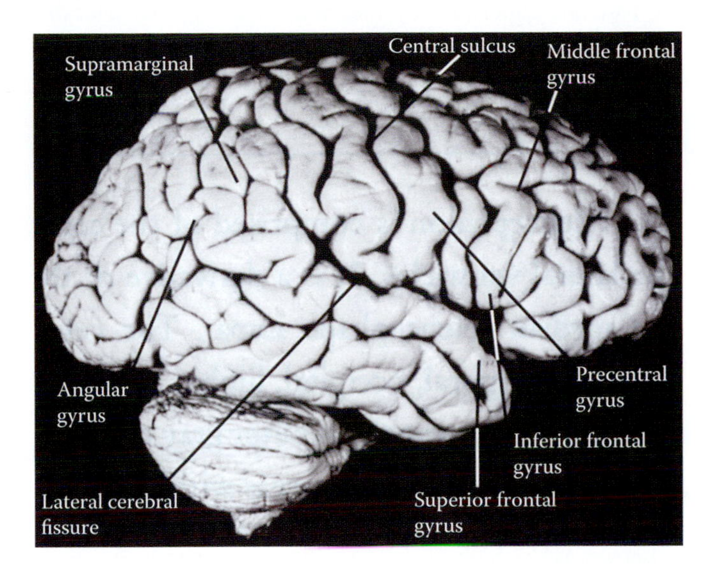

FIGURE 8.1 Photograph of the lateral surface of the brain showing prominent gyri. Lateral cerebral fissure and central sulcus are also shown.

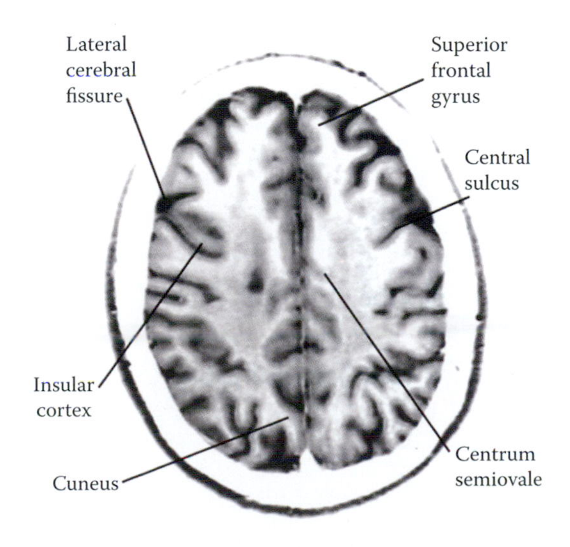

FIGURE 8.2 In this inverted MRI scan, cerebral cortex and central white matter are shown.

the corpus callosum and superior and lateral to the caudate nucleus, separated from each other by the septum pellucidum, which is crossed posteriorly by the fornix. This horn is bounded rostrally by the genu of the corpus callosum and the rostrum, superiorly by the callosal trunk, and laterally and inferiorly by the caudate nucleus. The central part (body) extends from foramen of Monro to the splenium of the corpus callosum. The floor of the central part is formed by the thalamus and caudate nucleus, which are separated. The caudate nucleus and thalamus also contribute to the lateral wall of the central part, separated *from each other* by the stria terminalis and the thalamostriate vein. Further inferiorly and medially, the fornix is separated from the thalamus by the choroidal fissure, which is occupied by the choroid plexus. The central part continues with the posterior horn in the occipital lobe and inferior horn in the temporal lobe. The *posterior horn* extends inside the occipital lobe medial to the calcarine fissure, with the latter forming a visible prominence on its medial wall known as the calcar avis. Above the calcar avis, a second protrusion becomes visible as a result of the medial course of the splenial fibers of the corpus callosum, which is known as the bulb of the posterior horn. The lateral wall and the roof of this horn are formed by the tapetal fibers of the trunk of the corpus callosum that separate this horn from the optic radiation. The transition zone between the posterior and inferior horns is termed the collateral trigone (Figures 8.3, 8.4, and 8.5).

The tail of the caudate nucleus continues in the floor of the posterior horn and the roof of the inferior horn. The amygdala, which is attached to the tail of the caudate nucleus, lies immediately rostral to the tip of the inferior horn. The inferior horn, the largest part, curves downward from the central part and extends laterally around the pulvinar and then rostrally ends near the temporal pole. The floor of this horn is formed by the collateral eminence (formed by the collateral sulcus), hippocampal gyrus, fimbria of the fornix, dentate gyrus, and choroid plexus. The roof and the lateral wall of this horn are formed by the tapetal fibers that emanate from the trunk of the corpus callosum. However the tail of the caudate nucleus, stria terminalis, and fimbria of the hippocampus also contribute to the roof of the inferior horn. Continuation of the choroid plexus into this horn is maintained via the choroid fissure between the stria terminalis and fimbria of the hippocampus. The fimbria of the hippocampus continues as the alveus and then with the crura of the fornix. This plexus is

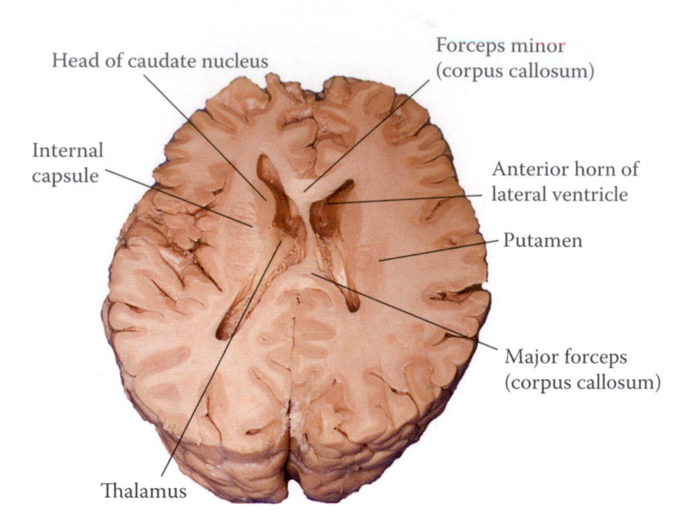

FIGURE 8.4 Horizontal section of the brain showing the location of the caudate nucleus and thalamus in relation to the anterior horn and central part of the lateral ventricle. The transition between the atrium and inferior horn of the lateral ventricle is visible.

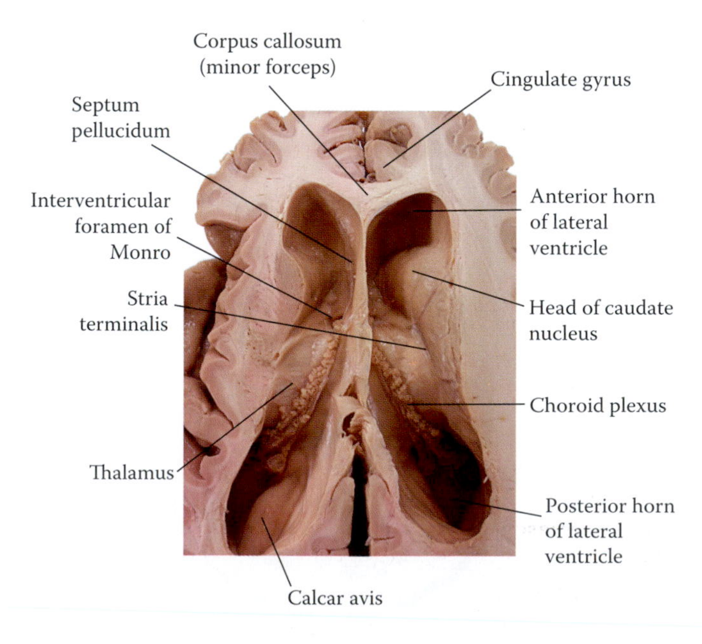

FIGURE 8.3 Horizontal section of the brain illustrating the boundaries of the anterior and posterior horns of the lateral ventricle and its central part. The connection of the lateral and third ventricles is maintained via the interventricular foramen of Monro.

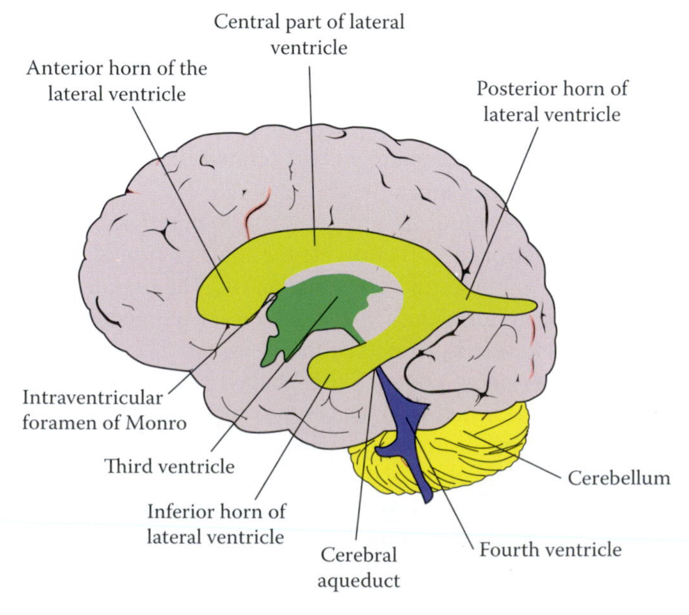

FIGURE 8.5 Diagram of a lateral view of the ventricular system.

supplied by the anterior and posterior choroidal arteries that originate from the internal carotid and posterior cerebral arteries, respectively.

FRONTAL LOBE

The frontal lobe (Figures 8.1, 8.2, 8.7, 8.8, and 8.9) lies rostral to the central sulcus and superior to the lateral cerebral (Sylvian) fissure. It occupies the anterior cranial fossa, superior to the olfactory bulb and tract, as well the cribriform plate of the ethmoid bone. It contains the superior and inferior frontal sulci, which separate the superior, middle, and inferior frontal gyri. Medially, the cingulate sulcus separates the medial part of the superior frontal gyrus (medial frontal gyrus) from the cingulate gyrus. The orbital gyri on the inferior surface of the frontal lobe are demarcated from the rectus gyrus by the olfactory sulcus that contains the olfactory tract. This lobe encompasses several functionally distinct areas including the primary motor (Brodmann area 4), the premotor (Brodmann area 6), supplementary motor (Brodmann 6), prefrontal (Brodmann areas 9, 8, 12, and 46), and anterior cingulate (Brodman areas 24 and 32) cortices. These frontal lobe areas are associated with certain gyri. The primary motor cortex is located in the precentral gyrus (Brodmann area 4) while the frontal eye field (Brodmann area 8) is contained in the middle frontal gyrus, which is responsible for conjugate deviation of the both eyes to the opposite side. The frontal lobe also contains Broca's motor speech center (Brodmann areas 44 and 45) in the inferior frontal gyrus. Within Broca's motor speech center, linguistic rules, grammar (syntax), and the template for phonation are formed, and melody and rhythm are regulated. The prefrontal cortex (PFC) occupies the area rostral to premotor cortex.

As discussed earlier, numerous gyri and cortical areas constitute the frontal lobe, including the precentral, superior, middle, and inferior frontal gyri and the premotor, prefrontal, and supplementary cortices. The *precentral gyrus* (Brodmann area 4) lies between the central and the precentral sulci and constitutes the primary motor cortex for the entire body, with the exception of the lower extremity. A distorted somatotopic organization of the body in this gyrus is known as the motor homunculus. Neurons of the precentral gyrus contribute to the formation of the corticospinal and corticobulbar fibers and maintain reciprocal connections with the ventral lateral nucleus of the thalamus (Figure 8.6).

> Damage to the motor cortex, as a result of trauma, tumor, hemorrhage, or thrombosis of the MCA, produces signs of upper motor neuron (UMN) spastic palsy (spasticity, hyperreflexia, Babinski sign, clonus), which are confined to the upper extremity, trunk, and head region (see Motor Neurons, Chapter 20).

Above the superior frontal sulcus lies the *superior frontal gyrus*, which continues medially with the medial frontal gyrus. The middle frontal gyrus, as discussed earlier, contains the frontal motor eye field (Brodmann area 8), which projects to the contralateral paramedian pontine reticular formation (PPRF). The PPRF, in turn, regulates conjugate horizontal eye movements through the ipsilateral abducens nucleus, which innervates the lateral rectus and controls the neurons of the oculomotor complex. This ocular movement is not regulated by the visual stimuli.

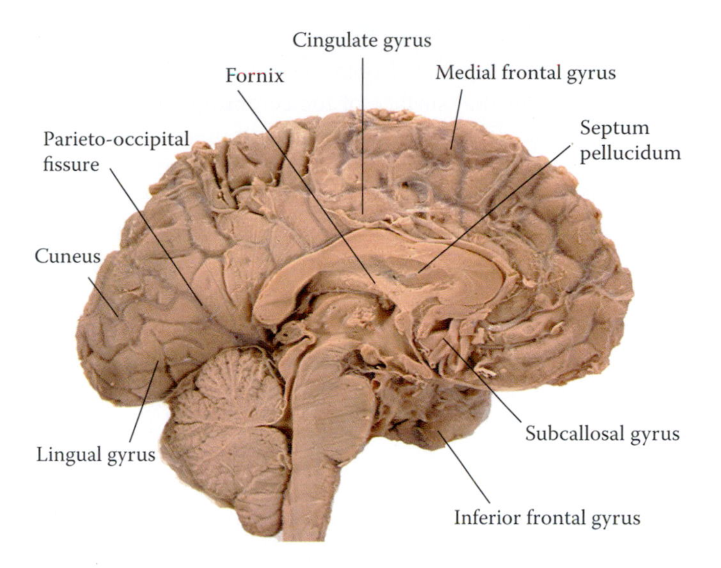

FIGURE 8.6 Midsagittal section of the brain showing the medial surface of the frontal, parietal, and occipital lobes.

> Frontal lobe seizures may activate the frontal eye field and produces conjugate eye movement to the opposite side. Damage to this area produces tonic deviation of both eyes to the side of the lesion (as if the patient is pointing to the affected side of the frontal lobe). This is due to the fact that a lesion in one frontal eye field leads to the unopposed action of the contralateral eye field. This ocular deficit gradually attenuates, and in a matter of hours or days, movement of the eyes as far as the midline becomes possible again. However, free gaze movements in all directions will be regained much later, with nystagmus being the final diagnostic indicator. The *inferior frontal gyrus* is separated from the middle frontal gyrus by the inferior frontal sulcus and is subdivided into orbital, triangular, and opercular parts. The orbital part lies anterior to the anterior ramus of the lateral cerebral fissure; the triangular part lies between the anterior and ascending rami of the fissure, whereas the opercular part is lodged posterior to the ascending ramus. Both the opercular and triangular parts of this gyrus form the Broca's motor speech center. Damage to Broca's center in the dominant hemisphere produces expressive aphasia, a condition that will be discussed in detail later. The *premotor cortex* corresponds to Brodmann areas 6 and occupies the area immediately rostral to the motor cortex, which

includes the superior frontal gyrus, which extends toward the medial surface of the cerebral hemisphere. Microscopically, it contains giant pyramidal cells of Betz, resembling the cytoarchitecture of the motor cortex. It is divided into area 6aα, which lies immediately rostral to the precentral gyrus (area 4); area 6aβ, which includes the superior frontal gyrus; and area 6b, which lodges the area at the lower ends of the precentral and postcentral gyri adjacent to the face area of the sensory and motor homunculus. A portion of 6aβ in the medial side of the brain belongs to the supplementary motor cortex. Most of the afferents to this cortical area are derived from the ventral anterior thalamic nucleus.

While stimulation of 6aα produces a motor response similar to that produced by the precentral gyrus, stimulation of Brodmann area 6aβ produces motor activities, which are characterized by attentive or orientative movements such as turning the head and eyes toward the contralateral side. Coordinated movements in the muscles that are derived from the branchial arches (facial, masticatory, laryngeal, and pharyngeal muscles) occur as a result of stimulation of area 6b. Isolated unilateral lesion of the entire cortical area 6 has been shown to result in contralateral motor apraxia (inability to perform familiar motor activity in the absence of any detectable motor or sensory deficits) and transient pathological grasp reflex. A bilateral lesion causes hypertonia in the muscles of the upper extremity. Paralysis or paresis is not seen in these lesions. Ablation of the motor and premotor cortices can cause spastic paralysis.

Prefrontal Cortex

The *PFC* is part of the frontal cortex rostral to the premotor cortex, a well-developed area in the human. It includes Brodmann areas 8, 9, 10, 12, 45 and the orbital gyri (Brodmann area 46). The PFC is considered a "silent" cortex, as it remains mute with electrical stimulation. It consists of dorsolateral (areas 8, 9, and 46), and ventrolateral (areas 12 and 45) portions as well as the orbitofrontal cortex, exhibiting differences relative to size, distribution, and density of their neurons. These areas within the PFC are well interconnected, allowing exchange and distribution of information. As a sensory multimodal converging cortical area (somatosensory, visual, auditory), it maintains connections with other cortical (occipital, temporal, parietal) and subcortical areas as well as areas that regulate reward, memory, and emotional responses such as mesencephalic and limbic structures. It establishes reciprocal connections with the dorsomedial nucleus of the thalamus. It is also connected to the temporal cortex via the uncinate fasciculus and to the parietal and occipitotemporal association cortices. Areas 8, 12, and 45 maintain multimodal connections, while areas 9, 12, and 46 are bimodal. It must be emphasized here that PFC is interconnected with association cortices and not with the primary sensory cortex. One area that deserves particular attention is the mechanism by which the PFC modulates memory, motivation, and emotion. This is accomplished by direct and indirect connections of the orbitofrontal part of PFC to the olfactory and gustatory cortices, as well as the hypothalamus, amygdala, and hippocampal gyrus.

In order for the PFC to influence behavior and motor activities, it must establish connections with certain areas of the frontal cortex. The dorsolateral area is interconnected to the supplementary motor and premotor cortices but not the primary motor cortex. It is also connected to the frontal eye field (area 8) rostral part of the cingulate gyrus as well as the cerebellum and the superior colliculus. With aging, the dorsolateral part of PFC, which oversees cognitive processing, shows a reduction in dendritic arborization and neuronal density. This causes cognitive activation that correlates with the decline in memory that occurs with aging. The PFC also influences stereotyped movements through projections to the striatum, which conveys the received information to the globus pallidus. Information received by the globus pallidus (GP) projects primarily to the VL nucleus of the thalamus, which, in turn, sends the impulses to the primary cortex. It has been suggested that basal ganglia may serve as a pathway for projections from the mesencephalic ventral tegmental areas to reach the PFC.

Through these connections, the PFC acts as an integrator of diverse sensory input, which eventually will be used to regulate cognition and conduct a goal-directed behavioral pattern. The PFC is important in the establishment of emotional response, programming, and intellectual functions through a controlled process. This includes controlling of thoughts; coordinated, sustained, willful, and socially acceptable behavior; and managing and pursuing long-term goals. However, these functions require complex multimodal sensory input and working memory. In order to perform nonreflexive functions a mechanism that places goal-related information at the disposal of the higher cortical PFC must be established through extensive and diverse interaction with other brain regions relative to goals and means of these actions. Selectivity in processing and storage of information enables proper and timely completion of the task yet adds a limitation to our ability to conduct multiple tasks simultaneously. Irrespective of these facts, behavior controlled by the PFC remains selective as to the environment, location, and anticipated tasks. It also exercises executive hierarchy over (inhibiting or enhancing) activities at subcortical levels.

Damage to the PFC has been documented in the case of Phineas Gage, who sustained a trauma to his head by an iron bar that pierced his skull and damaged the PFC. After surviving this horrible accident, as a skilled railroad foreman, he became disorganized, impulsive, aggressive, and easily irritable; was unable to control his mood; lacked concern for his position; and lost the ability to plan and execute tasks. Despite this horrible injury, his memory, speech, and intelligence remained intact. Prefrontal lobotomy is a surgical procedure, although very rarely used, that involves bilateral removal of the prefrontal cortices. It is utilized to modify the behavior of psychotic individuals and to relieve intractable chronic pain that is unresponsive to conventional analgesics. Patients who have undergone this type of operation no longer complain of pain and appear to be oblivious to it. They exhibit emotional lability and superficial affect. Alterations in personality, disposition, drive, and outlook are also prominent. Patients become disinhibited, lack initiative, and become less creative. They display an irresponsible attitude and seem indifferent and apathetic and have difficulty in planning and organizing their behavior, with no changes in intelligence. Lobotomized patients exhibit lack of restraint, absence of hostility, and boastful behavior. Reasoning, logical thinking, problem-solving abilities, and working memory may also be impaired. Considerable loss of drive and ambition is observed. These deficits indicate that the affected individual will be able to operate complex behavior through subcortical centers but loses the ability to coordinate these behaviors to produce reasonable actions that match the goals and associated limitations. Disinhibition, selective attention, cognition, and other executive dysfunction can be evaluated by Stroop test, which entails asking the patient to read the name of the color of a word written in black that defines another color (blue). The patient will read the written word instead of saying the color, indicating that he/she continues to conduct a previously learned behavior because of the fact that the request elicits a reflex-based response outside the control of PFC and within the domain of the subcortical areas.

Recent studies conducted on patients with depressive diseases have indicated that transcranial electromagnetic stimulation of the PFC, in which bouts of magnetic waves are passed through the brain, may be effective in the treatment of certain depressive illnesses. Corollary to this phenomenon is the depressive syndromes that are unresponsive to cyclic antidepressants exhibited by patients with infarction of the orbitofrontal branch of the MCA.

Lack of attention, indifference, and apathy seen in patients with PFC lesions may mimic manifestations of patients with attention deficit hyperactivity disorder (ADHD). This commonly diagnosed chronic childhood neurobehavioral disorder is characterized by significant attention deficits or restlessness and impulsiveness, or their combination. One aspect of this deficit may predominate, leading to classification into attention deficit hyperactivity disorder predominantly inattentive (ADHD-PI), where lack of attention prevails; attention deficit hyperactivity disorder hyperactive, impulsive type (ADHD-HI), in which impulsiveness and hyperactivity are prominent; and attention deficit hyperactivity disorder combined type (ADHD-C), where a combination of the previous two criteria are met. It is more commonly diagnosed in boys than girls and tends to be a chronic condition that continues into adulthood. Misdiagnosis is common as the dysfunctions seen in this condition are not clearly demarcated, and some of them are seen in one form or another in other neurobehavioral disorders, such as bipolar disorder, major depressive disorder, and obsessive–compulsive disorder (OCD), among many others. However, the length of the presentations and their extent compared to healthy classmates may help in the differential diagnosis. Patients experience academic difficulties and are perceived to be disruptive. In ADHD-PI, children become easily bored, amnestic, and easily confused; have a tendency to quickly alternate between various activities; and experience difficulty maintaining concentration, following step-by-step directions, or completing homework. In ADHD-HI, children are more talkative and always in motion; frequently fiddle; have a tendency to make inappropriate remarks; exhibit poor social skills; and are impatient and careless about the consequences of their action. As indicated earlier, ADHD-C patients show a combination of the above manifestations. It appears that executive functions that are associated with PFC, which regulates working memory, planning and initiation of action, vigilance, and social inhibition, are compromised. A reduction in brain volume, particularly the PFC of the left cerebral hemisphere, has been shown in individuals with ADHD. Others have suggested a possible role of the dorsal anterior cingulate gyrus and the neostriatum. Studies have shown a correlation between developmental dyspraxia (difficulty planning and coordinating movements, handwriting difficulty) including verbal dyspraxia (difficulty producing and sequencing speech sounds and poor language development) and ADHD. Children may experience depression, speech delay, and mood disorders, which can lead, in the absence of early intervention, to substance abuse, major depression, and possible criminal behavior. There is a strong genetic component to this condition.

Due to the important role that dopamine plays in the control of psychomotor activity—attention, motivation, and inhibition which are implicated in ADHD, frontostriatal dopamine has been extensively studied. Reduction in striatal dopamine in patients with ADHD, which is synthesized by the substantia nigra and ventral tegmental area and delivered by the nigrostriatal projection, and the efficacy of striatal dopamine transporter blockers such as methylphenidate in the treatment of this condition are consistent findings that support dopamine's role in the development of ADHD. Genes considered closely linked to this condition include dopamine active transporters genes (DAT1), DR4, DR5, and latrophilin-3 (LPHN3), as well as beta decarboxylase and monoamine oxidase A (MAO-A). These dopamine transporter genes are linked to the dopamine reuptake mechanism at the synaptic levels. MAO-A deaminates norepinephrine, epinephrine, serotonin, and dopamine. Thus, inhibition of MAO-A by monoamine oxidase inhibitor is the basis for therapy of clinical depression and anxiety.

The *supplementary motor area* (Brodmann area) primarily refers to the medial frontal gyrus, which forms the medial extension of the superior frontal gyrus above the cingulate gyrus. This area exhibits a small motor homunculus that functions independently from the primary motor cortex. Stimulation of this region results in bilateral synergistic movements of a postural nature of the axial and appendicular musculature; rapid uncoordinated movements; as well as a complex pattern of motor activities.

Frontal lobe hematoma can occur as a result of rupture of the MCA or its branches, producing *abulia*, which is characterized by a lack or impairment of verbal spontaneity and initiative as well as the inability to perform volitional acts or make decisions. Patients appear sedentary and withdrawn. Expansion of the hematoma to involve the frontal eye field may produce tonic conjugate deviation of both eyes toward the side of the lesion. It usually resolves after a few days as the intact contralateral frontal gaze center compensates for the deficit. Involvement of the precentral gyrus in this hematoma produces paralysis of the opposite half of the body.

Occlusion of the MCA can produce coma and forced deviation of the eyes toward the occluded side when the frontal eye field is involved. Occlusion of the *anterior cerebral artery* (ACA) proximal to the callosomarginal branch may cause a large infarction on the medial surface of the frontal lobe, resulting in intellectual deterioration, apraxia, visible primitive reflexes (such as sucking reflex), incontinence, and possible aphasia. Although rare, bilateral deficits associated with the anterior cerebral arteries may occur when both arteries arise from a common stem, which is occluded, producing significant infarctions in both frontal lobes. Patients with this extensive bilateral lesion exhibit a variety of symptoms including bilateral lower extremity paralysis or paresis; urinary and bowel incontinence; behavioral changes; development of primitive reflexes (e.g., grasp reflex and sucking reflex); and abulia (loss of initiative, drive, and reaction to stimuli with passive attitude). Frontal lobe lesions may also induce bow shift of the ACA and Bruns ataxia, a broad-based, short-stepped gait, increasing the risk of the affected individuals for falls.

Additionally, several pathological reflexes may be observed in individuals with frontal lobe lesions, such as grasp, snout, suck, and palmomental reflexes.

- The *grasp reflex* is an involuntary tonic grasp response, associated with slow flexion of the fingers. Attempts by the examiner to withdraw the grasped fingers will only augment the patient's grasp. It may be elicited by touching or stroking the anterior surface of the wrist, the center of palmar surface of the hand and digits, and between the thumb and index finger.
- The *snout reflex* is characterized by puckering and protrusion of the lips upon light percussion of the middle upper lip.
- The *suck reflex* is an involuntary sucking movement of the lips, produced by a blunt object stroked across the lips.
- The *palmomental reflex* exhibits contraction of the mentalis and the orbicularis oris muscles upon striking the palm of the infant's hand with a sharp object.

PARIETAL LOBE

The parietal lobe is bounded by a line that connects the preoccipital notch to the upper end of the parieto-occipital sulcus (Figures 8.7 and 8.8). On the lateral surface, this lobe consists of the *postcentral gyrus* (Brodmann areas 3, 1, and 2) and the *superior and inferior parietal lobules*. Medially, it contains the *precuneus* and the posterior part of the *paracentral lobule*. This lobe is concerned with mimicking (imitation of speech or action), pantomimicking actions (imitation of gestures without words), or copying.

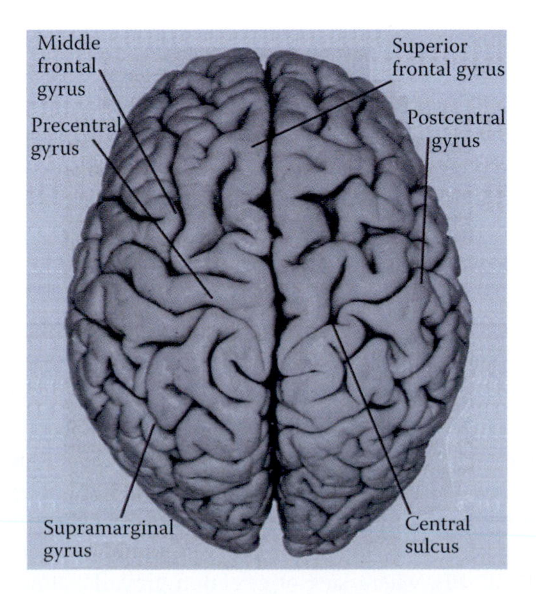

FIGURE 8.7 Dorsal surface of the brain with the precentral, postcentral, and supramarginal gyri clearly demarcated.

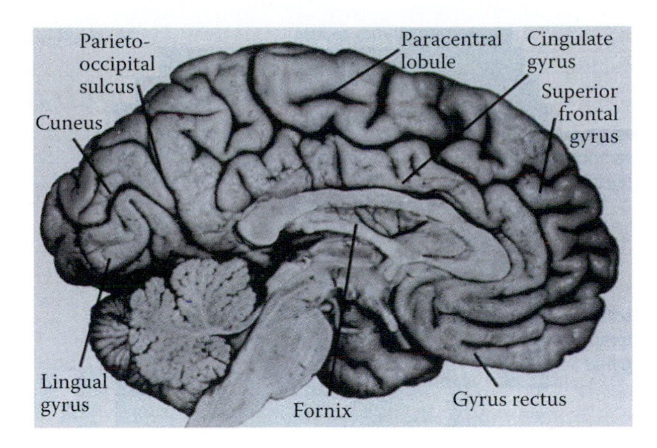

FIGURE 8.8 In this photograph of the medial surface of the brain, the individual gyri of the frontal, parietal, and occipital lobes are shown. The parieto-occipital and calcarine fissures are prominent.

Parietal lobe hematoma may produce loss of general sensations on the opposite side of the body. When the dominant hemisphere is involved, signs of neglect on the contralateral half of the visual field may also occur. Frequently, parietal lobe diseases may mimic certain manifestations of frontal lobe diseases.

The *postcentral gyrus* contains neurons that represent the primary sensory cortex (Brodmann areas 3, 1, and 2). The main input to this area arises from the ventral posterior nuclei of the thalamus (ventral posterolateral nucleus [VPL] and ventral posteromedial nucleus [VPM]), a somatotopically arranged projection in the form of a sensory homunculus (lips, tongue, and thumb have a larger representation). The body's distorted and disproportionate representation in this gyrus is also based upon the relative densities of the neuronal populations (sensory homunculus). Brodmann area 3 responds to joint sensations, Brodmann area 1 responds to cutaneous stimuli, and Brodmann areas 1 and 2 are excited by stimuli from joints. There are neurons in this gyrus that evoke a motor response upon stimulation. Some fibers of the postcentral gyrus descend in conjunction with the pyramidal tract and terminate in the gracilis and cuneatus nuclei.

The site of termination of the fibers that emanate from the postcentral gyrus may explain the deficit in kinesthetic perception that follows UMN lesions. Ablation of the postcentral gyrus may result in the inability to appreciate texture, weight, and slight changes in temperature. The ability to recognize painful, tactile, pressure, and proprioceptive sensations may not be significantly impaired, in contrast to the localization of noxious and tactile stimuli, which is severely affected.

The secondary somatosensory area is located at the base of the postcentral gyrus. However, certain parts of the body (e.g., face, mouth, and throat) are not represented in this area. Lesions of this area do not result in any recognizable deficit. The superior parietal lobule is represented by Brodmann areas 5 and 7. Damage to this lobule in the nondominant hemisphere impairs the ability to recognize certain parts of one's body on the contralateral side (hemispatial neglect). Initial hypotonia may also be observed. The inferior parietal lobule, which consists of the supramarginal and angular gyri, is separated from the superior parietal lobule by the interparietal sulcus. The supramarginal gyrus (Figure 8.7) encircles the posterior end of the lateral cerebral sulcus and is designated as Brodmann area 40.

Destruction of the supramarginal gyrus in the dominant hemisphere may disrupt its connection to other association cortices, resulting in tactile and proprioceptive agnosia or astereognosis, a condition that refers to the inability to recognize texture, form, and size by palpation; apraxia; left–right disorientation; and disturbances of body image.

The *angular gyrus* (Figure 8.1), designated as Brodmann area 39, surrounds the terminal part of the superior temporal sulcus. Due to its massive connections with the association cortices of the auditory, visual, and superior parietal lobe, the angular gyrus acts as a nodal point of integration for these modalities of sensations into symbols. It also signals the premotor cortex in the dominant hemisphere to initiate action via the superior longitudinal (arcuate) fasciculus within the same hemisphere and via the corpus callosum to the nondominant hemisphere. On the left side, it converts the written word to its auditory equivalent (graphemes to phonemes).

A lesion of the angular gyrus in the dominant hemisphere may produce visual agnosia, in which the ability to read and comprehend written words (alexia) and copy (agraphia) are lost. Bilateral damage to the angular gyri may result in the loss of ability to judge the location of objects and their relationship to each other in space.

- The *precuneus* is bounded by the marginal branch of the cingulate sulcus rostrally and the parieto-occipital sulcus caudally.
- The *paracentral lobule* is formed by the medial extension of the precentral and postcentral gyri. It contains neurons, which are responsible for the motor and sensory innervation of the lower extremity and the regulation of certain physiological functions, such as defecation and micturition.

Damage to the paracentral lobule can result from occlusion of the ACA, producing spastic paralysis of the muscles of the contralateral lower extremity with bowel and urinary incontinence. A lesion of the anterior portion of the parietal lobe results in *Dejerine cortical sensory syndrome* (Verger–Dejerine syndrome), which is characterized by contralateral loss of discriminative sensory modalities (joint sensation and stereognosis), including the face. Pain and temperature sensations remain intact, but the ability to localize these sensory impulses may be affected.

TEMPORAL LOBE

The temporal lobe (Figures 8.1 and 8.9) lies inferior to the lateral cerebral fissure, rostral to the preoccipital notch, and inferior and caudal to the frontal lobe. It contains the inferior horn of the lateral ventricle, the hippocampal and dentate gyri, the fimbria of the fornix, and the amygdala. The lateral surface of this lobe consists of the superior, middle, and inferior temporal gyri, which are separated by the temporal sulci. The superior temporal sulcus runs between the superior and middle temporal gyri and is surrounded caudally by the angular gyrus. Separation of the middle and inferior temporal gyri is maintained by the inferior temporal sulcus. Additional gyri exist on the inferior surface of the temporal lobe; these include the occipitotemporal and the parahippocampal gyri. Medial to the superior temporal gyrus, the *transverse gyri of Heschl* (Brodmann areas 41 and 42) make their appearance, constituting the primary auditory cortex.

Temporal lobe hematoma may produce herniation and brainstem compression and may lead to stupor and possible death. Selective destruction of parts of areas 41 and 42 produce sigoma contralaterally, a condition that is comparable to scotoma associated with lesions of the visual pathway.

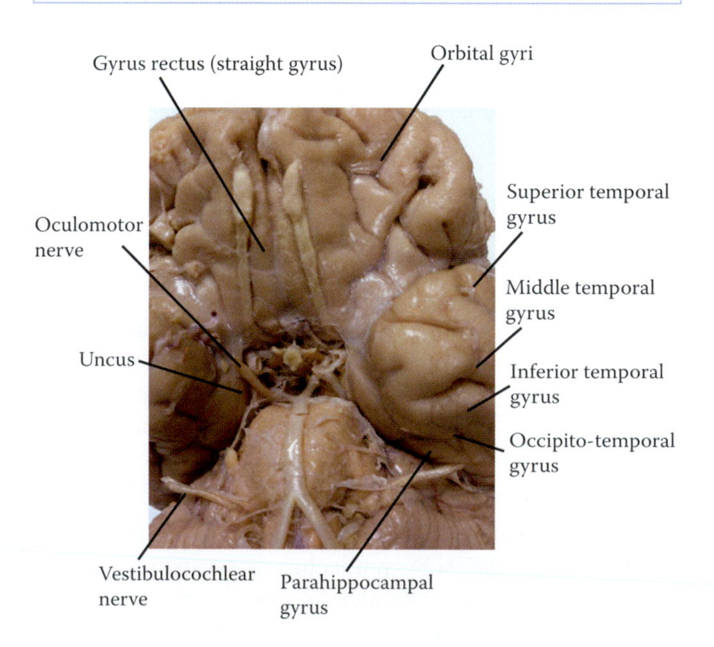

FIGURE 8.9 Inferior surface of the brain illustrating the gyri associated with the frontal and temporal lobes.

In the dominant hemisphere, the posterior part of the superior temporal gyrus contains the secondary auditory cortex (Brodmann areas 22, 51, and 52) or Wernicke's zone. *Wernicke's zone* is the site where templates for phonemes (speech sounds such as "f" or "ph" and words) are linked to primitive auditory sensations received from the primary auditory cortex. These templates are also linked to the auditory, visual, and olfactory systems, as well as to other sensory modalities in the angular gyrus. Due to these connections, Wernicke's area can regulate auditory perception and store visual imagery and verbal comprehension. Aside from the superior temporal gyrus, other gyri also exist on the lateral surface of the temporal lobe. Between the superior and inferior temporal sulci lies the *middle temporal gyrus*. The *inferior temporal gyrus* (Brodmann area 20) lies lateral to the occipitotemporal sulcus, making an appearance on both the inferior and lateral surfaces of this lobe. This cortical area is considered a higher visual association zone, which receives, in its posterior part, major input from the occipitotemporal cortex, representing the contralateral visual field.

Immediately lateral to the parahippocampal (fusiform) gyrus, the *occipitotemporal gyrus* makes its course. The *parahippocampal gyrus* (Figure 8.9) represents the inferior portion of the limbic lobe, expanding rostrally and medially into the *uncus*, a small tonguelike projection between the collateral and hippocampal sulci. This projection constitutes the primary olfactory center (Brodmann area 34), which receives and processes olfactory information received from the lateral olfactory stria. Uncus and olfactory pathways are considered integral parts of the limbic system.

Functional and Clinical Consideration

A hematoma that damages the superior frontal gyrus in the dominant (usually the left) hemisphere may produce receptive aphasia (inability to comprehend spoken language).

Uncal herniation is a condition in which the uncus as well as the medial portion of the temporal lobe is forced to herniate over the sharp margin of the tentorial notch into the posterior cranial fossa, usually as a result of unilateral supratentorial mass (hematoma or tumor). The mass often causes a substantial rise in the intracranial pressure (ICP) and displacement of the medial part of the temporal lobe (uncus) through the anterior part of the tentorium (tentorial notch) inferiorly. The swollen uncus may compress the oculomotor nerves and the posterior cerebral arteries and the crus cerebri on the opposite side. Patients with uncal herniation exhibit signs of ipsilateral or bilateral oculomotor palsy, which includes mydriasis (pupillary dilatation) followed by downward and lateral deviation of the eye, and eventual external ophthalmoplegia. Ipsilateral hemiplegia may occur as a result of compression of the corticospinal tract within the contralateral crus cerebri against the anterior edge of the tentorium (Kernohan notch phenomenon). Increased ICP may also be seen as a consequence of posterior cerebral artery compression, accompanied by occipital lobe infarction. As a result of this developing mass, the cingulate gyrus may also herniate through the inferior border of the falx cerebri, causing compression of the ipsilateral or contralateral ACA and infarction of the cingulate and medial frontal gyri, paracentral lobule, and precuneus. Irregularities in respiration ranging from Cheyne–Stokes breathing to respiratory arrest also occur. As in subdural hematomas, reduced states of consciousness and coma develop rapidly, as a result of compression of the ascending reticular activating system (ARAS) following oculomotor palsy. Signs of ipsilateral UMN paralysis, followed by decerebrate rigidity as a result of destruction of the inhibitory corticospinal tract in the crus cerebri and unopposed action of the excitatory pathways, are eventually concluded in total flaccidity.

Uncinate fits are seizures that produce involuntary movements of the mouth and tongue and perception of noxious hallucinatory odors, which may be accompanied by irrational fears of one's surroundings. They may occur as a result of a lesion, infarct, or tumor affecting the uncus, the amygdala, and possibly the gustatory area. Due to involvement of the temporal lobe, cognitive functions, such as memory, orientation, and attention may, also be impaired.

Pick's disease is another condition, although rare, that selectively produces maximal atrophy of the temporal and frontal lobes, sparing the posterior two-thirds of the superior temporal gyrus. It may also involve the occipitotemporal region (the silent cortical area), which is concerned with the storage of memories from the visual and auditory systems. The archipalium, which includes the dentate gyrus, pyramidal cells of the CA1 sector, and subiculum, are particular targets of this disease. Bilateral degeneration of the caudate nucleus, dorsomedial putamen, globus pallidus, and locus ceruleus are seen. Loss of the neuronal axons in the cortical white matter and degeneration of the lateral tuberal and dentate nuclei granular cell layer of the cerebellum are also visible. A striking dilation of the lateral ventricles and sparing of the substantia nigra are additional characteristic findings. Microscopically, neuronal loss in layers III and V and the appearance of swollen cells with pale cytoplasm and eccentric nuclei (Pick cells) occur. The presence of tau protein that aggregates into spherical argyrophilic inclusion (Pick) bodies in layers II and IV of the neocortex is considered a universal finding. Damage to this wide array of cortical areas may result in epileptic seizures combined with amnesia, nonfluent aphasia, auditory hallucinations, and the "déjà vu" phenomenon. Impaired social conduct including lack of appreciation of etiquette, disinhibition, and pacing are also observed. This slowly progressive disease is often fatal within 2–10 years.

In the central herniation (Central Syndrome) the diencephalon and parts of the temporal lobe are displaced through the tentorial notch inferiorly (descending tentorial displacement) into the posterior cranial fossa. This downward shift, a possibly fatal development, places an undue pull on the small pontine branches of the basilar artery and lacerates them, producing "Duret hemorrhage" in the ventral and paramedian areas of the upper pons and midbrain. This bleeding can be visible radiographically in the midline at the pontomesencephalic junction. A Kernohan notch may develop on the crus cerebri, compressing the corticospinal tract and producing ipsilateral hemiparesis, a false localizing sign known as Kernohan–Woltman syndrome. This bleeding can be visible radiographically in the midline at the pontomesencephalic junction. It is also accompanied by obliteration of the suprasellar cistern and compression of the cerebral peduncles that contain the corticospinal and corticobulbar tracts (CBTs). Venous obstruction may also develop as a result of this herniation. If the herniation is confined to the diencephalon, the condition may be reversible; however, herniation that involves the midbrain and upper pons leads to deep coma, posturing, Babinski sign with pupillary abnormality (nonresponsive pupil), dysconjugate eyes, and respiratory hyperventilation. In the late stage, deep coma continues and unresponsive and dilated pupils with flaccid palsy of the extremities develop. Ascending central herniation can be confirmed radiographically by an unusually small or even obliterated quadrigeminal cistern.

OCCIPITAL LOBE

The occipital lobe (Figures 8.6, 8.8, and 8.10) lies caudal to the imaginary line that connects the preoccipital notch to the upper end of the parieto-occipital fissure. On its lateral surface,

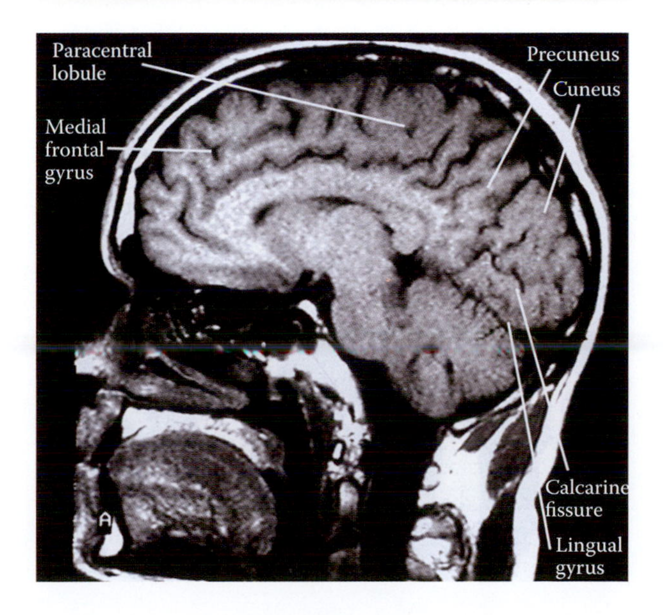

FIGURE 8.10 An inverted MRI scan of the medial surface of the brain showing some of the associated gyri.

the superior and inferior occipital gyri are separated by the lateral occipital sulcus. Caudal to the lateral occipital sulcus, the descending occipital gyrus makes its appearance. This lobe also consists of the cuneus and lingual gyrus, which are clearly visible medially. The *cuneus*, which is bounded rostrally by the parieto-occipital sulcus and caudally by the calcarine fissures, receives visual impulses from the lower quadrant of the opposite visual field. Inferior to the calcarine fissure and medial to the collateral sulcus, the lingual gyrus makes its appearance, receiving input from the upper quadrant of the opposite visual field.

Rostral to the occipital pole, the lunate sulcus runs vertically between the striate and peristriate cortices. The lunate sulcus, which contains the parastriate cortex, is crossed at its upper and lower ends by the superior and inferior polar sulci, respectively. Enclosed between the polar sulci is the macular area of the primary visual (striate) cortex.

CENTRAL LOBE (INSULAR CORTEX)

The insular cortex (Figure 8.11) or central lobe consists of long and short insular gyri, forming the floor of the lateral cerebral fossa. It continues anteriorly with the anterior perforated substance and is surrounded by an incomplete circular sulcus. This sulcus is deficient rostrally and inferiorly where the limen insula is located. Continuation of the insular cortex is marked by the opercular areas of the frontal, parietal, and temporal lobes. It receives input from the ventral posterior nucleus, medial geniculate body, dorsomedial thalamic nucleus, and pulvinar and intralaminar nuclei, maintaining ipsilateral connections with the primary and secondary somatosensory cortex, inferior parietal lobule, and the orbitofrontal cortex. It has been is suggested that the insular cortex plays an important modulatory role in the perception and recognition of fine touch, auditory impulses, and taste, and is thought to be associated with language function.

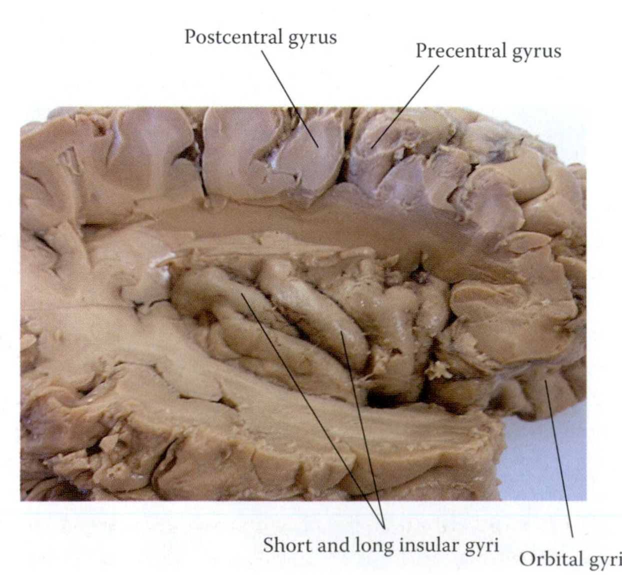

FIGURE 8.11 In this picture, the lateral surface of the brain is dissected to delineate clearly the boundaries of the insular gyri.

LIMBIC LOBE

The limbic lobe consists of the subcallosal, paraterminal (septal area), cingulate, and fasciolar gyri (Figures 8.8, 8.9, 8.10, and 8.12). A rostral region of this lobe, which lies between the lamina terminalis and the posterior olfactory sulcus, is termed the *subcallosal gyrus* (paraolfactory gyrus). It should also be noted that the *paraterminal gyrus* (precommissural septum gyrus) receives input from the olfactory system via the medial olfactory stria; the hippocampal gyrus through the precommissural column of the fornix; and the lateral hypothalamus via the medial forebrain bundle (MFB). It also receives impulses from the tegmental nuclei via the mammillary peduncle. Aside from its projection to the tegmental nuclei of the midbrain through the MFB, the paraterminal gyrus also sends fibers to the habenular nuclei contained in the stria medullaris thalami. It is also connected through the diagonal band of Broca to the periamygdaloid cortex.

The *cingulate gyrus* (Figures 8.8, 8.10, and 8.12) is part of the limbic lobe, maintaining reciprocal connections with the anterior nucleus of the thalamus. Through this thalamic nucleus, the hypothalamus influences the activities of the cingulate gyrus. This gyrus projects to the entorhinal cortex and influences the hypothalamus through fibers of the fornix. Due to these diverse connections, somatic and visceral responses may be elicited by stimulation of the anterior part of cingulate gyrus. Stimulation of the posterior cingulate gyrus elicits pleasurable reactions.

A unilateral cerebral mass or lesion due to traumatic brain injury, brain tumor, or intracranial bleeding can displace the cingulate gyrus toward the opposite hemisphere against the thin rostral part of the falx cerebri, resulting in cingulate or *subfalcial herniation* with

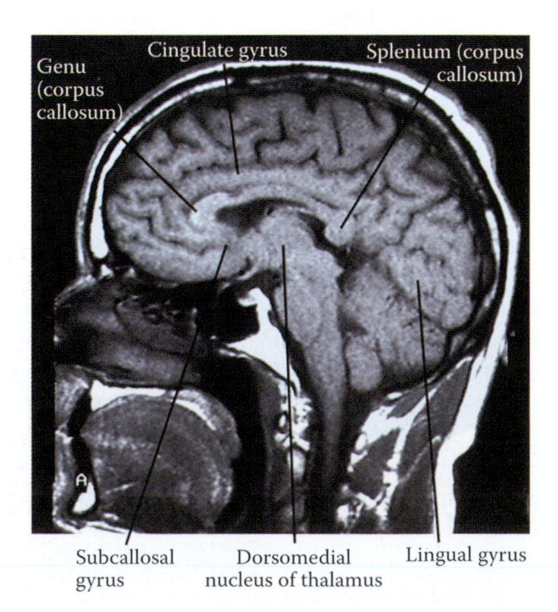

FIGURE 8.12 Inverted midsagittal MRI of the brain illustrating some components of the limbic lobe that surrounds the corpus callosum as well as adjacent structures.

secondary hemorrhagic infarcts. A subfalcine herniation is a common type of supratentorial herniation that occurs when the increase in ICP reaches such an extent that causes cingulate gyrus partial or complete herniation under the anteroinferior border of the flax cerebri. It must be remembered that herniation can occur in the absence of a marked increase in ICP, particularly if the mass lesions are confined to the margins of the brain compartments, causing local pressure that is not transmitted elsewhere in the brain. This is accompanied by compression and infarction of the ACA branches (callosomarginal, pericallosal, and frontopolar) and, when prolonged, can lead to necrosis of the cingulate gyrus. Compression of the ACA against the sharp edge of the lower border of the falx cerebri can lead to the development of aneurysm. This type of herniation also causes compression of the interventricular foramen of Monro and the lateral ventricle on the same side, leading to asymmetrical ventricles (dilated contralaterally and narrowed ipsilaterally). Obstruction of the ventricular system produces a noncommunicating type of hydrocephalus. When the subfalcine herniation occurs through the posteroinferior border of the flax cerebri, the internal cerebral veins and the cerebral vein of Galen are compressed, which exacerbates the progressing increase in ICP. Subfalcine herniation can be associated with coma and posturing and can be deadly it if progresses to central herniation. This condition is seen in conjunction with transtentorial herniation.

The *fasciolar gyrus* (retrosplenial gyrus) connects, just inferior to the splenium of the corpus callosum, the cingulate and the induseum griseum (stria Lancisi) to the dentate gyrus and Ammon's horn of the hippocampal formation. It lies on the lateral side of the gyrus Retzius, an inconstant protrusion from the posterior part of the parahippocampal gyrus inferior to the splenium.

CORPUS CALLOSUM

Midline interconnection of the cerebral hemispheres is maintained primarily by the *corpus callosum* (Figures 8.8, 8.12, and 8.13). This commissural structure, covered by the ventricular ependyma, forms the roof of the lateral ventricle. Superior to the corpus callosum, the anterior cerebral vessels and the falx cerebri can be seen. It consists of the rostrum, genu, trunk, and splenium. Fibers of the *genu* (forceps minor), which lies between the rostrum and trunk, connect the frontal lobes. Apart from its small size, the *rostrum* extends from the genu to the lamina terminalis; its superior surface is attached to the septum pellucidum. Wide cortical areas of the hemispheres are connected via the *trunk*, whereas the *splenium*, the thickest part of this commissure, forms the base of the longitudinal sagittal cerebral fissure, connecting the occipital lobes (forceps major). The splenium protrudes into the posterior horn of the lateral ventricle as the bulb of the posterior horn. The callosal trunk is covered by the induseum griseum (supracallosal gyrus), a thin lamina of gray matter, which is continuous with the cingulate, fasciolar, and paraterminal gyri, containing the medial and lateral longitudinal striae of Lancisi. Fibers of the trunk and splenium form the *tapetum*, which constitutes the roof and the lateral walls of the posterior and inferior horns of the lateral ventricle.

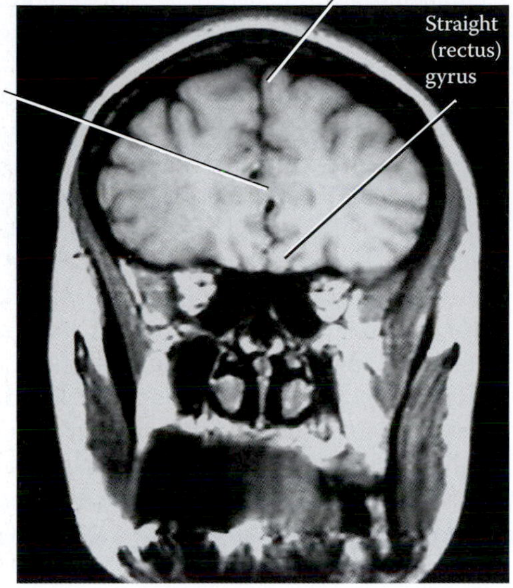

FIGURE 8.13 MRI scan through the genu of the corpus callosum. The frontal lobes are connected by the forceps minor, an extension of the genu of the corpus callosum.

Agenesis of the corpus callosum, a rare congenital condition, is characterized by a complete or partial failure of the corpus callosum to develop, resulting in the appearance of Probst bundles. These bundles run in a rostrocaudal direction within each half of the cerebral hemisphere and replace the fibers that cross the midline and connect the two hemispheres of the brain. Chromosomal abnormality, inherited genetic disorders, prenatal infections, toxic or traumatic injuries, and metabolic disorders are considered as possible causes of this condition. It has been proposed that a molecular dysfunction that undermines the signaling mechanism essential in the development of the components of the corpus callosum occurs in this condition, as well as other abnormalities collectively known as ciliopathies. Due to the variable nature of the manifestations of this anomaly, patients exhibit deficits that range from poor vision, hypotonia, and ataxia to spasticity, impaired motor development, seizure, and mental retardation. Radiographic imaging can complement the clinical findings and thus help in the establishment of a diagnosis of agenesis of the corpus callosum.

Disconnection syndrome (split brain) encompasses a constellation of symptoms, which results from interruption of the interhemispheric commissures or intrahemispheric connections. This syndrome can be produced be corpus callosotomy, which entails cutting the corpus callosum, a procedure commonly used in the past as a last resort treatment for intractable epilepsy. It can be seen when the corpus callosum together with parts of the white matter undergo necrosis in Marchiafava–Bignami syndrome, a condition that affects individuals who are heavy consumers of wine. Infarction of the pericallosal branch of the ACA may result in similar deficits, producing two hemispheres that function independently. This functional independence involves perception, cognition, mnemonic, learned, and volitional activities. Therefore, information generated in the nondominant hemisphere is expressed only in a nonverbal manner and could not be expressed in writing or speech, as the dominant hemisphere has a major role in linguistic expression. In individuals with a divided corpus callosum, most voluntary daily activities, native intelligence, memory, verbal reasoning, and temperament are not generally affected. Presence of bilateral sensory representation and the compensatory development of bilateral motor representation may explain the intactness of these functions. Various forms of agnosia may be associated with lesions of the corpus callosum. Due to interruption of the callosal fibers and the fact that information is not transferred from the right hemisphere to the speech center in the left hemisphere, patients may exhibit an inability to name objects presented into the left visual field or placed in the left hand. Affected individuals are also unable to carry out verbal commands with the left hand, but are capable of naming objects presented into the right visual field or hand. Thus, the inability to match an object in one hand (or seen with one eye) with one placed in the other hand (or seen with the other eye) becomes evident. These observations indicate that information received or generated by the nondominant hemisphere could not be expressed verbally or in writing and that visual information is not transferred between the two hemispheres. Comprehension of spoken and written languages is not usually affected due to its bilateral representation. Loss of emotional control and reasoning ability seen initially is followed by tremor and coma, which can lead to death.

A vertical partition, the *septum pellucidum* (supracommissural septum) extends between the corpus callosum and the fornix, forming the medial wall of the anterior and central parts of the corpus callosum. This septum (Figures 8.12 and 8.14) consists of two laminae of fibers, sparse gray matter, and neuroglia. Note that the *fornix* (Figures 8.6, 8.8, 8.12, and 8.14), a robust bundle at the inferior border of the septum pellucidum, is formed by the axons of the pyramidal layer of the hippocampal gyrus. This bundle starts as the alveus and converges into the fimbria, which ascends toward the splenium of the corpus callosum and, more rostrally, above

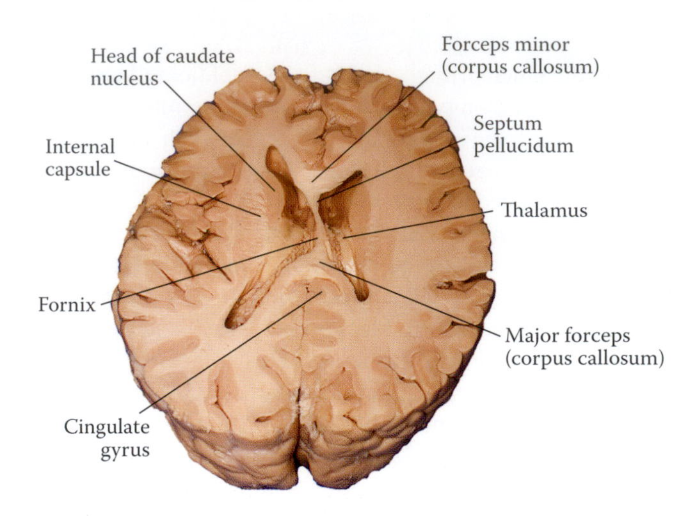

FIGURE 8.14 In this horizontal section, the septum pellucidum and its relation to the fornix are illustrated.

the thalamus as the crura of the fornix. These crura, interconnected by the *hippocampal commissure*, unite rostrally to form the body of the fornix, which attaches to the corpus callosum. Eventually, the body of the fornix divides into precommissural and postcommissural columns by the fibers of the anterior commissure. The precommissural column terminates in the lateral hypothalamus, while the postcommissural column projects primarily to the mammillary nuclei.

CEREBRAL CORTEX (GRAY MATTER)

In humans, the cerebral cortex (pallium) is the most highly evolved portion of the cerebral hemisphere and the central nervous system (CNS) in general. It consists of a thin shell of a large number of neuronal cell bodies, unmyelinated nerve fibers, glial cells, and blood vessels. The presence of the neuronal cell bodies and extensive capillary network is responsible for the gray appearance of the cerebral cortex. It is well documented that the cerebral cortex maintains a crucial role in the perception, fine discrimination, and integration of various modalities of sensation and in the regulation of visceral and somatic motor activities. It contains afferent (thalamocortical), efferent (projection), commissural, and association fibers. In order to increase the surface area of the brain, the cortex is thrown into convolutions (gyri) separated by sulci.

On a phylogenetic basis, the cerebral cortex is divided into the archipallium, paleopallium, and neopallium. The *archipallium* (oldest cortex) is comprised of the hippocampal, dentate, and fasciolar gyri and the subiculum. The hippocampal gyrus (*cornu ammonis*) is superior to the subiculum and the parahippocampal gyrus. It lies in the floor of the temporal horn of the lateral ventricle and forms the *pes hippocampi*, a rostral swelling that is covered by an ependymal layer. As mentioned earlier, the axons of the hippocampal neurons form the alveus, which continues with the fimbria of the fornix. The dentate gyrus, which lies between the hippocampal gyrus (cornu ammonis) superiorly and the subiculum inferiorly, is separated from the subiculum by the hippocampal sulcus. A transitional zone between the six-layered cortex of the parahippocampal gyrus and the three-layered cortex of the cornu ammonis is termed the *subiculum* (see also the limbic system). The *paleopallium* includes the olfactory cortex and the pyriform lobe, which are integral parts of the limbic system. Both the archipallium and paleopallium comprise the *allocortex* or heterogenetic cortex, consisting of three layers. Most of the cerebral cortex, approximately 90%, constitutes the *neopallium* (neocortex, or isocortex), which is formed by six distinct layers (homogenetic cortex). By the seventh month of intrauterine development, the six layers of the homogenetic cortical regions become distinct.

The cerebral cortex consists of pyramidal and nonpyramidal neurons such as the stellate (granule), fusiform neurons, horizontal cells of Cajal, and cells of Martinotti, which are arranged in horizontal and vertical layers. The *pyramidal cells* (Figures 8.15 and 8.16) have apical dendrites, which run perpendicular to the cortical surface, and basal dendrites and branch locally parallel to the cortical layer. Axons of the pyramidal neurons travel to the subcortical areas as projection fibers or course within the white mater as association fibers. These axons may give rise to recurrent collaterals and represent the primary output of the cerebral cortex. *Betz cells* are giant pyramidal cells, occupying the precentral gyrus. There are numerous *granule* cells, which function as interneurons, with their axons and many dendrites concentrated in lamina IV of the cerebral cortex (Golgi type II). *Fusiform* cells (Figure 8.16) occupy mainly the deep cortical layers, particularly layer V, and possess axons that form projection fibers. *Horizontal cells of Cajal* or horizontal cells (Figure 8.16) have axons that are confined to the superficial layers of the cerebral cortex, whereas *Martinotti cells* (Figure 8.16) are scattered diffusely throughout the cortical layers.

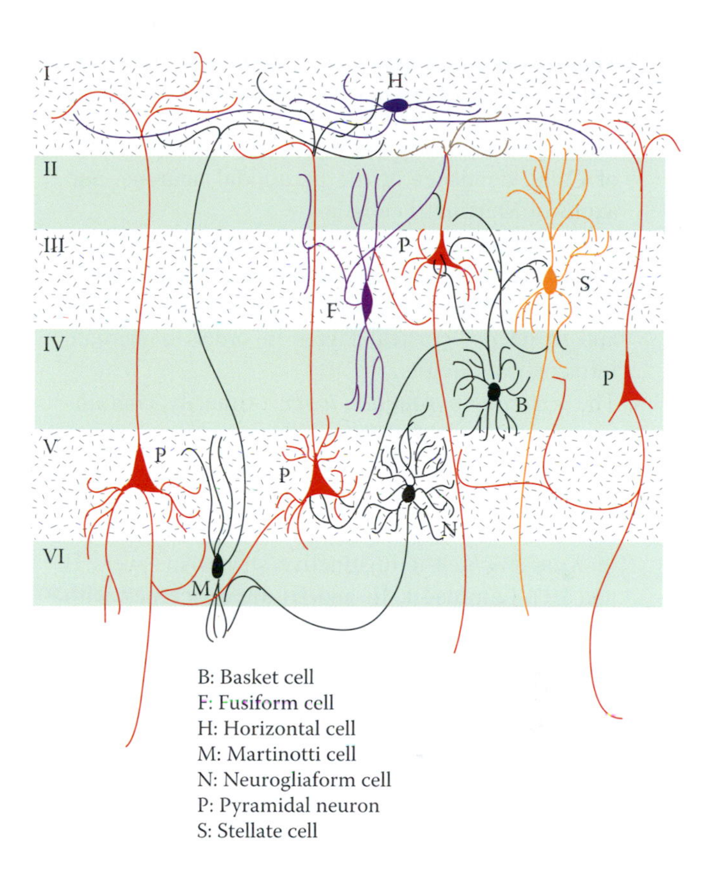

FIGURE 8.16 Schematic representation of the various neurons associated with the cerebral cortex.

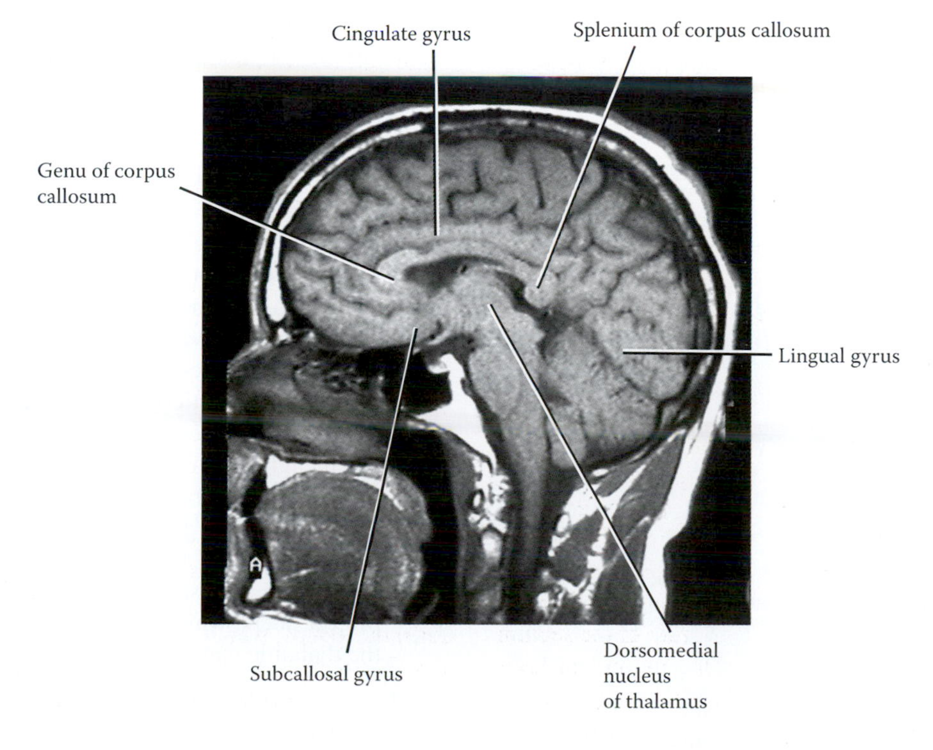

FIGURE 8.15 Typical pyramidal neuron showing the single axon and a distinct apical dendrite.

The *neuronal cytoarchitecture of the homotypical isocortex* reveals six layers or laminae (Figures 8.16, 8.17, and 8.18).

1. The *molecular layer* consists of horizontal cells of Cajal, dendrites of the pyramidal neurons, and axons of Martinotti cells.
2. The *external granular layer* is a receptive layer, consisting of small pyramidal and granule cells. Neurons of this layer project to the molecular layer and to deeper cortical layers in order to mediate intracortical circuits.
3. The *external pyramidal layer* primarily contains pyramidal neurons, projecting to the white matter as association fibers and to the opposite hemisphere as commissural fibers. This layer also contains granule and Martinotti cells, as well as the horizontal band of *Kaes-Bechterew* (distinctive stripes in layers II and III). Laminae I, II, and III are concerned with associative and receptive functions.
4. The *internal granular layer* receives all of the sensory projections from the thalamic relay nuclei via the thalamocortical fibers (e.g., optic radiation, auditory radiation, and primary sensory systems). It consists of densely packed stellate cells, containing the external myelinated *bands of Baillarger*. These outer and inner bands are formed by tangential thalamocortical fibers and may be visible to the naked eye. The outer band lies in lamina IV and is produced by afferents from the thalamic relay nuclei. The outer band is conspicuous in the striate cortex as the strip of Gennari, while the inner band is formed by the basal dendrites and the myelinated collaterals of the large pyramidal (Betz) cells. In the visual cortex, the internal granular layer consists almost exclusively of simple cells that respond to stimuli from only one eye (but not both). The cortical layers that lie superficial and deep to layer IV respond to visual inputs from both sides. The *infragranular (V and VI) layers* are the first to be formed during embryological development. Subsequent cell migration through the infragranular layers allows the neurons to form more superficial layers.
5. The *internal pyramidal layer* (ganglionic layer) contains the largest pyramidal (Betz) neurons, giving rise to the corticospinal and corticobulbar fibers. It is pierced by dendrites and axons from other layers, including association and commissural fibers.
6. The *multiform layer* contains small pyramidal and Martinotti cells, giving rise to projection fibers to the thalamus. It is crossed by commissural and association fibers. Cortical neurons of this layer project fibers, which maintain a feedback loop with the thalamic nuclei.

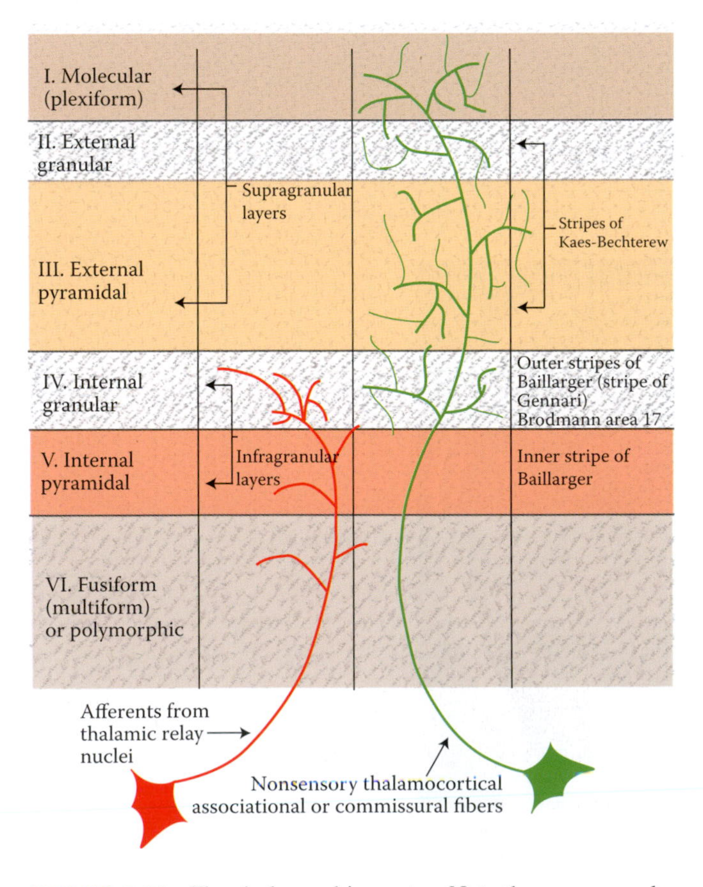

FIGURE 8.17 The six-layered isocortex. Note the supragranular and infragranular layers and the sites of termination of the specific thalamic afferent and nonsensory thalamocortical fibers.

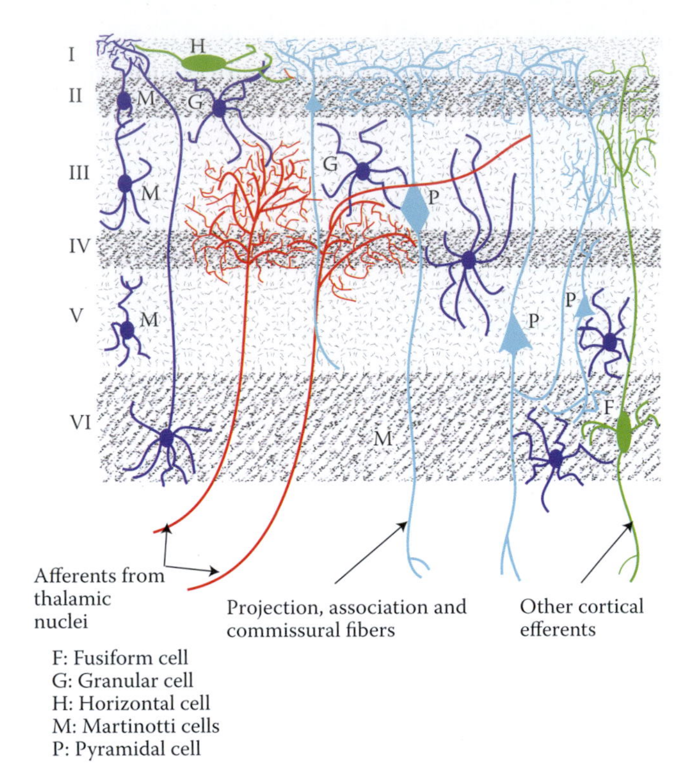

FIGURE 8.18 Cortical layers with their distinct neuronal population, showing areas of distribution of the afferents derived from the specific thalamic nuclei and areas of distribution of the commissural and association fibers.

The relative thickness of the pyramidal and granular layers may be used as a basis to classify the cerebral cortex into five distinct areas. This classification grades the cortex from a purely motor to a purely sensory cortex and from the thickest to the thinnest. These cortical layers also vary from a layer that contains the least number of granule cells to a layer that consists mostly of granule cells. According to this classification, the cerebral cortex is divided into agranular, frontal, parietal, polar, and granular cortices.

- The *agranular cortex* (motor cortex) is the thickest type, which lacks or contains only a few granule cells in layers II and IV. It is exemplified in the heterotypical motor cortex (Brodmann area 4) and partly in the premotor cortex (Brodmann areas 6 and 8) and paracentral lobule.
- The *frontal type of cortex* is a homotypical cortex with a very thin granular layer, which is represented in the superior frontal and inferior temporal gyri, superior parietal lobule, precuneus, and parts of the middle and inferior frontal gyri.
- Examination of the *parietal type of cortex* reveals six distinct layers with a particularly thin pyramidal layer, which is evident in the PFC, inferior parietal lobule, and superior temporal, occipitotemporal gyri.
- In the *polar cortex*, which is represented in the frontal and occipital poles, a well-developed granular layer is evident.
- The *granular cortex* (konicortex), the thinnest of all cerebral cortices, contains a granular layer that achieved maximum development and is symbolized in the homotypical cortices of cuneus, lingual, parahippocampal, and postcentral gyri as well as the transverse gyri of Heschl.

The cerebral cortex also exhibits vertical lamination, which represents the functional units of the cerebral cortex, extending through all cellular elements. This arrangement is absent in the frontal cortex but distinctly evident in the parietal, occipital, and temporal cortices. The neurons of the vertical columns in the sensory cortex establish contacts with each other via interneurons (Golgi type II cells). Each column receives impulses from the same receptors, is stimulated by the same modality of sensation, and discharges for the same duration, maintaining an identical temporal latency. The isocortex (neopallium) is also divided into sensory, motor, and association cortices. The sensory cortex is further subdivided into primary, secondary, and tertiary sensory cortices. The motor cortex is classified into primary motor, premotor, and supplementary motor cortices. On the other hand, the association cortices include parts of the parietal, temporal, and occipital cortices.

Sensory Cortex

The sensory cortex deals with the perception and recognition of sensory stimuli. It imparts unique characteristics to sensations, enabling their identification on the basis of both comparative and temporal (spatial) relationships. It includes primary and secondary sensory cortices.

Primary Sensory Cortex

The primary sensory cortices are modality and place specific, receiving information from the specific thalamic nuclei. Depending upon the modality, sensations from body and projections from the visual fields and auditory spectrum are represented topographically in the contralateral primary sensory cortices. They include the somesthetic cortex (Brodmann areas 3, 1, and 2); visual or striate cortex (Brodmann area 17); auditory cortex (Brodmann areas 41 and 42); gustatory cortex; and vestibular cortex.

The *primary somesthetic cortex* (Brodmann areas 3, 1, and 2) subserves general somatic afferents (deep and superficial), occupying the postcentral gyrus (Figures 8.19 and 8.32). It consists of three cytoarchitecturally distinct cortical stripes. Area 3 receives tactile sensation, and area 1 forms the apex of the postcentral gyrus, receiving deep and superficial sensation. Area 2 lies in the posterior surface of the postcentral gyrus, deals with deep sensation, and receives collaterals from the other two areas. The primary somesthetic cortex receives projections from the ventral posterolateral and the ventral posteromedial thalamic nuclei. Pain and thermal sensations are only minimally represented. The distorted representation of the body in this cortex is known as the sensory homunculus. This homunculus is formed according to the innervation density of the body part.

The *primary visual cortex* (striate cortex), which represents Brodmann area 17, lies on the banks of the calcarine fissure of the occipital lobe (Figure 8.23). It receives information from the third, fourth, fifth, and sixth layers of the lateral geniculate body. There is a visuotopic representation in which the peripheral part of the contralateral visual field is represented rostrally, while the macular visual field is delineated caudally in the occipital pole. No commissural fibers connect the striate cortices of both hemispheres. This cortical area receives blood supply from the internal carotid artery

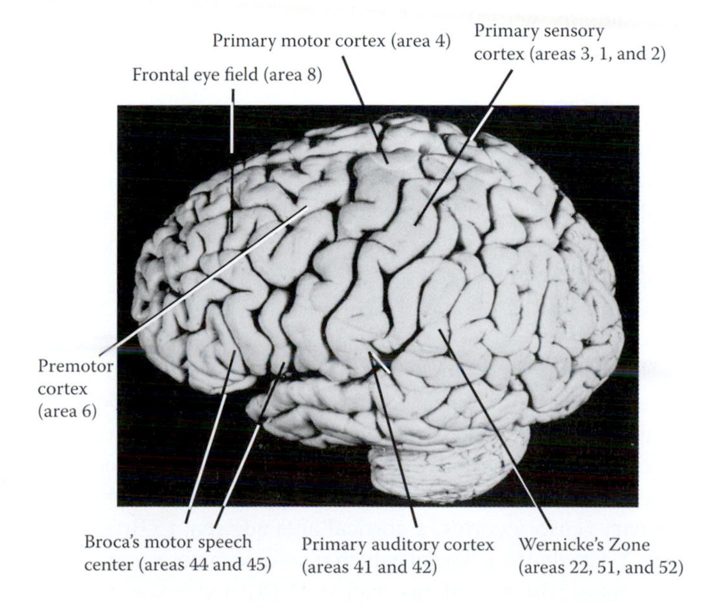

FIGURE 8.19 Lateral surface of the cerebral hemisphere. The main gyri of the frontal temporal and parietal lobes are indicated. Speech areas, frontal eye field, and auditory cortex are shown.

and the vertebrobasilar systems via the middle and posterior cerebral arteries, respectively. This cortex consists of vertical columns that discharge more or less as a unit, maintaining a topographical representation. It contains *simple cells* in the internal granular layer that respond to edges, rectangles of light, and bars presented in a particular receptive field axis of orientation to one eye. They possess "on" and "off" (inhibitory surrounding) centers. *Complex cells* have no "on" and "off" centers, spread in many layers of the striate cortex, and receive input from the simple cells. However, they do respond constantly to moving stimuli from both eyes. The vertical columns of the striate cortex may also be viewed as ocular dominance and orientation columns. *Ocular dominance columns* receive visual input from one eye only and are divided into alternate, independent, and superimposed stripes. Each column receives visual information from either the left or right eye. Development of these columns requires visual input. Deprivation of the binocular vision, as may be experienced in individuals with severe strabismus, may lead to unequal development of the ocular dominance columns and potential blindness. *Orientation columns* are much smaller than the dominance columns, are genetically determined and present at birth, and respond to a slit of light at a certain axis of orientation.

The *primary auditory cortex* (Brodmann areas 41 and 42) lies in the medial surface of the superior temporal gyrus (Figure 8.19). It is represented by the transverse gyri of Heschl, receiving the auditory radiation via the sublenticular part of the posterior limb of the internal capsule. Auditory radiation is formed by the axons of the ventral nucleus of the medial geniculate neurons. It should be noted that impulses reach the primary auditory cortex and emanate from the auditory receptors of both sides with contralateral predominance. In this cortex, a distinct tonotopic representation is exhibited in which higher frequencies are received medially and caudally, while lower frequencies occupy lateral and rostral areas.

> Due to the bilateral representation of the auditory impulses, a unilateral lesion of the auditory cortex produces impairment of hearing on the contralateral side but not total loss, with some degree of hearing loss on the ipsilateral side.

The *gustatory cortex* (Brodmann area 43), confined to the parietal operculum, receives afferents from the ventral posteromedial thalamic nucleus. The *vestibular cortex* is not a distinct cortical center, since the vestibular input is intermingled with other sensations. It is thought to occupy the distal part of the primary sensory cortex, a cortical area adjacent to Brodmann area 2. The thalamic nuclei, which receive somatosensory (VPL and ventral posterior inferior nucleus [VPI]) and motor impulses ventral lateral nucleus (VL), also receive vestibular input. This overlap may play a role in the regulation of conscious awareness of spatial relationships at the level of the thalamus.

Secondary Sensory Cortex

The secondary sensory cortices surround the primary sensory cortices, occupy a smaller area than the primary cortices, and receive input from the intralaminar and midline thalamic nuclei. Their topographic representation is either a mirror image or an inverted image, relative to the one perceived by the primary cortices. The *secondary somesthetic area* is primarily associated with noxious and painful stimuli. It occupies the superior lip of the lateral cerebral (Sylvian) fissure, distal to the postcentral gyrus. Large and diverse receptive areas convey a variety of sensory impulses to this cortex. Impulses are conveyed bilaterally, with a unilateral predominance, from the posterior thalamic zone and the ventral posterolateral thalamic nuclei. This cortex exhibits a distorted somatotopic arrangement in which the facial region lies adjacent to the corresponding area in the primary sensory cortex. Interestingly, anesthetics have a far greater effect on the secondary sensory area than on the primary sensory cortex.

The *secondary visual cortex* (Brodmann areas 18 and 19) surrounds the striate cortex and receives information from the primary visual cortex (Brodmann area 17) and the pulvinar. The visual impulses, which reach the superficial layers of the superior colliculus, project to the inferior and lateral part of the pulvinar. These impulses eventually terminate in the secondary visual cortex, constituting the *extrageniculate visual pathway*. Both secondary visual cortices are connected, subserving visual memory functions and other components of vision. In the *peristriate cortex* (Brodmann area 18), an inverted visual field receptive topography exists, as compared to the striate cortex. It receives input from Brodmann area 17, the pulvinar, and the lateral geniculate nucleus. It projects to the peristriate cortex of the opposite hemisphere. This cortical area is essential for visual depth perception (stereoscopic vision). The unique bilateral representation of the visual image is achieved by the interhemispheric connections of the peristriate cortices through the corpus callosum. This bilaterality, which is not mediated by the lateral geniculate nucleus, ensures that no gap exists between the single images generated by both eyes. The *parastriate cortex* (Brodmann area 19) surrounds Brodmann area 18, maintaining identical retinotopic representation. The *secondary auditory cortex* (Wernicke's zone—Brodmann area 22) surrounds the primary auditory cortex and receives afferents from the dorsal and medial subnuclei of the medial geniculate nucleus. This cortical area maintains reciprocal connections with the opposite hemisphere.

Motor Cortex

The motor cortex is also known as the agranular cortex because of the masking (attenuation) of the granular layers, particularly the inner granular layer. It occupies most of the frontal lobe and exerts control over the axial and appendicular muscles. It has a number of subclassifications, which include the primary and supplementary motor cortices, as well as premotor area and motor eye field.

Primary Motor Cortex

The *primary motor cortex* (Brodmann area 4), represented by the precentral gyrus and part of the paracentral lobule, contains giant pyramidal cells of Betz that project to the lumbosacral segments of the spinal cord (Figures 8.19 and 8.32). Like the primary sensory cortex, the body is also represented here in a distorted fashion and arranged according to the relative innervation density (motor homunculus). In this homunculus, the foot, leg, and thigh occupy the medial part of the paracentral lobule, whereas the gluteal region, trunk, and upper extremity, followed by the hand, digits, and head, in a descending manner, occupy the precentral gyrus. On the lower end of this homunculus, the tongue, muscles of mastication, and larynx are designated. A brief glance at the homunculus reveals disproportionately large areas for the hand and especially the thumb, as well as the face. Ablation of the precentral gyrus results in spastic palsy (increased muscle tone in the antigravity muscles) on the contralateral side. Approximately, up to 30% of corticospinal tract fibers arise from the primary motor cortex. The connection of the primary and secondary somatosensory cortices to the primary motor cortex enables the ventral posterolateral nucleus to convey information to the motor cortex. Specific projections to lamina V of the primary motor cortex arise from the VL nucleus of the thalamus. Other cortical areas such as Brodmann areas 5 and 6 also project to the motor cortex. Thus, the VL nucleus conveys the input received from the cerebellar nuclei to the motor cortex (Table 8.1).

> Contralateral flaccid palsy, followed by a gradual spasticity, Babinski sign, increased deep tendon reflexes, and clonus, are the prominent (upper motor) signs of lesions of the primary motor cortex. Recovery from such injury is only moderate.

Supplemental Motor Cortex

The *supplementary motor cortex* (Brodmann areas 8 and 9), a duplication of the primary motor cortex, occupies the medial frontal gyrus (medial part of the superior frontal gyrus) and overlaps with Brodmann areas 4 and 6. It mediates bilateral contraction of the postural muscles and plays an important role in the planning and initiation of movements. It becomes active even if the intended movement did not occur.

> Ablation of the supplementary motor cortex produces an increase in flexor muscle tone, leading to spasmodic contracture and pathologic grasp reflexes on both sides. The face and upper limb are represented anteriorly, the lower limb posteriorly, and the trunk occupies a more inferior position in this cortex.

TABLE 8.1
Primary Sensory Cortex

Cortical Areas	Major Afferents	Major Efferents	Function	Deficits
Postcentral gyrus; Brodmann areas 3, 1, and 2	Ventral posterolateral and ventral posteromedial nuclei, secondary somatosensory cortex along the upper bank of the lateral cerebral fissure, claustrum, basal nucleus of Meynert, and locus ceruleus	Superior parietal lobule (Brodmann areas 5, 7, and 6); ventral posteromedial and ventral posterolateral nuclei; and corticopontine fibers	Perception of somesthetic sensations in a somatotopic manner	Loss of discriminative sensations such as position of body parts, weight differences, texture, and two-point discrimination; slight reduction in pain and temperature perception
Inferior part of the postcentral gyrus adjacent to insular cortex	Visceral afferents	Not well defined	Integration of visceral sensations	Not well documented
Striate cortex—Brodmann area 17 (cuneate and lingual gyri)	Lateral geniculate nucleus	Lateral geniculate body and Brodmann area 18	Primary center for visual perception spatial representation	Contralateral homonymous hemianopsia (commonly with macular sparing; stimulation produces flash of light and visual hallucination
Transverse gyri of Heschl (Brodmann areas 41 and 42)	Medial geniculate body	Wernicke's zone and medial geniculate body	Perception of auditory impulses and tonotopic representation	Due to bilaterality of auditory projections, no noticeable auditory deficits will be detected
Olfactory cortex (pyriform lobe)	Olfactory bulb via olfactory tracts and striae	Dorsomedial nucleus, orbitofrontal and insular cortex, amygdala, entorhinal cortex, and hypothalamus	Perception of olfactory impulses	Ablation of the olfactory cortex produces anosmia; irritation, as in uncinate fits, produces olfactory aura that precedes seizures

Premotor Cortex

The *premotor cortex* (Brodmann area 6) occupies part of the superior frontal gyrus and maintains functional and topographical representations similar to the primary motor cortex (Figures 8.19 and 8.32). The motor programs in this cortex regulate the activities that are essential for any motor activity, such as the rhythm and strength of contraction of the muscles. The *frontal motor eye field* (Brodmann area 8) controls conjugate eye movement to the opposite side (Figure 8.19, Table 8.2).

TABLE 8.2
Motor Cortex

Cortex	Major Afferents	Major Efferents	Functions	Deficits
Precentral and postcentral gyri, and paracentral lobule	Premotor (Brodmann area 6); postcentral gyrus (areas 3, 1, and 2); contralateral (Brodmann area 4 and areas 5 and 7); ventral lateral and ventral anterior nuclei	Corticospinal, corticopontine, corticotegmental, corticothalamic to ventral lateral and ventral anterior nuclei	Provides motor control to the alpha motor neurons and regulates reflexes and muscle tone	Contralateral spastic palsy, Babinski sign, hyperreflexia, clonus, and loss of superficial abdominal and cremasteric reflexes
Premotor cortex (Brodmann area 6)	Precentral gyrus, postcentral gyrus, superior parietal lobule (areas 5 and 7), ventral anterior and ventral lateral nuclei	Corticostriate, ventral anterior, and ventral lateral nuclei	Motor for skeletal muscles	Stimulation of this area produces generalized movements; when area 4 is destroyed, it mimics the activities of area 4
Frontal eye field (Brodmann area 8)	Corticobulbar (e.g., pyramidal)	Other parts of the cortex	Voluntary eye movements conjugately to the opposite side	Both eyes conjugately move to the side of the lesion
Supplementary motor cortex	Postcentral gyrus (areas 3, 1, and 2); precentral gyrus (area 4); VL and VL nuclei	Area 4, striatum;, spinal cord via the corticospinal tract	Controls contraction of the postural muscles	Ablation produces spastic contracture of flexor muscles and grasp reflexes

TABLE 8.3
Association Cortex

Cortical Areas	Major Afferents	Major Efferents	Functions	Deficits
Superior parietal lobule (Brodmann areas 5 and 7)	Lateral posterior, lateral dorsal nuclei; Brodmann areas 3, 1, and 2 (postcentral gyrus	Corticotegmental and corticopontine pathways	Integration of general sensory and visual information	Contralateral astereognosis and inability to recognize writing on skin, although ability to feel and and appreciate general weight and temperature remain unaffected
Brodmann areas 18 and 19	Brodmann area 17 and pulvinar	Midbrain tegmentum, tectum, pontine nuclei, and pulvinar	Interpretation of visual information and regulation of optokinetics and accommodation reflexes	Inability to recognize an object in the opposite field of vision; deficits in optokinetic and accommodation reflexes; stimulation of this area produces visual hallucinations
Brodmann area 22 and Wernicke's speech center (Brodmann area 22)	Primary auditory cortex (Brodmann areas 41 and 42) and other cortical areas	Other association cortices	Integrates auditory and visual	Sensory aphasia may result from unilateral lesion but is most pronounced when the damage is bilateral
Prefrontal cortex	Dorsomedial nucleus and other cortical areas	Pontine nuclei, dorsomedial nucleus, and other cortical areas	Personality, drive, emotion, intellect; affective component of pain	Marked changes in personality, behavior, and judgment, most evident when lesions are bilateral; stimulation produces changes in blood pressure, respiratory rate, gastric motility, etc.
Cingulate gyrus	Anterior nucleus of thalamus	Anterior nucleus of thalamus	Autonomic manifestations and pain perception	Vascular changes and changes in temperature regulation
Broca's speech center (Brodmann areas 44 and 45)	Wernicke's zone and other cortical areas	Contralateral Broca's center and motor nuclei of cranial nerves	Phonation	Broca's aphasia
Inferior parietal lobule (Brodmann areas 39 and 40)	Pulvinar and secondary visual cortex	Prefrontal and premotor and insular cortices	Spatial and three-dimensional perception	Inability to draw, place blocks, orient spatially, or identify contralateral body parts

Association Cortex

Association cortices (Table 8.3) include cortical areas that are located between visual, auditory, and somatosensory cortices that integrate generated auditory, visual, gustatory, and general sensory impulses. This integration serves a variety of functions, including recognition of shape, form, and texture of objects; awareness of body image; and relationships of body parts to each other and their location. These cortical areas also regulate the conscious awareness of body scheme, physical being, and recognition and comprehension of language symbols. They may also be involved in planning of motor functions and modulation of sensory impulses. Association cortices encompass the superior parietal lobule (Brodmann areas 5 and 7) and inferior parietal lobule, which comprise the supramarginal (Brodmann area 40) and the angular gyri (Brodmann area 39), the posterior part of the superior temporal gyrus (Wernicke's area—Brodmann area 22), and the secondary visual cortex (Brodmann areas 18 and 19). Information that pertains to language processing resides in the perisylvian cortex, an association cortex that forms the periphery of the lateral (Sylvian) fissure.

CORTICAL AFFERENTS

Cortical afferents are derived primarily from the spinal cord, cerebellum, and basal nuclei. They project to the cerebral cortex via the thalamocortical radiations (peduncles). The cortical afferents that originate from the spinal cord include the dorsal column–medial lemniscus and the spinal lemniscus. These afferents convey impulses to the sensory cortex via thalamocortical fibers that emanate from the ventral posterolateral nucleus. Other afferents from the spinal trigeminal and principal sensory nuclei are conveyed through the trigeminal lenmnisci to the ventral posteromedial thalamic nucleus. These fibers also project to the sensory cortex via thalamocortical fibers. Afferents, which carry motor information, are conveyed to the motor and premotor cortices from the cerebellum and basal nuclei via neuronal axons of the ventral lateral and ventral anterior thalamic nuclei. The limbic lobe and PFC receive afferents, subserving emotion, behavior, memory, and mood, from the anterior and dorsomedial nuclei of the thalamus. Visual and auditory cortical afferents derived from the lateral and medial geniculate nuclei are carried by the optic and auditory radiations, respectively. The sleep–wake cycle is partly regulated by cortical afferents from the intralaminar thalamic nuclei within the ARAS. Noradrenergic afferents that arise from the locus ceruleus project to all the cortical areas and laminae (especially to lamina I) via the central tegmental tract and internal capsule, bypassing the thalamus. These noradrenergic fibers may inhibit the moderately active cortical neurons and enhance the signal-to-background ratio. They do not affect the cortical sensory neurons.

An important and consistent feature of the thalamocortical fibers is their organization into the superior, inferior, anterior, and posterior peduncles. The *superior peduncle* contains fibers that connect the ventral posterior nucleus to the postcentral gyrus and the ventral anterior nucleus to the premotor cortex. It also contains fibers that connect the ventral lateral nucleus to the motor cortex and the lateral dorsal and lateral posterior nuclei to association cortices of the parietal lobe. The *inferior peduncle* contains the auditory radiations, projecting from the medial geniculate nucleus to the transverse gyri of Heschl via the sublenticular part of the internal capsule. The *anterior peduncle* contains fibers connecting the dorsomedial and anterior thalamic nuclei to the cingulate and prefrontal cortices. The *posterior peduncle* runs in the retrolenticular part of the internal capsule, connecting the lateral geniculate nucleus to the primary visual cortex of the occipital lobe.

CORTICAL EFFERENTS

Several areas of the cerebral cortex provide massive cortical efferents to the spinal cord, brainstem, and thalamus. Some of these fibers are destined to the spinal cord, while others terminate in subcortical area (e.g., thalamus, brainstem, etc.).

- A prominent projection of the cortex is represented in the *corticospinal tract* (described in detail later with the UMNs), which originates from motor, premotor, and somesthetic areas of the cerebral cortex and descends in the ventral part of the brainstem. Most fibers of the corticospinal tract decussate at the level of the caudal medulla, forming the lateral corticospinal tract, while the remaining ipsilateral fibers constitute the anterior and anterolateral corticospinal tracts.
- The *CBT*, also described with UMNs, is a cortical pathway that acts upon the motor nuclei of the cranial nerves.
- The *corticothalamic tracts* form part of the reciprocal pathway that connects certain areas of the cerebral cortex to the thalamic nuclei. These corticofugal pathways include projections from the primary motor cortex (Brodmann area 4) to the ventral lateral and centromedian nuclei, as well as projections from the PFC to the dorsomedial nucleus. They also encompass projections from the cingulate gyrus to the anterior nucleus and from the premotor cortex (Brodmann area 6) to the ventral anterior and ventral lateral nuclei. In addition, cortical projections from premotor and motor cortical fibers to the intralaminar nuclei, and from the primary sensory cortex to the ventral posterolateral and ventral posteromedial nuclei, are also included. Projections from the primary auditory and visual cortices to the medial and lateral geniculate nuclei, respectively, constitute additional corticothalamic pathways. It is interesting to note that the thalamic reticular nucleus receives afferents from all areas of the cortex with no reciprocal projections to this area.
- Fibers of the *corticopontine tract* that project to the pontine nuclei maintain diverse origin from the

pyramidal neurons in layer V of the primary motor and prefrontal cortices (frontopontine), primary sensory cortex (parietopontine), temporal lobe (temporopontine), and visual cortex of the occipital lobe (occipitopontine). These fibers show a somatotopic arrangement in the crus cerebri of the midbrain, in which the frontopontine fibers occupy a medial position to the parieto-occipito-temporo-pontine fibers. This tract, the ipsilateral component of a major pathway known as the cortico-ponto-cerebellar, enables the cerebral cortex to influence cerebellar activity.

- A third group of fibers form the *corticoreticular tract*, which originates from the motor, premotor, visual, and auditory cortices and terminates in the nucleus reticularis gigantocellularis of the medulla and the nucleus reticularis pontis oralis of the pons. This pathway may be partially responsible for signs of decerebrate rigidity, which is seen upon transection of the brainstem at the intercollicular level.
- Another cortical projection, the *corticotectal tract*, consists of fibers from the secondary visual and primary auditory cortices as well as the frontal eye field. Visual tracking movements are regulated by fibers from the secondary visual cortex (Brodmann areas 18 and 19) that projects to the superficial layers of the ipsilateral superior colliculus, whereas control of saccadic eye movements is maintained by the projections of frontal eye field (Brodmann area 8) to the middle layers of the superior colliculus. In the same manner, movements of the eye and the head toward auditory stimuli are mediated by the projections of the primary auditory cortex to the deep layers of the superior colliculus.

The *corticorubral tract* is a component of the cortico-rubro-spinal tract, which originates from the motor and premotor cortices of both hemispheres and descends to terminate somatotopically in all areas of the red nucleus. Note that the ipsilateral corticorubral fibers from the motor cortex mainly terminate in the magnocellular part of the red nucleus, whereas the contralateral fibers from the premotor and motor cortices terminate in the parvocellular part of the red nucleus.

Finally, fibers from all areas of the cerebral cortex form the *corticostriate*, which project bilaterally to all regions of the caudate and putamen. Bilateral corticostriate fibers are primarily derived from the motor, premotor, and sensory cortices and are somatotopically arranged. In order for the contralateral fibers to reach the striatum, they cross in the corpus callosum and pursue their course in the subcallosal fasciculus. In particular, the caudate nucleus, a component of the striatum, receives input primarily from the PFC.

Cortical afferents accompanied by cortical efferents are contained in the *internal capsule*, a V-shaped bundle that consists of anterior the limb, genu, and posterior limb with retrolenticular and sublenticular parts (Figures 8.20, 8.21,

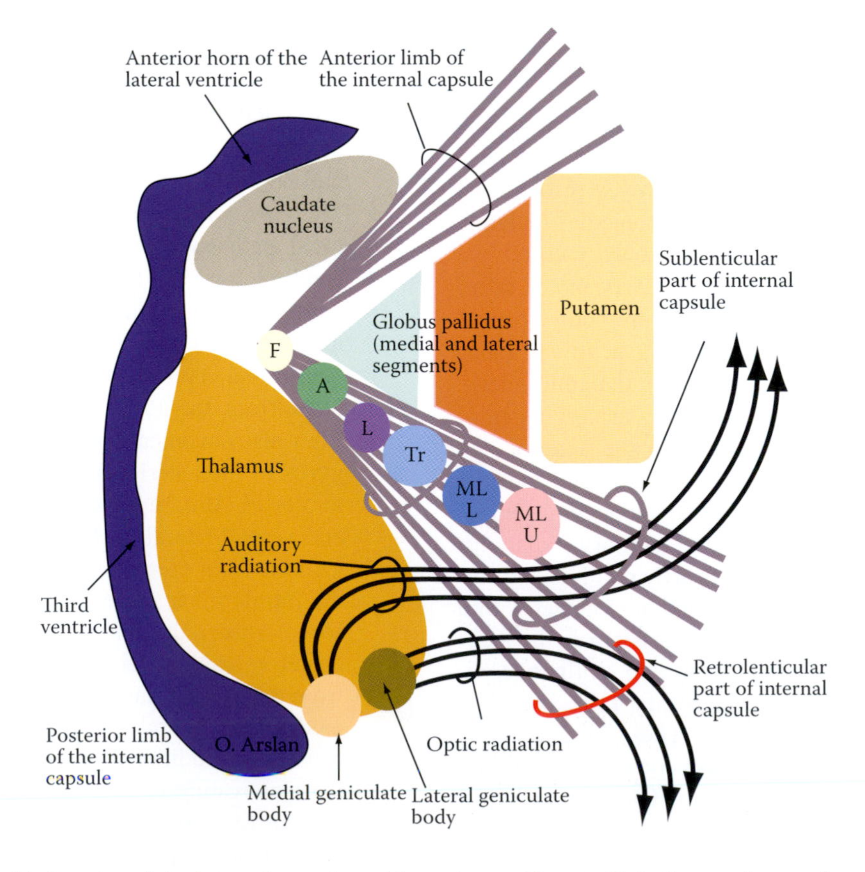

FIGURE 8.20 Topographic location of the internal capsule and its main constituents. Note the massive number of fibers crossing through the posterior limb of the internal capsule. A, arm; F, face (corticobulbar fibers); L, lower extremity; MLL, medial lemniscus fibers from the lower extremity; MLU, medial lemniscus fibers from the upper extremity; Tr, trunk (corticospinal fibers).

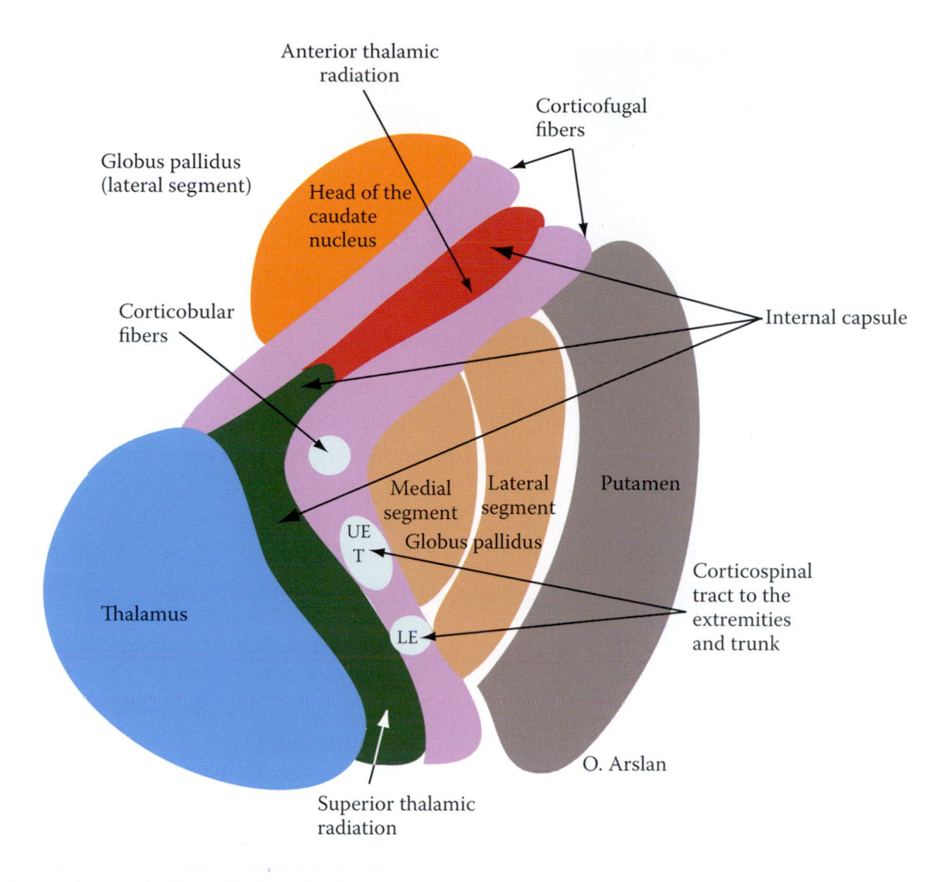

FIGURE 8.21 The internal capsule. Note its location between the thalamus, caudate nucleus, and the lentiform nucleus.

and 8.24). The corticothalamic and thalamocortical projections form the bulk of the internal capsule. An important aspect of the internal capsule is the somatotopic arrangement of its fibers. For instance, the anterior limb contains fibers that maintain reciprocal connections with the frontal lobe, the genu of the corpus callosum contains corticobulbar and corticospinal fibers whereas the posterior limb (lenticulothalamic) is much more massive, consisting of fibers of the corticospinal and frontopontine pathways and the superior thalamic radiation (spinothalamic tracts and medial lemniscus). It also contains the optic and auditory radiation in the retrolenticular and sublenticular parts, respectively. Due to this extensive concentration of sensory and motor fibers within, occlusion or hemorrhage associated with a small artery that supplies the posterior limb may lead to profound sensory or motor or combined deficits, as in lesions of the posterior limb. Visual and auditory deficits may also be observed if the retrolenticular and sublenticular portions are involved. Other deficits, such as spastic paralysis of the muscles of mastication, muscles of facial expression, and palatal muscles, may be observed if the corticobulbar fibers are involved.

Diverse arterial sources contribute to the blood supply of the internal capsule. In particular, the posterior communicating artery, which is most commonly hypoplastic, supplies the genu and a portion of the posterior limb of the internal capsule. Additional blood supply to the genu of the internal capsule may be derived from the internal carotid artery. The *medial striate artery* (recurrent artery of Heubner), which arises from the ACA, and sometimes from the MCA, pierces the anterior perforated substance to supply the rostral part of the anterior limb. The *anterior choroidal artery*, a branch of the internal carotid artery, runs above the uncus along the optic tract, and enters the inferior horn of the lateral ventricle via the choroidal fissure, to supply the posterior limb, including the retrolenticular part. The *lateral striate artery* (leticulostriate artery) supplies the anterior limb and the dorsal part of the posterior limb of the internal capsule.

Occlusion of the ACA proximal to the origin of the recurrent artery of Heubner may produce no detectable deficit if the collateral circulation from the corresponding artery of the opposite side is maintained. However, a lack of efficient anastomosis may result in infarction of the anterior limb of the internal capsule and destruction of the cortico-ponto-cerebellar tract, producing *frontal dystaxia* (partial ataxia in which the patient exhibits difficulty in controlling voluntary movements).

CEREBRAL WHITE MATTER

The white matter of the cerebral hemispheres lies deep to the gray matter of the cerebral cortex, consisting of nerve fibers and glial cells. It contains the basal nuclei and the central branches of the cerebral arteries. The fibers that course

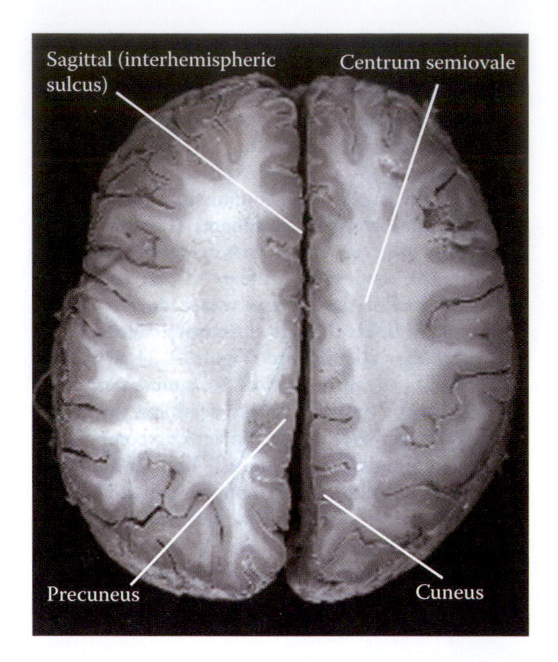

FIGURE 8.22 Horizontal section of the brain indicates the distinction between the central white matter and the peripheral cortical areas.

within the white matter are classified into commissural, association, and projection (Figure 8.22).

Commissural Fibers

The commissural fibers interconnect identical or nonidentical areas of the two cerebral hemispheres, forming the corpus callosum, anterior commissure, and hippocampal commissure. Interhemispheric communication that mediates the learning process is accomplished to a great extent by the *corpus callosum*, the largest of all commissures, which interconnects the cerebral hemispheres (Figures 8.23 and 8.25). However, exceptions exist in regard to this fact; for instance, the striate (primary visual) cortex and the hand area of the cerebral cortex do not project commissural fibers through the corpus callosum.

> Occlusion of the *posterior cerebral artery* may result in disruption of the splenium of the corpus callosum, preventing transfer of information from the right to the left visual cortices, a process that is essential for the visual recognition of objects. Therefore, patients with this type of lesion are able to read but unable to write or name colors. In normal individuals, the visual stimuli elicit nonverbal associations (tactile, taste, or smell), which are transmitted across the intact anterior part of the corpus callosum.
>
> Destruction of the anterior part of the corpus callosum (e.g., due to infarction of the ACA (*ACA syndrome*), distal to the origin of the anterior communicating artery, prevents verbal information from being conveyed from language centers of the left hemisphere to the appropriate centers of the right (mute) hemisphere. This may produce left arm apraxia in which the patient is unable to identify numbers or letters

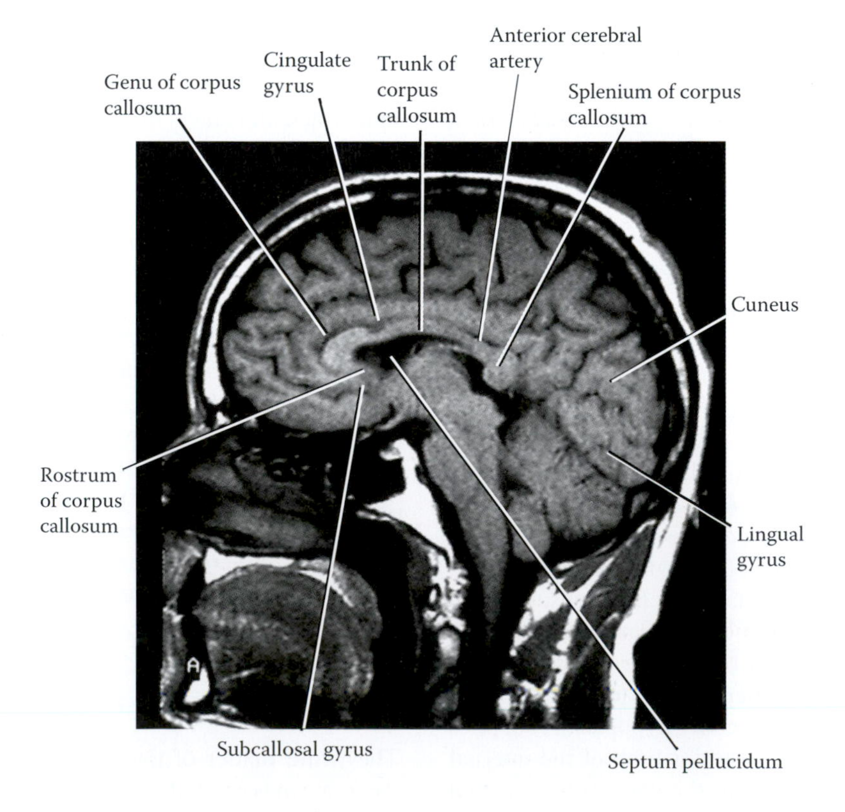

FIGURE 8.23 Inverted MRI showing components of the corpus callosum and their relationships to individual cerebral lobes, gyri, and the septum pellucidum.

written by a blunt object on the affected left extremity or perform movements using the left arm and leg upon verbal or written commands, although normal spontaneous movements are maintained. Due to infarction of the rostral part of the corpus callosum, the patient with left arm apraxia is also unable to name an object placed in the left hand or write or print with the left hand. Damage to the corpus callosum may also occur as a result of excessive consumption of red wine (Marchiafava–Bignami syndrome), producing subtle and variable manifestations.

Complete failure of the corpus callosum to develop, agenesis of the corpus callosum, may occur during the 4th to the 12th week of development. Agenesis of the corpus callosum, a congenital anomaly, is accompanied by a partial or complete absence of the cingulate gyrus and septum pellucidum, or the appearance of ipsilateral longitudinal fibers, which include some that fail to cross the midline. This form of agenesis is not commonly associated with neurological deficits, although mild mental retardation, seizure, and motor deficits may be seen. Another form of this condition is associated with development of small, multiple gyri (micropolygyria) and heterotopias of the gray matter. In general, agenesis of the corpus callosum is associated with inherited metabolic disorders (pyruvate dehydrogenase deficiency, glutaric aciduria type II) and with chromosomal abnormalities as in Dandy–Walker, Aicardi, and Cogan syndromes.

Dandy–Walker syndrome is another condition that is associated with agenesis (failure of development) of the corpus callosum. In this syndrome, the cerebellar vermis fails to develop, and the fourth ventricle is replaced by a cyst-like midline structure. There is thinning of walls of the posterior cranial fossa, and the roots of the cervical spinal nerves assume an ascending course in order to reach their corresponding foramina. Agenesis of the corpus callosum is also seen in Aicardi and Cogan syndromes. Patients with Aicardi syndrome exhibit seizure, microcephaly, and hemivertebra, while in Cogan syndrome, keratitis and vestibulo-auditory disorders are the predominant manifestations.

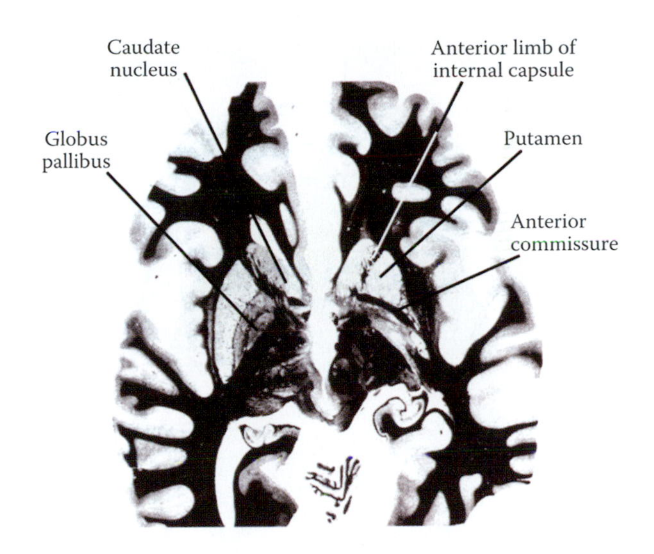

FIGURE 8.24 The course of the anterior commissure as a midline structure and its continuation inferior to the lentiform nucleus are illustrated.

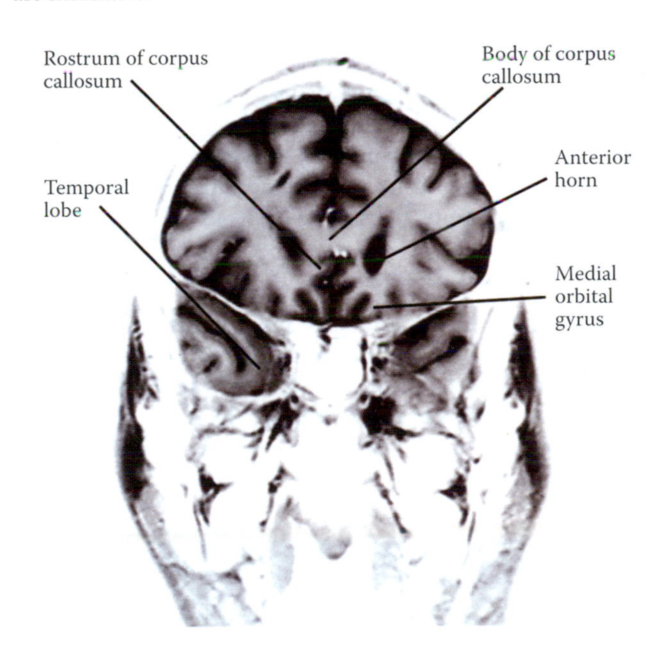

FIGURE 8.25 An inverted MRI scan (coronal section) of the brain through the anterior horn of the lateral ventricle. Observe the body and rostrum of the corpus callosum, individual cerebral lobes, gyri, and the septum pellucidum.

The *anterior commissure* (Figures 8.24, 8.25, and 8.26) is embedded in the upper part of the lamina terminalis, superior to the optic chiasma, resembling the handle of a bicycle. It divides the fornix into precommissural and postcommissural columns. This commissure exhibits an anterior smaller portion, which connects the olfactory tracts to the anterior perforated substances of both sides, and a posterior larger portion, interconnecting the parahippocampal and the middle and inferior temporal gyri, as well as the amygdaloid nuclei. In addition, the anterior commissure interconnects the anterior olfactory nuclei, diagonal bands of Broca, olfactory tubercles, prepyriform cortices, entorhinal areas, nucleus accumbens septi, bed nuclei of stria terminalis, and frontal lobes.

Another small yet important *posterior commissure* exists caudal to the pineal gland and rostral to the superior colliculi. This commissure contains fibers that arise from the habenular and pretectal nuclei and from the interstitial nucleus of Cajal, nucleus of Darkschewitsch, and the nucleus of the posterior commissure as well as from the superior colliculi. Additional interhemispheric linkage is secured by the hippocampal commissure, which extends between the crura of the fornix and hippocampal gyri, and inferior to the splenium of the corpus callosum.

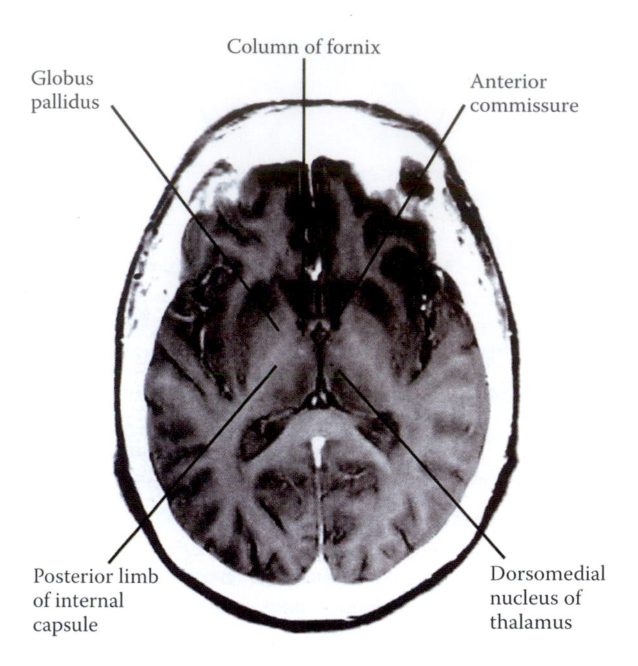

FIGURE 8.26 An inverted MRI scan (horizontal view) of the brain through the frontoparietal operculum. Note the curved midline anterior commissure rostral to the columns of the fornix. The splenium of the corpus callosum and the posterior horn of the lateral ventricle are also visible.

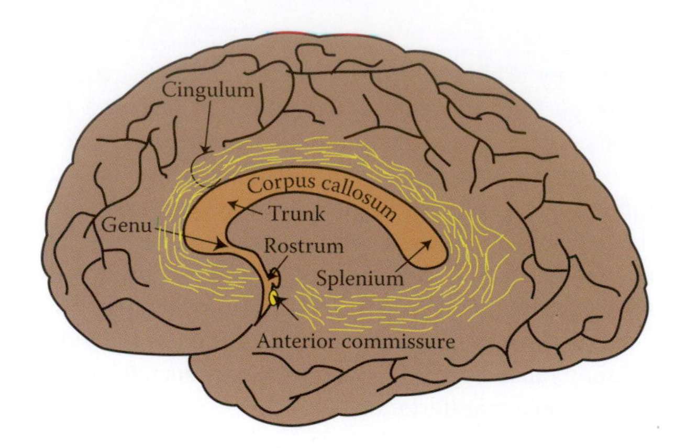

FIGURE 8.27 The cingulum within the cingulate gyrus. Observe its relationship to the corpus callosum.

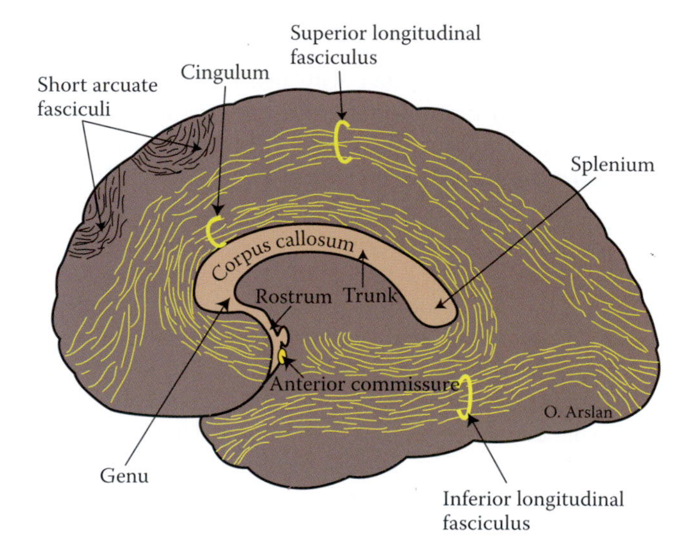

FIGURE 8.28 Midsagittal drawing of the brain. The superior and inferior longitudinal fasciculi are illustrated.

Association Fibers

Association fibers are axons within the white matter that establish linkage between areas within the same cerebral hemispheres, consisting of short and long fasciculi. Short fasciculi interconnect adjacent gyri within the same cerebral hemisphere. Long fasciculi, which include the cingulum, uncinate, superior longitudinal, inferior longitudinal, and superior and inferior occipitofrontal fasciculi, connect distant areas of the cerebral hemispheres.

Structures that comprise the limbic system, such as the subcallosal and paraterminal gyri rostrally and the parahippocampal and adjacent temporal gyri caudally, are interconnected by the cingulum, which is a curved bundle of association fibers that lies deep to and follows the course the cingulate gyrus (Figures 8.27 and 8.28). Broca's speech center is connected to the rostral parts of the temporal gyri via the *uncinate fasciculus*, enabling input generated or integrated in the temporal lobe to influence articulation (Figures 8.29, 8.30, and 8.31).

A group of large association fibers that course lateral to the corona radiata and the internal capsule form the *superior longitudinal fasciculus* (arcuate fasciculus) that connects the rostral parts of the frontal lobe to the secondary visual cortex (Brodmann areas 18 and 19) and the parietal and temporal gyri (Figures 8.28, 8.30, and 8.31).

Bilateral disruption of the superior longitudinal fasciculi may produce visual agnosia, which is characterized by the inability to name an object or describe its function.

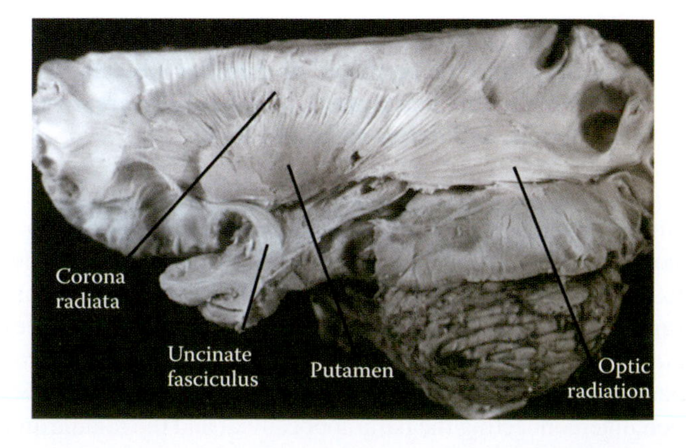

FIGURE 8.29 The lateral surface of the brain after removal of the frontal and temporal gyri. The fibers of the corona radiata are shown.

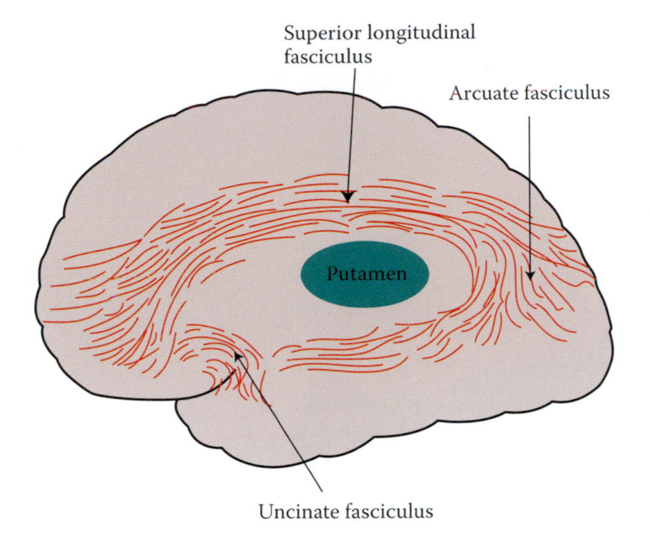

FIGURE 8.30 Schematic drawing of some of the association fibers within the cerebral hemisphere.

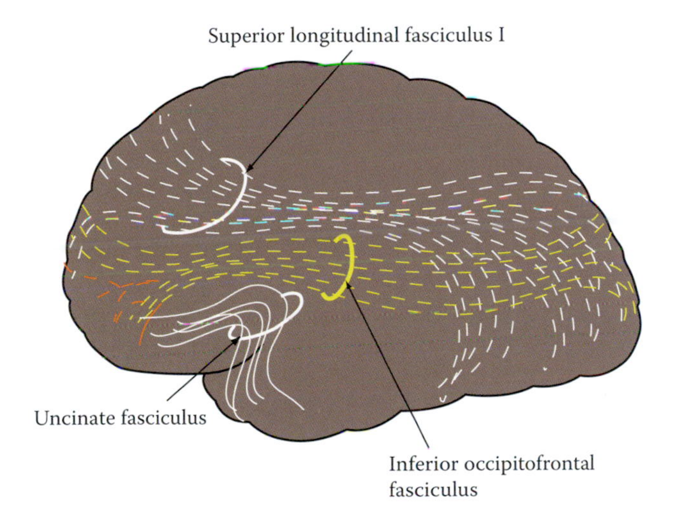

FIGURE 8.31 Diagrams of the inferior occipitofrontal, superior longitudinal, and uncinate fasciculi are illustrated to emphasize their relative locations.

Another group of association fibers forms the *inferior longitudinal fasciculus* (Figure 8.28), which extends from the secondary visual cortex to the inferotemporal cortex and courses lateral to the occipital horn of the lateral ventricle, separated by the optic radiation and tapetum.

Frontal and occipital lobes are also connected by the *superior occipitofrontal fasciculus* (subcallosal fasciculus) that lies inferior and lateral to the corpus callosum. This fasciculus is separated from the superior longitudinal fasciculus by the corona radiata. Additional bundle, the *inferior occipitofrontal fasciculus*, connects the frontal to the occipital lob and runs within the temporal lobe proximal to the uncinate fasciculus. It has been suggested that the uncinate fasciculus is part of the inferior occipitofrontal fasciculus (Figure 8.31).

Projection Fibers

Projection fibers are corticofugal axons that extend from the cerebral cortex to the subcortical nuclei, brainstem, and spinal cord. They form the corona radiata that runs between the superior longitudinal and the superior occipitofrontal fasciculi. They run within the internal capsule and may descend to terminate in the spinal cord or other subcortical areas. These projection fibers include the corticospinal, corticobulbar, corticostriate, corticopontine, and so forth.

CEREBRAL DYSFUNCTIONS

Cerebral lesions occur in a variety of diseases and conditions such as Alzheimer's disease (AD), encephalitis, Parkinsonism, pseudobulbar palsy (associated with destruction of the CBTs), and progressive supranuclear palsy. These diseases may produce speech and language disorders that may include speech derangement as in AD, palilalia (compulsive repetition of a phrase with increasing speed and decreasing volume), progressive supranuclear palsy, and *aphasia*. Other cerebral lesions produce *apraxia* (inability to perform familiar motor activity without sensory or motor damage) or *agnosia* (inability to recognize objects and symbols despite intactness of sensory pathways). Cortical dysfunctions may also produce *dementia* and *seizures*, which will be discussed later in this chapter.

Aphasia

Aphasia is the inability to comprehend the spoken and written language and express thoughts via words despite the fact that sensory systems, mechanisms of articulation, and the associated structures (Figures 8.32 and 8.33) are intact. This disorder should not be confused with other speech abnormalities that occur in cerebellar dysfunctions, hypoglossal or vagus nerve damage (changes in speech tone), or UMN palsy. Generally, the severity of aphasia depends upon the site of cortical damage and duration of the disorder. Although complete recovery does not always occur, the earlier the onset, the better the chance of recovery. Speech areas are comprised of the angular and opercular parts of the inferior frontal gyrus (Broca's motor speech center), posterior part of the superior temporal gyrus (Wernicke's receptive speech center), and the fibers that connect these centers in the ipsilateral and contralateral hemisphere. The following disorders involve cognitive functions (prepositional features) of language in the dominant hemisphere (usually the left).

Broca's aphasia (nonfluent, expressive, motor, verbal aphasia) (Figures 8.32 and 8.33) is seen in individuals with lesions of Broca's speech center (Brodmann areas 44 and 45). This condition involves disorders of speech and written language. It is characterized by

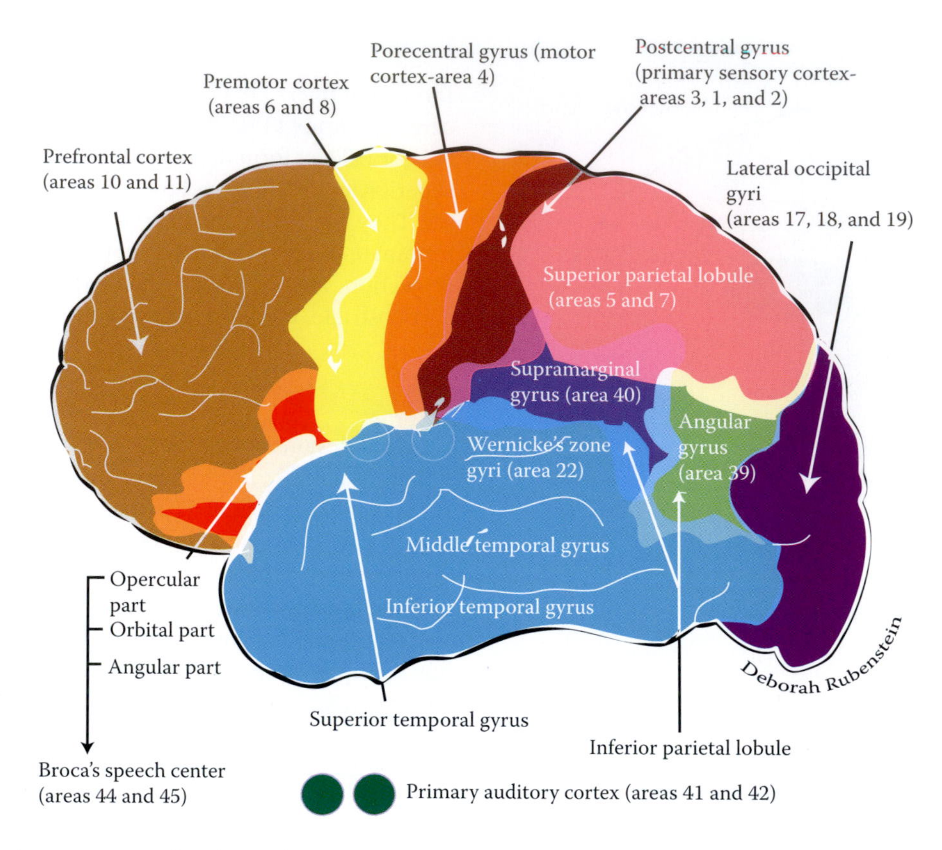

FIGURE 8.32 Prominent centers and gyri associated with the frontal, parietal, and temporal lobes. Broca's speech center and Wernicke's zone are distinctly illustrated.

difficulty initiating or repeating words with restricted vocabulary and grammar (agrammatic). Patients have difficulty in word retrieval and have a labored and awkward speech with a tendency to delete adverbs and adjectives as well as connecting words. Although comprehension of the spoken and written language is usually preserved, writing is severely affected. Due to disruption of the ipsilateral corticospinal, corticobulbar, and optic radiation, Broca's aphasia often is associated with right-sided hemiplegia, supranuclear

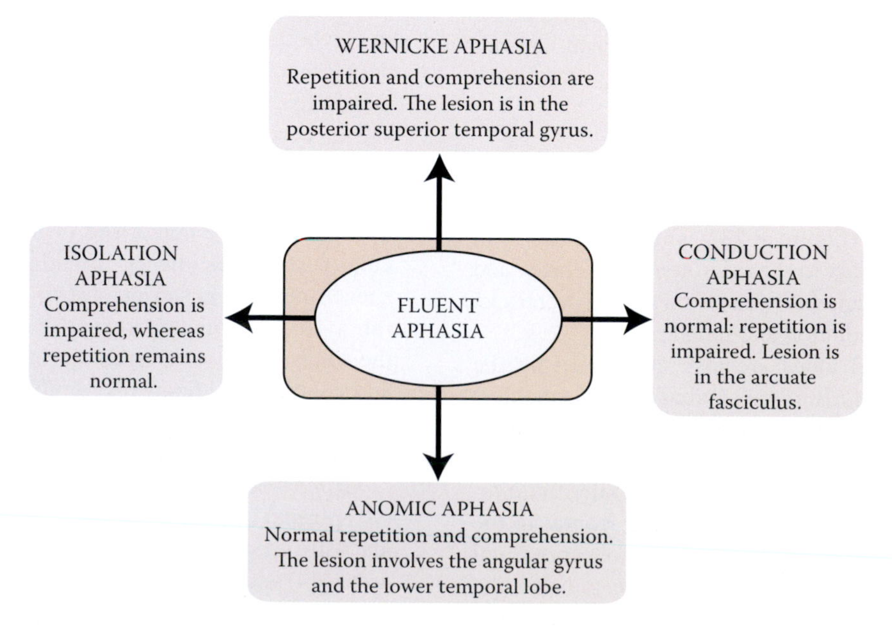

FIGURE 8.33 Schematic representations of various forms of fluent aphasia.

facial palsy (lower facial palsy), and right-sided homonymous hemianopsia. Tongue and lip movements are usually impaired. Involvement of the frontal eye field may result in conjugate deviation of both eyes to the left side.

Wernicke's aphasia (receptive, fluent, sensory, syntactic, acoustic aphasia) results from destruction of the association cortex that occupies the posterior portion of the superior temporal gyrus (Brodmann area 22) of the dominant hemisphere (Figures 8.32 and 8.33). Since Wernicke's area integrates verbal memory, visual, and sound patterns with the correct word phonemes that are essential elements for reading and writing, destruction of this area may impair the comprehension of spoken language and ability to read or write and find appropriate words (circumlocution). Patients also exhibit enormous verbal output, and a tendency to enhance speech by irrelevant word substitution, such as "fen for pen," known as *verbal paraphasia*. Sound transpositions (phonemic paraphasias) or substitution of phonemes (*literal aphasia*), and increased rate and pressure of speech (*logorrhea*), as well as unwillingness to terminate speech may also be observed. Although the grammar of the spoken language is not correct, the free usage of a variety of tenses in an unusual combination gives the impression of a speech with proper syntax. For an examiner who does not speak the patient's language, observing any speech dysfunctions may not be an easy task. Since patients remain unaware of the deficit, their speech conveys little meaning. For example, patients with Wernicke's aphasia have trouble repeating statements. Patients with Wernicke's aphasia may be labeled psychotic because of the similar nature of their speech to a thought disorder of frontal lobe origin. Patients may tend to develop a mania-like psychosis characterized by hyperactivity, rapid speech, euphoria, or irritable mood. Some patients with Wernicke's aphasia may exhibit *superior quadranopsia* (one-fourth blindness), indicating involvement of the geniculocalcarine pathway (Meyer's loop) as it courses through the temporal lobe. Individuals with Wernicke's aphasia may also exhibit a striking lack of concern (a finding that is not seen in individuals with Broca's aphasia) that may be replaced by paranoid behavior.

Anomic aphasia (semantic, amnesic aphasia, Figure 8.33) may occur at the end of Wernicke's aphasia or may be present as a distinct disorder. It is characterized by the inability to find words with relatively intact comprehension. Individuals with this condition do not exhibit paraphasia but lack substantive words in their speech. It is associated with alexia and agraphia, and occasionally with right superior quadranopsia. It can be caused by an injury to the parieto-occipital cortex, which may extend to involve the angular gyrus in the left dominant hemisphere. This condition may be an early language disturbance detected with expanding brain tumors.

A lesion that disrupts the arcuate fasciculus connecting Wernicke's zone to Broca's speech center causes conduction aphasia (central aphasia, Figure 8.33). The main deficit in this type of aphasia is the inability to repeat words and choose and sequence phonemes. Comprehension of silent reading remains intact, but the ability to read aloud is lost. Reading aloud requires intactness of the association visual cortex, splenium of the corpus callosum, angular gyrus, Wernicke's area, Broca's speech area, arcuate fasciculus, and motor cortex. Patients clearly become aware of their language deficit, especially when they entangle a key word. Some degree of apraxia may also be observed in the limb and facial muscles.

Transcortical aphasia (Figure 8.33) occurs as a result of compromise of the blood supply to the watershed areas surrounding the speech centers in the frontal and parietal lobes. These areas are supplied by the middle, anterior, and posterior cerebral arteries.

Transcortical sensory aphasia (isolated speech area syndrome) is a rare type of fluent aphasia, which may result from a cerebrovascular accident in the watershed areas of the parieto-occipital cortex and at the junctions of the middle, anterior, and posterior cerebral arteries. This vascular lesion disrupts the connection between Broca's and Wernicke's centers and other parts of the brain. Patients are unable to start conversation, and the response to a question generally contains confabulatory, automatic, irrelevant, and repetitious paraphasia. Repetition may take the form of echoing (parrotlike), word phrases, and melodies (*echolalia*). Although reading and writing abilities are abolished, articulation of memorized materials remains intact.

Transcortical motor aphasia (dynamic aphasia) (Figure 8.33) is a form of nonfluent aphasia (speech is reduced to less than 50 words per minute) in which repetition is unaffected and the speech is grammatically accurate. Inability to initiate conversation is the primary deficit, although comprehension of sounds and written language remains functional. Anoxia, or multiple infarctions, that disrupts the connections of Broca's area to the rest of the frontal lobe may be responsible for this disorder.

Transcortical motor and sensory aphasias are produced by lesions of the area surrounding the lateral cerebral fissure. Automatic repetition of words (*echolalia*) is the main speech-related function performed by these types of individuals.

Global aphasia results from a lesion that destroys Broca's and Wernicke's centers and the connecting arcuate fasciculus. There is a loss of ability to comprehend, articulate, read, write, and name a viewed object. It is usually associated with right homonymous hemianopsia, right hemiplegia, and right hemianesthesia. *Aphemia* (subcortical motor aphasia) is characterized by the inability to imitate, repeat, or produce sounds, with preservation of the ability to read and write. Auditory comprehension and word finding also remain intact. This condition may result from destruction of the output from Broca's speech center. *Subcortical sensory aphasia* (pure word deafness) results from destruction of the primary auditory cortex and the transcallosal fibers that carry information from the nondominant hemisphere. Since the Heschl gyrus is damaged on the left (dominant) hemisphere and no sound is able to reach Wernicke's zone from the opposite hemisphere, comprehension and repetition of the spoken word are not possible.

Alexia with agraphia is seen in individuals with a lesion involving the angular gyrus in the inferior parietal lobule. It is characterized by impairment of the ability to read and write and inability to comprehend symbols and words. Comprehension of sounds and the ability to articulate are not affected. Due to the proximity of the angular gyrus to the temporal lobe, as well as other areas of the parietal lobe, additional deficits may also be seen, including anomic aphasia, loss of right–left recognition, and ability to identify fingers.

Pure alexia or alexia without agraphia (pure word blindness) is a rare condition that occurs as a result of disruption in the visual association cortices (Brodmann areas 18 and 19) and the splenial fibers that convey visual information to the visual cortex and angular gyrus of the dominant hemisphere. It may be caused by occlusion of the left posterior cerebral artery. Thus, processing of auditory, visual and phonological aspects of language is disrupted. Patients exhibit right homonymous hemianopsia due to destruction of the left visual cortex. Disruption of the transcortical splenial fibers can result in the inability of patients to recognize words or letters, even their own. Since visually presented objects excite a variety of sensory systems in the mute hemisphere (tactile, taste, and olfactory) that are conveyed by the intact callosal fibers, naming these objects is possible. Generally, aphasics develop alexia because of the difficulty in comprehending the meaning of words, conversion of grapheme to phoneme (Wernicke's aphasia), or formulating grammar (Broca's aphasia).

Developmental dyslexia is a common CNS-based learning disability in which the patient exhibits impairment in reading and deficits that range from phonological awareness, decoding, processing, voice recognition, and orthographic coding (spelling mechanism) to verbal short-term memory, expressive language skills, and naming ability. Dyslexics have the ability to read and comprehend but at a peculiarly slow pace due to an impaired visual–verbal connection. However, they are poor spellers but with a quite-normal intelligence. They are unaware of the correlation between graphemes and phonemes and unable to link cognitive skills with visual processing. Dyslexics are easily distracted and, stressed, they experience difficulty piecing words into distinct sounds, utilizing sounds to create a word, or counting syllables in a word. They have difficulty naming and spelling. This condition may be seen with hyperactive attention deficit disorders, dysgraphia, and dyspraxia. Due to distinct neuroanatomical pathways for the spelling, reading, and writing in each language, a dyslexic in one language may not exhibit the deficit in another language. There is a body of evidence that points to unique structural defects in the brain of a dyslexic. A reduction in the sizes of Broca's speech center, Heschl gyri, insular cortex, magnocellular part of the lateral geniculate body (temporal visual processing), visual cortices, thalamus, inferior parietal lobule, and middle and inferior temporal lobe of the left cerebral hemisphere is seen in the brains of dyslexics. Therefore, brains of dyslexics exhibit reduced or lack of asymmetry between cerebral hemispheres and a decline of activity in the visual association cortices.

Pure agraphia (aphasic agraphia) is a rare condition seen in individuals with lesions of the angular gyrus. Reading remains intact, but writing and spelling are severely affected. This form of aphasia may also be produced by a lesion of the motor association cortex of the frontal lobe.

Disconnection syndromes, as discussed earlier, represent a constellation of deficits seen in complete transection of the corpus callosum. Patients with these syndromes may exhibit agraphia in the left hand, but not the right, as well as apraxia (ability to execute oral commands or perform familiar tasks is lost in the absence of sensory or motor deficits).

Hemioptic aphasia is characterized by the inability to name objects seen in the left visual field while maintaining the ability to recognize these objects via the left hand, which is guided by the right hemisphere. This syndrome is seen in individuals with bilateral epilepsy subsequent to surgical transection of the corpus.

Tactile aphasia is a condition in which the patient is unable to identify objects placed in the left hand but is able to do so when the object is placed in the right hand. As the right nondominant hemisphere, which has lost its connection to the left hemisphere (due to disruption of the callosal fibers), is responsible for the recognition of an object, identification of an object placed in the left hand will not be possible.

In summary, the following sequences may be of value in determining the role each center plays in naming a visual object:

Object → optic nerve and tract → lateral geniculate nucleus → Brodmann area 17 of the visual cortex → Brodmann area 18 → Angular gyrus (Brodmann area 39) → Wernicke's area (area 22)—pattern is formed → arcuate fasciculus → Broca's motor speech center (Brodmann areas 44 and 45) → facial area of the motor cortex → muscles of speech via certain cranial nerves.

Aprosodia is a condition in which the affective component of language, such as musical rhythm; facial expression; and comprehension of these gestures is lost or impaired. These disturbances accompany lesions of the nondominant hemisphere (usually the right). A lesion that involves the inferior frontal gyrus of the right hemisphere (mirror image of Broca's area) produces a speech that is monotonous; agestural (lacks gestures); and lacks melody, timing, and other affective content (*motor aprosodia*). These individuals tend to have difficulty in conveying tune, singing, and understanding emotional reactions. They exhibit hemiplegia and lower facial palsy on the left side. A lesion that destroys the posterior part of the superior temporal gyrus in the right hemisphere (mirror image of the Wernicke's zone) may result in the inability to recognize and comprehend the emotional content of the spoken language (*sensory aprosodia*). The emotional gestures produced will not be appropriate for the occasion and to the content of the speech.

Alexithymia, frequently seen in patients with psychosomatic disorders, is characterized by the inability to express emotion through words. Disruption of the connection between the affective (right) hemisphere and the expressive left hemisphere may account for this deficit.

Apraxia

Apraxia is a disorder characterized by the inability to execute learned voluntary functions without any detectable motor or sensory deficits, mental disability, or comprehension deficits. Apraxia is classified into kinetic, ideomotor, parietal, callosal, sympathetic, ideational, constructional, and buccofacial apraxia.

Kinetic apraxia is characterized by the inability to perform fine, skilled movements in one extremity, often due to a lesion of the contralateral primary motor cortex (Brodmann area 4).

Ideomotor apraxia is a disorder in which patients exhibit an inability to perform a given task upon command, despite retaining aptness to execute acts automatically, such as opening or closing the eyes. This deficit may be categorized into parietal, callosal, and sympathetic apraxia.

Parietal apraxia is due to destruction of the arcuate fasciculus that establishes a connection between the motor centers in the frontal lobe and the centers that formulate motor activities in the parietal lobe. This form of apraxia is bilateral and may be associated with conduction aphasia.

Callosal apraxia, as the name indicates, is produced by disruption of the genu of the corpus callosum that connects the premotor areas of both hemispheres. Thus, information generated in the upper extremity area of the premotor cortex of the left frontal lobe does not reach the corresponding area of the right frontal lobe. Since the motor activity is conveyed via the corticospinal tract, which is a crossed pathway, this deficit will be seen in the left arm.

Sympathetic apraxia is another form of ideomotor apraxia, resulting from damage to the premotor area of the left frontal lobe, an area adjacent to Broca's speech center. As discussed earlier, the premotor area of the left lobe provides motor commands to the corresponding area on the right side via the corpus callosum. Since the corticospinal tract is partly derived from the premotor area, the generated motor impulses will be conveyed to the spinal levels on the contralateral side. Therefore, destruction of the left premotor area produces a paralyzed right limb and an apractic left limb (in sympathy with the affected right limb). Patients with this lesion may also exhibit similar deficits in the buccofacial muscles and possible aphasia due to proximity of the lesion to Broca's center in the left hemisphere.

Constructional apraxia occurs as a result of a lesion of either the left or right parietal lobe or the connecting callosal fibers. Loss of perception of temporal relationships is manifested by the inability to copy simple figures. These patients do not have impairments in eating and chewing, and the ability to sing may be retained even though speech might be impaired.

Buccofacial apraxia, commonly associated with nonfluent aphasia, is characterized by the inability to move the tongue and facial muscles, including the muscles around the mouth, upon command.

Ideational apraxia is characterized by the inability to perform a complex task or series of acts in a purposeful manner and in a proper sequence. It is caused by a lesion in the parietal lobe of the dominant hemisphere or as a result of a diffuse brain disorder (dementia). It stems from the loss of ability to appreciate and formulate the idea necessary to carry out a complex task, although individual acts within the task can be executed without any difficulty.

Agnosia

Agnosia refers to the failure to recognize and understand the symbolic significance of sensory stimuli, despite the presence of the learned skill, intactness of the sensory receptors and pathways, and absence of mental disorders or dementia. This deficit is commonly associated with disruption of the connection of the primary, secondary, or tertiary sensory cortices with the association cortical areas that store memories for the stimulus. Agnosias are classified into visual, auditory, and tactile agnosia; simultanagnosia; anosognosia; Babinski agnosia; reduplicative paramnesia; asymbolia; and apractognosia.

Visual agnosia is the inability to recognize objects by vision, while retaining the capability to identify the same objects with other sensory modalities. Individuals with this condition do not have defects in the visual apparatus or pathway and can recognize people. This disability may change in severity from time to time. Lesions are generally confined to the visual association cortex of the temporal and parietal lobes and the connecting fibers to this cortex. Visual agnosia, a primary characteristic of *Klüver–Bucy syndrome*, may take the form of alexia, protopagnosia, facial and finger agnosia, graphagnosia, and color, auditory, and tactile agnosia.

Alexia is a form of visual agnosia that is characterized by the inability to read as a consequence of failure to recognize the written words.

Protopagnosia is a disability resulting from bilateral destruction of the occipitotemporal gyri. Patients with this condition are unable to identify familiar people, or they may identify them as impostors (*Capgras syndrome*), as strangers (*reverse Fergoli syndrome*), or as strangers mistaken for familiar people (*Fergoli syndrome*). Patients also are unable to identify familiar objects removed from their common visual context.

Facial agnosia is the inability to identify people by their faces, although identification of these individuals is possible via their voice, demeanor, or dress.

Finger agnosia (anomia) is a deficit characterized by the incapacity to identify and name fingers. This dysfunction is often seen in *Gertsmann syndrome* and is associated with anarithmetria (loss of arithmetic concept), agraphia or dysgraphia, and left–right disorientation of the body parts and objects. This syndrome is a combination of agnosia and apraxia and is produced by a lesion of the angular and supramarginal gyri of the dominant cerebral hemisphere.

Graphagnosia refers to the inability to recognize letters or numbers written on the palm.

Color agnosia refers to inability to identify the color of an object, despite retaining the capacity to match cards of different colors. It is not a sex-linked deficit, and lesions commonly affect the inferomedial parts of the occipital and temporal lobes. Patients with this deficit do not have color blindness.

Auditory agnosia occurs as a result of bilateral or unilateral destruction of the Wernicke's zone in the dominant hemisphere and is characterized by the inability to recognize familiar sounds with no detectable deficits in the auditory system or pathway.

Tactile agnosia is characterized by the inability to recognize objects through touch without impairment of the tactile receptors, spinal nerves, or ascending sensory pathways. The primary lesion is in the supramarginal gyrus of the dominant hemisphere (left), and to a lesser degree in the postcentral gyrus. *Astereognosis* refers to the loss of ability to recognize the texture, form, and shape of an object by tactile sensation (e.g., inability to recognize a coin placed in the palm of the hand).

Agnosias are also categorized into simultanagnosia, anosognosia, Babinski agnosia, reduplicative paramnesia, asymbolia, and apractognosia.

Simultanagnosia (spelling dyslexia) refers to the inability to recognize sensory stimuli received simultaneously and the inability to read except for the shortest words. This condition is a manifestation of a lesion of Brodmann area 18, or dysfunctions in scanning of visual images. It may be associated with visual deficits.

Anosognosia (asomatognosia), an important part of "hemispatial neglect," is characterized by indifference to or rationalization of the symptoms, or denial of an illness of serious nature. Patients appear inattentive and experience *allocheira*, which refers to an individual's experience of a stimulus, applied to one side of the body and being felt on the opposite side. Allocheira results from a lesion of the superior parietal lobule of the right (nondominant) hemisphere.

Babinski agnosia refers to the neglect or even the denial of existence of a paralyzed limb in a paralytic patient.

Reduplicative paramnesia is a striking delusional abnormality in which the patient perceives events, familiar places, and persons (time, place, and object) to have been duplicated. Patients may believe that their close relatives had been replaced by indistinguishable proxies. Capgras syndrome is believed to be a form of this syndrome. This syndrome has been reported in patients who have endured acute vascular lesion of the right frontemporo-occipital cortices. It may be accompanied by impairments of learning and memory, conceptualization, and executive nonverbal functions. However, language and visuospatial skills and motor speed may remain unaffected. It has been postulated that the distorted sense of familiarity and impaired ability to resolve the delusion may possibly result from temporal–limbic–frontal dysfunction and inability to integrate perceptual and memory-related information.

Asymbolia refers to the inability to comprehend familiar signs, symbols, gestures, hand positions, and finger configurations. This deficit translates into the inability to express concepts by means of learned signs. It may also encompass the failure to recognize the unpleasant component of a painful or threatening stimulus and the consequent lack of defense reaction. Patients with this disorder may experience impairment of language and drawing abilities while sparing the imitation of meaningless gestures.

Apractognosia denotes a variety of syndromes in which marked apraxic features are combined with the visual deficits. This condition, which may occur subsequent to right hemispheric lesion, may be associated with left and right disorientation and disturbance of the body scheme.

Dementia

Dementia refers to a progressive, diffuse, and multifocal decline in the intellectual and cognitive abilities that impairs daily functioning in the presence of normal consciousness. These include loss of memory accompanied by dysfunction in at least one other mental function such as memory, language, emotion, or behavior and cognition (judgment, abstract thinking, etc.). This may be classified into cortical and subcortical dementia. Cortical dementia is seen in AD (also discussed in Chapter 16), which accounts for approximately 50% of cases of dementia. Cortical dementia also occurs due to stroke, Pick's disease, Jakob–Creutzfeldt disease, subacute sclerosing panencephalitis (SSPE), neurosyphilis, normal pressure hydrocephalus (NPH, discussed later in this chapter), Lyme disease, and multiple sclerosis (MS, discussed in Chapter 1), and in individuals with frontal and temporoparietal lobe lesions.

AD is the most common cause of dementia that affects the elderly, and the risk of acquiring this disorder increases with the age. This neurodegenerative condition is associated with the deposition of β *amyloid* (42–amino acid chain) in the extracellular space, a fragment of amyloid precursor protein (APP), which is processed by presenilin 1 and 2. It is also associated with *neurofibrillary tangles.* β amyloid converts to an insoluble form that deposits slowly and causes destruction of the synaptic membrane and, ultimately, cell necrosis. The two types of plaques associated with amyloid are diffuse and senile plaques. Diffuse plaques are immature entities that occur normally with aging and are not associated with dementia. Senile plaques consist of βamyloid deposits surrounded with synaptic and inflammatory proteins as well as glial cells. These plaques are initially form in the hippocampal gyrus and basal forebrain (nucleus basalis of Mynert), and then expand to involve the entire isocortex (temporal, parietal, frontal, and then occipital), limbic, and subcortical areas. Neurofibrillary tangles are formed by a microtubule-associated protein, tau, which is found in the distal parts of the axons of the CNS but not the dendrites and, to a lesser degree, in the astrocytes and linked to the development of dementia. Tau is the product of alternative splicing exons 2, 3, and 10 of the *tau* gene from microtubule-associated protein tau (MAPT), which controls the stability and flexibility of the axonal microtubule. Hyperphosphorylation of tau results in its dissociation and the formation of helical filaments, collapse of the neuronal cytoskeleton, and neuronal loss. Neuronal loss occurs in the superior temporal gyrus, causing an inability to name objects and later expanding to cause global aphasia. Neurofibrillary tangles and senile plaques are pathognomic. Hirano bodies, which are refractile eosinophilic intracellular inclusion bodies, are also seen in AD. The acetylcholine level in the brain and CSF linked to cognitive impairment shows a dramatic decline as the disease progresses. Cholinergic neurons and particularly the nucleus basalis of Mynert undergo degeneration in AD. Advancing age, Apolipoprotein E (ApoE) genotype, and gene mutation that encodes APP constitute prominent risk factors for AD. Mutation seen with trisomy 21 (Down syndrome) predisposes patients with this syndrome to develop AD. ApoE genotype has five allelic forms, epsilon 1–5, which are encoded on chromosome 19. ApoE£4 in particular is associated with the amyloid deposition of AD. Patients with this disease present with anterograde amnesia and language disorders, which include inability to name and comprehend. Amnesia causes patients to ask, unaware of their disorder, the same question and tell the same story repeatedly. They experience difficulty locating familiar items such as keys, eyeglasses, pens, and so forth. They are disoriented, unable to navigate in an unfamiliar environment. Agitation, irritability, and anxiety are common, though some patients are pleasant and attempt to hide their symptoms. Paranoia and suspicion of others may also develop in these patients. A patient's condition may deteriorate to a degree that interferes with daily routine activities, and he/she requires assistance in simple tasks such as eating and dressing. Bradykinesia, hypokinesia, shuffling gait, and poor posture seen in Parkinson disease may also be seen. As the disease further progresses, seizures and bowel and urinary incontinence may develop. Death commonly occurs as a consequence of aspiration pneumonia.

Pick disease (frontotemporal dementia or lobar atrophy or sclerosis), as discussed earlier, is an extremely rare condition that shows severe signs of frontal or temporal lobe dysfunctions. It is characterized by spongiform degeneration of the superficial layers of the frontotemporal cortex with "knife edge" atrophy of the associated sulci accompanied by ballooning of the affected neurons (Pick cells). These changes are also accompanied by accumulation of the microglia, indicating a possible inflammatory basis for this disease. Pick bodies, which are pathognomic for Pick disease when they reside in neurons of the dentate gyrus, consist of intracytoplasmic argyrophilic inclusions. Despite the characteristic features of Pick bodies and Pick cells, they only exist in one-fourth of patients diagnosed with this disease. Formation of tau protein in the neuronal inclusions, astrocytes, and oligodendrocytes is also implicated in this disease, though the underlying mechanism appears to be unique for this disease. It is a slowly progressive disease that manifests initially in behavioral and personality disorders, impairment of affect, lack of insight, and poor mental function. Some patients become *disinhibited* and show distractibility, overactivity, inattention associated with socially inappropriate conduct, and signs of Klüver–Bucy syndrome (hypersexuality and hyperorality). This disinhibition is associated with degeneration of the orbitofrontal cortex. Others show *apathy* and lack of motivation and concern for self and to others, avoid expenditure, and appear depressed. They speak few words with prolonged intervals and without associated gestures (aprosody). Apathetic behavior is associated with lesions of the dorsolateral frontal cortex. Yet others exhibit *stereotypic* behaviors, such as repeating the same topic over and over within a setting, due to damage to the frontotempral and cingulate cortices. They become mentally inflexible, ritualistic, and compulsive but without anxiety. These deficits may be followed by the appearance of primitive reflexes such as grasp and sucking reflexes, akinesia, rigidity, motor neuron disease, progressive aphasia (fluent and nonfluent), mutism, and bowel and urinary incontinence. Dementia in this disease has a slow presenile course that resembles Alzheimer's dementia.

Vascular dementia (vascular cognitive impairment) is linked with reduction in or blockade of blood flow to the brain and is most commonly associated with hypertension. It commonly occurs after a stroke that could be abrupt in nature or follows a progressive course. It is considered as the most common cause of dementia after AD. This condition may be seen with other types of dementia, such as Lewy body dementia and AD. The severity and extent of manifestations and, ultimately, prognosis are linked to brain area affected, the degree of occlusion of the arteries, and the number and caliber of the affected arterial branches. Patients are confused and disoriented, have difficulty verbalizing their thoughts or comprehending spoken languages, and experience difficulty conducting daily activities. Hemiparesis, bradykinesia, hyperreflexia, Babinski sign, and gait and visual disorders may be observed. Occlusion of a large artery that supplies the frontal cortex can produce disinhibition, apathy, impaired planning and judgment, apraxia, and aphasia. If the occluded artery supplies the parietal lobe, signs of alexia or apraxia appear, while infarct of the medial temporal cortex is most likely to compromise memory function. Diseases of the small vessels can affect specific areas of the brain or well-defined pathways, such as the dorsolateral prefrontal, subcortical orbitofrontal, and medial frontal circuits, producing aphasia, inattentiveness, task instability, poor executive performance, behavioral changes that range from abulia (lack of will or initiative, passivity) and mania to compulsive conduct, mood swings, and depression with impairment of thought and physical activity.

Impairment of blood flow in the small arteries that supply deep areas of the brain produces manifestations of *Binswanger disease*. This progressive disease is a subcortical vascular dementia typically seen in chronically hypertensive and diabetic patients between the fifth and seventh decades of life due to severe atherosclerosis. It may also occur subsequent to autosomal dominant arteriopathy. The affected vessels supply the basal nuclei, internal capsule, and thalamus. This disease spares the capillaries, a fact that distinguishes its pathogenesis from that of AD. Patients are depressed, lack initiative, are withdrawn, and are ataxic, with a history of dizziness and syncope. They endure urinary incontinence and slow intellectual function.

Vascular dementia is most likely to shorten patient's lifespan to 3 years on average. Careful history, physical exam, neurocognitive testing, electrocardiography (ECG), chest radiograph, and magnetic resonance imaging (MRI) can be crucial in diagnosing this disease.

Jakob–Creutzfeldt disease, a progressive neurodegenerative disease, manifests as cortical dementia and commonly assumes sporadic or familial forms. In the *sporadic form*, the transmission of the disease may occur as a result of contaminated neurosurgical instruments, injection of human growth hormone, infected corneal transplant, and handling of cadaveric dura mater. Patients with this form of the disease may also show elevated levels of brain protein known as 14-3-3 in the CSF. In this form of neurodegenerative disorder, myoclonic jerking and periodic electroencephalography (EEG) complexes (sharp wave pattern superimposed on the slow background rhythm) accompany dementia. Patients may also develop a sleep disorder, ataxia, hemiparesis, aphasia, and hemianopsia. The *familial form* is thought to be an autosomal dominant condition with point mutations, deletions, or insertions in the coding sequence of the gene for PrP on the short arm of chromosome 20. The most common mutation that produces the clinical picture of this form of the disease is at codon 200. It has an earlier onset, and its course is more protracted. The EEG changes are often missing, and the 14-3-3 protein is not detected in the CSF, as is the case with the sporadic form. Different phenotypic manifestations may develop as a result of these mutations. Some may exhibit cerebellar ataxia, spastic paresis, and dementia at a later stage. Progressive insomnia, dementia, and dysautonomia that prove to be fatal may also be seen in association with mutation in the gene for PrP in this form of the disease.

Cortical dementia may also occur in subacute sclerosing panencephalitis (*SSPE*), a particularly fatal disease that affects children and young adults before the age of 20, months or years after a measles attack. The probable etiology is altered rubeola virus. It initially manifests as mood disorders, insomnia, hallucinations, and lack of concentration, which are eventually followed by myoclonic jerks and dementia. Patients may also exhibit choreiform movements, rigidity, dysphagia, and cortical blindness.

In *neurosyphilis*, which is caused by the spirochete *Treponema pallidum*, dementia develops as a result of brain infection and is seen as a late manifestation. It may be preceded by amnesia and certain personality changes. Patients with syphilitic dementia may also exhibit paresis, dysarthria, tremor, Argyll Robertson pupils, and locomotor ataxia.

Dementia pugilistica may also occur as a result of repeated head trauma or cerebral concussions in professional boxers, "punch-drunk state." Cortical dementia is manifested in the inability to recall events, but patients remain conscious and fully aware of their intellectual dysfunction.

Lyme disease, which is caused by the spirochete *Borrelia burgdorferi*, may manifest dementia in the second stage of the disease weeks or months after the onset of infection. Dementia may be detected in association with sleep and emotional disorders, slowing of memory, irritability, facial nerve palsy, and some degree of poor concentration.

AIDS dementia complex (ADC) is a common metabolic encephalopathy that occurs in the advanced stages of HIV infection and AIDS subsequent to meningitis and encephalitis. It is not caused by opportunistic infection, but patients' CD4+ T (T-helper) cell counts decrease to less than 200, and there is a high concentration of plasma viruses. The cardinal signs of this disease are gradual deterioration of intellectual capacity, poor motor function including loss of dexterity and coordination, amnesia, depression, aphasia and apathy, lack of initiative and spontaneity, irritability, and behavioral changes. If untreated with active antiretroviral medications, cognitive impairment later progresses in a variable course to dementia. Psychosis, profound mood swings, and bowel and urinary incontinence develop later in the course of the disease.

Evidence points toward the fact that HIV does not directly invade the neurons but, rather, infects and activates the macrophages and microglia within the brain that produce toxins, chemokines, and cytokines, triggering a cascade of events that end in neuronal death directly or via the astrocytes. Gp120 is an exterior membrane glycoprotein located on the cell plasma membrane (virion envelope). This glycoprotein interacts with CD4 receptor and chemokine co-receptor and binds noncovalently to gp41. The latter enables the virion to attach to the target cell membrane and traverse the lipid bilayer. The HIV viral protein gp120 induces mitochondrial-death proteins like caspases to influence the upregulation of the death receptor Fas, leading to apoptosis. Thus, inhibition of caspases prevents gp120 from causing apoptosis, a fact that can be utilized in therapeutic approaches.

Since the symptoms outlined earlier are also applicable to many other diseases, making the diagnosis of ADC remains a challenge. A mental status exam, computed tomography (CT) or MRI scans, and spinal tap can help to rule out other conditions. Treatment of ADC can be accomplished by highly active antiretroviral therapy (HAART).

Subcortical dementia, which is associated with lesions of the brainstem, cerebellum, or basal nuclei, may be observed in individuals with Parkinson's and Huntington's diseases (which are described with the extrapyramidal system in Chapter 21) and in Binswanger disease and progressive supranuclear palsy. In this type of dementia, speech dysfunction, motor deficits, and generalized slowing of cognitive function (mental akinesia) are prominent. Progressive supranuclear palsy (Steele–Richardson–Olszewski syndrome) exhibits Parkinsonian manifestations, including impairment of voluntary eye movements, prolongation of thought processes, irritability, and apathy.

Seizures

Seizures are irregular, short-lived, abrupt, disproportionate, and recurring depolarizations of cerebral neurons that elicit hypersynchrony and electrical storm. Seizures can be in the form of sensory disorders, disturbance of consciousness, convulsion, behavior, or a combination of these manifestations. Accumulation of extracellular potassium or reduction of the postsynaptic GABAergic inhibition usually accompanies heighetened neuronal ctivity seen in seizures. Seizures are viewed in two categories: partial and generalized.

Partial seizures manifest as abnormal movements or sensations and/or stereotyped behavioral patterns on one side. These focal seizures result from localized lesions such as scars, tumors or arteriovenous (AV) malformations, head trauma, infection, prenatal injury, fever, hypoglycemia/hyperglycemia, stroke, alcohol withdrawal, or congenital or metabolic disorders. They either remain localized or spread to adjacent cortical areas, depending on the extent of glutamate-mediated excitation combined with the decrease in GABAergic inhibition. Partial seizures are classified into simple and complex partial seizures.

A *simple partial seizure* is characterized by the relative localization of the abnormal discharge in the brain, usually to one hemisphere. It may arise from activation of foci in the primary motor, premotor, supplementary motor, or prefrontal cortices. Auras are very common and may be the only manifestation in this condition. It may involve the motor, sensory, and autonomic systems, and consciousness usually remains unaffected. Partial motor seizure may be manifested in turning the head and eyes to the contralateral side. *Jacksonian "march" seizure* is a motor seizure in which rhythmic and clonic twitching starts on the contralateral hand and "marches" up the arm, to the face, and down the leg in seconds, and the patient commonly becomes unconscious. However, it may be localized and may affect the foot, thumb, or mouth angle on the contralateral side. Partial sensory seizure is associated with a variety of sensations, depending upon the selective involvement of certain parts of the sensory homunculus. These sensations may include epigastric rising sensation and tingling in the lips, fingers, or toes that spreads to adjacent areas. It may also be associated with vertigo, olfactory hallucinations, or visual disorders such as sensations of darkness and light flashes.

Complex partial seizures occur infrequently and irregularly and are marked by impaired but not loss of consciousness, exhibiting widely varied clinical characteristics. Automatic behaviors such as chewing, lip smacking, fumbling with clothes, scratching genitalia, thrashing of arms or legs, or loss of postural tone may be seen in the complex partial seizure. "Aura" may precede this type of seizure. Dyscognitive states, which include increased familiarity (déjà vu) or unfamiliarity (jamais vu), autonomic responses, exhaustion, and illusions, may also be observed. However, these disorders are often more indicative of anxiety attacks than seizures.

Generalized seizures, which are mostly inherited, result from involvement of both cerebral hemispheres (mostly with no apparent structural damage). Many of the generalized seizures begin during childhood or adolescence. The two main categories of generalized seizures are the convulsive and nonconvulsive forms. The common convulsive type is the tonic–clonic (formerly known as grand mal) seizure, whereas the nonconvulsive forms include absence seizures (petit mal). Convulsive type may also include less common varieties such as purely tonic, atonic, or clonic generalized seizure. Nonconvulsive seizures also encompass atypical absence seizures. In summary, *convulsive generalized seizures include tonic–clonic seizures, status epilepticus, and tonic, atonic, and clonic seizures.*

Partial sensory seizure is associated with variety of sensations depending upon the selective involvement of certain parts of the sensory homunculus. These sensations may include the following:

The *tonic–clonic seizure* (grand mal epilepsy) occurs in 4%–10% of all cases of epilepsy and is viewed in two types: (1) awakening clonic–tonic–clonic seizures, which are provoked by sleep deprivation, excessive fatigue, and alcohol consumption, and (2) tonic–clonic seizures, which occur during both sleep and wake periods and maintain a better remission rate following drug therapy than the first type. In general, patients with tonic–clonic seizures may or may not sense the approach of convulsions (*prodrome phase*) in the forms of apathy or ecstasy, movement of the head, abdominal pain, headache, pallor or redness of the face, or unusual sensations. These experiences (aura) that last for few seconds, may represent an impending simple partial seizure. In the *ictal phase* (lasts 10–30 s), sudden loss of consciousness and falling to the ground and a brief period of flexion of the trunk and elbow, followed by a longer extension of the back and neck, jaw clamping, and stiffness of the limbs, may occur. Suspension of breathing, dilatation of the pupils, cyanosis, and possible urinary incontinence are also observed. This *tonic stage* is followed by a clonic phase in which a mild generalized tremor is followed by violent flexor spasm, facial grimaces, and tongue biting. At this stage, an increase of blood pressure and pulse rate, salivary secretion, and sweating occur.

When the clonic phase terminates, movements cease, and the patient remains apneic, lethargic, and confused and often falls asleep (postictal phase). The EEG shows characteristic initial desynchronization, lasting for few seconds, followed by a 10 s period of 10 Hz spikes. In the clonic phase, the spikes become mixed with slow waves, and then the EEG shows a polyspike-and-wave pattern. The EEG tracing becomes nearly flat when all movements have ceased.

Status epilepticus refers to the condition in which seizures last longer than 30 min or follow one another in a rapid fashion that a new wave of seizures begins before the previous one has ceased, with no recovery of consciousness or behavioral function. This condition is commonly caused by abrupt withdrawal of anticonvulsant medications, high fever, metabolic disorder, or cerebral lesions. There are two types of status epilepticus: convulsive and nonconvulsive. The *nonconvulsive type* affects behavior but does not produce tonic or clonic movements, which may be evident in the continuous lethargic and clouded state that a patient experiences with a prolonged absence seizure or a series of complex partial seizures. The *convulsive type* may be life threatening and may be fatal due to the sequence of events that may include circulatory collapse and hyperthermia.

The repeated nature of epileptic seizures may be explained experimentally on the basis of *kindling theory*. Induction of seizure activity in experimental animals increases in intensity subsequent to the increase in stimulus repetitions. Therefore, an intense seizure can be produced, after long period (months), by the application of relatively mild electrical stimulus. It has been suggested that prolonged electrical stimulation triggers a series of events in the neural circuitry that enables repeated seizure activity to be produced by an appropriate stimulus. This theory may be similar to long-term potentiation (LTP), in which a brief period of intense activity may elicit a persistent change in the synaptic property of the neuron.

Tonic seizures are characterized by sudden bilateral tonic (stiffening) muscle contractions of the entire body, accompanied by altered consciousness, without being followed by a clonic phase. They are brief and range from a few seconds to a minute and occur more commonly during sleep. This condition is usually caused by cerebral lesions and can occur at any age.

Atonic seizures exhibit several forms, but classically, they are characterized by loss of postural muscle tone and responsiveness and falling of the patient to the ground. Loss of muscle tone may be less severe, producing nodding of the head or drooping of the eyelids. These seizures occur at any age as brief episodes with sudden onset followed by immediate recovery.

Clonic seizures are characterized by generalized clonic (jerking) movements that are not preceded by a tonic phase. These movements are often asymmetric and irregular, and although rare, they usually occur in children.

Nonconvulsive generalized seizures comprise both absence and atypical absence seizures.

Absence seizures are brief episodes (2–15 s) of staring (simple absence), accompanied by impairment of awareness and responsiveness. During the episode, patients may stop talking or responding briefly. In a more complex absence seizure, blinking or synchronized mouth or hand movements accompany the staring episode. They occur suddenly and leave the patient postictally alert and responsive. These seizures may be so brief that patients themselves are sometimes unaware of them, and to the observer, it appears as a moment of absent-mindedness or daydreaming. They usually begin between the age of 4 and 14 and often resolve by the age of 18. Approximately 50% of individuals with this type of seizure develop generalized tonic–clonic convulsions by the end of the second decade.

Atypical absence seizures usually start and end gradually. They are not produced by hyperventilation, often last more than 10 s, and, like the typical absence, also begin at an early age. In contrast to the typical absence seizures, these episodes are often seen in individuals with mental retardation and other neurologic disorders. In this condition, staring is accompanied by partial diminution in responsiveness (i.e., able to respond to questions and remember events). Blinking or jerking movements and tonic or atonic seizures may also occur.

Coma

Coma refers to a prolonged state of unconsciousness associated with a lack of awareness of the surroundings and the inability to respond to repeated and constant stimulation. In this state, basic vital life support mechanisms such as respiratory and cardiovascular functions will continue unabated. The extent of consciousness and response to stimuli may vary according to site of the lesion and affected structures. This condition must be distinguished from obtundation, stupor, and delirium. Obtundation is a mild reduction in alertness and slowness in response to repeated stimuli, while stupor is a state in which brief arousal can only be achieved by using extreme stimulation. On the other hand, delirium is an acute hallucinatory confusional state associated with overactivation of the sympathetic system (palpitation, hypertension, sweating) that often shows fluctuation in intensity. Lack of motor activity in response to specific stimuli cannot by itself constitute a state of unconsciousness, as illustrated in "locked-in syndrome," which will be discussed later with the motor system. In this syndrome, a patient's hearing, perception of pain, vision, vertical eye movement, and sleep cycle remain unaffected, but he/she stays motionless, unable to respond.

A coma is a medical emergency that requires immediate intervention. It rarely last more than several weeks, and if it persists longer than 1 year, a patient's chances of recovery become remote, and he/she enters a persistent vegetative state and eventually dies. Decubitus ulcers and cystitis may develop during coma. The risk of asphyxiation may occur in deep coma, requiring careful respiratory support and management. Coma can be caused by primarily by traumatic brain injury, stroke, intoxication (such as drug abuse), and anoxic–ischemic brain injury but can also occur as a result of severe encephalitic or meningitic tumors, infection, seizures, and metabolic disorders (e.g., diabetic coma). The eyes of comatose patients are usually closed and are unresponsive to light (loss of pupillary light reflex). They lack the response to nociceptive stimuli but maintain some degree of reflex activity, exhibiting irregular breathing. Coma is classified into supratentorial-based (central, involving the cortical and diencephalic) and infratentorial-based (uncal, involving the brainstem) lesions. Coma assessment is done through measuring the response to verbal commands, eye opening in response to painful stimuli or voice, or spontaneous eye opening and motor movements to physical or verbal stimuli. These criteria constitute the Glasgow Coma Scale (GCS), which can accurately predict the outcome of a patient's condition. GCS of 7–8 or lower is considered comatose. A comatose patient's physical exam will be extremely important as it reveals information about a patient's respiration, posture, eye movements, and the pupil. Cheyne–Stokes breathing with alternating hyperventilation (crescendo–decrescendo pattern) and cyclical waxing and waning with recurrent episodes of apnea may indicate an extensive supratentorial (central herniation) lesion and can be associated with cardiac failure. This is due to the fact that cerebral cortical control of respiration is lost, and carbon dioxide–driven breathing takes effect, with accumulation of carbon dioxide leading to increased rate and depth of respiration (reactive hyperpnea). Midbrain and upper pontine lesions cause neurogenic hyperventilation, which exhibits in very deep and rapid respiration with a steady rate but without apnea. Apneustic breathing, characterized by rapid and shallow respiration and sudden deep gasping inspiration with a pause before expiration, is seen in comatose patients with infratentorial (uncal herniation with lower pontine) lesions. A medullary lesion produces ataxic breathing, which is characterized by irregular and shallow breathing with few gasps followed by absence of breathing.

Oculocephalogyric reflex (doll's eye), which is characterized by movements of both eyes in a direction opposite that of the head rotation, remains intact when the damage is caused by a supratentorial (central herniation with cerebral and diencephalic) lesion. A supratentorial lesion induced uncal herniation compresses the brainstem (midbrain) and produces partial or complete contralateral hemiparesis, ipsilateral oculomotor palsy, and contralateral lateral strabismus. Supratentorial (central herniation) lesion associated with a midbrain and bilateral medial longitudinal fasciculus (MLF) damage leads to the development of internuclear ophthalmoplegia, which is characterized by ipsilateral adductor paresis and contralateral nystagmus (dysconjugate eye movement) with intact vertical oculocephalic movements. Cortical and brainstem functions can be evaluated via the caloric test. In this test, instillation of cold water into one ear causes slow deviation of the eyes toward the stimulated side (slow phase) followed by a fast drift to the opposite side (fast phase). Failure of this ocular movement is indicative of a brainstem lesion and disruption of the vestibulo-ocular connections. Examination of posture can help locate the site of lesion in comatose patients.

Postural changes can be seen in traumatic brain injury, ICP, stroke, brain tumors, and encephalopathy. Unilateral posturing with spastic palsy occurs in stroke patients. Jakob–Creutzfeldt disease, brain abscess, and Reye syndrome in pediatric patients can induce posturing. Postural change in traumatic impact on the head among football players is known as "fencing response." Posturing can be induced by noxious stimuli in a mild form of coma, but not in deep coma, and may also occur without a stimulus. This is a reflection of relaxation of one group of muscles and the contraction of the antagonistic group of muscles induced by nociceptive stimuli. Posturing is an important indicator of the extent of brain damage and possible herniation and is thus utilized in determining the severity of a coma via the GCS for adults and the Pediatric GCS (for infants) and prognosis. Comatose patients commonly exhibit two stereotypical postures: decerebrate and decorticate posturing. Decerebrate posturing, which entails extended and adducted arms, extended and pronated forearms, extended thigh and legs, as well as plantar flexed foot and toes, indicates a lesion rostral to the red nucleus or at the intercollicular level of the midbrain. It develops as a result of disruption of the rubrospinal, corticospinal tracts and the cortically dependent medullary reticulospinal tract and the unopposed excitatory influences of the vestibulospinal and pontine reticulospinal tracts. This condition can be eliminated by rhizotomy. A patient with a decorticate posturing (mummy baby) exhibits adducted arms, flexed and supinated forearms, flexed wrists and digits, hands clenched into fists, extended thighs and legs, and everted and adducted feet. It results from the facilitatory effect of the rubrospinal tract on the flexor neurons of the cervical segments, which overcomes the excitatory effect of the medial and lateral vestibulospinal and pontine reticulospinal tracts on the extensor neurons. This is accompanied by disruption of the corticospinal tract that provides excitatory impulses to the flexor neurons of lumbosacral segments and the unopposed action of the pontine reticulospinal and lateral vestibulospinal tracts on the extensor neurons. This posture appears late in response to a noxious stimulus preceded by paratonia (hypertonus muscles that resist involuntary passive movements) when a supratentorial lesion occurs. However, if a patient's coma is due to infratentorial lesion, hemiplegia on the same side (seen in Kernohan notch phenomenon due to compression of the contralateral corticospinal tract in the crus cerebri) and paratonia on the opposite side are seen. No change in posturing is seen in pontine or medullary lesions, with the exception of flexor leg response.

Assessment of the functions of the cranial nerves through the evaluation of certain reflexes becomes challenging due to the unconscious status of the comatose patient. Corneal reflex, mediated by the ophthalmic (afferent) and the facial nerve (efferent), is associated with contraction of the orbicularis oculi (closure of the eyes) upon the application of a wisp of cotton to the cornea. This reflex is lost in patients with a dorsal pontine lesion.

Pupillary reaction to light stimulus is a significant part of the physical assessment of comatose patients. Normal pupils are equal in size and reactive to light. Application of light to one eye produces constriction of the pupil in the same eye (direct pupillary light reflex) and the contralateral eye (consensual or indirect light reflex). In supratentorial lesions involving the cerebral cortex or diencephalon, the pupils remain reactive but retain small size, while in supratentorial lesions that produce uncal herniation, the ipsilateral pupil becomes mydriatic and unresponsive to light stimulus due to compression of the parasympathetic fibers in the oculomotor nerve. In infratentorial lesions of the midbrain, the eyes remain unresponsive, assume a midline position, and show mydriasis. A lesion confined to the pons results in destruction of the sympathetic postsynaptic fibers within the oculomotor nerve, resulting in a miotic pupil that is reactive to light. In heroin or opiate intoxication–induced coma, the pupils are reactive to light. Hypothermia, anoxia, and overdose of serotonergic and cholinergic drugs can produce mydriatic and unreactive pupils.

The above changes should be viewed in their specific context as postural changes may not always be seen with ocular manifestations. For example, hypoglycemia can cause unconsciousness with decerebrate posturing but with intact oculocephalic and pupillary light reflexes. Similarly, in opioid toxicity, the pupils are miotic and reactive, but respiration is stunted or shallow and may be completely abolished.

Careful physical exam, EEG, and radiographic imaging help in the evaluation of a patient with coma and subsequent management. Better outcome is expected from drug intoxication and nonsevere head trauma than from cardiac arrest. These should be viewed with other signs, as extended bilateral pupillary abnormality, advancing age, extensive nature of the lesion, or absence of vestibulo-ocular reflex can be bad prognostic indicators for regaining function and recovery.

Cerebral Dominance

Anatomic cerebral asymmetry begins during embryonic development and as early as the second gestational trimester. These asymmetries are products of differences of size, cytoarchitecture, number of neurons, and dendritic arborization. These differences also translate into functional asymmetry relative to handedness, speech, memory, and so forth. Despite the similarity in the morphologic features of the two cerebral hemispheres and the symmetrical projections of the sensory pathways, each hemisphere remains specialized in certain higher cortical functions. In most individuals, the posterior part of the superior temporal gyrus (planum temporale), including the transverse gyri of Heschl, expands and shows greater length on the left cerebral hemisphere. As a result, the lateral cerebral (Sylvian) fissure is longer and more horizontal in the left hemisphere. A minority of brains exhibits this feature in the right hemisphere. Many brains may show a wider right frontal pole and a wider left occipital pole in a counter clockwise direction (Yakovlevian torque).

Identification of objects and comprehension of language may be accomplished by the *right hemisphere* (mute, *nondominant*, or creative hemisphere), utilizing visual and tactile information. This hemisphere also integrates visual impulses with spatial information and motor activities, as in drawing; interprets metaphors and tone of a dialogue; and mediates musical tones, facial recognition, construction, and other nonverbal activities. Thus, the nondominant hemisphere is holistic, concerned with perception of spatial information (superior parietal lobule), gesturing that accompanies speech (prosody), and recognition of familiar objects. In other words, this hemisphere is holistically creative, lacks details, and defies rules and logic. In 95% of males and 80% of females, the *dominant hemisphere*, usually the *left hemisphere*, is less creative and designed to carry out sequential analysis. It is conceived to comprehend spoken and written languages and to express thoughts into words. Additionally, sequencing of phonemic and syntactical characteristics of language, mathematical calculations, analytical functions, and fine-skilled motor activities are regulated by the left hemisphere. Right-handed individuals (dextrals) constitute 80% of the population, 10% are left-handed, and the remaining 10% of the population are ambidextrous. In approximately 90%–97% of individuals who use primarily their right hand, the left hemisphere is dominant for language. The other 3%–10% of right-handed persons have the speech center in the right hemisphere. In 60%–65% of left-handed individuals (sinistrals), the speech center is located in the left hemisphere; 20%–25% have the speech center in right hemisphere; and in 15%–20% of the population, the speech center is bilateral. In 60% of ambidextrous individuals, the speech center is located in the left hemisphere; in 10% of this population, it is positioned in the right hemisphere; and in 30% of these individuals, it lies in both hemispheres, particularly in women and individuals with homosexual orientation in both genders. Recovery from aphasia in sinistrals is more complete than in dextrals. The percentage of left-handed individuals in a family and the degree of right-handedness is likely to determine the extent of language dysfunction induced by a left-hemispheric lesion. Positron emission tomography (PET) may be used to detect the increased blood flow into the dominant hemisphere. The motor and sensory homunculus for the arm is larger in the left than the right hemisphere. The entorhinal cortex is more densely populated (large number) by the neurons, particularly in lamina II of the left hemisphere. There is also lateralization in regard to memory as verbal memory concentrates in the left hemisphere, in contrast to nonverbal memory, which resides in the right hemisphere.

There are indications that support the hypothesis that environmental and genetic determinants may alter brain asymmetry by suppressing the development of one hemisphere. The role of the corpus callosum in mediating cerebral asymmetry is evidenced by the fact that asymmetric cortical areas lack callosal connections. However, this anatomic and functional lateralization remains variable, and the magnitude of asymmetry may be less distinct. Asymmetric cerebral abnormalities are implicated in autism, Rasmussen's syndrome, schizophrenia, and dyslexia. Schizophrenia and dyslexia have been discussed earlier.

Autism

Autism is a neurodevelopment disorder that exhibits in impaired ability to communicate verbally (figurative language) and interact socially, the hallmark feature of this disease, accompanied by stereotyped, compulsive, ritualistic behavior that possibly becomes self-injurious (head-banging and/or hand flapping), with restricted interests and refusal to be interrupted from ongoing routine tasks. Impaired social interaction (autistics are loners) appears in different forms, such as avoiding face-to-face contact; ignoring or lack of intuition in that they have difficulty comprehending facial expressions, recognizing emotions, or gestures; and exhibiting profound proclivity to focus on one item and discuss restricted topics without empathy. They appear to be lonely. Patients are not responsive verbally when called upon and usually refer to themselves by their own names rather than using "I" and "me," and that may require joint attention. They have difficulty integrating words with gestures and are unable to translate symbols into words, exhibiting echolalia. Studies indicate that children this condition exhibit a destructive aggressive attitude, particularly those with mental retardation. When attempting to respond verbally, their communications remain unsynchronized, which may lead to unrealistic expectations of the extent of their comprehension.

These signs have a gradual onset, follow a steady course with no remission, and may be noticed first at the age of 6 months and become prominent by the age of 2 years, although continuation of the condition to older ages has been reported. It more commonly affects male than female children and is prevalent in all levels of the socioeconomic ladder and ethnic groups but with variable severity. Affected children may remain dependent on the care of others even in adulthood. Children's regression follows a normal initial development. This condition is considered part of autism spectrum disorders (ASD), which include Asperger syndrome, pervasive developmental disorder not otherwise specified (PDD-NOS), childhood disintegrative disorder, and Rett syndrome. Affected children may also exhibit signs and symptoms of attention deficit disorder, Tourette and Fragile X syndromes, and tuberous sclerosis. Some autistics may suffer from convulsions due to epileptic seizures in their adulthood. Autistics may exhibit mental retardation, anxiety, and sleep disorders. Autistics may be unusually gifted and talented in a limited area. Motor deficits that include incoordination and hypotonic muscles may also be seen. These signs may be seen in persons with mental disability as well as in highly functioning individuals.

The development of this disease has been attributed to genetics and environment factors. Studies have shown irregularities in several regions of the brain of autistics, for example, the cingulate gyrus, which are normally associated with timing of brain development. Autistics utilize different parts of the brain in performing social or nonsocial functions. As a result, the brain may tend to develop at a quicker pace subsequent to unusual neuronal growth that produces abnormal neuronal synaptic connectivity, impaired neuronal migration, and unbalanced interaction between neurons. Siblings of autistics are most likely to show signs of autism. Preliminary data indicate that genetic mutations, deletions, duplications, and/or impact of environmental factors (chemicals such as in heavy metals, bacteria or viral infections, smoking and pesticides) and teratogens on genes disrupt the normal early fetal brain development and growth and adversely affect the synaptic neuronal connectivity. Prevalence of autism appears significantly more commonly in individuals with 1q21.1. Indications that autistics show genetic alteration linked to chromosome 15 (15q13.3), chromosome 16 (16p13.1), and chromosome 17 (17p12) are inconclusive.

Abnormal levels of serotonin in the brains of autistic children have been reported. Data have been compiled regarding the role of metabotropic glutamate receptors and growth hormones in this disease. It has been reported that the brains of autistics have a structurally distorted "mirror neuron system" circuit, which regulates understanding and modeling of their reactions toward the gestures, emotion, intention of others. A poor connectivity and imbalance exist in autistics between parts of the brain that mediate social functions and those areas that regulate attention and goal-directed tasks. Thus, stimuli associated with audition, vision, language, and facial recognition are processed differently in autistics. There is also evidence that in autistics, the connection between the frontal cortex and the rest of the neocortex is weak and that this poor connectivity is brain hemispheric and predominantly confined to the association cortices (underconnectivity theory). The limitation of this theory is evident in the ability of autistics to execute some functions without deficits. Disturbances of social cognition are also explained on the basis of the empathizing–systemizing theory, which revolves around the fact that autistics have difficulty in assessing gestures and activities of others (empathizing) while remaining capable oft systemically controlling events generated by the brain (self). This type of selectivity in systemic versus empathic processing is further expanded to "extreme male brain theory," in which the male brain can perform systemization but is unable to produce empathic response. There are other theories that discuss disturbances of nonsocial "executive" functions in autistics.

Rasmussen's syndrome is a rare autoimmune chronic inflammatory condition associated with encephalitis of one cerebral hemisphere. Children under the age of 10 years, and rarely, adults, exhibit frequent seizures, loss of motor skills, aphasia, and hemiparesis that progressively worsens.

Wada test is a test conducted before brain tumor resection or surgical procedure utilized in the treatment of epilepsy. It is utilized to establish cerebral dominance and to determine the cerebral hemisphere that performs speech and memory functions. In this procedure, sodium amobarbital is injected into each internal carotid artery, one at a time. Therefore, the language and memory centers are blocked on one side in order to test the same functions in the opposite hemispheres. A brief aphasia, which accompanies injection into the internal carotid artery, may determine the dominant hemisphere. This procedure can produce a variety of complications, including contralateral hemiplegia and sensory neglect as well as manifestations of disinhibition.

BASAL (GANGLIA) NUCLEI

The basal (ganglia) nuclei are embedded in the white matter of the cerebral hemispheres. Together with the substantia nigra, red and subthalamic nuclei, reticular formation, and claustrum, they comprise the *subcortical motor system*. This system is concerned with stereotyped movements, suppression of cortically induced movements, and regulation of posture as well as adjustment of muscle tone. The basal nuclei influence the motor activity by projecting to specific nuclei of the thalamus, which, in turn, deliver the received impulses to the cerebral motor and premotor cortices. Specifically, the basal nuclei consist of the corpus striatum and amygdala. The amygdala is a structure that maintains connections with the striatum, thalamus, cerebral cortex, and structures that constitute the limbic system (see Limbic System, Chapter 17). The *corpus striatum* comprises the globus pallidus, caudate nucleus, and putamen. Both the caudate nucleus and putamen form the neostriatum, representing the afferent portion of the basal nuclei, whereas the lentiform nucleus refers to both the putamen and the globus pallidus. For further discussion of the basal nuclei, see Extrapyramidal Motor System, Chapter 21.

BLOOD SUPPLY OF THE CEREBRAL HEMISPHERES

Most of the neurological disorders, whether reversible or irreversible, are the result of vascular diseases or accidents. Knowledge of anatomy of the cerebral vessels is of utmost clinical importance, as it is essential in performing and interpreting angiographic imaging, understanding the deficits associated with vascular accidents, and developing a treatment plan. The blood supply to the brain tissue is maintained by the carotid and vertebrobasilar arterial systems. There are specific features of the capillaries associated with these two arterial systems. For instance, these capillaries are composed of nonfenestrated continuous endothelial cells, connected by tight junctions, and encased by perivascular end feet of astrocytes, possessing large number of mitochondria and pinocytotic vesicles. Approximately 15% of cardiac output, equivalent to 750 mL per minute, is directed to the brain. The gray matter of cerebral cortex receives far more blood than the white matter. The difference between the arterial and venous pressure may determine the cerebral blood flow. In addition, intrinsic factors such as the condition of the cerebral vessels, changes in the tension of carbon dioxide, and alteration in pH will also affect the blood flow to the brain. Due to the small size of frontal lobes in infants, the cerebral vessels are concentrated within the Sylvian triangle.

Vascular disorders are occlusive in nature, in which the dysfunction is a sequel to ischemia or hemorrhage. Cerebral emboli may occur at any time, but it is especially common during daytime. They may originate from the pulmonary veins, cardiac valves or chambers, and plaques in the aortic arch or its branches. Ulceration and occlusion increase the likelihood of stenosis in the affected vessels.

As mentioned earlier, the branches of the internal carotid and basilar arteries supply the brain and form the anterior and posterior circulation, respectively. The *internal carotid artery* (Figures 8.34, 8.35, 8.38, 8.39, and 8.40) arises from the common carotid artery at the level of the upper border

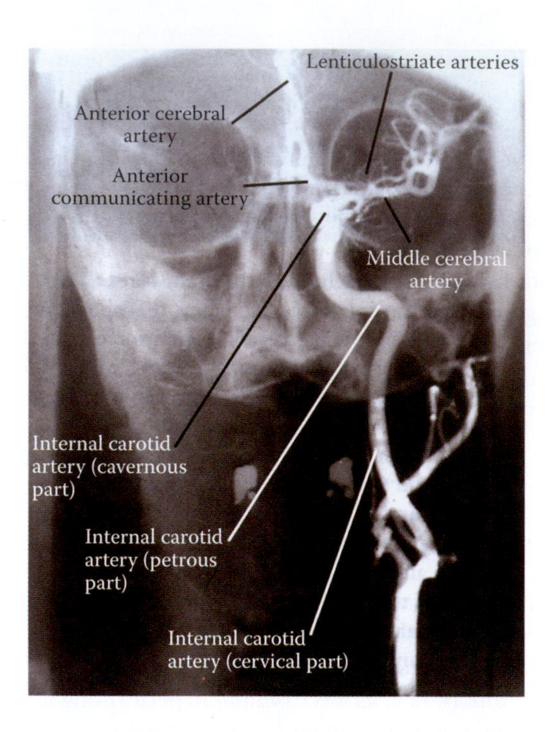

FIGURE 8.34 An anteroposterior arteriogram showing the common carotid artery and various parts of the internal carotid artery and its branches. The course of the internal carotid artery within the carotid sheath, petrous temporal bone, and cavernous sinus is illustrated.

of the thyroid cartilage and courses in close proximity to the sympathetic trunk in the neck. This vessel supplies the rostral two-thirds of the brain hemispheres and diencephalon (anterior circulation), as well as the nasal cavity, forehead, eye, and orbit. Initially, it ascends in the neck within the carotid sheath (*cervical part*) and then enters the carotid canal to gain access to the cranial cavity. Within the carotid canal (*petrous part*), it lies on the anterior wall of the tympanic cavity, separated by a thin bony lamella (Figure 8.36). This part courses near the trigeminal ganglion, separated from it by the roof of the carotid canal. During its intracranial course, the internal carotid artery passes over the cartilaginous plate of the foramen lacerum and continues through the carotid sulcus to the cavernous sinus, where it lies adjacent to the abducens, oculomotor, trochlear, ophthalmic, and maxillary nerves. Within this sinus, it gives rise to the cavernous, meningeal, and hypophyseal arteries. As it leaves the sinus, it turns anteriorly, ventral to the optic nerve and dorsal to the oculomotor nerve (*cerebral part*), giving rise to the ophthalmic, anterior cerebral, middle cerebral, posterior communicating, and anterior choroidal arteries.

The cavernous and cerebral parts of the internal carotid artery are collectively known as the carotid siphon (Figures 8.34, 8.35, and 8.36).

As the first branch of the internal carotid artery, the ophthalmic artery runs initially medial to the anterior clinoid

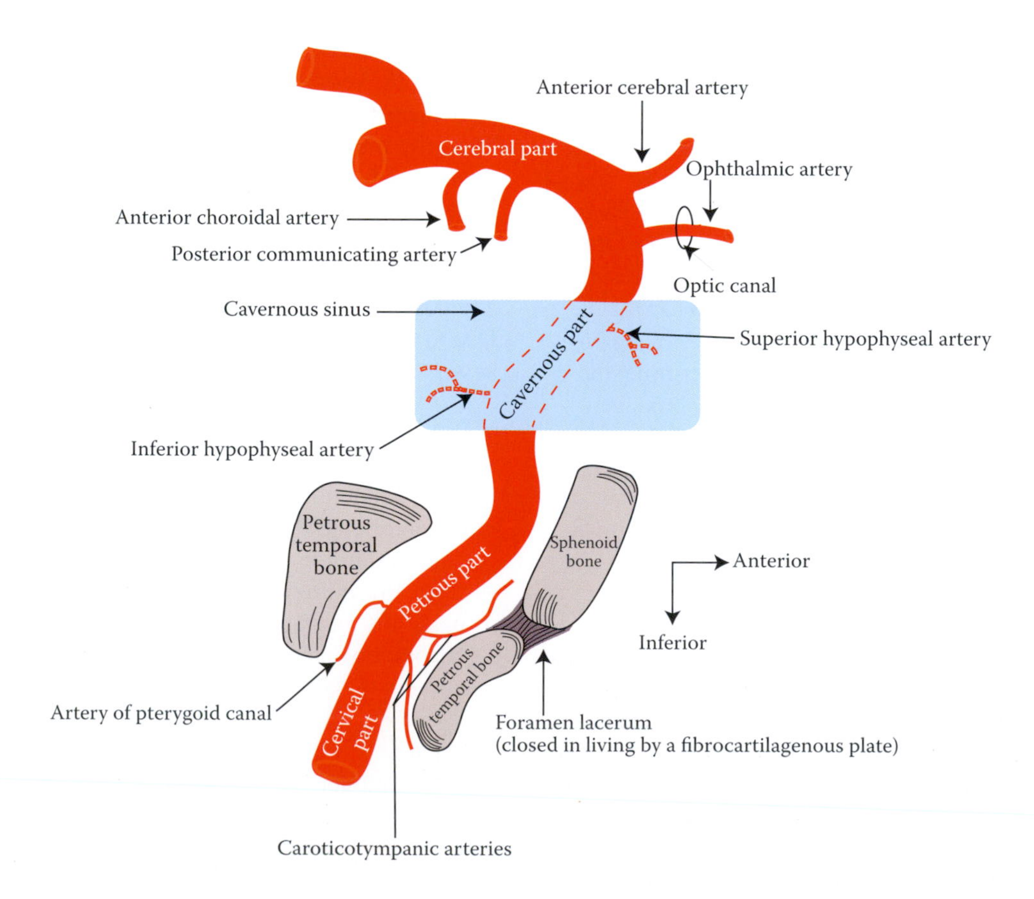

FIGURE 8.35 In this schematic drawing, the course of the internal carotid artery and principal branches is indicated.

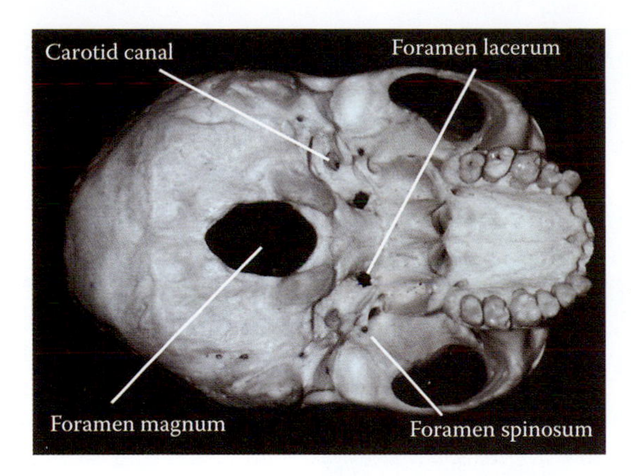

FIGURE 8.36 Inferior surface of the cranial base. The carotid canal and foramen lacerum are illustrated.

process and then enters the orbit through the optic canal. It courses superior and then lateral to the optic nerve toward the medial wall of the orbit passing between the superior rectus and the optic nerve. On the medial orbital wall, the ophthalmic artery travels with the nasociliary nerve between the superior oblique and the medial rectus. It proceeds to give rise to the central retinal, lacrimal, muscular, ciliary (long, short, anterior), supraorbital, ethmoidal (posterior and anterior), and meningeal branches. As it reaches the medial upper eyelid, it terminates as the supratrochlear and dorsal nasal arteries. Through these branches, the ophthalmic artery provides blood supply to the eyeball, extraocular and intraocular muscles, lacrimal gland, and eyelids. Despite the small caliber of the central retinal artery, it provides important blood supply to the retina. This branch pierces the optic nerve, accompanied by the corresponding vein, and supplies the four quadrants of the retina. It is considered as an end artery that lacks any anastomosis. Thus, occlusion of one branch of the central retinal artery produces quadranopsia. It also supplies the orbit, nasal cavity, ethmoidal sinuses, dura mater, and scalp (Figure 8.35).

One of the terminal branches of the internal carotid artery is the *ACA*, which runs medially and anteriorly toward the interhemispheric fissure, encircles the genu of the corpus callosum, and then courses above on the medial surface of the brain (Figures 8.34, 8.35, 8.37, 8.38, 8.39, and 8.42). The ACA divides into A1 and A2 segments. The A1 segment extends from its origin at the internal carotid artery to the anterior communicating artery. The A1 segments on both sides are connected via a short and transversely running vessel, the anterior communicating artery. The latter provides collateral circulation to the less perfused site in the event of vascular occlusion. A1 segments may arise from a single internal carotid artery, known as the azygos ACA. The A2 segment starts where A1 terminates, initially giving rise around the genu of corpus callosum to the orbitofrontal and frontopolar branches to the orbital gyri and the medial surface of the frontal lobe. It then gives rise to the callosomarginal and pericallosal branches. These two arteries, as their names indicate, run in close association with the corpus callosum and thus supply the cingulate gyrus, the caudal medial portion of the frontal lobe, and also the medial parietal lobe. As

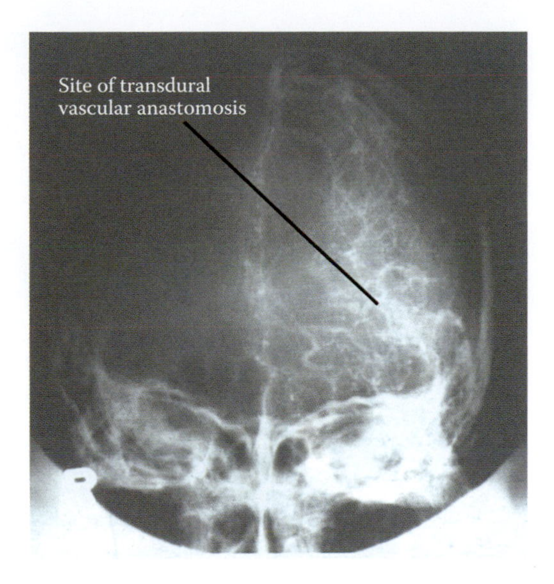

FIGURE 8.37 This angiogram of an individual with Moyamoya syndrome illustrates the site of transdural anastomosis.

you can see, through these multiple branches, the ACA supplies the medial surface of the brain hemisphere, including the cingulate, medial frontal gyrus; upper medial parts of the precentral and postcentral gyri; and the paracentral lobule (lower extremity motor and sensory cortices). Orbital and frontopolar branches supply the orbitofrontal cortex, and the rostral and medial parts of the frontal lobe, respectively. Within the cingulate sulcus, the cingulate branch makes its appearance, supplying the corresponding gyrus. Additionally, the callosomarginal branch supplies the paracentral lobule, whereas the pericallosal artery, a terminal branch of the ACA, supplies the precuneus, assuming a typical "sausage" configuration in syphilitic endarteritis. There are smaller, yet important, central branches that arise from the

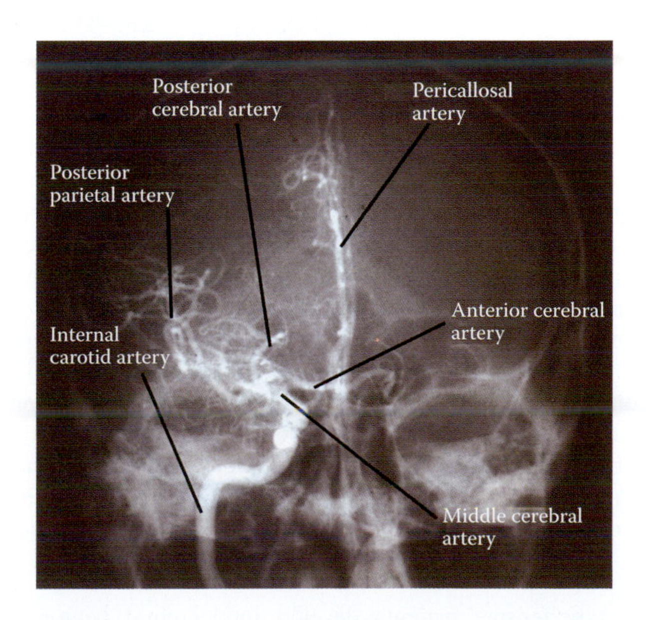

FIGURE 8.38 Another angiogram of the internal carotid artery showing branches of the middle and anterior cerebral arteries. A branch of the posterior circulation (posterior cerebral artery) is also shown.

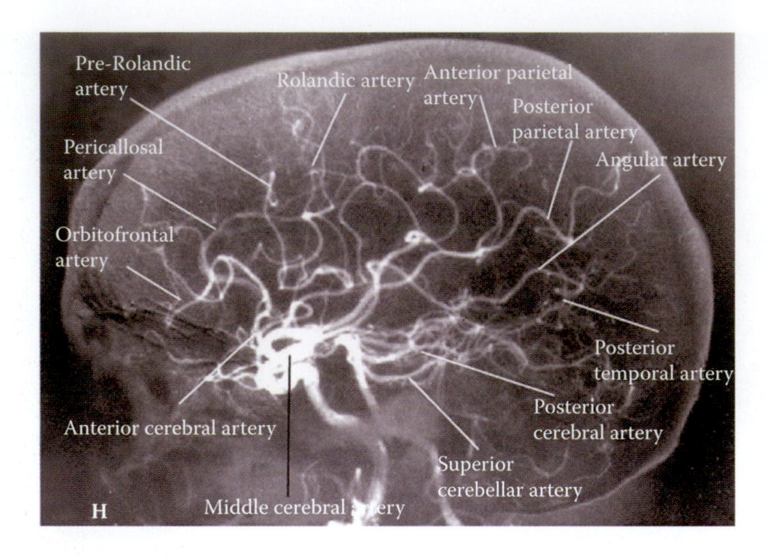

FIGURE 8.39 In this angiogram, branches of the internal carotid artery and branches of the middle cerebral artery are shown. Some branches of the posterior circulation are also visible.

ACA near its beginning and enter the anterior perforated substance. The most clinically significant branch is the medial striate (recurrent artery of Heubner), which supplies the head of the caudate nucleus, putamen, and anterior limb of the internal capsule. The latter branch is absent in about 3% of the population.

Most of the lateral surface of the brain is supplied by branches of the MCA, a larger terminal branch of the internal carotid artery above the anterior clinoid process (Figures 8.34, 8.35, 8.37, 8.38, and 8.39) where emboli are commonly lodged. The first segment, M1, stretches from the origin of MCA to the beginning of the lateral cerebral (Sylvian) fissure. M1 gives off the lateral striate and sometimes the medial striate branches, piercing the anterior perforated substance. The lateral striate arteries pierce the external capsule and the lentiform nucleus to supply the caudate nucleus. One of the lateral striate arteries, known as the lenticulostriate or Charcot artery of cerebral hemorrhage, is considered a major cause of parenchymal intracerebral bleeding in hypertensive patients. The M2 segment continues to run in the Sylvian fissure across the insular cortex. Here, it divides into the anterior temporal, orbitofrontal, pre-Rolandic, Rolandic, anterior and posterior parietal, angular, and posterior temporal. Areas of distribution of these branches include the precentral, postcentral, angular, supramarginal, superior temporal, middle frontal, and inferior frontal gyri, as well as the transverse gyri of Heschl. Brodmann area 8 (frontal eye field), which controls horizontal saccadic eye movements toward the contralateral visual field through projection to the PPRF, lies within the vascular domain of the ascending frontal branch "candelabra" of this artery, a branch formed by contribution of the pre-Rolandic, Rolandic, and anterior parietal branches.

Hypertensive patients develop intracerebral hemorrhage usually during strenuous activity, leading to the formation of hematoma that causes neurological manifestations within the first few hours. These symptoms may vary from severe headache and vomiting to contralateral hemiparesis or hemianesthesia, hemianopsia, and altered consciousness that can progress to coma and death, particularly if the ARAS in the pons and its projection to the thalamus is impacted. Seizures can also occur in patients with intracerebral bleeding, but less so with primary bleeding affecting the basal ganglia and thalamus. Edema or mass effect of the bleeding can produce manifestations based on location and affected structure. Anterior putaminal hemorrhage may produce hemiparesis of the opposite side of the body, while posteriorly located bleeding is more likely to cause reduction in sensory perception, but no change in consciousness occurs.

Due to the course of the MCA within the lateral cerebral fissure, tumors of the temporal lobe may displace this artery in an upward direction, which will be visible angiographically. Also, an interesting correlation exists between the extensive area of distribution of this artery and the increase in the vulnerability to infection. Lacunar infarcts and possible death due to rupture of the lenticulostriate artery may occur in hypertensive individuals. However, these lacunar infarcts do not usually cause headaches, and transient ischemic attacks (TIAs) prior to the infarction are infrequent.

Tumors of the temporal lobe may display the MCA upward. Also, an interesting correlation exists between the extensive area of distribution of this artery and the increase in the vulnerability to infection. Lacunar infarcts and possible death due to rupture of the lenticulostriate artery may occur in hypertensive individuals. However, these lacunar infarcts do not usually cause headaches, and TIAs, prior to the infarction, are infrequent.

In addition to cerebral branches, the internal carotid artery gives rise to the *anterior choroidal artery* (Figure 8.35), which arises in proximity of the posterior communicating artery and travels above the uncus and inferior to the optic tract. It supplies the basis pedunculi of the midbrain and the lateral geniculate nucleus. It continues rostrally to enter the inferior horn of the lateral ventricle via the choroid fissure and supply the choroid plexus. In addition, it provides blood supply to the hippocampal and dentate gyri, internal capsule, optic tract, optic radiation, central part of the midbrain, as well as the posterior limb of the internal capsule.

The posterior communicating artery arises from the internal carotid artery and contributes to the arterial circle of Willis and provides blood supply to the hypothalamus, optic tract, medial thalamus, walls of the third ventricle, and tuber cinereum. It runs superior to the oculomotor nerve, a relationship that bears clinical significance. An aneurysm that develops in this artery may produce signs of oculomotor palsy (see Cranial Nerves, Chapter 11). It connects the internal carotid artery to the posterior cerebral artery. It is usually a larger and dominant vessel on one side.

Occlusion of one or both internal carotid arteries (at the level of carotid siphon) activates the collateral circulation between the meningeal branches of the external and the internal carotid arteries. Collateral circulation also develops between branches of the lenticulostriate arteries. As a result of this anastomosis, one cerebral hemisphere is supplied with blood through a network of vessels (transdural anastomosis) resembling the "rete mirabile" of lower mammals. This anastomosis may be detected in Moyamoya syndrome via angiography as a puff of smoke.

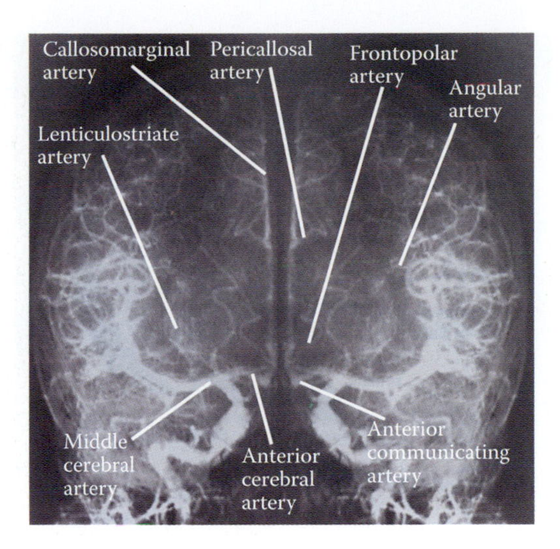

FIGURE 8.40 Angiogram of the internal carotid artery and its main cerebral branches. Branches of the middle cerebral artery are clearly visible.

Moyamoya syndrome (Figure 8.40) is a rare condition seen particularly in children and young female adults, between the second and third decades of life, as a result of unilateral acquired or congenital progressive occlusive or stenotic diseases of the major arteries that form the arterial circle of Willis, such as the internal carotid artery, and adjacent parts of the anterior and middle cerebral arteries. It was originally described in Japan and later in the West and is seen in association with neurofibromatosis, sickle-cell anemia, retinitis pigmentosa, Down syndrome, and Fanconi anemia. A hereditary component has been linked to q25.3, on chromosome 17. There is marked overgrowth of the tunica intima inward in addition to thrombotic plaques. Angiographically visible collateral arterial anastomosis will soon develop on the surface of the dura with occasional AV fistulas. Patients with this condition usually exhibit convulsions, severe headache associated with subarachnoid hemorrhage, alternating hemiplegia, aphasia, dyskinesia, visual and cognitive disorders, numbness, and other variable neurological disorders. Ischemic events, in the form of recurrent transient attacks (TIAs), are commonly observed in the younger individuals, whereas subarachnoid or cerebral hemorrhage is seen in the adult population. Surgical revascularization may be required to restore normal blood flow to the brain and meninges.

The *posterior cerebral artery* (Figures 8.37, 8.38, 8.41, and 8.42) is frequently double, which commonly emanates as a terminal branch of the basilar artery. It is separated from the superior cerebellar artery by the oculomotor nerve and more laterally by the trochlear nerve, a relation that has significant clinical value. It encircles the cerebral peduncle and travels on the upper surface of the tentorium cerebelli and also supplies the brain hemispheres. However, in 25% of individuals, one of the posterior cerebral arteries may arise from the internal carotid artery, replacing the posterior communicating branch. In 5% of individuals, the posterior

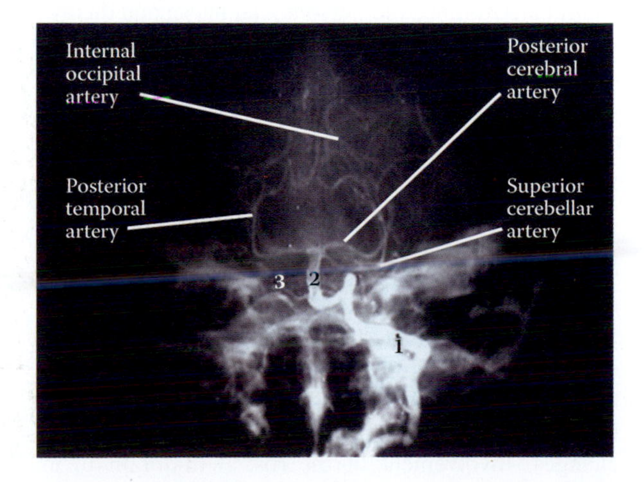

FIGURE 8.41 Angiogram of the posterior cerebral circulation. The vertebral artery and its branches are illustrated. (1) Vertebral artery; (2) basilar artery; (3) anterior inferior cerebellar artery.

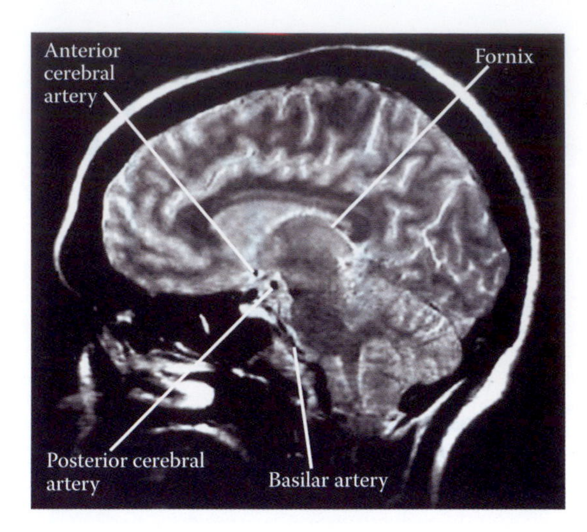

FIGURE 8.42 An MRI of the brain in the sagittal plane. Note the main branches of the anterior and posterior cerebral circulation.

cerebral arteries on both sides gain origin from the internal carotid arteries. This vessel connects to the internal carotid artery via the posterior communicating artery.

After providing blood to the midbrain and portions of the occipital and temporal lobes, the posterior cerebral artery divides into posterior temporal and internal occipital branches. The latter branch gives rise to the calcarine and parieto-occipital arteries. The calcarine artery runs within the calcarine fissure and supplies the primary visual cortex, establishing anastomosis with the MCA. Aside from its distribution to the cuneus and precuneus, the parieto-occipital branch also supplies the splenium of the corpus callosum (Figures 8.15 and 8.16). There are additional (posteromedial) branches from the posterior cerebral artery that pierce the posterior perforated substance and supply the globus pallidus, lateral wall of the third ventricle, and anterior thalamus. Posterior choroidal branches supply the inferior horn of the lateral ventricle and choroid plexus of the third ventricle. Posterolateral branches from the posterior cerebral artery supply the posterior and the paramedian thalamus, including the intralaminar, midline, and dorsomedial nuclei and the medial geniculate body. Macular sparing in individuals with cerebrovascular accident involving either the middle or posterior cerebral arteries may be attributed to the rich vascular anastomosis around the occipital pole between these two vessels.

The posterior cerebral artery, although rare, may be a single trunk (artery of Percheron) that supplies areas of distribution of the double cerebral arteries, including the paramedian thalamic regions and the rostral midbrain. Occlusion of this single artery can lead to bilateral paramedian thalamic infarcts. Patients exhibit disorder of consciousness and cognitive and behavioral changes. Involvement of the rostral midbrain in an infarct of the Percheron artery produces manifestations of oculomotor palsy.

Vulnerability to ischemia in hypertensive individuals is greater in *watershed areas*, which represent the cerebral zones supplied by the terminal branches of the anterior, posterior, and middle cerebral arteries that lack adequate blood supply and efficient perfusion pressure. When the posterior cerebral artery is derived from the internal carotid artery, TIA seen in occlusion of the internal carotid artery may also involve areas of distribution of the vertebrobasilar system (posterior circulation).

VENOUS DRAINAGE OF THE CEREBRAL HEMISPHERES

Venous drainage of the brain is maintained by superficial and deep cerebral veins. The superficial set of veins (Figures 8.43 and 8.44) includes the superior and inferior cerebral, superficial middle cerebral, and anastomotic (connecting) veins. Venous blood from the superior, upper lateral and medial surfaces of the cerebral hemispheres is drained via numerous *superior cerebral veins* that open into the superior sagittal sinus.

It is extremely significant to note that the anterior group of superior cerebral veins open at a right angle to the superior sagittal sinus, while the posterior group assumes a more oblique direction, against the current created by the blood in the dural sinuses. The *inferior cerebral veins* drain the orbital, frontal, and temporal gyri and open into the transverse sinus or the cavernous sinus, whereas the *superficial middle cerebral vein* runs across the lateral cerebral fissure to open into the cavernous or sphenoparietal sinus, draining the area around the lateral fissure. Two anastomotic veins interconnect the above-mentioned superficial cerebral veins.

These connecting veins are comprised of the *superior anastomotic vein of Trolard* connecting the superior cerebral and superficial middle cerebral veins to the superior sagittal sinus and the *inferior anastomotic vein of Labbe* that connects the middle cerebral vein to the transverse sinus.

Examination of the *deep cerebral veins* (Figures 8.43, 8.44, 8.45, 8.46, and 8.47) reveals the internal cerebral veins, great cerebral vein of Galen, and basal and occipital veins.

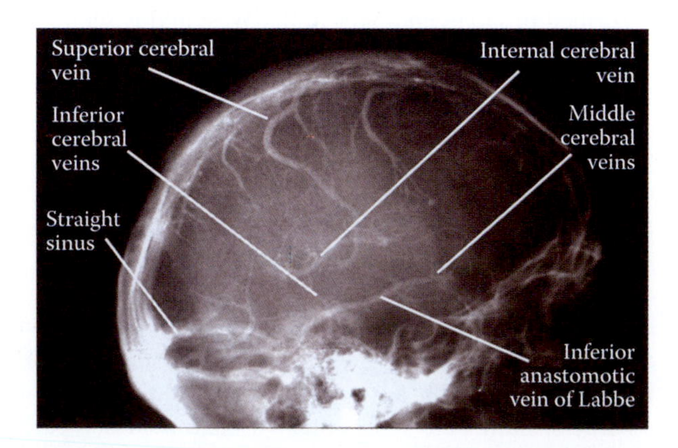

FIGURE 8.43 Angiogram of the superficial and cerebral deep veins and associated sinuses. Note the bridging veins and their drainage sites into the superior sagittal sinus.

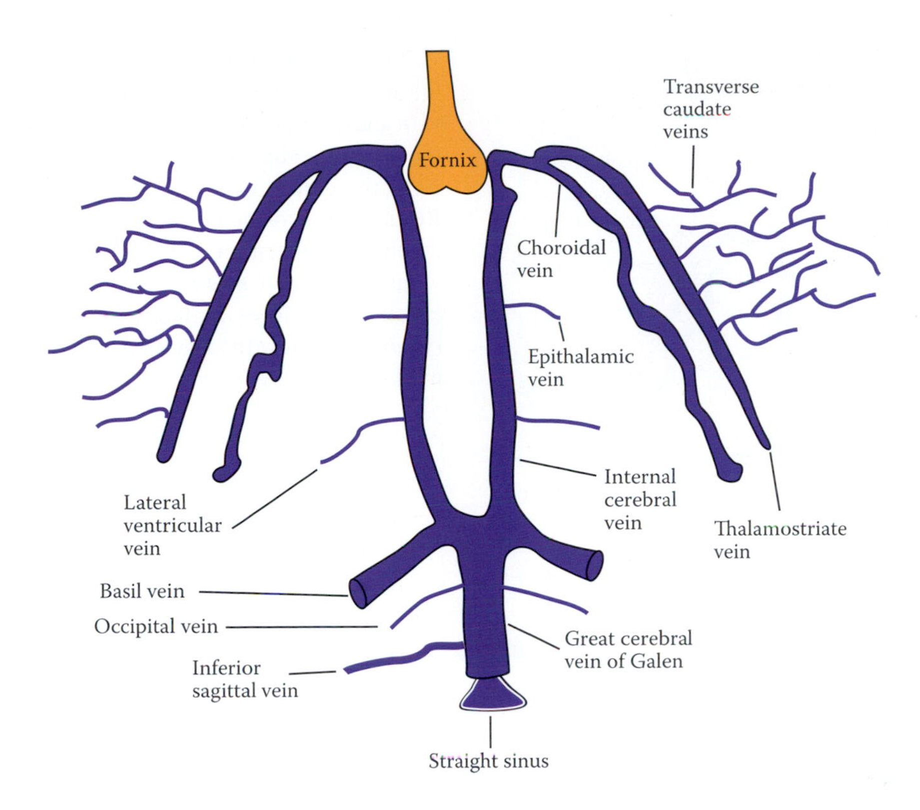

FIGURE 8.44 Schematic diagram of the deep cerebral vein. The great cerebral vein of Galen joins the venous blood from the inferior sagittal sinus to drain into the straight sinus.

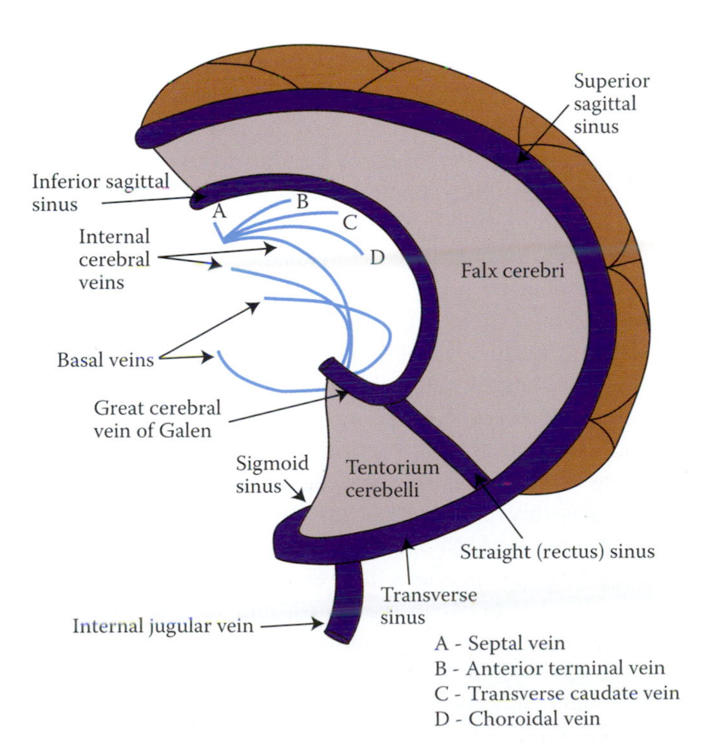

FIGURE 8.45 Tributaries of the internal cerebral vein and great cerebral vein of Galen are shown in this figure. The straight sinus, which receives the deep cerebral veins, is illustrated in relation to the tentorium cerebelli and falx cerebri.

The *internal cerebral vein* is formed by the *choroidal*, *septal*, *epithalamic*, *lateral ventricular*, and *thalamostriate* veins near the foramen of Monro. Each internal cerebral vein runs medial to the thalamus on the dorsal aspect of the third ventricle and courses inferior to the splenium of corpus callosum. The choroidal vein drains the venous blood of the

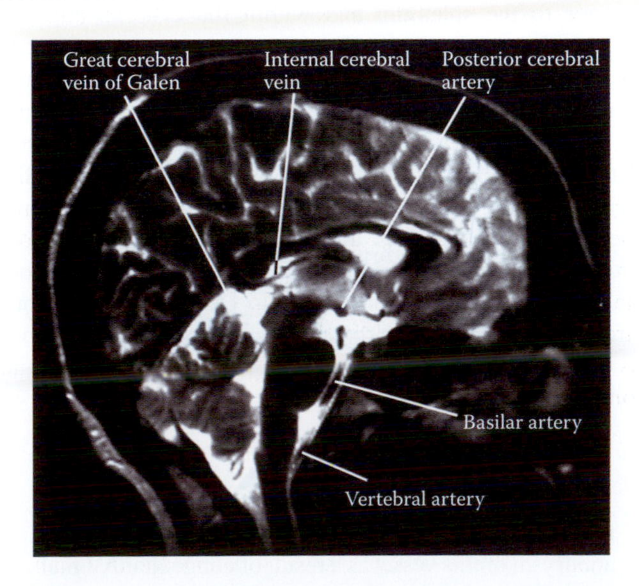

FIGURE 8.46 In this MRI scan, the internal cerebral vein and the great cerebral vein of Galen, as well as the straight sinus, are shown. Branches of the vertebrobasilar system are also illustrated.

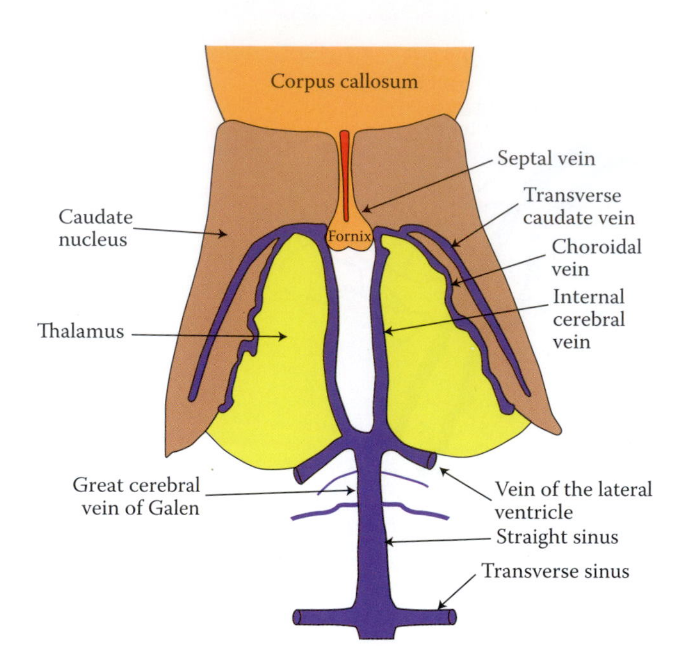

FIGURE 8.47 A detailed drawing of the various tributaries of the deep cerebral vein.

hippocampal gyrus, fornix, corpus callosum, and choroid plexus of the lateral ventricle. The septal vein drains the septum pellucidum and corpus callosum. The epithalamic vein receives venous blood from the habenula and pineal gland. The lateral ventricular vein receives venous tributaries from the caudal thalamus, parahippocampal gyrus, and lateral ventricle. The thalamostriate (terminal) vein courses caudally between the caudate nucleus and thalamus and drains both of these structures. It receives the choroidal vein posterior to the fornix, forming the internal cerebral vein.

The *great cerebral vein of Galen* (Figures 8.43, 8.44, 8.45, 8.46, and 8.47) is formed caudal to the pineal gland by the union of the two internal cerebral veins. This vein is located inferior to the splenium and within the cisterna ambiens (a dilatation of the subarachnoid space superior to the cerebellum). It joins the inferior sagittal sinus to open into the straight sinus at the junction of the tentorium cerebelli and the falx cerebri where it receives the occipital veins, posterior callosal veins, and basal vein of Rosenthal. Despite the rarity of thrombosis of the deep cerebral veins, it can produce seizures, headaches, and focal neurologic dysfunction, which are seen in superior sagittal thrombosis but without the increase in ICP or papilledema. Deep cerebral venous thrombosis associated with superior sagittal sinus thrombosis causes more serious disorders, such as increased ICP and coma with pupillary dysfunctions.

> Thinness of the wall of the great cerebral vein of Galen may predispose the vessel to rupture. Additionally, aneurysm of this vessel as a result of congenital AV malformation may produce hydrocephalus. The latter condition may be associated with congestive heart failure.

The *occipital vein* drains the venous blood of the occipital and the temporal lobes, whereas the *posterior callosal vein* carries the venous blood from the caudal portion of the corpus callosum to the great cerebral vein of Galen. Near the anterior perforated substance, the *basal vein* (of Rosenthal) is formed by contributions from the deep middle cerebral, anterior cerebral, inferior striate, and occipital veins. The *deep middle cerebral vein* drains the insular cortex and follows the lateral cerebral sulcus. The *anterior cerebral vein* drains the medial surface of brain accompanied by the ACA.

> A lesion of the parietal lobe may cause widening of the posterior part of falx cerebri and a "square shifting" of the anterior cerebral vein.

The *inferior striate vein* exits from the anterior perforated substance and drains the inferior surface of striatum.

> Thrombosis of the cerebral veins may occur as a result of mycotic or pyogenic infections or due to noninfectious conditions (e.g., malnutrition, hypercoagulable states). It may also be associated with trauma, administration of contraceptives, otitis media, and sinusitis or be a result of primary or secondary lesions elsewhere in the body. Clinical presentations are very variable depending on the location and extent of venous obstruction. Diffuse and constant headache is an early sign that worsens with recumbence. Visual deficits due to papilledema may be reversible unless it is prolonged. Perfusion of the optic nerves may be more compromised upon sudden changes in position, leading to episodes of transient blindness. Vomiting and focal or generalized convulsions due to cerebral edema may be seen with or without a focal motor deficit. Superficial cerebral veins in particular are prone to rupture at the points of their entry into the superior sagittal sinus (bridging veins), resulting in *subdural hematoma*. This is characterized by sudden and severe headache, vomiting, and dilatation of the facial veins, focal seizures, hemiplegia, dyskinesia, and possibly papilledema (discussed in detail later in this chapter). *AV malformations* (Figure 8.48) are congenital connections between arteries and veins that develop between the fourth and fifth weeks of embryonic life and are commonly seen in the cerebral vessels. The most frequently afflicted vessel with this malformation is the MCA. AV malformations are responsible for hemorrhage and focal or generalized seizures. They may mimic signs and symptoms of MS or tumors. Headache and intellectual deterioration are some of the manifestations of these anomalies.

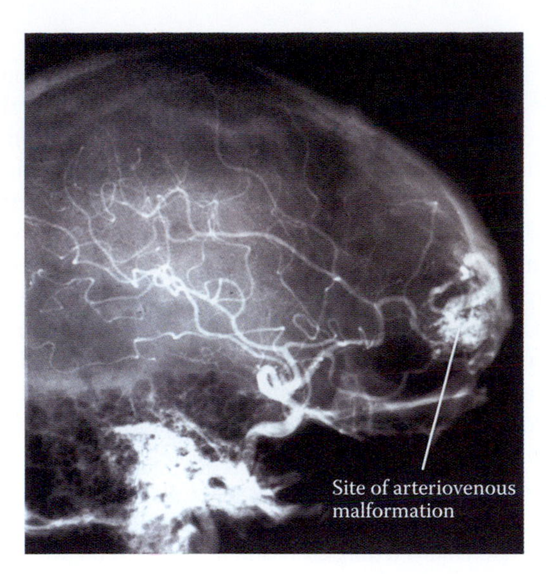

FIGURE 8.48 An internal carotid artery angiogram showing the general pattern of arteriovenous malformation involving the frontopolar branch of the middle cerebral artery.

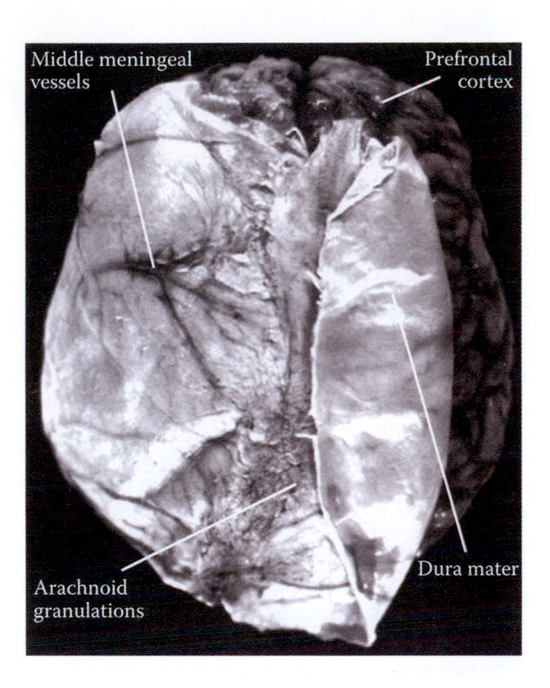

FIGURE 8.49 Photograph of the dura mater covering the brain. Notice the branches of the meningeal vessels.

MENINGES

Dura Mater

The brain is enveloped by the meninges that form coverings for the brain and contribute to the formation of the dural sinuses, brain barriers, and the CSF. These coverings send partitions that separate the cerebral hemispheres from each other and from the cerebellum and the brainstem. Meninges consist of the dura mater (*pachymeninx*), the arachnoid, and the pia mater (*leptomeninges*). Since meninges continue with the epineurium and perineurium of peripheral nerves, meningitis may produces irritation of the meninges around the brain, spinal cord, and peripheral nerves.

The *dura mater* (Latin for tough mother) is a collagenous membrane that covers the brain and spinal cord. It consists of meningeal and endosteal layers (Figure 8.49). At certain locations, the gap between these two layers results in the formation of the dural sinuses. This layer is the most sensitive of all meninges to painful stimuli. Dural innervation within the anterior cranial fossa is maintained by the meningeal branches of the ophthalmic and maxillary nerves. The meningeal branches of the maxillary nerve also contribute to the innervation of the dura mater in the middle cranial fossa. Within the posterior cranial fossa, the dura is supplied by the meningeal branches of the upper cervical spinal nerves. A recurrent branch of the ophthalmic nerve supplies the tentorium cerebelli. In the same manner, the blood supply of the dura mater is secured by branches of the ophthalmic and middle meningeal arteries in the anterior cranial fossa. Dura of the middle cranial fossa is supplied by the middle meningeal, accessory meningeal, ascending pharyngeal, and lacrimal arteries, whereas the meningeal branches of the occipital, vertebral, and ascending pharyngeal arteries provide blood supply to the dura in the posterior cranial fossa. Dural sinuses and, eventually, the internal jugular vein receive the venous blood of the dura mater.

Irritation of the epineurium and perineurium may account for the distinctive features of Kernig's sign, which is characterized by nuchal rigidity, pain in the hamstrings, and strong passive resistance upon attempt to extend the knee of a supine patient while the thigh is in flexed position. It may also account for automatic flexion of the hip and knee joints upon abrupt flexion of the neck in meningitic patients (Figure 8.50). In meningeal irritation, Brudzinski's sign can be elicited by attempting to flex the neck while the patient is in supine position, with the chest held down to the bed. This maneuver causes involuntary hip and knee flexion (Figure 8.51).

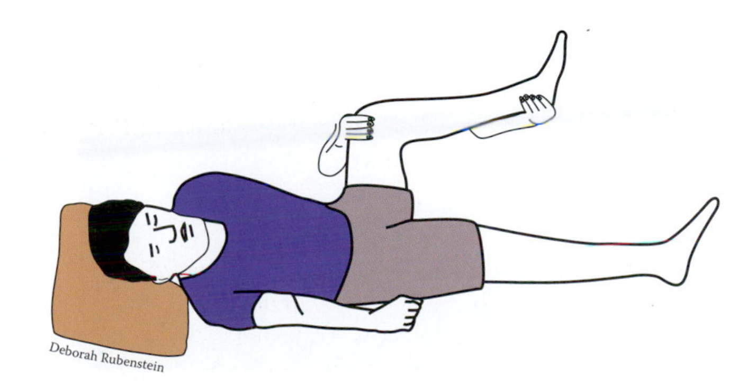

FIGURE 8.50 This diagram illustrates Kernig's sign.

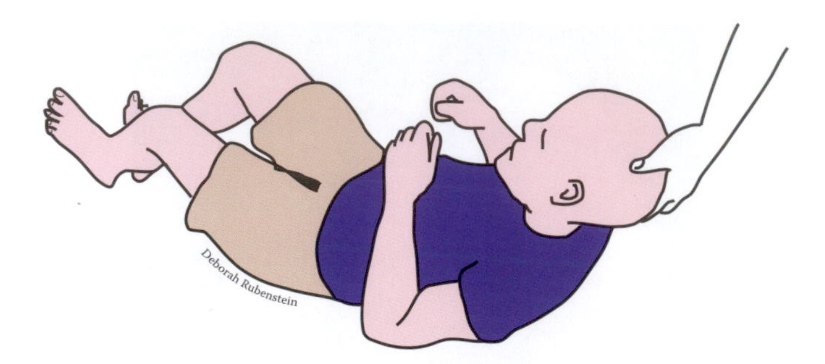

FIGURE 8.51 This schematic drawing shows Brudzinski's sign.

The epidural and subdural spaces, which lie superficial and deep to the dura mater, respectively, may be the sites of hematoma that produce serious manifestations. The clinical aspects of epidural and subdural hematomas are discussed below in detail.

Epidural hematoma (extradural hematoma) refers to a localized accumulation of blood between the bony skull and the endosteal layer of the dura (Figure 8.52). It is most commonly an acute unilateral process, most commonly of the temporal and parietal regions, occurring most commonly as a result of skull fracture-induced rupture of the anterior division of the middle meningeal artery or vein. This is a generally unilateral condition that results, in the majority of cases, from temporal bone fractures associated with mild head trauma. Epidural hematoma is less common in individuals older than 60 years as the dura mater becomes more firmly attached to the inner surface of the calvaria. It is a rapidly progressive condition, which may also involve the posterior cranial fossa, and classically presents lucid intervals, with initial loss of consciousness due to primary concussive injury followed by a return of wakefulness and final lapse into coma. Patients may also present with ipsilateral cerebellar dysfunctions, headache, and contralateral spastic UMN palsy. This condition is usually associated with transient loss of consciousness, requiring immediate surgical intervention. Due to the increased arterial blood pressure in epidural hematoma, MRI appears more rounded and isolated. This condition may lead to a rise in ICP, a condition that must be relieved before irreversible brain damage and herniation occur. Usually a hyperdense, biconvex collection between the calvaria and brain is observed in CT scan. Delayed epidural hematoma formation can occur in one-third of patients. Delayed diagnosis and failure to recognize the hematoma or unsuccessful treatment of this condition can lead to death.

A *subdural hematoma* refers to the accumulation of venous blood in the subdural space (Figure 8.53). It results from rupture of the bridging superficial cerebral veins as they drain into the dural sinuses. Most often, this condition occurs as a result of severe trauma to the front or back of the head, producing excessive anteroposterior displacement of the brain within the skull. Older patients are at greater risk due to the widened subdural space and stretched bridging veins subsequent to brain atrophy. It is much more common than epidural hematoma and tends to have a poorer prognosis. The progression of this condition is relatively slower than that of an epidural hematoma. Often,

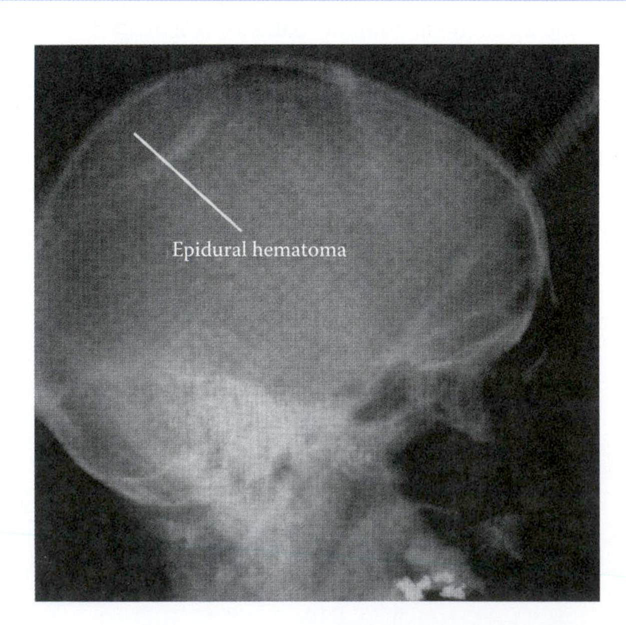

FIGURE 8.52 This radiographic image clearly illustrates sites of epidural hematoma.

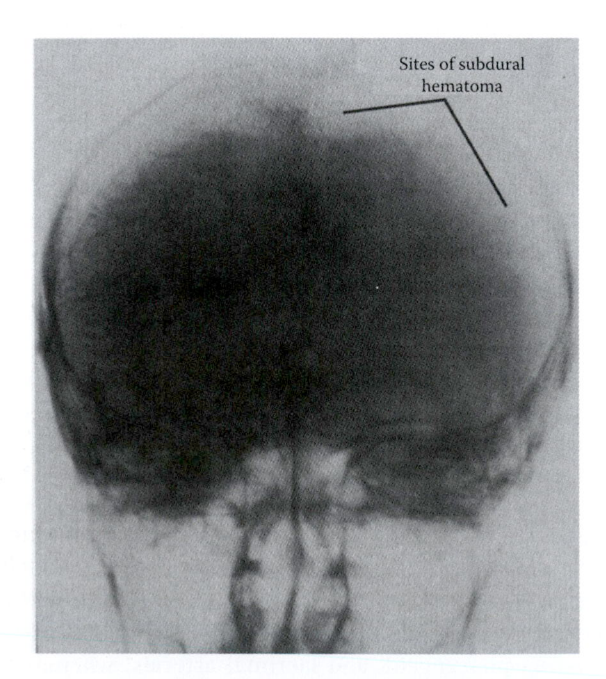

FIGURE 8.53 Radiographic image of the skull showing the extent of subdural hematoma.

symptoms do not appear until months after the initial head injury. Displacement of the pineal gland to the contralateral side, a radiographic finding, may be an initial sign of subdural hematoma. It is classified as an acute condition when it occurs within 3 days or a subacute or chronic condition when it shows manifestations between 3 days and 3 weeks. This condition is characterized by a loss or altered state of consciousness, unilateral pupillary dilatation (anisocoria—unequal pupils), and contralateral spastic palsy. The lucid interval is also commonly seen in acute subdural hematoma. However, spastic palsy may be seen ipsilaterally if the subdural hematoma is associated with compression of the midbrain (e.g., against the tentorial or Kernohan notch). Progression of this condition is more insidious, with symptoms appearing weeks or sometimes months after head injury. A sustained headache and mental obtundation are the most common symptoms. Imaging techniques that elucidate a brain shift may be useful in the diagnosis and assessment of asymptomatic cases. CT scans show hyperdense crescent-shaped images of the brain and the skull. They must be interpreted cautiously, since at certain stages (within weeks following injury), the hematoma itself becomes isodense with the surrounding brain tissue. MRI, which is a less cost-effective method of scanning, overcomes this diagnostic problem. Subdural bleeding, which follows the contour of the lateral surface of the brain, appears flat.

Chronic subdural hematoma is commonly seen in patients older than 50 years who have coagulopathy, often are chronic alcoholics, or have a history of mild injury that was neglected. It has been postulated that membranes that are prone to scanty bleeding are formed around the hematoma, leading to slow enlargement of the hematoma. Others implicate high osmotic pressure in the hematoma that forces the CSF into the hematoma. Patients often show symptoms that range from focal signs of cerebral dysfunction to possible signs of herniation.

Extensions of the dura mater in the form of dural partitions separate the two halves of the cerebral and cerebellar hemispheres, and also divides the cerebral hemispheres from the cerebellum. Part of these dural partition forms the diaphragm sella that covers the pituitary gland as it resides in the hypophyseal fossa. These dural partitions include the falx cerebri, tentorium cerebelli, and falx cerebelli.

The *falx cerebri* (Figure 8.54) is a dural partition that arches over the corpus callosum, extending between the two cerebral hemispheres. It stretches from the crista galli of the ethmoid bone, rostrally, to the tentorium cerebelli, caudally. Along the attachment of its superior border to the superior sagittal sulcus lies the superior sagittal sinus. This partition has a free inferior border that contains the inferior sagittal sinus.

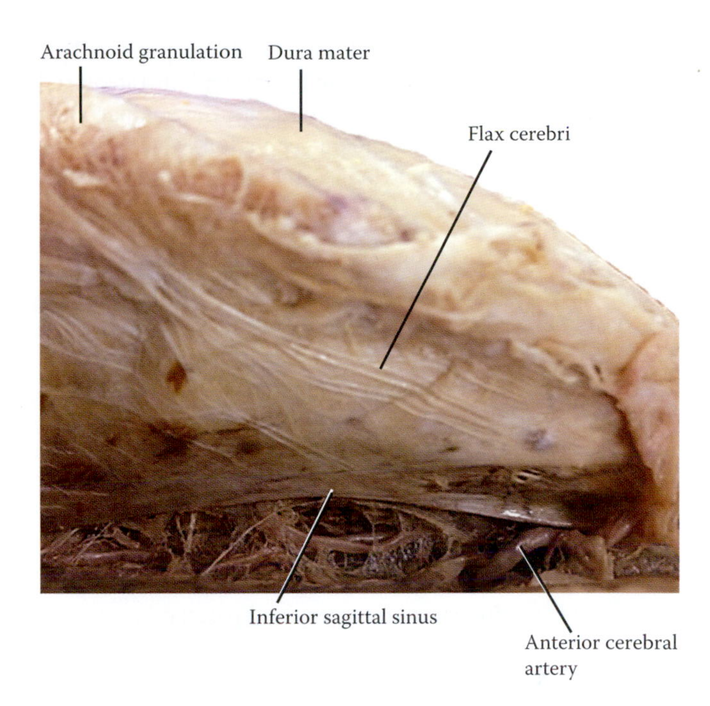

FIGURE 8.54 Falx cerebri.

The *tentorium cerebelli* (Figure 8.55) occupies the space between the occipital lobes of the brain and the cerebellum. It contains the transverse sinus in its attached border and the straight sinus at its junction with the falx cerebri (Figure 8.45). Both the great cerebral vein of Galen and the inferior sagittal sinus drain into the straight (rectus) sinus. Rostrally, the tentorium forms the tentorial notch, which allows the

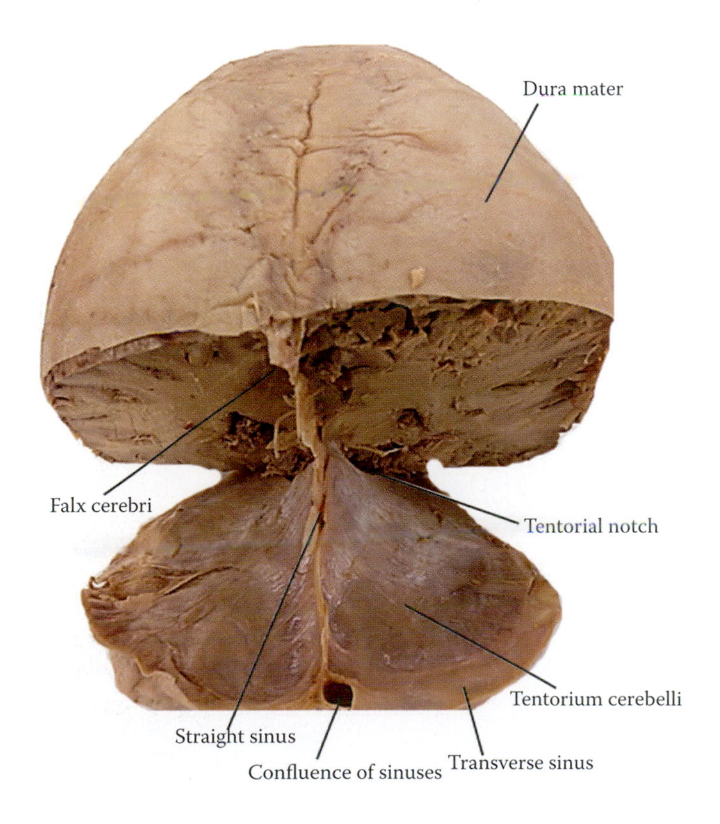

FIGURE 8.55 Tentorium cerebelli.

brainstem to pass through and join the diencephalon. This dural extension produces supratentorial and infratentorial compartments, respectively. The supratentorial compartment contains the occipital lobes and diencephalon, whereas in the infratentorial compartment lodges the cerebellum and part of the brainstem.

Increased pressure in a tentorial compartment, due to a developing mass, may eventually lead to herniation of part of the brain into the compartment with lower pressure. Bilateral supratentorial masses (Figure 8.56), which are produced by rostrocaudal displacement and compression of the midbrain, pons, and associated structures such as the tectum, reticular formation, oculomotor nerve, the diencephalon, may result in *transtentorial or central herniation*. In this condition, the undue pull of the displaced posterior cerebral artery upon the anterior choroidal artery and paramedian and pontine branches of the basilar artery produces stretching and shearing of these vessels. Stupor (due to compression of the fibers of the ARAS) and irregular and/or *Cheyne–Stokes* respirations (due to compression of the descending pathways to the respiratory center) are the main features of this condition. Due to its entrapment between the superior cerebellar and posterior cerebral arteries, the oculomotor nerve is most likely to be affected at the early stage of this disease. The supratentorial structures may also enlarge as a result of obstruction of the cerebral aqueduct. As a result of compression of the cerebral peduncle against the edge of the tentorium on the opposite side of herniation, the UMN dysfunctions will be seen on the side of herniation. Signs of decortication and decerebration will follow, leading to stiff posture and, later, rigidity. Additionally, paralysis of both vertical and horizontal (doll's eyes) movements and conjugate gaze to the contralateral side (due to destruction of the tectum, pontine tegmentum, and corticofugal fibers) may also be seen. *Uncal herniation* occurs as a result of a supratentorial mass and is characterized by displacement of the uncus into and inferior to the tentorial notch. A more detailed account on this condition is documented with the temporal lobe and uncus.

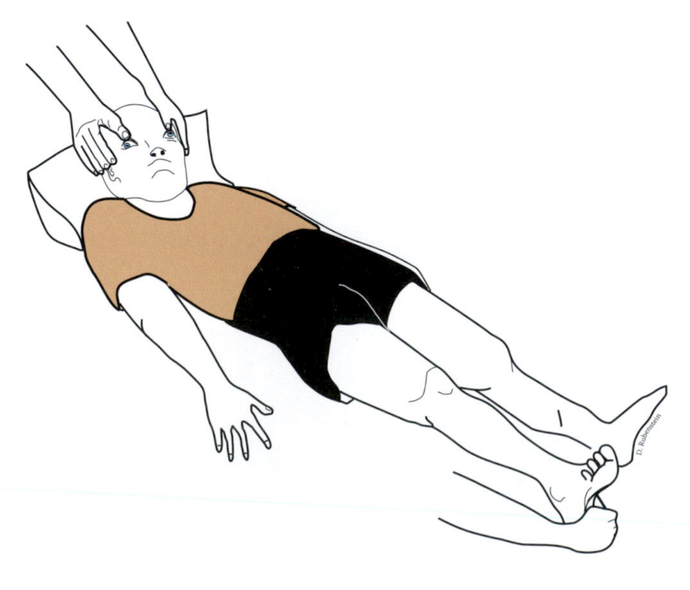

FIGURE 8.56 This is a depiction of an individual with transtentorial herniation.

The *falx cerebelli* is a sickle-shaped dural partition, located in the posterior cerebellar notch, separating the cerebellar hemispheres. It attaches posteriorly to the occipital crest, extending from the internal occipital protuberance to the foramen magnum, containing the occipital sinus.

Dural Sinuses

Dural sinuses (Figures 8.43, 8.45, 8.47, 8.49, and 8.54) are valveless venous channels, which are devoid of muscular tissue and are commonly located between the meningeal and endosteal layers of the dura mater. These sinuses, which drain intracranial structures, are classified into posterosuperior and anteroinferior groups. The posterosuperior group includes the superior and inferior sagittal sinuses, straight sinuses, confluence of sinuses, transverse (lateral) sinuses, sigmoid sinuses, and occipital sinus. Other sinuses such as the sphenoparietal, cavernous, intercavernous, and superior and inferior petrosal sinuses form the anteroinferior group.

Posterosuperior Group of Dural Sinuses

The *superior sagittal sinus* (Figures 8.43, 8.45, 8.47, 8.49, and 8.58) runs in the inner surface of the frontal bone from the crista galli, where it is connected to the emissary veins and to the veins of the nasal cavity, to the internal occipital protuberance, where it continues with the right transverse sinus. The connection to the nasal cavity is usually maintained via the foramen "cecum" if it is patent. The dilated posterior extremity of this sinus continues with the confluence of the sinuses. The superior sagittal sinus occupies the upper border of the falx cerebri and receives the *diploic veins*, arachnoid villi, and superior cerebral veins. The superior sagittal sinus receives venous lacunae, parietal emissary veins, and superior cerebral veins.

Venous lacunae are irregular venous pockets located on each side of the superior sagittal sinus, which receive the meningeal veins and the arachnoid granulations. They also communicate with each other as well as with the superior sagittal sinus.

Emissary veins are small veins, which pierce the skull, establishing a connection between dural sinuses (e.g., superior sagittal sinus) and the extracranial veins. They serve as a route by which infection may spread to the cranial cavity from areas outside the skull. Emissary veins are classified into frontal, parietal, temporal, and mastoid veins.

Diploic veins, which are contained in the cancellous tissue between the compact layers of the calvaria, are connected to the emissary veins. These veins are devoid of valves and are absent in the newborn and well developed in adults.

> The venous communications between the superior sagittal sinus and the extracranial veins may serve as a route for the spread of infection from the nose and scalp to the dural sinuses and, eventually, the systemic circulation. Obstruction of the superior sagittal sinus (Figure 8.57) as a result of thrombosis may produce focal neurologic signs and headaches exacerbated by sneezing, coughing, bending, or any activity that causes straining. Impediments to CSF absorption can lead to increased ICP and development of papilledema and transient ischemia, causing blurred vision. Like pseudotumor cerebri, changing position can produce visual obscuration, tinnitus, and vertigo. As the ICP increases with extension of the thrombus to the posterior end of the superior sagittal sinus, behavioral manifestations appear, and the patient enters into coma.

The *inferior sagittal sinus* (Figures 8.44 and 8.45) occupies the inferior margin of the falx cerebri dorsal to the corpus callosum. It continues with the straight sinus at the junction of the falx cerebri and tentorium cerebelli. The *straight sinus* (Figures 8.43, 8.45, 8.47, and 8.59) extends

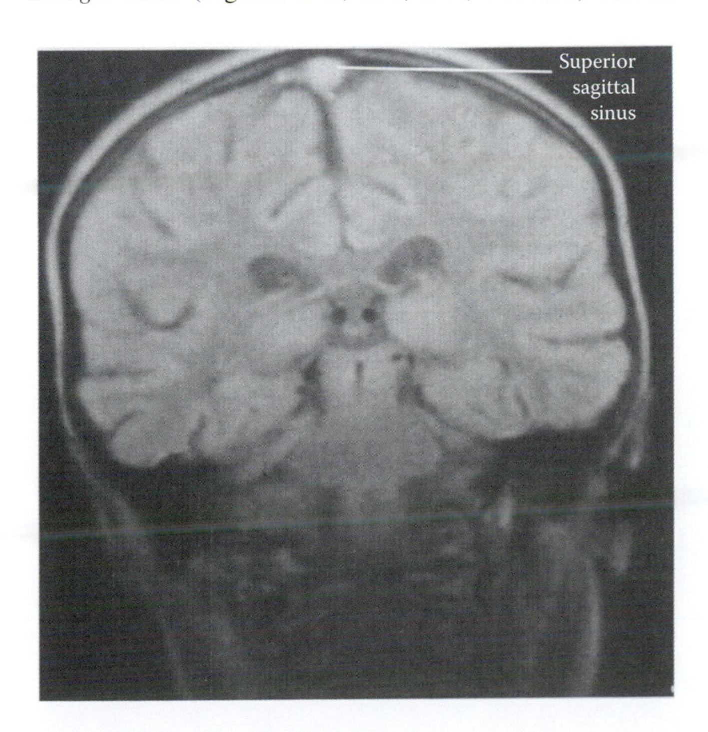

FIGURE 8.57 This MR image illustrates an obstructed superior sagittal sinus.

from the site of junction of the tentorium cerebelli and falx cerebri to the left transverse sinus. It receives the inferior sagittal sinus, superior cerebellar veins, and great cerebral vein of Galen. The *confluence of sinuses* (Figures 8.47 and 8.59) is formed at the site of union of the superior sagittal, straight, occipital and, transverse sinuses near the internal occipital protuberance.

The *transverse sinuses* (Figures 8.45, 8.47, 8.59, and 8.60) occupy the transverse sulci and the lateral and posterior borders of the tentorium cerebelli. It appears that the right transverse sinus is continuous with the superior sagittal sinus, while the left sinus continues with the straight sinus. As the transverse sinus drains into the sigmoid sinus, it receives the superficial middle cerebral, inferior cerebral, inferior cerebellar, and some of the diploic veins. Anastomotic veins of Labbe (inferior anastomotic vein) and Trolard (superior anastomotic vein) connect the superficial middle cerebral veins to the transverse and superior sagittal sinuses, respectively. Occasionally, the transverse sinus also receives arachnoid villi (Figure 8.58).

The *sigmoid sinuses* (Figure 8.59) form a curve on the medial surface of the mastoid part of the temporal bone and are separated from the mastoid antrum of the middle ear by a thin plate of bone. They are connected on both sides to the occipital and posterior auricular veins, as well as to the veins of the suboccipital triangle via the emissary veins. Thrombosis of this sinus, although rare, can occur subsequent to otitis media (otogenic sigmoid sinus thrombosis) and is seen in females using oral contraceptives.

The *occipital sinus* is contained in the posterior border of the falx cerebelli and lies anterior to the occipital crest, joining the confluence of sinuses near the occipital protuberance. It is connected to the occipital vein and the internal vertebral venous plexus, which serves as a collateral venous route upon blockage of the internal jugular vein. As the smallest of all dural sinuses, the occipital sinus begins near the margin of the foramen magnum and terminates in the confluence of sinuses.

Anteroinferior Group of Dural Sinuses

- The *sphenoparietal sinus* (Figure 8.59) runs along the inferior surface of the lesser wings of the sphenoid bone, joining the cavernous sinus on both sides. It receives the anterior temporal diploic veins and a branch of the middle meningeal vein.
- The *cavernous sinuses* (Figures 8.55 and 8.59) are located on both sides of the sella turcica, sphenoidal body, and pituitary gland. Each sinus extends from the superior orbital fissure to the apex of the petrous temporal bone and is connected to the pterygoid venous plexus via the emissary veins of the foramen ovale. It is interesting to note that this sinus contains the abducens, oculomotor, trochlear, ophthalmic, and maxillary nerves, as well as the internal carotid artery and associated sympathetic plexus.

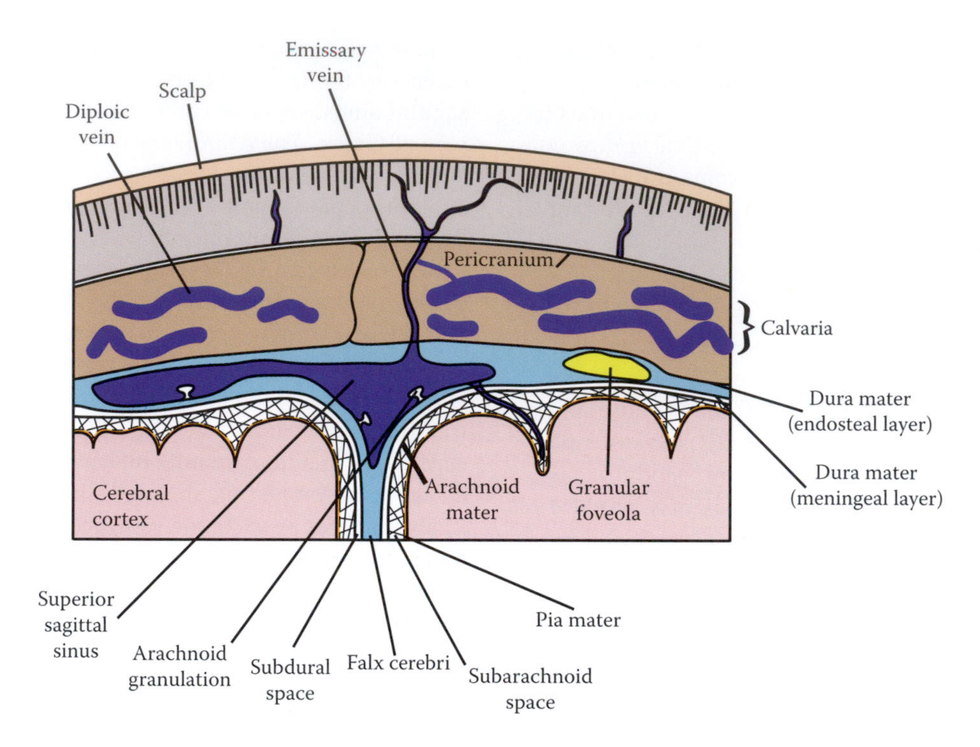

FIGURE 8.58 Schematic drawing of the scalp, calvaria, and superior sagittal sinus. The emissary veins and arachnoid granulations are shown to indicate the connection of the CSF and extracranial veins to the superior sagittal sinus.

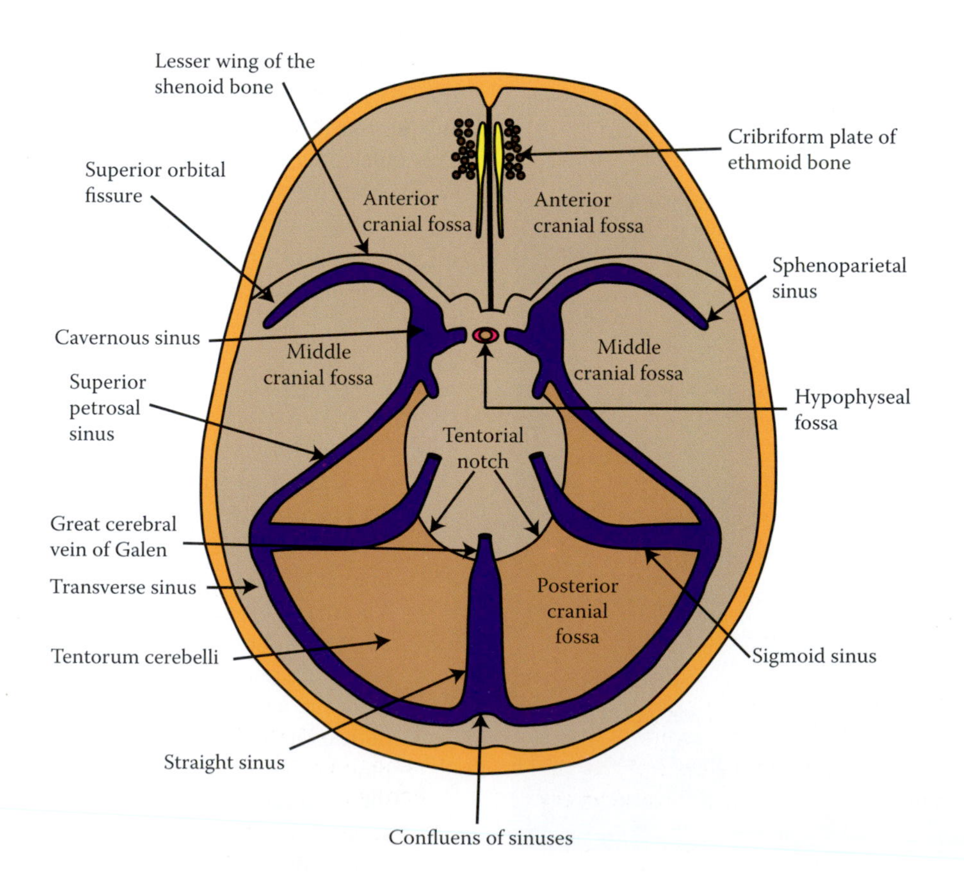

FIGURE 8.59 Diagram of the dural sinuses in the base of the cranium. Note the S-shaped configuration of certain dural sinuses on both sides of the cranial base.

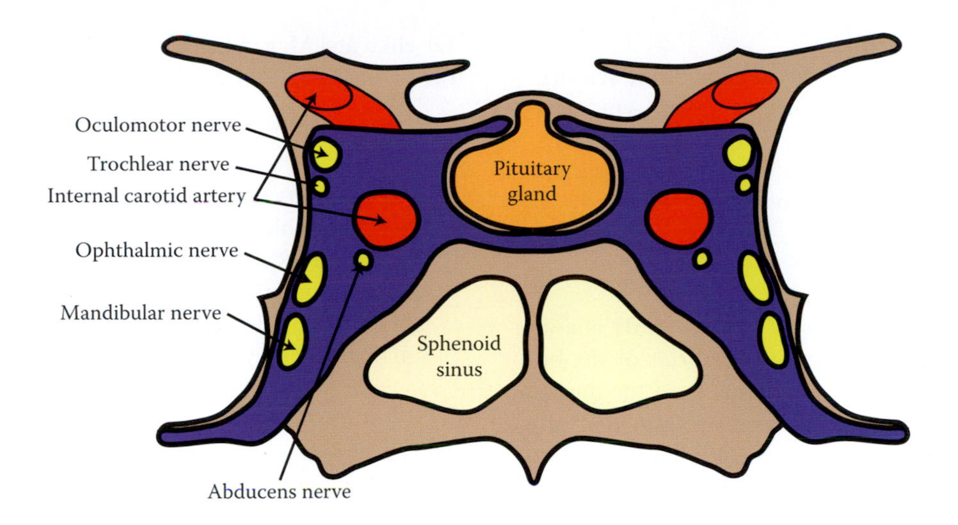

FIGURE 8.60 Content of the cavernous sinus. This dural sinus is unique in that it contains the internal carotid artery. The abducens nerve is the closest cranial nerve to the internal carotid artery.

The cavernous sinus communicates with the facial (angular) veins and pterygoid venous plexuses via the superior and inferior ophthalmic veins, respectively. Thus, thrombosis of the cavernous sinus can occur as a result of spread of infection from the face (e.g., infected acne near the medial canthus), infected scalp injuries, nasal furuncle, infected sphenoidal or ethmoidal sinuses, dental infection, orbital cellulitis, ear (otitis media), or infection localized in the infratemporal fossa. Cavernous sinus thrombosis can follow both acute and subacute courses and is associated with periorbital edema and swelling of the orbital soft tissue, which eventually leads to proptosis (forward displacement or bulging of the eyeball). Disturbances of consciousness, hypersomnia, and often headache, chills, diplopia, photophobia, visual impairment, and ocular pain occur. Exophthalmos is the most frequent presenting sign, while papilledema, facial edema, and subdural hematoma are less frequent findings. Retinal hemorrhage, chemosis, and varying degrees of ocular muscle palsy may also be seen due to compression of the oculomotor, trochlear, and abducens nerves. Sensory changes in the forehead and part of the face can be seen due to compression of the ophthalmic nerves. The abducens nerve is the most commonly affected nerve. Systemic manifestations such as fever and tachycardia are also seen. This condition may prove to be fatal if it is left untreated.

A caroticocavernous fistula is an abnormal AV shunt that develops inside the dural layers between the internal carotid artery (direct—type A) or its intracavernous branches of the internal carotid artery (indirect—type B) and cavernous sinus. However, meningeal branches of the external carotid artery (indirect—type C) may also be involved in this condition. Intracavernous branches of the internal carotid and meningeal branches of the external carotid arteries can also be involved in the same patient (type D). It may occur spontaneously or as a result of penetrating head trauma, basal skull fracture, complication of surgical procedures, aneurysmal rupture, or vascular disease. Patients with Ehler–Danlos syndrome and fibromuscular dysplasia are particularly prone to develop this condition. Clinical presentations of this condition include retrobulbar pain, pulsatile exophthalmos, double vision, progressive visual deficits, audible orbital or cranial bruit (a humming sound in the skull), and chemosis and dilatation of the subconjunctival vessels. CT scan, cerebral digital subtraction angiography (DSA), and arteriography can help diagnose this condition.

The *superior petrosal sinus* (Figure 8.59) follows the corresponding sulci, connecting the cavernous sinus to the transverse sinus. It receives the inferior cerebral as well as the superior cerebellar veins.

The *inferior petrosal sinus* occupies the inferior petrosal sulcus and connects the cavernous sinus to the internal jugular veins. It receives the superior cerebellar and labyrinthine veins and venous tributaries from the pons and medulla. It courses between the petrous temporal bone and the basilar part of the occipital bone.

Thrombosis of the inferior petrosal sinus, a complication of otitis media, causes retrobulbar pain due to compression of the ophthalmic nerve and abducens nerve palsy, leading to medial strabismus. Abducens nerve palsy is associated with Gradenigo syndrome or petrous apicitis, which exhibits also suppurative otitis media and pain in the distribution of the sensory branches of the trigeminal nerve. Other symptoms include fever, corneal hyporeflexia, and areflexia. This thrombosis may extend to the internal jugular vein caudally and the cavernous sinus rostrally.

The *arachnoid (spidery) mater* (Figure 8.61) is a delicate trabecular layer that covers the exterior of the brain without following the sulci and lies between the dura and pia mater. *Leptomeninges* is a common term that refers to the combined pia–arachnoid layer. It is separated from the pia and dura mater via the subarachnoid and subdural spaces, respectively. Dilatations of the subarachnoid space around the brain and brainstem that contains the cerebral vessels are termed *cisterns*. These cisterns occupy strategic locations around the brain and brainstem, which include the cerebellomedullary cistern, cisterna ambiens, pontine and interpeduncular cistern, cistern of the lamina terminalis, and supracallosal cistern.

Arachnoid Mater

The *cerebellomedullary cistern* (cisterna magna) establishes communication with the fourth ventricle via the foramina of Luschka and Magendie, containing the terminal branches of the posterior inferior cerebellar artery.

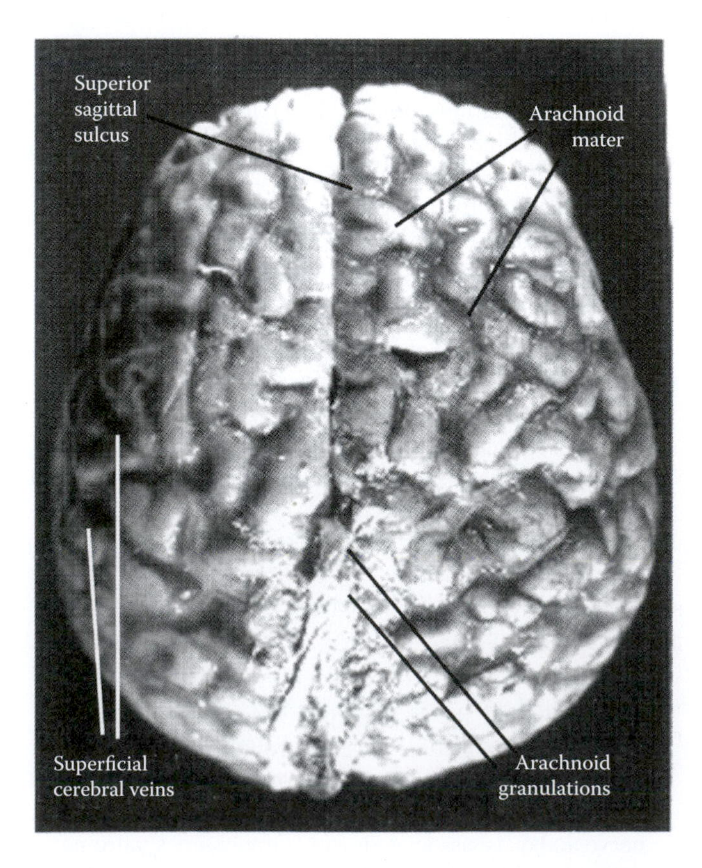

FIGURE 8.61 This photograph illustrates the arachnoid mater and arachnoid granulations.

Occlusion or atresia of the foramina of Luschka or Magendie may produce a noncommunicating hydrocephalus, which is characterized by enlargement of the ventricles, compression of the brain, and subsequent thinning of the cerebral cortex (cortical atrophy). Varying degrees of mental retardation and skull enlargement may also occur.

The *cisterna ambiens* is located posterior to the pineal gland and contains the great cerebral vein of Galen. The *pontine cistern* is located on the ventral surface of the pons, containing the basilar artery. The *interpeduncular cistern* encircles the arterial circle of Willis, mammillary bodies, and tuber cinereum. The *cistern of the lamina terminalis and supracallosal cistern* bridge the ACA.

Numerous minute arachnoid projections or villi, guarded by one-way valves, enter the superior sagittal and/or transverse sinuses. These one-way valves allow CSF to pass from the subarachnoid space to the systemic circulation, while preventing fluid movement in the opposite direction. It is important to note that the low pressure of the dural sinuses relative to the ICP also dictates the flow of CSF. Arachnoid villi, which are present in the fetus and newborn, form the macroscopic arachnoid granulations (Pacchionian bodies) at around 18 months of age that exert pressure on the inner surfaces of the calvaria, producing pressure atrophy and visible depressions. Extensive contact with the ependymal lining of the ventricles, pia mater, adjacent glial cells, and capillary endothelium of the choroid plexus may be secured by the long path followed by the CSF.

Meningiomas (Figure 8.62), which are the second most common benign primary intracranial neoplasms that affect primarily individuals in the fourth decade of their lives. They are believed to arise from the arachnoid villi along the course of the superior sagittal and sphenoparietal sinuses, but they could occur anywhere along the calvaria. They are easily resectable. Histologically, calcified psammoma bodies with whorls of fibrous tissue are seen. Hyperostosis, a reactive bone formation in the overlying calvaria, can occur in response to tumor and pose surgical risks. These tumors do not infiltrate, but they displace brain tissue and compress the brain and produce focal seizures, often as an early sign. Headaches are also common. Also, occlusion of the arachnoid villi due to thrombosis, infection, or tumors may produce a communicating hydrocephalus.

Subarachnoid hemorrhage and/or hematomas are commonly caused by rupture of the congenital berry aneurysm of the arterial circle of Willis. These saccular or oval-shaped and berrylike dilatations frequently occur

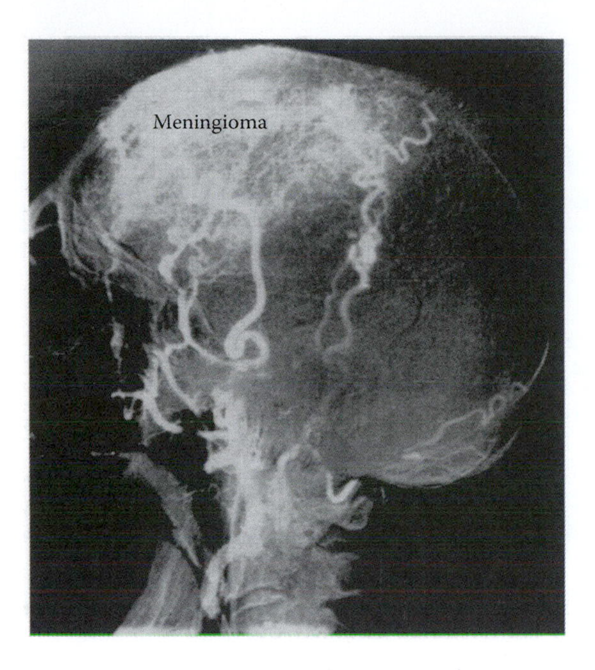

FIGURE 8.62 An angiogram of the internal carotid artery showing meningioma in the parietofrontal area.

near the proximal branches of the cerebral arteries (e.g., middle cerebral and posterior communicating arteries), where the muscular and elastic layers are deficient. They are associated with aortic coarctation and polycystic kidney. Compression of the optic, oculomotor, and trigeminal nerves by these aneurysms may occur prior to their rupture. Bleeding into the subarachnoid space produces symptoms that are frequently sudden in onset and include nuchal rigidity, severe ipsilateral headache, nausea, vomiting, transient vertigo, and occasional syncope (transient loss of consciousness that is preceded by lightheadedness). Hearing and visual impairment, as well as seizures (which occur in less than 20% of patients), may also be detected. Consciousness usually remains unaffected; however, disorders of concentration and attention, amnesia, and visual and hearing impairment may be observed. Signs of oculomotor nerve palsy and possible hemiplegia are also seen in patients with subarachnoid hemorrhage. CT scan and lumbar puncture (LP) may diagnose this condition. Neurologic complications may also include hydrocephalus, resulting from a clot that blocks the CSF pathway in the posterior fossa and vasospasm that produces cerebral infarctions.

Arachnoidal cysts (leptomeningeal cysts, Figure 8.63) are congenital lesions that develop from splitting of the arachnoid mater. Intrasellar cysts occupy extradural positions. They are divisible into simple and complex arachnoid cysts. The simple cysts maintain the ability to actively secrete CSF and are commonly found in the middle cranial fossa. Complex cysts may contain neuroglia and ependyma. Presentations of arachnoid cysts show variation according to their location. Increased ICP, visual deficits, developmental retardation, precocious puberty, hydrocephalus, and signs of bobble-head doll syndrome are seen in suprasellar arachnoid cysts. Cysts in the middle cranial fossa may manifest with hemiparesis, convulsions, and headache. Diffuse supratentorial or infratentorial cysts may also produce growth retardation, craniomegaly, and increased ICP.

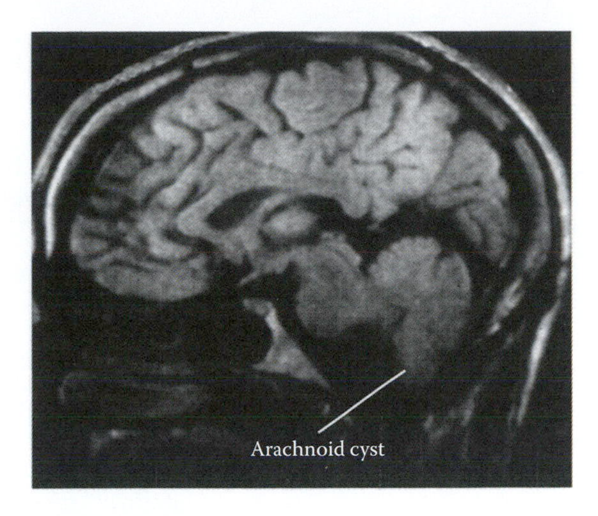

FIGURE 8.63 This MRI image illustrates an arachnoid cyst in the posterior cranial fossa, inferior and ventral to the cerebellum.

Pia Mater

The *pia (faithful) mater*, the innermost of the meninges, is a delicate layer that follows the architecture of the sulci and contributes to the formation of the blood–brain barrier (BBB). When the pia mater (containing small blood vessels) and ependyma join, they form the tela choroidea, which gives attachment to the choroid plexus. The pia mater consists of an outer epipial layer and an inner intima pia layer. As the cerebral arteries course within the subarachnoid space, they become isolated from the pia–arachnoid layer by a space that is continuous with the intrapial periarterial space, a gap between the smooth muscle of arterial capillary and pia mater. At the point of entrance of the arterial capillaries into the pia mater, one pial layer forms the pia–adventitial layer around the capillaries, which is separated from the external glial membrane by the periarterial subpial space. This layer follows the vessels into the brain, forming perivascular sheaths, which become discontinuous and eventually disappear as the vessels become capillaries. In this manner, leptomeningeal cells separate and form a regulatory interface between the arteries and the brain tissue, preventing the neurotransmitters and other pharmacologic agents released from nerves that supply the cerebral vessels from affecting the brain tissue. In this manner also, the subarachnoid space is separated by a layer of pia from the subpial and perivascular (Virchow–Robin) spaces.

The pia mater contributes to the formation of the brain barriers that maintain the ionic composition of CNS tissue

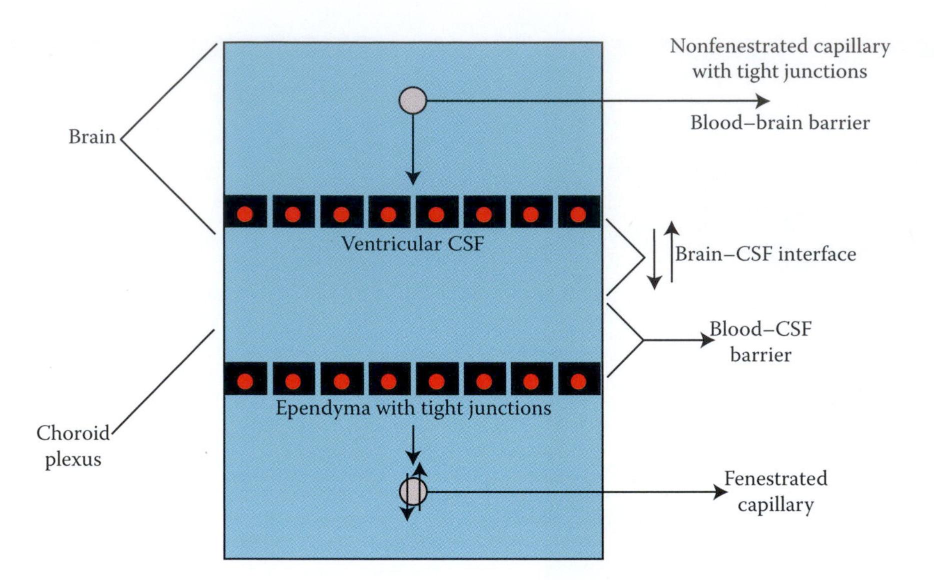

FIGURE 8.64 Diagram of the neural and vascular elements involved in the formation of the blood–brain and blood–CSF barriers.

by selectively allowing certain molecules to enter the brain or nerve tissue while excluding others. They (Figure 8.64) are classified into BBB, brain–CSF barrier, and blood–nerve barrier (BNB). These barriers are essential for the optimum functions of the central and peripheral nervous systems. However, small lipophilic molecules can gain access to the brain, utilizing specific transport systems, and are not affected by the barrier system. There are independent transporters for D-glucose and acidic, basic, and neutral amino acids.

BRAIN BARRIER

Brain barriers play an important role in maintaining chemistry of the brain for proper neuronal activity and have pivotal role in neurotoxicity and development of neurological disease. Changes in brain chemistry as a result of barrier dysfunction can produce disorders of memory, behavior, and learning. Certain chemicals tend to accumulate in certain parts of the brain, such as the nigral neurons, which are selectively affected by 1-methyl-4-phenyl-1,2,3,6-tetrahydropyridine, while lead (Pb) and manganese (Mn) accumulate distinctly in the hippocampus and basal ganglia, respectively. Cytoarchitecture of the endothelial cells and the changes associated with aging, cerebrovascular diseases, morphologic variations, blood flow differences, and barrier selectivity of neurotoxins need to be investigated to fully understand the mechanism of action of the barriers. The role of the brain barriers in developing the brain and the effect of chemical exposure during development needs also be explored.

The *BBB* is a system that maintains brain homeostasis, fibrinolysis, coagulation, vasomotor control, and activation as well as migration of leukocytes during pathologic processes. It is formed by the tight junctions of endothelial cells of the cerebral capillaries that isolate the brain from the blood and prevent oncotic and osmotic forces from influencing blood–tissue exchange. It controls transcapillary movements from blood to brain as well as in the opposite direction. Through this controlling mechanism, the BBB allows passage of substances of low molecular weight and small water-soluble molecules and macromolecules (O_2, CO_2, hormones) with high lipid solubility via specific symmetric and asymmetric carriers or by facilitated diffusion (Figure 8.53). The symmetric carriers may be located on the luminal or abluminal membrane. The asymmetry is evident in the presence of enzymes such as sodium; potassium ATPase on the abluminal membrane, which is responsible for the transport of Na from the endothelial cells into the brain; and potassium from the brain to into the endothelium. Since potassium has a critical impact on nerve impulse transmission and firing of neurons, asymmetry of potassium ATPase may be responsible for preserving a low level of potassium in the extracellular space. It has been suggested that this barrier system also depends upon the close interrelationship of the astrocytic end feet (glia limitans) with the basement membrane of the capillary endothelial cells. Astrocytes can also induce endothelial cells to form capillary-like structures, thus giving BBB-like properties to nonneuronal cells.

These endothelial cells contain large number of mitochondria and harbor enzymes that govern transport of ions to and from the CSF. They are noncontractile and do not contain actomysin filaments that respond to histamine. This fact may account for the unresponsiveness of brain capillaries to allergic disorders associated with histamine release, although local release of norepinephrine may result in reduction of blood flow into the brain capillaries via its action on the pericytes around these capillaries. Barrier-forming endothelial cells binding to each at the crossing point between blood and the brain occurs through transmembrane proteins such as occludin, claudins, junctional adhesion molecule (JAM), or endothelial cell–selective adhesion molecule (ESAM), which are anchored into the endothelial cells by a protein complex that includes zona occludens-1 (zo-1) and associated proteins. The extent of tightness of the interepithelial junctions of the choroid plexus is relatively limited, which explains the relative ease with which hydrophilic medications and toxicants can gain access to the brain through this plexus.

The BBB protects the brain against infections, particularly bacterial; however, when infection is followed by inflammation, permeability of the barrier changes, and more bacteria and/or viruses infiltrate, leading to serious consequences. However, *Borrelia bugdorferi*, that causes Lyme disease, and *Treponema pallidum*, which causes syphilis, can breach the BBB by creating tunnels through the blood vessel walls. Antibodies and certain neurotoxin such as *Clostridium botulinum* toxin have large molecules and are unable to cross the barrier system.

Therapeutically, many modalities of drug delivery through the BBB have been explored. These include the use of peptides, such as casomorphine, that cross the BBB; administration of vasoactive bradykinin; utilization of carrier-mediated transporters such as glucose and amino acid; disruption of active efflux transporters, such as P-glycoprotein (i.e., multidrug resistance proteins [MDR]), which is responsible for brain-to-blood efflux; and receptor-mediated transcytosis for insulin or transferrin. P-glycoproteins are ATP-binding cassette (ABC) transporters that belong to the ABCC carrier protein family and are expressed in cerebral capillaries. They control the efflux of a wide range of substances from the endothelium back into the blood circulation. As a result, the influx of hydrophobic drugs into the brain, such as vinca alkaloid antineoplastic drugs, ivermectin, cyclosporin A, digoxin, loperamide, or antiviral protease inhibitors, is relatively low.

Mannitol, intracerebral implantation, and high-intensity focused ultrasound (HIFU) can bypass the BBB. Liposomes loaded with nanoparticles, such as radiolabeled polyethylene glycol–coated hexadecylyanoacrylate nanospheres, are a promising method of delivery of drugs, particularly anticancer medications, across the BBB, although the cytoarchitecture of the barrier system may have already been disrupted by the tumor.

In the newborn, the endothelium of the cerebral capillaries of the BBB allows large molecules like albumin to be carried by their pinocytotic vesicles. This explains the high level of albumin in the CSF in neonates compared to infants. BBB may be disrupted as a result of brain tumors or as a sequel to bacterial meningitis. Stroke-induced cerebral edema may also impair the function of the BBB.

In general, the BBB prevents substances circulating in the bloodstream from gaining access to the brain and also plays a role in their modification and metabolism. Diffusion of a large concentration of D-glucose (2–3 times more than normally metabolized by the brain) through the BBB is facilitated by an insulin-dependent glucose transporter 1 (GLUT-1) glucose transporter. Reduced GLUT-1 transporter may be associated with seizures, impaired brain development, and mental retardation. Passage of structurally related essential amino acids (precursors of catecholamines and indolamines) through the cerebral capillaries to the brain is mediated by a single transporter. This allows an intense competition among neutral L-amino acids; thus, elevation in the plasma concentration of a rival amino acid may account for the inhibition of uptake of others. Therefore, high plasma levels of phenylalanine, as in phenylketonuria, may remarkably reduce the uptake of the competing essential amino acids. However, amino acids that are synthesized in the brain (such as amino acid neurotransmitter) are actively transported in a reverse direction outside the brain. CO_2, O_2, N_2O, and volatile anesthetics diffuse rapidly into the brain.

The BBB also controls the transport of calcium, magnesium, Mn, zinc, iron, cobalt, and molybdenum, which are essential CNS catalysts, second messengers, and gene expression factors. Enzyme stability and activation also require these elements, such as protein kinases, superoxide dismutase, and transcriptional factors. Mercury (Hg), Pb, and arsenic (As) can access the brain via the BBB. This transport is accomplished by the BBB vasculature through active or receptor-mediated transporters. Thus, damage to the endothelial cells can produce leakage of blood-borne material to the surrounding brain parenchyma. This is evident in Pb toxicity, which causes microvascular damage, widening of the interendothelial tight junctions, and enhanced pinocytotic activity, leading to cerebral edema, possible herniation, ventricular compression, and hemorrhage. The intracellular effect of Pb is also interesting as it sequesters in the same intramitochondrial compartment as Ca2, disrupts the Ca2 regulatory processes, and changes the phosphorylation chains by interfering with the protein kinase system. It may also interfere with astrocytic function and thus indirectly compromise BBB function. Methyl mercury (MeHg) can easily cross the BBB barriers, as it does not require a carrier system, and it can also cross the placenta. Consumption of high MeHg levels during pregnancy can cause encephalopathy in the offspring.

The *blood–CSF barrier (BCB)* is a system that is actively involved in the various early stages of brain development, brain maturation, CNS homeostasis, and neuroendocrine regulation. It is formed by the epithelial cells of the choroid plexus, which are connected by tight junctions, and the overlying pia–glial and ependyma–glial membranes. Thus, ventricular CSF diffuses into brain extracellular fluid and eventually into the subarachnoid space. The choroid plexus is located in the central part and the inferior horn of the lateral ventricle, roof of the third ventricle, and caudal roof of the lateral ventricle. This plexus regulates bidirectional transport and synthesizes transthyretin (TTR), transferrin, and ceruloplasmin that will be delivered to the CSF.

The tight junctions of the choroidal epithelial cells are relatively loose compared to the BBB to allow some degree of transport from the choroid plexus to the brain. The choroid epithelium forms a single layer that demarcates the blood in the choroid plexus from that of the ventricular CSF. It presents the apical microvilli and the basolateral enfoldings, which increase the surface area of exchange between the blood and CSF. The choroid plexus also has a high enzymatic activity for drugs and antioxidants such as glutathione peroxidase as well as high organic anion and cation transporters expression. The latter contains members that show selectivity relative to location; for example, solute carrier (SLC21) organic anion transporting polypeptides (oatp) is distributed on either the basolateral or the apical microvilli, whereas the SLC22, that is, organic anion transporters (OATs) and organic cation transporters (OCTs), are found on the apical villi. The multidrug resistance protein MRP1 is uniquely located at the basolateral membrane of the choroidal epithelium. These transporters either form a functional barrier to blood-borne toxic compounds and drugs or enhance the elimination of toxic endogenous metabolites. This barrier system is highly efficient, but functionally is suboptimal, as it allows drugs and xenobiotics out of the CSF. Pb's extent of accumulation in the choroid plexus is greater (nearly 100-fold) than in the endothelial cells of BBB. This accumulation may increase with age and exposure to a polluted environment. Pb accumulation in the choroid plexus changes TTR levels in the CSF by inhibiting its synthesis and disrupts the transepithelial transport. Disruption of this mechanism may affect thyroid hormone transport from the blood to the CSF, resulting in cognitive ability and irreversible mental deterioration. This is evident in Pb poisoning in children.

Functional and Clinical Significance

The significance of this barrier is well illustrated in patients with jaundice, where the bile is selectively excluded from entering the CSF or brain. The presence of a yellowish bile stain only in the stroma of the choroid capillaries substantiates the activity of this barrier. The sodium–potassium–ATPase system, which provides a pump allowing sodium into the CSF and potassium into the plasma, is contained in the choroid epithelium. This ionic movement assists the passage of large amount of plasma water from the capillary bed.

Structural Differences between BBB and BCB

BBB	Endothelial cells of cerebral capillary
	Basement membrane
	Perivascular feet of astrocytes
	Interstitial fluid
BCB	Endothelial cells of the choroid plexus capillaries
	Basement membrane
	Choroid epithelium
	CSF

The *BNB* comprises the perineurium and the capillaries of endoneurium. The wall of these capillaries is nonfenestrated, and the endothelial cells establish tight junctions. This barrier is functionally much more effective in the dorsal root ganglia and autonomic ganglion.

The clinical significance of barriers can be illustrated in (1) kernicterus, (2) cerebral edema, (3) brain scans, (4) the therapy of Parkinsonism, (5) epinephrine surge, (6) hyperglycemia/and or hypoglycemia, (7) MS, (8) AD, (9) meningitis, (10) HIV encephalitis, (11) rabies, (12) epilepsy, (13) brain abscess, (14) Progressive multifocal leukoencephalopathy (PML), and (15) trypanosomiasis. Other conditions that exhibit the role of the brain barrier are ischemia and certain viral or autoimmune diseases.

1. In infants, the immature liver cannot conjugate large amounts of bilirubin to serum albumin, leading to an increase of the unconjugated bilirubin. Inability of the BBB to block the unconjugated bilirubin from entering the CNS may be followed by its deposition in the basal and brainstem nuclei. Deposition of unconjugated bilirubin in the CNS is commonly detected in kernicterus or erythroblastosis (neonatorum) fetalis, a condition that develops as a consequence of Rh incompatibility between the mother and the fetus. It may also be associated with hemolytic diseases (low albumin reserve), reduced amount of glucuronyl transferase in the preterm infant, biliary atresia, low pH, and hepatitis. Infants with this predicament initially exhibit hypotonia followed by rigidity and gaze palsies. They are likely to die, and autopsy may reveal discoloration of the brain tissue, particularly of the basal nuclei. Survivors show signs of mental retardation and some forms of choreiform movements. Bilirubin toxicity may also occur as a result of a lack of or ineffective CNS bilirubin oxidase system due birth trauma or incomplete development.
2. Cerebral edema may be caused by bradykinin, which stimulates the influx of protein into the brain, and subsequent enlargement of the extracellular space of the white matter.
3. Brain scans also utilize the BBB concept, and contrast medium injected into the blood breaks down the BBB, allowing visualization and localization of lesions (gray matter contains a large number of capillaries that greatly exceeds the white matter) via various imaging techniques.
4. Administration of L-dopa, a precursor of dopamine, which is able to cross the BBB by a neutral amino acid carrier, may be used in the treatment of Parkinson's disease.
5. Neuronal function is maintained and released from the disastrous consequence of epinephrine surge by the BBB.
6. Neuronal activities may be inhibited, and coma may result in individuals with hyperglycemia, which eventually leads to accumulation of ketone bodies in the brain. Overactivity of the CNS and mental confusion may occur

in hypoglycemia. Lack of glucose in this condition may lead to insulin coma.

7. MS is a CNS demyelinating disease, which is discussed in detail in Chapter 2. It has been shown that in MS that the BBB is disrupted and the T lymphocytes cross over and start the demyelination process. As a result, MS is considered by some as a BBB-related condition in which the endothelial cells are damaged and P-glycoprotein becomes dysfunctional subsequent to oxidative stress. On this basis, antioxidants such as lipoic acid may be utilized to stabilize a disrupted BBB.
8. AD is the most common form of dementia that affects older people. It is an irreversible and progressive condition of the brain that causes impairment of memory and mental faculties that interfere with day-to-day life. Patients experience confusion, language disorders, irritability, and mood changes. It has been suggested that disruption of the BBB in Alzheimer patients allows plasma β amyloid (Aβ) and autoantibodies to enter the brain and bind to neurons and astrocytes, initiating a cascade of events that leads to the accumulation of cell surface–bound Aβ42 in the these structures.
9. Meningitis is an inflammation of the spinal/and or cerebral meninges caused commonly by *Streptococcus pneumoniae* and *Haemophilus influenzae*. Meningitis is associated with disruption of the BBB, which allows toxins and, for therapy, antibiotics to enter into the brain. Administration of antibiotics may aggravate the already-existing inflammatory response of the CNS by releasing neurotoxins from bacterial cell wall.
10. HIV encephalitis is an inflammation of the brain caused by HIV. It may be associated with headache, nuchal rigidity, confusion, and seizures. It has been proposed that in the initial stages of infection, HIV can manage to cross the BBB and circulate within monocytes ("Trojan horse theory"), which are later transformed into macrophages. Activated macrophages release viral particles known as virions into the brain tissue and initiate an inflammatory reaction involving the microglia. This cascade of events compromises the structural and functional integrity of the BBB.
11. Rabies is an acute viral CNS disease caused by an RNA virus transmitted mainly through the bite wounds contaminated by the saliva of an infected animal. The virus travels through the peripheral nerves to the spinal cord and then into the brain. Patients exhibit encephalitic and paralytic syndrome, which follows a prodrome stage with fever, headache, and paresthesia at the site of the bite. The BBB and BCB are considered the primary sites of rabies virus replication, where increased BBB and BCB permeability promotes viral clearance. As a result, the brain glucose and potassium concentration in the extracellular space, which is regulated by an active transport mechanism, as well as transmitters' levels in the CSF are adversely affected. A rise in the level of blood potassium renders the transport systems nonoperational, whereas excess of potassium in the CSF is removed via active transport into the blood.
12. Epilepsy is a common and potentially serious neurological disease that is characterized by recurrent seizures. Studies have linked the disruption of the BBB by either artificial or inflammatory mechanisms to the development of epileptic seizures. Additionally, effectiveness of antiepileptic drugs relates to expression of drug resistance molecules and transporters at the BBB.
13. Brain abscess is a rare but life-threatening condition, which may be indolent or fulminant, and occurs as a result of otitis media, paranasal sinus infection, and epidural abscess. It may be linked to heart anomalies such as atrial septal defect (ASD) or ventricular septal defect (VSD), pulmonary AV malformation, or endocarditis or may arise as a complication of surgery. It is characterized by a focal cerebral lesion with a hypodense center and a peripheral ring enhancement. Recent data indicate that the spread of bacterial infection and development of brain abscess occur as a consequence of brain barrier dysfunction. Patients present with relentless and progressive headache. Most patients, but not all, experience fever and may also show papilledema and signs of increased ICP. CT scan shows a focal cerebral lesion with a hypodense center and a peripheral ring enhancement. LP is contraindicated to prevent abscess dissemination into the ventricular system. As a treatment, the abscess can be aspirated, followed by cephalosporin or penicillin with metronidazole.
14. Progressive multifocal leukoencephalopathy (PML) is a rare usually fatal central nervous system demyelinating disease that is caused by JC polyomavirus. It can occur in AIDS and organ transplant patients subsequent to the administration of immunosuppressive

medications. It affects the white matter of the parietal and occipital lobes and destroys the oligodendrocytes, producing intranuclear inclusions. The virus infects the white mater after breaching the BBB.

15. Trypanosomiasis is a serious parasitic condition in which trypanosoma protozoa are found in brain tissue in the late stage of the disease. It occurs when the tsetse fly bite erupts into a wound and, as a result, the patient experiences fever, joint and muscle ache, and headache, and when advanced, the CNS will be involved. CNS involvement results in disruption of the circadian rhythm and sleep cycle (sleeping sickness), confusion, personality changes, seizures, and ataxia. The mode by which parasitic infection enters the brain from the blood is unclear; however, passage through the choroid epithelium (BCB), a circumventricular organ, has been suggested.

Circumventricular Organs

Not all areas of the brain harbor the barrier systems; in fact, structures that occupy positions near the ventricles, surrounding the brainstem and the diencephalon, known as the *circumventricular organs* (Figure 8.65), lack the barrier system and operate as points of contact between certain brain

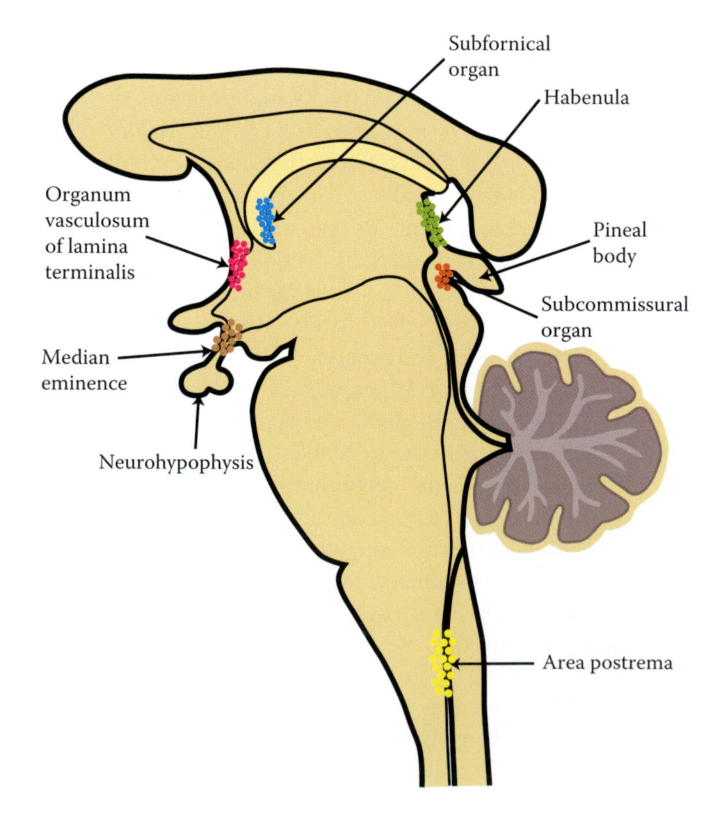

FIGURE 8.65 Drawing of the midsagittal brain illustrating the sites where blood–brain barriers are absent.

receptors and blood. These specialized areas, which may utilize the neurohumoral mechanism to exert their influence, include the subfornical organ, area postrema (AP), and organ vasculosum of the lamina terminalis; neurohypophysis and intermediate lobe of the pituitary gland; medial eminence; pineal gland; and subcommissural organ.

The *subfornical organ* is located near the anterior pole of thalamus and the interventricular foramen of Monro, maintaining massive projections to the supraoptic and paraventricular nuclei and the lateral hypothalamic area. Through these connections, it plays an important role in homeostasis, osmoregulation, and the circulation of blood in the choroid plexus. It binds angiotensin II and receives input from the hypothalamus, regulating water intake and inducing vasopressin secretion.

The *AP*, as discussed earlier with the fourth ventricle, is an emetic chemoreceptor center, located near the obex at the point of junction of the lateral walls of the caudal half of the fourth ventricle. It receives projection from the spinal cord and solitary nucleus. AP is intimately interconnected with the solitary tract nucleus. It is sensitive to apomorphine and digitalis glycosides, regulating food and water intake and cardiovascular functions.

The *vascular organ of the lamina terminalis* is a highly vascular structure, which lies superior and rostral to the optic chiasma and carries hypothalamic peptides (such as somatostatin, angiotensin II, and atrial natriuretics). Structurally, it is similar to the median eminence, maintaining a functional relationship with the preoptic area and preserving fluid balance.

The *neurohypophysis* receives terminals of the hypothalamic neurons that convey oxytocin, neurophysin, and vasopressin via the hypothalamohypophyseal tract.

The *medial eminence* serves as a link and transducer between the hypothalamic neurons that secrete the hormone-regulating factors and the portal–hypophyseal system. It also deals with the transduction of hypothalamic neuronal impulses.

The *pineal gland* is discussed in detail with the epithalamus. The *subcommissural organ* is located ventral to the posterior commissure and at the site of junction of the third ventricle and cerebral aqueduct. It secretes proteinaceous materials into the CSF; however, its role in salt and water balance is not yet confirmed.

VENTRICULAR SYSTEM AND CEREBROSPINAL FLUID

Since both the ventricular system and the subarachnoid space contain the *CSF*, a brief review of the ventricles with major emphasis on CSF, its pathway, and associated clinical conditions will be needed.

The *ventricular system* (Figures 8.66, 8.67, and 8.69), a derivative of the neural canal, is comprised of the lateral (see also the lateral, third, and fourth ventricles; cerebral aqueduct; and central canal). This system communicates with the venous blood via the subarachnoid space. The ependymal

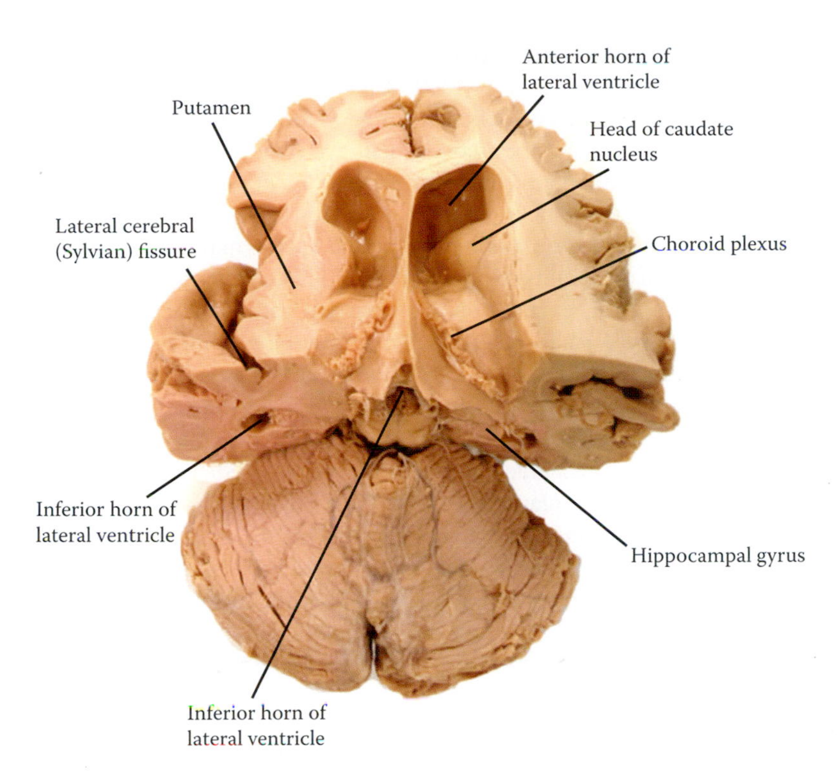

FIGURE 8.66 The floor of the lateral ventricle and the choroid plexus. The third ventricle and the floor of the fourth ventricle are also illustrated. The cerebral hemispheres, corpus callosum, and part of the cerebellum are removed.

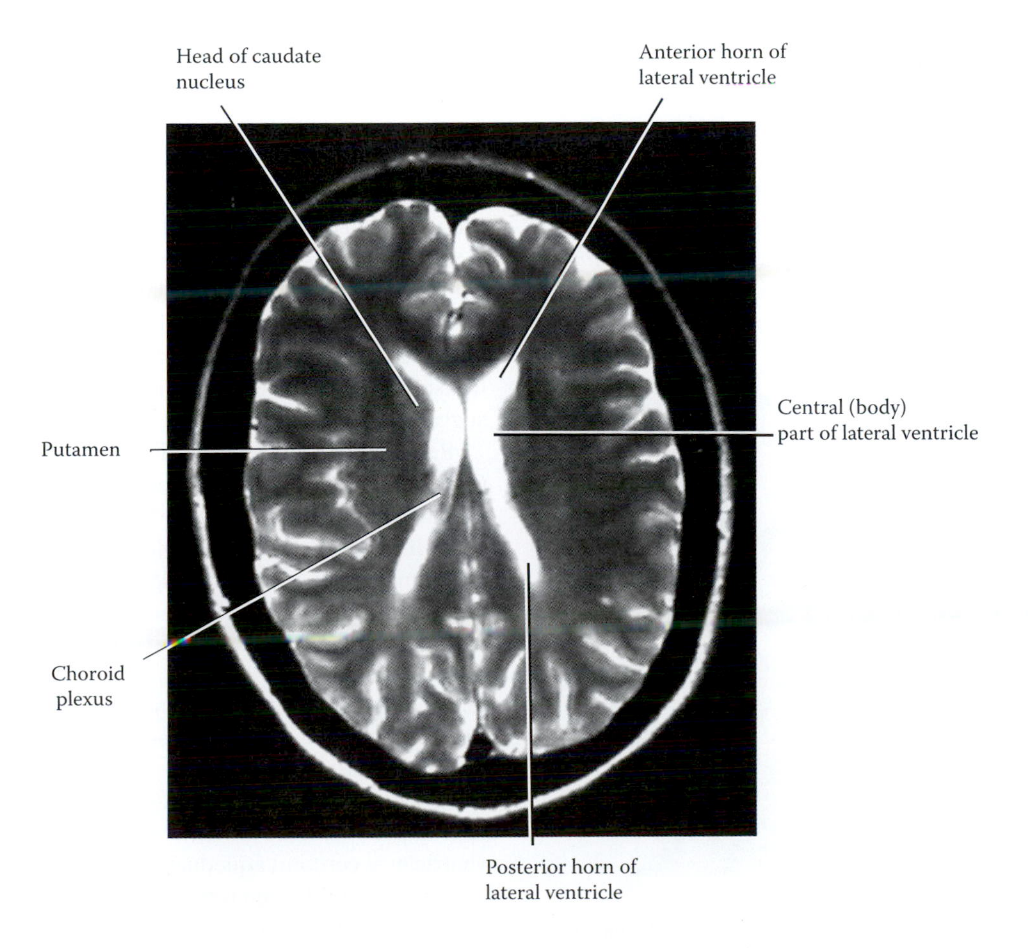

FIGURE 8.67 MRI scan of the brain showing some of the components of the lateral ventricles.

cells that selectively allow certain substances to gain access to the brain line the ventricles.

The *CSF* is a clear, colorless fluid that maintains an alkaline pH of 7.3. Turbidity or pinkish discoloration may be associated with the presence of fresh blood from subarachnoid hemorrhage; yellowish discoloration of CSF (xanthochromia) may result from disintegrated blood or an increased amount of protein and/or bilirubin. It contains trace amounts of protein, mainly immunoglobulin and, to a lesser, extent albumin. It also contains a few leukocytes and a small amount of glucose (80–120 mg/dL) and potassium (2.9 mEq/L), with a greater concentration of sodium and chloride. This is in contrast to serum, which contains a lower sodium concentration and higher potassium and calcium. The arachnoid mater, which forms part of the BCB, may also play a role in changing the composition of the CSF. Numerous peptides are also found in the CSF, which include luteinizing hormone–releasing factor, cholecystokinin, angiotensin II, substance P, somatostatin, thyroid releasing hormone, oxytocin, vasopressin, and. Alternations in the CSF concentrations of peptides may be used for the diagnosis of certain neurological diseases. The CSF maintains an alkaline pH of 7.3. The changes in the composition of the CSF may serve as a tool in the diagnosis of certain diseases. A low pH value may occur in conditions associated with acidosis and hypercapnia and in certain pulmonary diseases. Increased protein concentration is observed in spinal shock (complete transection of spinal cord) and in cases of extramedullary and intramedullary spinal cord tumors. Immunoglobulin G (IgG) is generally elevated in multiple sclerosis. Glucose levels decrease in bacterial, fungal, and viral meningitis. The CSF may contain neoplastic cells in primary and secondary neoplasms.

Secretion of the CSF is maintained at a rate of 0.3–0.4 mL/min by the choroid plexus (Figures 8.66 and 8.68) and, to a much lesser degree, by the ependyma. The *choroid plexus* is formed by the vascular epithelium of the pia mater and the ependyma that invaginates through the choroidal fissure, containing mostly simple cuboidal ependymal cells with microvilli of the brush border type. This plexus is innervated by the postsynaptic sympathetic fibers that emanate from the superior cervical ganglion of the sympathetic trunk. Selective prevention of the free entrance of protein and electrolytes from the blood to brain tissue is maintained by the ependymal cells that form tight junctions. Secretion and transport of certain hormones such as TTR, a carrier of thyroxin, retinol, and insulin-like growth factor-II into CSF may also be accomplished by the choroid plexus. Hardened bodies called *psammoma* (sand-like), which are composed of concentric rings of calcium carbonate, calcium, and magnesium phosphate, occur normally in the adult choroid plexus. CSF is circulated about three times in 24 h, and the absorption rate can be four to six times the normal rate of formation. Secretion of CSF may occur through a variety of processes that include filtration across the endothelial wall, hydrostatic pressure–dependent activity, and enzymatically controlled active process by the choroidal epithelium. The latter process is activated by ATPase and carbonic anhydrase. Cardiac glycosides produce inhibition of CSF secretion by uncoupling the mitochondrial oxidative phosphorylation. The steady secretion and absorption of the CSF is essential for maintaining a uniform pressure within the ventricles and the cranial vault. A vast number of 5-hydroxytryptamine receptors may influence the blood flow in the choroid plexus.

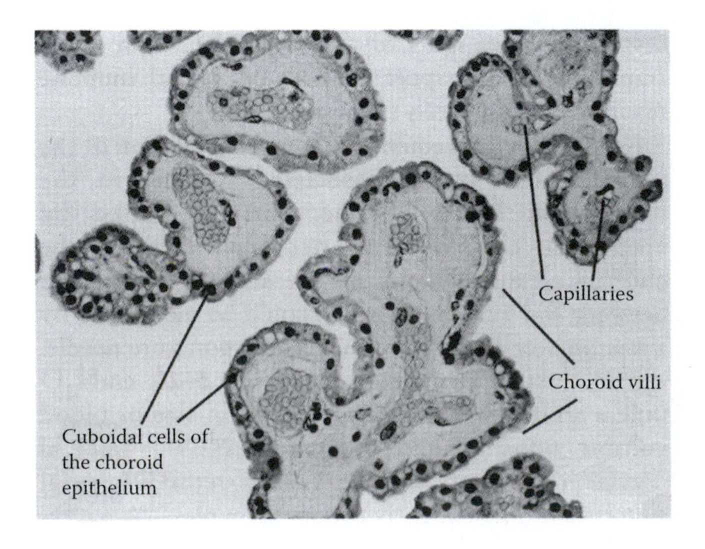

FIGURE 8.68 Photograph of the choroid plexus. Note the cuboidal cells of the choroid epithelium and associated capillary network.

CSF regulates ionic transport to and from the extracellular space (sink action), maintaining low concentrations of certain substances in the CSF and brain relative to plasma concentrations. This allows the creation of an efficient milieu for the conduction of nerve impulses. It also acts as a buffer to lessen the impact of head trauma (buoyancy effect), provides nutrients to the leptomeninges, and dramatically reduces the weight of the brain. Increased ICP and brain volume, as a result of vasodilatation of cerebral vessels or parenchymal swelling, may be counteracted by displacement of the CSF. The role of CSF in signal transduction, transport of hormones, and immune reaction has also been suggested.

Respiratory movements influence circulation of the CSF, cardiac systole-associated changes in the intracranial blood volume, current created by ependymal cilia, and arterial pulsation of the choroid plexus. The normal pressure of CSF ranges between 6 and 14 cm of water, which could be monitored by a manometer, attached to the LP needle. This pressure remains constant at 50–200 mm of water unless an increase or decrease in brain size or blood volume occurs (e.g., compression of the internal jugular veins increases the venous return of dural sinuses and subsequently produces increased ICP).

In general, the circulation of the CSF involves the following paths: lateral ventricle → foramen of Monro → third ventricle → cerebral aqueduct → fourth ventricle → foramen of Luschka and foramen of Magendie → cerebellomedulary cistern (cisterna magna) → superior sagittal. The CSF is also absorbed at the levels of spinal nerve root sheaths.

ICP increase may be associated with an abnormal increase in the CSF volume, cerebral edema, intracranial hemorrhage, or impairment of venous drainage. ICP due to space-occupying lesions such as tumors initially produces efflux of the CSF from the ventricular system and subarachnoid space, leading to compression of the brain surface against the rigid bony skull (Figure 8.69). These changes lead to flattening of the cerebral gyri. Rapid expansion of the space-occupying mass can raise the ICP to the level of the arterial pressure, leading to brain death prior to any possible herniation, and may produce papilledema, nausea and vomiting (due to activation of the vomiting center in the upper medulla), bradycardia (slowing of the heart due to vagal activation), and coma.

Idiopathic intracranial hypertension (pseudotumor cerebri) is a reversible condition in which ICP is increased in the absence of a tumor or underlying diseases. It occurs more often in women than men, especially in obese women, but children can also be affected. There is no specific etiology associated with this condition; however, oral contraceptives, anticonvulsive medications (phenytoin), tetracycline, steroids, tamoxifen, and high doses of vitamin A may increase the risk for development of this condition. The mechanism associated with this condition is not known; however, increased brain tissue volume due to increase in water content may contribute to the increase in pressure. Increased CSF volume and rate of blood flow have been advanced. Patients with hyperparathyroidism, Cushing and Addison disease, and iron-deficiency anemia and pregnant women are particularly prone to the development of this disease. Patients present with generalized throbbing headache that is worse in the morning and exacerbated with sneezing and coughing. Vomiting, diplopia due to abducens nerve palsy, generalized weakness, pulsatile tinnitus anosmia, and ataxia are also seen in patients with this disease. Other cranial nerve dysfunctions may also be observed, though rarely.

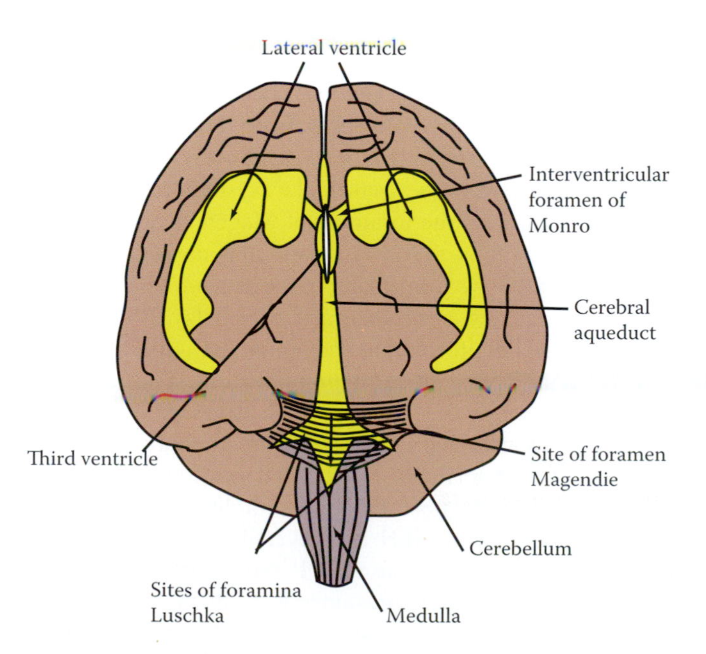

FIGURE 8.69 Schematic drawing of the ventricular system (posterior view).

Hydrocephalus

Accumulation of CSF in the ventricles produces *hydrocephalus* (Figures 8.70 and 8.71). This condition commonly results from obstruction of CSF pathways as in ependymoma, oversecretion of the CSF as in a papilloma of the choroid plexus, venous insufficiency as in dural sinus thrombosis, or defective absorption due to occlusion the arachnoid villi as in arachnoiditis. Hydrocephalus may also be produced as a result of atrophy or reduction in the total volume of the brain (*hydrocephalus ex vacuo*), as is seen in AD.

Hydrocephalus may be congenital or acquired:

Congenital hydrocephalus (overt-infantile hydrocephalus) occurs usually in the first few months of life or inside the uterus. It rarely extends to the fifth decade of life. It is commonly associated with stenosis of the cerebral aqueduct, obstruction of the foramina of Luschka and Magendie (Dandy–Walker syndrome), or obliteration of foramen of Magendie and cisterna magna (as in Arnold–Chiari Type II malformation). It may also result from asymmetrical fusion of cervical vertebrae, imperfect fusion or nonunion of the vertebral arches, meningomyelocele, and closure of the foramina of Magendie and Luschka (*Klippel–Feil*

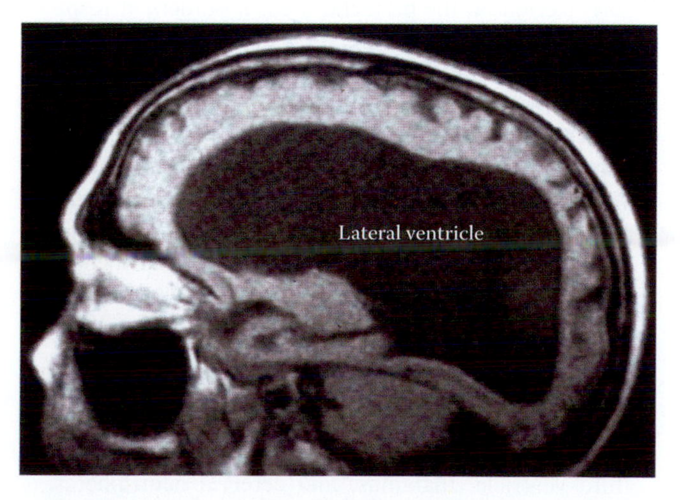

FIGURE 8.70 This MRI image shows the massive dilatation of the lateral ventricle in an individual with obstructive hydrocephalus. Note the remarkable thinning of the cerebral cortex.

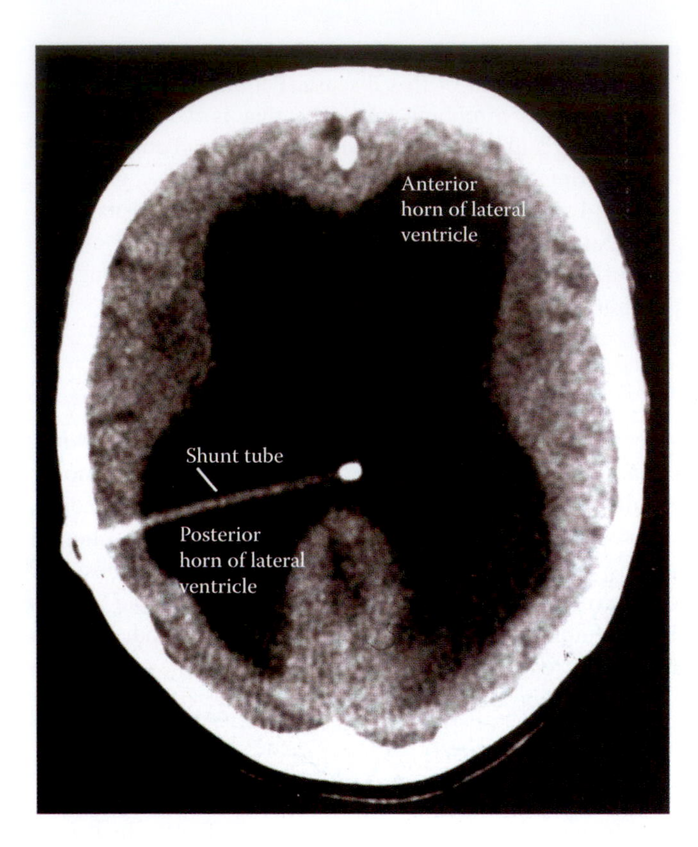

FIGURE 8.71 This image demonstrates a massive dilatation of the lateral ventricle in an individual with hydrocephalus. A surgical shunt is placed to drain the excess cerebrospinal fluid.

deformity). Congenital hydrocephalus commonly presents with meningomyelocele, increased ICP, separation of the sutures (diastasis), and protrusion of the fontanels. It is also associated with enlargement of the head, thinning of the scalp, visible superficial vessels, and downward (setting sun sign) position of the eyes. *Parinaud syndrome*, which exhibits vertical gaze palsy, can also be seen as a result of the pressure exerted on the pretectal area. Disorders of respiration (apneutic episodes), abducens nerve palsy, and "cracked-pot sound" upon percussion of the enlarged lateral ventricles (*Macewen's sign*) are also observed in hydrocephalic young children. It is interesting to note that the cerebral cortex is less affected than the white matter, and many patients survive this condition because of the capacity of the skull to expand. Signs of UMN palsy may also be seen. Although impairment of vision and paleness of the optic disc may be observed upon examination of the fundus, papilledema will not be seen in these individuals.

Acquired (occult) hydrocephalus may be caused by subarachnoid hemorrhage, postmeningitis, cysticercoids, vascular malformation, as well as tumors of the third ventricle, thalamus, and cerebral hemispheres. It is characterized by *visible papilledema* and less prominent frontal and occipital headache. It exhibits signs of frontal lobe dysfunctions such as inattentiveness, marked slowness in response, and easy distractibility. Memory impairment and reduction in mental and physical capacity are also noted. In older children and adults with rigid calvaria, gait apraxia and amnesia followed by dementia and slowing of thought process and, later, urinary incontinence may also occur. Imaging of the brain may reveal herniation of the third ventricle, erosion of the sella turcica, and atrophy of the corpus callosum. The onset of this condition is subacute and can cause intellectual deterioration followed by a restriction of movement. Manifestations of UMN palsy, which include deep tendon hypereflexia of the lower extremities, and Babinski sign, may be detected, along with gait disturbances. Surgical treatment involves ventriculoperitoneal shunting, which is the most commonly used technique in adults.

It may also involve subcutaneous shunting and draining the CSF into a systemic vein through the internal jugular vein. While improvement of gait is less likely, urinary incontinence is the most likely symptom to improve with shunting, while dementia is the least likely to undergo any changes. Hydrocephalus may also be classified into obstructive and nonobstructive forms:

Obstructive hydrocephalus, the most common type of hydrocephalus, results from obstruction of the flow of CSF inside the ventricular system or at its drainage site into the dural sinuses. It may occur as a result of stenosis of the cerebral aqueduct, atresia of foramina of Magendie and Luschka, obstruction of the fourth ventricle (as in Arnold–Chiari syndrome), or obstruction of the interventricular foramen of Monro. Obstructive hydrocephalus may be divided into communicating and noncommunicating types.

Communicating hydrocephalus occurs as a result of obstruction of the CSF pathway outside the ventricular system (e.g., in the cisterns and arachnoid villi), while normal connection between the ventricular system and the subarachnoid space via the foramina of Magendie and Luschka is maintained. It is a slowly developing condition, which is characterized by ventricular enlargement, deterioration of mental faculties, and bilateral UMN palsy. It is seen in Arnold–Chiari malformation and leptomeningitis.

NPH is one form of the communicating hydrocephalus commonly seen in the elderly, which develops over weeks or months upon disruption of the CSF circulation due to a subarachnoid hemorrhage, stroke, and posttraumatic events as a result of a fall, postmeningitic conditions, or posterior cranial fossa tumors. NPH is a reversible condition, which exhibits no signs of increased ICP, papilledema, or cranial nerves dysfunctions. It is presumed to be due to partial obliteration of the subarachnoid space around the cerebral hemispheres, combined with defective reabsorption of the CSF through the arachnoid villi. It is seen in 15% of Alzheimer's patients and may be associated with Parkinson disease, basilar artery ecstasies, and fluctuations in cerebrospinal pressure. Radiographic imaging (CT and MRI scans) may show expansion of the temporal lobe and ventricles with minimal or no cortical atrophy. The prominent clinical manifestations of this condition are *gait apraxia*, *dementia*, and *urinary incontinence*. Other manifestations include memory loss, speech disorders, behavioral changes, and less commonly, bowel incontinence. With treatment, gait apraxia is the first sign to improve. Urinary incontinence may involve urgency and frequency of urination, but in severe cases, patients are totally incontinent. Dementia, which includes psychomotor and cognitive impairment, is unusual. A careful clinical exam, imaging studies, and a spinal tap to remove excess CSF, measure CSF pressure, and perform lab analysis can aid in the diagnosis of this condition. A surgical shunt may not provide a complete cure but may relieve existing symptoms. Endoscopic third ventriculostomy may be performed to provide a drainage site for CSF in the third ventricle.

Noncommunicating hydrocephalus is caused by an obstructive process within the ventricular system, producing a loss of communication between the ventricular system and the subarachnoid space. Obstruction sites may include foramina of Monro, Magendie, and Luschka or the cerebral aqueduct. This form of hydrocephalus is often fatal and progresses very rapidly.

Nonobstructive hydrocephalus is a sequel to papilloma of the choroid plexus that results in oversecretion of CSF. It may also be caused by defective absorption at the arachnoid villi subsequent to hemorrhage, infections, dural fibrosis, or thrombosis of the superior sagittal sinus.

SUGGESTED READING

Aschner M, Gannon M. Manganese transport across the rat blood–brain barrier: Saturable and transferrin-dependent transport mechanisms. *Brain Res Bull* 1994;33:345–9.

Bertoni JM, Brown P, Goldfarb LG, Rubenstein R, Gajdusek DC. Familial Creutzfeldt–Jakob disease (codon 200 mutation) with supranuclear palsy. *JAMA* 1992;268:2413–15.

Bhatia KD, Wang L, Parkinson RJ, Wenderoth JD. Successful treatment of six cases of indirect carotid-cavernous fistula with ethylene vinyl alcohol copolymer (onyx) transvenous embolization. *J Neuro-ophthalmol* 2009;29(1):3–8.

Bogousslavsky J. In Donnan GA, Norrving B, Bamford JM, Bogousslavsky J, eds. *Lacunar and Other Subcortical Infarctions*. Oxford University Press, 1995;149–70.

Bonini F, Burle B, Liégeois-Chauvel C, Régis J, Chauvel P, Vidal F. Action monitoring and medial frontal cortex: Leading role of supplementary motor area. *Science* 2014;343(6173):888–91.

Bouchaud V, Bosler O. The circumventricular organs of the mammalian brain with special reference to monoaminergic innervation. *Int Rev Cytol* 1986;105:283–327.

Chang MC, Yeo SS, Jang SH. Callosal disconnection syndrome in a patient with corpus callosum hemorrhage: A diffusion tensor tractography study. *Arch Neurol* 2012;69(10):1374–5.

de Cock M, Maas YG, van de Bor M. Does perinatal exposure to endocrine disruptors induce autism spectrum and attention deficit hyperactivity disorders? *Acta Paediatr* 2012;101(8):811–8.

Endo M, Kawano N, Miyaska Y, Yada K. Cranial buff hole for revascularization in Moyamoya disease. *J Neurosurg* 1989;71:180–5.

Idro R, Otieno G, White S, Kahindi A, Fegan G, Ogutu B, Mithwani S, Maitland K, Neville BG, Newton CR. Decorticate, decerebrate and opisthotonic posturing and seizures in Kenyan children with cerebral malaria. *Malar J* 2005;4(57):57.

Kunii N, Morita A, Yoshikawa G, Kirino T. Subdural hematoma associated with dural metastasis—Case report. *Neurol Med Chir* 2005;45(10):519–22.

Liu F, McCullough LD. The middle cerebral artery occlusion model of transient focal cerebral ischemia. *Methods Mol Biol* 2014;1135:81–93.

Marchi N, Angelov L, Masaryk T, Fazio V, Granata T, Hernandez N, Hallene K, Diglaw T, Franic L, Najm I, Janigro D. Seizure-promoting effect of blood–brain barrier disruption. *Epilepsia* 2007;48(4):732–42.

Masliah E, Terry RD. Role of synaptic pathology in the mechanisms of dementia in Alzheimer's disease. *Clin Neurosci* 1993;1:192–8.

Moser DJ, Cohen RA, Paul RH, Paulsen JS, Ott BR, Gordon NM, Bell S, Stone WM. Executive function and magnetic resonance imaging subcortical hyperintensities in Vascular dementia. *Neuropsychiatry Neuropsychol Behav Neurol* 2001;14:89–9.

Percheron G. Arteries of the human thalamus: II. Arteries and paramedian thalamic territory of the communicating basilar artery. *Rev Neurol* 1976;132:309–24.

Rubia K. "Cool" inferior frontostriatal dysfunction in attention-deficit/hyperactivity disorder versus "hot" ventromedial orbitofrontal-limbic dysfunction in conduct disorder: A review. *Biol Psychiatry* 2011;69(12):e69–87.

Spiegel DR, Gorrepati P, Perkins KE, Williams A. A possible case of transient Anton's Syndrome status post bilateral occipital lobe infarct. *J Neuropsychiatry Clin Neurosci* 2013;25(3):E49.

Thapar A, Cooper M, Jefferies R, Stergiakouli E. What causes attention deficit hyperactivity disorder? *Arch Dis Child* 2012;97(3):260–5.

Vassilouthis J. The syndrome of normal-pressure hydrocephalus. *J Neurosurg* 1984;61:501–9.

Section III

Peripheral Neuroanatomy

9 Autonomic Nervous System (ANS)

The autonomic nervous system (ANS) regulates visceral motor, glandular secretion, contraction of the vessels and reflex activities, transmission of visceral sensations, as well as emotional behavior. It consists of neurons that are located in the central nervous system (CNS) but also extend in the peripheral nervous system (PNS) (Figures 9.1 and 9.2). It is not a fully autonomous entity, as the name may imply, but rather, an interdependent system that functions under massive input from the cerebral cortex and subcortical centers. The ability to voluntarily control the autonomic functions paved the way for biofeedback conditioning in which heart rate, body temperature, and blood pressure can be changed. This voluntary control is exemplified in the regulatory influences of the premotor cortex, cingulate gyrus, and hippocampal gyrus on the visceral motor nuclei of the cranial nerves and the intermediolateral columns of the thoracic and sacral spinal segments through relay neurons within the thalamus and brainstem reticular neurons. Through diverse connections with functionally diverse neurons at all levels of the CNS, the ANS maintains a stable homeostasis (internal stability), which is essential for normal physiological functions. The ANS consists of peripheral efferent and afferent fibers and central neurons in the spinal cord, brainstem, diencephalon, and brain that are closely associated with the somatic nervous system. The ANS is divisible into sympathetic and parasympathetic components. The afferent component transmits visceral pain and organic visceral sensations (e.g., hunger, malaise, nausea, libido, bladder, and rectal fullness). For the most part, visceral pain fibers initially accompany the sympathetic postganglionic and then the sympathetic preganglionic fibers. As briefly outlined, the efferent component innervates the smooth muscles, glandular tissue, and sweat glands.

AUTONOMIC NEURONS AND SYNAPTIC CONNECTIONS

The ANS acts through two sets of neurons: preganglionic (presynaptic) and postganglionic (postsynaptic) neurons, with the preganglionic neuronal cell bodies residing within the CNS (spinal cord and brainstem) whereas the postganglionic neurons occupying the paravertebral, prevertebral, or intramural ganglia. These sets of ganglia consist of multipolar neurons (Figure 9.3) that establish facilitatory or inhibitory synapses with the adjacent neurons, interneurons, or afferent cholinergic fibers. The multiplicity and pattern of the synaptic connectivity between the preganglionic and postganglionic neurons are aimed at intensifying autonomic response. In a slow and diffusely contacting smooth muscle, a single neuronal terminal branches markedly, innervates a large number of smooth muscle cells, maintaining a wide distance from the innervated structure. In rapidly contracting smooth muscles, the branching of the terminals is less pronounced, and the gap between the synaptic endings and the muscle cell is very narrow.

The axons of the preganglionic neurons consist of thinly myelinated (group B) fibers, whereas axons of the postganglionic neurons are of the unmyelinated type (C-group), rendering the conduction of generated response much slower. In contrast, axons of somatic nerves are principally myelinated. Terminals of the autonomic fibers in the smooth muscles contain varicosities and synaptic vesicles that resemble that of the somatic nerves. However, these terminals show diffuse and extensive branching in which a terminal nerve fiber may innervate several smooth muscle fibers. This branching is limited and well localized in the fast-acting muscles such as the dilator and constrictor pupillae. In the sympathetic nervous system, the vesicles are dense-cored and contain catecholamines, mainly norepinephrine and the enzymes involved in their synthesis and degradation (monoamine oxidases). In the parasympathetic system, the terminals have clear spherical vesicles and are cholinergic, resembling that of the neuromuscular junction in the somatic system. These cholinergic fibers can also cause the release of catecholamines from the sympathetic terminals. However, there are also purinergic terminals that utilize conjugated purine adenosine triphosphate (ATP). These are contained in dense vesicles within varicosities scattered in the myenteric plexus of the intestine (enteric system) as well as in the vessels, lungs, and reproductive organs. Their activity, evident in hyperpolarization of myocytes and subsequent peristaltic and sphincteric relaxation, is mediated by the cholinergic presynaptic sympathetic fibers.

The synaptic gaps in the ANS are much wider than that of the skeletal (somatic) muscles, which allows far greater diffusion capacity and accounts for the delayed response associated with this system relative to the SNS. The greater the synaptic gap, the slower the degradation processes of the neurotransmitter and the resultant prolonged action. Due to the direct and uninterrupted connections between the somatic neuronal axons and their targets (e.g., skeletal muscles), the generated response is immediate and rapid. Excision of the autonomic fibers may not produce atrophy of the denervated organ, as is the case with the somatic denervation. Autonomic afferents convey a variety of sensations such as visceral pain, thirst, hunger, cramps, and sensations of well-being or malaise, in contrast to somatic afferents that primarily transmit signals associated with tactile, thermal, painful, articular, and vibratory sensations. Afferents that convey visceral pain accompany the sympathetic presynaptic fibers and follow a reverse direction to that of the sympathetic

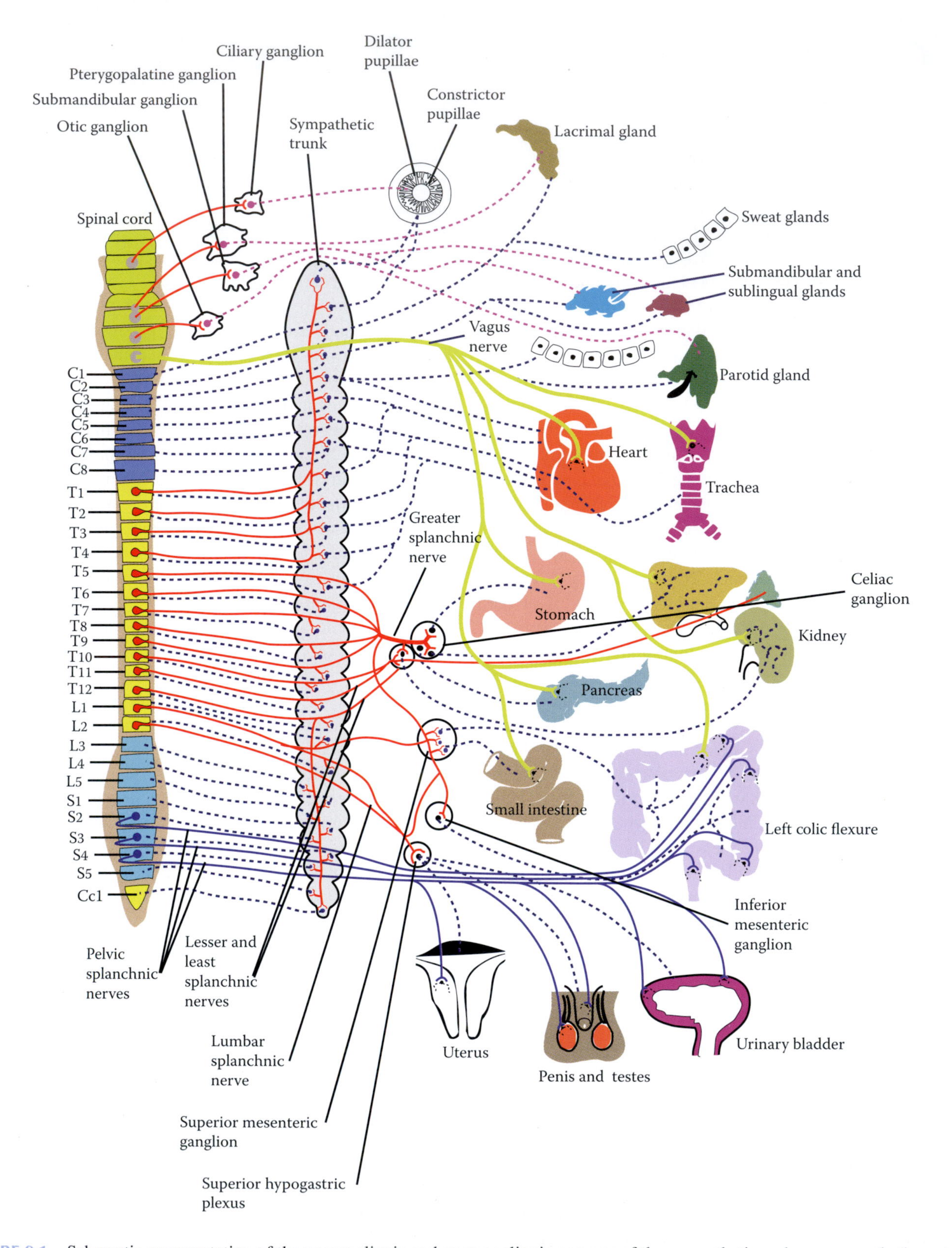

FIGURE 9.1 Schematic representation of the preganglionic and postganglionic neurons of the sympathetic and parasympathetic systems.

fibers, bypassing the prevertebral or paravertebral ganglia to reach the dorsal root via the white communicating rami. These afferents follow the dorsal roots and enter the spinal gray matter to establish synaptic connections and eventually join the anterolateral system. Additional differences between the ANS and SNS of the peripheral nervous system (PNS) also exist in regard to receptors and type of stimuli. Visceral receptors are mainly free nerve endings, whereas the somatic system comprises a variety of capsulated and non-encapsulated receptors. Conventional somatic stimuli may not elicit any

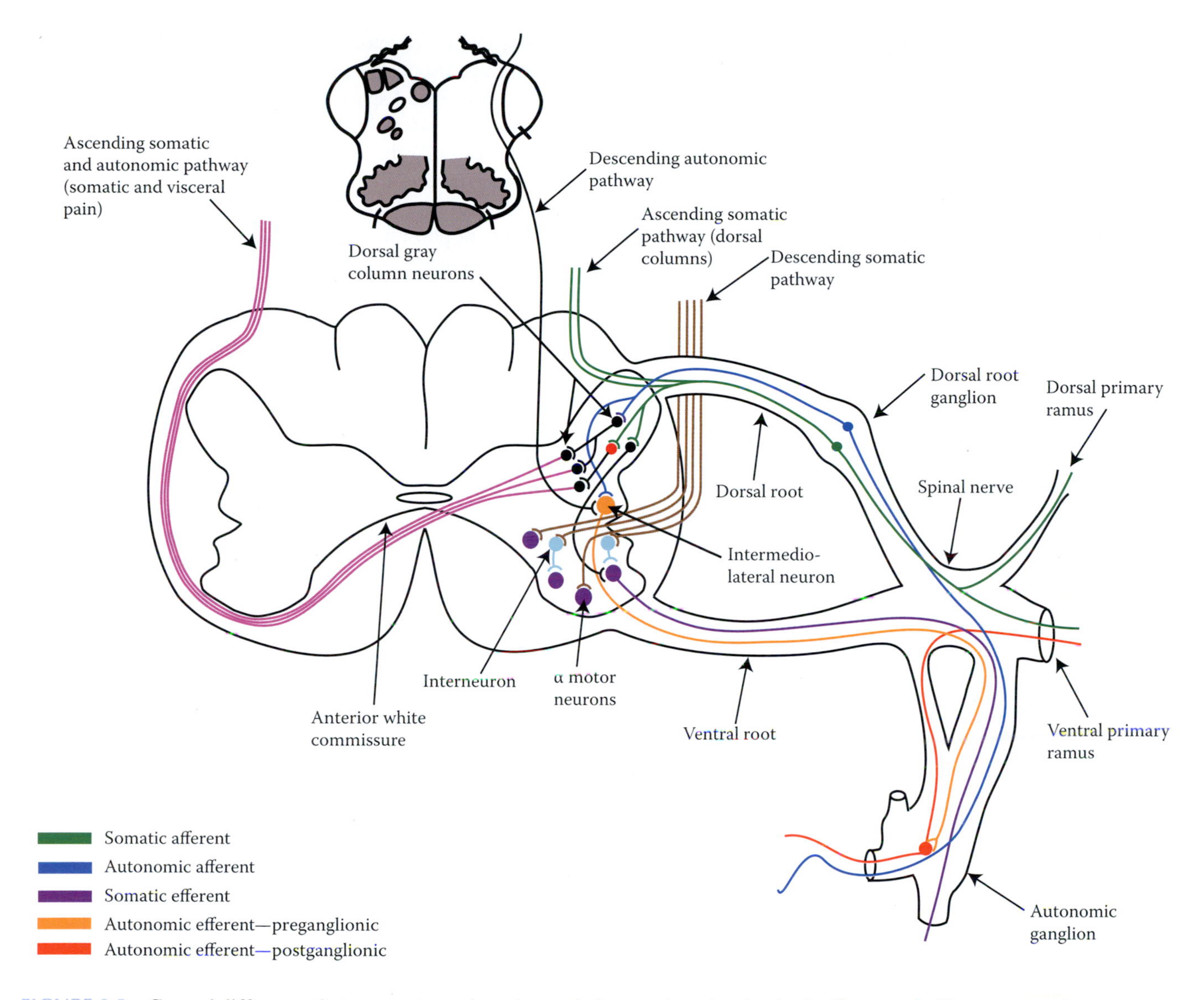

FIGURE 9.2 General differences between autonomic and somatic innervation, showing both efferent and afferent connections.

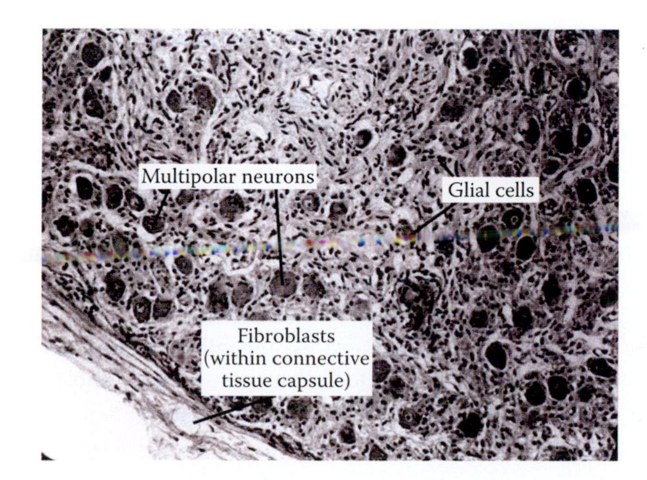

FIGURE 9.3 Photomicrograph of the multipolar autonomic neurons and associated glial cells.

visceral response; however, ischemia and tension within the wall of a visceral organ can activate the visceral receptors and elicit pain.

Despite these differences, both visceral and somatic afferents share neurons in the dorsal root ganglia. However, cranial somatic sensory neurons reside in the trigeminal ganglion, geniculate ganglion of the facial nerve, and the superior ganglia of the glossopharyngeal and vagus nerves, whereas the visceral sensory neurons are located in the geniculate ganglion and the inferior ganglia of the glossopharyngeal and vagus nerves. Additionally, both somatic and autonomic nerves form plexuses. Autonomic plexuses are scattered in the thoracic, abdominal, and pelvic cavities and surround the corresponding arteries to form the pulmonary, cardiac, celiac, aortic and hypogastric plexuses. Somatic nerve plexuses on the other hand are located in the neck, axilla, posterior abdominal and pelvic walls, forming the cervical, brachial, and lumbosacral plexuses. The ANS utilizes norepinephrine, dopamine, and acetylcholine as primary neurotransmitters, producing either an excitatory or inhibitory response. Modulators, such as hormones or tissue metabolites, may exert influences upon the transmitter release or their action.

As outlined above, there are two functionally diverse and antagonistic divisions within the ANS that operate in conjunction with the enteric nervous system. These are the sympathetic and parasympathetic systems.

SYMPATHETIC NERVOUS (THORACOLUMBAR) SYSTEM

The sympathetic is a mass response system brought into action during emergency and under stressful situations. It releases energy by increasing the blood pressure and blood glucose level as well as intensifying the rate of cardiac contractility and output (positive ionotropic and chronotropic effects). This system directs blood flow to the voluntary musculature at the expense of viscera and skin. Excitation of a disproportionately large number of postganglionic neurons (great degree of divergence) and high concentration of the circulating epinephrine in the blood stream are responsible for the mass response of the sympathetic system. It utilizes acetylcholine at the preganglionic level and norepinephrine at the postganglionic level. However, there are exceptions to this rule; acetylcholine may be utilized at the postganglionic level in the innervation of the blood vessels of skeletal muscle and the sweat glands. This may not be true in the case of sweat glands of the palm, which receive adrenergic fibers.

Other cotransmitters also play important roles in this system, such as ATP and neuropeptide Y (NPY). Sympathetic (adrenergic) receptors are integral membrane glycoproteins, which are classified into presynaptic and postsynaptic groups.

Presynaptic receptors are categorized into α_2 and β_2 groups. Additional groups and subtypes are discussed in detail in Chapter 12.

Group α_2 receptors are found on both cholinergic and adrenergic nerve terminals. They act on pancreatic islet (β) cells and on platelets, resulting in reduction of insulin secretion and platelet aggregation, respectively. The latter effect is due to inhibition of adenylate cyclase and activation of K^+ channels via Gi protein. On the adrenergic endings, α_2 (autoreceptors) receptors inhibit the release of norepinephrine by inhibiting neuronal Ca^{2+} channels.

Group β_2 (hormonal) receptors are found on the vascular, pupillary, ciliary, bronchial, gastrointestinal, and genitourinary tract smooth muscles, initiating relaxation of coronary vessels, bronchi, and skeletal arterioles, as well as the ciliary and constrictor pupilla muscles. These series of actions are produced by activation of adenyl cyclase, mediated by Gs protein. They also produce glycogenolysis in the skeletal muscles and liver. β_2 agonists relax the bronchial smooth muscles and therefore can be used in the treatment of asthma. β_2 antagonists are not essential.

Postsynaptic receptors are categorized into α_1 and β_1 groups; α agonists cause contraction of the smooth muscles of the arterioles and sphincters, whereas α antagonists may counteract this effect in individuals with hypertension and peripheral vascular disease.

α_1 group produces vasoconstriction and enhances glandular secretion (odoriferous apocrine sweat glands) via stimulation of phospholipase with formation of IP_3 (inositol-1,4,5-triphosphate) and diacylglycerol and increased cytosolic Ca^{2+}.

β_1 receptors act particularly on the cardiac muscles, producing increased rates and force of contraction and atrioventricular nodal velocity via activation of adenylyl cyclase and Ca^{2+} channels. The role of β_1 agonists in the stimulation of the nodal and ventricular muscles of the heart may be utilized in the treatment of heart failure. In the same manner, β_1 antagonists' (β blockers) role in reducing cardiac rate and force of contractility may be utilized in the treatment of angina pectoris.

Sympathetic innervation of the skin encompasses adrenergic and cholinergic fibers. Cholinergic fibers act upon the muscarinic receptors, enhancing the secretion of the eccrine sweat glands, while adrenergic fibers produce vasoconstriction of the cutaneous arterioles and activate the secretion of the odoriferous apocrine sweat glands.

Activation of the sympathetic nervous system, which may be mimicked by sympathomimetics, produces the following:

- Mydriasis (dilation of the pupils) by inducing contraction of the dilator pupillae muscles
- Relaxation of the ciliary muscle
- Increased contractility and rate of heartbeat (positive ionotropic and chronotropic effects)
- Vasodilatation of the skeletal and coronary arteries
- Vasoconstriction of the bronchial arteries, as well as arteries of the digestive system and skin
- Dilatation of bronchi and inhibition of bronchial secretion
- Inhibition of gastrointestinal motility and contraction of the sphincters
- Increased secretion of the sweat glands
- Contraction of the erector pilorum muscles
- Vasoconstriction of the genital arteries and contraction of the vas deferens, seminal vesicle, and prostate
- Inhibition of the detrusor muscle of the bladder and contraction of the urethral sphincters

The preganglionic sympathetic neurons (Figures 9.1 and 9.4) in the intermediolateral column of the thoracic and upper two or three lumbar spinal segments are connected to the postganglionic sympathetic neurons via the myelinated fibers of the white communicating rami. Postganglionic neurons form two sets of ganglia: paravertebral and prevertebral (Table 9.1).

PARAVERTEBRAL GANGLIA

The paravertebral ganglia (Figures 9.1, 9.4, 9.5, 9.8, and 9.9) form the sympathetic trunk, consisting of two symmetrical

TABLE 9.1

Ganglia of the Peripheral Nervous System

General Characteristics	Sensory Ganglia	Autonomic Ganglia
Location	Trigeminal, geniculate, superior, and inferior ganglia of the glossopharyngeal and vagus nerves Dorsal root ganglia	*Sympathetic* a. Paravertebral ganglia b. Prevertebral ganglia *Parasympathetic* a. Submandibular, otic, ciliary, and pterygopalatine ganglia b. Intramural ganglia
Type	Pseudounipolar or bipolar neurons	Multipolar neurons
Synaptic connections	None	Paravertebral, prevertebral, and intramural ganglia

chains parallel to the vertebral column that unite anterior to the coccyx at the ganglion impar. Multipolar neurons of the paravertebral ganglia receive presynaptic fibers from the intermediolateral columns of the thoracic and upper two or three lumbar segments via myelinated white communicating rami and provide postsynaptic fibers that join the spinal nerves via the unmyelinated gray communicating rami and also innervate visceral and vascular structures in the head, neck, and thoracic cavity. The paravertebral ganglia are divided into cervical, thoracic, lumbar, and sacral parts.

The cervical part of the sympathetic trunk (Figure 9.8) consists of the superior, middle, and inferior ganglia. Since these ganglia do not receive white communicating rami, the presynaptic fibers, which emerge from the upper thoracic spinal nerves, have to travel through the corresponding thoracic ganglia to reach their destination in the cervical ganglia. However, gray communicating rami do arise from these ganglia, supplying the upper four cervical spinal nerves.

The *superior cervical ganglion*, the largest cervical ganglion, is formed by fusion of the upper four cervical ganglia. It lies anterior to the longus capitis and posterior to the carotid sheath. This ganglion supplies fibers to the carotid body and the pharyngeal plexus and a cardiac branch (superior cardiac nerve), which contains efferent but not afferent nociceptive fibers. Most of the emerging postganglionic fibers form the external and internal carotid plexus around the corresponding arteries. Fibers of the internal carotid plexus enter the cranial cavity and supply the dura mater, and then enter the orbit to innervate the dilator pupilla, superior tarsal, and orbital muscles. Sympathetic postsynaptic fibers within the external carotid plexus to the face and neck are vasoconstrictor and sudomotor. While, those to the salivary glands are secretomotor mediated via the otic and submandibular ganglia. These fibers also run in the gray communicating rami to join the upper four cervical spinal nerves. Since the presynaptic fibers from the T1–T2 spinal segments project to and convert to postsynaptic fibers at the superior cervical ganglion and then innervate the structures in the head, removal of this ganglion deprives the ipsilateral side of the head of sympathetic innervation.

Ipsilateral disruption of the sympathetic fibers to the head produces manifestations of Horner's syndrome (Figure 9.6), which comprises ptosis (drooping of the upper eyelid), miosis (constriction of the pupil), enophthalmos (sunken eyeball), and anhidrosis (lack of sweating). Activation of the preganglionic sympathetic neurons in the intermediolateral columns at T1–T2 spinal segments that provide sympathetic fibers to the head requires descending autonomic input from the hypothalamus that travels in the lateral medulla and the lateral funiculus of the cervical spinal cord. Therefore, destruction of the lateral part of the medulla may also produce signs of Horner's syndrome, which are seen as a component of lateral medullary (Wallenberg's) syndrome.

The *middle cervical ganglion* (Figure 9.7), an inconstant ganglion and, when present, is located anterior to the inferior thyroid artery at the level of the sixth thoracic vertebra. It is formed by the fusion of the fifth and sixth cervical ganglia. It provides innervation to the heart and furnishes gray communicating rami to the fifth and sixth cervical ventral rami. It is connected to the cervicothoracic ganglion via the ansa subclavia (see below). It provides thyroid branches and the middle cardiac nerve, which is the largest sympathetic contribution to the deep cardiac plexus.

The *inferior cervical ganglion* joins the first thoracic ganglion to form the stellate ganglion. The cervicothoracic (*stellate*) ganglion (Figure 9.7) lies posterior to the initial part of the vertebral artery, apex of the lung, and cervical pleura, occupying the area between the transverse process of C7 and the neck of the first rib. It contributes gray communicating rami to the seventh and eighth cervical and first thoracic spinal nerves. It also supplies postganglionic branches to the subclavian artery and its branches and to the vertebral plexus, which extends into the cranial cavity. The preganglionic fibers that pass through the stellate ganglion, for the most part, project to the head and neck. However, vasomotor and sudomotor fibers are not contained in the white communicating ramus to the cervicothoracic ganglion. Postganglionic fibers from the stellate ganglion also travel within the inferior trunk of the brachial plexus and then within the ulnar, radial, and median nerves. In the hand, the postganglionic fibers leave these nerves and travel with the corresponding arteries. Occasionally, a vertebral ganglion may be present near the origin of the vertebral artery, which provides gray communicating rami to the fourth and fifth cervical spinal nerves.

The stellate and middle cervical ganglia are connected via the ansa subclavia, a nerve loop that encircles the subclavian

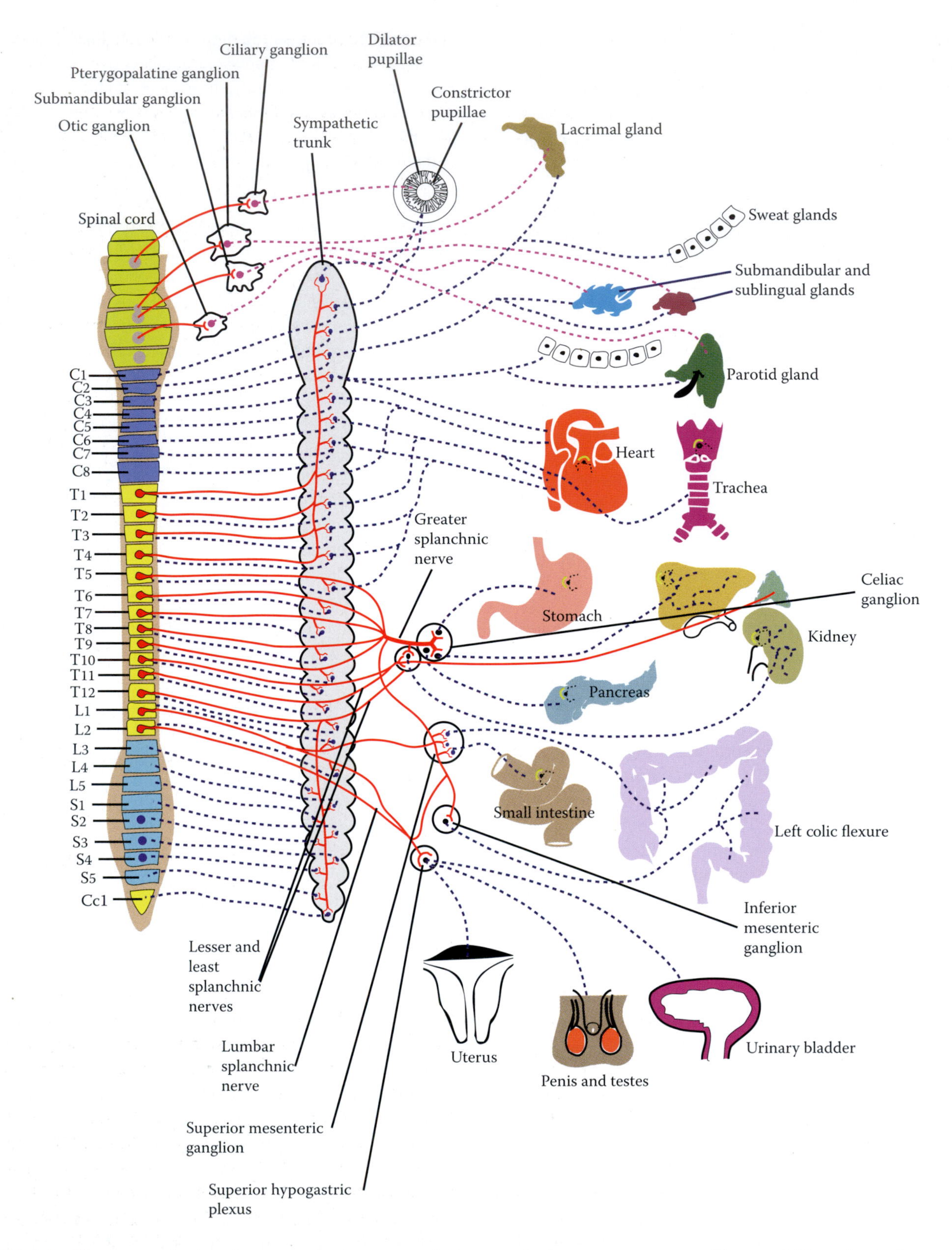

FIGURE 9.4 The preganglionic and postganglionic neurons of the sympathetic nervous system are illustrated. The splanchnic nerves and their associated ganglia are clearly marked.

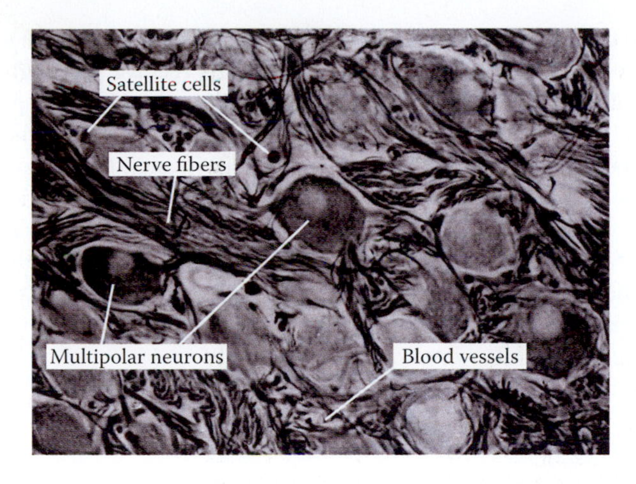

FIGURE 9.5 Photomicrograph of the multipolar neurons of the sympathetic ganglia and associate satellite cells.

FIGURE 9.6 Manifestations of left-sided Horner's syndrome.

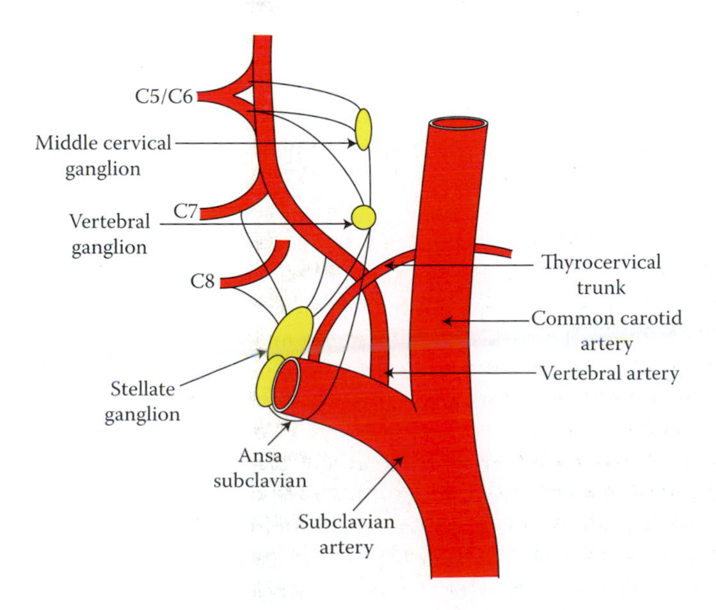

FIGURE 9.7 Drawing of the middle cervical and stellate ganglia and the connecting ansa subclavia.

artery on both sides and courses medial to the origins of the internal thoracic and vertebral arteries.

Pancoast tumor is a bronchogenic invasive carcinoma, squamous cell adenocarcinoma, or undifferentiated large-cell carcinoma of the superior pulmonary sulcus or apex of the lung, but it could also spreads to the adjacent ribs and vertebrae. It causes compression of the inferior trunk of the brachial plexus, producing pain and numbness in the shoulder and in the dermatomes C8–T1 (medial arm, forearm, and hand), with additional manifestations of Klumpke palsy manifested in atrophy mainly of intrinsic muscles of the hand and claw-hand configuration. In one-third of patients, compression of the stellate ganglion produces manifestations of Horner's syndrome, which include miosis, ptosis, and anhidrosis. Muscles of the arm can also be affected in this condition. The phrenic nerve may also be affected in this condition, producing diaphragmatic palsy. Additional manifestations such as respiratory disorders, cardiac arrhythmia, and facial swelling and engorgement of the facial and neck veins due to obstruction of the superior vena cava and brachiocephalic veins also occur. Hoarseness due to left recurrent laryngeal nerve palsy can also be seen in this syndrome. Signs of spinal cord compression can occasionally be seen due to erosion of the vertebral laminae by extension of the tumor.

Relief of upper extremity pain, alleviation of vascular spasm in the hands (seen in Raynaud's disease), or elimination of hyperhidrosis (excessive sweating) may be achieved by injection of anesthetic solution into the stellate ganglion (stellate ganglion block). Success of this procedure may be ascertained by the appearance of signs of Horner's syndrome and increased temperature of the ipsilateral upper extremity.

The thoracic part of the sympathetic trunk consists of 11 or 12 ganglia arranged anterior to the costal heads and covered by the costal pleura. These ganglia are connected to the thoracic spinal nerves via the white and gray communicating rami. As mentioned earlier, the first thoracic and the inferior cervical ganglia frequently join to form the cervicothoracic (stellate) ganglion. The second through the fifth thoracic ganglia provide sympathetic fibers to the posterior pulmonary and deep cardiac plexuses, while the upper five thoracic ganglia provide sympathetic fibers to the aortic plexus. Presynaptic fibers, mainly from T1–T6 (T7), which are destined to the cervical ganglia en route to the head, neck, and upper extremity, also travel within the cervical ganglia. Since vasoconstrictors to the upper extremity primarily emerge from the second and third thoracic spinal segments, excision of the corresponding thoracic ganglia (second and third) may denervate the vessels of the upper extremity. Presynaptic sympathetic fibers from the 5th through the 9th ganglia form the greater splanchnic nerve (Figure 9.4), presynaptic fibers from the 10th and 11th ganglia form the lesser splanchnic nerve, and fibers that emanate from the 12th thoracic ganglion form the least splanchnic nerve. Surgical removal of the paravertebral ganglia of the thoracic sympathetic trunk, sparing the intercostal nerves, produces circumscribed anhidrosis on the affected side without sensory deficits.

The lumbar and the thoracic parts of the sympathetic trunk is connected via a gap posterior to the medial arcuate ligament. The lumbar sympathetic ganglia receive white communicating rami from the upper four lumbar spinal nerves. These ganglia, which are located medial to the psoas major muscle and anterior to the lumbar vertebrae, give rise to the lumbar splanchnic nerves (preganglionic sympathetic fibers) that join the celiac, intermesenteric, and superior hypogastric plexuses. The first lumbar splanchnic nerve arises from the first lumbar ganglion and is destined to the celiac, renal, and aortic plexuses, with branches distributed to the cardia of the stomach, duodenum, and pancreas. The second lumbar splanchnic nerve arises from the corresponding ganglion and terminates in the aortic plexus, with some contribution to the innervation of the pancreas and duodenum. The third splanchnic nerve has its origin from both the third and fourth lumbar paravertebral ganglia and joins the superior hypogastric plexus anterior to the aortic bifurcation. The fourth lumbar splanchnic nerve is inconstant that may gain origin from the fourth or fifth lumbar ganglion and terminates in the superior and inferior (pelvic) hypogastric plexuses. Lumbar splanchnic nerves provide innervation to the aorta and inferior vena cava via the aortic plexus. Small branches are provided to the thoracic duct. Vasomotor fibers contained in the lumbar splanchnic nerves innervate the meninges in the vertebral column, joints, and adjacent muscles.

Postganglionic fibers from the lumbar paravertebral ganglia that travel in the femoral and obturator nerves, supply vasoconstrictor fibers to the femoral and obturator arteries and their branches. Therefore, surgical removal of the upper three or four lumbar ganglia or their preganglionic neurons may completely denervate the lower extremity vessels. Lesions confined to the lumbar part of the sympathetic trunk can occur as a result of Hodgkin's lymphoma and other invasive tumors of the para-aortic and retroperitoneal areas, producing diffuse leg pain, anhidrosis, and temperature elevation on the sole of the foot on the affected side without detectable sensory or motor impairment. Touching the plantar surfaces of the toes with the backs of the fingers and comparing the affected side, which will be warmer and anhidrotic, with the healthy side can have diagnostic value. Involvement of the lumbar plexus could be concluded if the sudomotor deficits are accompanied by sensory/and or motor dysfunctions.

The sacral part of the sympathetic trunk lies anterior to the sacrum and medial to the pelvic sacral foramina. It consists of four to five interconnected ganglia that join, anterior to the coccyx, the corresponding ganglia of the opposite side via the ganglion of impar. These ganglia receive preganglionic fibers from the lower thoracic and upper two lumbar spinal segments, giving rise to gray communicating rami that join the sacral and coccygeal plexuses. Postganglionic sympathetic fibers that run in the gray communicating rami provide vasomotor innervation to the gluteal and popliteal arteries by joining the gluteal and tibial nerves. The first two ganglia provide fibers to the inferior hypogastric (pelvic) plexus, while the rest of the sacral ganglia form a plexus around the median sacral artery.

Prevertebral Ganglia

The prevertebral ganglia (Figures 9.1 and 9.4) lie anterior to the lumbar part of the vertebral column, comprising the celiac, aorticorenal, superior mesenteric, and inferior mesenteric ganglia. The celiac ganglion, the largest prevertebral ganglion, is located around the celiac trunk and medial to the suprarenal gland. The caudal (lower) part of each celiac ganglion is known as the aorticorenal ganglion. Much smaller ganglia, such as the superior and inferior mesenteric, are lodged within the corresponding plexuses. Postganglionic fibers from the prevertebral plexuses travel in the femoral and obturator nerves, supplying vasoconstrictor fibers to the femoral and obturator arteries and their branches. Therefore, surgical removal of the upper three or four lumbar ganglia or their preganglionic neurons can completely denervate the vessels of the lower extremity.

PATTERN OF DISTRIBUTION OF THE SYMPATHETIC FIBERS

Axons of the sympathetic preganglionic neurons that synapse in the paravertebral ganglia usually follow an orderly course through the ventral root, spinal nerve, and then white communicating ramus to the reach the paravertebral. This is not true in the cervical and sacral segments, which lack the corresponding white communicating rami that connect the spinal nerves to the paravertebral ganglia. Axons of the postganglionic sympathetic neurons of the paravertebral ganglia follow varied courses (Figures 9.4 and 9.8). The sympathetic innervation of the sweat glands, erector pilorum muscles, and vessels of the extremities, thoracic and abdominal walls necessitates that the preganglionic fibers follow the course described earlier to reach the white communicating rami and then the paravertebral ganglia and continue beyond as postsynaptic fibers within the gray communicating rami to reach and distribute through the primary rami of the spinal nerves.

Fibers destined to the head and neck region (e.g., sweat glands of the face and the dilator pupillae muscles) ascend to synapse in the superior cervical ganglion, and then join plexuses around major blood vessels. The sympathetic postsynaptic fibers to the lower face arise from the external carotid plexus, a network of sympathetic fibers that encircle and follow the course of the corresponding artery.

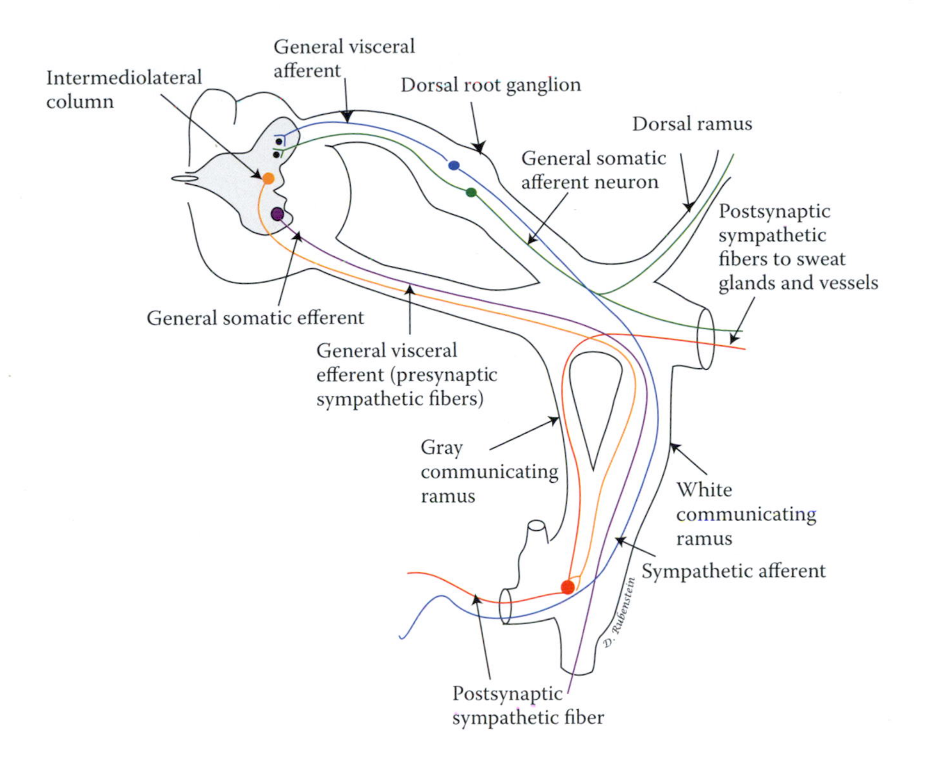

FIGURE 9.8 The functional components of a spinal nerve including general somatic afferent, general visceral afferent, general somatic efferent, and general visceral efferent are shown.

Sympathetic fibers to the sweat glands of the supraorbital region are contained within the supraorbital and supratrochlear branches of the frontal nerve. The latter, a branch of the ophthalmic nerve, receives its sympathetic fibers by communicating with the nasociliary nerve. Innervation of the dilator pupillae muscle is maintained by the postsynaptic sympathetic fibers that travel within the long ciliary branch of the nasociliary nerve. Sympathetic postsynaptic fibers to the superior tarsal muscle of the upper eyelid originate from the internal carotid plexus, as it travels within the cavernous sinus, and are contained within the oculomotor nerve. An interesting point to bear in mind is the fact that both the sympathetic postsynaptic fibers to the superior tarsal and the somatic fibers to the levator palpebrae muscles course within the oculomotor nerve. Sympathetic postganglionic fibers to the thoracic viscera originate from the cervical and upper five thoracic paravertebral ganglia. Fibers that are destined to the abdominal viscera bypass the sympathetic trunk to terminate in the prevertebral ganglia as the splanchnic nerves (Figures 9.1 and 9.4). The greater splanchnic nerve consists of preganglionic efferent and visceral afferent fibers that penetrate the crus of the diaphragm and enter the abdomen, establishing synaptic connections primarily with the celiac ganglion and partially with the aorticorenal ganglion. The lesser splanchnic nerve synapses in the aorticorenal ganglion, whereas the least splanchnic nerve (renal nerve) contributes to the renal plexus. Fibers that bypass both the paravertebral and prevertebral ganglia remain preganglionic and terminate in the chromaffin tissue of the adrenal medulla.

PARASYMPATHETIC NERVOUS (CRANIOSACRAL) SYSTEM

The parasympathetic is a local-response system, consisting of preganglionic and postganglionic neurons that act on the smooth muscles and viscera. The preganglionic parasympathetic fibers are contained in the pelvic splanchnic nerves and certain cranial nerves (Figures 9.1, 9.9, and 9.10). These preganglionic fibers establish connections with the postsynaptic parasympathetic neurons of the intramural ganglia on the pelvic and abdominal viscera or that of the parasympathetic ganglia in the head. Parasympathetic responses are manifested in miosis (constriction of the pupil), contraction of the ciliary muscle, decreased contractility and cardiac output (negative ionotropic and chronotropic effect), and constriction of the bronchi and bronchioles. Other manifestations of parasympathetic activation include increased gastrointestinal tract motility, constriction of the coronary arteries and vasodilatation of the vessels of the external genitalia and gastrointestinal tract, and contraction of the muscular wall of the urinary bladder. Dilation of cerebral vessels are primarily due to a change in CO2 concentration and not parasympathetic activation.

Acetylcholine, the main neurotransmitter at the parasympathetic terminals, is contained in the clear spherical vesicles, acting primarily in conjunction with cotransmitters such as vasoactive intestinal peptide (VIP) and, to a lesser degree, ATP. Due to the rapid degradation of acetylcholine and the low ratio of preganglionic to postganglionic neurons (less divergence), the action of the parasympathetic system remains localized and of short duration. Cholinergic

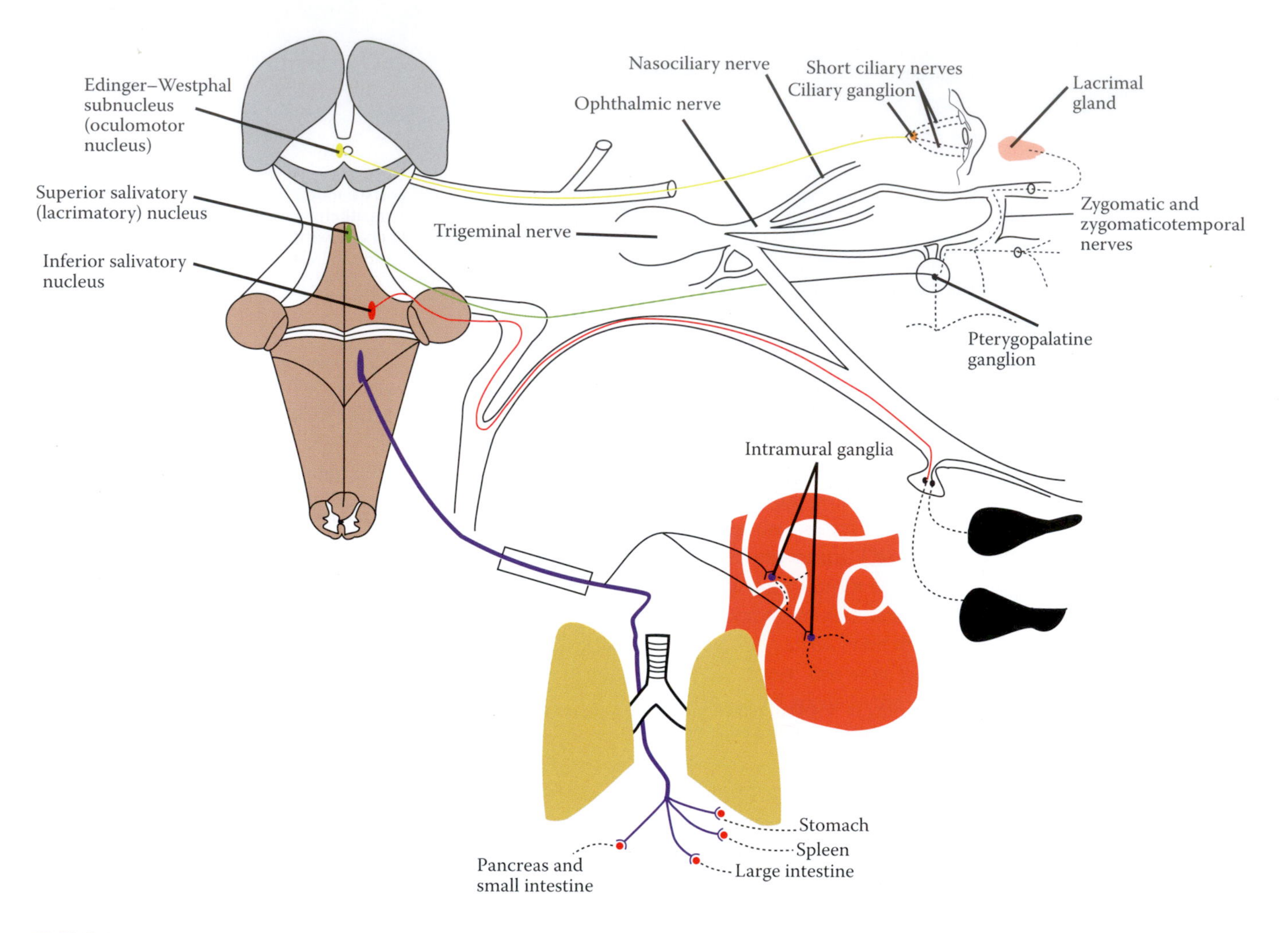

FIGURE 9.9 Cranial part of the parasympathetic nervous system. Note the associated nuclei, preganglionic fibers, and related ganglia.

receptors, which are activated by acetylcholine, are classified into nicotinic and muscarinic types.

Nicotinic receptors are further subdivided into nicotinic muscle receptor and nicotinic neuronal receptor. Nicotinic muscle receptors (C-10 receptor) are pentameric proteins, activation of which produces rapid increase in permeability of cells to sodium and calcium ions and subsequent depolarization and contraction of the skeletal muscle. Phosphorylation by cyclic adenosine monophosphate (cAMP) protein kinase, protein kinase C, or trypsin kinase increases the desensitization of these receptors. Muscle receptor contains α, β, γ, and δ or α, β, δ, and ε subunits in a pantameric complex. The reason for the difference is because the ε subunit replaces the γ in the adult. Subunit γ is particularly detected in the embryo or denervated muscle. Neuronal nicotinic receptors are categorized into two subunits, α and β, with the α occurring in at least seven different forms and β in three forms. They exist in the autonomic ganglia, adrenal medulla, and CNS. Neuronal nicotinic receptors are classified into bungarotoxin-insensitive (C-6) and bungarotoxin-sensitive nicotinic receptors. The former exist in the autonomic ganglia and produce depolarization and firing of the postganglionic neurons in the autonomic ganglia via opening of the cation channel. There are numerous agonists for these receptors, such as nicotine, phenyltrimethylammonium, methyl-isoarecolone, cytosine, and dimethylphenylpiperazinium, as well as a plethora of antagonists such as tubocurarine, lophotoxin, and dihydro-β-erythroidine.

Muscarinic receptors are coupled to G proteins and either act directly or indirectly on ion channels or are linked to second messenger systems. They are classified on a pharmacological basis into M_1–M_3 and on the basis of molecular cloning into M_4–M_5 subtypes. All five subtypes exist in the CNS. M_1 receptors show great affinity to pirenzepine and are found in the autonomic ganglia and glands. ll-[[2-[(diethylamino) methyl]-l-piperidinyl]acetyl]-5,11-dihydro-6H-pyrido[2,3-b][1,4]benzodiazepine-6-one (AFDX-116) shows high affinity to M_2 receptors in the myocardium and smooth muscles, whereas 4-diphenylacetoxy-*N*(2-chloroethyl)-piperdine hydrochloride (4-DAMP) displays high affinity to M_3 receptors in the smooth muscles and secretory glands. M_1, M_3, and M_5 are coupled to phosphoinositol (PI) hydrolysis, and M_2

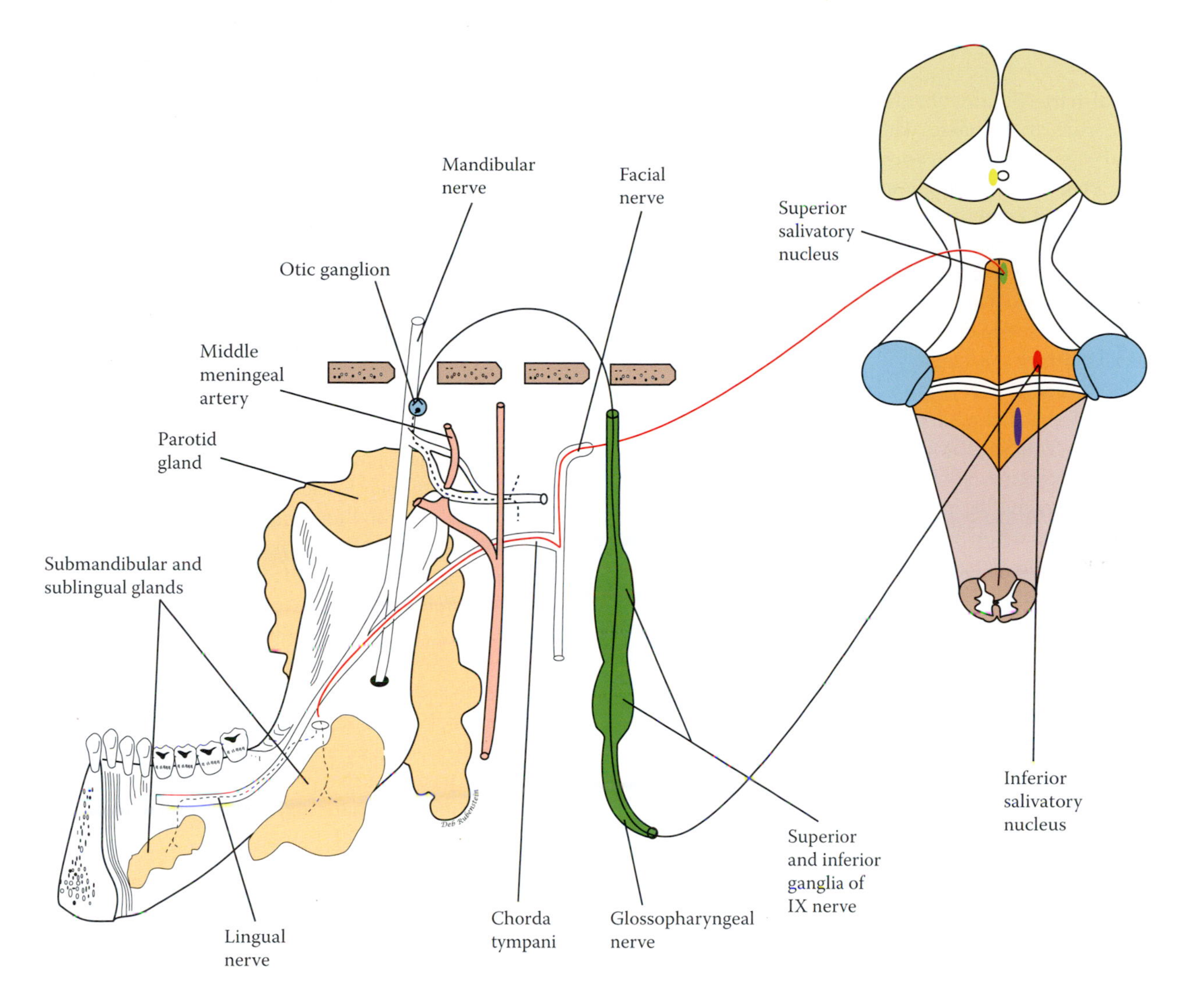

FIGURE 9.10 This diagram illustrates the parasympathetic neurons within the brainstem associated with innervation of the salivary glands.

and M_4 are coupled to cAMP. Activation of the muscarinic receptors produces depolarization or hyperpolarization by opening or closing the potassium, calcium, or chloride channels. Activation of M_1 receptors produces depolarization in the neurons of the autonomic ganglia. Stimulation of the M_2 receptors elicits hyperpolarization in the sinuatrial (SA) node, a decrease in the atrial contractile force and conduction velocity in the atrioventricular node, and a slight decrease in the ventricular contractile force. Activation of the M_3 receptors produces contraction of the smooth muscles and increased glandular secretion.

Acetylcholine acts upon the muscarinic receptors on the exocrine glands, heart, and smooth muscles. Cholinergic receptors are nicotinic in the autonomic ganglia and muscarinic at the postganglionic parasympathetic nerve endings. In the CNS, both muscarinic and nicotinic receptors exist. The combined effect of the muscarinic autoreceptors on the nerve endings (comparable to the α_2 autoreceptors of the sympathetic system) and acteylcholinesterase may prevent accumulation of acetylcholine in the synaptic cleft.

Cholinergic agents like carbachol (stimulates the bladder and bowel) and pilocarpine (produces constriction of the pupil) have similar effects to acetylcholine. Some of these agents act by inhibiting the enzyme cholinesterase and subsequently increasing the concentration of acetylcholine in the synaptic clefts. These cholinesterase inhibitors include physostigmine and di-isopropylfluorophosphate (DFP). Others, such as tubocurarine, act as antagonists by competing with natural mediators at the synaptic site. Anticholinergic medications may be used clinically to (1) induce dryness of the bronchi during surgery; (2) maintain dilatation of the pupil for in-depth ophthalmologic examination; (3) block the vagal inhibition in case of cardiac arrest; (4) prevent vomiting (antiemetic); (5) counteract the spastic effect of morphine on the gastrointestinal tract; (6) treat poisoning by overdose of cholinergic drugs; and (7) cause relaxation of the urinary bladder in individuals with cystitis. The parasympathetic system consists of cranial and sacral parts.

Cranial Part

The cranial part (Figures 9.9 and 9.10) consists of preganglionic neurons that course within the oculomotor, facial, glossopharyngeal, and vagus nerves. The oculomotor nerve (CN III) contains preganglionic parasympathetic fibers, which are derived from the Edinger–Westphal nucleus of the oculomotor nuclear complex. These fibers synapse in the ciliary ganglion, giving rise to postsynaptic fibers that eventually innervate the constrictor pupillae and the ciliary muscles.

CN III (inferior branch) → ciliary ganglion → constrictor pupillae and ciliary muscle.

The facial nerve (CN VII) contains preganglionic parasympathetic fibers that emanate from the neurons of the lacrimal and superior salivatory nuclei, establishing synapses in the pterygopalatine (sphenopalatine) and submandibular ganglia, respectively. The postsynaptic parasympathetic fibers from the pterygopalatine ganglion supply the lacrimal gland, mucus glands of the palate, nasal cavity, and pharynx via the greater petrosal nerve. On the other hand, the postsynaptic parasympathetic fibers from the submandibular ganglion innervate the submandibular and sublingual glands. CN VII → greater petrosal nerve → pterygopalatine ganglion → lacrimal gland, mucus glands of the palate and nasal cavity, and pharynx. CN VII → chorda tympani → submandibular ganglion → sublingual and submandibular glands.

The glossopharyngeal nerve (IX) contains preganglionic parasympathetic fibers in the lesser petrosal nerve that emanate from the medullary inferior salivatory nucleus that synapse in the otic ganglion. This ganglion sends postsynaptic secretomotor fibers to the parotid gland via the auriculotemporal nerve. CN IX → lesser petrosal nerve → otic ganglion → parotid gland.

The vagus nerve (CN X) contains preganglionic parasympathetic fibers, which are derived from the medullary dorsal motor nucleus of the vagus. These fibers synapse in the terminal (intramural) ganglia scattered along the thoracic and abdominal viscera (e.g., the pulmonary, myenteric, and submucosal plexuses). Intramural ganglia contain in abundance neurons that are nonadrenergic and noncholinergic. They may also contain excitatory transmitters such as serotonin and substance P or inhibitory transmitters such as VIP, ATP, or enkephalin. The vagal parasympathetic contributions to the abdominal viscera terminate at the junction of the right two-thirds and left one-third of the transverse colon.

Sacral Part

The sacral part (Figure 9.1) of the parasympathetic system includes the parasympathetic preganglionic axons, emanating from the intermediolateral column of the second, third, and fourth sacral spinal segments. The axons of these neurons leave through the ventral roots of the corresponding segments and form the pelvic splanchnic nerves, which supply the pelvic viscera and part of the abdominal viscera. The pelvic splanchnic nerves are excitatory to the muscular wall of the descending colon, sigmoid colon, rectum, and anal canal, as well as to a portion of the transverse colon. These splanchnic nerves are inhibitory to the urethral sphincters and vasodilatory to the erectile tissue of the external genitalia.

HIGHER AUTONOMIC CENTERS

These centers represent specific areas in the cerebral cortex, diencephalon, and brainstem that closely regulate the ANS. Stimulation or inhibition of these centers produces a variety of visceral changes. Autonomic centers in the cerebral cortex are scattered in the cingulate gyrus and hippocampal formation. In the diencephalon, the hypothalamus constitutes the principal area where parasympathetic (anteromedial) and sympathetic (posterolateral) centers are located. Hypothalamic control of the brainstem and spinal autonomic neurons is achieved via the dorsal longitudinal fasciculus (DLF), the mamillotegmental tract, and the medial forebrain bundle (MFB).

The DLF connects the medial hypothalamus to the dorsal motor nucleus of vagus, nucleus ambiguus, salivatory, and Edinger–Westphal nuclei, as well as the intermediolateral columns of the spinal cord. Medial hypothalamic neurons also send fibers to the dorsal motor nucleus of vagus, locus ceruleus, and raphe nuclei via the MFB. Brainstem raphe nuclei project to the prefrontal cortex, septal area, and cingulate gyrus also via the MFB. The mamillotegmental tract is formed by the axons of the mamillary neurons that project to the raphe nuclei and other nuclei of the mesencephalic and pontine reticular formation.

Transection of the lower pons disrupts the descending fibers from the pneumotaxic center, resulting in a deep respiratory cycle (apneustic breathing).

In the brainstem, the pontine autonomic is comprised of a cardiovascular center in the caudal pons between the superior olivary nucleus and the root of the facial nerve. The medullary autonomic (respiratory) is comprised of the inspiratory center around the solitary nucleus that contains opiate receptors upon which morphine acts as a depressant, and an expiratory center around the ambiguus nucleus. The latter projects to the motor neurons of the thoracic spinal segments and innervates the internal and innermost intercostals. The pneumotaxic center is comprised of the parabrachial nuclei of the pons, which influences the rate of breathing by shortening the respiratory cycle.

AUTONOMIC REFLEXES

A reflex is an innate and automatic response to a stimulus that occurs as a basic defense mechanism. It may be inherited or primitive, present at birth, and common to all human beings. It may also be conditioned, which is acquired as a result of experience. Intactness of the receptor, sensory (afferent) neuron, and a motor (efferent) neuron are essential for a typical reflex to occur. Afferent fibers deliver the generated impulses from a receptor to the CNS where it may be inhibited, facilitated, or modified, while the efferent fibers transmit the processed information to the effector organ. An interneuron may exist between the afferent and efferent neurons. Reflexes do not always operate independently; in fact, descending supraspinal pathways (somatic and visceral) modulate and regulate the neurons, which form the reflex arc. Conditions that affect the neural elements of a reflex arc or their supraspinal input may produce a variety of deficits.

A lesion that disrupts the reflex arc may result in circulatory, pupillary, and thermoregulatory abnormalities, disorders of sweating, denervation hypersensitivity, hyporeflexia, or areflexia, depending on the number of the affected segments. In peripheral neuropathy and poliomyelitis, the receptor, afferent, or efferent neurons of a reflex arc may be damaged, producing hyporeflexia or areflexia. Damage to the supraspinal pathways may produce hyperreflexia, hyporeflexia, or areflexia (e.g., upper motor neuron palsy manifests both deep tendon hyperreflexia and areflexia or hyporeflexia in the superficial abdominal reflexes). Due to the continuous generation of action potentials in the autonomic nerves and the role of reflexes in maintaining visceral functions, series of methodologies have been developed to assess reflex arc and the regulatory input from higher centers. These tests are designed to demonstrate the integrity of the ANS and define the location of the lesions that produce autonomic dysfunction.

Reflexes are categorized into superficial reflexes associated with the skin and mucus membrane and deep reflexes pertaining to the muscles and tendons. Reflexes may be mediated by cranial nerves (cranial reflexes) or spinal nerves (spinal reflexes). Additional classifications into visceral and somatic reflexes are based upon the nature of the innervated structure.

Visceral reflexes include viscero-visceral and viscerosomatic reflexes, while somatic reflexes comprise somato-somatic and somatovisceral reflexes. Visceral reflexes facilitate automatic adjustments of the entire organism to the internal and external environments. In order to promote digestion, some of these reflexes produce an increase in blood flow to the digestive tract following food ingestion, and decrease in absorption. Other reflexes may increase the rate and depth of respiration to meet the body's demand for oxygen in response to physical activity. Visceral reflexes are classified into viscero-visceral, viscerosomatic, and somatovisceral reflexes.

Viscero-visceral reflexes include the carotid sinus, Bainbridge, and carotid body reflexes.

Carotid sinus reflex is mediated by the carotid sinus (receptor), carotid sinus branch of the glossopharyngeal (afferent limb), reticular formation, and vagus nerve (efferent limb). Rapidly tilting the body from a horizontal position to standing position, produces a concomitant fall in blood pressure as the blood moves toward the lower part of the body. However, this arterial pressure change is immediately counteracted by the carotid sinus (baroreceptor) reflex and brought to a normal level by producing vasoconstriction in the abdominal viscera and lower extremities.

Any disruption in the afferent or efferent limb of this reflex can cause postural hypotension and possible syncope. Similarly, an increase in blood pressure stimulates the carotid sinus and activates the neural mechanism that adjusts the blood pressure to a normal level. Carotid sinus can be stimulated by manual massage, which elicits an increased discharge in the afferent and efferent fibers and subsequent changes in heart rate and blood pressure. This manual manipulation can induce carotid sinus syncope in which the patient loses consciousness, accompanied by convulsion. Death by strangulation and hanging has been attributed to complete disruption of the carotid reflex. Carotid sinus massage can be used to determine carotid sinus syncope and to distinguish between supraventricular tachycardia and ventricular tachycardia. The tonic activity of the vagus nerve in controlling heart rate can be tested by an intravenous administration of atropine, which blocks the vagal effect and induces tachycardia.

Bainbridge reflex monitors the central venous pressure through afferent nerve endings in the right atrium. These endings are represented by the peripheral processes of the neurons of the inferior ganglion of the vagus nerve. Distention of the right atrium produces reflex tachycardia due to vagal inhibition and sympathetic stimulation.

Carotid body reflex is initiated by an increase in carbon dioxide and a decrease in oxygen tensions of the blood. These changes stimulate the carotid body (chemoreceptor) and, eventually, the respiratory center through the vagus nerve.

Viscerosomatic reflexes comprise Hering–Breuer and vomiting reflexes.

Hering–Breuer reflex initiates expiration upon excitation of the terminals of the bronchial tree of the inflated lung. These excited terminals stimulate the expiratory center and the solitary nucleus. The expiratory center inhibits the inspiratory center, eliciting passive and elastic recoil of the lung.

Vomiting reflex is mediated by receptors that are located in the mucosa of the stomach, gallbladder, and duodenum. Activation of these receptors results in transmission of the generated impulses via the vagus nerve (afferent limb) to the solitary nucleus, medullary vomiting center, reticulospinal tracts, and neurons of the anterior horn and the intermediolateral columns of cervical and thoracic spinal segments.

Somatovisceral reflexes consist of the pupillary light, pupillary-skin (ciliospinal), accommodation, bladder, and rectal reflexes, and mass reflex of Riddoch.

The pupillary light reflex exhibits constriction of the pupil of both the stimulated eye (direct light reflex) and the contralateral eye (consensual light reflex) in response to direct light applied to one eye. This reflex is mediated by the optic and oculomotor nerves. The optic nerve, optic tract, and brachium of the superior colliculus form components of the afferent limb of this reflex. In this manner, the efferent limb consists of several components, which include the pretectal nucleus, Edinger–Westphal nucleus, oculomotor nerve, ciliary ganglion, and short ciliary nerves. Loss of light reflex is seen in diabetes mellitus, epidemic encephalitis, alcoholism, and neurosyphilis.

In Argyll Robertson pupil, pupillary construction is preserved in accommodation reflex but lost in light reflex. It is attributed to the disruption of the afferents of light reflex that pass medial to the lateral geniculate nucleus in their course to the pretectal nuclei, and preservation of the afferents of the accommodation reflex that continue to the lateral geniculate body in their path to the visual cortex. This type of lesion that selectively abolishes the afferents of light reflex but spares afferents of the accommodation reflex is seen in neurosyphilis. Patients with Argyll Robertson pupil remain unresponsive to atropine.

The pupillary-skin (ciliospinal) reflex is characterized by pupillary dilatation in response to painful stimuli. It may be elicited by a simple scratch, pinch, or cutaneous wound, especially involving the facial skin. The afferent limb (depending upon the site of the stimulus) may include neurons of the dorsal root ganglia or the trigeminal nerve and ganglion, as well as neurons of the posterior horn or the spinal trigeminal nucleus. The efferent limb includes the reticular formation, reticulospinal tracts, intermediolateral columns of the first thoracic spinal segment, and sympathetic pathway to the dilator pupilla muscle of the eye.

The accommodation reflex adjusts both eyes to near vision via convergence (adduction) of the eyes, constriction of the pupils, and increased curvature of the lenses. The afferent limb is formed by the retina, optic nerve, optic tract, lateral geniculate body, optic radiation and the visual cortex. The efferent limb comprises the pretectal and the Edinger Westphal nuclei, the presynaptic parasympathetic fibers to the ciliary ganglia and the postsynaptic fibers in the short ciliary nerves that innervate the ciliary and the constrictor pupilla muscles.

Marked pupillary constriction can be induced by the administration of methacholine if the lesion disrupts the preganglionic neurons of the parasympathetic nervous system. This response will be absent in healthy individuals. Instillation of cocaine does not elicit dilation of the pupil if the disruption is at the postganglionic neuronal level. This is due to the fact that cocaine acts by releasing and augmenting catecholamines, which will not be possible where sympathetic terminals are severed.

The bladder and rectal reflexes regulate the sphincteric control of micturition and defecation via the pelvic splanchnic nerves. Incontinence may occur as a result of disruption of this reflex arc. The urge to urinate or defecate may be lost upon interruption of the afferent fibers.

The mass reflex of Riddoch is elicited in an individual with spinal shock by stimulating the skin below the level of the spinal lesion. It is characterized by sudden evacuation of the bladder and bowel, flexion of the lower extremity, and sweating in response to an emotional stimulus.

In an individual who has endured spinal shock as a result of an injury to the spinal cord, the mass reflex can be elicited by stimulating the skin below the level of the injured spinal segment.

AUTONOMIC PLEXUSES

Autonomic plexuses represent a network of visceral nerve fibers that innervate structures in the thoracic, abdominal, and pelvic cavities. They are formed by sympathetic and parasympathetic nerve fibers and associated ganglia. Their names are derived from the corresponding arteries that they are associated with. These comprise the cardiac, pulmonary celiac, hepatic, gastric, suprarenal, renal, ureteric, superior mesenteric, aortic, inferior mesenteric, and superior and inferior hypogastric plexuses.

The cardiac plexus (Figures 9.4 and 9.10) provides innervation to the heart and the coronary arteries, consisting of superficial and deep parts. The superficial part of the cardiac plexus lies below the aortic arch and is formed by the cardiac branch of the left superior cervical ganglion and by the parasympathetic fibers of the vagus nerve. The deep part of this plexus lies anterior to the bifurcation of the trachea and is formed by the cardiac branches of the cervical (with the exception of the left superior cardiac branch) and upper four or five thoracic ganglia. In contrast, branches of the vagus and the recurrent laryngeal nerves form the parasympathetic component. Postganglionic fibers from the right vagus nerve establish synaptic connection with the sinoatrial node and with both atria. On the other hand, the postsynaptic fibers from the left vagus nerve act on the ventricular myocardium and the atrioventricular (His) bundle. Reduction of the contractile force of the heart and rate of contraction are achieved by stimulation of the vagus nerves that act upon the muscarinic receptors of the cardiac nodal tissue and atria.

Cardiac pain (e.g., due to myocardial ischemia) is transmitted by the C fibers of pseudounipolar neurons of the upper four or five thoracic spinal nerves that initially run in the middle and inferior cardiac branches of the sympathetic trunk. These fibers enter the dorsal horns of the corresponding spinal segments and synapse in certain laminae that form the anterolateral system. These connections may explain the referred pain to dermatomes of T1–T5, which is experienced by individuals with acute myocardial infarction.

The presynaptic muscarinic receptors on the sympathetic fibers are also inhibited by stimulation of the vagus nerves. Sympathetic postganglionic fibers act upon the β_1 receptors and, to a lesser degree, on the α_1 receptors in the sinoatrial and atrioventricular nodes, atrioventricular bundle, and ventricular myocardium. Activation of the β_2 receptors in the coronary arteries, by circulating epinephrine, produces relaxation of the vessels. Cholinergic presynaptic α_2 receptors on branches of the vagus nerve may also be inhibited by the sympathetic postganglionic fibers.

Subsidiaries of the cardiac plexus are the coronary plexuses that surround the coronary arteries.

The left coronary plexus is an extension of the deep cardiac plexus, supplying the left atrium and left ventricle. The right coronary plexus innervates the right chambers of the heart and is formed by the fibers of the deep and superficial parts of the cardiac plexus. The sympathetic fibers of this plexus, upon activation, produce coronary vasodilatation, while the parasympathetic fibers elicit vasoconstriction. The cardiac plexus continues with the pulmonary plexuses (Figure 9.11) around the corresponding arteries.

The pulmonary plexus, which lies partly anterior and partly posterior to the pulmonary hilus, receives parasympathetic fibers from the vagus nerve and sympathetic fibers from the second through the fifth thoracic spinal segments. This plexus innervates the pulmonary arteries, bronchi, and bronchial arteries.

The celiac plexus (Figures 9.4 and 9.12) surrounds the celiac trunk, and lies anterior to the diaphragmatic crura and medial to the suprarenal glands. It receives sympathetic fibers via the greater splanchnic (T5–T9 spinal segments) and the lesser splanchnic (T9–T11 spinal segments) nerve. The parasympathetic fibers are derived from the vagus nerve. This plexus, which also receives somatic fibers via the phrenic nerves, contains the celiac ganglia (visceral brain), where the greater and lesser splanchnic nerve establish synaptic connections. Due to the proximity of the lower part of the celiac ganglion (aorticorenal ganglion) to the renal artery, it contributes postsynaptic parasympathetic fibers to the renal plexus. The celiac, as the mother of all abdominal plexuses, has subsidiary plexuses, which innervate the liver, gallbladder, diaphragm, stomach, duodenum, spleen, adrenal glands, kidneys, and testes or ovaries (Figure 9.9).

The hepatic plexus, a continuation of the celiac plexus, surrounds the common hepatic artery and its branches and supplies the liver and gallbladder. Activation of the vagal parasympathetic fibers produces contraction of the gallbladder and bile duct and relaxation of the sphincter of Oddi. This plexus receives sympathetic contribution from the seventh through ninth spinal segments.

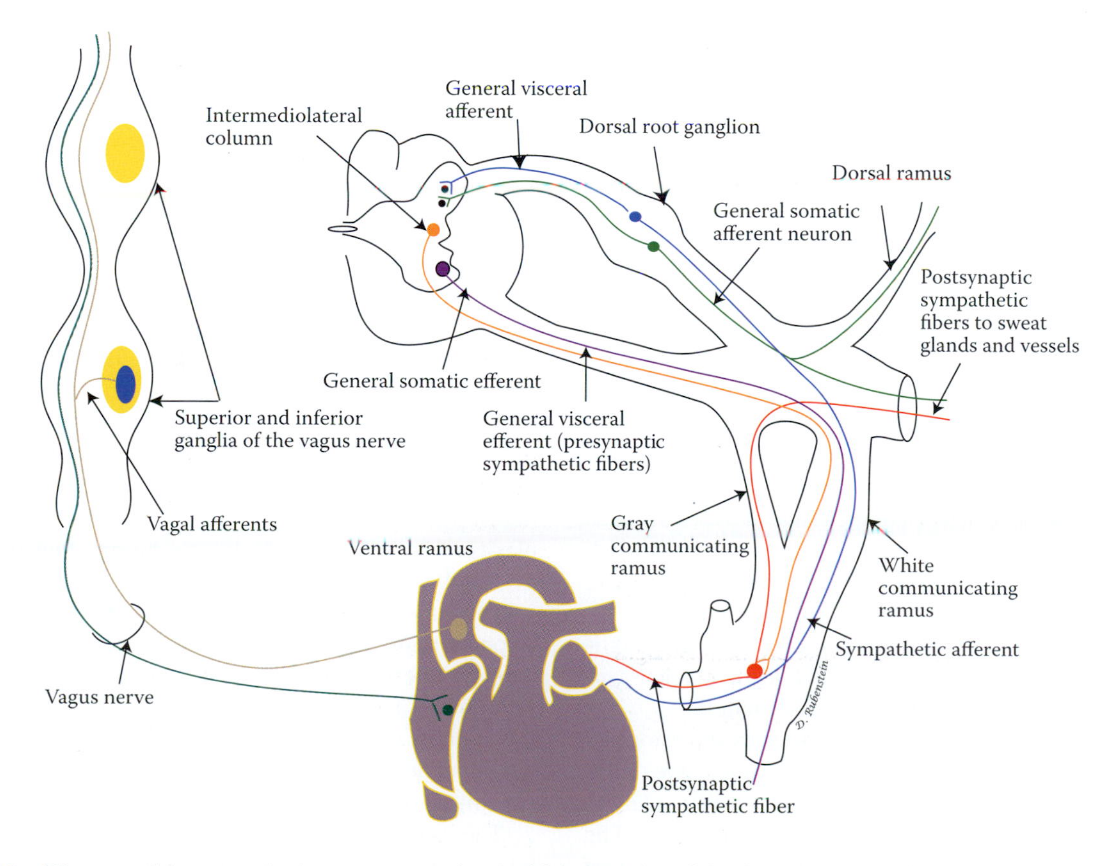

FIGURE 9.11 Diagram of the sympathetic neurons associated with innervation of the thoracic viscera (heart, lungs, and bronchi).

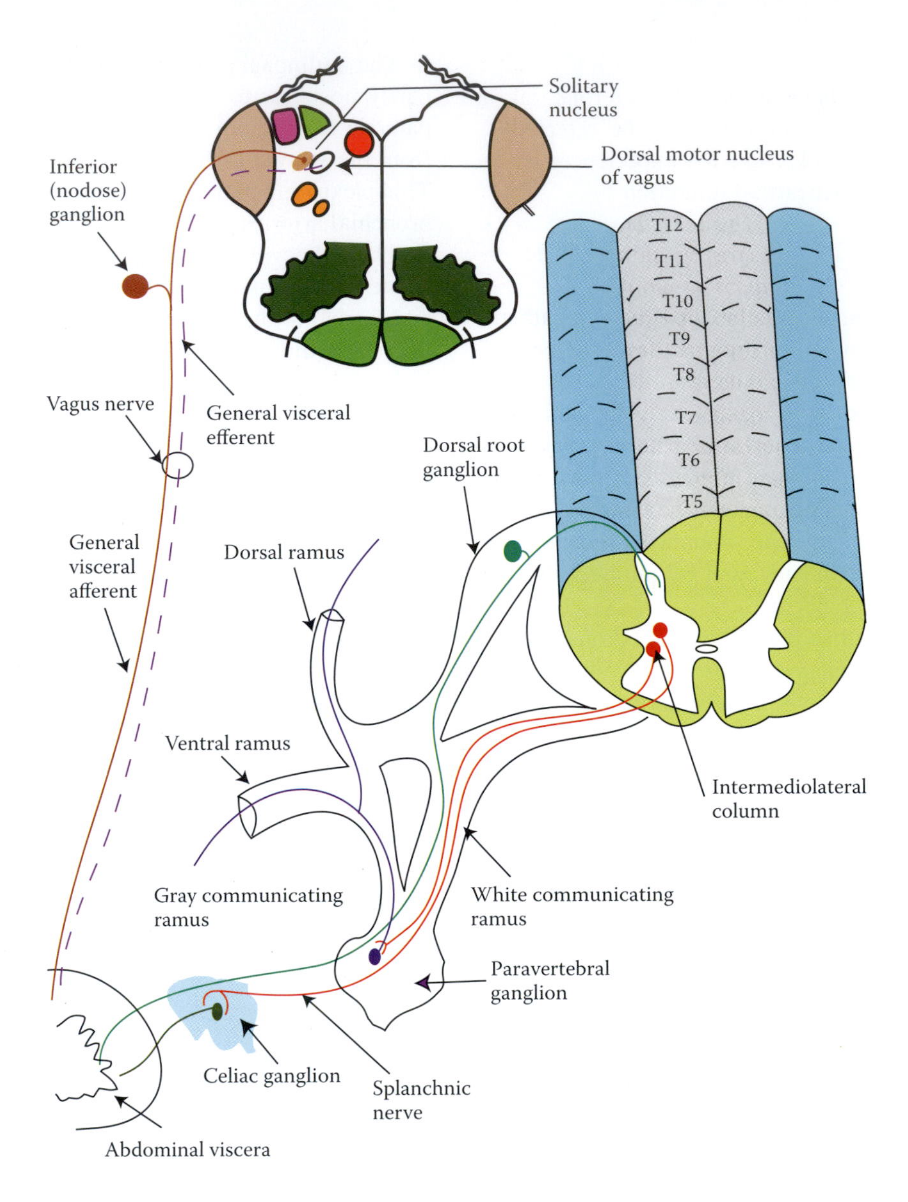

FIGURE 9.12 Schematic drawing of the sympathetic innervation of the abdominal viscera. Note the role of the celiac ganglia in mediating this innervation.

The gastric plexus consists of right and left plexuses; the right plexus, an extension of the hepatic plexus, innervates the pylorus. The sympathetic fibers produce contraction of the pyloric sphincter and inhibition of the gastric muscles, while the parasympathetic fibers maintain an opposing effect. The left gastric plexus, another extension of the celiac plexus, surrounds the left gastric artery. It exerts similar effects upon the stomach and pylorus. The sympathetic fibers, which supply the stomach, are derived from the T6–T10 thoracic spinal segments.

The suprarenal plexus is formed largely by the preganglionic sympathetic fibers of T8–L1 spinal segments that synapse in the chromaffin cells of the adrenal medulla.

The renal plexus surrounds the renal arteries and is formed by the sympathetic fibers of the lesser splanchnic nerve (T10–T11), least splanchnic nerve (T12), and first lumbar splanchnic nerve (L1), primarily exerting a vasomotor action. The vagal parasympathetic fibers serve as afferents, terminating in the wall of the kidney. This plexus also contributes to the ureteric and the gonadal plexuses.

The ureteric plexus receives sympathetic fibers from (T11–L1) spinal segments and parasympathetic fibers from the vagus and pelvic splanchnic nerves. Due to its close relationships to the abdominal and pelvic viscera, the ureteric plexus receives fibers from diverse sources including the aortic, renal, vesical, and hypogastric plexuses.

The superior mesenteric plexus (Figures 9.4 and 9.12) is a continuation of the celiac plexus, which is formed by sympathetic fibers of the 9th, 10th, 11th, and 12th thoracic and 1st lumbar (T10–L1) spinal segments and by the parasympathetic fibers of the vagus nerve. This plexus supplies part of the duodenum, jejunum, ileum, and approximately the right two-thirds of the large intestine.

The aortic (intermesenteric) plexus (Figures 9.4 and 9.12) encircles the aorta between the superior and inferior mesenteric arteries. It contributes to the testicular, inferior mesenteric, and hypogastric plexuses.

The inferior mesenteric plexus (Figures 9.4 and 9.12) surrounds the inferior mesenteric artery, containing sympathetic and parasympathetic fibers. The sympathetic fibers, which are inhibitory to the muscular walls of the descending colon, sigmoid colon, and upper part of the rectum, are derived from the first and second lumbar (L1 and L2) spinal segments. The parasympathetic fibers of the pelvic splanchnic nerves are excitatory, originating from the second, third, and fourth (S2–S4) sacral spinal segments.

The superior hypogastric (presacral nerve) plexus (Figure 9.4) is a continuation of the inferior mesenteric plexus. It receives sympathetic fibers from the 11th thoracic through the 2nd lumbar (T11–L2) spinal segments and parasympathetic fibers from the pelvic splanchnic nerves (S2–S4 spinal segments). It runs anterior to the sacrum, sacral promontory, and sacral plexus. It then divides into the right and left inferior hypogastric (pelvic) plexuses, supplying the pelvic structures.

Division of the sympathetic fibers of the superior hypogastric plexus (presacral neurectomy) is rarely performed in an attempt to relieve pain associated with diseased pelvic viscera. Dual transmission of pelvic pain via the sympathetic and parasympathetic fibers may render complete analgesia unattainable. In the male, removal of the superior hypogastric plexus may lead to loss of contraction of the seminal vesicles, prostate, and vas deferens, and eventual sterility.

The inferior hypogastric (pelvic) plexus (Figure 9.4) runs on both sides of the rectum, uterus, and bladder and gives rise to the vesical, middle rectal, prostatic, and uterovaginal plexuses. It contains parasympathetic fibers from the pelvic splanchnic nerves and sympathetic fibers from the lower thoracic and upper lumbar (T12–L1) spinal segments. The uterovaginal part of the pelvic plexus supplies the serosa, myometrium, and endometrium, as well as the vagina. The sympathetic fibers derived from the T12–L1 spinal segments produce uterine contraction and vasoconstriction, while the parasympathetic fibers produce relaxation of the myometrium and vasodilatation.

The vesical plexus, a subsidiary of the inferior hypogastric plexus, is associated with contraction of the detrusor muscles via the pelvic splanchnic nerves (parasympathetic), mediating micturition. The sympathetic component is derived from T11–L2 spinal segments, which also supply motor fibers to the vas deferens and the seminal vesicle.

The prostatic plexus, another subsidiary of the inferior hypogastric plexus, supplies the urethra, bulbourethral glands, corpora cavernosa, and corpus spongiosum via the lesser and greater cavernous nerves. The sympathetic part of the prostatic plexus controls ejaculation, inhibits the detrusor musculature of the bladder, and induces vasoconstriction. The parasympathetic part is formed by the pelvic splanchnic nerves, producing vasodilatation and erection.

ENTERIC NERVOUS SYSTEM

The enteric nervous system, an integral part of the ANS, consists of a group of neurons within the myenteric plexus of Auerbach's and the submucosal plexus of Meissner's and Henle's, as well as the pancreatic and cystic plexuses, which are derived from the neural crest cells. The myenteric plexus is located between the circular and longitudinal muscle layers, extending from the esophagus to the level of the internal anal sphincter. On the other hand, the submucosal plexus lies between the circular muscle and muscularis mucosa, stretching from the stomach to the anal canal. This system exerts a local reflex activity independent from the control of the brain and spinal cord. It is important to note that enteric nerves have more common features with the CNS than with the peripheral nerves. In fact, enteric nerves do not have collagenous coats, as is the case in the PNS. Furthermore, they lack the endoneurium and are supported by glial cells that resemble the astrocytes that contain glial fibrillary acidic protein (GFAP). Upon this network of neurons, the motility and the secretory functions of the gastrointestinal tract from the middle third of the esophagus to the anorectal junction remain dependent. The number of neurons associated with this system may be equivalent to or exceed the entire population of spinal neurons. This system of ganglia and plexuses is responsible for the induction of reflex peristalsis, independent of the direct commands of the brain. These ganglia maintain a blood–ganglion barrier and are not pierced by vessels or connective tissue septa. Some neurons of this system may subserve sensory function and respond to changes in the morphology of bowel shape. Others are simply interneurons that receive input from sensory neurons and project to the parasympathetic postganglionic neurons.

Numerous neuropeptides have been identified within this system of the neuronal network. These peptides may act to enhance or suppress the effects of transmitters or maintain a trophic role. Intrinsic motor neurons within this system may assume an excitatory role, utilizing acetylcholine and substance P as cotransmitters; others may project an inhibitory effect using ATP (cotransmitter in the large and small intestine), vasoactive intestinal polypeptide (VIP), and NO (nitric oxide) as cotransmitters.

Somatostatin is widely distributed in the gastrointestinal tract and the δ cells of the pancreas, where it inhibits the secretion of glucagon and insulin, a fact that may prove significant in diabetic patients. It is present in the dorsal root ganglia and autonomic plexuses. In the CNS, it is concentrated in the hypothalamus, amygdala, and neocortex, where it facilitates responsiveness to acetylcholine. In Alzheimer's disease, formation of somatostatin neuritic processes and depletion of its somatostatin-28 from the cortex are detected. Somatostatin-14, another form of this peptide, may show reduction upon the administration of cysteamine as a treatment for the metabolic disease known as cystinosis.

VIP is distributed in the pancreas, autonomic plexuses, and CNS. It is also contained in the parasympathetic cholinergic neurons of the salivary glands. Secretion of this peptide increases the glandular secretion (enhances the secretory function of acetylcholine) and blood flow to the intestine (as a result of vasodilatation). VIP-related peptides include the growth hormone–releasing hormone (GHRH) and pituitary adenylate cyclase–activating peptide (PACAP). GHRH is isolated from the intestine, and PACAP, as the name indicates, from the pituitary gland.

The lack of the parasympathetic ganglia in the myenteric and submucosal plexuses, as a result of failure of migration of the neural crest cells, is responsible for congenital megacolon of Hirschsprung's disease (also discussed at the end of this chapter), which is characterized by dilatation of the segment proximal to the affected segment. Patients exhibit constipation and abdominal pain. Other diseases that affect the enteric nervous system include herpes simplex, diabetes mellitus, amyloidosis, and Chagas disease.

AFFERENT COMPONENTS OF THE AUTONOMIC NERVOUS SYSTEM (ANS)

The post- and preganglionic ganglionic sympathetic fibers and less so the parasympathetic fibers are usually accompanied by afferent autonomic (visceral) fibers that originate from capsulated receptors in the visceral and vascular walls. Those that accompany sympathetic fibers use the white communicating rami to reach the dorsal roots of the spinal nerves. In addition to visceral pain, they also mediate visceral reflexes and transmit organic visceral sensations, libido, distention, hunger, and nausea. Stimuli that produce visceral pain include ischemia, distention and obstruction of the visceral wall. Cutting, burning, or crushing a viscera but do not induce visceral pain. Visceral afferents utilize mechanoreceptors, chemoreceptors, thermoreceptors, and osmoreceptors.

Visceral pain is transmitted by the visceral afferents that predominantly join and accompany the sympathetic efferents, in a reverse direction, and terminate in the same spinal segments that provided the sympathetic presynaptic efferents to the diseased visceral organ. This fact accounts for the phenomenon of referred pain in which pain from a diseased visceral organ is felt in the cutaneous areas of the spinal segments that originally provided the presynaptic sympathetic fibers to the affected organ. However, pain impulses from the bladder, anterior urethra, and uterine cervix pursue a course with the pelvic splanchnic nerves, in a reverse direction, to the second, third, and fourth sacral spinal segments. The superior hypogastric plexus and the lumbar splanchnic nerves also convey pain from the uterus, with the exception of the uterine cervix, to the lower thoracic and upper lumbar spinal segments. Therefore, dysmenorrhea (intractable pain associated with menses) can, for the most part, be alleviated by excision of the superior hypogastric plexus. Afferents from the testis and ovary run through the gonadal plexuses that terminate in the 10th and 11th spinal segments. General visceral afferents are also found in the glossopharyngeal and vagus nerves.

DISORDERS OF THE AUTONOMIC NERVOUS SYSTEM

Recognition and assessment of autonomic disorders can help in the diagnosis of clinical conditions, particularly those associated with the nervous and cardiovascular systems, and in the establishment of valid prognosis. The significance of autonomic dysfunctions in systemic diseases paved the way for the development of reliable, reproducible, and cost-effective objective assessment tests. These tests have become particularly critical in patients with diabetes mellitus. The impact of autonomic failure and its ramifications can be seen in a variety of disease processes and syndromes. These conditions can result from local or systemic lesions that compromise the integrity of the higher centers in the CNS, descending autonomic pathways, preganglionic neurons at spinal cord level, or postganglionic neurons in the paravertebral or prevertebral ganglia. ANS disorders can be associated with unusually large neurons that lie adjacent to the sympathetic ganglia and continue along their dendritic processes and axons. These disorders can also be linked to the presence of unmyelinated axonal masses of the Schwann cells around prevertebral ganglia and near abdominal viscera. Reduction in the number of cells in the intermediolateral column and severe demyelination of the vagus nerve are observed in diabetic or alcoholic neuropathy involving the ANS.

Autonomic dysfunction in tetanus may manifest excessive sweating with tachycardia and hypertension. Variations in autonomic manifestations correlate with lesion sites in the ganglia or individual autonomic nerve. For instance, when a lung tumor invades the costal pleura near the vertebral column, it disrupts the paravertebral ganglia and impairs autonomic function associated with the sweat glands and cutaneous vessels of paravertebral area; however, when the tumor expands to involve the intercostal nerves, hyperhidrosis will ensue as a result of irritation of these nerves. Autonomic dysfunctions may be exhibited as orthostatic hypotension, hypothermia, anhidrosis and heat stroke.

In spinal cord lesions, autonomic disturbances vary with the level of injury. Lesions of the cervical and upper thoracic spinal segments are most likely to produce combined sympathetic and parasympathetic dysfunctions, whereas damage to the lower thoracic segments are only associated with parasympathetic dysfunctions. Transection of the cervical part of the spinal cord may result in loss of all sensory and motor activities below the level of affected segment(s), as well as autonomic dysfunctions, including loss of sweating and piloerection, loss of micturition, impotence, and hypotension (spinal shock). Recovery of autonomic functions may occur as a result of the release from cortical and hypothalamic control. Since changes in blood pressure in individuals with cervical transection are no longer be mediated by autonomic centers in the brainstem, cutaneous stimulation below the level of the lesion may produce a rise in blood pressure, mydriasis, and sweating. Bladder function becomes automatic, and urination may occur when it is full. Following these changes, patients may manifest a triple or mass reflex in which a mild cutaneous stimulus produces flexion in all joints of the lower extremity (triple reflex), which disappears approximately 4 months following transection of the spinal cord.

Autonomic disorders can follow an acute or chronic course. Acute autonomic dysfunction is exemplified in sudden pandysautonomia, a self-limiting autoimmune condition in which all autonomic functions are disrupted and the patient exhibits anhidrosis with dry, hot skin, loss of salivary and mucus secretions in the oral and the nasal cavities, postural hypotension, nonreactive pupil, and lack of peristaltic movement with hypotonic urinary bladder. Sensation, motor coordination, deep tendon reflexes, and mental faculties are unaffected. Laboratory studies may show a high level of protein in the cerebrospinal fluid and serum glucose. Sural nerve biopsy that reveals demyelination may add additional diagnostic value. Chronic autonomic dysfunction can occur subsequent to prolonged immobilization or weightlessness or due to use of hypotensive medications or drugs that disturb thermoregulation.

Autonomic neuropathy produces changes that are variable and can involve the ganglia, unmyelinated visceral afferent and efferent fibers, vagus nerve, and associated smooth muscles. Autonomic neuropathy of the cardiac plexus produces a heart with resting rate (90–100 beats per minute) but without sinus arrhythmia or exertional increase in stroke volume. Heart rate usually shows great variation with deep breathing in healthy individuals, but these variations are absent in patients with autonomic dysfunction. Sinus arrhythmia (deep breathing–induced variation in heart rate) can be utilized in the determination of the effect of autonomic dysfunction on the heart. Physical activity increases heart rate, slowly reaching peak rates in a short period of time, while cessation of the activity produces delayed bradycardia. These changes in a denervated heart are attributed to the circulating catecholamines in the blood. Postural hypotension seen in autonomic neuropathy can adversely affect brain perfusion, resulting in transient visual blackout, vertigo, and syncope.

In individuals with mitral valve prolapse, autonomic hypersensitivity produces prolonged bradycardia following Valsalva maneuver, irregularities of heart rate, and ventricular fibrillation, particularly during invasive cardiac procedures.

In autonomic neuropathy involving the vagus nerve gastric secretion and intestinal motility cannot be induced by vagal stimulation or hypoglycemic conditions. Patients experience diarrhea or constipation, early satiety, nausea after meals, bloating, heartburn, and dysphagia. Lesions of the efferent limb of the baroreceptor reflex, which consists of sympathetic fibers to the blood vessels of the viscera, muscles, and skin, are thought to be responsible for postural hypotension seen in autonomic dysfunction. Reduction in the plasma levels of renin and norepinephrine is considered additionally contributory to this condition.

Severe diabetic diarrhea is uncommon, but when occurs, it is profuse, distressingly explosive in nature, watery, and usually postprandial or nocturnal. One-fifth of diabetic patients may also experience cardiovascular reflex disorders. Patients with diabetes may exhibit other autonomic manifestations including skin changes: atrophic, shiny, and red with hypohidrosis or cold whitish extremities with hyperhidrosis possibly due to denervation hypersensitivity.

The urinary bladder may also undergo dysfunction in autonomic neuropathy, resulting in neurogenic bladder. In order to understand neurogenic bladder, an overview of its innervation may be helpful. The urinary bladder receives somatic and autonomic innervation, which includes both sympathetic and parasympathetic fibers. Sympathetic presynaptic fibers primarily originate from the lower two thoracic and upper two lumbar spinal segments, convert into postsynaptic sympathetic fibers in ganglia within the superior and inferior hypogastric plexus, and then innervate the urinary bladder. These fibers inhibit the detrusor muscle and excite the muscles in the vesical trigone and internal urethral sphincter. An increase of bladder pressure allows the bladder to accommodate a larger volume of urine. Accordingly, demyelination of the sympathetic fibers in autonomic neuropathy reduces both the resistance to urethral outflow and the bladder's ability to accommodate urine and thus increases the frequency of micturition. The parasympathetic cholinergic fibers emanate from S2–S4 spinal segments as the pelvic splanchnic nerves and synapse on the intramural ganglia on the wall of the urinary bladder. Activation of the pelvic splanchnic nerves produces contraction of the detrusor muscle. For complete bladder emptying, supraspinal input must overcome these autonomically regulated mechanisms. Afferents that regulate reflex contraction of the detrusor muscle travel via both somatic and autonomic fibers and convey signals regarding distension from the bladder. Pain, in particular, travels in the pelvic and superior hypogastric plexus to the sacral and thoracolumbar spinal segments. Demyelination of these nerve fibers that supply the urinary bladder results in an increase in intervoid time with a limited number of urinations per day. Patients dribble (incontinent) and have difficulty initiating and maintaining voiding, and because of increased residual urine and decreased detrusor muscle activity, the urinary bladder is never completely empty.

Autonomic neuropathy can cause impotence in the male, which is the most frequent symptom in diabetic patients, by disrupting conduction in the nerves that mediate sexual function. Retrograde ejaculation due to failure of relaxation of the external urethral sphincter during orgasm can also occur in diabetic autonomic neuropathy. The pelvic splanchnic nerves are the vasodilators of the deep and dorsal arteries of the penis, which prolong erection through increased blood flow to the corpora cavernosa and corpus spongiosum. These nerves also have some role in stimulating the secretion of the bulbourethral (Cowper's) glands and prostate during the second phase of sexual response, in which blood flow and blood pressure, muscle tension, and respiratory rate are rapidly increased. This is followed by the third (orgasm) phase, in which sympathetic activation leads to emission by producing contraction of the vas deferens, seminal vesicles, and prostate. Contraction of the bulbospongiosus and ischiocavernosus muscles, which are innervated by the pudendal nerve, causes rhythmic contraction of these muscles and ejection of semen from the urethra and conclusion of the third phase of the sexual response.

Another aspect of autonomic dysfunction is the disorder of sweating. Environmental temperature changes and strenuous physical activity elicit thermoregulatory sweating, which diffusely occurs in the entire body, while stressful and emotional conditions produce sweating in specific areas of the body, such as the face, axilla, palm, and sole of the foot (emotional sweating). Sweating is associated with elevation of temperature in the cutaneous vessels and activation of cutaneous receptors that influence thermoregulatory centers. Sweating can be affected indirectly by postural changes that affect the temperature and, therefore, the blood flow in the cutaneous vessels. On this basis, there is a considerable increase of sweating in the exposed upper part of the body when a person lies on one side. Standing after lying down causes excitation of the sweat glands in the upper part of the body and inhibition in the lower part. Sweating can involve the entire body when the subject lies in a supine position. Despite the above facts, sweating, cutaneous temperature, and vasodilation may not be clearly interdependent as the above examples show. This becomes abundantly obvious in patients with Guillain–Barré syndrome who exhibit palm sweating in response to heat, but without cutaneous vasodilation. Sweating can be induced by the cutaneous administration of acetylcholine. This helps to determine the site of a lesion in patients with autonomic disorder. If the disruption is at the preganglionic level, thermal sweating will be absent, but sweating in response to this test is preserved.

Due to close association of the sudomotor fibers with the sensory fibers in the peripheral nerves, interruption of a peripheral nerve can lead to thermoregulatory and sweat secretion deficits that correspond to the area of distribution of the affected nerve. Thus, skin areas that show analgesia and anesthesia will be also anhidrotic. Corollary to this, regeneration of the affected nerve is associated with the return of sweat gland secretion. Impairment of sweat secretion in certain parts of the body may also indicate the degree of metastasis of a tumor. It could be an early sign of Pancoast tumor metastasis as it spreads posteriorly from the apex of the lung to disrupt the stellate ganglion, producing impairment of sweat secretion on the affected side of the head. In Adie syndrome, hypohidrotic patches on the trunk, extremities, or face are surrounded by hyperhidrotic areas, but without sensory impairment.

Hyperhidrosis (disorders of sweating) is characterized by increased sweating, in restful situations, due to overstimulation of the sympathetic postganglionic neurons that innervate the sweat glands. The anatomic structure of the sweat glands remains normal. It is categorized into essential hyperhidrosis of unknown etiology and secondary hyperhidrosis associated with fever and seen in a myriad of diseases, such as thyrotoxicosis, hypothalamic disorders due to ingestion of cholinergics, diabetes mellitus, gout, peripheral neuropathies, rheumatoid arthritis, polyarteritis nodosa, and thoracic outlet syndrome. Dumping syndrome, hypoglycemia, shock, syncope, intense pain, and also withdrawal from illicit drug or alcohol use can induce sweating but with cold skin. Hyperhidrosis, when it occurs, involves the palms of the hands, axilla, feet, and face. In familial dysautonomia, hyperhidrosis will be seen in the extremities, while in complex regional pain syndrome (CRPS) is confined to the area of distribution of the affected nerve. This socially uncomfortable condition, if left untreated, can last throughout patient's life. Palm sweating can be relieved by sympathectomy through removal of the second and third thoracic sympathetic ganglia. In autonomic neuropathy associated with diabetes mellitus, hyperhidrosis may involve the head, neck, trunk, and upper extremities, sparing the lower extremities. Symmetrical compensatory hyperhidrosis of the upper body is observed in lesions of the midthoracic spinal segments. Root lesions above T1 or below the L2 segment do not impair sweat secretion.

Facial sweating may occur unilaterally and involve the entire face or lower face and can be associated with lacrimation and nasal discharge. Facial sweating induced by eating (gustatory sweating) occurs as a consequence of aberrant cross-linkage between the sympathetic postsynaptic fibers that innervate the sweat glands and the postsynaptic parasympathetic fibers that innervate the parotid gland and run through the auriculotemporal nerve (Frey or auriculotemporal nerve syndrome). This reflex gustatory sweating is elicited by spicy or sour foods and can be treated by excision of the lesser petrosal branch of the glossopharyngeal nerve. Similarly, gustatory sweating in the submental area can occur as a result of cross-linkage between the chorda tympani that carries secretomotor (parasympathetic) fibers to the submandibular and sublingual glands and the sympathetic postsynaptic fibers that innervate the sweat glands in the area (chorda tympani syndrome).

Hyperhidrosis may be segmental in the upper extremity if the cause is a cervical rib. Despite the fact the sweat glands lose their innervation partially or completely in mononeuropathies, stress can induce hyperhidrosis by the possible effect of hormones and/or denervation hypersensitivity. Hyperhidrosis due to peripheral neuropathy of the sciatic or median nerve can be associated with CRPS in which the affected extremity is swollen, cyanotic, and extremely painful. Anhidrosis may follow hyperhidrosis if the peripheral nerve dysfunction is induced by progressive disease. In familial dysautonomia, sweating threshold is lowered, and excessive sweating occurs in the extremities despite peripheral nerve dysfunction.

Anhidrosis (lack of sweating) is the common and permanent deficit seen in the affected dermatomes following sympathectomy, a procedure utilized in the treatment of hyperhidrosis. However, this is not always true, particularly in one-third of patients, who exhibit hyperhidrosis as a result of excitatory sympathetic postsynaptic fibers that travel within branches of the trigeminal nerve. Presence of intact intermediate ganglia in the sympathetic trunk during sympathectomy may account for segmental sparing of anhidrosis. Sympathectomy may be responsible, though rarely, for causing alternating Horner's syndrome and hyperhidrosis. Anhidrosis is also seen in Horner's syndrome, leprosy, or diabetic neuropathy. In leprosy, anhidrotic areas are usually small and do not follow the course of the affected major nerve. Postsympathectomy, aberrant fibers that establish cross-linkage, may lead to the activation of sweat secretion and the development of alternating hyperhidrosis and anhidrosis. Anhidrosis induced by diabetic neuropathy, seen in the lower extremities or trunk, is associated with intolerance and excessive compensatory sweating on the head, neck, and face, particularly during meals.

In view of the extensive nature of the diseases and syndrome associated with autonomic dysfunction, our discussion will be limited to hereditary amyloidosis, CRPS, Guillain–Barré syndrome, Riley–Day syndrome, Fabry's disease, tetanus, Hirschsprung's disease, hyperhidrosis, Raynaud's disease or phenomenon, spinal cord lesions, Horner's syndrome, stellate ganglion syndrome, Shy–Drager syndrome, botulism, achalasia, Chagas disease, CRPS (reflex sympathetic dystrophy and causalgia), and autonomic ganglionopathy. It should be emphasized at this juncture that lesions of the hypothalamus, midbrain, reticular formation, and spinal cord may disrupt the central control of the higher autonomic regions. Hypothalamic and pontine lesions produce hyperpyrexia, while cortical lesions that produce paraplegia can also cause urinary incontinence and marked postural hypotension.

Hereditary amyloidosis is associated with a genetically variant protein that results in the production and deposition of amyloid in certain tissues. It mainly affects the heart, kidney, urinary bladder, and brain. Amyloid is a homogeneous, insoluble fibrillary and refractive protein that maintains affinity for certain dyes such as Congo red dye. It is characterized by sensory and motor neuropathy as well as autonomic dysfunction. Autonomic dysfunction occurs particularly in the Andrade and Rukavina types of this disease, in which sympathetic paravertebral and prevertebral ganglia are infiltrated with amyloid. Autonomic manifestations include explosive postprandial and nocturnal diarrhea; sexual dysfunction; postural hypotension; cardiac arrhythmia due to disorders of SA and atrioventricular (AV) nodes and His bundle; decreased peristalsis of the intestine; and impairment of urethral and anal sphincteric functions. Inanition (exhaustion) from diarrhea and chronic pyelonephritis due to urinary retention can lead to death of the patient.

CRPS is a chronic condition that exhibits pain and autonomic changes, usually in the extremities, and encompasses both reflex sympathetic dystrophy (type I) and causalgia (type II). Type I (reflex sympathetic dystrophy or Sudeck's atrophy) occurs as a result of bone fracture, trauma to soft tissue, or shoulder–hand syndrome after myocardial infarction. It is characterized by pain, which is disproportionate to the type of injury and aggravated by movements, with autonomic changes that include increased sweating and vasoconstriction. Type II (causalgia) occurs in partial lesions of a peripheral nerve and is associated with burning pain often accompanied by trophic cutaneous changes. Although type II can be associated with any nerve or plexus lesions, the median, sciatic nerves and certain roots of the brachial plexus (C7, C8, and T1) are particularly vulnerable due to the large number of associated autonomic nerve fibers. Burning pain, which is sympathetically induced, usually starts within a few hours of the injury and can be triggered by mild stimuli (allodynia), noise, or fright. Due to the severity of the burning sensation, the trivial nature of the stimuli, and overly protective attitude of the patients, a physician may assume that the condition is psychogenic. Edema, circulatory and massive trophic changes of the skin of the affected extremity, and disorders of joint movements usually accompany the burning pain sensation. Thickening of the skin under the nail in the form of ridges (Mees' nail bands) with pain confined to the territory of damaged nerve, elicited only by mild tactile stimuli with no burning quality, is known as algie diffusante. Transcutaneous electrical stimulation (TENS) seems effective in the treatment of burning pain associated with CRPS.

Guillain–Barré syndrome, which is also discussed in Chapter 2 is a serious autoimmune disorder seen commonly in patients with AIDS, herpes simplex, and mononucleosis, and after viral pulmonary or gastrointestinal infection. It occurs between the third and fifth decades and is common in both sexes. It produces demyelination of the nerves, leading commonly to ascending paresis and paralysis, pain, and paresthesia in the feet and hands, although ascending palsy may also be seen. Ataxia, visual disorders, dyspnea, and muscle spasm may also occur. Spontaneous episodes of hypotension, hypertension, tachycardia, cardiac dysrhythmia, syncope, dizziness, and orthostatic hypotension are seen in patients with this disease, possibly as the result of increased circulating catecholamines and disruption of the afferent and efferent limbs of the cardiovascular reflexes combined with denervation hypersensitivity.

Riley–Day syndrome (familial dysautonomia) is a congenital indolent autosomal recessive disorder of infants, which is often seen in children of Jewish descent. Reduction in the sensory neurons of the dorsal root ganglia and degeneration of the unmyelinated C fibers in the spinal nerves is most likely to account for impaired nociceptive and thermal perception. Loss of neurons within the trigeminal (Gasserian) ganglia and the postganglionic neurons of the autonomic ganglia also observed. Absence of propionyl coenzyme A carboxylase deficiency can produce hyperammonemia, resulting in reversible manifestations that mimic that of familial dysautonomia. Patients show a constellation of sensory and motor deficits, which include hypopathia (diminished response to painful stimuli), decreased taste sensation (due to reduced fungiform papillae), areflexia including corneal and spinal reflexes, hearing deficits, impaired vestibular reflexes, pulmonary infections, and recurrent episodes of severe vomiting. Hypopathia most likely develops as a consequence of marked reduction in the number of unmyelinated fibers. The autonomic disturbances in this syndrome include loss or decreased lacrimation and loss of the mechanisms that regulate blood pressure and temperature, blotching of the skin, episodes of hypertension and postural hypotension, and excessive sweating, particularly in stressful situations. Incomplete or failure of development and migration of the neural crest cells and, consequently, absence of the sensory and autonomic ganglia are linked to lack of nerve growth factors. Percutaneous administration of methacholine and histamine causes catecholamine secretion in the urine and triple response of Lewis, but without the flare component. The triple response of Lewis is a triphasic transient skin response that consists of a red line due to cutaneous vasodilation, spreading redness beyond the red line (flare), and wheal (swelling, edema) in the surrounding area, which occurs as a reaction to firm stroking or scratching of the skin.

In *Fabry's disease*, an X-linked lipid storage disease, there is a deficiency of *alpha-galactosidase-A* and accumulation of ceramide trihexoside in the autonomic ganglia, eyes, kidney, and cardiovascular system (see also Chapter 2). Symptoms usually start during childhood or adolescence. Patients show corneal opacity and burning sensations in the hands that exacerbates with physical activity and exposure to high temperature with elevated reddish-purple skin spots. Disorders of circulation and stroke can also occur. Autonomic dysfunctions include decreased sweat, saliva, and tear secretion, coupled with decreased circulating catecholamines and decreased sympathetic and parasympathetic activity.

Tetanus is an acute infectious exotoxin-mediating disease that occurs as a result of wound contamination by gram-positive, anaerobic *Clostridium tetani* or their spores that live in the soil and animal excretion and often involves a cut or deep wound caused by rusty nails, splinters, insect bites, burns, or IV drug administration sites. *C. tetani* releases tetanospasmin, a neurotoxin that spreads from the infected site to the CNS by binding to the neuromuscular junction and then following, in a retrograde direction, the peripheral nerve and their terminals, to the interneurons within the ventral horn of the spinal cord. Inhibition of the release of γ-aminobutyric acid (GABA) from the inhibitory interneurons results in excitation of the αγ and γ motor neurons, leading to hypertonicity and intermittent spasm of the muscles. Muscles in close proximity to the infection site show more severe spasms. The incubation period, which varies from 2 days to 2 weeks, can affect the extent and severity of the manifestations. The severe form of this disease occurs after an incubation period of 2–8 days. Typical early signs include trismus (locked jaw) and risus sardonicus, an involuntary spasmodic contracture that results from spasm of facial muscles. At a later stage, mild visual, tactile, or auditory stimuli produce painful spasm of all muscles. Rigidity of the back, neck, and abdominal muscles leads to opisthotonos posture, which exhibits hyperextended back, neck, and extremities, with a boardlike abdomen. This posture may be associated with bone fractures.

Autonomic dysfunction in tetanus is common, particularly with the severe form of this disease, and is a frequent cause of death. It can be seen in the second week after the onset of symptoms as the toxin spreads to the brainstem. Patients manifest hyperhidrosis, hypertension, tachycardia, and, less frequently resistant hypotension, bradyarrhythmia, and cardiac arrest. High levels of circulating catecholamines in these patients, up to 100-fold, may indicate hyperactivity of both the adrenal medulla and the sympathetic nervous system.

Hirschsprung's disease (congenital megacolon), as previously described, is a condition that results from failure of the neural crest cells to migrate, leading to loss of the parasympathetic ganglia in Auerbach's and Meissner's plexuses. This disorder, commonly seen in males, is characterized by the loss of the peristaltic movement, subsequent constriction of the affected segment of the intestine, and massive dilation of the intestinal segment proximal to the aganglionic as a result of retention of feces. It frequently involves the sigmoid colon and the rectum, but it may involve other parts of the colon or the entire colon and, rarely, the terminal ileum. It is more common in males (see also developmental aspects). Affected infants present with obstipation (intractable constipation or dyschezia), distension, anorexia, persistent urge to defecate, and vomiting. Since constipation in patients with Parkinson's disease may also exhibit deficiency in dopaminergic neurons of the enteric system, a possible correlation between the migration of neural crest cells and dopamine may require further investigation. It may also be possible to use these enteric dopaminergic neurons as donor grafts. If left untreated, toxic enterocolitis (toxic megacolon) may develop, leading to death. Rarely, the aganglionic segment is confined to the anus, which results in intermittent constipation with intervening episodes of diarrhea.

Raynaud's disease is a primary idiopathic vascular disorder, while Raynaud's phenomenon is secondary to conditions such as collagen vascular disease. It is characterized by spasmodic vasoconstriction of the digital arteries of the extremities in response to cold or emotional stress. This phenomenon may occur as a secondary condition to a cervical rib, scleroderma, thoracic outlet syndrome, atherosclerosis of the brachial artery, and connective tissue disease. It may be attributed to a lack of histamine-induced vasodilatation subsequent to a lack of the neural mechanism for histamine release in individuals with an intact hypothalamic sympathetic center. Emotional stimuli and cold may activate the sympathetic system, lowering the threshold for vasospastic response. Patients manifest intermittent pallor due to depletion of the blood in the capillary beds of the digits and cyanosis as a result of deoxygenation of the stagnant blood in the capillary beds. Color changes may involve redness of the affected digits (reactive hyperemia) as a result of dilation of the digital arteries, and engorgement of the capillary beds with oxygenated blood may also be observed. This will confer a ruddy complexion to the skin of the digits. Small painful ulcers may appear on the tips of the digits in the late course of this condition, particularly in patients with scleroderma. However, these ulcers do not occur proximal to the proximal interphalangeal joints, a distinguishing fact from trophic ulcers seen in synringomyelia and leprosy. Drug treatment should be reserved for severe cases. Oral administration of reserpine in doses of 0.25–0.5 mg once a day has been shown to increase blood flow to the fingers. Infusion of the brachial or radial artery with a single dose of reserpine has been reported to reduce pain and promote healing of ulceration. Mild cases may be controlled by protecting the body and extremities from cold and by using mild sedatives. Prazosin, the calcium antagonist nifedipine, phenoxybenzamine, and prostaglandins (thromboxane) are also effective medications for this condition.

Horner's syndrome is caused by a lesion of the intermediolateral column of the first thoracic spinal segment or emerging ventral root, degeneration of the lateral medulla, a lesion of the descending autonomic pathways from the hypothalamus, superior cervical gangliotomy, or syringomyelia. It may also be caused by percutaneous carotid puncture for cerebral angiography, a hematoma caused by a ruptured axillary artery, intracavernous lesions, birth trauma, enlargement of the cervical lymph nodes, thoracic tumors, destruction of the internal carotid plexus, or hypothalamic lesion. It is characterized by miosis (constriction of the pupil), ptosis (drooping of the upper eyelid due to paralysis of the superior tarsal muscle), anhidrosis (lack of sweating), dilation of the facial vessels, and apparent enophthalmos (sinking of the eyeball due to paralysis of the orbital muscle). Heterochromia, which refers to the diversity of colors in part or parts that should normally be one color, is a characteristic of congenital form of Horner's syndrome. In infants, Horner's syndrome may be associated with an unpigmented iris that assumes a bluish or mixed gray and blue appearance. *Stellate ganglion syndrome* is produced by compression of the stellate ganglion as seen in Pancoast tumor of the apical lobe of the lung, exhibiting signs of Horner's syndrome and reflex sympathetic dystrophy. The latter condition manifests vasodilatation and dryness of the skin of the upper extremity.

Achalasia refers to failure or incomplete relaxation of the lower esophageal sphincter, which is more common in males. In this condition, the normal peristalsis of the esophagus is replaced by abnormal contractions. It is classified into vigorous and classic achalasia. Vigorous achalasia resembles diffuse esophageal spasm, exhibiting simultaneous and repetitive contractions with large amplitude, whereas classic achalasia shows contractions of small amplitude. Secondary achalasia may result from infiltrating gastric carcinoma. It is characterize by dysphagia, chest pain, regurgitation and pulmonary aspiration, and projectile vomiting. Emotional disorders and hurried eating may predispose the individual to this condition. Although the esophageal myenteric plexus lack ganglia, the pathogenesis of this dysfunction is not well understood. Treatment may include administration of anticholinergics and calcium channel antagonists, or balloon dilatation. Surgical intervention in which the lower esophageal sphincter is incised may prove to be effective.

Chagas disease is an infectious and zoonotic disease caused by *Trypanosoma cruzi* and is transmitted from infected animals to humans by reduviid bugs. Chagoma, an inflammatory lesion, is often seen at the site of entry of the parasite. When the parasite enters through the conjunctiva, edema of the palpebrae and periocular tissue is a characteristic feature (Romana's sign). The heart is the most commonly affected organ, exhibiting cardiomyopathy, ventricular enlargement and thinning of the walls, mural thrombi, and apical aneurysm. The right branch of the His bundle is frequently damaged, producing atrioventricular block. Patients show signs of malaise, fever, and anorexia, which are associated with swelling of the face and lower extremities. This infectious parasitic agent may also cause destruction of the myenteric plexus in the esophageal, duodenal, colonic, and ureteric wall, producing megacolon, megaduodenum, and megaureter. Lymphadenopathy, meningoencephalitis, and increased incidence of esophageal varicosities are characteristics of this disease. This condition may be treated by nifurtimox, an effective drug against *T. cruzi* during the acute phase of the disease.

Shy-Drager syndrome (idiopathic orthostatic hypotension) is a multisystem disorder that includes autonomic dysfunctions, ataxia, and upper motor neuron palsy. Autonomic dysfunctions comprise anhidrosis (lack of sweating), impotence, postural hypotension, mydriasis and pupillary asymmetry, and bowel and bladder dysfunctions. The hallmark of this disease is postural hypotension, which is greater than 30/20 mm Hg on standing from a supine position. Patients also exhibit Parkinsonian manifestations in which rigidity and bradykinesia are very conspicuous. Neuronal loss has been shown in the intermediolateral column of the thoracic spinal segments, peripheral autonomic ganglia, substantia nigra, locus ceruleus, olivary nuclei, caudate nucleus, and dorsal motor nucleus of the vagus. These cellular losses are accompanied by gliosis and, in some cases, by Lewy bodies, which are typical of Parkinson's disease. Men are more frequently affected than women are, and the disease exhibits an insidious onset. Postural hypotension may be treated by medications that increase blood volume and by pressure (antigravity) stockings. Parkinsonian symptoms may be treated by the administration of Sinemet or bromocriptine as well as α agonists.

Botulism is caused by ingestion of food contaminated with *Clostridium botulinum* (anaerobic gram-positive organism), ingestion of spores, and production of toxin, or as a result of wound infection with the same bacteria. It is a paralytic disease, which initially affects the cranial nerves and expands to involve the limbs. Symptoms of botulism include autonomic disturbances such as nausea, vomiting, dysphagia, extremely dry throat, blurred vision, loss or diminished light reflex, and ptosis, in addition to skeletal muscle paralysis. Descending paralysis, which is symmetric, involving the head, neck, arm, and thorax is characteristic of this disease. Deep tendon reflexes are not generally affected, although the gag reflex may be depressed. Patients may die from respiratory failure. Patients may be given antitoxin (equine antitoxin) as well as cathartics and enemas to eliminate the toxin, supplemented with antibiotics.

Autonomic ganglionopathy is a rare idiopathic acquired neuromuscular disorder caused by autoantibodies directed against the ganglionic nicotinic cholinergic receptors in the sympathetic, parasympathetic, and enteric ganglia. Patients exhibit orthostatic hypotension, gastrointestinal motility disorder, tonic pupil, as well as impairment of working memory, attention, and executive functions.

SUGGESTED READING

Alexander MS, Biering-Sorensen F, Bodner D et al. International standards to document remaining autonomic function after spinal cord injury. *Spinal Cord* 2009;47(1):36–43.

Axelrod FB. Hereditary sensory and autonomic neuropathies. Familial dysautonomia and other HSANs. *Clin Auton Res* 2002;12 Suppl 1(7):12–14.

Baguley IJ, Heriseanu RE, Cameron ID, Nott MT, Slewa-Younan S. A critical review of the pathophysiology of dysautonomia following traumatic brain injury. *Neurocrit Care* 2008;8(2):293–300.

Djeddi DD, Kongolo G, Stéphan-Blanchard E, Ammari M, Léké A, Delanaud S, Bach V, Telliez F. Involvement of autonomic nervous activity changes in gastroesophageal reflux in neonates during sleep and wakefulness. *PLoS One* 2013;8(12):e83464.

Elliott J. Alpha-adrenoceptors in equine digital veins: Evidence for the presence of both alpha1 and alpha2-receptors mediating vasoconstriction. *J Vet Pharmacol Ther* (Blackwell Publishings) 1997;20(4):308–17.

Farrell KE, Keely S, Graham BA, Callister R, Callister RJ. A systematic review of the evidence for central nervous system plasticity in animal models of inflammatory-mediated gastrointestinal pain. *Bowel Dis* 2014;20(1):176–95.

Hiraba H, Inoue M, Gora K, Sato T, Nishimura S, Yamaoka M, Kumakura A, Ono S, Wakasa H, Nakayama E, Abe K, Ueda K. Facial vibrotactile stimulation activates the parasympathetic nervous system: Study of salivary secretion, heart rate, pupillary reflex, and functional near-infrared spectroscopy activity. *Biomed Res Int* 2014;2014:910812.

Kennedy RH, Bartley GB, Flanagan JC et al. Treatment of blepharospasm with botulinum toxin. *Mayo Clin Proc* 1989;64:1085–90.

Koike H, Sobue G. Autoimmune autonomic ganglionopathy and acute autonomic and sensory neuropathy. *Rinsho Shinkeigaku* 2013;53(11):1326–9.

MacDonald IA. The sympathic nervous system and its influence on metabolic function. In Bannister R, Mathias CJ, eds. *Autonomic Failure: A Textbook of Clinical Disorders of the Autonomic Nervous System.* Oxford: Oxford Medical Publishers, 1992, 197–211.

Miolan JP, Niel JP. The mammalian sympathetic prevertebral ganglia: Integratine properties and role in the nervous control of digestive tract motility. *J Auton Nerv Syst* 1996;58:125–38.

Mizeres N. The cardiac plexus in man. *Am J Anat* 1963;112:141–51.

Nathan PW, Smith MC. The location of descending fibers to sympathetic neurons supplying the eye and sudomotor neurons supplying the head and neck. *J Neurol Neurosurg Psychiatry* 1986;49:187–94.

Norvell JE. The aorticorenal ganglion and its role in renal innervation. *J Comp Neurol* 1968;133:101–12.

Pluta RM, Lynm C, Golub RM. Guillain–Barré Syndrome. *JAMA* 2011;305(3):319–319.

Schott GD. Visceral afferents: Their contribution to "sympathetic dependent" pain. *Brain* 1994;117:397–413.

Sinnreich Z, Nathan H. The ciliary ganglion in man. *Anat Anz* 1981;150:287–97.

Stark ME, Safir I, Wisco JJ. Probabilistic mapping of the cervical sympathetic trunk ganglia. *Auton Neurosci* 2014;181:79–84.

Valles M, Benito J, Portell E, Vidal J. Cerebral hemorrhage due to autonomic dysreflexia in a spinal cord injury patient. *Spinal Cord* 2005;43:738–40.

West CR, Wong SC, Krassioukov AV. Autonomic cardiovascular control in Paralympic athletes with spinal cord injury. *Med Sci Sports Exerc* 2014;46(1):60–8.

10 Spinal Nerves

Spinal nerves are formed by the union of the dorsal and ventral roots, which later divide into dorsal and ventral primary rami. The dorsal primary rami supply the skin and muscles of the back, while the ventral rami contribute to the formation of the cervical, brachial, and lumbosacral plexuses. Each plexus consists of ventral rami from a series of spinal segments, giving rise to branches that are derived from multiple spinal segments and supply a group of muscles and associated cutaneous areas. Some of the branches that arise from the plexuses are purely motor, others are sensory, and most carry both sensory and motor fibers. Damage to these branches may occur as a result of trauma or disease processes and may produce a constellation of muscular and cutaneous disorders. Because of the close relationship of some of the nerves to regional vessels, combined nerve and vascular disorders may ensue.

FORMATION, DISTRIBUTION, AND COMPONENTS OF THE SPINAL NERVES

Spinal nerves are formed by the union of the dorsal and ventral roots. Both of these roots run in the subarachnoid space to reach their points of exit at the intervertebral foramina. The central processes of the pseudounipolar neurons of the dorsal root ganglia (Figures 10.1 and 10.3) form the dorsal roots.

Dorsal root fibers enter the posterolateral sulcus of the spinal cord as medial and lateral bundles, receiving coverings from the pia mater, arachnoid mater, and the dural sheath. These roots consist of thickly and thinly myelinated, as well as unmyelinated, fibers. The thickly myelinated fibers comprise group Ia (annulospiral) and group II (flower spray endings) fibers that convey information from muscle spindles, as well as group Ib fibers of the Golgi tendon organs. The thickly myelinated fibers of the dorsal roots can selectively be blocked by the application of pressure on the dorsal roots. They also show selective degeneration in combined system degeneration disease (associated with pernicious anemia), tabes dorsalis, and arsenic poisoning. Smaller, thinly myelinated (Aδ) fibers and the unmyelinated C fibers that carry nociceptive impulses may be blocked more effectively by local anesthetics and can selectively be affected in beriberi disease (associated with vitamin B_1 deficiency). The skin area supplied by one dorsal root is known as a dermatome. Dermatomes (Figure 10.2) of successive dorsal roots show extensive overlap, which may be limited along the axial line.

The ventral roots are composed of axons of the α and δ motor neurons (GSE) of the ventral horn of the spinal cord, which supply the extrafusal and intrafusal muscle fibers, respectively. They also contain general visceral efferents (GVE) that emanate from the intermediolateral columns of the thoracic and upper lumbar segments (sympathetic fibers) or arise from the second through the fourth sacral spinal segments (parasympathetic fibers). In the intervertebral foramina, the dorsal and ventral roots unite to form the spinal nerves, which are accompanied by the meningeal and spinal branches of the segmental arteries.

There are 31 pairs of spinal nerves: 8 cervical, 12 thoracic, 5 lumbar, 5 sacral, and 1 coccygeal. All spinal nerves emerge via the intervertebral foramina (bounded anteriorly by the intervertebral disks and vertebral bodies, posteriorly by zygapophyseal joints, superiorly and inferiorly by the vertebral notches), with the exception of the first cervical (suboccipital) and the fifth sacral spinal nerves. The first cervical spinal (suboccipital) nerve leaves the vertebral column between the occiput and the atlas, whereas the fifth sacral spinal nerve exits through the sacral hiatus. In the same manner, the eighth cervical spinal nerve emerges inferior to the first thoracic vertebra. Sympathetic presynaptic fibers run through the gray communicating rami that connect the sympathetic ganglia to the spinal nerves. Due to the limited presence of the intermediolateral column that gives rise to the sympathetic presynaptic fibers in the thoracic and upper two or three lumbar segments, spinal nerves at these segments maintain additional connection to the sympathetic ganglia via the white communicating rami (Figure 10.3).

Proximity of the dorsal roots to the intervertebral disks can predispose them to compression injuries by the posterolaterally extruding nucleus pulposus of the intervertebral disks into the vertebral canal.

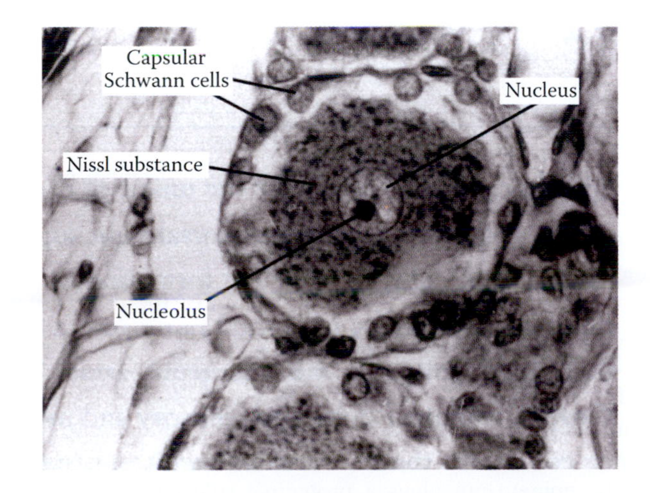

FIGURE 10.1 Photomicrograph of section of the dorsal root ganglion showing its main components.

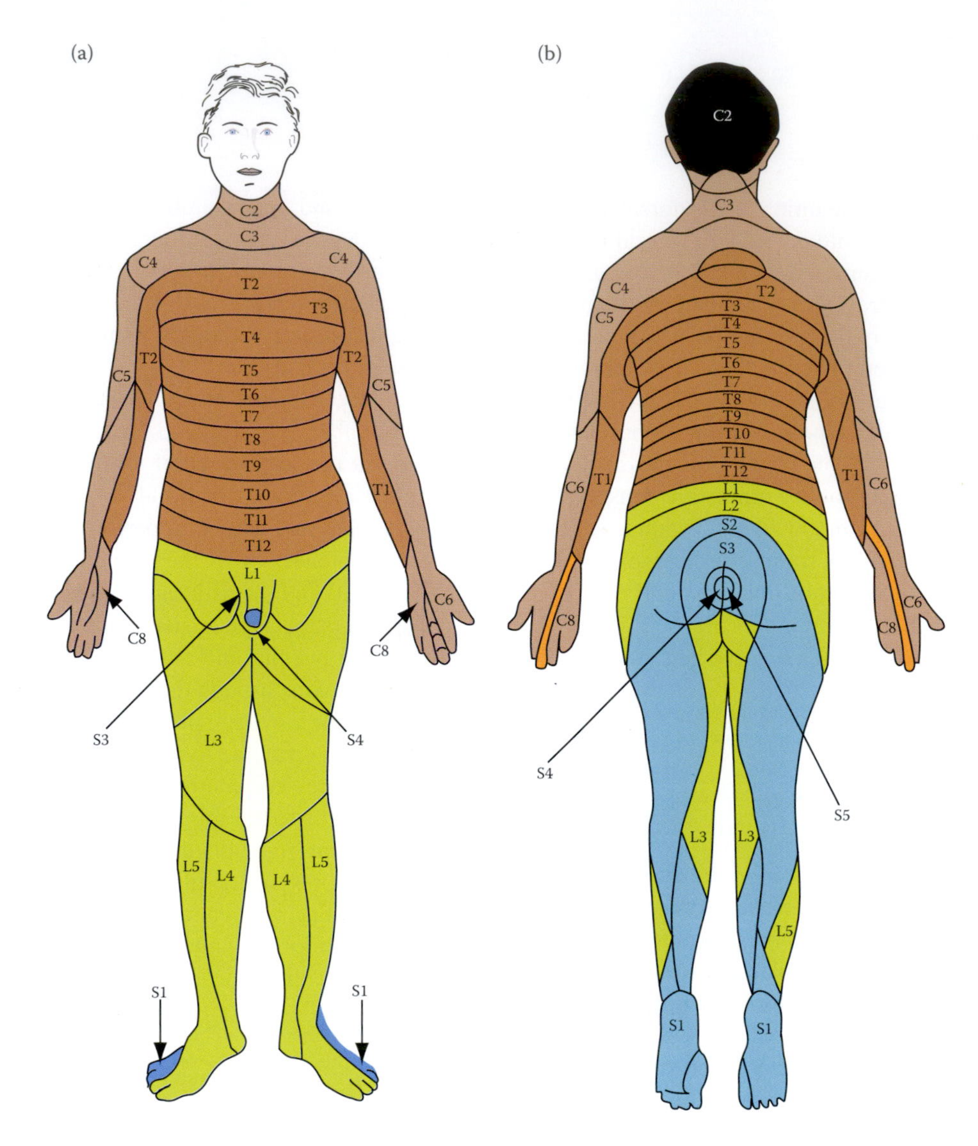

FIGURE 10.2 Dermatomes are mapped according to the band of skin innervated by dorsal roots of a single spinal segment. (a) Anterior view. (b) Posterior view.

Spinal nerves also give rise to recurrent meningeal branches that supply the spinal and cerebral dura mater, periosteum, blood vessels, posterior longitudinal ligament, as well as the intervertebral disks. Recurrence of pain in diseases associated with the vertebral column or spinal nerves can be attributed to irritation of these meningeal branches. Considerable overlap in the innervation of the spinal dura mater accounts for the perception of pain over several dermatomes upon irritation of a small area of the dura mater that receives a primary innervation via a single spinal nerve.

In general, the peripheral nerves are arranged in bundles or fasciculi that join together to form nerve trunks. The epineurium is a collagenous layer with variable amount of fat that covers and enwraps the nerve trunk. The fat content of the epineurium plays a protective role against injuries, and loss of this fatty layer may produce pressure palsies in bed-ridden chronic patients. Each fasciculus within a nerve trunk is encircled by the perineurium, a relatively thicker connective tissue sheath that exhibits epithelioid and myoid characteristics. The perineurium consists of collagen and cells derived from fibroblasts that continue with the coverings of the encapsulated receptors. Individual fibers are surrounded by the endoneurium, a loose connective tissue layer that is derived from the mesoderm. It consists of collagenous fibers, fibroblasts, Schwann cells, and endothelial cells that are immersed in a fluid, maintaining a higher pressure than the surrounding.

The pressure difference in the endoneurium, which is maintained by the perineurium, can be critical in preventing toxic contamination of the endoneurium. In addition, host neurons may send sprouts into the tubes formed by the endoneurium within a donor skin graft, allowing the reinnervation of the graft tissue to proceed.

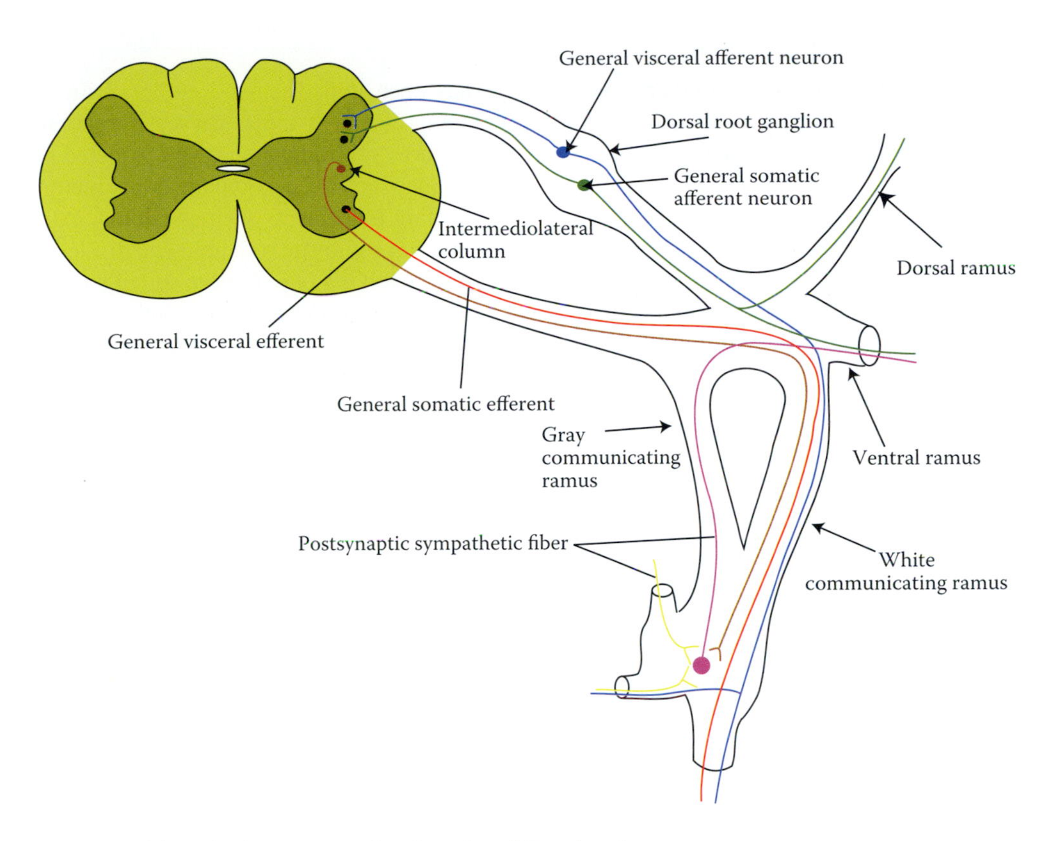

FIGURE 10.3 A diagram of the spinal nerve, associated rami, and functional components.

The nerve fibers and the connective tissue coverings are vascularized by intraneural capillaries called vasa nervorum. These vessels are categorized into extrinsic epineurial vessels and intrinsic longitudinal endoneurial microvessels. The unique course of these vessels may account for the relative resistance of peripheral nerves to ischemia. In addition to these coverings, nerve fibers may be ensheathed by myelin or remain unmyelinated. Thickness of the coverings also varies with the relative site of the nerve fiber and speed of conduction. Generally speaking, thickly myelinated fibers exist in high-pressure sites, innervating targets that require fast conduction. Unmyelinated nerve fibers are relatively thin and display a slower rate of conduction. They remain enfolded by the cytoplasm of the Schwann cells, in which each nerve fiber is enfolded by a single Schwann cell to form the Remak bundle. The unmyelinated fibers form the postganglionic autonomic fibers, the olfactory nerve, and the nociceptive C fibers. Following their formation, the spinal nerves divide into large ventral and smaller dorsal rami (Figure 10.3).

The ventral rami are mostly larger than the dorsal rami and supply the extremities and thoracic and abdominal walls. They form elaborate connections within the cervical, brachial, and lumbosacral plexuses. These connections enable one spinal segment to contribute to the formation of more than one spinal nerve and also make it possible for one muscle to be innervated by more than one spinal segment. The thoracic spinal nerves, for the most part, retain their segmental arrangement and form the intercostal nerves.

Pathological conditions involving the α motor neurons in the spinal cord, for example, poliomyelitis, or brainstem, or their axons (polyneuropathy), may produce spinal nerve dysfunctions via demyelination, followed by muscle denervation and paresis or paralysis. Paresis or paralysis may be preceded by visible involuntary contractions of the muscle fibers (fasciculation) involving a motor unit. Fibrillation involves a single muscle fiber, may not be visible through skin, and may only be detected through electromyography. The extent and severity of neuronal damage may determine the degree of dysfunction. On that basis, trauma or neurological diseases may produce disorders that are classified into neuropraxia, axonotmesis, and neurotmesis.

Neuropraxia is an incomplete, transient, and reversible loss of conduction without the loss of anatomic integrity. It results from transient ischemia or paranodal demyelination subsequent to severe compression. Neuropraxia may produce loss of deep tendon reflexes, and sensory dissociation with preservation of pain and thermal sensations, but with no detectable autonomic dysfunctions.

Axonotmesis refers to a complete interruption of an axon and its myelin sheath with preservation of the connective tissue stroma. It is characterized by immediate and complete loss of all sensory, motor, and autonomic functions. The degree of recovery of injured nerve fibers (commonly from a closed crushed injury) is dependent upon the length of the damaged segment and its distance from the innervated structure. The part of the axon distal to the site of injury undergoes Wallerian degeneration. After a latent period of approximately 1 month, downward-directed nerve regeneration may occur.

Neurotmesis refers to the complete anatomic disruption of neural and connective tissue elements of an axon. It is caused by injuries that penetrate nerves such as stab or gunshot wounds. Fibrosis and loss of continuity of endoneurial tubules render spontaneous recovery and regeneration almost impossible and neurosurgical repair a necessity.

The dorsal rami supply the skin and intrinsic muscles of the back and neck. They usually divide into medial and lateral branches (with the exception of the dorsal rami of the C1, C4, S5, and CC1 spinal nerves). In general, the medial branches of the dorsal rami of the cervical and upper thoracic spinal nerves provide primarily sensory while the lateral branches provide motor innervation. This pattern is reversed in the lower thoracic and lumbar spinal nerves. The dorsal rami of the sacral spinal nerves exit through the dorsal sacral foramina with the exception of the fifth sacral nerve, which divides into the medial and lateral branches. The dorsal ramus of the coccygeal nerve joins the lower two sacral dorsal rami to supply the coccygeal skin.

Spinal nerves may be affected in entrapment neuropathies, as a result of a localized injury or inflammation caused by mechanical irritation from impinging anatomical structure. Burning pain felt at rest associated with altered sensation is characteristic of these types of nerve injury. Injury to the spinal nerves or roots may also occur as a result of herniated intervertebral disks, tumors, osteoarthritis, spina bifida cystica, or cauda equina syndrome. These clinical conditions are generally dependent upon the extent of damage and the number of affected roots or nerves. Nerve root compression, for example, as a result of disk prolapse, commonly occurs at sites where the vertebral column is most mobile. The lower cervical and lower lumbar vertebrae are the frequent sites of root compression. Paresthesia or pain may result from compression of the dorsal roots.

Trauma to soft tissue or bony fracture that damages a spinal nerve can also injure the sympathetic fibers that accompany these nerves, resulting in burning pain sensation in a wider territory than the area of distribution of the affected spinal nerve (causalgia) accompanied by autonomic disturbances such as sweating and vasoconstriction (reflex sympathetic dystrophy). This topic has been will discussed earlier. Pain associated with compression of spinal nerves is generally confined to the area of distribution of the affected nerves and may or may not be accompanied by motor dysfunctions. Certain movements such as flexion, extension, or rotation can aggravate root pain due to a lesion or prolapsed disk involving one or more spinal roots. Since the ventral rami are the primary contributors to the cervical, brachial, lumbar, and sacral plexuses, a detailed discussion of these plexuses will be beneficial at this point.

CERVICAL SPINAL NERVES

There are eight cervical spinal nerves that divide into dorsal and ventral rami. The first cervical spinal nerve forms the suboccipital nerve, which runs in the suboccipital triangle and provides innervation to the rectus capitis posterior major and minor and the inferior and superior capitis oblique muscles. The medial branch of the dorsal ramus of the second cervical spinal nerve forms the greater occipital nerve, which encircles the inferior oblique muscle and ascends to supply the skin of the posterior scalp as far as the vertex. The ventral rami of the cervical spinal nerves form the cervical plexus and contribute partly to the brachial plexuses.

C3 and *C4 roots* contribute to the formation of the phrenic (primarily C4), supraclavicular, great auricular, and transverse cervical nerves. Lesions of C3 and C4 roots produce motor deficits associated with the diaphragm and pain or hypalgesia in the corresponding dermatomes. Referred pain in these dermatomes can reflect pathological processes involving the diaphragm (subphrenic abscess) and gallbladder (cholecystitis). Diaphragmatic motor deficits produce a paradoxical mobility during inspiration due to circumscribed relaxation of the affected area of the muscular diaphragm. Paralysis of the left diaphragm may trigger gastrocardiac manifestations (*Roemheld syndrome*). In this syndrome, irritation of the vagus nerve produces bradycardia and hypotension triggering activation of the autonomic system. The latter leads to constriction of the coronary arteries, causing angina pectoris and pain felt in the left arm, forearm, and fifth digit.

Injury to C5 root results in loss of biceps brachii reflex and pain over the upper part of the deltoid. Deltoid dysfunction and biceps brachii palsy become more evident when C6 is also involved in the injury.

Compression of C6 root can produce impairment of the biceps and brachioradialis muscle functions and biceps hyporeflexia or areflexia. Pain starts in the posterior arm from the deltoid muscle and continues downward to the radial side of the forearm and into the thumb. Manifestations of musculocutaneous nerve palsy may mimic these deficits; however, brachioradialis palsy and sensory deficit in the thumb will not be seen.

A lesion of C7 is most likely to cause pain that radiates to the middle finger, medial side of the index, and lateral side of the fourth digit on both the palmar and dorsal surfaces of the hand. Paresis of the triceps brachii and consequently diminished or absence of triceps reflex, atrophy of thenar muscles (C7 and C8), and pronator teres also occur. Some degree of weakness in the long flexors of the digits and midportion of the pectoralis muscle is also readily visible. Areas of pain, analgesia, or hypalgesia usually extend as a band to the mid dorsal surface of the forearm and lateral surface of the upper arm.

When C8 root is involved in a lesion, the deficits are confined to the fourth and fifth digits, hypothenar muscles, and sensory impairment on the ulnar side of the forearm and hand. Paresis of triceps brachii and associated hyporeflexia is also seen. Despite similarities of manifestations between ulnar nerve and C8 lesion, C8 radicular injury does not distinctly produce individual muscle paresis or atrophy as is the case with ulnar nerve. Horner syndrome is less likely to occur due to the fact that the sympathetic presynaptic fibers to the ipsilateral head emanate primarily from T1 spinal segment.

Cervical disk herniation is less common than the lumbar counterpart. It commonly occurs in the lower cervical disks between the fifth and sixth, and also between sixth and seventh cervical vertebrae. This is due to relative free mobility of the cervical part of the vertebral column, and that C6 acts as a fulcrum for cervical movements. Compression fractures of C5, C6, and C7 can occur as a result of diving in shallow waters. Degeneration of the intervertebral disks (cervical spondylosis) results in motor deficits and pain in the arm and shoulder as well as the neck. Arm pain is a common manifestation of cervical spine disorders and is associated with numbness and tingling that radiates to the tips of the digits. Pain can be exacerbated by provocative activity like sneezing or coughing. Prolapse of the intervertebral disk between the fifth and sixth vertebrae is most likely to compress the sixth cervical root. Prolapsed disk may protrude centrally and compresses the spinal cord, producing combined signs of upper motor neuron palsy and sensory deficits in the lower extremity. If the prolapse is large, it may also compress the anterior spinal artery resulting in ischemia and degeneration of the anterior 2/3 of the spinal cord (Beck syndrome). In Beck syndrome, the dorsal white columns are not affected, and as a result, the fine touch, vibratory sense, and conscious proprioception are preserved. Prolapse of the cervical intervertebral disk can occur in rear-end collisions or as a result of strong head turning when driving backward producing whiplash injuries. Using the arms in front of the body as in ironing and dish washing may also precipitate this type of herniation. Cervical disk prolapse can lead to bilateral radicular pain. Most herniation heals within weeks by the shrinkage and fibrosis of the protruding nucleus pulposus. The small diameter of the cervical intervertebral foramina and their locations near the zygapophyseal joints make them more prone to compression by the herniated disks.

CERVICAL PLEXUS

The cervical plexus (Figures 10.4 and 10.5) is formed by the ventral rami of the upper four cervical nerves, with a small contribution from the fifth cervical segment. It lies deep to the internal jugular vein and anterior to the middle scalene muscle. This plexus gives rise to sensory and motor nerves. It also provides segmental motor innervation to the geniohyoid, rectus capitis anterior and lateralis, longus capitis, and longus colli muscles. This plexus gives rise to ansa cervicalis, phrenic, lesser occipital, great auricular, transverse (colli) cervical, and supraclavicular nerves.

TERMINAL BRANCHES

The *ansa cervicalis* (C1, C2, C3) is a nerve loop formed by the union of the ventral ramus of the first cervical spinal nerve (superior root or descendens hypoglossi) and the ventral rami of the second and third cervical spinal nerves (inferior root or descendens cervicalis). The ansa cervicalis pierces the carotid sheath and runs superficial to the internal jugular vein, innervating the infrahyoid (strap) muscles (omohyoid, sternohyoid, and sternothyroid), with the exception of the thyrohyoid muscle, which is innervated by the ventral ramus of the first cervical spinal nerve.

The *phrenic nerve* (C3, C4, C5) is derived from the ventral rami of the third, fourth, and fifth cervical spinal nerves, with the largest contribution coming from the fourth cervical spinal segment. Frequently C3 and rarely C5 will be the main contributors to this nerve. This nerve descends posterior to the prevertebral fascia and anterior to the anterior scalene muscle. It courses within the superior and middle mediastina, between the mediastinal pleura and fibrous pericardium. This nerve runs anterior to the pulmonary root, separating it from the vagus nerve. In addition to providing motor innervation to the diaphragm, the phrenic nerve also contains sensory fibers that supply the central part of diaphragmatic pleura and diaphragmatic peritoneum, pericardium, mediastinal pleura, and the hepatic plexus.

Transection of the C4 segment, neuralgic amyotrophy, or injuries of the upper brachial or cervical plexus can compromise the function of the phrenic nerve. Other causes include surgical repair of an esophageal atresia or tracheoesophageal fistula, placement of drainage thoracic tube and, nerve block performed in the supraclavicular region. Phrenic nerve palsy can occur in Recklinghausen's disease of the apex of the lung, thymomas, lymphomas, and bronchial carcinoma. Pathological conditions involving the mediastinum such as diaphragmatic pleura and peritoneum or the gall bladder result in pain radiating to the dermatomes of third, fourth, and fifth cervical spinal nerves, which correspond to the back and upper part of the shoulder. Paralysis of the hemidiaphragm can result from excision

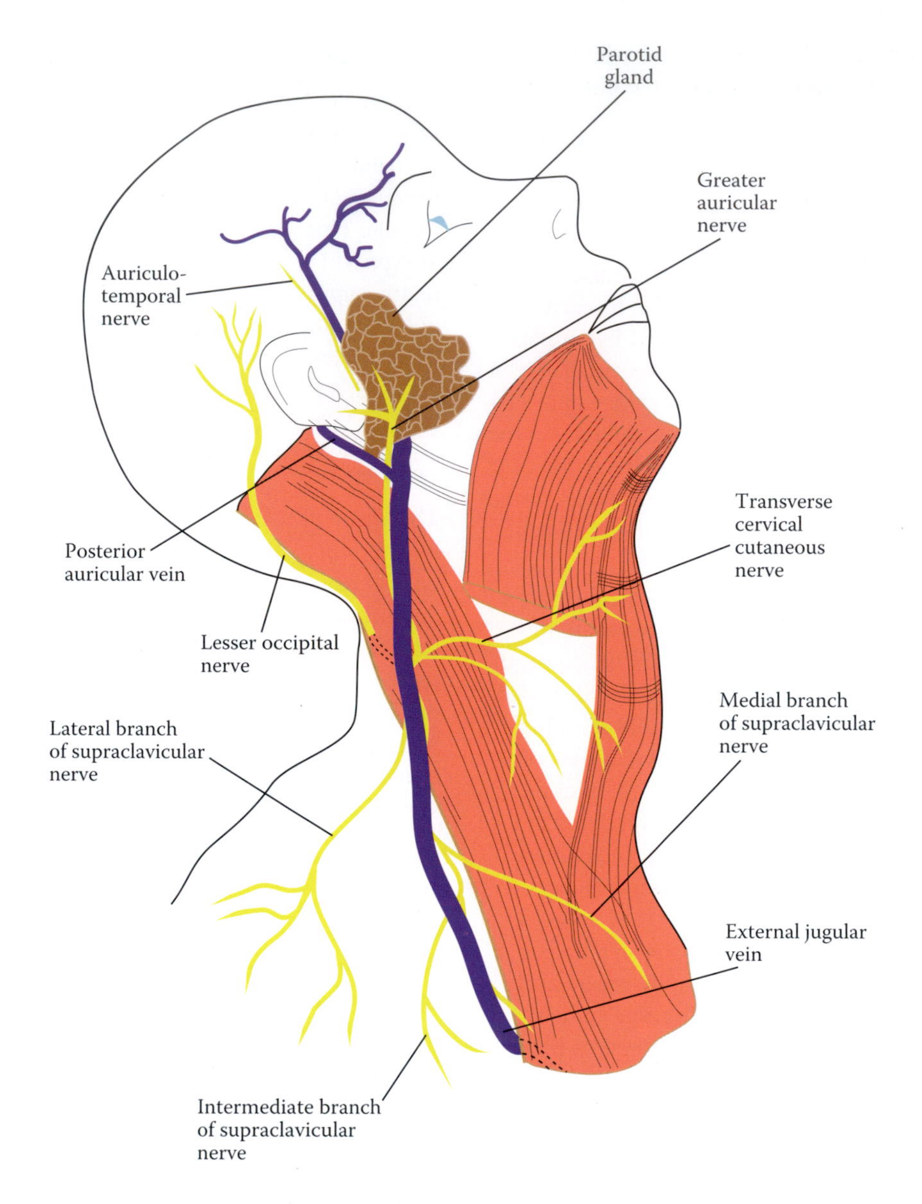

FIGURE 10.4 Schematic diagram of the superficial branches of the cervical plexus.

of the phrenic nerve in the neck unless an accessory phrenic nerve exists. Unilateral palsy may not be easily detected and require careful chest examination during expiration and inspiration supplemented by radiographic imaging. Sniff test and movement of the diaphragm during deep inspiration are also helpful in the diagnosis.

The *accessory phrenic nerve* is frequently derived from the fifth cervical spinal nerve. The superficial branches of the cervical plexus (listed below) exit at the midpoint of the posterior border of the sternocleidomastoid muscle (SCM) accompanied by the spinal accessory nerve.

The superficial branches of the cervical plexus (listed below) exit at the midpoint of the posterior border of the SCM accompanied by the spinal accessory nerve.

Taking this fact into consideration, complete cervical nerve block in radical neck dissection can be achieved by the administration of anesthetics into the midpoint of the posterior border of the SCM (Erb's point).

The *lesser occipital nerve* (C2) curves around the sternocleidomastoid muscle and supplies the upper part of the medial surface of the ear and the adjacent area of the posterior scalp.

The *greater auricular nerve* (C2–C3) ascends toward the parotid gland, accompanied by the external jugular vein, carrying sensation from the facial skin that covers the parotid gland, mastoid process, and the ear lobule. It is the only cutaneous nerve to the face that is not derived from the trigeminal nerve.

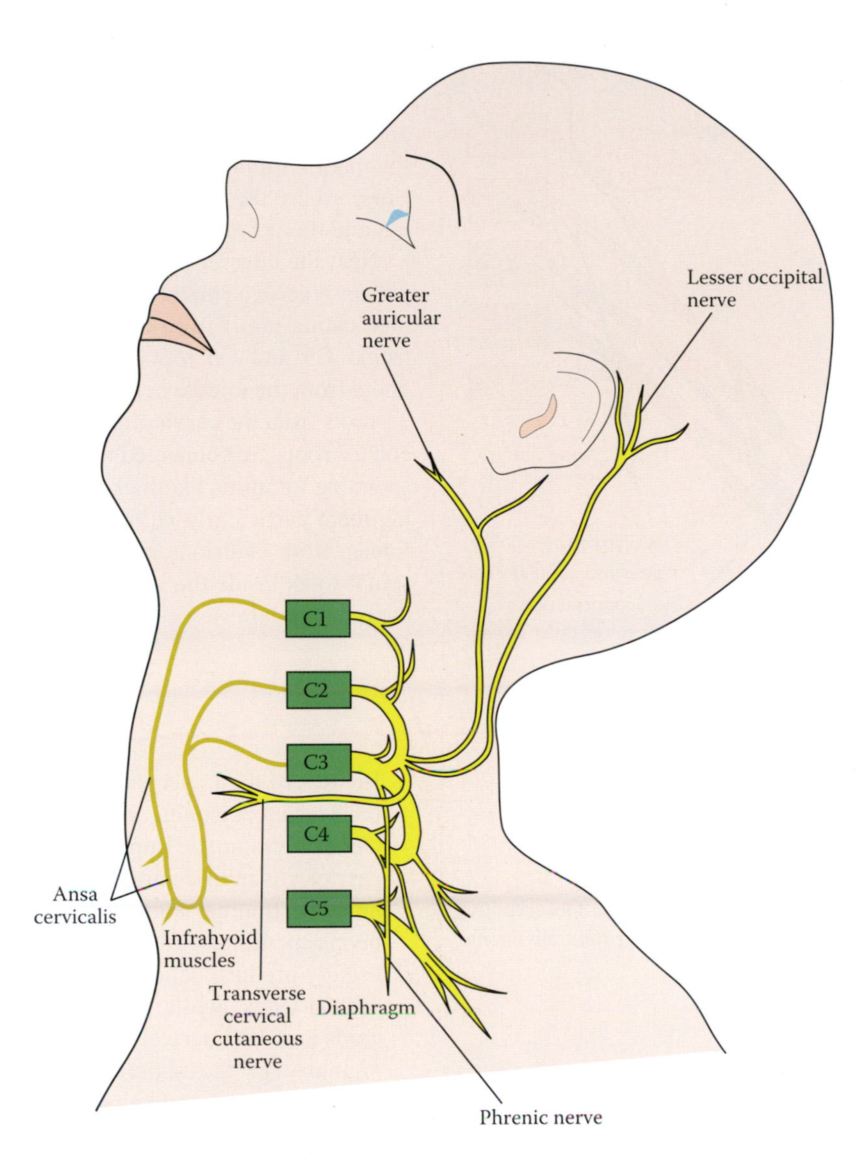

FIGURE 10.5 This diagram illustrates the components of the cervical plexus and its area of distribution.

The *transverse (colli) cervical nerve* (C2–C3) arises from the ventral rami of the second and third cervical spinal segments, supplying cutaneous fibers to the anterior and lateral neck.

The *supraclavicular nerves* (C3–C4) are derived from the ventral rami of the third and fourth cervical spinal nerves and descend deep to the platysma. They divide into lateral, intermediate, and medial branches, supplying the lower neck and the upper part of the anterior thorax.

BRACHIAL PLEXUS

The brachial plexus (Figure 10.6) is formed by the union of the ventral primary rami (roots) of the C5–T1 spinal nerves (with a small contribution from the ventral ramus of the C4 spinal nerve). This plexus lies in the posterior triangle of the neck, posterior to the clavicle, and between the anterior and middle scalene muscles. Because of the variable contributions of C4 and T2 ventral rami, the brachial plexus is classified into prefixed and postfixed types. Prefixed refers to a plexus formed by the ventral rami of C4–C8, and postfixed identifies the type of plexus in which C5 contribution is reduced and is formed by the ventral rami of C6–T2. Commonly, the ventral rami of C5 and C6 join at the lateral border of the middle scalene to form the superior trunk. The ventral ramus of C7 continues as the middle trunk, whereas the ventral rami of C8 and T1 spinal nerves unite behind the anterior scalene to form the inferior trunk. In the axilla, each trunk, posterior and superior to the clavicle, divides into anterior and posterior divisions. Union of the posterior divisions results in the formation of the posterior cord. Lateral to the axilla, the anterior divisions of the upper and middle trunks join and form the lateral cord, while the anterior division of the inferior trunk continues as the medial cord. In the axilla, the posterior and lateral cords lie on the lateral side of the axillary artery, while the medial cord posterior to the axillary artery. The roots of the brachial plexus give rise to the dorsal scapular and long thoracic nerves. The superior trunk gives origin to the suprascapular nerve; the lateral cord provides the musculocutaneous and lateral pectoral nerves, as well as the lateral root of the median nerve (Table 10.1).

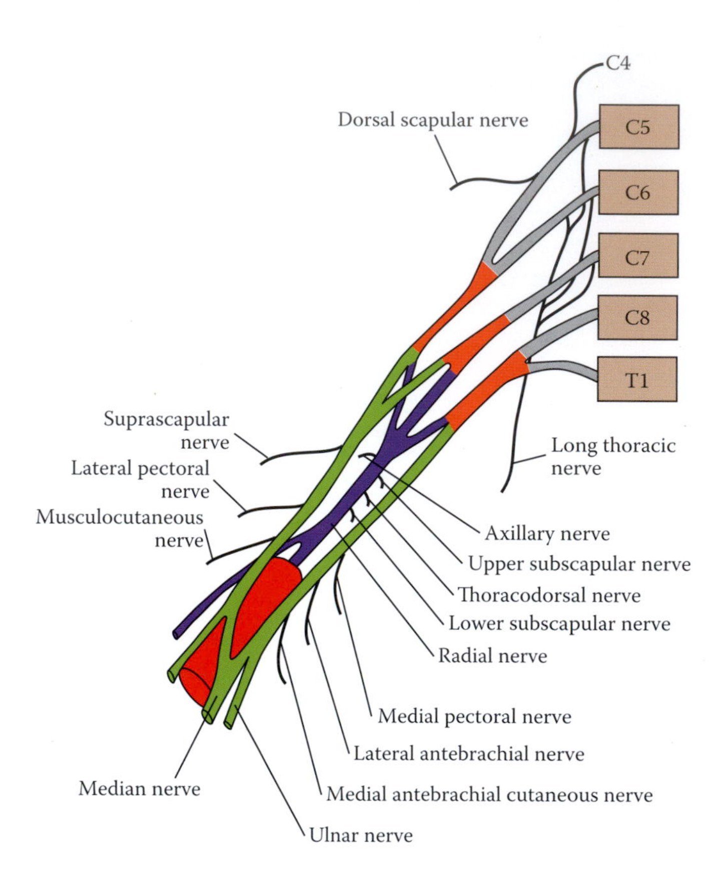

FIGURE 10.6 Formation of the brachial plexus. Observe the segmental contribution of the spinal cord to the trunks, divisions, cords, and peripheral branches.

TABLE 10.1
Summary

5	Roots	Ventral rami of C5–T1; give rise to the dorsal scapular and long thoracic nerves
3	Trunks	Superior trunk (C5 and C6); give rise to the suprascapular nerve and nerve to the subclavius, middle trunk (C7), and inferior trunk (C8 and Tl)
6	Divisions	Three anterior and three posterior
3	Cords	Lateral, posterior, medial
16	Branches	
	Posterior cord	Axillary, radial, thoracodorsal, upper, and lower subscapular nerves
	Lateral cord	Musculocutaneous and lateral pectoral nerves, and the lateral root of the median nerve
	Medial cord	Ulnar, medial brachial, medial antebrachial, medial pectoral, and medial root of the median nerve
		Medial pectoral, medial brachial and antebrachial cutaneous, and ulnar nerves, and medial root of the median nerve.

Note: The roots (ventral rami), trunks, divisions, cords, and branches of the brachial plexus may be simplified by the following mnemonic: Robert Taylor Drinks Cold Beer.

Numerous branches arise from the medial cord, which include the medial pectoral, ulnar, and medial brachial and antebrachial cutaneous nerves, the nerve to subclavius, and the medial root of the median nerve. Branches of the posterior cord are the axillary, radial, upper and lower subscapular, and thoracodorsal nerves.

Near the intervertebral foramina, the roots of the brachial plexus receive sympathetic postsynaptic fibers (gray communicating rami) from the cervical part of the sympathetic trunk. C5 and C6 roots receive sympathetic postsynaptic fibers from the middle cervical ganglion and the C7, C8, and T1 roots from the cervicothoracic (stellate) ganglion. C5, C6, and C7 roots are connected to the outer margins of the corresponding foramina via thickening of the epineurium, rendering them particularly vulnerable to avulsions due to traction forces. Root evulsions can produce ischemic injury of the spinal roots inside the intervertebral foramina by disrupting the radicular arteries.

Trunk Injuries

Injuries to the upper part of the plexus, depending on their location, extent, and types, are commonly associated with traumatic impact and gunshot wounds. Lower part plexus injuries are caused by regional lymphadenopathy, trauma, and fibrous tissue formation or by neoplastic lesions. Injuries sustained above the clavicle usually produce deficits associated with the upper and middle trunk, while infraclavicular injury affects the cords and their derivatives (individual nerves).

Injury to the superior trunk (Erb–Duchenne or Rucksack palsy) can be caused by an injury to C5–C6 due to hyperextension of the neck, which increases the angle between the shoulder and neck, or undue pull on the suprascapular nerve that anchors to the margins of the suprascapular notch. A fall from a motorcycle, a careless forceps delivery in breech position, heavy or maladjusted backpack traction on the head may also precipitate this type of injury. In cleidocranial synostosis (absence of presence of a rudimentary clavicle), the upper trunk is exposed and prone to injury.

Pressure from the rope that wraps around the shoulder in mountain climbers may also injure the upper trunk. Sustained pressure is generated by the strap of the backpack. Upper trunk injury may also occur in coma induced by the consumption of large doses of hypnotics. Patients with Parkinson's disease may also develop upper trunk palsy due to the effect of long-standing muscle rigidity and blockage. Abducted position of the arm during anesthesia, restrained abducted arms in restless patients, presence of shoulder support on the operating table while the patient is in Trendelenburg position (supine with the head lower than the feet), and use of curare drugs and muscle relaxants may predispose the patient to upper trunk injuries.

Patients present with an adducted (deltoid and supraspinatus are inoperative), extended, and medially rotated arm hanging limply at the side (coracobrachialis, infraspinatus, rhomboids, deltoid, and teres minor are impaired or weakened). The forearm is extended and pronated (biceps brachii and brachialis are nonfunctional), and the wrist is slightly flexed, forming the typical configuration of Waiter's tip hand (Figure 10.7). Due to the extensive overlap between the cutaneous fibers of the contiguous nerves, sensory loss associated with this type of injury will be confined to a small region of the shoulder. However, transection of the C5 and C6 roots can produce a larger area of sensory loss.

Hereditary neuralgic amyotrophy is a unilateral or bilateral autosomal dominant demyelinating disease associated with SEPT9 gene that affects more commonly males between the ages 20 and 30. It is considered a self-limiting condition, and it may take months to a year until symptoms completely disappear. Patients experience unilateral severe sharp pain in the outer sphere of the shoulder and the upper arm, commonly on the right side, unresponsive to conventional analgesics. Pain is usually followed shortly by selective muscular wasting in the shoulder muscles and less commonly the arm muscles. Abduction of the arm and flexion of the forearm can reduce the pain; conversely, extension of the elbow and lateral rotation and abduction of the arm exacerbate the pain. This episodic disease causes neuropathy commonly in C5 and C6 and some in the branches of the brachial plexus, particularly the long thoracic nerve. Axillary, radial, phrenic, and accessory nerves, though rarely, may be affected. This condition may be precipitated by vaccination, child delivery, exposure to low temperatures, and bacterial or viral infections. Other branches such as the median and radial nerve may also be affected. The disease shows relapse and remission, and complete recovery does not occur.

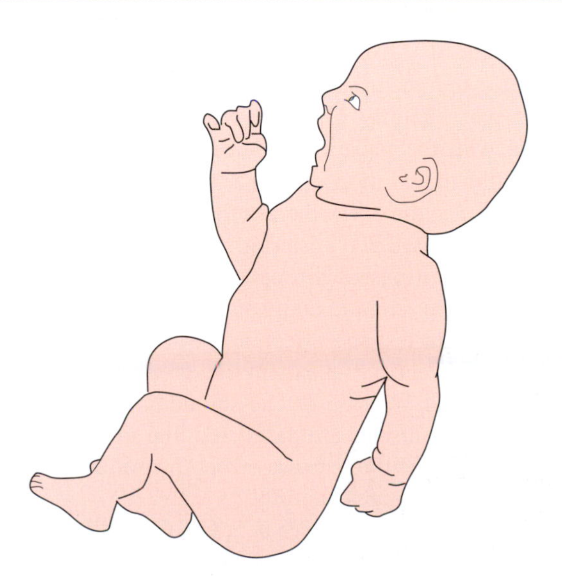

FIGURE 10.7 This diagram illustrates the manifestations of Erb–Duchenne palsy.

Injury of the middle trunk is uncommon, though when it occurs, it usually compromises the segments that contribute to the radial nerve with exception of the branch that supplies the brachioradialis (C5 and C6). As a result, very limited sensory diminution in the posterior surface of the forearm and radial part of the dorsum of the hand occurs.

Injury to the inferior trunk may be caused by excessive abduction of the arm, which may occur in an individual who clutches to an object while falling from a height (Dejerine–Klumpke palsy). It also occurs during a difficult breech delivery (birth palsy or obstetric paralysis), or upon a sharp angulation of the inferior trunk over the cervical rib (cervical rib syndrome), or when the person has been dragged by the arm. Abnormal insertion or spasm of the anterior and middle scalene muscles (scalene anterior syndrome), anomalous fibrous band that extends to the scalene tubercle of the first rib, and thoracic surgical procedures that require a sternal split can endanger the inferior trunk. Pressure exerted by expanding Pancoast tumor of the apex of the lung or metastatic breast cancer that involves the supraclavicular lymph nodes can produce lesion of the upper trunk. This condition is characterized by progressive weakness and eventual paralysis of the intrinsic muscles of the hand, especially the thenar muscles and first dorsal interosseous. Long flexors of the hand and the digits may or may not be affected. As a result, Klumpke palsy exhibits "claw-hand" configuration with hyperextension of the metacarpophalangeal joints and hyperflexion of the distal and proximal interphalangeal (IP) joints. Pain is usually severe and/or paresthesia is almost always felt along the medial border of the forearm that extends to the hand and medial two digits. Pain and paresthesia are exacerbated by carrying suitcases. Horner's syndrome (ptosis, miosis, and anhidrosis) may also be seen in this condition due to involvement of the cervicothoracic (stellate) ganglion.

When Klumpke palsy is caused by a fully developed or rudimentary cervical rib or by an anomalous fibrous band that extends from a long cervical transverse process of C7 to the first rib, the subclavian artery and vein are also compressed in conjunction with the inferior trunk. This results in combined neuronal and vascular disorders collectively known as thoracic outlet syndrome (TOS). Fully developed cervical rib with the associated neurovascular bundle produces a palpable protrusion in the supraclavicular fossa. The sharp edge of the anomalous fibrous band is most likely to cause symptoms than the mere presence of cervical rib even if it is fully developed. TOS may also be associated with constriction of the space between the clavicle and first rib subsequent to congenital or traumatic clavicular deformities, excessive abduction above 90°, downward pull of the scapula as a result of weakened or atrophic scapular and trapezius muscles, overweight, poor posture, and abnormalities in the insertion of the anterior and middle scalene muscles. In about one third of patients with this syndrome, a scalene minimus exists that originates from the transverse process of C7 and inserts into the first rib between the attachments of the anterior and middle scalene muscles.

TOS is commonly seen in women in their second to fourth decades of life. It usually exhibits paresthesia and/or pain in the medial forearm, medial hand, and the entire fifth digit that extends to the anterior and posterior thoracic walls. Pain is usually diffuse and vague in nature and position-dependent and can be triggered by carrying heaving objects. Paresthesia and hypesthesia can lead to complete anesthesia. Concomitantly, some degree of motor deficits (50% of patients) confined to the hand can also be seen gradually in these patients in the form of poor manual skills. There is some similarity between pronator teres syndrome and TOS, though in the former, pain and paresthesia are confined to the lateral side of the hand. Vascular deficits may not always be apparent until provocative maneuvers are employed, such as Adson's test and Wright's maneuver. Extending and turning the head toward the affected side on deep inspiration of a seated patient produce a weaker radial pulse (Adson's test). This maneuver induces contraction of the anterior and middle scalene muscles and produces narrowing of the interscalene space. Swelling of the arm subsequent to compression of the subclavian vein and reduced pulse, coldness, and cyanosis in the arm can be a sequela of subclavian artery compression. Reduction of blood flow within this vessel can also be attributed to irritation and then activation of the sympathetic plexus that surrounds the vessel, which mimics Raynaud syndrome. Rarely, a clot may form in the subclavian artery and travel to the digital arteries to cause pain and signs of ischemia. A stenotic bruit could be detected over the subclavian artery in the supraclavicular fossa due to kinking of this vessel over the rib, and the radial pulse is lost upon arm extension and abduction.

Some of the above manifestations are seen in costoclavicular syndrome, a rare condition in which the space between the clavicle and first rib is reduced, and the subclavian vessels and inferior trunk of the brachial plexus are affected. Sloping (droopy) shoulder, which increases with age, deformed clavicle or first rib, and widening of the superior thoracic aperture, can collectively produce manifestations of costoclavicular syndrome. This syndrome can be tested by asking the patient to move his/her shoulder backward and then downward while the examiner checks the radial pulse at the wrist and auscultates the subclavian vessels in the lower part of the posterior triangle of the neck.

Terminal Branches of the Roots

The dorsal scapular nerve (C5) (Figure 10.6) arises primarily from the ventral ramus of the fifth cervical spinal nerve and courses posteriorly through the posterior triangle of the neck toward the levator scapula and upper angle of the scapula after it pierces the middle scalene. En route, it innervates the levator scapula, as well as the major and minor rhomboids, accompanied by the deep branch of the dorsal scapular artery.

Traction injuries as a result of motorcycle accident, penetrating wounds, and avulsion of the brachial plexus can injure this nerve. Overactivity and fibrosis of the middle scalene muscle as a result of ischemic hypertrophy may damage the dorsal scapular nerve. Entrapment of this nerve within the middle scalene muscle hinders its ability to accommodate changes in position during movement of the head and arm. These movements exacerbate the preexisting weakness, atrophy, and pain in the rhomboids and levator scapula muscles. As a result of the muscular weakness, patients may exhibit impaired scapular retraction and weakened displacement of the flexed elbow when the hand is placed on the hip. The scapula deviates slightly laterally, and the medial border is pulled away from the thoracic wall (partial winging of the scapula). This inferior angle protrusion is abolished when the arm is abducted, which is in contrast to serratus anterior palsy. Attempts to correct the position of the shoulder as a result of this deficit produce deviation of medial border in that the upper part displaces medially and the lower part laterally. However, the motor deficit associated with the levator scapula is not prominent due to partial innervation of this muscle by the cervical plexus. Since the nerve pierces the middle scalene during its downward and posterior course, tenderness over the nerve as it passes through the muscle is a significant point that can aid confirmation of the injury. The examiner can test the affected muscles by pulling on the flexed forearm of a patient in a prone position with one hand and pressing on the shoulder with the other hand. If the muscles are paralyzed, the medial border of the scapula will move away from the midline and the inferior angle rotates laterally.

The *long thoracic nerve* (C5, C6, C7) (Figure 10.6) arises from the ventral rami of the C5, C6, and C7 spinal nerves, though C7 contribution may be absent. Initially, the upper two roots (C5, C6) pierce the middle scalene muscle and later unite with the lower root from the seventh (C7) cervical spinal segment that emerges between the anterior and middle scalene muscles. After the union of the contributing roots, the long thoracic enters the medial side of the axilla and then descends on the lateral surface of the serratus anterior muscle, which supplies, accompanied by the lateral thoracic vessels. This muscle works with the rhomboids to keep the medial edge of the scapula on the thoracic wall. The upper fibers pull the scapula anteriorly, whereas the lower fibers are particularly effective in bringing the scapula downward and anteriorly and the inferior angle laterally and inferiorly.

Proximity of the long thoracic nerve to the axillary lymph nodes, course within the middle scalene, and its vertical exposed location superficial to the serratus anterior muscle and lateral to the mammary gland account for the vulnerability of this nerve to injury in radical mastectomy. Carrying heavy objects (backpack) on the shoulder or entrapment within the middle scalene and violent shoulder movements as with using a heavy sledge hammer can compromise the integrity of this nerve. Men and particularly the dominant extremity are prone to this type of nerve injuries. Neuralgic amyotrophy and also Rocky Mountain spotted fever, diphtheria, and typhoid fever can produce long thoracic nerve palsy. Unlike the dorsal scapular nerve, entrapment of this nerve within the middle scalene muscle does not produce pain in the upper extremity. In this type of palsy, the deficits will not readily be obvious when the arms are at the side of the body. Because of serratus anterior muscle's role in keeping the scapula close to the thoracic wall, its atrophy results in displacement of the medial border of the scapula away from the thoracic wall, and protrusion of the inferior angle toward the midline (winged scapula), which becomes evident upon protraction and remains the main characteristic of long thoracic nerve dysfunction. Medial rotation of the scapula and slight posterior and downward displacement of the lateral end of the clavicle also occur. This is due to the inability of the muscle to hold the scapula against the thoracic wall and the unopposed action of the rhomboids. Weakened protraction and lateral rotation of the scapula are also common manifestations of this condition.

Terminal Branches of Superior Trunk

Suprascapular nerve (C5, C6) (Figure 10.6) arises from the ventral rami of C5 and C6 and sometimes from C4 and travels posterolaterally deep to the clavicle and trapezius and omohyoid muscles accompanied by the suprascapular vessels. It enters the supraspinous fossa via the suprascapular foramen, inferior to the supraspinous ligament, and supplies the supraspinatus muscle. Then, it descends through the spinoglenoid notch to gain access to the infraspinous fossa to innervate the infraspinatus and provide articular branches to the shoulder and acromioclavicular joints. In some individuals, the suprascapular nerve is both motor and sensory with cutaneous branches that supply the upper arm.

Although the suprascapular nerve is rarely damaged, injuries to this nerve may occur as a result of fibrosis and subsequent narrowing of the suprascapular foramen and/or in hereditary neuralgic amyotrophy (discussed earlier). Presence of cyst, mass, and ganglion near the suprascapular foramen can also compress the nerve. Entrapment can also be caused by the spinoglenoid ligament that extends from the lateral border of the spine of the scapula to the shoulder joint capsule or the scapular neck. In males, this ligament is predominant with extensive variations compared half of females who exhibit it. Shoulder trauma, traction, or compression injuries of this nerve can be seen in wrestlers, pitchers, volleyball players, and windsurfers. Rupture of the rotator cuff and shoulder dislocation is most likely to contribute to suprascapular nerve dysfunction. Fixation within the suprascapular foramen in the immobilized upper extremity of an individual with frozen shoulder and repeated compensatory motion of the scapula also endanger this nerve. Frozen shoulder, which commonly affects women over the ages of 50, occurring subsequent to postsurgical immobilization can progress to cause contracture of the muscles around the shoulder joint and possible rupture accompanied by trophic changes in the hand. Traction exerted on this nerve may eventually produce a pull on the upper trunk of the brachial plexus leading to Erb's palsy. This nerve can also be damaged in Colle's fracture as a result of downward movement of the thorax with the scapula fixed and the arm extended. It is seen in activities that require excessive arm abduction as in lifting weight or placing belongings to shelves and then bringing them down.

An injury to the suprascapular nerve produces atrophy of the supraspinatus and infraspinatus muscles and associated weakness of the lateral rotation (arm will be in a pronated position near the chest wall) and abduction of the arm up to 15° starting from the midline. Thus, initiation of abduction of the arm from the vertical position next to the chest becomes impossible. Inability to use the hand on the affected side to comb the back of the head and, when writing, a constant need to displace the medium being written upon away from the affected side toward the intact side are some of the signs that a patient experiences as a result of suprascapular nerve palsy. Pain sensation is confined to the posterior shoulder. Compression at the spinoglenoid notch spares the supraspinatus but affects the infraspinatus; thus, initiation of abduction is preserved. Atrophy and flattening of the supraspinous and infraspinous fossae in the elderly may not be diagnostic due to wasting of these muscles with aging. The supraspinatus muscle can be tested by abducting the arm against residence while the patient's neck is flexed to the ipsilateral side and head to the contralateral side. It can also be tested through cross-abduction, a maneuver performed by asking the patient to place his/her hand on the affected side on the opposite intact shoulder, abducting the affected arm up to 90°, followed by the examiner's pull on the affected elbow toward the intact side. This will elicit a severe pain if the suprascapular nerve is entrapped, particularly inside the suprascapular foramen.

The *nerve to the subclavius* (C5, C6), as the name implies, supplies the subclavius muscle, which acts as a cushion that prevents rupture of the subclavian artery in clavicular fracture.

Terminal Branches of Lateral Cord

The *musculocutaneous nerve* (C5, C6, C7) (Figures 10.6 and 10.8) is formed by the ventral rami of the fifth, sixth, and seventh spinal nerves near the lower border of the pectoralis minor. It pierces the coracobrachialis muscle, runs between the biceps brachii and brachialis where it lies close to the radial nerve, and continues to the forearm, lateral to the tendon of biceps brachii as the lateral antebrachial cutaneous nerve. The musculocutaneous nerve supplies the flexors of the elbow such as coracobrachialis, brachialis, and biceps brachii and provides articular branches to the elbow joint. It also provides cutaneous innervation to the lateral side of the forearm via the lateral antebrachial cutaneous nerve. The branch to the coracobrachialis arises earlier, whereas branches that supply the biceps brachii and brachialis emanate from the nerve after piercing the coracobrachialis.

Injury to the musculocutaneous nerve, although rare, results from a fracture of the humerus, shoulder dislocation, positioning of the arm during surgery, entrapment inside a hypertrophied coracobrachialis muscle, and from hereditary neuralgic amyotrophy. Common manifestations of this injury are atrophy of the biceps brachii and flattening of the anterior surface of the arm, weakened flexion of the arm, markedly weakened flexion of the forearm, weakened supination, and instability of the shoulder joint. When the forearm is supinated, flexion at the elbow becomes almost impossible. Instability of the shoulder joint may occur as a result of atrophy of the long head of biceps brachii and coracobrachialis, leading subsequently to separation of the head of the humerus from the glenoid cavity. Pain and paresthesia exacerbated by extension of the forearm at the elbow and eventual impairment of the cutaneous sensation in the lateral half of the forearm can also, though uncommonly, be observed in this injury. Heavy objects placed on the forearm and supported by the elbow may particularly compress the lateral antebrachial cutaneous branch of this nerve. Due to the overlap between sensory branches of the lateral antebrachial cutaneous nerve and the superficial branches of the radial nerve that supply the lateral forearm, sensory deficits may not be prominent. For the same reason, damaged lateral antebrachial nerve can be surgically removed and used as a graft. In order to test the biceps brachii and coracobrachialis, the patient's forearm is flexed against resistance while supinated. Coracobrachialis can be tested by flexing against resistance the laterally rotated arm.

The lateral pectoral nerve (C5, C6, C7) is larger than the medial pectoral nerve and passes inferior to the clavicle and across the subclavian vessels before it reaches the axillary fossa where it accompanies the thoracoacromial artery and supplies the pectoralis major muscle after piercing the clavipectoral fascia. It provides some nerve fibers that innervate the pectoralis minor (Figure 10.5). The lateral cord gives rise to the lateral root (C5, C6, C7), which joins the medial root from the medial cord and forms the median nerve.

Intactness of the lateral pectoral nerve is tested in the supine position by flexing the arm while the forearm extended and the contralateral shoulder is pressed against the examination table.

Terminal Branches of Medial Cord

The medial cord gives rise to the medial pectoral, medial brachial, medial antebrachial, and ulnar nerves, as well as to the medial root to the median nerve.

The medial pectoral nerve (C8, T1) (Figure 10.6) follows a course similar to that of the lateral pectoral nerve, and after passing between the axillary vein and artery, it pierces the pectoralis minor muscle to innervate both the pectoralis minor and major muscles. Integrity of the pectoralis minor can be tested by asking the patient to extend his arm and move his/her shoulder anteriorly while the examiner hand pushes the shoulder in the opposite direction.

The medial brachial cutaneous nerve (C8, T1) is the smallest branch of the brachial plexus that supplies the distal one third of the medial surface of the arm (Figure 10.9) after it joins the intercostobrachial nerve. It may be replaced by

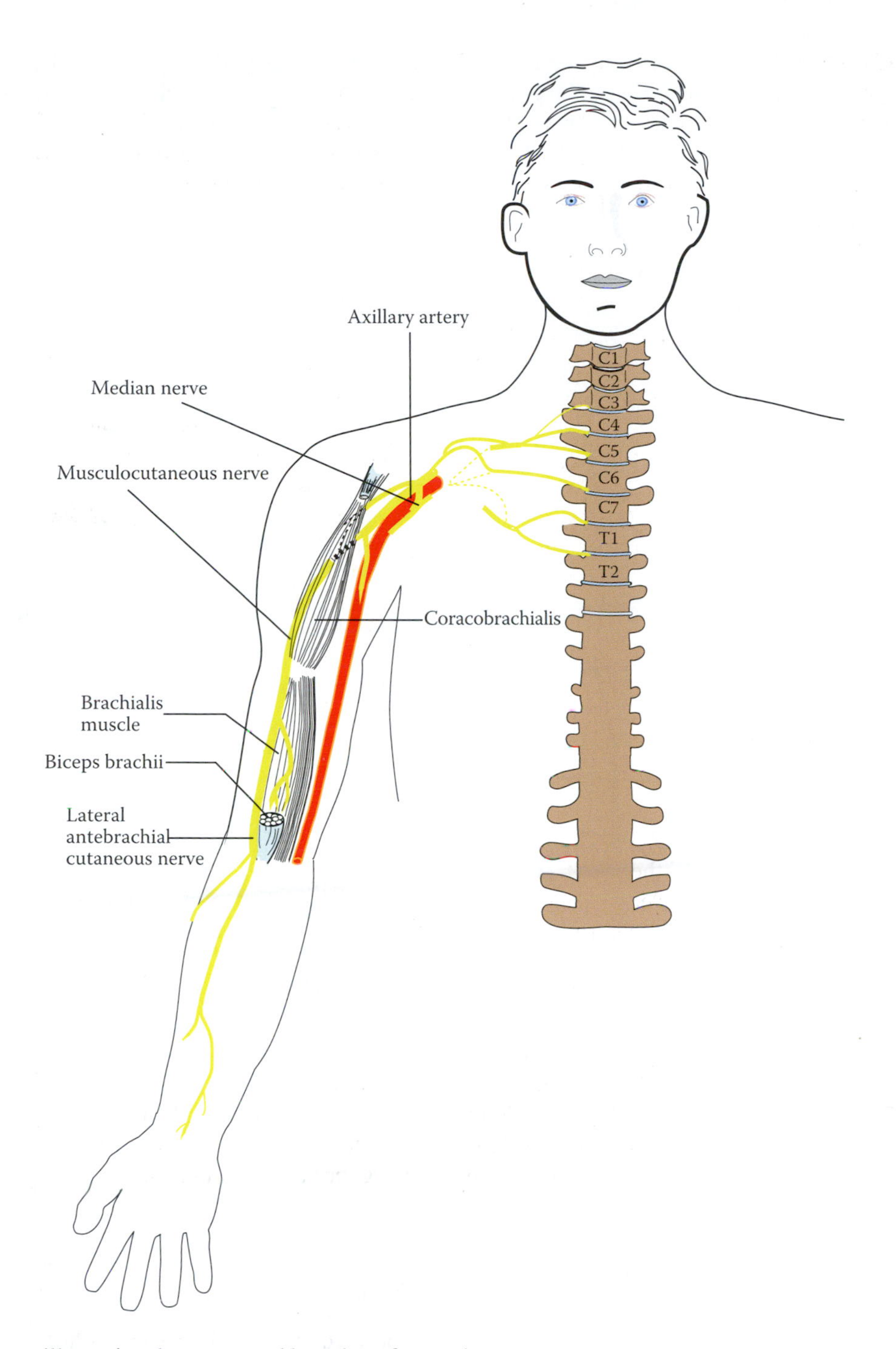

FIGURE 10.8 Diagram illustrating the course and branches of musculocutaneous nerve.

the combination of the intercostobrachial nerve and a branch from the third intercostal nerve. This nerve travels medial to the brachial artery and basilic vein and provides cutaneous fibers to the medial surface of the distal third of the arm.

The medial antebrachial cutaneous nerve (C8, T1) (Figure 10.9) courses between the axillary vessels and gives rise to a branch that supplies the anterior surface of the lower arm and then accompanies the brachial artery and basilic vein to the forearm where it divides into anterior and posterior branches. The anterior branch innervates the skin of the anterior surface of the medial forearm down to the wrist to connect with the cutaneous branches of the ulnar nerve. The posterior branch supplies the corresponding areas on the posterior surface of the medial forearm, establishing communication, in the same manner, with the sensory branches of ulnar and radial nerves.

The ulnar nerve (C8, T1) (Figure 10.10) branches off of the medial cord, but often receives contribution from ventral ramus of C7. It courses medial to the axillary and brachial arteries. In the lower medial arm, the ulnar nerve runs between the olecranon and medial epicondyle and then in the ulnar nerve sulcus on the medial epicondyle of the humerus accompanied by the superior ulnar collateral artery. At this site, the ulnar nerve is only covered by the skin and fascia.

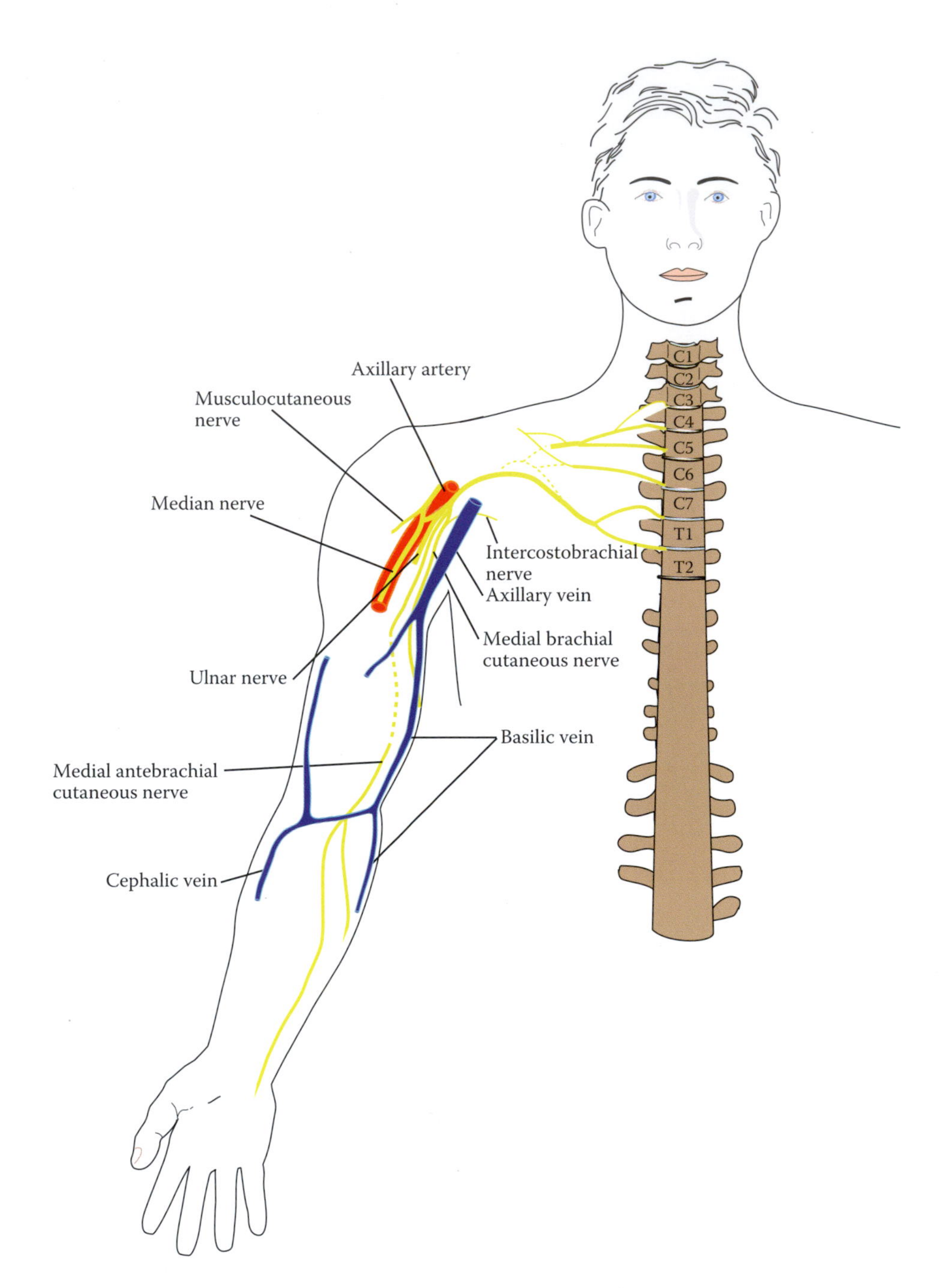

FIGURE 10.9 Schematic drawing of the course and areas of distribution of the medial brachial and antebrachial cutaneous nerves.

In the proximal and distal thirds of the arm, the ulnar nerve receives blood supply from the adjacent ulnar collateral arteries, making possible to use this segment of the nerve for transplant in evulsion injuries. It then enters the cubital tunnel, which is bounded anteriorly by the medial epicondyle and medially by the medial collateral ligament and fibrous capsule. The roof of the tunnel is formed by a fibrous band that connects the medial epicondyle to the olecranon and the tendinous origin of the flexor carpi ulnaris. It descends further between the two heads of the flexor carpi ulnaris and then between the flexor carpi ulnaris and medial half of flexor digitorum profundus to which it provides motor innervation. During its course in the forearm, the ulnar nerve is accompanied by the corresponding vessels.

Proximal to the wrist, the ulnar nerve gives rise to the dorsal and palmar cutaneous branches. The dorsal cutaneous branch runs deep to the lower part of the flexor carpi ulnaris and then courses forward to the dorsal surface along the medial border of the hand, superficial to the flexor retinaculum. It supplies the dorsal surfaces of the fifth digit, the adjacent sides of the fourth digit, and the medial side of the third digit with the exception of areas supplied by the median nerve. The palmar cutaneous branch arises from the mid-anterior forearm and supplies the skin of the medial side of the palm.

At the wrist, the ulnar nerve passes anterior to the flexor retinaculum and across the pisohamate ligament between the pisiform bone and the hook of the hamlet (Canal of Guyon), a

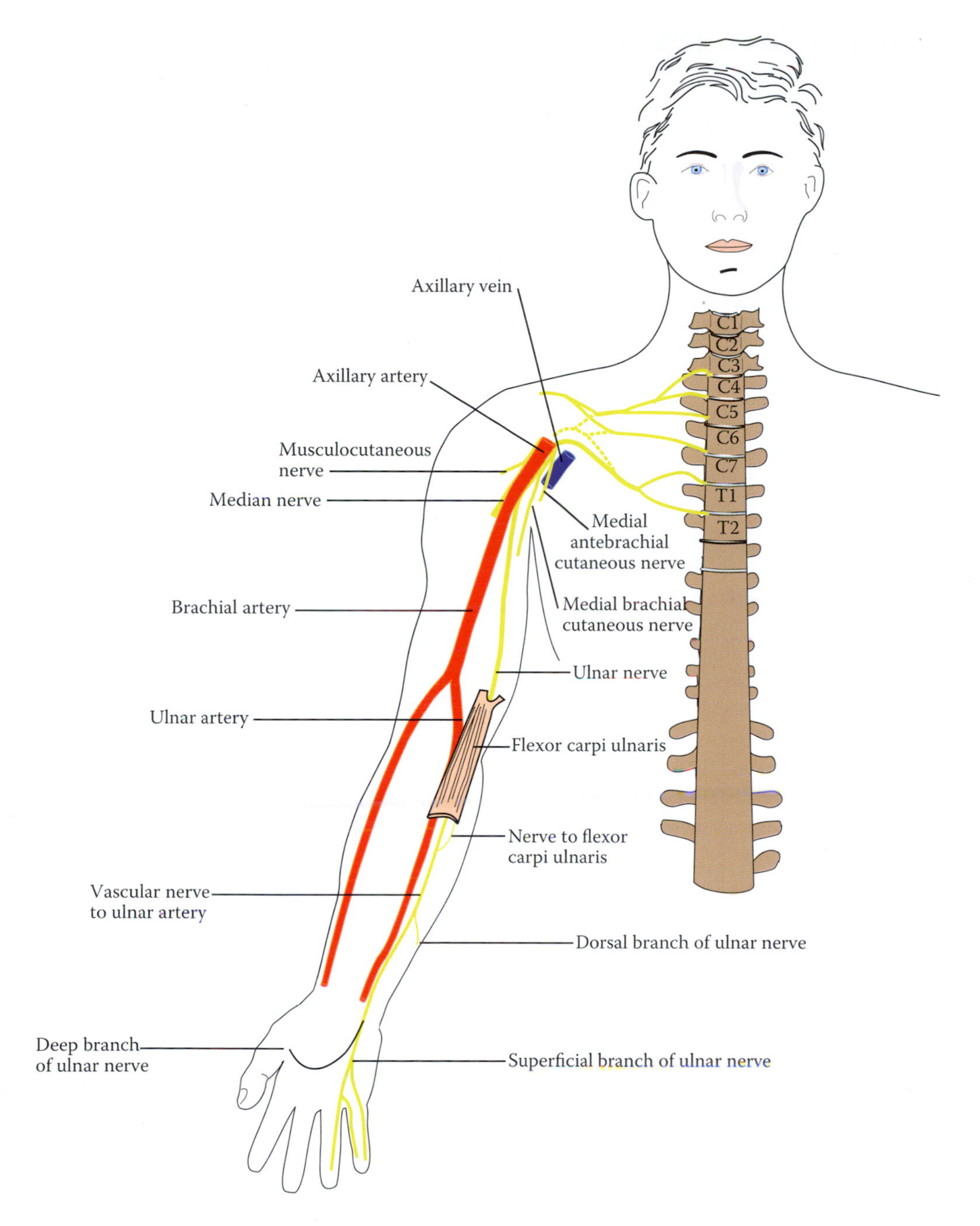

FIGURE 10.10 Ulnar nerve, its course, and areas of distribution.

common site of entrapment of the ulnar nerve. As it exits the canal, it divides into superficial and deep terminal branches. The superficial terminal branch supplies palmaris brevis and the skin of the palm, the fifth digit, and the medial side of the fourth digit. Through the deep terminal branch, the ulnar nerve innervates the hypothenar, interossei, two medial lumbricals, and the adductor pollicis.

Fibers of the ulnar nerve that supply the intrinsic muscles of the hand run within the median nerve in about 20% of individuals and then leave the median nerve distal to the elbow to join the ulnar nerve again (*Martin–Gruber anastomosis*). This anomalous connection, which is frequently seen on the right side, is believed to be an autosomal dominant condition. At the thenar eminence, the deep branch of the ulnar nerve may communicate with the recurrent branch of the median nerve to form the Riche–Cannieu anastomosis. This neural anastomosis allows the ulnar nerve to innervate the abductor pollicis longus, flexor pollicis brevis, opponens pollicis, and the lateral two lumbricals, and enables the ulnar nerve to innervate practically all hand muscles and cutaneous areas. The incidence of this variation points to hereditary basis.

After exiting the canal of Guyon, the ulnar nerve innervates the dorsal and palmar interossei, the adductor pollicis (in about 55% of individuals), the two medial lumbricals, and all hypothenar muscles: the abductor, flexor, and opponens digiti minimi. It also provides cutaneous innervation to one and a half of the medial portion of the palm and dorsum of the hand via the palmar and dorsal cutaneous branches.

However, these cutaneous branches to the hand may leave the ulnar nerve proximal to the canal of Guyon and flexor retinaculum, which requires local anesthetic be administered above the wrist when hand surgery is performed. In approximately 20% of the population, the ulnar nerve carries sensation from the skin of the entire fourth (ring) finger and medial half of the fifth finger.

Ulnar nerve injuries occur in fractures involving the medial epicondyle, dislocation of the elbow, and entrapment within the Guyon canal. It can also occur as a result of compression between heads of the flexor carpi ulnaris, prolonged leaning on the elbow (Vegas neuropathy), sustained flexion of the elbow, cubitus valgus deformity (tardy ulnar palsy), or entrapment within the cubital fossa (cubital tunnel syndrome). Application of tourniquets, pitching, inflammation and calcification of the medial collateral ligament, and positioning on the operating table may also contribute to the damage of the nerve. In order to test the functions of the muscles innervated by the ulnar nerve, movements of the digits and wrist controlled by the ulnar nerve are generated against resistance of the examiner. For instance, abduction and adduction of the digits reveal intactness of the interossei. Flexion of the wrist and ulnar deviation can test the integrity of the flexor carpi ulnaris. Flexion of the distal IP (DIP) joints of fifth and fourth digits against resistance with the hand and forearm held firmly by the examiner can test the function of the medial head of the flexor digitorum profundus.

Injury to the ulnar nerve at the axilla is rare. However, predisposition to injury is more common in the elbow due to the superficial position of the ulnar nerve in the condylar (ulnar nerve) sulcus of the medial epicondyle. Lesions of this nerve at the elbow occur as a result of recurrent trauma, subluxation of the ulnar nerve, excursion and subsequent traction or ulnar nerve elongation, elbow arthritis, or subluxation and subsequent displacement of the nerve anterior to the medial epicondyle. It can also occur in gouty tophus or as a result of entrapment in the aponeurotic tunnel between the flexor digitorum profundus and superficialis, producing cubital tunnel syndrome. Flexion of the elbow makes the ulnar nerve more prominent and increases its vulnerability.

In an injury of the ulnar nerve at the elbow, patients present with sensory disorders as the most common symptoms. These symptoms are not constant and show variability from day to day and last for months or years. Numbness, intermittent hypoesthesia, or paresthesia (abnormal sensations such as tingling, prickling, burning, and itching in the area of distribution of the ulnar nerve), radiating to the forearm with the possible involvement of the precondylar or intracapsular region is seen. Sensory disorders in the form of cramping, aching, or sharp pain may be felt in the elbow or hand and often triggered with elbow flexion and disappear upon extension. Pain may rarely be confined to the hypothenar skin. Patients may be awakened by nocturnal elbow pain. However, sensory deficits may not be apparent occasionally due to progressive nature of motor deficits. Loss of adduction and abduction of the fingers, loss of thumb adduction, and weakened flexion at the metacarpophalangeal (MP) joints occur. The fifth digit appears abducted because of the pull exerted by the extensor digiti minimi. There will also be weakened ulnar adduction and variable loss of flexion at the DIP joints of the ring and fifth digits and relatively mild form of claw hand. Patients with these deficits experience difficulty crossing or snapping fingers. Deficits in hand adduction (paralysis of flexor carpi ulnaris) and flexion of the DIP joints (atrophy of the flexor digitorum profundus) may not be apparent due to the fact that nerve fibers that supply these muscles are deeply located and thus protected compared to other fibers. Motor symptoms usually exacerbate with elbow flexion or wrist and hand movements. Elbow joint movements remain intact with normal carrying angle.

Cubital tunnel syndrome, an entrapment-related dysfunction, can result from cubitus valgus deformity induced by the malalignments of supracondylar fracture. The ulnar nerve can also be compressed in the tunnel formed by the tendinous arch that connects the ulnar and humeral heads of the flexor carpi ulnaris. The associated deficits mimic those of the elbow injury, though in this syndrome, the range of movements at the elbow joint is normal and the physiologic cubitus valgus remains unchanged.

Injury to the ulnar nerve at the wrist can occur as a result of entrapment within the canal of Guyon or compressed by a ganglion, lipoma, or rheumatoid synovial cyst. It can occur as a sequela of Colle's fracture, tumor, and sustained external pressure. Calcinosis caused by scleroderma, ulnar artery aneurysm, accessory abductor digiti minimi, fracture of the hook of hamate, and anomalous insertion of the flexor carpi ulnaris can endanger this nerve. Bicyclists, meat packers, gold and brass polishers, boot makers and cobblers, avid computer mouse users and users of code-sensitive machines for price check, oyster shuckers, and video game enthusiasts are particularly prone to ulnar damage at the wrist. If the injury is proximal to the hook of hamate in the canal of Guyon, atrophy of the hypothenar muscles and subsequent loss of thumb adduction occur. This makes scraping the thumb across the palm and formation of the letter "O" by the second digit and the thumb an impossible task. Loss of abduction and adduction of the digits is exemplified by the inability of a patient to hold on an object, usually a piece of paper, between the thumb and index or index and middle finger when the examiner pulls it away (Froment's sign). Due to the unopposed action of the extensor digiti minimi and the loose nature of the metacarpophalangeal joint, the fifth digit will assume an abducted position with prominent clawing posture (Wartenberg's sign). Since the dorsal and palmar cutaneous branches arise above the wrist, they are preserved and their cutaneous areas are spared. This makes compression of the ulnar nerve in the canal of Guyon a painless condition. If the lesion occurs either at the level or distal to the hook of hamate, hypothenar muscles will be spared. Ulnar claw hand can sometimes be seen in the adolescent females as a sign of developmental anomaly known as camptodactyly in which the lumbricals establish abnormal insertion to the affected digits. No neuronal sensory or motor deficits are seen, and the interossei and hypothenar muscles remain intact.

In the hand, if the terminal motor branch of the ulnar nerve is compressed against the hamate and pisiform, the sensory branches are spared and variable motor deficits will occur based on the level of injury. This type of injury is most likely to occur as a result of direct trauma (using the medial hand as a mallet) or prolonged pressure exerted by a pointed object like motorcycle handlebar against the medial edge of the hand. Damage to the deep branch at the wrist spares the hypothenar muscles and the third and fourth interossei and lumbricals but causes atrophy of the first dorsal interosseous. Therefore, claw hand with abducted fourth and fifth digits do not occur.

Ulnar claw hand (griffe cubitale), a characteristic configuration of the ulnar nerve damage *at the wrist*, is due to hyperextension at the metacarpophalangeal joints particularly of the ring and fifth digits and hyperflexion at the IP joints (Figures 10.11 and 10.12). Weakened extension and flexion at the IP joints of the fourth and fifth digits, hollowing of the palm (empty purse) due to hypothenar atrophy, and guttering of the grooves between the metacarpal bones due to atrophy of the interossei may also be seen in this type of injury. Impaired sensation, paresthesia, and nocturnal pain in the medial one third of the palm and dorsum of the hand may also be experienced by the patient.

Integrity of the ulnar nerve can be tested against resistance by spreading the fingers, flexing of the fourth and fifth digits at distal IP joints, and flexing of the metacarpophalangeal joints. Positive Froment's sign results from weakness of the grip between the thumb and index due to paresis of the adductor pollicis and the compensatory flexion of the distal IP joint of the thumb.

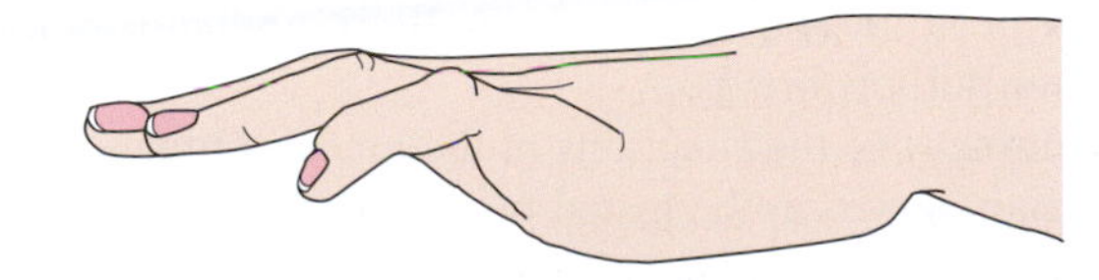

FIGURE 10.11 This drawing depicts the hand and digit disorders associated with ulnar nerve damage at the elbow. Note abnormal posture of the fourth and fifth digits and flattening of the dorsal interossei with normal hypothenar muscle.

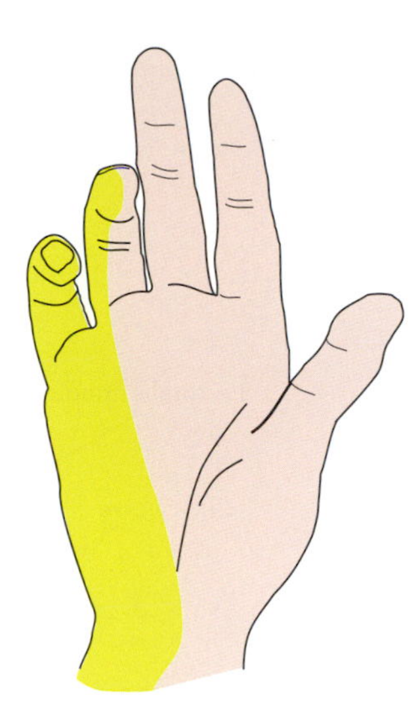

FIGURE 10.12 Manifestations of ulnar claw hand due to an injury to the ulnar nerve at the wrist. The shaded zone indicates the area of sensory loss.

The median nerve (C5, C6, C7, C8, T1) (Figures 10.13 and 10.14) originates from the ventral rami of C5–T1 and is formed by the union of the corresponding roots from the lateral and medial cords, anterior to the axillary artery. Sometimes, the musculocutaneous may also join the medial root when the lateral root is smaller than usual. Sensory fibers of this nerve are predominantly derived from C5 and C6 and descend within the lateral cord, while the motor fibers are given off mainly from C8–T1 segments and course within the medial cord. As it descends through the middle of the arm, it is accompanied by the brachial artery. In the cubital fossa, it lies posterior to the bicipital aponeurosis and anterior to the brachialis tendon. During its downward course toward the forearm between the humeral and ulnar heads of the pronator teres, it gives rise to the anterior interosseous nerve. This motor branch is thought to also carry proprioceptive fibers to the wrist and to the muscles innervated by it. A point of clinical significance to consider here is the fact that the anterior interosseous nerve may serve as a link between the ulnar and median nerves in Martin–Gruber anastomosis. Then, the nerve continues to descend on the posterior surface of the flexor digitorum superficialis. During this course in the anterior forearm, it innervates the pronator teres, flexor carpi radialis, palmaris longus (if present), and flexor digitorum superficialis. The anterior interosseous nerve supplies the pronator quadratus, the lateral half of flexor digitorum profundus, and the flexor pollicis longus. In its course to the hand, it passes within the carpal tunnel, accompanied by the tendons of the flexor digitorum superficialis, flexor digitorum profundus, and flexor pollicis longus. The carpal tunnel is a fibro-osseous tunnel formed anteriorly by the transverse carpal ligament, a deep layer of flexor retinaculum, and posteriorly deep to the flexor retinaculum. The latter extends between the scaphoid and trapezium on the radial side and the pisiform and hamlet on the ulnar side. Just distal to the inferior

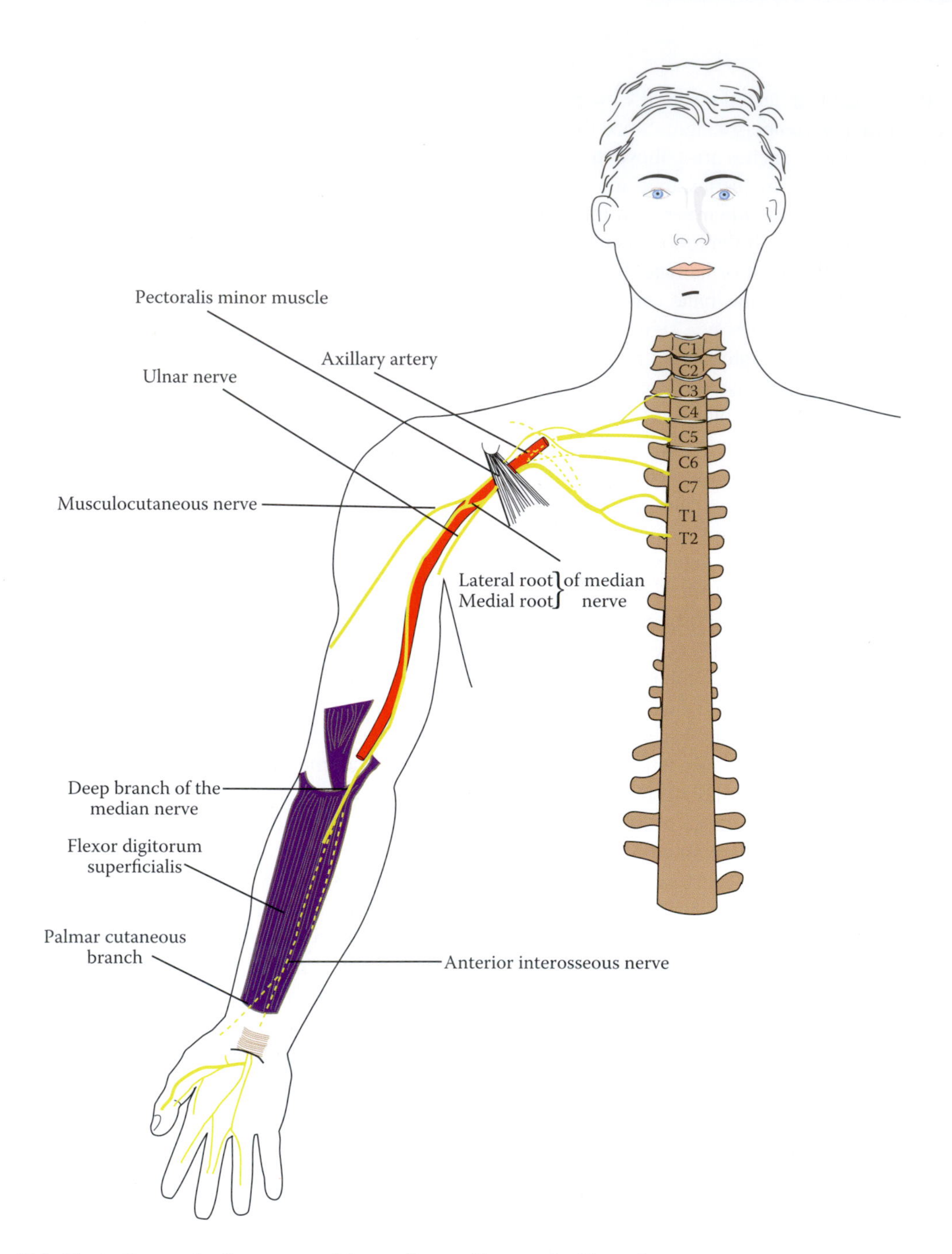

FIGURE 10.13 This illustration marks the course of the median and its terminal branches.

margin of the flexor retinaculum, the median nerve gives rise to a recurrent muscular branch from its lateral side that follows a recurrent course deep to the palmar aponeurosis to supply the abductor pollicis brevis, flexor pollicis brevis, and opponent pollicis.

Proximal to the flexor retinaculum, the median nerve gives rise to the palmar cutaneous branch to the skin of the thenar eminence and central palm. Finally, the median nerve provides the common palmar digital branches that divide into proper digital branches, which eventually supply the lateral two lumbricals and the skin of the lateral three and one-half digits. Approximately, in 50% of the patients, the median nerve also supplies the lumbrical of the fifth digit. It also supplies all muscles of the anterior forearm with the exception of the flexor carpi ulnaris and the medial half of the flexor digitorum profundus. The innervation of the median nerve can be abbreviated by this mnemonic, loaf, denoting one-half of lumbricals, opponens pollicis, abductor pollicis, and flexor pollicis brevis.

A summary of the functions of the muscles innervated by the median nerve can be tested by asking the patient to perform certain movements.

Pronator teres: flex the forearm at 30° then pronate it.

Pronator quadratus: pronate the completely flexed forearm.

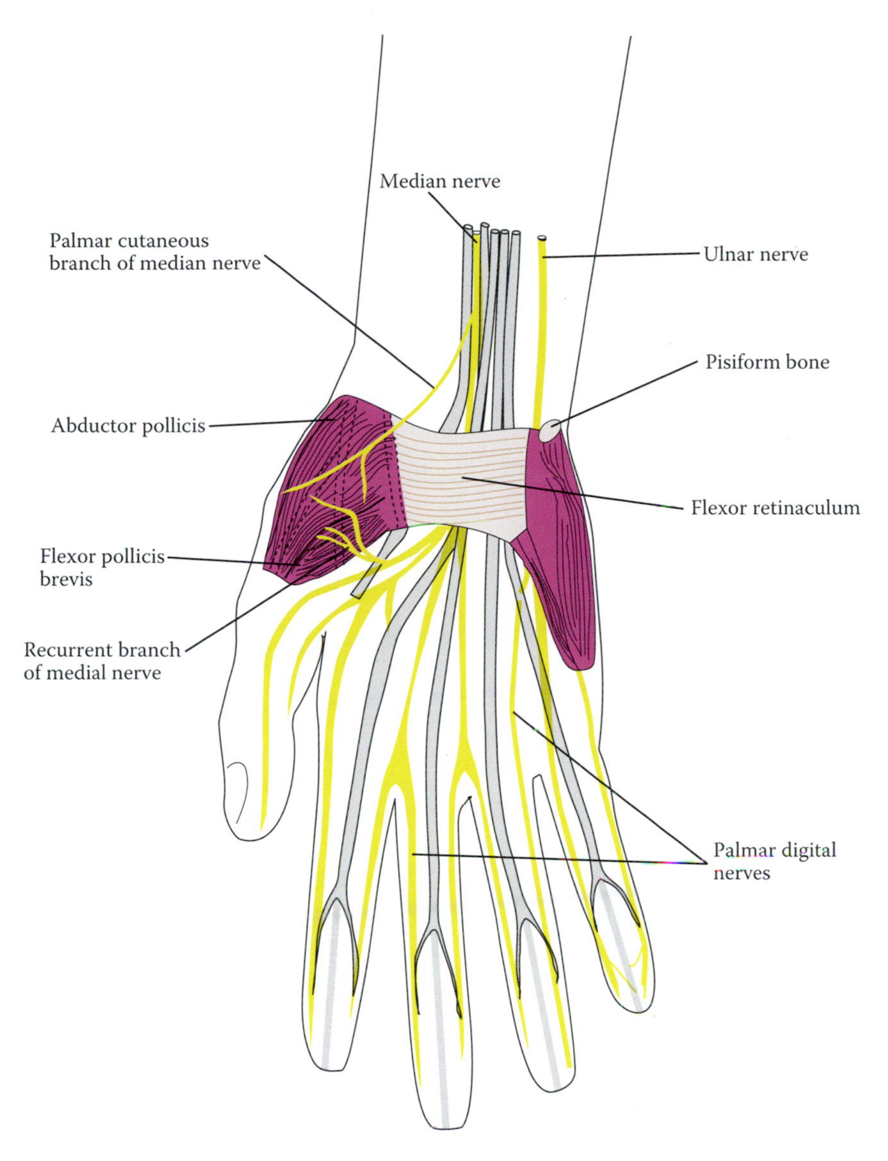

FIGURE 10.14 This drawing illustrates the course of the median nerve in the carpal tunnel, its branches in the hand. Notice the origin and course of the palmar branch of the median nerve.

Flexor pollicis longus: flex the thumb against resistance of the examiner's hand.

Flexor digitorum profundus (lateral part): flex the DIP joints while the MP and PIP joints are extended.

Lateral two lumbricals: flex the index and middle fingers at the MP joints while the DIP and PIP joints are held in extended position.

Abductor pollicis longus: abduct the thumb against resistance.

Opponens pollicis: oppose the thumb against resistance.

Median nerve is prone to damage in elbow dislocation, as a result of Volkmann's ischemic contracture, or is compressed during its course between the ulnar and humeral heads of the pronator teres. It can also be entrapped by the Struthers' ligament, an inconstant fibrous band that extends from the supracondylar ridge to the medial epicondyle of the humerus. Irrespective of the etiology, an injury to the median nerve at the elbow is most likely to produce loss of all functions associated with the median nerve including forearm pronation, thumb flexion, flexion at the IP joints of the index and middle fingers, and inability to make a fist. Inability to make a fist forces the patient to flex the (fourth and fifth) digits supplied by the ulnar nerve and produce preacher's hand, which also exhibits a slight flexion of metacarpophalangeal joints of the index and middle fingers by the interossei, and extension at the distal and proximal IP joints of the same fingers. Weakened flexion at the elbow and at the wrist and radial abduction of the hand are additional manifestations of this lesion.

The pronator syndrome, although uncommon, can occur as a result of its entrapment between the ulnar and humeral heads of pronator teres. It can also be entrapped within the fibrous arch formed by the bony attachments of the flexor digitorum superficialis and even as it passes between the biceps brachii tendon and fibrous band that extends to the antebrachial fascia. Throwing of a fast ball requires vigorous contraction of the pronator teres, and when sustained, it leads to hypertrophy of the muscle and compression of the median nerve. This syndrome, which may also be precipitated by repeated pronation and supination, manifests similar dysfunctions to those seen with median nerve lesions proximal to the elbow. Pronation may be weakened but not totally lost. Patients with this condition uniquely exhibit chronic forearm pain. Pain experienced by the patient can be exacerbated if the forearm is pronated against resistance or the flexed and pronated forearm is extended.

Pain in the palmar surface of the hand aggravated by either pronation or elbow flexion and paresthesia in the proximal forearm, elicited by forced supination of the forearm and extension of the hand at the wrist, are characteristics of this syndrome. Pain reproduction in the forearm as a result of certain movements against resistance may enable the examiner to pinpoint the site of entrapment of the median nerve in this condition. Volkmann's ischemic contracture may also lead, if untreated, to compression of the median nerve by the pressure of the swollen muscles of the forearm or pressure buildup by the accumulated fluid and blood under the flexor digitorum superficialis. Isolated compression of a branch of the median nerve may become evident in entrapment of the anterior interosseous nerve (AIN).

AIN palsy can occur as a result of distal humeral fracture, percutaneous puncture of the median cubital vein, and pressure exerted by fibrous bands formed by the flexor digitorum superficialis. A tight grip while the forearm is pronated and a pressure exerted by the tendinous origin of the ulnar head of pronator teres are additional possible causes of this nerve injury. Episodic nerve degeneration preceded by pain is seen in hereditary neuralgic amyotrophy. There are reported cases of spontaneous anterior interosseous nerve palsy with patients exhibiting severe and prolonged pain in the arm and proximal forearm, possibly precipitated by unusual motor activity. When the nerve palsy is induced by forearm fracture, loss of flexion at the distal IP joints of the thumb and index finger is noted without the initial pain. In addition to sensory manifestations, deficits in general include weakness or atrophy of the flexor pollicis longus and the lateral half of the flexor digitorum profundus and subsequent weakness of flexion at the distal IP joints of the thumb, index, and middle finger and impaired pinch maneuver (weakness of pinch grip due to paralysis of flexor pollicis longus and the lateral half of the flexor digitorum profundus). Attempt to make an even ring with the affected fingers, "O," the tip of the index finger touches the thumb at a more proximal point. Pronation will not significantly be affected in this syndrome due to the fact that pronator teres remains unaffected. Thumb opposition and sensation from the hand also remain intact. Weakened flexion and pronation and more importantly loss of flexion at the distal IP joints of the index and middle finger are the primary deficits of this syndrome.

Since the anterior interosseous nerve serves a variant link between the ulnar and median nerve trunks (Martin–Gruber anastomosis), damage to the AIN would most likely produce paralysis of the entire flexor digitorum profundus as well as other intrinsic hand (interossei, lumbrical, and hypothenar) muscles.

The median nerve may also be entrapped in the bicipital aponeurosis, producing signs of lacertus fibrosus syndrome. This type of entrapment produces pain upon forced pronation of the flexed and supinated forearm. Damage to the median nerve at the midpoint of the forearm produces partial paralysis of the flexor digitorum superficialis muscle, which results in pointing of the index finger due to the unopposed action of the extensors of the index finger (Figure 10.15.)

Median nerve damage at the wrist can occur subsequent to cuts across the wrist, anterior dislocation of the lunate bone, and compression in the carpal tunnel. This type of injury produces loss or weakened thumb opposition as in thumb pinching (thumb moves inward across the palm to oppose the fifth digit), atrophy of the thenar muscles, and pain and paresthesia or anesthesia in the lateral two-thirds of the palm. There may also be paresthesia or total loss of sensation in

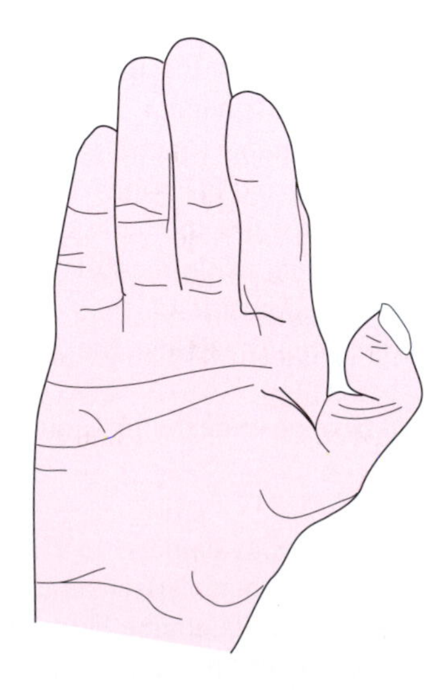

FIGURE 10.15 This is a drawing depicting the major manifestations of anterior interosseous nerve syndrome.

the palmar surfaces of the thumb, index and middle fingers, as well as the lateral one-half of the ring finger and the dorsal surfaces of the medial four digits as far as the middle phalanges. Weakened flexion at the metacarpophalangeal joints and weakened extension at the IP joints of the index and middle fingers are also observed. Weakened thumb abduction (inability to separate the thumb and index finger) and inadequate thumb rotation may produce "bottle sign," a deficit in which the subject is unable to maintain a grip around a bottle. Flattening and atrophy of the thenar eminence and the counter pull of the extensor and the abductor pollicis longus muscles of the thumb result in pulling the thumb to the same level with the other digits and producing the ape hand configuration.

Compression of the median nerve at the wrist may also occur inside the carpal tunnel producing signs of carpal tunnel syndrome. This syndrome, a unilateral and sometimes bilateral condition that often involves the dominant hand, is common in pregnancy and in middle-aged women during the premenstrual period. It may be caused by fluid retention, Colle's fracture, acromegaly, hypothyroidism, gout tophi, eosinophilic fasciitis, and congestive heart failure, tenosynovitis of the flexor tendons, mucopolysaccharidosis, or tuberculosis of the synovial sheaths. Since repetitive movements at the wrist displace the flexor tendons against the palmar side of the carpal tunnel, continuous and sustained flexion and extension of the wrist may be a predisposing factor for this condition. Congenital arteriovenous fistula or surgically induced Cimino–Bresica fistula and a persistent median artery can produce compression of the median nerve in the carpal tunnel. Diabetes mellitus, multiple myeloma, primary amyloidosis, and insect or snake bites can trigger manifestation of this condition. Transducers inserted into the canal may measure the increase in intracarpal pressure.

Carpal tunnel syndrome (Figure 10.16), the most common neuropathy of the hand, though variable, is characterized by acroparesthesia (tingling, numbness), pain in the radial three digits that increases at night or early morning and is frequently relieved by firm grasp of the hand (shaking hand). Demyelination of the median nerve, subsequent weakness or atrophy of the thenar muscles, and associated autonomic disturbances such as swelling and alteration in the texture of the skin in areas of distribution of the median nerve are also common manifestations of this disease. Opposition and, to a lesser degree, abduction of the thumb may eventually be affected, although variations of manifestations that range from numbness confined to the middle finger or atrophy of the muscle that act on the index finger without thenar involvement can also occur. The palmar cutaneous nerve that supplies the skin of the thenar skin, central palmar skin, is spared since it passes through the carpal tunnel.

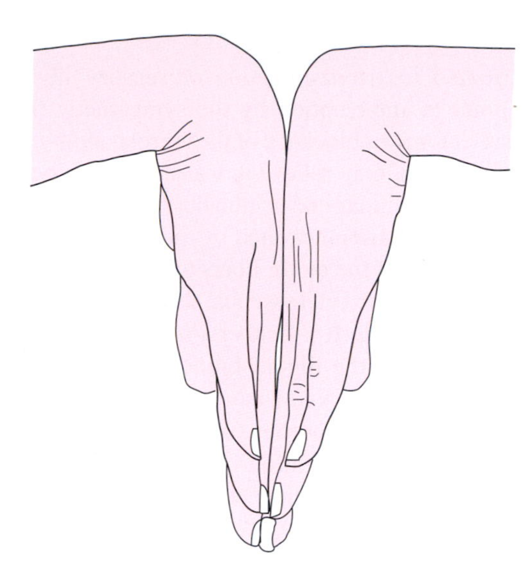

FIGURE 10.16 Manifestations of carpal tunnel syndrome. Note the wasting of the thenar muscles and ape-hand configuration.

Carpal tunnel syndrome is confirmed by the application of Tinel's sign and Phalen's maneuver. Tapping the wrist produces Tinel's sign, characterized by tingling and electrical sensation in the area of the sensory distribution of the median nerve. Forced flexion of the wrist (Phalen's maneuver) or forced extension of the wrist (reverse Phalen's maneuver) produces pain and tingling in the cutaneous distribution of the median nerve (Figure 10.17).

Anomalous innervation, as in Martin–Gruber anastomosis, must be ruled out in order to confirm the diagnosis of this condition. In general, injury to the median nerve is commonly associated with a burning and tearing pain sensation in the digits and palm of the hand, accompanied by vasomotor and sudomotor changes in the cutaneous areas wider than the area of distribution of the median nerve (*causalgia*). The affected part of the hand and digits becomes extremely sensitive to touch, including contact with clothes or air. Causalgia

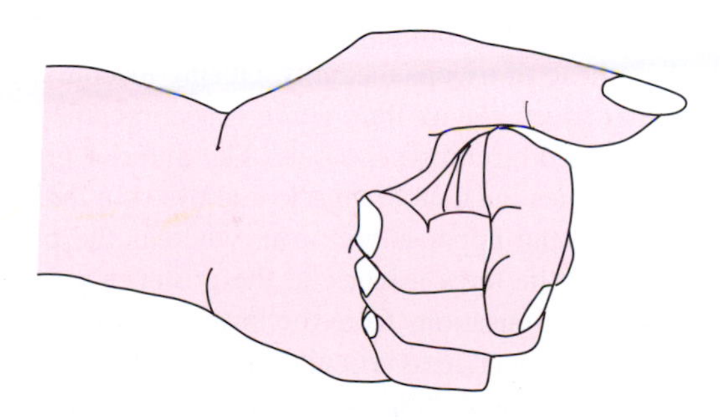

FIGURE 10.17 This diagram depicts the Phalen's test. Observe the acute flexion at the wrist.

is attributed to overstimulation of sensory fibers at their point of interruption by the sympathetic fibers. Sympathectomy or blockade of the corresponding sympathetic ganglia may relieve it. Causalgia may be also associated with ulnar and sciatic nerve injury. Muscles of the hand that are innervated by the ulnar nerve may also be affected if the motor fibers to these muscles run within the anterior interosseous nerve before joining the ulnar nerve (Martin–Gruber anastomosis).

A neuroma of the digital branches of the median nerve to the thumb is known as bowler's thumb, which exhibits skin atrophy, fibrous tissue accumulation, callus, and Tinel's sign. Similarly, digital branches can develop neuropathy in cheerleaders as a consequence of clapping and turning.

Combined damage of the ulnar and median nerves at the elbow or at a more proximal site produces loss of forearm pronation and flexion of the metacarpophalangeal and IP joints of the thumb and digits. Opposition of the thumb and fifth digit is lost. As a result, the hand becomes hyperextended (unopposed by the flexors), and the forearm assumes supinated position due to paralysis of the pronators. Fingers are hyperextended at the metacarpophalangeal joints due to paralysis of the interossei and lumbricals.

Combined damage to the median and ulnar nerves at the wrist results in *true claw hand*, which is characterized by hyperextension of the metacarpophalangeal joints and hyperflexion of the distal IP joints of the digits, unopposed by the action of interossei and lumbricals. Thumb opposition and adduction as well as atrophy of the thenar and hypothenar muscles are additional deficits of this injury. Atrophy of the interossei results in prominence of the long flexor tendons.

Terminal Branches of Posterior Cord

The axillary nerve (C5, C6) (Figure 10.6) is a considerably large branch that runs on the surface of the subscapular muscle posterior to the axillary artery and lateral to the radial nerve. At the lower border of the subscapular muscle, it pursues a curved course and enters the quadrangular space bounded by teres major distally, teres minor proximally, humerus laterally, and the long head of triceps medially. During its course in the quadrangular space, the axillary nerve is accompanied by the posterior humeral circumflex vessels. The anterior branch of this nerve supplies the deltoid muscle and the skin that covers the muscle and the upper lateral arm, whereas the posterior branch supplies the teres minor and the posterior part of the deltoid, carrying sensations from the lower lateral part of the arm. Further, it gives rise to articular branches to the shoulder joint. Function of the deltoid is checked by abduction of the arm against resistance above 90°. To test the function of the teres minor, the patient is placed on prone position and asked to laterally rotate his arm against resistance of the examiner's hand while the elbow is flexed at a 90° angle.

Damage to the axillary nerve may occur as a result of anterior inferior dislocation of the humeral head, manipulation to reduce dislocation, fracture of the surgical neck of the humerus, hereditary neuralgic amyotrophy, blunt trauma, though rarely, intramuscular injections, or pressure from the use of crutches. Sleeping in a prone position or lying on a surgical table with an abducted arm above 90° and forearm placed horizontally can traumatize this nerve. Axillary nerve palsy is characterized by atrophy of the deltoid and teres minor, loss of shoulder contour, and severe weakness of arm abduction up to 90°. Abduction is lost completely due to intactness of the supraspinatus, upper fibers of the trapezius, and serratus anterior muscles. Weakened lateral rotation (due to paralysis of the teres minor) and limited loss of cutaneous sensation from the shoulder are additional deficits of this condition. In *quadrangular space syndrome* seen in baseball pitchers, the axillary nerve may be affected in conjunction with the posterior humeral circumflex. Patients usually report anterior shoulder pain and paresthesia, which become intense upon abduction and external rotation. Deltoid weakness is rarely detected in this condition.

Radial nerve (C5, C6, C7, C8, and T1) is the largest branch of the brachial plexus formed by the ventral rami of the fifth, sixth, and seventh cervical and first thoracic spinal nerves (Figure 10.18). It derives from the posterior cord and descends posterior to the axillary artery and anterior to the latissimus dorsi and teres major. Before it reaches the posterior surface of the humerus, it maintains important relationships to the long and medial heads of the triceps brachii and further distally to medial and lateral heads of the same muscle. On the posterior surface of the humerus, it descends in the radial (spiral) sulcus accompanied by the deep brachial vessels. It innervates the triceps brachii by a branch that arises from the radial nerve inside the sulcus and above the mid-diaphysis of the humerus. The site of origin of this branch accounts for intactness of triceps brachii muscle in a fracture involving the radial nerve at the distal third of the humerus. During its course on the lateral side of the arm, the radial nerve passes between the brachialis and brachioradialis and then between brachioradialis and extensor carpi radialis longus. In the lower medial arm, the radial nerve gives rise to muscular, cutaneous, and articular branches. The muscular branches innervate the anconeus, brachioradialis, and extensor carpi radialis longus. Cutaneous branches distribute to the posterior arm (posterior brachial cutaneous nerve) and lower lateral arm and posterior forearm (posterior antebrachial cutaneous nerve). Elbow joint is the main recipient of the sensory articular branches of the radial nerve. As soon as it approaches the anterior surface of the lateral epicondyle, the

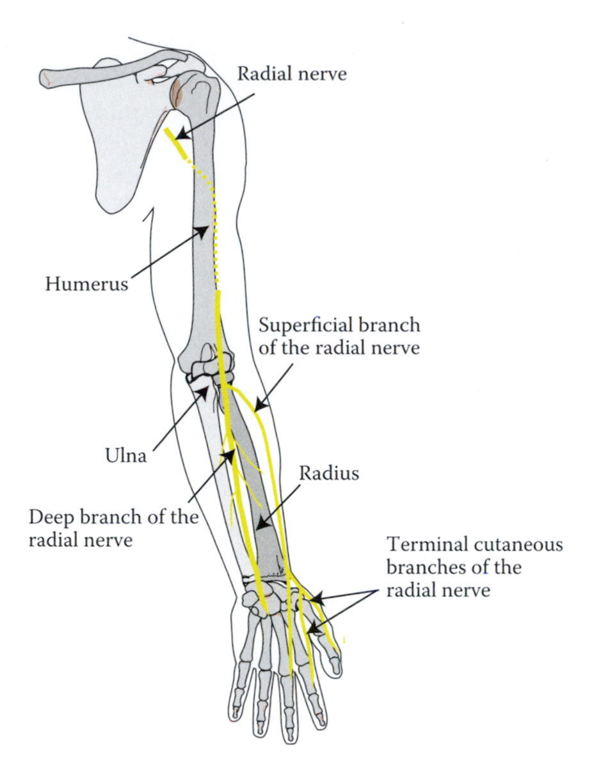

FIGURE 10.18 In this schematic drawing, the course, branches, and distribution of branches of radial nerve are shown.

radial nerve divides into superficial and deep terminal (posterior interosseous) branches.

The superficial terminal branch descends posterior (deep) to the brachioradialis and lateral to the radial artery and then anterior to the flexor digitorum superficialis and flexor pollicis longus. Proximal to the wrist, it encircles the distal end of the radius and courses deep to the tendon of brachioradialis to gain access to the dorsum of the hand. On the dorsum of the hand, the superficial branch of the radial nerve joins the posterior and lateral antebrachial cutaneous nerves and gives rise to branches that supply the radial side of the thumb and the adjacent area of the thenar eminence, as well as the three and a half digits of the dorsum of the hand as far as the midportions of the middle phalanges of the index, middle, and ring fingers. When the superficial branch of the radial nerve is absent, which rarely occurs, it is replaced by the dorsal branch of the ulnar nerve, and occasionally by the lateral antebrachial cutaneous nerve. These variations account for recurrence of pain in the radial nerve cutaneous area following a successful block of the superficial radial nerve.

The deep branch wraps around the radius and supplies the extensor carpi radialis brevis and supinator and then pierces the supinator muscle to emerge as the posterior interosseous nerve, innervating the extensor digitorum, extensor carpi ulnaris and extensor digiti minimi. In addition, the posterior interosseous nerve innervates the extensor pollicis longus, extensor indicis, extensor pollicis brevis, and abductor pollicis longus. This branch ends at the dorsal surfaces of the carpal bones as a pseudo ganglion, a swelling that provides articular branches to the wrist joint and ligaments. Anatomical studies have shown variations in the site of origin of various branches of the radial nerve. In some individuals, branches to the supinator and extensor carpi radialis brevis arise from the radial nerve proximal to the arcade of Frohse and before its division into superficial and deep branches directly from the posterior interosseous nerve.

Radial nerve is injured commonly in fractures that involve the midshaft of the humerus. Sleeping while inebriated with the arm hanging over the edge of a chair (sleep or Saturday night palsy), crutch misuse (crutch palsy), or misplaced pacemaker catheter compresses and injures the radial nerve. Arcade of Frohse, a fibrous band associated with the upper part of supinator muscle, can entrap the deep branch of the radial nerve. The radial nerve is also prone to compression as a result of fibrosis of the triceps brachii muscle seen in Parkinson's disease and Guillain–Barré syndrome. Damage to the radial nerve in the axilla results in loss of extension of the forearm at the elbow due to paralysis of the triceps brachii and anconeus muscle. Loss of extension of the hand at the wrist (wrist drop) and loss of extension of thumb and the metacarpophalangeal joints (finger drop), as well as impairment of cutaneous sensation or paresthesia in areas of distribution of the radial nerve (Figure 10.19), can also occur. Wrist drop in radial nerve palsy should be distinguished from the wrist drop seen in upper motor neuron palsy. In upper motor neuron palsy, the ability to grasp an object with the affected hand becomes possible by the synkinesia (involuntary muscle contractions that accompany voluntary movements), which will not occur in radial nerve palsy. Weakened flexion of the elbow, abduction of the thumb, radial and ulnar deviation of the hand, and weakened extension of the IP joints are additional manifestations of this type of injury. Since

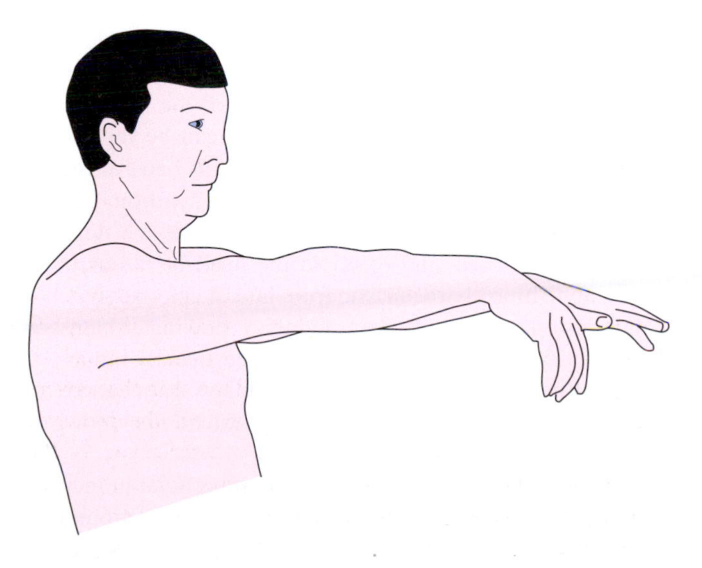

FIGURE 10.19 This drawing illustrates right hand wrist drop as a result of radial nerve damage.

the triceps muscle receives innervation proximal to the midpoint of the humerus, damage to the radial nerve immediately distal to this point will spare extension of the elbow.

Radial nerve lesions in the upper arm can be illustrated by testing the function of the triceps brachii and anconeus by extending the arm against resistance. The brachioradialis can be tested by flexing the mid-prone forearm, a maneuver that eliminates contraction of the biceps brachii and brachialis. Testing of the supinator function is accomplished by asking the patient to supinate the pronated forearm against the examiner's resistance. In the same manner, extensor digitorum muscle function can be checked by asking the patient to extend the MP joints of the second through the fifth digits and relax the IP joints.

Damage of the posterior interosseous nerve (PIN) occurs when it is entrapped within the extensor muscles of the forearm, particularly when the extended forearm is pronated while the hand flexed at the wrist. Wrestlers and individuals unaccustomed to the use of certain tools such as screwdriver where supinator (supinator syndrome) is overused are particularly prone to this nerve's palsy. Lipomas are more frequent cause of PIN injury, while rheumatoid arthritis, olecranon and bicipital bursitis, and presence of ganglia can cause, though very rarely, damage to this nerve. Posterior interosseous nerve palsy proximal to the supinator muscle produces initially weakened extensor digiti minimi followed by weakened wrist extension due to paralysis of the extensor carpi ulnaris, weakened forearm supination, and extension at the IP joints. The hand will be radially deviated upon extension because of intactness of the extensor carpi radialis longus and brevis, and brachioradialis. Abduction and extension of the thumb and extension of the digits are also weakened. Since the PIN is purely motor nerve, sensory changes are not seen; however, some radial paresthesia and deep ache in the forearm can be observed. Hyperextension, repeated pronation, and supination of the forearm may exacerbate these symptoms. PIN can be injured subsequent to the presence of a tendinous band at the arcade of Frohse producing radial tunnel syndrome. In this syndrome, lateral elbow aching pain is felt deep to brachioradialis and distal to the head of radius, a distinguishing characteristic from lateral epicondylitis. It is commonly experienced at night, following arduous activity or supination against resistance, usually radiating to the arm and forearm. Lateral elbow pain can be exacerbated by asking the patient to extend his elbow, wrist, and middle digit.

Due to the outward course of the superficial branch of the radial nerve, it is prone to traumatic injury around the wrist, which includes tight watchbands, bracelets or handcuffs, or bandage. Prolonged orchestral drumming has also been implicated in the entrapment of the posterior cutaneous nerve of the arm. Scaphoid fracture and cortisone injection to treat *tenosynovitis of de Quervain* can also traumatize this nerve. It is characterized by isolated pain or pain that radiates proximally to the elbow, dysesthesia, numbness, and positive Tinel's sign with no motor deficits and is associated with cheiralgia paresthetica or Wartenberg's disease. Trophic and vasomotor changes (reflex sympathetic dystrophy) may ensue as a sequela to compression of this nerve. The cutaneous deficit may vary due to considerable overlap with the terminal branches of the lateral antebrachial cutaneous nerve. Injury to the proximal part of the superficial radial nerve is less common due to its protected position deep to the brachioradialis; however, fractures of the radius can damage this nerve.

Subscapular nerve (C5, C6) consists of an upper branch (superior subscapular) and a lower branch (inferior subscapular). The former is much smaller in size and pierces the subscapular muscle at a higher point than the lower branch, while the latter branch enters and supplies the lower part of the subscapular muscle and then ends in the teres major muscle. The muscles innervated by this nerve cannot be tested independently due to the fact that latissimus dorsi, pectoralis major, and anterior fibers of deltoid muscle perform similar function.

Damage to the subscapular nerve may produce weakened medial rotation and adduction of the arm (due to paralysis of the subscapular and teres major muscles).

The thoracodorsal nerve (C6, C7, C8) (Figure 10.6) derives from the posterior cord and descends between the upper and lower branches of the subscapular nerve accompanied by the corresponding branch of the subscapular vessels (thoracodorsal artery and vein). Occasionally, it is given as a branch of the axillary nerve. It lies on the lateral side of the long thoracic nerve and supplies the latissimus dorsi muscle through the medial surface of the anterior border. Numerous activities can be performed by activation of this nerve such as climbing, pitching, and clutching to an object when falling from height.

The thoracodorsal nerve is most commonly damaged subsequent to radical mastectomy or fracture dislocation of the shoulder joint. It can also be affected as a result of an aneurysm of the subscapular or the thoracodorsal artery. Damage to this nerve produces weakened arm adduction and inability to do chin-ups. There will be an absence of the posterior axillary fold and inability to raise the extended and medially rotated arm posterior to the back while standing or in prone position. To test the latissimus dorsi, the extended and medially rotated arm of a prone patient is abducted against resistance of the examiner's hand.

THORACIC SPINAL NERVES

Due to the difference between the length of the vertebral canal and the length of the spinal cord, the lower thoracic spinal nerves pursue a longer course in order to reach their sites of exit at the intervertebral foramina. Accordingly, the thoracic spinal nerves emerge from the intervertebral foramina distal to the T1–T12 vertebrae. Following their exit, the thoracic spinal nerves divide into dorsal and ventral rami with the dorsal rami further splitting into medial and lateral branches near the zygapophyseal joints. The medial branches of the dorsal rami of the upper six thoracic spinal nerves pursue a 90° angled turn through the multifidi before they distribute primarily sensory fibers to the back, while the lateral branches provide principally motor fibers to the iliocostalis and levator costarum muscles. In the lower six thoracic spinal nerves, a reverse arrangement exists where the medial branches of the lower six thoracic spinal nerves are motor innervating the multifidi and the longissimus muscles, and the lateral branches are sensory to the corresponding dermatomes.

It has been documented that the sharp-angled turn of the medial branches of the dorsal rami of the upper six thoracic spinal nerves accounts for the paresthesia, burning sensation with diminution of the pinprick sensation at the medial border of the scapula observed in notalgia paresthetica.

Virtually all of the ventral rami of the thoracic spinal nerves, with one exception of the ventral ramus of T12, run in the intercostal sulci as the intercostal nerves (between the internal and innermost intercostal muscles), innervating the intercostal muscles, thoracic and abdominal walls, and the gluteal region, as well as the upper extremity. In addition to forming the first intercostal nerve, the ventral ramus of the first thoracic nerve contributes a large branch to the brachial plexus. The ventral ramus of the first thoracic spinal nerve lies dorsal to the stellate ganglion and pursues a course posterior to the cervical pleura (cupola) to reach the space between the anterior and middle scalene muscles. A particular branch, the *intercostobrachial nerve*, arises as the lateral cutaneous nerve from the second intercostal nerve (sometimes the third intercostal nerve) and joins the brachial plexus supplying the skin of the upper part of the medial arm. The upper six intercostal nerves supply the thoracic wall, costal pleura, the diaphragm, and the diaphragmatic pleura and peritoneum, while the lower five intercostal (thoracoabdominal) nerves course between the internal oblique and transverse abdominis muscles, piercing the anterior layer of rectus sheath, innervating the skin and muscles of the anterior abdomen, as well as the peritoneum. The ninth through the eleventh intercostal nerves pierce the diaphragm and then enter the pass through the internal oblique. The *tenth intercostal nerve* supplies the skin of the umbilicus, whereas the seventh, eighth, and ninth intercostal nerves innervate the skin of the supra-umbilical region. Lower two intercostal nerves supply the infra-umbilical region. The *subcostal nerve* is the largest, and it joins the ventral ramus of the first lumbar spinal nerve and courses posterior to the lateral arcuate ligament, anterior to the quadratus lumborum, and distal to the twelfth rib, accompanied by the subcostal vessels. It pierces the abdominal wall posterior to the anterior iliac spine and innervates the pyramidalis muscle, skin of the anterior gluteal region.

The origin, course, and area of distribution of the intercostal nerves can explain the mechanism of projected pain to the thoracic wall associated with inflamed costal or diaphragmatic pleura. It accounts for the pain sensation in the anterior abdominal wall as a result of subluxation of the interchondral joints of the lower costal cartilages or compression of the lower intercostal nerves subsequent to an inflammatory process, for example, tuberculosis of the lower thoracic vertebrae can produce constrictive or diffuse pain in the anterior abdominal as a result of compression of the intercostal nerves.

Lower five or six intercostal nerves that supply the abdominal muscles may be compressed as a result of entrapment within the rectus muscle and sheath, producing localized pain in the anterior abdominal wall, which is exacerbated by activities such as flexing the hip in supine position and placing pressure over the rectus abdominis. Injection of steroid can relieve manifestations of this disease.

Herpes zoster (shingles) is a systemic viral disease caused by *varicella zoster*, a group of DNA viruses that belong to the family of *Herpesviridae* that also produces chickenpox. This most common neurologic disease spreads through the respiratory route, by contact with secretions from skin lesions, or by vaccination. Immunocompromised, organ-transplant patients, patients with lymphoma, and elderly patients who experience waning of varicella-specific immunity are particularly prone to this disease. The virus establishes latent infection in perineuronal satellite cells of the dorsal root ganglia or sensory ganglia of the cranial nerves. Shingles occurs when the replication of the virus is reactivated commonly in the dorsal root ganglion cells and travels along the sensory root (trans-axonally) producing hemorrhagic inflammation and pain, followed by unilateral rashes that do not cross the midline and vesicular eruption in the course of the intercostal nerves. Vesicles that are filled with neutrophils erode to become shallow ulcers. Initially, patients exhibit myalgia, possibly fever, fatigue, and nuchal rigidity followed by unilateral dull, vague, and diffuse pain. Within a few days, well-localized lancinating burning pain becomes evident accompanied by herpetic blisters in the area of distribution of the sensory fibers of the ventral rami of the spinal nerves.

Lesions occur unilaterally, are mainly confined to the thoracic and lumbar segments, and usually heal within weeks. The cutaneous rashes may not always be visible particularly in the inguinal area and around the mammary gland. Herpetic blisters can be severe enough to cause lasting damage and even necrosis. Pain usually precedes the blisters by days. However, painless rashes also occur (zoster sine herpete). Inflammation and spread of virions may also involve the ventral and dorsal horns of the spinal cord and associated meninges producing occult focal poliomyelitis. It may also spread to the cerebral vasculature causing vasculopathy and vasculitis leading to stroke and meningoencephalitis. It can spread to the internal organs in immunocompromised individuals. When the neuropathic pain (combination of pain and numbness) becomes chronic and persists longer than 3 months, postherpetic neuralgia will develop. Postherpetic neuropathic pain may be accompanied by autonomic and motor deficits. Age, intensity of prodrome, and acute stages of this disease predispose patients to postherpetic neuralgia. However, overall recovery in young individuals is common with limited recurrence. Muscle weakness is expected, and when the virus affects the lower five or six thoracic spinal nerves, abdominal muscle palsy may ensue leading to loss of superficial abdominal reflex and possible hernia. Muscles weakness is usually accompanied by changes in the spinal reflexes. Herpes zoster can affect the geniculate and trigeminal ganglia, producing herpetic lesions in the skin of the concha of the ear and areas of distribution of the ophthalmic (ophthalmic zoster), maxillary, and mandibular nerves. Ophthalmic nerve involvement is usually indicated by rashes on the tip of the nose and can be accompanied by serious consequence such as keratitis, corneal ulcer, iritis, and even retinal cell necrosis. Rarely, a severe form of reflex sympathetic dystrophy with causalgia and manifestations of Horner syndrome are seen. Ramsay Hunt syndrome is a consequence of herpes zoster of the geniculate ganglion of the facial nerve (geniculate herpes), which causes vesicular eruption in the concha of the pinna and may produce vertigo, tinnitus, and neuronal deafness. Antivirals (acyclovir) are the mainstay medications in the treatment of the initial stage of this condition. Opioids and corticosteroids with tricyclic drugs may also be used for treatment of this condition.

Cardiac pain, commonly felt on the left side of the medial arm, forearm, and fifth digit, is attributed to activation of the sensory neurons within the first thoracic spinal segment. The neurons in the first thoracic spinal segment provide both sympathetic presynaptic fibers to the cardiac plexus and cutaneous fibers to the medial arm, forearm, and fifth digit via the medial antebrachial and brachial cutaneous nerves as well as the sensory branches of the ulnar nerve to the medial hand (T8–T1). It has been suggested that the pain-carrying afferent fibers from the heart to T1 spinal segment that accompany the sympathetic fibers lower the threshold of the sensory neurons within that particular segment. This change in neuronal threshold renders the normal cutaneous impulses that stream to T1 from corresponding dermatomes painful. Others suggest that convergence of pain-carrying afferents from the heart with that of afferents from skin dermatomes onto the same neuronal segment results in brain misinterpretation of the source of the pain as if it is emanating from the corresponding skin dermatomes rather than the heart itself.

Contraction of the abdominal muscles in response to cutaneous stimulation of the abdomen confirms the fact that the intercostal nerves subserve dual function of cutaneous and muscular innervation of the anterior abdominal wall. Rebound rigidity observed in the anterior abdomen of patients with appendicitis or diverticulitis is based on the fact that irritation of the parietal peritoneum adjacent to the inflamed appendix stimulates the intercostal nerves that innervate the peritoneum, skin, and abdominal muscles producing the observed rigidity.

Thoracic spinal nerve roots are rarely affected due to the restricted rotatory movement between the thoracic vertebrae. When it occurs, the prolapse is massive and it most likely to involve the mid and lower thoracic levels. However, direct trauma or cancer metastasis may cause collapse of the thoracic vertebrae and subsequent compression of the thoracic spinal nerve roots. Violent drawing of the entire body along the ground with one hand may specifically injure the dorsal root of the first thoracic spinal nerve, producing signs of Horner's syndrome and atrophy of the intrinsic muscles of the hand. Referred pain from internal organs and herpes zoster must be considered when dealing with thoracic pain. Due to involvement of the lower five or six thoracic spinal nerves in the innervation of the abdominal muscles, careful inspection of these muscles in thoracic nerve root injuries may be essential. Radicular injury to the thoracic spinal nerves causes pain in the corresponding dermatomes, which appears as band as is seen in shingles. These dermatomes maintain a dorsal and cranial course and with few exceptions do not reach the corresponding spinous processes. Due to the breadth of a thoracic root dermatome, the lateral and ventral parts are displaced cranially enabling the dorsal and ventral branches of a thoracic spinal nerve to meet.

Thoracic disk herniation is very rare and the incidence is between 0.2% and 5% of population. It is commonly seen between 30 and 50 years of age with male predominance. It involves the mid and lower thoracic part of the vertebral column, but commonly distal to T8 with T11/T12 as the most common level. Herniated disks calcify, and as a result, fragments spread to cause nerve root compression. Repetitive or prolonged bending forward, rotating the spine, poor posture (slouching), driving, and lifting can predispose an individual to this type of herniation. Most patients do not show symptoms, and finding a herniated thoracic disk on a radiographic image is mostly incidental. Patients experience a sudden severe onset of mid back pain in the upper back on one or both sides of the spine. It may show radiation following dermatomal pattern around the ribs, upper extremity, or abdominal wall, exacerbated by coughing or sneezing. Pain can be constant, sharp, shooting, or intermittent. Dysesthesia, paresthesia, and bladder and bowel incontinence can also occur. Lower thoracic root rupture can cause paralysis of the unilateral abdominal muscles. Muscle spasm and paresthesia can also be seen in this condition. However, cutaneous manifestations of herpes zoster and referred pain from diseased pulmonary, cardiovascular, gastrointestinal, or urogenital organs must be considered when evaluating this condition.

LUMBAR SPINAL NERVES

As in the case with other spinal nerves, these nerves divide into dorsal and ventral rami. The dorsal rami of the lumbar spinal nerves divide into medial branches that supply the multifidi and lateral branches, innervating the erector spinae. Lateral branches of the upper three dorsal rami form the superior cluneal nerves and supply the skin of the gluteal region. The ventral rami accompany the lumbar arteries, receiving, near their origins, the gray communicating rami from the sympathetic ganglia. Upper two or three ventral rami may also receive white communicating rami, conveying presynaptic sympathetic fibers. They form the lumbar plexus and innervate the muscles in the posterior abdominal wall and the lower extremity.

A lesion of the L3 root produces paresis of the quadriceps femoris muscle, diminution of patellar reflex, and pain or hypalgesia (decease sensitivity to nociceptive stimuli) in the thigh extending from the greater trochanter to the medial condyle of the femur. However, the sensation from the medial leg and medial border of the foot carried by the saphenous nerve remains unaffected, a fact that distinguishes femoral nerve palsy from L3 root dysfunction. Adductor paresis and hyporeflexia can occur if L3 and L4 roots are damaged.

Compression of L4 produces paresis of the vastus medialis and pain or impairment of sensation in the lower medial leg accompanied by the paresis of the tibialis anterior muscle. The latter motor dysfunction is a distinguishing characteristic of L4 root injury from femoral nerve palsy. Loss of sensation in the entire medial leg, impairment of sweating in the territory of the saphenous nerve, and patellar hyporeflexia are pathognomic for femoral nerve palsy.

Manifestations of L5 root lesion, which are common, include pain in the anterolateral leg and middle portion of the dorsum of the foot. The muscle within the anterior compartment of the leg, which is consistently affected, is the extensor hallucis longus with loss of posterior tibial reflex and preservation of patellar and ankle reflexes. Tibialis posterior reflex is elicited by tapping muscle tendon inferior and anterior to the medial malleolus producing inversion, adduction, and plantar flexion of the foot. However, when S1 is also involved, paresis of the extensor digitorum brevis is regularly observed. A combined lesion of L4 and L5 root can occur in a herniated disk between L4 and L5 producing paresis of the muscles in the anterior compartment of the leg (tibialis anterior and extensor hallucis longus) with intact muscular components of the lateral leg.

LUMBAR PLEXUS

Terminal Branches of the Lumbar Plexus

This plexus is formed by the ventral rami of the upper four lumbar spinal nerves, with small contribution from the twelfth thoracic spinal nerve. The contribution of the fourth lumbar ventral ramus is small. Each of the ventral rami of the first and second, and occasionally the third, is connected to the sympathetic ganglia via the white communicating rami. In the same manner, all ventral rami of the lumbar spinal nerves receive gray communicating rami. The ventral ramus of the fourth lumbar spinal segment contributes to both the lumbar and sacral plexuses (nervus furcalis). It is considered a prefixed plexus when the ventral rami of the third and fourth lumbar spinal nerves contribute to both the lumbar and sacral plexuses. However, when the fifth lumbar ventral ramus split between the lumbar and sacral plexuses, the lumbar plexus is considered a postfixed plexus. This plexus (Figure 10.20) runs anterior to the transverse processes of the lumbar vertebrae and posterior to the psoas major muscle, giving rise to the iliohypogastric, ilioinguinal, genitofemoral, lateral femoral cutaneous, femoral, obturator, and possibly accessory obturator nerves. This plexus has a simpler pattern than the brachial plexus in which a small branch from the twelfth thoracic ventral ramus joins the ventral ramus of the first lumbar spinal nerve and forms a trunk that splits into two branches. The larger branch gives rise to the iliohypogastric and ilioinguinal nerve, while the lower smaller branch unites with the second lumbar ventral ramus to form the genitofemoral nerve. Part of the ventral ramus of the second, the entire third, and part of the fourth lumbar ventral ramus unite and

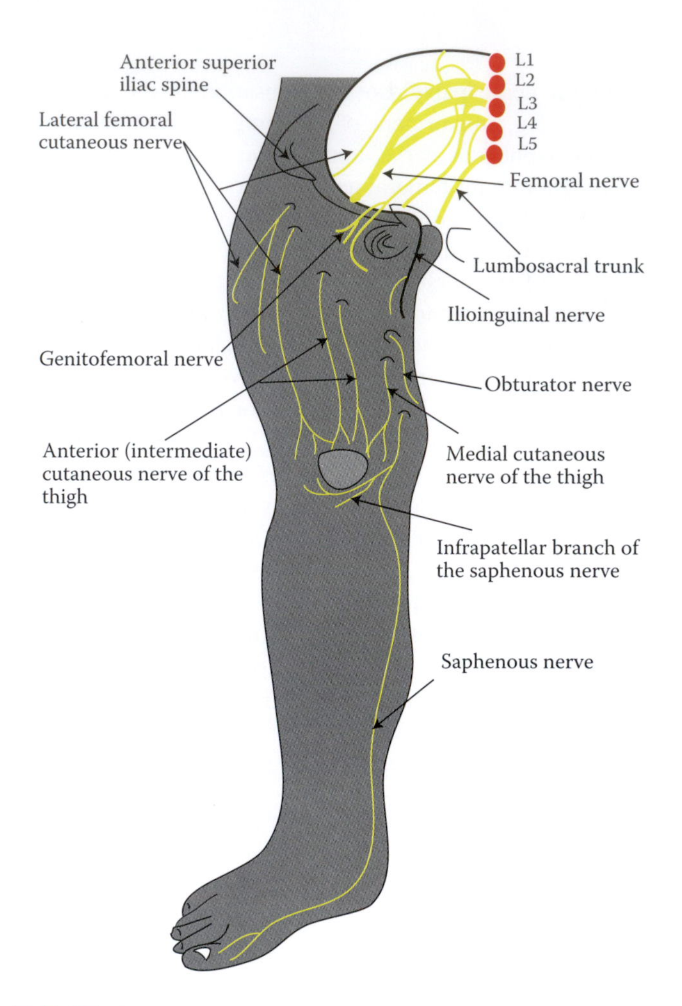

FIGURE 10.20 Diagram of the lumbar plexus showing segmental contributions to the individual branches.

then divide into anterior and posterior branches. Union of the anterior branches results in the formation of the obturator nerve, while union of the posterior branches produces the femoral nerve. Frequently, branches of the third and fourth ventral rami form the accessory obturator nerve. Parts of the posterior branches of the second and third lumbar ventral rami join to form the lateral femoral cutaneous nerve.

Iliohypogastric nerve (L1) (Figure 10.20) arises from the entire ventral ramus of the first lumbar spinal nerve with a smaller contribution from the subcostal nerve. It runs with the ilioinguinal nerve for a short distance, pierces the psoas major muscle, and courses between the kidney and the quadratus lumborum toward the iliac crest. It then pierces the transverse abdominis, running between the transverse and internal oblique muscles toward the deep ring of the inguinal canal. Analogous to the intercostal nerves, it divides into lateral and anterior branches, with the lateral branch supplying the skin of the posterolateral gluteal region and an anterior branch that runs superior to the superficial inguinal ring and innervates the skin of the suprapubic region. Both branches also supply the internal oblique and transverse abdominis muscles.

Damage to the iliohypogastric nerve can occur as a result of retroperitoneal tumors, surgical operation on the kidney, paranephrotic condition, entrapment in lower abdominal sutures, and pressure of a tight belt or jeans at the iliac crest. These conditions do not always produce significant clinical manifestations since other nerves also supply the muscles and skin innervated by this nerve. However, a surgical incision in the lower anterior abdominal wall that can compromise the function of the ilioinguinal nerve can weaken the anterior abdominal wall and predispose the patient to direct inguinal hernia. Damage to this nerve produces pain or very limited sensory loss in the area of distribution of this nerve due to overlap by the adjacent cutaneous branches.

The *ilioinguinal nerve* (L1) (Figure 10.20) resembles a typical intercostal nerve without a lateral branch. It maintains similar segmental origin to the iliohypogastric nerve and runs parallel but caudal to it across the iliacus muscle and then between the internal oblique and the transverse abdominis muscles. It continues through the inguinal canal inferior to the spermatic cord or the round ligament. This nerve emerges ventral and inferior to the anterior superior iliac spine through the superficial inguinal ring. It supplies the lower portions of the internal oblique and transverse abdominis muscles. Its anterior cutaneous branch pierces the aponeurosis of the external oblique muscle anterior to the anterior superior iliac spine and along the inguinal ligament and leaves the medial side of the superficial inguinal ring. This cutaneous branch supplies the skin of the upper medial thigh and the anterior part of the external genitalia. A strip of the skin parallel to inguinal ligament is supplied by a lateral recurrent branch.

The ilioinguinal nerve is prone to injury in surgical repair of an indirect inguinal hernia (herniorrhaphy), as a result of closure and dissection of the hernial sac, or subsequent to a low incision in the anterior abdominal wall while performing an appendectomy or intramuscular muscular injection. It may be compressed by constant and violent contraction of the transverse and internal oblique muscles of the anterior abdominal wall as a result of a fall from a height, abnormalities in the hip joints, or ligamentous disorders of the vertebral column. It can also be damaged as a result of nephrectomy or when an intramuscular injection is placed too far proximally above the iliac crest. Weakness in the abdominal muscles innervated by the ilioinguinal nerve can precipitate an inguinal hernia. Pain associated with entrapment of the ilioinguinal nerve may mimic urinary tract or gastrointestinal

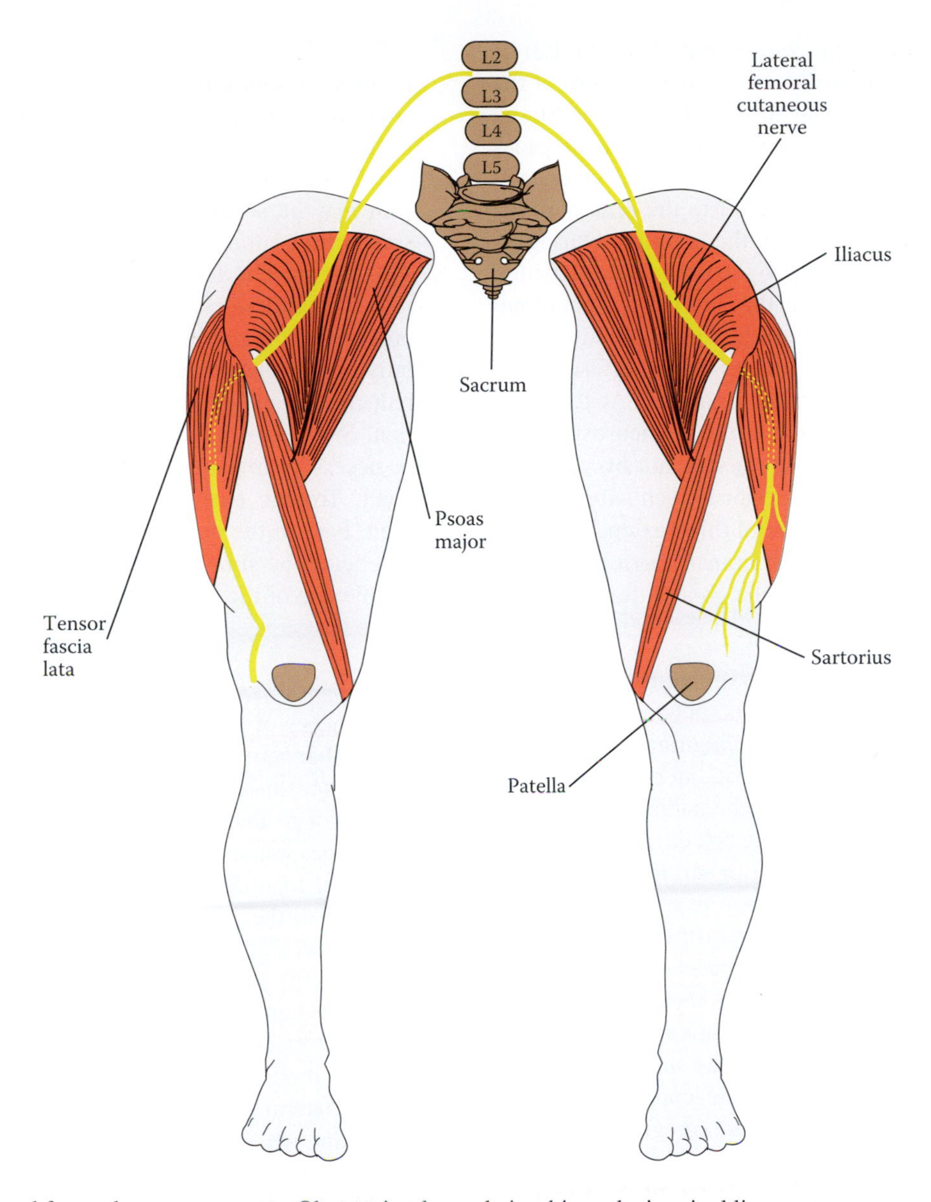

FIGURE 10.21 Lateral femoral cutaneous nerve. Observe its close relationship to the inguinal ligament.

disorders. Pain, which is of burning nature, can be felt over the lower abdomen, radiating in the mons pubis, root of the penis, and upper medial thigh. This burning sensation can limit the ability to extend and medially rotates the hip. Tenderness is felt medial to the anterior superior iliac spine. Patients walk with a stooped gait and a flexed hip and bent over (gait of a terrified rookie skier), and experience difficulty rising from a seated position.

The *genitofemoral nerve* (L1 and L2) (Figure 10.20) is derived from the anterior branches of the ventral rami of the first and second lumbar spinal nerves. It perforates the psoas major muscle, passes deep to the peritoneum, posterior to the ureter, and then divides into genital and femoral branches. These two branches course along the external iliac artery. The genital branch (L1) enters the deep inguinal ring, supplying the cremasteric muscle, skin of the scrotum, mons pubis, and the labia majora. The femoral branch (L2) descends lateral to the external iliac artery and posterior to the midpoint of the inguinal ligament, pierces the femoral sheath, and supplies the skin anterior to the upper part of the femoral (Scarpa) triangle.

Injury to this nerve can occur as a result of direct injury, herniorrhaphy, or during femoral vessel catheterization or appendectomy, particularly when the appendix is in a retrocecal position anterior to the psoas major muscle. Wearing tight jeans can also cause damage to this nerve. Sensory loss and/or pain (neuralgia of the spermatic cord) and absence of cremasteric loss occur.

The *lateral femoral cutaneous nerve* (L2 and L3) is formed by the posterior branches of the ventral rami of the second and third lumbar spinal nerves (Figures 10.20 and 10.21). This nerve leaves the posterior surface of the psoas major, runs in the double fascial layer of the iliacus, and continues a downward course posterior to the cecum on the right side and descending colon on the left side. During its course, it provides sensory fibers to the parietal peritoneum. This nerve passes posterior to or through the inguinal ligament, medial to the anterior superior iliac spine, and anterior to the sartorius and enters the space under the fascia lata, splitting into anterior and posterior branches. At the point of passage of the nerve to the thigh, the aponeurosis of the external oblique is anchored to the fascia lata by vertically running tendinous fibers. The anterior branch supplies part of the anterior and most of the lateral thigh extending as far downward to the knee. The lateral branch carries sensation from the upper lateral thigh below the greater trochanter.

The lateral femoral cutaneous nerve can be damaged in individuals with lumbar lordosis, pelvic cancer, as a result of entrapment within the inguinal ligament, bone marrow biopsy of the iliac crest, intertrochanteric osteotomies compression against the anterior superior iliac spine, pressure from ill-fitting belt during traction, retrocecal appendectomy, or anterior surgical approach to the hip joint. Constant adduction (e.g., sitting with crossed legs for prolonged period of time), compensatory stretching of the fascia and muscles around the nerve, and disorders in the ligaments that stabilize the vertebral column can contribute to damage of the lateral femoral cutaneous nerve. Obesity and prolonged standing to attention can precipitate compression of this nerve. Prolonged strenuous hike, pregnancy, confinement to a wooden bed in supine position, excessive angulation, and subsequent compression of the lateral femoral cutaneous nerve can occur as a sequela to extension of the leg while the thigh is abducted, particularly postpartum.

Compression of this nerve results in manifestations of *meralgia paresthetica*. This is commonly a unilateral condition seen predominantly in men at any age, particularly in obese individuals following a substantial weight loss or weight gain (pendulous abdomen). It is also seen as a result of wearing tight jeans or belt or prolonged standing, or subsequent to a developed pelvic tilt in muscular dystrophy patients. Occasionally, it is seen after abdominal operations. It has been observed following intertrochanteric osteotomy and in patients who spends long time lying in supine position on wooden floor. Despite its course behind the cecum on the right side and descending colon on the left side, no apparent abdominal disorders have been linked to these relationships. Patients exhibit intermittent painful paresthesia in the form of numbness and tingling or burning sensation in the anterolateral thigh extending from the groin to the knee. This sensation is not easily tolerable during the episodic attacks, which show relief when changing posture or movement. Affected skin may be so sensitive that the patient experiences severe discomfort to his or her own clothing (dysesthesia). The patient develops a habit of rubbing his thigh to alleviate these symptoms. No motor deficits will be observed. This condition can resolve itself and particularly in patients who successfully manage their weight reduction. Flexion of the hip may alleviate the sensation. Steroid therapy and/or surgical exploration may be necessary in patients resistant to conventional treatment. Trophic changes and hypotrichosis are rarely seen. Pain in the area of distribution of the nerve can be elicited by simultaneous hyperextension of the hip and flexion of the knee ("reverse Lasègue").

The *femoral nerve* (L2, L3, and L4) is formed by the posterior branches of the ventral rami of the second, third, and fourth lumbar spinal nerves (Figures 10.20 and 10.22). Early in its course, this nerve runs deep to the psoas major, and then anterior to the iliacus, descending between the iliacus and the psoas major covered by the iliac fascia. Proximal to the inguinal ligament, it provides branches to the iliopsoas muscle and to the pectineus, which runs deep to the femoral vessels. Within the femoral triangle, it supplies sensory fibers to the hip joint, and to the anterior thigh via the anterior (intermediate) femoral cutaneous branch, and to the skin of the lower medial thigh via the medial femoral cutaneous branch. As it descends posterior to the inguinal ligament, it lies on the lateral side of the femoral artery within the lacuna musculorum, where it divides into several terminal branches that fan out in the femoral triangle, innervating the sartorius, and quadriceps femoris.

The branch to the rectus femoris enters the muscle in its proximal third, the branch to the vastus lateralis follows a course posterior to the rectus femoris, while the branch that supplies the vastus medialis descends parallel to the saphenous nerve and femoral vessels. Similar course is pursued by the branch to the vastus intermedius. The saphenous nerve, the longest branch of the femoral nerve, accompanies the femoral artery in the femoral triangle and into the adductor canal, piercing the adductor membrane medial to the knee joint to supply the medial surface of the leg and the medial border of the foot (Figure 10.23). During its course in the medial leg, it is accompanied by the great (long) saphenous vein and gives rise, proximal to the medial femoral condyle, to the infrapatellar branch that carries sensation from medial aspect of the knee down to the tibial tuberosity.

The *obturator nerve* (L2, L3, L4) stems from the anterior branches of the ventral rami of the second, third, and fourth lumbar spinal nerves (Figure 10.24). Its proximal part is embedded in the psoas major muscle. Following a descending course on the medial border of the muscle, it crosses the

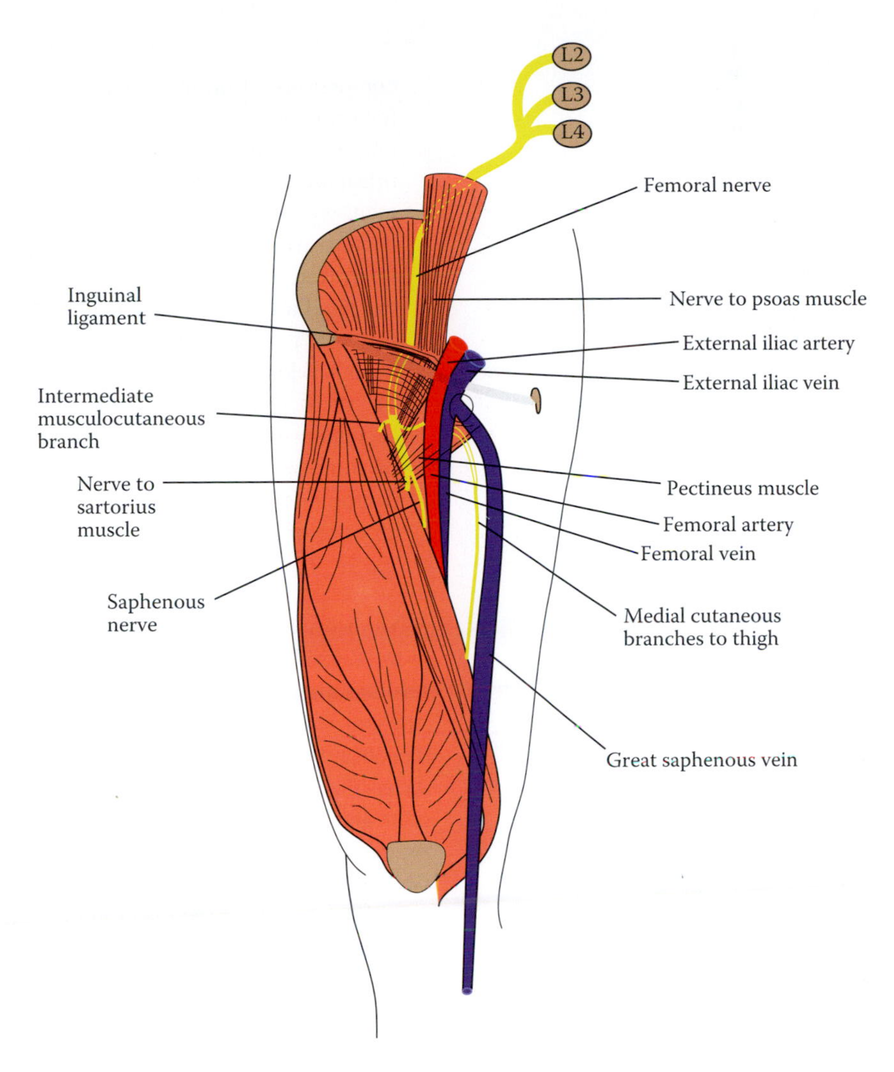

FIGURE 10.22 The origin, course, and distribution of the femoral nerve in the thigh.

Intactness of the femoral nerve can be tested by performing certain movement at the hip and knee joints. Hip flexors are tested by asking a seated patient to flex his/her hip against resistance from the examiner. External rotation of the flexed hip by the examiner (tailor's muscle) tests the integrity of the sartorius. With the patient supine in seated position and the leg hanging freely, knee extension will reveal the integrity of the rectus femoris and sartorius.

Injuries to the femoral nerve can occur in dislocation of the hip joint, hip surgery, as a result of stab or gunshot wound, or as a sequela to fractures of the coxa or proximal femur. Femoral nerve neuropathy in diabetic patient, retroperitoneal abscesses or tumors, pelvic surgery and in particular hysterectomy, vaginal hysterectomy, sustained abduction, and flexion and rotation of the leg in the lithotomy position can endanger the femoral nerve. Other causes include neurinoma of the L3 root, malignant lymphoma, neuralgic amyotrophy, and iliofemoral artery–induced ischemia of the supra-inguinal part of the femoral nerve. Additionally, retroperitoneal hematoma posterior to the psoas major sheath due to hemophilia or anticoagulant therapy, retrocecal appendectomy, or complication of femoral angiography may also cause painful femoral nerve palsy, which includes paralysis of the quadriceps femoris with loss of the patellar reflex and impairment of knee extension. Extension of the knee may still be possible via the iliotibial tract. Patients are unable to stand, buckle easily, fall frequently, and have difficulty walking on uneven surfaces and severe difficulty going up stairs; however, going downstairs, the affected leg is used to take the first step. They cannot climb stairs and are unable to swing the lower extremity forward during walking. Complete paralysis of the sartorius and rectus femoris and partial paralysis of pectineus muscle can also occur, leading to weakened thigh flexion. Knee extensors can be tested by extending the knee with the patient in supine position with the lower leg hanging freely. Sartorius muscle function can be tested

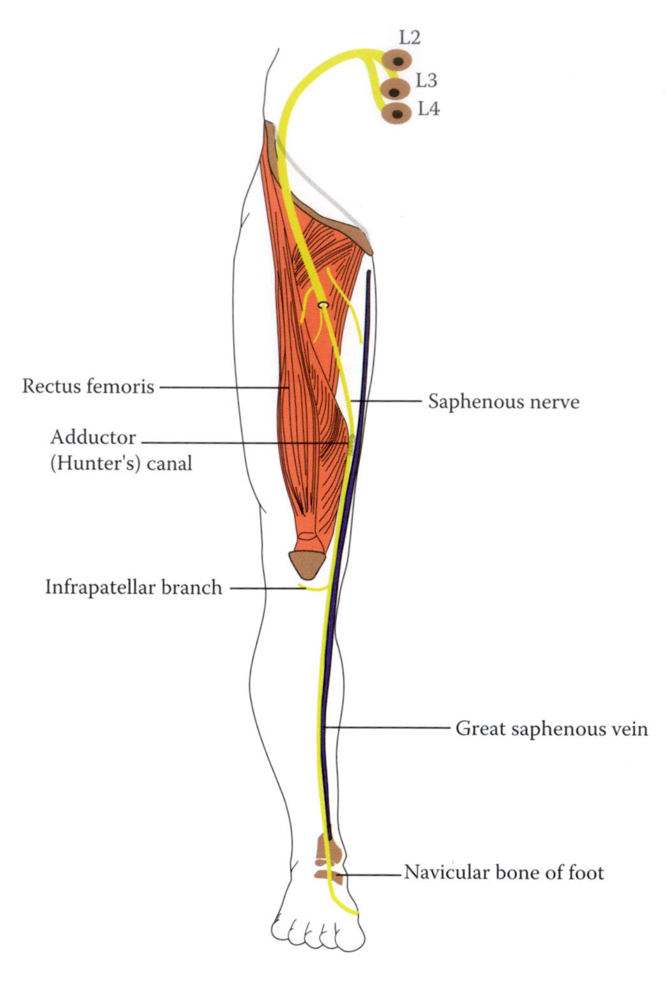

FIGURE 10.23 Course and termination of the saphenous branch of the femoral nerve is illustrated. Note the infrapatellar branch of this nerve.

by external rotation of flexed thigh. However, the iliopsoas, which is the main flexor of the thigh, remains intact or partially affected. Sensory loss on the anterior and lower medial thigh and the medial surface of the leg and foot is also observed in femoral nerve damage.

The *saphenous nerve*, in particular, can be entrapped as it exits the adductor canal. It is injured in surgical treatment of saphenous vein varicosities, as a result of insertion of aorticofemoral silicone prosthesis, excessive hyperextension of the hip joint, and irritation by phlebitis of the long saphenous vein or burns. It can be entrapped in the Hunter canal and injured in shunt surgery between the femoral artery and great saphenous vein in patients on chronic dialysis. Injured nerve is associated with numbness or anesthesia in the medial surface of the leg and medial border of the foot. Pain and heaviness may also develop in the medial leg, particularly when the thigh is hyperextended (reversed Lasègue sign). Tinel's sign can be elicited along the course of the nerve, typically at the site of compression. Entrapment can occur above the medial femoral condyle as the nerve pierces the fascia lata and pursues a superficial course. Accidental excision of infrapatellar branch of the saphenous nerve, its entrapment within the fascia medial to the knee, or its injury during arthroscopic knee surgery, can result in the formation of neuroma, eliciting excruciating pain over the patella (infrapatellar neuropathy). Sustained pressure that results from resting of the knee on a car door during a long trip or resting the medial knee against the surfboard (surfer's neuropathy) can compress the saphenous nerve, producing sharp pain. Manifestations of compression of this nerve (Gonyalgia paresthetica) can be in the form of numbness and paresthesia in the absence of any trauma.

pelvic brim, and along the lateral pelvic wall, it runs anterior to the sacroiliac joint and posterior to the common iliac vessels and ureters. On the lateral pelvis, it courses between the external iliac vein and internal iliac artery and then enters the obturator canal accompanied by the obturator vessels. Arterial anastomosis between the pubic branches of the obturator and inferior epigastric arteries occurs near the obturator canal in close proximity to the nerve. Lymph nodes are similarly located adjacent to the obturator nerve and vessels near the entrance of the obturator canal. Inside the canal, the obturator nerve is the most proximal, followed by the artery and then the vein with intervening adipose tissue. Following its exit from the obturator canal, the obturator nerve divides into anterior and posterior branches, separated by the adductor brevis muscle. The anterior branch supplies the adductor longus and brevis, gracilis, and pectineus muscles and then terminates as sensory branch that supplies the midportion of the medial thigh. The branch that innervates the obturator externus usually arises from the trunk of the nerve within the canal before bifurcation. The posterior branch passes deep to the adductor brevis and innervates the adductor magnus. There are articular branches provided by the posterior branch of the obturator nerve to the hip and posterior knee joint, which accounts for knee pain experienced by patients with arthritic condition of the hip joint.

The obturator nerve can be entrapped by an obturator hernial sac or within the obturator membrane, cut during hernial repair and operations involving the urogenital system. It can also be injured in pregnancy as a result of the pressure exerted by the head of the fetus, as a consequence of metastatic pelvic disease, pelvic fractures, hip replacement, or complicated labor. Damage to the obturator nerve has been reported in the obese elderly after exhausting long walk. Surgically, the obturator nerve may be excised in paraplegic patients due to cerebral palsy to relieve

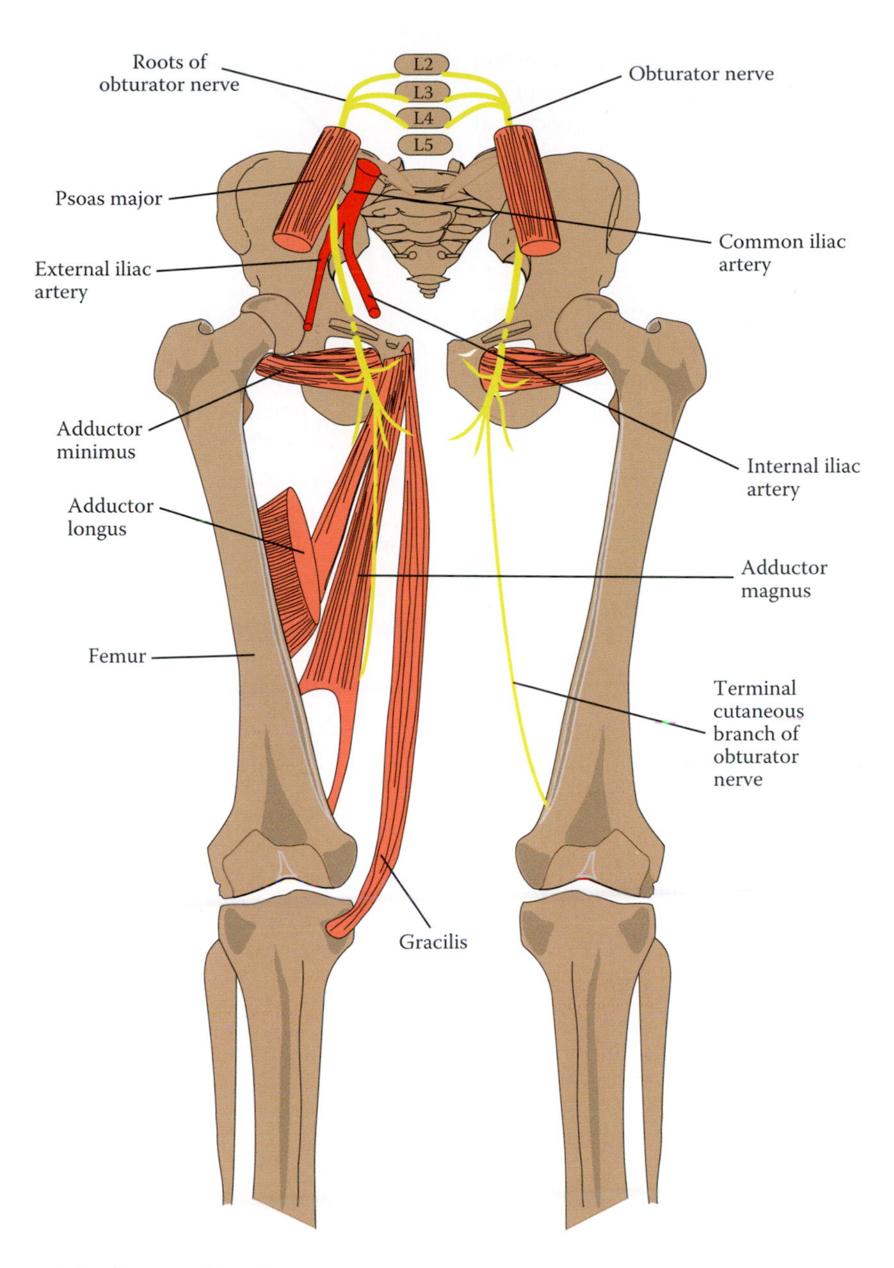

FIGURE 10.24 Course and distribution of the obturator nerve.

spasticity in the adductor muscles. The deficits associated with damaged obturator nerve are chiefly sensory that encompass paresthesia, sensory loss, or radiating pain over the middle part of the medial thigh that can be exacerbated when intra-abdominal pressure is increased during coughing or sneezing or by extension or lateral movement of the thigh. Deficits also include weakened or impaired ability to adduct the thigh or cross the legs, weakened medial rotation of the thigh, and tendency to swing the legs outward during walking. These may lead to motor disability during walking. The leg is held in abducted position producing a wide-steppage gait. Complete atrophy of the adductor muscles does not occur due to the dual innervation of the adductor magnus by the obturator and sciatic nerve. The affected individual experiences numbness or pain that radiates to the middle portion of the medial thigh. Diseases of the hip and knee joints produce pain that radiates to the cutaneous areas of the obturator nerve. Pubic bone osteitis and inflammatory edema following surgical operation of the pelvic organs or post-surgical scar can induce knee pain by causing compression or irritation of the articular rami of the posterior branch of the obturator nerve. Obturator neuralgia is an ill-defined and lesser-known condition, which produces neuropathic pain in the medial side of the thigh. The etiologies can be idiopathic, trauma-induced, or occurring subsequent to inguinal operation.

The accessory obturator nerve (L3, L4) is an inconstant nerve, which gains origin from the anterior branches of the ventral rami of the third and fourth lumbar spinal nerves. It travels posterior to the pectineus and supplies the obturator muscle and the hip joint.

SACRAL SPINAL NERVES

The dorsal rami of the sacral spinal nerves emerge from the dorsal sacral foramina with the exception of the fifth sacral nerve. Each dorsal ramus gives off medial branches that terminate in the multifidi and lateral branches that join the dorsal ramus of the fifth lumbar spinal nerve to form the middle cluneal nerves that supply the skin of the posterior gluteal region. Dorsal rami of the fourth and fifth sacral nerves also innervate cutaneous fibers to the skin overlying the coccyx. Visceral efferent fibers of the sacral plexus form the pelvic splanchnic nerves. These parasympathetic efferents arise from the intermediolateral columns of the second, third, and fourth sacral spinal segments and run in the ventral rami. Due to the vasodilator effect upon the penile arteries and the important role they play in erection, these parasympathetic fibers are also known as nervi erigentes. They also control micturition by inducing contraction of the detrusor muscles and relaxation of the urethral sphincters. Additionally, these nerves provide parasympathetic fibers to the left one-third of the transverse colon, descending colon, sigmoid colon, and the rectum.

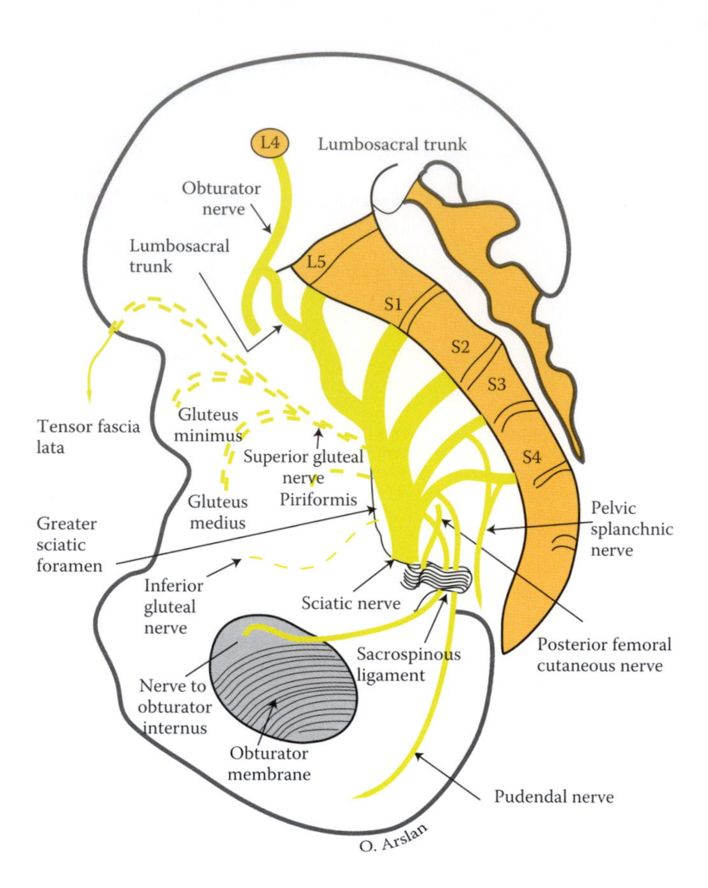

FIGURE 10.25 This is a schematic representation of the sacral plexus. Segmental contributions, divisions, and individual branches are illustrated.

> When S1 is traumatized, sensory and motor deficits ensue. The affected dermatome lies in close proximity to L5 dermatome with a dorsolateral disposition. It includes the posterior thigh and lower gluteal region proximally and the posterior malleolus, lateral border of the foot to the third and fifth toes distally. Paresis of the fibularis brevis, and consequently weakened eversion of the foot, is noted. Weakened posterior tibialis causes difficulty in pressing the foot against the floor when standing. Gastrocnemius and soleus are also affected with loss of calcaneal tendon (ankle jerk) reflex. Patients have difficulty standing on tip toes of the affected foot. You may also expect to see signs of gluteal muscle palsy and low positioned gluteal fold on the affected side and extremely rarely the positive Trendelenburg sign due to lesser gluteal muscle palsy. L5–S1 lesion affects the extensor digitorum longus and brevis muscles within the anterior compartment of the leg, sparing the tibialis anterior. It also produces ankle jerk hyporeflexia and paresis of the fibularis longus and brevis muscles.

SACRAL PLEXUS

The sacral plexus is embedded in the digitations of the piriformis muscle on the posterolateral wall of the pelvis, anterior to the sacrum and posterior to the rectum. It is formed by the union of the lumbosacral trunk with the ventral rami of the first, second, third, and part of the fourth sacral spinal nerves (Figures 10.25 and 10.26). The lumbosacral trunk results from the union of a portion of the ventral ramus of the fourth lumbar spinal nerve and the entire ventral ramus of the fifth lumbar spinal nerve. Branches of the sacral plexus leave through the greater sciatic foramen, proximal and/or distal to the piriformis muscle.

> A pregnant uterus, malignancies involving the rectum, and other pelvic structures can compress the sacral plexus. Disk herniation between L4 and L5 or L5 and S1 most often injures the L5 and S1 roots, respectively. Compression of this plexus produces pain that radiates to the posterior thigh and leg, and diminished or lost ankle or knee tendon reflexes. Aneurysm of the superior gluteal artery may specifically affect the lumbosacral trunk.

The sacral plexus gives rise to branches, which supply the gluteal region, posterior leg, and foot. These branches are the superior gluteal, inferior gluteal, posterior femoral cutaneous, pudendal, sciatic nerve, and pelvic splanchnic nerves. The latter has already been described earlier.

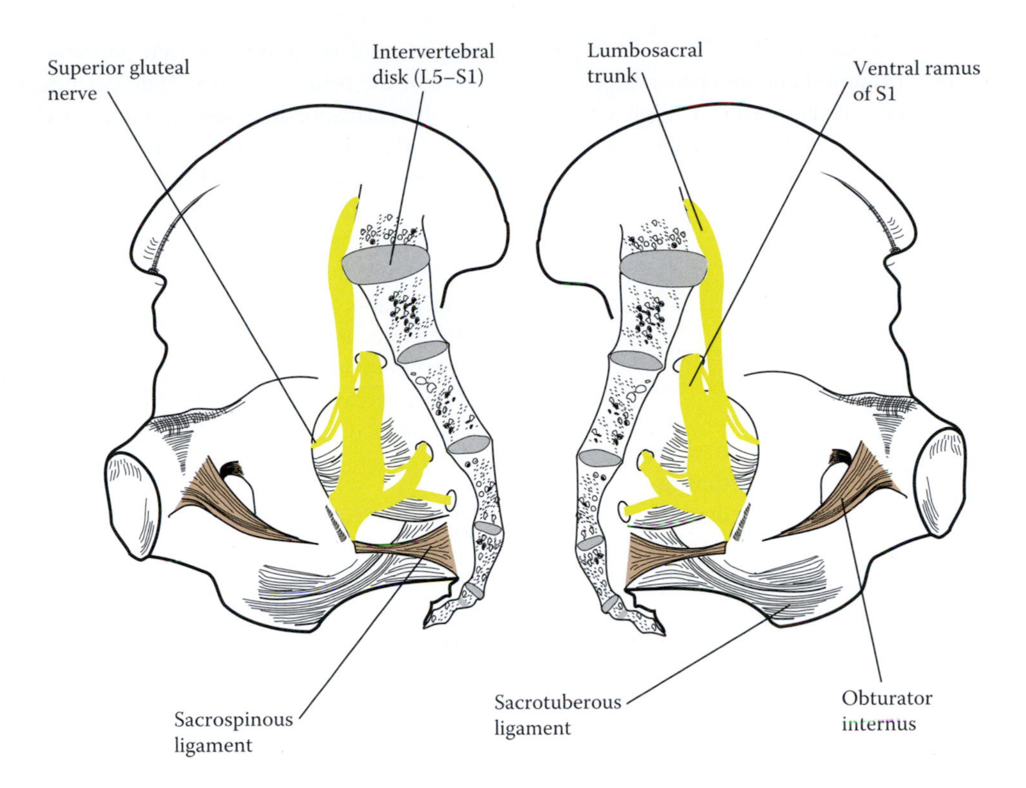

FIGURE 10.26 Roots of the sacral plexus is illustrated in relation to the piriformis muscle. Observe the lumbosacral trunk.

The superior gluteal nerve (L4, L5, S1) is formed by the posterior branches of the fourth and fifth lumbar and the first sacral ventral rami (Figures 10.25 and 10.26). It leaves the pelvis through the greater sciatic foramen and proximal to the piriformis muscle and enters the intergluteal space between the gluteus medius and minimus, accompanied by the corresponding vessels. It supplies the abductors and the medial rotators of the thigh (gluteus medius, gluteus minimus) and then continues forward to innervate the tensor fascia latae.

Injury to this nerve can result from intragluteal injections and surgical operation of the hip joint. Damage to the superior gluteal nerve, although rare, causes loss of abduction of the thigh and subsequent tilting of the pelvis toward the unsupported side when the foot is off the ground during walking, producing lurching gait as the paretic gluteus medius and minimus are unable to keep it in a horizontal position (Trendelenburg sign). As a compensatory mechanism, some patients exhibit "Duchenne sign," which is characterized by tilting pelvis toward the weight-bearing side, thus hindering the descent of the pelvis toward the swinging (unsupported) side.

Terminal Branches of the Sacral Plexus

The inferior gluteal nerve (L5, S1, S2) arises from the posterior branches of the fifth lumbar and first and second sacral ventral rami (Figures 10.25 and 10.26). This nerve exits the pelvis via the greater sciatic foramen, inferior to the piriformis muscle, accompanied by the corresponding vessels, sciatic nerve, posterior femoral cutaneous nerve, pudendal nerve, and internal pudendal vessels. It innervates the gluteus maximus muscle and provides articular branches to the hip joint.

Damage to the inferior gluteal nerve can result from gunshot wounds at the greater sciatic foramen, subsequent to child birth, a fracture of the neck of the femur, and surgical intervention to treat rectal cancer. Inferior gluteal nerve palsy produces atrophy and wasting of the gluteus maximus and impairment of extension of the thigh (although some extension remains possible by the hamstrings). Patients are unable to jump, climb stairs, or rise from a seated position.

The posterior femoral cutaneous nerve (S1, S2, S3) comes from the posterior branches of the first and second sacral and the ventral branches of the second and third sacral ventral rami (Figure 10.25). This nerve leaves the pelvis via the greater sciatic foramen, distal to the piriformis muscle. It supplies sensory fibers to the posterior thigh, as far down as the popliteal fossa. It gives rise to the inferior cluneal branches, which innervate the skin of the lower gluteal region. It also supplies the posterior part of the external genitalia via the perineal branch.

Damage to the posterior femoral cutaneous nerve produces anesthesia primarily in the posterior thigh with no motor deficits. Recurrence of pain after successful pudendal nerve block may be attributed to intactness (not affected by the anesthetic) of the perineal branch of the posterior femoral cutaneous nerve, which also supplies the external genitalia.

The *pudendal nerve* (S2, S3, S4) originates from the anterior divisions of the ventral rami of the second, third, and fourth sacral ventral rami (Figures 10.25 and 10.27). It leaves the pelvis via the greater sciatic foramen between the piriformis and the coccygeus muscles, medial to the sciatic nerve, and enters the gluteal region, accompanied by the internal pudendal vessels. Following its exit from the pelvis, the nerve crosses the ischial spine and the sacrospinous ligament and enters the ischiorectal fossa through the lesser sciatic foramen. It travels in the pudendal canal on the lateral wall of the ischiorectal fossa, giving rise to the inferior rectal branch. The inferior rectal nerve pierces the medial wall of the fascial pudendal canal and travels with the corresponding vessels medially to innervate the external anal sphincter and provide sensory fibers to the lining of the ectodermal part of the anal canal. It innervates the lower third of the vagina, interconnecting with the perineal branch of the posterior femoral cutaneous nerve and the posterior labial (scrotal) branches. After giving rise to the inferior rectal branch, the pudendal nerve divides into perineal branch and dorsal nerve of the penis or clitoris. The perineal branch accompanies the corresponding vessel and gives rise to the posterior scrotal (labial) branches, which are sensory to the scrotum or labia majora and possibly the lower third of the vagina, and to muscular branches to the urogenital muscles including the external urethral sphincter. The dorsal nerve of the penis or clitoris runs along the ischial ramus and then within the urogenital diaphragm, and then within the suspensory ligament of the penis or the clitoris, accompanied by the dorsal artery of the penis or clitoris. This nerve provides sensory fibers to the corpus cavernosum penis and clitoris and the skin and gland penis or clitoridis.

The pudendal nerve can be injured or compressed within the pudendal canal during horseback riding, as a result of a pressure from a mass or exudate in the ischiorectal fossa, pressure of a pregnant uterus, or fracture of the ischial spine. Damage to the pudendal nerve produces loss of sensation from the posterior part of the external genitalia and the ectodermal anal canal. It may also result in paralysis of the perineal muscles including the external urethral sphincter and the external anal sphincters.

Sciatic nerve (L4, L5, S1, S2, S3), the largest and the longest nerve in the body, is derived from the ventral rami of

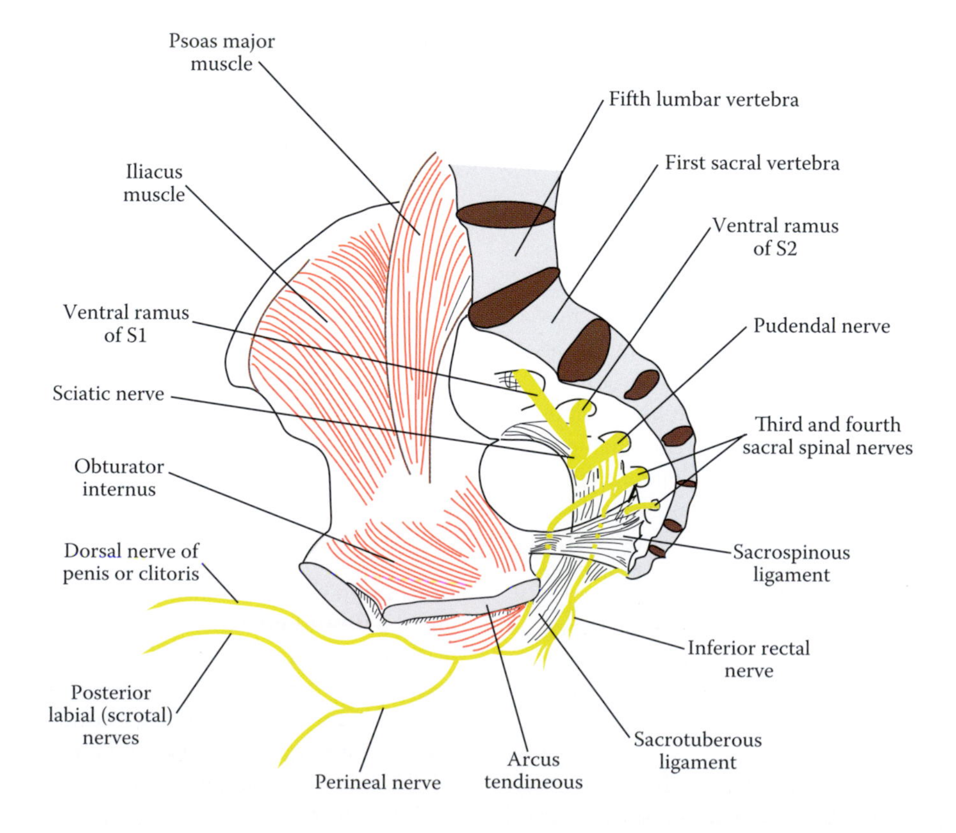

FIGURE 10.27 This diagram illustrates the course and branching of the pudendal nerve.

the fourth and fifth lumbar and the first, second, and third lumbar ventral rami. It emerges through the infra-piriform part of the greater sciatic foramen (distal to the piriformis (Figures 10.25, 10.28, and 10.29), descends between two prominent bony landmarks, the ischial tuberosity and greater trochanter, and then continues posterior to the gemelli and the tendon of the obturator internus muscle. It continues to descend posterior to the quadratus femoris that separates the nerve from the hip joint, margins of the acetabulum, and obturator externus. Along this entire course, the sciatic nerve is covered by the gluteus maximus and accompanied by the inferior gluteal vessels and more medially by the posterior femoral cutaneous nerve within the connective tissue of the subgluteal space. The latter connective tissue continues with the same through the greater sciatic foramen. During its downward course, it lies on the lateral side of the posterior cutaneous nerve of the thigh and branches of the inferior gluteal nerve. More distally, it lies anterior to the long head of the biceps femoris. At the lower third of the posterior thigh and near the adductor hiatus accompanied by the popliteal vessels, the sciatic nerve splits into two terminal branches: medially the tibial and laterally the common fibular nerve. The sciatic nerve supplies the hip joint and provides innervation to the long head of the biceps femoris, semitendinosus, and semimembranosus (hamstring) muscles and the part of the adductor magnus that originates from the ischial tuberosity. It receives blood supply from branches of the inferior gluteal artery and further down from branches of the medial and lateral femoral circumflex arteries that form the cruciate anastomosis. Occasionally the sciatic nerve pierces the piriformis muscle or splits earlier with one branch running superior and another branch descending inferior to the muscle.

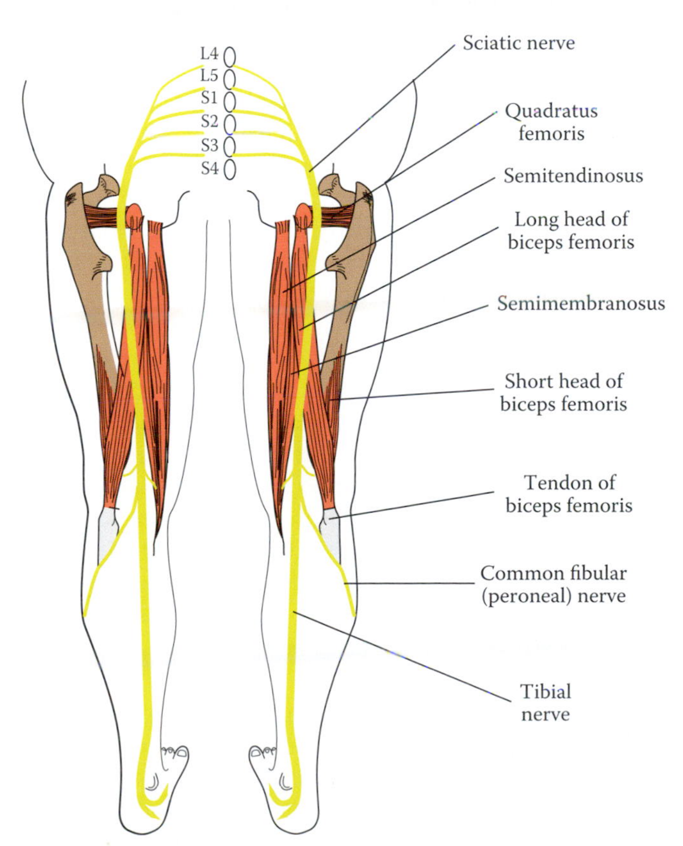

FIGURE 10.28 In this diagram, the segmental origin of the sciatic nerve, its course, and its division into the tibial and common fibular nerves are illustrated.

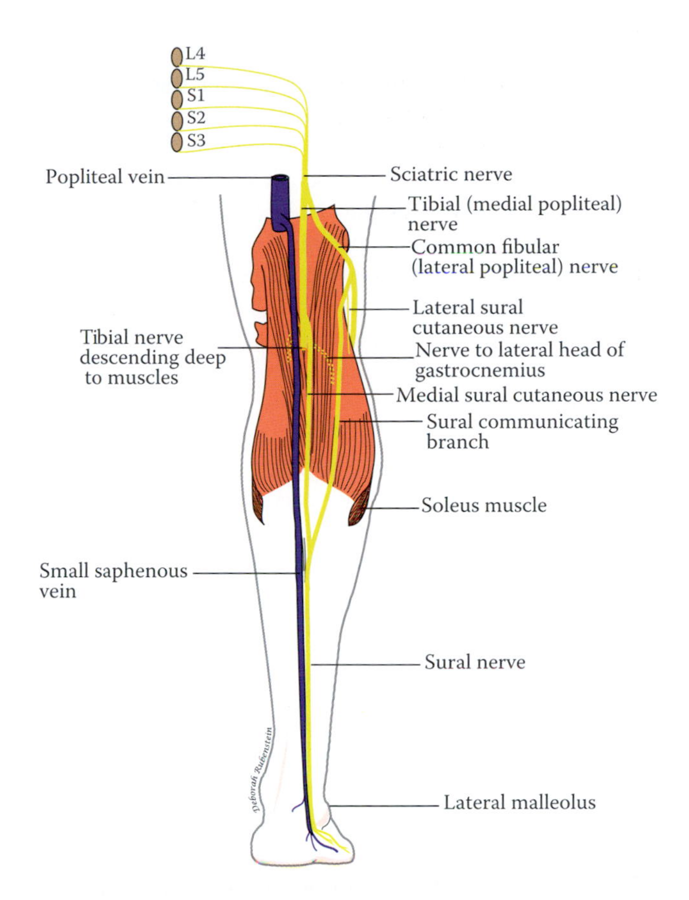

FIGURE 10.29 In this drawing, the common fibular and sural nerves, their relationships, and course are shown.

The sciatic nerve is subject to entrapment within the greater sciatic foramen, between parts of the piriformis muscle (piriformis syndrome), or compressed by an anomalous ligament or fibrous tissue within the greater sciatic foramen. Posterior dislocation of the hip joint, direct trauma, medially placed gluteal injection, prolonged recumbency after surgery or weight loss, hip surgery, fractures of the femur, and aneurysm of the internal iliac or inferior gluteal artery may also impair the function of this nerve. Presence of a myofascial band between the biceps femoris and the adductor magnus in the distal thigh can entrap the sciatic nerve and produce pain immediately proximal to the popliteal fossa. Anticoagulation therapy can produce retroperitoneal pelvic bleeding and as a consequence compression of the sciatic nerve. Sciatic nerve compression due to fibrosis of the gluteus maximus induced by intramuscular injection of pentazocine, an opioid analgesic has been reported. Endometriosis can produce cyclical progressive pain and weakness in the area of distribution of the sciatic nerve due to compression of the nerve fibers by the endometrial mass. Sacral meningocele may also produce compression of the sciatic nerve. Sciatic nerve compression against the ischial spine has been reported in lordotic patients due to compensatory hip flexion. Other causes include exostosis involving the gemelli and obturator internus, hamstring tears, stretching of the nerve during obstetrical intervention, and in Turner syndrome during surgical correction of the skin fold in the popliteal fossa that entraps the sciatic nerve.

Signs and symptoms associated with sciatic nerve damage are equivalent to a combination of damage to the tibial and common fibular nerves. When the sciatic nerve is affected near its origin, or as it crosses within the greater sciatic foramen, or at any site proximal to mid-posterior thigh, weakened knee flexion but not total loss due to intactness of the gracilis and Sartorius, and loss of plantar flexion of the foot, will ensue. Consequently, reflexes associated with the hamstring muscles will be diminished or lost. The patient is able to stand and walk but exhibits foot drop and toe drop and inability to move the foot. Trophic and vasomotor changes may also be seen. Sensations from the posterior leg and dorsum and plantar surfaces of the foot are lost, while sensations from the medial surface of the leg and medial border of the foot remain unaffected. Sensory innervation of the posterior thigh will not be affected unless a concomitant injury damages the posterior femoral cutaneous nerve. Projection of pain (usually of episodic nature) to the posterior thigh and posterior leg, as a result of overstretching of the irritated or inflamed sciatic nerve, is commonly referred to as *sciatica*. The patient may present with low back pain or sciatic pain or both. Backache may be acute, severe, and incapacitating. It may also be gradual in onset and diffuse in nature. Lumbar spasm and abnormalities of posture and restriction of spinal movement are usually seen. Complete recovery is possible; however, the tendency for recurrence of symptoms always exists. Lying down or standing may relieve pain, but it is aggravated by coughing, sneezing, or stooping. This condition is diagnosed by eliciting pain either through tapping the sciatic nerve or flexing the thigh at the hip (30°–70°) while the leg is extended and the patient is in a supine position (*Lasègue sign*). Patients attempt to relieve the pain by flexing the leg at the knee. Trauma to the sciatic nerve results in a severe persistent burning pain (*causalgia*), which may be accompanied by vasoconstriction and sweating in an area larger than the area of distribution of the nerve itself (reflex sympathetic dystrophy). This is due to possible involvement of the sympathetic nerves that accompany the neighboring arteries. When a direct trauma to the sciatic nerve occurs, the common fibular component is more likely to be affected than the tibial component.

Piriformis syndrome develops as a result of trauma to the gluteal region. It is characterized by severe pain in the gluteal region that radiates to the sacrum, the hip joint, and leg, exacerbated by flexing the trunk or carrying heavy weight. Tenderness will be felt at the site of exit of the sciatic nerve from the greater sciatic foramen accompanied by pain upon hip flexion and internal rotation.

Common fibular (peroneal) *nerve* (L4, L5, S1, S2) (Figures 10.25, 10.28, and 10.29) originates from the dorsal branches of the ventral rami of the fourth and fifth lumbar and the first two sacral spinal nerves. It descends posterior and lateral to the popliteal fossa and medial and then posterior to the biceps femoris tendon and the lateral collateral ligament. Then, it encircles the neck of the fibula and pierces the upper part of the fibularis longus, which forms a tendinous "fibular tunnel" over the nerve. Within this tunnel, the common fibular nerve divides into the superficial and the deep fibular (peroneal) nerves. Prior to its division, the common fibular nerve provides articular branches to the knee joint that accompany the superior and inferior lateral genicular arteries and the anterior recurrent tibial artery. It also gives rise to the lateral sural branch (lateral sural cutaneous nerve of the leg) that innervates the skin of the proximal leg. The sural communicating branch passes across the lateral head of the gastrocnemius to join the sural branch of the tibial nerve.

The common fibular nerve is the most predisposed branch of the sciatic nerve to injury even in the case of direct trauma to the sciatic nerve. The nerve is also prone to damage in gunshot wounds, subsequent to a spiral fracture of the neck of fibula, or as a result of the pressure exerted by a cyst on the lateral side of the popliteal fossa. Because of unusual sensitivity to the application of pressure, improperly fitting cast or prolonged squatting can also cause injury to this nerve. Rupture of the lateral collateral ligament, jogging, or sitting cross-legged after weight loss may also be associated with common fibular nerve injury. These conditions produce a painless *foot drop*, a common deficit that occurs due to paralysis and atrophy of the dorsiflexors (extensors) and evertors of the foot. Consequently, attempts to dorsiflex the foot or the toes will not be successful. It is associated with limited loss of sensation over the dorsum of the foot and the upper lateral leg. This limited sensory loss may be due to the overlap of the cutaneous innervation of the affected areas.

The superficial fibular (peroneal) nerve is a component of the lateral compartment of the leg. It begins immediately distal to the neck fibula and descends anterior to the fibula deep to the fibularis longus and then between the fibularis longus and brevis and extensor digitorum longus, which it innervates. It supplies the skin of the lower leg, pierces the crural fascia in the distal third of the leg, and then splits into medial and lateral branches. The larger medial branch provides sensory innervation on the medial side of the great and the adjacent sides of the second and third toes, communicating with the saphenous nerve on the medial border of the foot. The lateral branch supplies the lateral surface of the ankle joint and the adjacent sides of the third and fifth toes, connecting to the sural nerve.

> Interruption of the superficial fibular nerve occurs in lateral compartment syndrome, resulting in numbness and burning sensation on the dorsum of the foot. Impaired eversion, but not total loss, due to intactness of the extensor digitorum longus and fibularis tertius is also observed. Plantar flexion is also compromised due to paralysis of the fibularis longus and brevis; however, dorsiflexion remains possible. Patients walk with the foot in inverted position, and if prolonged, pes equinovarus may ensue. Injury to the superficial fibular nerve can also occur as the nerve pierces the deep fascia of the distal leg in its course to the dorsum of the foot. In this instance, the dysfunction is limited to a burning sensation in the area of distribution of the nerve on the dorsum of the foot.

The deep fibular (peroneal) nerve arises from the common fibular nerve between the upper part of fibularis longus and fibula and then descends deep to the extensor digitorum longus and anterior to the interosseous membrane. In the proximal part of the anterior compartment of the leg, the deep fibular nerve is accompanied by the anterior tibial vessels with varying relationship to the vessel (being lateral and then anterior). It innervates the tibialis anterior, extensor hallucis longus, extensor digitorum longus, and fibularis tertius. Immediately proximal to the ankle joint, the deep fibular nerve passes deep to the superior external retinaculum and then divides into medial and lateral branches. The medial branch, primarily sensory, accompanies the dorsalis pedis artery, coursing in the space between the tendon of the extensor hallucis longus and extensor hallucis brevis muscle. Continuing its descent posterior to the inferior extensor retinaculum, the medial branch passes deep to the extensor hallucis brevis on the dorsum of the foot, running between the tendons of the extensor hallucis brevis and extensor digitorum longus to supply the web of skin between the big toe and second toe. The lateral branch (motor) passes anterior to the ankle joint, crosses deep to the extensor digitorum brevis, which supplies, and follows an anterolateral course on the dorsum of the foot. Through its small twigs, the lateral branch supplies the extensor hallucis brevis, metatarsophalangeal joints of the middle three toes, and the second dorsal interosseous muscle. The trunk of the deep fibular nerve innervates the tibialis anterior, extensor digitorum longus, extensor hallucis longus, and the fibularis tertius muscle. It carries sensation from the dorsal surface of the skin web between the great toe and the second toe.

> Trauma to the dorsum of the foot, poorly fitting casts or shoes, or forceful plantar flexion or eversion of the foot may easily damage the deep fibular nerve. Since the anterior compartment is a confined space sealed by a bony wall and connective tissue septum, which allows no expansion, leg cast (shin splint) can produce compression of the associated vessels and nerves with resultant edema. The pressure from the developed edema may be sufficient to produce ischemic necrosis of the structures and signs of *anterior compartment syndrome*. An intense pain, redness, and swelling confined anterior to the tibia characterize this syndrome. Dorsiflexion of foot and toes becomes very painful. Paralysis of the tibialis anterior and extensor digitorum longus may also occur, producing *foot drop* (Figure 10.30). In addition to the above deficits, slight weakness of eversion and inversion of the foot may also be

FIGURE 10.30 This drawing is a depiction of left foot drop in an individual with damage to the common fibular nerve.

seen. The foot tends to evert upon walking leading, if prolonged, to pes valgus.

Injury to the terminal part of the deep fibular nerve at the ankle joint as it courses deep extensor retinaculum produces manifestation of *anterior tarsal tunnel syndrome*. Injuries to the deep fibular nerve may also occur subsequent to a ganglion cyst, osteophyte, hypertrophied extensor hallucis brevis, or surgical operation of the foot to reduce Lisfranc fracture (fractures of the cuboid, navicular bones, and the metatarsals bones and their subsequent displacement). This type of injury occurs when the foot is plantar flexed and the toes are dorsiflexed particularly in patients wearing high heels. Numbness and tingling in the dorsal web skin between the big and second toes and aching pain in the dorsum of the foot and the ankle joint are exacerbated by activity such as walking or inactivity as in sleeping. Extension or eversion of the foot can relieve the pain. Nocturnal foot pain, which may awaken the patient, paresthesia that radiates to the first web of skin, and positive Tinel's sign are specific manifestations that mimic that of carpal tunnel syndrome. Nocturnal foot pain can awaken the patient. Motor deficits are negligible because of the compensatory action of the extensor digitorum longus and the extensor halluces longus.

The tibial nerve (L4 to S3) is formed by the ventral branches of the ventral rami of the fourth and fifth lumbar and upper three sacral spinal nerves (Figures 10.25, 10.28, and 10.29). Following its division from the sciatic nerve, it descends vertically posterior to the popliteal fossa, superficial to the popliteal vein, where it gives rise to the sural nerve. It descends anterior to the triceps surae muscle (gastrocnemius and soleus), anterior to the fibrous band between the tibial and fibular origin of the soleus accompanied by the posterior tibial vessels. It innervates the popliteus muscle, the muscles of the posterior compartment of the leg, and the sole of the foot. It provides sensory branches to the knee joint, lower lateral surface of the leg, and the lateral border of the foot to the base of the fifth toe via the *sural nerve* (Figure 10.29).

The sural nerve joins a communicating branch from the common fibular nerve and later courses between the two heads of the gastrocnemius muscle, accompanied by the short saphenous nerve. In its downward course, it lies lateral to the calcaneal tendon and then between the lateral malleolus and calcaneus. During its course with the short saphenous vein in the upper leg, it connects to the terminal branches of the posterior cutaneous nerve of the thigh and later on the lateral side of the foot it connects to the superficial fibular nerve. The sural nerve provides sensory branches to the lower lateral and posterior leg and the lateral border of the foot and fifth toe.

Sural nerve biopsy is indicated to confirm diagnosis of certain diseases. Abnormalities detected in the sural nerve may help diagnosis of sarcoidosis, primary amyloidosis, and primary biliary cirrhosis. Disorders of the adrenocorticotropic and thyroid hormones may also produce changes in this nerve.

In the lower third of the leg, the tibial nerve becomes superficial and courses medial to the calcaneal tendon, posterior to the medial malleolus, and deep to the flexor retinaculum (laciniate ligament), where it bifurcates into the medial and lateral plantar nerves (Figures 10.31 and 10.32). Before its division deep to the flexor retinaculum, the tibial nerve gives off a medial calcaneal branch, which carries sensation from the skin of the heel. The medial plantar nerve, the largest branch, proceeds anteriorly lateral to the lateral plantar artery, deep to the abductor hallucis, and then between the latter muscle and the flexor digitorum brevis. It provides cutaneous branches that pierce the plantar aponeurosis and supply the skin of the medial two-thirds of the plantar surface of the foot.

At the bases of the metatarsal bones, the medial plantar nerve gives off three common plantar digital nerves. Each branch further divides into two branches to supply the adjacent sides of the hallux and the second toe, and second and third toes, as well as the third and fourth toes. It also

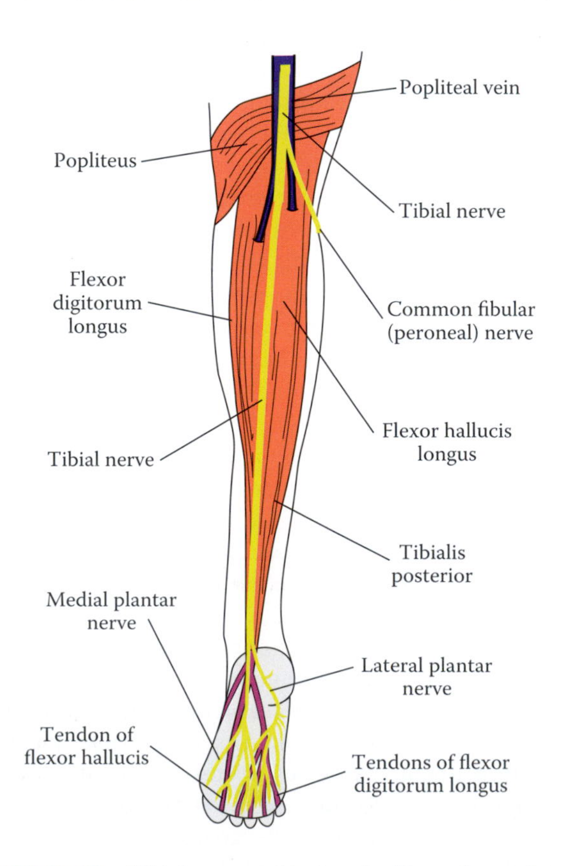

FIGURE 10.31 Tibial nerve and its terminal branches and medial and lateral plantar nerves.

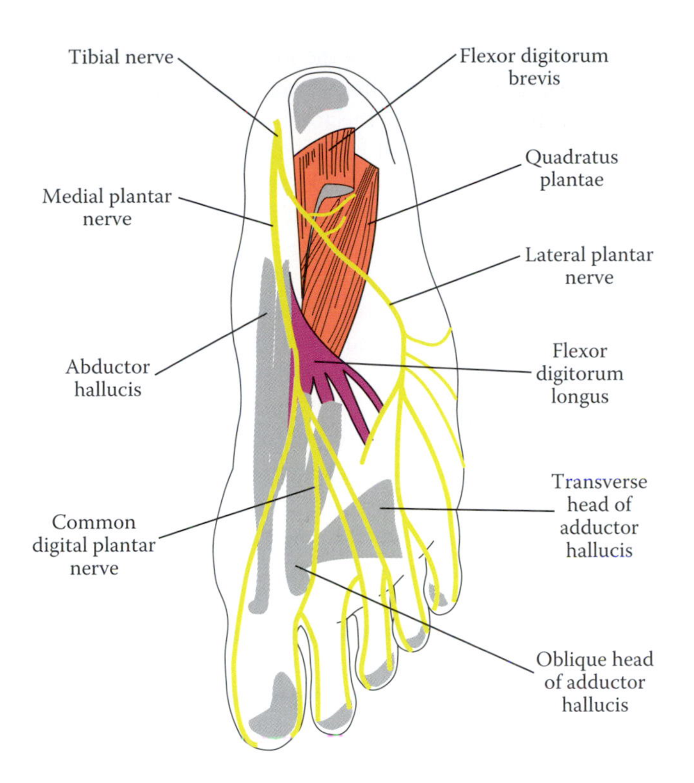

FIGURE 10.32 Medial and lateral plantar nerves.

innervates the abductor hallucis, flexor digitorum brevis, flexor hallucis brevis, and the first lumbrical muscles. The lateral plantar nerve courses anteriorly on the lateral side of the lateral plantar artery, between the flexor digitorum brevis and quadratus plantae. The nerve trunk supplies the quadratus plantae and abductor digiti minimi and carries sensation from the lateral one and a half of the plantar surface of the foot. The lateral plantar nerve then passes between the abductor digiti minimi and flexor digitorum brevis to divide into superficial and deep branches. Through the superficial branch, it supplies the skin of the lateral side of the fifth toe, the flexor digiti minimi, interossei in the fourth intermetatarsal space, and adjacent parts of the fourth and fifth toes. The deep branch of the lateral plantar nerve innervates all lumbricals with the exception of the first, the adductor halluces, and all interossei with the exception of those in the fourth intermetatarsal space.

Damage to the tibial nerve can occur in a fracture of the distal end of the femur, or as a result of trauma to the popliteal fossa, or subsequent to entrapment within the tarsal tunnel. Demyelination of fibers of this nerve can also be caused by thiamine deficiency as in *beriberi disease*. Tibial nerve palsy is associated with the loss of flexion in metatarsophalangeal and interphalangeal joints of the toes (due to paralysis and atrophy of the intrinsic and extrinsic plantar flexor muscles of the foot) with the resultant exaggerated plantar arch (pes cavus). Due to the unopposed actions of the extensor digitorum longus and brevis and the extensor hallucis and brevis muscles, a form of claw toe deformity can be seen. Loss of abduction and adduction of toes, weakened flexion of the leg at the knee, weakened inversion, and impaired plantar flexion may also occur. Due to the extensive area of distribution of the cutaneous branches of the tibial nerve, sensory deficit is striking, and numbness and burning pain may be felt in the sole of foot, especially upon standing. In general, compression of the tibial nerve may be suspected in individuals exhibiting a burning pain and paresthesia in the foot. Prolonged paralysis of the plantar flexors may cause shortening of the calcaneal tendon (Achilles tendon), producing *equinovarus deformity*, which is characterized by plantar hyperflexion, inversion of the foot, and medial rotation of the tibia. Often, the foot and the toes become dorsiflexed at the metatarsophalangeal and flexed at the IP joints (claw foot). The foot tends to strike the ground during walking as a result of dragging the lateral border and distal part of the *foot-slapping gait*. This condition can also arise from intrauterine compression of the spinal roots that contribute to the tibial nerve. Damage to the tibial nerve distal to the middle third of the leg can occur as a result of fractures of the medial malleolus, calcareous, or talus. It may also occur in dislocation of the ankle joint, posttraumatic edema, and compression within the flexor retinaculum (tarsal tunnel syndrome).

Tarsal tunnel syndrome, which is also known as posterior tarsal tunnel syndrome, is usually a unilateral condition associated with posttraumatic deformities of the knee, tight casts, ill-fitting shoes, and *Pott's* (*Dupuytren's*) fracture, which involves the distal end of the fibula and the medial malleolus. Ankle dislocation and twisting, heel varus, joint deformity, cysts, ganglia, anomalous accessory flexor digitorum, abnormally enlarged posterior tibial vein, or posterior tibial arterial aneurysm can all cause this syndrome. Acromegaly, hypothyroidism, rheumatoid arthritis, tenosynovitis, and diabetes mellitus can be associated with tarsal tunnel syndrome. Deficits depending upon the extent of nerve damage include radiating or shooting pain in the ankle and a burning sensation in the sole of the foot, particularly over the heads of the metatarsals, aggravated by walking. Many patients report that pain exacerbates upon rest after a long activity, particularly at night. Any or all the branches of the tibial nerve can be affected in the tarsal tunnel, including the medial plantar, lateral plantar, and calcaneal branches. Tinel's sign can be elicited by gently tapping the area immediately distal to the medial malleolus. Decreased vibratory sense and hypohidrosis are additional signs of this condition. Paresis of the plantar muscles of the foot with no detectable dysfunctions in the muscles of the leg is seen. The metacarpophalangeal joints of the lateral four toes may exhibit hyperextension (extensors are not counteracted by the lumbricals and interossei), whereas the IP joints show hyperflexion (flexors are not opposed by the lumbricals and interossei). This leads to conformational changes in the foot and to instability of the phalanges, which impair walking, particularly pushing-off phase.

Neuromas of the digital branches of the lateral plantar nerves cause a condition known as *Morton metatarsalgia* (Morton's toe neuroma), in which pain is felt in the anterior part of the sole of the foot. This painful neuropathy is caused by trauma and prolonged pressure as a result of jogging, long-distance walking, or wearing high-heeled shoes. It usually affects the digital branches between the third and fourth toes, producing degeneration of the axons and thickening of the perineurium. Damage to the endoneurial blood vessels also occurs. Patients exhibit intense burning pain over the head of the fourth and fifth metatarsals that radiates to the anterior parts of the corresponding toes. This is exacerbated by walking, squatting, kneeling, or standing, which necessitates extension of the toes, and is relieved by resting and lying down.

Damage to the sural branch of the tibial nerve can result from a *Baker's cyst* (a synovial cyst of the popliteal fossa) or fracture of the base of the fifth metatarsal bone. Impairment of sensation in the lower lateral leg and lateral border of the foot characterizes this injury.

Several tests can be performed to ascertain intactness of the muscles innervated by the tibial nerve. A standing patient can be asked by the examiner to stand on the toes of one while the other leg is in flexed position. This will test the soleus and gastrocnemius muscles. To test the tibialis posterior, the foot of a supine patient is plantar flexed and inverted by the examiner. In supine position, the tibialis anterior can be tested by asking the patient to plantar-flex his/her foot against the examiner's resistance. In order to check intactness of the muscles in the deep compartment of the posterior leg, a supine-positioned patient is asked to plantar-flex his/her toes against the examiner's resistance while the lower posterior leg is stabilized by the other hand of the examiner. Plantar flexor of the big toe can be assessed by asking a standing patient to press on a paper that is placed under both big toes and then pulled away by the examiner. On the intact side, the paper cannot easily be moved, while on the affected side, the paper will easily be released without tearing.

SPINAL REFLEXES

Spinal reflexes are locally mediated neuronal events, which are constantly modulated by the facilitatory and inhibitory influences of the descending supraspinal pathways. However, the ascending influences from the lower spinal segments are also exerted upon higher spinal levels. The dramatic intensity in the extensor rigidity of the forelimb muscles in a spinal animal, whose spinal cord has been transected at the level of the sixth thoracic spinal cord segment, is thought to be dependent upon the ascending inhibitory pathways (Shiff–Sherrington reflex). Spinal reflexes may be classified into superficial and deep reflexes.

Unilateral absence of the superficial abdominal reflex may be seen in both upper and lower motor neuron disorders. Upper motor neuron lesions that produce this reflex disorder usually involve the cerebral cortex and the descending autonomic pathways, whereas lower motor neuron lesions affect the lower three thoracic spinal segments in order to produce the loss of this reflex.

No significance is attached to the bilateral absence of this reflex; however, unilateral absence of it may indicate upper motor neuron palsy.

The cremasteric reflex is brisk in young adults and is usually absent in conus medullaris syndrome, varicocele, upper motor neuron palsy, and damages involving the upper lumbar roots.

Superficial Reflexes

The superficial reflexes are composed of the interscapular, superficial abdominal, cremasteric, gluteal superficial, plantar, anal, and bulbocavernous reflexes.

- Interscapular reflex refers to the reflex contraction of the rhomboid muscles and bilateral retraction of the scapula upon stroking the skin of the interscapular (T2–T4) area.
- Superficial abdominal reflex is elicited by stroking the skin of the abdomen from the periphery toward the umbilicus, stimulating the seventh through the twelfth thoracic (T7–T12) spinal segments. It results in contraction of the oblique abdominal muscles and movement of the umbilicus toward the side of the stimulus. However, obese and pregnant individuals usually do not exhibit this response.
- Cremasteric reflex, on the other hand, is characterized by contraction of the cremasteric muscle, followed by retraction of the ipsilateral testicle, upon a light stroke in a downward direction to the upper medial thigh. This reflex is mediated by the ilioinguinal nerve (LI) as the afferent limb and the genitofemoral nerve (LI, L2) as the efferent limb.
- Gluteal superficial reflex (L4–S1) is characterized by contraction of the gluteus maximus in response to the examiner's stroke of the skin of the buttock.
- Plantar reflex (L5–S2) is produced by stroking the lateral aspect of the foot, eliciting either plantar flexion of all the toes or no response at all.
- Anal reflex (S4, S5, CC1) is elicited by stroking the perianal region with a pinwheel, producing puckering of the anal orifice. It is abolished in tabes dorsalis, cauda equina, and conus medullaris syndromes.

- The bulbocavernous reflex (S2, S3, S4) may be utilized to reveal the intactness of the bladder and is particularly important in upper motor neuron diseases. This reflex is characterized by contraction of the bulbospongiosus muscle upon compressing the glans of the penis or clitoris or pinching the prepuce. Interruption of the efferent motor fibers of this reflex produces incontinence, while disruption of the afferent limb abolishes the urge to urinate and defecate.

Deep Reflexes

Deep reflexes include the stretch (myotatic), inverse myotatic (clasp knife), flexor, and crossed extension reflexes.

Stretch (myotatic) reflex (Figure 10.33) is elicited by tapping the tendon of a muscle, which produces increased length of the muscle fibers and subsequent activation of the muscle spindle. Contraction of the muscle spindle activates the annulospiral (Ia) fibers, which in turn monosynaptically stimulate the ipsilateral motor neurons, producing contraction of the stretched muscle. Annulospiral afferents establish excitatory monosynaptic connections with motor neurons of the synergists and disynaptic inhibitory connections to the motor neurons of the antagonistic muscle (reciprocal inhibition). Myotatic stretch reflexes are produced by tapping the patellar ligament in patellar reflex (L2, L3, and L4) and biceps brachii tendon in biceps reflex (C5, C6). It is also elicited upon tapping the tendon of the triceps in triceps reflex (C7, C8), tendon of the brachioradialis in radial reflex (C7, C8), gastrocnemius (Achilles) tendon in gastrocnemius or Achilles reflex (S1, S2), etc. Periosteoradial reflex produces flexion and supination of the forearm upon tapping the radial styloid process, while periosteoulnar reflex produces extension and ulnar abduction upon striking the ulnar styloid.

- Diminished or absence of myotatic reflex may occur in peripheral neuropathies, tabes dorsalis, poliomyelitis, diabetes mellitus, Holmes–Adie syndrome, sciatica, syringomyelia, and cervical spondylosis. Spinal shock, coma, and certain types of hydrocephalus can also diminish or abolish the myotatic reflexes. Upper motor neuron disorders such as stroke, multiple sclerosis, and spinal cord tumors or strychnine poisoning and anxiety disorders may render this reflex hyperactive (intensified). Westphal's sign, failure to produce the patellar (myotatic) reflex, may be reversed by the reinforcement of *Jendrassik* maneuver, which requires the patient to clench his/her hands while the patellar tendon is tapped.
- Contraction of a muscle can also be elicited by activation of the muscle spindle via the γ loop, without stretching the muscle. In the *y* loop, the contraction of the muscle spindle activates the primary Ia (annulospiral) endings, which in turn monosynaptically activate the motor neurons. The firing of these neurons results in contraction of the extrafusal muscle fibers. Thus, under normal conditions, the cerebral cortex can trigger muscle contraction and initiate postural changes and movement via two mechanisms: (1) directly by activating α motor

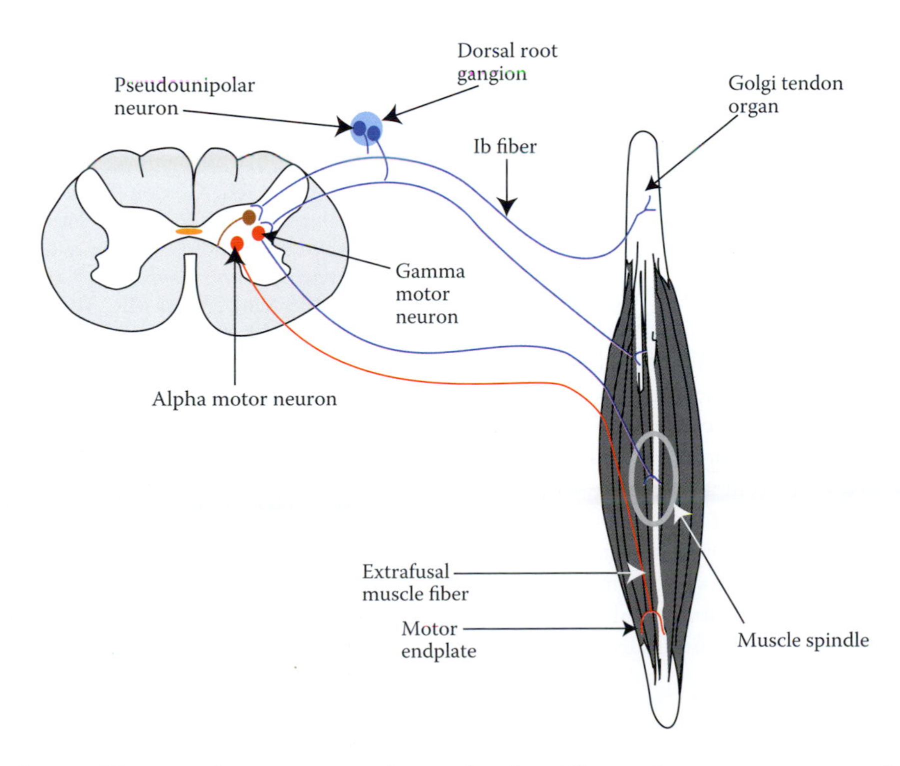

FIGURE 10.33 This diagram illustrates the components of myotatic reflex. Observe the gamma neurons and afferents of the muscle spindle.

neurons and (2) indirectly via the γ loop. The role of the γ loop can be illustrated when assuming an erect posture. Standing stretches the quadriceps muscle, which causes activation of the stretch receptor and subsequent contraction of the quadriceps femoris. However, the muscle begins to relax as soon as the tension in the muscle spindle is reduced and the rate of discharge from the motor neurons is diminished. In order to maintain erect posture, the *y* loop comes into action and activates the muscle spindle. Voluntary and precise movements are executed by the simultaneous activation of both systems, which are complementary. In general, activation of α motor system predominates when a quick response is desired, whereas activation of the γ system predominates when a smooth and precise movement is desired.

The inverse myotatic reflex is thought to underlie the mechanism of "clasp knife" phenomenon, which is observed in upper motor neuron palsy. In this reflex, passive stretching of the spastic muscle is met initially with great resistance to an extent, after which the muscle suddenly gives away. Sherrington named this phenomenon because of its similarity to the action of a jack or a switchblade knife.

- Inverse myotatic reflex comes into action upon stimulation of the Golgi tendon organ and the Ib fibers as a result of the tension developed in the contracted muscle. The Ib fibers establish disynaptic inhibitory (autogenic inhibition) contacts with the agonist neurons and excitatory synapses with the antagonistic neurons. The sum of these actions produces relaxation of the agonistic muscles.
- Flexor (withdrawal) reflex enables an individual to avoid harm by withdrawing from nociceptive or injurious stimuli. This reflex is mediated by the free nerve endings, to a lesser extent by the tactile receptors, as well as group III nerve fibers, conveying the impulses to the spinal cord. These afferent fibers establish polysynaptic excitatory and inhibitory connections with the motor neurons. The net effect of this circuitry is facilitation of ipsilateral flexor (agonists) motor neurons and inhibition of ipsilateral extensor (antagonist) motor neurons.
- Crossed extension reflex is characterized by flexion of the ipsilateral limb and extension of the contralateral limb in response to a strong nociceptive stimulus. This reflex is a byproduct of the flexion reflex, whereby the afferent fibers establish multisynapses at many levels of the spinal cord with the ipsilateral flexor neurons and. with the contralateral extensor neurons via the anterior white commissure.

SUGGESTED READING

Akgun H, Yucel M, Oz O, Demirkaya S. Differential diagnosis of carpal tunnel syndrome. *Turk Neurosurg* 2014;24(1):150.

Al-Qattan MM, El-Sayed AA. Obstetric brachial plexus palsy: The mallet grading system for shoulder function-revisited. *Biomed Res Int* 2014;2014:398121.

Callahan JD, Scully TB, Shapiro SA et al. Suprascapular nerve entrapment: A series of 27 cases. *J Neurosurg* 1991;74:893–6.

Cheatham SW, Kolber MJ, Salamh PA. Meralgia paresthetica: A review of the literature. *Int J Sports Phys Ther* 2013;8(6):883–93.

Kars HZ, Topaktas S, Dogan K. Aneurysmal peroneal nerve compression. *Neurosurgery* 1992;30:930–1.

Katirji B, Hardy RW Jr. Classic neurogenic thoracic outlet syndrome in a competitive swimmer: A true scalenus anticus syndrome. *Muscle Nerve* 1995;18:229–33.

Laha RK, Lunsford LD, Dujovny M. Lacertus fibrosus compression of the median nerve. *J Neurosurg* 1978;48:838–41.

Lee C-S, Tsai T-L. The relation of the sciatic nerve to the piriformis muscle. *J Formosan Med Assoc* 1974;73:75–80.

McKowen HC, Voorhies RM. Axillary nerve entrapment in the quadrilateral space: Case report. *J Neurosurg* 1987;66:932–4.

Meier C, Reulen HJ, Huber P, Mumenthaler M. Meningoradiculoneuritis mimicking vertebral disc herniation: A "neurosurgical" complication of Lyme-borreliosis. *Acta Neurochir* 1989;98:42–6.

Nakano KK, Lundergan C, Okihiro M. Anterior interosseous nerve syndromes: Diagnostic methods and alternative treatments. *Arch Neurol* 1977;34:477–80.

Roles NC, Maudsley RH. Radial tunnel syndrome: Resistant tennis elbow as a nerve entrapment. *J Bone Joint Surg* 1972;54B:499–508.

Seitz WH Jr, Matsuoka H, McAdoo J, Sherman G, Stickney DP. Acute compression of the median nerve at the elbow by the lacertus fibrosus. *J Shoulder Elbow Surg.* 2007;16(1):91–4.

Spangfort EV. The lumbar disc herniation. A computeraided analysis of 2,504 operations. *Acta Orthoped Scand* 1972;142 (Suppl):1–93.

Tai TW, Kuo LC, Chen WC, Wang LH, Chao SY, Huang CN, Jou IM. Anterior translation and morphologic changes of the ulnar nerve at the elbow in adolescent baseball players. *Ultrasound Med Biol* 2014;40(1):45–52.

Thomas K, Stein RB, Gordon T, Lee RG, Elleker MG. Patterns of reinnervation and motor unit recruitment in human hand muscles after complete ulnar and median nerve section and resuture. *J Neural Neurosurg Psychiat* 1987;50:259–68.

Thornton MW, Schweisthal MR. The phrenic nerve: Its terminal divisions and supply to the crura of the diaphragm. *Anat Rec* 1969;164:283–90.

van Alfen N, van Engelen BG, Reinders JW, Kremer H, Gabreëls FJ. The natural history of hereditary neuralgic amyotrophy in the Dutch population: Two distinct types? *Brain* 2000;123 (Pt 4):718–23.

Wiles CM, Whitehead S, Ward AB, Fletcher CDM. Not tarsal tunnel syndrome: A malignant "Triton" tumor of the tibial nerve. *J Neural Neurosurg Psychiat* 1987;50:479–81.

Wytrzes L, Marklex HG, Fisher M, Alfred HJ. Brachial neuropathy after brachial artery antecubital vein shunts for chronic hemodialysis. *Neurology* 1987;37:1398–400.

11 Cranial Nerves

The cranial nerves, as the name implies, lie within the cranial fossae, some may travel in the head, neck, while others continue their paths through the thoracic and abdominal cavities, innervating structures in these areas. They are classified according to their functional components and connections into purely sensory nerves (e.g., olfactory and optic); motor nerves (e.g., oculomotor, trochlear, abducens, accessory, and hypoglossal); and mixed nerves, containing both sensory and motor components (e.g., trigeminal, facial, glossopharyngeal, and vagus). The axons of the bipolar neurons in the olfactory mucosa form the olfactory nerve, while that of the Scarpa's and spiral ganglia form vestibulocochlear nerve. In the same manner the axons of the multipolar neurons of the retina make the optic nerve. Due to its origin from the retina, a telencephalic structure, the optic nerve, is considered as an extension of the central nervous system and is affected by central demyelinating diseases. The general and special sensory fibers are the central processes of the unipolar neurons of the geniculate ganglion of the facial nerve, and the bipolar neurons of the superior and inferior ganglia of the glossopharyngeal and vagus nerves. The motor fibers within the cranial nerves represent the axons of the multipolar neurons. A lesion of a cranial nerve or associated nucleus produces manifestations of lower motor neuron palsy: atrophy, flaccidity, and areflexia or hyporeflexia with the exception of the abducens nerve and nucleus. Affected structures usually deviate to the side of the lesion, with the exception a lesion of the vagus nerve, which produces deviation toward the intact side. Cranial nerve motor nuclei receive evenly distributed bilateral corticobulbar fibers with the exception of the facial motor neurons to the upper face which receives bilateral cortical input compared to neurons of the lower face which receive only contralateral cortical input. A lesion that disupts a cranial nerve and adjacent part of the corticospinal fibers produces alternating hemiplegia. Cranial nerves and associated nuclei can be affected in variety of conditions, such as the lateral medullary syndrome, Benedikt syndrome.

OLFACTORY NERVE

The olfactory receptors are formed by the dendrites of the bipolar neurons scattered in the mucosa of the superior concha and the opposing part of the nasal septum. The olfactory nerve (special visceral afferent [SVA]) represents the thinly unmyelinated axons of the bipolar neurons of the olfactory mucosa that form initially a plexiform network that bind together to establish well-defined filaments that pass through the cribriform plate of the ethmoid bone (Figures 11.2 and 11.3). These olfactory filaments are enfolded by Schwann cells and ensheathed by meninges that continue with the nasal periosteum. The subarachnoid space around the filaments also continues with the cerebral subarachnoid space.

These filaments enter the olfactory bulb, establishing synapses with the dendrites of the mitral cells (Figures 11.1 and 11.3). The olfactory bulb is an allocortex and consists of five layers, containing mitral and tufted cells (excitatory), granular and periglomerular (inhibitory) neurons. Periglomerular cells, which are GABAergic, receive excitatory fibers from the olfactory bipolar neurons and establish inhibitory connections with the surrounding mitral cells. Inhibition is also maintained by the dendrodendritic synaptic connections between a subset of periglomerular dopaminergic neurons and the GABAergic granule cells. The centrifugal fibers from the contralateral anterior olfactory nucleus also exert inhibitory influence through inhibitory internuncial neurons. Mitral cells also provide collateral fibers to the anterior olfactory nucleus. The axons of the mitral neurons (Figures 11.1 and 11.2) form the olfactory tract and project to the olfactory nucleus and then divides into medial, intermediate, and lateral olfactory striae. These striae bypass the thalamus in their course to the olfactory cortex. The lateral olfactory stria terminates primarily in the uncus (Figure 11.1), which constitutes the primary olfactory cortex, whereas the medial and intermediate olfactory striae terminate in the septal area and the anterior perforated substance, respectively.

The sense of olfaction can be impaired by lesions that affect the olfactory nerve, bulb, tract or the olfactory cortex (Neuronal olfactory disorders). It can also be compromised by conditions that affect the bony and soft tissue structures in the nasal cavity, nasopharynx, and paranasal air sinuses (conductive olfactory disorders). The olfactory disorder manifested in anosmia (loss of smell), is commonly acquired and rarely seen as a congenital anomaly. Bilateral anosmia induced by nasal infection (e.g., rhinitis sicca) or common cold, excessive smoking, and cocaine use, is commonly associated with ageusia (loss of taste). Anosmia due to viral influenza may completely is self- limiting and recovery is usually complete. Unilateral anosmia of nonrhinogenic origin may be a sign of a subfrontal lobe tumor, sphenoidal ridge meningiomas, or hypophysial tumors. Head trauma can lead to contusion of the olfactory bulb and separation of the olfactory filaments from the olfactory bulb, resulting in unilateral or bilateral *anosmia* or hyposmia. Reduction in smell sensation is observed in patients with Paget's disease,

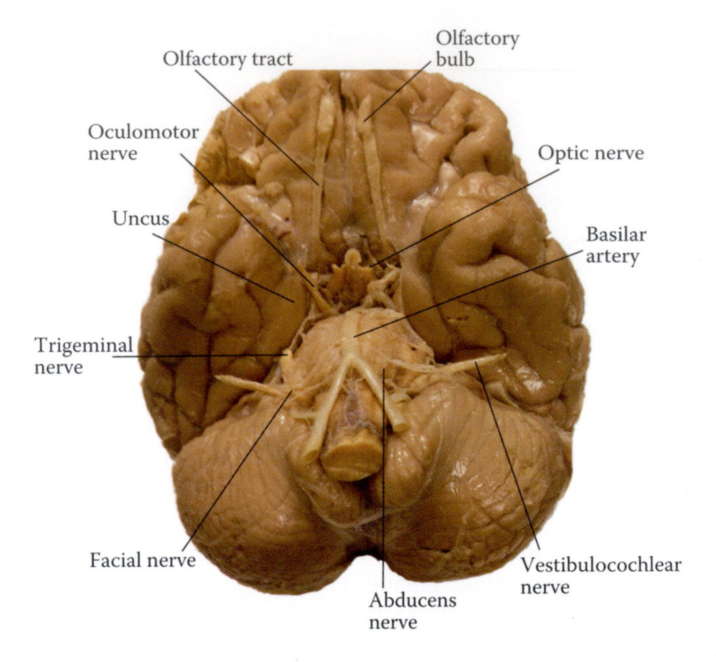

FIGURE 11.1 Inferior surface of the brain illustrating some of the associated cranial nerves.

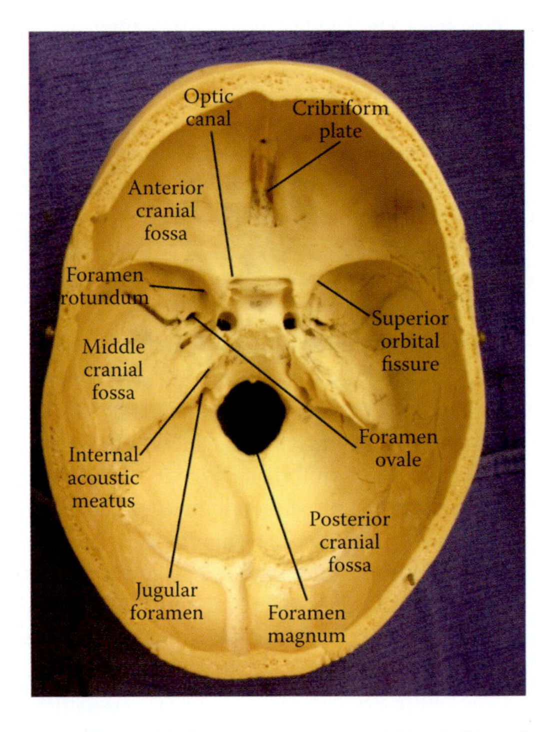

FIGURE 11.2 Cranial fossae with associated foramina and openings.

diabetes mellitus, and post-laryngotomy. Anosmia due to traumatic head injury, which accounts for one-fifth of all cases, may be detected weeks or months after the insult, lasting as long as the posttraumatic amnesia. Shearing of the olfactory filaments can be induced by occipital or temporal area impact. Anosmia induced by a fracture of the anterior cranial fossa and disruption of the olfactory filaments, is usually accompanied by leakage of cerebrospinal fluid through the nostrils (CSF rhinorrhea) and possible secondary infection transmitted from the nasal cavity. The latter is aided by continuation of the subarachnoid space around the olfactory filaments into the cranial cavity. Recovery from this type of injury may occur through replacement of the injured receptors. Olfactory hallucination with phantom smells however, is associated with a lesion or irritation of the uncus by mass.

Anosmia is also detected in Kallmann syndrome, a congenital disorder which is characterized by secondary hypogonadism and associated dwarfism with occasional color blindness. Anosmia can be caused by aplasia of the olfactory bulb and agenesis of the olfactory lobes. This random congenital condition which is seen predominantly in males occurs in conjunction with cleft palate. Anosmia is often accompanied by ageusia (impaired sense of taste). This combined disorder may be seen in an individual following head trauma and in patients with scleroderma who underwent treatment with histidine.

Hallucinatory disturbances of olfaction occur in psychotics and patients who suffer from clinical depression or alcohol withdrawal syndrome. Uncinate fits, which are also discussed in Chapter 8, occur in patients with temporal lobe epilepsy subsequent to overstimulation of the uncus. Patients perceive unpleasant olfactory aura and fear of the surroundings that precede the seizures. Cognitive functions, such as memory, orientation, and attention, may also be adversely affected.

Anosmia is also associated with olfactory groove meningiomas. These slow growing benign tumors develop in the midline of the anterior cranial fossa along the dura of the cribriform plate and planum sphenoidale. They may extend to involve the ethmoid sinuses, nasal cavity, frontal lobe, and optic tracts. Excessive growth (hyperostosis) of the underlying bone is commonly seen.

They represent 10% of intracranial meningiomas that predominantly affect women, with peak incidences between the fourth and sixth decades of life. Due to the slow growth of this tumor, anosmia develops gradually and patients become only aware when the tumor becomes large enough to produce headache, visual impairment due to compression of the optic tracts, and memory and personality changes due to the pressure exerted on the frontal lobe. Olfactory meningioma can, though rarely, produce ipsilateral optic nerve atrophy as a direct compressive effect of the tumor and papilledema contralaterally as a consequence of increased intracranial pressure. This combination of deficits represents manifestations of *Foster Kennedy* syndrome.

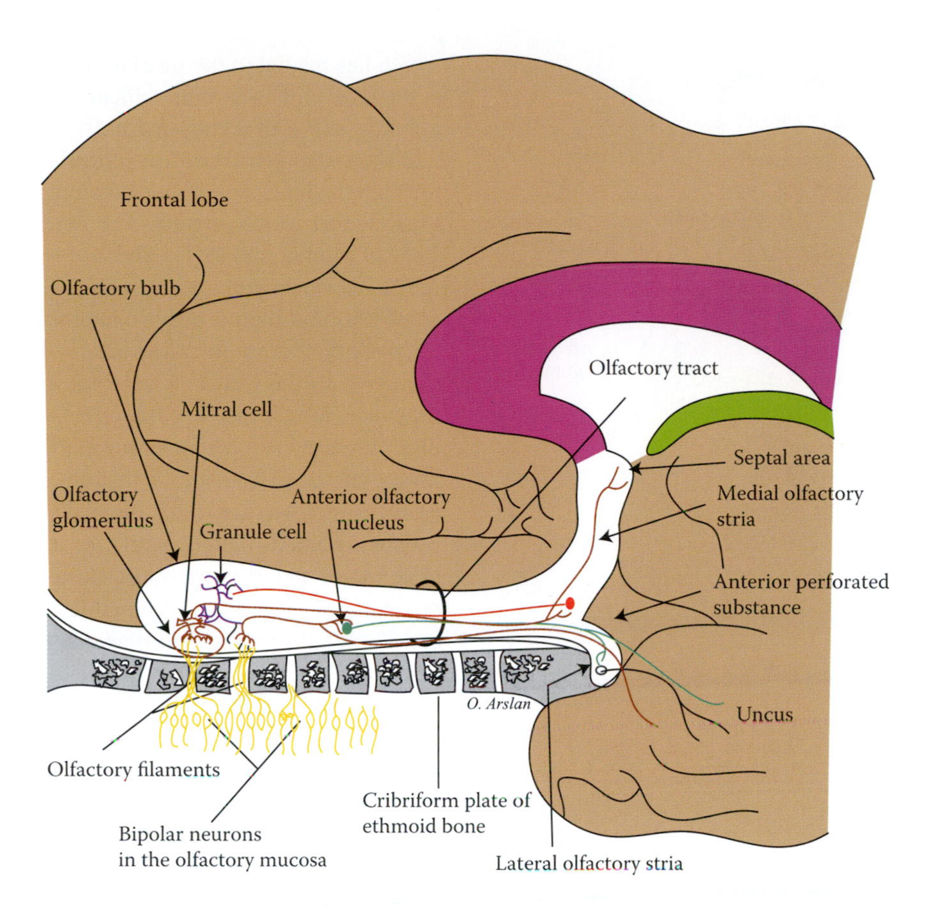

FIGURE 11.3 Olfactory nerve filaments, synaptic connection, and olfactory striae are followed in this diagram to their terminations in the septal area and uncus.

OPTIC NERVE

The optic nerve (special somatic afferent [SSA]), which is discussed in detail in Chapter 13, is formed by the axons of retinal ganglion cells. It is considered an extension of the brain for two main reasons; first, it is surrounded by myelin from the oligodendrocytes, and second, it is an embryological derivative of the forebrain diverticulum (Figures 11.1 and 11.4). Due to these reasons, patches of demyelination along the course of the optic nerve are seen in multiple sclerosis. The optic nerve acquires myelin in the orbit; otherwise, myelinated axons in the retina my cause light reflection and blurred vision. Visual information from the temporal and nasal halves of the corresponding retina, as well as impulses concerned with pupillary light and accommodation reflexes, is carried by the optic nerve. Fibers of the optic nerve leave the retina medial to the fovea centralis and converge on the optic disc, piercing the choroid layer and sclera and entering the orbit. In the orbit, it is crossed by the ophthalmic artery. Then, it leaves the orbit and gains access to the cranial cavity through the optic canal (Figure 11.2). Posterior to the optic canal, the nasal fibers decussate to form the optic chiasma (Figures 11.1 and 11.4). In the cranial cavity, the internal carotid artery lies on the lateral side of the optic nerve and ventral to the anterior cerebral artery. The presence of the central retinal artery and vein inside this nerve may account for the reduction of arterial and retardation of venous blood flow in these vessels upon compression of this nerve by a growing tumor or by increased intracranial pressure. Since cerebral meningeal coverings, associated subarachnoid space and CSF continue around the optic nerve, conditions that impede CSF circulation in the brain may eventually translate into pressure buildup that bilaterally compromises the integrity of this nerve.

The integrity of the optic nerve is determined by examination of the visual fields and visual acuity. This is accomplished by confrontational visual field testing, which involves closure of the examiner's right eye and patient's left eye while standing at eye level opposite each other. Visual acuity may be assessed by using the *Snellen eye chart*, positioned approximately 20 feet from the patient. Visual acuity may vary by environmental factors such as illumination and degree of contrast.

To detect the differences between both eyes in response to afferent stimuli, the *swinging light test* is employed. In this test, the patient is asked to look at a distant object while the examiner rapidly swings a light beam from one eye to the other. When directing the light into the blind eye, neither eye will show

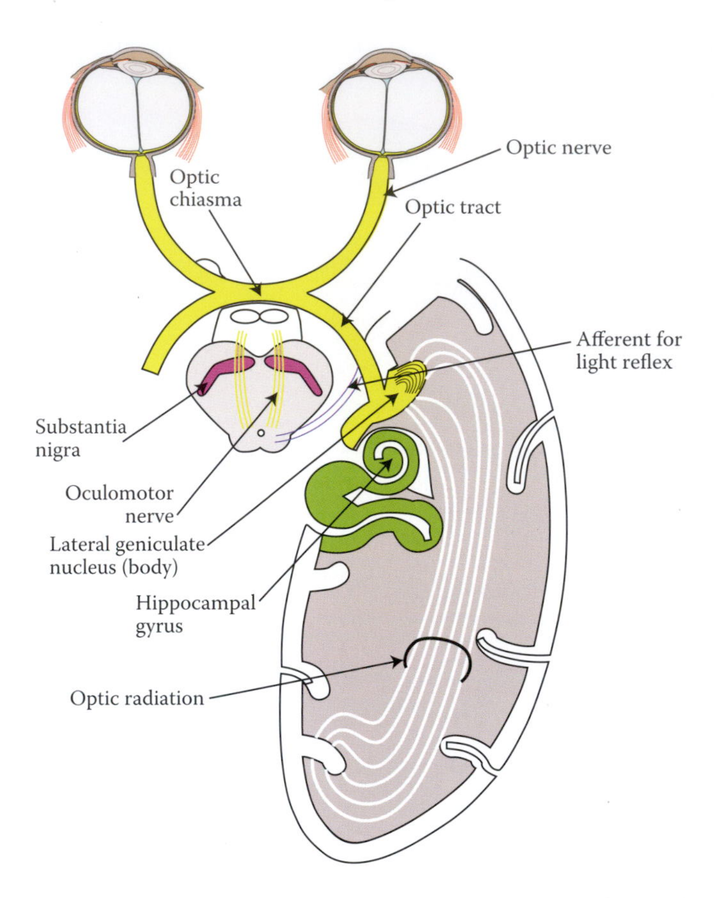

FIGURE 11.4 Impulses carried by the optic nerve are followed through the optic chiasma and optic tract to their final destination in the visual cortex.

constriction. However, upon moving the light back to the intact eye, the blind eye shows apparent pupillary dilatation due to the lack of afferents to the retina and optic nerve (*Marcus Gunn pupil*).

Complete destruction of the optic nerve results in total blindness on the affected side. Due to the bilateral connections of the visual fibers of the optic nerve, pupillary light and accommodation reflexes are lost on both sides when the affected eye is stimulated.

Prolonged elevation of intracranial pressure produces uniform compression of the optic nerve, leading to blindness. *Papilledema* (choked disc) is a condition in which the optic disc protrudes anteriorly as a result of dilatation of the subarachnoid space around the optic nerve, subsequent to increased intracranial pressure. This condition may also be associated with compression of the central retinal artery and vein and may be observed in hypertensive individuals.

OCULOMOTOR NERVE

The oculomotor nerve (general somatic efferent [GSE], general visceral efferent [GVE]) is formed by the neuronal axons of the oculomotor nucleus, a midline V-shaped nucleus, which lies medial to the medial longitudinal fasciculus (MLF) at the level of the superior colliculus of the midbrain. The oculomotor nerve descends medially, crosses the red nucleus, and then emerges from the ventral surface of the midbrain through the interpeduncular fossa, in close relationship to the crus cerebri which contains the corticospinal tract (Figures 11.1, 11.5, 11.6, and 11.7). This nerve continues in its course in the interpeduncular cistern and then between the superior cerebellar and the posterior cerebral arteries, and inferior to the posterior communicating artery. The close relationship of the nerve to the posterior cerebral artery bears clinical significance. When the cerebral edema and increased intracranial pressure leads to uncal herniation through the tentorial notch, the displaced uncus dislodges the posterior cerebral artery, which, in turn, pulls the oculomotor nerve downward, producing fixed dilated pupil. As it continues rostrally, the oculomotor nerve pierces the dura and enters the cavernous sinus, where it is accompanied by the trochlear, abducens, ophthalmic, and maxillary nerves as wells the internal carotid artery. It leaves the middle cranial fossa via the superior orbital fissure, and enters the orbit as a component of the tendinous annulus of *Zinn*, where it divides into superior and inferior rami. The superior ramus (GSE) supplies the superior rectus and levator palpebrae, whereas the inferior ramus provides innervation (GSE) to the medial and inferior recti and the inferior oblique and preganglionic parasympathetic fibers (GVE).

Selective ipsilateral palsy of the superior rectus and levator palpebrae muscles occurs as a result of an orbital mass or pathologic condition involving the anterior part of the cavernous sinus. This ipsilateral dysfunction of these muscles excludes oculomotor nuclear lesions, which also produce contralateral deficits.

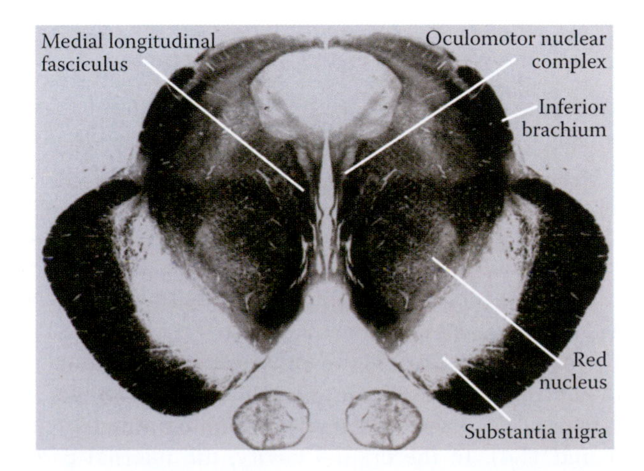

FIGURE 11.5 The oculomotor nuclear complex is illustrated, and its location is identified in reference to the medial longitudinal fasciculus and the periaqueductal gray matter.

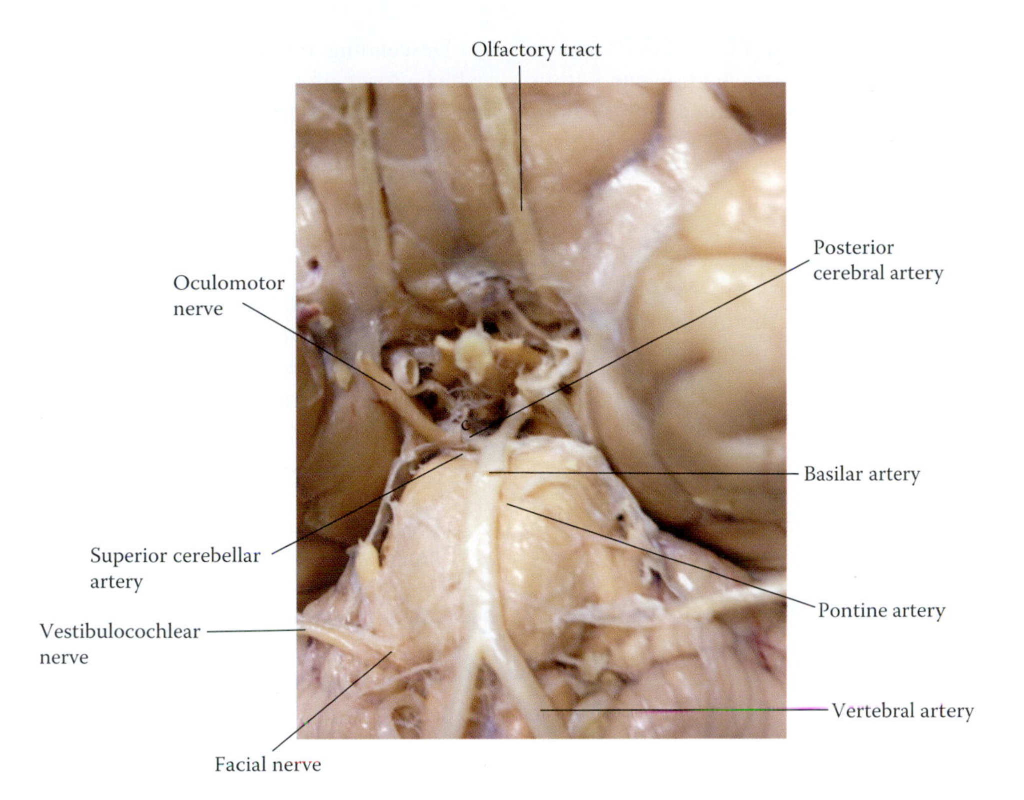

FIGURE 11.6 The course of the oculomotor nerve between the superior cerebellar and posterior cerebral arteries.

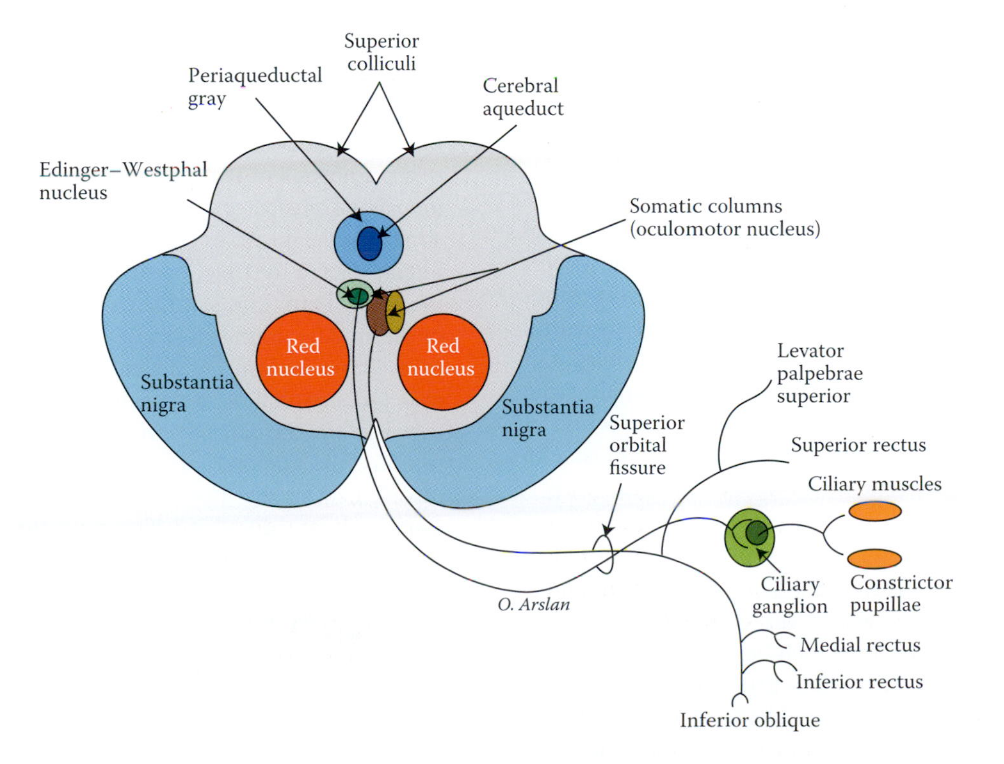

FIGURE 11.7 The functional components of the oculomotor nerve from their origin in the oculomotor nuclear complex to their site of innervation are shown in this drawing.

The preganglionic parasympathetic fibers (GVE) of the oculomtor nerve destined to the ciliary ganglion run within the inferior ramus and constitute the efferent limb of both pupillary light and accommodation reflexes.

Ipsilateral mydriasis and loss of accommodation without other signs of oculomotor palsy are seen in *Adie pupil* subsequent to disruption of the ipsilateral postsynaptic parasympathetic neurons of the ciliary ganglion due to viral or bacterial diseases. A random regenerative process enables these fibers to reinnervate the constrictor pupillae and ciliary muscles, producing a poorly reactive and mildly dilated pupil, which constricts slowly in response to near vision and then dilates slowly. This syndrome, mainly seen in young females, is associated with hyperreflexic deep tendons, disorders of sweating, photophobia, and hyperopia due to loss of accommodation.

The oculomotor nuclear complex consists of *somatic* and *visceral nuclear* columns. The somatic component of this nucleus (Figures 11.5 and 11.7) consists of the dorsal, intermediate, ventral, caudal central, and medial cellular columns. The dorsal column innervates the ipsilateral inferior rectus muscle; the intermediate column supplies the ipsilateral inferior oblique muscle; and the ventral column provides innervation to the ipsilateral medial rectus. The latter neuronal column is controlled by the internuclear neurons of the abducens nucleus in lateral gaze. The caudal central column, a midline subnucleus, provides bilateral innervation to the levator palpebrae muscle, whereas the medial column innervates the contralateral superior rectus muscle. Preganglionic parasympathetic fibers that synapse in the ciliary ganglion arise from the *Edinger–Westphal nucleus* (EWN), the visceral column of the oculomotor nucleus (Figure 11.7). These fibers synapse in the ciliary ganglion, and the emerging postganglionic fibers innervate the constrictor pupilla and ciliary muscles, mediating light and accommodation reflexes. EWN is the most superior subnucleus within the oculomotor complex located near the inferior margin of the periaqueductal gray matter. Fibers that emanate from this subnucleus are cholinergic and produce contraction of the constrictor pupillae in light reflex and ciliary muscle in accommodation.

The superficial and dorsal position of the preganglionic parasympathetic fibers within the oculomotor nerve explains the selectivity by which compressive lesions affect the pupillary reflex fibers first, while the same fibers are spared in diabetic oculomotor neuropathy. EWN is activated by input from the pretectal nuclei and the striate cortex.

Descending cortical fibers to the oculomotor nucleus, which form the corticobulbar fibers, are crossed and uncrossed and establish contact with the nucleus via the reticular formation. Input from the paramedian reticular formation projects to the ventral column of the oculmotor nucleus and mediate horizontal eye movements. Vestibular fibers also project to the oculomotor nucleus through the MLF, regulating movements of the head and fixation of gaze. Fibers from the accessory oculomotor nuclei (interstitial nucleus of Cajal, nucleus of Darkschewitsch, and nucleus of the posterior commissure) project to the oculomotor nuclei by crossing in the posterior commissure, mediating vertical and torsional eye movements.

Destruction of the oculomotor nerve may occur in combination with the corticospinal tract in *Weber's syndrome* or in conjunction with the red nucleus, spinothalamic tracts, medial lemniscus, and superior cerebellar peduncle in *Benedikt syndrome*. In Weber's syndrome, which will be also discussed with the motor system, ipsilateral signs of oculomotor palsy and contralateral signs of upper motor neuron palsy will be seen. In Benedikt syndrome, patients present with oculomotor palsy, tremor, and loss of proprioception on the same side and anesthesia contralaterally. The oculomotor nerve may also be damaged in conjunction with the red nucleus in *Claude's* (lower red nucleus) *syndrome*, exhibiting manifestations of oculomotor palsy, contralateral hemiataxia, but with no apparent hyperkinesia. Oculomotor nerve palsy with contralateral cerebellar ataxia, tremor, and signs of spastic palsy are seen in *Nothnagel's syndrome*. Since the oculomotor nerve courses immediately rostral to the superior cerebellar artery, caudal to the posterior cerebral artery, and inferior to the posterior communicating artery, an aneurysm of anyone of these vessels can produce oculomotor palsy. Transtentorial herniation may also pose undue stretch on the posterior cerebral artery, resulting in oculomotor nerve dysfunction that almost always involves the pupil. Thrombosis of the cavernous sinus, fractures of the middle cranial fossa or superior orbital fissure, parasellar neoplasm, and demyelinating diseases can also produce deficits associated with the oculomotor nerve. In amyotrophic lateral sclerosis (ALS) and poliomyelitis, the oculomotor nerve remains intact despite involvement of the motor neurons. Some of the above syndromes are discussed later with the motor system or under the heading of combined motor and sensory lesions.

Injury to the oculomotor nerve can be partial or complete, and the recovery may not be even. It may not show all the manifestations as evident in diabetic neuropathy, which spares the pupillary constriction mediated by the parasympathetic component of the EWN. These fibers may not be spared in brainstem lesions. Fibers, which mediate accommodation reflex, can selectively be damaged in diphtheria. Acute isolated painful oculomotor palsy is most commonly associated with aneurysm of the posterior communicating artery. Spontaneous intracranial hypotension may be associated with combined oculomotor and trochlear nerve palsies due the traction forces generated in this condition. Damage to the oculomotor nucleus may occur in multiple sclerosis and produces deficits similar to oculomotor nerve palsy with few exceptions. In complete oculomotor nucleus damage, contralateral superior rectus palsy will also be noticed, with inability to gaze vertically. Additionally, because of the bilaterally of innervation by the central caudal subnucleus of the oculomotor nuclear complex, the patient exhibits bilateral ptosis and unilateral oculomotor palsy.

A unilateral lesion of the lateral tegmentum of the midbrain may produce signs of oculomotor palsy on the same side and trochlear nerve palsy on the opposite side. In cavernous sinus thrombosis, oculomotor palsy can be accompanied by dysfunction of multiple other cranial nerves and possibly with signs of venous blockade. The cavernous sinus inflammatory process in Tolosa–Hunt syndrome causes a painful ocular motility disorder due to involvement of nerves III, IV, and VI and two branches of nerve V and severe headache accompanied by miotic pupil due to disruption of the sympathetic postsynaptic fibers around the internal carotid artery. The parasympathetic postganglionic fibers that mediate light reflex are spared in this condition. These disorders are accompanied by systemic manifestations such as fever.

A polyneuropathy of the third, fourth, and sixth cranial nerves and the ophthalmic and maxillary branches of the fifth cranial nerve may develop in cavernous sinus syndrome, a noninflammatory condition that can be attributed to a slow-growing mass. Diplopia with facial pain or numbness in the areas of distribution of the ophthalmic and maxillary nerves can be the initial presenting signs. Impairment of vision due compression of the optic nerve, hypesthesia in the area of the ophthalmic nerve, and ophthalmoplegia occur subsequent to middle cranial fossa tumor or inflammation and are seen in orbital apex (Jacod) syndrome.

Oculomotor palsy (Figure 11.8) is characterized by the following ipsilateral conditions.

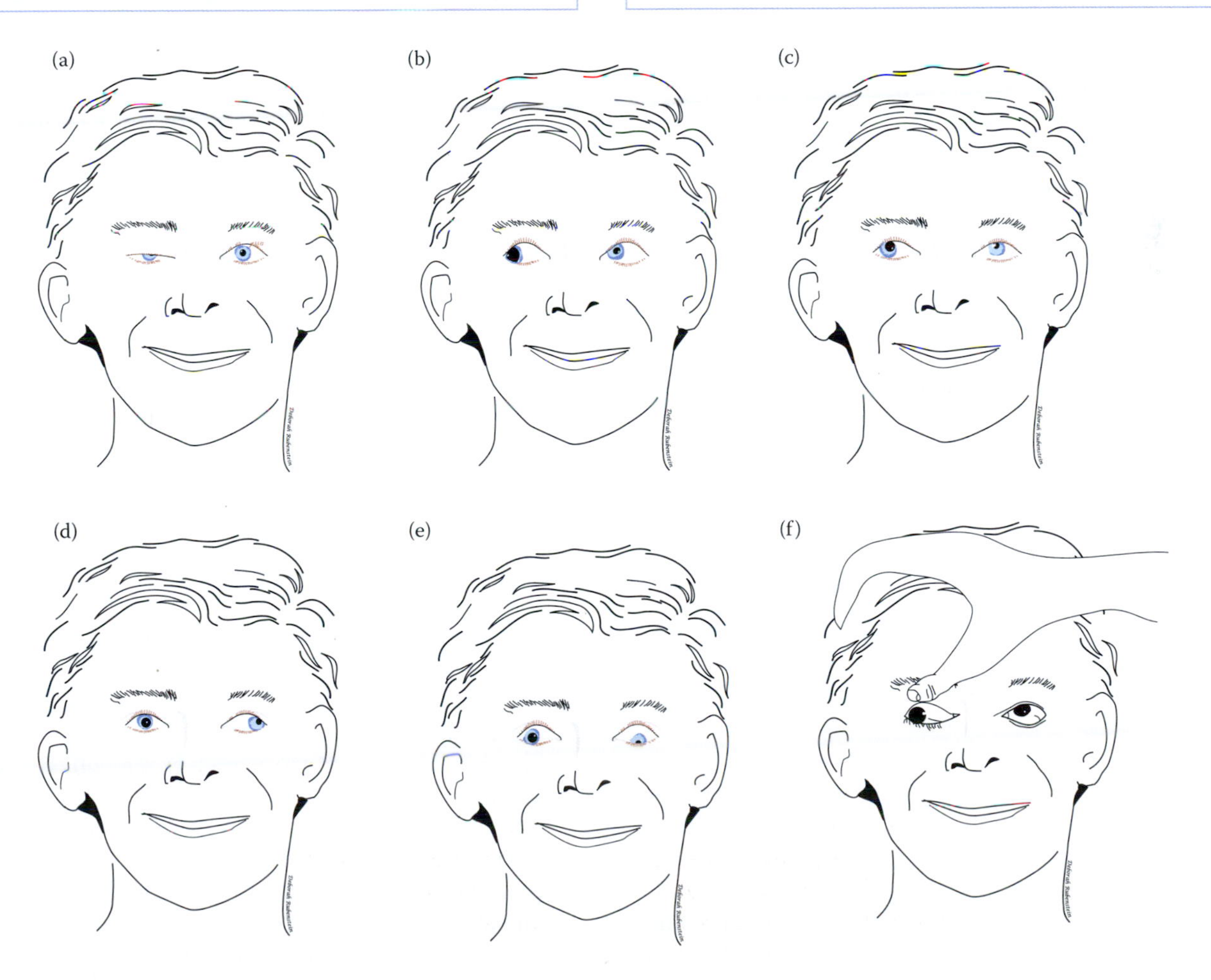

FIGURE 11.8 Manifestations of right oculomotor nerve palsy are depicted in diagrams (a) to (f). (a) Ptosis; (b) intact lateral gaze to the right side; (c) inability to gaze in an upward direction; (d) inability to adduct the right eye upon left lateral gaze; (e) inability to look downward on the right eye; and (f) paralyzed eyelid is raised, exhibiting lateral strabismus.

Ptosis (drooping of the upper eyelid due to paralysis of the levator palpebrae superior). This deficit should be distinguished from the ptosis observed in Horner's syndrome. In the latter condition, ptosis is less pronounced and occurs as a result of paralysis of the superior tarsal muscle, subsequent to disruption of the postsynaptic sympathetic fibers.

Mydriasis (dilatation of the pupil) is due to paralysis of the constrictor pupilla muscle and the unopposed action of the dilator pupillae muscle.

Diplopia (double vision) is seen in all directions except in lateral gaze, and the distance between true and false images is maximal in the direction of the gaze. The false image will always be peripheral to the true image.

Lateral strabismus (lateral deviation) and downward deviation of the eye due to activation of unopposed lateral rectus and the superior oblique muscles.

Enophthalmos (inward displacement of the eyeball), possibly an illusionary feature due to drooping of the upper eyelid.

Loss of pupillary and accommodation reflexes.

Examining the light reflex and observing movement of the eyes tests the integrity of the oculomotor nerve. To test the light reflex, the patient is asked to look at a distance, while the examiner shines a bright light into one eye of the patient; the examiner then observes the pupils in both eyes (direct and consensual light reflexes). In normal individuals, both pupils will constrict in response to the light applied to one eye. In lesions of the oculomotor nerve, the affected eye remains unreactive regardless of which eye is stimulated. Since both the optic and oculomotor nerves mediate accommodation reflex, evaluation of near vision may reveal the condition of these nerves. In order to accomplish this task, the patient is asked to look alternately at distant and close objects and observe convergence of both eyes. The functional integrity of the extraocular muscles that receive innervation from the oculomotor nerve may be tested by asking the patient to follow the examiner's finger as it traces an "H" configuration, first moving to patient's right, then up, then down, then back to the midline.

TROCHLEAR NERVE

The trochlear nerve (GSE) represents the axons of the neurons of the trochlear nucleus, which is located at the ventral edge of the periaqueductal gray at the level of the inferior colliculus. The trochlear nucleus receives ipsilateral vestibulo-ocular fibers and bilateral input from the accessory oculomotor nuclei. As it emerges from the nucleus, the trochlear nerve curves dorsolaterally toward the inferior colliculus medial to the superior cerebellar peduncle, decussates completely in the superior medullary velum, and exits from the dorsal part of the rostral pons caudal to the inferior colliculus (Figures 11.9 and 11.10).

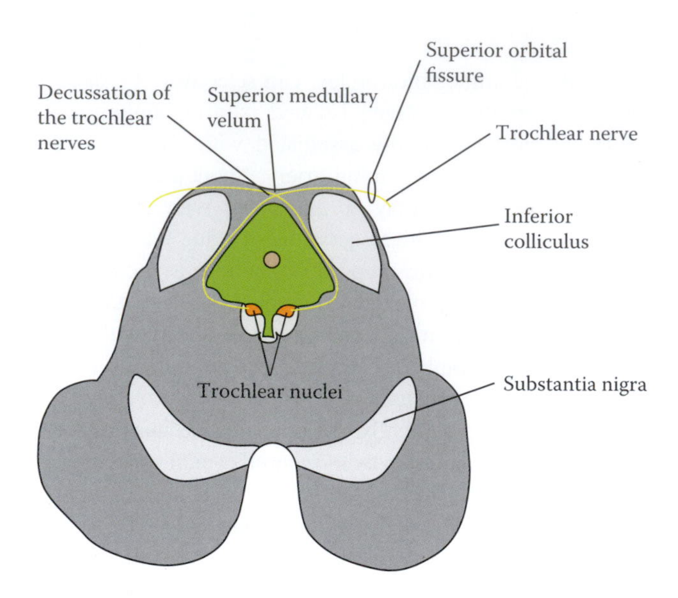

FIGURE 11.9 The course of the trochlear nerves and their decussation in the superior medullary velum is shown in this drawing.

The trochealr nerve then enters the cavernous sinus, and following its exit, it runs between the superior cerebellar and posterior cerebral arteries. Later, it enters the orbit via the superior orbital fissure and innervates the superior oblique muscle.

Prior to decussation and during its course in the midbrain within the central tegmental tract (CTT) and medial to the superior cerebellar peduncle, the trochlear nerve lies adjacent to the descending sympathetic fibers that are destined to activate the ciliospinal (Budge–Waller) center in the intermediolateral column of the thoracic and upper two lumbar spinal segments. This center sends preganglionic sympathetic fibers to the superior cervical ganglion, which, in turn, provides postganglionic fibers to the dilator pupilla muscle.

Trochlear nerve palsy can result from a traction injury on the nerve subsequent to frontal, occipital, or coccygeal impact or as a consequence of tectal compression against the tentorium cerebelli. It can occur in multiple sclerosis, cavernous sinus thrombosis, and superior orbital fissure syndrome. In the latter syndrome, the oculomotor (CN III) nerve is also damaged. It can also occur in orbital apex syndrome, which disrupts the functions of other nerves such as the optic (CN II), oculomotor (CN III), and abducens (CN VI) nerves in addition to the trochlear nerve. It can also develop in patients with pseudotumor cerebri and subsequent to spinal tap. Trochlear nerve palsy is characterized by impairment of downward gaze of the adducted eye (Figures 11.11 and 11.12). The eye on the affected side remains elevated and assumes a higher position when the eye is adducted than when it is abducted, and

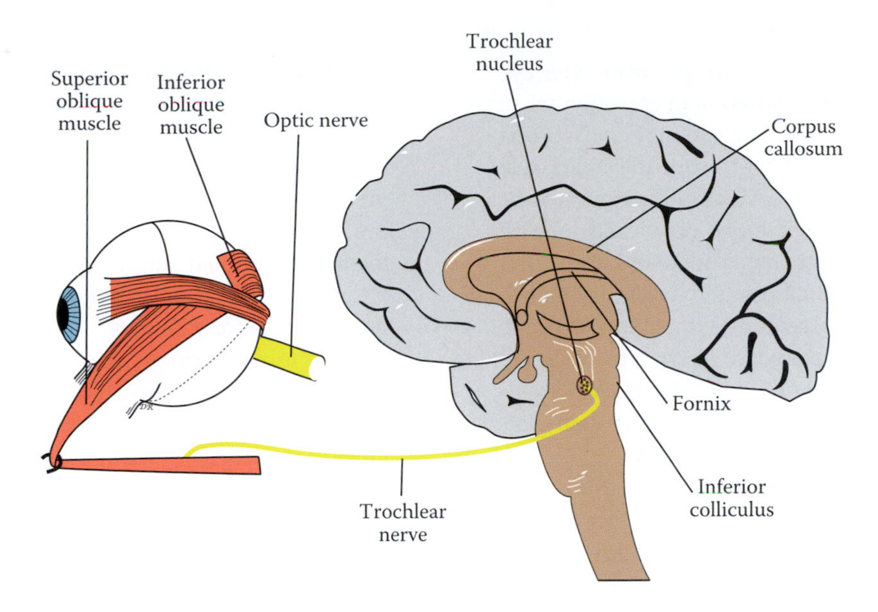

FIGURE 11.10 The origin, course, and final destination of the trochlear nerve.

FIGURE 11.11 The deficits associated with trochlear nerve palsy. (a) The right eye is elevated upon forward gaze. Elevation of the eye is (b) increased with adduction and (c) decreased with abduction.

FIGURE 11.12 Diagrams of manifestations of trochlear nerve palsy. (a) Maximal elevation upon tilting the head to the affected side. (b) Eye elevation disappears upon head tilting in the opposite direction.

decreases with abduction (hypertropic eye). The eye on the affected side assumes maximal elevation when the neck is bent toward the affected side and normal position when the neck is bent toward the intact side (Bielschowsky head-tilt test). Vertical diplopia will be more evident as the patient looks down and inward. Patients may develop asthenopia (ocular fatigue and pain, blurry vision, and diplopia) and torticollis. Due to close proximity of the descending sympathetic fibers to the predecussating trochear nerve fibers, a tectal lesion can produce manifestations of ipsilateral Horner's syndrome (miosis, ptosis, and anhidrosis) and contralateral trochlear nerve palsy.

Patients adopt a characteristic posture, tilting the head toward the opposite side so that the face will be directed toward the affected side (Beilschowsky sign). Maintenance of such a posture for an extended period will lead to *torticollis* (wry neck), which refers to the spasmodic contracture of the neck. Bilateral trochlear nerve palsy may occur in *superior medullary velum syndrome* as a result of a lesion that disrupts the decussating fibers of this nerve within the superior medullary.

Trigeminal gangliopathy can occur in systemic (scleroderma) sclerosis, an autoimmune disorders which causes overproduction of collagen and subsequent thickening of the skin and injury to the small digital arteries, and Sjögren syndrome, which produces dry eye and dry mouth. The basic pathophysiology involvement of trigeminal ganglion is the fact that the blood–brain barrier of this ganglion is more permeable to circulating autoantibodies.

Trigeminal neurinomas are rare benign intracranial tumors that involve the trigeminal ganglion and extend downward toward the posterior fossa or upward rostrally toward the middle cranial fossa and the cavernous sinus. Downward expansion can compress the cerebellum and facial and vestibulocochlear nerves, producing ataxia, vertigo, facial palsy, tinnitus, and neuronal deafness. Rostral expansion of the tumor may compress the lateral wall of the cavernous sinus, producing manifestations of oculomotor, trochlear, and abducens palsies and possibly optic nerve disruption. Nasopharyngeal carcinomas, lymphomas, chondromas, sarcomas, acoustic neuromas, and meningiomas can also involve the trigeminal nerve.

TRIGEMINAL NERVE

The trigeminal nerve (general somatic afferent [GSA], special visceral efferent [SVE]), the largest cranial nerve, exits the pons through the middle cerebellar peduncle (Figures 11.1 and 11.18). It has a sensory (trigeminal, Gasserian, semilunar) ganglion, which is formed by the unipolar neurons of the sensory fibers that transmit nociceptive, thermal, and tactile sensations, with the exception of proprioception. This ganglion lies anterior to the apex of the petrous temporal bone and resides in Meckel's cave. The trigeminal nerve gives off the ophthalmic (V1), maxillary (V2), and mandibular (V3) divisions. Functionally, the ophthalmic and maxillary branches constitute a large sensory root that terminates in the principal sensory and spinal trigeminal nuclei, while the mandibular division remains a mixed branch. The small sensory component of the mandibular branch conveys proprioception to the mesencephalic trigeminal nucleus. Postsynaptic sympathetic and postsynaptic parasympathetic fibers accompany the sensory fibers. Through these branches, the trigeminal nerve conveys sensation from the face, eyelids and eyeball, nose and nasal cavity, nasopharynx, external ear with exception of the earlobe, scalp as far as the lambdoid suture, paranasal sinuses, meninges, maxillary and mandibular teeth, hard and soft palate, temporomandibular joint, anterior two-thirds of the tongue, and floor of the mouth. It also provides motor fibers to the muscles of mastication, anterior belly of the digastric, tensor palatini, and tensor tympani through the mandibular branch.

The *ophthalmic division* (V1) runs in the cavernous sinus and reaches the orbit through the superior orbital fissure (Figure 11.15). It supplies the frontal and ethmoidal sinuses, eyeball, dura of the anterior cranial fossa, nasal cavity, upper eyelid, tip of the nose, and skin of the forehead region and scalp as far as the lambdoid suture. This division has frontal, nasociliary, and lacrimal branches. *The frontal nerve* provides sensory fibers to the forehead, upper eyelid, and scalp via its supratrochlear and supraorbital branches. The *lacrimal nerve* also transmits postsynaptic parasympathetic fibers to the lacrimal gland (Figure 11.18). The *nasociliary branch* of the ophthalmic nerve (Figure 11.13) runs with the ophthalmic artery, giving rise to sensory fibers to the lateral nose and the eyeball. It also carries presynaptic parasympathetic fibers that eventually run through the short ciliary branches to the ciliary body and the constrictor pupilla muscle.

Nasociliary neuralgia is an episodic or prolonged pain sensation in the medial canthus of the eye, eyeball, and external nose. Redness of the forehead, congestion of the nasal mucus membrane, lacrimation, and conjunctivitis usually accompany this condition. This may be triggered by stimulation of the medial canthus. Antibiotics, cortisone, and local anesthetic (5% solution of cocaine) may be used to treat this condition.

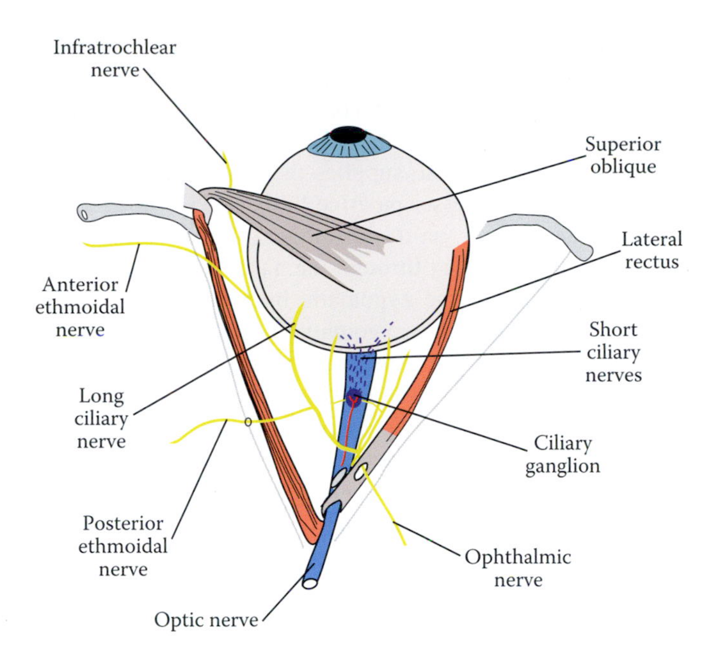

FIGURE 11.13 The course and branches of the nasociliary nerve.

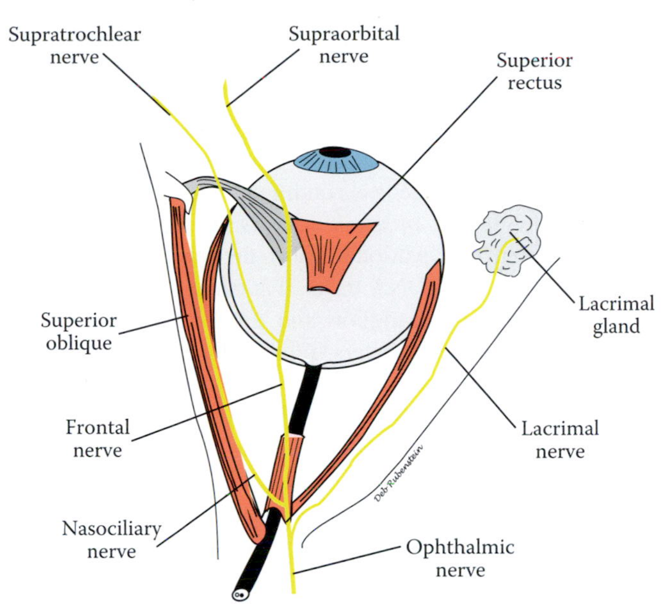

FIGURE 11.14 Frontal and lacrimal branches of the ophthalmic nerve.

The ophthalmic nerve mediates both corneal and lacrimal reflexes. The *corneal reflex* is a somatic reflex that is elicited by a light touch of the cornea or conjunctiva with a wisp of cotton, as the patient looks to the opposite side. Contraction of the orbicularis oculi muscles produces blinking in both eyes. This reflex is mediated by the indirect bilateral connections of the ophthalmic nerve (afferent limb) to the facial (efferent limb) motor neurons via interneurons of the pontine reticular formation. Loss of this reflex is a sign of neuropathy (Figures 11.14, 11.15, and 11.19).

Damage to the ophthalmic or the facial nerve or their central connections in the pons may produce loss of corneal reflex. Loss of the corneal reflex due to an ophthalmic nerve lesion may be observed in both eyes when the affected side is stimulated. However, this reflex may still be elicited in both eyes when the unaffected side is simulated (Figure 11.19). Damage to the facial nerve or nucleus produces loss of corneal reflex in the ipsilateral eye irrespective of which side is stimulated.

Lacrimation (tearing) *reflex* is mediated by the ophthalmic nerve (afferent limb) that conveys the signals to the superior salivatory nucleus (SSN) in the reticular formation of the pons. Axons of the presynaptic neurons of the SSN travel through the intermediate and the greater petrosal nerves to the pterygopalatine ganglion. The postsynaptic parasympathetic fibers from this ganglion reach the lacrimal gland through the zygomaticotemporal branch of the maxillary nerve and the lacrimal branches of the ophthalmic nerve.

Sluder's neuralgia (contact point headache) is a condition, which is associated with a lesion of the pterygopalatine ganglion. This unilateral disorder is characterized by unrelenting sharp or stabbing pain (resembles closely the cluster headache) in a single localized point associated the nasal, pharyngeal, or palatine branches of the maxillary nerve that pass through the pterygopalatine ganglion on their way to their sites of distribution. Pain is usually localized to the cheek, lower

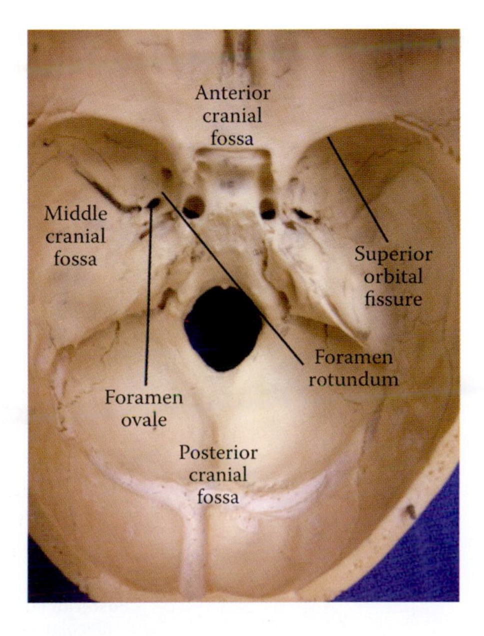

FIGURE 11.15 Main foramina and fissures that transmit branches of the trigeminal nerve.

eyelid, upper nose, palate, or maxillary teeth. Frequent sneezing is common in many cases. Photophobia and gustatory changes may also be experienced by the patients. Paranasal infection or other upper respiratory infection may induce this condition.

Herpes zoster ophthalmicus is a condition that results from reactivation of dormant varicella zoster, a double-stranded DNA virus, that causes an infection in the trigeminal ganglion and produces pain, rash, and then vesicular (blisters filled with serous exudate) eruptions in the course of the ophthalmic nerve (forehead, upper eyelid, and nose). The nasociliary branch may particularly be affected, causing perineuronal and intraneural inflammation on the surface of the eyeball and the skin of the nose down to its tip. The appearance of typical lesions at the tip of the nose indicates ocular involvement since the nasocilary nerve also innervates the cornea (Hutchinson sign.) This may lead to total visual loss and is rarely associated with middle cerebral artery infarct. The prodromal phase of this condition is associated with a low-grade fever and fatigue. Antiviral agents such as acyclovir are used as a principal therapy.

The *maxillary division* (V2) runs in the cavernous sinus, leaves the middle cranial fossa through the foramen rotundum (Figure 11.15), and enters the pterygopalatine fossa, where it is attached to the pterygopalatine ganglion (Figure 11.16). It provides sensory fibers to the skin overlying the maxilla, upper lip, lower eyelid, and side of the nose, as well as upper canine teeth via the infraorbital branch and the molar and premolar maxillary teeth and maxillary sinuses via the superior alveolar branches. It also supplies the mucosa of the palate via the greater and lesser palatine branches; the dura of the middle cranial fossa via the meningeal branches; the nasal cavity and nasopharynx through the nasal branches; and the temporal region via the zygomatic branch. The zygomatic nerve further divides into zygomaticofacial and zygomaticotemporal, branches. The latter branch gives rise to ganglionic branches to the pterygopalatine ganglion and communicates with the lacrimal nerve, conveying postganglionic parasympathetic fibers to the lacrimal gland. Irritation of the nasal mucosa produces *nasal (sneeze) reflex*, which is characterized by contraction of the muscles of the soft palate, pharynx and larynx, diaphragm, and intercostal muscles. The afferent limb of this reflex formed by the nasal fibers conveys these impulses to the spinal trigeminal nucleus, which transmit them, in turn, to the trigeminal motor and ambiguus nuclei, as well as the intercostal and motor neurons of the phrenic nerves to mediate contraction of the associated muscles.

The *mandibular branch* (V3) is sensory to the mandibular teeth; floor of the mouth; anterior two-thirds of the tongue; anterior wall of the auricle; and the external acoustic meatus, tympanic membrane, and skin overlying the lower jaw, chin, temporomandibular joint, lower lip, and lingual gingiva, (Figure 11.17). It innervates the muscles of mastication and the tensor palatini and tympani. It leaves

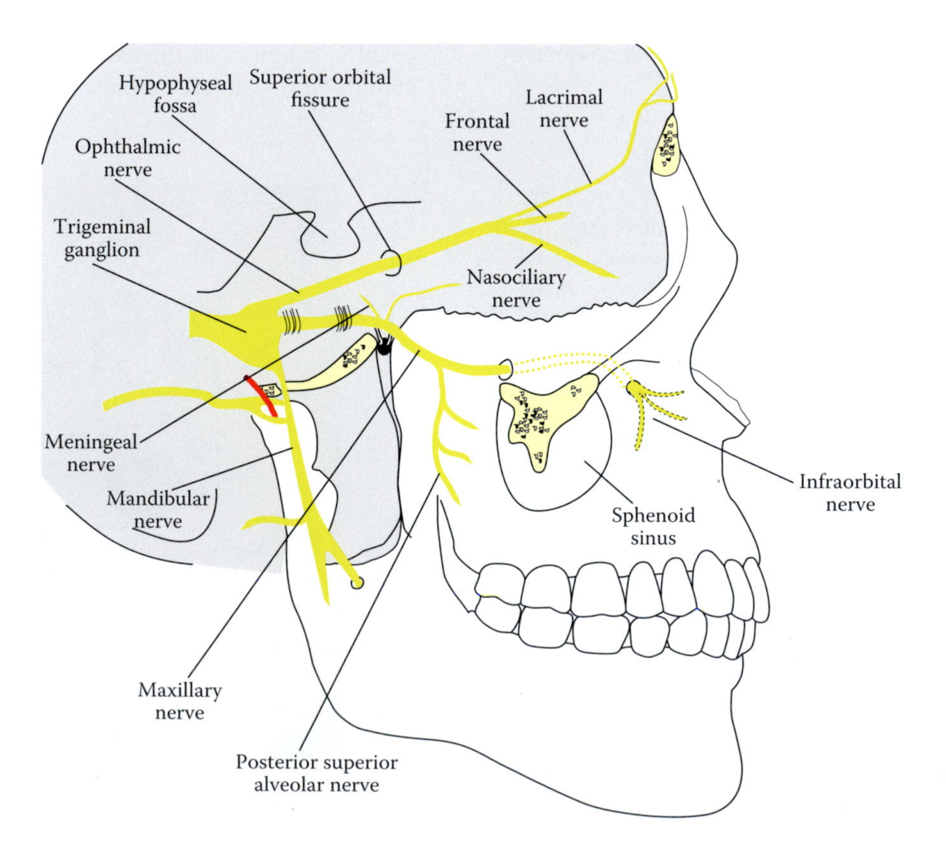

FIGURE 11.16 The origin, general course, and branches of the maxillary nerve.

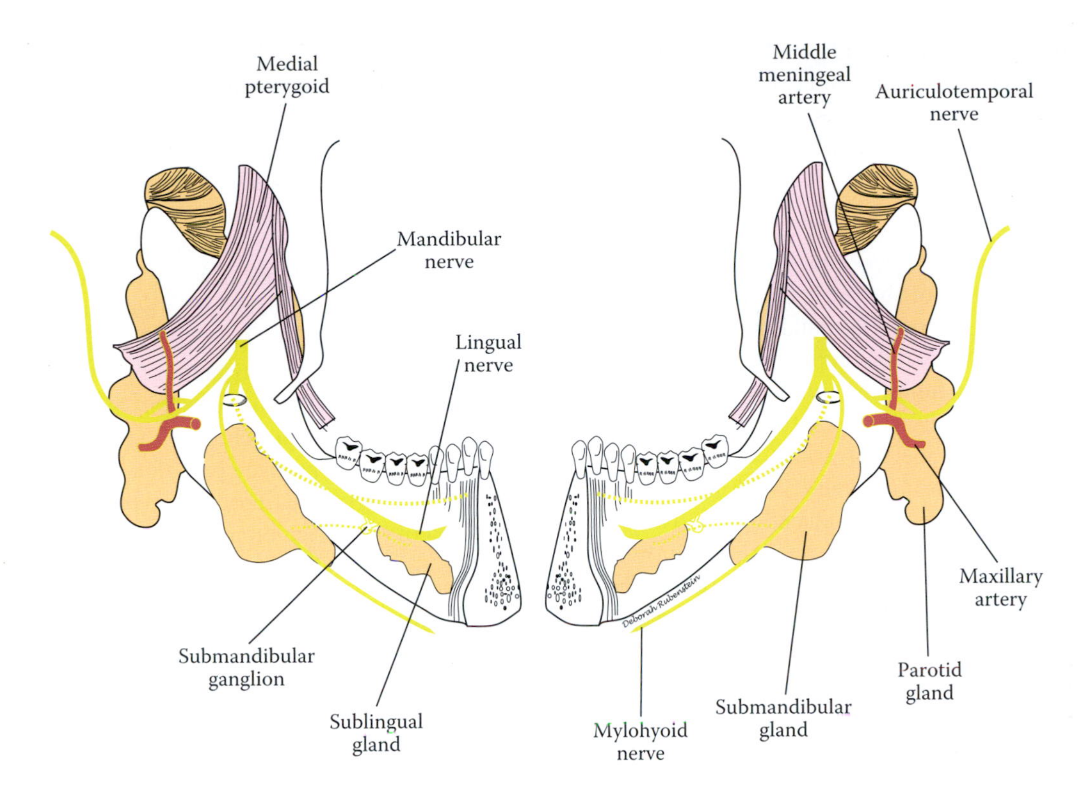

FIGURE 11.17 The principal branches of the mandibular nerve.

the middle cranial fossa via the foramen ovale (Figure 11.15) and enters the infratemporal fossa to course lateral to the otic ganglion. This branch divides into a primarily motor anterior trunk (with the exception of the buccal nerve, a sensory branch) and a principally sensory posterior trunk (with the exception of the mylohyoid nerve, a motor branch).

Major branches arise from the *posterior trunk* of the mandibular nerve, including the lingual, inferior alveolar, auriculotemporal, and mylohyoid branches.

The *lingual nerve* is sensory to the anterior two-thirds of the tongue, floor of the mouth, and lingual gingiva. During its course in the infratemporal fossa, it joins the chorda tympani branch of the facial nerve and proceeds to attach to the submandibular ganglion. The lingual nerve enables the chorda tympani to convey presynaptic parasympathetic fibers to the submandibular ganglion, mediating the secretion of the sublingual and submandibular glands. It also serves as a conduit for the fibers of the chorda tympani that convey taste sensation from the anterior two-thirds of the tongue to reach the brainstem and establish synaptic contact with the solitary nucleus.

Division of the lingual nerve distal to the site of union with the chorda tympani may produce loss of general and taste sensations from the anterior two-thirds of the tongue and impairment of salivary secretion from the submandibular and sublingual glands.

The *inferior alveolar nerve* (Figure 11.17) runs in the mandibular canal, supplies the mandibular teeth, and gives rise to the mental and incisive nerves. The mental nerve supplies the skin of the chin and is involved in the jaw jerk reflex. This monosynaptic reflex is characterized by sudden closure of the mouth (as a result of bilateral contraction of the masseter and temporalis muscles), following a downward tap on a finger placed on the jaw when the mouth is slightly open. It is mediated by the mandibular nerve, through the mesencephalic and the motor trigeminal nuclei.

Sudden onset of hypoesthesia, anesthesia, paresthesia, or pain over the area of distribution of the mental branch of the inferior alveolar nerve (numb chin syndrome) can be the primary manifestation of and forerunner to multiple sclerosis or progressive systemic malignancy. This uncommon unilateral condition occurs spontaneously with no history of dental diseases, trauma, or infection. It is caused by malignant infiltration of and neuropathy of the inferior alveolar nerve sheath.

Failure to elicit jaw jerk reflex may indicate a pontine lesion involving the trigeminal nerve or mesencephalic or motor trigeminal nuclei. Hyperactive jaw reflex is a manifestation of corticobulbar tract damage.

The *auriculotemporal nerve* (Figure 11.17) is the only branch that courses posteriorly, arising by two roots that encircle the middle meningeal artery. It carries postganglionic parasympathetic fibers to the parotid gland and sensory fibers to the temporomandibular joint, anterior temporal region, auricle, and external acoustic meatus. The *mylohyoid nerve* innervates the mylohyoid muscle and the anterior belly of the digastric muscle.

Anomalous connections between the postsynaptic parasympathetic fibers of the auriculotemporal nerve that are destined to the parotid gland and the sympathetic postsynaptic fibers that supply the sweat glands of the face may occur following infection and trauma, although rarely after surgical operation involving the parotid gland. These aberrant connections produce signs of *Frey syndrome* (gustatory sweating), which is characterized by sweating induced by salivatory stimuli. Patients with Frey syndrome exhibit flushing and sweating on the face, along the distribution of the auriculotemporal nerve, in response to tasting or eating. Auriculotemporal neuralgia may accompany this condition. Although a rare form of neuralgia, patients with this disorder may exhibit burning pain in the preauricular and temporal, regions triggered by chewing or tasting spicy food.

The *anterior trunk* of the mandibular nerve gives rise to nerves that supply the temporalis, lateral and medial pterygoid, masseter, tensor tympani, and tensor palatini muscles. The buccal nerve, the only sensory branch of the anterior trunk, supplies the skin and mucosa of the cheek.

Injury to the trigeminal nerve or its branches may occur as a result of a tumor of the cerebellopontine angle, otitis media, cavernous sinus thrombosis, fractures involving the facial bones or the middle cranial fossa, invasive dental procedures, or metastatic carcinomas. Multiple sclerosis may produce trigeminal neuralgia and transient facial anesthesia in young adults. In particular, the ophthalmic branch may be damaged as it courses within the superior orbital fissure in conjunction with the oculomotor, trochlear, and abducens nerves. It may also be injured in orbital apex syndrome together with the optic, oculomotor, trochlear, and abducens nerves. A fracture confined to the ramus of the mandible may put the mandibular nerve out of function.

A lesion of the trigeminal nerve produces combined sensory and motor as well as reflex disorders. These dysfunctions include unilateral anesthesia or pain in the area of distribution of the trigeminal nerve; loss of corneal reflex on both sides when the affected eye is stimulated; and atrophy of the muscles of mastication, tensor tympani, and palatini muscles. Additional deficits include numbness or pain in the facial region, palate, oral and nasal cavities, and anterior two-thirds of the tongue. Sensations from the temporomandibular joint, paranasal sinuses, and anterior part of the external acoustic meatus may also be impaired. The jaw jerk, oculocardiac (a reflex that mediates slowing of heart rate upon compression of the eyeball), sneezing, and lacrimation reflexes are impaired. Impairment of the postsynaptic parasympathetic innervation to the head region may also be noticed. Referred (projected) pain to the area of distribution of branches of the trigeminal nerve is common. For example, pain from carious tooth may project pain to the ear, or an ulcer of the tongue may produce pain that is felt in the ear and temporal region that corresponds to the area of distribution of the auriculotemporal nerve.

The trigeminal nerve has three sensory and a single motor nucleus (Figures 11.14, 11.15, and 11.18 through 11.20). The sensory nuclei of the trigeminal nerve are the spinal, principal sensory, and mesencephalic trigeminal nuclei.

The *spinal trigeminal nucleus* (GSA) extends from the midpons to the upper segments of the spinal cord, representing the rostral extension of the substantia gelatinosa. It receives thermal, painful, and tactile sensations from the head region within branches of the trigeminal, facial, glossopharyngeal, and vagus nerves. As they terminate in this nucleus, they form the spinal trigeminal tract, which are positioned lateral to the nucleus. Within the spinal trigeminal nucleus and tract, the ophthalmic fibers occupy a caudal position to the more rostral mandibular fibers, while the maxillary fibers maintain an intermediate position. Since pain and thermal fibers terminate in the most caudal portions of this nucleus and tract, within the caudal medulla, excision of the spinal trigeminal tract (tracteotomy) at this level may alleviate the intractable pain associated with trigeminal neuralgia.

In the midpons, the *principal* (pontine or chief) *sensory nucleus* (GSA) lies lateral to the trigeminal nerve fibers. It transmits tactile and pressure sensation, showing identical somatotopic distribution to that of the spinal trigeminal nucleus.

Neurons of the spinal trigeminal nucleus and the ventral part of the principal sensory nucleus give rise to axons regarded as secondary trigeminal fibers. These fibers cross the midline and run through the reticular formation, forming the *ventral trigeminal lemniscus* on the opposite side. Fibers from the dorsal part of the principal sensory nucleus are virtually derived from the mandibular nerve and ascend ipsilaterally as the *dorsal trigeminal lemniscus*. Both of the trigeminal lemnisci are associated topographically with the medial lemniscus and project to the ventral posteromedial (VPM) nucleus of the thalamus. The secondary trigeminal fibers have collaterals that establish contact with specific nuclei in the reticular formation, mediating various trigeminal reflexes.

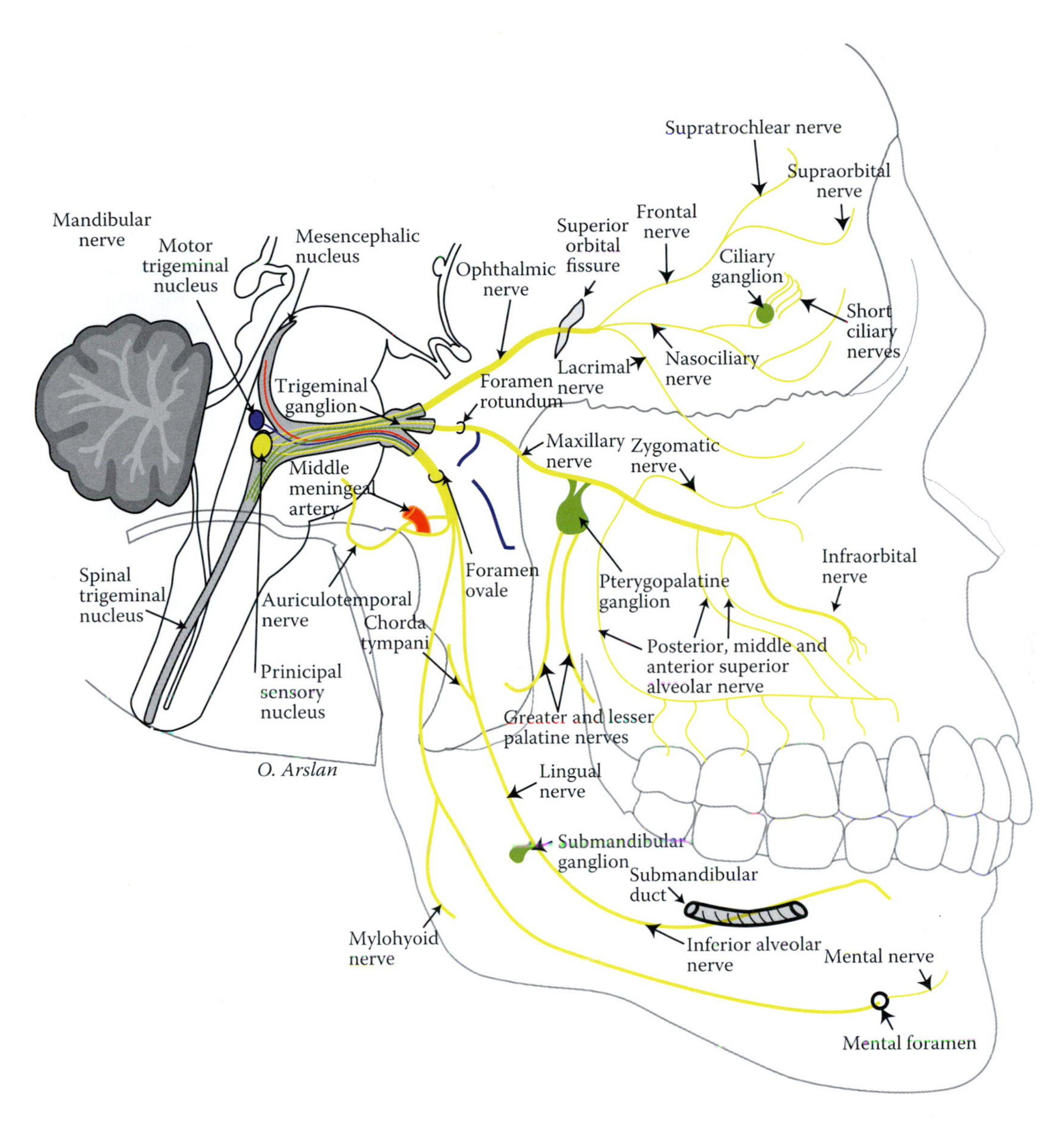

FIGURE 11.18 Functional components, branches, and associated nuclei of the trigeminal nerves. The course of these branches and their areas of distribution are documented to facilitate the understanding of the central and peripheral courses of this nerve.

The *mesencephalic nucleus* (GSA) is a unique nucleus that consists of unipolar neurons retained in the midbrain and receives fibers that transmit proprioceptive input from the temporomandibular joint, muscles of mastication, hard palate, mandibular and maxillary alveoli, and possibly the extraocular muscles. It projects to the cerebellum and superior colliculi, and mediates the jaw jerk reflex.

The *motor trigeminal nucleus* (SVE) is located medial to the entering trigeminal nerve fibers in the rostral pons (also discussed in Chapter 20). It exits medial to the larger sensory root and passes through the trigeminal ganglion and foramen ovale to be distributed to the muscles of mastication, anterior belly of the digastric, mylohyoid, tensor tympani, and tensor palatini muscles. It receives input primarily from the mesencephalic nucleus, mediating the jaw jerk reflex. This nucleus also receives bilateral corticobulbar fibers.

A lesion confined to the motor trigeminal nucleus may only result in atrophy of the muscles of mastication and deviation of the mandible toward the lesion side.

Trigeminal neuralgia (tic douloureux), a common idiopathic condition, exhibits a lightening or lancinating lower facial pain on the affected side, which lasts for a few seconds. This lifelong recurring disorder affects women twice as often as men and develops relatively late in life, between the ages of 50 and 60. The paroxysmal pain associated with this condition is typically intermittent and can be spontaneous but is frequently provoked by mild stimuli (e.g., touch, talking, shaving, application of cold, etc.) in one or more area of distribution of the trigeminal nerve, usually the maxillary or mandibular nerves. Patients may also exhibit signs

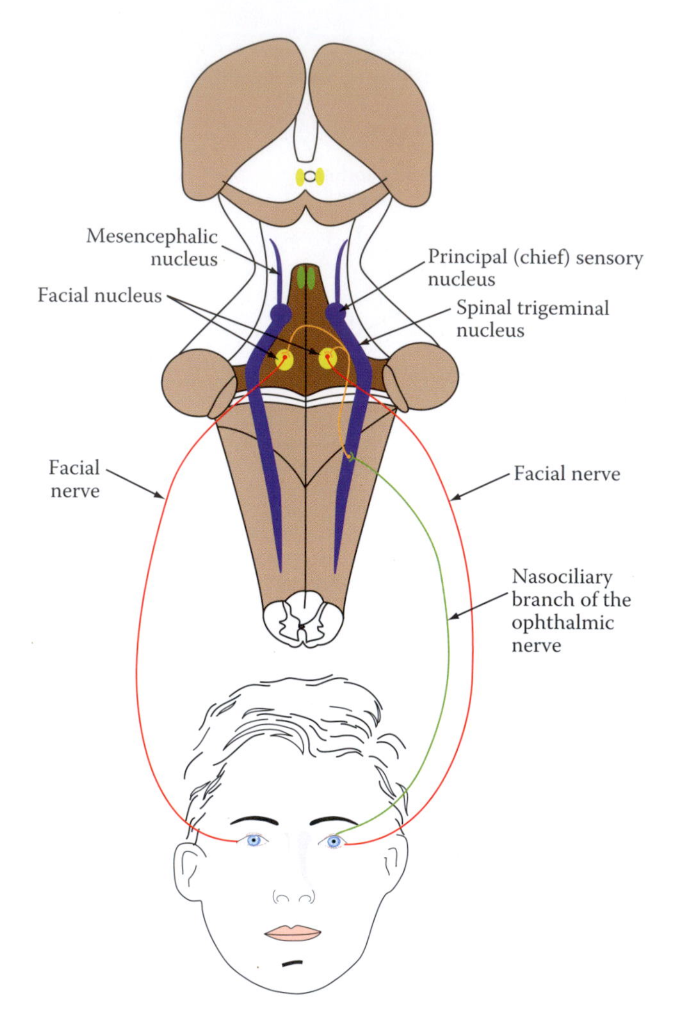

FIGURE 11.19 Corneal reflex. Both the trigeminal and facial nerves, as illustrated in this diagram, mediate this reflex.

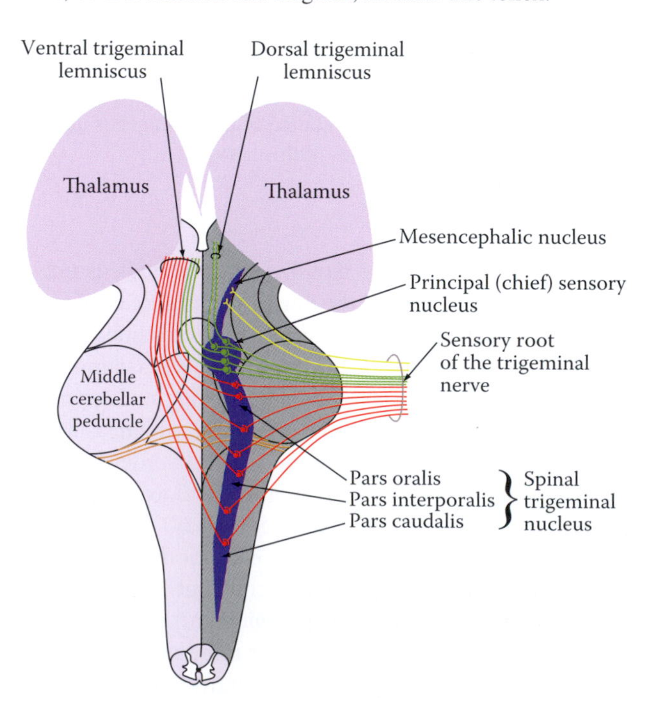

FIGURE 11.20 Trigeminal nuclei and their projections via the trigeminal lemnisci.

of autonomic disorders (lacrimation, salivation, and flushing of the face), reflex facial muscle spasm, and sensory loss. Aberrant superior cerebellar or cerebral arteries may produce this condition by compressing the root of the trigeminal nerve. Multiple sclerosis accounts for most cases of trigeminal neuralgia in young adults. Acoustic neuromas, arteriovenous malformations, and menigiomas can compress the trigeminal root and produce neuralgia. This condition may be treated by carbamezapine or dilantin or by injection of glycerol into the root of the trigeminal nerve. Decompression surgery can produce relief by placing a barrier between the trigeminal nerve and the anomalous vessel.

Branches of the trigeminal nerve may show port-wine discoloration in their area of distribution as a manifestation of *Sturge–Weber syndrome* (encephalotrigeminal angiomatosis). The areas of distribution of the ophthalmic nerve in the forehead, upper eyelid, and scalp are most commonly affected with this vascular malformation (Figures 11.21 and 11.22). These manifestations may be seen on one side or bilaterally. Calcified cortical atrophy may show characteristic railroad configuration. Calcified vascular abnormality may also be seen in the meninges. Patients with this condition may show mental retardation, convulsions, homonymous hemianopsia, and spastic hemiparesis if the cerebral atrophy and calcification is extensive.

ABDUCENS NERVE

The abducens nerve (GSE) is formed by the axons of the multipolar motor neurons of the abducens nucleus, which descends medially and passes through the paramedian reticular nuclei. During its course in the tegmentum and basilar pons, the abducens nerve runs adjacent to the corticospinal tract. It exits ventrally through the pontobulbar sulcus, maintaining a long intracranial course between the pons and the

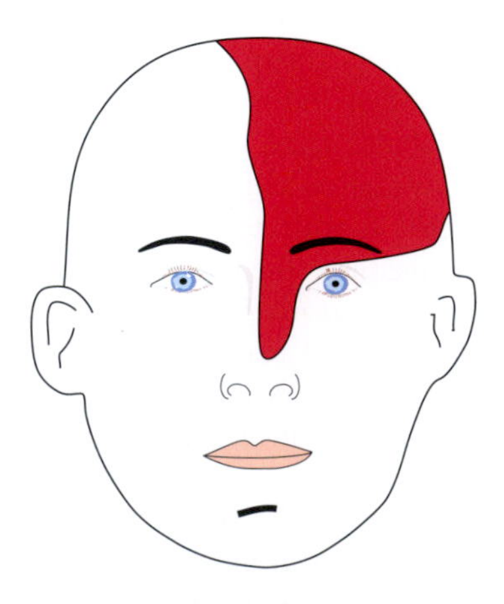

FIGURE 11.21 Port-wine discoloration (cutaneous vascular nevus) of the territory of the ophthalmic nerve. The patient is retarded and suffers from epileptic seizures.

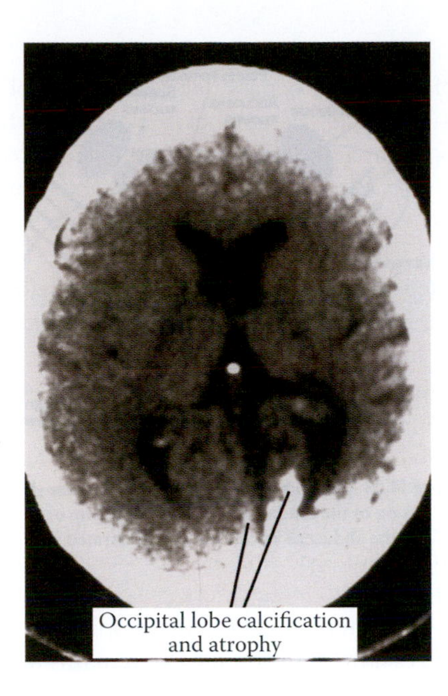

FIGURE 11.22 This image shows a prominent calcification in an individual with Sturge–Weber syndrome.

clivus within the pontine cistern. It courses below the inferior petrosal sinus, forms a bend over the apex of the petrous temporal bone, and enters the cavernous sinus through the *Dorello* canal just medial to Meckel's cave of the trigeminal ganglion. Within the cavernous sinus, the abducens nerve lies in close proximity to the internal carotid artery receiving sympathetic postsynaptic fibers from the carotid plexus that dilate the pupil. These sympathetic fibers join the nasociliary branch of the ophthalmic nerve and then the long ciliary nerve that passes through the ciliary ganglion to the dilator pupillae muscle of the iris. Then, it leaves the cavernous sinus to enter the orbit via the superior orbital fissure, innervating the lateral rectus muscle.

The abducens nucleus lies deep to the ependyma of pontine part of the fourth ventricle encircled by the motor fibers of the facial nerve, forming the facial colliculus (Figures 11.1 and 11.23). This nucleus is unique among all other cranial nerve motor nuclei as it contains a population of α motor neurons that give rise to the abducens nerve and a smaller population of interneurons that send axons through the contralateral MLF to the motor neurons of the oculomotor nucleus. These axons control the oculomotor neurons that innervate the medial rectus muscle. This functionally dual population of neurons may account for the distinct deficits produced by a lesion of the abducens nerve versus the abducens nucleus.

Proximity of the abducens nerve to the internal carotid artery may account for the initial signs of abducens nerve palsy in individuals with an aneurysm of the internal carotid artery. This intercavernous aneurysm also compresses the sympathetic fibers around the artery, producing Horner's syndrome in addition to painful abducens palsy. Also, due to its long intracranial course and sharp bend over the petrous temporal bone, this nerve is more prone to injury than any other cranial nerve. The sharp bend of the abducens nerve over the apex of the petrous temporal bone accounts for its undue stretch and damage when intracranial pressure is increased and the brain begins to displace

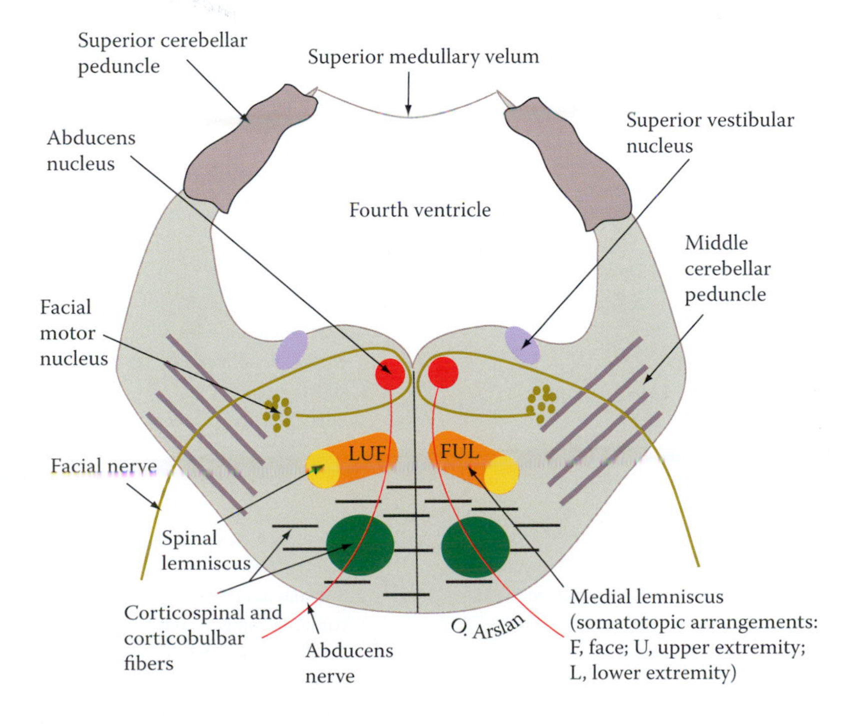

FIGURE 11.23 The abducens nucleus and nerve are shown in this section of the caudal pons. The relationship of the facial nerve to the abducens nucleus is also illustrated.

downward. Postlumbar puncture or spontaneous intracranial hypotension produces a brainstem shift that can also compromise the function of the abducens nerve function. Damage to this nerve may also occur in cavernous sinus thrombosis and fracture of the superior orbital fissure. Due to the close relationship of the abducens nerve to the petrous temporal bone, chronic otitis media can lead to mastoiditis and petrositis and then erosion of the bony wall, causing combined abducens nerve palsy, facial nerve palsy, and disruption of the trigeminal ganglion (*Gradenigo syndrome*). Medial strabismus or convergent squint in which the patient is unable to direct both eyes toward the same object characterizes abducens nerve palsy (Figure 11.24). This occurs due to paralysis of the lateral rectus muscle. Patients also experience horizontal diplopia (in acute stage) on attempted gaze to the affected side (Table 11.1). Chronic abducens nerve palsy may not exhibit diplopia as the image from the affected eye is suppressed psychologically. The abducens nucleus lesion is usually associated with ipsilateral facial nerve palsy. A lesion of abducens nucleus produces lateral gaze palsy toward the affected side due to disruption of the ipsilateral abducens nerve (loss of abduction) and internuclear neurons that control the ventral subnucleus of the oculomotor complex (adductor paresis) through the MLF. A lesion that damages the abducens nucleus and the ipsilateral MLF produces manifestations of "one-and-a-half" syndrome, which is characterized by conjugate horizontal gaze palsy to the same side and internuclear ophthalmoplegia when gaze is directed contralaterally. One eye remains motionless in the midline while the other eye can only abduct, but convergence is preserved.

Destruction of the abducens nerve and the adjacent fibers of the corticospinal tract on one side may produce signs of *middle alternating hemiplegia*, in which hemiplegia is manifested on the contralateral side, while signs of abducens nerve palsy remains ipsilateral. A lesion of the abducens nucleus results in disruption of the abducens nerve and the internuclear neurons that emanate from the abducens nucleus and project to the contralateral medial rectus, producing lateral gaze, which is characterized by adductor paresis on the lesion side and abductor palsy on the opposite side.

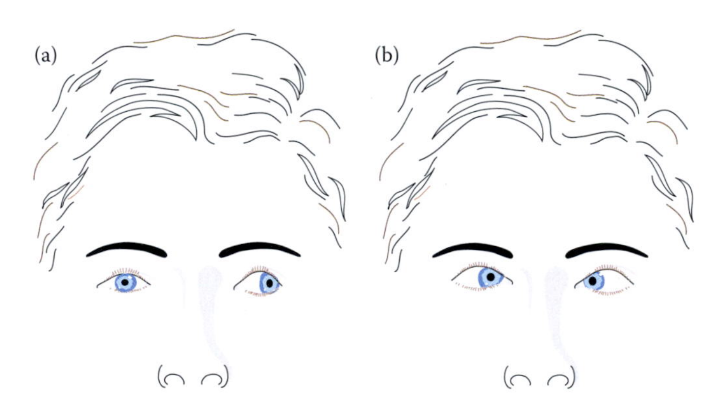

FIGURE 11.24 Deficits associated with right abducens nerve palsy. Note that the eye on the affected side is adducted at rest (a) and cannot abduct when attempting to look to the right (b).

TABLE 11.1
Cranial Nerves I–IV

Cranial Nerve	Location of Cell Bodies	Course	Distribution	Function
I. Olfactory	Neuroepithelial cells in nasal cavity	Cribriform plate of ethmoid	Olfactory mucus membrane	Olfaction
II. Optic	Ganglion cells in the retina	Optic canal	Retina	Vision and visual reflexes
III. Oculomotor	Somatic column of oculomotor nucleus Edinger–Westphal nucleus	Superior orbital fissure	Levator palpebrae and all extraocular muscles with the exception of lateral rectus and superior oblique Ciliary and constrictor pupillae muscles	Elevates the upper eyelid; adducts, elevates, or depresses the eyeball Mediates light and accommodation reflexes
IV. Trochlear	Trochlear nucleus	Superior orbital fissure	Superior oblique	Abducts, intorts, and depresses the eyeball
V. Trigeminal	Trigeminal (Gasserian) ganglion Trigeminal motor nucleus	V1—superior orbital fissure V2—foramen rotundum V3—foramen ovale Foramen ovale	Skin of face, scalp, gingiva, anterior two-thirds of tongue, oral and nasal cavities, eye, paranasal sinuses, and meanings Muscles of mastication, anterior belly of digastric, tensor tympani, and mylohyoid muscle	Cutaneous and proprioceptive information Mastication, mandibular movements, and deglutition
VI. Abducens	Abducens nucleus	Superior orbital fissure	Lateral rectus	Lateral deviation of the eye

FACIAL NERVE

The facial nerve (GSA, GVE, SVA, SVE) consists of a large somatic motor component, and a smaller intermediate root or nerve that occupies an intermediate position between the vestibulocochlear and facial motor nerve consisting of sensory and autonomic fibers. It leaves the brainstem at the cerebellopontine angle between the inferior cerebellar peduncle and the olivary eminence and enters the internal acoustic meatus (Figures 11.1, 11.2, 11.25, 11.26, and 11.28), accompanied by the vestibulocochlear and labyrinthine artery. Then, it follows a horizontal course dorsal to the cochlea (*labyrinthine segment*) and then runs vertically, forming the external genu, where the geniculate ganglion is located and the greater petrosal nerve arises. It descends along the posteromedial wall of the tympanic cavity (*tympanic segment*) and enters the facial canal within the mastoid part of the temporal bone (*mastoid segment*) and gives rise to the stapedius nerve and the chorda tympani. It leaves the skull through the stylomastoid foramen, innervating the auricular, occipitalis, posterior belly of the digastric, and stylohoid muscles. Then, it pierces the parotid gland (*parotid segment*), where it divides into an upper and a lower trunk. The upper trunk divides into the frontal, temporal, and zygomatic branches, whereas the lower trunk splits into the buccal, marginal mandibular, and cervical branches. The upper trunk supplies the frontalis, orbicularis oris, nasalis, procerus, corrugator supercilli, levator labii superioris, levator labii superioris alaque nasi, levator anguli oris, and zygomaticus major and minor, whereas the lower trunk innervates the levator anguli oris, levator labii superioris, buccinator, orbicularis oris, risorius, depressor labii, depressor anguli oris, mentalis, and platysma. These branches leave through the anterior border of the parotid gland to supply muscles of facial expression (*extracranial segment*). The facial nerve is supplied by the superficial petrosal branch of the middle meningeal artery and by the stylomastoid branch of the posterior auricular, occipital, superficial temporal, and transverse facial arteries and via the tympanic branch of the ascending pharyngeal arteries.

The facial nerve consists of the motor and intermediate (nerve) roots.

The *motor root* (SVE) is derived from neurons of the facial motor nucleus, which lies in close proximity to the spinal trigeminal nucleus and tract, ascends dorsomedially, and encircles the abducens nucleus (internal genu), forming the facial colliculus (intrapontine). It joins the intermediate nerve and leaves the caudal pons between the abducens and vestibulocochlear nerves and descends in the facial canal, exiting through the stylomastoid foramen. These motor fibers mediate both glabellar and corneal reflexes. *Corneal reflex* (also discussed with the trigeminal nerve (Figure 11.19) is characterized by contraction of the orbicularis oculi and the resultant blinking upon stimulation of the cornea by a wisp of cotton. Damage to the facial nerve also results in loss of this reflex on the side of lesion.

The *glabellar* (McCarthy's) *reflex* or *Myerson sign* exhibits forceful, persistent, involuntary, and repeated contraction of the orbicularis oculi muscle, which is elicited by repetitive finger taps on the forehead and supraorbital margin in a downward direction to the glabella. Damage to the facial nerve or its nucleus produces glabellar hyporeflexia, a disorder seen in Parkinson's disease and in patients with bilateral frontal lobe lesion, and occasionally in tense and overly stressed individuals.

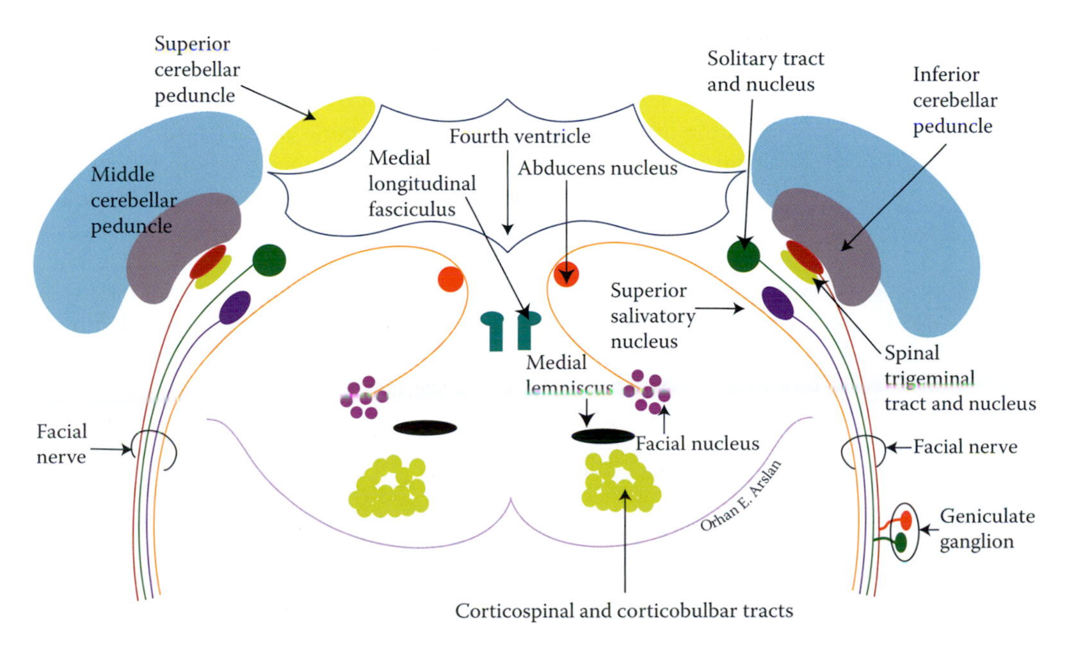

FIGURE 11.25 The functional components of the facial nerve, course, and associated nuclei are shown in this diagrammatic section of the pons.

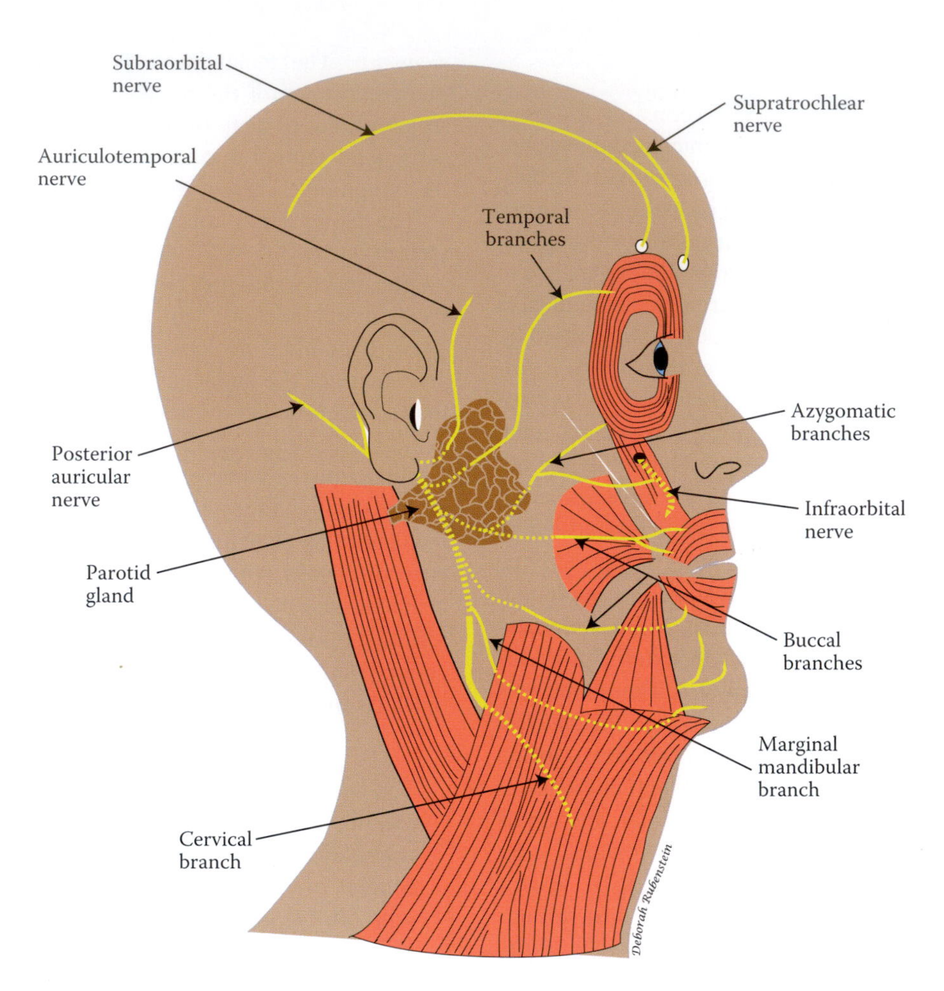

FIGURE 11.26 The course and branches of the facial nerve, outside the facial canal, are illustrated.

The *intermediate* (root) *nerve* contains general and special sensory as well as preganglionic parasympathetic fibers that run between the motor root and the vestibulocochlear nerve in close contact with the anterior inferior cerebellar artery. This root enters the facial canal accompanied by the labyrinthine branch of the anterior inferior cerebellar artery or basilar artery. The sensory neurons are located in the *geniculate ganglion*, a collection of unipolar neurons at the junction of the vertical and horizontal parts of the facial nerve located medial to the cochleariform process.

Ramsay Hunt syndrome, a condition that results from reactivation of the varicella zoster virus within the geniculate ganglion, produces ipsilateral infranuclear facial palsy and painful rash followed by vesicular eruptions in areas of distribution of branches of the facial nerve, which include the external acoustic meatus, anterior two-thirds of the tongue, and palate. Deafness and vertigo can also occur.

The somatic sensory fibers of the intermediate nerve terminate in the spinal trigeminal nucleus, while the SVA (taste) fibers establish synaptic connections with neurons of the solitary nucleus. This root also contains preganglionic parasympathetic fibers that run within the greater petrosal nerve and chorda tympani, providing secretomotor fibers to the submandibular, sublingual, and lacrimal glands. The facial nerve gives rise to the greater petrosal, chorda tympani, stapedius, posterior auricular nerve, and muscular branches.

The *greater petrosal nerve* (GVE, SVA) arises from the facial nerve at the level of the geniculate ganglion, carrying preganglionic parasympathetic fibers (GVE) from the SSN to the pterygopalatine ganglion (Figure 11.28). The postsynaptic parasympathetic fibers distribute to the lacrimal gland, mucus glands of the palate, and nasal cavity. It also carries taste (SVA) fibers from the palate to the solitary nucleus. This nerve should not be confused with the lesser petrosal branch of the glossopharyngeal nerve, which provides preganglionic parasympathetic fibers to the otic ganglion mediating parotid gland secretion.

The *chorda tympani* (SVE, GVE) runs in the upper quadrant of the tympanic membrane medial to the malleus and incus, leaves the skull via the petrotympanic fissure, and enters the infratemporal fossa, where it joins the lingual nerve. It carries preganglionic parasympathetic (GVE) fibers from the SSN, which are conveyed to the submandibular ganglion to regulate the secretion of the submandibular and sublingual glands. It also transmits taste sensation (SVA) from the anterior two-thirds of the tongue to the solitary nucleus (Figure 11.28).

Anomalous connections between the preganglionic parasympathetic fibers that emanate from the SSN, which include the chorda tympani and the greater petrosal nerve, result in salivation-induced lacrimation (*crocodile-tear syndrome*). This uncommon gustatory hyperhidrosis occurs on the same side of the face as a result of misdirected gustatory fiber nerve regeneration of an injured facial nerve.

The *stapedius nerve* (SVE) is formed by fibers of the facial motor nucleus that supplies the stapedius muscle, which contracts in response to high-frequency sound.

The *posterior auricular branch* (SVE) runs between the mastoid process and the external acoustic meatus and divides into auricular and occipital branches. The auricular branch innervates the posterior auricular muscles and other intrinsic muscles on the cranial aspect of the auricle, whereas the larger occipital branch supplies the occipital belly of the frontooccipitalis muscle (Figure 11.25). Other muscular branches of the facial nerve (SVE) supply the stylohyoid and the posterior belly of the digastric muscle. As discussed earlier, the muscular branches that arise from the trunk of the facial nerve within the parotid gland include the temporal, zygomatic, buccal, marginal mandibular, and cervical branches (Figure 11.26). The buccal branch of the facial nerve is motor mainly to the muscles around the mouth that accompany the parotid duct, while the buccal branch of the mandibular nerve is sensory to the skin and mucosa of the cheek. The auricular branch (GSA) of the facial nerve transmits general sensory impulses from the concha of the ear to the spinal trigeminal nucleus.

These nuclei are comprised of the facial motor, superior salivatory, solitary, and spinal trigeminal nerves (Figures 11.25 and 11.28).

The *facial motor nucleus* (SVE) is located in the tegmentum of the caudal pons, posterior to the dorsal trapezoid nucleus and ventromedial to the spinal trigeminal nucleus. It divides into lateral, intermediate and medial subnuclear groups with the lateral group projecting through the buccal branches, the intermediate through the temporal, orbital, and zygomatic branches while the medial group sends fibers to the posterior auricular, cervical, and stapedial branches. These subnuclei further subdivide into smaller nuclear groups that correspond to the muscles innervated by the facial nerve. It gives rise to the special visceral efferent fibers that encircle the abducens nucleus, forming the facial colliculus in the rostral part of the floor of the fourth ventricle. The uniqueness of this nucleus is illustrated in its highly distinctive corticobulbar connections. The part of the facial motor nucleus that provides innervation to the muscles around the mouth and lower face receives only crossed corticobulbar fibers, whereas the part of the nucleus that supplies the muscles around the orbit and the forehead region receives bilateral corticobulbar projections. This diverse cortical input accounts for the selective paralysis of the contralateral muscles around the mouth in individuals with unilateral corticobulbar damage, while sparing the muscles around the eye and orbit (Figure 11.27). Pyramidal input to this nucleus that runs in the medial lemniscus has been reported.

The *spinal trigeminal nucleus* (GSA) receives general somatic sensations from the concha, to be delivered to the ventral posteromedial nucleus of the thalamus.

The *superior salivatory nucleus* (GVE) lies adjacent to the caudal end of the facial motor nucleus in the caudal pons and provides preganglionic parasympathetic fibers to the pterygopalatine and the submandibular ganglia via the greater petrosal and chorda tympani branches, respectively. Postsynaptic fibers from these ganglia control the secretion of the lacrimal, submandibular, and sublingual glands and the mucus glands of the palate and pharynx.

The *solitary nucleus* (SVA) receives taste fibers via the central processes of the geniculate neurons that primarily originate from the anterior two-thirds of the tongue. Postsynaptic fibers from this nucleus terminate in the ventral posteromedial nucleus of the thalamus (Figure 11.28).

Facial palsy occurs as a result of damage to the corticobulbar fibers, facial motor nucleus, or facial nerve. These lesions, as documented below, are classified into two categories: supranuclear and infranuclear. In supranuclear lesions (Figure 11.29), the corticobulbar fibers that emanate from the cerebral motor cortex and project to the facial motor nucleus are disrupted, producing upper motor neuron signs in the muscles around the mouth on the contralateral side. Patients cannot voluntarily move the affected muscles (*voluntary facial palsy*) but remain responsive to emotional stimuli. Supranuclear lesions that affect the projections of the limbic system to the facial motor nucleus may result in *mimetic facial palsy*. In this condition, the patient remains unresponsive to emotional stimuli while preserving the ability to produce contraction of the affected muscles upon command.

Infranuclear lesions (Figure 11.30) are caused by damage to the facial nerve and/or the facial motor nucleus. These lesions may be the result of intrapontine infarct, mumps, acoustic neuroma, parotid tumors, Miller Fisher syndrome (a variant of Guillain–Barré syndrome), geniculate herpes (herpes zoster oticus), leprosy, leukemia, or sarcoidosis or may remain idiopathic. A lesion that affects the facial colliculus leads to concomitant damage to the facial nerve and abducens nucleus, resulting in ipsilateral facial nerve palsy and loss of lateral gaze toward the affected side. A brainstem glioma that disrupts the fibers of the abducens nerve can also affect the facial motor nucleus, causing concomitant ipsilateral facial palsy, medial strabismus, and diplopia. In the newborn, absence of the mastoid processes renders the facial nerves exposed on both sides and unusually vulnerable to damage by careless use of obstetrical forceps. The facial nerve and the trigeminal nerve can be damaged conjointly with the

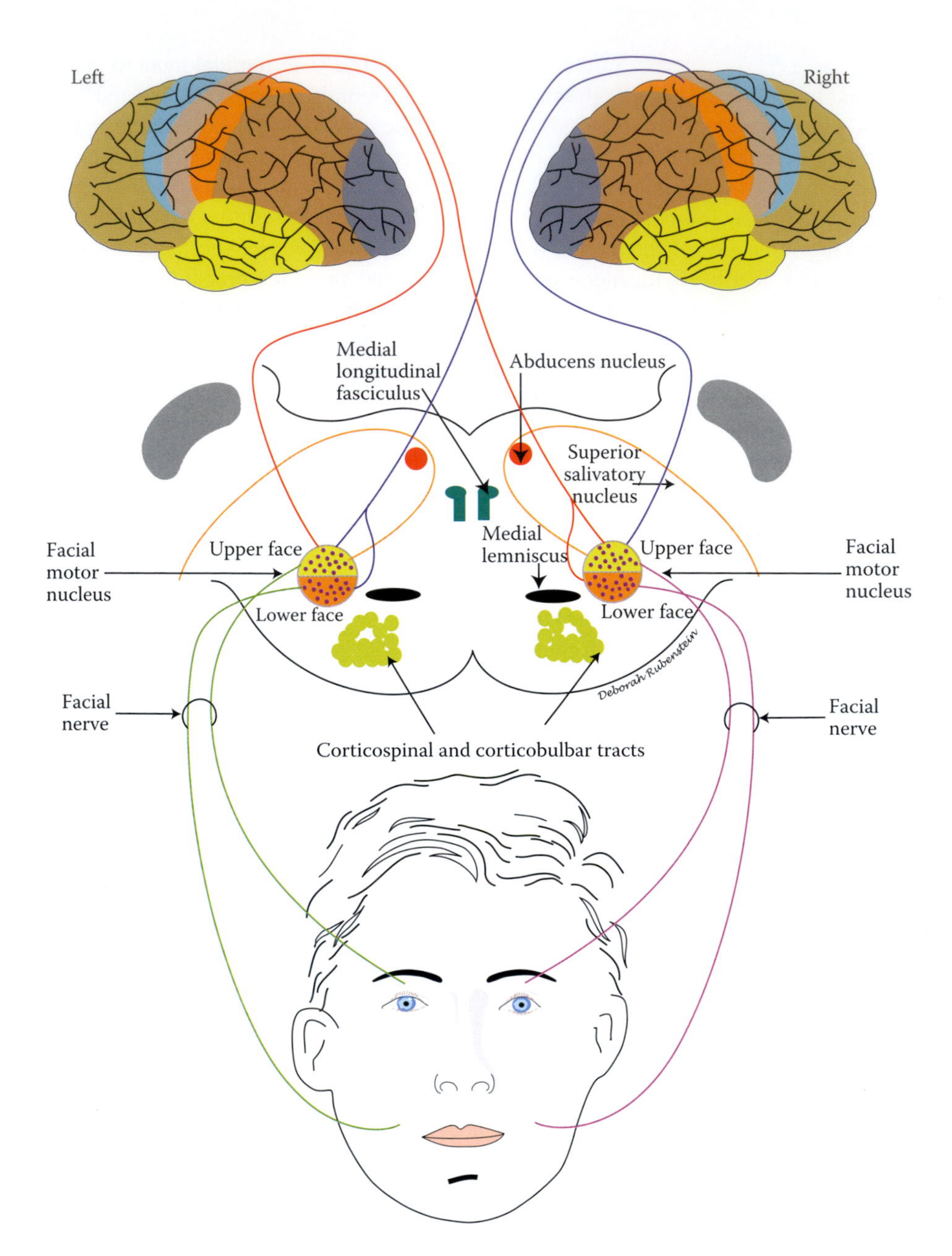

FIGURE 11.27 Schematic drawing of the preferential distribution of the corticobulbar tract in the facial motor nuclei. Note the bilateral distribution of the corticobulbar fibers to the neurons of the upper face and contralateral projection to the neurons of the lower face.

vestibulocochlear nerve as a result of compression by a large cerebellopontine angle tumor (acoustic neuroma). This type of compressive lesion produces, initially, tinnitus and then deafness, followed by vertigo and ipsilateral facial palsy and numbness.

Facial nerve damage produces ipsilateral symptoms, which vary dependent upon the site of the lesion. A lesion at or above the level of the geniculate ganglion (between the facial nucleus and the geniculate ganglion), as in tumors of the cerebellopontine angle or in fractures of the internal acoustic meatus, results in the ipsilateral deficits that include loss of lacrimation, taste, and general sensations from the palate due to disruption of the greater petrosal nerve. A lesion of the facial nerve at this site also results in the loss of taste sensation from the anterior two-thirds of the tongue and impairment of secretion of the submandibular and sublingual due to disruption of the chorda tympani, as does loss of sensation from the concha occur subsequent to damage to the auricular branch. Additionally, asymmetry of the face, widening of the palpebral fissure, inability to close the eye, loss of the corneal reflex, sagging of the angle of the mouth, and smoothing of facial sulci are also seen. Stapedius muscle palsy that produces hyperacusis (low hearing threshold) is also seen in a lesion of the facial nerve at this site.

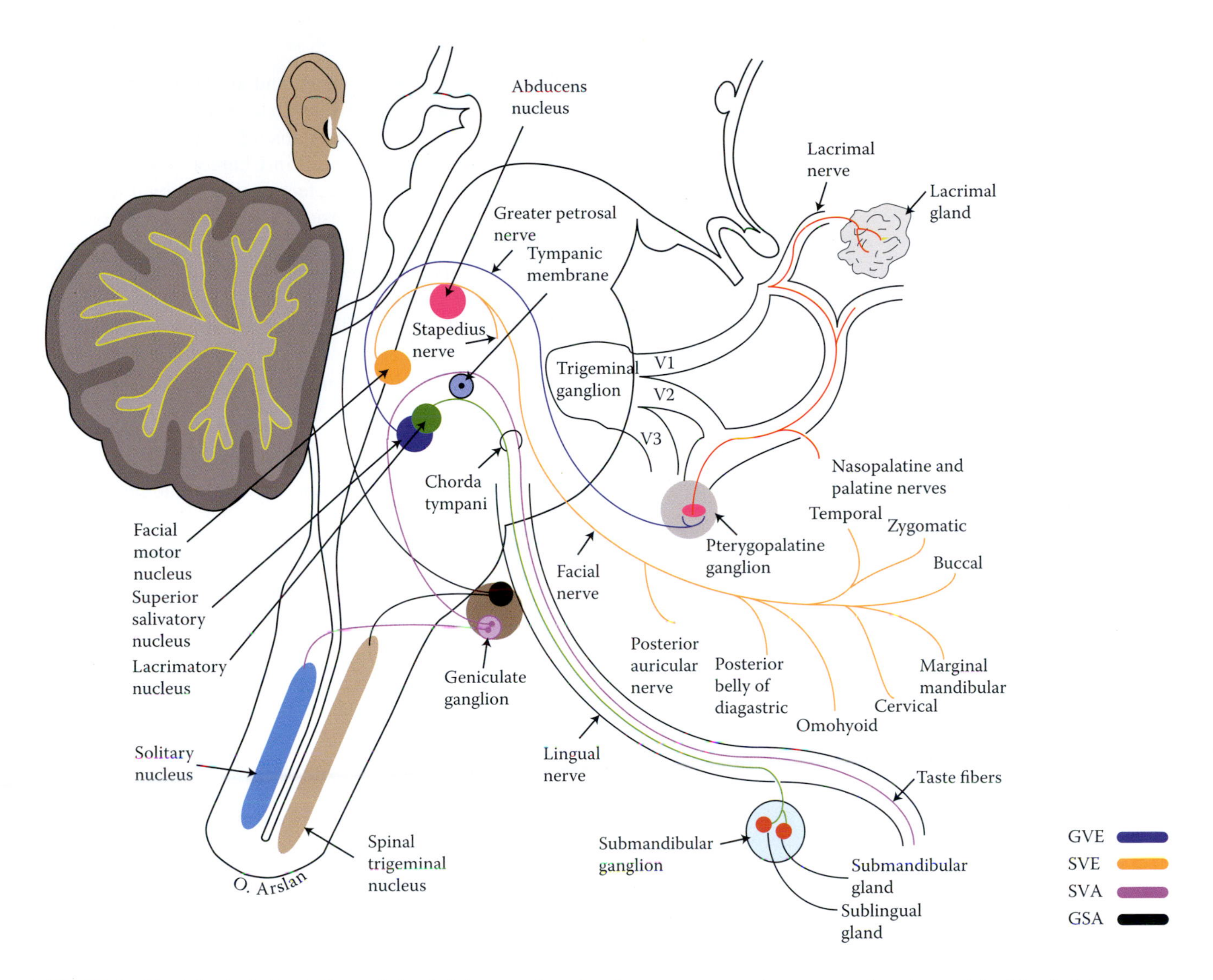

FIGURE 11.28 The individual fibers of the facial nerve, functional components, and sensory and motor nuclei of the facial nerve are illustrated.

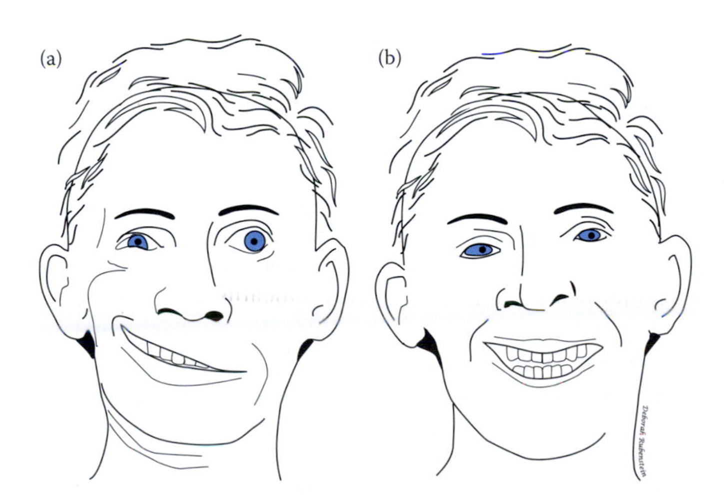

FIGURE 11.29 Manifestations of supranuclear facial palsy. Note the (a) prominent weakness in the muscles around the mouth when the patient is asked to open the mouth and (b) the retention of normal function of these muscles in response to emotional condition.

Unilateral destruction of the facial motor nucleus produces similar deficits as the lesion that disrupts the facial nerve at or proximal to the geniculate ganglion.

When the lesion occurs *immediately distal to the geniculate ganglion*, all of the above-listed deficits will be seen with the exception of lacrimation and taste sensation from the palate, which will be spared. Paralysis of the facial muscles of expression is the only deficit seen in individuals with a damaged facial nerve *at the stylomastoid foramen*. Impairment of corneal reflex due to facial nerve damage is observed only on the side of the lesion, regardless of which side is stimulated. Inability to close the eye by the unopposed action of the levator palpebrae superior and loss of blinking can increase the potential of corneal irritation or ulceration and may lead to keratitis and possible blindness. This complication can be avoided by placing a patch over the affected eye.

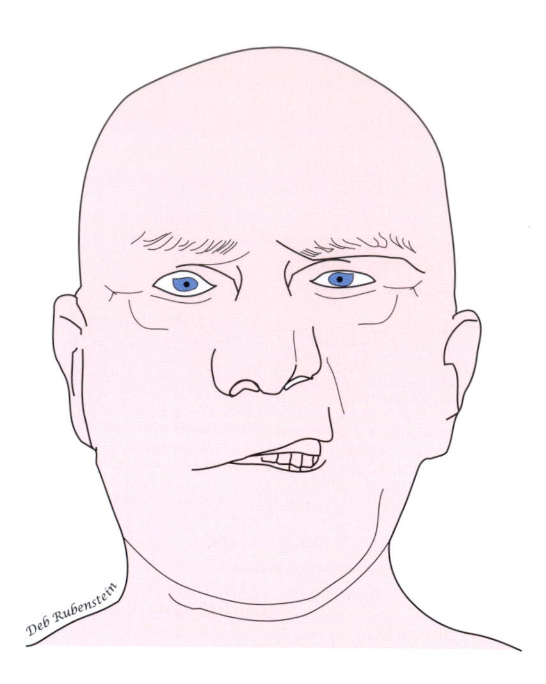

FIGURE 11.30 Right infranuclear facial palsy. Note the smoothing of the sulci, including the nasolabial sulcus, and sagging of the labial commissure on the affected side.

Bell's palsy is a benign inflammatory disease of unknown etiology, which results from compression or inflammation of the facial nerve. It may occur following exposure to cold or a viral infection (herpes simplex and varicella zoster) and is generally associated with edema, entrapment, and ischemia of the facial nerve within the narrow bony canal. This condition, which accounts for about 80% of all cases of facial palsy, may accompany otitis media, mastoiditis, and petrositis and may occur in diabetic patients and pregnant woman. Despite the benign nature of the disease, significant anxiety is endured by patients due to the fear of stroke or permanent facial disfigurement. Patients exhibit abrupt or progressive unilateral weakness of the facial muscles (facial asymmetry, depression of the angle of mouth), which may be preceded or accompanied by earache. Glandular secretion, stapedius muscle function, and taste sensation often remain unaffected. Epiphora (excessive tearing) due to ectropion and lagophthalmos (inability to close the eye completely) due to weakness of the orbicularis oculi muscle also occur. The latter can also be seen in comatose patients and individuals with blepharoplasty. The oculoauricular reflex, which is characterized by posterior movement of the ear when the patient directs his/her gaze as far laterally as possible, is lost in Bell's palsy. Patients may exhibit Bell's phenomenon, in which the eye turns upward and outward without accompanying eyelid closure. Recurrence is seen in approximately 10% of Bell's palsy patients. Unilateral recurrent facial palsy, folded and furrowed tongue, and swelling of the face occurs in Melkersson–Rosenthal syndrome, an autosomal dominant hereditary disorder. Recurrent acute painless facial palsy is seen in Charcot–Marie–Tooth IA neuropathy due to a defect of the neuronal myelin protein 22 gene. Synkinesis, an inappropriate involuntary facial muscular contraction that accompanies normal voluntary facial movement as in smiling, is seen with severe facial palsy as a sequel to nerve trauma. Prognosis of this condition depends on the severity of ear pain and the degree of paralysis, and it is poorer in individuals with ear pain and extensive facial palsy.

Facial myokymia, a unilateral subtle rippling movement of one or two facial muscles on one side of the face accompanied by contracture, is seen in brainstem glioma, multiple sclerosis, and Guillain–Barré syndrome.

VESTIBULOCOCHLEAR NERVE

The vestibulocochlear nerve (GSA) consists of a cochlear (auditory) and a vestibular component (Figure 11.1). This nerve leaves the brainstem at the cerebellopontine angle, accompanied by the facial nerve and the labyrinthine artery. For this reason, combined vestibular and auditory deficits, as well as facial palsy, may be seen in pathological conditions involving the internal acoustic meatus or the cerebellopontine angle (i.e., acoustic neuroma).

The vestibular nerve (SSA) represents the axons of the bipolar neurons of the superior and inferior vestibular (Scarpa's) ganglia (Figure 11.31) that establish a link between the vestibular receptors, brainstem, and cerebellum. This nerve conveys afferent and efferent impulses that mediate balance, postural control, spatial orientation, and gaze fixation. This encompasses recognition of position and movement via the visual system in relation to gravity and orientation of the body, and selection of suitable signals for postural adaptation (proprioceptive input). This expansive function requires close integration of the vestibular system with other sensory systems. The dendrites of the vestibular neurons pass through foramina in the internal acoustic meatus to reach the vestibular receptors (semicircular ducts, utricle, and saccule). The neurons of the superior vestibular ganglion receive information from the ampullary crests of the anterior and lateral semicircular ducts as well as the macula of the utricle, whereas the inferior vestibular ganglion receives vestibular input from the ampullary crests of the posterior semicircular duct and the macula of the saccule. See also the vestibular system, Chapter 15.

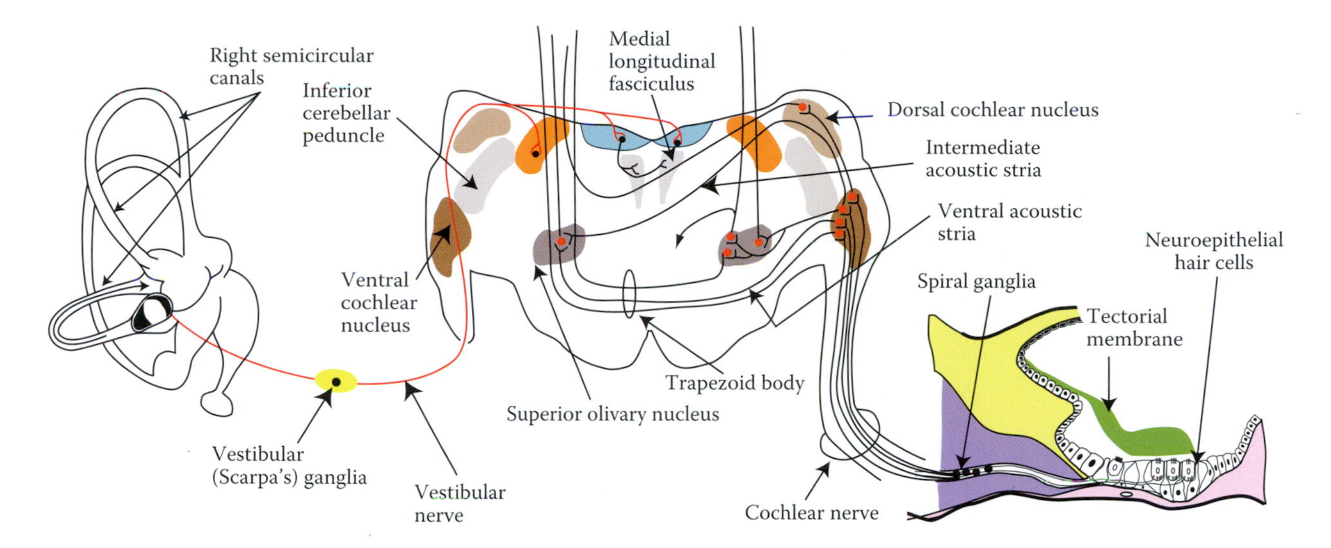

FIGURE 11.31 In this diagram, the vestibular and auditory receptors, associated ganglia, and course of individual nerves are illustrated. The connection of the vestibular nerve to the vestibular nuclei is also shown.

Intactness of the vestibular nerve and its connections may be evaluated by instilling cold or warm water (*caloric test*) or by rotating in the Barany chair. Dysfunction of the vestibular nerve produces ataxia, vertigo, and nystagmus, which are briefly explained below (see also the vestibular system, Chapter 15). Lesions of the vestibular nerve can be central or peripheral. A central infarct localized in the brainstem causes dysmetria, diplopia, dysarthria, dysphagia, and paresis or numbness around the mouth. Occlusion of the posterior inferior cerebellar artery (PICA) produces severe disequilibrium and inability to walk, nystagmus, and vertigo. Peripheral vestibular lesions are usually associated with a mild degree of disequilibrium. Vertigo and inability to walk, due to severe disequilibrium, indicate that central demyelinating or vascular (occlusion of PICA) diseases underlie these manifestations.

Ataxia (incoordination of motor activity) due to vestibular nerve dysfunction, is a severe and gravity dependent condition that often manifests as an intermittent incoordination of limb movements and becomes apparent during walking and standing.

Vertigo is a severe sense of rotation of the environment which is frequently intermittent and may be accompanied by oscillopsia (a back-and-forth movement of the visual objects with downbeat nystagmus), nausea, and vomiting. Vertigo may result in pallor, depression, and falling. In peripheral vestibular dysfunction, vertigo is caused by unilateral interruption of the tonic vestibular impulses. Head stability requires a reduction in the firing rate from one horizontal duct coupled with increased firing rate from the contralateral side. When this balance is disturbed with a peripheral vestibular lesion, the firing rate on the affected side is reduced, which is interpreted as head turning.

Nystagmus, an abnormal rhythmic oscillation of the eyeball, is produced visually by watching stationary targets from a moving vehicle (optokinetic nystagmus) or by extreme gaze to one side. It may also be produced iatrogenically by instilling cold or warm water into the ear (caloric test) or rotating in the Barany chair. Clinically, it may result from peripheral or central vestibular lesions (see also the vestibular system, Chapter 15). In a unilateral peripheral vestibular lesion, the firing rate of the horizontal semicircular canal is reduced, causing slow eye movement (slow phase of nystagmus) toward the affected side followed by corrective rapid eye movement (rapid phase of nystagmus) phase to the contralateral side. This nystagmus, which is gaze dependent, becomes more pronounced when the patient looks toward the affected side.

The *cochlear nerve* (SSA) is formed by the central processes of the bipolar neurons of the spiral ganglion, which are located in the modiolus of the cochlea (Figure 11.32). It contains both afferent and efferent fibers. The site of entry of the cochlear nerve into the pons and its course within the internal acoustic meatus are identical to that of the vestibular nerve. The dendrites of the bipolar neurons of the spiral ganglia receive auditory impulses from the organ of Corti (site of auditory neuroepithelial cells) within the cochlea and convey these impulses to the dorsal and ventral cochlear nuclei in the caudolateral part of the pons. Postsynaptic fibers synapse in the ipsilateral and contralateral superior olivary nuclei, while others continue to the opposite side, ascending in the lateral lemniscus. Postsynaptic neurons of the superior olivary nuclei project within the ipsilateral lateral lemniscus to the inferior colliculus. Fibers from the inferior colliculus travel rostrally to the medial geniculate nucleus and then through the auditory radiation to the transverse gyri of Heschl. For more details, see also the auditory system, Chapter 14.

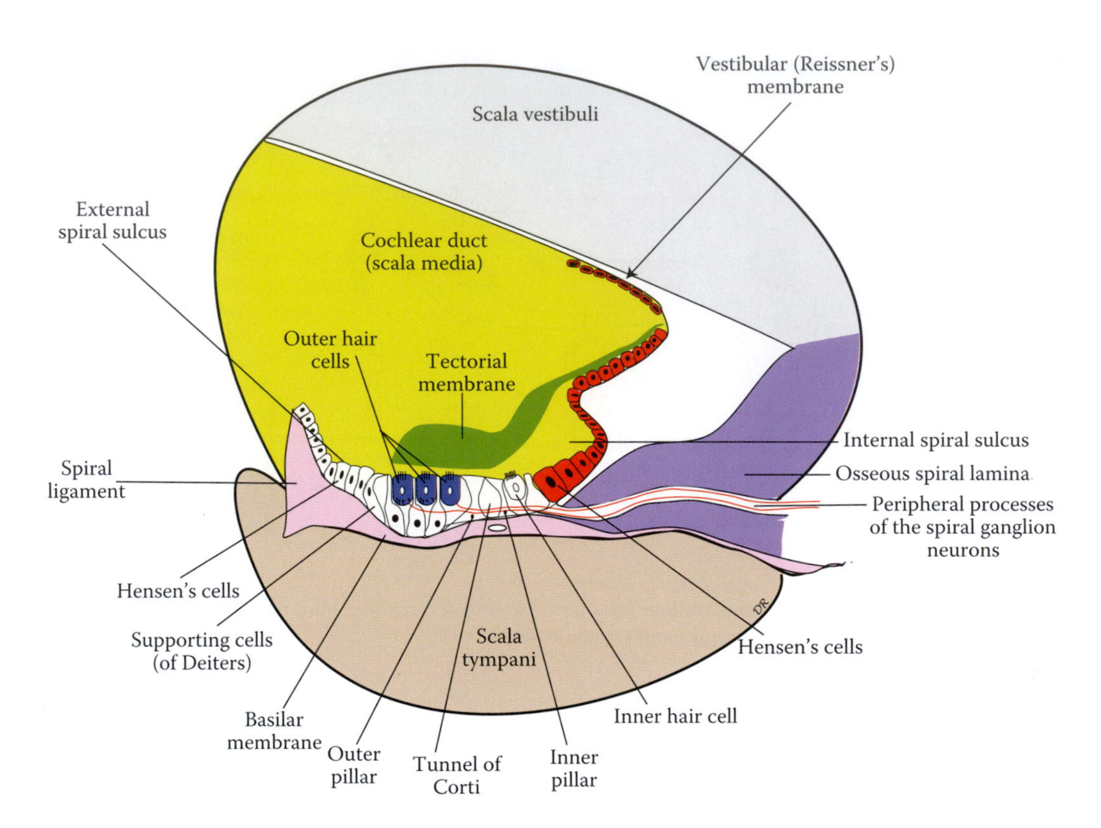

FIGURE 11.32 In this drawing, the neuroepithelial cells within the organ of Corti, spiral ganglia, and course of the cochlear nerve are illustrated.

Damage to the cochlear nerve, as a result of acoustic neuroma, or fracture involving the petrous temporal bone, produces sensorineuronal deafness on the side of the lesion. This type of deafness may be accompanied by tinnitus and is distinguished from conduction deafness by *Rinne* and *Weber* tests (see the auditory system, Chapter 14). Tinnitus refers to a unilateral or bilateral condition that ranges from a soft hissing sound or ringing noise to devastating loud constant roaring. It may be continuous or intermittent and is an important manifestation of damage to the cochlear nerve. It is often severe enough to interfere with normal daily activities.

Fractures of the posterior cranial fossa involving the internal acoustic meatus and tumors of the cerebellopontine angle (acoustic neuroma) may result in combined vestibular, cochlear, and facial nerve dysfunctions. This type of injury is characterized by deafness on the side of the lesion, vertigo (sense of rotation of the environment or self), nystagmus, ataxia, and signs of ipsilateral infranuclear facial palsy.

GLOSSOPHARYNGEAL NERVE

The glossopharyngeal nerve (general visceral afferent [GVA], GVE, GSA, SVA, SVE) exits the medulla from the rostral portion of the postolivary sulcus and leaves the skull through the jugular foramen accompanied by the vagus and accessory nerves (Figures 11.1, 11.33, and 11.34). It has two ganglia, a superior somatic sensory and an inferior visceral sensory ganglia. This nerve gives off tympanic, carotid sinus, lingual, pharyngeal, tonsillar, muscular, and auricular branches.

The *tympanic branch* (Jacobson's nerve—GSA) arises from the glossopharyngeal nerve in the jugular foramen and forms the tympanic plexus. This nerve conveys sensations from the mucosa of the middle ear, auditory tube, and mastoid air cells to the spinal trigeminal nucleus. The *lesser petrosal branch* (GVE) carries presynaptic parasympathetic fibers from the inferior salivatory nucleus of the medulla to the otic ganglionic. The *carotid sinus branch* (GVA) carries information from the carotid sinus, a baroreceptor, to the solitary nucleus, mediating the carotid sinus reflex, which is activated in response to fluctuation of blood pressure or carotid sinus massage. Visceral afferent fibers in the carotid sinus branch establish synaptic connections via interneurons in the reticular formation of the medulla with the dorsal motor nucleus of the vagus and, simultaneously, with the neurons of the reticulospinal tracts. In turn, the dorsal motor nucleus of the vagus sends preganglionic parasympathetic fibers to the cardiac plexus, reducing cardiac contractility (negative chronotropic effect) whereas the medullary reticulospinal tract inhibits or reduces the firing rate of the preganglionic sympathetic neurons that supply the cardiac plexus and the cutaneous arterioles. The decrease in the sympathetic output combined with the vagal inhibition results in a decrease of cardiac rate and output. The decrease in the peripheral vascular resistance leads to a decrease in blood pressure.

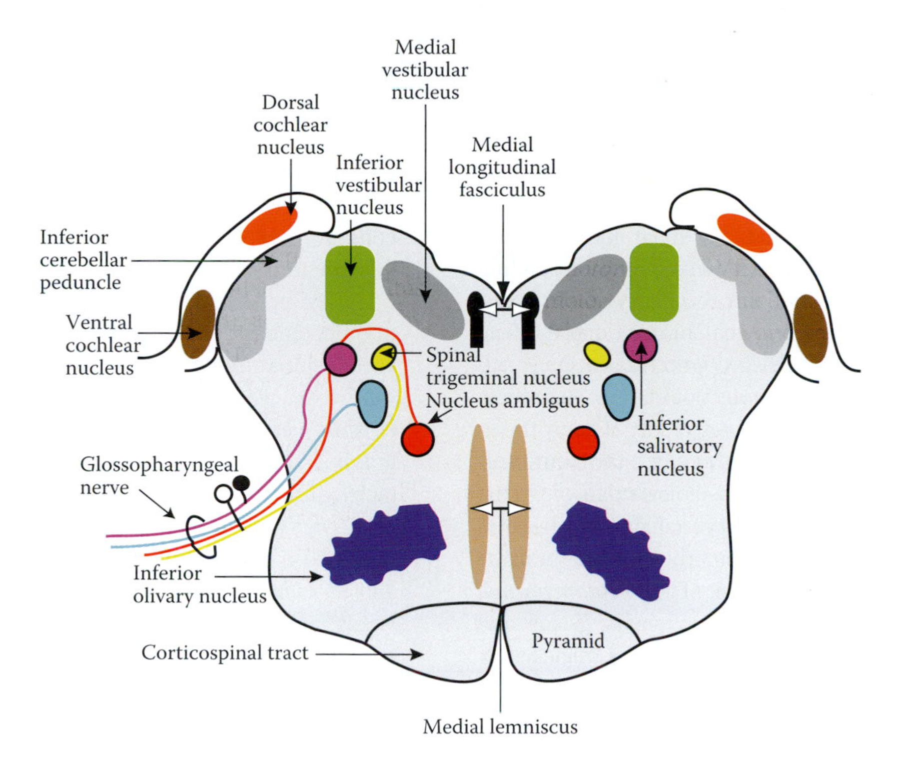

FIGURE 11.33 The functional components of the glossopharyngeal nerve and related nuclei are shown.

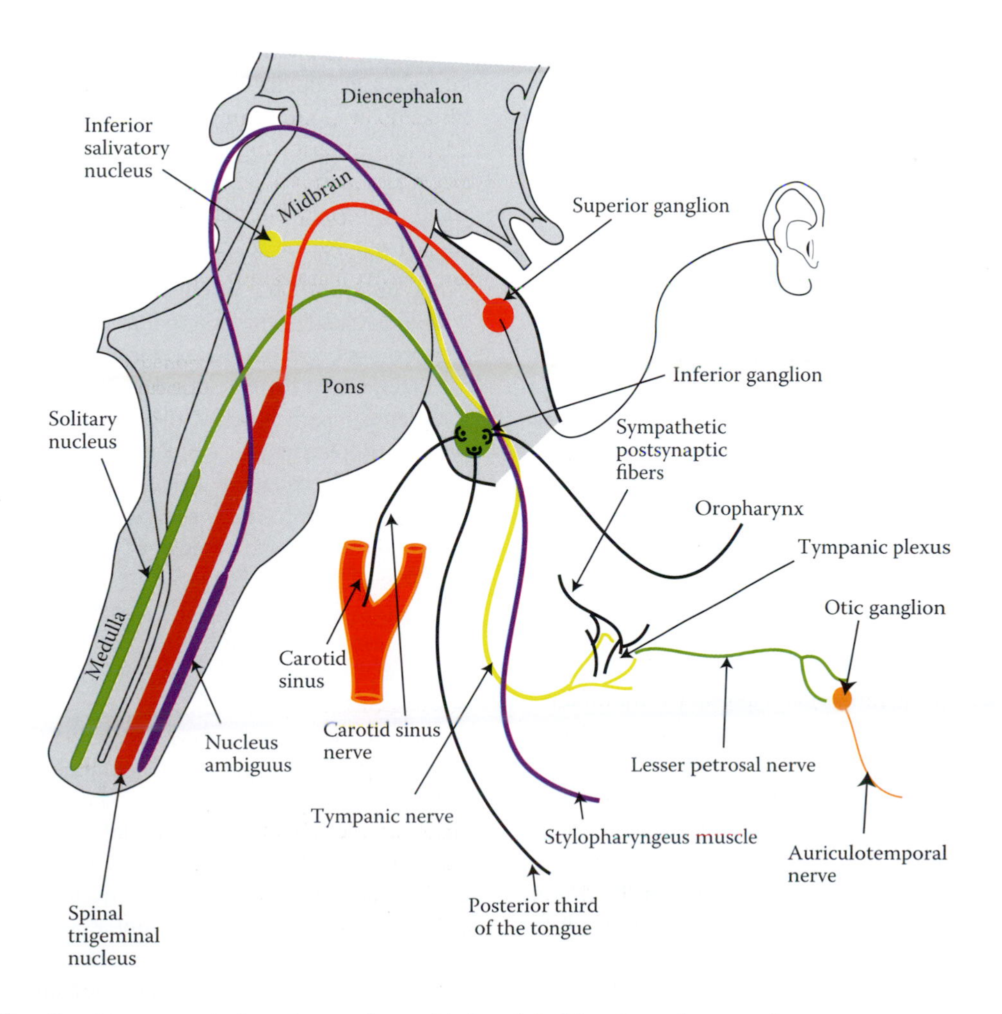

FIGURE 11.34 Functional components, branches, and associated nuclei of the glossopharyngeal nerve.

When this reflex becomes hyperactive as in individuals with vasomotor instability or in response to a mild stimulus, dizziness or syncope ensues (carotid sinus syncope). Lesions of the glossopharyngeal or the vagus nerves abolish this reflex.

The *lingual branch* carries special visceral and general sensory fibers (SVA, GSA) from the posterior one-third of the tongue to the solitary and spinal trigeminal nuclei, respectively. The *pharyngeal* (GSA) *branch* joins the corresponding branches of the vagus nerve to form the pharyngeal plexus. This plexus supplies the oropharyngeal mucosa and mediate the pharyngeal (gag) reflex through its connections to the spinal trigeminal, dorsal motor and ambiguus nuclei. The gag reflex, characterized by elevation of the stimulated part of the soft palate, is elicited by light stimulation of the oropharynx and soft palate. Unilateral loss of gag reflex may indicate damage to the glossopharyngeal nerve or the vagus nerve and may be observed in lateral medullary syndrome. In the elderly, this reflex may be reduced on both sides or may be absent. The *tonsillar branch* (GSA) carries sensations from the palatine tonsils, fauces, and soft palate to the spinal trigeminal nucleus. This branch may be overactivated in glossopharyngeal neuralgia. The *muscular branches* (SVE) arise from the ambiguus nucleus and supply the stylopharyngeus muscle, while the *auricular branch* (GSA) carries general sensation from the retroauricular area, terminating in the spinal trigeminal nucleus.

The *nuclei* associated with the glossopharyngeal nerve (Figures 11.33 and 11.34) are comprised of the nucleus ambiguus and solitary, spinal trigeminal, and inferior salivatory nuclei. The *nucleus ambiguus* (SVE) is located in the medulla and provides special visceral efferent fibers that supply the stylopharyngeus muscle. It also provides fibers to the cranial part of the accessory nerve and to the vagus nerve, innervating the laryngeal, pharyngeal, and palatine muscles. The *solitary nucleus* receives taste (SVA) sensations from the posterior one-third of the tongue; general visceral sensations (GVA) from the carotid sinus; and pain and temperature sensations from the middle ear, posterior one-third of the tongue, oropharynx, and tonsils. The *spinal trigeminal nucleus* (GSA) receives cutaneous sensation from the retroauricular area. The parasympathetic fibers from the *inferior salivatory nucleus* (GVE) are conveyed to the otic ganglion, which eventually innervate the parotid gland.

The glossopharyngeal nerve may be damaged in fractures of the posterior cranial fossa and stenosis of the jugular foramen. Demyelination caused by multiple sclerosis, tumors of the posterior cranial fossa, aneurysm of the internal carotid artery, and injuries involving the retroparotid space may also damage this nerve. Occlusion of the PICA may affect the associated nuclei in the medulla, producing deficits that may also be shared by lesions of the vagus and accessory nerves. Isolated glossopharyngeal nerve damage causes hypertension due to involvement of the carotid branch of the nerve, loss of sensation over the soft palate, fauces, oropharynx, and posterior third of the tongue on the side of the lesion as well as ipsilateral loss of parotid secretion and reduced or loss of gag and pharyngeal reflexes. Although isolated injury to the glossopharyngeal nerve is rare, the following are some of the conditions associated with irritation, compression, or damage to this nerve.

A lesion that disrupts the glossopharyngeal, vestibulocochlear, and vagus nerves, as well as the corticospinal tract may produce signs of *Bonnier's syndrome*, which is characterized by transient hypertension, vertigo, nystagmus, hearing deficits, dysphonia, hoarseness of voice, contralateral hemiplegia, and tachycardia.

Vernet's syndrome may occur as sequel to fractures of the base of the skull, involving the jugular foramen and its contents, which include the glossopharyngeal, vagus, and spinal accessory nerves. Disorders of this condition are analogous to the combined deficits associated with these individual nerves.

Villarreal's syndrome is a condition that results from injury to the retroparotid space, which involves the glossopharyngeal, vagus, accessory, and hypoglossal nerves. Sympathetic postganglionic fibers may also be disrupted in this condition, resulting in Horner's syndrome.

Glossopharyngeal neuralgia is a rare condition, which is characterized by spontaneous episodic attacks of excruciating pain in the tonsillar area, posterior third of the tongue, and external acoustic meatus, radiating to the throat, side of the neck, and back of the lower jaw. It is provoked by yawning, laughing, chewing, or swallowing of particularly cold liquid and may be associated with peritonsillar abscess, oropharyngeal carcinoma, and ossified stylohyoid ligament. It is rarely bilateral, and it may accompany trigeminal neuralgia. When ear pain is felt without signs of middle ear disease, oropharyngeal cancer must also be considered. Glossopharyngeal neuralgia may be associated with episodes of fainting, syncope, and reflex bradycardia as a result of involvement of the carotid sinus nerve (Table 11.2).

VAGUS NERVE

The vagus (GVA, GVE, GSA, SVA, SVE), as in the case of the glossopharyngeal nerve, is a composite nerve with diverse functional entities (Figures 11.1, 11.35, and 11.36). It travels ventrolaterally in the caudal medulla and passes through the spinal trigeminal tract and nucleus in close proximity to the nucleus ambiguus and spinal lemniscus. Due to this relationship, a lesion of the lateral medulla, as seen in *Wallenberg's (lateral medullar syndrome)*, is most likely to be associated with damage to the vagus nerve and associated nuclei. Patients with this syndrome exhibits dysphagia, dysphonia,

TABLE 11.2
Cranial Nerves VII–IX

Cranial Nerve	Component	Location of Cell Bodies	Course	Distribution	Function
VII. Facial nerve	SVE	Facial motor nucleus	Internal auditory meatus, facial canal, and stylomastoid foramen	Facial muscles of expression, stylohyoid, stapedius, and posterior belly of digastric	Facial expression and increase hearing threshold and elevation of hyoid bone
	GVE	Superior salivatory nucleus	Within the intermediate nerve via the internal auditory meatus, then via greater petrosal and pterygoid canal nerves to the pterygopalatine ganglion	Lacrimal gland, mucus glands of palate, pharynx, and nasal cavity	Parasympathetic
	SVA	Geniculate ganglion	Same as above	Via chorda tympani to anterior two-thirds of the tongue	Taste
	GSA	Geniculate ganglion	Via auricular branch	Concha of ear	General somatic sensation
VIII. Vestibulocochlear nerve					
Cochlear nerve	SSA	Spiral ganglion	Internal acoustic meatus	Organ of Corti	Audition
Vestibular nerve	SSA	Scarpa's ganglion	Same as above	Receptors in the semicircular canals, utricle, and saccule	Balance, orientation in three dimensions, and fixation of gaze
IX. Glossopharyngeal nerve	GVE	Inferior salivatory nucleus	Jugular foramen	Parotid gland	Parasympathetic
	SVE	Nucleus ambiguus	Jugular foramen	Stylopharyngeus	Swallowing
	GVA	Inferior ganglion	Same as above	Carotid sinus and body; posterior one-third of the tongue, oropharynx, palatine tonsils, and tympanic membrane	Baroreceptor and chemoreceptor; pain and temperature sensations from the mucosa of these areas
	SVA	Inferior ganglion	Same as above	Posterior one-third of the tongue	Taste
	GSA	Superior ganglion	Same as above	Retroauricular	General sensations

and alternating hemianesthesia. This nerve leaves the medulla via the postolivary sulcus, as a series of rootlets, and exits the skull through the jugular foramen, accompanied by the glossopharyngeal and accessory nerves. Then, it runs through the neck as the posterior component of the carotid sheath. During its course in the superior and posterior mediastina, it gives rise to branches to the cardiac and pulmonary plexuses. Later, the vagus nerves on both sides contribute to the formation of the anterior and posterior vagal trunks around the abdominal part of the esophagus, entering the abdomen through the esophageal hiatus of the diaphragm. In the abdomen, it contributes presynaptic parasympathetic fibers to the celiac, superior mesenteric, and aortic plexuses. The vagal (parasympathetic) contribution to the abdominal viscera terminates at the junction of the right two-thirds and left one-third of the transverse colon. The vagus nerve has a superior and inferior ganglion, containing neurons for somatic and visceral sensations, respectively. Within this nerve, the motor fibers belong to the cranial part of the accessory nerve, which distribute to the laryngeal, pharyngeal, and palatal muscles.

Through its course, the vagus nerve gives rise to branches in the cranial cavity, thorax, and abdomen, which include the meningeal, auricular, pharyngeal, carotid body, superior and inferior laryngeal, cardiac, pulmonary, esophageal, celiac, and superior mesenteric branches.

The *meningeal branch* (GSA) conveys sensations from the dura mater of the posterior cranial fossa to the spinal trigeminal nucleus. In the same manner, the *auricular branch* (GSA) carries somatic sensations from the external acoustic meatus to be conveyed to the spinal trigeminal nucleus.

The *pharyngeal branch* (SVE) consists of the cranial part of the accessory nerve that contributes to the formation of the pharyngeal plexus. Through this plexus, the vagus nerve innervates most of pharyngeal muscles (with the exception of the stylopharyngeus) and palatine muscles (with the exception of the tensor palatini). Branches that distribute to the epiglottic vallecula (SVA) convey taste sensations to the solitary nucleus.

Branches to the carotid body (GVE) arise from the inferior ganglion and course as a component of the pharyngeal

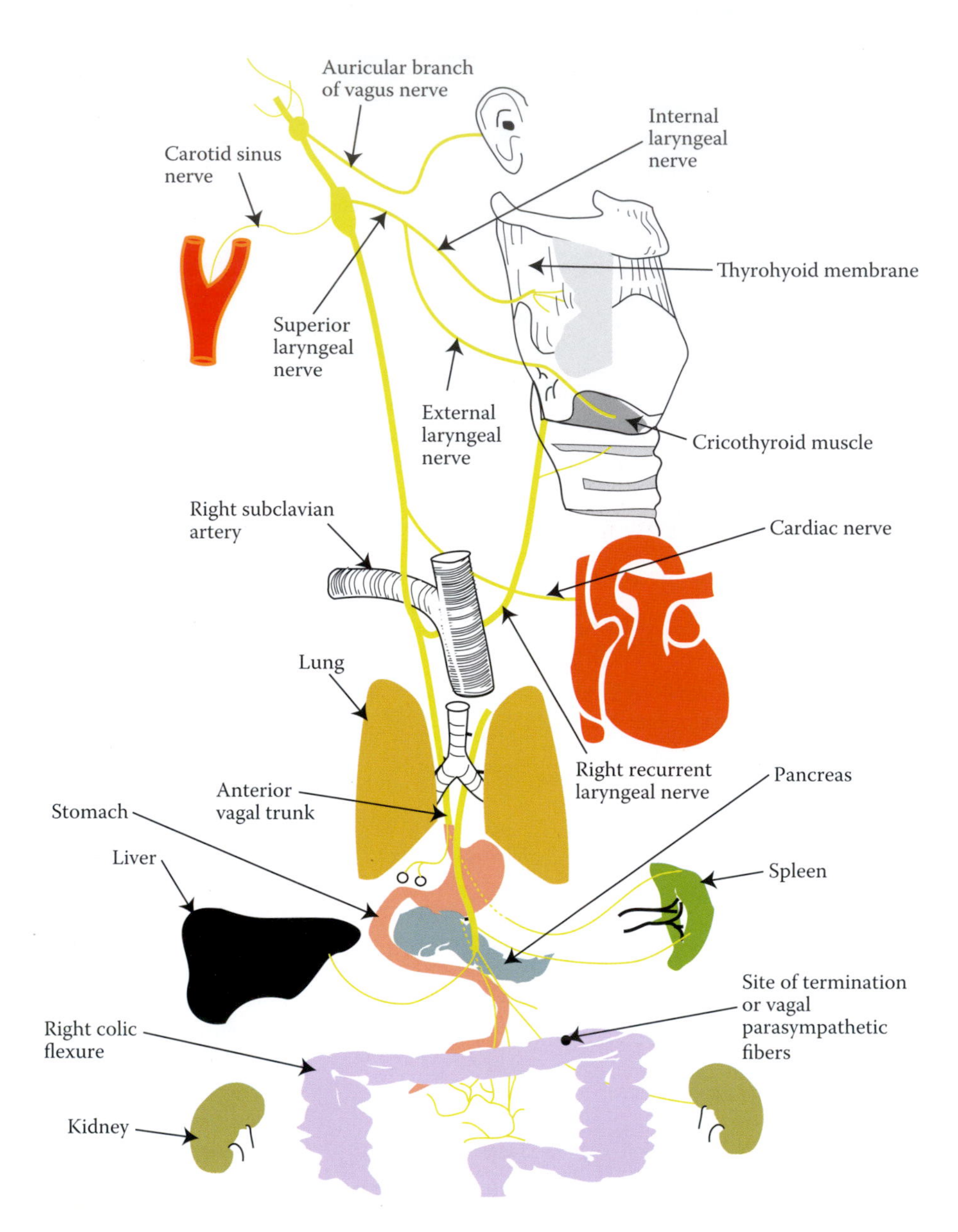

FIGURE 11.35 The origin, course, and distribution of branches of the vagus nerve.

branch and very rarely run within the superior laryngeal branch, transmitting signals about the changes in the level of carbon dioxide and oxygen tension. These fibers join the pharyngeal branches of the glossopharyngeal nerve and the cervical part of the sympathetic trunk, to form the pharyngeal plexus.

The *superior laryngeal branch* (GVA, SVE) divides into the internal laryngeal nerve (sensory) and external laryngeal nerve (motor). The *internal laryngeal nerve* (GVA, SVA) accompanies the superior laryngeal vessels in its course in the medial wall of the piriform recess and distributes branches to the laryngeal mucosa of the vestibule, laryngeal sinus, and epiglottic vallecula. This branch carries general sensations (GVA) from the laryngopharynx, piriform recess, and most of the laryngeal mucosa to the solitary nucleus. Taste sensation from the extreme posterior part of the tongue and the epiglottic vallecula is also conveyed by this nerve to the solitary nucleus. The *external laryngeal nerve* (SVE) emanates from the nucleus ambiguus and provides motor fibers to the cricothyroid muscle.

The *inferior* (recurrent) *laryngeal nerve* (GVA, SVE) encircles the subclavian artery on the right side and the aortic arch medial to the ligamentum arteriosum on the left side. This nerve carries general visceral sensation (GVA) from the infraglottic part of the larynx. During its course within the tracheoesophageal sulcus and medial to the thyroid gland, it runs in close proximity to the inferior thyroid artery, a relationship that bears important clinical significance in thyroidectomies.

The *cardiac branches* contribute parasympathetic fibers (GVE) to superficial and deep cardiac plexuses. The superficial plexus receives innervation from the left vagus nerve while the deep plexus receives contributions from the right vagus and the recurrent laryngeal nerves. These fibers slow the heart rate and produce constriction of the coronary arteries. They also carry general visceral sensation (GVA) from the heart.

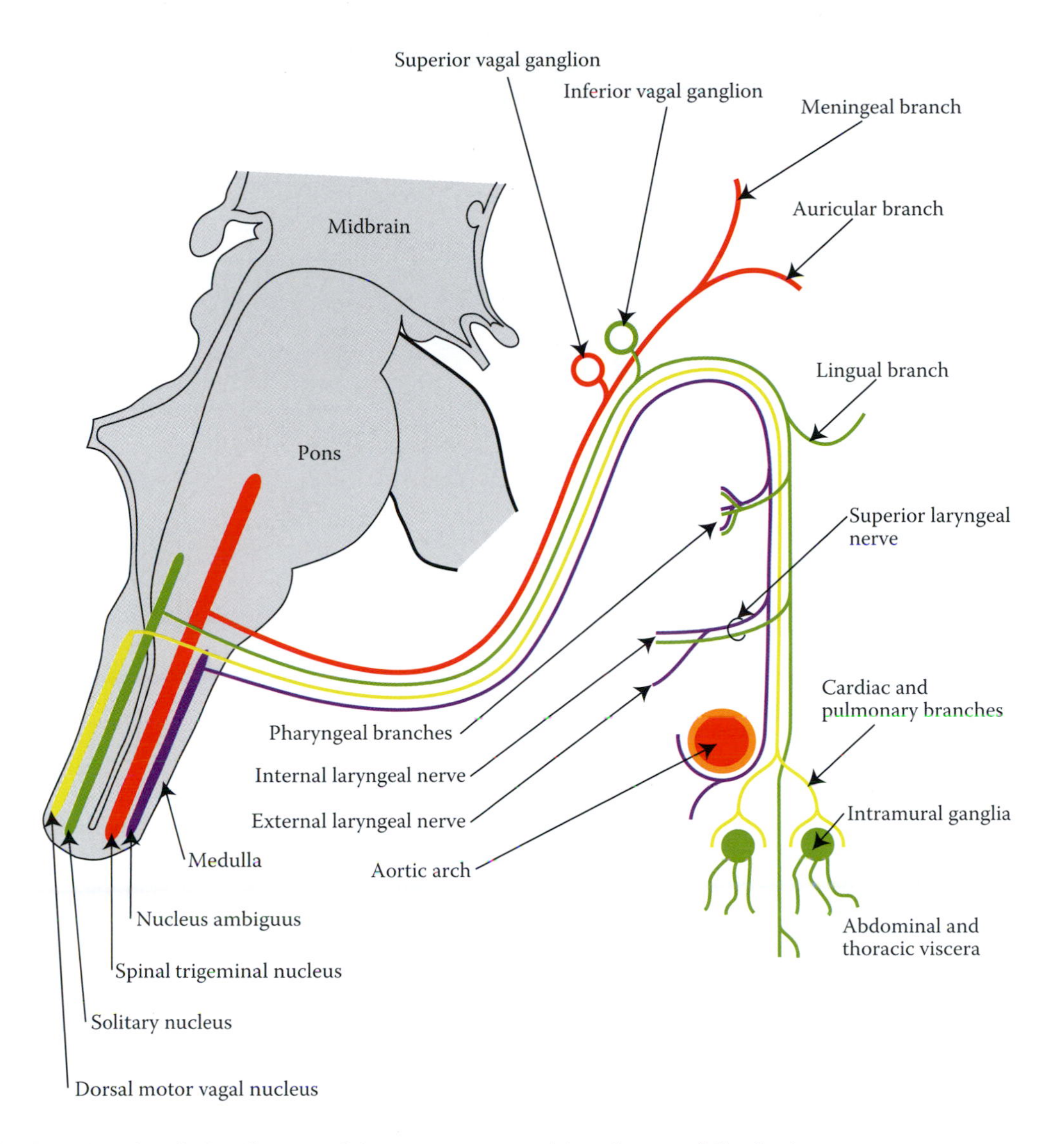

FIGURE 11.36 Complete descriptive diagram of the vagus nerve, nuclei, and areas of distribution.

The *pulmonary branches* are bronchoconstrictors and secretomotor (GVE) to the bronchial mucus glands, which run within the pulmonary plexus. They also carry information from stretch receptors (GVA) from the pulmonary bronchi.

The *esophageal branches* carry parasympathetic preganglionic (GVE) fibers that facilitate esophageal motility, as well as general visceral sensation (GVA) from the esophagus. These branches form the esophageal plexus and continue to the abdomen around the esophagus as the anterior and posterior vagal trunks. The anterior vagal trunk is formed primarily by the left vagus, while the posterior vagal trunk is principally derived from the right vagus.

Vagal trunks may selectively be severed in vagotomy, a surgical procedure used in the treatment of chronic gastric or duodenal ulcers. This approach is intended to greatly reduce hydrochloric acid secretion and thus enhance healing of the affected part of the gastrointestinal tract. It may be classified into truncal, selective, and high selective vagotomy. Truncal vagotomy may not desirable, due to the accompanied gastric stasis (dumping), atonia of the gallbladder, and impaired pancreatic secretion. In selective vagotomy, although the gastric branches of the vagus nerve, including nerves of *Latarget* to the antrum, are specifically cut, gastric dumping occurs, requiring surgical bypass. In high selective vagotomy, only the branches to the fundus and body of the stomach (acid-secreting areas) are cut, and gastric dumping is thus avoided. This procedure may also induce atrophy of the oxyntic cells, rendering it unresponsive to the circulating gastrin.

Branches to the celiac and superior mesenteric plexuses provide parasympathetic (GVE) fibers to the stomach, small intestine, cecum, ascending colon, and right two-thirds of the

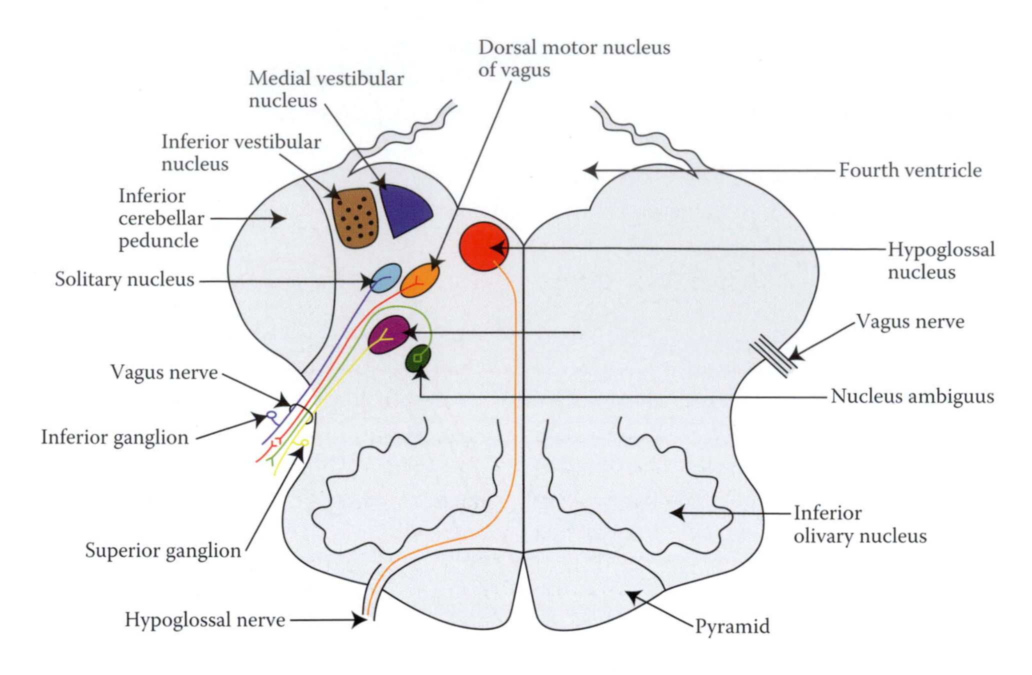

FIGURE 11.37 Drawing of the medulla at the level of the vagus nerve showing the associated nuclei and ganglia.

large intestine. It also provides GVA fibers, conveying sensory modalities of hunger, nausea, thirst, and bowel fullness to the solitary nucleus.

The vagal nerve nuclei include the dorsal motor nucleus of vagus, nucleus ambiguus, solitary nucleus, and spinal trigeminal nucleus (Figures 11.36 and 11.37). The *dorsal motor nucleus of the vagus* (GVE) occupies the area dorsolateral to the hypoglossal nucleus, forming the vagal trigone in the floor of the fourth ventricle. This nuclear column maintains similar dimensions to the hypoglossal nucleus both caudally and rostrally, providing parasympathetic preganglionic (GVE) fibers to the thoracic and abdominal viscera. The *nucleus ambiguus* (SVE) provides special visceral motor fibers that innervate the muscles of the pharynx, larynx, and soft palate. The *solitary nucleus* receives general visceral afferents (GVA), which constitute nearly 80% of the entire vagus nerve, from the bronchi, gastrointestinal tract, and carotid body. It also receives SVAs (taste sensation) from the root of the tongue and epiglottic vallecula. These afferents are conveyed to the ventral posteromedial nucleus of thalamus en route to the sensory cortex. The *spinal trigeminal nucleus* (GSA) receives general somatic sensations from the external acoustic meatus, concha of the ear, and dura mater of the posterior cranial fossa.

The vagus nerve is prone to damage in fractures of the posterior cranial fossa, involving the jugular foramen, or by an aneurysm of the common or internal carotid artery. A unilateral lesion of the vagus nerve results in slight difficulty in swallowing (dysphagia) and breathing (dyspnea), accompanied by regurgitation of food through the nasal cavity. It also produces hoarseness and a voice with nasal quality, transient tachycardia, loss of gag reflex, and deviation of the uvula toward the intact side on phonation. In unilateral vagal dysfunction, no consistent deficits are associated with the heart, lungs, or bowel functions. Bilateral disruption of the vagus nerves results in a serious condition that manifests cardiac arrhythmia, severe difficulty in breathing, dysphagia, dysphonia, abdominal pain, and stomach distention.

Emotional stress, crowded environment, and warmth often precipitate a common condition known as *vasovagal syncope*. It is characterized by sweating, an aura of nausea, and loss of consciousness. It is similar to the *reflex vasovagal syncope*, in which a fainting spell is caused by venipuncture. In order to test intactness of the vagus nerve, the patient will be asked to say “ah,” which produces elevation of the soft palate (uvular or palatal reflex). The *uvular* (palatal) *reflex* is characterized by equal and symmetrical movement of the soft palate and elevation of the uvula, upon stimulation of the mucosa of the soft palate or during phonation. It is mediated by the glossopharyngeal and vagus nerves.

A specific lesion of the *superior laryngeal nerve* produces anesthesia in the mucus membrane of the vestibule and the laryngeal sinus and paralysis of the cricothyroid muscle. This leads to a high risk of aspiration due to the inability to properly close the rima glottis and also due to loss of cough reflex subsequent to loss of sensation from the larynx. Consequently, the tension in the affected vocal folds will be lost, resulting in a monotonous voice and inability to raise the vocal pitch. Due to the close relationship of the inferior thyroid artery to the recurrent laryngeal nerve, ligation of this artery predisposes the *recurrent laryngeal nerve* to damage in thyroidectomy.

In particular, the left recurrent laryngeal nerve can also be damaged in an aneurysm of the aortic arch, as a result of bronchial and esophageal carcinoma, or in conditions that produce enlargement of the mediastinal lymph nodes. According to *Semon's law*, progressive lesions of the recurrent laryngeal nerve produce dysfunction in the abductors of the vocal folds before any significant deficits in the adductors. In contrast, the recovery involves the adductor muscles first followed by the abductors. Unilateral damage to the recurrent laryngeal nerve results in paralysis of the intrinsic laryngeal muscles, with the exception of the cricothyroid, as well as ipsilateral loss of sensation from the infraglottic part of the larynx. Initially, the voice is breathy, weak, and altered (like a whisper), with weak cough, but movement of the opposite vocal fold toward the midline may compensate for this deficit, rendering the voice fairly normal. The external laryngeal branch may be disrupted in ligation of the superior thyroid artery that accompanies the nerve, producing paralysis of the cricothyroid muscle and monotonous voice.

Voice can also be affected in spasmodic dysphonia, in which the laryngeal muscles undergo dystonia, resulting in voice breaks at irregular intervals, with the ability to sing normally. Regular breaks and tremulous voice occur in laryngeal tremor due to constant contraction of the intrinsic laryngeal muscles during phonation and at rest. Voice disorder in the form of weak and breathy voice is also seen in Parkinson's disease. Myasthenia gravis can also produce dysfunction of the laryngeal muscles and the resultant voice disorders.

ACCESSORY NERVE

The accessory nerve (Figure 11.38) has a cranial part (SVE) and a spinal part (GSE). The cranial part is derived from the neurons of the ambiguus nucleus (Figures 11.41 and 11.43) that joins the vagus nerve, distributing through the laryngeal and pharyngeal branches to the adductor muscles of the larynx, muscles of the palate, with the exception of the tensor palatini, and to muscles to the pharynx, with the exception of the stylopharyngeus (Figure 11.37). The spinal part originates from the lateral anterior gray column of the upper five spinal segments (accessory nucleus). A point of particular importance is the fact that the part of this nucleus that innervates the sternocleidomastoid receives fibers from the opposite corticospinal tract after their decussation in the medulla. The spinal accessory nerve enters the cranial cavity through the foramen magnum (Figures 11.15 and 11.37) and leaves the skull through the jugular foramen accompanied by the vagus and glossopharyngeal nerves. It joins the cranial part for a short distance in the jugular foramen and then leaves the foramen and descends to the neck in close proximity to the internal jugular vein and the internal carotid artery. During its course in the posterior triangle, it runs on the surface of the levator scapula, separated from it by the prevertebral fascia. Above the clavicle, it enters the medial surface of the trapezius. Here, it joins branches of C3 and C4 ventral rami. The spinal accessory nerve supplies the sternocleidomastoid and

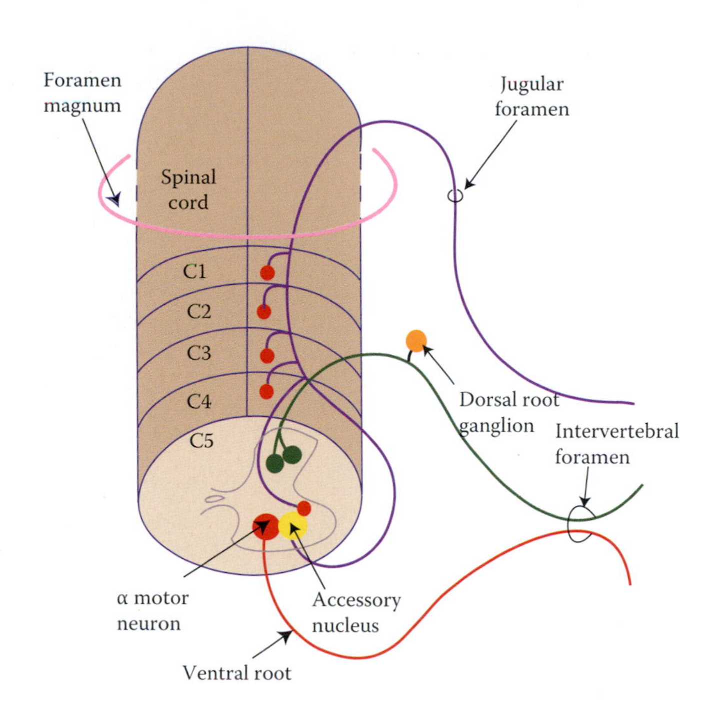

FIGURE 11.38 Drawing of the spinal accessory nerve; its origin, course, and areas of distribution are visible.

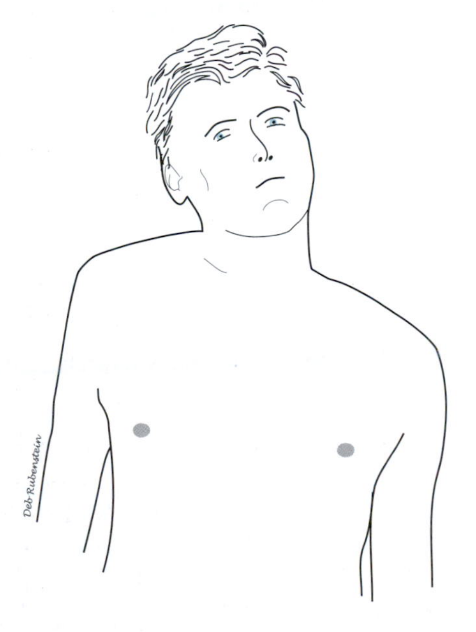

FIGURE 11.39 This is a depiction of a patient with right-side torticollis due to spasmodic contracture of the sternocleidomastoid muscle.

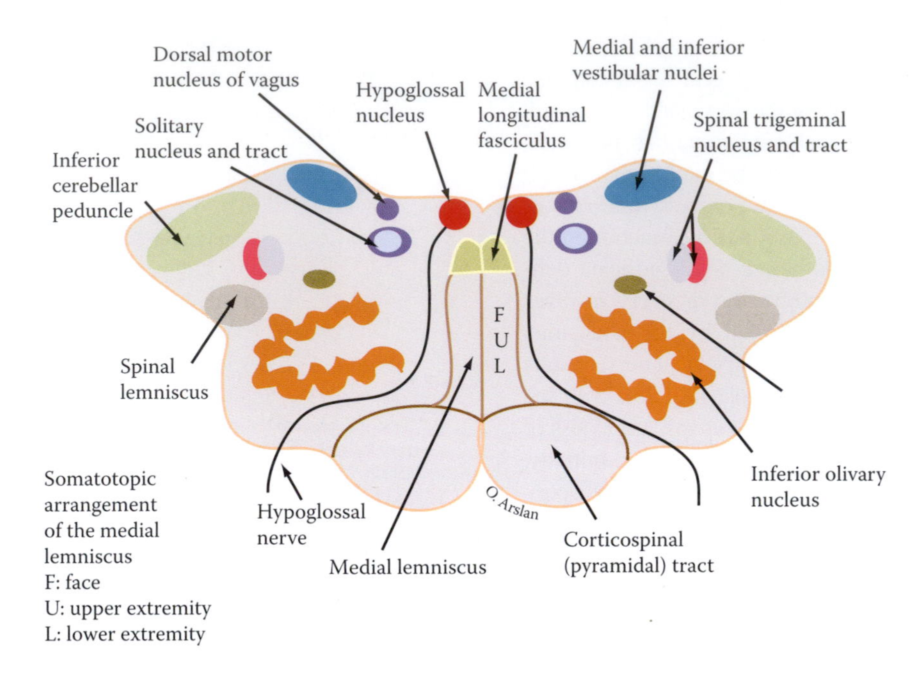

FIGURE 11.40 In this section, of the medulla at the midolivary level and the hypoglossal nuclei and nerves are shown. Observe the course of the hypoglossal nerve lateral to the medial lemniscus and between the pyramid and the inferior olivary nucleus.

upper trapezius, whereas the ventral rami of C3 and C4 innervate the caudal part of the trapezius. The second and third cervical nerves convey a sense of proprioception from the sternocleidomastoid muscle.

The accessory nerve is vulnerable to damage in radical cervical lymph dissection, a fairly common surgical procedure employed in metastatic carcinomas of the neck. Carotid endarterectomy, jugular venous cannulation, intraspinal meningioma, syringomyelia, or poliomyelitis can endanger this nerve. The pressure exerted by calcified tuberculous cervical lymph nodes or surgical attempts to remove these nodes may also damage the spinal accessory nerve. A stab wound in the neck or fractures of the foramen magnum or jugular foramen may also injure the accessory nerve. In addition, it may be damaged when the face mask of a football player is suddenly pulled laterally. Irritation of the spinal accessory nerve may lead to *torticollis* (Figure 11.39), a spasmodic contracture of the sternocleidomastoid muscle. Damage to the spinal accessory nerve may lead to paralysis of the trapezius and sternocleidomastoid muscles. A fracture of the jugular foramen, nasopharyngeal carcinoma, or rarely, radiation injury can produce combined deficits of the accessory, glossopharyngeal, and vagus nerves.

Paralysis of the trapezius muscle results in winging of the scapula, which becomes more prominent upon an attempt to abduct the arm on the affected side. This fact distinguishes winging of the scapula observed in long thoracic nerve damage from that of the accessory nerve damage. Paralysis of the sternocleidomastoid produces inability to turn the face to the opposite side. Due to the double crossing of the corticospinal tract fibers that project to the part of the accessory nucleus innervating the sternocleidomastoid, a lesion of the corticospinal tract rostral to the pons produces hemiplegia with weakness of the sternocleidomastoid muscle. Trapezius palsy due to this type of upper motor neuron lesion remains contralateral.

To test the integrity of the spinal accessory nerve, the trapezius and sternocleidomastoid functions are tested. To check intactness of the trapezius muscle, the patient is asked to shrug his/her shoulders against resistance. To test the sternocleidomastoid muscle, the patient is asked to turn his/her head to one side against resistance by the examiner.

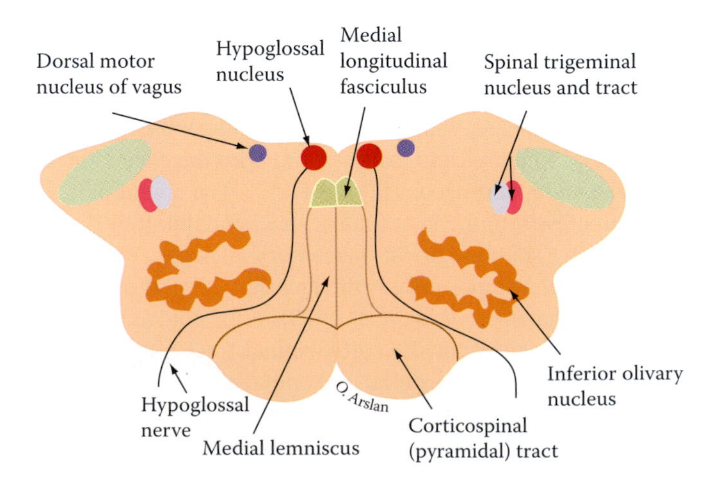

FIGURE 11.41 This drawing emphasizes some of the structures shown in Figure 11.40.

FIGURE 11.42 Diagram of the somatic and autonomic efferent nuclei of the cranial nerves.

HYPOGLOSSAL NERVE

The hypoglossal nerve (Figures 11.40 and 11.41), which is derived from neuronal axons of the hypoglossal nucleus (GSE) of the medulla, descends within the reticular formation, lateral to the medial lemniscus, and between the inferior olivary and pyramidal tract. The hypoglossal nucleus (Figures 11.40, 11.41, and 11.42) extends from the level of the stria medullaris to an area near the upper pole of the inferior olivary nucleus. Due to its proximity to the medial lemniscus and the pyramids, a single lesion, dependent upon the extent of the damage, may produce signs of medial medullary syndrome or inferior alternating hemiplegia.

It emerges from the medulla at the preolivary sulcus, leaves the skull through the hypoglossal canal, and runs in the occipital and carotid triangles, encircling the occipital branch of the external carotid artery. In the neck, a branch of the first cervical spinal segment joins the hypoglossal to later leave as the superior root of the ansa cervicalis. All intrinsic and extrinsic lingual muscles, with the exception of the palatoglossus (innervated by the pharyngeal plexus), receive motor innervation from the hypoglossal nerve. The hypoglossal nerve gives rise to meningeal, descending nerves to the thyrohyoid and geniohyoid muscles and other muscular branches. The meningeal branch,

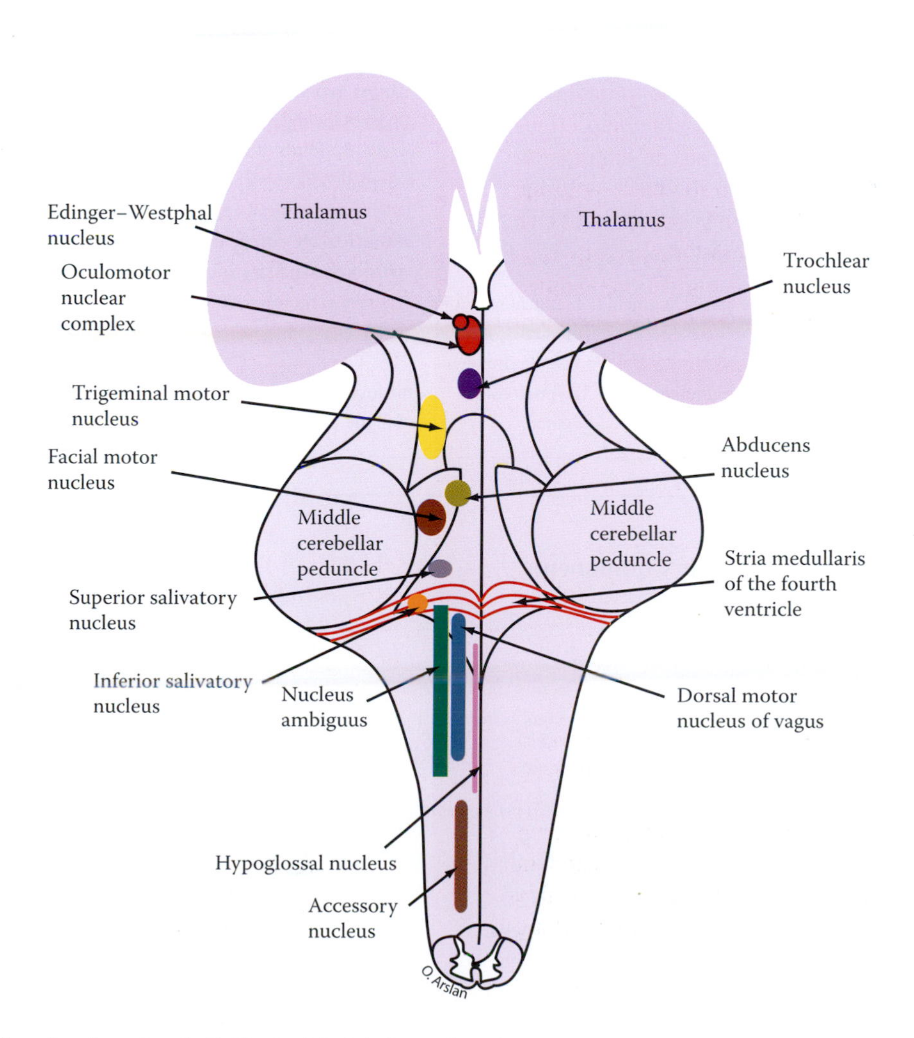

FIGURE 11.43 This is a drawing of an individual with right hypoglossal nerve palsy.

derived from the upper cervical spinal nerves, is sensory to the dura of the anterior cranial fossa and the dura that forms the occipital and inferior petrosal sinuses. A descending branch, derived from the ventral ramus of the first cervical spinal nerve, leaves the hypoglossal nerve as the superior root of the ansa cervicalis and innervates the geniohyoid and thyrohyoid.

A unilateral lesion of the hypoglossal nerve (Figure 11.43) may produce ipsilateral atrophy of the lingual muscles, fasciculation, furrowing of the affected side, and deviation of the tongue toward the lesion side upon protrusion. On retraction, the atrophied part of the tongue rises up higher than the other parts. Bilateral lesion of the hypoglossal nerves results in defective speech and difficulty in chewing and swallowing. The tongue lies motionless, and thus, swallowing becomes very difficult, forcing the patient to extend his/her head back and push the bolus of food into the pharynx with the aid of his/her fingers. Taste and tactile sensations are unaffected. The hypoglossal

TABLE 11.3

Cranial Nerves X–XII

Cranial Nerve	Components	Cell Bodies	Course	Distribution	Function
X. Vagus	GVE	Dorsal motor nucleus of vagus	Jugular foramen	Esophageal, bronchial, and muscles of the small intestine and right half of the large intestine	Secretomotor to the glandular tissue and motor to the intestinal wall
	SVE	Nucleus ambiguus	Same as above	Most muscles of the soft palate; pharynx and larynx	Movements of muscles of the soft palate, pharynx, and larynx during deglutition, respiration, and phonation
	GVA	Inferior (nodose) ganglion	Same as above	Visceral sensations from the pharynx, larynx, bronchi, aortic arch, and body, most of the digestive tract including the right two-thirds of large intestine	Visceral sensations; chemoreceptor and pressure changes
	GSA	Superior (jugular) ganglion	Same as above	External auditory meatus and auricle	Cutaneous sensibility
	SVA	Inferior (nodose) ganglion	Same as above	Taste sensation from epiglottic vallecula	Taste
XI. Accessory nerve	SVE	Nucleus ambiguus	Jugular foramen	Within branches of the vagus to pharyngeal and laryngeal muscles	Swallowing and phonation
	GSE	Upper five spinal segments	Foramen magnum rostrally; exit via jugular foramen	Trapezius and sternocleidomastoid muscles	Elevation of scapula and shoulder point; turning the face upward and to the opposite side
XII. Hypoglossal nerve	GSE	Hypoglossal nucleus	Hypoglossal canal	Intrinsic and extrinsic muscles of the tongue	Change the shape of the tongue and maintain its movement

TABLE 11. 4

Cranial Nerve Somatic and Autonomic Components

			Afferent	Efferent (Motor)		
				Parasympathetic Ganglia	Somatic	Branchial
I	Olfactory	+	Bipolar neurons in the olfactory mucosa	None	None	None
II	Optic	+	Extension of central nervous system	None	None	None
III	Oculomotor		None	Ciliary	+	None
IV	Trochlear		None	None	+	None
V	Trigeminal		Trigeminal (semilunar or Gasserian) ganglion	None	None	+
VI	Abducens		None	None	+	None
VII	Facial	+	Geniculate	Pterygopalatine and submandibular ganglia	None	+
VIII	Vestibulocochlear	+	Scarpa's and spiral ganglia	None	None	+
IX	Glossopharyngeal	+	Superior and inferior ganglia	Otic ganglion	None	+
X	Vagus	+	Superior and inferior ganglia	Intramural ganglia	None	+
XI	Accessory		None	None	None	+
XII	Hypoglossal	XII	None	None	+	None

and ambiguus nuclei may selectively be damaged in individuals with *Tapia syndrome*, which is characterized by paralysis of the muscles of the soft palate, posterior pharyngeal wall, vocal cords, and lingual muscles. Upper motor neuron lesions of the corticobulbar fibers that project to the hypoglossal nucleus produce deviation of the tongue toward the intact side away from the side of lesion. The hypoglossal nucleus is prone to damage in Arnold–Chiari malformation, neurinoma, meningioma, glomus jugulare tumor, syringobulbia, and inferior alternating hemiplegia (Tables 11.3 and 11.4). See also Chapter 20.

SUGGESTED READING

Arle JE, Abrahams JM, Zager EL, Taylor C, Galetta SL. Pupil-sparing third nerve palsy with preoperative improvement from a posterior communicating artery aneurysm. *Surg Neurol* 2002;57:423.

Bansal S, Khan J, Marsh IB. Unaugmented vertical muscle transposition surgery for chronic sixth nerve paralysis. *Strabismus* 2006;14(4):177–81.

Brazis PW. Palsies of the trochlear nerve: Diagnosis and localization-recent concepts. *Mayo Clin Proc* 1993;68:501–9.

Choi HG, Kwon SY, Won JY, Yoo SW, Lee MG, Kim SW, Park B. Comparisons of three indicators for Frey's syndrome: Subjective symptoms, minor's starch iodine test, and infrared thermography. *Clin Exp Otorhinolaryngol* 2013;6(4):249–53.

Gronseth GS, Paduga R. Evidence-based guideline update: Steroids and antivirals for Bell palsy: Report of the Guideline Development Subcommittee of the American Academy of Neurology. *Neurology* 2012;79(22):2209–13.

Kaddu S, Smolle J, Komericki P, Kerl H. Auriculotemporal (Frey) syndrome in late childhood: An unusual variant presenting as gustatory flushing mimicking food allergy. *Pediatr Dermatol* 2000;17:126–8.

Keller JT, van Loveren H. Pathophysiology of the pain of trigeminal neuralgia and atypical facial pain: A neuroanatomical perspective. *Clin Neurosurg* 1985;32:275–93.

Krammer EB, Rath T, Lischka MF. Somatotopic organization of the hypoglossal nucleus. A HRP study in the rat. *Brain Res* 1979;170:533–7.

Laine FJ, Smoker WR. Anatomy of the cranial nerves. *Neuroimaging Clin North Am* 1998;8:69–100.

Lossos A, Siegal T. Numb chin syndrome in cancer patients: Etiology, response to treatment, and prognostic significance. *Neurology* 1992;42:1181–4.

Marinella MA. Metastatic large cell lung cancer presenting with numb chin syndrome. *Respir Med* 1997;91:235–6.

Musani MA, Farooqui AN, Usman A, Atif S, Afaq S, Khambaty Y, Ahmed L. Association of herpes simplex virus infection and Bell's palsy. *J Pak Med Assoc* 2009;59(12):823–5.

Naraynsingh V, Cawich SO, Maharaj R, Dan D. Retrograde thyroidectomy: A technique for visualization and preservation of the external branch of superior laryngeal nerve. *Int J Surg Case Rep* 2014;5(3):122–25.

Richards BW, Jones FR Jr, Younge BR. Causes and prognosis in 4278 cases of paralysis of the oculomotor, trochlear, and abducens cranial nerves. *Am J Ophthalmol* 1992;113:489.

Shima F, Fukui M, Kitamura K, Kuromatsu C, Okamura T. Diagnosis and surgical treatment of spasmodic torticollis of 11th nerve origin. *Neurosurgery* 1988;22:358–63.

Socolovsky M, Di Masi G, Bonilla G, Malessy M. Spinal to accessory nerve transfer in traumatic brachial plexus palsy: Is body mass index a predictor of outcome? *Acta Neurochir (Wien)* 2014;156(1):159–63.

Spoendlin H, Schrott A. Analysis of the human auditory nerve. *Heart Res* 1989;43:25–38.

Tarlov EC. Microsurgical vestibular nerve section for intractable Meniere's disease. *Clin Neurosurg* 1985;33:667–84.

Uemura M, Matsuda K, Kume M et al. Topographical arrangement of hypoglossal motor neurons. An HRP study in the cat. *Neurosci Lett* 1979;13:99–104.

Wiegand DA, Fickel V. Acoustic neuromas. The patient's perspective. Subjective assessment of symptoms, diagnosis, therapy, and outcome in 541 patients. *Laryngoscope* 1989;99:179–87.

Section IV

Functional Neuroanatomy

12 Neurotransmitters

Classical neurotransmitters are small molecules of neuroactive agents actively involved in synaptic transmission and modulation. They are synthesized in the neurons and are released at the presynaptic terminals in sufficient amounts to affect the membrane potential or conductance of the postsynaptic neurons, producing inhibition or excitation. Their effect is commonly associated with the selective opening of specific ion channels in the postsynaptic membrane/and or phosphorylation of intracellular protein. They may bind directly to a receptor and cause second messenger–mediated changes in the neurotransmission. Exogenous administration of neurotransmitters may mimic the actions of the endogenous transmitters. Certain neurotransmitters may be released after a more sustained activation. Inactivation of these agents may occur locally at the terminals by enzymatic uptake and degradation, or diffusion and release. Some neurotransmitters do not act upon the postsynaptic membrane but affect its response to other neuromediators by enhancing or inhibiting their activities. Classical (small-molecule) neurotansmitters are comprised of amino acid neurotransmitters, acetylcholine, and biogenic amines.

Peptidergic neurotransmitters are scattered in the peripheral, central, and enteric nervous systems, and their synthesis is regulated by messenger RNA (mRNA) and ribosomes at the soma or dendrites. They are derived from an inactive prohormone, which is cleaved by certain proteolytic enzymes. Peptidergic neurotransmitters are organized structurally into families, and many are neurohormones that are synthesized in neurons and released into the blood circulation, cerebrospinal fluid (CSF), or the intercellular space by exocytosis.

AMINO ACID NEUROTRANSMITTERS

Amino acid neurotransmitters comprise gamma-aminobutyric acid (GABA) and glycine as proven inhibitory neurotransmitters, and glutamate and aspartate as putative stimulatory neurotransmitters. Others, such as proline, serine, and taurine, await further study to be considered as neurotransmitters. For our purposes, we will be dealing with the first group only.

Gamma-Amino Butyric Acid

GABA is the major inhibitory neurotransmitter in the brainstem, spinal cord, and Purkinje neurons of the cerebellum. It induces depolarization predominantly in the spinal cord and hyperpolarization in the cortical cells. It acts by increasing the permeability of the postsynaptic membrane to chloride. GABA is produced via an irreversible reaction of L-glutamic acid and glutamic acid decarboxylase (GAD), utilizing pyridoxal phosphate (a form of vitamin B6) as a cofactor. Substantial increase in the postmortem level of GABA may be attributed to the transient activation of GAD.

> Therefore, substances that decrease the amount of pyridoxine (hydrazides) or inhibit its action (e.g., sulfhydryl reagents, chloride, etc.) may repress the action of GABA and subsequently induce reversible epileptic seizures. Experimentally, localization of GAD may be of value in determining the concentration of GABA. Development of autoantibodies against GAD may be associated with stiff-man syndrome, a rare chronic neurological condition that exhibits progressive and fluctuating muscle spasm and atrophy.

GABA is metabolized principally by GABA transaminase (GABA-T), an extensively distributed enzyme, which bounds to pyridoxal phosphate, and may be inhibited by gabaculine. Transamination of GABA produces succinic semialdehyde, which is later reduced to γ-hydroxybutyrate (GHB).

> Increased levels of GHB subsequent to a deficiency of succinic semialdehyde dehydrogenase may occur as a manifestation of a congenital disorder of GABA metabolism, producing dementia and ataxia.

GABAergic neurons are scattered in high concentrations in many brain areas such as the lateral geniculate nucleus, Purkinje cell axons that project to the lateral vestibular and cerebellar nuclei, and also the striatal neurons that convey impulses to the substantia nigra and cortical neurons. Most interneurons are GABAergic, such as the amacrine and horizontal cells of the retina. GABAergic neurons are lacking or found in trace amounts in the peripheral nervous system. The highest concentrations of GABA are found in the diencephalon, whereas lower concentrations are localized in the cerebral hemispheres and brainstem.

Depolarization of the presynaptic neurons stimulates the release of GABA at the synaptic clefts. Reuptake into both presynaptic terminals and surrounding neuroglial cells terminates the action of GABA. Temperature and ion-dependent transport systems maintain this reuptake. In contrast to nerve terminals, GABA taken up into glial cells cannot be utilized, but instead, it may be metabolized to succinic semialdehyde by GABA-T. The semialdehyde is oxidized to succinate via succinic semialdehyde dehydrogenase.

Deficiency of succinic semialdehyde dehydrogenase may produce mental retardation and cerebellar disorders including hypotonia. These patients excrete copious amounts of both succinic semialdehyde and 4-hydroxybutyric acid. Deficiency of GABA-T produces deep tendon hyperreflexia, psychomotor retardation, and increased height. The latter effect may be attributed to the ability of GABA to enhance the release of growth hormones.

Glial GABA may be recovered via the Krebs cycle, where it is converted to glutamine. Glutamine is transferred to neurons, where it is converted by glutaminase to glutamate, which reenters the GABA shunt.

Overactivity of GABA produces an inhibitory effect on dopaminergic neurons by producing hyperpolarization at the postsynaptic level in the cerebral cortex and hyperpolarization at the presynaptic level in the spinal cord.

Blockage of GABAergic cortical neurons may be responsible for inducing convulsions and maintaining myoclonus. It has been suggested that lack of GABA in the substantia nigra, putamen, and caudate nucleus, subsequent to degeneration of the GABAergic neurons, may be associated with the involuntary choreiform movements observed in Huntington's disease. Additionally, overactivity of GABAergic neurons, which exert an inhibitory effect on dopaminergic nigral neurons, is thought to have a role in producing some of the signs and symptoms of Parkinson's disease. Also, the toxin of *Clostridium tetani* may bind to the presynaptic GABAergic cells of the α motor neurons of the spinal cord and brainstem, blocking the release of GABA. Inactivation of GABA blocks the inhibitory influences on the motor neurons, resulting in muscle spasm, rigidity, lockjaw, dysphagia, and opisthotonos.

GABA receptors, classified into GABA-a and GABA-b categories, are located in the neuronal cell membranes and astrocytes. The *GABA-a* receptors are more common than GABA-b and they belong to the same superfamily of ligand-activated receptors as the nicotinic acteylcholine receptors. They are G protein–coupled ionotropic receptors that consist of the α, β, γ, δ, and ρ subunits with additional subtypes. The ρ subunit, in particular, is abundant in the retina.

GABA-a receptors show binding sites for benzodiazepine and comprise a group of anxiolytic drugs that act particularly on GABA-a-γ_2. Barbiturates (e.g., phenobarbital) and antiepileptic drugs act on GABA-a-α and GABA-a-β receptors. These drugs increase chloride current and the duration of channel opening induced by GABA. GABA-a receptors are also the major molecular target for the volatile anesthetics and possibly ethanol. Neuroactive steroids (analogs of the progesterone and corticosterone derivatives) may exert antianxiety, sedative, and hypnotic effects via their potent modulatory effects of GABA-a receptors.

GABA-b receptors encompass two principal types of receptors that differ in regard to location. They are coupled indirectly to calcium and potassium channels via second messenger systems (G proteins). Inhibitory response of these receptors is produced, at both presynaptic and postsynaptic levels, by increased potassium or decreased calcium conductance and inhibition of cyclic adenosine monophosphate (cAMP) production. Receptor antagonists such as picrotoxin block GABA-b receptors, which mediate postsynaptic inhibitory potentials (IPSPs). They have selective affinity to baclofen, a GABA-analog (β-[4-chloro-phenyl-γ-aminobutyric acid]), which releases intracellular GABA, but not to bicuculline, and are not affected by benzodiazepine (e.g., Valium and Librium) or barbiturates (e.g., phenobarbital). Both groups of drugs increase GABA-induced chloride current by either intensifying the frequency of channels opening or prolonging its duration.

Some investigators suggest that improved cognitive ability may be achieved by blocking the GABA-b receptors that increase GABA and neuronal excitability of hippocampal neurons and thus improve memory encoding.

Glutamic Acid

Glutamic acid, the most abundant amino acid in the central nervous system (CNS), is a precursor for GABA and a fast-acting excitatory neurotransmitter. This neurotransmitter is involved in the formation of peptides and proteins as well as detoxification of ammonia in the cerebral cortex. The L-glutamate form of this neurotransmitter is synthesized in the nerve terminals via the Krebs cycle and transamination of α-oxyglutarate and also from glutamine in the glial cells. Glial cells and nerve endings release glutamic acid via a calcium-dependent exocytotic process. It has been found to have a powerful depolarizing effect on the neurons in all areas of the CNS. The sensitivity of glutamate receptors to glutamate NMDA receptor agonists is utilized as a basis for the classification of ionotropic glutamate receptors into NMDA and non-NMDA receptors. There are also metabotropic glutamate receptors, which exert effects via G protein. Neuronal dysfunctions associated with anoxia, seizures, or hypoglycemia may be due to the disproportionate inflow of calcium ions through the NMDA receptor channels and dramatic sustained increase in the level of glutamate (excitotoxicity). Anoxia impairs the sodium/potassium pump by reducing the ATP, which is followed by excessive increase in the level of potassium concentration in the extracellular space, promoting depolarization of neurons and inhibiting glutamate uptake and its release by reversing the glutamate transporter. This positive feedback system leads to a dramatic increase in the extracellular glutamate concentration. During this process, vast influx of sodium ions via both NMDA and non-NMDA receptor channels may enhance cellular necrosis by increasing the water content of the neuron (cytotoxic edema). Calcium ions gain access through NMDA receptors and voltage-dependent calcium channels and possibly via α-amino-3-hydroxy-5-methyl-4-isoxazole propionic acid (AMPA). Therefore, glutamate receptor antagonists, calcium channel blockers, and antioxidants can achieve suppression of calcium influx. Glutamate produces neuronal necrosis by utilizing calcium and free radicals as mediators. Calcium ions' role in neuronal death occurs initially upon activation of proteases and endonucleases, which lead to proteolysis of the microfilaments and eventual destruction of DNA. These ions also induce neuronal damage by enhancing the release of toxic hydroxyl free radicals via stimulation of NO synthetase (NOS) and phospholipase A_2. Glutamate is also released at synaptic sites where long-term potentiation (LTP), a sustained increase in the amplitude of excitatory potentials, occurs. LTP, which is thought to be associated with long-term memory, may be facilitated by the depolarization and calcium influx in the hippocampal neurons. It may also be associated with calcium/calmodulin–dependent protein kinase and protein kinase C.

Some scientific views indicate that the biochemical changes observed in Huntington's chorea may be directly or indirectly related to glutamate's actions as a neurotransmitter. This observation is based upon the experimental injection of NMDA agonists into the striatum and the subsequent neuronal loss that seems similar to that of Huntington's chorea. This is supported by the fact that NMDA receptor–mediated neurotoxicity may result from presynaptic and postsynaptic abnormalities that affect glutamate synapses and render some cells prone to toxicity by the normal glutamate levels. Lower concentrations of glutamate transporters may lead to toxic extracellular glutamate levels. Inhibition of glutamate release and blockage of NMDA- and kainate-mediated processes are the mechanisms utilized by medications for the treatment of amyotrophic lateral sclerosis (ALS), such as Riluzole.

A substance that acts like an AMPA agonist, stimulating its receptors, can selectively produce destruction of the upper and lower motor neurons. This is evident in individuals with neurolathyrism, an upper motor neuron disorder that occurs in the inhabitants of certain parts of Africa and Asia subsequent to a dietary reliance on chickpea (*Lathyrus sativus*). The toxin that produces this deficit is β-*N*-oxalylamino-L-alanine (BOAA).

In global ischemia, which occurs in individuals with cardiac arrest and neuronal death (e.g., in the CA1 pyramidal neurons of the hippocampal gyrus), may also be significantly reduced, when AMPA receptor antagonists are administered within the first 24 h of ischemic episodes. A combination of therapeutic hypothermia and the administration of an NMDA antagonist may achieve neural protection in instances of global ischemia. Neuronal protection of the penumbra (area in the immediate surroundings of the site of focal ischemia) can be accomplished to a remarkable degree by the administration of NMDA antagonists in the first few hours of the insult. Experimental evidence points to the significant role that AMPA antagonists can play in neuronal protection following focal ischemic insult.

Since NMDA receptor antagonists appear to prevent the induction of epileptic seizures, the role of NMDA receptor in induction of epilepsy has been strongly suggested.

Reproduction of the positive and negative symptoms of schizophrenia by 1-(1-phenylcyclohexyl)piperidine or phencyclidine (PCP) and other NMDA receptor antagonists and altered postmortem levels of glutamate in schizophrenic brains may support the role of glutamate in this psychiatric disorder.

AMPA and NMDA receptor agonists may become excitotoxic in the presence of bicarbonate. This observation is based on individuals with Guam disease, manifesting clinical signs of ALS, Parkinsonism, and senile dementia complex. This disease occurs in the inhabitants of the island of Guam (Chamorros), who consume seeds of cycad (*Cycas circinalis*, false sago) as a food source. The latter contains the neutral amino acid β-*N*-methyl-amino-L-alanine (BMAA), which exhibits an affinity for NMDA receptors but does not display in vitro toxicity or a direct excitatory effect.

Glycine

Glycine, the simplest of all amino acids in structure, is formed from serine in a reaction catalyzed by serine transhydroxymethylase. It may also be formed by reaction of transaminase with glutamate. Glycine is an essential component in the metabolism of peptides, proteins, nucleic acids, and porphyrins. Transmitter glycine potentiates synaptic activity by hyperpolarizing the membrane and increasing chloride permeability. Its inhibitory action is similar to that of GABA but mainly restricted to the spinal motor neurons and interneurons. It can be blocked by strychnine and not by bicuculline or picrotoxin. Glycine is found in high concentrations within the interneurons of the ventral horn of the spinal cord. It is released from the presynaptic terminals of primary afferent fibers of the retina, pons, and medulla. It has been shown that glycine has a role in both increasing the frequency of *N*-methyl-D-aspartate (NMDA) receptor channel opening and also preventing desensitization of these receptors, without involving the glycine receptors. β-alanine, taurine, L-alanine, L-serine, and proline activate glycine receptors.

The delay in muscle relaxation following voluntary contraction or percussion in *myotonia*, a condition that exhibits abnormally slow relaxation of the skeletal muscles following active contraction, is ascribed to a decrease in chloride conductance caused by reduction of glycine at synaptic clefts and its excretion in urine. Spasticity in antigravity muscles and hyperreflexia observed in upper motor neuron palsies are also thought to be mediated by glycine.

ACETYLCHOLINE

Acetylcholine is the first neurotransmitter to have been identified in the CNS. Extensive distribution of this neurotransmitter in the limbic system, interneurons of the neostriatum, α motor neurons of the spinal cord, and basal nucleus of Meynert has been subsequently confirmed. In the brainstem, the parabrachial nuclear complex, which is located dorsolateral to the superior cerebellar peduncle, contains high concentration of the cholinergic neurons. A prominent subnucleus of this complex, the pedunculopontine nucleus, is involved in the generation of rhythmic movements by projecting to the spinal cord. This neurotransmitter is also found in the retina, cornea, motor nuclei of the cranial nerves, and autonomic ganglia. In general, the function of acetylcholine varies with the site of its activity. It maintains an excitatory function in the central and the peripheral nervous systems, with the exception of its inhibitory effect upon the cardiac muscles. Acetylcholine is formed in the cytosol by the reversible reaction of acetyl coenzyme A (CoA) and choline and conversion of acetyl CoA to CoA, which is catalyzed by choline acetyl transferase (ChAT). Choline is derived from the degradation of acetylcholine by acetylcholinesterase within the synaptic cleft and also from the breakdown of phosphatidyl choline from other membrane sources. Lecithin produces a greater and longer-lasting rise in plasma choline levels. In general, the amount of acetylcholine at any given moment is dependent upon the amount of calcium influx and the duration of the action potential. Acetylcholinesterase, which is widely distributed in neuronal and nonneuronal tissues, exists in several molecular forms and catalyzes the hydrolysis of acetylcholine.

Cholinergic receptors, which are found in the brain and spinal cord, are classified into antagonistic muscarinic and nicotinic receptors.

The *muscarinic receptors* have several transmembrane sparing regions that are linked to guanine nucleotide triphosphate (GTP)–binding protein (G protein–coupled receptors), which respond slowly, stimulated by muscarine, and are blocked by atropine or scopolamine. They cause inhibition of adenyl cyclase, stimulation of phospholipase C, and regulation of ion channels. There are five muscarinic receptors, which display relatively slow response times to acetylcholine (ACh) binding. They are categorized into m1–m5. The m1 receptors are postsynaptic and excitatory and remain particularly unaffected in Alzheimer's disease. On the other hand, m2 receptors are presynaptically inhibitory, regulating the release of acteylcholine. This group of receptors is reduced in Alzheimer's disease, and their binding sites in the hippocampal gyrus, cerebral cortex, and striatum show appreciable age-related decrease. However, major impairments were detected with m2 control of dopamine (DA) release.

Nicotinic receptors respond quickly, are excitatory, and are activated by nicotine. These receptors are blocked by curare drugs or excess of nicotine or acetylcholine and are found in the neuromuscular junction, autonomic ganglia, cortex, and thalamus. They are desensitized by continued exposure to agonists. Nicotinic receptors have distinct α, β, γ, and δ subunits that contain four membrane-spanning α helices. It has been suggested that these subunits are arranged around an ion channel that remains closed at rest but opens when the ACh binds to an α subunit of these receptors. Neuronal nicotinic receptors that contain α2–α6 subunits are distinct from nonneuronal receptors, showing resistance to α-bungarotoxin and related α-neurotoxin.

Cholinergic neurons are thought to have crucial role in learning, memory, and cognitive ability. The role of cholinergic neurons in short-term memory is also assumed on the fact that centrally acting muscarinic blockers such as atropine and scopolamine may produce loss of memory and inability to execute learned tasks. On the contrary, chemicals that inhibit acetylcholinesterase (e.g., physostigmine) may elicit the opposite responses.

Cholinergic neurons within the nucleus basalis of Meynert, which project to wide areas of cerebral cortex, receive afferents from the limbic system and the hypothalamus and are involved in the ascending reticular activating system.

Cholinergic overactivation, which may be responsible for the clinical signs of Parkinson's disease, occurs as a result of either decreased dopaminergic activity or increased amount of acetylcholine. Administration of physostigmine increases the striatal acetylcholine concentration and often contributes to the exacerbation of Parkinson's disease.

Treatment of *preeclampsia* (hypertension-induced nervous disorders, e.g., seizure, coma, etc.) is based upon the fact that magnesium sulfate, the drug of choice, acts by inhibiting acetylcholine release.

Inhibition of the release of acetylcholine by the toxin of *Clostridium botulinum*, a calcium-dependent substance, occurs peripherally by binding to the external receptors at the synapse sites. Botulinum toxin cannot penetrate the CNS or exert any direct influence upon the central cholinergic neurons. Excessive inhibition of acetylcholinesterase produces a surplus of acetylcholine that binds to receptors, leading to exhaustion.

MONOAMINES

Monoamines (biogenic amines) are comprised of the *catecholamines* and the *indolamines.*

Monoamine transmitters are stored in synaptic vesicles; their release is a calcium-dependent processes, and they may be metabolized by monoamine oxidase (MAO). Plasma membrane transporter terminates synaptic action of the monoamine transmitters.

CATECHOLAMINES

Catecholamines are derivatives of beta-phenyl ethylalanine, with hydroxy groups on the third and fourth positions. They are synthesized by tyrosine, a common precursor for norepinephrine, epinephrine, and dopamine. In the central and peripheral nervous systems, transformation of tyrosine via a series of chemical changes may lead to the formation of norepinephrine, dopamine, or epinephrine, a process that is dependent upon tyrosine hydroxylase and dopamine β-hydroxylase. Release of catecholamines is stimulated by the influx of calcium.

Catecholamines are formed by L-tyrosine, which is converted to L-dopa (levodopa), via hydroxylation by tyrosine hydroxylase (rate-limiting enzyme). L-dopa is converted to dopamine following decarboxylation by an aromatic amino acid decarboxylase. Dopamine is either stored in the vesicles or hydroxylated to L-norepinephrine by dopamine β-hydroxylase. L-epinephrine is formed in the adrenal medulla from norepinephrine by the enzyme phenyl-ethanolamine-*N*-methyltransferase. Norepinephrine and dopamine are metabolized (inactivated) by MAO and catechol-*o*-methyltransferase (COMT). Inactivation of norepinephrine occurs by the reuptake mechanism into the presynaptic nerve terminals.

Catecholamines are found in the brain, chromaffin tissue of the adrenal medulla, and sympathetic nervous system, maintaining massive projections throughout the brain. Amine transmitters have slow modulatory influences, and most of their receptors are part of the G protein–coupled family. They may play a role in the regulation of visceral activities, emotion, and attention. Norepinephrine is the primary neurotransmitter in the sympathetic (peripheral nervous) system, while dopamine, serotonin, and norepinephrine act primarily in the CNS. Adrenergic receptors are classified into alpha (α) and beta (β) receptors. Although activation of both receptors produces inhibition of gastrointestinal tract motility, their classification in general is based on their respective excitatory and inhibitory effects on smooth muscles. These receptors are further subdivided into α_1 and α_2 and β_1, β_2, and β_3. Activation of the postsynaptic α_1 and presynaptic α_2 noradrenergic receptors produces vasoconstriction and inhibition of the release of norepinephrine, respectively. The latter process is unaffected by pertussis toxin, which inhibits G protein. α_1 receptors are inhibited by prazosin, a α_1-blocking substance. β *receptors* are closely linked to adenyl cyclase activation via G protein. Stimulation of the β receptors results in changes, which include vasodilatation of the coronary and abdominal arteries and relaxation of the ciliary, gastrointestinal, and detrusor muscles. They also produce activation of glycogenolysis, as well as dilatation of the bronchi. Activation of the β_1 and β_3 receptors causes an increase in the rate of both cardiac contractility and renin secretion (β_1) and enhances lipolysis (β_3), respectively. Low concentrations of epinephrine activate presynaptic β_2 receptors.

Enzymes involved in metabolic degradation of the catecholamines such as MAO and COMT maintain an important role in the expression of emotion. Lower levels of catecholamines may produce depression, while higher levels produce euphoria.

Norepinephrine

Norepinephrine in the CNS is concentrated in neurons of the locus ceruleus of the rostral pons and caudal medulla and the lateral tegmental nuclei. These noradrenergic neurons project to the entire cerebral cortex, hippocampus, cerebellum, and spinal cord as well as basilar pons and ventral medulla. It is released in small amounts from the adrenal gland during circulatory collapse. In dystonia, the level of norepinephrine drops in the red nucleus, hypothalamus, mammillary body, locus ceruleus, and subthalamic nucleus.

Norepinephrine regulates the degree of arousal, mood, memory, and learning and also modulates sound transmission (sharpening effect). An excess of norepinephrine has been shown to elicit euphoria, while its depletion may produce depression. The effects of mood-elevating drugs (antidepressants) such as MAO inhibitors and reuptake blockers such as reserpine (antidepressant) and amphetamines (sympathomimetics) are based upon the role of these agents in either increasing the concentration or depleting the endogenous norepinephrine. MAO action on norepinephrine results in the formation of vanilylmandelic acid (VMA), a readily detectable product in the urine. Measurement of VMA levels may bear diagnostic value in conditions such as pheochromocytoma and neuroblastoma.

Checking the levels of 3-methoxy-4-hydroxyphenethyleneglycol (MHPG), a major CNS metabolite of norepinephrine that is found in the urine, blood, and CSF, may help to assess the functional activity of the central adrenergic neurons.

Norepinephrine has potent excitatory but weak inhibitory effects on smooth muscles. It also has stronger affinity than epinephrine to β_3 receptors. Noradrenergic neurons in the brainstem reticular formation are classified, according to Dahlstrom and Füxe, into A_1 through A_8 groups. The A_1 group projects to the spinal cord, solitary nucleus, and hypothalamus and is located in the caudal lateral medulla. A_5 lies in the caudal pons near the superior olivary nucleus and projects to the intermediolateral column of the spinal cord. The locus ceruleus, which is designated as group A_6, occupies the rostral pons and caudal midbrain, projecting via the central tegmental tract, medial forebrain bundle, superior cerebellar peduncle, and tectospinal tract.

The extensive and global projection of the locus ceruleus to the intralaminar thalamic nuclei via the ascending reticular activating system may account for its role in paradoxical (REM) sleep.

Agonists and antagonists of α receptor exert variable influences on the firing rate of locus ceruleus neurons. The α_2 agonists such as clonidine inhibit the firing of locus neurons, in contrast to the α_2 antagonists such as yohimbine and iadazoxan that facilitate this neuronal activity. The A_7 group is located in the lateral pontine tegmentum of the isthmus sending projections to the spinal cord. In the peripheral nervous system, norepinephrine stimulates presynaptic and postsynaptic receptors.

Norepinephrine is thought to have a role in stiff-man syndrome, which is characterized by uncontrollable muscular spasm and stiffness. This suggestion is based upon the fact that increased level of 3-methoxy-4-hydroxyphenoglycol, a metabolite of norepinephrine, is detected in the urine of affected individuals.

Epinephrine

Epinephrine is released into the bloodstream from chromaffin cells of the adrenal medulla. This neurotransmitter stimulates the vascular smooth muscles and produces a rise in blood glucose concentration. Thus, arousal from insulin coma may be achieved by activation of glycogenolysis. Epinephrine has a stronger affinity for β adrenergic receptors in the smooth muscles of the vessels, bronchi, gastrointestinal tract, and urogenital system. Similarly, it also has stronger affinity to α_1 and α_2 receptors than norepinephrine. Epinephrine-containing neurons are classified into group C_1 in the lateral tegmentum, C_2 in the dorsal medulla, and C_3 in the medial longitudinal fasciculus. Certain noradrenergic neurons of the C_1 and C_2 groups project to the hypothalamus via the central tegmental tract and periventricular gray.

Dopamine

Dopamine, representing approximately 50% of the total catecholamines, is formed by conversion of tyrosine to L-dopa tyrosine hydroxylase and then to dopamine via L-aromatic amino acid decarboxylase. Orally administered tyrosine does not increase dopamine levels. However, orally administered levodopa (L-3,4-dihydroxyphenylalanine) is absorbed from the small intestine by active transport and later converted in the dopaminergic nigral neurons into dopamine. Dopaminergic neurons are concentrated in the tuberal nuclei of hypothalamus, nucleus accumbens, olfactory tubercle, midbrain, carotid body, and superior cervical ganglion. The latter also contains both cholinergic and noradrenergic neurons. In general, the distribution of dopamine parallels that of norepinephrine, with dopaminergic neurons outnumbering the noradrenergic neurons at the ratio of 3:1. High concentrations of dopamine and low concentrations of norepinephrine exist in the caudate nucleus and putamen, whereas the reverse occurs in the hypothalamus.

MAO-b oxidizes and selectively increases the level of dopamine. Inhibitors of this enzyme such as amphetamines have a profound mood-elevating effect and may even produce agitation.

Dopaminergic neurons in the pars compacta of the substantia nigra exert an inhibitory influence. The pars compacta synthesizes dopamine and delivers it to the neostriatum (caudate and putamen) via the nigrostriatal fibers.

Degeneration of the nigrostriatal fibers and depletion of dopamine accounts for the manifestations of Parkinson's disease. In *Parkinson's disease*, the mechanism of degradation of dopamine is maintained, but its synthetic machinery is impaired. Dopaminergic neurons in the ventral tegmental nuclei of the midbrain, which are localized medial to the substantia nigra, form the *mesolimbic system*, a group of neurons that maintain diffuse projections to the septal area, amygdala, entorhinal area, nucleus accumbens septi, olfactory tubercle, and pyriform cortex. The role of these projections in controlling mood and emotion accounts for the psychiatric disorders that accompany L-dopa therapy.

It has been suggested that overactivity of the mesolimbic dopaminergic neurons may be associated with schizophrenia. Antipsychotic drugs may increase dopaminergic neuronal activity and dopamine synthesis by blocking the postsynaptic dopaminergic receptors. Prolonged usage of these medications may lead to tolerance development. On the same basis, the inhibitory action of neuroleptic drugs upon these neurons may explain the improvement seen in patients with these disorders, as well as the unwanted side motor disorders.

Dopaminergic neurons in the ventral tegmentum and substantia nigra form the *mesocortical pathway* that projects to the prefrontal cortex (involved in motivation, attention, and social behavior) and entorhinal area. In contrast to the *mesolimbic system*, these neurons do not develop tolerance to continued usage of antipsychotic medications. Midbrain dopaminergic neurons, which are components of the mesocortical system, lack autoreceptors that regulate impulse traffic. They are generally presynaptic and respond to the same transmitter utilized by the neuron that contains them. They have a faster firing rate than the mesolimbic dopaminergic neurons and are less affected by the dopamine receptor–blocking agents such as haloperidol.

Dopaminergic neurons located in the tuberal nuclei of the hypothalamus project to the median eminence via the tuberinfundibular tract and then to the adenohypophysis via the hypophysial portal system. These projections are postulated to regulate the secretion of prolactin and melanocyte-stimulating hormone (MSH). Dopaminergic neurons in the retina and olfactory bulb may play a role in the phenomenon of lateral inhibition, which sharpens the visual and olfactory impulses and prevents neuronal cross talk.

Dopaminergic receptors are classified on an anatomical and biochemical basis into Dl and D2 receptors. Dl and D2 receptors in the caudate and putamen showed a detectable increase in density in Parkinsonism and a decrease in Huntington's chorea. Dl group of receptors which includes Dl and D5 activates $G\alpha_{s/olf}$ family of G proteins to stimulate cAMP production by AC (adenylyl cyclase) and are expressed postsynaptically on dopamine-receptive cells. Dl and D5 have similar concentration in the hypothalamus and temporal lobes. The D2-class dopamine receptors (D2, D3, and D4) couple to the $G\alpha_{i/o}$ family of G proteins and induce inhibition of AC, residing at both pre- and postsynaptic levels. The D4 subtype is particularly identified in the brains of schizophrenics.

Dopamine transporter (DAT), a sodium/calcium-dependent plasma protein, may bind to drugs like cocaine and amphetamine, producing behavioral and psychomotor changes. Thus, cocaine overdose may be treated by antagonists that prevent this binding. Substances that show high affinity for DAT such as 1-methyl,4-phenyl-1,2,3,4,6-tetrahydropyridine (MPTP) may prove to be toxic to dopaminergic neurons. D_1 receptors mediate the dopamine-stimulated increase of adenylate cyclase and, subsequently, intracellular cAMP. The role of D_2 receptors in motor activity, independent of adenylate cyclase, is clearly illustrated in the reduction of both motor and vocal tics upon administering drugs that act upon these receptors.

Additionally, the newer generation of antipsychotics such as clozapine (Clozaril) is thought to act upon D_2 receptors, resulting in selective inactivation of the dopaminergic neurons in the ventral tegmentum, but not in the neurons of the substantia nigra. These drugs may also have fewer side effects than the older generation of antipsychotics, which are known to produce tardive dyskinesia in some patients. D_3 and D_4 receptors maintain primary functional relationships to the limbic system and the telencephalon and secondary connections to the basal nuclei. The antipsychotic drug clozapine's strong affinity to D_3 and D_4 receptors may account for the suppression of unwanted subcortical motor side effects.

Dopamine's possible role in *Gilles de la Tourette syndrome* (hereditary multiple tic disorder), a childhood neurological condition that exhibits multiple motor and vocal tics and compulsive utterance, is based upon the improvement seen in patients with this disease following administration of dopaminergic antagonists. In psychiatric patients using certain neuroleptic drugs, a reduction of dopamine concentration in the substantia nigra may occur as a result of manganese intoxication or ingestion of levodopa, which produces hydrogen peroxide and hydroxyl free radicals, leading to involuntary motor activities.

INDOLAMINES

Indolamines are a family of biogenic amines that can act as neurotransmitters and share a common molecular structure, containing an indole ring with five members and an amine group (NH_2). They comprise serotonin and histamine.

SEROTONIN

Serotonin (5-hydroxytryptamine [5-HT]) is synthesized by hydroxylation of tryptophan and carboxylation of the product of the reaction by tryptophan hydroxylase. The latter enzyme also catalyzes the carboxylation process that forms serotonin and converts dopa to dopamine. The level of tissue oxygen, pteridine, and tryptophan (cofactors or substrate) may also influence the rate of 5-HT formation. Although presence of this neurotransmitter in the CNS represents a small fraction of its total concentrations in the whole body, serotonin is found in high concentration in the raphe nuclei, spinal cord, and hypophysis cerebri, where it is converted into melatonin by acetylation and methylation. Serotonin is actively transported from the cytoplasm to the storage vesicles. Vesicular transport involves vesicular transporter-$_1$ and -$_2$ ($VMAT_1$ and $VMAT_2$), which also function as antiporters to eliminate cytoplasmic toxic materials. For this very reason, vesicular transporters are known as toxin-extruding antiporters (TEXANs). Serotonin is stored in vesicles that do not contain ATP but, instead, contain a specific protein that binds to 5-HT (serotonin-binding protein) with high affinity in the presence of Fe^{2+}. Release of serotonin is thought to occur via exocytosis, and its rate is determined by the firing rate of serotoninergic soma in the raphe nuclei. Synaptic actions of 5-HT are terminated by binding of these molecules to specific transporter proteins on the serotoninergic neurons.

Changes in the 5-HT function have been implicated in schizophrenia, migraine, sleep, anxiety, and affective disorders. Neurotransmission at 5-HT receptors may be blocked by antidepressants (e.g., fluxetine); hallucinogens (e.g., LSD, anxiolytics [e.g., buspirone]); antiemetics (e.g., ondansetron), and antimigraine drugs (e.g., sumatriptan).

Serotonin receptors generally operate via a GTP-binding (G) protein and are classified, according to their coupling, to second messengers and their amino acid sequence homology. The $5\text{-}HT_1$ group of receptors are negatively coupled to adenylate cyclase via the G_i family of proteins, exhibiting high binding affinity to $[^3H]$-5-HT and mediating inhibition.

The 5-HT_1 family of receptors is further classified into 5-HT_{1A}, 5-HT_{1B}, 5-HT_{1D}, 5-HT_{1E}, and 5-HT_{1F}. The 5-HT_{1A} receptors, which mediate emotion, are localized mainly in the hippocampus (CA_1 sector), amygdala, neocortex, hypothalamus, and raphe nuclei. Neuronal hyperpolarization of this group of receptors is achieved by inhibiting adenylyl cyclase activity/and or opening of the potassium channels via G_i protein or by inhibition of calcium channels via G_O protein. 5-HT_{1A} autoreceptor agonists such as R(+)-8-hydroxy-2-(di-n-propylamino)-tetralin (R (1)8-OH-DPAT) are effective in inhibiting the neuronal activities of the midbrain raphe nuclei and the CA_1 pyramidal cells of the hippocampal gyrus. Other members of 5-HT_{1A} are also negatively coupled to adenylate cyclase. 5-HT_{1B} also utilizes G_i protein and mediates inhibition of adenylate cyclase. The same principles apply to 5-HT_{1C}, 5-HT_{1D}, and 5-HT_{1E}. Binding sites for [^{3}H]-5-HT in the choroid plexus are termed 5-HT_{1C} subtype, whereas the binding sites for [^{3}H]-5-HT in the bovine brain are designated as 5-HT_{1D} subtype. Inositol phosphate is liberated by the activation of 5-HT_{1C}, which leads to the opening of the calcium-dependent chloride channel.

The 5-HT_2 group contains receptors that maintain amino acid homology and are coupled to phospholipase C possibly via G_q. They can selectively be blocked by ketanserin and ritanserin and may produce excitatory effects, displaying high affinity to H^3-spiperone. 5-HT_2 antagonists may block the excitatory effects of glutamate and 5-HT receptors in the facial motor nucleus. This group of receptors is comprised of 5-HT_{2A}, 5-HT_{2B}, and 5-HT_{2C}. Prolonged stimulation of the 5-HT_{2B} and 5-HT_{2C} receptors may produce reduction in receptor density or sensitivity. Sustained administration of 5-HT antagonists may result in downregulation of both HT_{2A} and 5-HT_{2C}.

5-HT_3 is a member of ligand-gated ion channel that functions independently of G protein. They are densely populated at the nerve terminals in the entorhinal and frontal cortices, hippocampus, and area postrema, as well as the peripheral nervous system. These receptors are excitatory in the peripheral, enteric, and autonomic nervous systems, facilitating membrane depolarization to serotonin. They resemble the nicotinic cholinergic receptors, mediating fast synaptic transmission.

Additional receptors within the 5-HT family that are positively coupled to the adenylate cyclase include 5-HT_4, 5-HT_6, and 5-HT_7. 5-HT_4 receptor–binding sites are identified in the striatum, substantia nigra, olfactory tubercle, and atrium. They may mediate striatal dopamine release by s-hydroxytryptamine. However, 5-HT_5 receptors do not couple to adenylyl cyclase and consist of 5-HT_{5A} and 5-HT_{5B}. The 5-HT_{5A} mRNA transcripts are localized in the hippocampal gyrus, granule cells of the cerebellum, medial habenular nucleus, amygdala, thalamus, and olfactory bulb, whereas those of 5-HT_{5B} mRNA are found in the dorsal raphe nucleus, habenula, and hippocampus. 5-HT_6 receptor mRNA has been detected in the striatum, olfactory tubercle, and nucleus accumbens. 5-HT_7 is identified as a receptor in the vascular smooth muscle cells and astrocytes of the frontal cortex.

5-HT_6 receptors exhibit high affinity for LSD and antipsychotic and tricyclic antidepressants, such as clozapine, clomipramine, mianserin, and ritanserin, and positively couple to adenylate cyclase.

Uptake of 5-HT is accomplished by an active process that is temperature dependent and utilizes Cl^- and Na^+ and their remains. Therefore, inhibitors of Na/K ATPase may impair the uptake process of serotonin. Tricyclic antidepressants, such as imipramine and amitriptyline, inhibit reuptake of both serotonin and norepinephrine. Therefore, selective serotonin reuptake inhibitors (SSRIs) may also be used for the treatment of clinical depression. It has been suggested that serotonin reuptake inhibitors may alleviate symptoms of obsessive–compulsive disorder (OCD). Atypical antipsychotic drugs such as clozapine may inhibit central dopaminergic neurons but primarily maintain more powerful antagonistic action on 5-HT_{2A} and 5-HT_{2C} receptors. Agonists at 5-HT_{2A} and 5-HT_{2C} receptors may be responsible for the hallucinogenic activity of certain drugs, such as lysergic acid diethylamide (LSD). Sumatriptan, an effective medication in the treatment of migraine, is thought to derive its therapeutic role from the agonist action that displays for the 5-HT_{1D} and 5-HT_{1F} family of receptors. Depletion of the 5-HT content of serotonin is correlated with the tranquilizing action of reserpine, a hypotensive drug, which also depletes norepinephrine and dopamine contents.

Chemotherapeutic agents, such as cisplatin and dacarbazine, induce severe forms of nausea and vomiting. This results from a series of events that involve release of 5-HT from the chromaffin tissue of the gastrointestinal tract and the enteric nervous system. Released 5-HT specifically activates 5-HT_3 receptors, which are ligand-gated ion channel receptors, producing depolarization of the afferent nerves and increasing their firing rates. This eventually leads to activation of the chemoreceptor (emetic) trigger area. Thus, antagonistic agents that act on 5-HT_3 receptors in the GI tract, such as ondansteron and granisteron, and not on the emetic center, break these series of events and produce relief from nausea and vomiting.

Dahlstrom and Füxe described nine groups of serotoninergic neurons that are designated as B_1 through B_9. B_6 and B_7 represent the dorsal raphe nucleus; B_8 corresponds to the median raphe (superior central) nucleus, whereas B_9 forms part of the ventrolateral tegmentum of the pons and midbrain. B_1–B_4 groups are localized in the midpons through the caudal medulla, and they project mainly to the spinal cord. B_1, which is localized in the caudal part of the ventral medulla, has no known projections. B_3, which corresponds to the raphe magnus, projects to the spinal laminae I and II and the intermediolateral column. The B_2 group, also known as the nucleus raphe obscurus, projects to lamina IX of the spinal gray column, whereas B_6–B_9 nuclear groups project to the telencephalon and diencephalon. The B_8 group appears to project largely to the limbic system, whereas B_7 maintains specific projections to the neostriatum, thalamus, and cerebral and cerebellar cortices. Ascending serotoninergic projections from the rostral midbrain form the *dorsal periventricular tract* and the *ventral tegmental radiation,* which join the medial forebrain bundle and the dopaminergic and noradrenergic projections in the hypothalamus. The dorsal raphe nucleus projects to the striatum, whereas the median raphe nucleus sends fibers to the hippocampus, septal area, and hypothalamus. A somatotopic representation of the ascending serotoninergic projections exists in which the rostral and lateral parts of the dorsal raphe nucleus predominantly project to the frontal cortex. It is well established that serotoninergic neurons produce a combination of depolarization and increased membrane resistance of the neurons of the facial motor nucleus, which enhance the response of these neurons to other excitatory input.

Serotonin is implicated in the regulation of the slow phase of sleep, pituitary functions, and activities of the limbic system (behavior, thermoregulation, mood, and memory). It also plays an important role in the inhibition of pain transmission. Medullary raphe nuclei exert analgesic influence via projections to the spinal cord, whereas pontine and mesencephalic neurons contribute to the ascending reticular activating system via its rostral projections.

Destruction of the raphe nuclei and subsequent depletion of serotonin may produce reversible insomnia. The use of serotonin blockers such as ergotoxin, tricyclic antidepressants, MAO inhibitors, and sumatriptan in the treatment of migraine headache may support the role of serotonin in this condition.

Neurons of the dorsal raphe nucleus seem to be more prone to neurotoxicity generated by certain amphetamine derivatives, such as D-fenluramine, 3,4-methylenendioxymethamphetamine (MDMA or ecstasy), or parachloamphetamine (PCA). 5-hydroxyindoleacetaldehyde (5-HIAA), a serotonin metabolite, shows reduction in the CSF of MDMA users. Blocking the serotonin transport systems may avert this neurotoxicity. In contrast, the median raphe nucleus appears to be unaffected by these neurotoxic effects.

Histamine

Histamine acts both as a neurotransmitter and a neuromodulator in the CNS, mainly occupying the midbrain, tuberal and mammillary nuclei of the hypothalamus, and their extensions in the tuberoinfundibular and principal mammillary tracts, respectively. In the hypothalamus, it coexists with substance P (SP), met-enkephalin, and GAD. Nonneuronal histamine is contained in a substantial amount in the mast cells, where it is depleted by mast-degranulating drugs. Histamine exerts its influences on autonomic activity, temperature regulation, food and water intake (suppressant), vestibular function (may mediate motion sickness), sleep–wake cycles, and neurohumoral mechanisms (release of vasopressin, prolactin, adrenocorticotropic, etc.). It maintains the ability to excite CNS neurons and may be involved in locomotor and exploratory behavior, as well as diurnal changes in other CNS functions. Learning and retention of information may be enhanced by histamine. Histamine release is mostly nonsynaptic and widely diffuse, and release of neuronal histamine may be increased by stimulation of the D_2 dopaminergic and some serotoninergic and NMDA receptors. Histaminergic neurons project to glial cells, blood vessels, neurons, as well as capillary networks. Histamine receptors are identified as H_1, H_2, and H_3 according to their order of detection. H_1 receptors are shown to be involved in hormonal release, food intake, increased free calcium ion concentration, contraction of the smooth muscles, and increased capillary permeability. In the ventrolateral hypothalamus, H_1 receptor is involved in wakefulness. Both H_1 and H_2 are involved in regulation of pituitary gland function, whereas H_2 receptors may mediate endogenous analgesia.

It has been suggested that histamine may alter the blood–brain barrier, suppress the immune system, and produce certain vascular changes, contributing via these neurotoxic effects to the development of certain neurodegenerative diseases, such as multiple sclerosis, Alzheimer's disease, and Wernicke's encephalopathy. The H_1 receptor–mediated effect of histamine reduces seizure activity. Therefore, H_1 antagonists increase seizure onset and/or duration. H_1 antagonists that induce sedation include diphenhydramine and mepyramine, as well as meclizine (anti–motion sickness medication). H_2 receptors are involved in inhibition of gastric secretion, positive ionotropic and chronotropic effects upon the cardiac muscles, and inhibition of contraction of the smooth muscles. Therefore, H_2 antagonists (cimetidine and ranitidine) reduce gastric secretion and thus can be used for the treatment of gastric and duodenal ulcers. On the other hand, H_3 receptors may be involved in the regulation of histamine release and inhibition of acetylcholine, dopamine, norepinephrine, and other peptides.

NEUROPEPTIDES

Neuropeptides are substances that arise from inactive precursors. Their synthesis on ribosomes in the perikaryon or dendrites of a neuron is regulated by mRNA and packaged for release in the endoplasmic reticulum. Their eventual release is by a calcium-dependent process. Neuropeptides include the enkephalins, endorphins, SP, cholecystokinin (CCK), and hypothalamic peptides. Many peptides such as bradykinin, somatostatin, gastrin, secretin, and vasoactive polypeptide (VIP) are also shown to act upon the autonomic intestinal neurons and enteric nervous system, which will be discussed later with the autonomic nervous system. In view of the vast amount of information available in this area, the discussion will be restricted only to certain peptides.

Enkephalins

Enkephalins are pentapeptides that are present in the interneurons of the substantia gelatinosa, nucleus raphe magnus, and small intestine. Enkephalinergic neurons modulate pain via presynaptic inhibition upon afferents in the brainstem and spinal cord. They are classified into methionine enkephalin and leucine enkephalin.

Endorphin

Endorphins are peptides (naturally occurring opiates) consisting of C-terminally extended forms of Leu-enkephalin that bind to opiate receptors in the brain and induce analgesia similar to morphine. These receptors have the ability to bind to opiate agonists (e.g., morphine) or to antagonists (e.g., naloxone). Endorphins may be implicated in states of depression and generalized convulsions. They are derived from different genes, and they are classified into dynorphin-A, dynorphin-B, and α and β-endorphins. β-endorphins are contained in neurons within the diencephalon and pons and may be involved in the acquired intellectual deterioration in adults.

Substance P

Substance P is an 11-amino-acid oligopeptide that is present in the nerve endings of the unmyelinated class C or myelinated Aδ fibers, which carry nociceptive (painful) stimuli to the substantia gelatinosa. Therefore, axotomy may reduce the level of this peptide. SP is found in the dorsal root ganglia, gastrointestinal tract, and sensory ganglia of the cranial nerves, spinal trigeminal nucleus, basal nuclei, nucleus raphe magnus, periaqueductal gray, and hypothalamus. Other peptides that relate closely to SP are neurokinin A and neurokinin B with their specific receptors Nk_1 and NK_2. The neurokinin A gene is located on chromosome 7, whereas the gene for neurokinin B is positioned on chromosome 12. SP exerts a more powerful effect than both neurokinins at NK_1; however, it remains less potent at NK_2.

Cholecystokinin (CCK)

CCK is a neuromediator concentrated in the amygdala, hypothalamus, cerebral cortex, periaqueductal gray matter, and spinal dorsal gray columns. It coexists with other peptides in the substantia nigra, ventrotegmental area, and medulla. CCK_a and CCK_b are known receptors for CCK, with CCK_b being the predominant receptor in the brain. CCK_a is found in the nucleus accumbens septi, posterior hypothalamus, and area postrema of the medulla.

CCK is a neuropeptide that acts on the gastrointestinal tract and functions to stimulate the pancreatic secretions including insulin and glucagon. It has been shown that the level of CCK is low in bulemics and high in patients with celiac sprue. Because of its role in producing relaxation of the lower esophageal sphincter, CCK_a receptor antagonists may help reduce manifestations of gastroesophageal reflux disease (GERD). Due to CCK's role in digestion, CCK receptors theoretically can be blocked to reduce intestinal motility and alleviate some of the symptoms associated with chronic digestive disorders, such as irritable bowel syndrome. Control of obesity and increase of appetite may be achieved by use of CCK agonists and CCK receptor blockade respectively.

Hypothalamic Peptides

Hypothalamic peptides are 3- to 14-amino-acid peptidases, including thyrotropin-releasing hormone, somatostatin, corticotropin-releasing factor, melanocyte-stimulating factor, and luteinizing hormone–releasing factor.

SUGGESTED READING

Beckmann J, Lips KS. The non-neuronal cholinergic system in health and disease. *Pharmacology* 2013;92(5–6):286–302.

Cimino M, Marini P, Fomasafi D et al. Distribution of nicotinic receptors in cynomolgus monkey brain and ganglia: Localization of a3 sub-unit mRNA, a-bungarotoxin and nicotine binding sites. *Neuroscience* 1992;51:77–86.

Cohen SP, Mao J. Neuropathic pain: Mechanisms and their clinical implications. *BMJ* 2014;348:f7656.

de Castro-Neto EF, da Cunha RH, da Silveira DX, Yonamine M, Gouveia TL, Cavalheiro EA, Amado D, Naffah-Mazzacoratti Mda G. Changes in aminoacidergic and monoaminergic neurotransmission in the hippocampus and amygdala of rats after ayahuasca ingestion. *World J Biol Chem* 2013;4(4):141–7.

Dougherty PM, Paleck J, Zorn S. Combined application of excitatory amino acids and substance P produces long lasting changes in responses of primate spinothalamic tract neurons. *Brain Res Rev* 1993;18:227–46.

Finle PR. Selective serotonin reuptake inhibitors. *Ann Pharmacother* 1994;28:1359–69.

Foster AC, Kemp JA. Glutamate- and GABA-based CNS therapeutics. *Curr Opin Pharmacol* 2006;6(1):7–17.

Hills JM, Jessen KR. Transmission: Gamma-amino butyric acid (GABA), 5-hydroxytryptamine (5-HT) and dopamine. In Burnstock G, Hoyle CHV, eds. *Autonomic Neuroeffector Mechanisms*. Switzerland: Harwood Academic Publishers, 1992, 465–507.

Horder J, Lavender T, Mendez MA, O'Gorman R, Daly E, Craig MC, Lythgoe DJ, Barker GJ, Murphy DG. Reduced subcortical glutamate/glutamine in adults with autism spectrum disorders: A [(1)H]MRS study. *Transl Psychiatry* 2014 18;4:e364.

Li K, Xu E. The role and the mechanism of γ-aminobutyric acid during central nervous system development. *Neurosci Bull* 2008;24(3):195–200.

Marcondes Sari MH, Guerra Souza AC, Gonçalves Rosa S, Souza D, Dorneles Rodrigues OE, Wayne Nogueira C. Contribution of dopaminergic and adenosinergic systems in the antinociceptive effect of p-chloro-selenosteroid. *Eur J Pharmacol* 2014;725:79–86.

Pin JP, Duvoisin R. Neurotransmitter receptors I. The metabotropic glutamine receptors: Structure and function. *Neuropharmacology* 1995;34:1121.

Ramírez-León V, Ulfhake B, Arvidsson U, Verhofstad AA, Visser TJ, Hökfelt T. Serotoninergic, peptidergic and GABAergic innervation of the ventrolateral and dorsolateral motor nuclei in the cat S1/S2 segments: An immunofluorescence study. *J Chem Neuroanat* 1994;7(1–2):87–103.

Rheims S, Holmgren CD, Chazal G, Mulder J, Harkany T, Zilberter T, Zilberter Y. GABA action in immature neocortical neurons directly depends on the availability of ketone bodies. *J Neurochem* 2009;110(4):1330–8.

Strzelecki D, Szyburska J, Rabe-Jabłońska J. Two grams of sarcosine in schizophrenia - is it too much? A potential role of glutamate-serotonin interaction. *Neuropsychiatr Dis Treat* 2014;10:263–6.

Szabadics J, Varga C, Molnár G, Oláh S, Barzó P, Tamás G. Excitatory effect of GABAergic axo-axonic cells in cortical microcircuits. *Science* 2006;311(5758):233–5.

Valko PO, Gavrilov YV, Yamamoto M, Reddy H, Haybaeck J, Mignot E, Baumann CR, Scammell TE. Increase of histaminergic tuberomammillary neurons in narcolepsy. *Ann Neurol* 2013;74(6):794–804.

van Ree JM, Matthysse S. Psychiatric disorders: Neurotransmitters and neuropeptides. *Prog Brain Res* 1986;65:1–228.

Yadav V, Ryu J-H, Suda N, Tanaka KF, Gingrich JA, Schütz G, Glorieux FH, Chiang CY et al. Lrp5 Controls bone formation by inhibiting serotonin synthesis in the duodenum. *Cell* 2008;135(5):825–37.

Section V

Special Somatic Sensations

13 Visual System

The visual system is a special somatic afferent system, which receives, processes, and recognizes visual impulses with the associated memories. It forms binocular images and regulates associated reflexes. It is the only sensory system that is totally dependent upon the integrity of the cerebral cortex. In order for the visual images and associated memory to be constructed, visual impulses must pass through a chain of structures and neurons that are located in the eye and the visual pathway, encompassing the cornea, iris, anterior and posterior chambers of the eye, vitreous body, retina, optic nerve, optic tract, thalamus, visual radiation, and visual cortex. The visual system works closely with the vestibular neurons and gaze centers and cortical eye fields to ensure that eye movements are coordinated and that steady, proper binocular images are formed. Pathologic processes produces visual deficit that may be total or partial, confined to one or both visual fields.

PERIPHERAL VISUAL APPARATUS

The eyeball is a visual organ that is situated in the bony orbital cavity. It is surrounded by the orbital fat and separated by a thin fibrous (Tenon's) capsule.

EYEBALL

The eyeball (Figure 13.1) consists of an outer fibrous, an intermediate vascular, and an inner neuronal layer.

Tunica Fibrosa

The tunica fibrosa, the outermost layer of the eyeball, consists of the cornea and retina.

Cornea

The cornea (Figure 13.1) is an avascular structure that forms one-sixth of the fibrous tunic and represents the main refractive medium of the eyeball. It has no lymphatics, receives a rich nerve supply from the long ciliary nerves, and is highly resistant to infection. It forms the anterior wall of the anterior chamber of the eye and joins the sclera at the corneal–scleral junction.

> A Kayser–Fleischer ring, a greenish-brown or golden brown pigmentation around the corneoscleral junction, is formed by the deposition of copper in the Descemet's membrane and is seen in Wilson's disease (hepatolenticular degeneration). See also the subcortical motor system.

> The corneoscleral junction is marked by the Schlemm canal, a venous channel that receives the aqueous humor from the anterior chamber of the eye. Occlusion of this canal may lead to accumulation of aqueous humor, increased intraocular pressure, and the resultant glaucoma. This is discussed in detail later in this chapter.

Sclera

The sclera (Figure 13.1) is a fibrous structure that preserves the shape of the eyeball, resists intraocular pressure, and provides a smooth surface for eye movements, giving attachment to the extraocular muscles. It is continuous anteriorly at the limbus with the connective tissue stroma of the cornea and posteriorly with the dural sheath of the optic nerve. At the lamina cribrosa sclera (weakest part of this layer), the sclera is pierced by fibers of the optic nerve as well as the posterior ciliary vessels.

Tunica Vasculosa

The tunica vasculosa represents the intermediate layer of the eyeball, consisting of the choroid layer, ciliary body, and iris (Figure 13.1).

Choroid Layer

The choroid is a vascular layer that consists mainly of arteries and veins with the veins joining together to form four or five vorticose veins that drain into the anterior ciliary veins. The arteries within this layer are formed by the ciliary branches of the ophthalmic artery that extend to the iris, form the major and minor arterial iridal circles, supplementing the central retinal artery in providing blood supply to the retina.

Ciliary Body

The intermediate part of the tunica vasculosa is known as the ciliary body, which extends from the lateral end of the iris to the ora serrata (site of junction of the light-sensitive and nonsensitive parts of the retina). It consists of ciliary muscles and processes, giving attachment to the suspensory ligaments of the lens (zonular fibers). When viewing distant objects, the ciliary muscles are relaxed, while the zonular ligaments are stretched and taut. Viewing near objects is enabled by the contraction of the ciliary muscles accompanied by movement

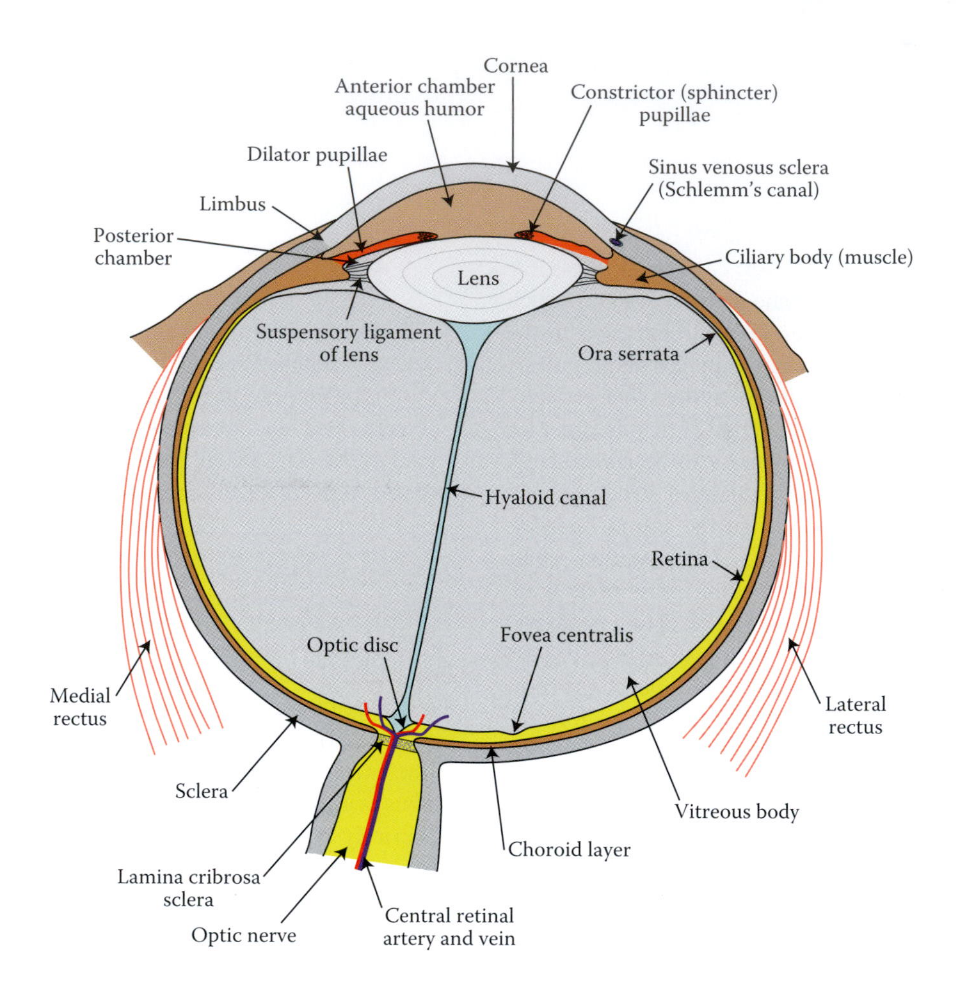

FIGURE 13.1 Section of the eyeball and associated layers. The neuronal organization of the retina is also illustrated.

of the ciliary bodies toward the iris and relaxation of the suspensory ligaments, resulting in an increase in lens curvature. The ciliary processes that give attachments to the zonular fibers also secrete the aqueous humor into the posterior chamber by active transport and diffusion from the capillaries.

Lens

The lens (Figure 13.1), a main component of the eye chambers, is a biconvex, colorless, avascular structure, derived from the surface ectoderm, and positioned between the iris and the anterior chamber. It lies posterior to the iris, embedded in the hyaloid fossa, receiving the suspensory ligaments of the lens. It is functionally similar to the cornea with a less refractive (diopteric) power. It has a transparent elastic capsule, a cortical zone, and a nucleus. The anterior and posterior poles represent the most convex parts of the lens. Changes in the lens curvature are regulated by the suspensory ligaments of the lens and the muscles of the ciliary body.

In some individuals, the lens may be absent as a developmental anomaly (*primary aphakia*) or as a result of degeneration (*secondary aphakia*). Corneal opacity and cataracts of the anterior lens are seen in *congenital anomaly of Peter*, in which gradual impairment of vision (misty appearance of visual objects) and diplopia are observed. The opacity may be confined to the nucleus of the lens (central cataract), producing myopia and poor vision during the day and better vision in dim light. Peripheral cataracts result in poor vision in dim light and better vision in bright daylight. Congenital cataracts may be seen at birth due to metabolic or chromosomal abnormalities, infection, or maternal diseases.

The lens may be affected by a myriad of clinical conditions, which include presbyopia and sunflower cataract.

Presbyopia develops as a result of aging and by the conversion of the lens into a less pliable structure, rendering it less reactive to contraction of the ciliary muscles. Presbyopic patients are hyperopic, exhibit difficulty in reading fine print, and endure the inconvenience of holding reading materials farther away to achieve optimum vision.

Sunflower cataract (chalcosis lentis) is a condition that is seen in Wilson's disease due to the impregnation of the subcapsular area of the lens with the radiating metallic greenish-brown opacity.

Iris and Pupil

The iris (Figure 13.1) forms the adjustable diaphragm of the eye that encircles the pupil, consisting of circular muscle fibers (constrictor pupilla) and radial fibers (dilator pupilla muscle), pigment cells, and epithelium. The constrictor pupilla, which acts as a pupillary sphincter, is innervated by the postganglionic parasympathetic fibers, whereas the dilator pupilla receives innervation from the postganglionic sympathetic fibers. The iridal pigment is the same in all individuals; however, the amount of pigment and the pattern of its distribution determine the eye color. Failure of closure of intraocular fissure during development due to genetic abnormality results in coloboma (defect or absence) of the iris, choroid layer, retina, and optic disc. The pupil, an opening in the center of the iris, may exhibit abnormalities such as miosis, mydriasis, aniscoria, Argyll Robertson pupil, hippus, Adie (tonic) pupil, and Marcus Gunn pupil.

Miosis refers to constriction of the pupil that is commonly seen in Horner's syndrome. Bilateral miotic pupils may result from metabolic encephalopathies, destructive pontine lesions, or opiate use. Miosis of pontine origin may be due to disruption of the descending sympathetic pathways. Disruption of the efferent sympathetic fibers in the carotid sheath, near the apex of the lung or base of the neck, may also produce unilateral miosis.

Mydriasis (dilatation of the pupil) is manifested in oculomotor palsy as a result of the unopposed action of the dilator pupilla muscle. This sign may be seen subsequent to aneurysms of the posterior communicating, superior cerebellar, and posterior cerebral arteries, and also in uncal herniation. Traumatic mydriasis is usually unilateral, occurs in response to direct trauma, and may not be accompanied by ocular muscle dysfunction. Bilateral mydriasis may be seen in trauma patients with poor vascular perfusion subsequent to hypotension or increased intracranial pressure. Prompt restoration of pupillary response may follow adequate vascular perfusion. Patients with Cheyne–Stokes respiration may exhibit mydriasis in hyperventilation phase and miosis in apneustic phase.

Anisocoria refers to unequal pupils, in which one eye may show constriction. Physiological anisocoria occurs in 20% of the population, exhibiting mild difference (up to 2 mm) in pupil size. Determination of whether the smaller or larger pupil is abnormal may require comparing pupil size in the dark and the room light. Sympathetic anisocoria will be more marked in dim light due to the subnormal constriction of the affected (small) pupil. This may result from iritis, disruption of the cervical sympathetics, or application of miotic (miosis-inducing) medications. Parasympathetic anisocoria will be evident in room light since the affected (larger) pupil constricts subnormally. This type of aniscoria may be seen in individuals with oculomotor palsy or glaucoma or as a result of application of mydriatic (mydriasis-inducing) drugs such as atropine.

Argyll Robertson pupil is characterized by the inability of the pupil to constrict in response to light, while remaining briskly responsive (constricts) in accommodation (light-near dissociation). This condition occurs in neurosyphilis, diabetes mellitus, and severe vitamin B deficiency. The lesion is thought to be located in the rostral midbrain medial to the lateral geniculate nucleus (LGN), disrupting the afferent limb of the pupillary light reflex, while preserving the afferent limb of the accommodation reflex. Reverse Argyll Robertson pupil is seen in syphilis and Parkinson's disease.

Hippus is a phenomenon in which the pupil exhibits spontaneous, intermittent rhythmical constriction and dilatation. Although the diagnostic value is questionable, this condition may be associated with hysteria, multiple sclerosis (MS), brain abscess, and Cheyne–Stokes respiration.

Adie (tonic) pupil is characterized by ipsilateral mydriasis and loss of constriction in accommodation without other signs of oculomotor palsy. It occurs subsequent to viral or bacterial diseases that disrupt the ipsilateral postsynaptic parasympathetic neurons of the ciliary ganglion. Random regeneration of these fibers produces a poorly reactive and mildly dilated pupil, which constricts slowly in response to near vision (accommodation) and then dilates also in a slow manner. This syndrome, mainly seen in young females, is also associated with hyperreflexic deep tendons, disorders of sweating, photophobia, and hyperopia due to loss of accommodation.

Marcus Gunn pupil (*relative afferent pupillary* defect (RAPD) manifests normal consensual reflex upon application of flashlight to the intact eye, but show dilatation when the light is quickly passed from the intact to the affected eye (*positive swinging flashlight test*). It has been suggested that the visual pathway on the affected side mistakenly responds to this reduction in light stimulation by pupillary dilation as if the light itself is less luminous. It is observed in individuals with ipsilateral retrobulbar neuropathy or a lesion of the optic nerve or retina. The consensual reflex is preserved, but the depth of perception of moving and colored objects is lost. The visual deficits are exacerbated by exercise or by any efforts that increase the body temperature (Uhthoff's sign). The latter sign is due to the possible change in the conduction of the affected nerve or variation in the sodium and potassium concentration around the myelin of the optic nerve following physical activity. Ocular abnormality, such as ptosis and retraction (winking) of the eyelid on the affected side is also seen in Marcus Gunn jaw-winking syndrome in response to opening the mouth or deviation of the jaw. This condition should not be confused with Marcus Gunn pupil.

Anterior Chamber of the Eye

The anterior chamber (Figure 13.1) of the eye is bounded anteriorly by the cornea, and posteriorly by the iris and the lens. It contains the aqueous humor, which crosses the trabecular network to gain access to the irideocorneal angle and the canal of Schlemm. The area between the iris anteriorly and the lens and zonular fibers posteriorly represents the posterior chamber. Both eye chambers communicate via the pupil, containing the aqueous humor, which maintains the intraocular pressure. The aqueous humor also serves as a path for metabolites from the cornea and lens, carries nutrients such as glucose, and plays a role in respiratory gaseous exchange. Failure of this communication may result in glaucoma.

Glaucoma is a condition in which the intraocular pressure is elevated independent from any other diseases of the eye (primary glaucoma) or as a result of ocular diseases (secondary glaucoma). Primary (chronic or open-angle) glaucoma may be congenital or acquired and may result from obstruction at the canal of Schlemm, aqueous veins, or trabecular meshwork at the irideocorneal angle.

Open-angle glaucoma is the most common form of glaucoma, which is produced by a gradual increase in intraocular pressure, accompanied by a gradual loss of peripheral vision, ending in total blindness. It usually begins in the fourth or fifth decades of life in individuals with familial history of a variety of glaucoma. Symptoms are absent at the onset, and its diagnosis may be confirmed by examination of the fundus of the eye and detection of increased intraocular pressure. In later stages of this disease, the optic cup is abnormally deep and permanently put out of function. A lack of clear symptoms in the initial stages of this disease is an important indication that regular ophthalmologic examination is highly recommended for individuals over the age of 40. Pilocarpine, which constricts the pupil and increases the outflow of the aqueous humor through the irideocorneal angle, may be used topically to treat this condition. Timolol (Timoptic) may reduce the production of aqueous humor, but its side effects on the cardiovascular system may make it less desirable medication. Marijuana may also lower intraocular pressure in this type of glaucoma.

Closed-angle glaucoma results from increased intraocular pressure subsequent to adhesion of the iris to the cornea and closure of the *irideocorneal* angle. It may not always be spontaneous, but iatrogenic, resulting from the application of medications that dilate the pupil and block the irideocorneal (filtration) angle by the iris itself. *Tricyclic* antidepressants with anticholinergic properties may precipitate this condition. Patients are usually older than 40 years, with a family history of glaucoma. Few patients may complain of seeing halos around lights. This condition may be acute or chronic.

Acute (*closed*) angle glaucoma is produced by the sudden obstruction of aqueous humor circulation, which produces pain and visual impairment of the affected eye. In this condition, the eye appears red, the cornea seems hazy, and the blood vessels are dilated.

Chronic (*closed*) angle glaucoma occurs as a result of gradual obstruction of the *irideocorneal angle*, showing similar signs to acute closed-angle glaucoma. Topical and systemic treatments and even laser iridectomy may have to be promptly applied to reduce production of aqueous humor. Glaucomas may be associated with lesions involving the peripheral retina or the optic nerve, producing *peripheral scotoma* (focal blindness in the form of dark or colored spot).

The vitreous body (Figure 13.1) is a clear and colorless substance, occupying most of the eyeball posterior to the lens. It primarily consists of water, hyaluronic acid, trace amounts of mucoproteins, and some salts. It contains fibrils, which may be visible as floating objects. The hyaloid membrane surrounds the vitreous body and thickens at the ora serrata of the retina to form the ciliary zonule. The ciliary zonule forms the suspensory ligaments of the lens. The hyaloid canal pierces the vitreous body and stretches from the optic disc to the posterior pole of the lens.

REFRACTIVE DISORDERS

In a normally relaxed eye (emmetropia), no optical defects exist, and it is adapted to the far vision. The lens is flat, and the suspensory ligaments of the lens (zonular fibers) are tense (Figures 13.1 and 13.3). However, near vision requires the use of accommodative power that includes an increase in lens curvature, constriction of the pupil, and convergence. The increase in lens curvature shortens the focal distance and allows images from closer objects to fall on the retina. The changes in near vision occur when the eyeball has a normal anteroposterior dimension and functional refractive media within normal range. An abnormally long or short eyeball, an uneven curvature of the refractive media, or a nonfunctional or overly functional accommodative apparatus may produce a variety of disorders, including myopia, hyperopia, astigmatism, presbyopia, and anisometropia.

- Myopia (nearsightedness) is an optical disorder characterized by the inability to see far objects. It may be due to an abnormally long eyeball axis or a refractive power that is too strong (Figure 13.3). Therefore, the image from a distant object falls anterior to the retina. As the object moves closer, the focal point moves back to the same degree until the image is spotted on the retina and clear vision is achieved. This condition may be associated with rhegmatogenous retinal detachment, in which the retina is detached and broken up into pieces. Myopia is corrected by a concave lens.

- Hyperopia (farsightedness) is the most common optical disorder associated with an abnormally short axis of the eyeball or weak refractive power (Figure 13.2). Due to shortness of the optical axis, images from distant objects fall behind the retina in the relaxed eye. However, accommodation focuses the images on the retina, allowing clear vision of distant objects. A convex lens corrects hyperopia.
- Astigmatism results from an uneven curvature of the refractive surfaces (egg-shaped), leading to a change in the angle of refraction of the horizontal and perpendicular light rays and subsequent focusing of these rays on different spots on the retina. As a result, blurred vision ensues. A cylindrical lens, which is concave or convex on one axis and flat on other, corrects this disorder.
- Presbyopia, a condition that develops with aging, is characterized by an inelastic, hard, and less pliable lens, which lacks or has limited power of accommodation. Individuals with this disorder are hyperopes (farsighted).
- Anisometropia is a rare disorder in which the refractive powers between both eyes remain different.

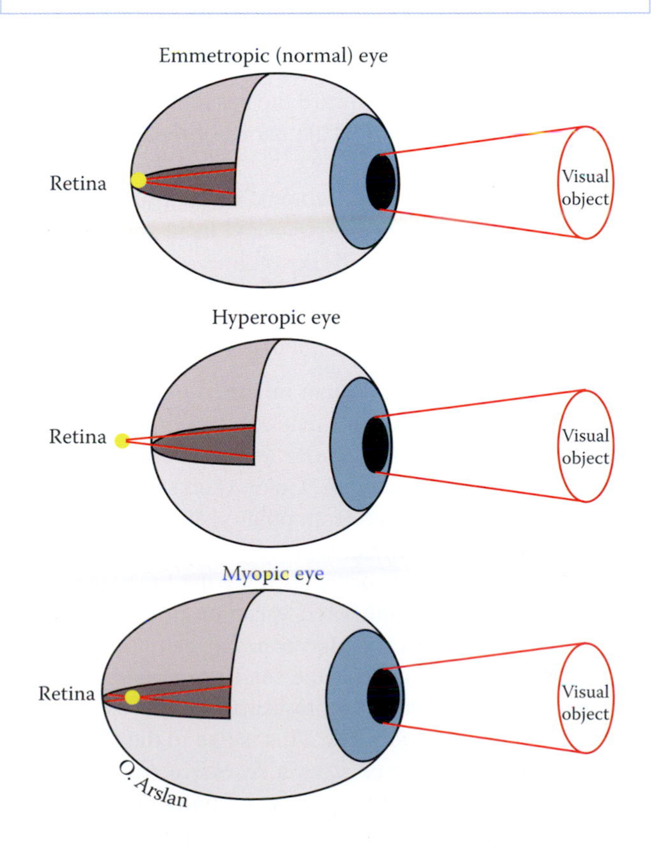

FIGURE 13.2 Refractive defects associated with vision.

Tunica Nervosa

Tunica nervosa consists of the pigment epithelial layer and retina (Figure 13.1). The pigment epithelium is loosely bound to the *retina*, which contains the photoreceptor layer. The pigment epithelial layer offers mechanical support, absorbs excess light, and provides nutrients for photoreceptors.

Retina

The retina (Figures 13.1 and 13.2) is the inner layer of the tunica nervosa that develops as the optic vesicle from the diencephalon, consisting of the pars optica, pars ciliaris, and pars iridica. The par optica joins the par ciliaris at the ora serrata. The pars optica is the only light-sensitive part of the retina, which contains photoreceptors, and it is loosely bound to the pigment epithelium. It designed to detect and transduce images into neuronal signals and further analyze the visual input as it continues through the visual pathways. It consists of 10 layers and 9 types of cells. The retina is incompletely fused to the pigment epithelium and is separated by a potential space.

Detachment of the retina from the pigment epithelium may be complete or focal, occurring as a result of trauma or disease processes. It may be associated with a breakup of the retina (rhegmatogenous detachment) subsequent to direct trauma. Retinal detachment, as is seen in diabetic vitreoretinopathy, is associated with an intact retina that has undergone undue traction by the fibrovascular bundles between the vitreous body and the retina. Retinal detachment may also occur when pathological processes allow exudate derived from the choroid layer to enter the subretinal space (exudative retinal detachment). Retinal detachment may cause blurred vision, light flashes, or the appearance of floating bodies. Cherry-red spots may be seen on fundoscopic examination of the retina in individuals with Tay–Sachs disease.

The peripheral retina and ciliary body join at the ora serrata. At the ora serrata, the sensory layers of the retina and the retinal pigment epithelium fuse, thus limiting the spread of any pathological subretinal fluid. The tunica nervosa contains photoreceptors, which are divided into cones and rods. Cones vary in number from 6–7 million and are distributed among the rods, except in the fovea centralis. They occupy a central position, whereas the rods are localized in the periphery. The optic disc is the site where axons of the ganglionic layer leave the eyeball, the optic nerve is formed, and the photoreceptors are absent. The physiologic cup is the lighter-colored central part of the disc, which is penetrated by retinal vessels. The normal cup-to-disc (c/d) ratio of 1:5 may be lost in glaucoma. The macula lutea, a yellowish spot on the temporal side of the optic disc, contains the fovea centralis. The latter is the site for acute vision occupied only by cones. The cones possess a higher threshold of excitability and have a 1:1 synapse ratio with the dendrites of the bipolar neurons. Cones are specialized for daylight vision (phototopic) and color discrimination.

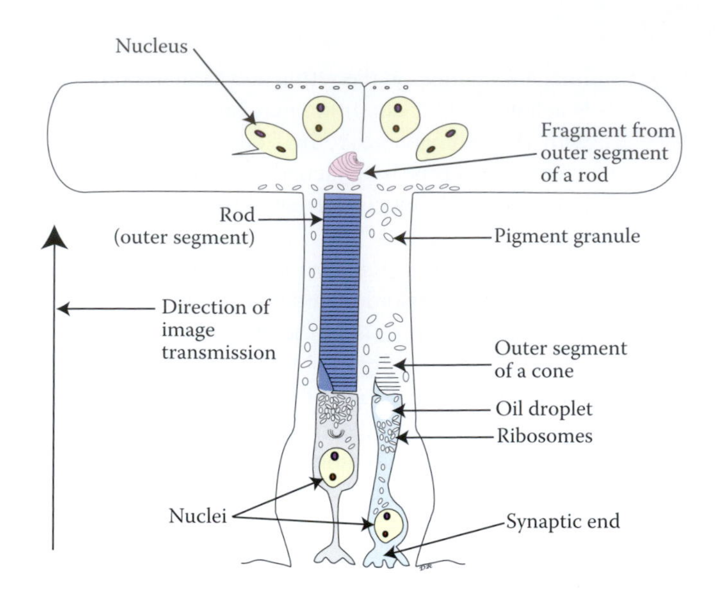

FIGURE 13.3 Structural organization of the photoreceptors.

Macular degeneration is a painless condition that affects persons over the age of 50 in which there is metamorphosia (image distortion), blurred vision, color blindness, loss of contrast sensitivity (contours and shadows are not distinct), central scotomas, and drastic decrease or loss of central vision. The painless nature of this condition makes it unnoticed for some time. It is categorized into nonexudative (dry) and exudative (wet) types. In the nonexudative (dry) type, small, discrete yellowish clusters of cellular debris, "drusen," accumulate around the macula, causing ophthalmoscopically visible lesions that coalesce with advancing condition to form larger lesions that cause atrophy of the retinal pigment epithelial layer and compromise central (acute color discriminating) vision though loss of photoreceptors. Cellular debris also deposits between the retinal pigment epithelium and choroid layer, compromising the nourishment of the photoreceptors and causing retinal detachment. Drusen, which are large and soft, can be associated with high levels of cholesterol deposits. In the exudative (wet) type, which is the less common form of this condition, the abnormal choroidal neovascularization in the choriocapillaries through Bruch's membrane causes leakage of blood and protein below the macula. Scar formation and bleeding can induce irreversible retinal damage. In the exudative type, there will be changes in the visual hyperacuity (ability to identify misalignment of visual objects) before any changes in the visual acuity, and this can be assessed by preferential hyperactivity perimetry. In the advanced stage of the disease, irreversible damage occurs to the photoreceptors and leads, although rarely, to blindness. Advancing age, family history with a relative affected with this condition, and the presence of genes for factor H (CFH), factor B (CFB), and factor 3 (C3) are considered predisposing factors for this disease. Early intervention with anti-vascular endothelial growth factor (VEGF) and angiogenesis inhibitors, such as *ranibizumab* (*Lucentis*) or *bevacizumab* (Avastin), may help avert blindness (Figure 13.3).

A lesion rostral to the optic chiasma involving the foveal part of the retina and the corresponding part of the optic nerve are commonly seen in MS. These lesions may be associated with optic neuritis (inflammation of the optic disc) or retrobulbar neuritis (inflammation of the optic nerve).

Dark or colored spot in the center of the visual field is referred to as central scotoma. The point of exit of axons of the optic nerve from the eyeball marks the optic disc of the retina, commonly known as the blind spot. The focal blind spot attributed to the optic disc is termed physiological scotoma.

Arcuate scotoma is a pathological focal visual deficit, which results from a lesion in the retina or optic nerve fibers. It occurs near the optic disc and arches superiorly or inferiorly toward the nasal field of the retina and in the direction of the axons of the ganglionic multipolar neurons.

Scintillating (flittering) scotoma (teichopsia) is characterized by floating of irregular and lucid spot, sometimes with a zigzag or wall-like outline, which may last for up to 20 to 25 min. It usually occurs secondary to an occipital lobe lesion. It may also be associated with migraine (migraine aura).

Color blindness may be an inherited (sex-linked) or acquired condition. Patients may exhibit blindness to all colors (*achromatopia*) or to one (*monochromatopia*) or two colors (*dichromatopia*). Color vision is mediated by the cones, segregated from other visual information in the retina, and eventually processed in a specialized pathway in the visual cortex utilizing the LGN and the optic radiation. The inherited variation in the amount of photopigments in the blue cones, green cones, and red cones may account for the sex-linked condition of color blindness. This condition affects 8% of males compared to 2% of females. This is due to the fact that red and green genes exist as a recessive trait on the male X chromosome. One percent of males lack the red gene (protanopes—lack the long-wave mechanism), and 2%–3% lack the green gene (deuteranopes—lack the medium-length mechanism). The gene for the blue color is present on an autosome on the seventh chromosome and is rarely affected by mutation. It is thought that all three cone genes maintain a common ancestral red gene. The red gene may have given rise to the blue cone pigment, which, in turn, has given origin to the red and green cone pigments. Trichromats are individuals with normal three-color vision or with one normal color vision and two feeble red vision (protanomaly). Trichromats may also have three feeble green vision (deuteranomaly) or four weak blue vision (tritanomaly). These weaknesses are due to the reduction in the amount of cone pigment and are unrelated to neuronal circuitry associated with processing of the color vision. Dichromats, individuals with two-color vision (color blind) and who lack one of the pigments, may not perceive red (protanopes) due to a lack of erythrolabe, green (deuteranopes) as a result of absence of chlorolabe, or blue (tritanopes—lack the short-wave length mechanism) due to the absence of cyanolabe. Deuteranopia are much more common than *protanopia at a ratio of 3:1.* Gene loss or recombination between genes, which produces a hybrid gene on the X chromosome, may occur in individuals with red–green color blindness. However, disorders of color vision may also be acquired. Pathological conditions that affect the outer layer of the retina may produce blindness to blue color (tritanopia) as a result of loss of the processing mechanism of short-wave length. In this manner, pathological elements that affect the optic nerve and the inner retinal layer may cause loss of red–green color vision.

Rods (Figures 13.2 and 13.3) are the most numerous, averaging between 100 and 130 million per retina. They are peripherally located and activated by lower illumination. Rods visualize black, white, and gray colors under twilight or scotopic (achromatic) vision. In dim light, rods contract to maximize the surface area exposed to the limited light. The outer segments of rods contain discs, which are sloughed and removed by pigment cells. Dendrites of horizontal cells in the inner nuclear layer interconnect cones and rods. The processes of the large glial (Müller) cells, which hold the retinal layers together, constitute the outer limiting membrane. The photoreceptors form synaptic linkage with the dendrites of the bipolar neurons at the outer plexiform layer.

Accumulation of pigment cells and floating discs of the outer segments may account for the retinal degeneration and black-colored lesions in the fundus seen in retinitis pigmentosa. In this disease, which occurs often as a result of genetic mutation for rhodopsin and peripherin, a glycoprotein of the outer segments, disc debris and black clumps of pigment, "bone specules," in the peripheral part of the retina may hinder the diffusion of nutrients from capillaries of the choroid layer to the photoreceptors, thus accounting for the retinal degeneration observed in this blinding disease. Progressive nyctalopia (night blindness) and ring scotoma are conditions in which the center and extreme peripheral part of the retina are spared while the mid portion of the periphery is affected to a great extent.

A small lesion or petechial hemorrhage in the retina near the optic disc produces a focal blindness or scotoma in which central visual acuity is impaired. Vitamin A plays a significant role in vision. Darkness causes vitamin A to undergo reverse changes into *retinin*, which bonds with opsin to form rhodopsin. Nyctalopia (night blindness) is associated with vitamin A deficiency.

The bipolar neurons (Figure 13.1) are depolarizing or hyperpolarizing neurons that represent the primary (first-order) neurons in the visual pathway. Depolarizing (invaginating) biploar neurons are inhibited by darkness. They stimulate the "on"-type ganglionic cells and are released from inhibition by illumination. Hyperpolarizing (flat) neurons, which are inhibited by light, excite the "off"-type ganglionic neurons, maintaining different receptors. The nuclei of these neurons are located in the inner nuclear layer, while the axons are spread in the inner plexiform layer, establishing contacts with the dendrites of the ganglion cells.

The amacrine cells (Figure 13.1), which resemble the granule cells of the olfactory bulb, form inhibitory synapses upon the dendrites of the ganglion cells and maintain reciprocal connections with the bipolar neurons. These cells contain different transmitters and, together with the ganglionic neurons, are the only excitable (produce action potentials) neurons of the retina. The horizontal cells establish inhibitory dendrodendritic synapses with the bipolar neurons, intensifying contrast by inactivating the bipolar and ganglionic neurons.

Ganglionic multipolar neurons (Figures 13.1 and 13.5) form the second-order neurons in the visual pathway, which have the capacity to fire at a fairly steady rate even in the absence of visual stimuli. They are either "on" type or "off" type, dependent upon their synaptic connections with the bipolar neurons. They are classified into sustained X, transient Y, and intermediate W cells. Sustained X cells subserve constant on or off response, analyzing the field in regard to their shapes and colors of objects in the visual field. Transient Y cells are relatively few in number, giving rise to momentary response to rapidly moving objects. Y cells project to the superior colliculus (SC) and thalamus to detect the movement of objects in the visual field. W cells are small and project to the pretectum to mediate the pupillary light reflex. The dendrites of the ganglionic neurons are connected with the axons of the bipolar neurons in the inner plexiform layer. Some ganglionic neurons in the nasal halves of the retina may fixate the visual image on the fovea centralis, preventing the image from slipping off. This occurs when axons of these specific ganglionic neurons project via the midbrain reticular formation to the inferior olivary nucleus. This is followed by projection of the olivocerebellar fibers to the same Purkinje neurons that receive input from the medial vestibular nucleus.

The axons of the ganglionic neurons leave the eyeball through the lamina cribrosa sclera, forming the optic nerve. The retina is divided by the optic axis into a medial (nasal) half and a lateral (temporal) half; a horizontal plane further divides the retina into upper and lower nasal and upper and lower temporal quadrants. The nasal half of the retina of each eye receives visual impulses from the temporal half of the visual field and vice versa. The upper quadrant of the retina of one eye sees the lower quadrant of the contralateral visual field and vice versa.

The retina is dependent upon the arterial supply of the ophthalmic artery (Figure 13.4). This artery arises from the cerebral part of the internal carotid artery medial to the anterior clinoid process. It enters the orbit via the optic canal accompanied by the optic nerve. It supplies the eye, forehead, dura matter, ethmoidal sinuses, and nasal cavity. In the orbit, it frequently runs superior and then medial to the optic nerve, accompanied by the nasociliary nerve. It gives rise to the central retinal, anterior ciliary, and the long and short posterior ciliary arteries.

The central retinal artery (Figure 13.4) penetrates the optic nerve near the eyeball and divides into four branches. These branches are end arteries that supply the four quadrants of the retina. The branches appear thinner and brighter red than the corresponding vein, with a normal artery-to-vein ratio of 2:3. In hypertensive individuals, the central retinal artery may exhibit narrowing or spasm and become thickened or sclerotic,

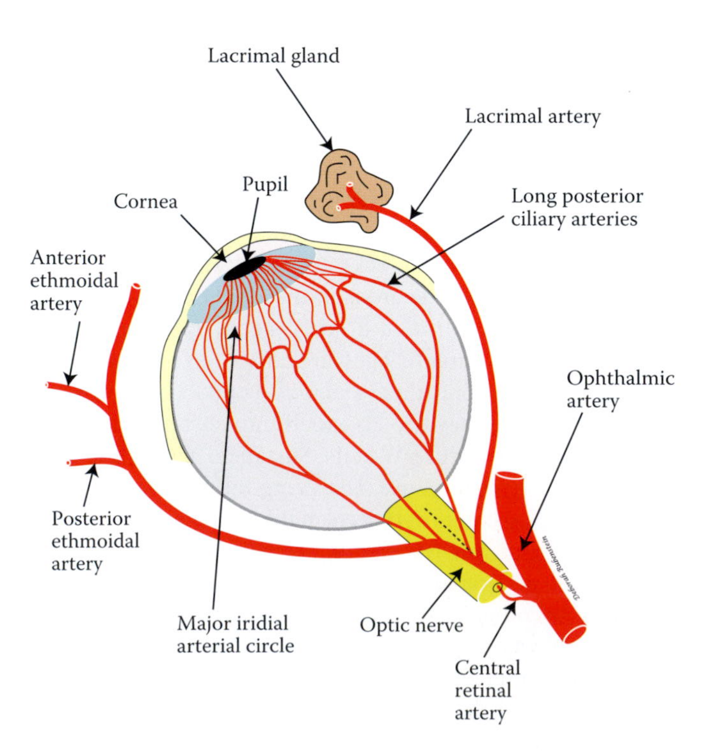

FIGURE 13.4 The ciliary and central retinal branches of the ophthalmic artery.

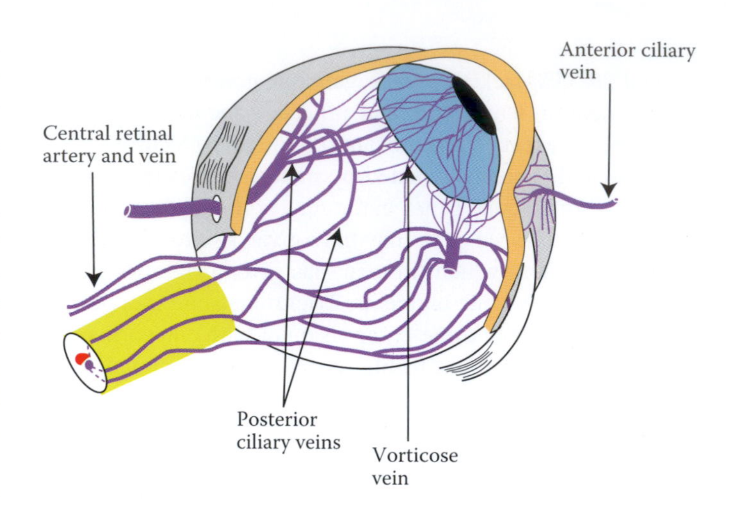

FIGURE 13.5 The central retinal vein and the anterior and posterior ciliary veins are illustrated.

changing color to orange-metallic. The central retinal vein (Figure 13.5) may be concealed by the more superficial and widened arterial wall branches that give the appearance of a discontinuous venous column. The long and short posterior ciliary arteries supply the choroid and ciliary processes and establish anastomosis with branches of the central retinal artery.

Occlusion of one branch of the central retinal artery may produce quadranopsia. A decrease of blood flow in the central retinal artery may indicate possible occlusion of the internal carotid artery. Incomplete occlusion of the internal carotid artery exhibits sudden transient monocular blindness in the form of a blackout or misty vision appearing as a shade or curtain that covers the visual field from side to side or from above without permanent visual loss (amaurosis fugax). This transient visual attack, which lasts from seconds to minutes, may also result from compression of the ophthalmic artery by the intraocular pressure subsequent to reduction in the pressure of the carotid system. Bilateral reduction in the blood pressure of the ophthalmic artery relative to the pressure of the brachial artery may indicate bilateral carotid disease.

Examination of the fundus of the eye may reveal, in a normal person, a more sharply defined temporal edge than the nasal edge. The optic disc appears pinkish in light-skinned persons and yellowish-orange in dark-skinned individuals. Pallor of the disc may suggest atrophy of the optic nerve. Retinal vessels radiate from the center of the optic disc and divide into branches that distribute to the retinal quadrants. Hypertension and arteriosclerosis alter the morphology of these vessels. Essential or malignant hypertension may produce retinal exudate, hemorrhage into the plexiform layer of the retina, cotton wool patches, a lipid star in the macula, and irregular narrowing of the retinal arteries.

The physiologic cup is the lighter-colored central part of the disc, which is penetrated by retinal vessels. The normal c/d ratio of 1:5 is genetically determined. Only 2% of normal eyes have a ratio more than 0.7. Unequal c/d ratios, in which the difference between the two eyes is more than 0.1, is seen in 8% of normal individuals and in 70% of patients with early glaucoma. A changing c/d ratio is significant because glaucomatous expansion of the optic cup is superimposed upon the amount of physiological cupping present before the onset of raised intraocular pressure. During the early stages of glaucoma, the increase in size of a small cup may not be detected because its dimensions may still be smaller than the physiological cup. Therefore, estimation of the cup size does not by itself carry diagnostic value, unless the increase is profound. Glaucomatous cups are usually larger than physiological cups, although a large cup may not be pathological.

OPTIC NERVE

The optic nerve (Figures 13.1, 13.6, and 13.11) is formed by the unmyelinated axons of the ganglionic layer of the retina, which acquire myelin outside the eyeball. Embryologically, it develops with the retina as an extension of the telencephalon. It is invested by the meninges and surrounded by the cerebrospinal fluid (CSF). Fibers arising from the fovea centralis follow a straight course to the temporal quadrant of the optic disc, forming the spindle-shaped papillomacular bundle. The fibers of this bundle pursue a central position inside the optic nerve. Fibers that arise from superior and inferior parts of the macula meet at the upper and lower poles of the optic disc, respectively. Those fibers that arise from the nasal retina follow a radial course, while the fibers of the temporal retina pursue an arcuate path around the papillomacular bundle to reach the optic disc. The upper and lower nasal fibers also contain fibers from the peripheral temporal parts of the retina. The optic nerve is crossed by the ophthalmic artery and pierced by a branch of it, the central retinal artery, as it leaves

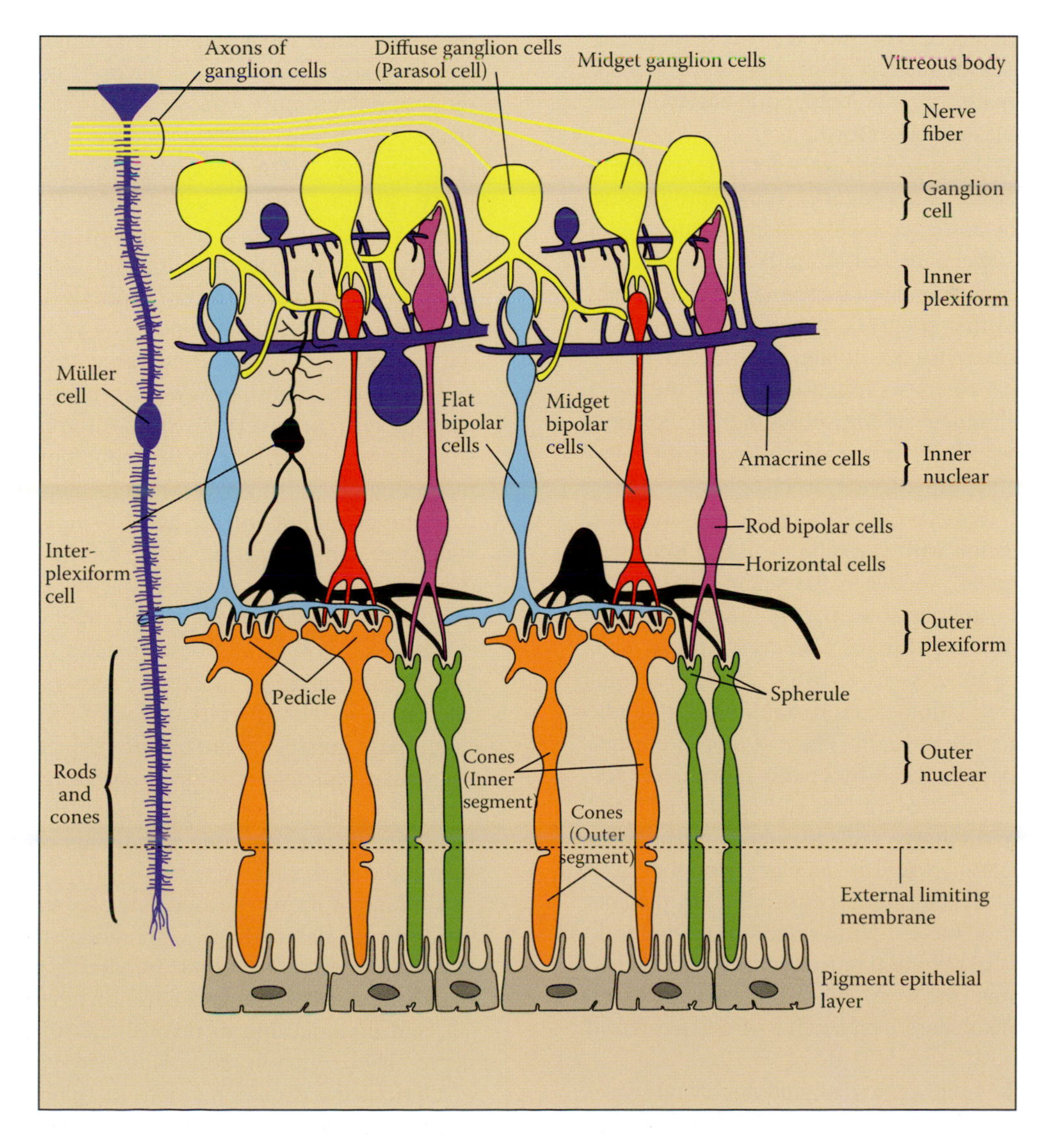

FIGURE 13.6 Neuronal series associated with transmission of visual image to the optic nerve.

the eye. It then travels further in the orbit and enters the optic canal. After exiting the canal, it pursues a course inferior to the frontal lobe. The optic nerve also forms the common afferent limb for both the pupillary light and accommodation reflexes.

The central position of the papillomacular bundle fibers within the optic nerve, their high metabolic activity, and their dependence on the blood supply of externally located ciliary arteries make them more prone to metabolic, toxic, and compressive injuries. Consequently, central scotomata or those that extend further to the optic disc can occur as a result of these insults. The arcuate fibers reaching the superotemporal and inferotemporal aspects of the optic disc are most vulnerable to glaucomatous insult, and the fibers of the papillomacular bundle are the most resistant. MS, which affects the temporal part of the optic disc, disrupts this bundle, producing central scotoma. Paget's disease, suprasellar tumors, or fractures involving the optic canal may damage the optic nerve. The optic nerve can be the target of many diseases such as glaucoma, optic neuritis, ischemic neuropathy, papilledema, optic nerve drusen, congenital dysplasia, and so forth.

Glaucoma, as discussed earlier, refers an increase in intraocular pressure that produces slowly progressive optic neuropathy associated with "cupping" atrophy of the optic disc. Patients present initially with scotomata in the form of arches or circles confined to the peripheral visual field that progress to complete peripheral visual loss. Central vision will later be affected.

Optic neuritis refers to inflammation of the optic nerve and development of demyelinating patches posterior to the eyeball, or intraocularly (papillitis) subsequent to MS, syphilis, Lyme disease, herpes zoster, vasculitis, and diabetes. It also occurs in measles, mumps, or infection with varicella viruses. Optic neuritis is seen typically between the second and fourth decades of life and episodically in nearly 50% of patients with MS. One-third of patients with optic neuritis can develop other signs of MS. It may be accompanied by demyelination of the papillomacular bundle and white matter of the brain. There is a brief episode of monocular pain induced by ocular movements followed by sudden and severe loss of visual acuity and color vision, particularly red. In adults, this deficit is usually unilateral, while in pediatric patients, it is often a bilateral deficit. Optic neuritis in MS patients produces pallor of the temporal quadrant of the optic disc and defects in temporal halves of the visual fields. The spontaneous recovery of vision usually seen after few months occurs irrespective of corticosteroid administration.

An unusually small optic disc and nocturnal hypotension may be responsible for ischemia of the optic disc (*anterior ischemic optic neuropathy*), a condition associated with abrupt, painless, and unilateral visual loss noticed in the early morning. This ischemia may be associated with giant cell (temporal) arteritis, a systemic vasculitis that involves branches of the external and also internal carotid arteries. Giant cell arteritis is seen in patients in the sixth decade of their life and presents with monocular or binocular visual loss, scalp tenderness, fever, jaw and tongue claudication, diplopia, and tinnitus. Visual deficit may result from involvement of the ophthalmic artery and its branches. In one-half of the patients, polymyalgia rheumatica may be a preexisting condition. Physical exam reveals tender temporal scalp and generalized reduced pulsation. Presence of giant cells in the biopsy specimens of affected arteries may not be a uniform finding due to a segmental pattern of vasculitis.

Papilledema (choked disc) is a condition characterized by bilateral passive elevation of the margins of the optic discs as a result of increased intracranial pressure. Since the subarachnoid and subdural spaces of the brain also extend around the optic nerve, increased intracranial pressure can be transmitted along these spaces, producing edema around the nerve and retardation of venous drainage. Tumors that involve the optic nerve sheath, tectum, cerebellum, fourth ventricle (ependymoma), cerebral hemisphere, and corpus callosum may also produce papilledema. Cerebellar tumors (e.g., *medulloblastoma*) may protrude into the fourth ventricle and obstruct the pathway of the cerebrospinal fluid, producing increased intracranial pressure and papilledema earlier than any other tumors of the central nervous system. Cavernous sinus thrombosis as well as sinusitis may contribute to this condition by impeding the venous blood flow. It is rarely seen in congenital cyanotic conditions of the heart or in Guillain–Barré syndrome (an idiopathic acute febrile inflammatory disease that produces polyneuropathy). It may occur in pseudotumor cerebri (idiopathic intracranial hypertension), which is seen in obese females of childbearing age. Since pontine or medullary tumors do not generally interfere with the circulation of the cerebrospinal fluid, these masses do not usually induce papilledema. Therefore, patients may die from brainstem compression before developing papilledema. Papilledema can be detected by examining the dilated retinal veins in the fundus of the eye.

Inflammation of the optic disc is suspected when exudate and hemorrhage with moderate elevation of the disc margin are present. Headache, nausea, vomiting, hemiparesis, visual obscuration or even hemianopsia, and diplopia due to involvement of the abducens nerve may be seen in individuals with papilledema.

Optic nerve drusen refers to the globules of mucopolysaccharides and proteinaceous material (hyaline bodies) from axonal degeneration of retinal neuronal axons or stasis of the axoplasmic transport within these axons that progressively accumulate anterior to the lamina cribrosa and within the optic nerve disc. This insidious condition is bilateral in the majority of patients and can be inherited (autosomal dominant) and affects males and females to an equal degree. Drusen is not always visible and may become more prominent after the first decade of life or with atrophy of the optic nerve axons. Presence of these globules may lead to compression of the optic nerve axons and the central retinal vessels, causing retinal scaring. One complication of this condition is the development of choroidal neovascular membrane and abnormal blood vessel growth beneath the retina and subsequent loss of central acute vision. It may be associated with retinitis pigmentosa and Noonan syndrome and can mimic papilledema.

Integrity of the optic nerve is determined by examination of the visual fields and visual acuity. This is accomplished by confrontational visual field testing, which involves closure of the examiner's right eye and patient's left eye while standing at eye level opposite each other. This is followed by the examiner's simultaneous show of one or two fingers on each hand and his request that the patient ascertain the fingers that he has seen. The other eye will be tested the same way from upper to lower quadrants. In normal individuals, the fingers will be seen at the same time by the examiner and patient. *Scotoma* (focal blindness), which occurs in glaucoma and tumors of the central nervous system, may be detected upon widening of the visual field of a patient by pulling the examiner's hand away from the patient. A flashing light beam or a pencil may also be used, and the patient is asked to state the timing of its appearance and direction. Visual acuity may be assessed by using the *Snellen eye chart*, positioned approximately 20 feet from the patient. Each eye is tested separately, and the first number in the standard ratio 20/20 denotes the actual distance of the patient from the chart, while the second number represents the distance at which a person with normal vision can read the chart. Visual acuity of each eye, which reflects the macular function, should be tested independently with and without glasses. For this purpose, the examiner may use a newspaper article or attempt to present a picture or small objects to be identified by the patient. Visual acuity may vary by environmental factors such as illumination and degree of contrast.

To detect the differences between both eyes in response to afferent stimuli, the *swinging light test* is employed. In this test, the patient is asked to look at a distant object while the examiner rapidly swings a light beam from one eye to the other. When directing the light into the blind eye, neither eye will show constriction. However, upon moving the light quickly from the intact to the affected eye, the affected eye shows apparent pupillary dilatation due to the lack of afferents to the retina and optic nerve (Marcus Gunn pupil).

OPTIC CHIASMA

The optic chiasma (Figures 13.7, 13.12, 13.14, and 13.16) is formed rostral to the hypothalamus by the decussation of the nasal fibers of the optic nerves, including the fibers from the macula, infundibulum, and tuber cinereum. It is formed inferior and anterior to the third ventricle and superior and rostral to the pituitary gland and sella turcica. It lies medial to the internal carotid and the site of its bifurcation into the anterior and middle cerebral arteries. It is also caudal to the anterior communicating and medial to the posterior communicating arteries. Through this vascular network, the optic chiasma receives its arterial blood supply. Corollary to this, the venous drainage pursues a similar course into the adjacent anterior cerebral vein and also to the basal vein of Rosenthal.

The decussating inferior nasal fibers occupy a rostral position to that of the superior fibers within the optic chiasma. Fibers from the lower nasal quadrant form a short loop into the medial part of the contralateral optic nerve prior to joining the optic tract as the *anterior knee fibers of von Willebrand.* This accounts for superior temporal quadranopsia in the contralateral eye, which accompanies optic nerve lesion. Similarly, the fibers from the superior nasal quadrant of the retina form a short loop backward into the ipsilateral optic tract known as the *posterior knee fibers of von Willebrand.* Macular fibers and the nearby central retinal area occupy the central chiasma. Beyond the optic chiasma, the temporal fibers continue caudally to join the ipsilateral optic tract, whereas the crossed fibers join the contralateral optic tract.

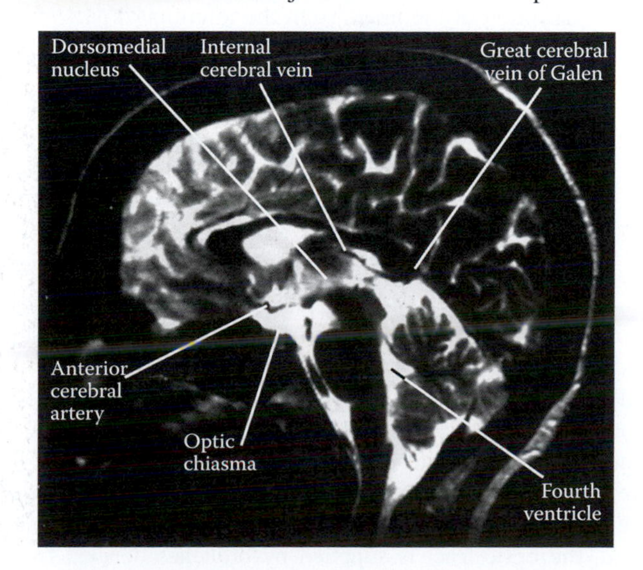

FIGURE 13.7 Magnetic resonance imaging (MRI) scan of the brain. Observe the course of the optic chiasma and its relationship to the anterior cerebral artery.

Due to the close relationship of the optic chiasma to the adenohypophysis, adenomas of the pituitary or, rarely, craniopharyngioma may compress the center of the optic chiasma and completely disrupt the nasal fibers of the retina, producing bitemporal heteronymous hemianopsia (*tunnel vision*). Bitemporal heteronymous hemianopsia (Figure 13.14) may be suspected if a patient can read left-hand letters only with the right eye and right-hand letters with the left eye, or if the patient reports frequent bumping into people or electric poles while walking. However, due to the benign nature of the expanding pituitary adenoma and the pattern of arrangement of the crossing nasal fibers, the disruption and the associated visual deficits occur gradually. Initially, the visual deficit is confined to the superior temporal quadrant and visual acuity and color vision, with central vision being spared. As the tumor expands, the inferior temporal quadranopsia develops, and then complete visual loss in the temporal visual half becomes evident. A lesion that affects the posterior chiasma and beginning of the optic tract produces posterior junctional scotoma, which is characterized by incongruous contralateral hemianopsia due to disruption of the nascent optic tract fibers and ipsilateral inferior temporal quadranosia as a result of damage to the crossing superior nasal fibers of the optic chiasma. Binasal heteronymous hemianopsia may be produced by aneurysms of both internal carotid arteries. Unilateral nasal hemianopsia may possibly occur if the aneurysm of the internal carotid artery is ipsilateral, which may mimic a pituitary tumor, producing visual deficits and radiographically detectable sellar enlargement. A chiasmal lesion near its junction with the optic nerve may also produce junctional scotoma, which is characterized by superior temporal quadranopsia (due to disruption of the fibers from the inferonasal part of the retina located in the rostral part of the optic chiasma) and loss of ipsilateral central vision. Based on the pattern of fiber arrangement, a chiasmal lesion also produces superior or inferior bitemporal quadranopsia or monocular temporal hemianopsia.

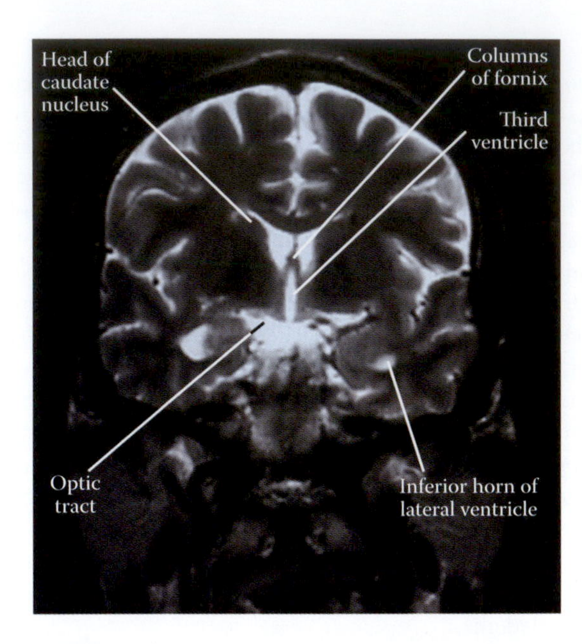

FIGURE 13.8 MRI scan of the brain (coronal view). Note the course of the optic tract in relation to the third ventricle.

OPTIC TRACT

The optic tract (Figures 13.8, 13.9, 13.10, 13.11, 13.13, 13.14, and 13.15) is formed by the crossed nasal fibers and ipsilateral temporal fibers of the optic nerve that carries impulses from the opposite visual field. It courses between the tuber cinereum and the anterior perforated substance, adjacent to the internal carotid artery, and then continues around the crus cerebri and thalamus to the LGN. In its anterior one-third, the optic tract is supplied by branches derived from the internal carotid, middle cerebral, and posterior communicating arteries, and in its posterior two-thirds, by a single anterior choroidal artery. The fibers that originate from the macula lutea, occupy an intermediate position, fibers from the upper quadrant of the retina reside in an anterior and medial location, and the fibers from the lower quadrant of the retina occupy a more lateral and posterior position in the optic tract. Most of the fibers of the optic tract project to the lateral geniculate body (LGB), a visual relay nucleus of the thalamus, where they establish synaptic linkage with its neurons. Some fibers of the optic tract bypass the LGB and terminate in the pretectum and tectum, containing the efferent neurons for the pupillary light reflex. The fibers that bypass the LGB enter the SC and the pretectum, activating the Edinger–Westphal nucleus of both sides via crossed fibers of the posterior commissure. The Edinger–Westphal nucleus

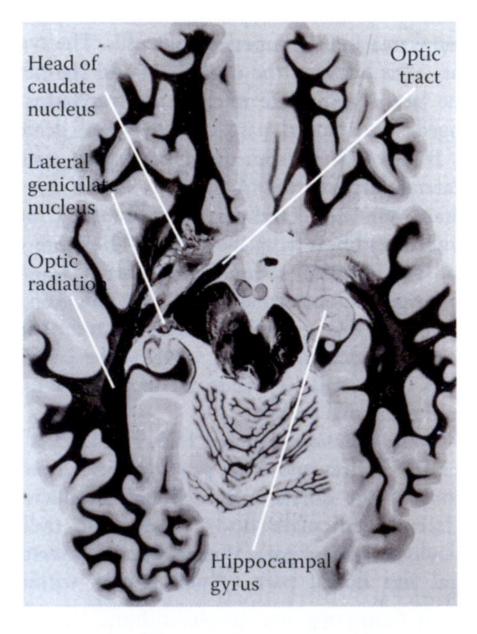

FIGURE 13.9 Horizontal section of the brain. The optic tract, lateral geniculate body, and optic radiation are prominently displayed.

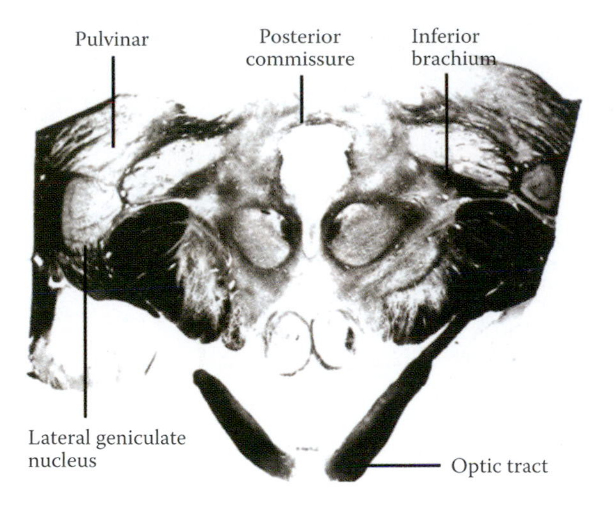

FIGURE 13.10 In this section, the optic tract and the lateral geniculate bodies are also clearly visible.

provides preganglionic parasympathetic fibers to the ciliary ganglion that control the curvature of the lens and the contraction of the sphincter pupillae muscle through the short ciliary nerves.

A lesion, such as caused by a tumor, aneurysm, or trauma that disrupts all the fibers of the optic tract, produces homonymous hemianopsia on the contralateral visual field (Figure 13.12). This will be accompanied by a mild afferent pupillary defect in contralateral side as more fibers have entered the optic tract from the opposite than the ipsilateral side. Optic disc pallor is another manifestation of optic tract lesion in which disruption of the ipsilateral temporal retinal fibers causes upper and lower pole disc atrophy, while disruption of the contralateral nasal fibers and the contralateral nasal papillomacular bundle produces a "band or bowtie" disc characterized by atrophy in the temporal and nasal poles. Due to proximity of the internal carotid artery to the optic tract, aneurysm of this vessel may compress the ipsilateral (temporal) fibers of the optic tract, resulting in nasal hemianopsia (Figure 13.10). Partial damage to the optic tract produces the characteristic configuration of wedge-shaped loss of visual field.

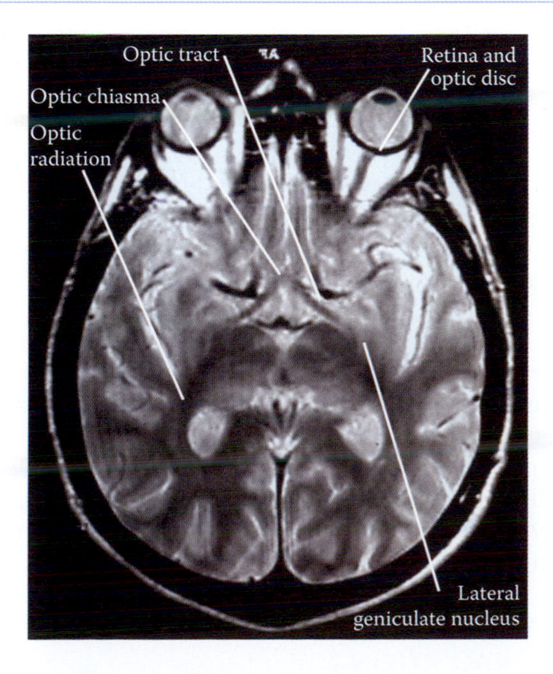

FIGURE 13.11 This MRI scan illustrates some of the elements associated with the visual system. Note the optic nerve, optic chiasma, optic tract, lateral geniculate nucleus, and optic radiation.

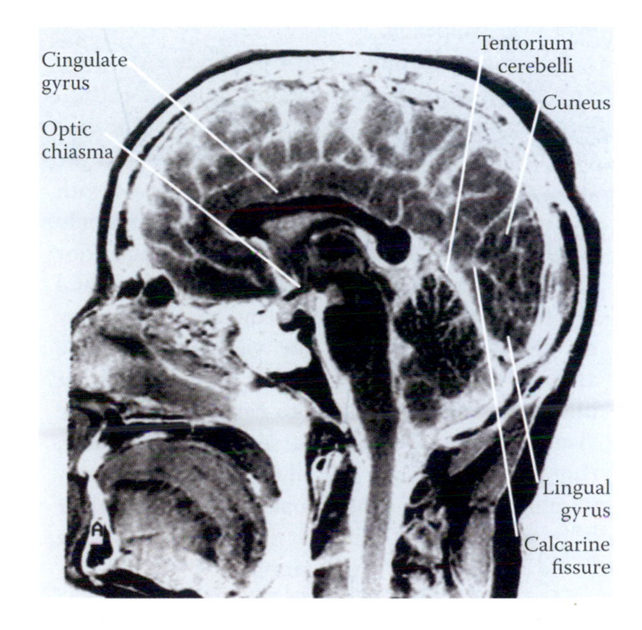

FIGURE 13.12 MRI scan (midsagittal view) of the brain illustrating the visual cortex within the lingual gyrus and cuneus.

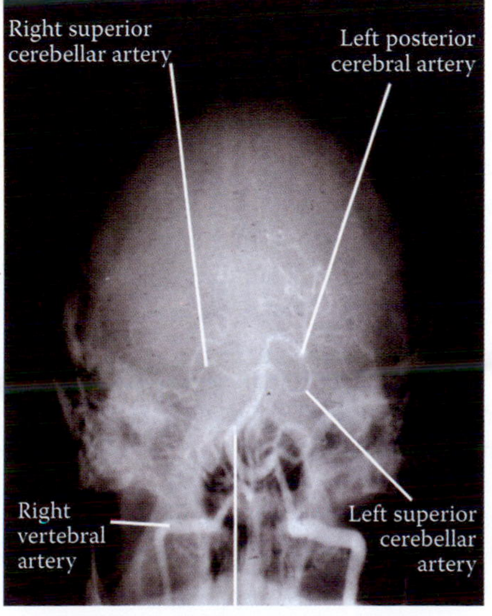

FIGURE 13.13 In this angiogram of the vertebrobasilar system, the right posterior cerebral artery is obstructed.

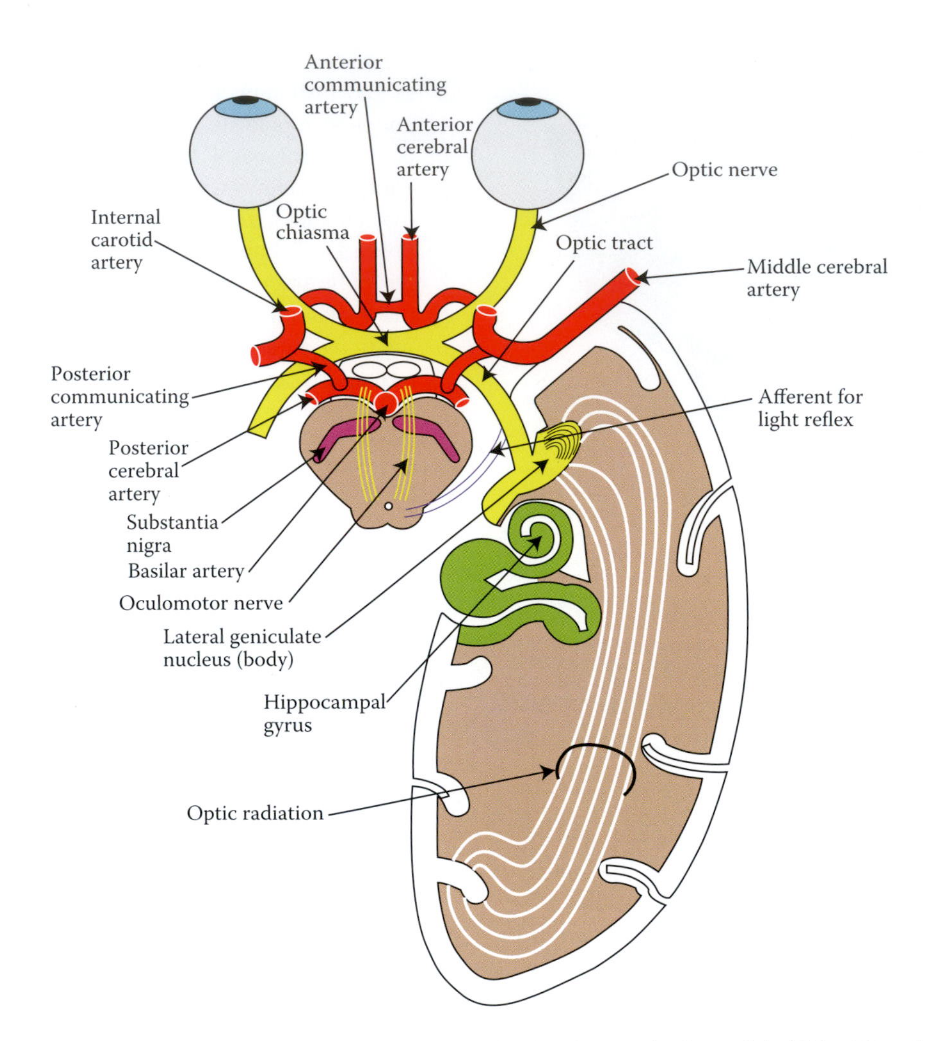

FIGURE 13.14 Drawing of the visual pathway from the optic nerve to the visual cortex in the occipital lobe. The relationship of the arterial circle of Willis to the optic nerve, optic chiasma, and optic tract is shown.

LATERAL GENICULATE NUCLEUS (LGN)

The LGN is a visual thalamic relay nucleus for the retinal fibers that are contained in the optic tract. It consists of six layers intervened by white matter bands. There is a retinotopic arrangement associated with these layers with a 1:1 synapse ratio, which enables the creation of a contralateral half of the visual field, in which layers 1, 4, and 6 receive fibers from the contralateral retina, while layers 2, 3, and 5 receive fibers from the ipsilateral retina (Figures 13.9, 13.10, 13.11, 13.14, 13.15, and 13.16). Layers 1 and 2 form the ventral (magnocellular) subnucleus, and layers 3 to 6 comprise the dorsal (parvicellular) subnucleus. Most fibers of the optic tract terminate in the LGN. Despite the fact that LGN is a visual synaptic station, retinal input constitutes only 20% afferents, while the posterior parietal and occipital lobes as well as the midbrain reticular formation form the rest of the input to the LGN. Fibers bypass the LGB to synapse in the pretectal area and the SC. The synaptic connections between the optic tract and the neurons of the LGN are somatotopically arranged. The medial part of the LGN receives fibers from the upper retinal quadrant, the lateral part receives fibers from the lower retinal quadrant, and the central part of the LGN receives fibers from the macula. The axons of dorsal (parvicellular) subnucleus of the LGB neurons form the geniculocalcarine tract (optic radiation). The medial part of this projection terminates in the superior bank of the calcarine fissure, whereas the lateral part projects in a similarly precise manner to the inferior bank of the visual cortex. The LGN receives blood supply from branches of the posterior cerebral and posterior communicating artery.

A lesion that involves the area medial to the LGN, as is seen in neurosyphilis, may selectively disrupt the fibers that mediate constriction of the pupil in light reflex, sparing the afferent limb of accommodation reflex. Individuals with this type of lesion may exhibit pupillary constriction in accommodation but not in response to light (*Argyll Robertson pupil.*)

OPTIC RADIATION

The optic radiation (Figures 13.9, 13.11, 13.12, 13.13, 13.14, and 13.15) represents the myelinated postsynaptic fibers of

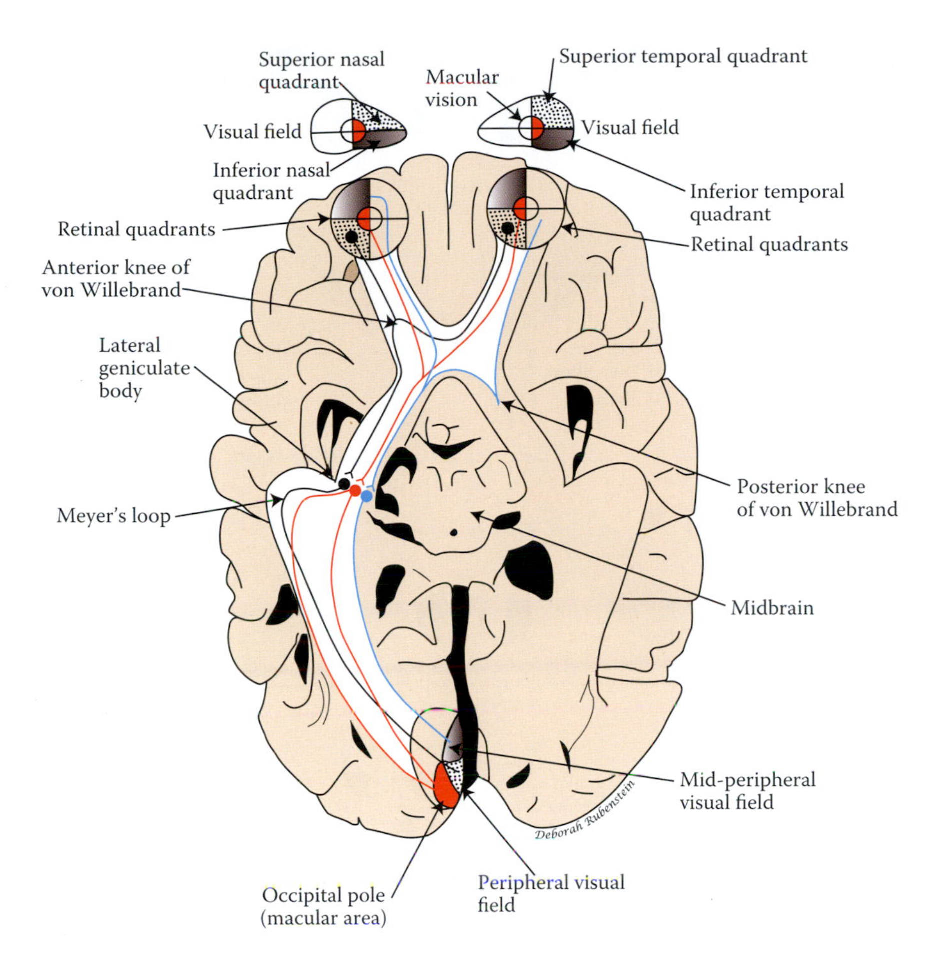

FIGURE 13.15 The course of the visual image from the retina through the optic nerve, optic tract, lateral geniculate body, and optic radiation, to the visual cortex is shown.

neurons from the dorsal (parvicellular) subnucleus of the LGN. The optic radiation (geniculocalcarine tract), shaped like a crescent, has superior and inferior parts that course within the retrolenticular part of the internal capsule en route to the visual cortex. Each part represents one-fourth of the visual field of the contralateral side. Fibers derived from the upper retinal quadrant run in the superior part of the optic radiation, and fibers from the lower retinal quadrant are shifted to the lower part of the optic radiation. The foveal fibers occupy the most lateral position of the optic radiation. The superior (upper) fibers of the optic radiation follow a direct path through the parietal lobe to end in the cuneus in the upper bank of the calcarine fissure. The inferior (lower) part of the optic radiation, known as *Meyer's loop*, forms a curve into the temporal lobe running adjacent to the tip of the inferior horn and then continues lateral to the posterior horn of the lateral ventricle. Then, it turns back to rejoin the rest of the optic radiation. Meyer's loop conveys visual impulses from the contralateral superior visual field and terminates in the lower bank of the calcarine fissure (lingual gyrus). The anterior choroidal and, to a lesser degree, the posterior choroidal supply the rostral part of the optic radiation, whereas the mid portion is provided blood supply by the middle cerebral artery. The calcarine artery is the chief source of blood supply to the posterior part of the optic radiation.

Complete disruption of the optic radiation produces contralateral homonymous hemianopsia that will not easily be distinguished from parietal or occipital lesions or from lesions that disrupt the optic tract and LGN. However, the presence of RAPD and "bowtie" optic disc atrophy may serve as distinguishing characteristics of a lesion of the crossed nasal fibers in the contralateral optic tract derived from the nasal papillomacular area. Similarly, in a parietal lesion that disrupts the entire optic radiation, visual field defects are bilateral and homonymous, and the smooth pursuit ocular movement to the side of the lesion is lost. Ischemic changes due to hypotension, occlusion of the posterior cerebral or basilar arteries, contrecoup head trauma, or cardiac arrest can cause a bilateral occipital lobe infarction with a unique keyhole defect seen in the vertical meridian.

Since the inferior portion of the optic radiation (*Meyer's loop*) follows a separate course within the temporal lobe before joining the bulk of the geniculocalcarine tract, selective damage to the optic radiation in the temporal lobe, a frequently occurring lesion, may produce mildly

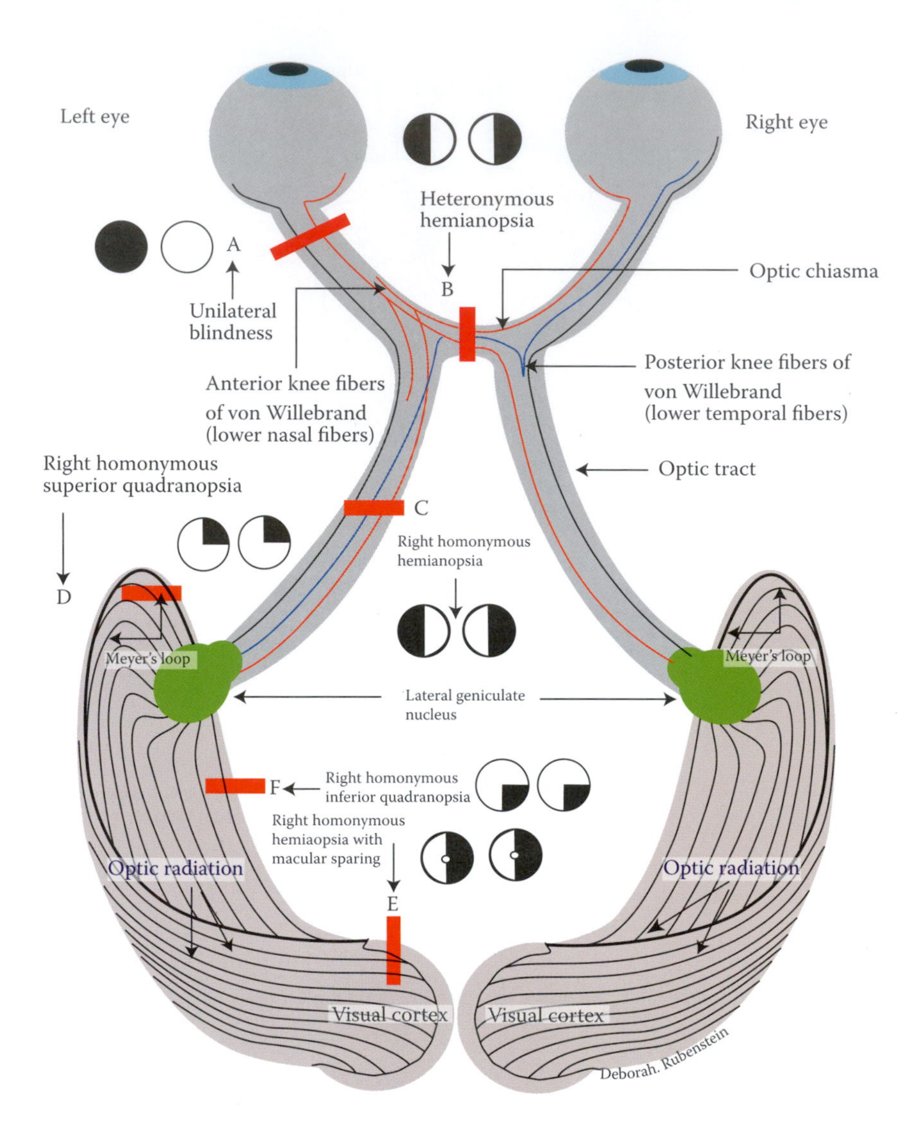

FIGURE 13.16 A detailed view of the lesions associated with the visual system and pertinent dysfunctions.

incongruous superior quadranopsia in the opposite visual field. This visual field defect resembles a wedge or "slice" removed from the superior visual field and is commonly termed a "pie in the sky." Edema caused by bleeding from the medial striate artery (a branch of the middle cerebral artery) may also compress the optic radiation, resulting in transient homonymous hemianopsia, which lasts until the edema subsides. Vascular lesions affecting the optic radiation may also be caused by occlusion of the anterior choroidal and posterior cerebral arteries. An abscess that develops in the temporal lobe, above the level of the auditory meatus, may compress and disrupt the fibers of Meyer's loop, producing quadranopsia in the contralateral visual field. Homonymous visual field defects due to lesions of Meyer's loop tend to be incongruous. Those defects that result from damage to the optic radiation near the visual cortex are congruous (edges of the visual field defect in each eye are identical in shape). Damage to the superior fibers of the optic radiation, which rarely occurs in parietal lobe lesion, produces a contralateral inferior quadranopsia. The most lateral part of the temporal visual field, which measure 25°, is seen only by one eye on the same side and projects to the rostral-most part of the striate cortex adjacent to the parieto-occipital cortex. As a result of this unique visual field representation, a lesion of the most rostral part of the calcarine fissure produces monocular homonymous defect. Unlike the optic tract, a lesion of the striate cortex produces congruent visual deficits without accompanying optic disc atrophy, central vision loss, or RAPD.

VISUAL CORTEX

Primary Visual Cortex

The primary visual or striate cortex (Brodmann area 17) is the principal cortical area for visual perception, integration, and formation of a binocular image. It is mainly confined medially to the banks of the calcarine fissure, although part of this cortex also extends slightly around the occipital pole to the lateral surface. The six-layer organization of the visual cortex is discussed in detail in Chapter 8. It has a point-to-point connection with the LGN. Due to this precise connection, a small lesion in the visual cortex may result in scotoma (focal blindness). Area 17 includes portions of the lingual and cuneate gyri, extending to the lateral surface of the occipital lobe. It consists of a very thin granular cortex, in which layer IV is divided into densely packed upper and lower sublayers and a lighter middle layer with fewer small cells between the giant stellate cells. The light middle layer has a thickened outer myelin-rich band, is visible to the naked eye in sections of the fresh brain, and is known as the band of Gennari. Area 17 receives information from all neurons of the LGB that project to Brodmann areas 18 and 19 (Figures 13.12, 13.15, and 13.16).

The interconnection between area 17 of both cerebral hemispheres is not well developed. Visual fibers that reach the pulvinar deal with the contralateral visual field and project to layers I, III, and IV of cortical areas 18 and 19 and to the supragranular layers of Brodmann area 17. The latter projection constitutes the extrageniculate visual pathway.

Perception of visual images (e.g., individual may be able to read an article if brought into focus) remains intact even with bilateral damage to the striate cortex as long as the occipital pole is spared. Bilateral destruction of the occipital poles, on the other hand, markedly impairs the ability to clearly and accurately observe visual fields.

> Macular sparing a phenomenon in which a lesion involving the occipital lobe or occipital pole results a visual defect that spares central vision. This may be due to incomplete lesion of the striate cortex and sparing of the posterolateral part around the occipital pole, efficient arterial anastomosis between the middle and posterior cerebral arteries, presence of a separate blood supply, patient's ability to shift ocular fixation, or bilateral representation of the macular area in both cerebral hemispheres. Macular sparing is seen with cortical blindness, accompanied by contralateral incongruous homonymous hemianopsia.

Dark bars against a light background, and straight edges separating areas of different degrees of brightness effectively stimulate the visual cortex. The primary visual cortex consists of functional units that are arranged in columns of cells exhibiting different receptive fields. These functional units include the ocular dominance and orientation columns that are arranged perpendicular to the cortical surface.

The ocular dominance columns, partially formed at birth, which run at a right angle to the cortical surface, receive visual input from both eyes. However, they are arranged in such a manner that the visual input to a cortical column is only derived from one eye (dominant). The close proximity of the ocular dominance columns of the right and left eyes renders selective disruption of the input from a single eye very difficult. Segregation of the visual impulses into right and left laminae of the dominance column occurs in layer IV. However, no ocular dominance columns exist in the parts of the striate cortex, which receive impulses from the optic disc and also from the peripheral most temporal visual field of the ipsilateral eye. Occlusion of one eye during postnatal development may permanently hinder growth of the associated column. The *orientation columns* are smaller than the dominance columns and extend from the white matter to the pial surface of the cerebral cortex. They contain cells that possess the same receptive field axis of orientation and have "on" and "off" centers. The visual cortex is primarily supplied by the calcarine branch of the posterior cerebral artery, although the middle cerebral artery also contributes through its anastomotic connections. The striate cortex consists of simple, complex, and hypercomplex cells. Simple cells have similar characteristics to those of the retina and LGN.

> Amblyopia (lazy eye) is a disorder that develops from a prolonged suppression of an image in one eye between the second and fourth years of life. It may be the result of congenital strabismus and the inadequate stimulation of one eye by visual image. It occurs in children who exhibit diplopia as a sequel to functional imbalance between the extraocular muscles and subsequent attempts to eliminate the image in one eye by constantly utilizing the other eye. As the cross-eyed child favors one eye over another, the unused eye eventually loses visual acuity and may permanently be blind. In this condition, no deficits are recorded in the refractive media or ocular apparatus. Amblyopia may also occur as a result of nutritional deficiency and in alcoholics. This condition may be associated with damage to the optic nerves and bilateral scotoma. Blurred vision and optic atrophy may also occur in this condition.
>
> Unilateral or bilateral occlusion of the posterior cerebral artery (Figure 13.13) is commonly associated with a variety of deficits and syndromes. Infarction of the posterior cerebral artery is the most common etiology of visual deficits of occipital lobe origin (Figure 13.11). Transient occlusion of the vertebral arteries on both sides, which may occur as a result of cervical spondylosis and subsequent narrowing of the transverse foramina, may dramatically reduce the blood flow in the labyrinthine and posterior cerebral arteries. A patient with this condition may experience vertigo and transient blindness, which last for few seconds, without remembering that these disorders ever have happened.
>
> Anton's syndrome is an expression of the psychological ramification of cortical blindness, which is caused by disruption of the corticothalamic connection between area 17 and the thalamus, and it is also observed in individuals with nondominant hemispheric damage. It commonly results from bilateral occlusion of the posterior cerebral arteries. Patients have normal and reactive pupils but may show indifference or pay no attention to half of

the visual field of the affected side. They are generally unaware of their blindness and attempt to name objects and describe the surrounding objects in the visual field, though they cannot tell illuminated from nonilluminated areas. Patients consistently deny that they are blind and insist that poor lighting or disinterest is the cause for their visual deficits.

Occlusion of the posterior cerebral arteries can cause bilateral degeneration of the parieto-occipital cortex between Brodmann areas 19 and 7, producing signs and symptoms of Balint syndrome. This syndrome is characterized by the inability to appreciate or scan the peripheral visual field (due to lack of coordination with the oculomotor system) or use visual cues to grasp an object. Infarction of the posterior cerebral artery may also produce a combination of hemianopsia or quadranopsia, macular sparing, and hemianesthesia with no muscle paralysis. If the infarct involves the dominant hemisphere, Charcot–Wilbrand syndrome may develop, which is characterized by visual agnosia. Gertsmann syndrome, transcortical sensory aphasia, and alexia without agraphia are also seen in posterior cerebral artery infarcts.

Secondary Visual Cortex

The secondary visual cortex (Brodmann area 18) adjoins the striate cortex, deals with visual memories, and receives visual impulses from Brodmann area 17. It is a mirror-image representation of Brodmann area 17, which consists of a six-layered granular cortex. It interconnects Brodmann areas 17 and 19 and does not contain the band of Gennari. This cortex, as in the case of area 17, responds best to dark bars and edges. The majority of cells in Brodmann area 18 are complex cells arranged in columns. Usually, in the dominant hemisphere, the upper lateral portion of Brodmann area 18 deals with memories for inanimate objects, while the lower medial portion is concerned with memories for living parts or individuals. In order for the visual object to be recognized, information must project to Brodmann area 18 of the dominant hemisphere (for fine feature analysis) via the splenium of the corpus callosum.

Disruption of the connection between Brodmann *areas 18 of both* cerebral hemispheres may occur upon excision of the corpus callosum, producing unilateral visual agnosia. Patients with this type of deficit are unable to recognize images received by the right (nondominant) hemisphere of the brain. Bilateral visual agnosia results from a lesion of Brodmann area 18 in the dominant cerebral hemisphere. Visually agnostic patients cannot recognize objects without using tactile, auditory, gustatory, or olfactory clues.

Lesions that damage the upper lateral or lower medial parts of the secondary visual cortex in the left dominant hemisphere may result in autotopagnosia, which is characterized by failure of the patient to distinguish living people from objects. Lesions of the dominant hemisphere confined to the upper part of Brodmann area 17 and the occipital association cortex adjacent to the angular gyrus produce finger agnosia. This condition manifests inability to name objects, identify fingers, write, do arithmetical calculations, or recognize left from right. Achromotopsia, the inability to recognize color in only one-half of the visual field, may occur independently. Stimulation of Brodmann area 18 results in visual hallucinations in the form of sparkling lights.

Tertiary Visual Cortex

The tertiary visual cortex (Brodmann area 19), a mirror image of Brodmann area 18, occupies the area lateral to the secondary visual cortex. It is responsible for recalling (revisualizing) formerly seen images. The hypercomplex cells are the primary neurons in Brodmann area 19 that receive visual input from both eyes. Stimulation of this area produces colorful visual images of moving events and objects. The middle part of area 19 relates to the macula and object sizes, whereas the inferior part of this area responds exclusively to color. Movement activates a small area anterior to the macular zone of area 19. In order for the images to be recalled, visual information must project from Brodmann area 18 to Brodmann area 19, where they are activated by various types of stimuli (e.g., auditory, tactile, olfactory, etc.). It is important to note that recalling symbols is a function of the angular gyrus.

Lesions involving the parietal lobe and Brodmann area 19 of the occipital lobe may cause dysfunction similar to astereognosis. Therefore, the ability to recall objects by using tactile stimuli may be lost. Bilateral disruption of the connections between the association visual cortices and the entorhinal cortex (Brodmann area 28) may occur as a result of basilar artery insufficiency that extends to involve the posterior cerebral arteries. This disconnection, which is associated with degeneration of the occipitotemporal area, result in anterograde visual amnesia and difficulty in visually adapting to new and unfamiliar territory despite intactness of the visual apparatus. Visual changes in migraine headaches include blurred vision, flashing lights, wavy lines, and scotoma. Hemiparesis, ophthalmoplegia, or aphasia may accompany these symptoms.

It is worth noting that the inferior temporal gyrus serves as a visual association cortex, contains visual memory stores, and receives input from the entorhinal cortex (Brodmann area 22) and Brodmann areas 7, 18, and 19. These connections may explain the visual hallucination associated with temporal lobe epilepsy, as well as the vivid scenes experienced by patients undergoing brain operation. By the fifth postnatal month, the visual association cortices (Brodmann areas 18 and 19) become involved in stereopsis, a mechanism that enables the brain to measure the incongruity between the two retinal images, thus constructing a complete three-dimensional image.

OCULAR MOVEMENTS

Eye movements provide a significant index of the functional activity of the motor nuclei of the extraocular muscles and the neurons within the brainstem reticular formation. They enable image of the object of interest to be held simultaneously on both fovea centralis despite the head rotation of the observer or movements of the target. Diverting the gaze toward the target to be seen is accomplished through rapid (saccadic) eye movement; stabilization of the moving image on the fovea is mediated by smooth pursuit movements, whereas fixation of gaze during head rotation is performed by the vestibulo-ocular and optokinetic system. Despite the divergence of the neural mechanisms of the ocular movements, the common pathway that links them remains primarily confined to the pons (horizontal movement) and midbrain (vertical movement). There are two primary motor gaze centers: a horizontal gaze center located in the paramedian pontine reticular formation (PPRF) that extends from the pontobulbar sulcus to the pontomesencephalic junction and a vertical gaze center in the rostral interstitial nucleus of the MLF (riMLF) at the superior collicular level of the midbrain. The PPRF contains excitatory burst neurons, which discharge at high frequencies prior to and during ipsilateral saccadic movement, providing the pulse or the signals that control eye movement velocity. Extraneous saccadic movements are prevented by a distinct set of pause neurons scattered in the midline of the caudal pons within the nucleus raphe interpositus that exert tonic inhibitory influence.

The vestibular nuclei and nucleus prepositus hypoglossi project to the abducens nuclei as well as to the PPRF. The projection to the abducens nuclei carries vestibular input that plays a role in the fixation of gaze, but it also conveys signals from the nucleus prepositus hypoglossi through the cerebellum to regulate smooth pursuit movement. The reciprocal connection with the PPRF and the neurons of the abducens nuclei enable the eye to maintain a desired (abducted) position despite the antagonistic viscoelastic forces of the orbit that tend to pull the eye (adducts) in the direction of the primary (gazing straight ahead) position following an ipsilateral horizontal saccade. Thus the abducens neurons discharge during a horizontal saccadic movement to deviate the eye rapidly to a desired new position (pulse) and also retain this new position (step). The vertical gaze center in the riMLF regulates the vertical saccadic movements. Bilaterality of up-and-down movements is maintained by the fibers that run in the posterior commissure and connect the vertical gaze centers on both sides. The riMLF also projects to the oculomtor nucleus.

A lesion of the posterior commissure and dorsal riMLF as seen in Parinaud syndrome produces primarily upward (vertical) gaze palsy. A lesion of the ventral part of the riMLF is most likely to result principally in downward (vertical) gaze palsy. Another center that mediates vertical pursuit movement is located in the interstitial nucleus of Cajal.

Ocular movements are classified into conjugate (version) movements, where the visual axes of both eyes remain

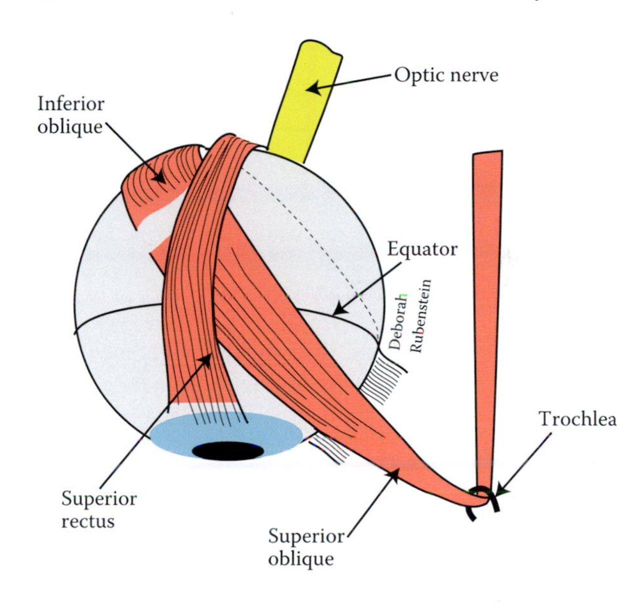

FIGURE 13.17 Diagram of the eyeball demonstrating the sites of attachment of the oblique muscles in relation to the optic axis.

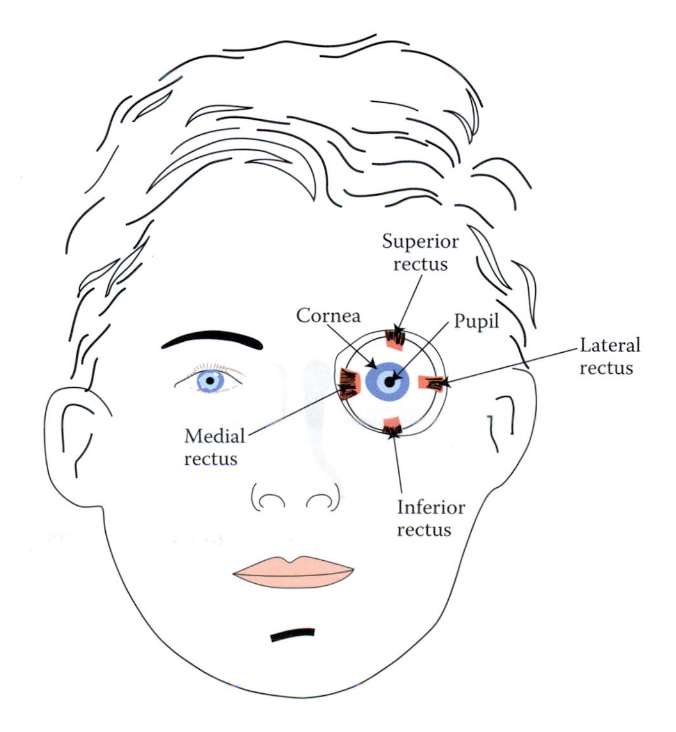

FIGURE 13.18 Diagram of the rectus muscle.

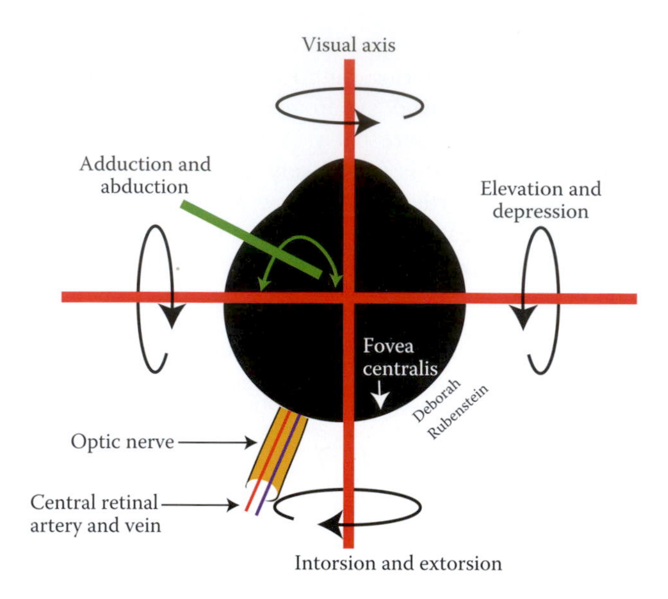

FIGURE 13.19 Diagram of the ocular movement around the visual axis. These movements encompass adduction, abduction, elevation, depression, intorsion, and extorsion.

parallel (for far vision), and disconjugate (vergence) movement, in which the visual axes intersect (for near vision) by the contraction of the medial rectus muscles (Figures 13.17, 13.18, and 13.19).

Disconjugate (Vergence) Movement

In this type of movement the visual axes intersect (for near vision) by the contraction of the medial recti. It deals with tracking of approaching (converging) or receding (diverging) objects that require slow movements of the eyes in opposite directions.

Conjugate (Version) Movement

In this type of ocular movement the visual axes of both eyes remain parallel (for far vision), and depend upon the integrity of certain gaze centers and the medial longitudinal fasciculus (MLF). The MLF is the principal internuclear pathway that interconnects the motor nuclei that innervate the extraocular muscles, coordinates conjugate eye movements, and ensures binocular vision.

Conjugate eye movements are further categorized into saccadic, smooth pursuit, and vestibulo-ocular movements.

Saccadic Eye Movement

Saccadic movements of the eye are involuntary, ballistic, and rapid movements that include successive jumps of the eye from one point of visual fixation to another. It is the only conjugate movement that could be produced voluntarily, for example, when reading or visualizing the items in a room. Saccades occur virtually in all voluntary eye movements with the exception of smooth pursuit eye movements. In contrast to smooth pursuit movement, visual acuity is diminished during saccades. Saccadic movements are used to improve reading speed by increasing the numbers of words read in a single fixation.

Several cortical areas are involved in saccadic movements, which include the frontal eye field in the premotor cortex (Brodmann area 8), the supplementary eye field in the rostral part of the supplementary motor cortex (Brodmann area 6) and the dorsolateral prefrontal cortex, which occupies an area rostral to the frontal eye field (Brodmann area 46); and a posterior eye field in the superior part of the angular gyrus (Brodmann area 39). These cortical areas project to the PPRF and the riMLF. The frontal eye field initiates saccades that predict the appearance of an expected target, when a previously seen target appears, or when scanning is require to search for an image of interest. The posterior eye field in the parietal lobe is associated with the integration of visual and spatial stimuli that triggers saccadic movement in response to the sudden appearance of visual or auditory targets. The dorsolateral prefrontal cortex is believed to provide a spatial map for saccadic movements that govern a previously seen target and eliminate misdirected saccades, whereas the supplementary eye field helps to produce the sequence of the saccades. The superior colliculus (SC), which receives visual, auditory, and somatic sensory terminals, projects to the horizontal and vertical gaze centers and joins with cortical input to play an important role in the initiation and accurate targeting of the saccades. Like the cortical projections, these tectal inputs to the gaze centers discharge very shortly before the saccadic movement begins and is linked to the presentation of the visual image.

Saccadic movements are mediated by the cortical input from the frontal eye field indirectly to the ipsilateral SC and via the SC projection to the contralateral PPRF. The activated PPRF stimulates the ipsilateral abducens nucleus, and through the internuclear neurons of the abducens, it activates the contralateral neurons of the medial rectus of the oculomotor nuclear complex. The activated PPRF also projects to the medullary reticular formation, which provides inhibitory commands to the local neurons, which, in turn, send axons to the contralateral abducens nucleus, causing a reduction in its activity. Thus, through this connection, the ipsilateral abducens nucleus is activated, whereas the contralateral abducens nucleus is inhibited. Direct cortical input from the frontal lobe to the contralateral PPRF regulates saccadic eye movements independently. The direct and indirect pathways constitute the anterior system. However, tectal projection is dependent upon the activation of selected neurons in the SC by corticotectal fibers. Cortical and tectal input project in a similar manner to the riMLF to mediate vertical gaze. There is a second pathway (posterior system) that originates from the posterior eye field to the PPRF and riMLF via the SC. A third pathway is associated with memory-guided saccadic movement generation that begins from the frontal lobe and projects to the SC via the basal nuclei. This entails projection from the frontal cortex to the caudate nucleus (corticostriate) and then via the striatonigral fibers to the pars reticulata (SNpr) of the substantia nigra, which, in turn, projects to the SC and, eventually, to the gaze centers.

Saccadic movements are controlled by the contralateral frontal cortex and are not affected by sedatives or analgesics. They are lost in Huntington's chorea and ophthalmoplegia of supranuclear origin. Cerebellar diseases may produce overshooting and undershooting of saccadic movements. Damage to the frontal eye field results in loss of ability to produce saccadic movements to the contralateral side and deviations of the eyes toward the lesion side (see also cortical dysfunctions, Chapter 8). A lesion that disrupts the SC does not eliminate saccades but changes the velocity, frequency, accuracy, and latency of these ocular movements.

Vestibulo-Ocular Eye Movement

Vestibulo-ocular movement (reflex) is designed to fixate gaze during rapid head movement. Slow movement, movement with eyes closed or head movement in darkness usually elicits minimal transient vestibulo-ocular response. The latter type of movement activates the optokinetic system, which may follow the vestibulo-ocular reflex when the head movement begins to dissipate. It is a conjugate ocular movement that involves compensatory eye movement the same distance as the head but in the opposite direction, mediated by the frontal eye fields that maintain connections with the parietotemporal cortex, and also with the primary, secondary, and tertiary visual cortices that initiate and guide the smooth pursuit movement. The mechanical stimuli generated in the vestibular receptors are transduced to impulses that travel in the primary vestibular fibers (axons of the bipolar neurons of the Scarpa ganglia) terminating in the cerebellum and the vestibular nuclei. Vestibular impulses are then transmitted to the abducens nucleus and, eventually, to the PPRF. Visual impulses from the neocortex are thought to be integrated with the vestibular impulses at the PPRF directly or via a relay in the SC.

Smooth Pursuit Eye Movement

Smooth pursuit movement is a slow conjugate eye movement that becomes active in tracking moving targets. Visual acuity is maintained during this movement. In order for the smooth pursuit movement to occur the, middle temporal visual area (MT), also known as the V5, must be activated to identify and register the velocity and direction of the moving target. MT lies caudal to the ascending limb of the inferior temporal sulcus at the junction of areas 19 and 37 near the occipitotemporal border. MT conveys the signal to the medial superior temporal visual area (MST), which lies in the inferior parietal lobule rostral and superior to MT. The MST and frontal eye field project to specific nuclei in the basilar pons that, in turn, project bilaterally to the posterior part of the cerebellar vermis and to the contralateral flocculus and fastigial nuclei. Signals received by the cerebellum are conveyed to the nucleus prepositus hypoglossi and medial vestibular nucleus and, through these nuclei, to the PPRF and riMLF.

Damage to MST produces ipsilateral impairment of smooth pursuit movement. Cerebellar dysfunctions and administration of sedatives and analgesics produce fragmentation of this movement into a series of saccades.

DISORDERS OF OCULAR MOVEMENTS

These disorders include nystagmus, conjugate gaze palsy, ocular dysmetria, oculogyric reflex, opsoclonus, ocular flutter, ocular bobbing ocular myoclonus, oscillopsia, and congenital ocular motor apraxia.

Nystagmus is an involuntary, rhythmic oscillation of the eye in response to an imbalance in the vestibular impulses (see also the vestibular system).

Conjugate gaze palsy includes lateral gaze and vertical gaze palsies. *Lateral gaze palsy* refers to the inability to look to the side of the lesion resulting from destruction of the abducens nucleus. *Vertical gaze palsy* is characterized by the inability to look up-or downward and is associated with lesions of the vertical gaze center in the rostral midbrain.

Ocular dysmetria denotes an error in ocular fixation, producing overshooting of the intended target followed by oscillation of the eyeball. This is commonly seen in cerebellar *vermian* lesion.

The oculogyric reflex is characterized by upward or side-to-side rolling movements of the eyes accompanied by abnormal contractions of the facial muscles. It is an expression of an acute dysgenic condition induced by neuroleptics. This reflex may be the result of metabolic disorders of dopamine and may be alleviated with anticholinergic medications.

Opsoclonus (dancing eyes in infants) is another ocular disorder that exhibits a random, conjugate saccadic movement of the eyes in all directions with unequal amplitudes. It is a manifestation of pretectal lesions or viral encephalitis.

Ocular flutter, seen in individuals with cerebellar lesions, is characterized by sudden, rapid, and spontaneous to-and-fro oscillations of the eyes. It is associated with blurred vision and may be seen with changes in fixation regardless of the direction of the gaze.

Ocular bobbing refers to the fast, spontaneous (not rhythmic) downward deviation of both eyes, followed by slow synchronous return of the eyes to the original position. This phenomenon may be seen in comatose individuals with lesions of the pons, cerebellum, or cerebral cortex.

Ocular myoclonus is a term used to describe the rhythmic, rotatory, or pendular movements of the eyes synchronously with similar movements of the palatal, pharyngeal, laryngeal, lingual, and diaphragmatic muscles.

Oscillopsia refers to a bilateral illusionary oscillation of objects in the visual field during head movement which shows variations that range from poor visual acuity and blurring to bobbing, in which the ability to hold images steady on the retina is lost. Oscillopsia may be vertical or horizontal or exhibit multiple forms of ocular movements. It is typically seen in the Arnold–Chiari malformation, superior canal dehiscence syndrome, and vestibular dysfunction. Patients with this deficit also exhibit a constant sense of nausea and vertigo. Unilateral oscillopsia in the form of shimmering and shaking of vision in one eye could be due to myokymia, a condition associated with spontaneous and involuntary contraction of the muscles due to vascular compression of the ocular muscles, particularly the superior oblique (superior oblique myokymyia), or high alcohol or caffeine intake.

Congenital ocular motor apraxia (Cogan syndrome) is a disorder of conjugate deviation of the eyes in which voluntary saccades are absent. The eye movements only occur when the head is in motion. The head abruptly turns to the side to visualize the object while the eyes move in the opposite direction of the movement.

OCULAR REFLEXES

Ocular reflexes are comprised of the direct pupillary, consensual pupillary, accommodation, ciliospinal, oculocardiac, and oculoauricular reflexes, as well as blink reflex of Descartes.

The direct pupillary light reflex (Figure 13.20) is produced by shining a beam of light into the eye and observing the pupillary constriction on the stimulated eye. This reflex is mediated by the optic nerve (afferent limb) and the oculomotor nerve (efferent limb). Information, which is carried by the optic nerve, is delivered to the optic tract and bilaterally to the oculomotor nuclei. This reflex is lost in Argyll Robertson pupil, a pupillary disorder that occurs in neurosyphilis, diabetes mellitus, and epidemic encephalitis and alcoholism.

The consensual pupillary light reflex (Figure 13.20) is characterized by constriction of both pupils in response to application of light to one eye. It is mediated by the bilateral connection of the optic tract to the oculomotor neurons via the central commissural connections. Disruption of the optic tract fibers that are destined to the oculomotor nuclei, as a result of a lesion medial to the LGB, produces manifestations of *Argyll Robertson pupil.* The latter exhibits loss of pupillary constriction in light reflex, while maintaining it in accommodation.

The accommodation reflex (Figure 13.20) exhibits certain changes in the eye that are associated with near vision. These changes include convergence (adduction of the eyes), miosis,

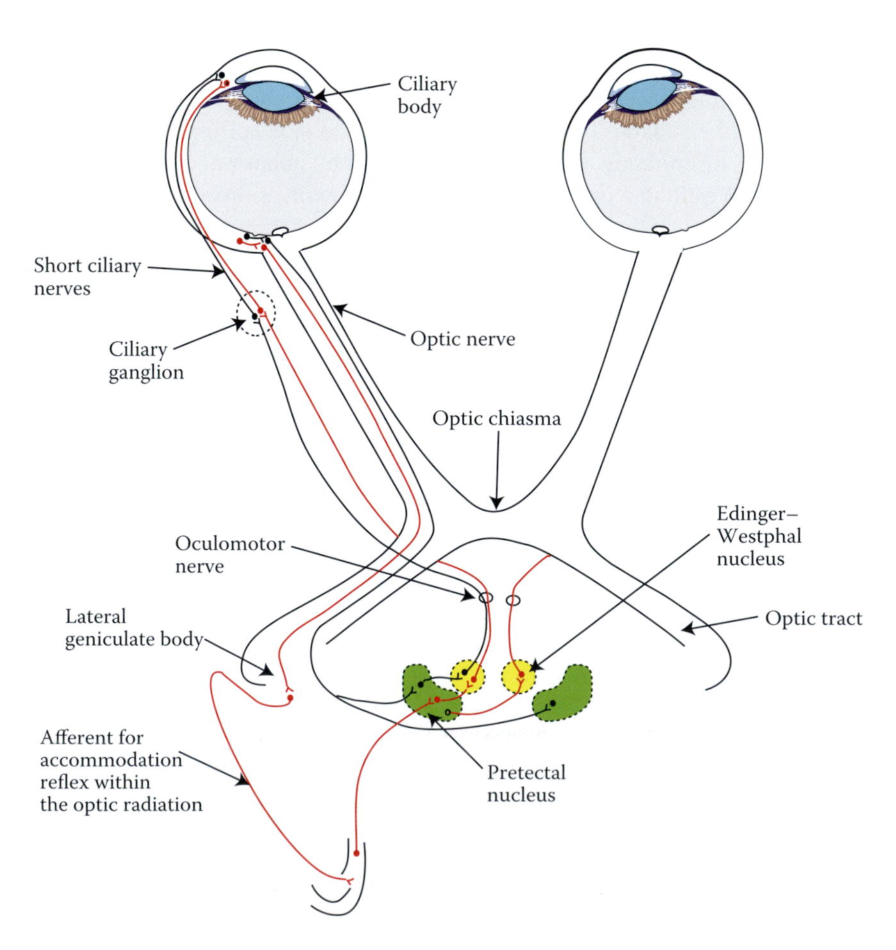

FIGURE 13.20 The reflex arcs of accommodation and pupillary light reflex.

and increased curvature of the lens. It requires the utilization of the visual cortex as well as the optic nerve, optic tract, and oculomotor nuclei.

The ciliospinal reflex exhibits pupillary dilatation in response to painful stimulation of a dermatomal area (e.g., pinching the neck or face). This reflex is dependent upon the integrity of the cervical postsynaptic sympathetic fibers as well as the presynaptic neurons of the first and second thoracic spinal segments.

The oculocardiac reflex is characterized by bradycardia (slowing of heart rate) in response to a pressure applied on the eyeball. It is mediated in the medulla by the ophthalmic nerve's (afferent limb) connections, via interneurons, to the spinal trigeminal nucleus, dorsal motor nucleus of the vagus, and the cardiovascular center (efferent limb).

The oculocephalic reflex (doll's eye movement) is typically produced in unconscious patients by passively rotating the head to one side, which results in initial movements of the eyes contralaterally and then toward the midline, irrespective of the direction of rotation. Raising the head lowers the eyes and the reverse is true (Cantelli sign). This reflex is dependent upon the integrity of the vestibular, oculomotor, and abducens nerves and nuclei as well as the MLF, which is inhibited in the awake individual by the descending cortical influences. Closure of the eyelids facilitates this reflex by eliminating the cortical input. Patients who have bilateral cortical lesions, as in comatose individuals, with intact brainstem connections between the oculomotor nerve and the vestibulocochlear nuclei, exhibit a brisk doll's eye movement.

Loss of the oculocephalic reflex is an ominous finding, which indicates metabolic depression or a lesion in the brainstem that disrupts the connection between the third and eighth cranial nerves. Suppression of the ascending reticular activating system and loss of consciousness occur when the lesion is located rostral to the pontine and midbrain gaze centers. Therefore, loss of this reflex in comatose patients may indicate that the trauma has damaged the caudal pons and did not spare the lateral gaze center, requiring urgent intervention. Impaired oculocephalic response may also occur as a result of malpositioning or inadequate head rotation.

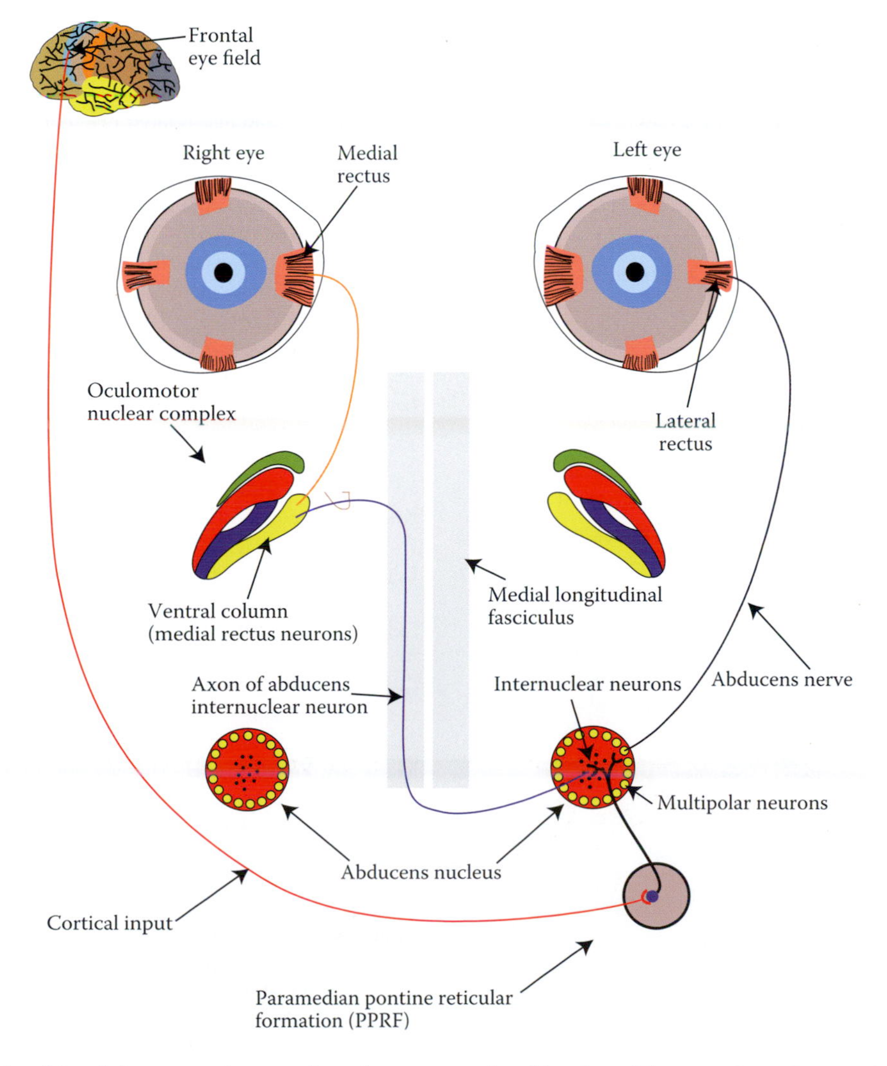

FIGURE 13.21 The role of the abducens nucleus as a lateral gaze center in adduction of the contralateral eye and abduction of the ipsilateral eye. The role of the cerebral cortex in influencing eye movement is also illustrated.

The oculoauricular reflex is elicited by asking the patient to look to the extreme temporal side. It is characterized by contraction of the posterior auricular muscles and the subsequent movement of the ear posteriorly, contralateral to the stimulated side. This reflex is absent in Bell's palsy.

The blink reflex of Descartes is produced by an object that abruptly and unexpectedly approaches the eye. This reflex is mediated by the optic and facial nerves and is characterized by contraction of the orbicularis oculi in response to this stimulus.

GAZE CENTERS

Gaze centers are represented by the lateral and vertical gaze centers in the pons and midbrain, respectively. The lateral gaze (horizontal) center (Figure 13.21) is located in the abducens nucleus and the adjacent PPRF. This region includes a pulse generator for fast eye movements and an integrator that determines the ultimate resting position of the eye. It projects to the ipsilateral abducens nucleus, which controls the contralateral medial rectus muscle, and the ipsilateral lateral rectus muscle. The corticotectal tract, which is derived from the frontal eye field (Brodmann area 8), carries information that projects to the contralateral gaze center and regulates contralateral voluntary conjugate eye movements. Corticotectal fibers that are derived from the occipital lobe (Brodmann areas 17, 18, and 19) control involuntary smooth pursuit eye movement.

> Lesions of the contralateral pontine lateral gaze, the ipsilateral frontal eye field, or the ipsilateral corticomesencephalic tract may produce gaze palsy to the opposite side.

The vertical gaze center is located in the riMLF adjacent to the SC of the midbrain. This premotor center is damaged in Parinaud syndrome (Figure 13.22), which is characterized by the inability to gaze upward, with weakness of convergence and sometimes loss of pupillary light reflex and mydriasis. It primarily occurs as a result of a lesion in the superior colliculi or the posterior commissure, subsequent to a pineal gland tumor. The supranuclear mechanism for upward gaze is situated closer to the third ventricle than the center for downward gaze.

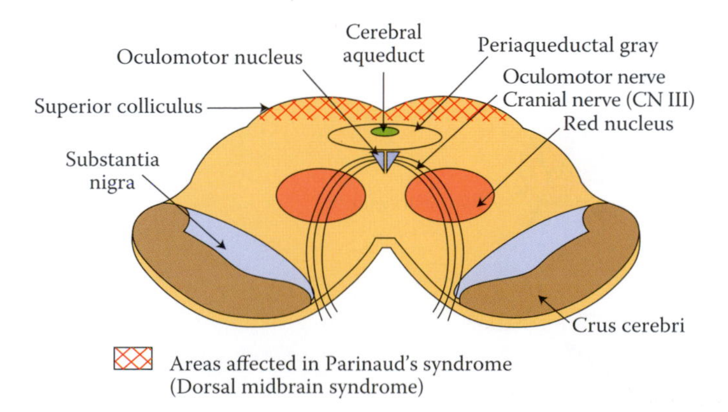

FIGURE 13.22 Section of the midbrain showing the lesion associated with Parinaud syndrome. In this syndrome, the vertical gaze center, which is represented in the superior colliculus, is disrupted.

> Posterior tumors of the third ventricle may result specifically in upward gaze palsy. Upward gaze palsy may also be seen in individuals with subdural hemorrhage or hydrocephalus. Selective downward gaze palsy can result from damage to the interstitial nucleus of Cajal, which lies dorsomedial to the MLF. Posterior thalamic hemorrhage is associated with downward deviation of the eye. Pretectal syndrome, which occurs as a consequence of vascular occlusion or neoplasms that are confined to the pretectum or the tectum, exhibits bilateral paralysis or paresis of vertical gaze, nystagmus, and lid retraction.

SUGGESTED READING

Betti V, Della Penna S, de Pasquale F, Mantini D, Marzetti L, Romani GL, Corbetta M. Natural scenes viewing alters the dynamics of functional connectivity in the human brain. *Neuron* 2013;79(4):782–97.

Carpentier S, Knaus M, Suh M. Associations between lutein, zeaxanthin, and age-related macular degeneration: An overview. *Crit Rev Food Sci Nutr* 2009;49(4):313–26.

De Pasquale R, Sherman SM. A modulatory effect of the feedback from higher visual areas to V1 in the mouse. *J Neurophysiol* 2013;109(10):2618–31.

Gaymard B, Pierrot-Deseilligny C, Rivaud S, Velut S. Smooth pursuit eye movement deficits after pontine nuclei lesions in humans. *J Neurol Neurosurg Psychiatry* 1993;56:799–807.

Hogan MJ, Alvarado IA, Weddell JE. *Histology of the Human Eye*. Philadelphia, PA: W.B. Saunders Co., 1971.

Huerta MD, Harting IK. The mammalian superior colliculus studies of its morphology and connections. In Vanegas H, ed. *Comparative Neurology of the Optic Tectum*. New York: Plenum, 1984, 687–773.

Kaas JH. Theories of visual cortex organization in primates. *Cereb Cortex* 1997;12:91–125.

Legothetis NK, Sheinberg DL. Visual object recognition. *Annu Rev Neurosci* 1996;19:577–621.

Leigh RJ, Zee DS. *The Neurology of Eye Movements*, 2nd ed. Philadelphia, PA: F.A. Davis, 1991.

Lopes-Ferreira D, Neves H, Queiros A, Faria-Ribeiro M, Peixoto-de-Matos SC, González-Méijome JM. Ocular dominance and visual function testing. *Biomed Res Int* 2013;2013:238943.

Marik SA, Yamahachi H, Meyer Zum Alten Borgloh S, Gilbert CD. Large-Scale Axonal Reorganization of Inhibitory Neurons following Retinal Lesions. *J Neurosci* 2014;34(5):1625–32.

Mays LE. Neural control of vergence eye movements: Convergence and divergence neurons in the midbrain. *J Neurophysiol* 1984;51:1091–108.

Merle H, Donnio A, Ayeboua L, Plumelle Y, Smadja D, Thomas L. Occipital infarction revealed by quadranopsia following snakebite by Bothrops lanceolatus. *Am J Trop Med Hyg* 2005;73(3):583–5.

Merriam EP, Gardner JL, Movshon JA, Heeger DJ. Modulation of visual responses by gaze direction in human visual cortex. *J Neurosci* 2013;33(24):9879–89.

Metitieri T, Barba C, Pellacani S, Viggiano MP, Guerrini R. Making memories: The development of long-term visual knowledge in children with visual agnosia. *Neural Plast* 2013; 2013:306432.

Peters A, Payne BR, Budd J. A numerical analysis of the geniculocortical input to striate cortex in the monkey. *Cereb Cortex* 1994;4:215–29.

Sengpiel F, Blakemore C. The neural basis of suppression and amblyopia in strabismus, part 2. *Eye* 1996;10:250–8.

Sparks DL, Mays L. Signal transformations required for the generation of saccadic eye movements. *Annu Rev Neurosci* 1990;13:309–36.

Straube A, Leigh RJ, Bronstein A, Heide W, Riordan-Eva P, Tijssen CC, Dehaene I, Straumann D. EFNS task force—Therapy of nystagmus and oscillopsia. *Eur J Neurol* 2004; 11:83–9.

Westheimer G. Editorial: Visual acuity and hyperacuity. *Invest Ophthalmol* 1975;145(8):570–2.

14 Auditory System

The auditory system is a special somatic afferent (SSA) system that transmits airborne vibrations via the external acoustic meatus, middle ear cavity and the perilymph to the organ of Corti. Mechanical displacement of the neuroepithelial hair cells is eventually converted into auditory impulses and conveyed via the cochlear nerve and ascending auditory pathways primarily to the contralateral primary and secondary auditory cortices via a chain of neurons in the brainstem and diencephalon. Auditory dysfunctions can occur as a result of a lesion of the cochlear nerve, auditory pathways, or auditory cortex. Auditory dysfunctions can occur in conjunction with vestibular dysfunctions.

PERIPHERAL AUDITORY APPARATUS

On developmental, structural, and functional basis, the peripheral auditory apparatus consists of the external, middle, and inner ear.

External Ear

The *auricle* is a funnel-like structure that collects and directs air vibrations through the external acoustic meatus (Figure 14.1). It is primarily a cartilaginous structure, which consists of numerous irregular curvatures and eminences. These eminences include the helix and antihelix separated by the scaphoid fossa. The antihelix surrounds the concha, a depression that leads into the external acoustic meatus. The upper part of the antihelix divides into two crura by the triangular fossa. The part of the concha above the anterior end of the helix is known as the cymba concha. The latter overlies a triangle area superior to the external acoustic meatus, which marks the lateral wall of the mastoid antrum. The concha is guarded anteriorly and posteriorly by a cartilaginous projection known as the tragus and antitragus, respectively. These two projections are separated by the intertragic notch. The lobule (ear lobe), a noncartilaginous part of the auricle, consists of fibrous and fatty tissues. The auricle gives attachment to a group of primitive muscles of facial expression, which are innervated by the facial nerve. The auricle receives sensory innervation from the great auricular, lesser occipital, facial, auriculotemporal, and vagus nerves.

The *external acoustic meatus* (Figure 14.1) is an S-shaped tube, which is cartilaginous in its lateral one-third and bony in its medial two-thirds, extending from the concha to the lateral wall of the tympanic cavity.

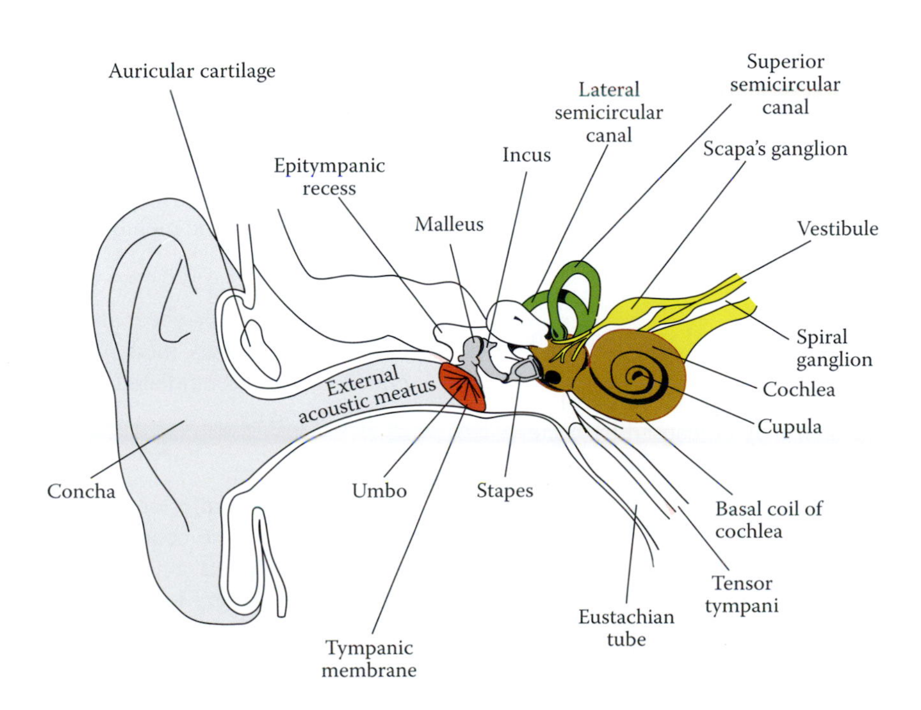

FIGURE 14.1 Diagram of the external, middle, and inner ear.

In the adult, the S-shaped curve of the external acoustic meatus is corrected, and proper visualization of the canal during otoscopic examination may be maintained by pulling the auricle upward and posteriorly. In children, shortness of the canal and the equal length of the bony and cartilaginous parts render the tympanic membrane more vulnerable to injuries during examination. The angle of junction between the bony and cartilaginous parts is a common site of entrapment of foreign bodies.

In the newborn, the anteroinferior wall of the bony external acoustic meatus contains the foramen of Huschke, which persists until approximately 5 years of age. The cartilaginous part continues laterally with the concha. The bony part forms a sulcus medially for the insertion of the tympanic membrane. The external acoustic meatus (Figure 14.1) lies superior to the parotid gland, anterior to the mastoid air cells, and inferior to the middle cranial fossa.

The firm adherence of skin to the underlying cartilage and bone of the external acoustic meatus accounts for the overstimulation of the nociceptors and the resultant excruciating pain associated with inflammatory conditions of the this canal.

Cerumen (earwax) is a secretion of the subcutaneous glands in the medial part of the external acoustic meatus. These glands receive sensory innervation from the vagus nerve and the auriculotemporal branch of the mandibular nerve. The adhesive qualities of the cerumen may help to protect the ear canal from foreign bodies. The auriculotemporal nerve is responsible for the referred earache associated with tooth decay or lingual ulcer. Vagal innervation of the external acoustic meatus explains the coughing and sneezing reflexes and bradycardia associated with excessive irrigation of the external acoustic meatus.

Aberrant connection between the auricular branch of the vagus nerve with the chorda tympani following ear surgery produces *gustatory otolgia-wet ear syndrome* which is characterized by taste induced secretion in the external acoustic meatus.

Inflammation of the perichondrium of the cartilaginous auricle (perichondritis) may be caused by bacterial infection subsequent to traumatic injury, insect bites, or incised superficial abscess. It is characterized by accumulation of pus between the cartilage and perichondrium, occasionally leading to avascular necrosis and a deformed external ear. Suction drainage and systemic antibiotics are required for treatment.

External otitis (*Pseudomonas* osteomyleitis of the temporal bone) is another condition that affects the external ear and commonly occurs in diabetic patients, particularly the elderly. It begins as Pseudomonas aeroginosus infection, which progresses to become a *Pseudomonas* osteomyelitis.

Persistent and severe earache and development of granulation tissue that blocks the external canal are some of the symptoms of this condition. Conductive hearing loss and facial palsy may also occur. This condition may spread to the entire temporal bone, if it is not controlled. Surgical intervention and intravenous antibiotic therapy may be required.

Middle Ear

The middle ear (Figures 14.1, 14.2, 14.3, 14.4, and 14.5) consists of irregular air-filled cavities within the temporal bone, containing the ossicles (malleus, incus, and stapes). This cavity serves a mechanical function in transmitting mechanical energy, in the form of the airborne vibrations of sound waves, from the external environment to the inner ear. This cavity is connected anteriorly to the nasopharynx via the pharyngotympanic (Eustachian) tube and posteriorly to the mastoid antrum. It has lateral, medial, anterior, and posterior walls in addition to the roof and floor. The *lateral wall* of this cavity is formed by the tympanic membrane within the tympanic sulcus, and the epitympanic recess. The *tympanic membrane* (Figures 14.1, 14.2, 14.3, and 14.5) in the adult has a fibrocartilaginous periphery lodged in the tympanic sulcus at a 45° angle with the horizontal plane. In children, this membrane maintains a horizontal plane.

Structurally, it consists of a connective tissue layer, covered externally by the skin and lined by mucosa. It has a large, taut distal part (pars tensa) and a smaller, proximal, and triangular part known as the pars flaccida. The flaccid part lies above the malleolar folds, extending between the lateral process of the malleus and the deficient edges of the tympanic sulcus. The handle of the malleus, which attaches to the medial surface of the pars tensa of the tympanic membrane, forms the umbo, a central depression on the lateral surface of the tympanic membrane.

The umbo may disappear in individuals with middle ear infections (otitis media) due to the pressure generated by the accumulated inflammatory fluid in the tympanic cavity.

The anterior and inferior quadrant of the tympanic membrane is known as the cone of light, or *triangle of Politzer.* The chorda tympani, a branch of the facial nerve, runs between the inner and intermediate layers of the tympanic membrane, medial to the handle of the malleus.

Drainage of inflammatory exudate from the tympanic cavity is commonly performed through an opening in

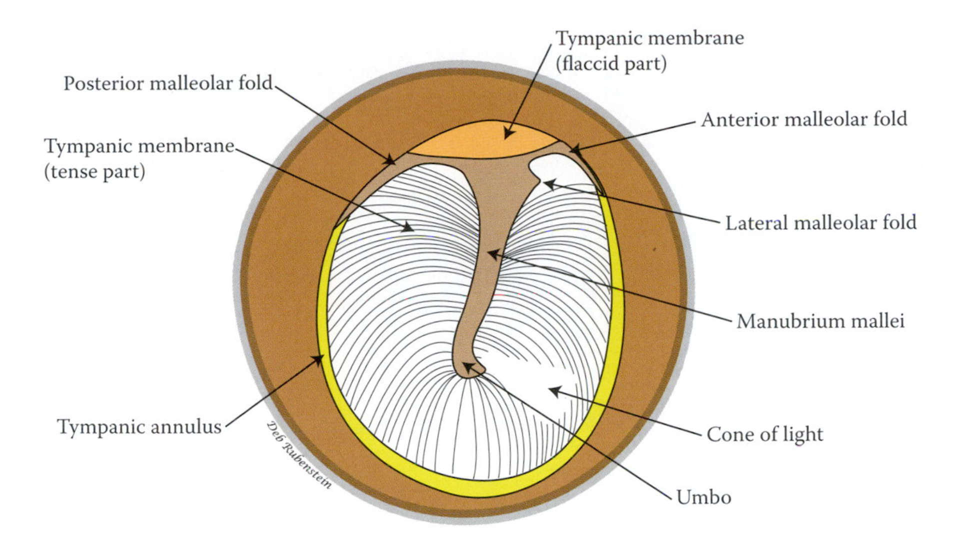

FIGURE 14.2 The tympanic membrane and associated ligaments.

the posteroinferior quadrant of the tympanic membrane. This quadrant is less vascular and contains no prominent nerves or ossicles.

Barotitis media (aerotitis) is a condition that results from sudden change in the atmospheric pressure relative to the pressure in the tympanic cavity. Descent of an airplane or deep sea diving usually brings about this abrupt ambient pressure change. This is mediated by reflex swallowing and widening of the auditory (Eustachian) tube. Partial or complete occlusion of the auditory (Eustachian) tube due to allergy, upper respiratory tract infection, or enlarged tubal tonsils may render the pressure in the tympanic cavity lower than the atmospheric pressure. This leads to retraction of the tympanic membrane and the transudation of blood from the blood vessels of the lamina propria of the mucus membrane. Bleeding into the tympanic cavity and rupture of the tympanic membrane may occur in severe pressure differentials. An individual with allergy or respiratory tract infection may be advised not fly or apply nasal vasoconstrictors when flying. Conductive deafness and severe pain usually accompany sudden pressure changes. Sometimes, a perilymphatic fistula from the oval or round window may accompany the bleeding and is generally accompanied by sensorineuronal hearing loss and vertigo.

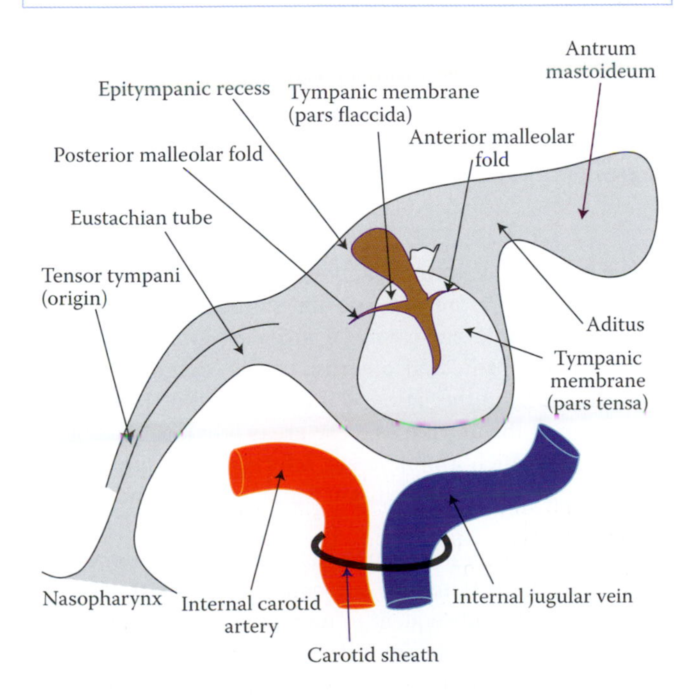

FIGURE 14.3 Diagram of the floor, lateral and anterior walls of the middle ear (tympanic cavity).

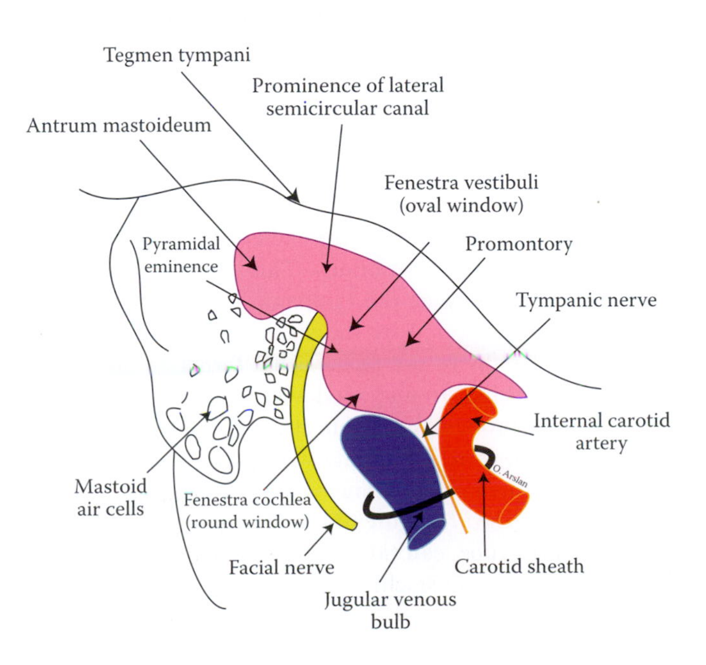

FIGURE 14.4 Schematic drawing of the medial and anterior walls, the roof and floor of the middle ear (tympanic cavity).

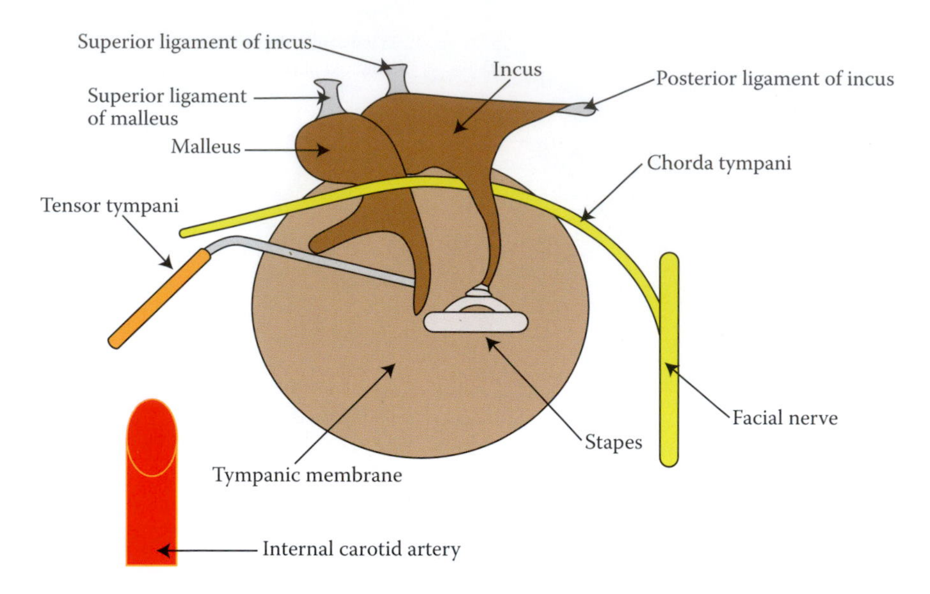

FIGURE 14.5 Diagram of the Rinne test.

Bulbous (infectious) *myringitis* is another condition that affects the tympanic membrane as a result of viral or bacterial infections. It is characterized by the formation of small fluid-filled vesicles on the tympanic membrane. This may progress to produce otitis media with fever and hearing loss. Infectious myringitis persists for 2 days and is usually caused by *Streptococcus pneumoniae* or mycoplasma infections. Antibiotics, analgesics, and induced rupture of the vesicles are common therapeutic measures for this condition.

The *medial wall* (Figure 14.4) contains the promontory, a bony prominence formed by the basilar part of the cochlea. The promontory contains the tympanic plexus, which is formed by the tympanic branch of the glossopharyngeal nerve and the carotid–tympanic nerve (sympathetic) fibers. This plexus supplies sensory fibers to the tympanic cavity, auditory tube, and mastoid air cells. The opening that lies posterior and superior to the promontory is known as the oval window (fenestra vestibuli). This opening establishes communication between the tympanic cavity and the scala vestibuli and is covered by the stapes. Immediately above the fenestra vestibuli, the facial canal forms an eminence. The fenestra cochlea (round window) is located posterior and inferior to the promontory, connecting the middle ear cavity to the scala tympani. The round window is occupied by the secondary tympanic membrane.

The *anterior wall* (Figures 14.1, 14.3, and 14.4) is formed by the auditory tube, carotid canal, and tensor tympani muscle. The auditory (Eustachian) tube connects the tympanic cavity to the nasopharynx, equalizing the pressure between these two cavities. Thus, this tube may serve as a route for the spread of infection from the pharynx to the middle ear. The auditory tube consists of lateral bony and medial cartilaginous parts. The medial cartilaginous part forms the tubal torus, a mucosal eminence in the lateral wall of the nasopharynx, which continues inferiorly with the salpingopharyngeal fold. The cartilaginous part gives attachment to the tensor palatini, salpingopharyngeus, and levator palatini muscles. The tensor palatini may be responsible for opening of the pharyngeal opening of the tube during swallowing. The carotid canal, which forms the anterior wall of the middle ear, is located within the petrous temporal bone and transmits the internal carotid artery.

The *posterior wall* (Figure 14.3) of the tympanic cavity bears several features, which include the aditus for the mastoid antrum, pyramidal eminence, and the fossa for the short process of the incus. The aditus is the gate to the mastoid antrum, containing on its medial wall a prominence formed by the lateral semicircular canal. The mastoid antrum is an air sinus, which is connected anteriorly to the epitympanic recess (via the mastoid antrum) and inferiorly to the mastoid air cells.

The antrum lies immediately medial to the suprameatal triangle, a common site for surgical intervention into the middle ear cavity. It also lies inferior to the tegmen tympani, and anterior to the sigmoid sinus, separated from the latter by a bony lamella. The relationship of the antrum to these areas provides possible routes by which a middle ear infection may spread to the temporal lobe of the brain, sigmoid sinus, and mastoid air cells.

Acute mastoiditis is an inflammatory condition that involves the soft tissues surrounding the mastoid air cells, usually, subsequent to untreated or inadequately treated otitis media (middle ear infection). Surgical mastoiditis, a medical and surgical emergency, encompasses osteitis and periosteitis of the mastoid bone, accompanied by transverse and sigmoid sinus thrombosis. Swelling, pitting edema, erythema, and percussion tenderness are some of the clinical signs of this condition. Downward and anterior displacement of the auricle may also occur. Abscess formed in the mastoid bone may occasionally involve the facial nerve, producing facial nerve palsy. Intravenous administration of antibiotics and surgical intervention that involves myringotomy, drainage of any abscess, and cortical mastoidectomy may be required. If treatment of mastoiditis is inadequate, sepsis, meningitis, brain abscess, and even death may ensue. Antibiotics may convert temporarily an acute mastoiditis into masked mastoiditis.

The pyramidal eminence contains the stapedius muscle and lies between the facial canal posteriorly and the oval window anteriorly.

The *roof* (Figure 14.3) of this cavity is formed by the tegmen tympani, which separates it from the meninges and the temporal lobe of the brain. It contains the superior petrosal sinus giving passage to veins that serve as a conduit for the spread of infection from the tympanic cavity to the temporal lobe and the meninges. In infants, unossified areas of the roof may also serve as a route for the spread of infection.

The *floor* (Figures 14.3 and 14.4) of the middle ear cavity is formed by a thin bony lamina, which lodges the internal jugular vein. Spread of infection to the systemic circulation from the middle ear cavity may occur if the separating bony lamina is unossified.

The middle ear cavity also contains *bony ossicles* (Figures 14.1 and 14.5), which transmit the vibration from the tympanic membrane to the perilymph of the inner ear. These ossicles include the malleus, incus, and stapes. These ossicles form synovial joints with each other to facilitate their movements.

The *malleus* has a head, neck, manubrium, and anterior and lateral processes. The head forms a sellar-type joint with the incus in the incudomalleolar articulation. Both the body of the incus and the head of the malleus are located in the epitympanic recess, proximal to the tympanic membrane. The manubrium is embedded in the medial surface of the tympanic membrane, forming the umbo. The tendon of the tensor tympani muscle inserts into the upper end of the manubrium. The anterior process is short and is connected to the petrotympanic fissure, while the lateral process is longer and gives attachments to the malleolar folds.

The *incus* resembles the premolar tooth, having two processes (long and short) and a body. The body articulates with the head of the malleus, and the long process articulates in a spheroidal joint with the stapes. The short process attaches to the fossa incudis.

The *stapes* resembles a stirrup consisting of a head, neck, and oval base that covers the oval window via the annular ligament. The head articulates with the lenticular process of the incus. The neck gives attachment to the tendon of the stapedius muscle. Movement of the stapes is pistonlike, producing waves within the perilymph of the scala vestibuli.

Paralysis of the stapedius muscle may directly affect the movement of the stapes, thereby lowering the hearing threshold and resulting in hyperacusis. *Otosclerosis* is the most common cause of progressive conductive hearing loss in the adult. This hereditary condition is characterized by excessive bony growth and the formation of irregularly arranged immature bones and ankylosing of the joints between the ossicles, particularly the stapedial base. When otosclerotic plaques impinge upon the scala media, neuronal deafness occurs. Microsurgery and replacement of the affected stapes with prosthesis may be required for treatment. Hearing aids may serve the same purpose.

Otitis media refers to a bacterial or viral infection of the middle ear, usually secondary to upper respiratory tract infections. This condition, which is more common in children, may be acute or chronic. Acute otitis media may be suppurative or serous. *Acute suppurative otitis media* most commonly develops as a result of bacterial contamination via the Eustachian tube in the presence of preexisting inflammation in the middle ear. Eustachian tube dysfunction results in absorption of oxygen and its replacement by carbon dioxide, which initiates an inflammatory response followed by accumulation of transudate. Persistence of this condition produces exudate that may be contaminated by the infected nasopharyngeal content. Parainfluenza, coxsackieviruses, and adenoviruses are the most frequent viruses involved in this condition. Bacterial otitis media is most commonly caused by *S. pneumoniae*, *Staphylococcus aureus*, *Haemophilus influenzae*, *Mycoplasma pneumonia,* and group A *Streptococcus pyogenes*. Persistent and severe earache and temporary hearing loss are some of the initial symptoms. Fever, nausea, and diarrhea may occur in young children. Additional signs include bulging of the tympanic membrane or its severe retraction. Perforation of the tympanic membrane may occur, and pulsatile discharge may be seen. Mastoiditis, labyrinthitis, and meningitis are some of the complications of the disease. Therapy may include antibiotics and myringotomy (a surgical opening in the eardrum) to drain the accumulated pus.

Acute and serous otitis media is a disorder in which a middle ear effusion develops subsequent to persistent occlusion of the auditory (Eustachian) tube. Upper respiratory infection, chronic rhinosinusitis, cleft palate, and nasopharyngeal adenoids may lead to persistent occlusion of the Eustachian tube. The serous transudate may become mucoid and glue-like exudate after few days, containing bacterial contaminants.

This condition is more common in young children because of the presence of a narrow Eustachian tube, enlarged adenoids, and frequency of nose and throat inflammation. This also occurs when maximum acquisition of speech skills is needed. In this condition, the tympanic membrane thickens and frequently is retracted by the negative pressure of the tympanic cavity. The pars flaccida of the tympanic membrane may also be retracted, leading to cholesteatoma (keratoma). The latter is a growth of normal stratified squamous epithelium that enlarges and eventually destroys the ossicles and even the inner ear and cranial cavity. Cholesteatoma becomes an excellent site for the growth of bacteria. Tympanometry may be employed to measure the pressure on both sides of the tympanic membrane. Effusion, in general, is a self-limiting process, which resolves in a period of 2 weeks. Antibiotics, nasal decongestants, and antihistaminic medications may be used. A blocked Eustachian tube may be forced open by asking the patient to breathe out through his/her mouth while his/her nostrils are pinched shut. Drainage of fluid from the tympanic cavity may be accomplished by myringotomy, a surgical procedure that involves making an incision in the tympanic membrane.

Chronic suppurative otitis media is always associated with central or peripheral perforations of the tympanic membrane with chronic purulent otorrhea. These perforations result in conductive hearing loss. Exacerbations of this condition due to upper respiratory tract infection may lead to the formation of aural polyps and destructive changes in the middle ear cavity. Peripheral perforations usually occur in the posterior superior quadrant of the pars tensa, destroying large areas of the tympanic membrane, including the annulus tympanicus and the mucus membrane. Labyrinthitis, facial palsy, and intracranial suppuration are more likely complications of peripheral than central perforations. Pars flaccida (attic) perforations may extend to the epitympanic recess. Peripheral and pars flaccida perforations are frequently associated with cholesteatoma. In central perforation, chronic otitis media may be exacerbated after upper respiratory infection, resulting in painless discharge from the ear. Persistence of infection may lead to conductive hearing loss. Topical antibiotics are an initial step in the treatment of chronic suppurative otitis media, although neomycin-containing preparations may be contraindicated in patients with perforated tympanic membrane because of the possible otoxicity-induced neuronal deafness. When otorrhea (ear discharge) exists, daily irrigation with Burrow's solution may be advised. Surgical closure of the tympanic membrane may be required.

Inner Ear

The inner ear (Figures 14.4 and 14.6) contains the receptors for the auditory and vestibular systems, consisting of bony and membranous labyrinths. The bony labyrinth encloses the membranous labyrinth. The bony labyrinth is lined by periosteum and is filled with perilymph, consisting of the cochlea, vestibule, and semicircular canals. The vestibule represents the middle portion of the bony labyrinth, connected to the tympanic cavity via the oval window (fenestra vestibulae) and the round window (fenestra cochlea). The membranous labyrinth in the vestibule forms the utricle and saccule. The utricle responds to linear acceleration and deceleration and gravitational pull, while the saccule is thought to deal with vibratory sense.

Abrupt compression or decompression injuries, head trauma, heavy lifting, or straining my result in the formation of round or oval window fistulas. These fistulas may produce fluctuating hearing loss and/or tinnitus, which may improve overnight and worsen during the day.

The medial wall of the *vestibule* (Figure 14.6) contains the aqueduct of the vestibule, containing the endolymphatic duct. The latter duct connects the utricle to the saccule. The posterior part of the bony labyrinth is formed by the semicircular canals, which enclose the membranous semicircular ducts. The semicircular canals are organized perpendicular to each other, representing the three dimensions in space. They comprise the anterior (superior), lateral (horizontal), and posterior semicircular canals. The anterior canal is parallel to the posterior canal of the contralateral side. Each canal has a dilated lower part called the ampulla, which contains the ampullary crest and the neuroepithelial receptor cells.

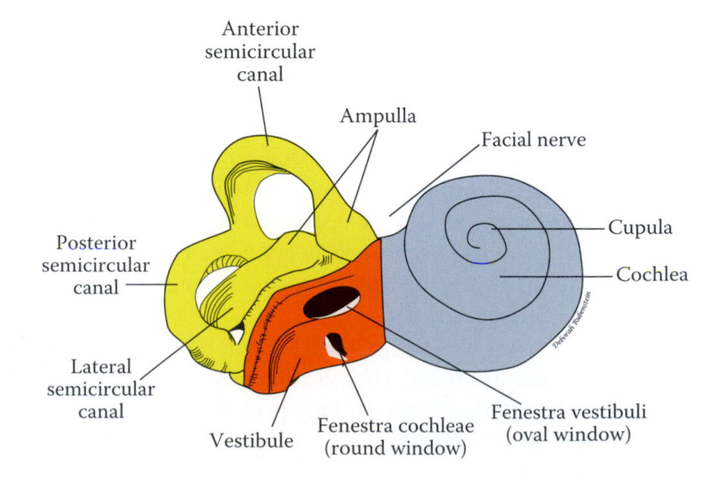

FIGURE 14.6 In this schematic diagram, the main components of the inner ear are illustrated. Observe the close relationship of the facial canal and nerve to the vestibule.

The anterior part of the bony labyrinth, the *cochlea* (Figures 14.1 and 14.6), forms two and one half turns, which end at the apex or cupula. The basilar part of the cochlea forms the promontory on the medial wall of the tympanic cavity. The cochlea is connected to the middle ear cavity by the fenestra vestibuli (oval window), which is covered by the stapes and by the fenestra cochlea (round window). The fenestra cochlea is covered by the secondary tympanic membrane. The bony cochlea encloses the membranous cochlear duct (scala media) containing the auditory receptors. It also contains the organ of Corti, the scala vestibuli, and the scala tympani. The scala tympani is connected to the subarachnoid space by the cochlear canaliculus.

The cochlea consists of a bony shell, a central axis (modiolus), and a spiral lamina that protrudes into the auditory canal. The modiolus, which is comprised of trabecular tissue, forms the pillar around which the cochlea and spiral lamina make two and half turns, containing foramina for the cochlear nerve branches. The spiral bony lamina protrudes inside the cochlear canal and is connected to the spiral ligament of the cochlear bony wall by the basilar and vestibular membranes. The *basilar membrane* (Figure 14.7), which contains the outer and inner hair cells and rods, is an integral part of the mechanism that transduces the mechanical energy created by the sound waves to chemoelectric potentials. The tips of the hair cells are embedded in the tectorial membrane. High frequencies activate the neuroepithelial hair cells in the basal turn of the cochlea, whereas low-frequency sounds stimulate the corresponding neuroepithelial cells in the apical turn.

The *vestibular membrane* of Reissner extends from the upper lip of the spiral lamina to the stria vascularis (Figure 14.7). The latter is a vascular area on the inner wall of the cochlea. The vestibular and the basilar membranes and the cochlear wall constitute the boundaries of the endolymph-filled cochlear duct or scala media. The cochlear duct is separated from the perilymph-filled scala tympani and scala vestibuli via the basilar and vestibular membranes, respectively. The scala vestibuli and tympani are separated from each other throughout the entire length of cochlear duct except at the cupula, where a connection is formed via the helicotrema. The cochlear canaliculus interconnects the scala tympani with the subarachnoid space.

Tears of the intracochlear (Reissner's) membrane have been identified frequently in patients with cochlear hydrops. This condition may produce potassium poisoning of the sensorineural structures of the cochlea, leading to fluctuating hearing loss.

The *perilymph* resembles cerebrospinal fluid and extracellular space and occupies the perilymphatic space between the bony and membranous labyrinths. It is connected to the subarachnoid space via the cochlear canaliculus. It is believed that the perilymph is an ultrafiltrate of blood or, possibly, that it may be a derivative of the cerebrospinal fluid surrounding the vestibulocochlear nerve or contained within the cochlear canaliculus. The perilymphatic spaces of the semicircular canals and vestibule are connected to the scala vestibuli, and through the helicotrema, to the scala tympani.

The *endolymph* fills the membranous labyrinth and resembles the intracellular fluid in its ionic composition. In contrast to the perilymph, the endolymph contains a high concentration of potassium ions and a lower concentration of sodium ions. The endolymph may be derived from the stria vascularis, a specialized epithelium with rich blood capillaries; from the dark epithelial cells of the inner layer of the utricle and semicircular ducts; and also from the platinum semilunatum (a specialized inner epithelium near the ampullary crests and macula).

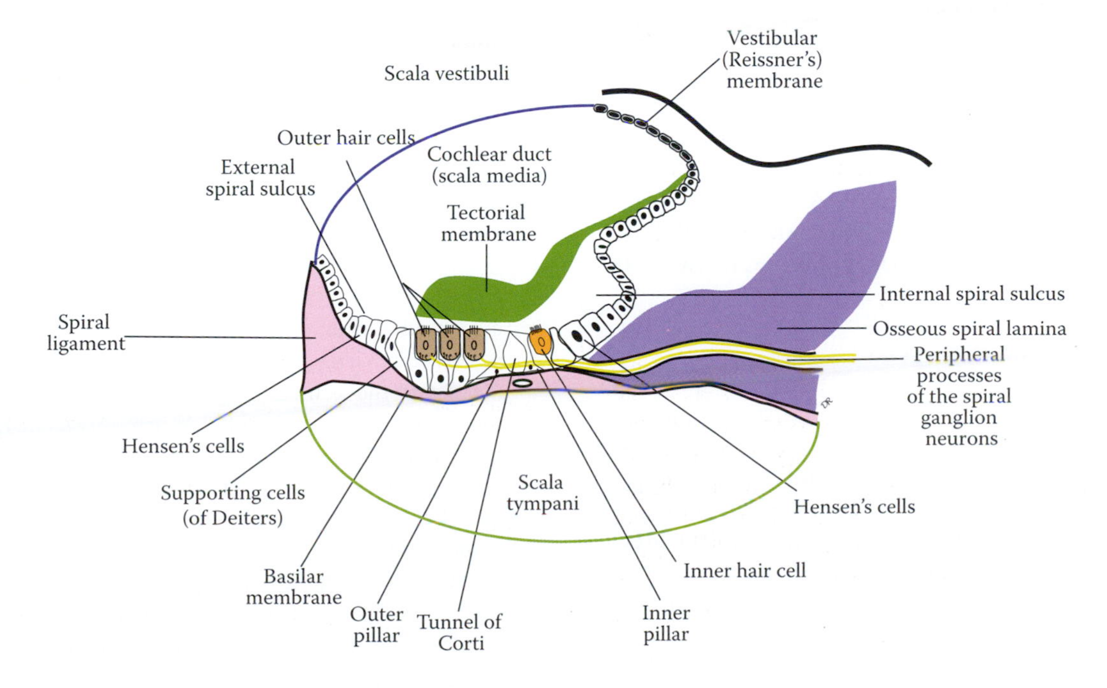

FIGURE 14.7 The cochlear duct and the organ of Corti.

Airborne vibration is initially collected by the auricle, where binaural audition is enhanced via curves and depressions of this cartilaginous structure. Sound is then funneled through the external acoustic meatus, which creates vibrations on the tympanic membrane and subsequent movement of the malleus and incus. Movement of the incus produces a pistonlike movement of the stapes on the oval window, creating a current in the perilymph of the scala vestibuli. The traveling current, which induces pressure changes, passes through the helicotrema to the scala tympani and then to the compliant basilar membrane. Vibration of the basilar membrane, upon which the neuroepithelial cells (outer and inner hair cells) rest, results in mechanical displacement of the hair cells within the tectorial membrane accompanied by a shearing force. This results in the propagation of a wave of electrochemical energy and its conversion via transductive process into nerve impulses, which are carried by the peripheral and then the central processes of the spiral ganglion neurons. High-frequency sounds generate a current in the basilar membrane near the base of the cochlea, whereas low-frequency sounds with peak amplitude of wave in the basilar membrane near cupola. This supports the concept that tonotopic localization of sound frequencies may be determined by the mechanical features of the basilar membrane.

The organ of Corti may be damaged by a variety of agents and conditions. Degeneration of the hair cells in the basal coil of the cochlea may occur in the elderly, producing loss of high tone. Midfrequency hearing loss may occur as a result of degeneration of the hair cells in the middle cochlea. The latter deficit is seen in industrial workers and rock band performers who are exposed to loud noises. Prolonged treatment with certain antibiotics such as streptomycin, neomycin, kanamycin and dihydrostreptomycin, or aspirin may produce ototoxicity and degeneration of the outer hair cells nearest to the tunnel of Corti. The trend of degeneration will continue to involve the lateral outer hair cells and then the inner hair cells. At first, the hearing loss will be confined to higher frequencies, and then, as the degeneration progresses, lower frequencies will also be affected.

Spiral Ganglion

The spiral ganglia (Figure 14.7) are located in the modiolus of the cochlea and consist of bipolar neurons, representing the first-order neurons of the auditory system. The cochlear nerve, which is formed by the central process of the neurons of the spiral ganglion neurons, constitutes the largest component of the vestibulocochlear nerve. This nerve exits at the cerebellopontine angle, traveling within the internal acoustic meatus accompanied by the vestibular and facial nerves and the labyrinthine artery. The internal acoustic meatus lies within the temporal bone, anterior and superior to the jugular foramen. It is divided into upper and lower parts by a transverse crest. The upper part of the meatus gives passage to the facial nerve and branches of vestibular nerve to the utricle and the lateral semicircular ducts, while the lower part of the meatus transmits the cochlear nerve as well as branches of the vestibular nerve to the posterior semicircular duct. The labyrinthine artery also pierces the internal acoustic meatus to supply the structures in the inner ear. This artery, which arises frequently from the anterior inferior cerebellar artery and less often from the lower basilar artery, runs with the facial and vestibulocochlear nerves.

Cochlear Nerve

The cochlear nerve, as discussed earlier, is formed mainly by the axons of the neurons (type I and II) of the spiral ganglia, which are located in the spiral osseous laminae, but it also contains efferent fibers. Type I cells convey information from the inner hair cells via myelinated axons and represent the vast majority of the neurons, whereas the type II cells transmit information from the inner hair cells through unmyelinated axons. The fibers of the cochlear nerve accompany the vestibular nerve in their course toward the brainstem by initially connecting to and then joining the vestibular nerve (via Oort's anastomosis) in the internal acoustic meatus before separating again. It enters the brainstem at the cerebellopontine angle and courses on the lateral side of the inferior cerebellar peduncle to distribute in the cochlear nuclei in a manner that a single cochlear nerve fiber stimulates several neurons of the cochlear nuclei. Virtually all *cochlear nerve* fibers (Figures 14.7 and 14.8) distribute to the ventral and dorsal cochlear nucleus. The efferent component of the cochlear nerve forms the olivocochlear bundle. Fibers from the basal part of the cochlea that receives high-frequency sound terminate mainly in the ventral portions of both cochlear nuclei. The cochlear nerve fibers that convey low-frequency sound from the apical turn of the cochlea terminate in the dorsal parts of the cochlear nuclei.

Herpes zoster oticus (Ramsay Hunt syndrome), as discussed earlier, is a viral disease caused by the herpes zoster virus that invades the vestibulocochlear nerve and the geniculate ganglion of the facial nerve. Symptoms include neuronal hearing loss, vertigo, and facial palsy. Vesicular eruptions along the course of the cutaneous branches of the facial nerve in the concha can also be seen. These symptoms may be transient or complicated by the involvement of other cranial nerves. Corticosteroid therapy and antiviral medication (acyclovir) may be used in the treatment of this condition. Analgesics for pain and diazepam for the treatment of vertigo may also be used.

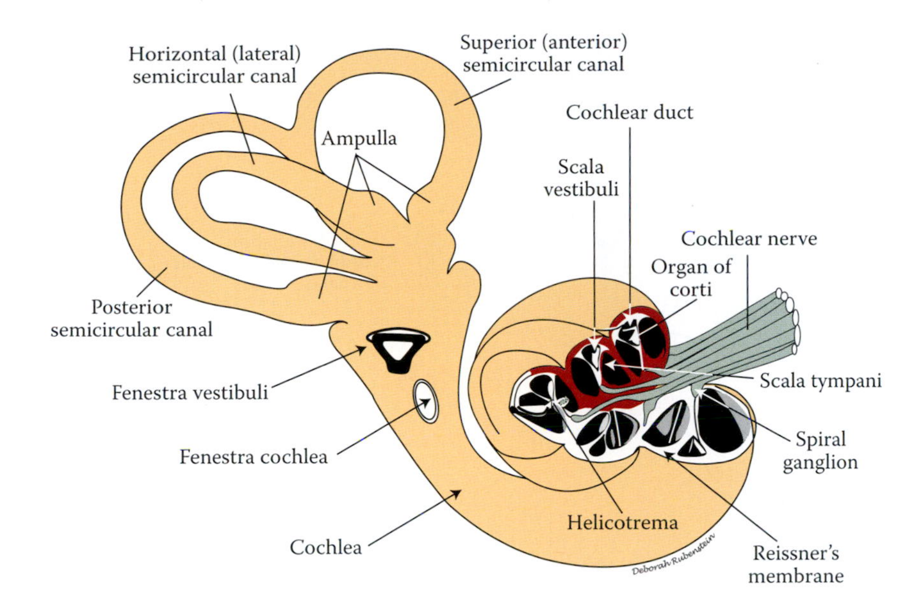

FIGURE 14.8 Section through the cochlea showing the scalae vestibuli, tympani, and media, as well as the organ of Corti.

CENTRAL AUDITORY PATHWAYS

Cochlear Nuclei

The *cochlear nuclei* are located dorsolateral to the inferior cerebellar peduncle, with the dorsal nucleus forming the acoustic tubercle in the rhomboid fossa on the posterior part of the inferior cerebellar peduncle. This nucleus appears to be continuous with the ventral cochlear nucleus. Many types of neurons are contained in a more cytoarchitecturally complex ventral cochlear nucleus that receives, in an orderly topographic manner, the ascending fibers of the cochlear nerve. The postsynaptic fibers from the cochlear nuclei form three acoustic striae, contributing to the lateral lemniscus.

Acoustic Striae

The *ventral acoustic stria* (Figures 14.9 and 14.10) is the largest acoustic stria, which arises from the ventral portion of the ventral cochlear nucleus and forms the trapezoid body by decussating with the corresponding stria of the opposite side. The trapezoid body, which contains raphe nuclei, lies ventral to the medial lemniscus in the ventral tegmentum of the pons, adjacent to the median raphe, its fibers continue on the contralateral side as the lateral lemniscus dorsal to the superior olivary nucleus. Some fibers of the ventral acoustic stria terminate bilaterally in the superior olivary nuclei. The superior olivary nucleus receives bilateral input from the ventral cochlear nuclei, which facilitate detection of the

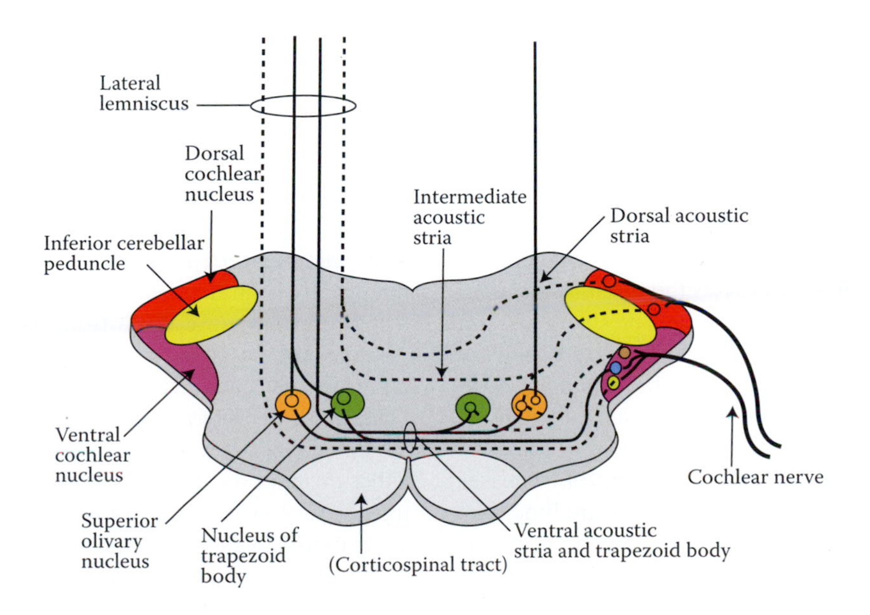

FIGURE 14.9 The auditory pathway from the cochlear nerve to the lateral lemniscus.

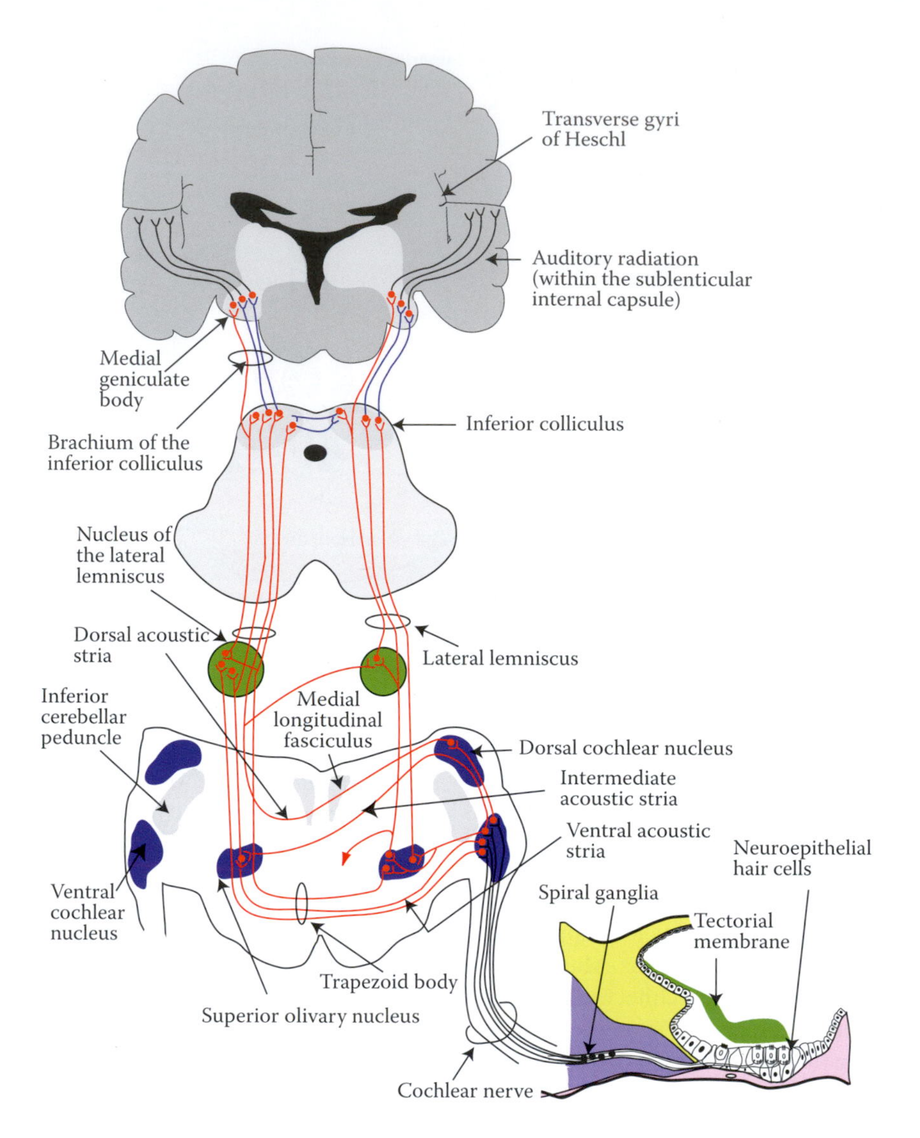

FIGURE 14.10 Complete neuronal sequence associated with the transmission of the auditory impulses from the external ear to primary auditory cortex in the temporal lobe.

difference in auditory impulse arrival between the two ears, enabling localization of sound. Postsynaptic fibers from the superior olivary nuclei project predominantly to the lateral lemniscus of the same side.

The *intermediate acoustic stria* (Figures 14.9 and 14.10) is the smallest stria, which originates from the dorsal portion of the ventral acoustic nucleus, crosses the midline, and contributes to the lateral lemniscus of the opposite side. This stria also projects bilaterally to the periolivary and retro-olivary nuclei, which form the inhibitory olivocochlear bundle. However, there are no fibers from this stria that terminate in the superior olivary nuclei.

The *dorsal acoustic stria* (Figures 14.9 and 14.10) arises from the dorsal cochlear nucleus, crosses the midline, and becomes part of the contralateral lateral lemniscus. Some fibers of this stria project to the superior olivary nuclei, which consist of bipolar neurons that play an important role in sound localization. These nuclei contribute fibers only to the ipsilateral lateral lemniscus. All acoustic striae eventually contribute fibers to the contralateral lateral lemniscus.

Lateral Lemniscus

The *lateral lemniscus* (Figures 14.9 and 14.10) contains predominantly contralateral and some ipsilateral secondary auditory fibers that are derived from the cochlear nuclei. It also contains tertiary fibers originating from the superior olivary and trapezoid nuclei and the nuclei of the lateral lemniscus. Damage to the lateral lemniscus primarily results in contralateral auditory deficits. Complete deafness will not occur in either ear since the lateral lemniscus contains bilateral ascending auditory fibers. The vast majority of the lateral lemniscal fibers terminate in the central nucleus of the inferior colliculus; however, some fibers may project to the contralateral central nucleus via the inferior collicular commissure, and others may directly reach the medial geniculate nucleus (MGN).

Inferior Colliculus

The *inferior colliculus* (Figure 14.11) is an auditory relay and reflex center that consists of a central nucleus continuous with the adjacent periaqueductal gray matter, which receives bilateral ascending fibers of the lateral lemniscus, and a pericentral nucleus that is concerned with the auditory impulses from one ear. The central nucleus is further subdivided into a dorsomedial and ventromedial part covered by a four-layered dorsal cortex that shows diversity in reference to the shape of neurons and their dendritic organization. The laminar arrangement of this nucleus enables detection of various frequencies and tonal discrimination since the dorsal and ventral parts of receive high- and low-frequency auditory impulses, respectively. Studies show that neurons of the central nucleus exhibit preferential connection relative to the source of incoming sounds in that the great majority of neurons respond to bilateral auditory input, while the other group of neurons are excited by contralateral input and inhibited by ipsilateral auditory signals. Neurons of the central nucleus of the inferior colliculus project through the brachium of the inferior colliculus to the ventral laminated part and the magnocellular divisions of the MGN. Experimental studies demonstrated that lesions of the inferior colliculus and its brachium produce deficits associated with localization, tone detection, auditory learning, and reflexes. The projections of the inferior colliculus to the spinal and brainstem neurons cross the superior colliculus in their course to their targets, establishing synaptic linkage with the neurons that form the tectospinal and tectotegmental pathways, accounting for the reflexive turning of the head and neck toward auditory and visual stimuli. This process is augmented by the presence of the intercollicular commissure.

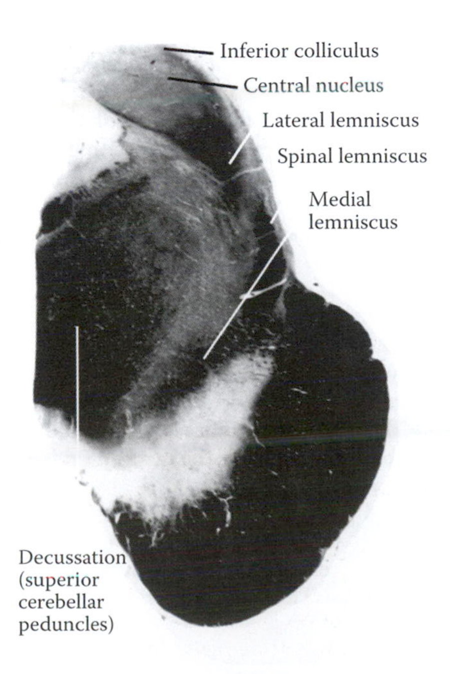

FIGURE 14.11 Image of the midbrain at the inferior collicular level showing the central nucleus and the termination of the lateral lemniscus.

As indicated earlier, the postsynaptic fibers of central nucleus neurons of the inferior colliculus run through the inferior brachium to terminate in the MGN, while some other fibers bypass the inferior colliculus to reach the MGN.

Medial Geniculate Nucleus

The medial geniculate nucleus (MGN), an auditory relay nucleus between the inferior colliculus and the primary auditory cortex, is located ventrolateral to the pulvinar, separated by the brachium of the superior colliculus. It consists of medial, ventral, and dorsal nuclei, which are distinguished by their morphologic characteristics, connections, and density. The medial (magnocellular) nucleus receives fibers from the inferior colliculus for the detection of intensity and duration of sound and from the deep part of the superior colliculus, indicating the possible role of this nucleus in the mediation of modalities other than sound. Neurons of this part of the MGB are preferentially tuned to certain frequencies and show reduction in response proportional to the sound intensity. It projects diffusely to lamina VI of the auditory, insular, and opercular cortices. The dorsal (posterior) nucleus overlies the ventral nucleus and consists of principal cells and interneurons but without frequency-based layering. This nucleus receives afferents from the pericentral nucleus of the inferior colliculus and from other auditory relay nuclei. A broad range of frequencies is regulated in the dorsal nucleus, which account for the lack of tonotopic organization. Projection of the dorsal nucleus is limited to the secondary auditory cortex (Brodmann area 22). Neurons of the ventral nucleus that receive afferents from the ipsilateral inferior colliculus via the brachium of the inferior brachium, are responsible for relaying intensity, binaural information, and frequency to the auditory cortex. It exhibits tonotopic arrangement in which low frequencies project laterally, whereas high-pitched sounds are conveyed medially. This ventral nucleus projects primarily to layer IV of the primary auditory cortex. Commissural neurons do not exist between the medial geniculate bodies.

Auditory Radiation

Axons from the MGN (Figure 14.12) form the *auditory radiation* that projects to the transverse gyrus of Heschl (Brodmann areas 41 and 42) via the sublenticular portion of the posterior limb of the internal capsule, which also contains the optic radiation, as well as sensory and motor fibers. The primary auditory cortex provides descending projection to the inferior colliculus through the MGN by synapsing within this nucleus or bypassing it. These descending fibers and the olivocochlear bundle may be part of a system that modulates the auditory impulses.

AUDITORY CORTICES

The *primary auditory cortex* (Brodmann area 41, 42) is located medial to the superior temporal gyrus, where low frequencies (low-pitched tones) are received laterally and high frequencies

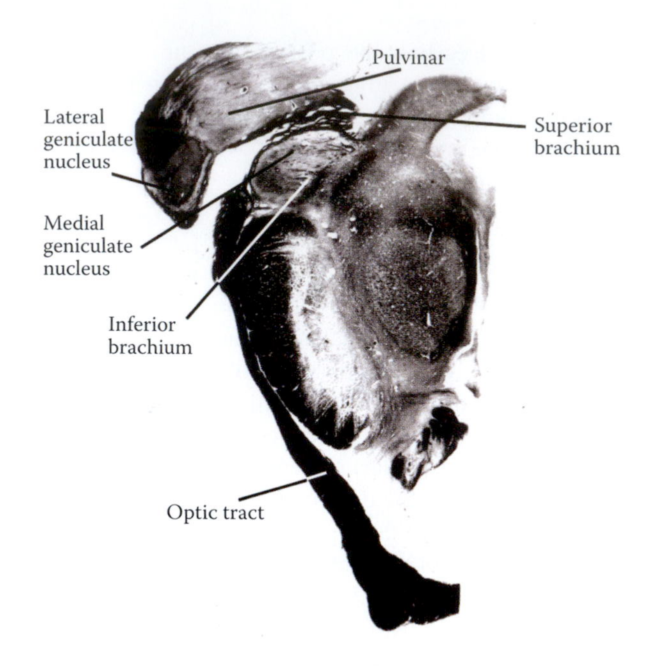

FIGURE 14.12 Section through the diencephalomesencephalic junction to show the medial and lateral geniculate bodies.

(high-pitched tones) are perceived caudally and medially in close proximity to the insula. The auditory association cortex (Brodmann area 22), which occupies the posterior part of the superior temporal gyrus, regulates spoken language through connections with the entorhinal cortex.

> Disruption of the connection between the auditory association and entorhinal cortices may produce anterograde auditory amnesia, in which a patient is unable to retain spoken language. Bilateral damage to the auditory cortex may result in impairment of the ability to distinguish and decipher tones in specific models. Perception of sound and discrimination of tones remain intact in these individuals.

Mechanisms that regulate the transmission of auditory impulses include the descending cortical and subcortical projection, olivocochlear bundle, and stapedius reflex. The descending cortical projection encompasses cortical input to the MGN. Descending fibers from subcortical nuclei, such as the MGN, form the geniculotectal tract. Other descending fibers that act upon the lower auditory neurons are also included in this regulating mechanism. These descending fibers maintain reverse course to that of the lateral lemniscus, exerting an inhibitory influence upon background noise. Through this inhibition, the desired auditory signals are sharpened. The olivocochlear bundle (Figure 14.13) contains crossed and uncrossed fibers originating from the retro-olivary and periolivary nuclei. It is formed by the myelinated medial olivocochlear fibers that project to the outer hair cells and by the unmyelinated fibers that target the inner hair cells and form the lateral olivocochlear bundle. These inhibitory efferent fibers that initially run through the vestibular nerve and later project to the outer hair cells via the cochlear nerve fibers may be utilized in the fine auditory (speech) discrimination in the presence of background noise.

Dampening of the impact of high-frequency sound is mediated by the stapedius reflex (Figure 14.13), which involves activation of the facial and trigeminal motor nuclei in response to loud noises. It comprises bilateral projection of the auditory impulses to the facial and trigeminal motor nuclei via the superior olivary nuclei. Activation of the facial neurons results in contraction of the stapedius and tensor tympani muscles. Paralysis of the stapedius muscle in facial nerve palsy may produce hyperacusis (low hearing threshold). Paralysis of the tensor tympani muscle as a result of mandibular nerve damage produces hypacusis (high hearing threshold).

In summary, neurons associated with the auditory system include the spiral ganglia, cochlear nuclei, superior olivary nuclei, trapezoid body, lateral lemniscus, inferior colliculus, MGN, and the transverse gyrus of Heschl. Most of the auditory fibers are contralateral, but some remain ipsilateral.

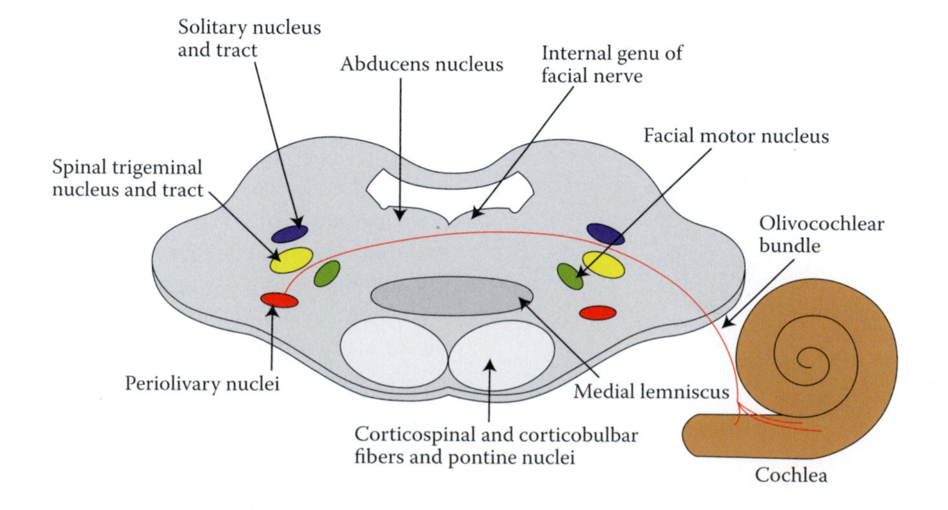

FIGURE 14.13 The origin of the inhibitory olivochochlear bundle and its course to the outer hair cells of the cochlea.

AUDITORY DYSFUNCTIONS

Sudden hearing loss may develop over a period of few hours or less, usually in only one ear. One of the initial signs of this condition is the loud sound heard in the affected side. Viral disease such as mumps, measles, influenza, chicken pox, or infectious mononucleosis may be the culprit. Strenuous activities such as heavy weight lifting may exert undue pressure on the inner ear, producing sudden deafness. Most patients recover their hearing within 2 weeks. Hearing impairment refers to a defect in the identification of acoustic information. It may involve any portion of the transducer mechanism of the ear that includes a defect in the mechanical conduction of sound waves, an abnormality in sensorineuronal coding, a disturbance in the sensory transmission to the CNS, or a combination of these deficits. Inflammatory conditions can cause hearing loss associated with otalgia, otorrhea, aural fullness, and headache. Involvement of the inner and neural elements is associated with hearing deficit, tinnitus, and vertigo. These concomitant deficits are seen with both ototoxicity and Meniére disease. Multiple sclerosis can produce both ocular and hearing problems. Facial trigeminal nerve dysfunction concomitant with hearing loss may occur in patients with expanding tumors, such as acoustic neuroma. Hearing loss can be viewed in two categories: conductive and sensory neuronal.

CONDUCTIVE DEAFNESS

Conductive deafness occurs as a result of either mechanical defects or inefficiencies. These may include occlusion of the external acoustic meatus by the cerumen, infection of the soft tissue associated with this canal (exostosis), or nodular overgrowth in the external canal, which is seen in long-distance swimmers as a result of exposure to cold water. Congenital atresia and stenosis of the external canal due to fibrosis or cicatrix are some additional causes. Middle ear infection or trauma and buildup of purulent exudate in the tympanic cavity that impair the mobility and the mechanical efficiency of the tympanic membrane may also produce conductive hearing loss. Tympanosclerosis as a result of scar tissue formation and deposition of calcium in the membrane or healed perforation may also impair the elasticity of the tympanic membrane and lead to hearing loss. Bony fusion of the ossicles (otosclerosis), formation of a ball sac within the tympanic cavity (cholestatoma) following perforation or a retraction pocket in the tympanic membrane, perilymphatic fistula of the round window, or the vestibule that extends through the oval window to the middle ear cavity can also causes conductive deafness.

SENSORINEURONAL DEAFNESS

Sensorineuronal hearing loss (nerve deafness) may occur as a result of lesions of the cochlea, cochlear nerve, or its nuclei. Noise trauma as a result of prolonged exposure to excessive levels of noise can produce deafness by damaging the hair cells of the basal turn of the cochlea, which receive high-frequency sound waves. However, this type of deafness should not be confused with *temporary threshold shift*, a temporary hearing loss resulting from sudden exposure to loud noise. Deafness due to this condition improves within 24–48 h. *Presbycusis*, the second most common etiology of neuronal deafness, is an age-related gradual degeneration of the neuroepithelial cells of the organ of Corti that manifests deafness mainly to high frequencies, is a slowly progressive, bilateral, and symmetric condition with genetic predisposition. It can be associated with the administration of ototoxic medication and history of exposure to high frequency sounds. Sensory neuronal hearing loss may also occur in otosyphilis as a late manifestation of the disease, affecting higher frequencies first and progressing to involve lower frequencies as the auditory and vestibular dysfunctions become bilateral. Temporal bone osteitis, microvascular infarction, and *Mycobacterium tuberculosis* infiltration are considered the underlying causes of deafness. Flulike conditions are a common cause of sudden deafness without vestibular dysfunctions.

Destruction of the organ of Corti as is seen in Meniére disease, congenital atresia of the labyrinth (deaf mutism), acquired atresia of the labyrinth subsequent to meningitis, loud noise, administration of ototoxic medications (i.e., streptomycin, neomycin, or quinone), or acoustic neuroma may also produce nerve deafness. Rapid, progressive, and total hearing loss may occur subsequent to viral or bacterial infections of the inner ear. Malarial infections may produce permanent neuronal deafness. Neuronal deafness due to labyrinthitis is usually associated with vertigo and ataxia.

Fluctuating hearing loss and ataxia and gradual progression of hearing loss occur in syphilitic labyrinthitis. Meniére disease, a condition that develops as a result of a disorder of endolympahtic circulation and formation of endolymphatic hydrops, produces fluctuating hearing loss, vertigo, ataxia, and tinnitus. Infarct as a result of occlusion of the anterior inferior cerebellar artery that supplies the inferior cerebellum, lateral medulla, and inferolateral part of the pons produces ipsilateral deafness, facial palsy, gait ataxia, and contralateral hemianesthesia and ipsilateral facial anesthesia. Occlusion of the labyrinthine branch of the anterior inferior cerebellar artery or basilar artery due to a thrombotic or embolic episode or subsequent to compression by an acoustic neuroma may also cause hearing loss. Otoxic effects of medications such as salicylates and aminoglycosides may also produce sensorineuronal deafness. Salicylates, in particular, produce reversible symptoms, which include tinnitus, deafness to all frequencies, vertigo, and ataxia. Reversible ototoxic effects may result from ethacrynic acid and furosemide. Chemotherapeutic agents such as nitrogen mustard and *cis*-platinum may also produce profound hearing loss. Streptomycin and gentamicin produce greater adverse effects on the vestibular system and to some degree on the organ of Corti. Kanamycin, tobramycin, and neomycin produce significant damage to the cochlear system. Aminoglycosides damage the inner hair cells, producing neuronal deafness to high frequencies, which progresses to involve all frequencies.

Congenital disorders, such as *Mondini deformity*, in which most of the cochlea fails to develop in both ears, produces neuronal deafness. This may affect acquisition of language skills at an early age. It may sometimes be misdiagnosed as mental retardation. Retrocochlear deafness due to lesions the cochlear nerve or the central auditory pathways is seen as a result of cerebrovascular diseases, stroke, intracranial hemorrhage, demyelinating diseases such as multiple sclerosis (4%–10%), or head trauma. If multiple sclerosis (MS) is suspected, cerebrospinal fluid (CSF) evaluation for increased immunoglobulin-G (IgG) index and oligoclonal bands in gel electrophoresis can be done to confirm the condition. Perventricular white matter lesions on the inferior colliculus and/or the cochlear nuclei may also be seen in MS patients. Cerebellopontine angle tumors and acoustic neuroma have an insidious onset.

Acoustic neuroma is a common tumor that arises from the perineurium of the vestibular nerve (vestibular schwannoma) and extends to involve the cochlear nerve. It accounts for 10% of intracranial tumors and may be idiopathic or associated with neurofibromatosis (*von Recklinghausen's disease*). Type I neurofibromatosis is sporadic and usually unilateral. Neurofibromatosis type II exhibits bilateral neuroma and shows autosomal dominant inheritance. Genetic causes are rarely associated with the bilateral form of this disease. The tumor is usually solitary and involves the vestibular nerve within the internal acoustic meatus and expands to the posterior cranial fossa and lodges in the cerebellopontine angle, as the name of tumor indicates. Erosion and widening of the internal acoustic meatus may be observed radiographically in this condition. It produces dysfunction of the vestibulocochlear, facial, and trigeminal nerves. It may expand and disrupt the lateral part of the pons. Symptoms include early hearing impairment, which is progressive and sometimes associated with tinnitus (80% patients), facial numbness, headache, gustatory changes, ataxia, vertigo, vomiting, involuntary movements, and rarely, facial pain. Rarely the glossopharyngeal nerve is involved, leading to dysphagia.

Tinnitus

Tinnitus is classified into subjective tinnitus, the most common and objective tinnitus. *Subjective* tinnitus is a continuous or intermittent unilateral or bilateral noise heard, unilaterally or bilaterally, by the patient alone in the form of hissing, humming, electric buzzing, whooshing, whistling, or machinery-like roaring.

It can be presented concomitantly with sensorineuronal hearing loss and may be severe enough to interfere with normal sleep. Unilateral tinnitus may be a sign of Meniére disease, acoustic neuroma, or otosclerosis. Meniére disease may produce a low-pitched continuous tinnitus similar to an ocean roar that intensifies before the attack of vertigo. Otosclerosis also produces low-pitched and continuous tinnitus. High-pitched tinnitus heard over external sounds may result from physical trauma. Chronic exposure to noise and toxicity with drugs such as streptomycin, salicylates, and quinine may also produce bilateral high-pitched tinnitus. However, tinnitus induced by drugs such as indomethacin, quinidine, levodopa, propranolol, carbamezapine, salicylates, aminophylline, and caffeine is not commonly associated with hearing loss.

Objective tinnitus can be heard by the patient and the examiner when he/she places a stethoscope (with bell removed) into the patient's external acoustic meatus, or with the bell around the ear. An abnormally patent (patulous) Eustachian tube may produce a blowing sound that coincides with inspiration and expiration. Extensive weight loss due to dieting or as a consequence of debilitating disease may result in objective tinnitus. It may also be heard in an individual with tetany and as a result of contracture of the palatal muscles and tensor tympani. Objective tinnitus may be audible in individuals with arteriovenous malformation, carotid artery aneurysm, and vascular tumors in the form of bruit. In these cases, the tinnitus remains pulsatile and in synchrony with the heartbeat. Pulsatile tinnitus may indicate vasculitis, idiopathic intracranial hypotension, or giant cell arteritis. Turbulence within the internal jugular vein produces a venous hum, which is heard by the examiner in the form of a "whistling" or continuous machinelike sound.

AUDITORY TESTS

Audiometry

Deafness may be assessed by *audiometry*, a sophisticated procedure used to measure hearing ability and determine the threshold of audibility. In this test, a manual or automatic electronic device that produces auditory stimuli of known frequency and intensity is used with headphones fitted over the ears. In *Békésy audiometry*, continuous and interrupted tones are presented to the patient by pressing a signal button. Monaural tracings are made, which measure the increments by which the patient must increase the volume in order to hear the continuous and interrupted tones just above the threshold. The intensity of the tone decreases as long as the button is depressed and increases when it is released. Four basic configurations have been established based on analysis of tracings. Type I is considered normal, type II indicates a lesion in the cochlea, and types III or IV usually point to a retrocochlear lesion. An objective method of determining auditory acuity may also be achieved by recording and averaging the electrical potentials elicited by the cortex in response to stimulation by pure tones (cortical audiometry). Electrocochleographic audiometry is another procedure that measures the electrical potentials from the inner ear in response to auditory stimuli. Hearing threshold may also be determined by electrodermal audiometry, in which a patient is conditioned to pure tones by harmless electric shock. The anticipation results in brief electrodermal response, which is recorded. Hearing threshold is determined by the lowest intensity at which electrodermal response is attained. On the other hand, *speech audiometry* measures, in decibels, the threshold of speech perception (speech reception threshold [SRT]) and the ability to understand approximately 50 monosyllabic words (speech discrimination). Efficacy of audiometry is restricted by the accuracy of the results, cooperation of the patient, differences in the locations of the headphones, and development of proficiency at detecting threshold.

SRT is done by presenting a standardized series of words to a patient and then measuring the lowest intensity at which these words are comprehended. Pure tone audiometry, as the name indicates, measures hearing at various tones and frequencies. In normal individuals, the threshold varies with frequencies, and the ability to hear below 45 Hz is poor.

Brainstem Auditory Evoked Response

One of the newest methods that ascertain the integrity of primary and secondary neural pathways is the *brainstem auditory evoked response* (BAER). In this method, a large number of clicks are delivered to each ear at one time. The series of waves are recorded via scalp electrodes and maximized by computers. Within 10 ms after each stimulus, a series of seven waves appear. The first five waves relate to the auditory neurons and pathways. Reduction in amplitude and voltage or delay in the appearance of the waves indicates a structural lesion within the auditory system.

Weber Test

In the *Weber test* (Figure 14.14), a vibrating tuning fork is placed on the vertex (top) of the patient's head or on the nasal bone and the patient is asked to tell whether the sound is heard equally on both sides (*positive Weber test*) or whether the sound is greater on one side. In conduction deafness, the sound is lateralized (heard better) on the affected side, while in nerve deafness, the sound is heard more loudly in the unaffected ear.

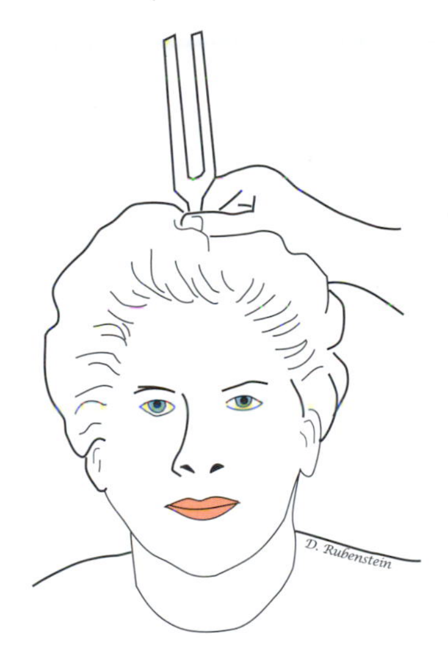

FIGURE 14.14 This is a schematic diagram of the Weber test. The sound is lateralized to the impaired side.

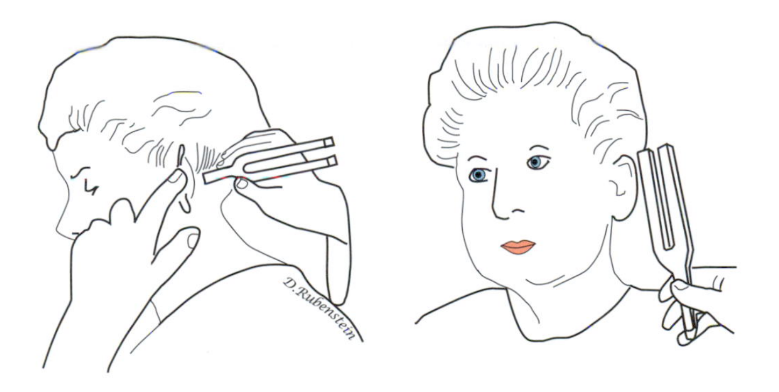

FIGURE 14.15 This diagram illustrates the Rinne test.

Rinne Test

In the *Rinne test* (Figure 14.15), air conduction is compared to bone conduction by using a tuning fork. The examiner first applies a tuning fork of 256–512 Hz frequency to the mastoid process, and when the patient can no longer hear the sound, the tuning fork is placed in front of the ear, and lateralization of the sound is observed. Because air conduction lasts longer than bone conduction, the sound is heard through the air longer and louder than bone (positive Rinne test). In conduction deafness, sound will be perceived longer and louder through the bone than through the air (negative Rinne test). In neuronal deafness, both air and bone conduction are compromised; however, air conduction remains longer and louder than bone conduction.

SUGGESTED READING

Bajo VM, Merchan MA, Lopex DE, Rouiller ER. Neuronal morphology and efferent projections of the dorsal nucleus of the lateral lemniscus in the rat. *J Comp Neurol* 1993;334:241–62.

Barkat TR, Polley DB, Hensch TK. A critical period for auditory thalamocortical connectivity. *Nat Neurosci* 2011;14(9):1189–94.

Boscariol M, Garcia VL, Guimarães CA, Montenegro MA, Hage SR, Cendes F, Guerreiro MM. Auditory processing disorder in perisylvian syndrome. *Brain Dev* 2010;32(4):299–304.

Cant NB, Benson CG. Parallel auditory pathways: Projection patterns of the different neuronal populations in the dorsal and ventral cochlear nuclei. *Brain Res Bull* 2003;60(5–6):457–74.

Faye-Lund H, Osen KK. Anatomy of the inferior colliculus in rat. *Anal Embryol* 1985;171:1–20.

Fettiplace R et al. The sensory and motor roles of auditory hair cells. *Nat Rev* 2006;7:19–29.

Griffiths TD. Central auditory pathologies. *Br Med Bull* 2002; 63(1):107–20.

Hartley DE, Moore DR. Effects of otitis media with effusion on auditory temporal resolution. *Int J Pediatr Otorhinolaryngol* 2005;69(6):757–69.

Heffner HE, Heffner RS. Hearing loss in Japanese macaques following bilateral auditory cortex lesions. *J Neurophysiol* 1986;55(2):256–71.

Hudde H, Weistenhöfer C. Key features of the human middle ear. *J Otorhinolaryngol* 2006;324–9.

Kandler K, Clause A, Noh J. Tonotopic reorganization of developing auditory brainstem circuits. *Nat Neurosci* 2009;12:711–7.

Kass JH et al. Auditory processing in primate cerebral cortex. *Curr Opin Neurol* 1999;9:500.

MacKechnie CA, Greenberg JJ, Gerkin RC, McCall AA, Hirsch BE, Durrant JD, Raz Y. Rinne revisited: Steel versus aluminum tuning forks. *Otolaryngol Head Neck Surg* 2013;149(6): 907–13.

Pandya DN, Rosene DL, Doolittle AM. Corticothalamic connections of auditory-related areas of the temporal lobe in the rhesus monkey. *J Comp Neurol* 1994;345:447–71.

Robertson D, Anderson CJ. Acute and chronic effects of unilateral elimination of auditory nerve activity on susceptibilityto temporary deafness induced by loud sound in the guinea pig. *Brain Res* 1994;646:37–43.

Shmonga Y, Takada M, Mizuno N. Direct projections from the nonlaminated divisions of the medial geniculate nucleus to the temporal polar cortex and amygdala in the cat. *J Comp Neurol* 1994;340:405–26.

Sichel JY, Eliashar R, Dano I. Explaining the Weber test. *Otolaryngol Head Neck Surg* 2000;122:465–6.

Spoendlin H, Schrott A. Analysis of the human auditory nerve. *Hearing Res* 1989;43:25–38.

Tanaka Y, Kamo T, Yoshida M, Yamadori A. "So-called" cortical deafness. *Brain* 1991;114:2385–401.

Wynne DP, Zeng FG, Bhatt S, Michalewski HJ, Dimitrijevic A, Starr A. Loudness adaptation accompanying ribbon synapse and auditory nerve disorders. *Brain* 2013;136(Pt 5):1626–38.

15 Vestibular System

The vestibular system is a special somatic afferent system that maintains the orientation in the three dimensions, fixation of gaze during head rotation, and control of muscle tone. In order to accomplish these diverse activities, generated impulses at the receptor level are carried by the vesibular nerve to the cerebellum and the vestibular nuclei. Projections from these nuclei eventually reach the extraocular motor nuclei, cerebellum, and spinal cord.

PERIPHERAL VESTIBULAR APPARATUS

Vestibular receptors are contained in the inner ear within the semicircular canals and the vestibule. These receptors convey vestibular inputs through the dendrites of the bipolar neurons of Scarpa's ganglia to the cerebellum and brainstem. Massive projections from the brainstem to subcortical and spinal neurons allow the generated impulses to exert influences on vast areas of the nervous system.

Semicircular Canals

Semicircular canals (Figure 15.1) are three in number, consisting of the anterior (superior), posterior, and lateral (horizontal) canals. They are unequal in length, form an incomplete circle, and are perpendicular to each other, enclosing the endolymph-filled membranous semicircular ducts. They contain the perilymph-filled perilymphatic spaces separated from the endolymph-filled semicircular ducts. The anterior (superior) semicircular, a vertical canal, lies inferior to the arcuate eminence and is oriented horizontally to the axis of the petrous temporal bone. The ampulla of this canal opens into the upper and lateral part of the vestibule. The lateral (horizontal) canal is directed horizontally posteriorly and laterally, joining the vestibule just inferior to the opening of the ampullary end of the superior semicircular canal and superior to the fenestra vestibuli (oval window). The posterior semicircular, a vertical canal, is positioned parallel to the posterior surface of the petrous temporal bone. The horizontal semicircular canals are in the same plane, whereas the anterior canal of one side lies parallel to the posterior semicircular canal of the opposite side.

The crista ampullaris, a dilatation at each end of the semicircular ducts, contains neuroepithelial cells. These neuroepithelial (hair) cells of the ampullary crest have stereocilia embedded in a gelatinous substance known as the cupula. A single cilium, the kinocilium, exists at one edge of the stereocilia. The stereocilia and the kinocilium receive the peripheral processes of the neurons of the vestibular ganglia. During depolarization, the cilia move toward the kinocilium, whereas in hyperpolarization, the reverse movement occurs. When a current develops in the endolymph subsequent to head movement, the cupula deflects to one side. The cupula

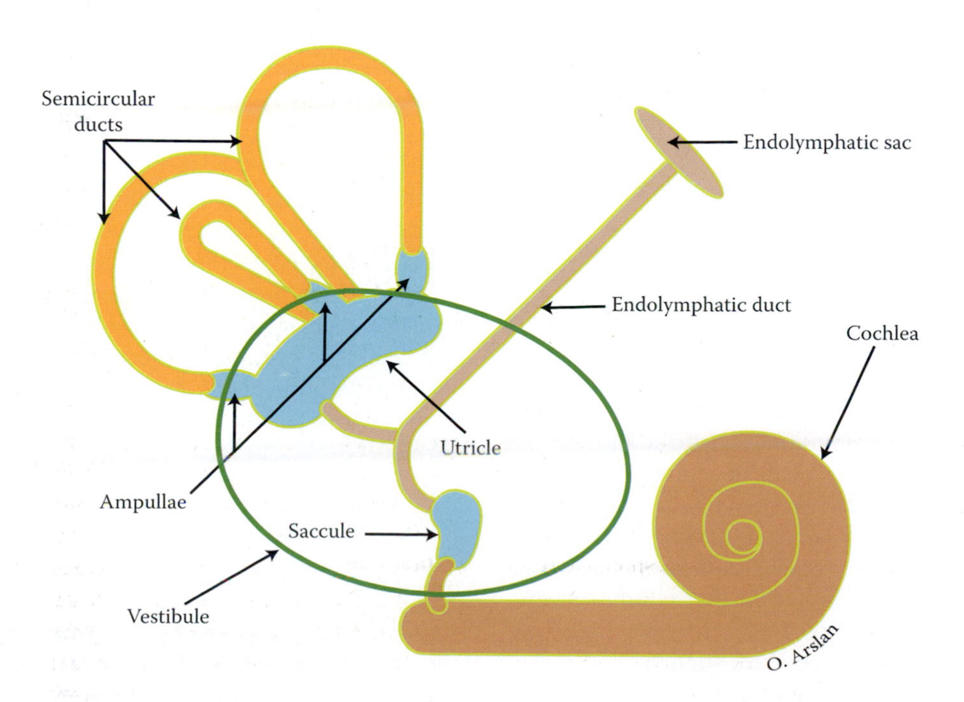

FIGURE 15.1 Various components of the vestibular receptors.

and the enclosed ampullary crests of the semicircular ducts are kinetic receptors due to their sensitivity to angular acceleration and deceleration.

Movement of the endolymph in the semicircular ducts occurs during head rotation. Due to the inertia created by head rotation, the endolymph lags behind, and the cupula deflects in the direction opposite to that of the head rotation. Movement of the cupula, followed by movement of the stereocilia toward the kinocilium, produces depolarization of the hair cells. As the rotation continues, the endolymph catches up with the head rotation and moves in the same direction. When the head rotation suddenly stops, the endolymph continues to move in the original direction, and the cupula and stereocilia move in the reverse direction. In general, the lateral ampullary crista is excited by movement of the endolymph toward the utricle from the lateral semicircular duct, whereas excitation of the superior (anterior) and posterior ampullary cristae occurs upon movement of the endolymph out of the utricle. Deflection of the hair cells in these structures results in excitation of the vestibular nerve endings. The generated nerve impulses in the excited neuroepithelial cells are transmitted to the neurons of the vestibular ganglia.

Vestibule

The *vestibule* (Figure 15.1) is the central part of the osseous labyrinth that lies medial to the middle ear cavity, anterior to the semicircular canals and posterior to the cochlea. It contains the endolymph-filled membranous utricle and saccule, and the fenestra vestibuli (oval window) in its lateral wall, which is covered by the stapes and the annular ligament.

Saccule

The *saccule* is located in a recess on the anteromedial wall of the vestibule adjacent to the inferior vestibular area. Posterior to the saccule, the cochlear recess can be seen, which contains the fibers of the vestibulocochlear nerve destined to the cochlear duct.

Utricle

The *utricle* is located posterosuperiorly in an elliptical depression in the roof and medial wall of the vestibule. The neuroepithelial cells of the utricle rest on the macula, a horizontal structure in the erect posture, which allows changes in its firing pattern when the head is flexed or extended. The saccular macula contains neuroepithelial cells on its medial wall, which assume a vertical position and are able to respond to head tilt to one side. The stereocilia of neuroepithelial cells of both maculae are embedded in a gelatinous material covered with otoliths consisting of crystals of calcium carbonate particles. The kinocilia in the utricular and saccular maculae are located in different positions, which expands the sensitivity of the receptor neurons to head tilts in different directions. In general, the utricle and saccule are static receptors that respond to gravitational pull, head tilt, and linear acceleration and deceleration. The macula of the saccule is thought to be sensitive to vibration.

Deflection of the stereocilia of the macula utriculi or sacculi upon head tilting, induced by the pull on the underlying cilia, is transduced into vestibular impulses, which are transmitted by the dendrites of the bipolar neurons of Scarpa's ganglia. The neuroepithelial hair cells are facilitated in one half of the macula and inhibited in the other half. Excessive stimulation of the utricle may produce signs of motion sickness. The maculae project to the lateral vestibular nuclei, which form the lateral vestibulospinal tracts. Thus, an increase in extensor muscle tone via the ipsilateral vestibulospinal tract may occur upon tilting the head to one side. Flexion or extension of the head may activate the maculae on both sides and, eventually, the vestibulospinal tracts bilaterally. Contraction of the extensor muscles is coordinated by cerebellar cortical input or directly via the fastigiovestibular fibers.

Vestibular Ganglion

These vestibular receptors receive the peripheral processes (dendrites) of the bipolar neurons of the *vestibular (Scarpa's) ganglia* that consist of the superior and inferior vestibular ganglia. The superior vestibular ganglion is comprised of neurons that receive information from the anterior (superior) and lateral semicircular ducts and the utricle. The inferior vestibular ganglion is formed by neurons that innervate the posterior semicircular ducts and the saccule. Central processes of the bipolar neurons of the vestibular ganglia (the primary vestibular fibers) form the vestibular nerve.

Vestibular Nerve

The *vestibular nerve* runs in close association with the cochlear and facial nerves and the labyrinthine artery after leaving the internal acoustic meatus (Figure 15.2). It enters the brainstem through the pontocerebellar angle between the pons, cerebellum, and medulla. Then, it passes between the inferior cerebellar peduncle and the spinal trigeminal tract, bifurcating into ascending and descending branches that primarily distribute to the vestibular nuclei (Figure 15.2). Some of the fibers bypass the vestibular nuclei and terminate in the cerebellum as primary vestibulocerebellar fibers, projecting to the fastigial nuclei, ipsilateral uvula, and flocculonodular lobe of the cerebellum. The primary vestibulocerebellar projections run through the superior vestibular nucleus and then the juxtarestiform body (medial smaller part of the inferior cerebellar peduncle) to terminate ipsilaterally as mossy fibers in the granular layer of the nodule and caudal uvula in the ventral part of the anterior lobe and deep part of the vermis.

CENTRAL VESTIBULAR PATHWAYS

Vestibular Nuclei

Vestibular nuclei are located in the vestibular area of the rhomboid fossa, consisting of the superior, inferior, medial, and lateral nuclei. They receive afferents from the interstitial nucleus of Cajal, the flocculonodular lobe, the fastigial nuclei, and the vermal part of the cerebellum. No known

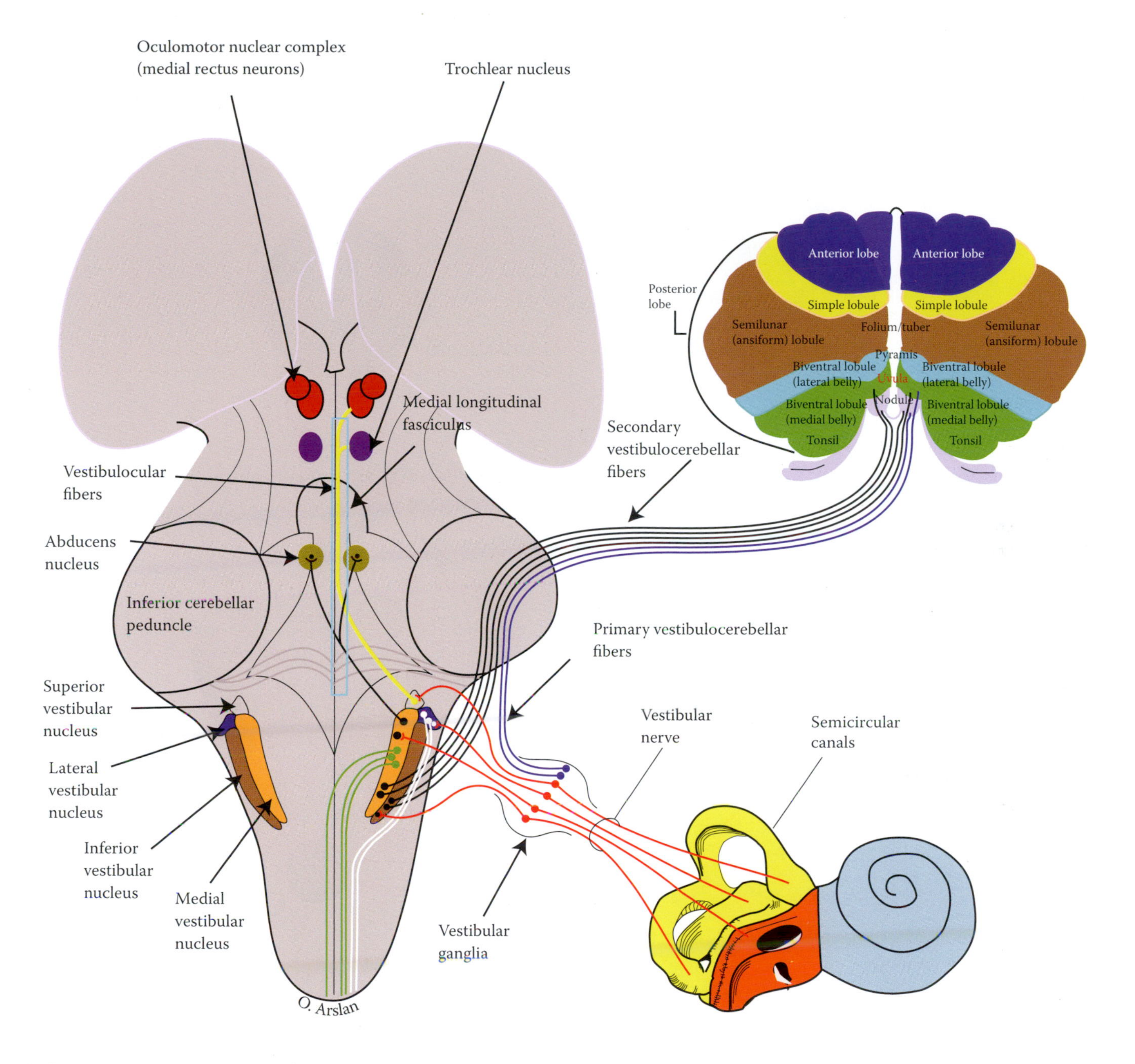

FIGURE 15.2 Schematic diagram of the vestibulocochlear apparatus and associated nerves and nuclei.

afferents to these nuclei are derived from the cerebral cortex or basal nuclei. The vestibular nuclei (Figure 15.3) project extensively to the cerebellum, and also to the motor nuclei of the extraocular muscles, interstitial and parasolitary nuclei, cervical segments of the spinal cord, and thalamus.

The *superior vestibular nucleus* is the most rostral vestibular nucleus, which lies ventral and medial to the superior cerebellar peduncle, adjacent to the mesencephalic nucleus and the principal sensory nucleus.

The *inferior vestibular nucleus* has a speckled appearance as it is crossed by the descending vestibular fibers. It is the smallest nucleus in this group located on the lateral side of the medial vestibular nucleus, extending from the caudal medulla (at the level of the lateral cuneate nucleus) to the level of the superior vestibular nucleus near the pontomedullary junction. It receives input from the vestibular nerve, giving rise to the bilateral ascending and descending projections through the medial longitudinal fasciculus (MLF) to the motor nuclei of the extraocular muscles.

The *medial vestibular nucleus* is the longest of the vestibular nuclei, extending from the level of the hypoglossal nucleus to the level of the abducens nucleus. It separates the dorsal vagal motor nucleus from the rhomboid fossa and continues caudally with the nucleus intercalatus. It is crossed by the fibers of the stria medullaris and receives input from the uvula and nodule, consisting of small and

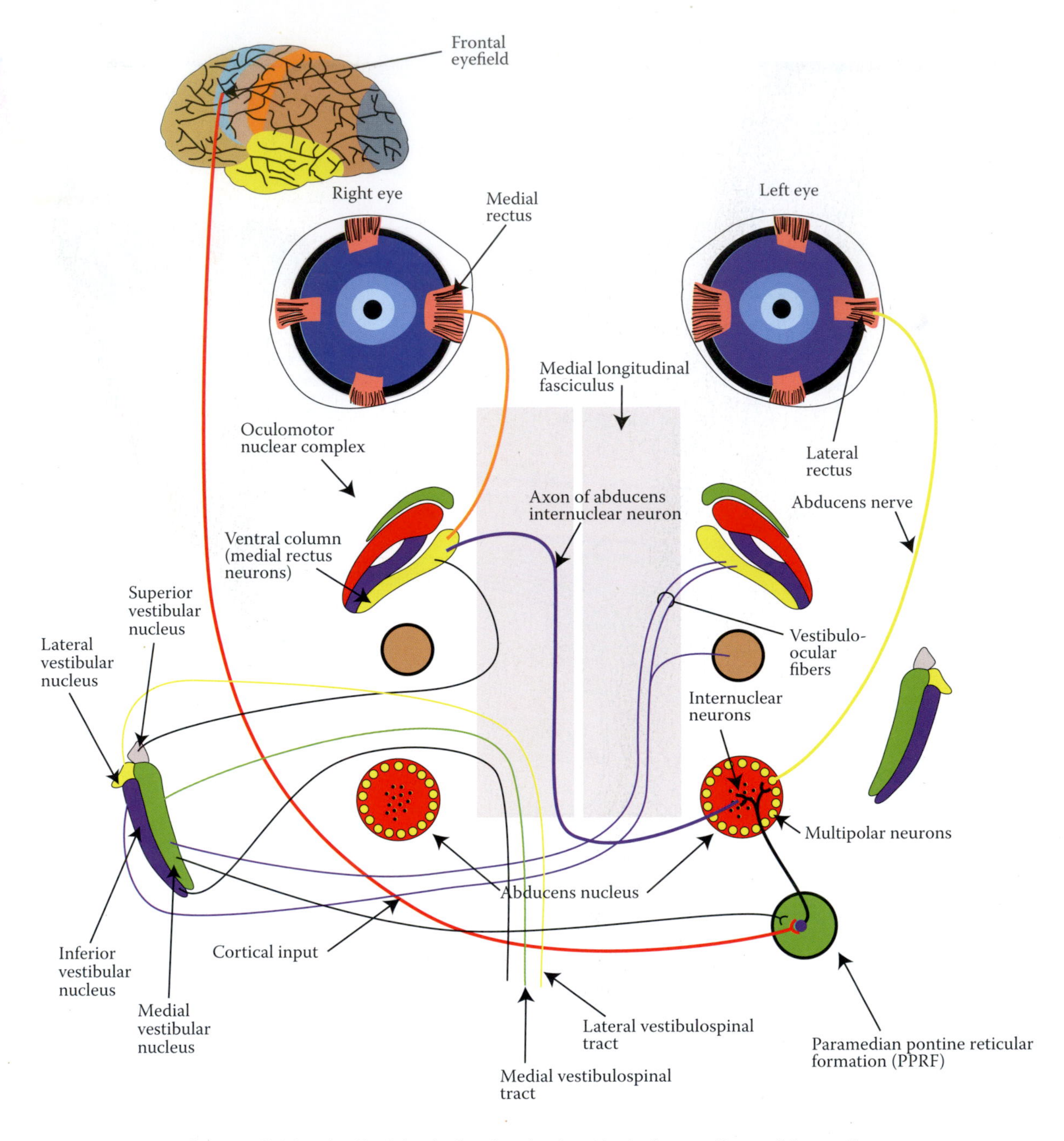

FIGURE 15.3 Diagram of the medial longitudinal fasciculus showing its principal ascending and descending components.

medium-sized cells, which are the source of the medial vestibulospinal tract. This pathway contains some axons from the inferior and lateral vestibular nuclei and is incorporated in the MLF.

The *lateral vestibular* (Deiters) *nucleus* contains giant neurons, which are the source of the main excitatory lateral vestibulospinal tract. It lies ventrolateral to the rostral part of the medial vestibular nucleus, extending rostrally from the level of the abducens nucleus and caudal end of the superior vestibular nucleus caudally to the site of entrance of the primary vestibular fibers. It receives input from the B zone of the anterior vermal cortex (lateral parallel strip of the anterior lobe and the simple lobule) but not from the fastigial nucleus or labyrinth. Damage to the B zone may produce disinhibition of Dieter's nucleus and eventual extensor hypotonia. It mediates cerebellar influences on postural and labyrinthine reflexes.

Secondary Vestibulocerebellar Fibers

Secondary vestibulocerebellar fibers originate from the vestibular nuclei and distribute to the cerebellum, spinal cord, and motor nuclei of the extraocular muscles and thalamus. Those fibers derived from the medial and inferior vestibular nuclei project to the flocculonodular lobe of the cerebellum through the juxtarestiform body.

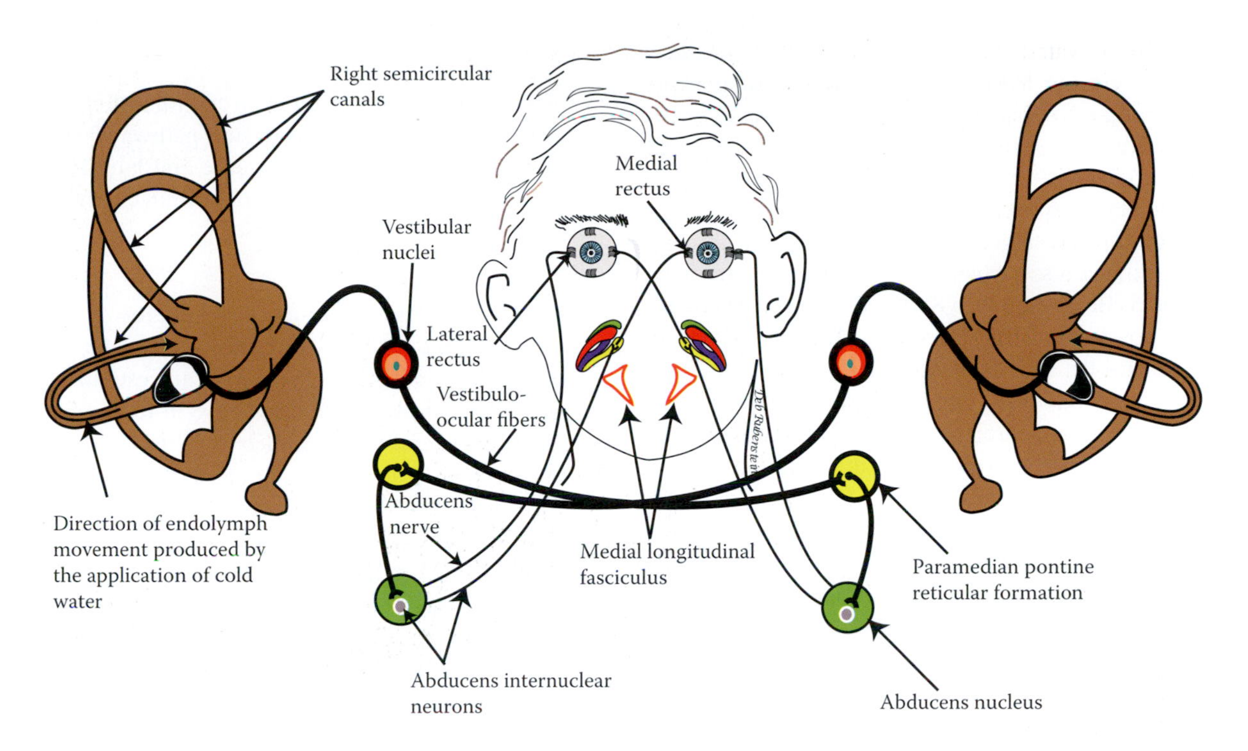

FIGURE 15.4 This drawing illustrates the mechanism by which vestibular impulses, generated by caloric test (e.g., application of cold water), produce eye movements. Note the vestibular input to the abducens nucleus (lateral gaze center) and ensuing fibers that act upon the medial and lateral rectus muscles.

Vestibulospinal Tracts

Secondary vestibular fibers to the spinal cord (Figures 15.3 and 15.4) form the medial and lateral vestibulospinal tracts. The medial vestibulospinal tract, a monosynaptically inhibitory pathway to the axial muscles of the trunk and neck, originates primarily from the medial vestibular nucleus. It descends in the sulcomarginal fasciculus within the anterior funiculus to reach the midthoracic segments. It is a bilateral tract with an ipsilateral predominance contained within the MLF. The *lateral vestibulospinal tract* is the principal vestibular projection to the spinal cord. It originates from the lateral vestibular nucleus and runs ipsilaterally through the entire length of the spinal cord, terminating in laminae VII and VIII. It courses initially in the outer part of the anterolateral funiculus and then in the anterior funiculus. The origin and distribution of this pathway is somatotopically organized with fibers emanating from the rostroventral, central and dorsocaudal parts of the lateral vestibular nucleus projecting to the cervical, thoracic and the lumbosacral segments, respectively. It is monosynaptically and polysynaptically excitatory to the alpha motor neurons of the extensor muscles in the extremities, trunk, and neck.

Vestibulo-Ocular Fibers

Secondary vestibular fibers to the motor nuclei of the extraocular muscles (Figure 15.3) enable the vestibular system to coordinate eye movements via the MLF. The MLF (Figure 15.3) is a composite bundle, containing ascending and descending components. The ascending component includes the vestibulo-ocular fibers, which project bilaterally to the motor nuclei of the oculomotor, trochlear, and abducens nerves as well as the axons of the internuclear neurons of the contralateral abducens nucleus. These fibers mediate the vestibulo-ocular reflex, coordinate the contraction of the extraocular muscles, and fixate our gaze during head rotation. In the vestibulo-ocular reflex, impulses generated from the vestibular nuclei and mainly the medial vestibular nucleus (e.g., left) as a result of head movement toward the left side are transmitted to the (right) paramedian pontine reticular formation (PPRF), which, in turn, projects to the ipsilateral (right) abducens nucleus. The activated (right) abducens nucleus sends commands via the internuclear neurons to the neurons of the medial rectus muscle of the contralateral (left) oculomotor nuclear complex that produces adduction of the left eye and to the ipsilateral motor neurons of the (right) abducens nucleus that causes abduction of the right eye. The net result is movements of the eyes to the side opposite to head movement.

Activation of the abducens internuclear and abducens motor neurons produces movement of both eyes to the side of the original stimulus. As mentioned earlier, abducens internuclear neurons mediate conjugate adduction of one eye with the abduction of the opposite eye in lateral gaze. The vertical gaze center is also activated by the input generated from the medial vestibular nucleus. This occurs when bilateral movement of the endolymph into the superior ampullae (during neck flexion) and inferior ampullae (during neck extension) activate the corresponding ampullae, and subsequently, the medial vestibular nuclei. Although the superior vestibular nucleus is also activated during head movement, it exerts inhibitory role via the MLF. The descending components contain the medial vestibulospinal tract, the pontine reticulospinal tract, the interstitiospinal tract, and the tectospinal tract (Figure 15.3).

To view an object of interest, both eyes have to be directed conjugately toward the target. This is accomplished through saccadic eye movements in which impulses generated from the frontal eye field (Brodmann area 8) of the left hemisphere, for example, are conveyed to right the contralateral PPRF, which, in turn, stimulates the ipsilateral right abducens nucleus. This mechanism is also discussed above. Activation of the internuclear and multipolar neurons within the abducens nucleus produces deviation of the eyes toward the right side or the side of the stimulated abducens nucleus. A lesion of the frontal eye field disrupts the neural circuit that mediates conjugate deviation of the eyes toward the opposite side. As a result the contralateral frontal eye field will be dominant, and both eyes conjugately deviate toward the side of the lesion. If the cortical lesion is extensive enough to also involve the precentral gyrus and the ensuing corticospinal tract, the patient exhibit manifestations of contralateral upper motor neuron palsy and conjugate deviation of both eyes toward the same side, as seen in comatose patients. Saccades toward the opposite side may resume as recovery is attained.

Note that the vestibular input to the abducens nucleus through lateral gaze center and ensuing impulses act upon the medial and lateral rectus muscles through the associated neurons.

In summary: vestibular receptors → peripheral processes of the vestibular neurons → vestibular ganglia → primary vestibular fibers → vestibular nuclei → cerebellum. The postsynaptic fibers from the vestibular nuclei also form the secondary vestibular fibers that project to the spinal cord, motor nuclei, and cerebellum.

VESTIBULAR DYSFUNCTIONS

Vestibular deficits may result from central lesions that involve the vestibular nuclei and associated pathways or peripheral dysfunctions due to vestibular nerve and labyrinthine damage. Below are some of the most important conditions associated with vestibular dysfunctions.

A unilateral lesion that disrupts the vestibular nerve produces an imbalance of the vestibular input, causing overstimulation of the contralateral vestibular nuclei accompanied by an equal increase in the discharge of the contralateral vestibular nerve. This leads to activation of the ipsilateral (side of the lesion) PPRF that projects to the ipsilateral abducens nucleus and eventual slow deviation of the eyes toward the side of the lesion (slow phase of nystagmus). The slow phase, which is of vestibular origin, is followed and counteracted by rapid adjustment movement in which both eyes move toward the opposite side (rapid phase of nystagmus). Since nystagmus is named on the basis of the fast phase, a lesion of the right vestibular nerve, for example, produces a right nystagmus.

Internuclear ophthalmoplegia (*MLF syndrome*) is commonly seen in multiple sclerosis as a result of degeneration of the medial longitudinal fasciculi (MLF) rostral to the abducens nuclei (Figures 15.5 and 15.6). It may also result from occlusion of the paramedian pontine branches of the basilar artery (brainstem stroke). This condition is characterized by paresis of adduction on attempted lateral gaze to the opposite side. Adductor paresis is due to disruption of the axons of the internuclear neurons, which project to the oculomotor neurons and innervate the medial rectus.

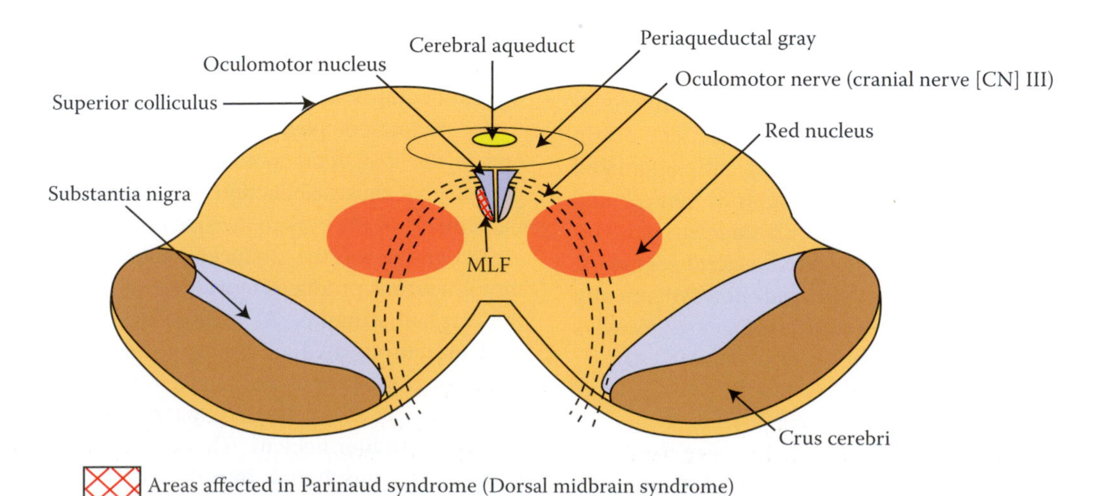

FIGURE 15.5 Section of the midbrain at the level of the superior colliculus. Observe the lesion that involves the medial longitudinal fasciculus producing internuclear ophthalmoplegia.

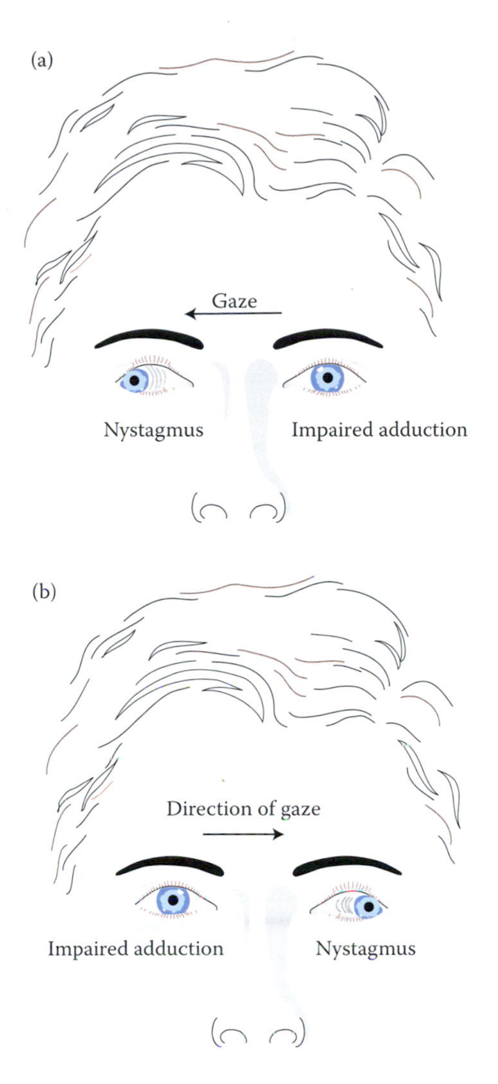

FIGURE 15.6 Manifestations of internuclear ophthalmoplegia (medial longitudinal fasciculus [MLF] syndrome). (a) Left MLF lesion with adductor paresis of the left eye and monocular nystagmus in the right eye. (b) Right MLF lesion with impaired adduction of the right eye and nystagmus in the left eye.

Ataxic nystagmus of the abducting eye (*Harris's sign*), due to compensatory augmentation of the vestibular output, is seen as a result of damage to the vestibular fibers within the affected MLF. Horizontal diplopia is marked upon gazing to the contralateral side; however, convergence remains unaffected due to intactness of the oculomotor nuclei and the associated pathway. In young adults, bilateral internuclear ophthalmoplegia is virtually pathognomonic of multiple sclerosis.

One-and-a-half syndrome is a rare condition characterized by anterior internuclear ophthalmoplegia on the opposite side of the lesion and conjugate horizontal gaze palsy toward the lesion side. In other words, one eye remains motionless in the midline, while the other eye can only abduct. Nystagmus is seen in the contralateral abducting eye with intact convergence. This syndrome results from a combined lesion of the ipsilateral caudal MLF and the ipsilateral paramedian reticular formation (PPRF). Ipsilateral damage to the PPRF produces lateral gaze palsy toward the opposite side (inability to abduct the ipsilateral eye or adduct the contralateral eye). Disruption of the MLF produces adductor paresis on the same side and nystagmus on the opposite side. It occurs in pontine hemorrhage, tumor, ischemia, and multiple sclerosis.

Labrynthitis refers to inflammation of the inner ear that is typically caused by a virus and, less commonly, by bacteria. It can be a consequence of upper respiratory infection, but it can also occur as a result of head trauma or otitis media or as an adverse effect of medication. This condition is associated with inflammation of the vestibular nerve that results in the generation of impulses interpreted by the brain as head turning signals. The main symptom of this disease is incapacitating vertigo that starts weeks after respiratory infection. Patients also exhibit nystagmus, nausea, vomiting, tinnitus, possible deafness, and general malaise.

Vestibular neuritis is a benign, a self-limiting mononeuropathy of the vestibular nerve of unknown etiology, although some evidence indicates a viral etiology. Patients with this condition may exhibit a single prostrating paroxysmal attack of vertigo or a series of prolonged vertiginous bouts that lasts for 2 to 3 days or weeks that rarely wane in weeks. It is associated with spontaneous horizontal nystagmus (fast phase or saccades toward the intact side) evoked by gaze in the direction of the fast phase (Alexander's law), nausea, truncal ataxia, and vomiting. A unique feature of the condition is the absence of any hearing deficit. Minor episodes of ataxia may be experienced by the patient after the acute episode has subsided. Symptomatic treatment with vestibular sedatives may be essential. Several etiologies are attributed to the disease, such as previous upper respiratory infection, common cold, ischemia, or infection with herpes simplex.

Meniére disease is a commonly unilateral condition that primarily affects middle-aged individuals. It is seen in both sexes equally, with a peak onset between 20 and 50 years of age. It is thought to result from abnormal circulation of the endolymph within the vestibular apparatus, as well as progressive distension of the endolymphatic sac, cochlear duct, and saccule. The etiology of this distension has yet to be determined, but theories usually suggest either overproduction or decreased resorption of the endolymph. Some evidence suggests that individuals with Meniére disease have abnormal temporal bone and skull features along with an abnormal endolymphatic sac from birth. It is associated with recurrent episodes of severe and explosive vertigo, vomiting, aural pressure, prostration, and also tinnitus, which last for several minutes to hours.

In this condition, a low-frequency hearing deficit that progresses with time may also be observed. Stress, premenstrual fluid retention, and specific food may precipitate these symptoms. The most serious symptom of this condition is vertigo, which may last from 20 min to several hours. Hearing loss and tinnitus may precede the first episode of vertigo by several months or years in *Lermoyez's* variant of this disease. In this disease, spontaneous nystagmus may be observed during the episode; however, caloric-induced nystagmus is lost on the affected side. This disease should be differentiated from basilar artery ischemia, which exhibits additional manifestations such as visual deficits and diplopia.

Meniére disease is classified into five stages, which determine the appropriate management plan. In *stage I*, only the cochlea is involved, and the patient experiences tinnitus, fullness, and low-tone hearing loss. This stage responds best to treatment, which includes diuretics, vasodilatory drugs (e.g., promethazine), and a low-salt diet. *Stage II* involves more widespread endolympahtic hydrops of the saccule and other parts of the labyrinth. Hearing loss for low tones, tinnitus, fullness, and dizzy spells are experienced by the patient at this stage. Treatment may include diuretics, a low-salt diet, and dexamethasone. Persistence of symptoms may require insertion of a shunt tube into the endolymphatic duct. If the shunt is not successful, streptomycin perfusion may be essential to destroy the end organ. This is achieved by aminoglycoside, which replaces the calcium ions in the cell membrane and produces the loss of stereocilia. In *stage III*, the endolymphatic sac becomes smaller and more displaced. The symptoms also include tinnitus, vertigo, and fullness. At this stage, vestibular neurectomy, which carries the complication of hearing loss as a side effect, may be required. In *stage IV*, the obstruction to the flow of the endolymph is complete, and the hydrops fills the vestibule. The patient experiences no vertigo since the hydrostatic pressure cannot increase. If the disease is bilateral, the patient cannot get motion sickness. There is no known treatment for the disease at this stage. In *stage V*, multiple obstruction and/or rupture of the membranous labyrinth may occur. No treatment is available, although cochlear implants may be used when the patient's biggest concern is bilateral deafness and not vertigo.

Combined vestibular and auditory deficits are seen ipsilaterally in lesions at the internal acoustic meatus or cerebellopontine angle. Fractures of the temporal bone may result in disruption of the vestibular and cochlear nerves.

Acoustic neuroma (tumor of the pontocerebellar angle) is a benign slow-growing schwannoma that arises in the vestibular nerve and may expand to involve the cochlear nerve, compressing the adjacent vestibular nuclei, inferior cerebellar peduncle, trigeminal nerve, glossopharyngeal nerve, and spinal trigeminal tract. Middle-aged or elderly patients are more prone to this disease than the rest of the population. Despite its vestibular origin, acoustic neuroma initially presents with unilateral tinnitus followed by vertigo and gradual deafness that may span months or years. Patients present with hypoactive caloric response on electronystagmogram and rarely exhibit vertigo. Intactness of the facial nerve may be ascertained by checking the corneal reflex. Patients may also feel pain in the area of distribution of one or more branches of the trigeminal nerve. Radiating pain in the oropharynx, dysphagia, and hoarseness of voice may also occur if the tumor expands further downward. Deficits associated with the lesions of the vestibular apparatus, nuclei, and pathways include vertigo (sense of rotation of the environment or self), ataxia, past-pointing, nystagmus, and some other deficits depending upon the affected structure. These deficits include the following.

Vertigo refers to a complete or partial loss of spatial orientation (should be distinguished from dizziness and blackouts). Patients may have a feeling of moving in space or of objects moving around them. This deficit is associated with otogenic disorders such as Meniére disease, neurologic disorders as in multiple sclerosis, visual disorders such as diplopia, or ischemic attacks of the vertebrobasilar system. Other etiologies of vertigo may include hysteria and blood disorders such as leukemia. A peripheral lesion–induced vertigo produces severe, episodic, and paroxysmal attacks. Central vertigo, a persistent condition, is generally accompanied by nystagmus and ataxia. These manifestations are due to inhibition of one labyrinth and overactivation of the unaffected labyrinth. A patient falls to the side of the dysfunction or swerves to that side during walking. Vertigo subsides within hours when the patient assumes a recumbent position due to the inhibitory action of the flocculonodular lobe upon the intact vestibular nuclei. A sufficient sense of position is maintained, when standing, via the visual cues and utilization of proprioceptive stimuli in the dorsal columns. Darkness may impede this compensatory system. It is not usually seen in lesions of the vestibular nuclei or in cold water caloric testing of the affected side. Vertigo is rarely seen in patients with acoustic neuroma and is absent in patients with bilateral vestibular dysfunction, although bilateral dysfunction can cause disruption of the vestibulo-ocular reflex and disturbance of visual fixation during head movement.

Intense, brief, head maneuver–induced and reliably reproduced vertigo is known as benign paroxysmal positional vertigo (BPPV). This common type of vertigo is an inner ear disorder that is often sudden; lasts for few seconds; and usually is triggered by head tilting, lying down, bending or looking up. It is caused by otolith calcium crystals dislodged from the utricle and entering into the semicircular ducts (commonly posterior), causing abnormal endolymph displacement, making them more sensitive to gravitational pull.

Patients may experience the symptoms days before snow or rain, when under stress, or when deprived of normal sleep. This condition affects individuals in certain professions, such as car mechanics, yoga instructors, and plumbers, because of the habitual head positioning. Patients who are typically over the age of 60 also exhibit nausea, vomiting, syncope, visual disorder (inability to see or read during the episode), and rotatory torsional nystagmus toward the affected ear. BPPV can be tested by Dix–Hallpike maneuver, which aims at inducing fatigue to nystagmus and eventually eliminate it. In this maneuver, the examiner asks the patient to quickly lie backward over the examination table with the neck extended 30° from the horizontal while holding the patient's head at 45° angle to one side. The examiner observes the nystagmus for at least 45 s. The Epley maneuver is used to provide relief by repositioning the otolith particles from the semicircular ducts to the utricle using gravity. Following the therapy, patients are provided with a collar to avoid any position that may move the particles back into the semicircular ducts.

Ataxia (incoordination) is characterized by a wide-based, cautious, and unsteady gait, especially during turns and when walking on uneven surfaces. It is severe, gravity dependent, and generally associated with a lack of coordination of voluntary movements.

Past-pointing refers to overshooting of the patient's finger when attempting to reach a target. This is used to check the integrity of the vestibular system by asking the patient to touch the examiner's finger with his/her right index finger following rotation in a revolving (Barany) chair. In individuals with an intact vestibular apparatus, the patient's index finger is placed to the right of the examiner's finger, following rotation to the right. The reverse is true of rotation to the left.

Nystagmus is an alternating, rhythmic oscillation of the eye in response to an imbalance in the vestibular impulses. This condition may be produced artificially by visual (optokinetic nystagmus) or thermal (caloric nystagmus) stimuli. A lesion of the temporal or parietal cortex may abolish visually induced (optokinetic) nystagmus. It may be seen with eyes at rest or may be accentuated by ocular movement. Congenital nystagmus, which may be familial in origin, is characterized by pendular movement of the eyes at rest, triggered by head or eye movements. It is generally described as having fast and slow phases (phasic or jerk nystagmus). Occasionally, nystagmus may develop in normal individuals when they attempt to maintain extreme gaze to one side.

Pathological nystagmus is generally seen at rest. Clinically, it is described according to the direction of the rapid phase. The slow phase of the jerk nystagmus is due to the tonic vestibular stimuli, while the rapid phase (saccadic movement) is the result of the corrective action of the frontal cortex. Nystagmus can occur as a result of peripheral (labyrinthine) disorders, cerebellar diseases, ocular dysfunction (amblyopia), or circulatory disturbances of the endolymph (Meniére disease). It can be gaze dependent or an independent movement. It may also be observed in individuals with central lesions involving the ascending vestibular pathways (e.g., MLF). Impairment of ocular fixation associated with myasthenia gravis, optic atrophy, and poor illumination (miner's nystagmus), may generate pendular nystagmus with two phases of equal velocity. Nystagmus caused by peripheral lesions is generally horizontal, has a rapid phase toward the intact side, and is commonly associated with diplopia and dysarthria. Diseases of the inner ear may produce positional nystagmus. The latter may also be elicited upon sudden movement of the head 30° backward over the edge of the bed and simultaneously rotating to one side or the other. A left turn results in nystagmus in a clockwise direction, whereas a right turn produces nystagmus in a counterclockwise direction.

Brainstem lesions may cause gaze-dependent, coarse, and unidirectional nystagmus, which may be horizontal or vertical. The fast saccadic phase of nystagmus is lost in parietal lobe lesions which becomes evident when the striped color in optokinetic test is moved toward the affected side. Central vestibular lesions produce nystagmus with the rapid phase toward the opposite side. Vertical nystagmus with downward gaze may be observed in individuals with lesions involving the structures within the posterior cranial fossa and foramen magnum, as in Arnold–Chiari malformation and also in tegmental diseases. Vertical nystagmus with upward gaze is observed in lesions of the tegmentum and ipsilateral cerebellum. Lesions of the ascending component of the MLF (rostral to the abducens nuclei) result in *anterior internuclear ophthalmoplegia*, which is characterized by the inability to adduct the ipsilateral eye in lateral gaze and by nystagmus in the opposite eye.

VESTIBULAR TESTS

An *optokinetic test* is performed by observing the displacement of the entire visual field (e.g., watching striped colors on a revolving drum). The fast phase of this nystagmus is in the direction opposite to the movement of the object. This test may be of value in the determination of the validity of malingerers' claim that they cannot see and seems of particular value in the examination of hysterical patients.

A *caloric test* is based upon initiating a current within the endolymph by cooling or warming the inner ear through irrigation of the external acoustic meatus. The patient will be asked to tilt his/her head 60° from the vertical plane to bring the horizontal semicircular canals to a vertical position. Vertical gaze can be tested by rotating the head in the vertical plane, which produces compensatory upward and downward gaze. Instillation of cold water in both ears produces tonic upward movement of the eyes, while bilateral warm water tests produce tonic downward gaze. The effect of angular acceleration on movement of the endolymph is shown in Figures 15.4 and 15.7. The response to caloric test is decreased on the affected side in Meniére disease. In comatose patients, with a suppressed reticular activating system, the caloric test elicits tonic deviation of the eye only. Supratentorial lesions produce loss of oculocephalic reflex with preservation of caloric response. Frontal lesions do not interfere with either the caloric test or the oculocephalic reflex. Intactness of the vestibular system is determined by iatrogenic tests, which include optokinetic, caloric, and/or positional tests. An easy mnemonic to remember is COWS (cold water opposite, warm water same side).

The *Romberg test* determines the integrity of the system that mainly mediates proprioception; however, vestibular dysfunction can produce a positive Romberg test. In this test, the patient is asked by the examiner, who stands close by as a precaution, to stand with the feet together, hands to the sides, looking straight ahead, first with the eyes open and then with the eyes closed, and observe excessive swaying.

The *timed up and go test* (TUG) uses the time that a person takes to rise from a seated position, walk 10 feet, turn around, walk back, and return to the original position. TUG scores of 10 s or less indicate neurologic intactness. Fourteen seconds is within normal limits for frail elderly patients; however, the same score in a young individual may suggest that the person is prone to falls. Greater than 20 s and up to 30 s is a normal range for a dependent or disabled patient who needs assistance. The average practical cutoff value for the TUG is 12 s. This testing modality helps to determine the integrity of the vestibular, visual, and somatosensory systems utilized in postural regulation.

Dynamic gait index is produced by a patient's response to eight functional walking tests that measure gait and balance and fall risk. Each type of walking is given a score of 0–3, with 24 being the maximum possible points. An abnormal index is seen in the elderly and patients with stroke and vestibular disorders. A

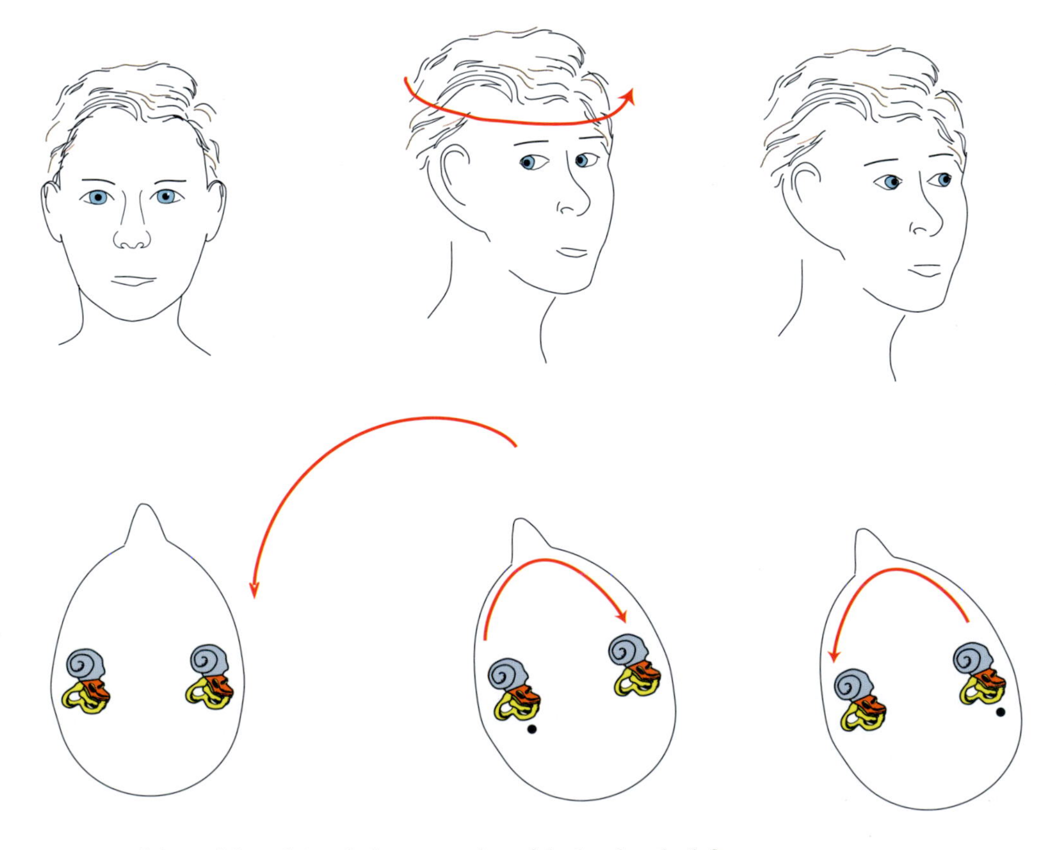

FIGURE 15.7 Movement of the endolymph in relation to rotation of the head to the left.

score of 19 or less is associated with increased risk of falling with no neurologic impairment. A score of 7 is found in older adults with a history of falls but without neurologic deficits.

The *Berg balance scale* is a testing modality used to determine the static and dynamic balance, consisting of a set of 14 tasks. This test takes 15–20 min and includes rising from a seated position and standing on one foot. The degree of postural control is rated from 0 (unable) to 4 (independent) and also by the final total of all of the scores.

SUGGESTED READING

Boyle R. Activity of medial vestibulospinal tract cells during rotation and ocular movement in the alert squirrel monkey. *J Neurophysiol* 1993;70:2176–80.

Buttner-Ennever JA. A review of otolith pathways to brainstem and cerebellum. *Ann N Y Acad Sci* 1999;871:51–64.

Buttner U, Lang W. The vestibulocortical pathway: Neurophysiological and anatomical studies in the monkey. *Prog Brain Res* 1979;50:581–8.

Buttner U, Waespe W. Purkinje cell activity in the primate flocculus during optokinetic stimulation, smooth pursuit eye movements and VOR-suppression. *Exp Brain Res* 1984;55:97–104.

Della Santina CC, Cremer PD, Carey JP, Minor LB. The vestibulo-ocular reflex during self-generated head movements by human subjects with unilateral vestibular hypofunction: Improved gain, latency, and alignment provide evidence for preprogramming. *Ann N Y Acad Sci* 2001;942:465–6.

Dickman JD, Angelaki DE. Dynamics of vestibular neurons during rotational motion in alert rhesus monkeys. *Exp Brain Res* 2004;155:91–101.

Gu Y, Deangelis GC, Angelaki DE. A functional link between area MSTd and heading perception based on vestibular signals. *Nat Neurosci* 2007;10:1038–47.

Horak FB. Postural orientation and equilibrium: What do we need to know about neural control of balance to prevent falls? *Age Ageing* 2006;35(Suppl 2):ii7–11.

Klam F, Graf W. Vestibular signals of posterior parietal cortex neurons during active and passive head movements in macaque monkeys. *Ann N Y Acad Sci* 2003;1004:271–82.

Le P, Pfeifer C, Homan J, Lasak J. Vertigo and sudden hearing loss at 35,000 feet. *JAMA Otolaryngol Head Neck Surg* 2014;140(1):79–80.

Liu S, Tong D, Liu M, Lv D, Li Y. The safe zone of posterior semicircular canal resection in suboccipital retrosigmoid sinus approach for acoustic neuroma surgery. *J Craniofac Surg* 2013;24(6):2103–5.

McCrea RA, Gdowski GT. Firing behaviour of squirrel monkey eye movement-related vestibular nucleus neurons during gaze saccades. *J Physiol* 2003;546:207–24.

Nakamagoe K, Fujizuka N, Koganezawa T, Yamaguchi T, Tamaoka A. Downbeat nystagmus associated with damage to the medial longitudinal fasciculus of the pons: A vestibular balance control mechanism via the lower brainstem paramedian tract neurons. *J Neurol Sci* 2013;328(1–2):98–101.

Reisine H, Simpson JI, Henn V. A geometric analysis of semicircular canals and induced activity in their peripheral afferents in the rhesus monkey. *Ann N Y Acad Sci* 1988;545:10–20.

Roy JE, Cullen KE. A neural correlate for vestibulo-ocular reflex suppression during voluntary eye-head gaze shifts. *Nat Neurosci* 1998;1:404–10.

Roy JE, Cullen KE. Vestibuloocular reflex signal modulation during voluntary and passive head movements. *J Neurophysiol* 2002;87:2337–57.

Roy JE, Cullen KE. Dissociating self-generated from passively applied head motion: Neural mechanisms in the vestibular nuclei. *J Neurosci* 2004;24:2102–11.

Sadeghi SG, Goldberg JM, Minor LB, Cullen KE. Efferent-mediated responses in vestibular nerve afferents of the alert macaque. *J Neurophysiol* 2009;101(2):988–1001.

Sadeghi SG, Goldberg JM, Minor LB, Cullen KE. Efferent mediated responses in vestibular nerve afferents of the alert macaque. *J Neurophysiol* 2009;101:988–1001.

Sara SA, Teh BM, Friedland P. Bilateral sudden sensorineural hearing loss: Review. *J Laryngol Otol* 2014;128 (Suppl 1):S8–15.

Section VI

Special Visceral Sensations

16 Olfactory System

Chemicals that dissolve in liquid stimulate both olfactory and taste receptors in a similar way. The perception of food flavor is dependent upon the close interaction between the olfactory and gustatory (taste) systems. The olfactory system detects, recognizes, and finely discriminate airborne odorants. This special visceral sensation becomes functional at birth and effectively operates even at low molecular concentration of the odorants. Odorants bind reversibly to the receptor membrane proteins in the olfactory epithelium, generating depolarization and action potentials through a series of interactions. Propagated action potentials converge on the olfactory bulb and course through the olfactory tract and the striae to the pyriform cortex. Association cortices also receive olfactory projection indirectly from the pyriform cortex via the thalamus. Olfaction is considered one of the oldest systems that directly influences feeding behavior, emotion, social attitude, protective reaction, and sexual desire. This is achieved through extensive connections to diverse areas of the brain including the septal area, entorhinal cortex, and amygdala. Pheromones can elicit responses that affect the reproductive cycles, possibly through a hypothalamic connection; however, the precise mechanism that underlies the processing of these chemosensory signals and the role of the vomeronasal organ (VNO) remain at best preliminary.

PERIPHERAL OLFACTORY APPARATUS

The olfactory apparatus consists of the sensory epithelium with supporting cells and chemoreceptor cells that detect and discriminate between different airborne odorants. Olfactory information is conveyed, in the form of electrical signals, via the olfactory nerves to the olfactory bulb and eventually via the olfactory tract and striae to the primary olfactory cortex. The olfactory mucosa is pigmented and covers the posterior part of the superior nasal concha, the upper posterior part of the lateral nasal wall, the roof of the nasal cavity, the upper part of the nasal septum, the inferior surface of the cribriform plate of the ethmoid, and the sphenoethmoidal recess. It is much thicker than the respiratory epithelium and consists of pseudostratified epithelium that contains olfactory receptors underlined by the lamina propria. The latter contains olfactory filaments and the subepithelial Bowman glands, which secrete mucus containing microvilli and sensory cilia. The olfactory epithelium consists of the olfactory receptors, sustentacular cells, proper basal cells, and globose cells.

Olfactory Receptors

The olfactory receptor cells are bipolar cells with soma, apical dendrites, and basal axons. The location of the cell bodies in close proximity to the sensory surface is a unique feature that distinguishes them from other receptors, though they do not respond to odorants. The cell body of each receptor appears ellipsoidal with a single apical dendrite that extends to the mucus layer on the epithelial surface and forms rounded extensions receiving centrioles that eventually become the olfactory cilia. Olfactory cilia are unique in that they elicit a strong electrical response upon stimulation by the odorants, contain an abundant amount of actin, and have close resemblance to the microvilli. Due to these characteristics, olfactory cilia maximize the sensory area of each cell for detection of odorants through trailing ends that travel toward the basal layer of the epithelium, where they join other axons to form the olfactory filaments. Olfactory filaments are ensheathed by glial cells as far as the olfactory bulb. These glial cells that continue around the olfactory filaments to their termination inside the olfactory bulb are unique cells that gain origin from the olfactory placode. They contain abundant glial fibrillary acidic protein (GFAP) intermediate filaments, which are also present in glial cells of the central nervous system, such as astrocytes.

The olfactory filaments pass through the cribriform plate as medial and lateral groups to enter the olfactory bulb and synapse with the mitral, basket, and periglomerular cells in the olfactory glomeruli. Plasma membrane contains concentrations of intramembranous particles that cover the exposed surfaces of the olfactory dendrites. These particles are thought to be the site of olfactory reception and ion channels related to sensory transduction; their concentrations accord with their ability to detect low concentrations of odorants. The placode from which the neuroepithelial receptor cells, sustentacular cells, and glial cells are derived also gives rise to the VNO and to a group of cells that enter the medial eminence of the hypothalamus where the cells of the luteinizing hormone–releasing hormone are located. The path of migration of these cells to the hypothalamus is thought to form the nervi terminalis. These unmyelinated fibers of the bipolar and multipolar neurons extend along the medial sides of the olfactory tracts and pass through the cribriform plate to reach the nasal mucosa. They enter the brain, establishing contacts with the neurons of the septal area, hypothalamus, lamina terminalis, and anterior perforated substance (APS).

Olfactory receptor cells show diversity at the cellular and molecular levels. Their axons contain a number of immunohistochemically demonstrable characteristic proteins, such as 19 kDa olfactory marker protein and carnosine. They also express surface molecules that cross-react with the ABO blood group antigens and carbohydrates that bind several lectins. A large number of olfactory receptor genes that code for olfactory receptor protein molecules have been isolated. However, expression of these genes remains very limited as each olfactory receptor neuron expresses one copy of a single (the other copy is mute) or very few odorant receptor genes. This clearly points in the direction that only one copy of these genes is activated in transduction and that different odorants stimulate individual subsets of the olfactory receptor neurons.

Olfactory receptors undergo periodic replacement and show regeneration following excision of the olfactory nerve or when they are directly damaged. Regeneration may be delayed when scar tissue is formed along the pathway of the interrupted axon, which initiates a surge of the mitosis in the globose cells. Severe respiratory infection with influenza virus can also impede the regenerative process. Periodic slow and continuous replacement of receptor cells throughout life occurs as a result of mitotic division of the globose cells, which are the stem cells at the base of the epithelium, followed by the growth of the dendrites to the olfactory mucosa and axons that enter the olfactory bulb. The newly formed cells assume a location below the nuclei of the supporting cells, and then they shed or are removed by phagocytosis. This process is balanced by the degeneration of the existing cells and the continuous addition of new receptors. The turnover does not occur in an orderly fashion and can be affected by external factors, such as the quality of inhaled air. Although some receptors may have a long life span, the overall cycle of replacement is considered a form of apoptosis, which is time and position dependent. This process of replacement of receptors is accompanied by regeneration of the axons and formation of new synaptic connections within the olfactory bulb and sometimes in the frontal lobe of the brain. This unique process, a characteristic of fetal development, seems to continue in the adult olfactory bulb. The presence of fetal neural cell adhesion molecules (N-CAMs), intermediate filaments, and fetal tubulin in the olfactory receptors supports the notion of the fetal pattern of cell turnover. An impaired sense of smell in the elderly is accompanied by loss of olfactory receptors, which are replaced by ciliated epithelium of the respiratory system.

Olfactory secretion, which is produced by the Bowman glands, allows odorants to dissolve and diffuse in a sufficient concentration to stimulate the sensory receptor. It regulates the flow of ions essential for transduction and elimination of harmful substances and used odorants. The latter function is governed by the odorant-binding proteins (OBPs), which are carried by the cilia of the respiratory epithelium. Olfactory secretion also contains substances that act as a first line of defense against antigens such as the sulfated proteoglycans, IgA, lysozyme, and lactoferrin. Increased thickness of the mucus as a result of nasal congestion can negatively influence the ability to detect olfactory stimuli.

The microvilli of the sustentacular (supporting) cells pass between the olfactory cilia and proceed toward the overlying mucus, while their nuclei form a layer on the exterior of the receptors. They have a long life span, are very slowly replaced, and remain stable, unlike the receptor cells, which undergo rapid turnover. Their role in the insulation of the receptors, elimination of debris and excess or used odorants and toxic substances, and in establishing structural support to the epithelium has been reported. Further, the supporting cells maintain the ionic environment essential for olfaction, serve as an anchoring point for the receptors, play a role in receptor cell maturation and turnover, and also ensheath and isolate the receptor cells. The presence of high levels of cytochrome-P450, endocytotic vesicles, and other detoxifying agents in the supporting cells of the olfactory epithelium may suggest the possible role of these cells in the elimination of used and unwanted odorants, degradation of the mucus from the epithelium, and subsequent intensification of olfaction. The phagocytic activities of secondary lysosomal (residual) bodies in the bases of the supporting cells are responsible for the pigmentation of the olfactory area. Subsequent to phagocytosis, lysosomal extensions in lamellated dense bodies at the basal lamina are formed that resemble lipofuscin granules. These bodies are responsible for the pigmentation of the olfactory epithelium, which becomes darker as their number increase with age. The sustentacular cells play an important role in regulation of the ionic environment of the receptors, removal of debris and toxic substances, maturation of the receptors, and their insulation. In contrast to the rapid turnover of the receptors cells, the supporting cells remain fairly stable.

Basal cell proper lies in contact with the basal lamina, exhibiting numerous intermediate filaments embedded in the desmosomes that establish contact with the sustentacular cells. They are flat cells with darkly stained cytoplasm and condensed nuclei. Globose (blastema) cells are elliptical with pale nuclei and cytoplasm containing a large number of ribosomes. They are considered as neuroblasts undergoing active mitoses (neurogenesis) that closely relate to the extent of turnover of receptor cells. Mitosis continues throughout life. Microvillous cells with characteristic apical long microvilli may be considered as olfactory receptor cells, although they lack olfactory marker protein and do not show degeneration subsequent to disruption of the olfactory cells. Olfactory epithelium also contains free terminals of the trigeminal nerve, which convey noxious stimuli from inhaled air.

Olfactory transduction involves odorant (stimulus) dissolution in the thin layer of the olfactory mucosa and its diffusion to the exposed receptor membrane in the distal segments of the olfactory cilia and olfactory knob, where the olfactory receptor G protein (Golf) is present. This is followed by activation of the cilial secondary messenger systems, which utilize the adenylate cyclase pathway through the guanosine triphosphate (GTP)–dependent adenylate cyclase III. This leads to the formation of cAMP, activation and then opening of voltage-sensitive calcium/sodium channels. This is followed by depolarization of the cilia of receptor cells and then the axon hillock and initiation of action potential in the olfactory nerve. The cilial secondary messenger systems may also utilize the phosphoinositide pathway through the GTP-dependent phospholipase C (PLC) to produce depolarization and action potential production. PLC, a membrane-bound enzyme activated through protein receptors, cleaves the phospholipid phosphatidylinositol-4,5-bisphosphate (PIP2), producing diacylglycerol (DAG) and inositol-1,4,5-trisphosphate (IP3). IP3 enters the cytoplasm and binds to IP3 receptors, which causes opening of calcium channels, leading to an increase in the concentration of calcium. Calcium and DAG activate protein kinase C (PKC), which phosphorylates cytosolic protein molecules, leading to additional intracellular changes.

There is an indication that denseness of the olfactory receptor proteins is related to the ability to acutely distinguish various odorants. A single receptor may be activated by a single odorant or multiple odorant molecules irrespective of the extent of the affinity of the odorants to bind to a receptor molecule. The variation in the concentration of the odorants and length of exposure to these stimuli can alter, desensitize, and possibly attenuate individual receptor neuronal response. In mammals and, to a lesser degree, in humans, auxiliary olfactory tubes develop in the anterior edge of the base of the nasal septum and are enclosed by a bony or cartilaginous shell. These structures, which are known as the VNOs, are divided by the nasal septum and open into the nasal cavity or indirectly into the oral cavity. They are chiefly concerned with identifying pheromones that regulate sexual reaction, behavior, and temperature. The medially located sensory component of the organ, a pseudostratified epithelium, consists of relatively large cell bodies with microvilli but without cilia or globose cells. While the nonsensory cells occupy the cavernous tissue in the lateral wall, enabling odorants to enter and leave these organs in a liquid phase (secreted by the VNO glands) via a pumping mechanism regulated by changes in blood pressure. Axons of the sensory cells within VNO project to the accessory olfactory bulb that sends impulses to the amygdala and the bed nucleus of the stria terminalis and, eventually, to the anterior hypothalamus. These primitive entities usually undergo postnatal involution in the human.

Olfactory Nerve

The *olfactory nerve* or olfactory filaments represent the central processes of the bipolar neurons located in the olfactory mucosa of the superior nasal concha and the corresponding part of the nasal septum. These unmyelinated filaments are among the smallest and slowest in conduction. They are enclosed by glial cells, form medial and lateral fasciculi, and enter the olfactory bulb in the anterior cranial fossa by traversing the cribriform plate of the ethmoid bone. Extension of the dura around the olfactory nerve fibers continues with the periosteum of the nasal cavity, and the arachnoid–pia layer merges with the perineurium. The olfactory filaments form synaptic linkage with the mitral periglomerular and tufted cells.

The subarachnoid space continues around the olfactory filaments, serving as a conduit for the spread of infections to the meninges. Due to the close relationship of the olfactory filaments to the anterior cranial fossa, fractures of this area may lead to anosmia accompanied by leakage of cerebrospinal fluid (CSF) from the nasal cavity (CSF rhinorrhea). Anosmia can also be caused by inflammatory processes that affect the nasal cavity, including the conchae and the paranasal sinuses. Nasal polyps caused by allergy or associated with cystic fibrosis can lead to anosmia. Intracranial pressure increase induced by neurosyphilis, frontal lobe tumors, meningiomas of the floor of the anterior cranial fossa, abscesses as a result of meningitis, hypophysial tumors, hypothyroidism, cadmium toxicity, pernicious anemia, and meningiomas of the sella turcica may lead to unilateral anosmia. Aneurysms of the internal carotid artery, Refsum's disease, schizophrenia, Paget's disease, and sarcoidosis can also cause anosmia. Medications that are used for the treatment of cardiac arrhythmia such as amiadarone, alpha blockers such as dibenzyline, and the cold remedy drug Zicam have been reported to cause anosmia. Anosmia can be an early sign of Parkinson's disease and Alzheimer's disease (AD). Patients with acquired anosmia may experience reduction in libido and appetite.

Congenital anosmia follows an autosomal dominant pattern.

OLFACTORY BULB

The *olfactory bulb* (Figure 16.1) develops from the floor of the primitive cerebral hemispheres that elongates to form the olfactory tract. It is located between the inferior surface of the frontal lobe and the cribriform plate of the ethmoid bone and lies in the anterior part of the olfactory sulcus. The embryonic olfactory bulb develops as a ventricle-containing diverticulum, which forms the basis of the laminar cellular organization of the bulb. Fusion of the walls of the olfactory bulb ventricle renders it a solid structure. The olfactory bulb is comprised of the olfactory nerve layer, molecular layer, mitral cell layer, internal granular layer, and layer of olfactory tract fibers. As the name indicates, the olfactory nerve layer consists of the unmyelinated axons in various stages of development or degeneration as well as laminin. Presence of the latter protein, which influences cell differentiation, migration, and adhesion, as well as phenotype and survival, may be linked to the development of the neuroblasts that form a pool for the olfactory receptor cell replacement. The olfactory nerves divide in

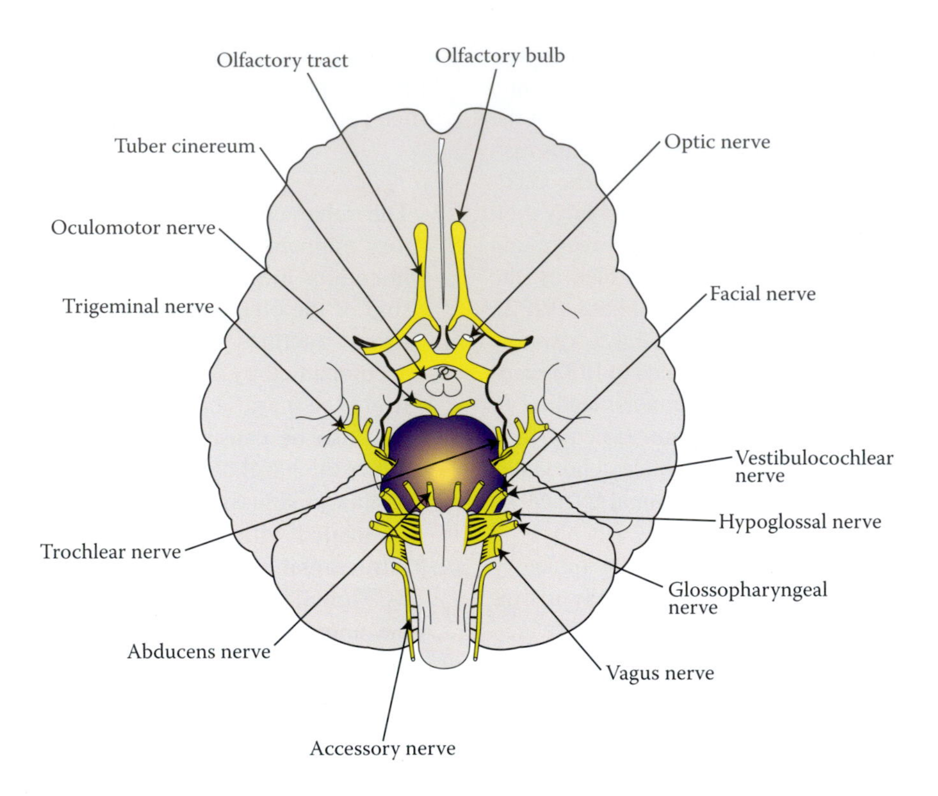

FIGURE 16.1 General scheme of the cranial nerves. Note the olfactory bulb and tract.

the glomerular layer and, through their branches, establish synaptic contact with the dendrites of the mitral, internal tufted, and periglomerular cells. Examination of the external plexiform layer (EPL) reveals a superficial part that is formed by the bodies of the tufted cells and a deep part that harbors the dendrites of the mitral and tufted cells. Bodies of the large mitral cells primarily form a thin sheath with their dendrites, contributing to the olfactory glomeruli and EPL, whereas the axons of the mitral cells proceed to the deep layers. EPL receives cholinergic and monoaminergic input from the basal forebrain and the brainstem. Tufted cells with few granule cell bodies and axons as well as collaterals of the mitral cells constitute a separate layer known as the internal plexiform layer. The granule cell layer (GCL) consists of closely packed cells with their processes and afferent and efferent axons. The olfactory synaptic glomeruli are formed by the dendrites of the mitral, internal tufted, and periglomerular cells as well as axons of the mitral cells. GCL receives robust serotoninergic, adrenergic, and noradrenergic afferent input.

The olfactory bulb acts as a relay, integration, and feedback center for complex pathways. It is a synaptic site of the olfactory filaments with the mitral cells, and with the axons originating from the anterior olfactory nucleus (AON) and the contralateral olfactory bulb. The AON consists of pyramidal neurons, which are located in the posterior portion of the olfactory bulb and may extend into the olfactory trigone and the striae. These neurons receive axonal collaterals of mitral and tufted cells and provide recurrent collaterals to synapse with the dendrites of the ipsilateral tufted and granule cells. The pyramidal neurons also project to the contralateral olfactory bulb and AON through the anterior commissure (Figure 16.2). The granular and glomerular cell layers of the olfactory bulb receive input from the pyramidal neurons of the olfactory cortex, AON, cholinergic neurons of the horizontal limb nucleus of the diagonal band of Broca, serotoninergic neurons of the raphe nuclei, and the noradrenergic neurons of the locus ceruleus. Efferents of the olfactory bulb are formed mainly by the axons of the mitral and tufted cells. Mitral and tufted cells are the pillar cells in the olfactory bulb. As discussed earlier, the mitral cell possesses a primary apical dendrite that bypasses the outer plexiform layer and enters the olfactory glomerulus via a cluster of branches that spread across the entire glomerulus, and secondary dendrites with few branches that terminate in the EPL. Axons of the mitral cells form the bulk of the olfactory tract, giving rise to recurrent collaterals that diffuse in the granule cell and internal plexiform layers. Through this arrangement, the mitral cells can influence olfactory input through the glomeruli in the superficial layer and modulate output in the deep layers. It is presumed that glutamate and aspartate are utilized as neurotransmitters in the connections of the dendrites and axons of the mitral and tufted cells.

Despite the morphological similarities of the mitral and tufted cells, the presence of certain characteristics enabled the categorization of the tufted cells into the external, middle, and internal groups of cells. The external tufted cells are considered intrinsic, as their dendrites and their collaterals remain confined to the internal plexiform and GCL. Those cells that are located in close proximity to and

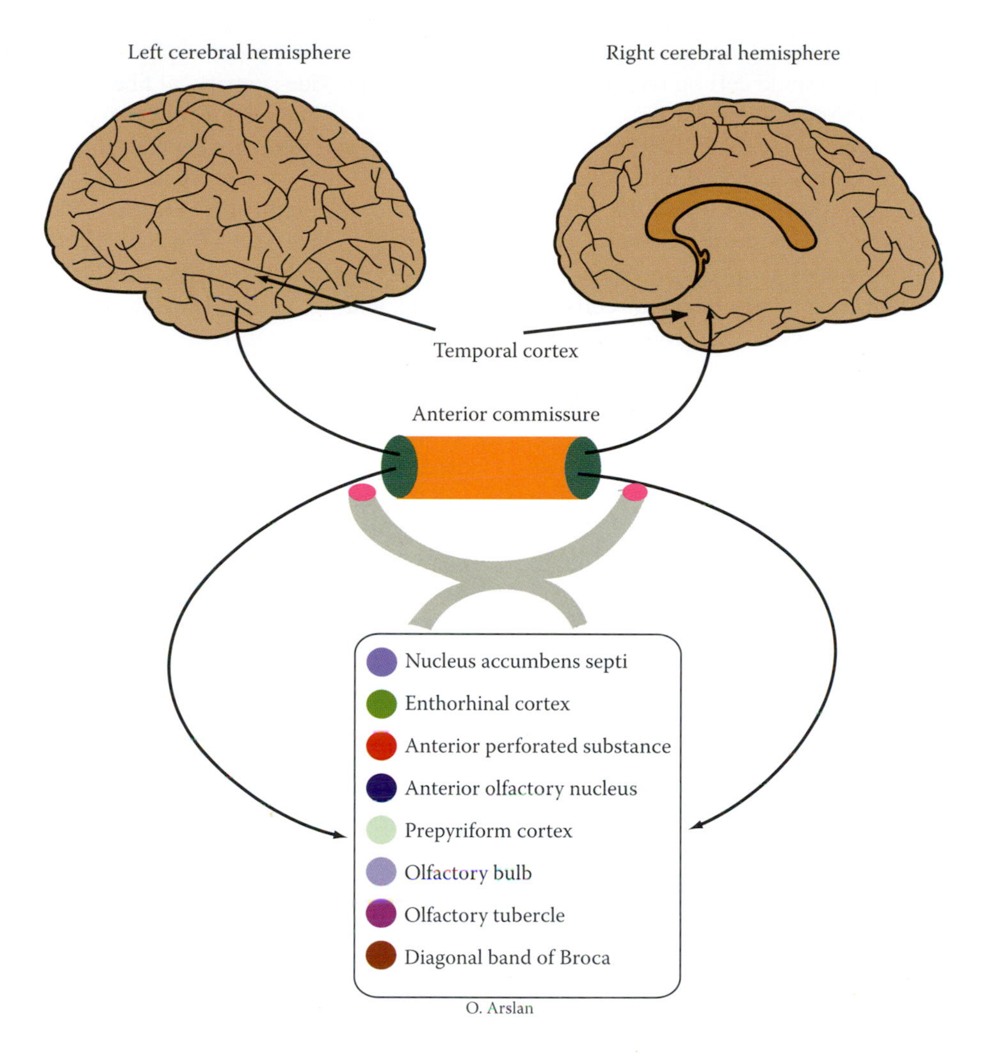

FIGURE 16.2 Schematic drawing of the anterior commissure. Observe the identical structures in both hemispheres connected by this commissure.

resemble the mitral cells form the internal tufted cells. In view of their number and axonal contribution to olfactory tract, the middle tufted cells are considered the main group of tufted cells that provide axons to the olfactory tract, with collaterals to the internal plexiform layer and dendritic branches that join the olfactory glomeruli. Tufted cells utilize substance P in addition to the classical neurotransmitters. The axons of the mitral and tufted cells form a parallel output within the olfactory tract but with selective projections to the piriform and entorhinal cortices, AON, and also the amygdala. This selectivity of connection extends to the granule cells, which differentially influence the output of the mitral and tufted cells.

Granule cells are anaxonic, utilize gamma aminobutyric acid (GABA) as a neurotransmitter, and resemble the amacrine cells of the retina. Their deep dendritic processes immerse in the GCL, while the superficial dendritic rami spread in to the EPL. Through their extensive connections with the mitral and tufted cells and terminals of afferents, the spines of the granule cell dendrites are considered important sites where olfactory input is regulated. Superficially placed granule cells provide dendrites that establish linkage with the tufted cell dendrites in the superficial part of the EPL, whereas the deeply placed granule cells establish synaptic connections with the dendrites of the mitral cells in the deep part of the same layer. Those granule cells that are positioned between the superficial and deep groups remain within the confines of the same layer, unable to project dendritic processes to the adjacent parts of the olfactory bulb.

Periglomerular cells are primarily dopaminergic (Dahlström cell group A15); however, others are GABAergic or utilize a combination of both of these neurotransmitters as well as enkephalin. Within the olfactory glomeruli, their dendrites connect with the dendrites of the mitral and tufted cells as well as with the terminals of the olfactory filaments, whereas their short axons extend outside the glomerulus and establish interglomerular connection, enabling activities within one glomerulus to affect the output of neighboring glomeruli.

The olfactory glomerulus is a site of excitatory and inhibitory axodendritic synaptic connection of the olfactory nerve axon with the dendrite of the mitral, tufted, and periglomerular cells as well as dendrodendritic synapses involving the mitral, tufted, and periglomerular cells. The synaptic linkage between the granule cell on one hand and the mitral/tufted cell on the other hand is thought to be inhibitory, a fact that may account for the inhibitory control of this connection on the olfactory bulb output. In contrast, the synapse between the mitral/tufted cell and the granule cell appears to be excitatory. The synaptic connectivity within these glomeruli continues to maintain specificity despite the continued production of new receptor cells and elimination of the degenerated cells. Serotonergic fibers that emanate from the mesencephalic raphe nuclei enhance the centrifugal influence of the glomeruli on the processing of olfactory information.

ANTERIOR COMMISSURE

The *anterior commissure* (Figure 16.2) is a robust bundle of myelinated fibers shaped like the handle of a bicycle that lies superior to the optic chiasma. It crosses the lamina terminalis and divides the fornix into precommissural and postcommissural columns. This commissure consists of anterior and posterior bundles. The anterior olfactory bundle courses forward toward the APS and olfactory tracts. It interconnects the olfactory bulbs, tubercles and tracts, containing the centrifugal axons of the AON of one side that project to the corresponding nucleus on the contralateral side and to the granule cells. The granule and tufted cells activate other parts of the olfactory system through this connection. These axons act as a feedback loop, representing an efferent system that regulates the incoming olfactory pathways. The larger, posterior bundle courses inferior to the lenticular nucleus, running through the external capsule and terminating in the anterior temporal and parahippocampal gyri. It interconnects the medial and, to a lesser degree, the middle and inferior temporal gyri, entorhinal area, diagonal band of Broca, APS, prepyriform cortex, amygdala, and bed nucleus of the stria terminalis of both sides.

OLFACTORY PATHWAYS

The *olfactory tract* (Figure 16.2) runs in the olfactory sulcus inferior to the frontal lobe and lateral to the gyrus rectus. It is formed by the centripetal axons of the mitral and tufted cells as well as the centrifugal contralateral axons of the AON neurons, horizontal limb nucleus of the diagonal band of Broca, serotonergic neurons of the mesencephalic dorsal raphe nucleus, and noradrenergic neurons of the locus ceruleus. The cellular and laminar organization of the olfactory bulb disappears in the olfactory tract, including the individual mitral, tufted, and periglomerular cells, with the exception of the granule cells that form the AON, a poorly laminated rostral cortical structure that receives collaterals from the centripetal axons of the mitral and tufted cells. The olfactory tract a trilaminar cortex that consists of a superficial plexiform (layer I) that receives the centripetal input from the olfactory bulb, an intermediate pyramidal (layer II), as well as a deeper polymorphic cell (layer III) layers. Layers II and III provide centrifugal fibers to the olfactory bulb and projections to the piriform cortex and send axons that continue with the main centripetal fibers of the olfactory tract into the olfactory striae. Neurons of the AON continue caudally with the prepiriform cortex, APS, and precommissural septal area. Near the APS, the olfactory tract presents the olfactory (pyramid) trigone, which diverge into the lateral, medial, and smaller intermediate olfactory striae.

The *olfactory tubercle*, although harder to delineate, is a prominence that lies caudal to the olfactory trigone and adjacent to the APS, separated from the lentiform nucleus by several structures such as the substantia inominata, fibers of the anterior commissure, and ansa lenticularis. It is a trilaminar structure that receives afferents from the dorsomedial nucleus of the thalamus and establishes reciprocal connection with the amygdala, while efferents project to the entorhinal area, hippocampal formation, and septal area via the medial forebrain bundle and stria medullaris thalami.

As the *lateral olfactory stria* continues with the semilunar gyrus at the rostral end of the uncus, it is covered by the lateral olfactory gyrus that blends with the gyrus ambiens of the limen insulae, forming the prepiriform cortex that further continues caudally with the entorhinal areas (Brodmann area 28). The entorhinal area and prepiriform and periamygdaloid cortices form the piriform lobe, which lies medial to the rhinal sulcus. During its course toward the uncus, the lateral olfactory stria wraps around the APS. The *medial olfactory stria* runs anterior to the lamina terminalis, accompanied by the diagonal band of Broca, terminating ipsilaterally in the paraterminal (septal area) and subcallosal (parolfactory) gyri. The subcallosal gyrus lies between the anterior and posterior olfactory sulci, separated from the paraterminal gyrus (precommissural septum) by the posterior olfactory sulcus. Olfactory impulses received by the septal area travel through the medial forebrain bundle to the lateral hypothalamus, preoptic area, and tegmentum of the midbrain. This pathway enables the olfactory impulses to influence the autonomic centers in the brainstem. Additionally, olfactory impulses from the septal area also project to the ipsilateral habenula via the stria medullaris thalami. The habenula, in turn, projects to the interpeduncular and tegmental nuclei via the fasciculus retroflexus, and to the opposite habenula through the habenular commissure. The autonomic (visceral) nuclei of the brainstem and the spinal cord receive projections from the reticular (tegmental) nuclei of the midbrain via the dorsal longitudinal fasciculus of Schütz and the reticulospinal tracts, respectively. These massive connections explain the emotional reaction as well as the somatic and visceral reflexes (such as tongue, eye, and neck movements; salivation; increased gastric secretion; retching; and vomiting) in response to odors. The intermediate stria terminates in the APS (Figure 16.3).

The *APS* lies between the optic chiasma and optic tract and demarcates the site of branching of the internal carotid artery into the middle and anterior cerebral arteries. This area, pierced by central branches of the internal carotid,

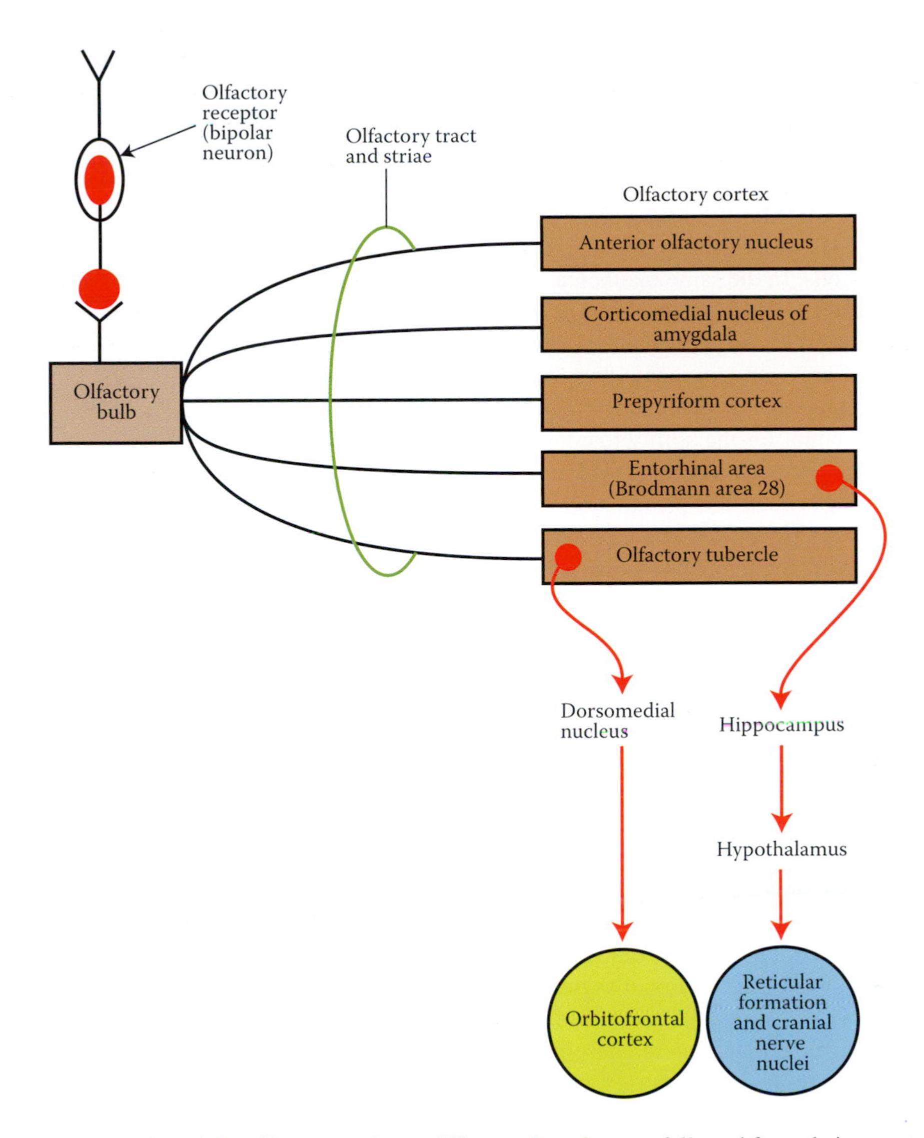

FIGURE 16.3 Neuronal organization of the olfactory pathway. Olfactory impulses are followed from their receptor to the site of perception and recognition at the olfactory cortex. Projection of the olfactory impulses to the thalamus, hypothalamus, and reticular formation are also shown.

anterior, and middle cerebral arteries, is located between the diverging medial and lateral olfactory striae. The APS lies medial to the uncus and lateral to the optic chiasma and continues with the prepyriform cortex and the semilunar gyrus (periamygdaloid area) laterally and the septal area rostrally. Superiorly, it continues with the claustrum and corpus striatum through the substantia innominata, which contains the cholinergic neurons of the nucleus basalis of Meynert, ventral globus pallidus, and extension of the central nucleus of amygdala that merges with the bed nucleus of stria terminalis. The fibers of the ansa lenticularis and anterior commissure separate the APS from the globus pallidus. The granule cells located in the pyramidal layer of the APS form the islands of Calleja, which contain the nucleus accumbens septi. The APS forms the olfactory tubercle posterior to the olfactory trigone and receives, if present, the intermediate olfactory striae. The diagonal band of Broca is a cortical structure located caudal to the APS that continues with the periamygdaloid area caudally and the paraterminal gyrus rostrally.

OLFACTORY CORTICES

The *primary olfactory cortex* includes the piriform (prepiriform) cortex and the periamygdaloid cortex, which consists of the corticomedial nucleus of the amygdala and the associated semilunar gyrus. The piriform cortex comprises the lateral olfactory and ambiens gyri, which maintain reciprocal connection with the entorhinal cortex, basolateral amygdaloid nucleus, preoptic area, septal area, and dorsomedial nucleus of the thalamus. The stria terminalis conveys olfactory information from the corticomedial nucleus of the amygdala to the preoptic area and the ventromedial nucleus of the hypothalamus. The piriform lobe represents the uncus (intralimbic and uncinate gyri), olfactory tubercle, lateral olfactory gyrus, and entorhinal cortex (Brodmann area 28).

The *piriform cortex* is a trilaminar paleocortex that consists of a superficial plexiform layer (layer I), a superficial compact layer (layer II), and a deeper sporadically arranged cell layer (layer III). These layers show specificity relative to their connections, where the axons of the mitral cell are received by the superficial part of layer I, terminating in the deep parts of layers II and III by synapsing with the distal dendrites of the pyramidal cells. In the same manner, association olfactory cortices project to the deep part of layer I and establish synaptic connections with the proximal dendrites of the pyramidal cells. Lack of precise topographic distribution of olfactory input and the presence of an intrinsic afferent synaptic connection may account for the ability to construct memory relative to previous odorants. As discussed earlier, extensive connections exist between the primary olfactory cortex through the fibers of the pyramidal cells and the posterolateral orbitofrontal cortex, hypothalamus, magnocellular part of the dorsomedial nucleus, and basolateral amygdala that enable the olfactory impulses to exert significant influences on various activities regulated by these structures. The lateroposterior and centroposterior orbitofrontal cortex are the main targets of piriform cortical projection to the frontal lobe. The centroposterior part of the orbitofrontal cortex represents the primary site of projection of the fibers of the magnocellular part of the dorsomedial thalamic nucleus, and when damaged, the ability to detect and discriminate odorants will be lost. It appears that bilateral olfactory stimulation of the piriform cortices causes stimulation of the orbitofrontal cortex on one side. This fact demonstrates the lateralization of the olfactory sensation and the functional asymmetry of the cerebral hemisphere with regard to olfaction. The overlap between the pyramidal cell projections of the piriform cortex and the gustatory fibers to the insular cortex may shed light on the olfactory deficit–induced taste disorders.

The *entorhinal cortex* (Brodmann area 28) is a six-layered cortex that lacks the internal granular layer with medial (28a) and lateral (28b) parts as well as rostrocaudal cellular arrangement. Layer 1 is acellular plexiform, layer 2 marks the boundary of the entorhinal cortex through visible small protrusions known as the verrucae hippocampae formed by the large pyramidal and stellate cells, and layer 3 is formed by medium-sized pyramidal cells. Layer 4, known as lamina dessicans, lacks cell bodies; layer 5 consists of pyramidal cells; and layer 6 lies beneath the presubiculum and parasubiculum and around the perforant pathway. The entorhinal cortex lies caudal to the amygdala, rostral to the hippocampal gyrus, medial to the collateral sulcus, and adjacent to Brodmann areas 35 and 36. Primary olfactory cortical projections reach the rostral part of area 28 but not the caudal part. Area 28 provides massive projections to the dentate gyrus, hippocampal gyrus, and also the frontal lobe through the uncinate fasciculus, receiving profuse input from diffuse areas of the neocortex, as well as from the olfactory bulb and piriform and periamygdaloid cortices. Superficial layers of area 28 are rich with vasoactive intestinal polypeptide and cholecystokinin immunoreactive cells. Area 28 gives rise to fibers that contain encephalin, as well as to the perforant pathway that utilizes glutamate in its projection to the dentate gyrus.

The *basal forebrain nucleus of Meynert*, which consists predominantly of cholinergic neurons in the substantia innominata, is implicated in *AD*. These cholinergic neurons project extensively to the cerebral cortex. Numerous neurofibrillary tangles (NFTs) and neuritic plaques are also seen in the CA1 zone of the hippocampal gyrus. The *tau* protein that cross-links the microtubules in the normal perikaryon becomes excessively phosphorylated, unable to cross-link microtubules, leading to the formation of the paired helical filaments as an early stage of NFT formation. Hyperphosphorylation of *tau* proteins occurs as a result of activation of kinases and/or deactivation of phosphatases. As AD progresses, the NFT and its associated apical dendrites become filled with abnormally phosphorylated *tau* protein and helical filaments. In the end stages of the disease, the cytoskeleton of the pyramidal neuron disintegrates, and the remaining *tau* protein and paired helical filaments form a "ghost NFT." NFTs, often referred to as "intracellular amyloid," also contain ubiquitin and glycosaminoglycans (GAGs). When aluminum has been demonstrated within NFT in cortical areas of AD-affected individuals, some investigators suggested the role of this metal as a cause of this disease. Presence of NFTs is commonly associated with dystrophic neurites (neuropil threads or curly fibers) in the cortical neuropil. Although dystrophic neurites are not specific to AD, their abundant presence may correlate with clinical dementia and presence of NFT. MAP-2 is another microtubule-associated protein that is found in healthy cells and becomes abnormally phosphorylated. Additionally, A-68, an early marker protein for AD, is suspected to be a modified form of *tau*.

OLFACTORY CORTICAL DYSFUNCTIONS

Neuronal loss in the nucleus basalis of Meynert, a characteristic of Alzheimer disease (AD), is closely related to the severity of dementia. This is followed by involvement of the association cortices of the frontal, parietal, and temporal lobes. Cortical neurons most affected are the pyramidal-shaped neurons, which project to other cortical layers such as I, II, III, and VI. Other limbic structures such as the amygdala, locus ceruleus, median raphe nuclei, hypothalamus, and cingulate and orbitofrontal cortex may also be involved. Some investigators have suggested that *Hirano bodies*, which are eosinophilic rods composed of alpha-actinin, vinculin, and tropomyosin epitopes, are possible products of the degradative process that the cytoskeletal elements undergo. However, no clear-cut evidence has been documented to support this suggestion.

The role of cholinergic neurons in normal intellectual activity is confirmed by studies in which central anticholinergic medications produced cognitive dysfunctions that mimic Alzheimer's dementia. In AD, other biochemical imbalances, morphological abnormality, or degeneration is also noticed in neurons associated with somatostatin, norepinephrine, substance P, and vasopressin. New evidence has supported the concept that Aβ (diffuse plaques) deposition may occur prior to Alzheimer's-type neuronal or glial changes. This assumption that Aβ arises from aberrant (mutant) proteolysis of the βAPP gene (on the long arm of chromosome 21) following membrane injury is held because Aβ is derived from an integral membrane sequence and it is highly insoluble when it is derived from senile plaques and meningovascular deposits. Molecular evidence suggests that mutations in the βAPP gene may initiate β-amyloidosis prior to any existing pathological changes. Some have attributed familial AD to mutations in the βAPP gene. Drugs that inhibit the protease, which produces the C terminus of Aβ (γ-secretase), may be used as therapeutic agents in the treatment of AD. Detection of soluble Aβ in the CSF and plasma of normal and AD patients, as well as in a variety of cultured cells, has led to the development of enzyme-linked immunosorbent assay (ELISA). These assays have established that many AD patients show lower CSF concentrations of the soluble $A\beta_{42}$ peptide than do normal elderly persons.

In addition to βAPP gene mutations, presenilin 1 and 2 (PS1 and PS2) genes, which are required for embryogenesis, have also been implicated in familial AD. The toxic effects of mutant PS1 and PS2 genes are manifested in the dysregulation of γ-secretase(s) that selectively intensifies the proteolysis of βAPP at amyloid-beta 42 ($A\beta_{42}$) and results in dramatic increase in $A\beta_{42}$ level. Another major risk factor for the development of AD is the excess of natural ε4 polymorphism of the ApoE gene. Inheritance of apolipoprotein E (ApoE) ε3 alleles, which is the most common ApoE gene pool, has been shown to significantly increase the likelihood of developing late onset of AD, while the inheritance of the ApoE ε4-allele strongly increases the risk for developing AD at an earlier age. The increase in the likelihood of AD may also occur upon inheritance of the ApoE4 protein on chromosome 19, which lacks cysteins and is thus unable to undergo disulfide cross-linking.

It has been shown that AD subjects with two ε4 alleles have higher concentrations of Aβ peptide deposits (particularly the highly aggregation-prone, 42-residue form) in the brain. In fact 30–40% of individuals with ε4 alleles are at risk to develop this disease. However, it should also be emphasized that NFTs may form in a variety of other diseases, indicating that its presence is a mere response to brain insults. The linkage between the development of AD and Down syndrome has been reported. Nonfibrillar, diffuse, and amorphous forms of Aβ deposits have been found in the limbic, striatum, and association cortices of trisomic patients. This early accumulation of diffuse plaques may be resulted from the elevated APP gene and its expression and eventual increase in Aβ concentration. Synaptic losses near the sites of these deposits have also been observed.

Clinically, *AD* (see also Chapters 16 and 18) exhibits progressive intellectual dementia, which includes impairment of judgment and memory. In the initial stages, patients may seem sociable and alert but show signs of confusion and depression in a new and unaccustomed environment. Patients also exhibit the inability to recall of a list of items after a few minutes' delay without cues or hints (delayed free recall) while maintaining the ability to recall items from the same list with the aid of cues or multiple cues (recognition memory). Language impairment, memory loss, depression, and anomia may be seen following the initial stage of this disease. Comprehension of both written and verbal communication (lack of spontaneity in speech) also occurs. Babinski sign and hyperreflexic jaw jerk may also be seen at this stage. Patients do not exhibit hemiparesis or visual deficits (homonymous hemianopsia). The profound deficits in memory and other cognitive therapy are striking in the late stages of this disease. At this stage, patients become unresponsive and incapacitated and curl into a fetal posture. Cholinergic agonist Cognex (tacrine), which increases the level of acetylcholine by inhibiting degradation of acetylcholinesterase and *selegeline*, a monoamine oxidase-B (MAO-B) inhibitor and powerful antioxidant, may be effective in the treatment of AD.

The reciprocal connections between the dorsomedial nucleus and the posterior prefrontal and orbitofrontal cortices (tertiary olfactory cortex) may explain the possible role of the prefrontal cortex in odor discrimination. A tumor or lesion involving the uncus, parahippocampal gyrus, interpeduncular fossa, or amygdala may cause uncinate fits. The gustatory cortical area (inferior parts of the postcentral and precentral gyri) may also be involved in this condition. Patients with uncinate fits may have periods of hallucinatory olfactory perception (olfactory aura) or cacosmia, a perception of foul odors accompanied by minor or major seizures and by fear of the unreality of the environment. Chewing or lip-smacking may also be observed. Cacosmia is a form of parosmia in which the patient experiences heightened sensitivity and a lower threshold toward olfactory stimuli in the form of unpleasant odor such as the odors of putrid egg or garbage that a person with a normal olfactory system would not perceive. It can result from upper respiratory infection, prolonged and recurrent exposure to volatile chemicals, head trauma, and rarely, brain tumor or epileptic seizure, as discussed above. Dopamine deficiency in Parkinson's disease is also implicated in cacosmia due to disturbances in the function of the olfactory system.

SUGGESTED READING

Ankel-Simons F. Chapter 9: Sense organs and viscera. *Primate Anatomy*, 3rd ed. Academic Press, 2007, 392–514.

Chen S, Tan HY, Wu ZH, Sun CP, He JX, Li XC, Shao M. Imaging of olfactory bulb and gray matter volumes in brain areas associated with olfactory function in patients with Parkinson's disease and multiple system atrophy. *Eur J Radiol* 2014;83(3):564–70.

Collet S, Grulois V, Bertrand B, Rombaux P. Post-traumatic olfactory dysfunction: A cohort study and update. *B-ENT* 2009;5(Suppl 13):97–107. Review.

DeVries SH, Baylor DA. Synaptic circuitry of the retina and olfactory bulb. *Neuron* 1993;10(Suppl):139–49.

Dionne VE. How do you smell? Principle in question. *Trends Neurosci* 1988;11:188–99.

Franselli J, Landis BN, Heilmann S, Hauswald B, Huttenbrink KB, Lacroix JS, Leopold DA, Hummel T. Clinical presentation of qualitative olfactory dysfunction. *Eur Arch Otohinolaryngol* 2004;261:411–5.

Fukazawa K. A local steroid injection method for olfactory loss due to upper respiratory infection. *Chem Senses* 2005;30: 1212–3.

Holbrook EH, Leopold DA. Anosmia: Diagnosis and management. *Curr Opin Otolaryngol Head Neck Surg* 2003;11:54–60.

Jacek S, Stevenson RJ, Miller LA. Olfactory dysfunction in temporal lobe epilepsy: A case of ictus-related parosmia. *Epilepsy Behav* 2007;11(3):466–70.

Keller M, Baum MJ, Brock O, Brennan PA, Bakker J. The main and the accessory olfactory systems interact in the control of mate recognition and sexual behavior. *Behav Brain Res* 2009;200(2):268–76.

Kotan D, Tatar A, Aygul R, Ulvi H. Assessment of nasal parameters in determination of olfactory dysfunction in Parkinson's disease. *J Int Med Res* 2013;41(2):334–9.

Landis BN, Frasnelli J, Hummel T. Euosmia: A rare form of parosmia. *Acta Oto-Laryngologica* 2006;126(1):101–3.

Nakamagoe K, Fujizuka N, Koganezawa T, Yamaguchi T, Tamaoka A. Downbeat nystagmus associated with damage to the medial longitudinal fasciculus of the pons: A vestibular balance control mechanism via the lower brainstem paramedian tract neurons. *J Neurol Sci* 2013;328(1–2):98–101.

Neundorfer B, Valdivieso T. Parosmia and anosmia under L-Dopa therapy. *Nervenarzt* 1977;48(5):283–4.

Patel RM, Pinto JM. Olfaction: Anatomy, physiology, and disease. *Clin Anat* 2014;27(1):54–60.

Tirindelli R, Dibattista M, Pifferi S, Menini A. From pheromones to behavior. *Physiol Rev* 2009;89(3):921–56.

Witt M, Hummel T. Vomeronasal versus olfactory epithelium: Is there a cellular basis for human vomeronasal perception? *Int Rev Cytol* 2006;248:209–59.

Wyatt TD. *Pheromones and Animal Behaviour: Communication by Smell and Taste*. Cambridge: Cambridge University Press, 2003, 295.

Wysocki CJ, Preti G. Facts, fallacies, fears, and frustrations with human pheromones. *Anat Rec A Discov Mol Cell Evol Biol* 2004;281(1):1201–11.

Zatorre RJ, Jones-Gotman M, Evans AC, Meyer E. Functional localization and lateralization of human olfactory cortex. *Nature* 1992;360:339–40.

17 Limbic System

The limbic system (visceral brain) directly or indirectly affects somatic and visceral motor (autonomic) functions by modulating the activities of the brainstem reticular formation, spinal cord, and the hypothalamus. It maintains homeostasis, integrates the olfactory impulses, changes the behavioral pattern, and modifies or inhibits reactions to stimuli via rewarding and punishing centers. It also influences the activities of the pituitary gland. This system plays an important role in encoding and establishing memory patterns. Virtually all sensory systems including olfactory, visual, and auditory impulses maintain connections to the limbic system. It encompasses the entorhinal cortex and associated connections, hippocampal formation (cornu ammonis, subiculum, dentate gyrus), thalamus including the anterior and dorsomedial nuclei, hypothalamic areas and nuclei, septal area, induseum griseum, amygdala, prefrontal cortex, cingulate and parahippocampal gyri, epithalamus including the pineal gland, habenula, and stria medullaris thalami.

Thus, the limbic system regulates emotion, behavior, drive, long-term memory, maintenance of cognitive maps for navigation, and learning processes. This system is also involved the formation of spatial memory, processing of the signals that convey motivationally significant stimuli related to reward and fear as well as social behavior. In summary, it is a system that integrates the activities that preserve the individual and species. The limbic lobe includes Brodmann areas 35a, 29, 24a, b, c, and Brodmann area 25, which are represented by the medial part of the perirhinal, retrosplenial, ventral parts of the anterior cingulate, and infralimbic cortices, respectively.

HIPPOCAMPAL FORMATION

The *hippocampal formation* represents phylogenetically the oldest part of the cortex (archipallium) and includes the hippocampal gyrus (cornu ammonis), fornix, dentate gyrus, subicular complex (subiculum, presubiculum, parasubiculum), and the entorhinal cortex. The *hippocampal gyrus* (cornu ammonis) develops as an enfolding of the cerebral cortex into the inferior horn of the lateral ventricle (Figures 17.1, 17.2, and 17.3). It is one of the simplest and most primitive structures of the human brain that stretches from the foramen of Monro to the tip of the inferior horn of the lateral ventricle. It lies superior to the parahippocampal gyrus, expanding anteriorly to form the pes hippocampi, which is crossed by several shallow grooves. The parahippocampal gyrus lies medial to the collateral sulcus and continues with the subiculum, a juxtallocortex. The subiculum extends with the inferior surface of the dentate gyrus and further laterally with the cornu ammonis, thus forming a curve that points toward the middle of the dentate gyrus.

HIPPOCAMPAL GYRUS

The hippocampal gyrus or cornu ammonis (CA) is acellular and plexiform structure for the most part with the exception of the pyramidal layer. It is divided into three zones designated as CA1, CA2, and CA3. Zone CA4 has also been proposed but lack of cytoarchitectural delineation from C3 makes this distinction untenable. In the CA1 zone (Sommer's sector), considered the most complex of all hippocampal layers, the pyramidal neurons are characteristically small and are located near the subiculum, distant from the dentate gyrus. This zone is particularly susceptible to anoxia, as it occurs in an epileptic seizure involving the temporal lobe. Pyramidal cells in fields CA3 and CA2 that project to CA1 zone and constitute Schaffer collaterals utilize gamma aminobutyric acid (GABA) as a neurotransmitter, establishing synaptic contacts with the apical dendrites of adjacent pyramidal neurons in CA3. Stimulation of Schaffer's collaterals of CA3 pyramidal cells, which uses glutamate and/or aspartate as a neurotransmitter, produces dramatic increase in the number of their synapses on the CA1 pyramidal neurons, while prolonged stimulation may reduce the formation of new synapses. Mossy fibers from the dentate granule cells that project to the pyramidal cell layer of CA3 utilize glutamate and/or aspartate as a neurotransmitter, synapsing on their proximal dendrites. CA1 and CA2 are not clearly demarcated. The same is true with regard to CA2 and CA3, which are also not easily distinguishable as the pyramidal layer of one field continues with the

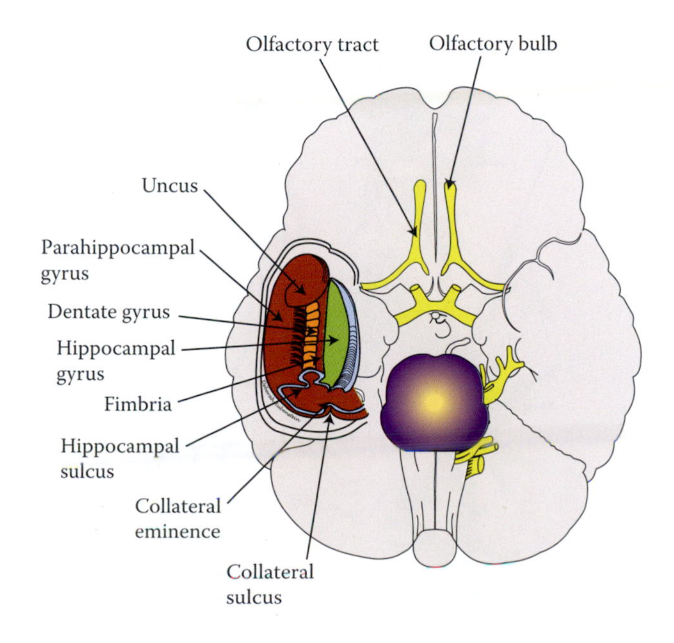

FIGURE 17.1 Schematic drawing of the hippocampal formation and its various constituents.

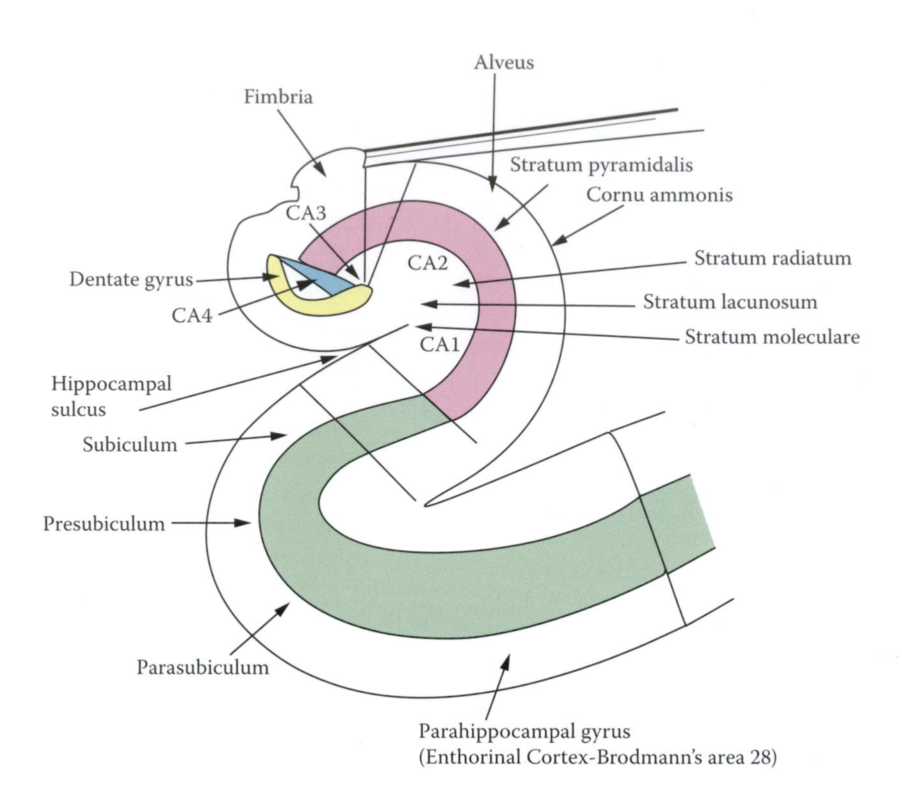

FIGURE 17.2 Hippocampal formation and its main components are shown. Note also the various sectors of the hippocampal gyrus (cornu ammonis).

other, and are sandwiched between CA1 and CA4. Zone CA2 receives massive input from the hypothalamus but lacks any afferents from the mossy fibers of the dentate granule cells.

CA is a trilaminar archipallium, which is composed of ependyma, alveus, stratum oriens, stratum pyramidalis, stratum radiatum, stratum lacunosum, and stratum moleculare. Alveus is formed by the axons of the pyramidal neurons of the hippocampus and the subiculum that converge onto the fimbria of the fornix. Pyramidal neuronal layer, as the name indicates, is composed of the pyramidal cells. Stratum lucidum, which remains less distinct in human and absent in CA1 and CA2 fields, represents a site where mossy fibers make synaptic contacts with the proximal dendrites of the pyramidal cells in CA3. Virtually all layers of the cornu ammonis contain vasoactive intestinal peptide (VIP) and cholecystokinin reactive neurons. Strata lacunosum and moleculare also contain somatostatin. Cells in CA3 and CA2 in stratum radiatum and stratum oriens receive afferents from the septal nuclei, hypothalamus, and Schaffer's collaterals from CA3 and associational fibers from the hippocampus. The perforant pathway pursued by the efferents of the entorhinal cortex (area 28) that project to the dentate gyrus runs through the stratum radiatum and lacunosum–moleculare and establishes en route synapses using glutamate and aspartate, with the distal dendrites of the hippocampal pyramidal neurons.

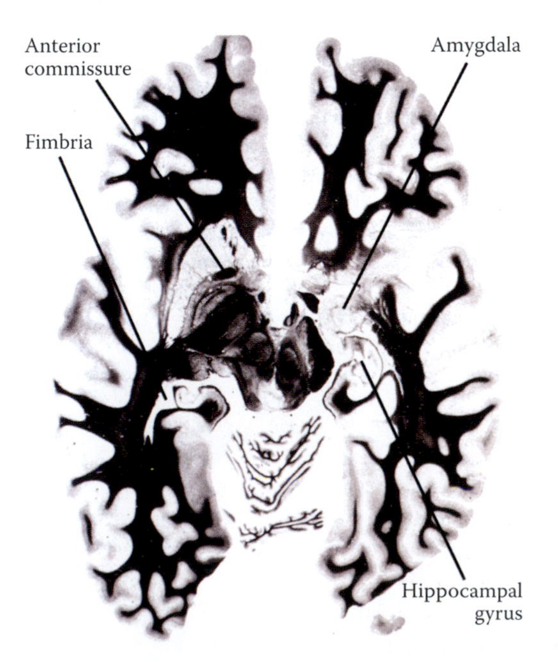

FIGURE 17.3 In this horizontal section of the brain, the hippocampal gyrus, the fimbria of fornix, and the amygdala are indicated. Other structures are also shown to emphasize their relations to the adjacent structures within this section.

> Memory traces are believed to be scattered in the stellate cells of association cortices via the entorhinal cortex and fornix.

CA4 is the region of the cornu ammonis adjacent to the hilus and the dentate gyrus. The CA4 and CA3 regions contain the largest neurons, which form synapses with the mossy fibers of the granule cell layer of the dentate gyrus.

Schizophrenia

It has been suggested that damage to the hippocampal gyrus may account for serious deficits associated with schizophrenia. *Schizophrenia* (a term that replaced "dementia precox" in the psychiatric literature) is a disease that affects as high as 1%–1.5% of the adult population and involves a group of disturbances that share a common phenotype. It equally affects men and women with some differences in course and onset between the sexes. Men have an earlier onset of schizophrenia than women do. Greater than half of all male schizophrenics, but less than one-third of female schizophrenics, may show manifestations before the age of 25. Studies have shown that the peak age of onset for males is 15–25 years, whereas for women, it ranges between the ages of 25 and 35. In general, the outcome for female schizophrenics may be better than the outcome for male schizophrenics. A particularly interesting finding is that individuals who later develop schizophrenia are more likely to have been born in the winter and early spring months and less likely to have been born in late spring or summer. Monozygotic twins have a 50% concordance rate.

The disturbances associated with schizophrenia include incoherence of thought (hallucinations or false perception), feeling, and behavior. Patients may have difficulty establishing social relationships and experience delusions, mood disturbances, auditory hallucination, and even motor overactivity and violent or bizarre behavior. These symptoms, which are variable in severity, may be categorized into positive, negative, and disorganized symptoms. Positive manifestations principally comprise delusions (paranoia, grandiosity, bizarre thoughts, and tactile delusions) and hallucinations (mainly auditory). Withdrawal, lethargy, loss of spontaneity and initiative, motivational impairment, and indifference to emotional stimuli constitute the *negative manifestations* of this psychotic disorder. Schizoid individuals present with the negative symptoms. Lack of thought content, illogical ideas, and incoherence are the primary disorganizational symptoms of this disease. Some patients are classified as schizotypal when mild positive and disorganizational symptoms are present. Patients are also depressed and exhibit anxiety. When mood-related changes are predominant in this disease, a schizoaffective disorder is produced.

Genetic, environmental, and neurophysiological factors may be responsible for this disease. One model for the integration of biological, social, and environmental factors is the stress-diathesis model. This model states that a person with a specific vulnerability (diathesis) may develop symptoms of schizophrenia when subjected to particular environmental stress factors. The stress may be biological or environmental or both. Recent research has implicated a pathophysiological role for the limbic system, prefrontal cortex, and basal nuclei.

According to the ontogenic hypothesis of schizophrenia, the dendrites of the pyramidal cells of the hippocampal gyrus appear to undergo disorientation subsequent to deranged embryological development. The degree of deviation of the apical shafts of the pyramidal cells may remain proportional to the severity of symptoms and multiple hospitalizations. The abnormal deviation varied from 70° to 180° from normal. Due to this directional deviation, abnormal afferents converge on these dendrites. Additionally, during development, primitive neurons migrate from the neuroepithelial zone to the hippocampus via the radial glial cells. These radial glial cells function both as directional guides and a structural support for the migrating neuroblasts. Without this support and guidance, the neuroblasts fail to develop and migrate properly. During embryogenesis, the close proximity or adhesion of the migrating neuroblasts to the radial glial cells is maintained by the neuronal cell adhesion molecules (NCAMs). This mechanism is essential for proper migration, alignment, and lamination resulting in cluster of cells lined up side by side with particular polar orientation. Uncoupling of the NCAM from the neuronal–glial complex allows the migratory neurons to leave the radial glia within the hippocampus.

Since neuronal migration occurs in the second trimester of gestation, cellular derangement, subsequent to insult by maternal illness, must occur during this stage of development. This theory is supported by the observation that the number of cases of schizophrenia in offspring of mothers infected with influenza virus showed significant increase. The amount of available sialic acid may determine the cell binding capacity of the NCAM. Since capsular neuromidase-producing viruses affect the sialic acid, changing the binding property of NCAM eventually affects the migratory pattern of neurons during embryogenesis. Other investigators challenge the ontogenic theory on the basis of the fact that schizophrenia occurs in teenagers and that the symptoms wax and wane with progressive deterioration and remission. They claim that infection or injury may result in miswiring and aberrant regeneration, leading to abnormal sprouting and synaptic reorganization of projection sites and increased excitability and abnormal behavioral pattern of the affected neurons. This reorganization may also lead to abnormal functioning of the structures that receive input from the hippocampal gyrus.

An interesting observation that needs to be pointed out in regard to development of schizophrenia is that the myelination of the neuronal pathways of the limbic system occurs during late adolescence under the influence of the neurotropic gonadal hormones. These hormones are involved in the development and lengthening of the dendritic spines in CA1 and CA3. In fact, numerous glucocorticoid receptors exist in the hippocampal gyrus, and hormonal role in the growth of dendrites of the granule cells seems obvious. During adolescence, the frequency of hippocampal neuronal firing is proportional to the hormonal release. This massive firing predisposes the cells to damage, initiating sprouting and cellular reorganization.

Dopamine is considered the primary neurotransmitter involved in the development of schizophrenia. The simplest form of dopamine hypothesis is based upon the overactivity of the *mesolimbic dopaminergic system*. This theory relies upon the fact that, first, most antipsychotics are antagonists to the dopamine type (D2) receptors, and second, drugs that increase dopaminergic neuronal activity, such as amphetamine, are psychotomimetic. The theory also suggests that dopamine type (D1) receptor may be responsible for the so-called negative symptoms of schizophrenia. Another part of dopamine theory is the significant increase in plasma levels of the dopamine metabolite, homovanillic acid, in schizophrenics. The problems with dopamine hypothesis are twofold: the fact that antipsychotic medications are useful in the treatment of virtually all psychotic states suggests that dopaminergic hyperactivity is unique to schizophrenia; and some electrophysiological data suggest that the firing rate of dopaminergic neurons may actually increase in response to long-term treatment with antipsychotics (see also dopamine in Chapter 12). Excitatory amino acids may also play a role in the pathophysiology of schizophrenia.

Treatment of schizophrenia may be accomplished by dopamine receptor antagonists (D2 receptors that are coupled to adenylyl cyclase) such as risperidone clozapine, melperone, sertindole, and ziprasidone. Even with treatment, only 50% of patients are not severely debilitated. Also, dopamine receptor antagonists are associated with side effects such as akathesia (restlessness), rigidity, tremor, tardive dyskinesia (tongue darting), and neuroleptic malignant syndrome. Risperidone, an antipsychotic drug, is antagonistic to serotonin type 2 (5-HT2) as well as to the dopamine type 2 receptors. Both positive and negative symptoms are improved by this medication. Clozapine is also an effective antipsychotic drug, which primarily antagonizes the D4 receptor and, to a lesser degree, the D2 and serotonin receptors. Since clozapine is associated with agranulocytosis in 1%–2% of patients, careful monitoring of the blood may be required.

Dentate Gyrus

The *dentate gyrus* (Figures 17.1 and 17.2) lies medial and inferior to the cornu ammonis and alveus, superior to the subiculum and lateral to the fimbria of the fornix. It is separated from the subiculum by the hippocampal sulcus. The induseum griseum and the fasciolar gyrus are considered as posterior extensions of the dentate gyrus. Anteriorly and inferiorly, it continues into the uncus as the band of Giacomini or the tail of the dentate gyrus. It lacks the pyramidal neuronal layer, consisting of an external molecular layer that contains the highest concentration of GABAergic neurons, which is adjacent to the molecular layer of the cornu ammonis, a dense granule cell layer, and an innermost polymorphic layer. The latter also contains somatostatin. Cholecystokinin is found in the hilar region of the dentate gyrus.

The axons of the granular cell neurons project to the hippocampal gyrus through the polymorphic layer, whereas their dendrites remain confined to the overlying molecular layer, a primary site of projection of the entorhinal cortex. Granule cell layer contains deeply located GABAergic neurons and also opioid peptide dynorphin.

The CA4 field is considered as part of the pyramidal layer surrounded by the granule cell layer. The molecular and the granule cell layers form the dentate fascia. Association fibers confined to the same side are the main component of the polymorphic layer of this gyrus. Examining the subiculum (Figure 17.2) reveals a transitional area that shows graded variation between a three-layered cortex of the cornu ammonis and the adjacent entorhinal cortex as well as the parahippocampal cortex. It is characterized by a marked thickening of the pyramidal layer. The subiculum is divided into the prosubiculum (which is the closest to and merges with CA1); presubiculum, a region adjacent to the entorhinal cortex, containing large neurons; and the subiculum proper. Glutamate and/or aspartate are utilized by the afferents to the dentate gyrus that emanate from the entorhinal cortex.

SUBICULAR COMPLEX

The subicular complex consists of the subiculum, presubiculum, and parasubiculum and contains cholecystokinin (CCK) neurons. The subiculum consists of a middle pyramidal neuronal layer that sends dendrites to form the superficial molecular layer and a deep polymorphic layer. Major projections arise from the pyramidal layer of the subiculum destined to the septal area, anterior thalamus, nucleus accumbens septi, entorhinal area, and the mammillary nuclei of the hypothalamus. The presubiculum lies medial to the subiculum and consists of a middle pyramidal cell layer, a superficial plexiform layer, and a deep layer that continues with the subiculum or entorhinal cortex. The parasubiculum is a transition between the entorhinal cortex and the rest of the subiculum as the deep cell layers of these cortical areas are not easily discernible.

Synaptic activation of the hippocampus entails initial activation of the dendritic spines of the granule cells in the external part of the molecular layer of the dentate gyrus through projection from layers II and III of the entorhinal cortex that pursues the perforant path. Activated granule cells send massive mossy fibers to the proximal dendrites of CA3 pyramidal cells, which, in turn, project heavily to the stratum radiatum of the CA1 hippocampal field. Then CA1 sends profuse projection to the subiculum and through the subiculum it reaches the entorhinal cortex. This intrinsic loop is interrupted by projections of the pyramidal neurons to the lateral septal nucleus and subicular efferents to the amygdala, ventral striatum, mammillary bodies, and anterior thalamus. Cornu ammonis projects to the septal nuclear complex, while the subicular projection is mainly directed toward the mammillary body as well as the striatum and parahippocampal gyrus, orbitofrontal cortex, and perirhinal cortex. Subcortical efferents project to all zones of the hippocampal gyrus. The subiculum and the entorhinal cortex receive projection from most areas of the neocortex including the insula contrary to the dentate gyrus and cornu ammonis, which receive no neocortical afferents.

CONNECTIONS OF THE HIPPOCAMPAL FORMATION

Afferents of the Hippocampal Formation

Afferents of the hippocampal formation (Figure 17.8) arise virtually from all sareas of the limbic system (with the exception of the olfactory cortex) and are topographically organized but not modality specific. These afferents are specifically derived from the entorhinal area, the medial septal nucleus, supramamillary area of the posterior hypothalamus, cingulate gyrus, and from the contralateral hippocampal formation. Cholinergic neurons of the septal area and the diagonal band of Broca project in a diffuse manner to all layers of the hippocampal formation via the fornix. In particular, the medial septal nucleus projects to the dentate gyrus and to certain parts of the cornu ammonis (CA3, CA4) via the fornix. The entorhinal area (secondary olfactory cortex—Brodmann's area 28) forms the main afferents to the hippocampal and dentate gyri, which arise from the medial and lateral parts of the entorhinal area, and pursue alvear (superficial) or perforant (deep) paths. Alvear fibers originate from the medial part of the entorhinal area, run toward the ventricular surface, and terminate in the subiculum at CA1. Perforant axons derive from the lateral part of the entorhinal area and traverse the subiculum, terminating in the dentate gyrus and virtually all the layers of the cornu ammonis (except CA4). The cingulate gyrus projects indirectly to the hippocampal gyrus via relays in the presubiculum and the entorhinal cortex. These fibers run within the cingulum and some of the cortical association fibers, dorsal to the corpus callosum. Noradrenergic input to the hippocampal gyrus is derived from the locus coeruleus. Anterior thalamic nucleus specifically projects to the presubiculum. Pyriform and prefrontal cortices also provide input to the hippocampal formation. Visual, auditory, and somatosensory cortical projections reaches the hippocampal formation via the entorhinal cortex. Projections of the hippocampal formation to corresponding areas of the both hemispheres are maintained via the hippocampal commissure that connects the crura of the fornix.

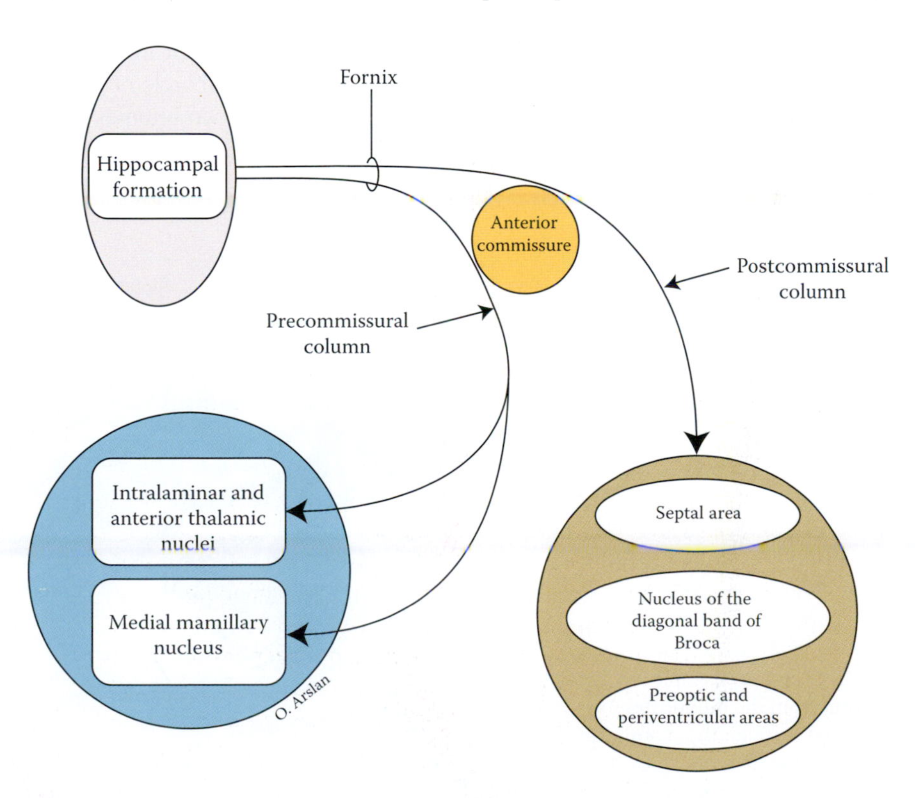

FIGURE 17.4 This diagram depicts the main efferent pathway from the hippocampal formation, which is represented in the fornix. The fornix divides into precommissural and postcommissural columns.

Efferents of Hippocampal Formation

The fornix is the main output of the hippocampal formation, originating from the hippocampal gyrus (Figures 17.4, 17.5, and 17.8). Through this tract, the hippocampal formation influences the activities of the anterior thalamic nucleus and midbrain reticular formation. The fornix is formed primarily by the axons of the pyramidal neurons of the cornu ammonis and, to a lesser degree, from the subiculum (with the exception of the presubiculum). The subicular fibers of the fornix project to the cingulate gyrus as well as to the entorhinal area. Through this massive projection, the hippocampal gyrus reaches the septal area, the preoptic area, and the hypothalamus. Some efferents of the hippocampal formation are also derived from the dentate gyrus.

Afferents to the hippocampus are also carried by the fornix. The axons, which comprise the fornix, converge on the ventricular surface of the hippocampal gyrus as the alveus and then continue as the fimbria. The fimbria of the fornix, the longitudinal stria, and the induseum griseum, which cross the corpus callosum, are collectively known as the *dorsal fornix*. The latter conveys information from the hippocampal gyrus to the fasciolar, the cingulate gyri, and the septum pellucidum. The fimbria extends rostrally to the uncus, constitute the inferior border of the choroidal Fissure, stretching caudally with the crura of the fornix around the thalamus, inferior to the splenium of the corpus callosum. The crura of fornix on both sides are then connected by the hippocampal commissure (psalterium), which also connects CA3 and CA4 zones. Between the psalterium and the corpus callosum, an anomalous cyst (cavum vergae) may exist, which communicates with lateral ventricle.

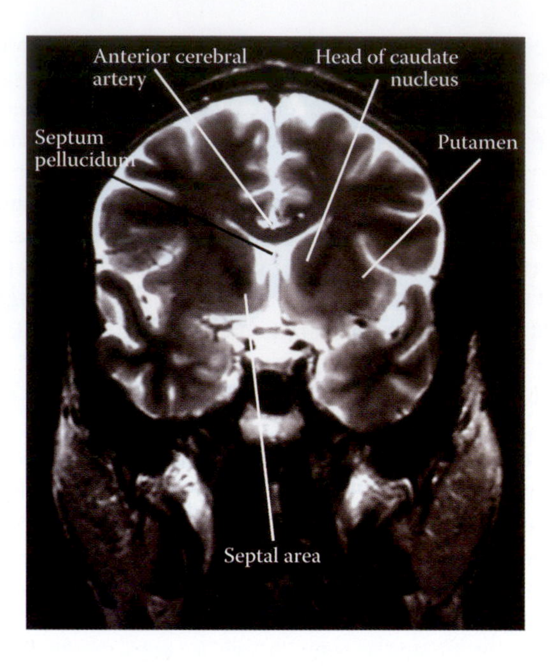

FIGURE 17.6 This MRI scan shows the septal area adjacent to the striatum and columns of the fornix. The septal area represents the nodal point of the limbic system, allowing impulses to be dispersed to diffuse areas of the brainstem and cerebral cortex.

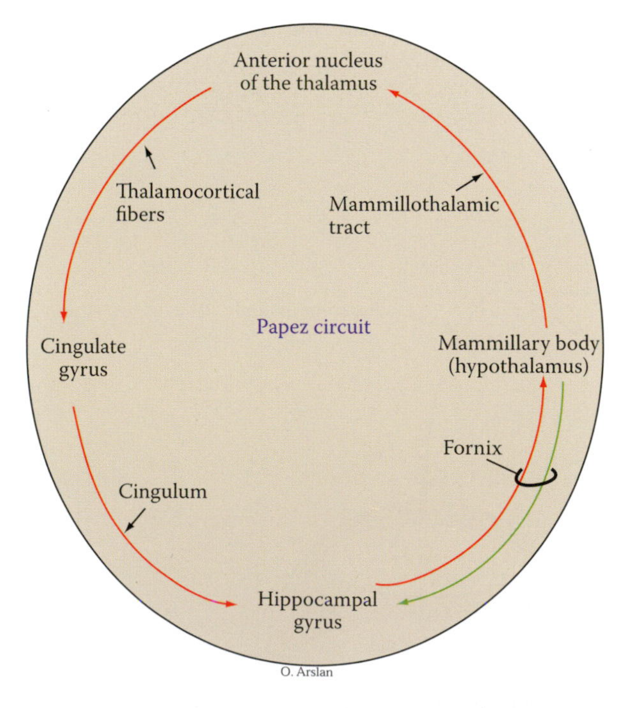

FIGURE 17.5 This feedback circuit represents the Papez circuit of emotion in which the efferent from the hippocampus gyrus projects to the mammillary body and through the mammillo-thalamic tract to the anterior nucleus of the thalamus. Cortico-thalamic fibers enable impulses received by the anterior nucleus of thalamus to project to the cingulate gyrus and via the cingulum back to the hippocampal gyrus where the circuit is completed.

The union of the crura, above the roof of the third ventricle, leads to the formation of the body of the fornix, which is attached to the inferior margin of the septum pellucidum. Above the anterior tubercle of the thalamus, the body of the fornix divides into two bundles and later into two columns. The precommissural column arises from the cornu ammonis and the subicular fields (with the exception of the presubiculum) and terminates in the septal area, preoptic nuclei,

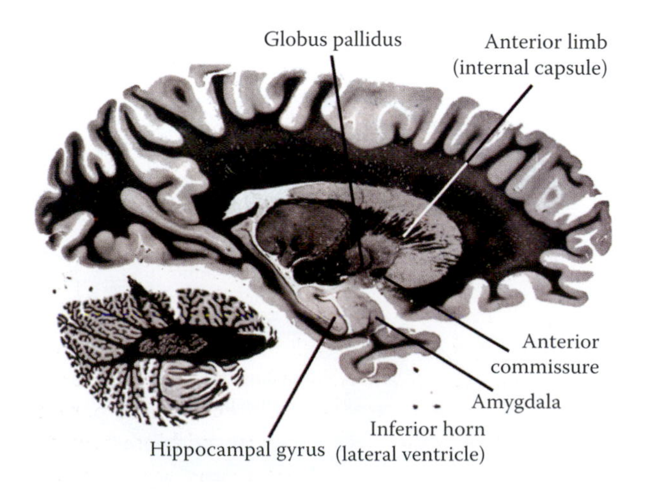

FIGURE 17.7 Coronal section of the brain through the anterior commissure. The amygdala, a significant structure in the make-up of the limbic system, occupies a prominent area rostral to the inferior horn of the lateral ventricle, adjacent to the uncus.

anterior hypothalamus, and cingulate gyrus. The postcommissural column arises from the subiculum and terminates in the mammillary body, intralaminar thalamic nuclei, lateral hypothalamus, habenula, and midbrain tegmentum (Figure 17.4). It can be concluded from the above data that the hippocampal formation serves as a closed feedback circuit involving the fornix, mammillary body, mammillothalamic tract, anterior nucleus of the thalamus, cingulate gyrus, and cingulum. These connections form the basis for the *Papez circuit* of emotion (Figures 17.5 and 17.6).

FUNCTIONAL AND CLINICAL CONSIDERATION

Hippocampal linkage to the hypothalamus via the septal area may account for the important role that the hippocampal formation plays in the control of behavior. Expression of aggressive behavior is also induced by the effect of the hippocampus gyrus upon the supplementary motor areas of the cerebral cortex. The corticosteroid and estradiol containing hippocampal neurons that act upon the hypothalamus mediate endocrine function. These connections to the hypothalamus is also responsible for the respiratory and cardiovascular changes observed upon stimulation of the hippocampus.

Memory and Amnesia

Many areas of the brain, including the hippocampus, are implicated in encoding, storage, and retrieval of learned information such as the amygdala and mammillary bodies. Long-term potentiation (LTP) and depression are thought to cause changes in neuronal connectivity that forms the basis for learning and memory. For the *memory* of an experience to be encoded, it has to be received, registered, and processed, followed by storage of the recorded information and then its retrieval (recalling). In other words, memory encompasses the retention, reactivation, and reconstruction of a given event, which can be exhibited through thought and behavior. The hippocampus is believed to mediate declarative and spatial memory and consolidation of newly acquired information from short-term to long-term memory, while the amygdala is involved in emotionally charged memory. Thus, surgical removal of the hippocampal gyri produces short attention span and distractibility.

Memory is formed in infants as young as 6 months of age, and with advancing age, the ability to quickly recall information that occurred over a longer period of time increases. Since the dentate and hippocampal gyrus as well as the frontal cortex develop after the age of 6 months, recall of temporal order of information (sequence of two-action process) does not occur in infants younger than 6 months. Construction of memory is a dynamic process that involves induction, maintenance, and recall of information.

Memory and accuracy of recall can be affected through repetition, stress, odors, and verbal requests. The latter is enhanced by excitement and curtailed by excessive and sustained stress. Heightened emotional state is associated with strong memory that causes it. The strength and longevity of memories are directly proportional to the intensity of the emotion experienced during the event, a fact that can underlie the neurologic basis of posttraumatic stress disorder and therapeutic approach. Memory encoding process in the hippocampal gyrus and recall are negatively influenced by glucocorticoids and catecholamines released in stressful situations. A clear impairment of memory performance under the stress has been shown, which is attributed to the distraction experienced during the memory encoding process. This may not always be true as the memory may be enhanced if it is linked to a learning context even in the presence of stressful situation. Studies have shown that when learning and retrieval contexts are congruent (similar), performance of a task seems to enhance irrespective of the presence or absence of stress. This finding may possibly be extended further to student test performance or eyewitness account of event when it is conducted in a familiar setting rather than an unfamiliar environment.

Memory can be broadly classified into short and long term. Short-term memory is believed to be based primarily on an acoustic code for storing information and less so on visual code, although this may not always be applicable. It revolves around the limited capacity and duration to retain (4–5) names or numbers for several seconds to a minute. It depends on transient patterns of neuronal communication within the dorsolateral prefrontal cortex and the parietal lab. Chunking, a maneuver through which a long number can be divided into meaningful groups, can enhance short-term memory.

Aspects of short-term memory cannot easily be distinguished from working memory model, and they are regarded by some investigators as one system. Working memory involves the mechanism that underlies performance of dissimilar visual tasks than similar ones. It enables the person to conduct processes that entail reasoning and comprehension by performing verbal and nonverbal tasks. It has been proposed that the anterior cingulate gyrus, basal nuclei, parietal cortex, and frontal lobe play significant role in working memory. The model is based on four proposed pillars: the central executive, the phonological loop, the visuospatial sketchpad, and the episodic buffer. The central executive is tasked with funneling the received information and coordinating it among various domains. The cognitive processes that perform monitoring task, when performing simultaneous tasks, and completing goal-directed actions (suppression of irrelevant information and tasks and bringing attention to relevant information and tasks) are also accomplished by the central executive (attention) domain. This model also entails transient storage of auditory information (sound of language) and its maintenance and refreshment through continuous and repetitive articulation in silence within a phonological loop. Further, visuospatial sketchpad is tasked with encoding of information relative to visual (shape, texture and color) and spatial (location) tasks, such as constructing visual images and forming mental pictures of the constructed images. Later, spatial, visual, and verbal information funneled to different domains is linked, temporarily stored, and possibly enriched with semantic information in the episodic buffer to establish an integrated linear and unified system.

Others propose that working memory can hold a limited number of concepts, which serve as a guide for retrieval of related information through retrieval structures. Encoding for working memory entails sensory input that activates and causes prolonged spiking of the individual neurons that persist even after cessation of the stimulus. Both prefrontal cortex and the medial part of the temporal lobe play an important role in working memory; although, the role of the former appears to be more substantial than the latter.

It is thought that synaptic consolidation and system consolidation are the processes that mediate the transformation of a short-term into a long-term memory. Synaptic consolidation is associated with a protein synthesis process within the medial temporal lobe, whereas systemic consolidation transforms medial temporal lobe–dependent memory into an independent memory over the span of months or years. Research studies have demonstrated that prevention after retrieval can influence subsequent retrieval of the memory and that postretrieval treatment with protein synthesis inhibitors can lead to an amnestic state. The latter finding is supported by the fact that memories are regularly restructured during the retrieval process and that a retrieved memory is not a replica of the initial experiences.

Long-term memory maintains larger capacity to recall a large number of items for prolonged period of time through semantic encoding. It is enhanced through repetition and episodic memory, which relies on attaining information regarding the time, location, and nature of the experience. This requires widespread and durable changes in synaptic connectivity between neurons of different regions of the brain. Consolidation of recently acquired information is thought to be enhanced by sleep and that the brain activity during sleep may mimic the activity that occurs during the acquisition of new information. Atkinson–Shiffrin memory model proposes a multilevel model that includes episodic and procedural memory and that silent repetition is the sole mechanism by which information can be consolidated for long-term storage. This model has been challenged through research and accumulated clinical data. Long-term memory has been classified into declarative and procedural memories.

Declarative memory entails conscious recall of the information, which is explicitly stored and retrieved. This is a fast system but with limited capacity that plays a crucial function in establishing long-term memories. This requires connecting multiple areas of neocortex that subserve perception and short-term memory of the events. The neocortex slowly joins the long-term memory storage irrespective of the activity of the medial temporal lobe and diencephalon. This type of memory is further subdivided into semantic and episodic memories. Semantic memory is a concept-based memory that does not involve or tie to any particular event that gives meaning to otherwise meaningless words or sentences. Semantic memory task is associated with activation of the hippocampal gyrus, inferior prefrontal cortex, and posterior temporal cortices in the left hemisphere, as well as with activation of the inferior temporal and middle frontal gyri in the right hemisphere. Frontotemporal lobar degeneration, Alzheimer's disease, and encephalitis due to herpes simplex virus can impair semantic memory differently. Semantic dementia, a neurodegenerative condition of the frontotemporal lobe, tends to involve general categories of semantic memory including verbal and nonverbal domain and is associated with worsening fluent aphasia, anomia, loss of ability to comprehend the meaning of words, and also dyslexia. It commonly results from atrophy of the left inferior temporal lobe. Patients experience behavioral changes, difficulty finding words, and inability to match meanings to pictures or objects. Certain categories of semantic memory impairment are specifically affected in viral encephalitis. In semantic memory impairment associated with Alzheimer's disease, the ability to recognize, describe, and name objects is lost.

Episodic memory is specific to a particular contextual knowledge associated with locations, times, and related emotions. It is considered a collection of past experiences that occurred at a given time and location. It is the summation of process that includes affect, sensation, perception, and conceptual recollection. Encoding for episodic memory includes sustained changes in the molecular structure in the form of LTP and spike-timing–dependent plasticity (STDP), which lead to subsequent changes in synaptic connections. LTP, as discussed earlier, refers to augmentation of signal transmission through synchronous stimulation of the neurons, whereas STDP is a process through which neuronal synaptic connections are adjusted based on the relative timing of generated spikes (input and output). This type of memory requires visual imagery of the event, subjective sense of time, and the consciousness of presence (autonoetic) in a particular, though subjective, time. Other than visual imagery, familiarity, recollection of semantic information, and descriptions also constitute main components of this type of memory. In summary, episodic memory pertains to a perspective of visual imagery and is based on the episode of a short-lived past experience that it can be relived though it is forgettable. It exhibits temporal feature and undergoes prolonged stimulation and inhibition. In a way, episodic memory connects together items in semantic memory and that episodic and semantic memories are parts of the overall declarative memory.

New episodic memories require the involvement of primarily the prefrontal cortex and the hippocampal gyrus. Patients with damage to the prefrontal cortex exhibit lack of ability to remember the time and location of an event though remain capable of recognizing the visual imagery of the event seen in the past. It has also been reported that prefrontal cortex enhances encoding by adding semantics (meaning) into the processed information. A view that is supported by evidence indicates that the hippocampal gyrus acts as a temporary storage center for memories followed by their consolidation in the isocortex.

Procedural (implicit) memory is repetition-based long-term memory without newly acquired explicit memory formation and is transformed into motor skills. This type of memory does not allow conscious recall of information, and accessing previous experiences is done unconsciously. It is retrieved and put into use in the development of motor and cognitive skills. It is linked to the dorsolateral striatum, the cerebellum, and the limbic system. The dorsolateral striatum mediates motor activity through direct and indirect pathways in the form of feedback loop. These pathways are formed by GABA-related medium spiny neurons that also contain dopaminergic (DRD1, DRD2), muscarinic (M4), and purinergic adenosine (A_{2A}) receptors and cholinergic interneurons. The cerebellum coordinates motor activity, adjusts muscle tone, and fine-tunes skills needed to perform procedural motor functions. In particular, the cerebellar cortex contains the engram for initial memory trace, which is distributed via the Purkinje neurons to other areas of the brain. The part of the limbic cortex that continues with the neostriatum caudomedially forms the marginal division zone (MrD), an area thought to be linked to procedural memory. MrD consists of spindle-shaped neurons with specific connections and reactivity to monoamines and neuropeptides. These neurons, which link the limbic system and the basal nucleus of Meynert, show activity mainly on the left side during the performance of memory (auditory digital working) task. It has been suggested that MrD may play a significant role in the execution of digital working memory.

Studies have identified several areas of the brain important in memory including the hippocampus, dentate gyrus, subiculum, amygdala, parahippocampal, entorhinal, and perirhinal cortices. The important diencephalic structures involved in memory includes the anterior thalamic, dorsomedial nucleus, and midline nuclei with their efferents and afferents that traverse the internal medullary lamina. The nucleus basalis of Meynert appears to be involved more in attention-related functions than in memory functions. The hippocampus is believed to be associated with spatial learning and declarative memory, whereas the amygdala is involved in emotional memory. Midline diencephalic region, specifically the dorsomedial nucleus of the thalamus and the mammillary bodies of the hypothalamus, plays an important role in memory.

Anterograde amnesia, which results from destruction of the hippocampal gyrus, may explain the role of this gyrus in encoding of short-term memory. Seizures of hippocampal origin exhibit unique low threshold activity that remains generally localized with no behavioral change or loss of consciousness. This is generally true unless other areas of the limbic system are involved. Individuals with these seizures appear to be confused and may show signs of aggressive behavior as well as auditory and gustatory hallucinations. Bilateral removal of the hippocampal gyrus causes short-term memory loss, confusion, and compensatory confabulation (tendency to fabricate, recite imaginary experiences to fill the gaps in memory, and give irrelevant answers to reasonable questions).

Alcoholism and subsequent thiamine deficiency, failure to get food rich with thiamine, and continued ingestion of carbohydrate can be predisposing factors to Wernicke's encephalopathy. Patients exhibit confusional state, nystagmus, ataxia, ophthalmoplegia, and sometimes stupor and fatal autonomic insufficiency. If untreated, this condition may lead to *Korsakoff's psychosis* (amnestic confabulatory syndrome) in which the patient's conversation becomes unintelligible, accompanied by disturbance of orientation, agnosia or apraxia, susceptibility to external stimulation and suggestion, amnesia, confabulation, and hallucination. It is commonly associated with bilateral degeneration of the hippocampus, mammillary body, and possibly the dorsomedial nucleus of the thalamus. Patients with this condition exhibit a striking difficulty in remembering events after the onset of the disease and difficulty in retaining newly acquired information and skills.

Wernicke–Korsakoff's syndrome refers to the combined Wernicke's encephalopathy and Korsakoff's psychosis, which results from the inability to encode the semantic component of information at the initial stage of learning. This condition can follow an acute, subacute, or chronic course. Mental confusion is also obvious, but consciousness and intellect apparently are preserved. Patients are able to learn but do so at a much slower rate, yet they appear to forget retained information over a period of time similar to healthy individuals. Other signs and symptoms of this disease may include nystagmus, medial strabismus, gaze palsy, and ataxic gait. Temporal lobotomy, a rarely performed neurosurgical operation for the treatment of certain types of epilepsy (psychomotor type), may cause similar deficits to Korsakoff's syndrome. It is important to note that transient global amnesia, which results from bilateral temporal lobe ischemia subsequent to atherosclerosis, presents with a sudden impairment of recent memory (retrograde amnesia often taking part) lasting hours, days, or weeks.

Hyperthymesic syndrome is a disorder that is characterized by an unusual ability to remember events that consist of long stream of details of personal, sematic autobiographical accounts of important and mundane past experiences that are encoded involuntarily and retrieved automatically. Individuals with this condition have both semantic and episodic memory (visual and facts) and tend to exhibit remarkable obsession with use of dates as mnemonic devices and that one regained memory triggers another. This unconscious process does not entail experiences outside the realm of a person's life. It can be burdensome and exhausting and may significantly curtail cognitive capacity of the individual. Imaging studies of hyperthymesic brains show enlargement in the caudate nucleus, frontal lobe, and temporal lobe, which contains the hippocampal gyrus. The caudate nucleus is associated with procedural memory as well as obsessive compulsive disorder, the frontal lobe is associated with executive function and facts, whereas the hippocampal gyrus is active in declarative memory.

SEPTAL AREA

The *septal area* lies rostral and superior to the lamina terminalis and anterior commissure, representing the nodal point for integration of impulses associated with the limbic system (Figures 17.6 and 17.8). It is composed of the precommissural and supracommissural parts. The supracommissural part corresponds to the septum pellucidum, which consists of laminae of fibers with gray matter and neuroglia. The precommissural part corresponds to the paraterminal gyrus lodged between the posterior paraolfactory sulcus and the lamina terminalis, continuing with the diagonal band of Broca, medial olfactory stria, and induseum griseum. Its anterior end is known as the prehippocampal rudiment. It contains the septal nuclei, corresponding to the cortical area between the lamina terminalis and the paraolfactory sulcus. The anterior extension of the paraterminal gyrus merges with the prehippocampal rudiment, whereas the superior end joins the induseum griseum. Septal nuclei are found in the vicinity of the columns of the fornix, which includes the lateral and medial septal nuclei, the nucleus of the diagonal band of Broca (septadiagonal complex), and the nucleus accumbens septi. The septal nuclear complex consists of the medial, lateral, dorsal, ventral, and caudal nuclear subdivisions. Some authors include the bed nucleus of the stria terminalis as a component of the caudal nucleus. Due to proximity of the medial septal nucleus to the diagonal band of Broca, their overwhelming cholinergic function and shared connection, a medial septal/diagonal band complex is viewed as a single entity. This complex sends cholinergic projections to the hippocampal formation via the fornix and the cingulate gyrus (Figure 17.11).

These neuronal projections also contain the amino acid GABA, galanin, and nitric oxide synthase. The same is true with regard to cholinergic nucleus basalis of Meynert. This complex receives projections through the medial forebrain bundle (MFB) from the dorsal tegmental and medial mammillary nuclei, lateral preoptic, and anterior hypothalamic area. Adrenergic neurons of the locus coeruleus and the medulla also send fibers to the medial septal/diagonal band complex.

Afferents to the lateral septal nucleus emanate from zones CA1 and CA3 of the cornu ammonis, from the subiculum, and from the preoptic and ventromedial hypothalamic nuclei. This nucleus also receives input from the paraventricular nuclei of the thalamus that contain arginine and vasopressin, which respond to changes in the blood concentrations of the sex hormones. Kölliker–Fuse and dorsal vagal motor nuclei also provide afferents to this part of the septal nuclear complex. Adrenergic efferents from the locus coeruleus and A1 and A2 medullary neurons, serotonergic input from the mesencephalic raphe nuclei, and dopaminergic projection from the ventral tegmental area (A10) converge on the lateral septal nucleus. The MFB enables the lateral septal nucleus to influence the activities of the hypothalamus and midbrain reticular formation through projections to the anterior hypothalamic area, preoptic and supramammillary areas, and the mesencephalic ventral tegmentum.

Due to the ascending projections of the adrenergic and serotonergic neurons of the brainstem reticular formation to the cortical components of the limbic system and the hypothalamus, mood-elevating drugs may have antagonistic effects upon these pathways. Through its massive connections, the septal area may be implicated in the regulation of behavior-related autonomic activities. This area may also be involved in theta rhythm detected in EEG during rapid eye movement (REM) sleep and motor activity. Lesions involving the septal area cause heightened reactions including

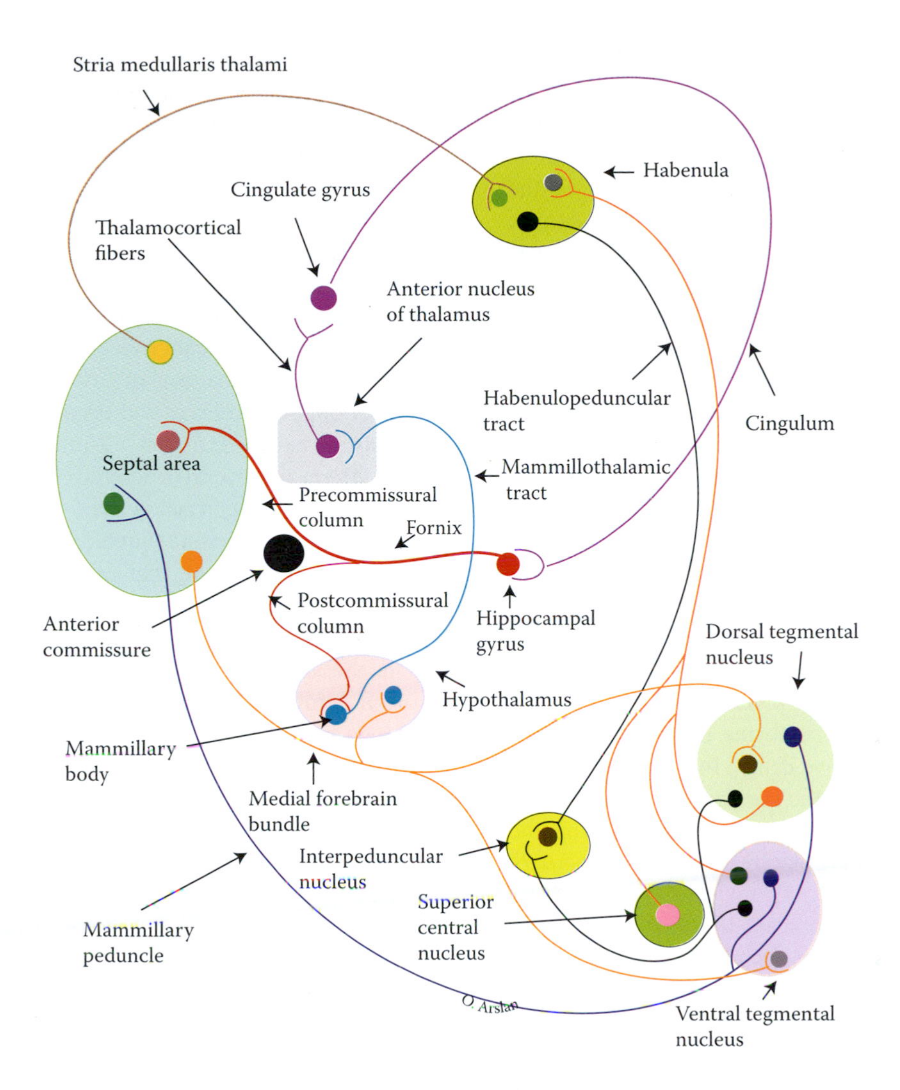

FIGURE 17.8 This diagram illustrates the major structures associated with the limbic system and their connections. The septal area projects through the habenula to the reticular formation and receives input from the hippocampal gyrus and from the tegmental nuclei via the MFB and the mammillary peduncle.

hyper-emotionality and an increase in aggressive behavior. Stimulation of this area reduces the aggressive behavior.

The septal area sends inhibitory commands to the hypothalamus, preoptic areas, and the tegmentum of the midbrain via the MFB. This connection may account for the hyperactivity and septal rage associated with septal lesion, although amygdalar fibers may also be involved in this condition.

Information received by the hypothalamus is delivered to the midbrain tegmentum via the mammillotegmental tract and the dorsal longitudinal fasciculus, and to the anterior nucleus of the thalamus via the mammillothalamic tract. Through these connections, the septal area forms a closed feedback loop, and together with the hypothalamus, it constitutes an integral part of the limbic system. It exerts a significant influence upon neuroendocrine, somatic, and visceral motor functions. Reciprocal connections exist between the amygdala, via the stria terminalis, with the septal area and diagonal band of Broca.

Transfer of memory traces from the hippocampus to the prefrontal cortex may involve the septal projection to the dorsomedial nucleus of the thalamus.

The *septal area* conveys impulses to the medial habenular nuclei and certain midline thalamic nuclei via the stria medullaris thalami, a grossly visible pathway that courses dorsomedial to the thalamus. The latter connection is reciprocal. Habenular projection to the interpeduncular nucleus and the midbrain ventral tegmentum via the fasciculus retroflexus enables the septal area to affect the activities of the reticular formation. The nucleus of the diagonal band of Broca establishes connections with the olfactory, cingulate, and prefrontal cortices as well as the amygdala, habenula, and DM nucleus of the thalamus.

The *nucleus accumbens septi* is located between the head of the caudate nucleus and the anterior part of the putamen and to the lateral side of the septal area. Together with the olfactory tubercle, it forms the ventral striatum. It is implicated in reward, pleasure, addiction, laughter, and placebo effects. It consists primarily of the medium spiny GABAergic neurons that constitute the principal efferent from the nucleus. Axonal projection of the nucleus is destined to the ventral pallidum, which, in turn, projects to the dorsomedial nucleus of the thalamus. The latter nucleus projects to the prefrontal cortex and the neostriatum. Efferents from the nucleus accumbens also reach the substantia nigra and the pontine reticular formation. Afferents to the nucleus accumbens originate from the CA1 zone and the ventral subiculum of the hippocampus, prefrontal cortex, basolateral amygdala, and dopaminergic fibers from the ventral tegmental area of the midbrain, which constitutes the mesolimbic (cortico-striato-thalamo-cortical) loop. The site of termination of the tegmental input is also the center of action of self-administered addictive drugs. In order for the hippocampal input to excite the medium spiny neurons of the nucleus accumbens and thus maintain priming, the subicular neurons hyperpolarize, whereas neurons of the CA1 zone depolarize.

Dopaminergic receptors in the nucleus accumbens together with the mesencephalic tegmental area projections (cell group A10) are implicated in neural mechanism of reward and addiction (incentive motivation). The addictive role of psychomotor stimulants in enhancing dopamine release (e.g., amphetamine) or blocking its reuptake (e.g., cocaine) may be attributed to their interaction with the dopaminergic system. Therefore, dopamine receptor antagonists may reduce the effect of cocaine. The mesolimbic dopamine system, which is formed by the dopaminergic neurons of the nucleus accumbens, is considered a major factor in determining the addictive potential of drugs. Administration of dopamine receptor blockers or a lesion of the dopaminergic afferents of this nucleus derived from the ventral tegmental area within the MFB diminishes reward-based activity. The elevated level of dopamine in the nucleus accumbens when reward-based activity is conducted and the fact that dopamine transmission in this nucleus mediates psychomotor drug-induced motor activity strengthen the notion that the mesolimbic–dopamine system also mediates rewarding features of these drugs. The mesolimbic system also enhances appetitive behavior in the presence of conditioned and unconditioned stimuli. However, dopamine transmission in the dorsal striatum mediates the stereotyped and flexible responses that accompany direct contact with the intended target when the motivational stimuli terminate. This is evident in the fact that dopamine depletion in the ventral striatum or administration of dopamine receptor antagonists produces a marked reduction in appetitive responses and motor activity in the presence of a conditioned stimulus. Studies have shown that perceptible changes in the activity of the dopaminergic neurons of the nucleus accumbens occur in response to reward or to a conditioned stimulus that signals a sense of reward. Due to this significant role of the nucleus accumbens in the neural mechanism of reward, drugs that augment the dopamine release and transmission in this nucleus may serve as a basis for self-inflicted drug abuse.

There are several drugs and chemical substances that influence the neuronal activity of the accumbens septi. They include cocaine that blocks the reuptake of dopamine, amphetamine that promotes dopamine release, and heroin that interacts with receptors in the nucleus accumbens and mesencephalic dopaminergic neurons in the ventral tegmental area. Thus, when cocaine is self-administered intravenously in conjunction with a dopamine receptor antagonist, the effect of cocaine is reduced. Increase in the rate of administration of the dopamine receptor antagonist requires a proportional increase in the rate of intravenous administration of the drug up to a point that blockade of the receptors is complete and the addictive effect of the drug is eliminated. Heroin primarily increases dopamine release in the nucleus accumbens, but this is also coupled with some increase in the neuronal activity of the mesencephalic dopaminergic neurons through their interaction with the μ-opiate receptors in the ventral tegmental area. Again, the reinforcing effect of heroin remains primarily dependent on the dopaminergic neuronal activity of the nucleus accumbens. This is evident in the fact that dose-dependent self-administration of heroin increases proportionally with the administration of methyl naloxinium into the nucleus accumbens. However, systemic infusion of the dopamine antagonists remains ineffective in countering the self-administered heroin, which is attributed to duality of opiate receptor-dependent mechanism that activates the mesolimbic system in the nucleus accumbens independent from its stimulation of the mesencephalic ventral tegmentum.

Experimental data indicate that the dopaminergic function of the ventral striatum modulates the effects of conditioned as well as nonconditioned aspects of rewards. In order for the ventral striatum to exert its influence on behavior controlled by reward-related stimuli or conditioned reinforces through dopamine release, the basolateral amygdala projects heavily to the ventral striatum, completing the "limbic-ventral striatal loop." Due to the connection of the nucleus accumbens to the prefrontal cortex and dorsomedial thalamic nucleus, this loop is also known as the cortico-striato-thalamo-cortical loop. It appears that this loop is a significant circuit in the regulation of the emotional control on behavior, which includes the control of the neuroadaptive processes that mediate the ability of drug-paired stimuli to acquire re-enforcing feature and control of the drug-seeking behavior (craving). It has been argued that D1 receptor–dependent neural mechanisms within the medial prefrontal cortex and basolateral amygdala form substrates for the compelling effects of drug-related stimuli. Opioid receptors' role in the mediation of conditioned alcohol-seeking behavior has been suggested. Additionally, conditioning factors (i.e., exposure to drug-associated stimuli) and stress can intensify predisposition to relapse. In view of the above, simultaneous effect of environmental triggers on relapse must be considered in the development of therapeutic approaches. In addition to its involvement in addiction, the nucleus accumbens plays a role in rewards such as sex and food. It is stimulated during emotional scenes induced by music, pictures, or mental imagery of pleasant and serene situation. It serves an interface center between the limbic system and the motor cortex. Attempts have been made to induce deep brain stimulation of the nucleus accumbens through surgical placement of electrodes as a treatment for severe depression. Attempts have also been made to treat alcoholism through the ablation of the nucleus accumbens. Research findings indicate that during placebo administration, expectation of benefit from the placebo induces changes in the neural circuitry and neurotransmitter systems that correlate with the observed physiologic changes.

INDUSEUM GRISEUM

The *induseum griseum* (supracallosal gyrus), located above the corpus callosum, is continuous rostrally with the septal area, the diagonal band of Broca, and the anterior perforated substance. The fasciolar (splenial) gyrus, which is continuous with the dentate gyrus, represents the caudal connections of the induseum griseum. The induseum griseum contains the medial and lateral longitudinal striae of Lancisi. Some fibers of the longitudinal stria intersect the corpus callosum, contributing to the dorsal fornix, while other fibers continue rostrally with the paraterminal gyrus and caudally with the fasciolar gyrus.

AMYGDALA

The *amygdala* (Figures 17.3, 17.7, 17.9, 17.10, 17.11 and 17.14) is a nuclear complex that is embedded within the dorsomedial part of the anterior pole of the temporal lobe deep to the uncus and rostral to the tip of the inferior horn of the lateral ventricle. It bounds the tip of the inferior horn medially, ventrally, and superiorly. It continues with the dorsally located claustrum, the external capsule, and the magnocellular nucleus basalis of Meynert, which separates it from the lentiform nucleus. It lies inferior to the semilunar, ambiens, and uncinate gyri. It is medial to the optic tract and fused with the tail of the caudate nucleus. The amygdala is adjacent to the pyriform lobe, which consists of the prepyriform (cortical area near the lateral olfactory stria) and periamygdaloid cortex. Caudally, the amygdala continues with the hippocampal formation, forming the amygdalohippocampal area. Amygdalae are larger in the male than the female.

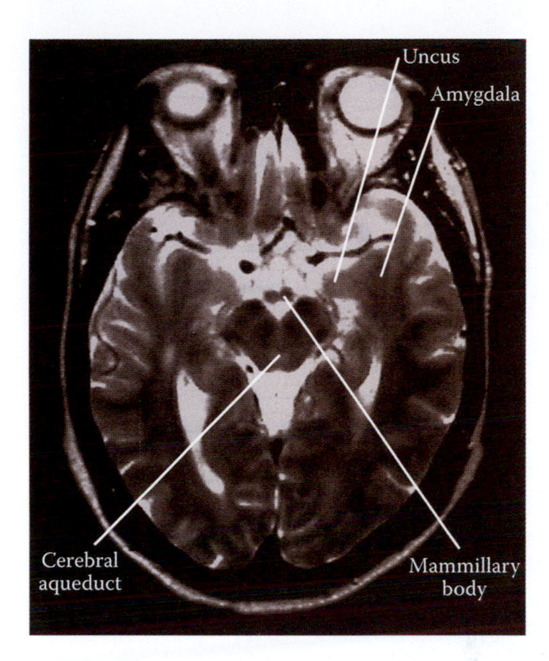

FIGURE 17.9 In this MRI scan, some of the components of the limbic system, for example, amygdala and hippocampal gyrus, are shown.

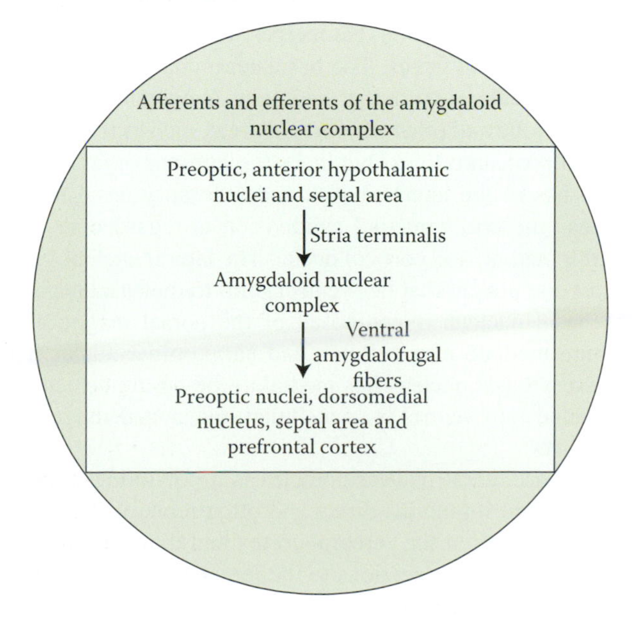

FIGURE 17.10 Diagram of the afferent and efferent connections of the amygdala.

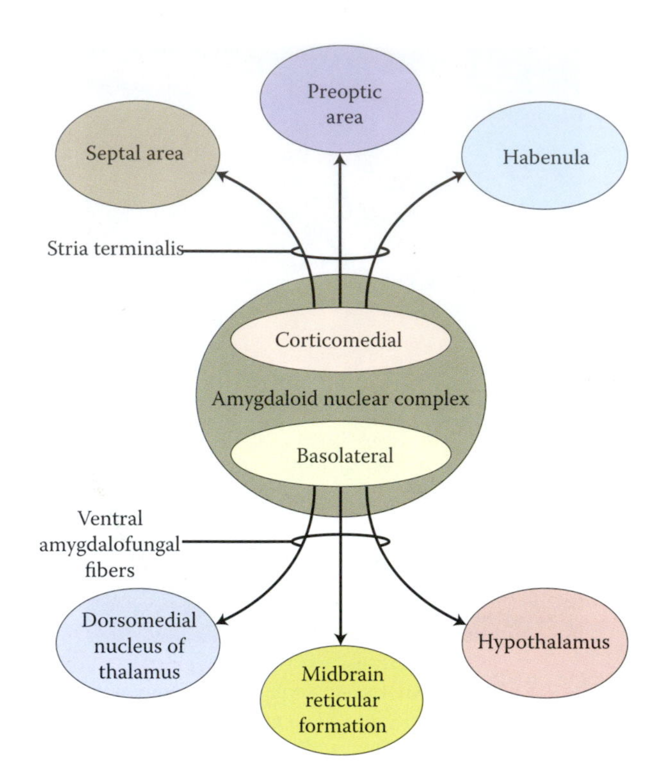

FIGURE 17.11 Schematic diagram of the outputs of the corticomedial and basolateral subnuclei of the amygydaloid nuclear complex.

This nuclear complex, which underlies the mechanism of emotional behavior through connections with the hypothalamus, reticular formation, cranial nerve nuclei, ventral tegmental area, adrenergic neurons of locus coeruleus, and cholinergic neurons of the laterodorsal tegmental nucleus, also acts on the dopaminergic and noradrenergic neurons, mediating consciousness and REM sleep.

The amygdala encompasses a corticomedial subdivision, which continues with the anterior perforated substance, and a basolateral subdivision that merges with the claustrum and parahippocampal gyrus. The basolateral complex is considered a cortical structure that maintains connections with the temporal lobe and other neocortical areas such as the precentral and postcentral gyri but lacks the laminar organization. It consists of the lateral, basal, and accessory basal nuclei, whereas the corticomedial subdivision is regarded to have central, medial, and cortical nuclei. The lateral nucleus is the largest component that lies ventrolateral to the basal nucleus. The basal nucleus is composed of the dorsal magnocellular, intermediate parvicellular, and paralaminar nuclei. The accessory basal nucleus lies medial to the basal nucleus and is divided into ventral, parvicellular, and dorsal magnocellular parts.

The basolateral nuclear complex is a polymodal cortical structure that maintains direct and often reciprocal connections with areas of the cerebral cortex and thalamus, as well as unidirectional projections to the motor and premotor cortices. Cortical characteristic is also evident in the manner that the neuropeptide Y (NY), somatostatin, and cholecystokinin are distributed. Somatostatin and NY are found in large concentration in the lateral nucleus. Excitatory amino acid neurotransmitters such as aspartate and glutamate are also used by this nuclear subdivision. The lateral nucleus, the largest subdivision, lies ventrolateral to the basal nucleus. The basal nucleus consists of dorsal magnocellular, intermediate parvicellular, and paralaminar subnuclei. The accessory basal nucleus located medial to the basal nucleus consists of similar subdivisions to that of the basal nucleus.

The basolateral nuclear complex receives serotoninergic projections from the raphe nuclei that ascend through the MFB. Dopaminergic projections from the midbrain ventral tegmental area (A10) are mainly received by the lateral and central nuclei and the parvicellular part of the basal nucleus. Dense cholinergic projections from the magnocellular division of nucleus basalis of Mynert reach the basal and parvicellular accessory basal nuclei. The high concentration of benzodiazepine binding sites, particularly $GABA_A$ binding sites in the lateral and accessory basal nuclei, may explain the anxiolytic action of benzodiazepine on the amygdala and form a possible neural basis of fear and anxiety. This subdivision contains the highest concentration of opiate receptors in the entire brain, which may be activated under stressful situation.

The corticomedial nuclear complex consists of the central, medial, and cortical nuclei as well as the associated subnuclei including the periamygdaloid nucleus. The central nucleus, which divides into medial and lateral parts, is located dorsomedial to the basal nucleus, adjacent to the putamen, occupying the caudal portion of the amygdaloid nuclear complex. It continues with the anterior amygdaloid area. Extension of the central and medial nucleus with the bed nucleus of the stria terminalis, component of the substantia innominate inferior to the lentiform nucleus and basal forebrain, constitutes the "extended amygdala." The periamygdaloid nucleus appears to merge with the basal nucleus of the basolateral complex. The central and medial nuclei are considered by some as an extension of the basal nuclei.

The nuclei within the corticomedial nuclear complex receive noradrenergic projections from the locus coeruleus and lateral tegmental nucleus via the MFB. The central nucleus lies dorsal and medial to the basal nucleus, and together with the medial nucleus, it merges with the bed nucleus of stria terminalis and periamygdaloid cortex, forming the centromedial amygdaloid nuclear complex. It retains chemical and cellular continuity with the bed nucleus via the stria terminalis and the sublenticular basal forebrain.

The central nucleus projects to the periaqueductal gray matter, substantia nigra, ventral tegmental area, parabrachial nuclei, dorsal motor nucleus of vagus, and solitary nucleus. Projections to the central nucleus arise directly from the parabrachial, and many other nuclei convey their impulses to the central nucleus via the parabrachial nucleus. These connections account for the important role that amygdala plays in the regulation of cardiovascular, respiratory, and gustatory systems. The lateral, basal, and accessory basal nuclei project (not reciprocated) mainly to the magnocellular part of the dorsomedial thalamic nucleus that conveys impulses to the prefrontal cortex, enabling the amygdala to influence the activities of this part of the cerebral cortex. On the other hand, the central and medial amygdaloid nuclei maintain reciprocal connections with the midline thalamic nuclei. Some experimental data also suggest projections from the ventral posteromedial nucleus to the lateral nucleus of the amygdala.

The paraventricular nucleus of hypothalamus sends oxytocin and vasopressin-immunoreactive terminals to the central amygdaloid nucleus. Vasopressin-immunoreactive neurons are also found in the medial nucleus, which also receives projections from the suprachiasmatic nucleus of hypothalamus. The medial and central nuclei contain β-endorphin and enkephalin immunoreactive neurons. Processing of olfactory stimuli and pheromones occurs in the cortical nucleus. Certain nuclei of both subdivisions of the amygdaloid nuclear complex contain high concentrations of steroid hormones and their binding sites. Estrogen-concentrating neurons are found in the medial, accessory basal, and posterior cortical nuclei, whereas dihydrotestosterone-concentrating neurons are abundant in the lateral, parvicellular basal, and accessory basal nuclei. Enzymes, such as reductase, that convert testosterone into nonaromatizable 5a-dihydrotestosterone, and aromatase, which convert testosterone and erostenedione to estradiol, are abundant in the amygdala. Thus, changes in the hormonal levels that accompany the menstrual cycle may affect the functions of amygdaloid neurons.

Unidirectional intrinsic connections exist among amygdaloid nuclei in which the lateral nucleus projects primarily to all divisions of the basal, accessory basal, paralaminar, and anterior cortical nuclei and less densely to the central nucleus. Divisions of the basal nucleus project to the accessory basal nuclei, and the accessory basal nucleus maintains profuse connections to the central nucleus. The central nucleus projects to the anterior cortical nucleus. In general, impulses that reach the corticomedial nucleus of the amygdala are derived from the olfactory bulb and the anterior olfactory nucleus of both sides. The basolateral nuclear complex (Figures 17.10, 17.11, and 17.14) receives input from the pyriform lobe, inferior temporal gyrus, nucleus of the diagonal band of Broca, and the orbitofrontal cortex. The orbitofrontal, cingulate, and temporal neocortex (the largest contributor) maintain reciprocal connection with the amygdala. The robust connections of the individual nuclei of the amygdala, which are generally reciprocal with the exception of the connections with the striatum, thalamus, and parts of the cerebral cortex, have to be taken into consideration when overall function of the amygdala is examined. Substantial cholinergic projections from the magnocellular nucleus basalis of Mynert terminate in the basal nucleus (magnocellular part) and accessory basal nucleus. In turn, the central nucleus, the parvicellular part of the basal nucleus, and the magnocellular accessory basal nucleus project back to the nucleus basalis of Mynert and the nucleus of the diagonal band of Broca. Striatal projections form a substantial component of the amygdaloid output from the basal and accessory basal nuclei to the nucleus accumbens septi, ventromedial parts of the caudate nucleus, and putamen. The hippocampal gyrus, midline thalamic nuclei, and the prefrontal cortex provide efferents to the amygdala that interdigitate with efferents that originate from the nucleus accumbens septi. Amygdala also has direct connections with the hippocampal formation, and also with the temporal, occipital, frontal, cingulate, and insular cortices.

Noradrenergic projections from the locus coeruleus and the lateral tegmental nucleus and serotonergic projections dorsal and medial to the amygdala run through the MFB. There is also a high concentration of dopamine β-hydroxylase in the central, lateral, and intercalated nuclei that is derived from the ventral tegmental area (A10) with some adrenergic input. However, serotonergic input is far greater than the noradrenergic and dopaminergic counterparts within the amygdala with the densest concentrations located in the amygdala–hippocampal area, and in the central, lateral, parvicellular basal, and magnocellular accessory basal nuclei. The anterolateral part of the magnocellular nucleus basalis of Meynert provides cholinergic innervation to the nucleus of the lateral olfactory tract and to the basal and parvicellular accessory nuclei as well as to the amygdala–hippocampal area.

Amygdalar neurons in the accessory basal and basal nuclei that project to the striatum utilize glutamate/aspartate as a neurotransmitter. These neurotransmitters are also contained in the lateral, anterior cortical, and periamygdaloid nuclei. β-endorphin and cholecystokinin, VIP, and corticotropin releasing hormone from the hypothalamic paraventricular nucleus are found in a considerable concentration in the medial and central nuclei. Lateral and basal nuclei contain high concentration of somatostatin and NY. Vasopressin, oxytocin, and encephalin are found in the central nucleus, whereas vasopressin is primarily concentrated in the medial nucleus, which receives afferents that emanate from the suprachiasmatic nucleus of the thalamus. Amygdaloid nuclear complex and associated receptors contain steroid hormones. Neurons of the accessory basal, posterior cortical, and medial nuclei contain high concentrations of estrogenic neurons. The medial, lateral, accessory, and parvicellular basal nuclei contain dihydrotestosterone as well as 5-alpha reductase.

Amygdaloid projections are maintained primarily via the stria terminalis and the ventral amygdalofugal pathways. The stria terminalis runs parallel to the fornix, containing bilateral afferent and efferent fibers, to and from the amygdala, which cross in the anterior commissure. It initially courses in the roof of the inferior horn of the lateral ventricle medial to the tail of the caudate nucleus. Then, it travels in the floor of the central part of the lateral ventricle between the caudate nucleus and the thalamus, adjacent to the thalamostriate vein, dividing into supracommissural, commissural, and subcommissural components. Most of the fibers of the supracommissural and subcommissural components are amygdalofugal, distributing to the septal and preoptic areas as well as to the hypothalamus. Others terminate in the pyriform lobe, the anterior perforated substance, and the dorsomedial nucleus of the thalamus. Fibers of the stria terminalis that joins columns of the fornix and the stria medullaris are primarily subcommissural. Commissural fibers interconnect the amygdala of both sides via the anterior commissure.

The *ventral amygdalofugal fibers* (Figures 17.10, 17.11, and 17.14), the largest output from the amygdala, originate from the basolateral nucleus, course ventral to the globus pallidus, and distribute to the hypothalamus and preoptic area. These fibers reach the brainstem reticular formation via the MFB. Both nuclear components of the amygdala project to the hypothalamus, the dorsomedial nucleus, and the midline nuclei of the thalamus. Stimulation of the amygdala may produce activities associated with food intake, behavioral and visceral changes, and affect. It may also evoke feelings of relief, relaxation, pleasant sensations, and autonomic response such as bradycardia and pupillary dilatation. As with any other constituent of the limbic system, the amygdala poses antagonistic effects upon behavior and endocrine functions. Stimulation of the corticomedial and basolateral nuclei inhibits and facilitates aggressive behavior, respectively. Activation of the basolateral nucleus enhances the release of the growth and adrenocorticotropic hormones.

Estrogen containing neurons of the amygdala (estradiol receptors) may induce ovulation if the corticomedial component is stimulated, whereas transection of the stria terminalis abolishes ovulation. The stria terminalis may exert both excitatory and inhibitory effects on the secretion of the gonadotropic hormones. This hormonal effect may account for the hypersexuality that occurs in individuals with bilateral amygdalectomy.

There is a primary role of the amygdala in the formation and storage of emotional memories. The lateral nucleus of the basolateral nuclear complex is involved in the emotional response to aversive stimuli through LTP (a phenomenon of neuronal plasticity that also enhances behavioral learning in which prolonged enhancement of signal transmission between neurons that received simultaneous and synchronous activation). Synaptic reactions in the lateral nucleus associated with imprinting of emotional experiences are translated into fear through its connections with the central nucleus and the bed nucleus of stria terminalis. The central nucleus generates reaction to fearful stimuli through tachycardia, mydriasis, sweating, tachypnea, immobility, and release of stress hormones. This function becomes evident in lesions of the amygdala, which produces impairment of acquisition and expression of fearful condition of Pavlov.

The amygdala also regulates long-term memory consolidation and the memory associated with fear conditioning (inhibitory avoidance), possibly via LTP. Consolidation allows modulation of the stored memory traces. Emotional arousal, associated with a learning event, as occurs in stressful situation, augments the extent of retention of that event. Thus, a correlation exists between the emotional-arousing nature of the learned event, ability to retain it, and the amygdalar activity evident through the theta rhythm. The latter is an oscillatory pattern of the electroencephalography signals, which is thought to indicate a neuronal synchronized event that enhances memory retention and synaptic plasticity by increasing connectivity between neocortical memory storage sites and the temporal lobe. "Karuna" meditation has been demonstrated to involve heightened neuronal activity in the amygdala, insular, and temporoparietal cortices.

Experimental ablation of the anterior temporal lobes, including the amygdala and uncus in monkey, produces the commonly known *Klüver–Bucy syndrome*. Manifestations of this disease include sexual hyperactivity accompanied by change in appetite and dietary habits, including hyperphagia (overeating) and hypoemotionality (loss of fear and reduction in maternal behavior). Changes in behavior from aggressive to placid are also seen in this syndrome. Temporary memory loss may also occur. Damage to the visual association areas of the temporal cortex causes visual hallucinations and visual agnosia with inability to recognize familiar objects or members of the species. Inability to visually identify objects leads to a persistent tendency and a strong compulsion to orally examine, on repetitive basis, edible and nonedible objects (hypermetamorphosis). The human Klüver–Bucy syndrome following traumatic injury to the temporal lobes, Alzheimer's disease or Pick's disease, Rett syndrome, porphyria, carbon monoxide poisoning, stroke, and herpes simplex encephalitis may exhibit manifestations similar to those of the monkeys; however, sexual activity is not prominent. Dementia, aphasia, amnesia, and excessive eating are common manifestations. Signs of this condition may also be observed in *posttraumatic apallic syndrome* (coma vigil), which is also characterized by reactionless gaze, lack of spontaneous movements, and appearance of primitive reflexes, hyperkinesia, and rigidity.

Imaging studies have shed more light on the functions of the amygdala in humans. Research data have been shown that the left amygdala in particular is associated with obsessive-compulsive personality, posttraumatic stress, and anxiety disorders. Smaller amygdala has been reported in people with bipolar disorder. Amygdalar activity has been demonstrated to dramatically increase in individuals with borderline personality disorder who experience difficulty distinguishing between actual threatening and neutral visual images. This activity is also heightened when confronted with frightening images and situations where social phobia develops. However, since damage to the amygdalae does not eliminate fear but enhances reaction to fearful stimuli, the role of the amygdala in processing of these stimuli cannot be clearly stated. It appears that emotionally charged event or experience has memory enhancement effect that activates the amygdala.

Many psychological disorders have been linked to the size of the amygdala. Anxiety disorders have been shown to occur more frequently in individuals with a smaller left amygdala, and that psychotherapy and the administration of antidepressants tend to increase its size. As one of the predilection site of toxoplasma cysts, amygdalar damage may be responsible for the paranoid manifestations observed in patients with this parasitic disease. Despite the negative emotional conditions mediated by the amygdala, its role in positive emotion has been documented.

Pattern of activity of the amygdala has been shown to exhibit differences according to sexual orientation in homosexual and heterosexual males and females. Homosexual men show more female-like patterns of activity in the amygdala than heterosexual males do. Similarly, homosexual females tend to exhibit more male-like patterns of activity. It has been reported that the connection of the amygdala was extensive in the left amygdala in homosexual males and heterosexual females, whereas amygdalar connections appear to be more widespread from the right amygdala in homosexual females and heterosexual males.

Larger amygdala has been shown to correlate with social networking, accurate social judgment about other person's faces, and possibly emotional intelligence and aspects of social cooperation. Respect to personal space is also mediated by the amygdala as it is activated in individuals when put in close physical contact with others. Modulation and demonstration of aggressive behavior and stimulation of the amygdala increase both sexual and aggressive conduct. Left amygdala has been implicated in the reward system. Stimulation of the right and left amygdalae produces functionally different reactions. Positive (pleasurable) and negative (fear and sadness) emotions are produced by stimulation of the left amygdala, whereas activation of the right amygdala induces fear and anxiety. The difference between the right and left amygdalae also extend to lateralization in males and females. Activity was shown to lateralize to the right amygdala mediating reaction to stressful situation in the males and to the left amygdala that also mediates thought-based reaction in females. There is some indication that females are more prone to anxiety disorders than males, and that females lose serotonin receptors in the amygdala in response to unusually stressful situation while males display an increase in these receptors.

Decortication (excision of the cerebral cortex including the limbic lobe) may produce sham rage, which is characterized by combative, violent reactions to mild stimuli. Dementia, a cardinal manifestation of limbic encephalitis, exhibits profound neuronal loss as a result of gliosis, perivascular cuffing, and infiltration of microglia in the cerebral cortex. These deficits are reversible as long as the treatment is not delayed for a long period of time. Patients with limbic system disease may also exhibit "florid delirium" and overreacting to a situation or acting inappropriately.

Due to the important role of the amygdala in initiating a state of fear, bilateral lesions confined to the amygdala produce a behavior that lacks the reaction to fearful situations. These lesions appear in patients with Urbach–Wiethe disease, a rare autosomal recessive disorder that shows variable manifestations that include thickening of the basement membrane of the skin, palpebral beaded papules, neuropsychiatric disorders, and epileptic seizures. This condition is attributed to mutations of the extracellular matrix protein 1 gene contained in chromosome 1 at 1q21. In this disease, bilateral symmetrical calcification of the amygdala, periamygdaloid cortex, and medial temporal gyri commonly occur. Due to the important role of the amygdala in emotional memory, calcification within the vessels that supply it produces anxiety, psychotic and mood disorders, schizophrenia-like symptoms, and epileptic seizure. Protein kinase C-epsilon located in the amygdala regulates the behavioral response to alcohol consumption as alcohol intoxication and binge drinking have been linked to amygdalar damage.

LIMBIC LOBE

The limbic lobe is the cortical ring, which surrounds the corpus callosum and consists of the cingulate, parahippocampal, paraterminal gyri, and the isthmus (which connects the cingulate to the parahippocampal gyrus).

CINGULATE GYRUS

The *cingulate gyrus* borders the corpus callosum, bounded superiorly by the cingulate sulcus and inferiorly by the callosal sulcus. This gyrus continues with the parahippocampal gyrus through the isthmus. It is connected to the parahippocampal gyrus via the cingulum and to the dorsomedial nucleus of the thalamus through the thalamocortical and corticothalamic fibers. The dorsomedial nucleus acts as an indirect pathway between the hypothalamus and the cingulate gyrus. The cingulate gyrus establishes connections with the temporal lobe via the uncinate fasciculus and with the ipsilateral corpus striatum via the internal capsule. It is also connected to other limbic structures, receiving afferents from the hippocampal gyrus, anterior nucleus of the thalamus, nucleus of the diagonal band of Broca, and dopaminergic input from the midbrain ventral tegmentum. Despite the fact that no output from the cingulate gyrus projects to the hypothalamus, autonomic activities including changes in blood pressure, cardiac output, respiration, and motility of the digestive tract may be seen upon stimulation of the cingulate gyrus. Tumors of the cingulate gyrus are associated with changes in behavior. The

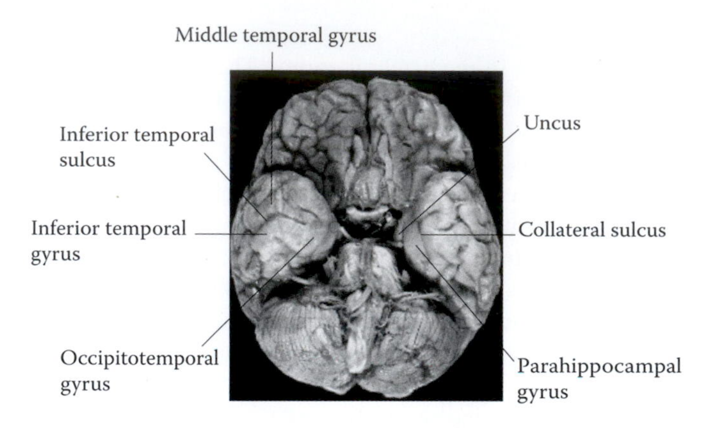

FIGURE 17.12 Photograph of the inferior surface of the brain illustrating the temporal gyri and the uncus. Note the proximity of the uncus to the ocular motor nerve.

parahippocampal gyrus (Figure 17.12) represents the inferior portion of the limbic lobe, expanding rostrally into the uncus between the collateral and hippocampal sulci.

PREFRONTAL CORTEX

The *prefrontal cortex* (Figure 17.12) develops to a great extent in humans and is linked to both intellectual and emotional processes. It includes the orbital gyri and Brodmann's areas 9 and 10. The prefrontal cortex is connected to the parietal, temporal, and occipital association areas. It receives information from the DM nucleus of the thalamus, other parts of cerebral cortex, septal area, monoamine neurons, and the hypothalamus. It conveys this information to the hippocampus, the amygdala, and the subiculum. Stimulation of the prefrontal cortex inhibits aggressive behavior. There is no direct projection from the prefrontal cortex to the hypothalamus.

The role of the prefrontal cortex in emotion and behavior is best demonstrated in individuals who have undergone bilateral prefrontal lobotomy, a surgical procedure that was once commonly employed to treat psychotic patients. Bilateral lobotomy alters aggressive behavior and tends to alleviate the emotional distress associated with chronic intractable pain unresponsive to conventional analgesics. In humans who have sustained frontal lobe damage as a result of tumors or wounds, perceptual and intellectual deficits as well as derangement of behavioral programming and abstract thinking may become evident. Patients lose the ability to identify the goal of their intended action, and their reactions change between gloom and elation in a superficial and abrupt manner. Patients also exhibit signs of indifference to their position in society and become less concerned with monetary matters. Patients may acknowledge pain, but they show little emotional distress.

HYPOTHALAMUS

The *hypothalamus* contains centers that deal with feeding, behavior, drinking habits, expression of emotion, and hormonal control, maintaining homeostasis (see hypothalamus-Chapter 7). Certain *thalamic nuclei* also maintain connections with the limbic lobe as well as the prefrontal cortex. For instance, the dorsomedial and anterior thalamic nuclei maintain bilateral connections with the cingulate and the prefrontal cortices. Lesions of the dorsomedial thalamic nucleus have been associated with anterograde amnesia. Another component of the diencephalon, the *epithalamus*, contains neurons that mediate activities of the limbic system via massive connections to the septal area, the hypothalamus, and the brainstem reticular formation. In summary, the limbic system operates via the Papez circuit, a feedback loop that utilizes the septal area, and the circuit that revolves around the amygdaloid nuclear complex.

Papez circuit of emotion (Figures 17.5 and 17.8) is formed by fibers of the fornix that carry impulses from the hippocampal gyrus to the mammillary body. Information received by the mammillary body is conveyed via the fornix to the anterior nucleus of the thalamus and later to the cingulate gyrus within the thalamocortical fibers. The final connection between the cingulate gyrus and the hippocampal gyrus, which is maintained via the cingulum, completes this feedback loop.

The *septal area* (Figures 17.8 and 17.13) serves as a nodal point for the afferents and efferents that converge on this area, mediating limbic system functions. As can be seen (Figure 17.8), the septal area receives input from the midbrain reticular formation via the MFB and conveys this information via the stria medullaris thalami to the habenula and via the fasciculus retroflexus to the interpeduncular nucleus. The connections of the interpeduncular nucleus to the habenula and reticular formation complete this circuit.

The *amygdaloid nuclear complex* (Figures 17.9, 17.10, 17.11, and 17.14) also serves through a feedback circuit to promote the function of the limbic system. As discussed earlier, the amygdala projects to the septal area, which through various connections links to the reticular formation. The feedback loop is completed by direct amygdalar projection to the reticular formation via the amygdalofugal fibers and through the latter to the prefrontal cortex.

In order for the limbic system to respond to stress, different neurochemical signals interact and diverse responses have to be elicited in various structures associated with this system. However, virtually all limbic system structures express glucocorticoid and mineralocorticoid receptors, which allow the glucocorticoid to modulate limbic signaling patterns. The role of given limbic structures is both region- and stimulus-specific and that each limbic structure projects to various other brain structures that possess distinct subcortical targets. It has been suggested that disorders of the limbic and hypothalamo-pituitary-adrenocortical (HPA) axis may be responsible for producing affective disorders. In particular, the hippocampus and anterior cingulate gyrus inhibit the stress-induced HPA

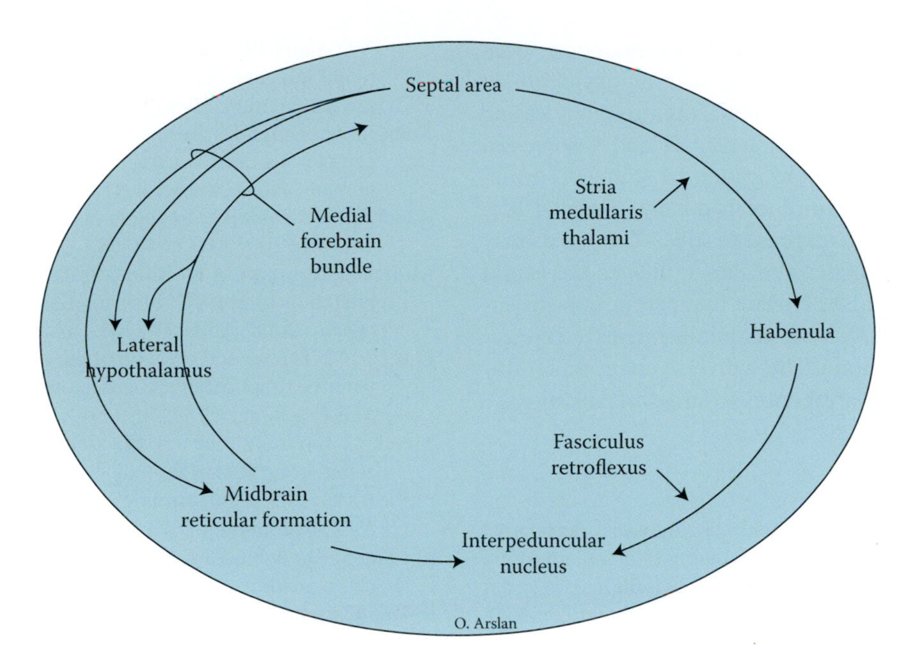

FIGURE 17.13 This feedback circuit (Papez circuit of emotion) is initiated by impulses that are generated by the hippocampal gyrus, delivered to the mammillary body, and via the mammillo-thalamic tract to the anterior thalamic nucleus. In order to complete this feedback loop, impulses received by the anterior thalamic nucleus are conveyed back to the hippocampal gyrus via the cingulate gyrus and cingulum, respectively.

activation and glucocorticoid secretion, whereas the amygdalar nuclear complex may increase glucocorticoid secretion. To access corticotropin-releasing hormone neurons, limbic structures, such as the hippocampal gyrus, neocortex, and amygdala ordinarily do not project to HPA effector neurons of the paraventricular nucleus (PVN) but rather project, in an overlapping fashion, to the glucocorticoid neurons of PVN indirectly via neurons of the bed nucleus of the stria terminalis, hypothalamus, and brainstem. This may indicate that input to the PVN is integrated at subcortical levels. The issue of intrinsic neurochemical signaling in response deserves some analysis. It appears that peptides that are rich in the limbic system form the chemical basis of stress response. The chemical structure multitude types of the peptides enable a single molecule to activate the mechanisms that produce the autonomic, endocrine, and behavioral responses. Stress can be induced

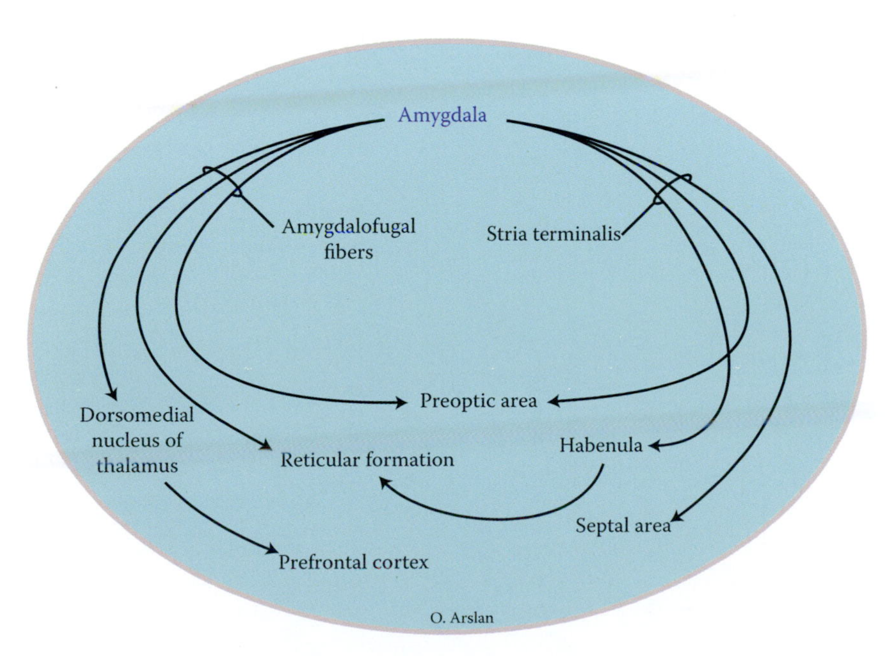

FIGURE 17.14 Summary diagram of the efferent projections of the amygdala. As indicated, the stria terminalis projects to the septal and preoptic areas as well as the habenula. The amygdalofugal fibers enable the amygdala to project to the thalamus, hypothalamus, cerebral cortex, and reticular formation.

through hypoglycemia, dehydration, cold, or anxiety, or even by the demands of reproductive cycle, such as oxytocin and endorphin. NY responds to hypoglycemia, thyrotropin-releasing factor responds to cold, and angiotensin II and vasopressin are activated in response to dehydration to stimulate drinking and increase blood pressure, respectively. Gonadal hormones undergo reduction in response to stress. Corticoids may modify the actions of peptides and thus influence the neural response to stress. Stress also activates adrenergic, dopaminergic, and serotonergic neurons that maintain extensive connection to other areas of the brain. These diverse connections may explain the equally diverse responses to stressful conditions.

SUGGESTED READING

Adolphs R, Cahill L, Schul R, Babinsky R. Impaired declarative memory for emotional material following bilateral amygdala damage in humans. *Learn Mem* 1997;4:291–300.

Ally B, Hussey E, Donahue M. A case of hyperthymesia: Rethinking the role of the amygdala in autobiographical memory. *Neurocase* 2012;1–16.

Asari T, Konishi S, Jimura K, Chikazoe J, Nakamura N, Miyashita Y. Amygdalar enlargement associated with unique perception. *Cortex* 2010;46(1):94–9.

Blair RJR. The amygdala and ventromedial prefrontal cortex: Functional contributions and dysfunction in psychopathy. *Philos Trans R Soc Lond B Biol Sci* 2008;363(1503):2557–65.

Botvinick MM, Cohen JD, Carter CS. Conflict monitoring and anterior cingulate cortex: An update. *Trends Cogn Sci* 2004; 8:539–46.

Bzdok D, Laird A, Zilles K, Fox PT, Eickhoff S. An investigation of the structural, connectional and functional sub-specialization in the human amygdala. *Hum Brain Mapp* 2012;34(12):3247–66.

Cavanna AE, Cauda F, D'Agata F, Sacco K, Duca S, Geminiani G. Mapping pleasure pathways: The functional connectivity of the nucleus accumbens. *J Neuropsychiatry Clin Neurosci* 2011;23(2):30–30.

Costa VD, Lang PJ, Sabatinelli D, Bradley MM, Versace F. Emotional imagery: Assessing pleasure and arousal in the brain's reward circuitry. *Hum Brain Mapp* 2010;31(9):1446–57.

Kahn I, Shohamy D. Intrinsic connectivity between the hippocampus, nucleus accumbens, and ventral tegmental area in humans. *Hippocampus* 2013;23(3):187–92.

Kauer JA, Malenka RC. Synaptic plasticity and addiction. *Nat Rev Neurosci* 2007;8(11):844–58.

Kullmann DM, Lamsa K. Roles of distinct glutamate receptors in induction of anti-Hebbian long-term potentiation. *J Physiol (Lond.)* 2008;586(6):1481–6.

Laruelle M. The second revision of the dopamine theory of schizophrenia: Implications for treatment and drug development. *Biol Psychiatry* 2013;74(2):80–1.

Merck C, Jonin PY, Laisney M, Vichard H, Belliard S. When the zebra loses its stripes but is still in the savannah: Results from a semantic priming paradigm in semantic dementia. *Neuropsychologia* 2014;53:221–32.

O'Mara S. The subiculum: What it does, what it might do, and what neuroanatomy has yet to tell us. *J Anat* 2005;207:271–82.

Solano-Castiella E, Anwander A, Lohmann G, Weiss M, Docherty C, Geyer S, Reimer E, Friederici AD, Turner R. Diffusion tensor imaging segments the human amygdala in vivo. *Neuroimage* 2010;49(4):2958–65.

Schacter DL, Gilbert DT, Wegner DM. *Explicit and Implicit Memory*. Psychology, 2nd ed. New York: Worth, Incorporated, 2011, 238.

Schwabe L, Wolf OT. Learning under stress impairs memory formation. *Neurobiol Learn Mem* 2010;93(2):183–8.

Shu SY, Bao R, Bao XM, Zheng ZC, Nui DB. Synaptic connections between the projection from the marginal division of the striatum to the Meynert's basal nucleus and its relationship to learning and memory behavior of the rat. *Chin J Histochem Cytochem* 1998;7:1–11.

Volk LJ, Bachman JL, Johnson R, Yu Y, Huganir RL. PKM-ζ is not required for hippocampal synaptic plasticity, learning and memory. *Nature* 2013;493(7432):420–3.

White FJ. Synaptic regulation of mesocorticolimbic dopamine neurons. *Annu Rev Neurosci* 1996;19:405–36.

18 Gustatory System

The gustatory system regulates the perception, transduction, and transmission of the gustatory impulses. It recognizes the concentration and the pleasurable or harmful qualities of the consumed substances. It responds to a diverse type of taste sensations. Taste, as a chemical sensation, shares common connections and characteristics with the olfactory system, entailing the stimulation of the gustatory modified epithelial receptors, primarily G protein coupled and, to lesser degree, ion channel receptors. The taste receptors are scattered on the distinct papillae of the dorsum of the tongue, epiglottic vallecula, and soft palate. Transduction of the chemical stimuli by the gustatory receptors is eventually transmitted via fibers of the facial, glossopharyngeal, and vagus nerves, which link with the neurons of the solitary nucleus and, ultimately, with the gustatory cortex through the ventromedial thalamic nucleus. A parallel link enables the gustatory impulses to travel to the hypothalamus and amygdala to regulate autonomic activities induced by taste. The quality of taste is determined by activation of several clusters of well-defined central and peripheral gustatory neurons. Due to the multisynaptic connection of the gustatory pathway, continuous replacement of the taste receptors and the reflexive taste-elicited behavior are factors that enable this system to resist potential injuries.

Taste, together with the olfactory and thermal sensations as well as information that pertain to texture form integral components of the chemical sensations and the subjective perception of flavor. This is evident in the fact that airborne odorants pass through the oral and nasal cavities and are received by the taste receptors and olfactory epithelium, respectively. Obstruction of the nasal cavity or anosmia can block the transmission of the airborne odorants and impair taste sensation, while allowing the tastants to be conveyed in the aqueous solution. Similarly, the gustatory system is influenced by the sight and even the sound surrounding the environment. Since taste sensation is much simpler and maintains a higher threshold than olfaction, appreciation of flavor may be intensified by olfactory impulses as is exemplified by appreciation of coffee and chocolate flavors. The ability to taste may vary according to gender, age, and fluctuation in metabolic activities, as well as changes in body temperature. Taste also determines our eating habits, weight, and indirectly, the electrolyte balance within the gastrointestinal system. It initiates salivation, swallowing, secretions of digestive enzymes including insulin as well as activation of gagging reflex in response to unpleasant and inedible substances.

PERIPHERAL GUSTATORY RECEPTORS

Tastants are categorized into salty, sweet, sour, and umami taste (Japanese for delicious taste). Other types of taste sensations such as metallic, starchy, and astringent are also exist. Despite the diversity of chemical stimulants, all parts of the tongue respond to all tastants. Though some taste buds may respond to all categories of stimulants, other may respond to one or very few tastants. Receptors for a particular class of taste are selectively responsive to a wider range of similar chemicals. It has been suggested that perception of taste is a culmination of a complex array of responses from a specific area of the tongue irrespective of the number of receptors.

Classification of tastants bears metabolic, nutritional, and protective significance. Solubility in saliva and the presence of specialized carrier molecules, such as von Ebner gland (VEG) protein for more lipophilic molecules, can determine the intensity of taste. The mechanism of transduction can occur either by the passage of ions directly through the membranes or by stimulation of G protein after reception of specific proteins and, finally, by the second messenger-mediated ion gating.

Sweet taste is acquired from carbohydrates and amino acids and generates needed energy (glucose) and storage energy (glycogen). It has been proposed that sweet taste transduction follows two models. One model follows the G-protein coupled receptor-G ($GPCRG_s$)-cAMP route in which natural sweeteners such as sucrose stimulate the G protein–coupled receptors (GPCR) T1R2 (human taste receptor family 1, member 2) and T2R3 (human taste receptor family 2, member 3) located on chromosome 5, which releases gustducin (transducer of intracellular signals) activating adenylyl cyclase. The latter catalyzes the conversion of ATP to 3′,5′-cyclic AMP (cAMP) and pyrophosphate. Generated cAMP may act either directly by producing an influx of cations through cAMP-gated channels or indirectly by stimulating protein kinase A, which causes phosphorylation and potassium efflux in the apical membrane. This subsequently leads to opening of the voltage-gated calcium channels, depolarization, and neurotransmitter release. The other model follows the $GPCRG_q/G\beta\gamma\text{-}IP_3$ pathway used by artificial sweeteners such as saccharin. Sacharin binds and activates GPCRs coupled to phospholipase C ($PLC\beta_2$) by either $\alpha\text{-}G_q$ or $G\beta\gamma$. Subsequent activation of $PLC\beta_2$ generates inositol-1,4,5-trisphosphate (IP_3) and diacylglycerol (DAG). IP_3 and DAG cause calcium release from intracellular stores. Accumulation of calcium ions leads to cellular depolarization and neurotransmitter release.

Salty taste of sodium chloride depends on the existence of monovalent cations that act as an osmotically active compound, playing a critical role in ion and water homeostasis. Sodium chloride receptors are the most primitive of taste receptors, which contain Na^+ ion channels (epithelial sodium channel [EnaC]) that depolarize the taste cells and open calcium gates regulated by voltage, allowing ions to enter the cells and cause release of neurotransmitters. Interaction between tastes does occur, particularly the large anions that inhibit the monovalent cation-mediated salty tastes as well as sweet and bitter substances.

Sour taste indicates the presence of an acidic compound (H^+ ion) and is associated primarily with the proton concentration and, to a lesser extent, with the particular anion involved. It is mediated by different receptors. One of the receptors is the ENaC, which allows the entrance of H^+ ions into the cell. This protein is involved in the salty taste transductive mechanism, a fact that accounts for the variation of salty taste when sour stimuli are present. There is also mammalian degenerin (MDEG1) of the ENaC family receptors, which permits potassium to leave the cells, counteracted by the H^+ ions that block and trap the potassium ions. Another group of receptors attach to H^+ ions and allow sodium ion influx down the concentration gradient into the cell leading, to the opening of a voltage-regulated calcium gate, depolarization, and eventual release of neurotransmitter.

Substances that are harmful generally taste bitter, and the taste is elicited by divalent cations and alkaloids (strychnine, nicotine, and atropine). Action of bitter-tasting compounds is mediated by GPCRs of the T2R family, particularly T2Rs that respond specifically to bitter tastants. Stimulation of the GPCR causes the release gustducin, which activates phosphodiesterase. The latter, through a cascade of events, forms a secondary messenger that blocks potassium ion channels and also stimulates the endoplasmic reticulum to release calcium, leading to accumulation of potassium ions in the cell, initiation of depolarization, and eventual neurotransmitter release.

Umami taste, typical of monosodium L-glutamate, triggers a pleasurable response, leading to the intake of peptides and protein synthesis essential for elements in enzymes, hemoglobin, and antibodies. It has been suggested that monosodium L-glutamate bonds to receptor mGluR4, causing the G protein complex to activate a secondary receptor, which ultimately leads to neurotransmitter release. The umami taste also uses T1R1 and T1R3 receptors that respond to most of the 20 amino acids encountered in the diet. Absence of the T1R3 receptor causes reduced sensitivity for this taste, but not total loss.

Connection of the solitary neurons to the motor and parasympathetic neurons of the facial nerve enables gustatory impulses to produce contraction of the facial and laryngeal muscles in the form of facial grimaces, puckering, transient hoarseness as a response to a taste that is bitter or offensive. Thus, suitability of the substance as a nutrient, and the protective steps that can be taken before ingestion are determined. Both peripheral and central neurons associated with gustatory impulses respond broadly to more than one modality of taste sensation as well as nongustatory stimuli (thermal and tactile). This variation is intensity bound and has a perceptual quality-dependent characteristic. Gustatory stimulation primarily occurs in response to molecules that are soluble, hydrophilic, and nonvolatile. It requires variable threshold concentrations of substances, low for bitter and noxious-tasting compounds and higher for good-tasting substances.

Despite the lack of conclusive evidence relative to the presence of topographic arrangement of taste fibers and projections to the solitary nucleus and the gustatory cortex, segregation of various taste-conveying cranial nerve fibers does exist within the solitary nucleus and in the primary gustatory cortical center. Corollary to this, sensitivity of gustatory cells in the taste buds shows a difference according to these nerves. Frequency of neuronal impulses and the number of responding neurons may constitute a basis for the intense taste experience. The pleasurable aspect of taste can be attributed to activation of the neurons of the amygdala. Neuronal responses are broadly tuned across stimuli, and no single group of neurons is able to differentiate between different taste qualities. Taste quality is thought to be represented by the extent of neural activity across a number of different afferents and is thus represented in the population response. Discrimination among stimuli or different qualities of tastes occurs via activation of a group of receptors and not by a single cell.

Taste buds or canaliculus gustatorius are renewed periodically because of their short life span and their number is around 5000, although the range can vary between 500 and 20,000. The taste buds are sparsely scattered and in a variable manner on the fungiform (mushroomlike) papillae, numbering a total of 250, which are concentrated in the anterior two-thirds and primarily on the tip of the tongue. There is an average of 3 taste buds for each papilla, which respond to umami taste, and salty substances. The circumvallate papillae are arranged as a single row anterior to the sulcus terminalis that demarcates the anterior two-thirds and posterior one-third of the tongue, with a large number of taste buds (250 taste buds for each papilla). Taste buds are also abundant on the foliate papillae, which are arranged on the posterolateral tongue, with one papilla on each side. Foliate papillae contain a total of 1200 taste buds that respond to bitter and sour taste. Central tongue and filiform papillae lack taste buds. This anteroposterior localization of the papillae accounts for the corresponding and gradual pattern of taste bud stimulation. Taste buds are concentrated on the posterior and central part of the soft palate near its junction with the hard palate and are embedded in keratinized epithelium. They are most numerous in fetal life and undergo atrophy in postnatal life.

Each taste bud is composed of a cluster of fusiform neuroepithelial cells located on the surface of the tongue around taste pores on the mucosal surface, where they are covered by a dense glycoprotein-rich extracellular material. Dissolved chemical substances in the oral mucosa diffuse through the taste pores and then through dense extracellular layers within the apices to reach the taste receptors, where they initiate depolarization and action potential. This will activate the synaptic connections at their bases, initiating an action potential in the terminal of the afferents of the associated cranial nerves.

Taste buds contain parietal (precursors), sustentacular, basal (longer life span), and gustatory cells (short life span), which are modified neuroepithelial cells with epithelial and neuronal characteristics. Through electron micrograph, five types of cells are recognized (types I–V), which show varying staining characteristics and location (dark and light cells arranged in a concentric manner with apical microvilli projecting through the epithelium and basal domains). Type I cells are the most abundant, are peripherally located, contain high concentrations of chromatin, and are thin. Type III cells exhibit prominent vesicles and multiple mitochondria and contain neural cell adhesion molecule (N-CAM) molecules. Types II and III are considered receptor cells that establish contact with the afferent neuronal fibers. They are separated at their bases from the lamina propria by basal lamina where contact with afferent fibers occurs. Type IV cells are lodged at the posterior end of the taste bud.

Taste buds do not show structural variation, and the neuroepithelial cells within these buds respond to multiple stimuli. Each neuroepithelial cell is associated with synaptic transmission and action potential, and they are therefore also termed "para neurons." It receives terminals at its base from several neurons, and several cells may be innervated by one neuron. Therefore, the 50 cells of the taste buds receive, correspondingly, 50 nerve fibers. Nerve fibers have many terminals that supply widely distributed taste buds and multiple sensory cells in each bud. Some of the nerve fibers divide before reaching the taste buds (perigemmal plexus), while others branch inside the bud (intragemmal plexus). Fibers that constitute the perigemmal plexus are free nerve endings and contain substance P and calcitonin gene-related peptide (CGRP). Intragemmal plexus fibers form a synaptic linkage with either type I or II and type III cells.

Interaction of taste stimuli and receptors and transduction occur in the apical microvilli, whereas the bases of these cells produce graded electrical potentials that activate the neurons and contain specialized areas that interact with afferents from the associated ganglia of facial, glossopharyngeal, and vagus nerves. This trophic input influences the maintenance and differentiation of taste cells and sensitivity of the sensory cells, but not their receptor expression. They decline in number at a rate of 1% per year, increasing after the age of 40. This fact may explain the tendency for elderly people to eat more spicy food than the average population. Denervation causes rapid dedifferentiation and death of the taste buds.

Regeneration of nerve terminals leads to the formation of new and functional taste buds. Taste cells regenerate from the basal cells. Several factors affect the perception of taste. Zinc, an essential mineral, modulates the concentration of gustin through carbonic anhydrase VI and indirectly influences taste bud regeneration. It is an important component of parotid gland salivary protein, responsible for the development of normal taste buds. Zinc also is a cofactor for alkaline phosphatase abundantly present in taste bud membranes. It also causes elevation of calcium concentration in saliva, an essential step in taste transduction.

Gustatory cells which are located on the tongue respond primarily but not exclusively to sweet and salty modalities, while those cells on the laryngopharynx and the palate perceives principally bitter and sour taste sensation). Lingual papillae that subserve gustatory functions contain group of neuroepithelial and supporting sustentacular cells, which provide, on a weekly basis, the taste cells upon their natural death. A variety of cell surface molecules exist that may play a role in maintaining the structural integrity of the receptors and mediation of transduction though receptor–afferent interaction. These surface molecules include the N-CAM, α-gustducin (a gustatory G protein), serotonin, blood group antigens (A, B, H, and Lewis^{b}), vasoactive intestinal peptide (VIP), cytokeratins, and neural enolases.

It has been proposed that information coding of taste quality utilizes both the labeled-line and across-fiber pattern coding models. In the labeled-line model, an individual taste receptor responds to only a single tastant and is transmitted by distinct single afferent pathway. The across-fiber pattern model is based on the fact that individual taste cells respond to different tastants and taste quality is conveyed to the gustatory cortex by an afferent fiber system with overlapping response spectra. The pattern of activity across all of the afferent nerve fibers, as the name indicates, codes for a particular quality. Since multiple taste qualities are conveyed by the same fiber, a process of pattern recognition must occur, but with a preferential response to a particular taste quality.

GUSTATORY NEURONS AND ASSOCIATED GANGLIA

Neurons (Figure 18.1) that receive taste sensation are unipolar, contained in the geniculate and the inferior ganglia of the glossopharyngeal and vagus nerves. Each afferent fiber divides into branches that connect to several taste buds that are located on different areas of the tongue. Peripheral processes of these neurons run within the chorda tympani, which joins the lingual branch of the trigeminal nerve; the superficial greater petrosal branch of the facial nerve, which runs in the pterygoid canal nerve and distributes via the middle and posterior palatine branches; the lingual branch of the glossopharyngeal nerve; and the internal laryngeal branch of the vagus nerve. It appears that some of these nerves, particularly the facial nerve, exhibit discriminative capability for different taste qualities, a characteristic that glossopharyngeal and internal laryngeal nerves lack. Other information relative to temperature and texture of substances is also conveyed by branches of the trigeminal and glossopharyngeal nerves.

Sweet-tasting stimuli act on the receptors in the anterior two-thirds of the tongue and the soft palate, which receive innervation from branches of the facial nerve. Bitter-tasting stimuli activate receptors innervated by the glossopharyngeal nerve. Central processes of these neurons form the solitary tract, and their terminals synapse in the rostral third of the solitary (gustatory) nucleus in the medulla, whereas the general visceral afferents including input about gastrointestinal motility terminate in the more caudal half of the nucleus. Thus, the integration of general visceral and special visceral afferent (taste) and associated reflexes is achieved at this level.

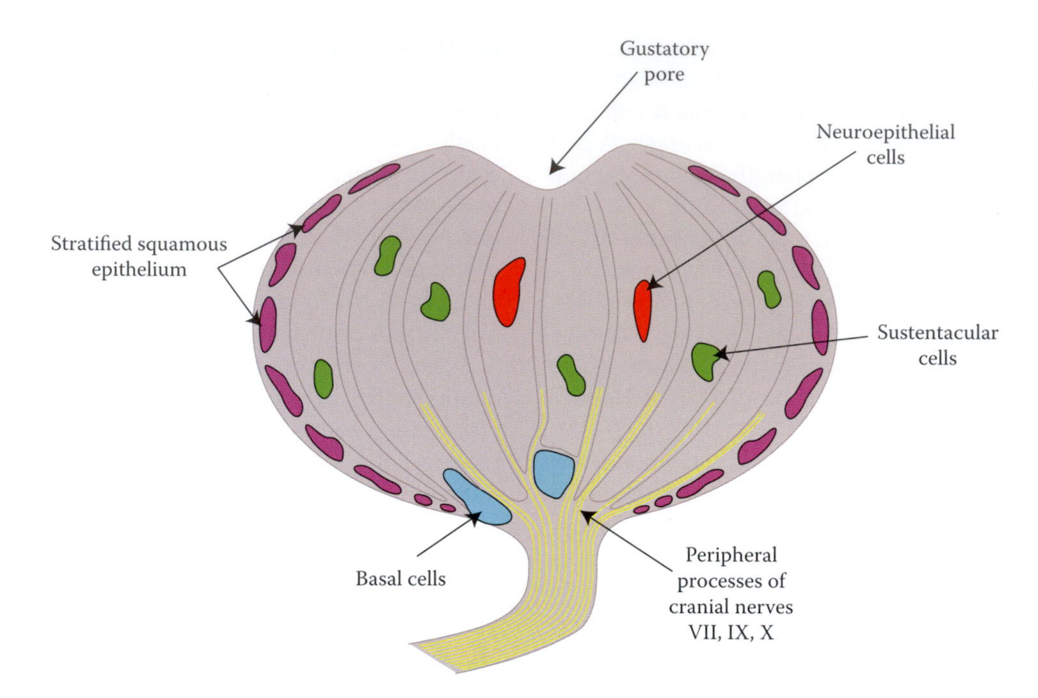

FIGURE 18.1 Schematic drawing of taste bud showing the gustatory pore, sustentacular, basal, and nueroepithelial cells as well as the peripheral processes of the facial (VII), glossopharyngeal (IX) and vagus (X) nerves.

GUSTATORY PATHWAYS

Neurons of the solitary nucleus also establish connections directly or through interneurons with the motor nuclei of the trigeminal and facial and hypoglossal nerves as well as the ambiguous nucleus to mediate reflexes that pertain to ingestion and reaction to bitter- and noxious-tasting substances. Termination within the gustatory nucleus follows an orderly manner in which the facial nerve terminates rostrally, the vagus nerve caudally, and the glossopharyngeal nerve projects to the intermediate part of the nucleus.

Postsynaptic fibers from the gustatory part of the solitary nucleus cross the midline and form the solitariothalamic tract, which ascends as a component of the ventral tegmental tract dorsomedial to the medial lemniscus. This tract, which terminates in the medial part of the ventral posteromedial nucleus of the thalamus, ascends in conjunction with the medial lemniscus. It projects to the parvocellular part of the ventromedial nucleus of the thalamus (accessory arcuate nucleus). From the thalamus, the taste fibers radiate through the internal capsule and project to the primary gustatory cortex located adjacent to the anteroinferior part (tongue area) of the primary sensory cortex and limen insula and around the lateral cerebral fissure (parietofrontal operculum and anterior insular cortex).

A second prominent pathway conveys gustatory afferents to the lateral hypothalamus that regulates feeding and autonomic function, as well as to the bed nucleus of the stria terminalis and the central nucleus of the amygdala to mediate emotional reactions associated with taste sensation. The orbitofrontal cortex also receives gustatory input in conjunction with visual, olfactory, and somatic afferents, which account for the motivation to eat or not to eat. Some taste fibers leave the solitary nucleus to establish synaptic linkages with the ambiguus and hypoglossal nuclei, modulating reflex activities, satiety, and visceral responses to pleasurable and offensive experiences of eating a particular food. There are descending fibers from the insular gustatory cortex to the solitary nucleus that regulate (inhibit or excite) neurons of this nucleus in a manner that provokes appropriate cortical response to taste sensation. Inhibitory descending input is mediated by GABA through $GABA_A$ receptors, whereas excitatory input to the solitary nucleus utilizes substance P, which is inhibited by met-enkephalin.

GUSTATORY DYSFUNCTIONS

Taste disorder can have serious health consequences. Combined loss of taste and smell sensations may be a sign of neurodegenerative diseases such as Alzheimer's and Parkinson's. A distorted gustatory system can be a risk factor for a variety of diseases including heart disease, diabetes, and stroke, as it directly affects eating habits and the desire to consume excessive or extremely limited amounts of food.

Dysgeusia refers to a distortion of the sense of taste, usually a metallic taste, which can be associated with complete loss of taste (ageusia) and decrease in taste sensitivity (hypogeusia). Dysgeusia can occur as a result of a reduction of type III cell microvilli and intracellular vesicles. It can also occur due to chemotherapy (cyclophosphamide, cisplatin, and etoposide), which can cause mucositis in the oral cavity, as well as inhibition of regeneration of new taste buds and salivary secretion. Amiloride directly blocks ENaC and inhibits sodium channels linked to taste receptors. Tetracycline and lithium carbonate leave traces in the saliva that induce metallic flavor. Sulfahydryl groups including penicillamine that cause zinc deficiency can also produce dysgeusia. Angiotensin II receptor antagonists, such as eprosartan, have been linked to dysgeusia. This medication is used for the treatment of hypertension, acts on the renin–angiotensin system, and produces vasodilation and inhibition of the sympathetic neurons.

Other conditions associated with dysgeusia include diabetes mellitus, hypothyroidism, periodontal disease, and asthma. Dysgeusia can occur subsequent to xerostomia (dry mouth syndrome) because of the change in the composition of the saliva, insufficient concentration, and reduced salivary flow. Xerostomia is associated with nasal congestion and mouth breathing and can occur subsequent to increased use of anticholinergic or sympathomimetic drugs or diuretics. Dysguesia can result from damage to individual nerve fibers that convey taste sensation from the tongue and palate, or occurs subsequent to lesions of the gustatory pathways in the pons, mesencephalon, or thalamus.

Close proximity of the gustatory and micturition neurons in the pons and cerebral cortex may possibly explain the dysgeusia seen in individuals with urinary obstruction. Other factors that can lead to dysgeusia also include gastric reflux, lead poisoning, laryngoscopy, tonsillectomy, and exposure to pesticides. Hypogeusia (reduction in gustatory sense) may be associated with Sheehan's syndrome, hypothyroidism, and sarcoidosis. Giant cell arteritis may exhibit gustatory disorders in the elderly.

Phantogeusia refers to a persistent spontaneous abnormal taste with no external stimulus, which is attributed to damage of the afferent fibers subsequent to oral infection or overuse of certain medications. Research shows that some T2R taste receptor genes were increased significantly in patients with phantogeusia, indicating that increased expression of taste receptor genes is involved in the pathogenesis of phantogeusia.

SUGGESTED READING

Bahar A, Dudai Y, Ahissar E. Neural signature of taste familiarity in the gustatory cortex of the freely behaving rat. *J Neurophysiol* 2004;92:3298–308.

Bradbury J. Taste perception: Cracking the code. *PLoS Biol* 2004;2(3):E64.

Cheon BK, Hill DL. Nerve cut induced decrease of chorda tympani terminal fields in the NTS of adult control and sodium-restricted rats. *Chem Senses* 2003;28:A62.

Feng P, Huang L, Wang H. Taste bud homeostasis in health, disease, and aging. *Chem Senses* 2014;39(1):3–16.

Hong JH, Omur-Ozbek P, Stanek BT, Dietrich AM, Duncan SE, Lee YW, Lesser G. Taste and odor abnormalities in cancer patients. *J Support Oncol* 2009;59–64.

King AB, Menon RS, Hachinski V, Cechetto DF. Human forebrain activation by visceral stimuli. *J Comp Neurol* 1999;413:572–82.

Kobayashi M. Functional organization of the human gustatory cortex. *J Oral Biosci* 2006;48(4):244–60.

Leonard NL, Renehan WE, Schweitzer L. Structure and function of gustatory neurons in the nucleus of the solitary tract. IV. The morphology and synaptology of GABA-immunoreactive terminals. *Neuroscience* 1999;92:151–62.

Lundy RF Jr, Contreras RJ. Gustatory neuron types in rat geniculate ganglion. *J Neurophysiol* 1999;82:2970–88.

Malaty J, Malaty IA. Smell and taste disorders in primary care. 2013;88(12):852–9. Review.

Ohishi Y, Komiyama S, Shiba Y. Predominant role of the chorda tympani nerve in the maintenance of the taste pores: The influence of gustatory denervation in ear surgery. *J Laryngol Otol* 2000;114:576–80.

Pittman DW, Contreras RJ. Dietary NaCl influences the organization of chorda tympani neurons projecting to the nucleus of the solitary tract in rats. *Chem Senses* 2002;27:333–41.

Pritchard TC, Macaluso DA, Eslinger PJ. Taste perception in patients with insular cortex lesions. *Behav Neurosci* 1999; 113:663–71.

Saito H. Gustatory otalgia and wet ear syndrome: A possible cross-innervation after ear surgery. *Laryngoscope* 1999;109:569–72.

Sbarbati A, Crescimanno C, Benati D, Osculati F. Solitary chemosensory cells in the developing chemoreceptorial epithelium of the vallate papilla. *J Neurocytol* 1998;27:631–5.

Scott TR, Giza BK. Issues of gustatory neural coding: Where they stand today. *Physiol Behav* 2000;69:65–76.

Shuler MG, Krimm RF, Hill DL. Neuron/target plasticity in the peripheral gustatory system. *J Comp Neurol* 2004;472: 183–92.

Streefland C, Jansen K. Intramedullary projections of the rostral nucleus of the solitary tract in the rat: Gustatory influences on autonomic output. *Chem Senses* 1999;24:655–64.

Section VII

General Somatic Sensations

19 Cortical and Subcortical Sensory Systems

The ascending pathways convey conscious (cortical) and unconscious (subcortical) sensory information to the higher levels of the central nervous system. These pathways are concerned with transmission of a variety of sensory modalities, regulation of muscle tone and mediation of intersegmental reflexes. They may exhibit monosynaptic connections or utilize an extensive network of neurons. The modalities of the general somatic sensations include pain, temperature, tactile, joint, vibration, and pressure sensations. These modalities are categorized into epicritic (discriminative) and protopathic sensations. The *epicritic* (discriminative) modalities include sensations such as fine touch, two-point discrimination (the ability to distinguish two blunt points from one another), joint sensation, and vibratory sense. These sensations are received by the encapsulated receptors and transmitted by the thickly myelinated and fast conducting fibers. The *protopathic* sensations, which include pain, temperature, and crude touch, are received by the free nerve endings (uncapsulated receptors) and conveyed to the spinal cord by small, thinly myelinated and/or unmyelinated fibers. These fibers represent the peripheral processes of the pseudounipolar neurons of the dorsal root ganglia (DRG), which are contained in the peripheral nerves. Visceral pain also projects to the cerebral cortex; however, its peripheral transmission is maintained by fibers that predominantly accompany the sympathetic fibers. The central transmission of this modality is identical to the somatic sensations. Subcortical sensations emanate from Golgi tendon organs, muscle spindle, and other exteroceptive receptors, conveying information to the cerebellum, inferior olivary nucleus, and tectum via the spinocerebellar, trigeminocerebellar, spino-olivary, and spinotectal tracts.

Ascending sensations utilize the peripheral and central processes of the neurons of the spinal dorsal root and the cranial nerve ganglia to convey impulses from receptors to the spinal cord and brainstem, respectively. Activated spinal neurons that transmit cortical sensations relay in the ventral posterolateral thalamic nucleus, whereas the neurons that conduct impulses from spinal gray columns to the cerebellum, tectum, and inferior olivary nuclei form direct pathways to the site of terminations (subcortical sensations). Sensory impulses received by brainstem sensory nuclei from cranial nerves project to the sensory cortex via relay in the ventral posteromedial thalamic nucleus or to the cerebellum directly without intervening relay nuclei.

SENSORY FIBERS

Peripheral nerve fibers that transmit sensory impulses are classified by *Erlanger and Gasser* on the basis of the fiber diameter, rate of conduction, and degree of myelination into groups A, B, and C fibers. Olfactory and optic nerves are not included in this classification. The "A" fibers, the largest and the fastest in conduction, are classified into Aα, Aβ, Aγ, and Aδ fibers. Each group of fibers, with a few exceptions, is composed of efferent and afferent fibers. Afferents are confined to groups A and C. Components of group "A" afferents include Aα, Aβ, and Aγ. Group Aα consists of sensory fibers from the muscle spindles, which terminate in the deeper layers of the dorsal horn and also in the ventral horn; group Aβ fibers carry sensations from the Golgi tendon organs (GTOs) and Meissner's and Pacinian corpuscles ending in laminae III–VI. Class Aδ fibers carry nociceptive and visceral stimuli to laminae I, IV, and V. Class C fibers are nonmyelinated and considered to be the slowest in conduction, conveying nociceptive and olfactory stimuli. *Efferents* are classified into classes A, B, and C group fibers. Group A includes Aα, Aβ, and Aγ. Class Aα fibers supply fast-twitching extrafusal muscles, Aβ fibers form the plaque endings of the muscle spindle fibers, and collaterals of the Aα axons and Aγ fibers constitute the plate and trail endings of intrafusal muscles. Class B fibers comprise the preganglionic sympathetic and parasympathetic fibers. Group C fibers are nonmyelinated and of small-caliber fibers, which convey postganglionic autonomic fibers.

Lloyd also classifies sensory fibers of the peripheral nerve into *groups I, II, III, and IV. Group I* fibers consist of the primary sensory Ia fibers associated with the muscle spindle (equivalent to Aα) and Ib fibers that convey impulses from GTOs. The *Ia group* consists of thickly myelinated and fast conducting fibers that receive information from the muscle spindle and establish excitatory monosynaptic connections with the spinal motor neurons of the synergists and inhibitory connections with the antagonists. They are sensitive to stretch stimuli, mediating the monosynaptic myotatic reflex. Tapping a deep tendon (e.g., patellar tendon) with a reflex hammer produces a stretch in the attached muscle, activating the Ia fibers and subsequently the spinal motor neurons. Activation of the spinal motor neurons results in contraction of the innervated muscle. The *Ib* afferents consist of smaller diameter fibers, which mediate the *inverse myotatic* (clasp-knife) *reflex* and convey inhibitory impulses to the synergist motor neurons of the spinal cord. It has been shown that interneurons in the reflex pathways of Ib afferents receive short-latency excitation from low threshold cutaneous and joint afferents. When a limb movement is initiated and suddenly meets an obstacle, inputs from cutaneous and joint receptors will trigger the Ib afferent inhibitory system, producing reduction in the muscle tension. *Group II* fibers include afferents from the pacinian corpuscles, hair follicles, and secondary fibers from muscle spindles. These fibers are inhibitory and play a major role in protecting a muscle when it experiences undue tension. *Group III* (comparable to Aδ) and *group IV* (comparable to C fibers) fibers carry nociceptive stimuli, including the free nerve endings in the walls of the blood vessels and in the follicles of the finer hair.

SENSORY RECEPTORS

All sensory systems use receptors to transduce various forms of energy into neuronal activity, which, if of sufficient intensity, results in the generation of nerve impulses whose frequency and pattern may determine the sensory experience. Receptors convey the generated stimuli, through a series of neurons—primary, secondary, or tertiary—to the cortical or subcortical areas of the central nervous system. For the most part, the type of modality of sensation mediated by a particular axon is specific for each axon, and furthermore, not all activities in the sensory receptors are consciously perceived. Receptors, activated by one or more modalities, are classified into mechanoreceptors, chemoreceptors, photoreceptors, and osmoreceptors. The mechanoreceptors convey tactile, pressure, and auditory impulses, whereas chemoreceptors are concerned with taste, smell, and chemical changes, such as the fluctuation in levels of carbon dioxide. Photoreceptors, on the other hand, react to electromagnetic waves, whereas osmoreceptors are sensitive to the changes in osmotic pressure. Regulation of blood flow and pressure is accomplished by receptors in the adventitia of the blood vessels. Receptors are also classified according to their areas of distribution into exteroceptors, proprioceptors, and interoceptors. Exteroceptors are superficial receptors that respond to general cutaneous stimuli or to special sensory stimuli. General exteroceptors are represented by the free nerve endings and encapsulated terminals. Exteroceptors also respond to special sensory stimuli such as the olfactory, visual, acoustic, and taste sensations. Receptors that respond to stimuli associated with the direction, position of the body, and extent of movement, such as the GTO, muscle spindle, Pacinian corpuscles, and vestibular receptors, are known as proprioceptors. Proprioceptors maintain balance and control muscle contraction and grading. Receptors, primarily free nerve endings, located in the walls of the viscera and the adventitia of blood vessels are known as interoceptors. They show sensitivity to excessive tension and stretch of the visceral wall.

General sensory receptors are also categorized into capsulated and encapsulated receptors. The *encapsulated receptors* consist of the muscle spindles, GTOs, pacinian and Messiness' corpuscles, and Ruffini endings. Less distinct, and not totally recognized by anatomists, are the Krause'e end bulb and Golgi-Mazzoni corpuscles.

Muscle (neuromuscular) spindles are numerous in the intrinsic hand muscles, antigravity axial muscles, and flexor muscles of the upper extremity and extensors of the lower extremity (Figures 19.1 and 19.2). They are few in numbers (4–16) in the

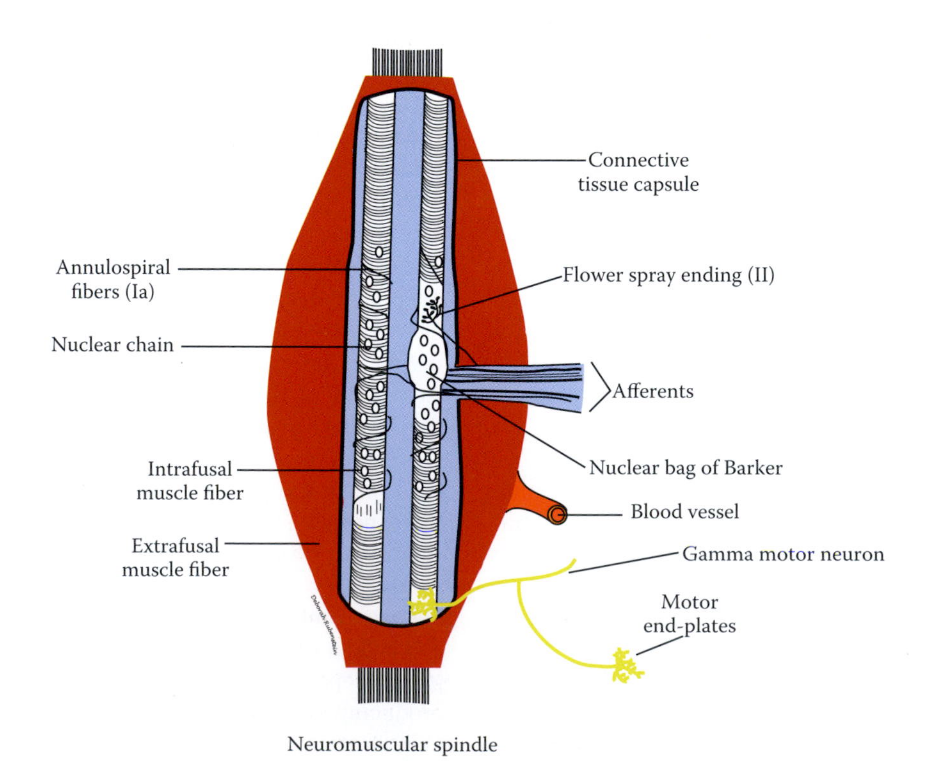

FIGURE 19.1 This diagram illustrates the two forms of the intrafusal muscle fibers, Ia and II afferents, and the terminals of the gamma fibers.

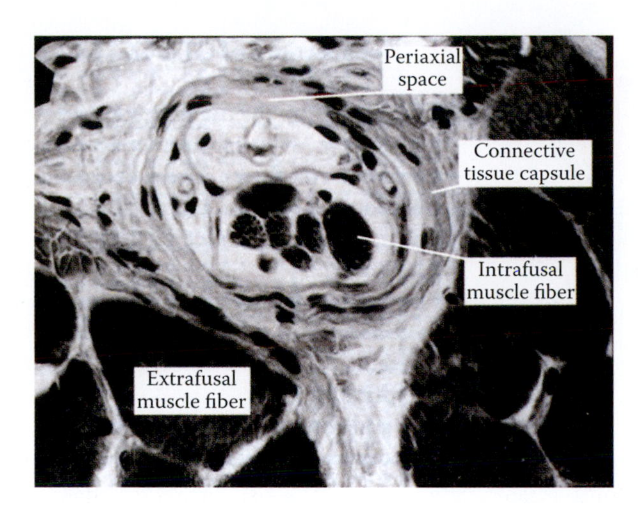

FIGURE 19.2 Photomicrograph of a section through the muscle spindle. The connective tissue capsule and intrafusal muscle fibers are clearly visible.

muscles that contain predominantly fast glycolytic (white) fibers. Each muscle spindle contains a few specialized intrafusal muscle fibers surrounded by an external capsule that consists of fibroblasts and collagen and an internal capsule that invests individual fibers as tubes. The external and internal capsules are separated by the gelatinous substance formed by glycosaminoglycans. The intrafusal muscle fibers are classified based on the arrangement of the nuclei in their equatorial sarcoplasm into nuclear bag (nuclei are equatorial and form protrusion) and nuclear chain fibers (single row of axially oriented nuclei). Differences are also exemplified in the fact that nuclear bag fibers appear to be thicker, extending to the extrafusal fibers compared to the relatively thinner nuclear chain fibers that attach to the polar zones. On histochemical, ultrastructural, and physiologic basis, the nuclear bag fibers are further categorized into dynamic nuclear bag$_1$ (lack M lines and elastic fibers, contain scanty sarcoplasmic reticulum and ATPase enriched with mitochondria and oxidative enzymes) and static bag$_2$ (exhibit prominent M lines, elastic fibers around the polar region, high level of ATPase, and lower oxidative enzymes). The muscle spindle receives two types of thickly myelinated fibers: primary annulospiral (Ib) and secondary flower spray (type II) endings. The primary annulospiral fibers spiral around the equators of the intrafusal fibers and are rapidly adapting, whereas the secondary flower spray endings are slowly adapting, respond to stretch, spread in a beaded manner, and are associated with the nuclear chain fiber.

The muscle spindle acts as a length detector excited in response to stretch, mediating the monosynaptic myotatic reflex. This passive stretch may be produced when a reflex hammer strikes the patellar ligament. It receives γ efferents as well as Ia and II afferents. Group Ia and II fibers that convey impulses generated by passive stretch of a muscle establish excitatory synaptic contacts upon the spinal α motor neurons. The rate of firing of the Group Ia fibers depends upon the rate of stretch; the more rapid the stretch, the more impulses generated (phasic response). The plate ending motor neurons intensify the latter response. Impulses generated by the secondary (II) fibers depend upon the extent of maintained stretch (tonic response). Tonic response correlates with the intensity of the trail ending motor neurons. Excitation of α motor neurons produces contraction of the quadriceps muscle and extension of the leg, which is accompanied by inhibition of α motor neurons of the antagonist muscles via the inhibitory synaptic connections of the Ia and II fibers.

The *Golgi tendon organs* are found at the junction of a muscle and its tendon, mostly in the tendinous part. They consist of small loosely arranged bundles of collagenous fibers (intrafusal fasciculi) that parallel the extrafusal muscle fibers. They receive unmyelinated nerve terminals and are enclosed by a capsule that consists of concentric cytoplasmic sheets. The capsule receives one or several thickly myelinated Ib fibers that ramify and form leaf-like configuration, rich in vesicles and mitochondria. GTOs are activated by excessive tension placed on the muscle as a result of active contraction or passive stretch. They are slow adapting high threshold (type III) endings that provide cortical and subcortical proprioceptive information to the cerebral cortex and cerebellum, respectively. They reflexively inhibit the development of excessive tension, mediating inverse myotatic reflex. In contrast, a decrease in muscle tension, as a result of fatigue, produces the opposite effect of reducing activation of these receptors. The signals from GTOs are carried by the Ib fibers to the spinal cord, establishing disynaptic connections with neurons of the agonist and antagonist muscles. Central connections of GTOs are studied later in this chapter.

The largest of all receptors are the Pacinian corpuscles (Figure 19.3) that consist of a capsule, concentric lamella separated by fluid that adds to their turgidity, and terminals of the thickly myelinated Aα fibers. They are abundant in the genital organs, mesentery, nipple, periosteum, ligaments, deep part of joint capsules, palmar aspect of the hand and digits, pancreas, and mesenteries. These receptors (type II endings), activated by vibration, tension, and movement, as well as sudden stress-related changes, are considered rapidly adapting and low threshold mechanoreceptors.

The tactile corpuscles of Meissner (Figure 19.3) are a group of receptors that consist of a central core surrounded by a capsule, receiving terminals of tactile sensory nerve endings. They are found in the glabrous skin of the hand and foot, lips, prepuce, anterior forearm, palpebral conjunctiva, and tongue of infants and young adults. These receptors, which respond best to movement across a textured surface, as when touching cloth or reading Braille, usually disappear with aging.

Ruffini endings (Figure 19.3) are cylindrical sensory organs that respond to pressure and warmth, as well as to simultaneously applied two blunt points. They are type II slowly adapting mechanoreceptors, each consisting of a granular matrix surrounded by a thin capsule, and pierced by Aβ nerve fibers. These receptors are found in the glabrous and hairy skin (dermal stretch receptors), gland penis, joint capsules (provide awareness of joint position and movement),

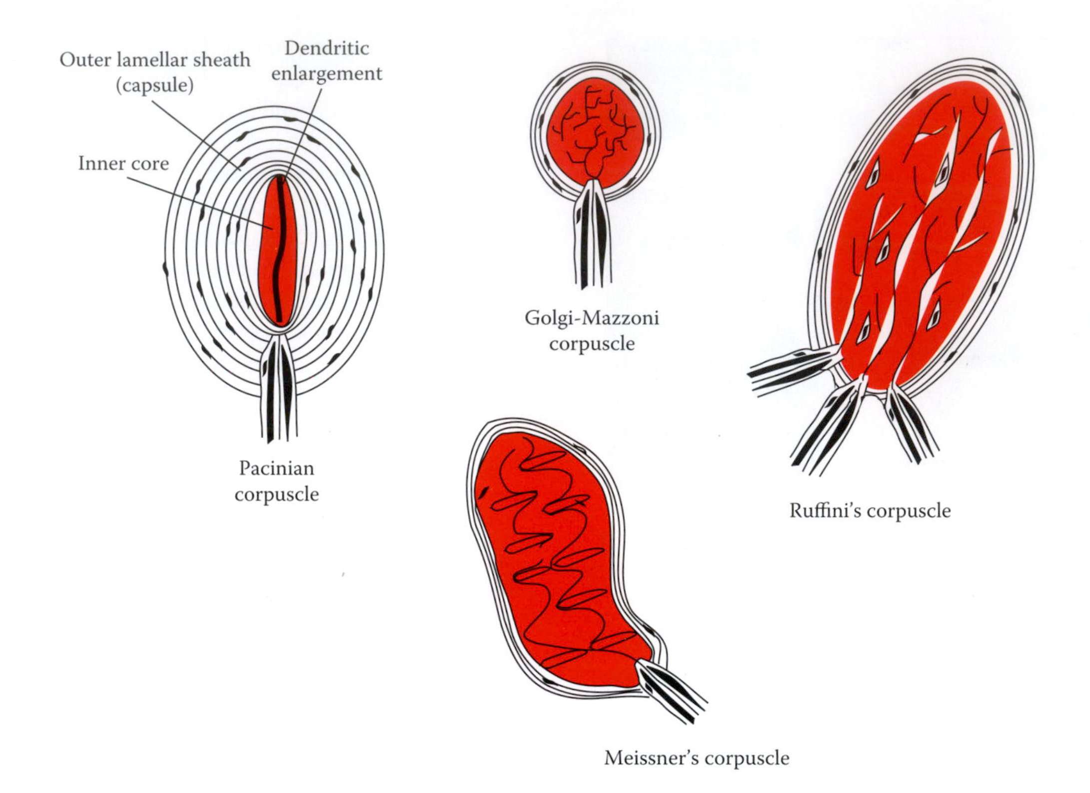

FIGURE 19.3 Various forms of capsulated and noncapsulated receptors.

and tendinous insertions activated upon grasping heavy objects, carrying a luggage, and when scratching.

The nonencapsulated receptors are composed of the free nerve endings, and Merkel's disks. The free nerve endings are sensory endings that branch to form a plexus and are abundant in all connective tissue including the epidermis, cornea, mucus membranes, meninges, joint capsules, fasciae, Haversian canal, oral cavity, respiratory tract, and tooth pulp. These receptors that carry primarily pain and thermal sensations receive thinly myelinated or unmyelinated type IV fibers and respond to light mechanical tactile, thermal stimuli, and articular pain.

Merkel's disks (tactile menisci) are slowly adapting epidermal cells in close apposition to expanded (bowl-shaped) nerve terminals of Aβ (type II) fibers. They are located between the epidermal keratinocytes or bulbar epithelium of hair follicles and are activated by tactile sensations. They are receptors that respond to bending hairs and to vertical indentation of the skin. Merkel's disks are considered free nerve endings when not engaged in synaptic linkage subserving trophic function.

As discussed earlier, sensory systems utilize series of neurons, in an orderly fashion, to convey impulses to higher centers in the central nervous system. The *first-order* neurons, for all general somatic afferents from the body, are located in the dorsal root ganglia (DRG) of the spinal nerves. The second-order neurons may be positioned in the gray matter of the spinal cord or in the brainstem, whereas the third-order neurons are formed by the neuronal cell bodies of the ventral posterolateral (VPL) nucleus of the

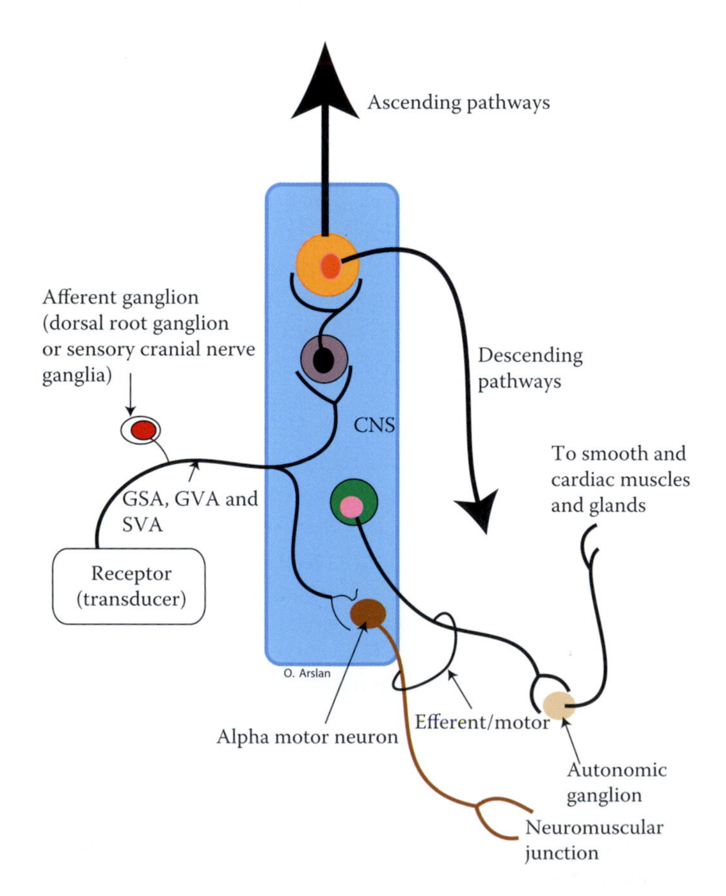

FIGURE 19.4 In this schematic diagram, the peripheral course of the somatic and visceral sensory fibers and their central connections are explored. Relevant reflex connections and somatic and visceral innervation are also depicted.

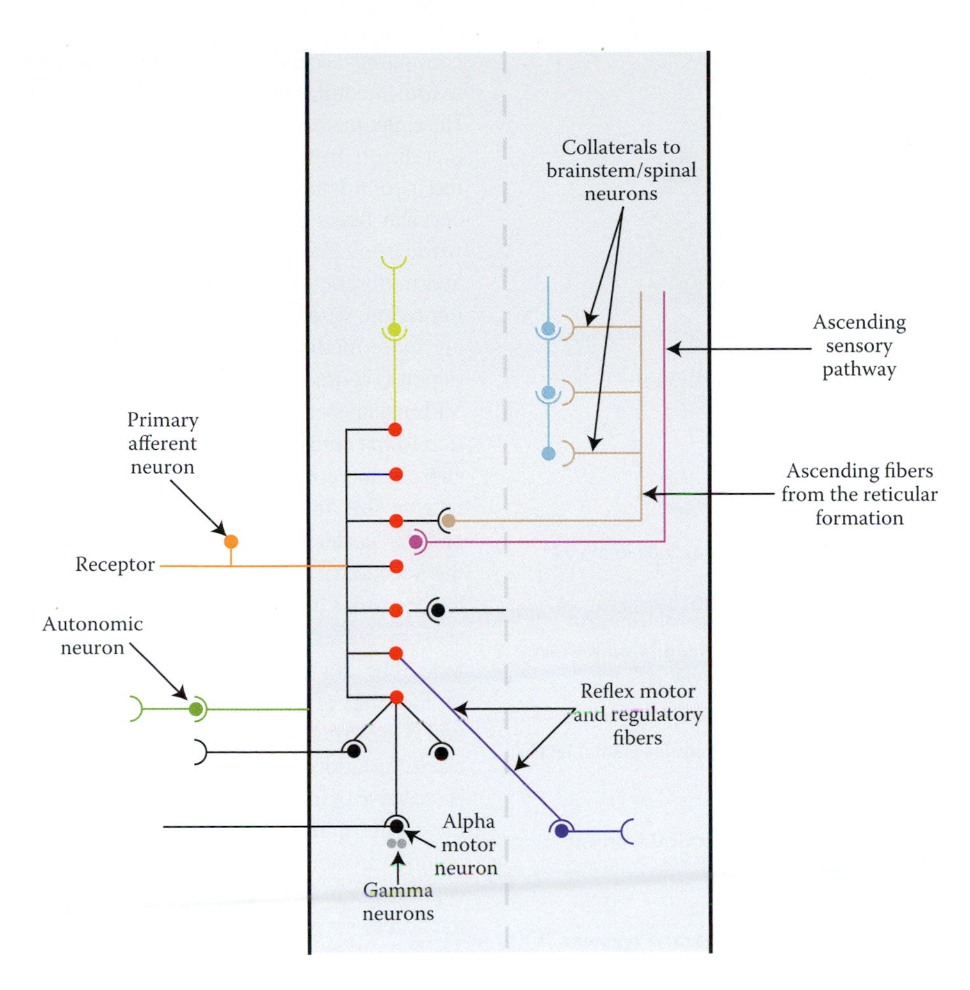

FIGURE 19.5 This diagram traces the connection of the primary afferent neurons with spinal neurons. The role of these neurons in the mediation of reflex activity on the ipsilateral and contralateral side as well as in the formation of ascending sensory pathways is illustrated.

thalamus (Figures 19.4 and 19.5). There are numerous pathways that convey general somatic afferents to the cerebral cortex (ascending cortical pathways). These pathways ascend in the dorsal, lateral, and anterior funiculi of the spinal cord and carry joint, tactile, vibratory, nociceptive, and thermal sensations. They are mainly crossed pathways and project to thalamus and then to the cerebral cortex. Cortical pathways are composed of the dorsal column–medial lemniscus, spinothalamic tracts, and the trigeminal lemnisci. Pathways that subserve special cortical sensations such as visual and auditory have been described in previous chapters.

CORTICAL SENSATIONS FROM THE BODY

As discussed above the sensory modalities that emanate from receptors in the body and destined to the cerebral cortex are transmitted via the dorsal white columns and anterolateral system, whereas those sensations that derive from the head are conveyed via the trigeminal lemniscus.

The *dorsal white columns* (Figures 19.6, 19.7, 19.8, and 19.9) transmit fine tactile sensation, vibratory sense, position and movement sense (kinesthesia), two-point discrimination of simultaneously applied blunt pressure points, and stereognosis (ability to recognize form, size, texture, and weight of objects). The receptive fields for this pathway are small, primarily scattered in the fingers and tongue, and less numerous in the skin of the back. These sensations are conveyed by the thickly myelinated, fast conducting fibers, which represent the central processes of the pseudounipolar neurons of the DRG that occupy the medial portion of the dorsal roots, bifurcating into ascending and descending branches within the spinal cord. The descending branches of the dorsal root fibers form the fasciculus interfascicularis (comma tract of Schultze) in the cervical and thoracic spinal segments and the fasciculus septomarginalis (located near the posterior median septum) in the lumbar and sacral spinal segments. The long ascending branches travel ipsilaterally in the posterior funiculus and are somatotopically organized with the most medial fibers conveying discriminative sensations from the lower extremities and lower half of the trunk and extending the entire length of the spinal cord as the gracilis fasciculus. Despite the modality specificity, the fasciculus gracilis also contains fibers from cutaneous receptors of the lower extremity and fibers from stretch receptors derived particularly from the second and third lumbar spinal segments. Virtually all fibers from proprioceptive receptors, conveying position and movement senses from the lower extremity and lower half of the body, travel within the gracilis fasciculus. Long fibers

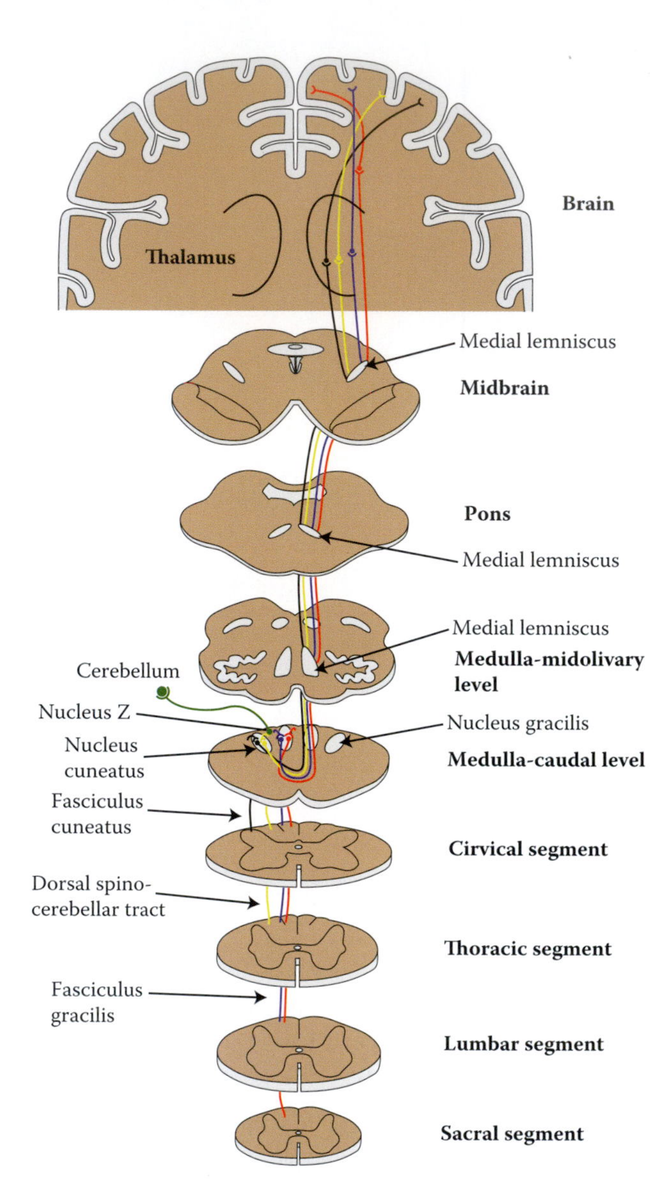

FIGURE 19.6 This diagram shows the course of most of the fibers that carry conscious proprioception from the lower extremity via the dorsal column–medial lemniscus pathway.

of this fasciculus terminate in the gracilis nucleus. Shorter fibers, which constitute the majority of fibers, synapse in the Clarke's nucleus and then in the nucleus Z of Brodal and Pompeiano, before reaching the upper cervical segments and continuing through the internal arcuate fibers to the thalamus (Figures 19.6 and 19.7).

Since proprioceptive fibers from the lower extremity leave earlier, the gracilis fasciculus in the cervical spinal segments contain, for the most part, cutaneous impulses, terminating in the nucleus gracilis. Fibers that transmit joint and fine tactile sensations from the upper extremities and upper half of the trunk are positioned lateral to the fasciculus gracilis and form the fasciculus cuneatus. Fasciculus cuneatus extends in the cervical and upper six thoracic spinal segments, terminating in the ipsilateral cuneate nucleus. Postsynaptic axons from cuneate and gracilis nuclei and nucleus Z constitute the internal arcuate fibers, which decussate in the caudal medulla, continuing ascendingly on the contralateral side through the rostral medulla, pons, and midbrain as the medial lemniscus. Here, the medial lemniscus exhibits a somatotopic arrangement that differs from the spinal level (Figure 19.10). In the medulla, the medial lemniscus assumes a vertical position, in which the cervical fibers are located dorsally, the sacral fibers ventrally, while trunk fibers maintain an intermediate position. In the pons and midbrain, the medial lemniscus shifts to a horizontal configuration, where the cervical fibers are lateral, the sacral fibers are more medial, and the trunk fibers occupy an intermediate region (Figure 19.10). The medial lemniscus terminates in the VPL nucleus of the thalamus. The VPL then relays this information to the cerebral cortex via the posterior limb of the internal capsule and corona radiata. Areas of the cerebral cortex, which receive this information, include the primary somatosensory and the posterior insular cortices of the contralateral side and the secondary somatosensory cortices of both sides. The posterior insular cortex receives input from the secondary somatosensory cortex and Brodmann's areas 5 and 7 of the parietal lobe. The dorsal column–medial lemniscus pathway operates on the basis of a one-to-one synapse ratio and lateral inhibition and is governed by the regulatory influence of the corticospinal tract. Variations in the sensory experiences when the individual is attentive or alert may be attributed to the influences exerted by the corticospinal tract. Rostral to the eight cervical spinal segments, the cuneatus fasciculus transmits unconscious proprioception fibers from the upper extremities that terminate upon the accessory cuneate nucleus. This nucleus forms the ipsilateral cuneocerebellar tract (will be discussed shortly with the subcortical sensory pathways).

Phylogenetically older fibers of the dorsal column pursue a deeper course, synapse in the most peripheral neurons of the medullary nuclei of the dorsal column and ascend within the medial lemniscus. These fibers establish synaptic contacts with the reticular formation and eventually become part of the spinoreticulothalamic pathway. This connection may explain the pain sensation felt upon electrical stimulation of the dorsal column–medial lemniscus pathway.

Dorsal columns show degeneration in *tabes dorsalis* (Figure 19.11), a slowly progressive condition that is commonly seen as a late manifestation of neurosyphilis and also in diabetes mellitus. It results from degeneration of the dorsal root fibers of the lumbosacral and sometimes cervical spinal nerves and the posterior white columns. Patients with tabes dorsalis may present with early signs of incontinence (atonic bladder) and sexual impotence. They may also experience, formication, bouts of lightning pain, identical in severity to trigeminal neuralgia, particularly in the eyes (tabetic ocular crises). Pain may not be uniform and may be reduced or lost later in certain regions. Argyll–Robertson pupil, in which a patient's eye can accommodate far vision but remains unresponsive to light, is another feature of the disease.

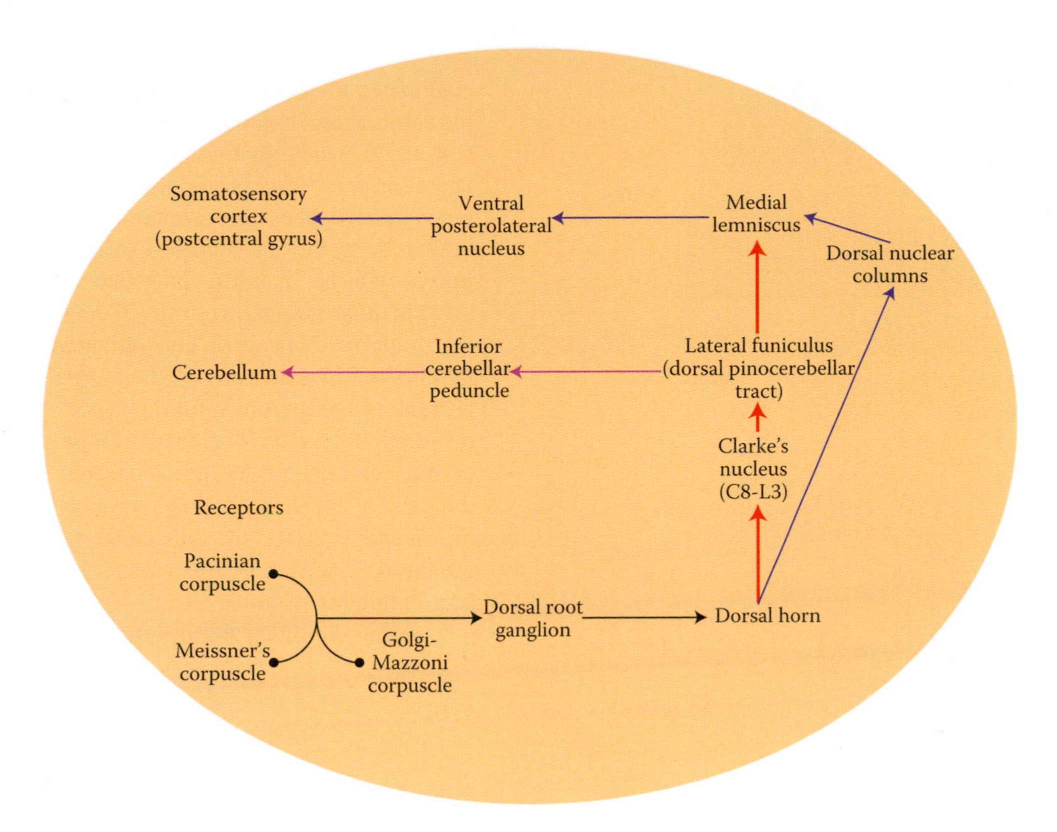

FIGURE 19.7 This is a more simplified schematic drawing of Figure 19.6. Dorsal column–medial lemniscal pathway and its association with the dorsal spinocerebellar tract in the transmission of proprioceptive fibers from the lower extremity are shown. Most of the fibers that convey proprioceptive impulses from the lower extremity travel in the dorsal spinocerebellar tract before terminating in the nucleus Z.

Due to the loss of position and vibratory senses in the lower limb, patients may exhibit positive Romberg's sign and a high steppage-broad gait, compensating for the loss of proprioception from the feet. *Romberg's sign* may be utilized to determine the patient's ability to maintain balance while standing in an erect position. The patient's feet must be approximated with his/her eyes gazing straight ahead and arms by his/her sides. Some degree of swaying is possible; however, tendency to fall must be reassessed by repeating the test. Anxiety-oriented individuals and hysterical patients are naturally prone to test positive. Patients usually fall down in poorly illuminated areas due to the lack of visual cues as well as joint sensation. This test may be repeated with eyes closed to exclude the utilization of visual cues in maintaining equilibrium and preventing the patient from falling. Degenerative changes may cause swelling and pain in the joints, loss of sensation from the bladder, and optic atrophy. Individuals also exhibit loss of two-point discrimination (healthy individuals should be able to distinguish two blunt points applied simultaneously to the finger tips with a range of 4–5 mm). It is important to note that in cerebellar lesions, the swaying is very marked, occurring with open or closed eyes.

Note that in some patients with severely damaged dorsal columns, joint and vibratory senses and the two-point discrimination remain preserved. This may be explained on the basis that the posterior spinocerebellar tract transmits joint sensation from the lower extremity. Vibratory sensation may be shared by the spinothalamic tract. The position sense from the upper extremity may be mediated via collaterals of the primary afferents (Ia fibers), which are given to the spinothalamic tract.

The *anterolateral system* (Figures 19.12 and 19.13) is a morphological designation for the combined lateral, ventral spinothalamic and spinoreticulothalamic tracts. The *lateral spinothalamic tract* (neo-spinothalamic-lateral system) is a crossed pathway, which conveys thermal and painful sensations from somatic and visceral structures. Painful and thermal stimuli activate the free nerve endings and generate impulses that enter the spinal cord via the lateral bundle of the dorsal root as C and Aδ fibers. Group C fibers, which carry slow, diffuse, and aching (visceral) pain, as well as thermal sensations, are unmyelinated and terminate primarily in the substantia gelatinosa (lamina II), establishing indirect contacts with neurons of the anterolateral system. The Aδ fibers are thinly myelinated, convey sharp and fast pain in addition to cold stimuli, and establish synapse with neurons of spinal laminae I–V. These two types of fibers may be inhibited by stimulation of the adjacent thickly myelinated and fast conducting (Aβ) fibers. Organic visceral sensations signal a bodily need such as hunger, thirst, libido, nausea,

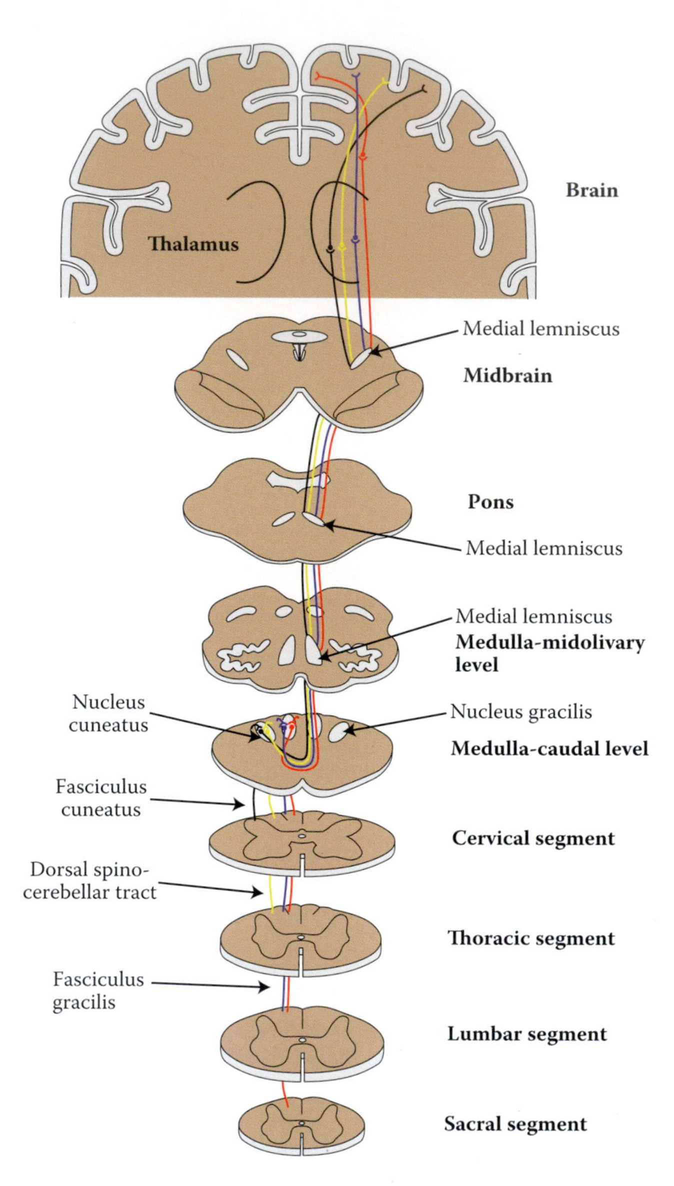

FIGURE 19.8 Diagram illustrates the course of the discriminative sensations from the upper extremity and upper half of the trunk and discriminative cutaneous sensations from the lower extremity. Note the ipsilateral course of this pathway in the spinal cord and contralateral course at supraspinal levels.

and bladder and bowel fullness and are generally conveyed to the spinal cord via the parasympathetic fibers.

Transmission of pain occurs as a result of imbalance between the afferent impulses that reach the spinal gray matter in the presence of massive afferents through the thinly myelinated fibers. The *gate theory* of analgesia and the rationale behind using transcutaneous electrical nerve stimulation (TENS) is based upon the inhibitory role that thickly myelinated and large-diameter fibers (from tactile corpuscles) exert, when stimulated, upon the unmyelinated C and thinly myelinated Aδ fibers. The large diameter and thickly myelinated fibers are excitatory to the interneurons of the substantia gelatinosa and similarly to the T cells in lamina IV, which give rise to the fibers of the anterolateral system. In contrast, the thinly myelinated and unmyelinated fibers are excitatory to the T cells and inhibitory to the interneurons of the substantia gelatinosa, which, in turn, presynaptically inhibit the afferents that synapse with the T cells. Stimulation of the thinly myelinated or unmyelinated afferent fibers inhibits the interneurons of the substantia gelatinosa, preventing them from inhibiting the T cells and thus opening the gate for nociceptive impulses that ascend to the ventrolateral thalamic nucleus. Excitation of the thickly myelinated fibers produces excitation of the interneurons of the substantia gelatinosa, which inhibit the T cells in lamina IV and thus pain transmission. Inhibition of transmission is aided by supraspinal descending inhibitory pathways. TENS is employed in the treatment of intractable pain associated with rheumatoid arthritis and postherpetic neuralgia.

Neuronal sensitization is associated with transmission of noxious stimuli that seems to occur at the peripheral terminals and at the central synaptic sites in the dorsal horn of the spinal cord. At the peripheral or receptor endings, increased sensitivity does not only occur at the traditional nociceptor endings but also at those mechanoreceptors and thermoceptors that are activated subsequent to inflammatory process. These receptors remain silent until a traumatic injury or inflammatory mechanism is initiated, and they are recruited by chemical mediators such as bradykinin, prostaglandins, serotonin, cytokines, tumor necrosis factor (TNF), and nerve growth factor (NGF). At the synaptic sites in the dorsal horn, processing of the nociceptive information is accomplished by lowering the threshold of spinal neurons that receive the afferent through prolonged repetitive facilitatory effect of incoming afferent fibers after a brief stimulation. The latter mechanism accounts for hyperalgesia observed after tissue injury. Lowering the threshold in central sensitization involves certain excitatory neurotransmitters, such as glutamate and aspartate, and the N-methyl-D-aspartate (NMDA) receptors. These receptors come to action upon repetitive stimulation of the unmyelinated fibers allowing calcium to enter the postsynaptic sites and initiate the activation of the nitric oxide (NO), an important neurotransmitter associated with lowering neuronal threshold and maintaining hyperexcitable state. The above facts about central sensitization and neuronal plasticity may enable the development of medications that address the issue of hyperalgesia at the spinal level and hyperexcitable neurons and thus maintain normal signaling pattern.

The nucleus raphe magnus in the medulla, the neurons of the medial reticular zone, and the periaqueductal gray matter constitute a system that provides endogenous analgesia. The nucleus raphe magnus and the medial reticular zone contain

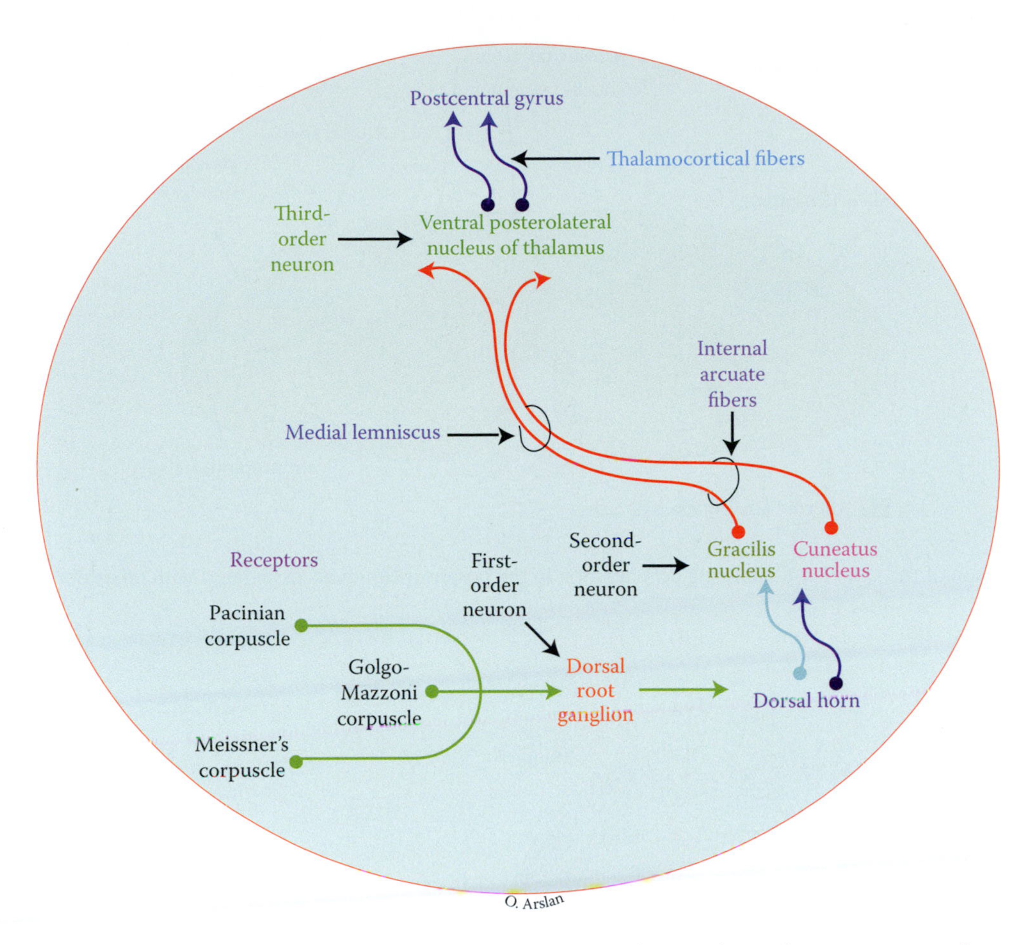

FIGURE 19.9 This is a simplified diagram for the dorsal column–medial lemniscus pathway that conveys conscious proprioception and discriminative cutaneous sensations from the upper extremity and discriminative cutaneous sensations from the lower extremity as illustrated in Figure 19.8.

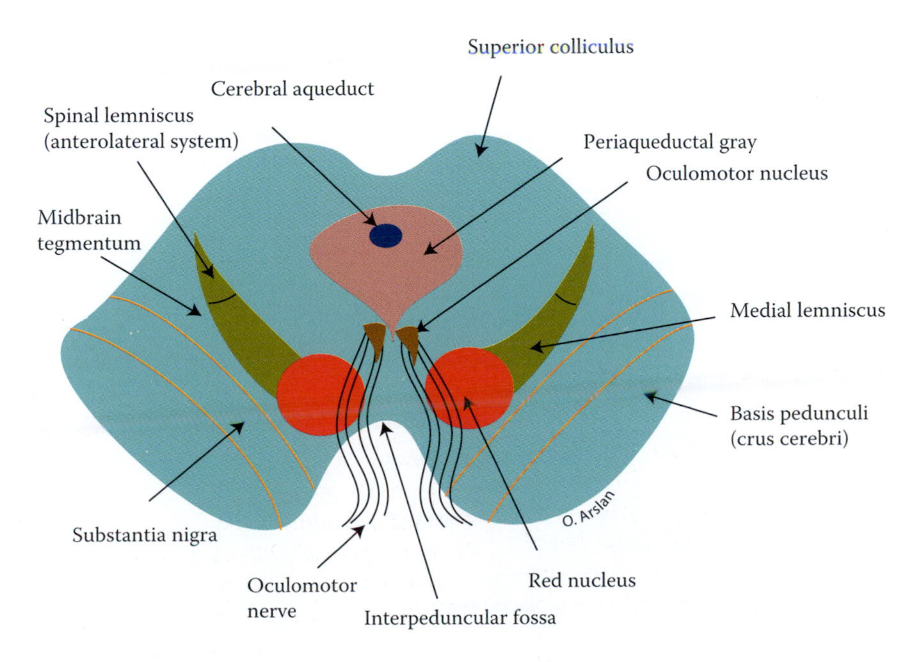

FIGURE 19.10 This diagram illustrates the somatotopic arrangement of the fibers within the medial lemniscus in which the lower extremity fibers occupy lateral and dorsal positions to the fibers that emanate from the upper extremity.

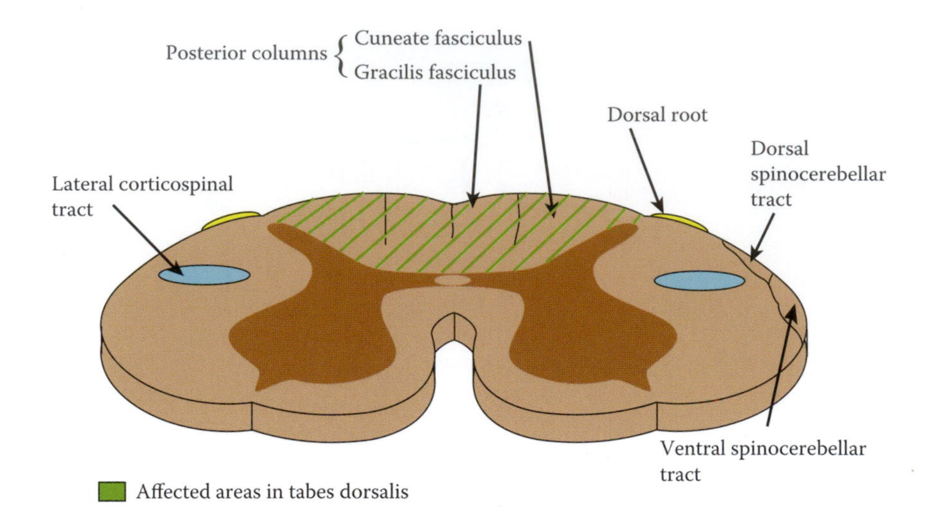

FIGURE 19.11 Section of the spinal cord showing degeneration in the posterior funiculus associated with tabes dorsalis.

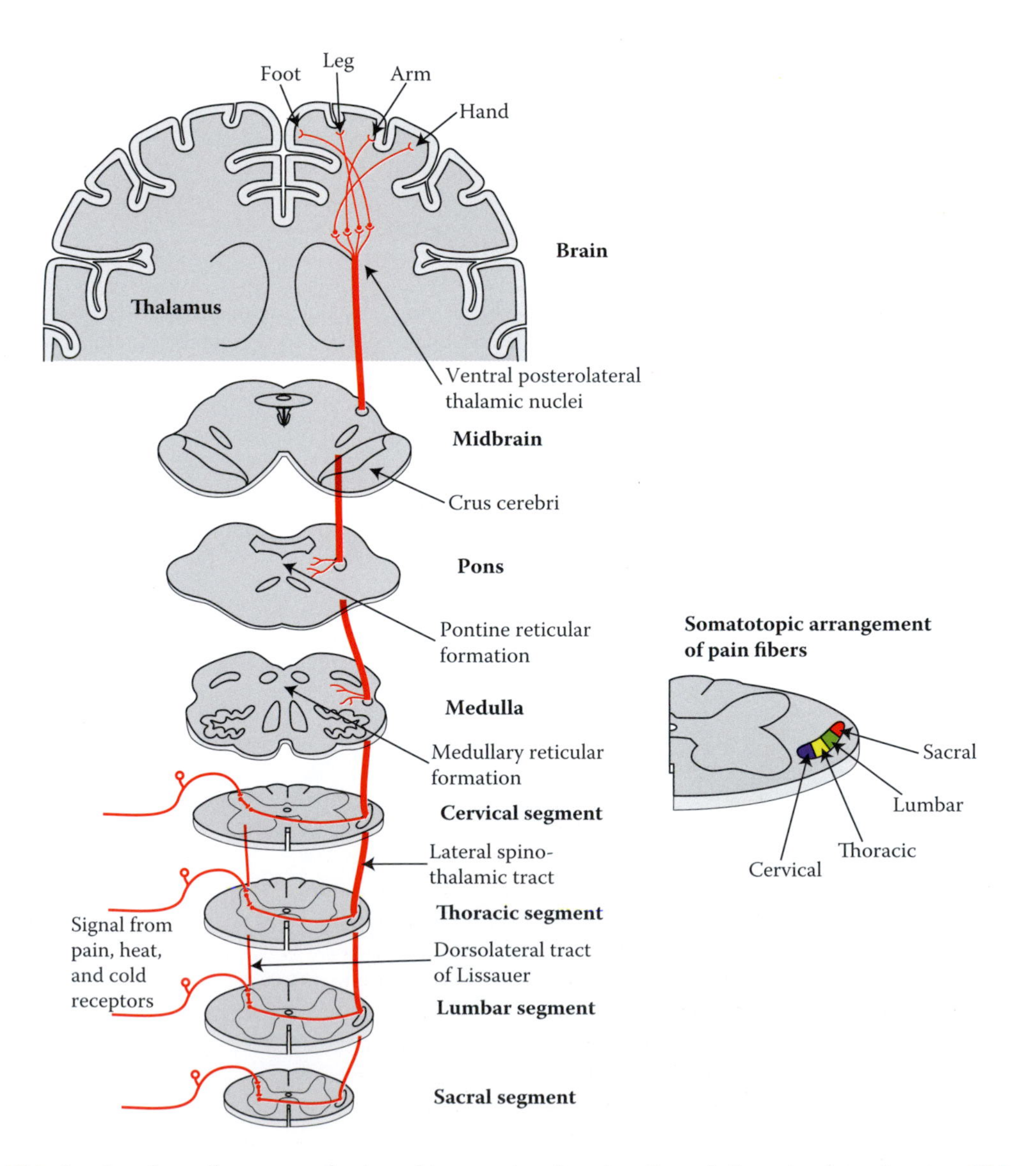

FIGURE 19.12 This drawing shows the course of pain and temperature impulses through the anterolateral system. This is a contralateral pathway that shows reverse somatotopic arrangement to that of the dorsal column.

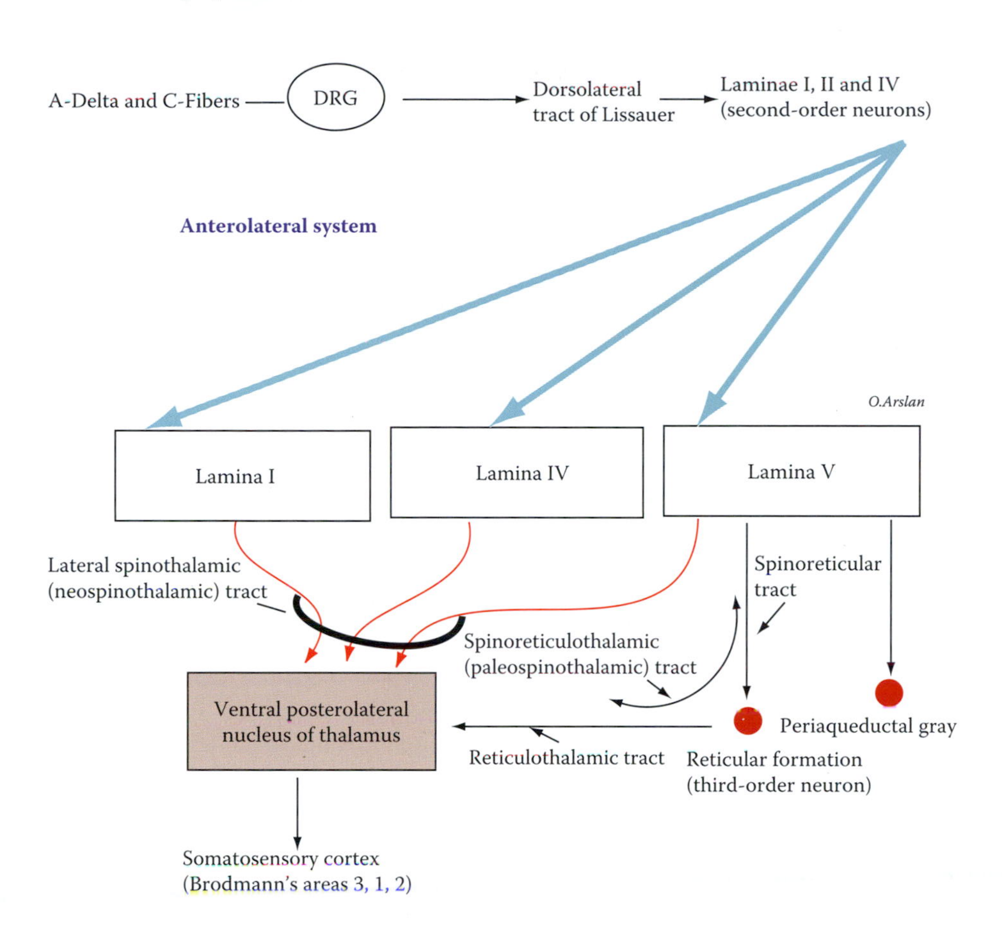

FIGURE 19.13 This is a simplified diagram of the organization of the anterolateral system. Other pathways associated with transmission of pain such as spinoreticular are also illustrated.

serotonin, cholecystokinin, thyrotropin-releasing hormone (TRH), enkephalin, GABA, neurotensin, and dynorphin with some neurons containing multiple neuropeptides. These neurons receive input from the cerebellum, the periaqueductal gray matter, and the spinal cord, conveying efferent fibers to the spinal trigeminal nucleus at the brainstem level and to the substantia gelatinosa at the spinal level to modulate nociception. At a higher level, the neurons in the mesencephalic periaqueductal gray matter also contain similar neuromediators such as serotonin, GABA, substance P, cholecystokinin, enkephalin, etc. It receives hypothalamic afferent that transmits histamine, luteinizing hormone releasing hormone, oxytocin, vasopressin, adrenocorticotrophic hormone, angiotensin II, and neurotensin, among many others. Similarly, efferent fibers from the periaqueductal gray matter project to the spinal neurons as well as the ponds and medulla.

Nociceptive impulses for the most part are carried by the Aδ and C fibers that ascend one or two segments above the level of their entry in the dorsolateral tract of Lissauer and then synapse in certain Rexed laminae of the spinal cord; others, although few, may establish synaptic connections at the same level of their entry into the spinal cord. The postsynaptic fibers from laminae I, IV, and V cross the midline in the anterior white commissure to form the lateral spinothalamic tract, a component of the anterolateral system, on the contralateral side. However, not all nociceptive fibers contribute to the lateral spinothalamic tract; in fact, some may synapse on interneurons that send processes to inhibit the C and Aδ fibers by secreting enkephalins. Information within the lateral spinothalamic tract is somatotopically arranged. The dorsal portion of this tract transmits temperature, whereas the ventral portion conveys pain sensation. Visceral pain fibers occupy the most medial part of the lateral spinothalamic tract, adjacent to the gray matter. Fibers that originate from the lower half of the body occupy a lateral position to the fibers from the upper half of the body (Figure 19.14).

Due to the somatotopic arrangement of pain fibers, an expanding spinal intramedullary tumor may disrupt the fibers from the cervical, thoracic, and lumbar regions while sparing the sacral fibers (*sacral sparing*). Patients with this type of tumor may exhibit pain and thermal anesthesia in the body without any detectable loss in the dermatomes of the sacral segments. Projections of painful stimuli to the orbitofrontal cortex may account for the indifference patients with damage to the prefrontal cortex exhibit toward pain despite acknowledging its existence. The secondary sensory cortex plays an important role in the distinction of sharp pain from dull pain.

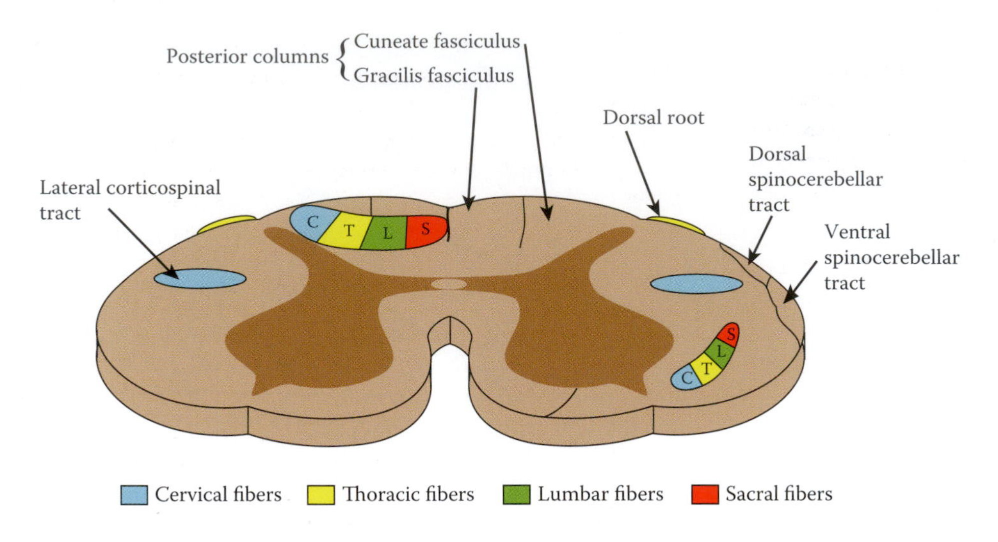

FIGURE 19.14 In this drawing, the somatotopic arrangement of the anterolateral system is compared to that of the dorsal column. Note that the sacral fibers are the most lateral, whereas the cervical fibers occupy the most medial portion of the pain pathway.

The lateral spinothalamic tract runs in the lateral part of the brainstem and intermingles with the fibers of the ventral spinothalamic tract, forming the anterolateral system (spinal lemniscus). As it ascends, it provides collaterals to the reticular nuclei of the brainstem. It then projects to the VPL nucleus of the thalamus, intralaminar thalamic nuclei of both sides (with the exception of the centromedian nucleus), and posterior thalamic zone. Experiments on decorticated animals have confirmed the role of the thalamus in pain recognition. The VPL nucleus projects to the postcentral gyrus, which is believed to deal with the localization of painful stimuli. Information from the posterior thalamic zone and the intralaminar nuclei of the thalamus is transmitted to the secondary sensory and orbitofrontal cortices, respectively. All these thalamocortical projections are contained in the internal capsule and the corona radiata. Through these thalamocortical connections, the anterolateral system excites various cortical centers, producing pleasurable or displeasurable (affective) sensations associated with pain. Unlike the dorsal column-medial lemniscus, the lateral spinothalamic tract does not function on the basis of a one-to-one synapse ratio or lateral inhibition. However, it is regulated by the corticospinal tract, accounting for variation in the sensory experiences and perception by an attentive or alert individual. In fact, studies show that the response of cells that gives rise to the lateral spinothalamic tract to noxious stimuli can be inhibited by stimulation of the primary sensory cortex and posterior parietal lobe. The cortical inhibitory pathway is part of the endogenous analgesic mechanism that also includes the periventricular area, VPL nucleus, nucleus raphe magnus, periaqueductal gray, parabrachial area, certain parts of the pontine and mesencephalic reticular formation, and associated connections. Endogenous analgesia can also be achieved by stimulation of the Aδ fibers (see also gate control theory).

Pain transmission is modified not only by the corticospinal tract but also by the descending serotonergic and noradrenergic neurons and neurotransmitters such as endorphins, adrenocorticotrophic hormone, naloxane, and enkephalin. These polypeptides, which are synthesized in the central nervous system, maintain similar analgesic effect to morphine, develop tolerance, and may induce euphoria, mood changes, and possible respiratory depression. They are located in the spinal cord, periaqueductal gray matter, and limbic system. Ability of certain individuals to tolerate pain under extreme physical and emotional conditions, as in a professional boxer's tolerance to repeated blows or a soldier's emotional indifference to the pain produced by a shrapnel wound, is attributed to the dampening effect of the pyramidal tract fibers upon the internuncial neurons of the posterior gray columns and the analgesic role of endogenous opioids (β-endorphin) released by activation of the tuberal (arcuate) nucleus of the hypothalamus that project to the periaqueductal gray matter. Diffuse noxious inhibitory control (DNIC) is a phenomenon in which noxious stimuli applied to one part of the body produce analgesia in another pain-stricken area. This phenomenon may have practical application in relief of pain associated with tooth decay by application of ice to the skin of the dorsum of the hand. In the same manner, DNIC may explain the analgesic effect of acupuncture produced by insertion of a needle into the skin. Transmission of pain may also be suppressed, and profound analgesia may be achieved by the descending fibers of the periaqueductal gray matter that act upon the primary afferent neurons. Pain may be reduced or abolished from the thorax and abdomen by the injection of alcohol to a peripheral nerve, although complete blocking of a nerve may produce paralysis, a side effect that renders this procedure of minimal clinical value.

Visceral pain is perceived as a dull, aching, slow, diffuse, and deeply seated pain. The site of actual visceral pain sensation is referred to a site distant from the location of the diseased organ (referred pain). Referred pain should be distinguished from projected somatic pain (sciatica), which is induced by irritation of a nerve trunk. The visceral pain may be elicited by overdistention of the walls of the viscera (ureteral or biliary stone) or ischemia due to thrombosis or emboli, impairing the blood supply to the affected organ. In other words, conventional stimuli, such as cutting or burning of a viscus, do not elicit pain. Visceral pain may be also produced by palpation of the abdominal wall, in search of the diseased organ, through the parietal peritoneum.

The *spinoreticular tract* (Figure 19.13) is an integral part of the ascending reticular activating system, maintaining an important role in the synchronization and desynchronization of the electrocortical activity of the brain, as well as transmission of pain sensation. It establishes a direct link between the spinal cord and the brainstem reticular formation, extending the entire length of the spinal cord. This tract terminates primarily upon cells of the nucleus reticularis gigantocellularis in the ipsilateral medulla. However, some fibers may terminate bilaterally upon cells of the nucleus reticularis pontine oralis and caudalis. This pathway further extends from the brainstem reticular formation to the thalamus, constituting the *spinoreticulothalamic* (paleo-spinothalamic) tract. The spinoreticulothalamic tract is a multisynaptic, multineuronal pathway, which terminates in the intralaminar nuclei of the thalamus, establishing synaptic connections with the hypothalamus.

The spinocervical tract originates from laminae III–V and ascends in the dorsolateral funiculus of the spinal cord, terminating in the ipsilateral lateral cervical nucleus. The latter nucleus lies in the lateral funiculus of the upper two cervical segments, conveying the received information via the medial lemniscus to the contralateral VPL nucleus and posterior thalamic zone completing the spinocervicothalamic tract. Some fibers also terminate in the contralateral midbrain. This pathway conveys thermal, noxious stimuli as well as stimuli relative to hair displacement and remains under the inhibitory control of the descending pathways.

The spinoreticulothalamic, a phylogenetically old system, carries poorly localized, diffuse, and vague pain from visceral organs. The connection of the spinoreticulothalamic tract to the hypothalamus accounts for the visceral activity and emotional responses associated with pain (e.g., facial expression, GI activity in response to pain).

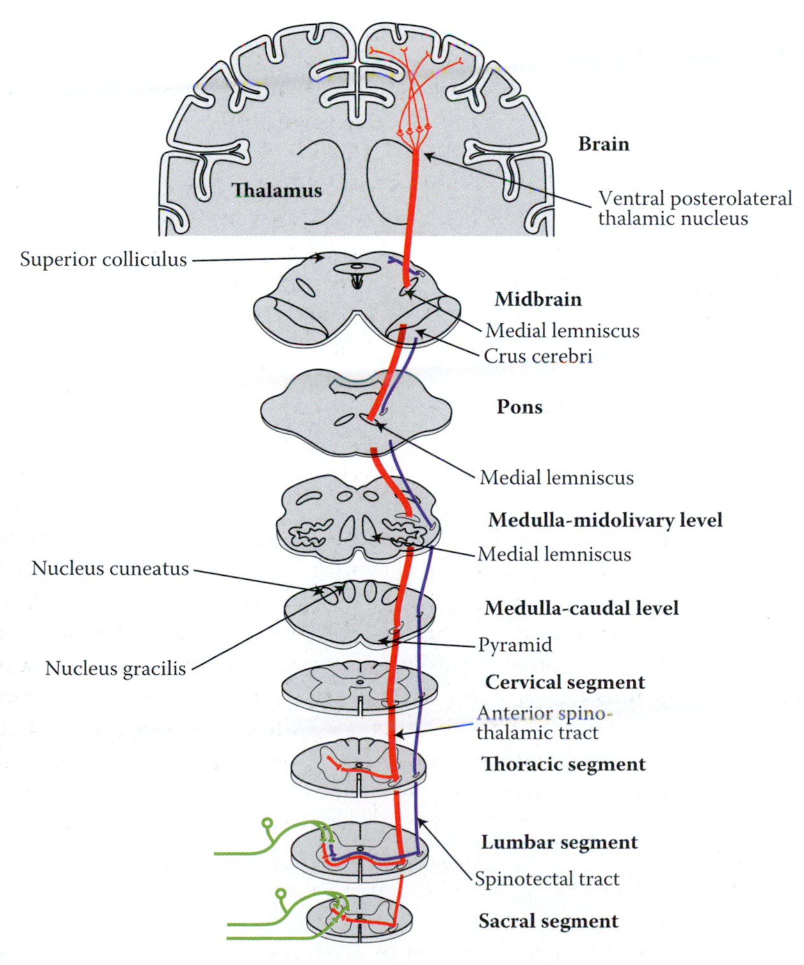

FIGURE 19.15 Transmission of crude touch is maintained by the ventral spinothalamic tract, which is depicted in this picture. This pathway, which forms the anterior component of the anterolateral system, is also thought to transmit itching, tickling, and possibly libidinous sensation. Note that the spinotectal tract, an integral part of the anterolateral system, is also illustrated.

The *ventral spinothalamic tract* (paleospinothalamic–anterior system) runs in the ventral funiculus as a component of the anterolateral system, medial to the fibers of the ventral roots and dorsal to the vestibulospinal tract, transmitting signals associated with crude light touch, pressure sensation, possibly tickling, itching (via the Merkel's disks and Meissner's corpuscles), as well as libidinous sensations (Figure 19.15). This pathway is intermingled with fibers of the spinoreticular and reticulospinal tracts. Since fine touch and discriminative tactile sensations are also conveyed in the dorsal columns, this pathway remains clinically insignificant. This tract is formed by axons of second-order neurons located in laminae III and IV, which send axons that decussate in the anterior white commissure in the segment proximal to the site of entrance of the dorsal roots and ascend contralaterally as part of the anterolateral system (spinal lemniscus), although some fibers ascend ipsilaterally. It shows somatotopic arrangement, where fibers carrying information from the caudal parts of the body are located lateral to those originating from the cranial parts, and this arrangement continues through the spinal cord, medulla, and pons. Then the lateral fibers assume more dorsal position, whereas the medial fibers position ventrally in the midbrain during their course to the third order neurons of the thalamus. This pathway projects to the VPL nucleus of the thalamus (third-order neuron), which, in turn, conveys the stimuli to the primary sensory cortex. During its course toward the thalamus, it gives off numerous collaterals to the medullary reticular formation.

Lesions localized in the spinal cord or brainstem may disrupt the lateral spinothalamic (neospinothalamic) tract or its fibers. In the spinal cord, this pathway may be damaged in the lateral funiculus or at the anterior white commissure. Unilateral destruction of the lateral spinothalamic tract results in analgesia and thermal anesthesia on the contralateral side of the body, extending one or two segments below the level of the lesion. Due to the bilateral representation of the painful stimuli from the perineal region and abdominal and pelvic viscera, destruction of this tract on one side may not be sufficient to produce complete anesthesia in these regions. It is worth noting that the return of some pain sensibility following a lesion may be attributed to the presence of uncrossed intersegmental spinothalamic tracts. Destruction of the posterior part of the lateral spinothalamic tract may extend to involve the posterior spinocerebellar tract, producing also incoordination of motor activity (cerebellar ataxia).

Disruption of the decussating pain fibers in the anterior white commissure produces bilateral analgesia. Destruction of the anterior white commissure at the cervical and the upper thoracic segments results in a tuxedo jacket-type anesthesia, a common manifestation of syringomyelia (commissural syndrome), which involves the upper extremities and upper half of the trunk (Table 19.1).

Syringomyelia (Figure 19.16) is a rare neurologic condition with insidious onset, which results from cavitation (syrinx formation) around the central canal (hydromyelia). These cyst-like tubular cavities are irregular in shape derived from a diverticulum and contain yellowish fluid. They may communicate with the central canal or remain isolated and are commonly seen maximally in the cervical segments (cape distribution), extending to the thoracic segments. The anterior gray columns and the lateral spinothalamic tract will also be damaged when the cavitation expands anteriorly and laterally. Enlarged cavities expand further to involve the lateral corticospinal tract and the dorsal columns. The syrinx often extends rostrally into the brainstem producing syringobulbia. Clinical features associated with syringomyelia may appear before the age of 30 and are characterized by dissociated

TABLE 19.1
General Sensory Dysfunctions

Conditions	Affected Structures	Deficits
Syringomyelia	1. Anterior commissure	1. Bilateral loss of pain and temperature sensations
	2. Ventral gray column	2. Bilateral or unilateral flaccid paralysis
	3. Intermediolateral column	3. Signs of Horner's syndrome
Syringobulbia	1. Vestibular nuclei	1. Vertigo and nystagmus
	2. Nucleus ambiguus	2. Dysphagia and hoarseness of voice
	3. Spinal trigeminal nucleus	3. Analgesia and thermal anesthesia on ipsilateral face
	4. Hypoglossal nucleus	4. Weakness of lingual muscles and dysarthria
Tabes dorsalis	1. Dorsal columns	1. Ataxia due to loss of proprioception
	2. Dorsal roots	2. Bilateral discriminative touch
Friedreich's ataxia (spinocerebellar ataxia)	Clarke's nucleus, ipsilateral dorsal spinocerebellar tract, contralateral ventral spinocerebellar tract, and dorsal columns	Gait ataxia involving all extremities due to bilateral impairment of proprioception and cerebellar deficits

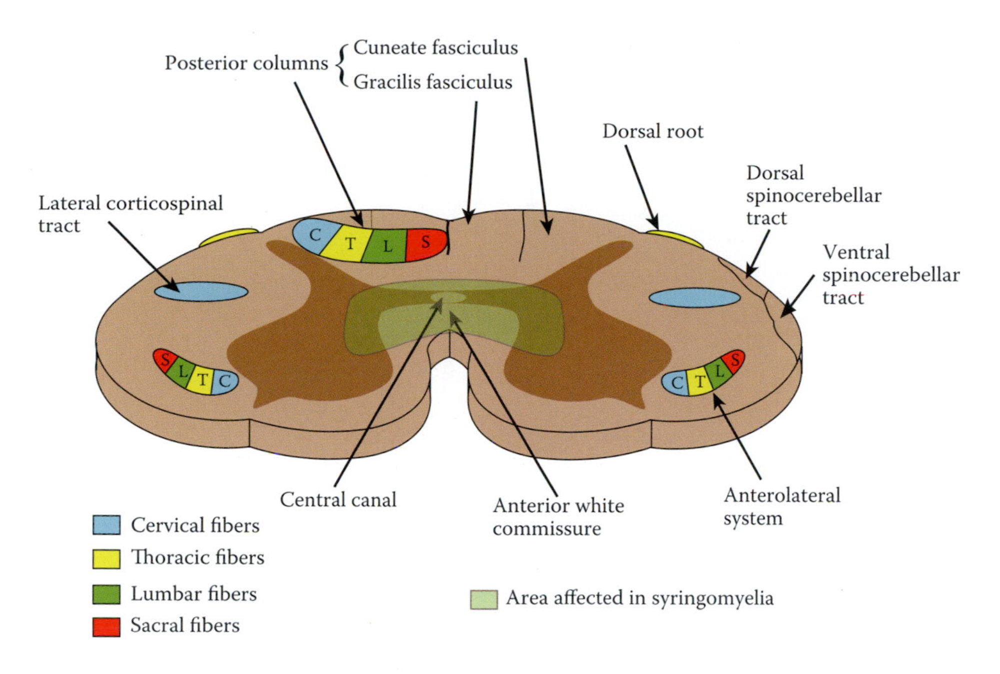

FIGURE 19.16 In this diagram, the lesion is caused by cavitation around the central canal disrupting the decussating fibers of the anterolateral system. Pain and temperature sensations are lost bilaterally.

sensory loss, which refers to the bilateral loss of pain (analgesia) and thermal sensation (thermoanesthesia) while preserving fine touch and proprioception. Failure to appreciate thermal and painful stimuli may result in cigarette burns and ulceration of the digits. Analgesia may follow signs of upper motor neuron paralysis after years. Loss of pain, proprioception, and fine motor control can result in Charcot joint, which is characterized by joint swelling with degenerative changes in the presence of intact skin. Wasting of the intrinsic muscles of hand unilaterally and bilaterally is commonly seen as a result of damage to the α motor neurons of C8–T1 spinal segments and may mimic features of ulnar nerve palsy. Rarely severe kyphoscoliosis occurs as a consequence of degeneration of the α motor neurons and paresis of the paravertebral and extremity muscles. Symptoms may worsen in some patients with straining or Valsalva maneuver. Syringomyelia is thought to have a congenital basis as it develops in patients with Arnold–Chiari malformation. Trauma or intramedullary spinal tumors such as ependymoma and astrocytoma can predispose the patient to the development of spinal syrinx. Imaging, particularly MRI, and also myelography and CT will be helpful in the diagnosis of this condition.

Horner's syndrome, the most frequently seen feature of syringomyelia, is the result of involvement of the intermediolateral column. Autonomic dysfunction together with loss of pain sensation may cause mutilation of the fingers.

Syringobulbia is a condition that affects structures in the medulla and pons, producing damage to the nucleus ambiguus, hypoglossal and vestibular nuclei, and the spinal trigeminal tract and nucleus. The symptoms and signs of this condition are characteristically ipsilateral and consist of dysphagia, nystagmus, analgesia and thermoanesthesia of the face, dysarthria, and hoarseness of voice. Patients may exhibit increased curvature of the vertebral column among other signs of poor physical condition.

As mentioned earlier, due to the somatotopic arrangement of the fibers within the anterolateral system, an expanding intramedullary tumor may disrupt the cervical, thoracic, and lumbar fibers leaving the sacral fibers unaffected. Individuals with this type of lesion may exhibit anesthesia for painful and thermal stimuli in the entire body, leaving the sacral area intact. This phenomenon, which manifests total anesthesia in body with the exception of the sacral region, is known as *sacral sparing* (Figure 19.17).

Phantom limb pain or sensation may be experienced by individuals subsequent to amputation of a limb weeks, months, or years after a traumatic accident. Patients experience the feeling of movements in that particular limb. It commonly develops in patients in their third or fourth decade of life. Patients will be aware of the distal parts of the phantom limb more markedly than the proximal parts. Body parts that maintain larger areas in the sensory cortical homunculus may have more lasting phantom sensation than areas with smaller cortical representation. Phantom perception has some psychogenic basis, although activation of hypersensitive thalamic and cortical neurons

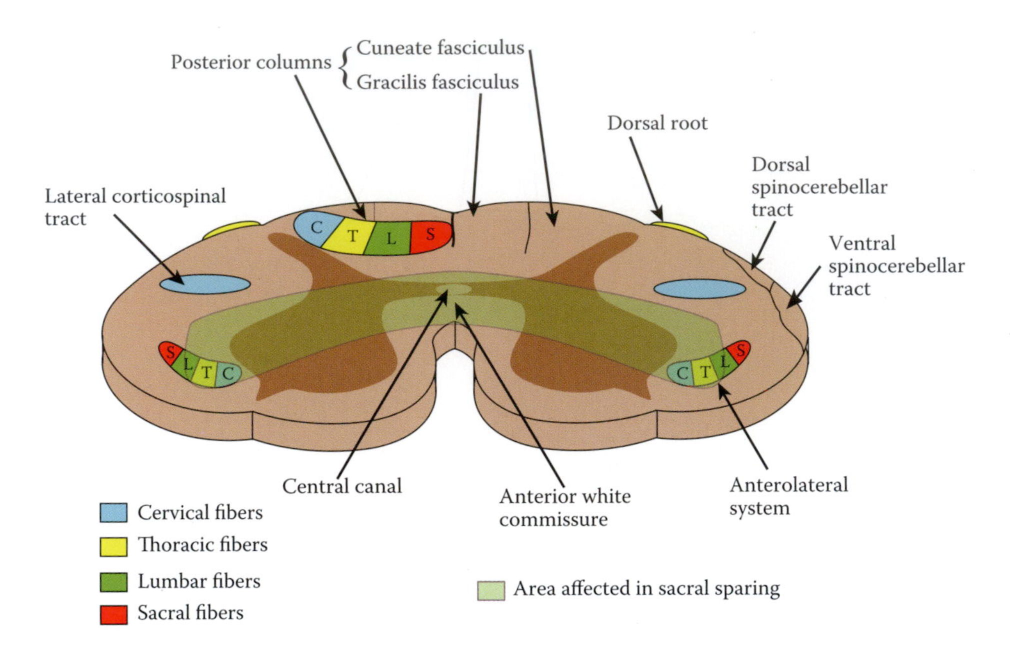

FIGURE 19.17 In this picture, an expanding intramedullary lesion disrupted the pain fibers from the entire body with the exception of the sacral region, resulting in the phenomena of sacral sparing.

may have a role in this phenomenon. The phantom limb may be abolished by excision of the sympathetic nerve fibers to the amputated part. Visual input from mirror therapy has been reported to diminish phantom limb pain. In this module of therapy amputees place a mirror between their missing and intact limbs, and then simultaneously move both the phantom and intact limb. During this exercise the patient continues to view the reflected image of the intact limb moving in the mirror. It has been proposed that removal of sensory input from the amputated limb leads to reorganization in the somatosensory cortex, revealing visual responses, which may contribute to phantom limb pain, and that mirror therapy helps to reverse these responses. *Stump pain*, as a result of a surgical procedure, may be attributed to development of neuroma or irritation by the scar tissue. Neuroma may occur as a result of regenerating nerve sprouts within the scar tissue of the amputated limb.

Low back pain ranks second only to common cold as a reason for primary care visits. It is the second leading cause of work absenteeism. Most low back pain is self-limiting, and patients recover without specific treatment. A small number of cases may exhibit persistent or progressive neurologic dysfunction associated with radiographic findings. Low back pain may be associated with sciatica, in which the pain follows a dermatomal distribution, radiating to the buttock and down the leg often with numbness stretching into the lower leg and foot. Very few patients with low back pain exhibit urinary retention, saddle anesthesia, or bilateral neurological dysfunctions. Serious conditions that produce low back pain, such as abdominal aortic aneurysm, tumors, abscesses, emboli, and cauda equina syndrome, should be excluded before initiating any treatment. Several methods may be employed to confirm the positive findings of low back pain: (1) reduced *Schober index*, which measures the distance between L5 spinous process and a point 10 cm proximal to it in both erect and maximally flexed lumbar vertebrae (normal distance ranges 10–15 cm); (2) forward bending that produces reflex flexion of the knee on the affected side (*Neri's sign*); and (3) well-leg, straight leg raising test. Hematological tests such as erythrocyte sedimentation rate may be used to determine the medical etiology of back pain from the mechanically induced pain. Malignancies, systemic lupus erythematosus, and rheumatoid diseases may show elevation of ESR. C-reactive protein may show an increase in individuals with back pain associated with tuberculosis and rheumatoid arthritis. Hematocrit value, rheumatoid factor, platelet count, level of serum alkaline phosphatase, and uric acid may also be used to differentiate the medical causes of this condition. Computed tomography and MRI techniques and myelogram proved to be useful in the evaluation of abnormalities of lumbosacral spine. Back pain can occur as a result of herniation of the nucleus pulposus of the intervertebral disk. Most common sites of herniation are L4–L5 and L5–S1, accounting for 95% of the total disorders. A third of all patients with disk herniation develop sciatica approximately 6–10 years after the onset of pain. Disk herniation at L4–L5 causes pain and numbness, which radiate to the posterior thigh, anterolateral leg, medial foot, and great toe. Foot drop and weakness of dorsiflexion may be associated with this herniation.

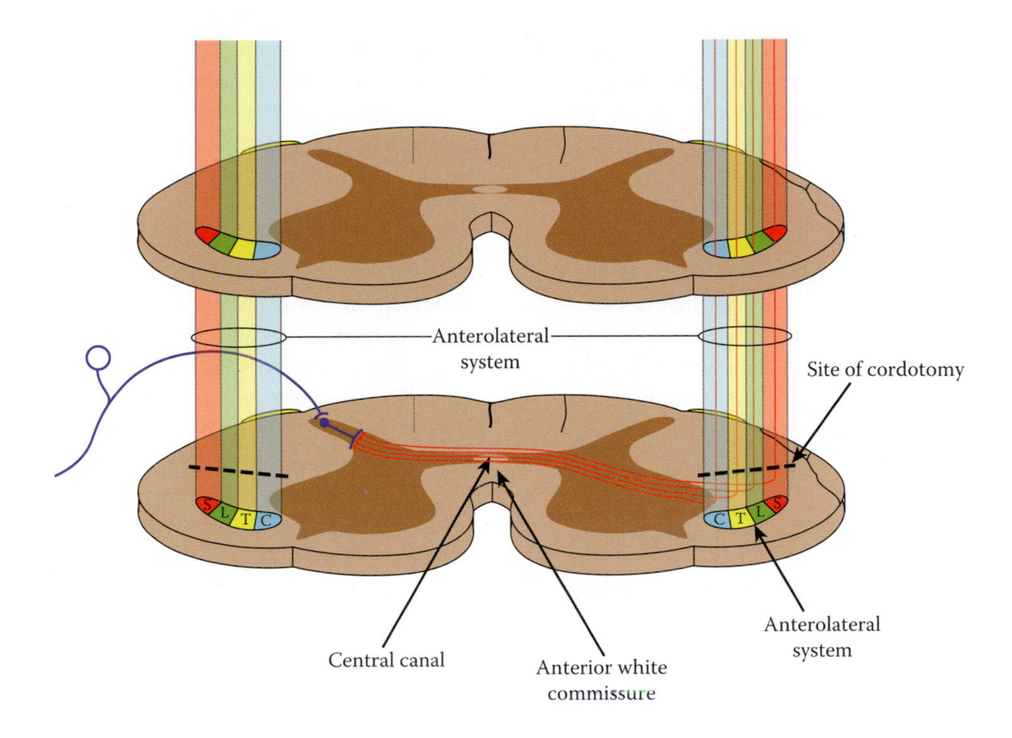

FIGURE 19.18 Cordotomy, a surgical procedure by which pain pathway is excised, is depicted in this drawing. Pain and temperature sensations are lost on the contralateral side.

L5–S1 disk herniation presents with pain and numbness of the posterior thigh and leg, posterolateral foot, and lateral toes. Plantar flexion of the foot and toes may also be weakened. Crossed *Lasègue's sign*, in which the pain radiates down the back of the affected leg when the contralateral leg is passively elevated, may also be used to confirm this condition. Pain may be felt along the course of the sciatic nerve as far distal as the calcaneal tendon (*Valleix's pressure points*). High lumbar herniation at the L2–L4 level is less frequent and may present with pain and numbness of the posterolateral or anterior thigh radiating to the anteromedial leg. Additional deficits include weakness of knee extension and hyporeflexic knee joint. Osteoarthritis, spondylolisthesis spondylitis, chronic inflammatory disease that involves the sacroiliac and shoulder joints, Reiter's disease (manifests triad of urethritis, arthritis, and conjunctivitis), and vertebral osteomyelitis may also cause back pain.

Rest, back exercise, TENS, acupuncture, and injection of corticosteroid may be of value. Acetaminophen is a standard treatment for this condition, although morphine and meperidine may be essential in individuals with osteoporosis or malignancies. Nonsteroidal anti-inflammatory drugs (NSAIDs) may be used as analgesic and to counteract the inflammatory process. Surgery may be an option if other therapeutic measures fail.

Analgesia may also be attained by *cordotomy* (Figure 19.18), a surgical technique that requires excision of the lateral spinothalamic tract in the lateral funiculus of the spinal cord. In this procedure, the dentate ligaments, which are the pial extensions that attach to the spinal cord at sites demarcating the dorsally located lateral corticospinal tract from the ventrally positioned lateral spinothalamic tract, are utilized. Cordotomy may be used in cases of malignant intractable pain, which is unresponsive to conventional analgesics. It is more useful in conditions where the pain is restricted to one side of the body or one limb.

Bilateral cordotomy, which may be employed for more generalized and diffuse pain, carries the risk of causing incontinence and fatal respiratory impairment (*Ondine's curse*). The latter refers to the death during sleep as a result of cessation of voluntary and involuntary breathing. Due to these reasons and to avoid damage to the phrenic nucleus, a surgical approach at the T1 spinal segment is more preferable. In this procedure, a needle is introduced into the anterolateral quadrant of the spinal cord through the intervertebral space between the atlas and the axis. To prevent any serious adverse effects, an electrode may be passed through the needle. The path of the needle and the stimulating electrode are monitored via imaging techniques. Any motor response from the ipsilateral upper or lower extremity should be considered a serious sign that the needle has entered the lateral corticospinal tract and that it must be rerouted. Tingling sensations are usually felt as the needle enters the lateral spinothalamic tract, which is the site of cordotomy.

Reduction in the ability to appreciate pain and thermal sensations will be detected upon excision of the lateral spinothalamic tract (LSTT) via the surgical needle. Postoperative anesthesia will not be permanent, and it may disappear completely after a period of 1 year due to existence of intact uncrossed pain fibers. Since fibers of the lateral spinothalamic tract ascend two to three segments in the dorsolateral tract of Lissauer before entering the dorsal gray column, the anesthesia induced by this procedure will start four to eight segments below the level of cordotomy, and therefore, the pain from the upper extremity may not be abolished completely.

CORTICAL SENSATIONS FROM THE HEAD

The *trigeminal lemniscus* (see also the trigeminal nerve) conveys general sensation (GSA) from the head region to the thalamus. The first-order neurons for these pathways are located in the sensory ganglia of the trigeminal, facial, glossopharyngeal, and vagus nerves. Depending upon the modality of sensation, the central processes of the ganglionic neurons enter the brainstem, establishing synapses with the sensory nuclei of the trigeminal nerve. Fibers conveying nociceptive, thermal and some tactile sensations form the spinal trigeminal tract and synapse in the spinal trigeminal nucleus. Pressure and tactile fibers terminate in the principal sensory nucleus, whereas proprioceptive fibers are transmitted to the mesencephalic nucleus. The axons of the neurons of the spinal trigeminal nucleus and the principal sensory nuclei form the dorsal and ventral trigeminal tracts. The *ventral trigeminal lemniscus* (tract) is a crossed tract, which originates from the spinal trigeminal nucleus and the ventral part of the principal sensory nucleus. It ascends with the medial lemniscus and terminates in the ventral posteromedial nucleus of the thalamus (*VPM*). The *VPM* conveys this information through the thalamocortical fibers to the postcentral gyrus (primary sensory cortex). These ascending thalamocortical fibers occupy the posterior limb of the internal capsule. Damage to the ventral trigeminal tract may occur in lesions of the pontine and midbrain tegmentum, resulting in contralateral loss of pain, thermal, and, to some degree, tactile sensation from the facial region. Axons, which are derived from the dorsal part of the principal sensory nucleus, where the mandibular nerve fibers terminate, form the ipsilateral *dorsal trigeminal lemniscus* (tract). This pathway projects to the ventral posteromedial (VPM) nucleus, conveying this information to the postcentral gyrus via the thalami-cortical fibers. Destruction of the dorsal trigeminal tract results in ipsilateral loss of pressure and tactile sensation. However, sensory loss will not be significant since these sensations are also carried by the ventral trigeminal tract.

SUBCORTICAL SENSATIONS

Ascending subcortical pathways terminate in the cerebellum, tectum, reticular formation, and the olivary nuclei and include the spinocerebellar, spinotectal, spinoreticular (discussed earlier), and spino-olivary tracts.

The *spinocerebellar tracts* project to the spinocerebellum (anterior lobe of the cerebellum), carrying proprioceptive, stretch, and tactile sensations. They consist of the dorsal and ventral spinocerebellar, cuneocerebellar, and rostral spinocerebellar tracts.

The *dorsal spinocerebellar tract* represents the axons (the largest in the entire CNS) of the Clarke's column, which extends in the entire thoracic and upper two or three lumbar spinal segments. It is an ipsilateral tract that ascends close to the surface of the dorsolateral funiculus and enters the spinocerebellum (anterior lobe of the cerebellum) through the inferior cerebellar peduncle. This tract carries proprioceptive, tactile, and pressure impulses from individual muscles and joints of the lower limb and lower half of the trunk. The upper limb equivalent of the dorsal spinocerebellar tract is the *cuneocerebellar tract*, which originates from the accessory cuneate nucleus and terminates ipsilaterally in the spinocerebellum and pontocerebellum (anterior and posterior lobes of the cerebellum) via the inferior cerebellar peduncle.

The *ventral spinocerebellar tract* is derived from the gray columns of the intermediate and the border cells of the ventral gray column in the thoracic, lumbar, and sacral segments. It is a crossed tract, conveying proprioceptive information from the entire lower extremity, particularly from the internuncial neurons that mediate flexor reflexes and provide input to the cerebellum. It also conveys information from the collateral of primary afferents emanating from the muscles and joints of the lower extremity. This pathway reaches the contralateral anterior lobe of the cerebellum through the superior cerebellar peduncle. The upper limb equivalent of this tract is the ipsilateral *rostral spinocerebellar tract*, which arises from laminae Vll of the cervical enlargement and upper thoracic spinal segments. It enters the cerebellum through the inferior and superior cerebellar peduncles to be distributed to the anterior lobe of the cerebellum. The ventral and rostral spinocerebellar tracts relay information about neuronal activities of the descending motor pathways. Proprioception from the muscles of mastication, alveoli of the maxillary and mandibular bones, and extraocular muscles is delivered to the mesencephalic nucleus of the trigeminal nerve and later conveyed via the *trigeminocerebellar tract* to the ipsilateral spinocerebellum coursing within the superior cerebellar peduncle.

Degeneration of the spinocerebellar tracts, dorsal columns, and Clarke's nucleus occurs in *Friedreich's* (hereditary) *ataxia*, an autosomal recessive disease that is seen in childhood and before the end of puberty. It is characterized by gait ataxia (high-arched steppage), disturbances of speech, scoliosis (exaggerated lateral curvature of the vertebral column), and paralysis of the muscles of the lower extremities. This disease is often associated with hypertrophic cardiomyopathy. It also manifests areflexia, pes cavus (exaggerated longitudinal arch of the foot), nystagmus, vertigo, hearing impairment, visual deficits, and intention tremor.

The *spino-olivary tract* originates from laminae of the spinal cord, ascends in the anterolateral funiculus, and terminates in the accessory olivary nuclei. This pathway carries information from GTOs, muscle spindles, and some other exteroceptive receptors. A corresponding *dorsal spino-olivary tract* also exists in the dorsal funiculus, which terminates in the inferior olivary nucleus.

The *spinotectal tract* (Figures 19.13 and 19.15) is composed of axons that are derived from lamina VII of the spinal gray matter. These axons cross the midline in the anterior white commissure and ascend on the contralateral side as a component of the anterolateral system (spinal lemniscus), terminating in the superior colliculus and the periaqueductal gray matter. The periaqueductal gray, which contains enkephalin and β endorphins, receives input from the frontal lobe and hypothalamus, projecting to the serotinergic tegmental nuclei of the medulla. Although the functional significance of this pathway is not clear, its role in modulating the transmission of pain, thermal, and tactile sensations is currently being studied.

SUGGESTED READING

Almeida TF, Roizenblatt S, Tufik S. Afferent pain pathways: A neuroanatomical review. *Brain Res* 2004;1000(1–2):40–56.

Apkarian AV, Bushnell MC, Treede RD, Zubieta JK. Human brain mechanisms of pain perception and regulation in health and disease. *Eur J Pain* 2005;9(4):463–84.

Apkarian AV, Hodge CJ. Primate spinothalamic pathways. I. A quantitative study of the cells of origin of the spinothalamic pathway. *J Comp Neurol* 1989;288:447–73.

Apkarian AV, Hodge CJ. Primate spinothalamic pathways. II. The cells of origin of the dorsolateral and ventral spinothalamic pathways. *J Comp Neurol* 1989;288:474–92.

Apkarian AV, Hodge CJ. Primate spinothalamic pathways. III. Thalamic termination of the dorsolateral and ventral spinothalamic pathways. *J Comp Neurol* 1989;288:493–511.

Diers M, Christmann C, Koeppe C, Ruf M, Flor H. Mirrored, imagined, and executed movements differentially activate sensorimotor cortex in amputees with and without phantom limb pain. *Pain* 2010;149(2):296–304.

Green AL, Pereira EA, Aziz TZ. Deep brain stimulation and pleasure. In Kringelbach ML, Berridge KC, eds. *Pleasures of the Brain*. New York: Oxford University Press, 2010, 302–19.

Hellman KM, Mendelson SJ, Mendez-Duarte MA, Russell JL, Mason P. Opioid microinjection into raphe magnus modulates cardiorespiratory function in mice and rats. *Am J Physiol Regul Integr Comp Physiol* 2009;297(5):R1400–8.

Keszthelyi D, Troost FJ, Simrén M, Ludidi S, Kruimel JW, Conchillo JM, Masclee AA. Revisiting concepts of visceral nociception in irritable bowel syndrome. *Eur J Pain* 2012;16(10):1444–54.

Lamm C, Decety J, Singer T. Meta-analytic evidence for common and distinct neural networks associated with directly experienced pain and empathy for pain. *Neuroimage* 2011;54(3):2492–502.

Landi A, Nigro L, Marotta N, Mancarella C, Donnarumma P, Delfini R. Syringomyelia associated with cervical spondylosis: A rare condition. *World J Clin Cases*. 2013;1(3):111–5.

Moffie D. Spinothalamic fibers' pain conduction and cordotomy. *Clin Neurol Neurosurg* 1975;78:261–8.

Osborn CE, Poppele RE. Sensory integration by the dorsal spinocerebellar tract circuitry. *Neuroscience* 1993;54:945–56.

Sarnthein J, Jeanmonod D. High thalamocortical theta coherence in patients with neurogenic pain. *Neuroimage* 2008; 39(4):1910–7.

Schott GD. Visceral afferents: Their contribution to "sympathetic dependent" pain. *Brain* 1994;117:397–413.

Sidall PJ, Cousins MJ. Neurobiology of pain. *Int Anesthesiol Clin* 1997;35:1–26.

Stevens RT, London SM, Apkarian AV. Spinothalamocortical projections to the secondary somatosensory cortex (SII) in squirrel monkey. *Brain Res* 1993;631:241–6.

Van Oosterwijck J, Meeus M, Paul L, De Schryver M, Pascal A, Lambrecht L, Nijs J. Pain physiology education improves health status and endogenous pain inhibition in fibromyalgia: A double-blind randomized controlled trial. *Clin J Pain* 2013;29(10):873–82.

Wall PD. The gate control theory of pain mechanisms. A reexamination and re-statement. *Brain* 1978;101:1–18.

Willis WD, Westlund KN. Neuroanatomy of the pain system and of the pathways that modulate pain. *J Clin Neurophys* 1997;14:2–31.

Section VIII

Motor Systems

20 Upper and Lower Motor Neuron Systems

Motor neurons are comprised of upper and lower neurons and their axons. Upper motor neurons (UMNs) are represented by the corticospinal tract (CST) and corticobulbar tract (CBT), and also by the descending autonomic pathways which emanate from higher centers in the cerebral cortex and diencephalon. Lower motor neurons (LMNs) are formed by the α motor neurons of the ventral horn of the spinal gray columns and by the motor nuclei of cranial nerves. UMNs regulate the activities of α motor neurons and as a result they also control the reflex activities at spinal and brainstem levels. Damage to the UMNs produces manifestations of spastic palsy whereas a lesion that involves the LMNs results in flaccid palsy.

UMNs are comprised of neurons of the primary motor, premotor, and supplementary motor cortices (Figure 20.1). The premotor cortex (Brodmann area 6) occupies the frontal lobe rostral to the motor cortex, while the supplementary motor area lies on the medial side of the superior frontal gyrus, controlling patterning and initiation of motor activities. The primary motor cortex (Brodmann area 4) is arranged somatotopically (Figure 20.2), in which parts of the contralateral body are represented in a distorted fashion (*motor homunculus*). In this homunculus, the thumb in particular, and the hand, in general, occupies a larger area than other parts of the body. This is attributed to the relative density of the neurons associated with movements of the digits. The thumb and index finger occupy an area near the face, while the body lies superior to the head between neurons of the shoulder and hip. The head occupies the lower part of the homunculus near the lateral cerebral (Sylvian) fissure, whereas the lower extremity is confined to the paracentral lobule, which lies rostral to the precuneus on the medial surface of the cerebral hemisphere. These neurons project to the spinal cord and brainstem via the CST and the CBT, respectively. The CST and CBT exert a wide range of influences on motor and reflex activities and sensory transmission. These effects are facilitated by their projections to the spinal and

FIGURE 20.2 Diagram of the motor homunculus in which the entire body is represented in a distorted fashion on the motor cortex. The motor areas for the head and hand are considerably larger than other areas due to neuronal density.

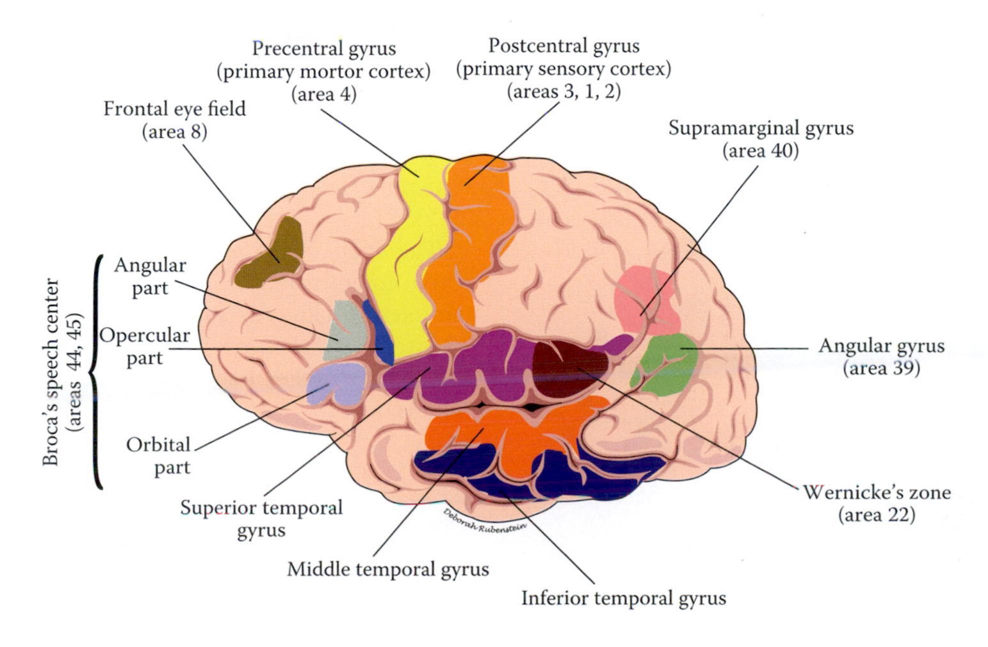

FIGURE 20.1 The motor and premotor cortices are illustrated.

cranial motor neurons, as well as thalamic, striatal, and other brainstem neurons.

The *CST* (*pyramidal*) is a phylogenetically new pathway that exists in man and other mammals. It continues to develop throughout the first 2 years of life. This principal motor pathway extends the entire length of the spinal cord, intermingling with fibers of the rubrospinal tract in the lateral funiculus of the spinal cord (Figure 20.4). It regulates the voluntary motor activities, especially the skilled and fine movements of the digits. It exerts facilitatory influences on the flexor neurons and inhibitory effects on neurons of antigravity muscles, executing antagonistic function to that of the vestibulospinal tract. This major pathway is derived from the precentral gyrus (Brodmann areas 4), premotor area (Brodmann areas 6 and 8), and postcentral gyrus (Brodmann areas 3, 1, and 2). The largest fibers originate from the giant pyramidal cells of Betz. The primary motor cortex contributes 80% of the total fibers of this tract, premotor supplies 10%, and the remaining fibers originate from Brodmann areas 3, 1, 2, and 5. Therefore, the CST maintains motor and sensory components. The motor component sends collaterals to the striatum, primary sensory cortex, thalamus, red nucleus, inferior olivary nucleus, pontine nuclei, and reticular nuclei in the medulla and pons. It also conveys impulses to the motor nuclei of the cranial nerves as it proceeds to α and γ spinal motor neurons. The sensory component, which is derived from the postcentral gyrus and paracentral lobule, sends a substantial number of inhibitory fibers to the GABA neurons of the striatum, gracilis and cuneatus nuclei, locus ceruleus, and substantia gelatinosa. This tract descends through the corona radiata (Figures 20.3, 20.4, and 20.5) and the posterior limb of the internal capsule, continuing through the basis pedunculi of the midbrain, basilar pons, and pyramids of the ventral medulla. During its descending course, it gives collaterals to the striatum.

As the CST travels in the brainstem, it sends collateral fibers to the locus ceruleus; the red, pontine, and dorsal

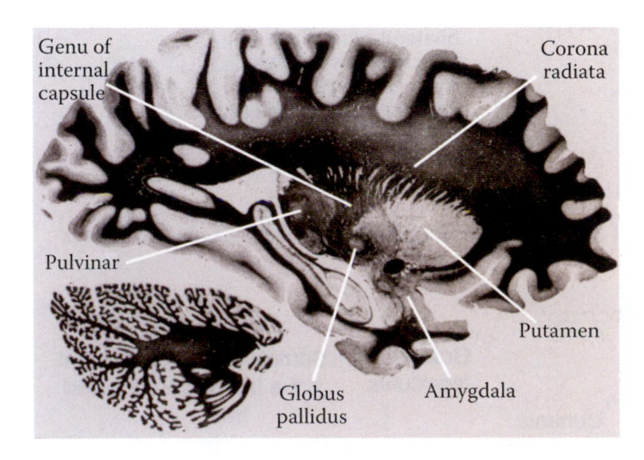

FIGURE 20.3 Photograph of a sagittal section of the brain showing the corona radiata and the basal nuclei. The frontal and temporal gyri are removed in this picture.

> The CST travels in close proximity to the oculomotor nerve in the midbrain, abducens nerve in the pons, and hypoglossal nerve in the medulla. Therefore, a single lesion that involves both the CST and a cranial nerve may produce various forms of *alternating hemiplegia* that combine both of these motor deficits.

column nuclei; as well as the inferior olivary nuclei. Other descending fibers may terminate in the dorsal column nuclei and neurons of the dorsal horn. This pathway is also known as the pyramidal tract due to its course within the pyramids of the medulla. Within the CST, the sacral fibers occupy a lateral position to the thoracic and lumbar fibers, while the cervical fibers remain most medial.

Most of the fibers of the CST (80%–85%) decussate in the caudal part of the medulla, forming the *lateral CST* (Figure 20.6). The remaining ipsilateral fibers form the anterior and anterolateral CSTs. The lateral CST primarily terminates in the cervical segments (55%), other fibers end in the lumbosacral segments (25%), and the remaining terminate in the thoracic segments (20%). The ventral fibers of this tract are derived from the motor cortex, whereas the dorsal fibers originate from the neurons of the primary sensory cortex. Additionally, fibers that have a longer course assume a superficial position relative to the shorter fibers. Sacral fibers remain the most lateral; the cervical fibers most medial, whereas the thoracic and lumbar fibers occupy an intermediate position within the lateral CST.

Fibers of this tract that originate from the postcentral gyrus terminate in the proper sensory nucleus, modulating the transmission of sensory impulses. For the most part, the CST acts upon the α motor neurons through interneurons of laminae VII and VIII of the spinal cord gray matter. Some fibers exert direct influence upon the α motor neurons in lamina IX, innervating muscles, which are involved in fine and skilled movements. The remaining uncrossed fibers (10%–15%) of the CST descend in the ipsilateral anterior funiculus, forming the *anterior CST* (Figures 20.4 and 20.5).

The fibers of the anterior CST decussate in the anterior white commissure before terminating in the neurons of the intermediate gray and central parts of the ventral horn. This tract, which exerts a facilitatory effect on the cervical and axial muscles, may only extend to the upper thoracic segments.

> Some ipsilateral corticospinal fibers may form the anterolateral CST that joins the lateral CST in the lateral funiculus. The presence of these ipsilateral fibers may hasten the recovery from severe motor weakness that accompanies extensive cordotomy (surgical

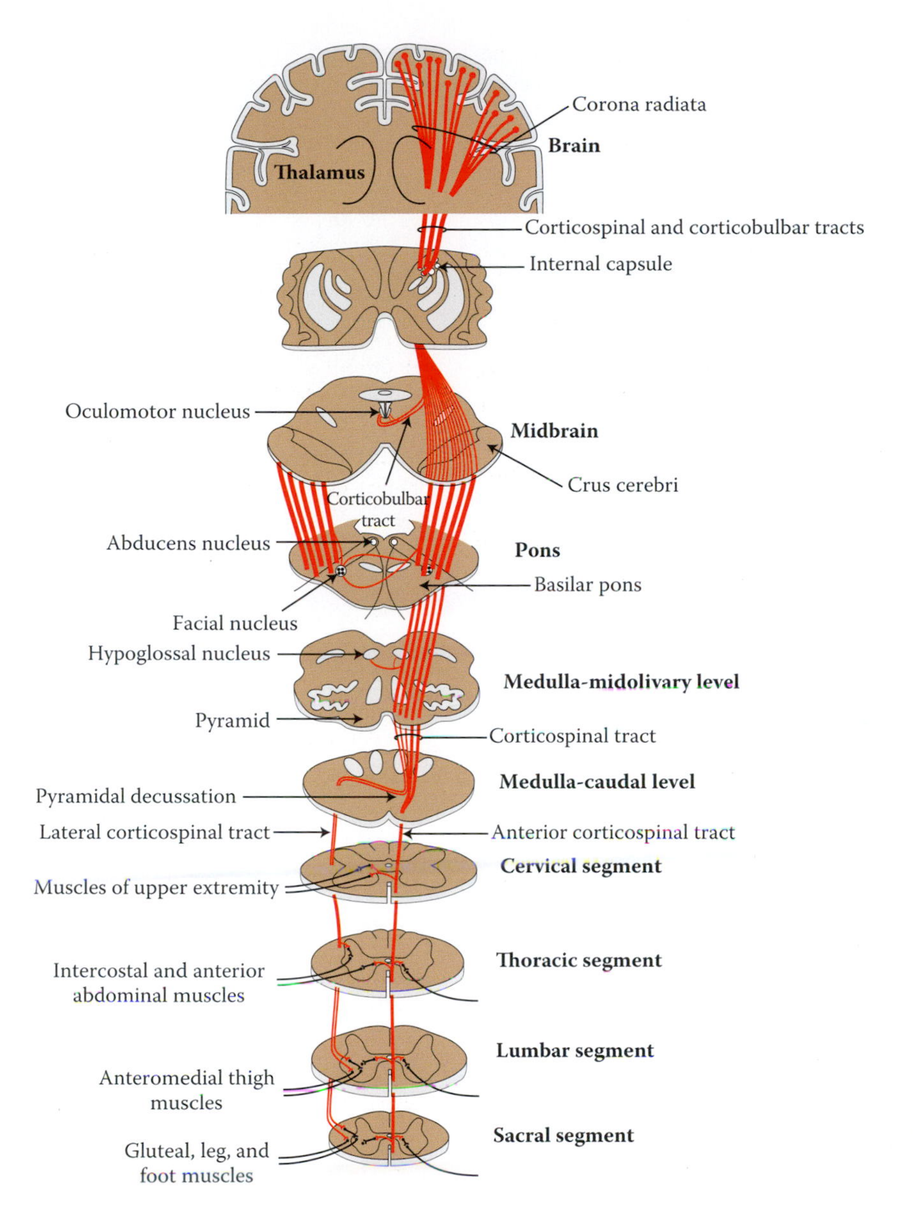

FIGURE 20.4 Diagram of the corticospinal and corticobulbar tracts. The red fibers represent the lateral corticospinal tract; the green fibers show the course of the ventral neurons of the intermediate gray and central parts of the ventral horn.

removal of the lateral spinothalamic tract), damaging the lateral CST. Motor recovery may involve early return of flexion and adduction of the arm and extension and adduction of the thigh. Recovery of hand functions remains very poor. Compensatory activation of the ipsilateral fibers of the CST and their role in motor recovery may be illustrated in individuals with progressive glioma who have suffered stroke or undergone surgical removal of one cerebral hemisphere. Ipsilateral corticospinal fibers may also play an important role in execution of certain motor activities upon command with one hand in individuals with an excised corpus callosum.

UPPER MOTOR NEURON (UMN) PALSIES

Damage to the motor cortex or the CST may occur as a result of cerebrovascular or degenerative lesions. These lesions produce a constellation of signs of UMN palsy (Figures 20.7 and 20.8). When the adjacent cranial nerve is also involved, tract signs of *alternating hemiplegia* are produced, including spastic palsy (UMN) on the contralateral side and cranial nerve dysfunction (LMNs) on the same side (Table 20.1).

Occlusive lesion of the basilar artery may produce bilateral flaccid palsy followed by spasticity, and possible coma. Bilateral occlusion or stenosis of the carotid

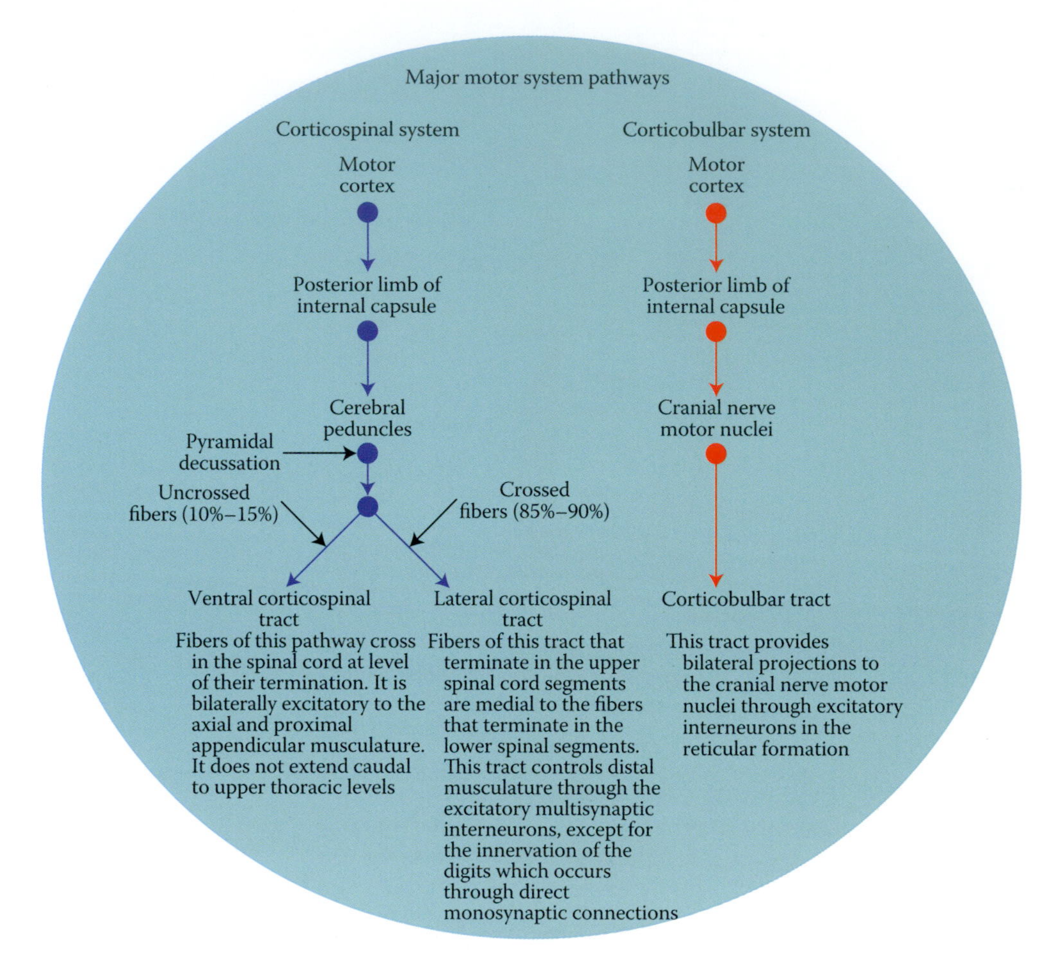

FIGURE 20.5 This is a schematic drawing of the course and termination of the corticospinal and corticobulbar tracts. As noted, the cerebral cortex also influences motor activity via the red nucleus and the rubrospinal tract.

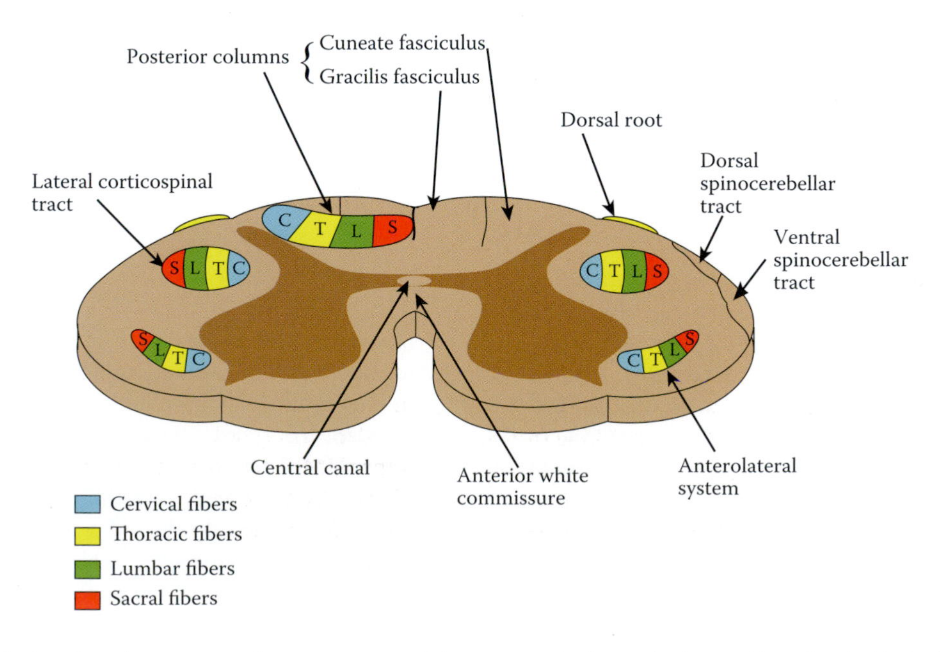

FIGURE 20.6 In this drawing, the somatotopic arrangement of the corticospinal tract is shown. Note the similarity between the arrangement of the fibers of the corticospinal tract and the anterolateral system.

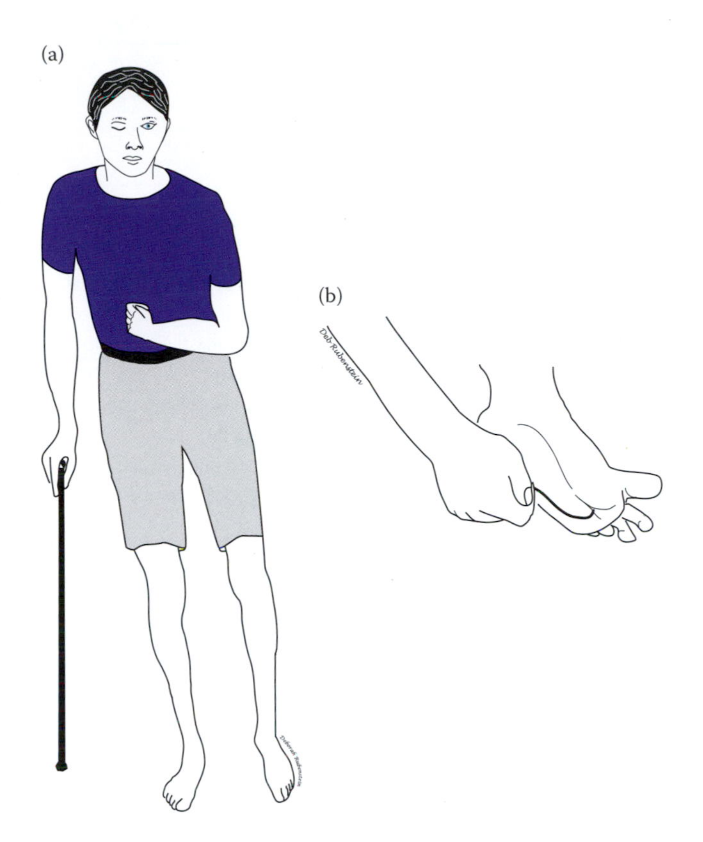

FIGURE 20.7 (a) A patient with signs of upper motor neuron palsy. Note the characteristic gait and the flexed upper extremity. (b) Babinski sign.

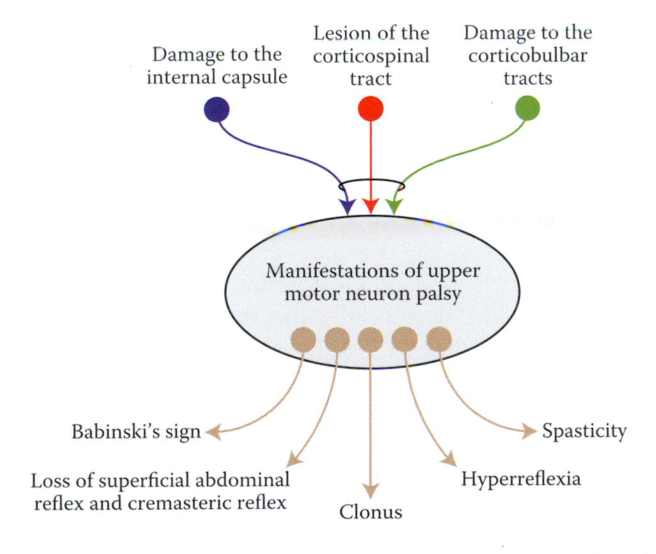

FIGURE 20.8 Summary of the deficits associated with the lesions of the corticospinal tract (upper motor neuron palsy). Spasticity and hyperreflexia are some of the characteristics of this type of dysfunction.

arteries may produce quadriplegia and conjugate horizontal deviation of both eyes and, eventually, coma. Transient episodes of hemiparesis, which may lead to hemiparalysis, aphasia, light-headedness, carotidynia (pain over the occluded artery), and confusion may also be detected in some individuals, although presence of good collateral circulation can prevent these deficits from appearing in young individuals. Occlusion of the middle cerebral artery, which supplies the motor areas of the head and upper extremity, may produce signs of UMN palsy that vary considerably with the affected arterial branch. In this manner, motor disorders may be restricted to the head and/or upper extremity, sparing the lower extremity. On the other hand, occlusion of the anterior cerebral artery, which provides blood supply to the medial surface of the brain and the paracentral lobule (motor centers for the lower extremity), may result in motor dysfunctions restricted to the lower extremity. Irritation of the precentral gyrus by a trauma-induced scar is a common culprit that causes an orderly and progressive or marching type of convulsions (e.g., twitching) of the distal muscle groups that advances proximally (*Jacksonian seizure*).

Signs of *UMN palsy* (Figures 20.7 and 20.8) include spasticity, Babinski sign, Gordon's leg and finger signs, Oppenheim's and Hoffman's signs, Chaddock response, inverse myotatic reflex, clonus, hyperreflexia, and loss of superficial abdominal reflex. *Spasticity* refers to the gradual resistance to a passive movement of a joint that predominates in the antigravity muscles, which are the flexors of the upper extremity and the extensors of the lower extremity. It is observed within a matter of weeks following the insult. *Babinski sign* (extensor plantar response) or *upgoing toe reflex* (Figure 20.7) is elicited by a gradual stimulation of the lateral plantar surface of the foot with a blunt object. Despite some variation, this reflex is characterized by dorsiflexion of the big toe and abduction (fanning) of the other toes. It is part of a generalized nociceptive reflex, which also includes flexion of the lower extremity. It is important to note that presence of this reflex may indicate UMN disease only in adults. Children under 2 years of age may normally exhibit this reflex due to incomplete development of the CST. It may also be seen in unconscious individuals, during seizures and sleep, following anesthesia, in fatigue states, as a result of intoxication or following an epileptic seizure. Despite the appearance of UMN deficits, Babinski sign may be absent in chronic paraplegic or hemiplegic patients and in individuals with paralyzed extensor hallucis longus subsequent to common fibular (peroneal) nerve damage.

Babinski reflex is absent in 25% of individuals with other UMN disorders. Equivocal Babinski may be confirmed by Gordon's leg, Gordon's finger, Oppenheim's and Hoffman's signs, Chaddock response, inverse myotatic response, and clonus.

Gordon's leg sign is a Babinski-like sign produced in response to squeezing of the calf muscle in an anteroposterior direction.

TABLE 20.1
Motor Dysfunctions

Conditions	Affected Structures	Deficits
Ventromedian syndrome	Corticospinal tracts	Cruciate hemiplegia
Dorsolateral syndrome	Corticospinal tracts and spinal accessory nucleus	Cruciate hemiplegia and paralysis of the trapezius and sternocleidomastoid muscles
Cerebral palsy	Prenatal or postnatal brain damage	Upper motor neuron palsy (hemiplegia or diplegia), mental retardation, and visual deficits
Pseudobulbar palsy	Corticobulbar tracts	Spastic palsy of the lingual, pharyngeal, masticatory, and palatal muscles and emotional outbursts
Bulbar palsy	Ambiguus and hypoglossal nuclei	Dysphagia, dysarthria, loss of gag and jaw reflexes, and deviation of the uvula to the affected side
Poliomyelitis	α motor neurons of the lumbar and cervical spinal segments; trigeminal motor nucleus may also be affected	Bilateral flaccid paralysis of the affected muscles and Kernig's sign; masticatory and respiratory muscles may also be involved

Gordon's finger sign is characterized by extension of the flexed thumb and index finger in response to application of pressure on the pisiform bone.

Oppenheim's sign is another Babinski-like response produced by a firm downward stroking of the medial surface of the tibia and tibialis anterior muscle.

Hoffman's sign is characterized by clawing movement of the fingers, which includes a sharp and sudden flexion and adduction of the thumb and flexion of the fingers upon flicking the distal phalanx of the middle finger of the pronated hand while grasping it between the index and the thumb. Although this reflex may be indicative of an increase of muscle tone, it should not be used as the only diagnostic sign for UMN palsy.

Chaddock response is a Babinski-like response, which is elicited by the application of a stimulus in a downward direction on the lateral malleolus and the lateral side of the foot extending toward the fifth toe. Since frontal lobe lesions generally produce inhibition of the extensor plantar response and excitation of the grasp reflexes, this response may be used to confirm frontal lobe damage.

Inverse myotatic reflex (clasp-knife reflex) is elicited by passive flexion of the spastic upper or lower extremity. It is mediated by the disynaptic inhibitory action of the Golgi tendon organs on the agonist muscles and the excitatory influence upon the antagonist muscles.

Clonus is a repetitive series of myotatic reflexes elicited by passive dorsiflexion of the foot at the ankle joint (ankle clonus) or by pulling the patella downward (patellar clonus). It may also be elicited by an abrupt and sustained application of a passive stretch as in tapping the quadriceps tendon. In *ankle clonus*, forcible and rapid dorsiflexion of the foot stretches the gastrocnemius muscle, activating the muscle spindle and leading to the firing of the annulospiral (1a) fibers. Group 1a fibers convey the generated impulses from the annulospiral fibers to the α motor neurons of the spinal cord. These activated motor neurons produce contraction of the gastrocnemius muscle, resulting in plantar flexion of the foot. Plantar flexion of the foot at the ankle joint stretches the tibialis anterior muscle, stimulating the muscle spindle. This repetitive series of rapid dorsiflexion and plantar flexion of the foot continues for a prolonged period of time in UMN palsy. Ankle clonus seen in tetanus may also be a quickly exhaustible deficit. Patellar clonus (trepidation sign) exhibits clonic contraction of the quadriceps muscle in response to tapping the quadriceps tendon or upon a downward pull on the patella and stretching of the quadriceps muscle.

Hyperreflexia denotes an increase in reflex activity (as in the patellar reflex) due to a lowered threshold of the deep tendon reflexes. *Hyperreflexia* is an increase in reflex activity due to a lower threshold of the deep tendon reflexes. The presence of myotatic appendages (pseudo-after-discharge) may result in limb falling by its own weight "dead beat" rather than pendular termination of the reflex.

Superficial abdominal reflex is elicited in normal individuals by scratching the abdominal wall and observing the contraction of abdominal muscles and deviation of the umbilicus to the side of the stimulus. In UMN palsy, the umbilicus remains stationary, and the reflex will be absent.

Damage to the CSTs may also occur in ventromedian and dorsolateral syndromes, as well as in cerebral palsy.

In *ventromedian syndrome* (cruciate hemiplegia), signs of spastic palsy are manifested in the ipsilateral upper extremity and the contralateral lower extremity (Figure 20.9). This syndrome results from a lesion at the site of pyramidal decussation that disrupts the fibers of the lower extremity superior (before) to the site of decussation and the fibers to the upper extremity neurons inferior to (after) the level of the decussation.

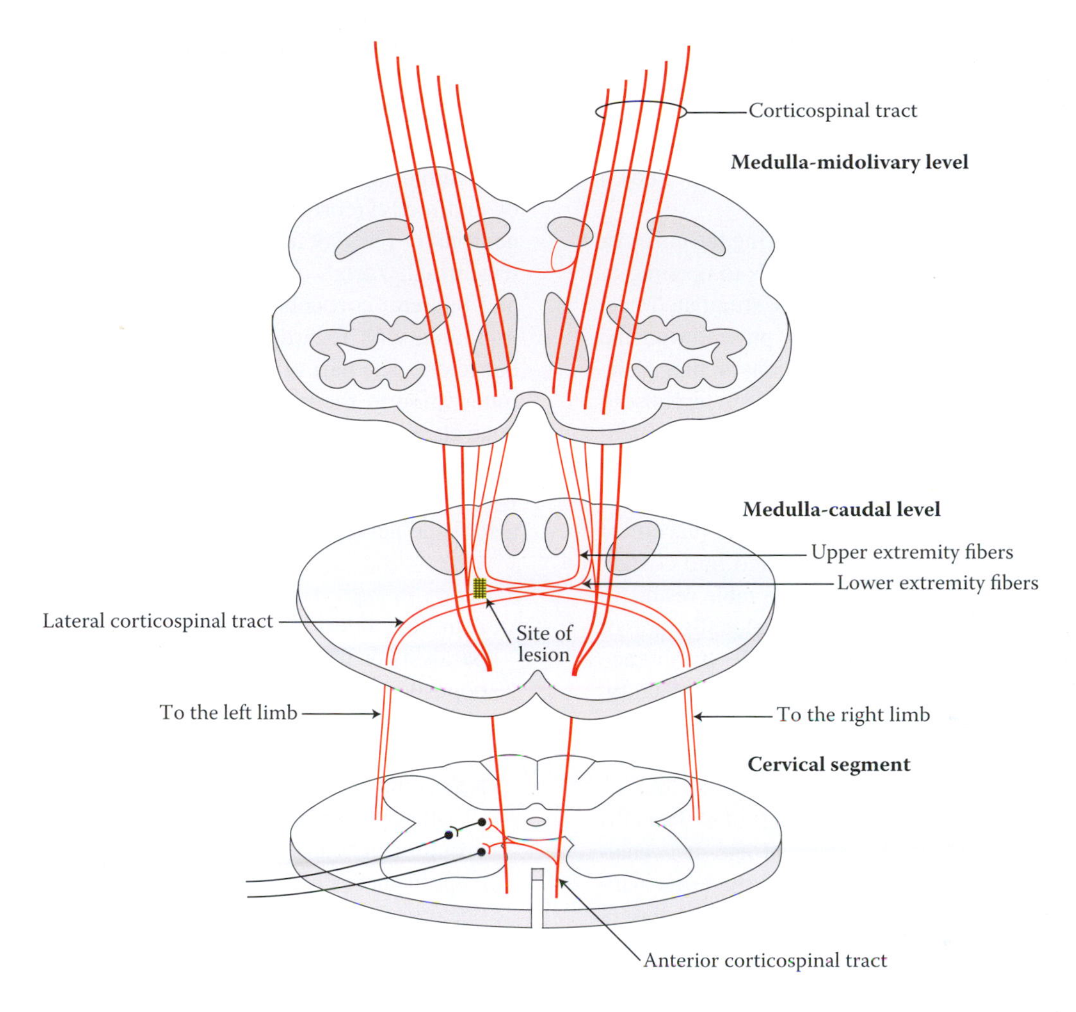

FIGURE 20.9 Diagram of a lesion of the corticospinal tracts at the site of their decussation in the caudal medulla. Due to the somatotopic arrangements of these fibers, upper motor dysfunction will be seen in the ipsilateral upper extremity and contralateral lower extremity.

In *dorsolateral syndrome*, the lesion is at the pyramidal decussation and may also involve the spinal accessory nucleus, producing paralysis of the trapezius and sternocleidomastoid muscles and spastic palsy in the ipsilateral upper extremity and contralateral lower extremity.

Another important condition that exhibits UMN deficits is *cerebral palsy*, which refers to a group of prenatal or postnatal heterogeneous neuromotor dysfunctions that occur as a result of periventricular hemorrhage, malformation of the brain, or asphyxia. It may be the result of hypoxia or ischemia following infection or trauma. Cerebral palsy may be of a progressive or nonprogressive nature. Most infants have normal social and mental development, although one-third of these individuals exhibit varying degrees of mental retardation. Children with minimal cerebral palsy are clumsy and exhibit some visual deficits and signs of UMN palsy. Others may show dyskinesia, ataxia, or combinations of spastic and dyskinetic forms of cerebral palsy. Patients with the spastic form may develop hemiplegia, diplegia, diparesis, tetraplegia, or tetraparesis, and it is occasionally associated with epileptic seizures.

In the dyskinetic form of this disease, the least common type of cerebral palsy, abnormal involuntary movements such as athetosis, choreiform movements, or a combination of choreoathetoid movements and speech disturbances may occur. The ataxic form presents hypotonia, unsteady gait, and loss of coordination.

In determining the integrity of the cerebral motor cortex in *newborns* and *infants*, certain maneuvers may be employed to elicit specific reflexes and signs. These infantile reflexes are primitive and present at birth or appear during the first 6 to 8 months, and they disappear later in life. They include Moro, grasp, rooting, incurvation, placing, crossed extension, parachute, Landau, and tonic neck reflexes.

The *Moro* (startle) *reflex* is evoked by sound or by sudden extension of the neck, which may persist up to 3 years of age. It is characterized by symmetric abduction and extension of the arms, extension of the lower extremity, flexion of the thumb and great toe, and abduction and extension of the digits. Asymmetry in these movements may indicate UMN palsy or brachial plexus injury.

An infant's forceful grasp of the examiner's finger upon stroking the palm of the infant's hand is known as *grasp reflex*.

The *rooting reflex* is elicited by stroking the cheek or perioral region of the infant, which leads to opening of the mouth and turning of the head to the stimulated side.

All normally developed newborns present with the *incurvation* (Galant's) *reflex*, which exhibits arching of the legs and head of the infant on the same side in response to stroking the back of a prone-positioned infant. The arching in this reflex is produced by unilateral contraction of the muscular columns of the erector spinae muscle.

In the *placing reflex or reaction*, passive placement of the dorsum of an infant's foot or hand into contact with the under surface of the edge of a table results in flexion of the hip and knee joints and relocalization of the stimulated foot on the top of the table. Its absence or presence on one side indicates a motor abnormality.

Crossed extension reflex is characterized by flexion followed by extension and abduction of the contralateral leg upon stimulation of the plantar aspect of the foot.

The *parachute reflex* appears at the age of 6–9 months and commonly persists until 2 years of age. It is characterized by abduction and extension of the arms symmetrically upon tilting the head forward from an upright position while holding the infant by the waist. Asymmetry indicates maldevelopment of the motor system.

The *Landau reflex* is characterized by extension of the neck and spine and, to some degree, the lower extremities upon suspending the infant in prone and horizontal positions.

Assessing the degree of brain development can be done via the *tonic neck reflex*, which is elicited by passively turning the head to one side. It is characterized by extension of the upper and lower extremities on the side to which the head is turned and flexion on the joints of the opposite side.

In *adult* patients with UMN palsy, other reflexes may be elicited such as grasp reflex and mass reflex of Riddoch. Generally, these reflexes are expressions of the release of intact structures from the inhibitory and regulatory effects of the supraspinal pathways.

The adult *grasp reflex* is characterized by forcible flexion of the toes and the reluctance to release the grasp in response to stroking the center of the plantar surface of the foot by a blunt object. Presence of this reflex indicates a lesion of the frontal lobe on the contralateral side.

The *mass reflex of Riddoch*, although normal in infants, may occur in adults under stressful conditions. It is characterized by abrupt evacuation of the bladder and sometimes bowel and sweating, accompanied by flexion of the lower extremity.

The *CBT* conveys impulses from the cerebral cortex to the motor nuclei of the cranial nerves (Figures 20.4 and 20.5). It is primarily a bilateral tract with a contralateral predominance. The fibers of this tract project to the motor nuclei of the cranial nerves either directly via a monosynaptic route or indirectly (cortico-reticulo-bulbar tract) through interneurons in the reticular formation. The motor nuclei of the trigeminal, facial, and hypoglossal nerves receive direct and bilateral corticobulbar fibers. The facial motor nucleus is unique with regard to its connections to the corticobulbar fibers. The part of the facial motor nucleus that provides innervation to the muscles around the mouth receives only contralateral corticobulbar projections, while the part of the nucleus that innervates the forehead and upper facial muscles receives bilateral corticobulbar fibers from both sides (see also facial nerve).

A unilateral lesion of the CBT is most likely to result in paralysis or paresis of the lower facial muscles on the opposite side, while preserving or minimally affecting all other motor nuclei. Interestingly, an ipsilateral lesion of the CBT in the pons may produce glabellar hyperreflexia.

Bilateral destruction of the CBT may result in a group of deficits collectively known as *pseudobulbar* (supranuclear) *palsy*. This form of dysfunction may occur in amyotrophic lateral sclerosis (ALS, which may also produce bulbar palsy), multiple cerebral infarcts, Alzheimer's disease, bilateral cortical atrophy, cerebral palsy, bilateral lesions of the frontal lobes, and rostral brainstem damage. It is also seen in occlusion of the posterior cerebral and basilar arteries. It is characterized by spastic paralysis of the lingual, pharyngeal, laryngeal, masticatory, and palatal muscles. Dysphagia due to spastic palsies of the pharyngeal and palatal muscles may lead to accumulation of food particles in the nasopharynx or trachea. Patients may develop a tendency to choke when swallowing. Therefore, it is advisable that patients be given a diet in the form of paste. Sensory deficits and cerebellar dysfunctions are absent. Stimulation of the posterior pharynx and palate results in strong gag sensation and forceful contraction of the palatal muscles and elevation the soft palate (gag hyperreflexia). Severe dysarthria (speech difficulty) in the form of explosive speech with nasal sound is also seen as a result of weakness (paresis) of the facial and buccal muscles.

Articulation is not commonly impaired in regard to syntax, grammar, or comprehension of the spoken language, but aphasia may be a presenting problem in some patients. The inability to voluntarily control breathing (e.g., coughing and blowing out) may cause death as a sequel to aspiration pneumonia. Due to disruption of the limbic system's connections to the motor nuclei of the affected muscles, individuals with this disorder may also exhibit signs of emotional disturbances including expressionless gaze, sudden and unrestrained (pathological) crying, or laughing or giggling in response to minimal stimuli. These emotional outbursts may last for a few minutes or until exhaustion. Some cannot smile or express sadness by frowning. Patients with pseudobulbar palsy also exhibit jaw jerk hyperreflexia, dementia, and intellectual deterioration.

On the other hand, *bulbar* (medullary) *palsy* is the result of the damage to the motor nuclei of the 9th, 10th, 11th, and 12th cranial nerves (IX–XII). This condition may be seen in poliomyelitis, ALS (Lou Gehrig's disease), and occlusion of the vertebral artery. Patients may also exhibit dysphagia (difficulty swallowing) and dysarthria (thickened speech with heavy nasal intonation). However, the gag and jaw jerk reflexes remain depressed, and the uvula deviates toward the intact side. Emotional and intellectual deficits are absent.

LOWER MOTOR NEURONS (LMNs)

LMNs are group of neurons that lie in the spinal cord and brainstem. They are represented by the α and γ motor neurons of the ventral horn, intermediolateral column neurons of the lateral horn, and motor nuclei of the cranial nerves (III, IV, VI, and XII). These neurons provide general somatic, general visceral, and/or special visceral fibers, and they constitute the final common pathway for the motor system (Figure 20.10). They occupy Rexed lamina IX, exhibiting a somatotopic arrangement within the ventral horn of the spinal cord in which neurons that innervate the hand and forearm occupy a lateral position to those that supply the shoulder and arm. Neurons for the foot and leg are located lateral to the neurons of the hip and trunk. Additionally, the flexor neurons are dorsal to the extensor neurons, and the adductor neurons are dorsal to the abductor neurons. The axons of α and γ motor neurons, which receive descending motor fibers either directly or indirectly, run through the ventral roots and then the spinal nerves. Both α and γ neurons are involved in voluntary movements via simultaneous contraction of the intrafusal and extrafusal muscles, respectively. These activities are mediated via gamma loop, which enables the brain to continuously monitor the state of contraction of the agonists and relaxation of the antagonists.

The *general somatic motor efferent* (GSE) component of the oculomotor nucleus provides innervation to most of the extraocular muscles, whereas the GSE of the trochlear nucleus innervates the superior oblique muscle. GSE fibers in the abducens nucleus innervate the lateral rectus muscle and also provide fibers to the contralateral medial rectus neurons of the oculomotor nucleus through the medial longitudinal fasciculus (MLF). These two sets of GSE fibers from the abducens nucleus become functional in lateral gaze, exhibiting abduction of the ipsilateral eye and adduction of the contralateral eye. GSE fibers of the hypoglossal nucleus innervate the lingual muscles.

The *general visceral motor neurons* consist of a spinal part contained in the intermediolateral columns of the thoracolumbar (presynaptic sympathetic) and sacral spinal (presynaptic parasympathetic) segments and a cranial part

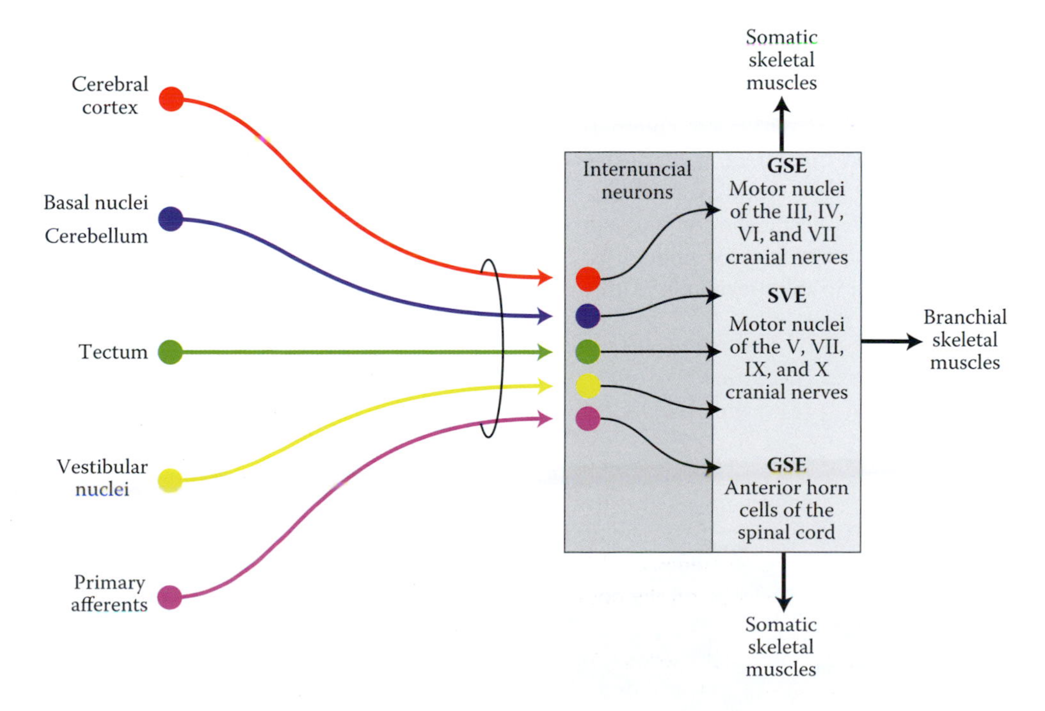

FIGURE 20.10 Schematic drawing of the somatic and visceral lower motor neurons within the spinal and cranial nerves. Note that the *general somatic efferent* neurons are represented in the spinal α motor neurons and the motor nuclei of the oculomotor, trochlear, abducens, and hypoglossal nerves, whereas the *special visceral motor* neurons innervate muscles of branchial origin. As is seen in this diagram, these lower motor neurons are influenced by activities of the cerebral cortex, basal nuclei, vestibular system, and tectum.

contained in the motor nuclei of cranial nerves III, VII, IX, and X (see autonomic nervous system). Presynaptic parasympathetic neurons form the Edinger–Westphal nucleus of the oculomotor nerve, superior salivatory nucleus of the facial nerve, inferior salivatory nucleus of the glossopharyngeal nerve, and dorsal motor nucleus of vagus.

The *special visceral efferent neurons* (SVE) constitute the motor nuclei of the trigeminal, facial, glossopharyngeal, vagus, and accessory nerves that provide innervation to muscles of branchial arch origin. Neurons of the trigeminal motor nucleus innervate the muscles of mastication, tensor tympani, tensor palatini, mylohyoid and anterior belly of digastric. In a similar fashion, the facial motor nucleus (SVE) is responsible for the innervation of the facial muscles of expression, stapedius, stylohyoid, and posterior belly of the digastric. Nucleus ambiguus (SVE) sends fibers through the glossopharyngeal, vagus, and accessory nerves to the muscles of pharynx, larynx, and soft palate.

LOWER MOTOR NEURON (LMN) PALSIES

Lesions of the LMNs may involve the anterior horn, ventral root, spinal nerve, or motor nuclei of cranial nerves, producing signs of *LMN palsy*. These deficits consist of ipsilateral flaccid paralysis, atrophy (wasting of the affected muscles), fasciculation, and areflexia. Fasciculation results from the action potentials in the terminal branches of the nerve fibers prior to their degeneration. They are visible through the skin and may be elicited by strenuous exercise. These deficits are exemplified at spinal level in poliomyelitis.

Poliomyelitis is a viral disease caused by poliovirus, a member of the family of Picornaviridae of enteroviruses that inhabit the gastrointestinal tract. It nearly has been eradicated from the western world, yet temperate climates of the Eastern Mediterranean, Africa, and Indian subcontinent remain a hot bed for poliovirus. Despite the successful record in the prevention of this disease, cases of inactivated poliovirus vaccine-induced paralytic polio, after the first immunization, have been reported. Vaccine-associated viruses can be excreted for 2 months. Humans act as a reservoir for this highly infectious virus, and it spreads during summer predominantly through the fecal–oral route. The virus enters through the oral route and resides in the pharynx and gastrointestinal tract even before any symptoms appear. The incubation period for poliomyelitis ranges between 1 and 3 weeks. It attacks the lymphoid tissue and enters blood circulation and then invades the central nervous system (CNS). The lesions caused by this virus are random, and the degeneration is followed by gliosis. Some neurons that escape the disease may increase in size and form relatively giant motor units. It affects the anterior horn neurons of the lumbar and cervical enlargements and, to a lesser degree, the brainstem motor neurons. Polio infections, for the most part, remain asymptomatic, and a small percentage of patients exhibit systemic symptoms such as fever, sore throat, nausea, vomiting, and abdominal pain. In small population of patients, signs of aseptic meningitis may be seen in the form of nuchal rigidity followed by recovery. The neurological symptoms usually appear late in the course of the disease, including flaccid paralysis, areflexia, and atrophy of the affected muscles. Involuntary spasm of the hamstring muscles and pain (*Kernig's sign*) can be elicited in this disease when the patient's hip and knee are flexed at 90° initially in a supine position and then the knee extended while the hip maintains a flexed position (Figure 20.11). The *paralytic stage* of this disease is characterized by tenderness and painful contractions of muscles, which may be elicited by movement or exposure to cold. Usually, the appendicular muscles are asymmetrically affected (spinal polio), although muscles of mastication, respiration, and swallowing may also be involved (bulbar polio). The facial muscles are less frequently involved. Paralysis of these muscles may be severe and generalized, and

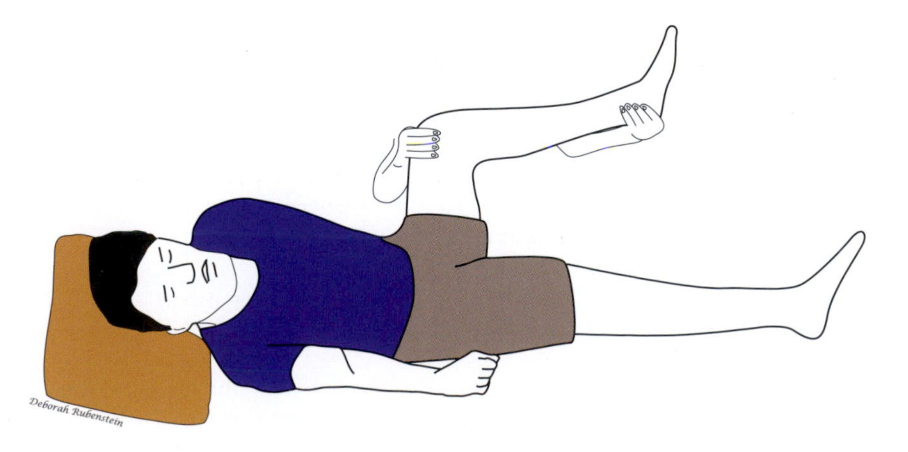

FIGURE 20.11 Kernig's sign. Observe the examiner's attempt to extend the knee.

the recovery remains slow and lasts for weeks. Urine retention may be observed, but it disappears in a short period of time. No sensory disturbances or deficits will be detected in this disease. In infants, bone growth will also be adversely affected. Recovery is possible if the respiratory muscles and diaphragm remain unaffected. The Cerebrospinal Fluid shows an increase in protein and cellular components. In one-fourth of the patients, increased paresis may occur years after recovery from paralytic poliomyelitis.

Combined UMN and LMN deficits may occur as a result of lesions in the cerebral cortex, brainstem, or spinal cord (Table 20.2). *Spinal lesions* that produce combined motor deficits are manifested in *ALS* (Lou Gehrig's disease), a relatively uncommon and progressive degenerative disease (Figure 20.12), which is characterized by bilateral degeneration of both the CST and CBT (UMNs) and the anterior gray columns (LMNs). Signs of LMN palsy are seen at the levels of the involved segments, whereas UMN deficits are manifested contralaterally below the levels of the affected segments. ALS affects both the proximal and distal musculature, and its distribution is asymmetrical and may remain confined to one side for a period of time. The spasticity may or may not precede the muscular atrophy. No sensory loss will be detected. This disease may be fatal when it involves the anterior horns of the cervical and thoracic segments of the spinal cord due to respiratory depression. Despite the diffuse nature of destruction of CNS motor neurons, mental faculties remain unaffected. Surprisingly, deep tendon reflexes show hyperactivity with muscle atrophy. However, motor impairment does not involve bladder and bowel functions or ocular movements. Involvement of the motor nuclei of the trigeminal, facial, and hypoglossal nerves in *bulbar amyotrophic lateral sclerosis* may result in facial asymmetry, dysarthria, and dysphagia.

Brainstem lesions may also produce combined UMN and LMN dysfunctions, which are seen in the medulla, pons, and midbrain.

TABLE 20.2
Combined Upper and Lower Motor Neuron Dysfunctions

Clinical Conditions	Affected Structures	Deficits
Amyotrophic lateral sclerosis (ALS)	Ventral horns (LMN); lateral corticospinal and corticobulbar tracts; as well as trigeminal, facial, and hypoglossal motor nuclei	Bilateral flaccid and spastic palsies of the affected muscles; bilateral paralysis of the masticatory, pharyngeal, and laryngeal muscles
Bulbar amyotrophic lateral sclerosis	Trigeminal motor, facial, hypoglossal, and ambiguus nuclei	Bilateral atrophy of masticatory, facial, and lingual muscles; cardiac arrhythmia may also be seen
Cestan–Chenais syndrome	Nucleus ambiguus, corticospinal tract, and descending autonomic fibers	Dysphagia, dysphonia, contralateral hemiplegia, and Horner's syndrome
Inferior alternating hemiplegia	Ipsilateral corticospinal tract and hypoglossal nerve	Flaccid palsy of the lingual muscles and contralateral hemiplegia
Jackson's syndrome	Ambiguus and hypoglossal nuclei and the corticospinal tract	Signs of Avellis syndrome and ipsilateral paralysis of the lingual muscles
Schmidt's syndrome	Ambiguus nucleus and corticospinal tract	Ipsilateral paralysis of the vocal cord, palate, trapezius, and sternocleidomastoid muscles; contralateral hemiplegia may also be observed
Foville's syndrome	Abducens nucleus, facial nerve, and corticospinal tract	Ipsilateral facial palsy, contralateral hemiparesis, and lateral gaze palsy
Millard–Gubler syndrome	Facial and abducens nerves and corticospinal tract	Facial palsy, medial strabismus, diplopia, and contralateral hemiplegia
Locked-in syndrome	Corticospinal and corticobulbar tracts on both sides	Quadriplegia and inability to articulate; eye movements remain intact
Raymond–Cestan syndrome	Abducens nerve and corticospinal tract, sometimes the trigeminal nerve	Signs of abducens nerve palsy, contralateral hemiparesis, and sometimes sensory loss in the face
Interpeduncular syndrome	Corticospinal tracts and oculomotor nerves on both sides	Oculomotor palsy and quadriplegia
Superior alternating hemiplegia (Weber's syndrome)	Oculomotor nerve and corticospinal tract	Ipsilateral oculomotor palsy and contralateral hemiplegia
Decerebrate rigidity	Corticospinal and medullary reticulospinal tracts	Rigid posture with extended arms, legs, and feet and flexed hand
Decorticate rigidity	Cerebral cortex	Flexion of the elbow, wrist, and fingers, and extension of the hip, knee, and ankle joints

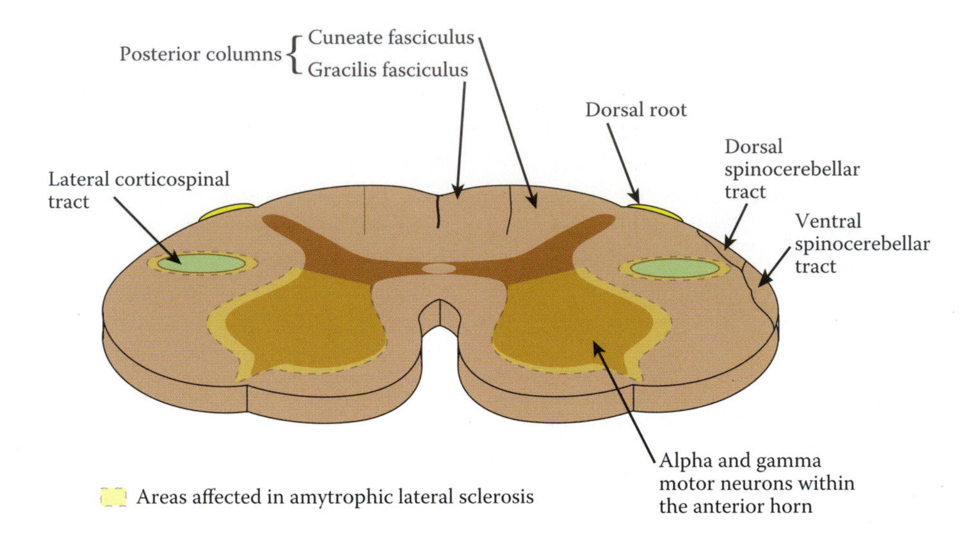

FIGURE 20.12 Section of the spinal cord shows bilateral degeneration of the lateral corticospinal tracts and anterior horns in amyotrophic lateral sclerosis.

Medullary lesions produce combined upper and LMN deficits that are manifested in inferior alternating hemiplegia, Jackson's syndrome, and Schmidt's syndrome.

Inferior alternating hemiplegia (Figure 20.13) results from a single lesion involving both the pyramidal tract and the hypoglossal nerve on the same side. Flaccid paralysis of the ipsilateral lingual muscles and contralateral spastic paralysis of the upper and lower extremities characterize this condition.

Jackson's syndrome is caused by a lesion that damages the nucleus ambiguus, hypoglossal nucleus, and the pyramidal tract. Patients exhibit manifestations of the inferior alternating hemiplegia as well paralysis of the muscle of pharynx, larynx and soft palate.

Schmidt's syndrome, characterized by contralateral spastic hemiplegia and ipsilateral flaccid palsy of the trapezius and sternocleidomastoid muscles and muscles of pharynx, larynx and soft palate, is produced by a lesion that disrupts the ipsilateral nucleus ambiguus and the pyramidal tract.

Pontine lesions also produce combined motor deficits that are seen in Millard–Gubler, Foville's, Raymond's, and locked-in syndromes.

Foville's syndrome results from a combined lesion of the abducens nucleus, facial nerve, solitary and spinal trigeminal nuclei, central tegmental tract, and CST. In addition to the ipsilateral facial palsy and contralateral hemiplegia, patients also exhibit ipsilateral facial

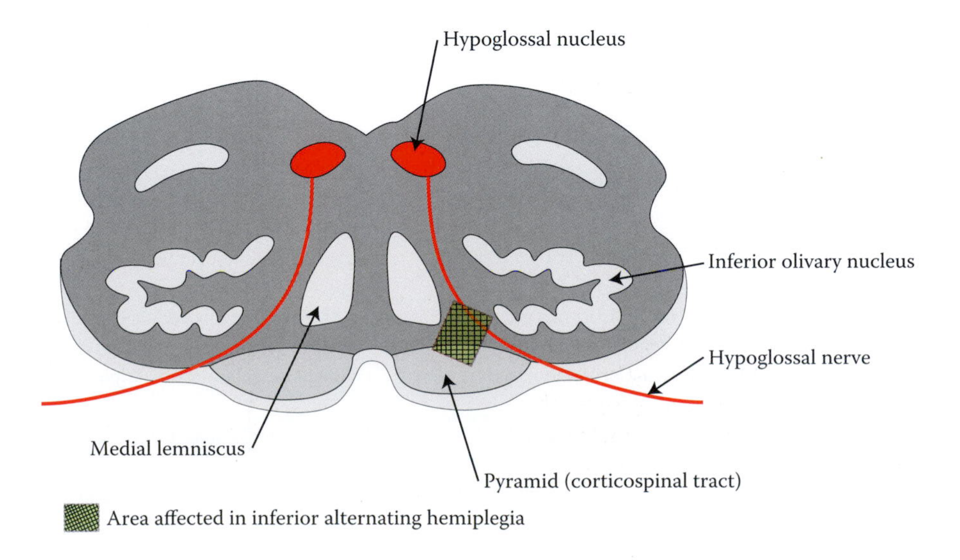

FIGURE 20.13 Image of the medulla with a lesion involving the corticospinal tract and hypoglossal nerve in inferior (hypoglossal) alternating hemiplegia.

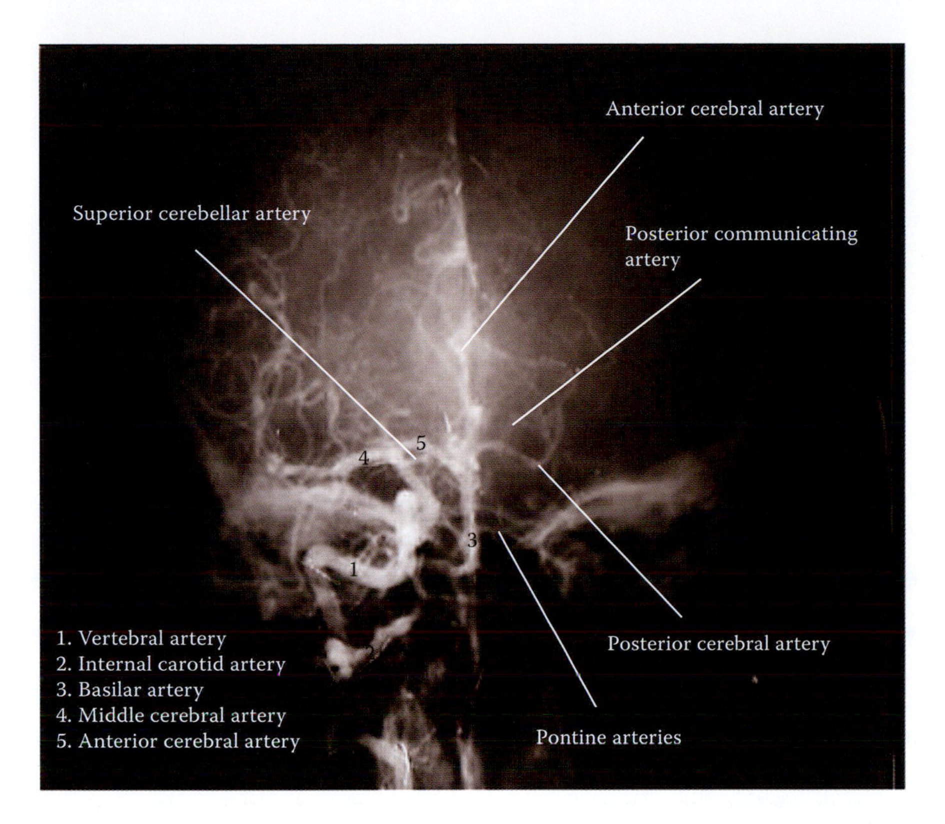

FIGURE 20.14 In this drawing of the caudal pons at the level of the abducens and facial nerves, observe the lesion that disrupts the abducens nerve and the corticospinal tract, responsible for Raymond's syndrome.

anesthesia and dysgeusia, Horner's syndrome and lateral gaze palsy. The latter is characterized by loss of conjugate horizontal movement to the ipsilateral side and forced deviation of both eyes to the contralateral side.

Raymond's syndrome (middle or abducens alternating hemiplegia) is another pontine disorder that results from a lesion of the abducens nerve and the CST. Patients exhibit ipsilateral signs of abducens nerve palsy and contralateral hemiplegia (Figure 20.14).

Millard–Gubler syndrome is caused by intrapontine hemorrhage, thrombosis, or tumor, which results in damage to the facial and abducens nerves, as well as the CST. Patients present with diplopia, facial palsy, and medial strabismus on the ipsilateral side and hemiplegia on the contralateral side.

In a similar fashion, *midbrain* lesions produce combined motor deficits, which are clearly seen in Weber's and interpeduncular syndromes.

Weber's syndrome (Figures 20.15 and 20.16) is a fairly common condition, which results from damage to the oculomotor nerve and the CST. It is characterized by signs of ipsilateral oculomotor palsy and contralateral hemiplegia. Hemorrhage due to a ruptured aneurysm of the posterior communicating artery, unilateral thrombosis of paramedian branches of the posterior cerebral artery, and metastatic lesions are common etiologies of this condition.

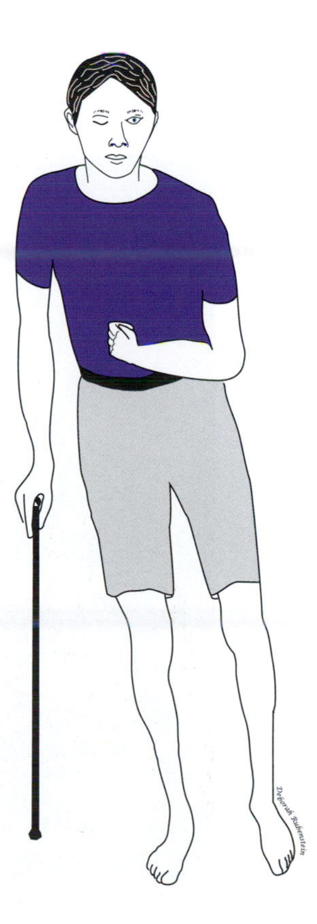

FIGURE 20.15 This drawing depicts a patient with Weber's syndrome. Oculomotor dysfunctions (e.g., ptosis) on the right side and spastic palsy on the left half of the body are evident.

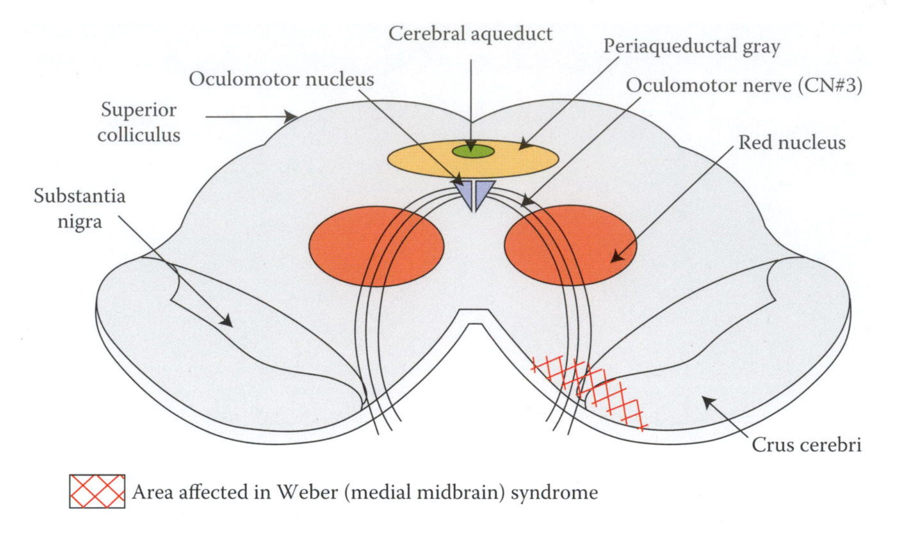

FIGURE 20.16 Drawing of the midbrain at the level of the superior colliculus. The hatched area represents the lesion site associated with Weber's syndrome, involving the CST and the oculomotor nerve.

Interpeduncular syndrome is usually of vascular origin and results from bilateral damage to the CSTs and the oculomotor nerves. It is characterized by quadriplegia, bilateral loss of convergence and vertical eye movements, ptosis, and mydriasis.

Locked-in syndrome (pseudocoma) is secondary to sustained multiple vascular infarcts caused by occlusion of the pontine branches of the basilar artery (Figure 20.17). This syndrome is associated with bilateral destruction of the CBT and CST in the basilar pons, while preserving the dorsal tegmentum. Patients remain conscious and aware of their surroundings (intact cognitive and affective functions), but suffer from mutism (inability to articulate) and quadriplegia. They are unresponsive to external stimuli and unable to move their trunk or limbs, while

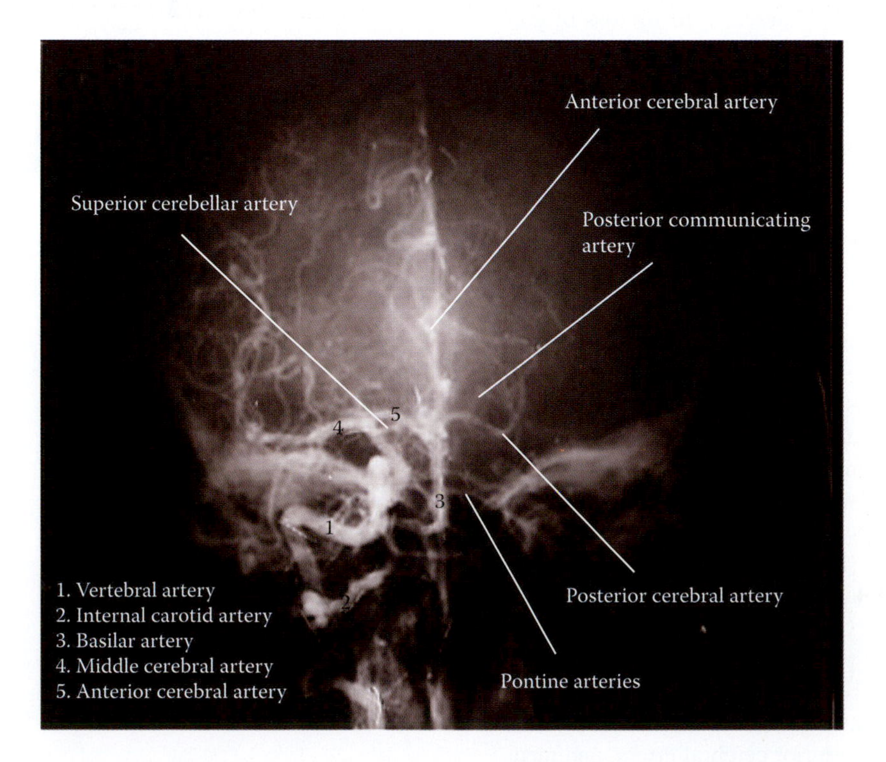

FIGURE 20.17 In this angiogram, both the internal carotid artery and vertebrobasilar system are indicated. The connection between these two arterial systems is maintained by the posterior communicating artery, which is also shown in this angiogram. Note the pontine arteries from that supply the CBT and CST and are involved in locked-in syndrome.

sparing movements of the eye and eyelid. This condition may mimic signs and symptoms of severe cases of Guillain–Barré syndrome and myasthenia gravis. Due to intactness of the midbrain and cerebrum, cerebral cortical activity remains normal despite the physical impairment, and partial recovery is possible. One of the important aspects of this condition is the ability of patients to comprehend the events surrounding them and understand peoples' comments and reactions.

Cerebral lesions may also produce combined upper and lower motor dysfunctions that are evident in decerebrate rigidity.

Decerebrate (gamma) *rigidity* (Figure 20.18) may be attributed to the facilitatory effect of the pontine reticulospinal and the vestibulospinal tracts and to the lack of inhibitory effect of the cortically dependent medullary reticulospinal tract. It is caused by heightened activities of the γ neurons that result from transection of the midbrain at the intercollicular level or from a pathological condition anywhere between the midbrain and the first cervical spinal segment. It may occur spontaneously or in response to mild or noxious stimuli. Decerebrate rigidity may be seen as a result of vascular occlusion, compression, metabolic disorders such as hypoxia or hypoglycemia, or inflammatory process. The antigravity muscles show increased tone as an expression of the facilitation of the γ motor neurons, augmenting the firing rate of the muscle spindles and the α motor neurons. Patients with this condition exhibit a rigid posture in which the jaw is clenched, the extremities are fully extended, the feet are plantar flexed, the hands are extended or clenched, and the wrists are usually flexed and facing forward. Some patients may exhibit tonic neck reflex, which is characterized by flexion of the left elbow and extension of the right elbow upon passive turning of the head to the right side with no detectable movements of the lower extremity. Signs of decerebrate rigidity can be reduced or abolished by destruction of the vestibular nuclei, vestibulospinal tract, or inner ear labyrinth. Segmental transection of the dorsal roots may prove beneficial in relieving the patient from signs of rigidity. This condition should not be confused with decorticate rigidity.

Patients with *decorticate rigidity* (Figure 20.18) exhibits flexed elbow, wrist, and fingers with the arms and forearms held tight to the chest, extended thigh, extended and internally rotated leg; and plantar flexed foot. It is seen in comatose individuals with a diffuse lesion of the cerebral hemispheres.

Decorticate posturing can results from supratentorial lesions rostral to the red nucleus. This posture is produced subsequent to the facilitatory effect of the rubrospinal tract on the flexor neurons of the cervical segments that overcomes the excitatory effect of the medial and lateral vestibulospinal and pontine reticulospinal tracts on the extensor neurons producing the flexed posture of the upper extremities. While disruption of the inhibitory corticospinal tract leads to the unopposed excitatory action of the pontine reticulospinal and lateral vestibulospinal tracts on the extensor

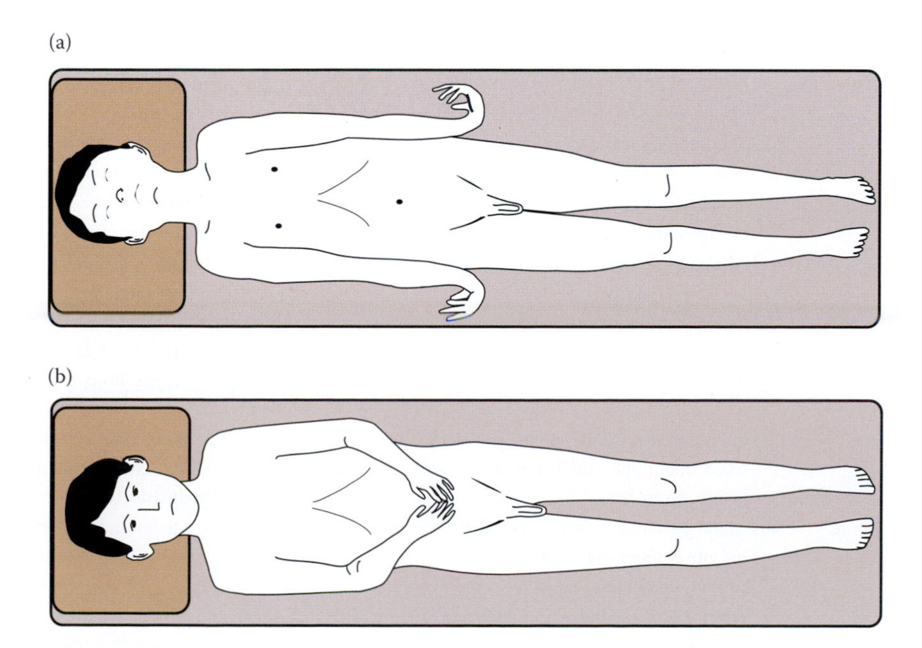

FIGURE 20.18 These diagrams illustrate (a) decerebrate rigidity and (b) decorticate rigidity.

neurons producing the extended posture in the lower extremities seen in these patients. Plantar flexion of the feet is most likely to be attributed to the unopposed facilitatory input to the flexor neurons of the sacral spinal segments that control flexion at the ankle joint.

Combined sensory and motor dysfunctions occur as a result of lesions involving the spinal cord, brainstem, or cerebral cortex (Table 20.3).

Spinal lesions that produce combined system deficits may be seen in anterior spinal artery syndrome, Brown-Séquard syndrome (spinal cord hemisection), cauda equina syndrome, conus medullaris syndrome, transection of the spinal cord, ependymoma, multiple sclerosis (MS), spina bifida, pellagra, subacute combined system degeneration, and syringomyelia.

Anterior spinal artery (Beck) *syndrome* (Figure 20.19) is characterized by softening of the anterior two-thirds of the spinal cord and the resultant destruction of the spinothalamic and pyramidal tracts. It may be produced by occlusion of the anterior spinal artery as

TABLE 20.3
Combined Motor and Sensory Dysfunctions

Clinical Conditions	Affected Structures	Deficits
Anterior spinal artery (Beck) syndrome	Spinothalamic and corticospinal tracts	Bilateral anesthesia and upper motor neuron palsy
Brown-Séquard Syndrome	Anterolateral system, dorsal column, corticospinal tract, descending autonomic fibers, and neurons of the ventral and dorsal horns	Contralateral hemianesthesia of the body, ipsilateral loss of fine touch and proprioception, ipsilateral spastic palsy, signs of Horner's syndrome, loss of all sensations, flaccid palsy, atrophy, and areflexia on the side of the affected segment
Cauda equina syndrome	Lumbosacral roots	Pain in the involved dermatomes and flaccid palsy
Conus medullaris syndrome	Sacral segments	Bladder, bowel, and sexual disturbances, and anesthesia in the gluteal region
Ependymoma	Ependyma of sacral segments of spinal cord and fourth ventricle	Muscle paresis, incontinence, hydrocephalus, and paraplegia
Multiple sclerosis	Dorsal columns, corticospinal tracts, optic nerve, lateral spinothalamic tract, and ventral horn of the spinal cord	Locomotor ataxia, upper motor neuron palsy, paresthesia or anesthesia, visual deficits, and incontinence
Spina bifida (myeloschisis)	Lumbosacral roots and/or segments	Incontinence, sexual dysfunction, sensory and motor loss, and back pain
Pellagra	Anterior horn, cortical neurons, dorsal columns, and spinal nerves	Locomotor ataxia, cogwheel rigidity, paresis, and subcortical motor signs
Refsum disease	Dorsal columns, ventral horn, and spinal nerves	Locomotor ataxia, polyneuritis, visual deficits, and deafness
Subacute combined system degeneration	Dorsal columns and corticospinal tract	Impairment of proprioception and vibratory sensations, and signs of upper motor neuron palsy
Transection of the spinal cord	All ascending and descending tracts, dorsal and ventral horns	Loss of all sensations, muscle tone, and reflexes below the level of transaction followed by stages of recovery
Avellis syndrome	Nucleus ambiguus, lateral spinothalamic, and corticospinal tracts	Dysphagia, dysphonia, contralateral hemianesthesia, and hemiparesis
Medial medullary syndrome	Hypoglossal nucleus, medial lemniscus and corticospinal tract	Ipsilateral atrophy of lingual muscles, contralateral loss of proprioception and signs of spastic paralysis
Lateral medullary (PICA/Wallenberg's) syndrome	Nucleus ambiguous spinal trigeminal nucleus anterolateral system descending sympathetic fibers	Dysphagia, ipsilateral loss of pain and temperature sensations from the face, contralateral loss of pain and temperature sensations from the body, Horner's syndrome
Anterior inferior cerebellar artery (AICA) syndrome	Facial motor nucleus, spinal lemniscus, spinal trigeminal tract, middle cerebellar peduncle, and vestibular nuclei	Ipsilateral facial paralysis, contralateral hemianesthesia, ipsilateral hemianesthesia of face, and signs of cerebellar dysfunctions
Superior cerebellar artery syndrome	Lateral gaze center, anterolateral system, and middle cerebellar peduncle	Lateral gaze palsy, cerebellar dysfunctions, and contralateral hemianesthesia
Benedikt syndrome	Superior cerebellar peduncle, oculomotor nerve, red nucleus, medial, spinal lemnisci, and possibly the corticospinal tract	Ataxia, oculomotor palsy, contralateral loss of position sense and hemianesthesia, and signs of cerebellar dysfunction. This Syndrome may also exhibit hemiparesis

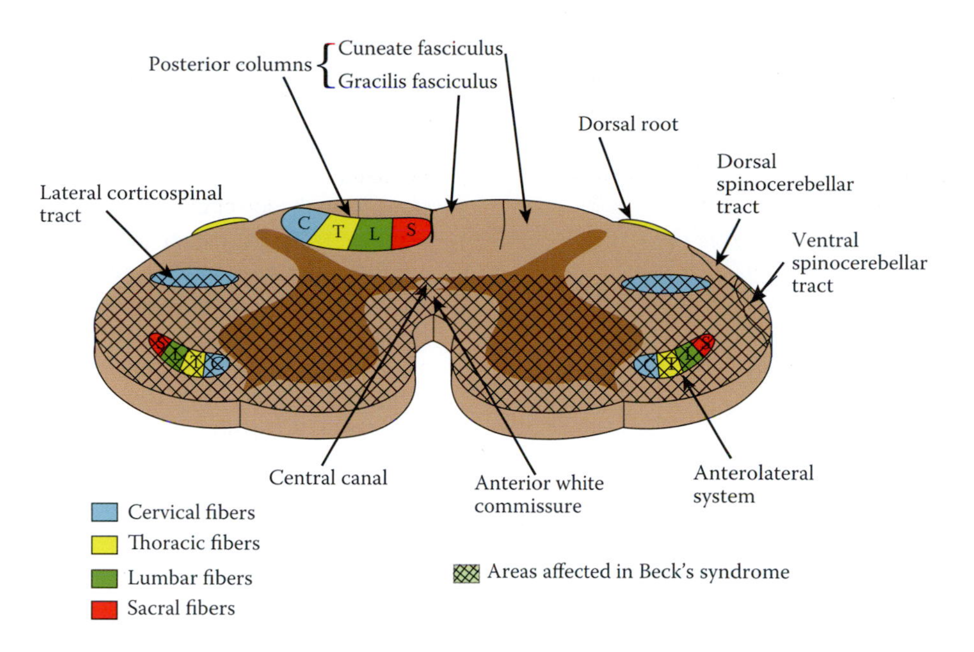

FIGURE 20.19 Drawing of the section of the spinal cord showing degeneration of its anterior two-thirds associated with occlusion of the anterior spinal artery.

a result of thrombosis secondary to atherosclerosis or developing tumor. There is bilateral loss of pain and temperature sensations, which extend one or two segments below the level of the lesion. Tactile, vibratory, and position senses remain unaffected.

Brown-Séquard syndrome (hemisection of the spinal cord) is produced by disruption of the dorsal root, dorsal horn, ventral gray column, and ascending and descending pathways on one side of the spinal cord (Figures 20.20 and 20.21). This condition may result from shrapnel wounds or fracture dislocation of the vertebrae. It is characterized by ipsilateral loss of motor functions (UMN and LMN) and discriminative tactile as well as joint sensations below the level of the lesion. All sensory and motor activities are impaired at the level of the affected segment(s) on the lesion side. Pain and temperature sensations are lost on the contralateral side, extending one or two segments below the level of the lesion. Signs of UMN palsy (e.g., hyperactive tendon reflexes, spasticity, positive Babinski sign, Hoffman's sign, clonus, etc.) are seen ipsilaterally below the level of the lesion. Symptoms of LMN palsy, such as flaccidity, areflexia, and muscle atrophy, may be observed at the level of the affected segment.

Cauda equina syndrome is a condition produced by compression of the lumbosacral roots as a result of a tumor or prolapse of the intervertebral disc below the first lumbar vertebra. Pain, as a first sign of this condition, is constant, radiating across the dermatomes of the involved roots, followed by flaccid paralysis (wasting of the tibialis anterior muscle), areflexia, and saddle-shaped anesthesia in the gluteal region (when the lower sacral roots are involved). Pain may be felt in the lumbar area or in both legs. It may be aggravated by physical activity and coughing, and in contrast to the pain generated by herniated disc, it remains relatively unaffected by bed rest. Disturbance of micturition may appear late in the course of this condition, but no pyramidal signs may be observed. Numbness, tingling, or burning sensation may be felt long before any objective findings of sensory loss are detected. Both knee and ankle jerk reflexes may be lost. Sensory loss over the anterior thigh may indicate compression of the upper lumbar roots, whereas compression of the lower lumbar roots produces sensory loss in the anterior and lateral leg.

Conus medullaris syndrome is a condition that may result from aortic aneurysm, prolapse of the L1/L2 intervertebral disc, abdominal operations such as nephrectomy, and sympathectomy, neoplasm that damage all or many of the sacral spinal segments. It is characterized by impaired bladder (atonic bladder that has lost the ability to initiate or inhibit urination) and bowel functions, as well as sexual disturbances (impotence). Sensory loss will be detected in the gluteal region but not in the posterior thigh or leg due to extensive sensory overlap. No major motor deficits are observed, although weakness of movement in the feet may occur in individuals with a lesion involving the ventral and dorsal spinal gray columns.

Transection of the spinal cord may result from shrapnel wounds, expanding intramedullary and/or extramedullary tumors, trauma, or occlusion of the artery of lumbar enlargement. This condition produces

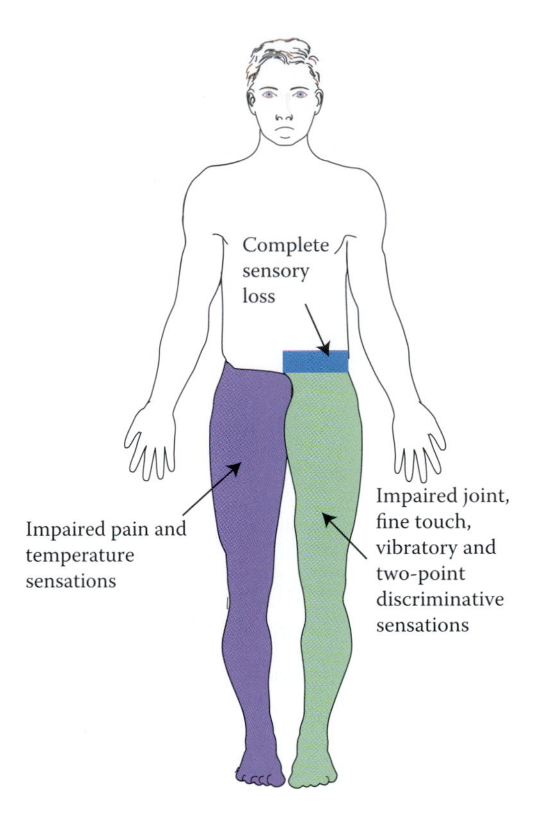

FIGURE 20.20 This schematic drawing illustrates the deficits associated with hemisection of the spinal cord at the left 10th thoracic spinal segment (Brown-Séquard syndrome).

manifestations of *spinal shock*, which includes loss of all somatic and visceral sensations, muscle tone, and reflexes below the level of the lesion. There may be a narrow zone of hyperesthesia at the upper margin of the anesthetic region. Sensory loss may not correspond to the level of the lesion since a lesion that starts from the periphery inward is most likely to initially affect the outermost fibers that carry pain and temperature sensations from the lower extremity. In contrast, a lesion that expands from the center in an outward direction disrupts these sensations in a reverse way. The recovery stages in humans may last more than 6 months and terminate by the appearance of Babinski sign. Recovery phases are orderly, starting with the appearance of minimal reflexes, followed by flexor muscle spasm and alternate flexor–extensor muscle spasm and eventual extensor spasm. Exaggerated flexor reflex activity such as triple flexor reflex (flexion of the hip, knee, and thigh) in response to a mild stimulus may occur during the stage of flexor spasm. Hyperreflexia may be due to hypersensitivity of spinal motor neurons and interneurons and/or release from the inhibitory influences of the UMNs. Bladder and bowel dysfunctions are the most distressing symptoms associated with this condition.

Ependymoma, a rare glioma of the ependymal lining of the ventricular system and the spinal cord, occurs typically in the vicinity of the fourth ventricle in the first two decades of life and in the sacral spinal segments in middle age. In some patients, the onset of the condition is associated with trauma. Symptoms usually include progressive muscle paresis with insidious onset and pain. Incontinence, hydrocephalus, paraplegia, sensory disorders, and priapism are also seen. Patients may be relieved of pain upon standing or walking and show a positive *Lasegue's test* (pain and limitation of

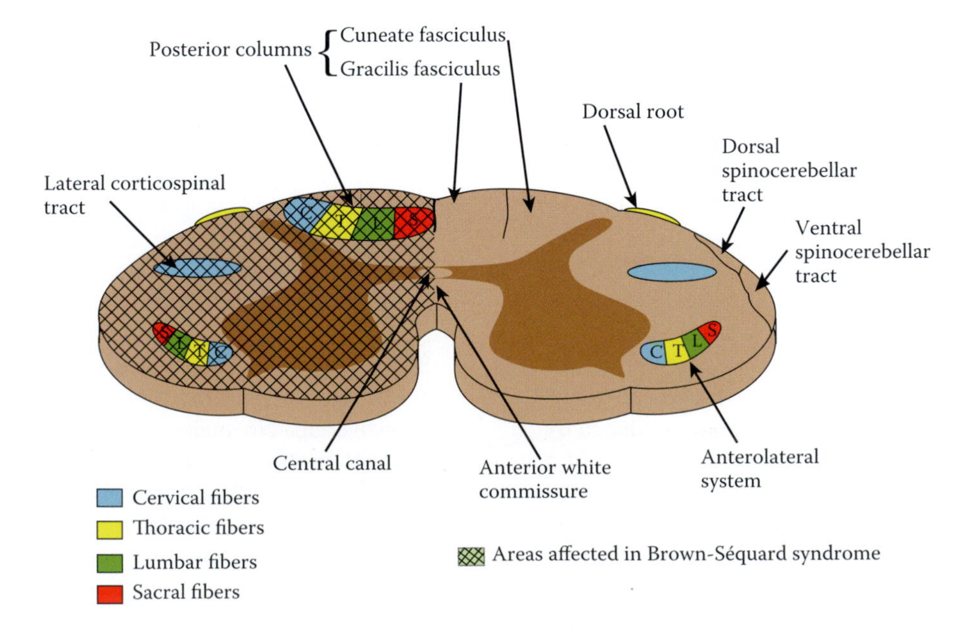

FIGURE 20.21 In this section, degeneration in the left half of the spinal cord is shown. This type of lesion disrupts the fibers of the dorsal columns; the lateral corticospinal, reticulospinal, spinothalamic, spinocerebellar, and rubrospinal tracts; the ventral horn, and the anterior commissure.

movements upon flexion of the thigh at the hip when the knee is extended). The average survival of these patients is 12 years.

Spina bifida (myeloschisis), as was explained earlier in relation to the development of the nervous system, is a congenital abnormality that results from failure of closure of the posterior neuropores. It is commonly seen in the lumbosacral region and may also be associated with Arnold–Chiari syndrome. This condition is classified into spina bifida occulta and spina bifida cystica. *Spina bifida occulta* is usually asymptomatic, unless a mass develops and compresses the spinal cord at the site of anomaly, producing bladder dysfunction and retardation of lower limb development. Back pain and sciatica may accompany this condition. *Spina bifida cystica*, seen with syringomyelia, is associated with herniation of the meninges (meningocele) or with herniation of spinal cord tissue, nerve roots, and meninges (meningomyelocele). Bladder and bowel incontinence, sexual dysfunction, sensory and motor loss, or combined manifestations of spinal root and spinal cord lesions may be observed in this condition.

Multiple sclerosis is an autoimmune disease that produces demyelination in the CNS (see also Chapter 2). Demyelination in the CNS shows a predilection for the optic nerve, spinal cord, periventricular area, brainstem, and cerebellum. The most common initial sign is visual deficits (monocular visual loss and diplopia) with retrobulbar pain. Bilateral anterior internuclear ophthalmoplegia (MLF syndrome), resulting from plaque formation in the MLF, is almost pathognomic for this disease. Due to ipsilateral adductor paresis, MS patients with MLF syndrome may exhibit diplopia on lateral gaze. However, unilateral anterior ophthalmoplegia is not unique for MS as it also occurs in young patients with systemic lupus and in older individuals with basilar artery infarctions. In a third of patients, bladder dysfunction, bed-wetting and the frequent urge to urinate are initially observed. Involvement of the posterior columns of the spinal cord leads to sensory ataxia, resulting in disturbances of gait and equilibrium and a positive Romberg's sign (patients have wide-based gait and lurch forward).

Pellagra is caused by a deficiency of nicotinic acid or tryptophan and is seen in individuals who are dependent upon corn as the main component of their diet. Acute encephalopathy, cogwheel rigidity, locomotor ataxia, confusion, depression, polyneuropathy, subcortical motor signs, and paresis are seen in this disease. Further, degeneration of the neurons of the anterior horn and cerebral cortex, dorsal columns, and spinal nerves is also evident in this disease.

Subacute combined system degeneration occurs as a result of lack of the intrinsic factor, which facilitates the absorption of vitamin B_{12} and is commonly associated with pernicious anemia (Figure 20.22). A megaloblastic anemia, a late sign of this disease, may be detected a few weeks prior to death. This disease initially presents with acroparesthesia, an abnormal sensation in the form of tingling and numbness in the toes and, later, in the digits of the hand, that progresses to an extent that the patient avoids

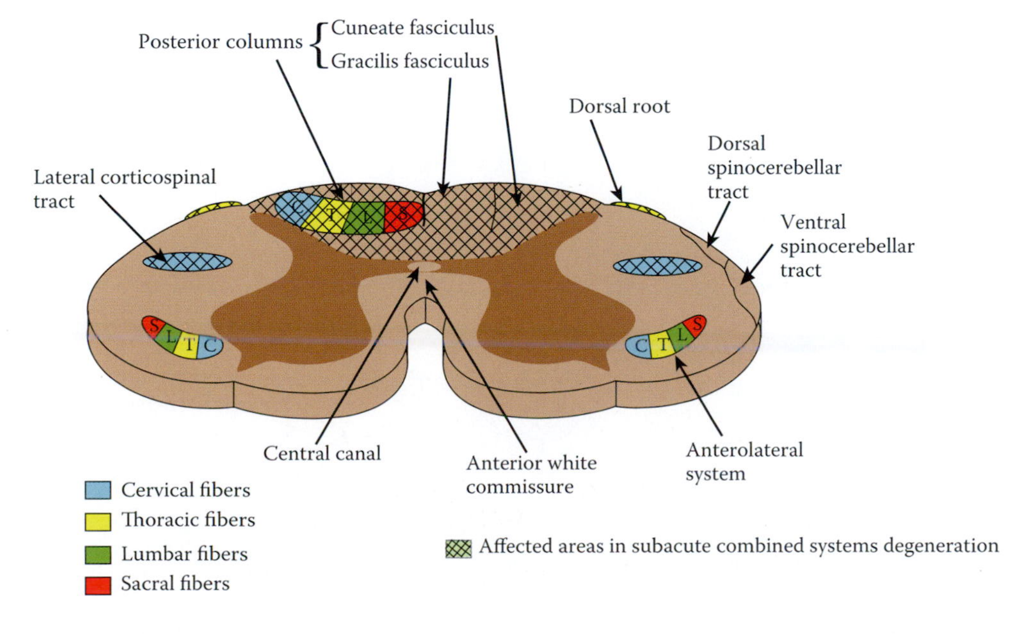

FIGURE 20.22 This is a view of a section of the spinal cord demonstrating degeneration in the dorsal column–medial lemniscus pathway and the lateral corticospinal tracts on both sides in subacute combined degeneration. The dorsal root fibers may also be involved. These lesions produce combined sensory and motor deficits commonly associated with deficiency of vitamin B_{12}.

hand-to-hand contact. Others may exhibit granular foot sensations (as in standing on gravel), or sciatica on the same side, becoming eventually bilateral. *Lhermitte sign*, an electrical sensation that travels down the vertebral column to the lower extremity, is seen when a patient flexes his/her neck. A positive *Romberg's sign*, which refers to exaggerated swaying of the trunk and body when the eyes are closed, is also seen in this condition. This sign is negative in cerebellar and labyrinthine diseases. Degeneration of the pyramidal tract and lateral and dorsal columns occur bilaterally, following no particular order. Disruption of the pyramidal tracts produces signs of UMN palsy, while degeneration of the dorsal columns results in locomotor ataxia and loss of sense of vibration and two-point discrimination. Other sensory changes are attributed to degeneration of the lateral columns and the dorsal root fibers (Table 20.2).

Combined system dysfunctions may be associated with brainstem lesions affecting the medulla, pons, or midbrain. Medullary lesions elicit signs of Avellis, medial medullary, and lateral medullary syndromes.

Avellis syndrome is caused by a lesion that destroys the nucleus ambiguus, lateral spinothalamic tract, and CST. Patients present with ipsilateral paralysis of the soft palate and vocal cords, contralateral hemianesthesia, and spastic paralysis of the extremities. If the lesion is extensive enough to involve the sympathetic fibers, this condition is called the *Cestan–Chenais syndrome*.

Medial medullary syndrome (Figure 20.23) results from occlusion of the paramedian (bulbar) branches of the anterior spinal artery, producing degeneration of the medial lemniscus, hypoglossal nerve, and pyramidal fibers on the ipsilateral side. Signs and symptoms of this condition include contralateral spastic paralysis in the muscles of the extremities and ipsilateral flaccid palsy of the intrinsic lingual muscles. Paralysis of the lingual muscles leads to atrophy of the ipsilateral side of the tongue and its deviation to the affected side upon protrusion. Patients with this disease may also exhibit loss of positional sense, discriminative tactile sensation, vibratory sense, and two-point discrimination.

Lateral medullary (Wallenberg) *syndrome* results from occlusion of the *posterior inferior cerebellar artery* (PICA), producing, depending on the extent of the lesion, degeneration of the spinal trigeminal tract and nucleus, spinal lemniscus, nucleus ambiguus, vestibular and cochlear nuclei, as well as the inferior cerebellar peduncle. (Figures 20.24, 20.25, and 20.29). It is characterized by dysphagia, dysphonia (hoarseness of voice), Horner's syndrome, and alternating hemianesthesia (anesthesia of the ipsilateral face and contralateral body). Ataxia, vertigo, loss of the gag reflex, and loss of taste sensation from the posterior one-third of the tongue are additional manifestations of this disease. Its onset is sudden, exhibiting violent rotatory vertigo (may mimic labyrinthitis), nausea, vomiting, nystagmus, and hiccups. Consciousness and UMNs remain intact, and the prognosis for a complete recovery is good. Death may occur suddenly in some patients with lateral medullary syndrome due to possible overactivation of the dorsal motor nucleus of the vagus. Although recovery from this syndrome is generally good, large ischemia involving the inferior cerebellum may cause compression of the medulla and increased pressure of

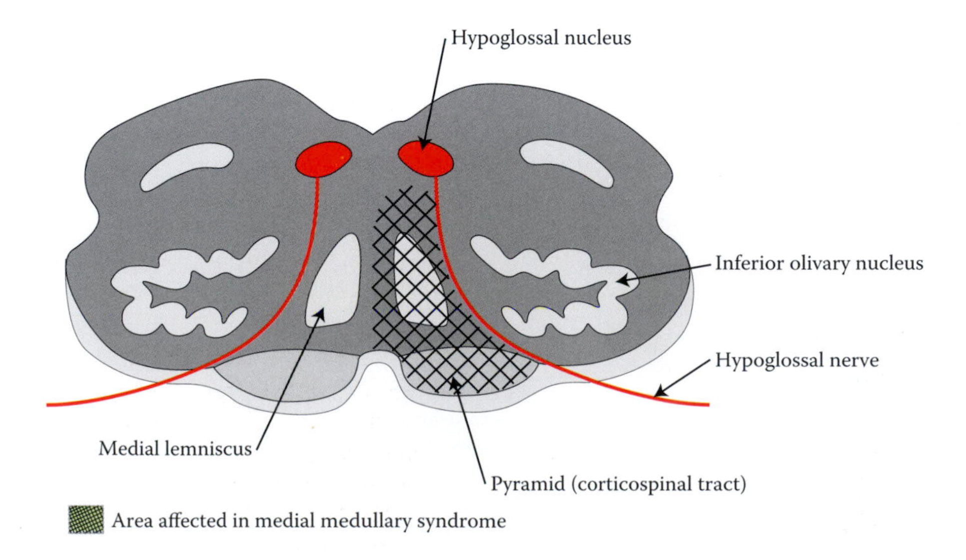

FIGURE 20.23 Medulla at midolivary level illustrating an ipsilateral lesion that involves the medial lemniscus, hypoglossal nerve, and corticospinal tract.

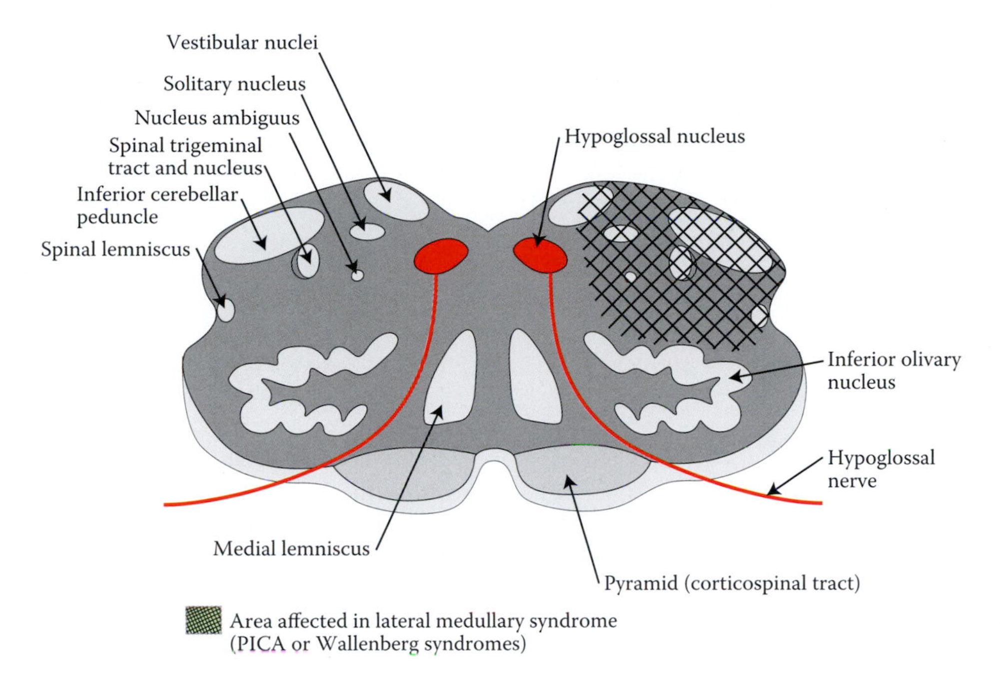

FIGURE 20.24 Drawing of the site of degeneration in the lateral medulla involving the anterolateral system, spinal trigeminal tract and nucleus, nucleus ambiguus, inferior cerebellar peduncle, and descending autonomic fibers.

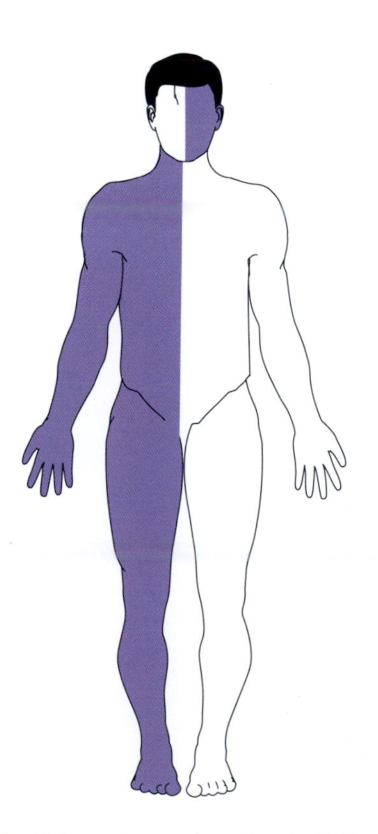

FIGURE 20.25 Schematic drawing of an individual with lateral medullary syndrome.

the posterior cranial fossa, dramatically changing the prognosis. Bilateral lateral medullary ischemia may result in an inability to initiate breathing during sleep and subsequent death (*Ondine's curse*).

Pontine lesions that produce combined sensory and motor deficits are seen in the anterior inferior cerebellar artery (AICA) and superior cerebellar artery syndromes, as well as medial pontine syndromes.

AICA (inferior lateral pontine) *syndrome* (Figure 20.26) results from occlusion of the anterior inferior cerebellar artery. It is characterized by anesthesia of the ipsilateral face and contralateral body, paralysis of the lower facial muscles, lateral gaze palsy, possible deafness, vertigo, nausea, and cerebellar deficits. Cerebellar deficits include ataxia, hypertonia, nystagmus, and intention tremor.

Superior cerebellar artery syndrome results from occlusion of the corresponding vessel, producing paralysis or paresis of conjugate eye movements, hemianesthesia on the contralateral side, and signs of cerebellar dysfunctions. Loss of fine touch, vibratory, and position senses is also seen. Impaired optokinetic nystagmus or skew deviation of the eyes may occur.

Medial pontine syndrome (Figure 20.27) may be associated with occlusion of the paramedian and short circumferential branches of the basilar artery,

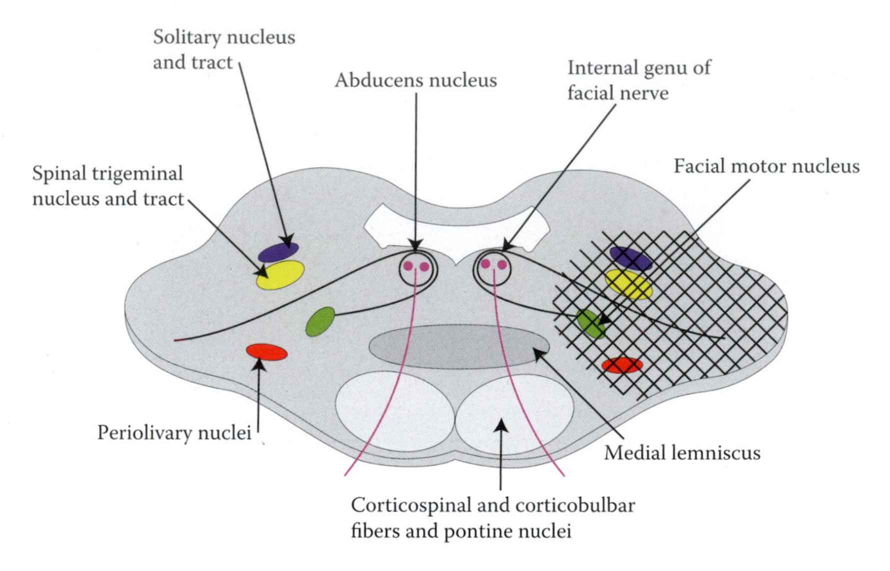

FIGURE 20.26 Areas of degeneration in the lateral pons associated with occlusion of the anterior inferior cerebellar artery.

producing destruction of the abducens and facial nerves, medial lemniscus, CST, and possibly MLF. This syndrome combines signs of abducens alternating hemiplegia, contralateral ataxia, and diplopia (caused by paralysis of the lateral rectus muscle). Involvement of the MLF may produce manifestations of anterior internuclear ophthalmoplegia, which is characterized by ipsilateral adductor paresis and contralateral monocular nystagmus.

Midbrain lesions are primarily manifested in *Benedikt syndrome* (Figure 20.28), which results from occlusion of the circumferential branches of the posterior cerebral artery (Figure 20.29) and subsequent destruction of the red nucleus, superior cerebellar peduncle, medial lemniscus, and spinothalamic tracts. It is characterized by signs of oculomotor palsy (ipsilateral ptosis, mydriasis, lateral strabismus, diplopia, loss of light and accommodation reflexes); intention tremor; ataxia chiefly in the upper extremity; nystagmus; vertigo; and loss of pain and temperature, position, and vibratory senses on the contralateral side. If the CST is involved, hemiparesis without Babinski sign and sensory loss may be detected in conjunction with UMN deficits.

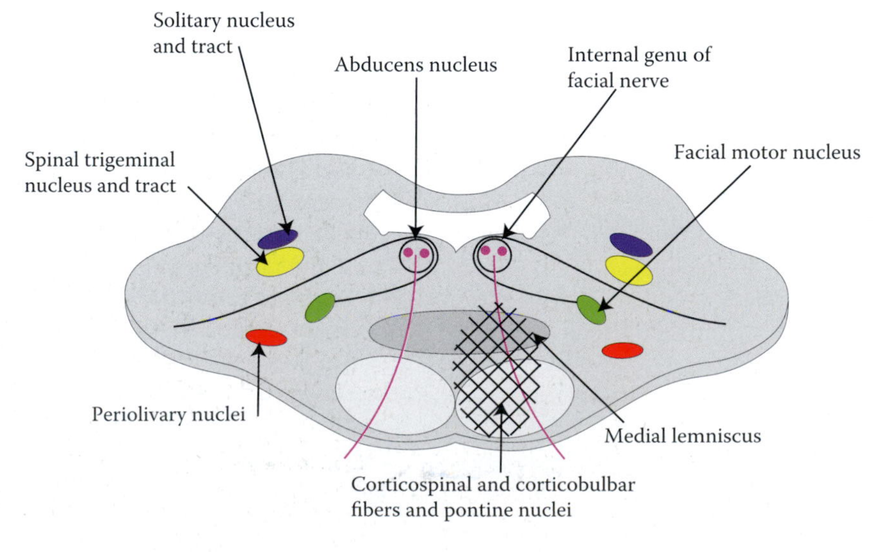

FIGURE 20.27 View of the medial pons showing degeneration in the corticospinal tract, abducens and facial nerves, medial lemniscus, and facial motor nucleus. This lesion is associated with occlusion of the paramedian and circumferential branches of the basilar artery.

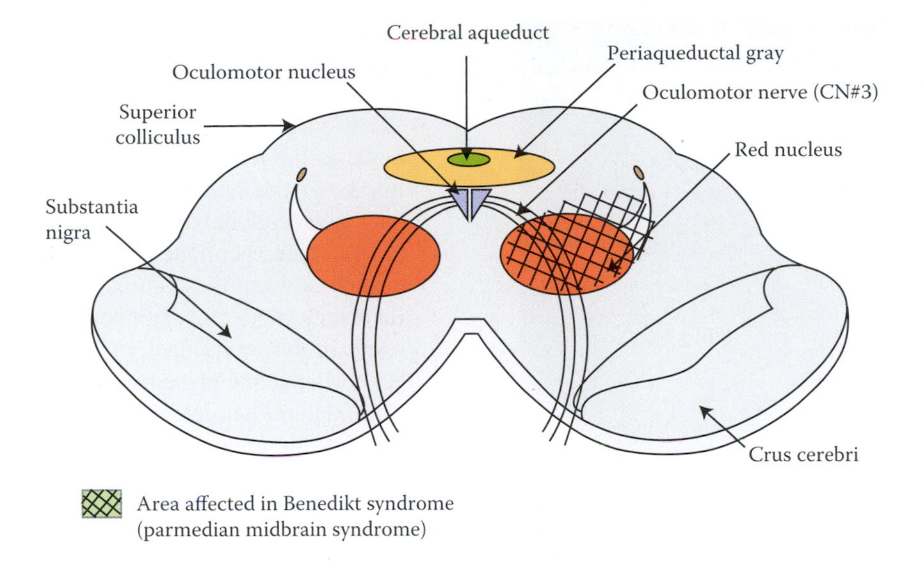

FIGURE 20.28 This drawing represents sites of degeneration in Benedikt syndrome. Note that the oculomotor nerve, medial lemniscus, red nucleus, and possibly the spinal lemniscus are affected.

Cerebral lesions that produce combined motor and sensory disturbances may result from trauma or cerebrovascular accidents that involve a main cerebral artery or its tributaries, including occlusion of the internal carotid or its cerebral branches (anterior and middle cerebral) or the lenticulostriate arteries.

Obstruction of the anterior cerebral artery distal to the anterior communicating branch may result in contralateral spastic palsy and sensory disturbances in the lower extremity and possibly mental confusion. In the absence of an efficient anastomosis with the middle cerebral artery, occlusive diseases affecting the callosomarginal branch may produce an infarction of the paracentral lobule, leading to paralysis of the contralateral leg, cortical sensory loss, and urinary and bowel incontinence. It may also lead to an isolated infarction of the head of the caudate nucleus with no detectable deficits. Unilateral occlusion of the trunk of the anterior cerebral artery is less likely to result in a significant deficit since the anterior communicating artery allows blood to flow to the affected side from the intact side. Obstruction of the terminal portion of the anterior cerebral artery distal to the origin of the callosomarginal artery may not produce any clinically recognizable symptoms. Aneurysm of the pericallosal branch of the anterior cerebral artery (Figure 20.30) may produce headache, mental confusion, and seizure.

Occlusion middle cerebral artery (Figures 20.29, 20.30, and 20.31), although infrequent, may be secondary to thrombi within the internal carotid artery. Depending upon the site of occlusion in the dominant or nondominant hemisphere and the affected arterial branch, a constellation of deficits, which range from spastic palsy and supranuclear facial palsy to sensory disturbances, visual deficits, and speech disorders (aphasia), may be seen. These deficits are commonly seen in the morning, showing progression and fluctuation in the subsequent days. In general, middle cerebral artery syndrome exhibits contralateral hemiplegia (especially of the face and arm), contralateral cortical hemianesthesia, contralateral hemianopsia, aphasia, agraphia, alexia (if it involves the dominant

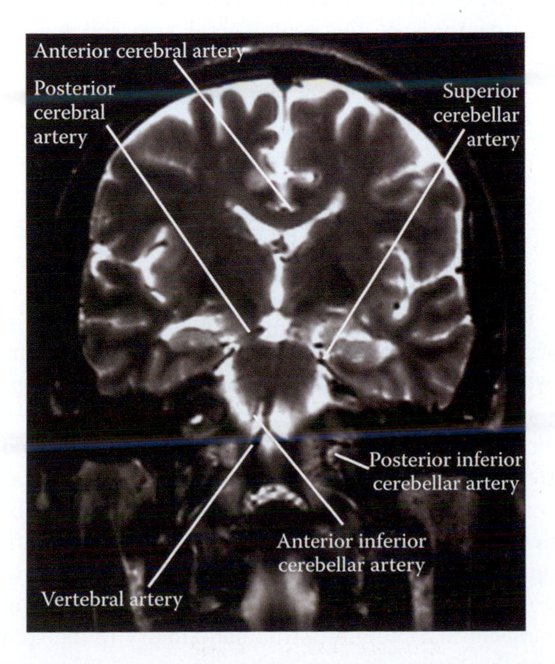

FIGURE 20.29 In this magnetic resonance image, branches of the internal carotid artery and vertebrobasilar system are shown. Anterior inferior cerebellar, superior cerebellar and the posterior cerebral branches of the basilar artery are clearly visible.

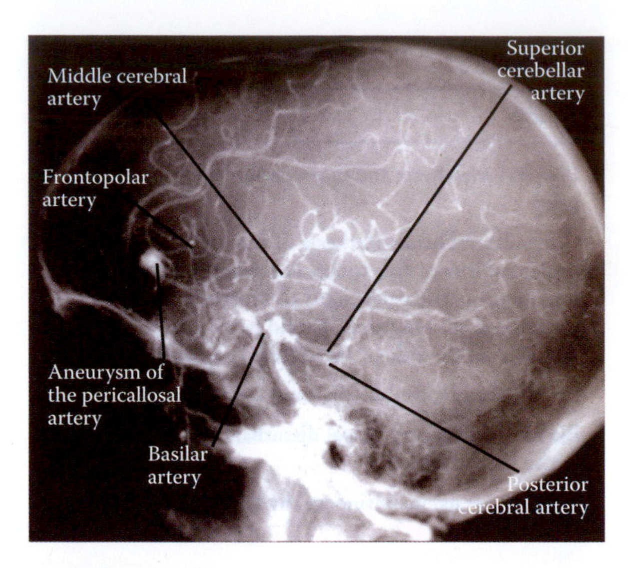

FIGURE 20.30 This angiogram shows an aneurysm of the pericallosal branch of the anterior cerebral artery. Note the middle cerebral artery and branches of the basilar artery.

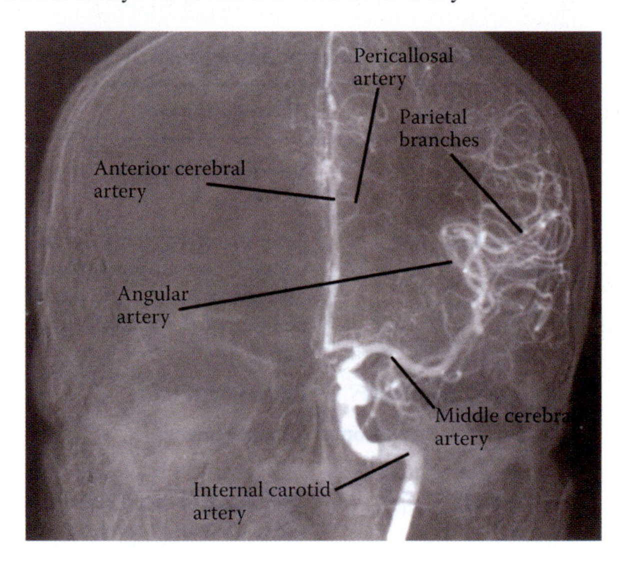

FIGURE 20.31 This internal carotid artery angiogram (arterial phase) is shown as an overview of the distribution of the middle cerebral and anterior cerebral arteries.

hemisphere), apraxia, and anosognosia (when the nondominant hemisphere is affected).

Occlusion of the middle cerebral artery proximal to the origin of the lenticulostriate branches, the most serious occlusion, produces extensive damage to the anterior and posterior limbs of the internal capsule. This type of occlusion results in sensory and motor deficits on the contralateral half of the face, hand, arm, and leg; homonymous hemianopsia; and global aphasia (if the dominant hemisphere is involved). It is important to note that a deep infarction of the middle cerebral artery at or above the described site may produce lacunar infarcts restricted to the internal capsule, sparing the cerebral cortex. This selectivity may occur as a result of the poor circulation in the lenticulostriate arteries and relatively good collateral circulation on the lateral surface of the brain. When the occlusion occurs distal to the lenticulostriate artery, it may produce an infarct of the opercular cortex. The cortical deficits may include global dysphasia, right–left disorientation, homonymous hemianopsia, and graphic language disturbances when the dominant hemisphere is involved. Involvement of the nondominant hemisphere may result in dyspraxia, lack of initiative, and a failure to acknowledge the presence of any neurological deficits. Contralateral hemiplegia and hemisensory loss, which involves the upper extremity and face but not the leg, may also be observed.

- Occlusion of the superior branch of the middle cerebral artery that supplies the precentral and postcentral gyri as well as Broca's center may produce contralateral hemiplegia and hemisensory loss in the upper extremity and face, and expressive (Broca's) aphasia. Infarcts associated with the inferior division of the middle cerebral artery results in contralateral homonymous hemianopsia, disorders of spatial thoughts, and stereognosis. It is also associated with failure to recognize the contralateral extremity, anosognosia (loss of ability to recognize bodily deficits subsequent to a lesion of the nondominant hemisphere), and receptive aphasia (when the dominant hemisphere is involved). The rolandic branch of the middle cerebral artery may also be occluded, producing sensory loss and motor paralysis of the contralateral arm and lower face. Infarcts due to occlusion of the ascending frontal branch in the dominant hemisphere produce Broca's aphasia, supranuclear facial palsy, and tonic deviation of the eyes to the side of the lesion. Occlusion of the angular artery produces receptive (Wernicke's) or global aphasia and apractognosia.

Rupture of the anterior choroidal branch of the internal carotid artery causes contralateral hemianesthesia and hemiplegia with homonymous hemianopsia or superior quadranopsia, but without cerebral edema. The relatively small size of the anterior choroidal artery and its long subarachnoid course may account for its susceptibility to thrombosis. Absence of cerebral edema may be an important feature of this condition compared to occlusion of the middle cerebral artery.

An expanding tumor in the frontoparietal area (paracentral lobule) or a traumatic injury to the spinal cord, brainstem, or cerebral cortex may damage the sensory and motor cortical areas, producing deficits similar to that of the vascular lesions. Excessive and constant stimulation of the temporal lobe by a tumor may lead to generalized tonic–clonic seizure, disorders of consciousness, urinary incontinence, tongue biting (clonic phase), aura, and fear sensation, followed by fatigue and confusion.

SUGGESTED READING

Bäumer D, Talbot K, Turner MR. Advances in motor neurone disease. *J R Soc Med.* 2014;107(1):14–21.

Caccia MR, Ubiali E, Schieroni F. Axonal excitability and motor propagation velocity of peripheral nerves in patients with acute vascular lesions of the brain. *J Neurol Neurosurg Psychiatry*. 1976;39(9):900–4.

Everhart DE, Harrison DW, Crews WD Jr. Hemispheric asymmetry as a function of handedness: Perception of facial affect stimuli. *Percept Mot Skills*. 1996;82(1):264–6.

Flament D, Onsott D, Fu Q-G, Ebner TJ. Distance- and error-related discharge of cells in premotor cortex of rhesus monkeys. *Neurosci Lett.* 1993;153:144–18.

Gopalakrishnan CV, Dhakoji A, Nair S. Giant vertebral artery aneurysm presenting with 'hemiplegia cruciata'. *Clin Neurol Neurosurg*. 2013;115(9):1908–10.

Halsband U, Ito N, Tanji 3, Freund H-J. The role of premotor cortex and the supplementary motor area in the temporal control of movement in man. *Brain*. 1993;116:243–66.

Higashigawa M, Maegawa K, Honma H, Yoshino A, Onozato K, Nashida Y, Fujiwara T, Inoue M. Vaccine-associated paralytic poliomyelitis in an infant with perianal abscesses. *J Infect Chemother.* 2010;16(5):356–9.

Jin XL, Li XH, Zhang LM, Zhao J. The interaction of leukocytes and adhesion molecules in mesenteric microvessel endothelial cells after internal capsule hemorrhage. *Microcirculation*. 2012;19(6):539–46.

Kennedy, PR. Corticospinal, rubrospinal and rubro-olivary projections: A unifying hypothesis. *Trends Neurosci.* 1990;13:474–8.

Kim SG, Ashe J, Hendrich K, Ellermann JM, Merkle H, Ugurbil K, Georgopoulos AP. Functional magnetic resonance imaging of motor cortex: Hemispheric asymmetry and handedness. *Science*. 1993;261(5121):615–7.

Maffini F, Cocorocchio E, Pruneri G, Bonomo G, Peccatori F, Chiapparini L, Vincenzo SD, Martinelli G, Viale G. Locked-in syndrome after basilary artery thrombosis by mucormycosis masquerading as meningoencephalitis in a lymphoma patient. *Ecancermedicalscience.* 2013;7:382.

Marín-Padilla M. The mammalian neocortex new pyramidal neuron: A new conception. *Front Neuroanat.* 2014;7:51.

Menon P, Kiernan MC, Vucic S. Cortical dysfunction underlies the development of the split-hand in amyotrophic lateral sclerosis. *PLoS One*. 2014;9(1):e87124.

Sevencan F, Ertem M, Öner H, Aras Kılınç E, Köse OÖ, Demircioğlu S, Dilmen F, Eldemir R, Öncül M. Acute flaccid paralysis surveillance in southeastern Turkey, 1999–2010. *Turk J Pediatr.* 2013;55(3):283–91.

Spaccavento S, Del Prete M, Craca A, Loverre A. A case of atypical progressive supranuclear palsy. *Clin Interv Aging*. 2014;9:31–9.

Tan AM, Chakrabarty S, Kimura H, Martin JH. Selective corticospinal tract injury in the rat induces primary afferent fiber sprouting in the spinal cord and hyperreflexia. *J Neurosci.* 2012;32(37):12896–908.

Tanji J. The supplementary motor area in the cerebral cortex. *Neurosci. Res*. 1994;19:251–68.

Thevenon A, Serafi R, Fontaine C, Grauwin MY, Buisset N, Tiffreau V. An unusual cause of foot **clonus**: Spasticity of fibularis longus muscle. *Ann Phys Rehabil Med.* 2013;56(6):482–8.

Yavuz SU, Mrachacz-Kersting N, Sebik O, Berna Ünver M, Farina D, Türker KS. Human stretch reflex pathways reexamined. *J Neurophysiol.* 2014;111(3):602–12.

21 Extrapyramidal Motor System

The extrapyramidal system comprises a group of structures that are concerned with stereotyped movements, complex motor activities, suppression of cortically induced movements, regulation of posture, and adjustment of muscle tone. It consists of the basal nuclei, claustrum, substantia nigra, red nucleus, subthalamic nucleus, and reticular formation. These neurons influence motor activity by projecting to specific thalamic nuclei en route to the motor and premotor cortices. The cortical influence on subcortical and spinal motor neurons is mediated via the corticofugal fibers. Therefore, projection of the basal nuclei, a component of the extrapyramidal system, to the motor cortex via the thalamus and thalamocortical fibers may account for the important role that this system plays in initiating and planning movements.

BASAL (GANGLIA) NUCLEI

Extrapyramidal motor system consists of the basal nuclei, substantia nigra, red nucleus, subthalamic nucleus, and reticular formation. The *basal nuclei* (Figures 21.1 through 21.6) are a collection of subcortical nuclei embedded in the white matter of the cerebral hemispheres, which include the globus pallidus, caudate nucleus, and putamen that collectively form the corpus striatum. Some authors have extended the conventional definition of the basal nuclei to also include the subthalamic nucleus and the substantia nigra. The basal nuclei regulate stereotyped movements and mediate control of saccadic eye movements via reciprocal connections to the frontal eyefield and via their projections to the superior colliculus. They also coordinate orientation-associated memory and behavior via the bidirectional connections to the prefrontal and orbitofrontal cortices. This diverse origin and connections may account for the important role that the basal nuclei play in regulating motor activity and higher cognitive functions.

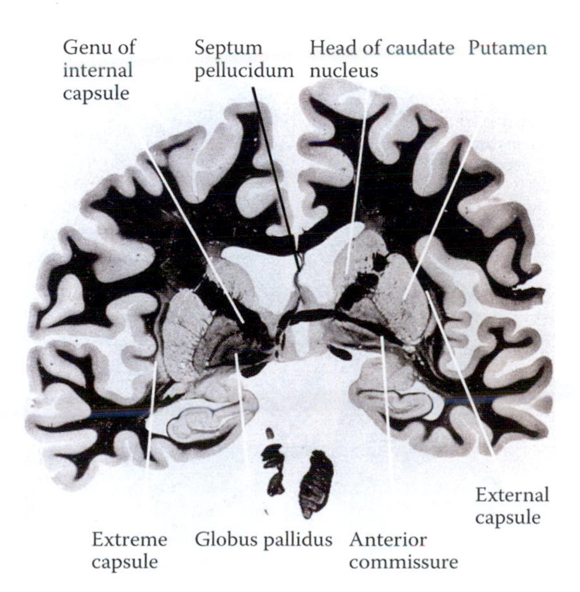

FIGURE 21.1 Coronal section of the brain. Observe the various components of the basal nuclei in relation to the internal capsule and the lateral ventricle.

CORPUS STRIATUM

The *corpus striatum* is divisible into a dorsal striatum, which consists of the caudate and lentiform nuclei, a smaller ventral component consisting of the ventral striatum and ventral pallidum.

Dorsal Striatum

The *dorsal striatum* is derived from the telencephalon, maintaining identical cellular structures. In general, the dorsal striatum modulates complex motor responses. The *caudate nucleus* is a comma-shaped structure that lies on the lateral side of the thalamus, separated from it by sulcus terminalis, which contains the stria terminalis and the thalamostriate vein. The stria terminalis together with the hippocampal fimbria and fornix form the boundaries of the choroid fissure. The caudate nucleus is composed of a large rostral portion,

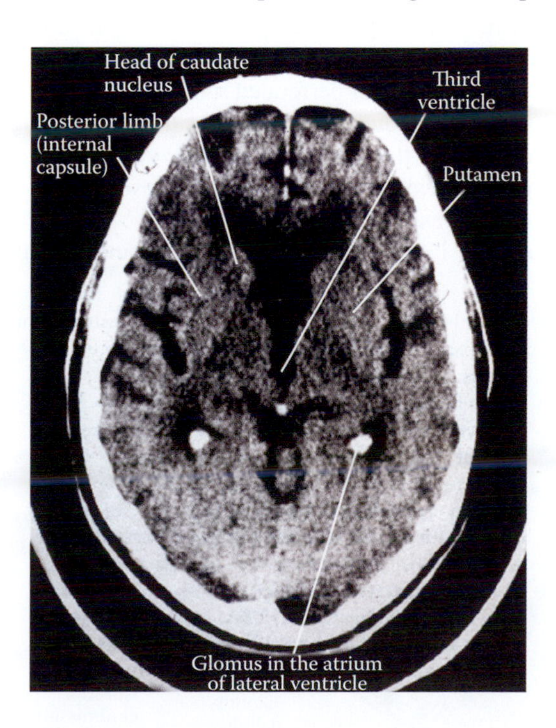

FIGURE 21.2 Computed tomography scan of the brain. The putamen and caudate nucleus are clearly visible.

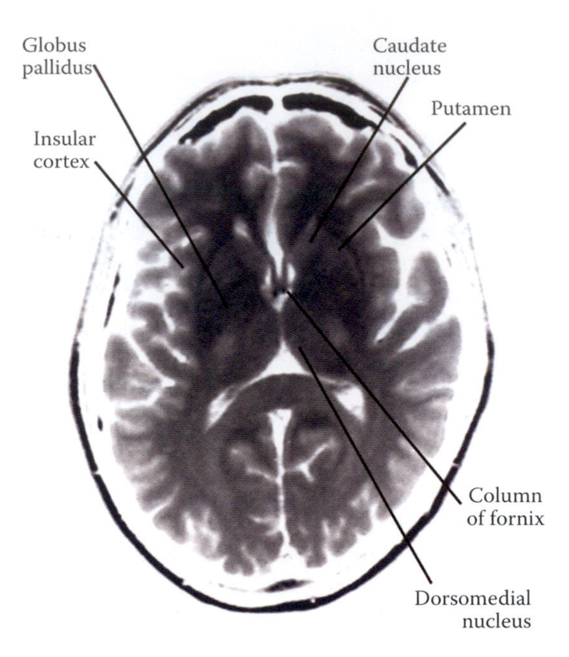

FIGURE 21.3 MRI scan showing a more elaborate structural organization of the basal nuclei.

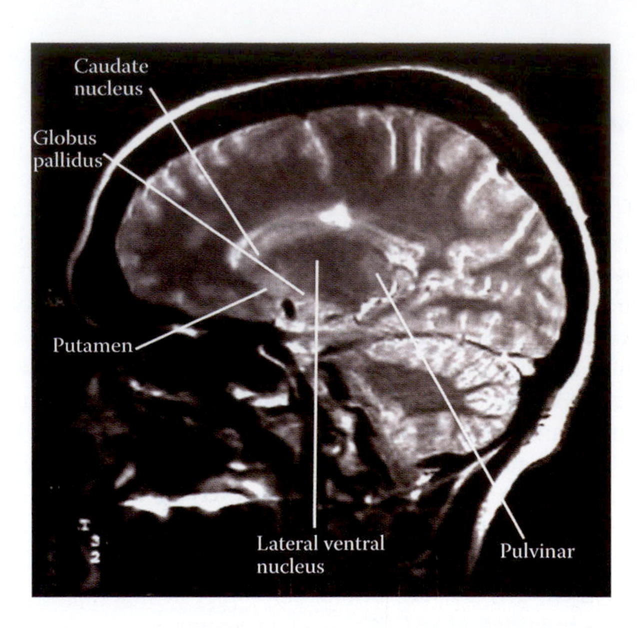

FIGURE 21.5 MRI scan of the brain demonstrates the relationship of the basal nuclei to the thalamus and lateral ventricle.

in which the head narrows to a body and then continues with a tail that curves in a downward direction. The head forms the floor of the anterior horn of the lateral ventricle, whereas the body is located in the floor of the central portion (body) of the lateral ventricle, dorsolateral to the thalamus. The tail curves downward and runs into the roof of the inferior horn of the lateral ventricle, terminating at the posterior end of the amygdala. The fronto-occipitalis and the subcallosal fasciculi separate the lateral and medial parts of the superior surface of the caudate nucleus from the corpus callosum, respectively. Superior to the anterior perforated substance, the inferior part of the head of the caudate nucleus joins the inferior part of the putamen to form the fundus *striati of Brockhaus* that spreads within the anterior limb of the internal capsule.

The *putamen*, the most lateral component of the corpus striatum, lies lateral to the globus pallidus and medial to the external capsule and is crossed inferiorly by the fibers of the anterior commissure. It joins the head of the caudate nucleus rostrally, separated from the globus pallidus by the external medullary lamina. The putamen and the globus pallidus form the lentiform nucleus, a triangular-shaped nucleus with an apex faced medially and a base directed laterally. It is located medial to the insular cortex separated from it by the claustrum, external and extreme capsules. The internal capsule separates the lentiform nucleus from the caudate nucleus.

Cytoarchitecturally, the striatum consists mainly of the GABAergic medium spiny neurons, and to a lesser degree, of the leptodendritic (Deiter's) neurons that resemble the pallidal neurons, also of the tonically active cholinergic (spidery)

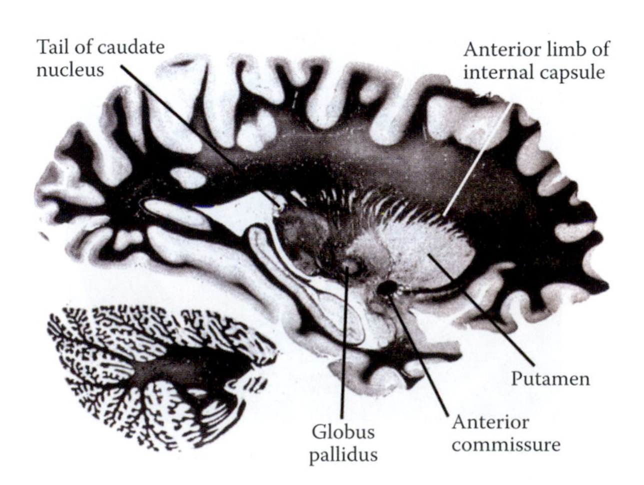

FIGURE 21.4 Midsagittal view of the brain. Observe the massive putamen, globus pallidus, and the fibers of the internal capsule.

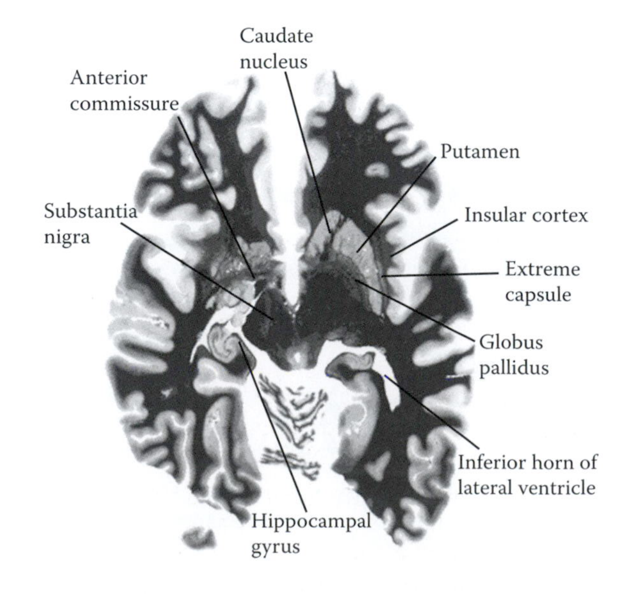

FIGURE 21.6 Horizontal section of the brain. Observe the continuation of the putamen with the caudate nucleus as well as the medial and lateral segments of the globus pallidus.

neurons that exhibit large perikaryon with short dendrites and axons, and the microneurons, which are mainly GABAergic interneurons that mediate local circuits. Dendritic spines allow synaptic connections with the cortical fibers. The dorsal striatal neurons maintain low hyperpolarized excitable state and require robust synchronized cortical input. The dorsal striatum receives blood supply from the lateral striate (lenticulostriate) artery, anterior choroidal artery and the medial striate artery (recurrent artery of Heubner). The latter vessel is a branch of the anterior cerebral artery which provides blood supply to the head of the caudate and the adjacent part of the putamen.

Occlusion of the anterior cerebral artery proximal to the origin of the medial striate branch may produce no detectable deficits if the collateral circulation with the corresponding artery of the opposite side is maintained. The relative small size of the anterior choroidal artery, which supplies the tail of the caudate and its long subarachnoid course, may account for its susceptibility to thrombosis. A hematoma that develops on the putamen as a result of hypertension or cerebrovascular malformation may quickly expand to involve the lateral ventricle and the adjacent internal capsule. Headache, vomiting, stiff neck, confusion, and diminished consciousness, which are manifested in putaminal hemorrhage, are attributed to accumulation of blood in the lateral ventricle. Destruction of the internal capsule by this hematoma may also produce contralateral hemiparesis, contralateral conjugate gaze palsy, and Horner's syndrome.

Ventral Striatum

The *ventral striatum* consists of the nucleus accumbens, great portion of the anterior perforated substance, and the olfactory tubercle. The paraolfactory cortex, septal nuclei, and the diagonal band of Broca merge with the nucleus accumbens. Inferior and posterior to the ventral striatum and separated by the anterior commissure is the ventral pallidum. The latter is located superior and lateral to the nucleus basalis of Meynert and caudal to the anterior perforated substance.

The striated appearance of the corpus striatum is mainly attributed to the radiated manner by which the myelinated and unmyelinated striatal afferents and efferents run through it and form the "Wilson's pencils." Small multipolar neurons constitute the main cellular component of both the ventral and dorsal striatum, establishing through its spiny dendrites synaptic connections (symmetric (type I) with the majority of striatal afferents. Intrinsic synaptic connections within the striatum are, for the most part, of asymmetric type (type II).

In general, most striatal neurons contain γ-aminobutyric acid (GABA), enkephalin, substance P (SP), as well as dopamine. The distribution of the neuroactive chemicals in the striatum shows variation in concentrations in different parts of the striatum irrespective of their origin. Rostral striatum contains high concentration of dopamine, substance P, and acetylcholine acid, while glutamic decarboxylase (GAD) and serotonin show high concentrations in the caudal striatum. The enkephalinergic neurons have D2 dopamine receptors, whereas SP neurons have D1 receptors. Somatostatin, acetylcholinesterase, choline acetyltransferase, and pancreatic polypeptide containing neurons also exist. Striosomes and matrix are not clearly evident in humans, although the matrix seems to be the main part in putamen. Striosomes, patches within the matrix of the caudate nucleus and less so in putamen, contain higher concentration of dopamine than the matrix.

The matrix contains acetylcholine and somatostatin receptors that receive projections from the thalamus. Striosomes also contain enkephalin, dopamine, substance P, and opiate receptors. The rest of the striatal matrix holds acetylcholine and somatostatin that receive input from the intralaminar thalamic nuclei and superficial layers of the neocortex. However, the distribution of all afferents to the striatum does not follow a uniform pattern with regard to the matrix and striosomes as seen in nigrostriatal and thalamostriatal afferent distribution. Unlike efferent fibers from the superficial neocortical layers, which terminate principally in the striatal matrix, efferent fibers from deeper cortical layers project to striosomes. Striatal efferents are not uniformly arranged either as the striatopallidal and striatonigral efferents form clusters emanating from the striosomes. Ventral striatum consists mainly of striosomes and less-defined matrix.

Dorsal Pallidum

The *dorsal pallidum* or the globus pallidus is located lateral to the posterior limb of the external capsule and medial to the putamen (Figures 21.1, 21.4, and 21.5). It has a very limited number of neurons and is crossed by a dense bundle of myelinated fibers (striato-pallido-nigral fibers), which give it a pale color as the name indicates. It consists of medial and lateral segments separated by the internal medullary lamina. The neuronal density is higher in the external segment than in the internal segment. Striatal afferents contained in the Wilson's pencils connect with the dendrites of the multipolar neurons of this segment forming right-angle synaptic connections that maximize contacts. Its main connections remain essentially confined to other basal nuclei. The fast spontaneous activity is interrupted by several seconds of inactivity.

The medial segment, in particular, exhibits similar cytological, morphological, and functional characteristics to that of the pars reticulata of the substantia nigra, containing substance P. Phylogenetically, it develops earlier than the lateral segment. It is a derivative of the diencephalon and forms the efferent (output) portion of the basal nuclei. It is further divided by the intermediate lamina into outer and inner parts. Neurons of the lateral segment are charged spontaneously and continuously without long intervals of inactivity (fast-spiking pacemaker). This segment receives robust input from the striatum as well as dopaminergic afferent fibers from the substantia nigra pars compacta.

GABA, distributed evenly in both the internal and external segments of the pallidum, appears to be primarily utilized by its neurons compared to a lesser degree of utilization of acetylcholine. Substance P is concentrated in the internal segment and enkephalin in the external segment. GABA and substance P are primarily utilized by the pallidonigral projection to the pars reticulate of the substantia nigra.

Ventral Pallidum

The *ventral pallidum* represents the area superolateral to the nucleus basalis of Meynert and caudal to the anterior perforated substance (substantia innominata), separated from the dorsal pallidum by the anterior commissure. The ventral pallidum also uses GABA and, to a greater extent, acetylcholine as neurotransmitters. In addition, the ventral striatum projects to the ventral pallidum using substance P and enkephalin as neurotransmitters.

CLAUSTRUM

The *claustrum*, a gray matter mass that resembles the thalamus, is located between the external and extreme capsules. It continues with the amygdala, the prepiriform cortex, and the anterior perforated substance. It has temporal (prepiriform) and insular parts, maintaining reciprocal connections with most areas of the isocortex. It receives retinotopically organized visual fibers from both eyes, as well as sensory fibers from the postcentral gyrus. The claustrum, superior colliculus, and pulvinar may mediate activities associated with visual attention.

SUBSTANTIA NIGRA

The *substantia nigra* (Figures 21.6 and 21.7) represents the largest nucleus of the brainstem that lies dorsal to the crus cerebri and ventral to the mesencephalic tegmentum. It is a bilaterally represented pigmented nucleus that extends through the entire length of the mesencephalon from the pons to the subthalamic nucleus. It is crossed by the medial fibers of the oculomotor nerve that exit through the interpeduncular fossa. It maintains reciprocal connection with the basal nuclei and receives corticonigral fibers from the precentral and postcentral gyri, which partly terminate in the pars reticularis while others continue into the red nucleus and the reticular formation. Ascending sensory pathways also provide collateral to the substantia nigra.

It consists of two adjoining but contrasting components with cell-rich pars compacta that exhibits larger and thicker dendritic arborizations and sparsely cell-populated pars

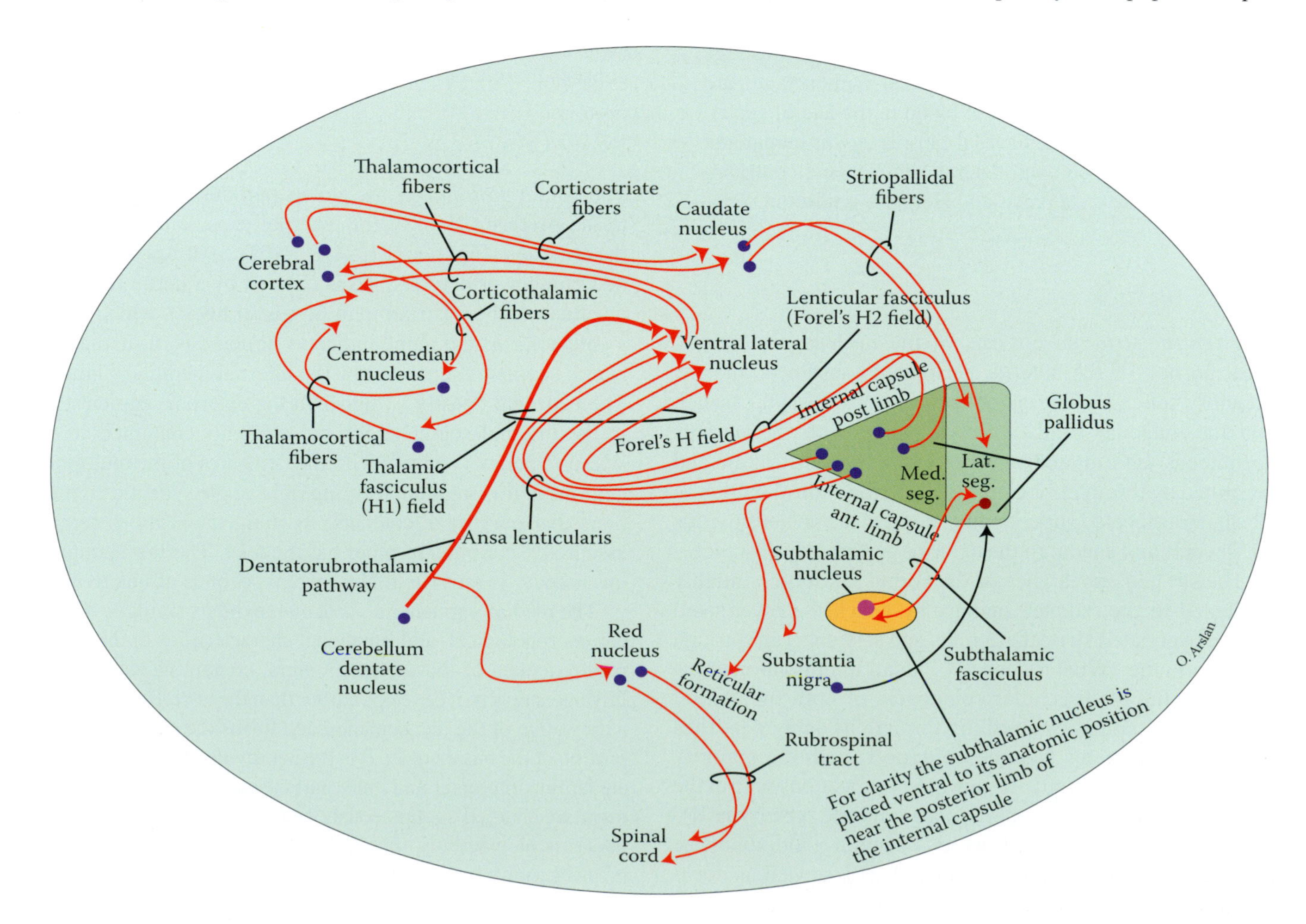

FIGURE 21.7 Section through the subthalamic area. Observe the subthalamic nucleus, lenticular and thalamic fasciculi, caudate nucleus, putamen, and substantia nigra.

reticulata. The pigment granules reside mainly in the *pars compacta* and increase with age through the deposition of melanin. In contrast to the pars reticulata, dopaminergic neurons of the pars compact are "slow-spiking pacemakers," and their activity is believed to be linked to reward and prediction of reward. It has been reported that dopamine release is associated with phasic responses of these neurons to reward-related activity, particularly reward-prediction behavioral inaccuracies. The pars compacta on both sides interconnects in the midline through the paranigral nucleus, which is also known as the ventral tegmental dopaminergic A10 cell group of Tsai. The latter forms the mesolimbic dopamine system, which provides dopaminergic projections to the prefrontal and anterior cingulate cortices as well as the dorsal striatum.

Clinical depression in parkinsonian patients is accompanied by marked dysfunction and reduced neuronal activity of the prefrontal cortex and the striatal neurons as well as the dopaminergic, noradrenergic, and serotonergic neurons within the brainstem. Parkinsonian depression is thought to be an expression of the disruption of the direct connections of the basal nuclei to the frontal and limbic cortices as well as a reflection of the disruption of the indirect ascending connection to the prefrontal cortex via the substantia nigra pars reticulata, dorsomedial and ventromedial thalamic nuclei. It has been proposed that patients with compulsive-obsessive disorders and hyperactive child syndrome manifest depression as a result of this disruption.

The *pedunculopontine tegmental nucleus* (cholinergic cell group 5) and the lateral dorsal tegmental nucleus (cholinergic cell group 6) lie in close proximity to group cell A10. The A8 (retrorubral nucleus or nucleus parabrachialis pigmentosus) and dopaminergic A9 cell groups of Dählstrom and Fuxe together constitute the mesostriatal dopamine system. Cell groups A10 and A8 that spread inside the crus cerebri do not receive striatal afferents. Collectively, cell groups A8, A9, and A10 form the ventral tegmental (VT) system that regulates adaptive behavior through projections to the cerebellum and monoaminergic neurons of the brainstem. The VT system utilizes the medial forebrain bundle (MFB) to project to the prefrontal, orbitofrontal, and cingulate cortices (mesocortical system) as well as to the amygdala, entorhinal cortex, nucleus accumbens, and the interstitial nucleus of the stria terminalis (mesorhombic system). It also utilizes MFB to send signals to the striatum (mesostriatal system) and to the diencephalon (mesodiencephalic system). In addition to the above, the cells in the lateral parts of A9 and A10 cell groups also contain somatostatin, while the medial parts of these cell groups contain cholecystokinin. The thin neuronal axons of the pars compacta that carry dopamine travel dorsally, encircle the medial border of the subthalamic nucleus to enter the H2 field, and then cross the internal capsule to project to the upper part of the medial pallidum and then the striatum. Their fibers end in a nonhomogeneous manner in the striatum, but not in the matrix or in the striosomes.

Pars reticulata contains large multipolar neurons that are GABAergic and resemble the pallidal neurons, and represents the efferent fibers of the substantia nigra. But it also harbors dopaminergic dendrites of the pars compacta with small interneurons. Ventrally, it continues with the subthalamic region, intermingles with the crus cerebri, and extends with the globus pallidus, containing unusual concentration of iron. The dendrites of the multipolar neurons receive the striatonigral comb fiber system and are oriented at a right angle to enable maximal exposure to these afferents. Striatonigral fibers utilize GABA and differentially substance P throughout and enkephalin in the medial part.

The pars reticulata is considered a fast-spiking pacemaker, and its stimulation may not elicit motor activity as a few neurons are capable of responding to any form of movements but are responsive to signals that relate to memory, attention, or the mechanism that prepares for movements. It shows degeneration in Parkinson's disease (see also the midbrain). Despite the high concentration of dopamine in these neurons, cholinergic neurons represent one-fourth of neuronal population in the pars compacta of the substantia nigra. GABAergic neurons of the pars reticularis influence the activities of the ventral anterior and dorsomedial thalamic nuclei via the nigrothalamic pathway. These neurons also send projections to the superior colliculus of the ipsilateral side via the nigrotectal fibers to mediate saccadic eye movements (fast-spiking pacemaker). The area that contains nigrotectal neurons is considered by some as the substantia nigra pars lateralis. This nigrotectal connection is part of the striato-nigra-tectal circuit that influences ocular motor activity. Disruption of this circuit is responsible for the fixed gaze observed in Parkinson's disease, Huntington chorea, and supranuclear gaze palsy. Similarly, pars reticularis influences the spinal alpha and gamma motor indirectly via projection to the reticular formation and the pedunculopontine nucleus (nigrotegmental tract). There exist dopaminergic projections from the pars compacta and glutaminergic projection from pars reticularis.

RED NUCLEUS

The *red nucleus* (Figures 21.7 and 21.9) occupies the center of the midbrain tegmentum dorsomedial to the substantia nigra. The pinkish tinge of the red nucleus is attributed to an iron pigment in the multipolar neurons. It is encircled by the superior cerebellar peduncle, which consists of the dentato-rubrothalamic fibers and is crossed and partially surrounded by the fibers of the oculomotor nerve en route to the interpeduncular fossa. It merges superiorly with the reticular formation and the interstitial nucleus. It consists of magnocellular and parvocellular components. The magnocellular part of this nucleus gives rise to the contralateral rubrospinal tract, a less distinct pathway in humans that thought to regulate flexor muscle tone. This nucleus receives contralateral cerebellar projections from the dentate and globose and emboliform nuclei, and provides ipsilateral projection to the inferior olivary nucleus via the rubro-olivary projection, which is contained in the central tegmental tract. It also receives ipsilateral corticorubral fibers from the precentral and postcentral gyri, and bilateral input from the supplementary motor cortex. In turn, the red nucleus projects to the ipsilateral motor cortex via relay in the ventral lateral (VL) nucleus of thalamus. Therefore, it is conceivable that the rubro-olivary tract may switch the control of movements from the corticospinal tract to the rubrospinal tract for programmed automation. In the same manner, the rubro-olivary tract may allow the corticospinal tract to intervene in response to changes during ongoing automated movements by the rubrospinal pathway. It appears that the rubrospinal tract delivers cerebellar and cerebral input to the upper three cervical spinal segments in humans. Recovery of motor control (in monkeys), following corticospinal tract dysfunction, may be attributed to the compensatory role that the rubro-olivary tract plays in rerouting the motor commands to the cerebellum and back to the rubrospinal tract. In addition, afferent fibers from the substantia nigra, globus pallidus, hypothalamus, and subthalamic nucleus also terminate in the red nucleus. There is bidirectional connection between the red nucleus and the superior colliculus on both sides.

In *Benedikt's syndrome*, a condition associated with occlusion of the posterior cerebral artery, the *red nucleus* is damaged in conjunction with the oculomotor nerve, medial and spinal lemnisci, superior cerebellar peduncle, and possibly the corticospinal tract. This syndrome exhibits signs of oculomotor nerve palsy, ataxia (chiefly in the upper extremity), coarse intention tremor, adiadochokinesis, and contralateral hypotonia. Hemiparesis may be seen without Babinski sign, and sensory loss may be detected in conjunction with upper motor neuron deficits. Destruction of the spinal and trigeminal lemnisci, usually of vascular origin, produces loss of pain, temperature and light touch, and vibratory senses on the contralateral side of the body and face, respectively.

SUBTHALAMIC NUCLEUS

Subthalamic nucleus (Figures 21.7 and 21.8), a biconvex lenticular nucleus in the subthalamus, is separated by the zona incerta from a medially located internal capsule and a ventrally positioned ventral thalamic nuclear group. It lies medial and dorsal to the crus cerebra, encapsulated by the fibers of the subthalamic fasciculus. It consists of multipolar neurons with small interneurons with long ellipsoid dendritic arborizations that lack spines. These multipolar neurons are "fast-spiking pacemakers." The small number of interneurons constitutes nearly 10% of the GABAergic of the intrinsic circuitry. Afferents to this nucleus emanate primarily from the lateral pallidal segment, but cortically derived glutaminergic excitatory afferents also project to the subthalamic nucleus that leads to inhibition of pallidal neurons. While striatopallidal and the pallido-subthalamic fibers are inhibitory GABAergic, the subthalamic nucleus utilizes the excitatory neurotransmitter glutamate. Prolonged stimulation of the subthalamic nucleus suppresses dyskinetic motor activity seen in Parkinson particularly as it relates to dopamine therapy. In contrast to the medial pallidal segment and substantia nigra pars reticulata, the subthalamic nucleus and the lateral pallidal segment are fast-firing pacemakers (central pacemaker of the basal nuclei with synchronous bursts) that lack direct thalamic projection. There is a dense striatal projection to the lateral pallidal segment but not to the subthalamic nucleus, while the cortical fibers project to the subthalamic nucleus but not the lateral pallidum. The connection between the subthalamus and lateral pallidum is both excitatory and inhibitory; the subthalamopallidal fibers subserve excitatory influence compared to the pallidosubthalamic fibers which exert inhibitory effect. Through these connections, the subthalamic nucleus integrates motor activities through its connections with the basal nuclei, substantia nigra, and the tegmentum of the midbrain. It is thought to have an inhibitory influence upon the globus pallidus.

A lesion of the subthalamic nucleus produces *hemiballism*, a rare disorder with sudden onset that affects the proximal part of the contralateral extremities. It is commonly seen in hypertensive individuals as a sequel to a vascular accident. It manifests forceful and uncontrollable movements commonly on the contralateral extremities, resembling a baseball pitcher's windup. This condition exacerbates under stressful conditions.

RETICULAR FORMATION

Reticular formation, another component of the extrapyramidal system, occupies the central core of the brainstem, consisting of the paramedian, medial, and lateral nuclear columns. It is concerned with the regulation of somatic and visceral (autonomic) motor activities and reflexes. It also modulates the electrocortical activities of the brain, mediates pain transmission, and regulates emotional expression.

The connections of the ventral striatum and ventral pallidum are mainly with the limbic system and temporal and orbitofrontal cortices. Efferent fibers from the corpus striatum arise mainly from the pallidum and the pars reticulata of the substantia nigra. The pattern of connection starts with corticostriate fibers terminating in the striatum (dorsal and ventral), which, in turn, projects to the pallidum and substantia nigra pars reticulata. The pallidal and nigral efferent fibers then convey the information to the dorsomedial supplementary motor cortex or to the orbitofrontal, cingulate gyrus, and the superior colliculus. This connection provides

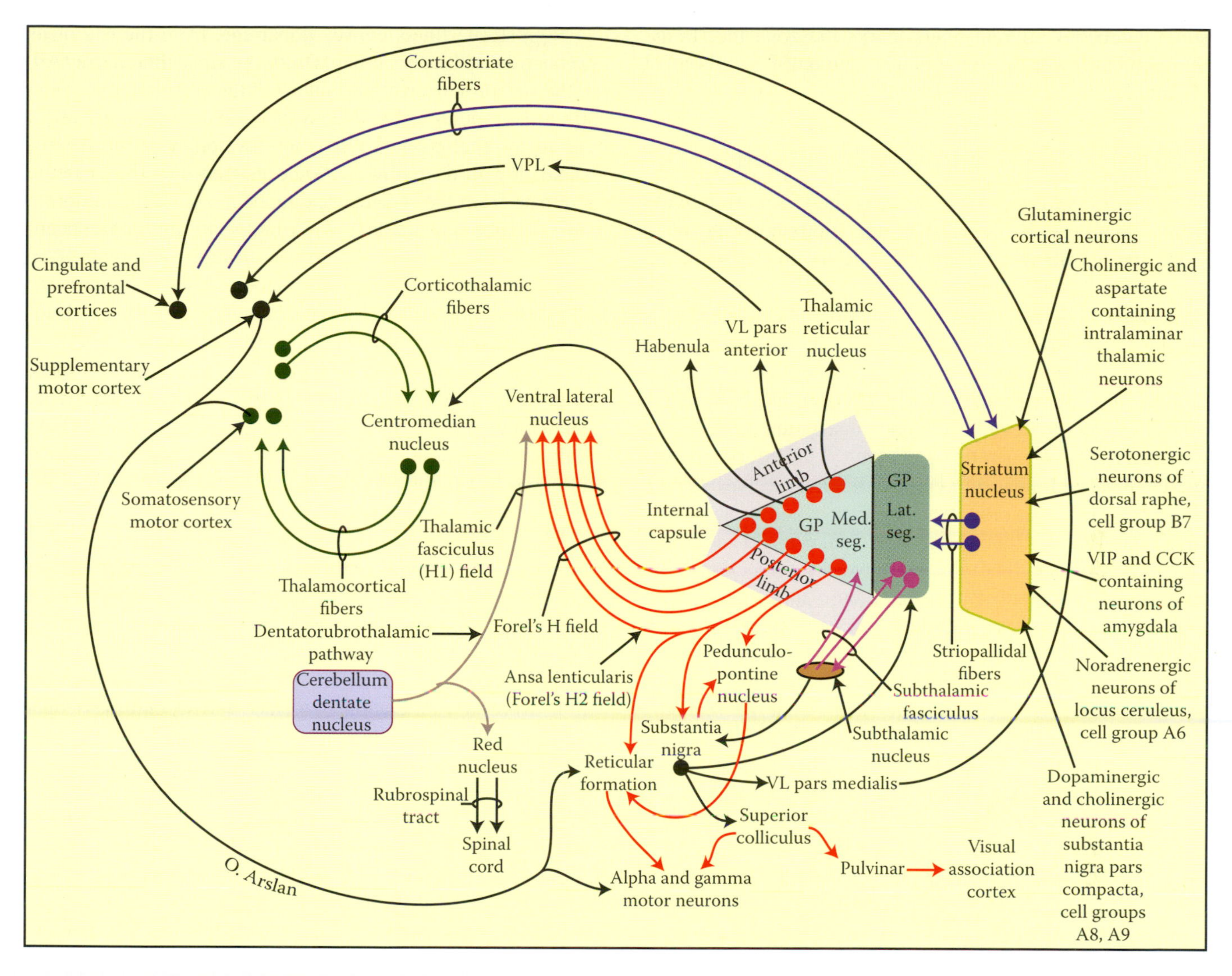

FIGURE 21.8 This schematic drawing shows the afferents and efferents of the basal nuclei. Note the massive output from the medial segment of the globus pallidus to VL nucleus of thalamus.

an avenue for the cognitive, affective, symbolic, and sensory input to reach the supplementary motor cortex, which is an essential part of the neocortex in planning purposeful movements. Disruption of this link that includes the mesostriatal and mesolimbic systems accounts for the limited movement ability (bradykinesia) observed in Parkinson's disease as well as *N*-methyl 4-phenyl 1,2,3,6-tetrahydropyridine (MPTP) poisoning. Despite the limitation of movement and speech, patients report, following therapy, that they were aware and are planning to perform movements without the ability to do so. Presence of projection from the visual cortex to the premotor cortex may explain the reliance of a patient with Parkinson's disease on visual guides for movement control in the absence of involvement of the supplementary motor cortex.

Afferent fibers to the basal nuclei terminate in the caudate, putamen, and the lateral segment of the dorsal pallidum. Nigrostriatal fibers that convey dopamine from the pars compacta to the striatum are affected in Parkinson's disease leading to the depletion of dopamine in the striatum and substantia nigra. There is a topographical representation of the striatonigral projection from the dorsal striatum to the substantia nigra where the fibers that emanate from the head of the caudate terminate rostrally compared to the putaminal fibers that distribute diffusely throughout the substantia nigra. Lateral segment of the dorsal pallidum sends GABAergic efferent fibers to the pars compacta but mostly to the pars reticulata of the substantia nigra. Subthalamic nucleus also projects to the pars reticulata. Efferent fibers of the corpus striatum project exclusively to the nuclei in the midbrain such as the substantia nigra or to diencephalic structures such as the thalamus, hypothalamus, and subthalamic nucleus. Due to the strategic location of the internal capsule between the basal nuclei and the diencephalon, the emanating pallidal axons must either pass through the internal capsule (fasciculus lenticularis and fasciculus subthalamicus) or run around it (ansa lenticularis). Similarly, because of the location of the basis pedunculi between the basal nuclei and the midbrain, the strionigral comb fiber system must pass through the basis pedunculi en route to the substantia nigra. The path of efferent fibers from the corpus striatum with some cerebellar efferent fibers marks the zona incerta and the "Haubenregionen" or hat plume region of Forel (Forel H, H_1, and H_2).

CONNECTIONS OF THE CORPUS STRIATUM

The *dorsal striatum* receives corticostriate fibers from widespread areas of the cerebral cortex, thalamostriate fibers from the intralaminar thalamic nuclei, and nigrostriatal fibers from the substantia nigra (Figures 21.7 and 21.8). The *corticostriate* input consists of a massive fiber system comparable to the corticopontine fibers, which originate from the entire neocortex, with the exception of the olfactory, visual, and auditory cortices and particularly from layers V and VI and generally project to the ipsilateral striatum, utilizing the excitatory neurotransmitter glutamate. The inferior part of the caudate and adjacent ventral striatum receives specific projections from the orbitofrontal cortex. Caudate nucleus, in particular, receives input from the frontal eyefield and frontal association cortex. The putamen receives projections from the ipsilateral somatosensory, premotor, and motor cortex and also from the contralateral motor cortex via the corpus callosum. Additional projections to the caudate nucleus arise from the association areas of the frontal, parietal, and temporal cortices. The corticostriate fibers undergo reduction through numerical convergence between the source of these projection and their targets leading to dampened input map. Since the corticostriatal fibers can systematically terminate together or independently from each other, cortical information conveyed forms the basis of separation and recombination. These fibers also exhibit a spatial "remapping."

The thalamostriate fibers emanate from the centromedian and central lateral intralaminar thalamic nuclei and convey cerebellar and pallidal input through the striatothalamic fibers to the supplementary motor and orbitofrontal cortices via the thalamostriate fibers, completing a closed feedback circuit. Through this circuit, the basal nuclei affects not only the supplementary motor cortex but also the behavioral and cognitive functions. The nucleus centralis pars paralateralis, pars parafascicularis, and pars media constitute components of the CM-PF complex, containing a high concentration of parvalbumin, a calcium-binding albumin, which is related structurally to calmodulin and troponin C. Parvalbumin is contained in the GABAergic interneurons of the reticular thalamic nuclei and CM-PF nuclear complex and is expressed in the basket cells of the cerebral cortex and in the Purkinje neurons of the cerebellum. Interneurons that contain this type of albumin protein are fast-spiking and produce gamma waves of the electroencephalogram. Nuclei centralis pars parafascicularis and pars media are the major source of thalamostriatal fibers that utilize glutamate as a neurotransmitter. There is some degree of selectivity of termination regarding their input, as the neuronal axons of the pars parafascicularis project to the nonmatrix and associative part of the striatum, whereas the pars media is specifically destined to the matrix compartment and forms the striatal–medial pallidal–pars media striatum (Nauta–Mehler's) circuit.

Nigrostriate fibers convey dopamine, from the site of its production at the pars compacta of the substantia nigra (A9) to the striatum, exerting inhibitory influence upon the dorsal striatum through a direct path or possible excitatory effect via an indirect path. These afferents comprise the mesostriatal dopamine pathway. Some afferents to the striatum originate from the serotonergic raphe nuclei of the midbrain (dorsal raphe nucleus B7), contributing to the high concentration of serotonin in the dorsal striatum. Afferents from adrenergic neurons of the locus coeruleus (cell group A6) also terminate in the striatum. These dopaminergic, serotonergic, and adrenergic projections to the striatum modulate otherwise inactive striatal neurons to interact with the thalamic and cortical input. Although tremor is seen in conjunction with disruption of the nigrostriatal fibers in patients with Parkinson's disease, destruction of the nigrostriatal connection does not induce tremor without also disruption of the dentatorubral and dentatothalamic fibers.

Efferent fibers from the dorsal striatum project to both segments of the globus pallidus (striatopallidal) and the pars reticulata of the substantia nigra (striatonigral fibers). Both striatopallidal and striatonigral fibers deliver inhibitory GABA and excitatory substance P. Substance P may enhance the production and utilization of the neurotransmitter dopamine. Receptor/ion channel interface and second messenger may determine the variation in the inhibitory and excitatory effects of GABA, substance P, and dopamine. Striatonigral fibers project to the pars reticulata of the substantia nigra directly or via the pallidum. A disproportionately large number of striatal axons converge on a given single pallidal or nigral neurons with a ratio of 100:1. This is evident by the fact that 90% of pallidal and nigral dendrites establish synapses with the striatal neurons. This convergence is thought to enable the pallidal neurons to focus their response through a center-surround pattern of activity (central inhibition and peripheral excitation). Direct striato-pallido-nigral fibers consist of thinly myelinated axons of the striatal spiny neurons grouped into pencils "converging like the spokes of a wheel" forming the comb system. The indirect striatopallidal pathway involves projection from the external pallidal segment to the subthalamic nucleus through the subthalamic fasciculus and through the same fasciculus back to the internal pallidal segment. The internal pallidal segment, in turn, projects to the mesencephalic reticular formation and the pedunculopontine nuclei. Modulated input received by these nuclei is conveyed to the pars reticulata of the substantia nigra through relay in the tectum, brainstem, spinal cord, as well as the medial part of the VL nucleus of thalamus. The latter sends projection to the prefrontal and anterior cingulate cortices.

Afferents to the ventral striatum arise from the hippocampal gyrus, entorhinal cortex, olfactory tubercle, anterior cingulate, nucleus accumbens, orbitofrontal, anterior cingulate, and temporal cortices. It also receives serotonergic projections from the dorsal raphe (group B7), noradrenergic from the locus coeruleus (A6), and dopaminergic from the substantia nigra (medial part of pars compacta) and the paranigral nucleus (A10). The latter dopaminergic projections form the *mesolimbic dopamine* pathway, which utilizes the MFB to also send fibers to the entorhinal area, amygdala, septal area, and hippocampus.

Efferent fibers of the ventral striatum project to the ventral pallidum and the substantia nigra pars reticulata. Projection to the pars reticulata occurs directly or through utilization of the subthalamic nucleus as a relay nucleus.

As discussed earlier, the globus pallidus or dorsal pallidum is divided into lateral and medial segments by the internal medullary lamina. It receives GABAergic fibers primarily from the striatum (striopallidal fibers) and to a lesser extent from the subthalamic nucleus. There are no direct spinal projections from the pallidum. The striopallidal fibers produce an initial inhibition, followed by excitation, an arrangement thought to control the extent of their inhibitory signal and spatially focus on the restricted number of pallidal neurons.

The output from the medial segment of dorsal pallidum is represented primarily by the lenticular fasciculus, ansa lenticularis, and thalamic fasciculus (Figures 21.7 and 21.8). The lenticular fasciculus (Forel's field H2) traverses the posterior limb of the internal capsule and then joins the ansa lenticularis, whereas the ansa lenticularis (Figures 21.8 and 21.9) loops around the posterior limb of the internal capsule. When lenticular fasciculus and ansa lenticularis merge rostral to the midbrain in Forel's field H, they are joined by the dentatorubrothalamic fibers to form the thalamic fasciculus (Forel's field H1). The latter delivers impulses to the ipsilateral VL, ventral anterior, and the centromedian thalamic nuclei. The VL and ventral anterior thalamic nuclei influence, through the thalamocortical radiations, the ipsilateral motor and premotor cortices, respectively. The centromedian nucleus acts as a feedback loop, delivering generated impulses back to the striatum. It appears that tremor in Parkinson's disease occurs as a result of neuronal firing in bursts in the VL nucleus of thalamus that corresponds closely to the involuntary movement,

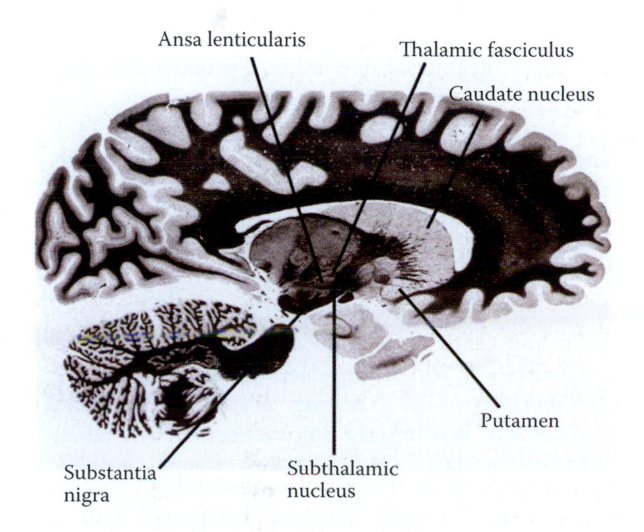

FIGURE 21.9 This diagram shows the feedback circuit formed by the efferents and afferents of the centromedian nucleus of the thalamus, which involves reciprocal connection with the cerebral cortex and projections from the medial segment of the globus pallidus.

which is mediated by cerebellar input to the posterior part of the VL nucleus and eventual projection to the primary motor cortex. This is supported by studies indicating that thalamotomy and, in particular, extirpation of the VL nucleus abolish parkinsonian tremor. A few fibers may project to the tegmentum of the midbrain as the *pallidotegmental tract*, constituting the principal link between the basal nuclei and the reticular formation. Some pallidal projections terminate in the habenular nucleus (10%), thus establishing a link between the basal nuclei and the limbic system. Projections from the medial segment are also directed posteriorly and through collaterals to the lateral thalamus, pedunculopontine, and retrorubral nuclei.

The lateral segment of the globus pallidus projects to the subthalamic nucleus via the subthalamic fasciculus (Figure 21.7), utilizing the neurotransmitter GABA. Information received by the ventral pallidum, which originates from the ventral striatum, projects to the dorsomedial and midline thalamic nuclei and via thalamocortical fibers to the cingulate and prefrontal cortices and hippocampal gyrus.

It is evident from above that the function of basal nuclei and associated neurons operates in a reciprocal open and closed feedback circuits. In the closed feedback circuit, the connection is set in a way that a particular cortical area that receives input would feed into this circuit that leads to projection back to the same cortical area. This type of feedback is demonstrated in several connections between the prefrontal and anterior cingulate cortices via the dorsomedial and ventromedial thalamic nuclei, somatosensory motor cortex via the centromedian nucleus, and the supplementary motor cortex through the VL nucleus pars anterior. In the open type of the circuit, the loop starts with a cortical area that feeds into this circuit and through series of relay nuclei convey the impulses to another cortical area outside the loop. There is mixed loop that entails open and closed components of the corticocortical feedback circuit. The cortical connections to different parts of the striatum form the basis for the ability of the cerebral cortex to transform mental assessments to motor behavior and the ability to introduce adjustments for ongoing motor activity. The anterior caudate nucleus through its connection with the prefrontal cortex mediates working memory and ability to plan. This arrangement also holds true relative to the striatal projection to the pallidum.

FUNCTIONAL LOOPS AND PATHWAYS

It has been suggested that several loops, with closed and open components, are involved in the connections of the basal nuclei. The closed corticocortical feedback circuit through parts of basal nuclei is reinforced by additional afferent fibers from related but distinct cortical areas to specific parts of the basal nuclei involved in the open component of the loop. This is evident in the supplementary motor loop via the part of the putamen that also receives input from the premotor, primary motor, and somatosensory cortices. Similarly, the frontal eyefield loop occurs via the part of the caudate nucleus that is also the site of termination of the posterior parietal and the dorsolateral prefrontal cortices. It has been further reported that this closed-/open-circuit arrangement may have structural (matrix and striosomes) roots and is related to the differential distribution of the neuroactive elements in parts of the basal nuclei. Thus, striatal efferent fibers that project to the pallidum and substantia nigra pars reticulata can be categorized on the basis of dopaminergic receptor types and also on the origin from the striosomes (found mainly in the striatum) which are controlled by the hippocampal and olfactory cortices and deep part of layer V of the neocortex. It is also classified relative to their origin from the matrix, mainly the putaminal matrix, that controls the superior colliculus via the pars reticulata and receives input from the superficial part of layer V of the neocortex. Therefore, distinct yet overlapping loops appear to function within the structures that constitute the extrapyramidal system. One of the loops associated with the striosomes of the dorsal and ventral striatum utilizes mesostriatal and mesolimbic projections to the limbic system and deeper layers of the neocortex. Whereas the loop associated with the matrix conveys cortical input from the superficial layer V of the neocortex to the pallidum to influence the activities of the supplementary cortex and also to the superior colliculus via the pars reticulata substantia nigra to mediate head and eye movements. It is not surprising to see disease of the extrapyramidal system to manifest disorders that parallel the affected functional feedback.

As evident from the above, the tonic inhibitory output from the internal pallidal segment and the substantia nigra pars reticulata is modulated by two parallel pathways. The *direct pathway* permits activation of the VL and VA nuclei of thalamus as well as the premotor and motor cortices by exerting transient suppression on the tonically active inhibitory neurons of the internal pallidal segment, thus promoting disinhibition that allows inputs to be funneled to the thalamus and cortex. Therefore, direct pathways when activated release tonic inhibition.

The *indirect pathway* promotes focusing the gating by increasing the tonic inhibitory impulses to thalamus generated by the internal pallidal segment. In the indirect pathway, another population of striatal neurons project to the lateral pallidal segment of the dorsal pallidum, which, in turn, sends excitatory impulses to the subthalamic nucleus and the internal pallidal segment. The subthalamic nucleus sends impulses to the internal pallidal segment and the substantia nigra pars reticularis. In the direct pathway, the tonically facilitated GABAergic neurons of the lateral pallidal segment are inhibited by the striatal projections. Subsequently, the subthalamic nucleus released from this inhibition conveys excitatory impulses to the internal pallidal segment, which augments the inhibitory outflow to the VL and VA nuclei of thalamus and through their connections increases inhibition of the motor and premotor cortices. Thus, the indirect pathway exerts control over, or acts as a "brake" modulating, the disinhibitory influences of the direct pathway.

Disturbance of balance in the inhibitory influence exerted within the direct and indirect pathways can account for the hypokinetic disorders observed in Parkinson's disease and the hyperkinetic manifestations that occur in Huntington's disease. Disruption of the circuitry within the direct pathway accounts for the hypokinetic disorders seen in Parkinson's disease. Hypokinesis is a reflection of lack of disinhibition of the basal nuclei, which reduces the excitability of the thalamus and ultimately the motor cortex. This is also accompanied by heightened discharge of the inhibitory neurons in the substantia nigra pars reticulata. In the same manner, reduction in the excitability of the tectal neurons produces low frequency and amplitude saccades. Understanding the circuits involved in the indirect pathway can explain the dysfunctions seen in Huntington's chorea in which the medium spiny striatal neurons that exert inhibitory influence on the neurons of the lateral pallidal segment undergo degeneration. Thus, the neurons of the external pallidal segment become highly activated, which, in turn, inhibit the excitatory projections of the subthalamic nucleus to the internal pallidal segment. As a consequence, the inhibitory outflow of the basal nuclei to thalamus is reduced leading to inappropriate excitation of the motor and premotor cortices and the development of involuntary movements. Release of this inhibition from the thalamus may also be responsible for the development of other related manifestations, which are not movement-related. Circuitry of the indirect pathway can also shed light on the manifestations of Parkinson's disease, as a lesion of the subthalamic nucleus can alleviate some of the movement disorders.

In summary, the corticostriate fibers activate the striatum to release the inhibitory GABA, which act upon the globus pallidus and substantia nigra. The striopallidal fibers block the release of inhibitory GABA from globus pallidus, which act upon the thalamus, resulting in activation of the thalamocortical fibers. This activation is thought to facilitate movement by exciting the motor cortex. Influences exerted by the subthalamic nucleus upon motor activity follow different logic. The corticostriate fibers produce excitation of the striatum and subsequent inhibition of the lateral segment of the globus pallidus, leading to disinhibition of the subthalamic nucleus and ultimate activation of the pallidal output.

DYSFUNCTIONS OF THE EXTRAPYRAMIDAL SYSTEM

Lesions associated with the extrapyramidal system may produce abnormal involuntary movements (dyskinesia) contralateral to the lesion side. Dyskinesia may be due to the excess of neural output and the imbalance in the dopamino-cholinergic-GABAergic systems. These motor disturbances can simply be explained as intact structures released from the inhibitory influences of the damaged structures associated with this motor system. Motor deficits may also be expressed as a reduction in movements (hypokinesia) and alteration in the residual contraction of the muscles (e.g., rigidity seen in Parkinson's disease). Dyskinesia may assume the forms of tremor, rigidity, choreiform movements, athetosis, dystonia, myoclonus, tardive dyskinesia, akathisia, and tics.

Tremor (tremor at rest) is an involuntary rhythmic, sinusoidal, and alternating bilateral movement in a group of joints seen at rest that may disappear during voluntary action. It is rarely seen unilaterally and rarely involves the legs. Tremor may be aggravated under stress and in situations that require maintaining a constant posture. It often diminishes with alcohol intake and during sleep and voluntary movements.

Midbrain or rubral tremor is an extremely disabling condition, which may also be seen at rest, exhibiting exacerbation during movement. This form of tremor is seen in multiple sclerosis resulting from destruction of the superior cerebellar peduncle that surrounds the red nucleus. Tremor may also be observed in neonates as a result of low level of serum magnesium, epileptic seizures, and individuals with hepatolenticular degeneration (Wilson's disease).

Rigidity refers to the generalized increase in the tone of all muscle groups (agonists and antagonists) to the same degree without significant change in reflex activity. In contrast, spasticity primarily affects the antigravity muscles. Rigidity may also be responsible for the fragmentation of a movement and for the slow regaining of the resting position subsequent to sudden release of passive resistance from the affected part. This is in contrast to the rebound phenomenon observed in cerebellar dysfunction. In rigidity, joints exhibit wax-like resistance to passive movements in all directions (during flexion and extension of the wrist or elbow). Conversely, spasticity exhibits resistance only in the initial stage of passive movement. Although rigidity is generally accompanied by tremor, congenital tremor observed in erythroblastosis fetalis (kernicterus) remains an exception.

Rigidity and tremor are pathognomic to *Parkinson's disease* (Figure 21.10), a slow progressive condition that develops as a consequence of encephalitis, manganese or carbon dioxide poisoning, trauma, neurosyphilis, cerebrovascular accidents, or cerebral arteriosclerosis. Phenothiazine and catecholamine-depleting drugs like reserpine and metoclopramide (Reglan), which is useful for esophageal reflux and gastric stasis, may also produce characteristics of reversible form of Parkinson's disease. The similarity between medication-induced Parkinsonism and Parkinson' disease is so great that the distinction is almost impossible.

True Parkinson's (*paralysis agitans*) disease is a familial disorder most frequently seen in males. In this disease, the melanin-containing cells of pars compacta of the substantia nigra, nigrostriatal fibers, locus coeruleus (noradrenergic neurons), raphe nuclei (serotoninergic neurons), dorsal motor nucleus of vagus, and the globus pallidus undergo degeneration and some depigmentation. Pathological examination of the pars compacta of the substantia nigra may reveal eosinophilic inclusion or Lewy bodies. Deficits seen in Parkinson's disease are attributed to the abnormal reduction or depletion of dopamine in the striatum and the pars compacta of the substantia nigra and loss of its inhibitory effect but with normal dopamine receptors. Glucose level and striatal neurons remain within normal limits unlike the loss of striatal neurons and the dopamine receptor sites that accompany changes in glucose metabolism (evident in PET scan studies) seen in Huntington chorea, progressive supranuclear gaze palsy, and Wilson's disease. This is followed by overactivity of the cholinergic intrastriatal neurons and alteration in the output mechanism of the GABAergic neurons. It has been proposed that Parkinson's disease is an abnormal acceleration of the aging process based upon the fact that dopamine shows an increase in concentration very early in life, followed by a rapid decrease in the first two decades of life, and then by prolonged decline of dopamine, a process that may even accelerate further by infection or toxicity.

FIGURE 21.10 This is a depiction of the expressionless (masked) facies of an individual with Parkinsonism.

In Parkinson's disease, the resting tremor is intermittent and slow (4–6 Hz) and shows characteristics of "pill rolling" in which the cupped hand appears as though it was shaking pills. Tremor involves extension of the interphalangeal joints of the thumb and digits and flexion of the metacarpophalangeal joints. It usually begins unilaterally, most frequently involving the right hand, and tends to disappear with action, intense concentration, and during sleep. It re-emerges as the extremity assumes a resting position. Rhythmic, regular, and involuntary contractions of the labial muscles without tongue movements (*Rabbit syndrome*) may be seen as a manifestation of Parkinson's disease or as a side effect of antipsychotic drugs, which respond to anticholinergic medications. The rigidity seen in Parkinson's disease produces lead-pipe or cog-wheel phenomenon (ratchet-like features) as a result of the gradual and alternate increase and decrease in muscle resistance to passive movement.

Patients with Parkinson's disease exhibit flexed and stooped postures, which may be corrected upon command. A marked decrease or absence of movements (bradykinesia or akinesia) is manifested in small handwriting (micrographia). Akinesia may become so severe, resulting in immobile posture. The movements may become slow to an extent that conveying the food to a patient's mouth may require an unusually long time. Patients exhibit difficulty in initiating or stopping movements, as well as difficulty in performing a simple task, such as buttoning a shirt or feeding. Habitual automatic movements become more difficult than unusual, or they may assume the pattern of reflex activity. Thus, running may be easier than walking. Shuffling gait with short and rapid steps (festination) may also be seen in individuals with Parkinson's disease. Patients may also lack the associated movements during walking (i.e., arm swing) and are unable to bend forward or flex the knee in an attempt to sit on a chair (sitting en bloc) or stand up. This is attributed to the possible lack of control mechanism exerted by the basal nuclei over the descending dopaminergic and aminergic neurons of the reticular formation on the movement generating interneurons of the spinal cord. Patients may have difficulty in maintaining the standing posture if a sudden move has occurred. Slow monotonous repetitive speech and masked expressionless face are also common symptoms of this disease. The head, shoulder, and body move together as a block when making turns, leading to entrapment in corners. Righting reflex, which involves smooth turns from the supine to prone position, is severely impaired. Intellect remains intact, although some indications of slow thinking and response may be detected. Clinical depression in these patients is accompanied by marked reduced firing in the prefrontal and the striatal neurons as well as the aminergic neurons within the brainstem. This is thought to be an expression of a disorder caused by disruption of the direct connection between the basal nuclei, frontal lobe, and limbic cortex, as well as the indirect connection via the substantia nigra pars reticulate with the dorsomedial and ventromedial thalamic nuclei.

Patients with Parkinsonism usually prefer cold weather and are easily kept warm. They may be comfortable with light cover even in cold weather. Emptying of the bladder may be inadequate, and constipation is a major and common complaint. Erectile function is impaired and blood pressure may be low. Patients commonly exhibit seborrheic dermatitis, which can be controlled with adequate hygiene. Tendon reflexes are usually unaffected, and abnormal extensor plantar reflex suggests a Parkinson-like syndrome. However, uninhibited glabellar reflex (Myerson sign), snout reflex, and palmomental reflexes are common to Parkinson's disease, even early in the course of this condition.

L-dopa, a precursor of dopamine, which permeates the blood-brain barrier, has been proven effective in ameliorating akinesia and rigidity but is less effective in reducing tremor. Initially, one-fourth of the nigrostriatal neurons must be intact to convert L-dopa to dopamine. Dopamine agonists such as bromocriptine may be used as a supplement to L-dopa. L-dopa may be supplemented with medications such as amantadine (Symmetrel) that enhance the release of dopamine and norepinephrine from presynaptic neurons. Tocopherol (vitamin E) is proven to have an important role in protecting dopamine from the effects of toxins and free radicals. Combined with Deprenyl, an antidepressant and antioxidant drug that blocks dopamine metabolism, tocopherol may prove to be of value in the treatment of this disease. Carbidopa, a decarboxylase inhibitor that does not cross the blood-brain barrier, is usually given in combination with L-dopa to raise the L-dopa concentration in the brain and minimize its side effects. Overdosage of L-dopa produces reversible choreiform movements of the extremities and perioral muscles. Since dopamine depletion is associated with excess of acetylcholine, anticholinergic medications are relatively effective in treating Parkinson's disease and drug-induced Parkinsonism and dystonia. It is important to bear in mind that anticholinergic therapy may accelerate the development of tardive dyskinesia and may contribute to mental retardation. Neurosurgical procedure involving the contralateral thalamus (thalamotomy) used to be a successful method of treatment until the introduction of L-dopa. This surgical procedure also carried the risk of damaging the corticospinal and corticobulbar tracts within the posterior limb of the internal capsule, producing various forms of upper motor neuron palsies. Some patients developed dystonia and athetoid movements as complications of neurosurgical intervention.

Artificially induced lesions in the globus pallidus and/or the VL nucleus of the thalamus may alleviate tremor and rigidity. Posteroventral pallidotomy reemerged as a method of treatment developed by Leksell, which achieved good relief from tremor, rigidity, and akinesia. However, central homonymous visual field deficit, facial weakness, and dysphasia are the reported complications of this procedure. Innovative approaches to the treatment of Parkinson's disease included transplantation of dopamine producing adrenal medullary cells into the basal nuclei or ventricular system. In view of the fact that the brain is an immunological privileged site, attempts to experimentally transplant mesencephalic fetal cells were carried out, though with no clear-cut success.

Chorea is a brisk, rapid, graceful, and purposeless movement of short duration, unpredictable in direction, timing, and location, which starts suddenly and shows no rhythm. These movements are parakinetic, which appear to be meaningful, but in reality, they do not serve any rational purpose. They may occur in isolation or may be accompanied with other involuntary movements. They are seen primarily in the distal appendicular musculature (e.g., hand) and appear to move from one part of the body to another in a random and continuous fashion. Chorea may be confined to one side of the body (hemichorea) or specific muscles, for example, respiratory muscles (respiratory chorea). Diffuse degenerative process in the neostriatum may result in choreiform movement intermixed with tics, parkinsonism, and dystonia. These movements show exacerbation under stressful situations and during walking, persist as long as the patient is awake, and may even continue during sleep. Twitching in the face, lip-smacking or pouting, cheek puffing, tongue rolling, and jaw protrusion usually accompany the involuntary movements. Patients may not be aware of these involuntary movements and may be thought of as fidgeting or being clumsy. Hypotonia, a decrease in the muscle tone or atonia, which occurs at the end of each involuntary movement, may lead to a delay in the relaxation of the contracted muscles. On neurological exam, the patient exhibits "milkmaid grip," a sign in which the patient squeezes the examiner's hand, upon request, by performing "milking motion" of contraction and relaxation. This is also considered an important Jones' criterion in diagnosing rheumatic fever. Ocular disorders (disruption of saccades and gaze abnormalities) and jerky finger-to-nose testing also occur. Chorea is considered as a manifestation of impairment of the modulatory effect of the basal nuclei upon thalamocortical projections and receptor blockade of the neurotransmitter. The putamen, the globus pallidus, and the subthalamic nucleus play an important role in the development of chorea. In healthy individuals, efferents from the subthalamic nucleus utilize a glutamate excitatory pathway in its excitatory projection to the globus pallidus and the substantia nigra. This pathway excites the pallidal neurons that project to the thalamus and thus inhibits corticothalamic projection to the motor cortex. In patients with chorea, the excitatory subthalamic projection is reduced or abolished, producing disinhibition of the pallido thalamic projection. Two classic diseases exhibit this type of movements: Sydenham's and Huntington's chorea.

Sydenham's chorea (St. Vitus' dance) is a rare reversible disorder, which occurs in children between the ages of 5 and 15 years and commonly follows endocarditis as a complication of A beta hemolytic streptococci-induced rheumatic fever (Figure 21.11). The pathogenesis of this disease is based on the fact that streptococcal M proteins induce the production of antibodies (IgG) against neurons of the caudate nucleus and subthalamic nucleus, disrupting their function. It usually manifests months after the acute infection and can last up to 2 years. It should be differentiated from dystonia, tic disorders, and dyskinesia associated with drug withdrawal. The involuntary movements are uncoordinated, are more rapid than those of Huntington chorea, and may have a lightning character. Involuntary choreiform movements in this disorder involve primarily the feet and hand and also the face in the form of grimaces, but can affect all extremities. The affected hand undergoes alternating change in tension (milking sign) with progressively worsening handwriting. Patients show speech difficulty due to disorders of the tongue (lingual fasciculation) and masticatory and facial muscles. Children develop unusual behavior and remain irritable.

Huntington chorea (HD), a devastating progressive hereditary illness that can be fatal and produces a slowly developing dementia. It is an autosomal dominant disease, and the mutated gene is located on chromosome number 4. The location of the mutated gene may allow prenatal diagnosis of this disease, eventual cloning, and replacement of the gene. Children of

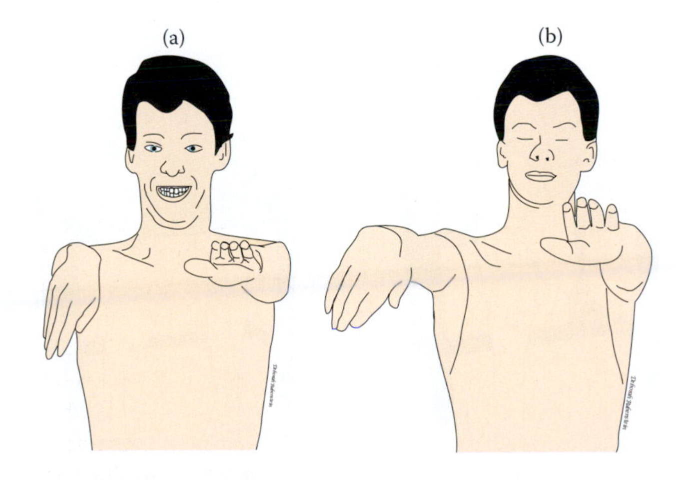

FIGURE 21.11 These two diagrams illustrate the choreiform movements associated with Sydenham's chorea. (a) The patient's forearms are pronated and his right hand is flexed at the wrist, while he is grimacing. (b) The same individual with eyes closed and having a hypotonic right arm.

patients with this disease have a 50% chance of being affected. It is characterized by degeneration of the neostriatum and cerebral cortex.

In this disease, neurodegenerative changes and cellular loss occur in the striatal spiny "medium-sized" neurons that utilize inhibitory GABA as a neurotransmitter in their projections to the globus pallidus and substantia nigra. The reduction of the glutamic acid decarboxylase, an enzyme responsible for the conversion of L-glutamic acid to GABA, is thought to be crucial in the pathogenesis of this disease. Loss of the striatal neurons may cause an increase in thalamic inhibition and thus increases its excitatory impact on the cerebral motor cortex and the resultant disorganized hyperkinetic movement. Changes in the levels of substance P, enkephalin, dynorphin, and the cholecystokinin may also be noticed. Neurodegenerative changes also occur in the frontal and temporal cortices. Mental deterioration, which includes paranoid delusions and eventual dementia, may occur prior to the onset of involuntary movements or immediately afterward. Symptoms typically appear late in the fourth or fifth decade after the affected individual have had children and passed on the gene (Figure 21.12). Unfortunately, one of the saddest travesties of this illness in the past was people's perception that afflicted patients were witches, creating the environment for their cruel and unfair treatment, torture, and sometimes execution. Severity and length of symptoms of HD may vary according to the age of onset. Symptoms in the adults last nearly a decade longer than that in the juvenile population. Early presentations may include inattentiveness, lack of concentration, difficulty expressing thoughts, gradual development of behavioral and emotional disorders, impulsiveness, and depression anhedonia (inability to appreciate enjoyable experience). These may, though infrequently, be associated with visual and auditory hallucinations. Gradual decline of mental faculty, memory disorder, and judgment and reasoning abnormalities are some of the signs of cognitive impairment and dementia.

Ataxia, which is manifested as lurching, dance-like walking pattern, is seen. Choreiform movements are seen in the fingers and toes and then spread to other parts of the body accompanied by fidgeting and clumsiness. Atypical (*Westphal subtype*) Huntington's chorea may occur in children and occasionally in adults, exhibiting rigidity and akinesia without choreiform movements.

Athetosis (Figure 21.13) is a slow, irregular, writhing, and worm-like movement involving the distal muscles of the extremities, neck, face, and tongue.

FIGURE 21.12 These diagrams show manifestations of Huntington's chorea. Observe the bizarre posture and uncoordinated movements of the upper and lower extremities. Affected individuals are disoriented but alert, exhibiting speech disorder and disorientation.

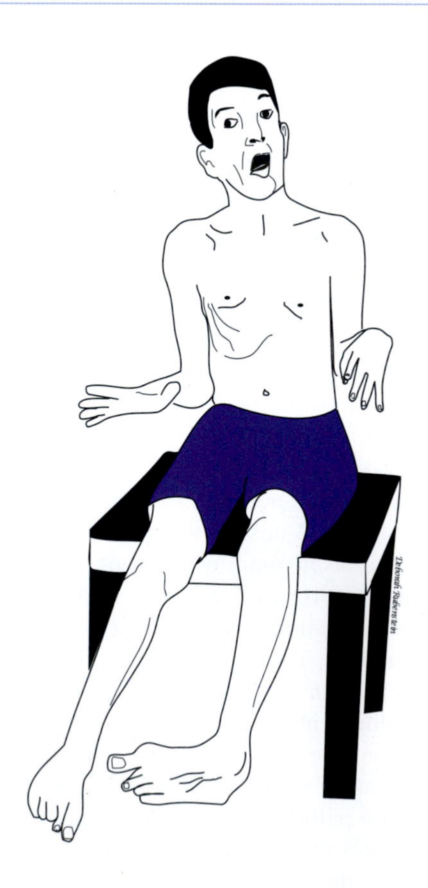

FIGURE 21.13 This drawing depicts an individual with athetosis affecting all muscles of the limbs, neck, and face. Note the flexion in the upper extremities and extension of the lower extremities.

These movements are cramp like, characterized by hyperflexion or hyperextension of the joints of the hand with bizarre configuration accompanied by adduction and abduction of the shoulder joint. They alternate between extension–pronation and flexion–supination of the arm, flexion–extension of the fingers, and eversion–inversion of the foot. Sometimes dorsiflexion of the great toe may be erroneously diagnosed as sign of Babinski. Athetoid movements occur in birth trauma–related anoxia. They are also seen in kernicterus, a neurological condition caused by Rh or other blood incompatibility between the mother and the fetus, which is associated with hemolysis, and accumulation of nonconjugated bilirubin in the brain. Here, athetosis exhibits faster rate and blends with chorea forming choreoathetoid movement.

Dystonia (torsion dystonia) is an idiopathic hereditary disorder characterized by contracture of the axial or appendicular musculature, resulting in severe torsion and deformed rigid posture (Figure 21.14). It is a physically impairing condition with no detectable impact on the mental status. Dystonia is also seen in hepatolenticular degeneration (Wilson's disease) and Huntington's disease and in Lesch–Nyhan syndrome. In contrast to athetosis where the distal muscles are affected, dystonia primarily involves the proximal muscles. The twisted and fixed posture of a dystonic patient can be very striking and repulsive. It manifests itself in childhood or teenage years. Overactivity of the striatal dopaminergic system, which increases dopamine concentration, is thought to be a factor in this disorder. This hypothesis is based on the observation that drugs that induce this condition block dopaminergic receptors with the resultant increase in dopamine concentration. It is furthermore based on the fact that administration of dopaminergic drugs exacerbates the associated movement disorders. Some patients may show minimal signs collectively known as *writer's cramp*, which is characterized by tonic contractions of the hand muscles during writing.

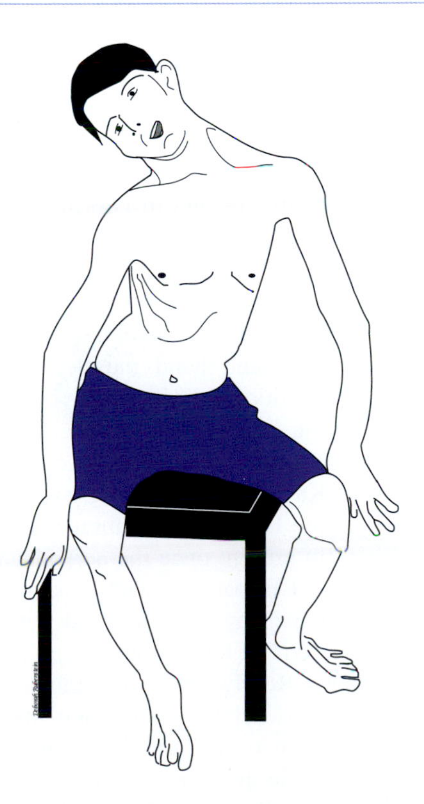

FIGURE 21.14 This is a depiction of a patient with torsion dystonia. Observe the torsions of the limbs, neck, and spine, which are produced by the slow, yet powerful, involuntary movements.

Myoclonus is a short, abrupt, irregular, uncoordinated, and lightning-like involuntary contraction of a muscle or a group of muscles, which may be induced by voluntary action or by a stimulus from the examiner. This is most likely due to decreased cortical inhibition, a common basis shared by epileptic seizures. It may or may not be synchronous or symmetrical, and may be focal or generalized. The repetitive fashion of the myoclonic jerks involving one or more sites is characteristic of this condition. It can be classified into negative and positive myoclonus. Negative myoclonus (asterixis) is a bilateral arrhythmic interruption of sustained muscle contraction (relaxation) causing lapses of posture, which is seen in Wilson's disease and most often in stuporous patients due to hepatic failure-induced encephalopathies. Positive myoclonus refers to involuntary muscle contractions. Physiologic myoclonus is exemplified in diaphragmatic and hypnic myoclonus. Diaphragmatic myoclonus involves the diaphragmatic muscles producing constant hiccup, whereas nocturnal myoclonus (hypnic jerk) occurs at the onset of sleep or upon awakening. Essential myoclonus is a pathologic myoclonus that is induced by voluntary movements, but is not associated with epileptic seizures, dementia, or ataxia and remains nonprogressive. Most of the essential myoclonus has familial basis (autosomal dominant) and commonly accompanied tremor or dystonia. It can be associated with the damage sustained by the cerebral cortex and abnormal neuronal discharge (cortical myoclonus) as a result of infectious diseases (spongioform encephalopathy, i.e., Jakob–Creutzfeldt disease) or metabolic disorders such as toxemia or uremia. Seizures and dementia may accompany cortical myoclonus. Subcortical pathologic myoclonus involves the brainstem. Hypertrophic degeneration of the inferior olivary nuclear complex in the medulla and associated neuronal feedback circuit produces *palatal myoclonus*, a unilateral or bilateral rhythmic and rapid contraction of the soft palate, which occurs regularly during sleep. It is not usually associated with mental disorder and generally results from infarction of the inferior olivary nucleus. Bilateral stereotypic and repeated extension of the foot and big toes, flexion of the hip, knee and ankle joints during non-REM sleep is a form of myoclonus seen in restless leg syndrome. A generalized or segmental myoclonus associated with stress is known as psychogenic myoclonus.

Tardive dyskinesia is a very common symptom that develops in a great number of patients as a side effect of antipsychotic medications (e.g., phenothiazines) administered for a prolonged period of time. Choreoathetoid movements of the limbs and trunk, and tongue protrusion, particularly lip smacking and facial grimacing, characterize this irreversible and persistent condition. Patients with this condition may exhibit loud labored respiration similar to that seen in Tourette's syndrome. Supersensitivity of the dopaminergic striatal neurons and depletion of GABA and glutamic acid decarboxylase may be attributed to the development of this condition. Several structures such as substantia nigra and inferior olivary nucleus may show some degree of degeneration in this condition.

Akathesia is characterized by a state of restlessness and jitters and continuous compulsion to move in order to relieve the uncomfortable sensation in the leg and thigh. These movements may mimic those seen in Parkinson's disease and other disorders of the basal nuclei. It may also be seen as a consequence of usage of neuroleptics, resembling the dyskinesia produced by excessive L-dopa. Akathesia may also be seen in other conditions such as parietal lobe lesions or portacaval encephalopathy. It has some resemblance to choreiform movements but is restricted to the lower extremity and is generally accompanied by subjective restlessness that exceeds leg movements. The latter is an important factor that distinguishes akathesia from agitation. One form of this condition is *tardive akathesia*, which may mimic signs of manic depression or anxiety states. It is treated with opioid analgesic such as propoxyphene.

Wilson's disease (hepatolenticular degeneration) is a rare autosomal recessive disorder of copper metabolism, which commonly occurs in the offspring of blood-related (consanguineous) marriages. The gene is located on chromosome 13, which shows a large number of mutations, while the gene for ceruloplasmin is located on chromosome 3. Ceruloplasmin contains 95% of the copper in human serum and has a significant role in iron efflux from the cells. In aceruloplasminimia, normal copper level is maintained while the iron gradually accumulates in the brain and pancreatic islets of Langerhans. Initial manifestations usually appear in the third or fourth decade of life. This serious condition is caused by a low level of circulating copper binding $\alpha 2$ globulin (ceruloplasmin) leading to a high level of free copper in the serum, which deposits in the basal nuclei, kidney, and liver. Over 30 proteins depend on copper to perform their functions including cytochrome c oxidase, superoxide dismutase, ceruloplasmin, dopamine-β hydroxylase, and lysyl oxidase.

Copper is an important element in a variety of cellular activities, including neurotransmitters, connective tissue and peptide biosynthesis, cellular respiration, antioxidants and iron, and metabolism. In fact, the metabolic fates of copper and iron are related. Deficiency of copper generates cellular iron deficiency, which results in growth retardation, impairment of immune response, and disorders of bone mineralization.

Copper is a key component of cytochrome c oxidase and dopamine-β hydroxylase and is found in high concentrations in the adrenergic neurons. Deposition of copper ultimately produces neuronal degeneration in the putamen and changes in the liver that range from elevation of serum transaminase level and chronic active hepatitis to multilobular cirrhosis and jaundice. However, release of excess copper suddenly into the blood can cause hemolytic anemia and liver failure. The cerebellum, skeletal muscles, and spinal and cranial nerves are spared. Rigidity, dysphagia, difficulty in speech, loss of facial expression, behavioral changes, and dementia are distinctive signs of this disease. Copper level is determined by the balance between gastrointestinal absorption and excretion through the bile. The latter is facilitated by a copper-transporting ATPase located in the Golgi apparatus near the hepatic canaliculi. Most of the plasma copper binds to ceruloplasmin, a glycoprotein synthesized in the liver.

Dysarthria (speech disorder), dysphagia, and bradykinesia are commonly seen as initial neurologic presentations. Tremor, a manifestation of this disease, is accompanied by dystonia and involves the hands, arms, legs, and trunk, a fact that distinguishes it from the tremor that occurs in Parkinson's disease. In the limbs, it produces flapping tremor that resembles "wing or chest beating." Dystonia and chorioathetoid movements are also seen with dystonia affecting and producing spasm of the facial (involuntary grimacing), cervical (torticollis), and hand muscles. Irritability, aggression, and emotional lability can occur in this disease. Deposition of the copper at the limbus (corneoscleral junction) in the upper and lower poles of the cornea produces Kayser–Fleischer ring, a yellow-brown pigmentation in the Descemet membrane. This ocular sign is universally present once the nervous system is involved but absent in patients with hepatic dysfunction. Ocular deficits may be associated with disruption of the neural circuit for smooth pursuit movement and slow saccades. Rarely, and as a late manifestation, sunflower cataract develops. Glomerular and tubular dysfunctions and glucosuria are also seen as a result of involvement of the urinary system. This disease is treated with D-penicillamine, a copper chelator. A low copper diet and potassium sulfide supplement, which retards copper absorption, is also needed. This disease has a prevalence of 1/200,000 individuals.

Tics are rapid, repetitive, irregular, and nonrhythmic stereotyped movements (motor tics) or abnormal vocalizations (vocal or phonemic tics) that occur suddenly, that can be mimicked by the observer, and that often can be willfully held under control by the patient over a variable period of time, usually at the expense of mounting inner psychic tension. The inner tension is often relieved when the tic is allowed to occur. These movements usually involve several muscle groups simultaneously and may assume the form of a complex motor act. Tics may be intensified or preceded by excitement, anxiety, rage, or exhaustion and are suppressed by intense concentration. They may persist during sleep and may be accompanied by sleepwalking. Tics occur in response to external stimuli, are not mediated by the corticospinal tract, and have no premovement (premonitory) recorded potential in the motor cortex. Dopamine receptor hypersensitivity, disturbances of frontal cortical-basal ganglia and limbic pathways, and impaired cerebral lateralization have been implicated in the underlying mechanism of this disorder.

Simple tics are repetitive, often without rhythm, and unlike myoclonus, they are associated with other tics and may be in the form of shoulder jerk, prolonged eye blink, or head toss. *Complex motor tics* are coordinated movements that resemble voluntary actions intended to conceal tic disorders through variety of activities such as rubbing, tapping, jumping, stamping, and adjusting hair (parakinesia). Compulsive tics are complex tics that are ritualistic (time of the day and number of times), performed to reduce anxiety or obsession or to ward off harm or horrible event. They are preceded by uncomfortable sensation, such as itching (sensory tics). On the contrary, impulsive compulsive tics are performed without any regard to consequences and are intended to obtain arousal and gratification. *Vocal tics* may be in the form of throat clearing, grunting, sniffing, or vocalization of words or fractions of words. Patients may mimic words (echolalia) or movements (echopraxia or echokinesis).

Tic disorders refer to childhood disorders that affect one-fifth of school-age children, and 1% of children may present with tic syndrome before the age of puberty. They are rapid, nonrhythmical, and stereotyped with sudden onset, which includes transient and chronic tic disorders, Tourette's syndrome, and tic disorder not otherwise specified (NOS). However, all types of tic disorders can occur in the same family. *Transient tic disorder* is the most common of idiopathic tic disorder that exhibits multiple motor and/or phonemic tics that occur multiple times during the day and last for a minimum of 1 month, but do not exceed 1 year. Chronic tic disorder can be single or multiple with motor tic being the most common form of the disease. However, phonemic tics, which lasts on and off for more than a year is less common.

Tourette's syndrome is a familial (autosomal dominant) chronic disorder, which begins as simple tics in childhood between the age of 3 and 10 years and lasts for more than a year. It is the more extreme form of the tic disorder spectrum. It later advances into multiple motor and vocal tics. It affects boys three to four times more frequently than girls (3–4 males/1 female ratio). Most cases develop by the age 7, and most severe tic symptoms are seen in early puberty. It exhibits stereotyped, repetitive, and involuntary chronic, simple, complex, and dystonic motor, and simple and complex vocal tics. Tics in this syndrome are suppressible voluntarily or by alcohol intake or through focused activity and are exacerbated by stress, anxiety, and certain physical experiences. This syndrome starts with *simple tics* such as eye blinking and advances to a more complex form. Complex movements are distinct and coordinated, involving several muscle groups, and might include facial grimacing combined with a head twist and a shoulder shrug. Other complex motor tics may actually appear purposeful, including sniffing, spitting, licking or touching objects, or twisting. Tics continue during sleep, though to a lesser extent. Vocal tics may assume the form of repetitive throat clearing, sniffing, grunting, or barking noises and may start 1–2 years later, and evolve into complex vocal tics, which include unprovoked outburst of *coprolalia* (utterance of obscene or socially inappropriate phrases or words) or echolalia (repeating the words or phrases of others), although the percentage of patients with coprolalia remains relatively low.

Most of the obscene words are fragmented, containing only fractions of the actual words. The latter phenomenon may be associated with indecent gestures (*copropraxia*) or lewd thoughts. Secondary tics or tourettism occurs secondary to use of neuroleptics or is associated with congenital or neuropsychiatric disorders. CT scan and MRI do not show any abnormality, although electroencephalogram may not be normal. Tics in this lifelong disease change in distribution and intensity and may show remission. Anxiety and tension associated with school year tend to aggravate this condition. Patients may develop depression, anxiety disorder, obsessive-compulsive disorder (OCD), or attention-deficit hyperactivity disorder (ADHD). Intelligence remains unaffected. Methylphenidate (Ritalin), a medication used for the treatment of ADHD, and cocaine, which blocks the reuptake of norepinephrine and dopamine, may predispose children for Tourette's syndrome. This syndrome may be treated by haloperidol, which suppresses vocal and motor tics, but it may also cause gynecomastia. Clonidine, fluphenazine, pimozide, and tetrabenazine are also commonly used.

SUGGESTED READING

Alcaro A, Huber R, Panksepp J. Behavioral functions of the mesolimbic dopaminergic system: An affective neuroethological perspective. *Brain Res Rev* 2007;56(2):283–321.

Alm PA. Stuttering and the basal ganglia circuits: A critical review of possible relations. *J Commun Disord* 2004;37(4):325–69.

Cameron IG, Watanabe M, Pari G, Munoz DP. Executive impairment in Parkinson's disease: Response automaticity and task switching. *Neuropsychologia* 2010;48(7):1948–57.

Chakravarthy VS, Joseph D, Bapi RS. What do the basal ganglia do? A modeling perspective. *Biol Cybern* 2010;103(3):237–53.

Christopher L, Marras C, Duff-Canning S, Koshimori Y, Chen R, Boileau I, Segura B, Monchi O, Lang AE, Rusjan P, Houle S, Strafella AP. Combined insular and striatal dopamine dysfunction are associated with executive deficits in Parkinson's disease with mild cognitive impairment. *Brain* 2014;137(Pt 2):565–75.

François C, Tande D, Yelnik J, Hirsh EC. Distribution and morphology of nigral axons projecting to the thalamus in primates. *J Comp Neurol* 2002;447:249–60.

Geisler S, Derst C, Veh RW, Zahm DS. Glutamatergic afferents of the ventral tegmental area in the rat. *J Neurosci* 2007;27(21):5730–43.

Ikemoto S. Dopamine reward circuitry: Two projection systems from the ventral midbrain to the nucleus accumbensolfactory tubercle complex. *Brain Res Rev* 2007;56(1):27–78.

Janezic S, Threlfell S, Dodson PD, Dowie MJ, Taylor TN, Potgieter D, Parkkinen L, Senior SL, Anwar S, Ryan B, Deltheil T, Kosillo P, Cioroch M, Wagner K, Ansorge O, Bannerman DM, Bolam JP, Magill PJ, Cragg SJ, Wade-Martins R. Deficits in dopaminergic transmission precede neuron loss and dysfunction in a new Parkinson model. *Proc Natl Acad Sci USA* 2013;110(42):E4016–25.

Jenkinson N, Nandi D, Oram R, Stein JF, Aziz TZ. Pedunculopontine nucleus electric stimulation alleviates akinesia independently of dopaminergic mechanisms. *Neuroreport* 2006;17:639–41.

Kamishina H, Yurcisin G, Corwin J, Reep R. Striatal projections from the rat lateral posterior thalamic nucleus. *Brain Res* 2008;1204:24–39.

Lammel S, Hetzel A, Haeckel O, Jones I, Liss B, Roeper J. Unique properties of mesoprefrontal neurons within a dual mesocorticolimbic dopamine system. *Neuron* 2008;57(5):760–73.

Levesque JC, Parent A. GABAergic interneurons in human subthalamic nucleus. *Mov Disord* 2005;20:574–84.

Menasegovia J, Bolam J, Magill P. Pedunculopontine nucleus and basal ganglia: Distant relatives or part of the same family? *Trends Neurosci* 2004;27(10):585–8.

Middleton FA, Strick PL. Basal ganglia "projections" to the prefrontal cortex of the primate. *Cereb Cortex* 2002;12:926–35.

Olson VG, Nestler EJ. Topographical organization of GABAergic neurons within the ventral tegmental area of the rat. *Synapse* 2007;61(2):87–95.

Parent M, Parent A. Single-axon tracing and three-dimensional reconstruction of centre median-parafascicular thalamic neurons in primates. *J Comp Neurol* 2005;481(1):127–44.

Parent M, Parent M. Single-axon tracing study of corticostriatal projections arising from primary motor cortex in primates. *J Comp Neuro* 2006;496:202–13.

Swain JE, Scahill L, Lombroso PJ, King RA, Leckman JF. Tourette syndrome and tic disorders: A decade of progress. *J Am Acad Child Adolesc Psychiatry* 2007;46(8):947–68.

Walker K, Lawrenson J, Wilmshurst JM. Sydenham's choreaclinical and therapeutic update 320 years down the line. *SAMJ* 2006;96(9):906–12.

Index

Page numbers followed by f and t indicate figures and tables, respectively.

A

α-amino-3-hydroxy-5-methyl-4-isolaxone propionic acid (AMP), 85
Abdominal reflex, superficial, 264
Abducens nerve, 282–284, 283f, 284f
Abducens nucleus, 15, 64
Abscesses
  spinal epidural, 47
  subdural, 48
Absence seizures, 163
Accessory (lateral) cuneate nucleus, 60
Accessory nerve, 299–300, 299f, 300f
Accessory phrenic nerve, 228
Accommodation reflex, 210, 342–343, 342f
Acetaminophen, 431
Acetylcholine (ACh), 5, 32, 37, 38, 39, 41, 68, 205, 207, 310, 465, 466
  distribution of, 310
  function of, 310
Acetylcholine receptor (AChR), 40
Acetylcholinesterase, 310
Acetyl coenzyme A (CoA), 310
Achalasia, 217, 220
AChR. *See* Acetylcholine receptor (AChR)
Achromatopia, 326. *See also* Color blindness
Achromotopsia, 338
Acoustic neuroma, 98, 360, 370
Acoustic stria, 355
  dorsal, 355f, 356, 356f
  intermediate, 355f, 356, 356f
  ventral, 355–356, 355f, 356f
Acquired (occult) hydrocephalus, 192
Acquired immune deficiency syndrome (AIDS), 31
ACTH. *See* Adrenocorticotropic hormone (ACTH)
Actin, 12, 25, 27
Actin-binding proteins, 12
Acute autonomic dysfunction, 215
Acute mastoiditis, 350
Acute motor axonal neuropathy (AMAN), 29
Acute suppurative otitis media, 351
AD. *See* Alzheimer's disease (AD)
Adenohypophysis, 123
Adenoma
  of anterior pituitary, 123
Adenosine triphosphatase (ATPase), 25
Adenosine triphosphate (ATP), 197, 200
Adenylate cyclase, 25
ADHD. *See* Attention deficit hyperactivity disorder (ADHD)
ADHD-C, 133
ADHD-PI, 133
Adiadochokinesis, 97
Adie pupil, 272, 323
Adrenal medulla, 5
Adrenergic neurons, 32
Adrenergic receptors, 311
  alpha receptors, 311
  beta receptors, 311
Adrenocorticotropic hormone (ACTH), 31
Adson's test, 232
Afferent fibers, 468, 469
  of ANS, 214
  cortical, 147
  hypothalamic, 119–120, 119f–120f
Afferents
  cerebellar, 87–91, 88f–90f
  from ventral striatum, 470
AFP. *See* Alpha-fetoprotein (AFP)
Agnosia, 158–159
Agranular cortex, 143
Agrin signals, 13
AICA. *See* Anterior inferior cerebellar artery (AICA)
AICA syndrome. *See* Anterior inferior cerebellar artery syndrome
AIDS. *See* Acquired immune deficiency syndrome (AIDS)
AIDS dementia complex (ADC), 161
AIN. *See* Anterior interosseous nerve (AIN)
Akathesia, 478
Alar plate, 3, 11f, 12, 13, 14, 14f, 15
  derivatives of, 15
  of mesencephalon, 16, 17f
  rhombic lip of, cerebellum from, 15–16, 16f
Alcoholism, nucleus accumbens and, 399
Alexia, 158
  with agraphia, 156
  pure, 156
Alexithymia, 157
Alkaloids, 408
Allocortex, 141
α-amino-3-hydroxy-5-methyl-4-isoxazole propionic acid (AMPA), 309
Alpha-fetoprotein (AFP), 7
Alpha motor neurons, 64
ALS. *See* Amyotrophic lateral sclerosis (ALS)
Alzheimer's disease (AD), 121, 153, 159–160, 213, 384
  anosmia and, 379
  development of, risk factor for, 385
  muscarinic receptors and, 310
Amacrine cells, retina, 327
AMAN. *See* Acute motor axonal neuropathy (AMAN)
Amaurosis fugax, 328
Amblyopia, 337
Amiculum olivae, 61
Amiloride, 411
Aminergic cerebellar afferents, 91
Amino acid neurotransmitters, 37, 307
  gamma-aminobutyric acid, 307–308
  glutamic acid, 308–309
  glycine, 309–310
Amino acids, 407
Amniocentesis, 7
Amniotic cavity, 4
α–motor neurons, 52
AMPA receptor antagonists, 309
Amphetamines, 78, 311, 312
Amygdala, 101, 119–120, 122, 130, 399–403, 399f
  afferent and efferent connections of, 399f
  basolateral nuclear complex, 400
  corticomedial nuclear complex, 400
  disorders linked with, 402–403
  extended amygdala, 400
  in formation and storage of emotional memories, role of, 402
  role of, 400–401
  size of, psychological disorders with, 402–403
  stimulation of, 402
  stria terminalis and ventral amygdalofugal pathways, 401
  ventral amygdalofugal fibers, 399f, 400f, 402
Amygdaloid nuclear complex, 404, 405f
Amyloidosis, hereditary, 217
Amyotrophic lateral sclerosis (ALS), 272, 309, 447t, 448f
Analgesia, 431
Anal reflex, 264
Anaxonic cells, 36
Anemia, megaloblastic, 455
Anencephaly (exencephaly), 3, 6
Anesthesia
  caudal, 47
  epidural, 47
  spinal, 48
Angiokeratoma corporis diffusum (Fabry's disease), 34
Angiotensin, 118
Angiotensin II receptor antagonists, 411
Angular gyrus, 135
  lesion of, 135
Anhidrosis (lack of sweating), 217
Aniridia, 12
Anisocoria, 323
Anisometropia, 325
Ankle clonus, 442
Anomic aphasia, 155
Anosmia, 267, 379
  bilateral, 267
  due to traumatic head injury, 268
  in Kallmann syndrome, 268
  olfactory groove meningiomas and, 268
  unilateral, 267
Anosognosia, 158
ANS. *See* Autonomic nervous system (ANS)
Ansa cervicalis, 227
Ansa peduncularis, 108
Antergrade (Wallerian) degeneration, 28
Anterior cerebral artery (ACA)
  obstruction of, 459
  occlusion of, 134
Anterior cerebral vein, 174
Anterior choroidal artery, 149, 171
Anterior commissure, 151, 381f, 382
  formation of, 18

Anterior compartment syndrome, 261
Anterior corticospinal tract, 55, 55f
Anterior inferior cerebellar artery (AICA), 83–84
occlusion of, 84
Anterior inferior cerebellar artery (AICA) syndrome, 452t, 457
Anterior internuclear ophthalmoplegia, 371
Anterior interosseous nerve (AIN), 242
Anterior interosseous nerve (AIN) palsy, 242
Anterior ischemic optic neuropathy, 330
Anterior knee fibers of von Willebrand, 331
Anterior nucleus, 101f–103f, 103
Anterior olfactory nucleus (AON), 380, 382
Anterior perforated substance (APS), 377, 382–383
Anterior spinal artery, 49, 49f–50f
occlusion of, 49–50, 51
Anterior spinal artery (Beck) syndrome, 452–453, 452t
Anterior tarsal tunnel syndrome, 262
Anterograde amnesia, 103, 395
Anterolateral system, 421, 424f, 425f
Anticholinergic therapy, 474
Antidepressants, 311
Antidiuretic hormone (ADH) neurons, 114
Antiepileptic drugs, GABA-a receptors and, 308
Anti-ganglioside3 (GD3) antibodies, 29
Antigravity muscles, 451
Antipsychotic drugs, 312
dopaminergic receptors and, 313
prolonged usage of, 312
Anton's syndrome, 337–338
Antrum, 350
Anxiety disorders, amygdala size and, 403
AON. *See* Anterior olfactory nucleus (AON)
Aortic (intermesenteric) plexus, 202f, 212–213, 212f
Aphasia, 153–157, 154f
anomic, 155
Broca's, 153–155, 154f
global, 155–156
hemioptic, 156
subcortical sensory, 156
tactile, 156
transcortical, 155
Wernicke's, 155
Aphemia, 156
Apolipoprotein E (ApoE), 385
Apractognosia, 159
Apraxia, 157
Aprosodia, 157
A-68 protein, 384
APS. *See* Anterior perforated substance (APS)
Arachnoidal cysts, 184
Arachnoid cysts, 48
Arachnoiditis, 48
Arachnoid mater, 5, 182–183, 182f
spinal cord, 46f, 47–48
ARAS. *See* Ascending reticular activating system (ARAS)
Archicerebellar lesions, 96
Archicerebellum, 84
Archipallium, 140
Arcuate nuclei, 61
Arcuate (tuberal) nucleus, 112
Arcuate scotoma, 326
Arcuatocerebellar fibers, 89, 90f
Area postrema, 60
A_1 receptors, 200
Argyll Robertson pupil, 323, 334, 342, 420
Arm pain, cervical spine disorders, 227
Arnold–Chiari malformation, 7, 429
Arnold–Chiari syndrome, 9, 455
type I, 9
type II, 9
type III, 9
type IV, 9
Arteriovenous malformation (AVMA), 47
Artery of Adamkiewicz, 49–50
Articulation, 444
Ascending pathways, spinal, 54f, 54t
in lateral funiculus, 54
in posterior funiculus, 54
in ventral funiculus, 54–55
Ascending reticular activating system (ARAS), 74–75, 102, 118, 137
ASD. *See* Autism spectrum disorders (ASD)
L-aspartate, 87
Asperger syndrome, 166
Association cortex, 146t, 147
Association fibers, 152–153, 152f–153f
Asterixis, 477
Astigmatism, 325
Astrocytes, 21–22, 21f–22f, 23f, 28, 85
characteristics, 21–22, 22f
fibrous, 22
functions, 22
as glycogen storage site in CNS, 22
modified, 22
perivascular end feet, 22f
protoplasmic, 22
role in blood–brain barrier formation, 21, 22
Astrocytoma, 22
Astrotactin, 11
A_6 subunit, 85
Asymbolia, 159
Asymmetric synapses, 38
Asynergy, 96–97
Ataxia, 29, 291, 307, 371, 476
Ataxia (posture and gait abnormalities), 96
Athetosis, 476, 476f
Atonic bladder, 453
Atonic seizures, 163
ATP. *See* Adenosine triphosphate (ATP)
ATPase. *See* Adenosine triphosphatase (ATPase)
Attention deficit hyperactivity disorder (ADHD), 133
Atypical absence seizures, 163
Audiometry, 361
Auditory agnosia, 158
Auditory cortex, 357–358
bilateral damage to, 357
primary, 357
Auditory dysfunctions, 347
Auditory nerve, 5
Auditory radiation, 357, 358f
Auditory system, 347
central auditory pathways
acoustic stria, 355–356, 356f
auditory radiation, 357, 358f
cochlear nuclei, 355
inferior colliculus, 357, 357f
lateral lemniscus, 355f, 356, 356f
medial geniculate nucleus, 357
cochlear nerve, 354, 354f, 355f
cortices, 357–358
dysfunctions, 359
conductive deafness, 359
sensorineuronal deafness, 359–360
tinnitus, 360
external ear, 347–348, 347f
inner ear, 349f, 352–354
middle ear, 347f, 348–352, 348f
spiral ganglia, 353f, 354
Auditory tests, 361. *See also* Auditory system
audiometry, 361
brainstem auditory evoked response, 361
Rinne test, 361f, 362
Weber test, 361, 361f
Auriculotemporal neuralgia, 280
Autism, 166–167
Autism spectrum disorders (ASD), 166
Autonomic centers, higher, 208
Autonomic ganglia, 5
Autonomic ganglionopathy, 217, 220
Autonomic nervous system (ANS), 197–220
afferent components, 214
autonomic neurons and synaptic connections, 197–200, 198f–199f
disorders, 214–220
enteric nervous system, 213–214
higher autonomic centers, 208
overview, 197
parasympathetic nervous (craniosacral) system, 198f, 205–208, 206f, 207f
cranial part, 208
sacral part, 208
plexuses, 210–213, 211f–212f
reflexes, 208–210
sympathetic fibers, distribution pattern of, 204–205, 205f
sympathetic nervous (thoracolumbar) system, 200
paravertebral ganglia, 198f, 200–204, 201t, 202f–203f
prevertebral ganglia, 198f, 202f, 204
visceral *vs.* somatic afferents, 197–199, 199f
Autonomic neuropathy, 215–216
impotence in male in, 216
severe diabetic diarrhea in, 215
urinary bladder dysfunction in, 215
Autonomic plexuses, 210–213, 211f–212f
Autonomic reflexes, 208–210
Autotopagnosia, 338
Avastin. *See* Bevacizumab
Avellis syndrome, 452t, 456
AVMA. *See* Arteriovenous malformation (AVMA)
Axillary nerve, 244
injury to, 244
Axo-axonic synapses, 38
Axodendritic synapses, 38
Axolemma, 25
Axonal guidance, 3
Axonotmesis, 225–226
Axons, 15, 25–26
Axoplasmic transport, 25–26
fast phase, 25
intermediate phase, 25
slow phase, 25
Axosomatic synapse, 37
Azathioprine, 31

B

Babinski agnosia, 158
Babinski sign, 137, 441, 441f
Back pain, 430, 455
Baclofen, 308
Bacterial otitis media, 351
BAER. *See* Brainstem auditory evoked response (BAER)

Bainbridge reflex, 209
Baker's cyst, 264
Balint syndrome, 338
Band of Gennari, 337
Barbiturates, GABA-a receptors and, 308
Barotitis media (aerotitis), 349
Basal forebrain nucleus of Meynert, 384
Basal (ganglia) nuclei, 167, 463, 463f, 464f
  embryonic developmental, 17–19, 18f
Basal plate, 3, 10, 11f, 14, 14f, 15
  of mesencephalon, 16, 17f
Basilar artery, 49, 49f–50f, 63, 63f–64f, 83, 102
  stenosis of, 97
Basis pedunculi, 67
Basivertebral veins, 51
Basket cells, 86
Batson's venous plexus, 51
BBB. *See* Blood-brain barrier (BBB)
BCB. *See* Blood-CSF barrier (BCB)
BDNF. *See* Brain-derived neurotrophic factor (BDNF)
Beck syndrome, 49, 227, 452, 452t
Bed-wetting (enuresis), 77
Beilschowsky sign, 276
Bêkêsy audiometry, 361
Bell's palsy, 290
Benedikt syndrome, 98, 272, 452t, 458, 459f, 468
Benign paroxysmal positional vertigo (BPPV), 370–371
Benzodiazepine, 119
Bergmann cells, 87–88
Bergmann glial cells, 11, 16, 22, 85
Beri-beri disease, 263
β-*N*-methyl-amino-l-alanine (BMAA), 309
β-*N*-oxalylamino-l-alanine (BOAA), 309
Betz cells, 141
Bevacizumab, 326
Bielschowsky head-tilt test, 276
Bielschowsky sign, 69
Bilateral analgesia, 428
Bilateral anterior internuclear ophthalmoplegia (MLF syndrome), 455
Bilateral cerebellar dysfunctions, 9
Bilateral cordotomy, 431
Bilateral destruction of CBT, 444
Bilateral lobotomy, 404
Bilateral occlusion, 439
Bilateral trochlear nerve palsy (superior medullary velum syndrome), 66
Bilateral visual agnosia, 338
Binasal heteronymous hemianopsia, 332
Binswanger disease, 160
Biogenic amines. *See* Monoamines
Biological clock, 114
Bipolar neurons, 32, 36, 36f
Bitemporal heteronymous hemianopsia, 332, 334f
Bitter-tasting compounds, action of, 408
Bitter-tasting stimuli, 409
Black widow spider *(Latrodectus mactans)* toxicity, 39
Bladder reflex, 210
Blind spot, 326
Blink reflex of Descartes, 344
Blood-brain barrier (BBB), 183, 184–186
  astrocytes and, 21, 22
  functional and clinical significance, 186–188
  vs. BCB, 186t
Blood-CSF barrier (BCB), 185
  vs. BBB, 186t
Blood-nerve barrier (BNB), 184, 186
Blood supply
  cerebellum, 83–84
  cerebral hemispheres, 167–172, 168f–172f
  spinal cord, 48–51, 49f, 50f
  thalamus, 102
BMP4. *See* Bone morphogenetic protein 4 (BMP4)
BNB. *See* Blood–nerve barrier (BNB)
Bone morphogenetic protein 4 (BMP4), 3, 4
Bonnier's syndrome, 294
Borderline personality disorder, 402
Botulism, 39, 217, 220
Bowman glands, 378
BPPV. *See* Benign paroxysmal positional vertigo (BPPV)
Brachial plexus, 229–230, 230f, 230t
  terminal branches of
    lateral cord, 234
    medial cord, 230f, 234–244, 235f–237f, 239f–243f
    posterior cord, 230f, 244–246
    roots, 232–233
    superior trunk, 233–234
  trunk injuries, 230–232, 231f
Brachydactyly syndrome, 6
Bradykinesia, 469
Brain
  computed tomography scan of, 463
  coronal section, 463f
  horizontal section of, 464f
Brain barriers, 184–188, 184f. *See also* Blood-brain barrier (BBB); Blood-CSF barrier (BCB); Blood-nerve barrier (BNB)
  BBB *vs.* BCB, 186t
  circumventricular organs, 188, 188f
  functional and clinical significance, 186–188
Brain-derived neurotrophic factor (BDNF), 31
Brainstem, 57–71
  fourth ventricle, 57–62
    caudal medulla, 59–62
  medulla, 57, 57f
  midbrain, 66–71, 67f
    caudal, 68–69, 69f
    rostral, 70–71, 70f
  overview, 57
  pons, 62–66, 63f–64f
    caudal, 64–65, 64f
    midpons, 65–66, 65f
    rostral, 66, 66f
  vertebrobasilar arterial system and, 58–59, 58f
Brainstem auditory evoked response (BAER), 361
Brainstem lesions, 447
Brain tumor (glioma)
  astrocytoma, 22
  neurofibromas, 23
  Schwannomas, 23
Brain vesicles
  derivatives of, 11t
  embryonic development, 10, 10f
  primary, 10, 10f, 11t
  secondary, 10, 10f, 11t
Branched-chain ketoaciduria (maple syrup urine disease), 33–34, 34f, 35f
B_1 receptors, 200
Broca's aphasia, 153–155, 154f
Brown-Séquard syndrome, 452t, 453, 454f
Brudzinski's sign, 175, 176f
Buccofacial apraxia, 157
Bulbar amyotrophic lateral sclerosis, 447t
Bulbar palsy, 442t, 445
Bulbocavernous reflex, 265
Bulbous myringitis, 350
Bungaro multicinctus, 39
Bungarotoxin, 39
Burrow's solution, 352

C

Cacosmia, 385–386
Cajal–Retzius cells, 11
Calbindin, 86
Calcinosis, 238
Calcitonin gene-related peptide (CGRP), 409
Calcium-binding proteins, 85, 86
Callosal apraxia, 157
Calmodulin, 86
Calmodulin-binding membrane-associated phosphoprotein, 25
Caloric test, 291, 372
CAMP. *See* Cyclic adenosine monophosphate (cAMP)
Canaliculus gustatorius, 408
Canal of Guyon, 237, 238, 239
Cantelli sign, 343
Capgras syndrome, 158
Capsulated receptors, 416, 418f
Carbachol, 207
Carbidopa, 474
Carbohydrates, 407
Cardiac pain, 210–211, 248
Cardiac plexus, 207f, 210–211, 211f
Carotid bodies, 5
Carotid body reflex, 209
Carotid sinus reflex, 209
Carotid sinus syncope, 294
Carpal tunnel, 239, 241f
Carpal tunnel syndrome, 243, 243f
Cataplexy, 78
Catecholamines, 5, 37, 197, 311
  dopamine, 312–313
  epinephrine, 312
  norepinephrine, 311–312
  synthesis of, 311
Catechol-*o*-methyltransferase (COMT), 311
Cauda equina, 45–46, 45f–46f
  formation of, 13
Cauda equina syndrome, 45–46, 452t, 453
Caudal anesthesia (saddle block), 47
Caudal medulla, 59–62, 75, 76
  midolivary level, 60–62, 60f
  motor decussation level, 59, 59f
  pontomedullary junction, 62
  rostral medulla, 62, 62f
  sensory decussation level, 59–60, 59f–60f
  venous drainage, 59
Caudal midbrain, 68–69, 69f
  dorsal raphe nucleus, 69
  dorsal tegmental (supratrochlear) nucleus, 69
  inferior colliculus, 68
  lateral lemniscus, 69
  locus ceruleus nucleus, 69
  parabigeminal nucleus, 69
  pedunculopontine nucleus, 69
  superior cerebellar peduncles, 69
  trochlear nucleus, 68
  ventral tegmental nucleus, 69
Caudal pons, 64–65, 64f
Caudate nucleus, 130, 463, 470

Causalgia, 243–244, 260
Cavernous sinus syndrome, 273
Cavernous sinus thrombosis, 181
CBT. *See* Corticobulbar tract (CBT)
CD4. *See* Cluster designation-4 (CD4)
Celiac plexus, 211, 212f
Cellular adhesion molecules (CAMs), 21
Cellular differentiation, 3
Central canal (spinal cord), 51
Central gray, 60
Central lobe (insular cortex), 138, 138f
Central nervous system (CNS), 3, 21, 23, 101, 140, 197
  brainstem. *See* Brainstem
  MAG, 26, 27
  myelin, 27
  P_1 and P_2 proteins, 26
  presynaptic differentiation, 13
  regeneration in, 31–32
  retrograde (indirect Wallerian) degeneration in, 28
  spinal cord. *See* Spinal cord
  synapses in, 37–38
  synaptogenesis, 13
  transynaptic (transneuronal) degeneration in, 28
Central retinal artery, 328, 328f
Central retinal vein, 328, 328f
Central scotoma, 326
Central tegmental tract (CTT), 68, 74, 75
Centrioles, 25
Centromedian nucleus, 471
Centromedian–parafascicular nuclear (CM–PF) complex, 109
Cephalic flexure, 10, 10f
Cerebellar aplasia, 99
Cerebellar cognitive affective syndrome, 96
Cerebellar hemispheres, 81
Cerebellar herniation, 98
Cerebellar ischemia, 97
Cerebellar peduncle, 63, 63f
Cerebellar tumors, 330
Cerebellar veins, 84
Cerebellomedullary cistern, 182
Cerebellopontine angle, 62
Cerebellovestibular circuit, 92f, 93
Cerebellum, 57, 62, 81–99
  afferents, 87–91, 88f–90f
  blood supply and venous drainage, 83–84
  central lobule, 82
  circuits
    cerebellovestibular, 92f, 93
    cortico-cerebro-cerebellar, 94, 94f
    intracerebellar, 95
    reticulocerebellar, 93, 93f
    rubrocerebellar, 93, 94f
  classification, 84
  cortex, 84–87, 85f–86f
    Golgi neurons, 84, 85, 85f
    granular layer, 84–85, 85f, 86f
    LTD, 85
    molecular layer, 84, 86–87, 86f
    mossy fiber rosettes, 84, 85, 86f
    Purkinje cell layer, 84, 85–86, 85f, 86f
  efferents, 91–93, 91f
  embryonic developmental, 15–16, 16f
  fissures, 81
  functional and clinical consideration, 81, 95–98
    feedforward processing, 95
    modularity, 95
    motor coordination, 95–96
    motor learning, 95
    neuronal plasticity, 95
  hemispheres, 81–82
  inferior view, 83f
  lesions and associated diseases, 98–99, 99f
    signs and symptoms, 96
  lesions and associated dysfunctions/diseases
    acoustic neuroma, 98
    adiadochokinesis, 97
    archicerebellar lesions, 96
    asynergy, 96–97
    Benedikt syndrome, 98
    cerebellar aplasia, 99
    cerebellar herniation, 98
    Dandy–Walker syndrome, 98
    dementia, 98
    dysmetria, 97
    Foix's syndrome, 98
    Friedreich's ataxia, 99, 99f
    frontal lobe tumors, 99
    headache, 98
    hemorrhage, 99
    hyperkinesia, 97
    hypotonia, 97
    MS, 98
    nausea and vomiting, 98
    neocerebellar lesions, 96
    Nothnagel's syndrome, 98–99
    ocular and visual disorders, 97–98
    paleocerebellar lesions, 96
    posture and gait abnormalities (ataxia), 96
    rebound phenomenon of Holmes, 97
  morphologic characteristics, 81–83, 81f–83f
  nuclei, 87
  overview, 81
  superior surface, 82f
  vermal segment, 82, 84f
  vermis and associated cortical parts, 82–83, 83f
Cerebral aqueduct, 67, 101
  obstruction of, 67
Cerebral artery(ies)
  anterior, 134
  middle, 129
  posterior, 171–172
Cerebral cortex (gray matter), 140–147, 141f–142f
  agranular, 143
  association cortex, 146t, 147
  embryonic development, 11
  external granular layer, 142
  external pyramidal layer, 142
  frontal type of, 143
  granular, 143
  internal granular layer, 142
  molecular layer, 142
  motor cortex, 143f, 144–146, 146t
  multiform layer, 142
  parietal type of, 143
  polar, 143
  sensory cortex, 143–144, 143f
Cerebral dominance, 165–166
Cerebral hemispheres, 129–131, 129f–130f
  anterior horn, 129–130
  blood supply, 167–172, 168f–172f
  cerebral cortex. *See* Cerebral cortex (gray matter)
  characteristics, 129–131, 129f–131f
  embryonic developmental, 17–19, 18f
  posterior horn, 130
  venous drainage, 172–174, 172f–175f
  white matter of, 149–153, 150f
    association fibers, 152–153, 152f–153f
    commissural fibers, 150–151, 150f–151f
    projection fibers, 153
Cerebral lesion, 451
Cerebral motor cortex in newborns/infants, 443
Cerebral palsy, 442t, 443
Cerebral peduncle, 67
Cerebral veins. *See also specific* types
  thrombosis of, 174
Cerebrospinal fluid (CSF), 7, 23, 47, 57, 129, 188–193, 189f–191f, 447
  hydrocephalus, 191–193, 191f–192f
Ceruloplasmin, 478
Cerumen, 348
Cervical disk herniation, 227
Cervical flexure, 10, 10f
Cervical ganglion, 201, 203f
Cervical plexus, 226–227, 228f, 229f
  accessory phrenic nerve, 227
  ansa cervicalis, 227
  greater auricular nerve, 227
  lesser occipital nerve, 227
  phrenic nerve, 227
  supraclavicular nerves, 228
  terminal branches of, 227–229
  transverse (colli) cervical nerve, 228
Cervical rib syndrome, 231
Cervical spinal nerves, 223, 226–227
Cervical spinal segments, 53, 53f
Cestan–Chenais syndrome, 447t, 456
Chaddock response, 442
Chagas disease, 217, 220
Chalcosis lentis. *See* Sunflower cataract
Charcot–Marie–Tooth IA neuropathy, 290
Charcot–Marie–Tooth neuropathy, 26
Charcot–Wilbrand syndrome, 338
Chemical sensations, 407
Chemical synapses, 38
Chemoreceptors, 416
Chemotherapy, 410
Cherry-red spots, 325
Childhood disintegrative disorder, 166
Cholecystokinin (CCK), 316
Choline, 310
Choline acetyl transferase, 85
Cholinergic neurons, 32, 310
Cholinergic receptors, 205–206, 207, 310
  muscarinic receptors, 310
  nicotinic receptors, 310
Cholinergic synapses, 38
Cholinesterase inhibitors, 207
Chordin, 4
Chorea, 475
Choreiform movements, 475
Choroid fissure
  formation of, 18
Choroid plexus, 15, 18, 58
Chromatin bodies, 24
Chronic autonomic dysfunction, 215
Chronic suppurative otitis media, 352
Chunking, 393
Ciliary body, 5
Ciliary neurotrophic factor (CNTF), 31
Ciliospinal reflex, 343
Cimino–Bresica fistula, 243
Cingulate gyrus, 102, 138, 403–404
Circadian rhythm
  nonphotic cues role in, 114

Circuits, cerebellar
cerebellovestibular, 92f, 93
cortico-cerebro-cerebellar, 94, 94f
intracerebellar, 95
reticulocerebellar, 93, 93f
rubrocerebellar, 93, 94f
Circumvallate papillae, 408
Circumventricular organs, 188, 188f
Cisterna ambiens, 182
Clarke's nuclear column, 52, 53
Clarke's nucleus, 66, 420
Clasp-knife reflex, 442
Claude's syndrome, 272
Claustrum, 466
Clinical depression, 467, 474
Clonic seizures, 163
Clonidine, 312
Clonus, 442
Closed-angle glaucoma, 324
Clostridium botulinum, 39, 220
toxin, 310
Clostridium tetani, 218
toxin, 308
Clozapine, 313, 390
Clozaril. *See* Clozapine
Cluster designation-4 (CD4), 28–29
CM–PF complex. *See* Centromedian–parafascicular nuclear (CM–PF) complex
CNP. *See* 2´,3´cyclic nucleotide 3´phosphodiesterase (CNP)
CNS. *See* Central nervous system (CNS)
CNTF. *See* Ciliary neurotrophic factor (CNTF)
Cocaine overdose, 313
Coccygeal spinal nerve, 223, 226
Cochlear hydrops, 353
Cochlear nerve, 291, 354, 355f
Cochlear nuclei, 355
Cogan syndrome. *See* Congenital ocular motor apraxia
Cognex (tacrine), 385
Colchicine, 25
Color agnosia, 158
Color blindness, 326–327
Colpocephaly, 9
Coma, 163–165
Combined sensory and motor dysfunctions, 452, 452t
Combined UMN and LMN deficits, 447, 447f
Commissural fibers, 150–151, 150f–151f
Commissural syndrome, 428
Common fibular (peroneal) nerve, 256f, 259f, 260
injury to, 260
Communicating hydrocephalus, 192
Complex motor tics, 479
Complex partial seizures, 162
Complex regional pain syndrome (CRPS), 216, 217–218
Compound granular corpuscles, 24
Compulsive tics, 479
Computed tomography, 463f
Conductive deafness, 359
Confluence of sinuses, 179, 180f
Congenital anomaly of Peter, 322
Congenital hydrocephalus, 191–192
Congenital ocular motor apraxia, 342
Conjugate gaze palsy, 341
Constructional apraxia, 157
Contact point headache. *See* Sluder's neuralgia
Contralateral flaccid palsy, 145
Contralateral hemiplegia, 460
Contralateral homonymous hemianopsia, 333, 335
Contralateral pathway, 424f
Conus medullaris syndrome, 452t, 453
Copper, 479
Coprolalia, 479
Cordotomy, 431
Corneal reflex, 277, 282f, 285
Cornu ammonis (CA). *See* Hippocampal gyrus
Coronal section of brain, 463f
Coronary plexuses, 211
Corpus callosum, 129, 136f, 138f–139f, 139–140, 150
agenesis of, 139
failure of, 151
formation of, 18–19
Corpus cerebelli, 81
Corpus striatum. *See also* Extrapyramidal motor system
connections of, 470–471
dorsal pallidum, 465–466
dorsal striatum, 463–465
ventral pallidum, 466
ventral striatum, 465
Cortical afferents, 147
Cortical dementia, 159
Cortical efferents, 147–149, 148f–149f
Cortical pathways, 419
Cortical sensations. *See also* Extrapyramidal system
from body, 419–432, 420f–427f, 428t, 429f, 430f, 431f
anterolateral system, 421, 424f, 425f
dorsal white columns, 419, 420f, 421f, 422f, 423f
Horner's syndrome, 429
low back pain, 430
phantom limb pain, 429
sensory dysfunctions, 428t
spinoreticular tract, 425f, 427
stump pain, 430
syringobulbia, 429
syringomyelia, 428, 429f
ventral spinothalamic tract, 428
from head
dorsal spinocerebellar tract, 432
spinocerebellar tracts, 432
spino-olivary tract, 433
spinotectal tract, 433
trigeminal lemniscus, 432
ventral spinocerebellar tract, 432
Corticobulbar tract (CBT), 437, 440f
articulation, 444
bilateral destruction of, 444
bulbar palsy, 445
unilateral lesion of, 444
Cortico-cerebro-cerebellar circuit, 94, 94f
Corticonucleo-cerebellar fibers, 91–92, 91f
Corticopontine tract, 147–148
Corticoreticular tract, 148
Corticorubral tract, 148
Corticospinal and corticobulbar tracts (CBTs), 137, 147
Corticospinal tract (CST), 147, 437, 438–439, 439f, 440f
Corticostriate, 148
Cortico-striato-thalamo-cortical loop, 398
Corticotectal tract, 148
Corticothalamic fibers, 102
Corticothalamic tracts, 147
Corticotropin-releasing factor, 87
Costoclavicular syndrome, 232
Cranial meningocele, 7
Cranial nerves, 267, 268f, 284t, 295t, 302t
abducens nerve, 282–284, 283f, 284f
accessory nerve, 299–300, 299f, 300f
classification of, 267
facial nerve, 285–290, 285f, 286f, 288f–290f
glossopharyngeal nerve, 292–294, 293f
hypoglossal nerve, 301–303
oculomotor nerve, 270–274, 271f, 273f
olfactory nerve, 267–268, 269f
optic nerve, 269–270, 270f
somatic and autonomic components, 302t
trigeminal nerve, 276–282, 277f
trochlear nerve, 274–276, 275f
vagus nerve, 294–299, 296f–298f
vestibulocochlear nerve, 290–292, 291f, 292f
Cranial part, parasympathetic nervous system, 208
Craniopharyngioma, 122
Cranioschisis, 6
C-reactive protein, 430
Cremasteric reflex, 264
Crocodile-tear syndrome, 287
Crossed extension reflex, 266, 444
CRPS. *See* Complex regional pain syndrome (CRPS)
Crude touch transmission, 427f
Crus cerebri, 68
Crutch palsy, 245
CSF. *See* Cerebrospinal fluid (CSF)
CTT. *See* Central tegmental tract (CTT)
Cubclavian steal syndrome, 97
Cubital tunnel syndrome, 238
Cuneate fasciculi, 59
Cuneate nucleus, 60, 60f
Cuneatus fasciculus, 53
Cuneocerebellar tract, 88, 90f, 432
Cuneus, 138
Currarino syndrome, 8
Cyclic adenosine monophosphate (cAMP), 27, 407
Cyclic adenosine monophosphate (cAMP) protein kinase, 206
Cyclic guanosine 3´,5´-monophosphate (cGMP)–dependent-protein kinase, 85
2´,3´cyclic nucleotide 3´phosphodiesterase (CNP), 27
Cyclophosphamide, 31
Cyclosporine, 31
Cytochalasin B, 12
Cytokines, 28

D

Dalton myelin proteolipid protein (DM20), 26
Dandy–Walker syndrome, 9, 98, 151
Death, 456
Decerebrate rigidity, 447t, 451f
Decorticate posturing, 451
Decorticate rigidity, 447t, 451f
Deep cerebral veins, 172, 173f, 174f
Deep cerebral venous thrombosis, 174
Deep fibular (peroneal) nerve, 261
injury to, 261–262
Deep middle cerebral vein, 174
Deep reflexes, 265–266, 265f
Dejerine cortical sensory syndrome (Verger–Dejerine syndrome), 135–136
Dejerine–Klumpke palsy, 231

Dejerine–Roussy syndrome, 110–111
Delta sleep/deep sleep, 77
Dementia, 97, 98, 159–161, 307, 403
Dementia pugilistica, 161
Demyelination, 27, 28, 30, 455
metabolic disorders, 32–35, 32t
Dendrites, 25, 467
Dendritic spines, 465
Dendrosomatic synapses, 38
Denervation, 409
Dentate gyrus, 387f, 388f, 390
Dentate nucleus, 87
Dentatorubrothalamic tract, 92
Deprenyl, 474
Depression
catecholamines and, 311
norepinephrine depletion and, 311
Dermatome, 223, 224f
Dermoid sinuses, in spina bifida occulta, 8
Descemet membrane, 478
Descending pathways, spinal, 54t, 55f
in lateral funiculus, 55
in ventral funiculus, 55
Deuteranopia, 327
Developmental dyslexia, 156
Devic syndrome, 30
DFP. *See* Di-isopropylfluorophosphate (DFP)
Diabetic neuropathy, 29
Diacylglycerol (DAG), 379, 407
Diaphragmatic myoclonus, 477
Diastematomyelia
in spina bifida cystica, 9
Dichromatopia, 326. *See also* Color blindness
Dichromats, 327
Diencephalon, 10, 10f, 11t, 101–126
dorsal surface of, 101f
embryonic developmental, 16–17
epithalamus, 123–124, 124f
functional and clinical consideration, 123–124
functional and clinical consideration, 110–111, 111t, 115–119, 123
habenula, 124f–126f, 125–126
hypothalamic afferents, 119–120, 119f–120f
hypothalamic efferents, 120–122, 120f
hypothalamus, 111–115, 111f–113f
nuclei and associated areas, 112–115, 112f–113f
intralaminar nuclear group, 102f, 105f–107f, 109–110, 109f
lateral nuclear group
lateral dorsal, 102f, 105f, 106f, 107, 107f
lateral posterior, 102f, 105f, 106f, 107, 107f
pulvinar, 101f–102f, 104f, 106f, 107–108, 108f
medial nuclear group
dorsomedial nucleus, 101f–104f, 106f–109f, 108–109, 110f
midline nuclear group, 110
overview, 101
pineal gland, 124–125
pituitary gland (hypophysis cerebri), 122–123, 122f
posterior commissure, 126
reticular nuclear group, 110
stria medullaris, 125
subthalamus, 125f–126f, 126
thalamic nuclear group, 102–106, 102f
anterior nucleus, 101f–103f, 103
lateral geniculate body, 101f, 102f, 103–104, 104f
medial geniculate body, 101f, 102f, 104, 104f
ventral anterior nucleus, 102f–104f, 105–106, 106f
ventral lateral nucleus, 102f–104f, 106, 106f
ventral nuclear group, 103
ventral posterior nucleus, 102f, 104–105, 104f, 106f
thalamus, 101–102, 101f–102f
Diffuse noxious inhibitory control (DNIC), 426
Di-isopropylfluorophosphate (DFP), 207
Diploic veins, 179
Diplopia, 97–98, 274
Direct hypothalamospinal tract, 122
Direct pupillary light reflex, 342, 342f
Disconnection syndrome, 139–140, 156
Distal IP (DIP) joints, flexion at, 238
Dix–Hallpike maneuver, 371
DLF. *See* Dorsal longitudinal fasciculus (DLF)
Doll's eye movement. *See* Oculocephalic reflex
Dopamine, 311, 312–313, 465, 473, 474, 479
role in psychomotor activity, 133
Dopamine active transporters genes (DAT1), 134
Dopamine β-hydroxylase, 311
Dopamine receptor antagonists, in schizophrenia treatment, 390
Dopaminergic neurons, 32, 312–313, 467
Dopamine transporter (DAT), 313
Dorsal cochlear nucleus, 64
Dorsal fornix, 392
Dorsal funiculus, 53, 53f
Dorsalin genes, 12
Dorsal longitudinal fasciculus (DLF), 68, 120, 122, 208
Dorsal motor nucleus of the vagus, 61
Dorsal pallidum, 465–466
Dorsal periventricular tract, 315
Dorsal raphe nucleus, 66, 69, 73, 74f
Dorsal root fibers, 223
Dorsal root ganglion (DRG), 5, 13, 14, 24, 54, 223, 223f, 415
Dorsal scapular nerve, 230f, 232
Dorsal spinocerebellar tract, 52, 432
Dorsal spino-olivary tract, 433
Dorsal striatum, 463–465
Dorsal tegmental (supratrochlear) nucleus, 69
Dorsal trigeminal lemniscus (tract), 432
Dorsal white columns, 419, 420f, 421f, 422f, 423f
Dorsiflexion, of foot and toes, 261
Dorsiflexion of toe, 477
Dorsolateral syndrome, 442t, 443
Dorsomedial nucleus, 101f–104f, 106f–109f, 108–109, 110f
removal of, 109
Dorsomedial subdivision, hypothalamus, 113
Down syndrome, 171
D-penicillamine, 478
DRG. *See* Dorsal root ganglion (DRG)
Drusen, 326
D-tubocurarine, 41
"Duchenne sign," 257
Duperficial reflexes, 264–265
Dural sinuses, 172f–175f, 177f, 178, 179–182, 180f–182f
posterosuperior group of, 172f–175f, 177f, 178–179, 180f
Dura mater, 175–178, 175f–178f
Dura mater (pachymeninx), spinal cord, 46–47, 46f
"Duret hemorrhage," 137
Dynamic gait index, 372–373
Dynein, 25
Dysarthria (speech disorder), 444, 478
Dysgeusia, 410, 411. *See also* Gustatory dysfunctions
Dyskinesia, 472, 473
Dyslexia, developmental, 156
Dysmetria, 97
Dysphagia, 9, 444
Dysphonia, 9
Dystonia, 311, 477, 478

E

Ear. *See also* Auditory system
external, 347–348, 347f
auricle, 347
concha, 347
external acoustic meatus, 347–348
inner, 349f, 352–354
ampulla, 352
basilar membrane, 353, 353f
bony labyrinth, 352
cochlea, 352f, 353
endolymph, 353
membranous labyrinth, 352
organ of Corti, 353, 353f, 354
perilymph, 353
semicircular canals, 352
vestibular membrane of Reissner, 353, 353f
vestibule, 352, 352f
middle, 347f, 348–352, 348f
anterior wall of cavity, 350
auditory tube, 350
bony ossicles, 351
carotid canal, 350
chorda tympani, 348
floor, 351
incus, 351
lateral wall of cavity, 348
malleus, 351
medial wall of cavity, 350
oval window, 350
posterior wall of cavity, 350
pyramidal eminence, 351
roof, 351
stapes, 351
triangle of Politzer, 348
tympanic membrane, 348–350, 349f
tympanic plexus, 350
Earwax. *See* Cerumen
E-CAMs. *See* Extracellular matrix adhesion molecules (E-CAMs)
Edinger–Westphal nucleus (EWN), 16, 272
Edrophonium chloride, 41
Efferent fibers
cortical, 147–149, 148f–149f
from dorsal striatum, 470
hypothalamic, 119f, 120–122, 120f
of ventral striatum, 471
Efferents, 415
cerebellar, 91–93, 91f
Ehlers–Danlos syndrome, 9
Electrical synapses, 38
Electrocochleographic audiometry, 361
Electron micrograph, 408
Emboliform nucleus, 87
Emissary veins, 178–179
Encapsulated receptors, 416
Encephalitis, 153

Encephalocele, 7
Encephalotrigeminal angiomatosis. *See* Sturge–Weber syndrome
Endocrine-active tumors, 123
Endocrine-inactive tumors, 123
Endogenous opioids, 426
Endoneurium, 224
pressure difference in, 224
Endorphins, 316
Endosteal layer, spinal cord, 46, 46f
Energy (glucose), 407
Enkephalinergic neurons, 465
Enkephalins, 37, 85, 316
Enophthalmos, 274
Enteric nervous system, 213–214
Entopeduncular nucleus, 126
Entrapment neuropathies, spinal nerves in, 226
Enzyme-linked immunosorbent assay (ELISA), 385
Ependymal cells, 23
Ependymoma, 452, 454
Epicritic (discriminative) modalities, 415
Epidural anesthesia, 47
Epidural hematoma, 176, 176f
Epidural space, 176
spinal cord, 46, 46f
Epidural venous plexus, 51
Epinephrine, 312
Epineurium, 224
Epithalamus, 101, 123–124, 124f , 404
embryonic developmental, 16–17
functional and clinical consideration, 123–124
Epley maneuver, 371
Equinovarus deformity, 263
Erb–Duchenne palsy, 230–231, 231f
Erb–Goldflam disease. *See* Myasthenia gravis
Essential myoclonus, 477
Euphoria
catecholamines and, 311
norepinephrine in, level of, 311
Eustachian tube, 350, 351
blocked, 352
dysfunction of, 351
Exogenous myelomonocytic cells, 31
External otitis, 348
External vertebral (epidural) plexus, 51
Exteroceptors, 416
Extracellular matrix adhesion molecules (E-CAMs), 12
Extrageniculate visual pathway, 70, 108, 144
Extrapyramidal motor system
basal (ganglia) nuclei, 463, 463f
claustrum, 466
corpus striatum
connections of, 470–471
dorsal pallidum, 465–466
dorsal striatum, 463–465
ventral pallidum, 466
ventral striatum, 465
dysfunctions of, 479
functional loop and pathways, 471–472
red nucleus, 467–468
reticular formation, 468–470
substantia nigra, 466–467, 466f
subthalamic nucleus, 468, 469f
Extrapyramidal system
cortical sensations
from body, 419–432, 420f–427f, 428t, 429f, 430f, 431f
from head, 432
features of, 415
sensory fibers, 415–416
sensory receptors, 416–419, 416f, 418f, 419f
subcortical sensations, 432–433
Eyeball. *See* Visual system
Eye movements, 463. *See also* Rapid eye movements (REM)

F

Fabry's disease, 32, 34, 217, 218
Facial agnosia, 158
Facial colliculi, 57
Facial motor nucleus, 15, 64
Facial myokymia, 290
Facial nerves, 5, 285–290, 285f, 286f, 288f–290f, 410f
chorda tympani, 286
corneal reflex, 285
extracranial segment, 285
facial motor nucleus, 287
geniculate ganglion, 286
greater petrosal nerve, 286
intermediate (root) nerve, 286
labyrinthine segment, 285
mastoid segment, 285
motor root, 285
parotid segment, 285
posterior auricular branch, 287
solitary nucleus, 287
spinal trigeminal nucleus, 287
stapedius nerve, 287
superior salivatory nucleus, 287
tympanic segment, 285
Facial palsy, 287
Facial sweating, 217
F-actin, 25, 27
Falx cerebri, 177, 177f, 178
Fanconi anemia, 171
Fasciculation, 446
Fasciculus cuneatus, 420
Fasciolar gyrus (retrosplenial gyrus), 139
Fastigial nucleus, 87
Fastigiovestibular pathway, 92–93, 92f
Fastigium, 81
Fast-spiking pacemaker, 467
Fatty acids
myelin, 27
Feedback circuit bt efferents/afferents, 471f
Feedforward processing, 95
Femoral nerve, 252, 253f
injuries to, 253
Femoxetine, 78
Fergoli syndrome, 158
FGF. *See* Fibroblast growth factor (FGF)
Fibroblast growth factor (FGF), 12
Fibronectin, 12, 21, 31
Fibrous astrocytes, 22
Finger agnosia, 158
First-order neurons, 318
Fissures, cerebellar, 81–82, 82f–83f
Flexor (withdrawal) reflex, 266
Flocculonodular lobe, 84
Floor plate, 14
formation of, 4
Fluxetine, 78
Foix's syndrome, 98
Foliate papillae, 408
Follistatin, 4
Foot drop, 261–262, 261f, 430
Foot-slapping gait, 263
Foramen of Huschke, 348
Foramen of Monro, 17, 101
Foramina of Luschka, 57, 58
Foster Kennedy syndrome, 268
Fourth ventricle, 57–62, 101
caudal medulla, 59–62
midolivary level, 60–62, 60f
motor decussation level, 59, 59f
pontomedullary junction, 62
rostral medulla, 62, 62f
sensory decussation level, 59–60, 59f–60f
isthmus rhombencephali, 57
metencephalon, 57
myelencephalon, 57
Fovea centralis, 326
Foville's syndrome, 447t, 448
Frey syndrome, 280
Friedreich's ataxia, 99, 99f, 432
Frölich syndrome, 118
Froment's sign, 238, 239
Frontal dystaxia, 149
Frontal eyefield loop, 471
Frontal lobe, 131–132
hematoma, 134
prefrontal cortex, 132–134
seizures, 131–32
tumors, 99
Frontal motor eye field, 145
Functional loop and pathways, 471–472, 471–4732
Fusiform cells, 141, 142f
Fusiform neurons, 36

G

GABA. *See* γ-aminobutyric acid (GABA)
GABAergic neurons, 32, 87, 114, 307
GABA transaminase (GABA-T), 307
GAD. *See* Glutamic acid decarboxylase (GAD)
Gag reflex, 294
Galactocerebroside, 27
Galactosyl glycerides, 27
Galanin, 112
γ-aminobutyric acid (GABA), 21, 37, 85, 112, 114, 115, 307–308, 381
action mechanism, 307
glial, 308
inactivation of, 308
overactivity of, 308
production of, 307
receptors
GABA-a receptors, 308
GABA-b receptors, 308
transamination of, 307
γ-hydroxybutyrate (GHB), 307
Gamma interferon (IFNγ), 28
Ganglia
of PNS, 201t
Ganglionic multipolar neurons, in visual pathway, 327–328
Ganglioside monosialic2 (GM2), 34
Gangliosides, 27, 34
GAP-43. *See* Growth-associated protein 43 (GAP-43)
Gap junctions, 38
Gastric plexus, 212
Gate theory, 422
Gaucher's disease, 32, 33
Gaze centers, 343f, 344
GBS. *See* Guillain–Barré syndrome (GBS)

GDN. *See* Glial-derived nexin (GDN)
Generalized seizures, 162
General somatic afferents, 14, 15
General somatic efferents (GSEs), 61
fibers, 14, 15
General visceral afferents (GVAs), 14, 61
General visceral efferents (GVEs), 61, 223
fibers, 14, 15
Genetic aspects, nervous system developmental, 12–13
Genitofemoral nerve, 250f, 251
injury to, 251
Germinoma, 125
GFAP. *See* Glial fibrillary acidic protein (GFAP)
GGF. *See* Glial growth factor (GGF)
Ghrelin, 118
GHRH. *See* Growth hormone–releasing hormone (GHRH)
Giant cell arteritis, 330
Gilles de la Tourette syndrome, 313
Glabellar hyporeflexia, 285
Glabellar reflex, 285
Glasgow Coma Scale (GCS), 164
Glaucoma, 324, 330
acute angle, 324
chronic angle, 324
closed-angle, 324
open-angle, 324
and peripheral scotoma, 324
primary, 324
secondary, 324
Glaucomatous cups, 329
Glial acidic fibrillary protein (GFAP), 31
Glial cells, 3, 21
Glial-derived neurite-promoting factor (GNPF), 21
Glial-derived nexin (GDN), 21
Glial fibrillary acidic protein (GFAP), 213, 377
Glial growth factor (GGF), 31
Glioblasts, 18
Global aphasia, 155–156
Global ischemia, 309
Globoid cell leukodystrophy (Krabbe's disease), 32, 33
Globose nucleus, 87
Globus pallidus, 474
Glomerular neurons, 36
Glomerulonephritis, 29
Glossopharyngeal nerve, 5, 292–294, 293f, 410f
auricular branch, 294
carotid sinus branch, 292
damage to, 294
inferior salivatory nucleus, 294
lesser petrosal branch, 292
lingual branch, 294
muscular branches, 294
nucleus ambiguus, 294
pharyngeal branch, 294
solitary nucleus, 294
spinal trigeminal nucleus, 294
tonsillar branch, 294
tympanic branch, 292
Glossopharyngeal neuralgia, 294
Glutamate, 21
L-glutamate, 87, 88
Glutamate receptor antagonists, 309
Glutamic acid, 308–309
Glutamic acid decarboxylase (GAD), 68, 307
Glutamic decarboxylase (GAD), 465
Gluteal superficial reflex, 264
Glycine, 309–310
Glycogen, 407
astrocytes as storage site in CNS of, 22
Glycosaminoglycans, 417
GM2. *See* Ganglioside monosialic2 (GM2)
GM-4 (sialosylgalactosylceramide-4), 27
γ motor neurons, 52
GNPF. *See* Glial-derived neurite-promoting factor (GNPF)
Golgi neurons, 95
cerebellar cortex, 84, 85, 85f
Golgi tendon organs (GTO), 415, 417
Golgi type I, 36
Golgi type II, 16, 37
Gordon's finger sign, 442
G protein, 407
G protein-coupled receptors (GPCR), 407
Gracilis fasciculus, 53, 59, 419, 420
Gracilis nucleus, 59–60, 59f–60f
Gradenigo syndrome, 181, 284
Granular cortex, 143
Granular layer, cerebellar cortex, 84–85, 85f, 86f
Granule cells, 36
Graphagnosia, 158
Grasp reflex, 134, 444
Gray commissures, 51
Gray matter, 45, 51–52, 52f
cerebral cortex. *See* Cerebral cortex (gray matter)
lamina I, 51
lamina II (substantia gelatinosa), 51–52
lamina III and IV, 52
lamina IX, 52
lamina V, 52
lamina VI, 52
lamina VII, 52
lamina VIII, 52
lamina X (area X), 52
lissauer zone (dorsolateral fasciculus), 52
periaqueductal, 120
Gray's type I synapse, 38
Gray's type II synapse, 38
Great cerebral vein of Galen, 174
Greater auricular nerve, 227
Griffe cubitale (ulnar claw hand), 239, 239f
Group α_2 receptors, 200
Group β_2 (hormonal) receptors, 200
Growth-associated protein 43 (GAP-43), 25, 31
Growth hormone–releasing hormone (GHRH), 214
GSEs. *See* General somatic efferents (GSEs)
Guam disease, 309
Guanine nucleotide triphosphate (GTP)–binding protein, 310
Guidepost cells, 12
Guillain–Barré syndrome (GBS), 26, 29, 216, 217, 218, 245, 330, 451
Gustatory cortex, 144
Gustatory dysfunctions, 410–411
dysgeusia, 410–411
hypogeusia, 411
phantogeusia, 411
taste disorder, 410
Gustatory neurons and associated ganglia, 409, 410f
Gustatory otolgia-wet ear syndrome, 348
Gustatory pathways, 410
Gustatory receptors, peripheral. *See* Peripheral gustatory receptors
Gutter cells, 24
GVAs. *See* General visceral afferents (GVAs)
GVEs. *See* General visceral efferents (GVEs)

H

Habenula, 123, 124f–126f, 125–126
Hallucination, hypnagogic, 78
H_1 antagonists, 315
H_2 antagonists, 315
Harris's sign, 369
Headache
as cerebellar dysfunction, 98
Hearing impairment, 359
Hearing loss, 359. *See also* Auditory system
Hematological tests, 430
Hematoma, 465
epidural, 176, 176f
frontal lobe, 134
parietal lobe, 135
spinal epidural, 47
subdural, 48, 176–177, 176f
temporal lobe, 136
thalamic, 111
Hemiballism, 468
Hemicholinium, 39
Hemioptic aphasia, 156
Hemiparesis, 441, 468
Hemorrhage, 449
cerebellar, 99
subarachnoid, 182–184
Hepatic plexus, 211
Hepatolenticular degeneration, 477, 478
Hereditary amyloidosis, 217
Hereditary multiple tic disorder. *See* Gilles de la Tourette syndrome
Hereditary neuralgic amyotrophy, 231
Hering–Breuer reflex, 209
Herniation, cerebellar, 98
Herpes zoster (shingles), 247–248
Herpes zoster ophthalmicus, 278
Herpes zoster oticus, 354
Herring bodies, 123
Hexosaminidase A, 34
Higher autonomic centers, 208
Hippocampal commissure, 140
formation of, 18–19
Hippocampal formation, 119, 387, 387f, 388f
afferents of, 391, 397f
efferents of, 391f, 392–393, 392f, 397f
Hippocampal gyrus (cornu ammonis), 25, 140, 387–389, 387f, 388f. *See also* Schizophrenia
Hippus, 323
Hirano bodies, 384
Hirschsprung's disease, 6, 217, 219
Histamine, 37, 112, 315
HLA. *See* Human leukocyte antigen (HLA)
Hodgkin's lymphoma, 204
Hoffman's sign, 442
Holotelencephaly (holoprosencephaly), 7
Homeobox genes, 12
Homonymous hemianopsia, 110, 333, 333f
Homonymous visual field defects, 336
Horizontal cells of Cajal, 141, 142f
Hormone regulating factors (HRFs), 114
Horner's syndrome, 201, 203, 203f, 217, 219–220, 323, 429
HOX-A, 12
HOX-B, 12
HOX-C, 12
HOX-D, 12
Human leukocyte antigen (HLA), 30
Human leukocyte antigen (HLA) DR2, 78
Huntington chorea (HD), 309, 467, 475, 476f

Huntington's disease, GABAergic neurons and, 308
Hydrocephalus, 7, 9, 191–193, 191f–192f
communicating, 192
congenital, 191–192
noncommunicating, 193
nonobstructive, 193
NPH, 192–193
obstructive, 192
5-hydroxytryptamine (5-HT). *See* Serotonin
Hyperhidrosis (disorder of sweating), 216–217
Hyperkinesia, 97
Hyperopia (farsightedness), 325, 325f
Hyperphagia (excessive eating), 118
Hyperreflexia, 441f, 442, 454
Hyperthymesic syndrome, 396
Hypertrophic degeneration, 477
Hypnagogic hallucination, 78
Hypogeusia, 411. *See also* Gustatory dysfunctions
Hypoglossal nerve, 300f, 301–303, 301f
Hypoglossal nucleus, 57, 60, 61
Hypokinesis, 472
Hyporeflexia, 97
Hypothalamic peptides, 316
Hypothalamic sulcus, 101
Hypothalamo-hypophyseal tracts, 121
Hypothalamo-pituitary-adrenocortical (HPA) axis, 404
Hypothalamus, 101, 102, 111–115, 111f–113f, 404–406
afferents, 119–120, 119f–120f
efferents, 120–122, 120f
embryonic developmental, 16–17
functional and clinical consideration, 115–119
lesions of, 121
nuclei and associated areas, 112–115, 112f–113f
temperature regulation and, 118
tumors, 121
Hypotonia, 97, 475

I

Iadazoxan, 312
Ideational apraxia, 157
Ideomotor apraxia, 157
Idiopathic intracranial hypertension (pseudotumor cerebri), 191
IFNβ-1b, 31
IFNγ. *See* Gamma interferon (IFNγ)
IgG. *See* ImmunoglobulinG (IgG)
IGL. *See* Intergeniculate thalamic leaflet (IGL)
Iliohypogastric nerve, 250, 250f
injury to, 250
Ilioinguinal nerve, 250, 250f
injury to, 250–251
ImmunoglobulinG (IgG), 31
Incurvation (Galant's) reflex, 444
Indirect pathways, 472
Indirect Wallerian (retrograde) degeneration, 28
Indolamines, 313
histamine, 315
serotonin, 313–315
structure of, 313
Induseum griseum, 399
Infectious myringitis. *See* Bulbous myringitis
Inferior alternating hemiplegia, 447t, 448, 448f
Inferior anastomotic vein of Labbe, 172
Inferior cerebellar peduncle, 62, 65, 81
Inferior cerebral veins, 172
Inferior cervical ganglion, 201, 203f
Inferior colliculus, 67, 68, 69f, 357, 357f
Inferior frontal gyrus, 131
Inferior gluteal nerve, 257
injury to, 257
Inferior hypogastric (pelvic) plexus, 202f, 213
Inferior longitudinal fasciculus, 153
Inferior mesenteric plexus, 202f, 212f, 213
Inferior occipitofrontal fasciculus, 153
Inferior olivary nuclear complex, 61, 62
Inferior olivary nucleus, 61
Inferior petrosal sinus, 181
thrombosis of, 181–182
Inferior sagittal sinus, 179
Inferior salivatory nucleus, 62
Inferior striate vein, 174
Inferior temporal gyrus, 136
Inferior temporal quadranopsia, 332
Inferior vestibular nucleus, 61
Infranuclear lesions, 287, 290f
Infundibulum, 17
Injury(ies)
to axillary nerve, 244
to common fibular (peroneal) nerve, 260
to deep fibular nerve, 261–262
to femoral nerve, 253
to genitofemoral nerve, 251
to iliohypogastric nerve, 250
to ilioinguinal nerve, 250–251
to inferior gluteal nerve, 257
to lateral femoral cutaneous nerve, 252
to median nerve, 241–244
to musculocutaneous nerve, 234
to obturator nerve, 254–255
to posterior femoral cutaneous nerve, 258
to pudendal nerve, 258
to radial nerve, 245–246
to sciatic nerve, 259–260
spinal nerves, 226
to superficial fibular (peroneal) nerve, 261
to superior gluteal nerve, 257
to tibial nerve, 263–264
trunk, 230–232, 231f
ulnar nerve, 238–239, 239f
Insular cortex (central lobe), 138, 138f
Integrin, 3, 12
Intercavernous aneurysm, 283
Intercostobrachial nerve, 247
Interfascicular oligodendrocytes, 22, 22f
Intergeniculate thalamic leaflet (IGL), 113
Intermediolateral nucleus, 52, 53
Internal carotid artery, 167–168
unilateral obstruction of, 115
Internal vertebral (epidural) plexus, 51
Interneurons, 470
Internuclear ophthalmoplegia, 368–369, 368f, 369f
Interpeduncular syndrome, 447t, 450
Interscapular reflex, 264
Intracerebellar circuit, 95
Intracochlear (Reissner's) membrane, tears of, 353
Intrafusal muscle fibers, 416f, 417
Intralaminar nuclear group, 102f, 105f–107f, 109–110, 109f
Intramedullary spinal tumors, 429
Intraocular muscles, 5
Intraocular pressure, 324. *See also* Glaucoma
Inverse myotatic reflex, 266, 415, 442
Involuntary choreiform movements, 475
Ipsilateral corticospinal fibers, 439
Irideocorneal angle, 324
Ischemia of optic disc, 330
Isthmus rhombencephali, 57

J

Jacksonian "march" seizure, 162
Jacksonian seizure, 441
Jackson's syndrome, 447t, 448
Jakob–Creutzfeldt disease, 159, 160–161
Jaw reflex, hyperactive, 279

K

Kallmann syndrome, anosmia in, 268
Kayser–Fleischer ring, 321, 478
Kernig's sign, 175, 175f, 446, 446f
Kernohan–Woltman syndrome, 137
Kinesin, 25
Kinetic apraxia, 157
Klippel–Feil deformity, 7
Klumpke palsy, 203, 231–232
Klüver–Bucy syndrome, 158, 402
Kölliker-Fuse nucleus, 75, 76
Korsakoff's psychosis, 395
Korsakoff's syndrome, 103
Korsakoff–Wernicke syndrome, 32
Krabbe's disease, 27, 33
Krox20, transcription factor, 12
Kyphoscoliosis, 429

L

Labrynthitis, 369
Labyrinthine artery, 354
Lacrimation (tearing) reflex, 277
Lambert–Eaton syndrome, 41
"warm-up" phenomenon in, 41
Lamina terminalis, 6, 111
Laminin, 3, 12, 21
Landau reflex, 444
Laryngopharynx, 409
Lasègue's sign, 260, 431
Lasegue's test, 454
Lateral cord
terminal branches of, 234
Lateral corticospinal tract, 55, 55f
Lateral dorsal (LD) nucleus, 102f, 105f, 106f, 107, 107f
Lateral femoral cutaneous nerve, 251f, 252
injury to, 252
Lateral funiculus, 53
Lateral gaze palsy, 341
Lateral geniculate body (LGB), 101f, 102f, 103–104, 104f, 332, 333f
Lateral geniculate nucleus (LGN), 334
Lateral lemniscus, 65, 69, 355f, 356, 356f
Lateral medullar syndrome, *Wallenberg's*, 294
Lateral medullary syndrome, *Wallenberg's*, 267, 294, 456, 457f
Lateral nuclear group
lateral dorsal, 102f, 105f, 106f, 107, 107f
lateral posterior, 102f, 105f, 106f, 107, 107f
pulvinar, 101f–102f, 104f, 106f, 107–108, 108f
Lateral olfactory stria, 382
Lateral pectoral nerve, 234
Lateral posterior (LP) nucleus, 102f, 105f, 106f, 107, 107f
Lateral reticular nucleus, 62, 75

Lateral reticular zone, 75–77, 76f–77f
Lateral spinothalamic tract (LSTT), 54, 54f, 421, 432
Lateral strabismus, 274
Lateral striate artery, 149
Lateral ventricle, 129
Lateral vestibular (Deiters) nucleus special somatic afferent (SSA), 64
Lateral vestibulospinal tract, 367
Latrodectus mactans (black widow spider) toxicity, 39
α-latrotoxin, 39
Lazy eye. *See* Amblyopia
L-dopa, 474, 478
Lemniscal trigone, 66
Lenticular fasciculus, 471
Leptin, 118, 119
Leptodendritic (Deiter's) neurons, 464
Leptomeninges, 182
Lermoyez's disease, 370
Lesch–Nyhan syndrome, 477
Lesions
- of angular gyrus, 135
- cerebellar
  - archicerebellar, 96
  - neocerebellar, 96
  - paleocerebellar, 96
- of cerebellar efferent fibers, 93
- of hypothalamus, 121
- of locus ceruleus, 78
- of raphe nuclei, 73
- reflexes and, 209
- spinal cord
  - autonomic disturbances and, 214–215
- of substantia nigra, 68
- of trochlear nerve, 69

Lesser occipital nerve, 227
Leukocyte function–associated antigen-1 (LFA-I), 28
Levator palpebrae muscle, ipsilateral palsy of, 270f
LFA-I. *See* Leukocyte function–associated antigen-1 (LFA-I)
LGB. *See* Lateral geniculate body (LGB)
L-glutamic acid, 476
LGN. *See* Lateral geniculate nucleus (LGN)
Lhermitte sign, 456
Libidinous sensations, 428
Limbic lobe, 135f–138f, 138–139, 403
Limbic system, 387. *See also* Memory
- amygdala, 399–403, 399f
- cingulate gyrus, 403–404, 404f
- dentate gyrus, 387f, 388f, 390
- hippocampal formation, 387, 387f, 388f
  - afferents of, 391, 397f
  - in control of behavior, role of, 393
  - efferents of, 391f, 392–393, 392f, 397f
- hippocampal gyrus, 387–389, 387f
- hypothalamus, 404–406, 405f
- induseum griseum, 399
- limbic lobe, 403
- memory and amnesia, 393–396
- prefrontal cortex, 404, 404f
- role of, 387
- septal area, 396–399, 397f
- subicular complex, 390–391

Limbic system circuitry
- anterior nucleus and, 103

Limbic-ventral striatal loop, 398
Lingual papillae, 409
Lipofuscin granules (corporal amylacea), 25
Lipoma-covered defect, in spina bifida occulta, 8
Lipomyelomeningocele, 9
Lipoprotein, 25
Lissauer zone (dorsolateral fasciculus), 52
Lissencephaly, 9
Lithium carbonate, 411
LMN. *See* Lower motor neurons (LMN)
Locked-in syndrome, 63, 447t, 450, 450f
Locus ceruleus, 25, 57, 77, 120, 311, 312
- lesions of, 78
- role in REM, 77

Locus ceruleus nucleus, 69
Long-term depression (LTD), 85
Long-term potentiation (LTP), 309, 393
Long thoracic nerve, 230f, 232–233
Lou Gehrig's disease, 447
Low back pain, 430
Lower motor neurons (LMN), 437
- general somatic motor efferent (GSE), 445
- general visceral motor neurons, 445
- special visceral efferent neurons (SVE), 445, 446

Lower motor neurons (LMN) palsies, 446–461. *See also* Upper motor neurons (UMN) palsies
- AICA syndrome, 457
- anterior spinal artery (Beck) syndrome, 452–453
- Avellis syndrome, 456
- brainstem lesions, 447
- Brown-Séquard syndrome, 453
- Cauda equina syndrome, 453
- cerebral lesion, 451, 459
- combined sensory and motor dysfunctions, 452, 452t
- combined system dysfunctions, 456
- combined UMN and LMN deficits, 447, 447f
- Conus medullaris syndrome, 453
- decerebrate (gamma) rigidity, 451, 451f
- Ependymoma, 454
- Foville's syndrome, 448
- inferior alternating hemiplegia, 448, 448f
- interpeduncular syndrome, 450
- Jackson's syndrome, 448
- lateral medullary (Wallenberg) syndrome, 456
- locked-in syndrome (pseudocoma), 450, 450f
- medial medullary syndrome, 456
- medial pontine syndrome, 457–458
- medullary lesions, 448
- midbrain lesions, 449, 458
- Millard–Gubler syndrome, 449
- multiple sclerosis, 455
- myeloschisis, 455
- obstruction of anterior cerebral artery, 459
- occlusion middle cerebral artery, 459, 460f
- pellagra, 455
- poliomyelitis, 446
- Pontine lesions, 448, 457
- Raymond's syndrome, 449
- Schmidt's syndrome, 448
- spina bifida occulta, 455
- spinal lesions, 452
- subacute combined system degeneration, 455
- superior cerebellar artery syndrome, 457
- symptoms of, 453
- transection of spinal cord, 453
- Weber's syndrome, 449

LP. *See* Lumbar puncture (LP)
LTD. *See* Long-term depression (LTD)
Lucentis. *See* Ranibizumab
Lumbar central disc herniation, 45
Lumbar plexus
- terminal branches of, 249–256, 250f–251f, 253f–255f

Lumbar puncture (LP), 45, 48
Lumbar spinal nerves, 223, 249
Lumbar spinal segments, 53, 53f
Lyme disease, 159, 161
Lysosomes, 25

M

Macrocephaly, 7
Macroglia, 21–23
- astrocytes, 21–22, 21f–22f, 23f
- ependymal cells, 23
- oligodendrocytes, 22, 22f–23f
- Schwann cells, 22–23

Macrographia, 97
Macrophages, 28
Macula lutea, 326
Macular degeneration, 326
- exudative, 326
- nonexudative, 326

Macular sparing, 337
Magnesium sulfate, in preeclampsia, 310
Magnetic resonance imaging (MRI), 31
Magnocellular neurons, 114
Major histocompatibility complex (MHC I/II) antigen, 28–29
Mamillotegmental tract, 208
Mammalian degenerin (MDEG1), 408
Mammillary body(ies), 101, 112
Mammillotegmental tract, 121
Mammillothalamic tract, 121
Mantle layer, 10, 11f
- division of, 3

MAP-2. *See* Microtubule-associated protein 2 (MAP-2)
Maple syrup urine disease, 28, 33–34, 34f, 35f
MAP-2 protein, 384
MAPs. *See* Microtubule-associated proteins (MAPs)
Marchiafava–Bignami syndrome, 140, 151
Marcus Gunn jaw-winking syndrome, 323
Marcus Gunn pupil, 270, 323, 331
Marfan syndrome, 9
Marginal division zone (MrD), 395
Marijuana, 324
Martin–Gruber anastomosis, 237, 239, 242, 243
Martinotti cells, 141, 142f
Mass reflex of Riddoch, 210, 444
Mastoid antrum, 350
MBP. *See* Myelin basic protein (MBP)
McCarthy's reflex, 285
Mdial forebrain bundle (MFB), 118, 119, 122
Mechanoreceptors, 416
Meckel– Gruber syndrome, 6
Meclizine, 315
Medial antebrachial cutaneous nerve, 235, 236f
Medial brachial cutaneous nerve, 234–235, 236f
Medial cord
- terminal branches of, 230f, 234–244, 235f–237f, 239f–243f

Medial forebrain bundle (MFB), 73, 74, 75, 138, 208, 396, 467
Medial geniculate body (MGB), 101f, 102f, 104, 104f
Medial geniculate nucleus (MGN), 356, 357
Medial lemniscus, 68

Medial longitudinal fasciculus (MLF), 55, 59, 64, 65, 68, 71, 93, 272, 340, 365, 366f, 367, 368, 368f, 369f
Medial medullary syndrome, 452t, 456
Medial nuclear group
  dorsomedial nucleus, 101f–104f, 106f–109f, 108–109, 110f
Medial olfactory stria, 382
Medial parabrachial nucleus, 75, 76
Medial pectoral nerve, 230f, 234
Medial reticular nuclei, 74, 76f
Medial striate artery, 149
Medial superior temporal visual area (MST), 341
Medial vestibular nucleus (MVN), 61
Median nerve, 239–244, 240f
  injury to, 241–244
Medulla, 57, 57f
  caudal. *See* Caudal medulla
  rostral, 62, 62f
Medullary lesions, 448
Medulla spinalis. *See* Spinal cord
Melanocytes, 5
Melkersson–Rosenthal syndrome, 290
Memory, 393. *See also* Limbic system
  Atkinson–Shiffrin memory model, 394
  brain areas in, 395
  construction of, 393
  declarative, 394
  episodic, 394–395
  hippocampal gyrus in, role of, 395
  hyperthymesic syndrome and, 396
  long-term, 394
  procedural, 395
  semantic, 394
  short-term, 393
  Wernicke's encephalopathy and, 395
  working, 393–394
Meniére disease, 359, 369–370
Meninges
  anteroinferior group of dural sinuses, 179–182, 180f–182f
  arachnoid mater, 182–183
  dural sinuses, 172f–175f, 177f, 178
  dura mater, 175–178, 175f–178f
  pia mater, 183–184
  posterosuperior group of dural sinuses, 172f–175f, 177f, 178–179, 180f
Meningiomas, 182, 183f
Meningocele
  in spina bifida cystica, 8
Meningoencephalocele, 7
Meningomyelocele
  in spina bifida cystica, 9
Meperidine, 431
Meralgia paresthetica, 252
Merkel's disks (tactile menisci), 418
Mesencephalic nucleus, 66
Mesencephalon (midbrain) , 10, 10f, 11t
  embryonic developmental, 16, 17f
Mesocortical pathway, 312
Mesolimbic dopaminergic neurons, 120
Mesolimbic dopaminergic system, 390
Mesolimbic system, 312
Metabolic disorders
  demyelinating, 32–35, 32t
Metabotropic glutamate receptors, 85
Metacarpophalangeal (MP) joints, flexion at, 238
Metachromatic leukodystrophy (MLD), 27, 32, 35
Metathalamus, 103–104
Metencephalon, 10, 10f, 11t, 57
  embryonic developmental
    cerebellum, 15–16, 16f
    pons, 15, 15f–16f
3-methoxy-4-hydroxy-phenethyleneglycol (MHPG), 311
3,4-methylenendioxymethamphetamine (MDMA), 315
Methylphenidate (Ritalin), 78, 479
1-methyl,4-phenyl-1,2,3,4,6-tetrahydropyridine (MPTP), 313
N-methyl 4-phenyl 1,2,3,6-tetrahydropyridine (MPTP), 469
Methylprednisone, 31
Meyer's loop, 335
MFB. *See* Medial forebrain bundle (MFB)
MGB. *See* Medial geniculate body (MGB)
MGN. *See* Medial geniculate nucleus (MGN)
MHC I/II antigen. *See* Major histocompatibility complex (MHC I/II) antigen
Microcephaly, 7
Microfilaments, 25
Microglia, 23–24, 23f
  phagocytosis of cellular debris, 23
Microtubule-associated protein 2 (MAP-2), 25
Microtubule-associated proteins (MAPs), 25
Microtubules, 25
Midbrain, 66–71, 67f
  basis pedunculi, 67
  caudal, 68–69, 69f
  central tegmental tract, 68
  cerebral aqueduct, 67
  cerebral peduncle, 67
  classification, 67
  crus cerebri, 68
  dopaminergic neurons, 312
  dorsal longitudinal fasciculus, 68
  inferior colliculi, 67
  lesions, 449, 458
  reticular formation, 119
  rostral, 70–71, 70f
  substantia nigra, 68
  superior colliculi, 67
  tectum, 67
  tegmentum, 67
  tremor, 473
Middle alternating hemiplegia, 284
Middle cerebellar peduncle, 65, 66, 81
Middle cerebral artery (MCA), 129
  occlusion of, 134
Middle cervical ganglion, 201, 203f
Middle frontal gyrus, 131
Middle temporal gyrus, 136
Midline nuclear group, 110
Midolivary level, caudal medulla, 60–62, 60f
Midpons, 65–66, 65f
Migraine headache, serotonin blockers in, 315
Migration, 3
Millard–Gubler syndrome, 447t, 449
Miller Fisher syndrome, 287
Mimetic facial palsy, 287
Minor dense line, 27
Miosis, 323
Mitaxantrone, 31
Mitochondria, 25
MLD. *See* Metachromatic leukodystrophy (MLD)
MLF. *See* Medial longitudinal fasciculus (MLF)
MLF syndrome. *See* Internuclear ophthalmoplegia
Modafinil, 78
Molecular aspects, nervous system developmental, 12–13
Molecular layer, cerebellar cortex, 84, 86–87, 86f
Mondini deformity, 360
Monoamine oxidase (MAO), 311
Monoamine oxidase A (MAO-A), 134
Monoamines, 311
  catecholamines, 311–313
  indolamines, 313–315
Monochromatopia, 326. *See also* Color blindness
Monosodium l-glutamate, 408
Moro (startle) reflex, 443–444
Morphine, 431
Morton metatarsalgia (Morton's toe neuroma), 264
Mossy fiber rosettes, cerebellar cortex, 84, 85, 86f, 88
Motor coordination
  cerebellum role in, 95–96
Motor cortex, 143f, 144–146, 146t
  premotor, 146, 146t
  primary, 145, 145t
  supplementary, 145
Motor decussation level, caudal medulla, 59, 59f
Motor deficits, 472
Motor end plate, 38, 39
Motor homunculus, 437, 437f
Motor learning
  cerebellum role in, 95
Motor recovery, 439
Motor tics, 479
Motor trigeminal nucleus, 66
Moyamoya syndrome, 171, 171f
MRI. *See* Magnetic Resonance Imaging (MRI)
MS. *See* Multiple sclerosis (MS)
MSK1 genes, 6
MSK3 genes, 6
Müller cells, 22
Multiple sclerosis (MS), 26, 29–31, 30f, 159, 282, 359, 360, 452, 455
Multiple sleep latency test (MSLT), 78
Multipolar autonomic neurons, 197, 199f
Multipolar neurons, 11, 12, 32, 36–37, 37f, 465
Multisynapses, 37
Muscarinic receptors, 206, 310
Muscle spindle, 417, 417f
Muscle (neuromuscular) spindles, 416, 417
Musculocutaneous nerve, 230f, 234, 235f
  injury to, 234
MVN. *See* Medial vestibular nucleus (MVN)
Myasthenia gravis, 40–41, 40f
Mydriasis, 274, 323
Myelencephalon, 10, 10f, 11t, 57
  embryonic developmental, 14–15, 14f, 15f
Myelin, 21, 26–28, 26f. *See also* Demyelination; Myelination
  CNS, 27
  fatty acids, 27
  MAG, 26, 27
  PMP22, 27
  PNS, 27
  P_0 proteins, 26, 27
  P_1 proteins, 26
  P_2 proteins, 26
Myelin-associated protein (MAG), 26, 27
Myelination, 27
Myelin basic protein (MBP), 27
Myelography, 48
Myeloschisis, 455. *See also* Spina bifida
Myenteric plexus, 213

Myerson sign, 285
Myoclonus, 477
Myopia, 324–325, 325f
Myotonia, 310
Myringotomy, 352

N

Narcolepsy, 78
Narcoleptics, 78
Nasal cavity, obstruction of, 407
Nasal hemianopsia, 333
Nasal polyps, 379
Nasal (sneeze) reflex, 278
Nasociliary nerve, 276, 277f
Nasociliary neuralgia, 276
Nausea and vomiting
  as cerebellar dysfunction, 98
N-cadherin, 4, 12
N-CAM. *See* Nerve cell adhesion molecule (N-CAM)
N-CAM genes. *See* Neural cell adhesion molecule (N-CAM) genes
Nearsightedness. *See* Myopia
Necrobiosis, 14
Negative myoclonus (asterixis), 477
Neimann–Pick disease, 27, 32, 34–35
Neocerebellar lesions, 96
Neocerebellum, 84
Neopallium, 141
Neo-spinothalamic-lateral system, 421
Neri's sign, 430
Nerve cell adhesion molecule (N-CAM), 4, 12
Nerve deafness, 359. *See also* Sensorineuronal hearing loss
Nerve fibers, 409
Nerve gas (di-isopropyl fluorophosphate), 41
Nerve growth factor (NGF), 21, 31
Nerve growth factor receptor (NGF-R), 31
Nerve root compression, 226
Nerve to the subclavius, 234
Nervous system, basic elements, 21–41
  demyelinating metabolic disorders, 32–35
  functional and clinical consideration, 29–31
  myelin, 26–28, 26f
  neuroglia
    macroglia, 21–23
    microglia, 23–24, 23f
  neuronal degeneration
    antergrade, 28
    retrograde, 28
    transynaptic, 28–29
  neurons, 24–26
    classification, 32, 36–37
  overview, 21
  regeneration, 31–32
  synaptic connectivity, 37–41
Nervous system, developmental aspects, 3–19
  brain vesicles development, 10, 10f, 11t
  diencephalon (thalamus, hypothalamus, epithalamus and subthalamus), 16–17
  genetic aspects, 12–13
  medulla spinalis (spinal cord), 13–14, 13f
  mesencephalon (midbrain), 16, 17f
  metencephalon
    cerebellum, 15–16, 16f
    pons, 15, 15f–16f
  molecular aspects, 12–13
  myelencephalon, 14–15, 14f, 15f
  neural tube
    defects, 4–10, 4f–5f, 7f, 8f
    differentiation, 10–12, 11f
    formation, 3–4, 3f–4f
  overview, 3
  telencephalon (cerebral hemispheres, basal nuclei and ventricular system), 17–19, 18f
*Neu*ERBB2, 31
Neural canal, 3
  formation of, 5f
Neural cell adhesion molecule (N-CAM), 409
  genes, 31
Neural crest cells, 3
  derivatives of, 5, 6t
  formation of, 4–5, 5f
  migration of, 4, 4f
    failure of, 5–6. *See also* Hirschsprung's disease (congenital aganglionic megacolon)
Neural ectoderm (cephalocaudal), 3
Neural folds, 3f, 5f
Neural groove, 4
Neural induction, 3
Neural plate, 3, 3f
  formation of, 4, 5f
Neural sulcus
  formation of, 4, 5f
Neural tube, 3
  defects, 4–10, 4f–5f, 7f, 8f
    anencephaly (exencephaly), 6
    cranioschisis, 6
    encephalocele, 7
    Hirschsprung's disease (congenital aganglionic megacolon), 6
    macrocephaly, 7
    meningoencephalocele, 7
    microcephaly, 7
    spina bifida, 7–9, 8f. *See also* Spina bifida
    spinal dysraphism (defective fusion), 7
  differentiation, 10–12, 11f
    rostral neural tube, 10f
  formation of, 3–4, 3f–4f, 5f, 7f
    primary neurulation, 4
    secondary neurulation, 4
  transformation into spinal cord, 14
  ventrodorsal patterning, 12
Neuroactive steroids, 308
Neuroblasts, 18
  migration of, 10–11, 14
Neurodegenerative diseases, sign of, 410
Neuroectodermal cells, 3
Neuroepithelial cell, 17, 409
Neurofibrillary tangles (NFTs), 384
Neurofibromas, 23
Neurofibromatosis, 171
  type I and II, 360
Neurofilaments, 25
Neuroglia
  features, 24f
  macroglia, 21–23
    astrocytes, 21–22, 21f–22f, 23f
    ependymal cells, 23
    oligodendrocytes, 22, 22f–23f
    Schwann cells, 22–23
  microglia, 23–24, 23f
Neuroglial adhesion molecules (NgCAMs), 12
Neurohypophysis, 17, 123
Neurolathyrism, 309
Neurolemma, 5
Neuroligins (NLGNs), 13
Neuroma, 430
Neuromuscular junction, 38–39, 38f
Neuronal cell adhesion molecules (NCAMs), 389
Neuronal degeneration
  antergrade, 28
  retrograde, 28
  transynaptic, 28–29
Neuronal olfactory disorders, 267
Neuronal plasticity (sensitization), 25, 95, 422
Neuronal responses, 408
Neuronal sprouting, 21
Neurons, 3, 24–26, 409, 410f
  adrenergic, 32
  anaxonic, 37
  autonomic, 197–200, 198f–199f
  bipolar, 32, 36, 36f
  cholinergic, 32
  classification, 32, 36–37
  dopaminergic, 32
  features, 24, 24f
  fusiform, 36
  GABAergic, 32
  glomerular, 36
  multipolar, 11, 12, 32, 36–37, 37f
  noradrenergic, 32
  processes
    axons, 25–26
    dendrites, 25
  pseudounipolar, 32, 36, 36f
  Purkinje, 36
  pyramidal, 36
  regeneration, 31–35
  serotoninergic, 32
  soma (perikaryon), 24–25
  stellate (star), 36
  unipolar, 36
Neuropeptides, 315–316
  cholecystokinin, 316
  endorphins, 316
  enkephalins, 316
  hypothalamic peptides, 316
  substance P, 316
Neuropeptide Y (NPY), 200
Neuropores
  caudal, 4
  closure of, 3, 4
    failure of, 6. *See also* Rachischisis
  rostral, 4
Neuropraxia, 225
Neurosyphilis, 159, 161
Neurotmesis, 226
α-neurotoxin, 41
Neurotransmitters, 307, 422. *See also specific* types
  acetylcholine, 310
  acetylcholine as, 466
  action mechanism of, 307
  amino acid, 307–310
  classical, 307
  indolamines, 313–315
  monoamines, 311–313
  neuropeptides, 315–316
  peptidergic, 307
  release of, 407, 408
Neurotrophins, 13
Neurotubules, 25
Neurulation
  primary, 4
  secondary, 4
NgCAMs. *See* Neuroglial adhesion molecules (NgCAMs)
NGF. *See* Nerve growth factor (NGF)

NGF-R. *See* Nerve growth factor receptor (NGF-R)
NGF receptor, 23
Nicotinic receptors, 206, 310
  muscle (C-10 receptor), 206
  neuronal receptor, 206
Night terrors, 77
Nigrostriatal fibers, 469
Nigrostriate fibers, 470
Nissl bodies, 24, 28
Nitric oxide (NO), 213, 422
NMDA. *See* N-methyl-D-aspartatereceptors (NMDA)
NMDA receptor antagonists, 309
NMDA receptor–mediated neurotoxicity, 309
N-methyl-D-aspartate receptors (NMDA), 85, 87, 113
Nociceptive impulses, 425
Nodes of Ranvier, 25, 27
Noggin, 4
Noncommunicating hydrocephalus, 193
Nonconvulsive generalized seizures, 163
Nonencapsulated receptors, 418, 418f
Non-HOX homeobox genes, 12
Nonneuronal histamine, 315
Non-NMDA glutamate receptors, 85
Nonobstructive hydrocephalus, 193
Non–rapid eye movement (NREM), 73
  characterization, 77
  delta sleep/deep sleep, 77
  slow-wave sleep, 77
  stages, 77–78
Nonsteroidal anti-inflammatory drugs (NSAID), 431
Noradrenergic neurons, 32
Norepinephrine, 78, 311–312
Normal pressure hydrocephalus (NPH), 159, 192–193
NO synthetase (NOS), 309
Nothnagel's syndrome, 98–99, 272
Notochord (chordamesoderm), 3
NPH. *See* Normal pressure hydrocephalus (NPH)
NPY. *See* Neuropeptide Y (NPY)
NREM. *See* Non-rapid eye movement (NREM)
NREM–REM cycle, 78
Nuclei centralis pars parafascicularis, 470
Nucleus accumbens septi, 398–399
Nucleus ambiguus, 61, 62, 75
Nucleus basalis of Meynert, 395
Nucleus raphe magnus, 62, 73, 74f
Nucleus raphe obscurus, 62, 73, 74f
Nucleus raphe pallidus, 62, 73, 74f
Nucleus reticularis gigantocellularis, 62, 74, 76f
Nucleus reticularis parvocellularis, 62, 75
Nucleus reticularis pontine caudalis, 74, 76f
Nucleus reticularis pontine oralis, 74, 76f
Numb chin syndrome, 279
Nyctalopia, 327
Nystagmus, 96, 97, 291, 341, 371
  congenital, 371
  horizontal, 371
  jerk, 371
  by peripheral lesions, 371
  vertical, 371

O

Obex, 58, 60
Obsessive–compulsive disorder (OCD), 133
Obstructive hydrocephalus, 192
Obturator nerve, 252–256, 254f
  injuries to, 254–255
Occipital lobe, 131f, 135f, 137–138, 137f
Occipital sinus, 179
Occipital vein, 174
Occipitotemporal gyrus, 136
Occlusion
  of ACA, 134
  of AICA, 84
  of MCA, 134
  middle cerebral artery, 459, 460f
  of pontine arteries, 63
  of subclavian artery, 97
  of vertebral artery, 97
OCD. *See* Obsessive–compulsive disorder (OCD)
Ocular bobbing, 341
Ocular deficits, 478
Ocular disorders, 97–98, 475
Ocular dominance columns, 337
Ocular dysmetria, 341
Ocular flutter, 341
Ocular movements, 339–340, 339f, 340f
  conjugate movement, 340
    saccadic, 340–341
    smooth pursuit, 341
    vestibulo-ocular, 341
  disconjugate movement, 340
  disorders of, 341–342
Ocular myoclonus, 341
Ocular reflexes, 342–344
  accommodation reflex, 342–343, 342f
  blink reflex of Descartes, 344
  ciliospinal reflex, 343
  consensual pupillary light reflex, 342, 342f
  direct pupillary light reflex, 342, 342f
  oculoauricular reflex, 344
  oculocardiac reflex, 343
  oculocephalic reflex, 343
Oculoauricular reflex, 344
Oculocardiac reflex, 343
Oculocephalic reflex, 343
Oculocephalogyric reflex, 164
Oculogyric reflex, 341
Oculomotor dysfunctions, 449f
Oculomotor nerve, 270–274, 271f, 273f
  injury to, 272–273
Oculomotor nuclear complex, 71, 270f, 272
Oculomotor nucleus, 25
Oculomotor palsy, 273–274, 273f
Odorant-binding proteins (OBPs), 378
Odorants, 377. *See also* Olfactory system
Olfactory bulb, 267, 379–382, 380f, 381f
  anterior olfactory nucleus (AON), 380
  external plexiform layer (EPL), 380
  granule cell layer (GCL), 380
  granule cells, 381
  internal plexiform layer, 380
  mitral and tufted cells, 380–381
  olfactory glomerulus, 381–382
  olfactory nerve layer, 379
  periglomerular cells, 381
  pyramidal neurons, 380
Olfactory cilia, 377
Olfactory cortex, 101
Olfactory meningioma, 268
Olfactory nerve, 267–268, 269f
Olfactory receptor G protein (Golf), 378
Olfactory system, 377
  anterior commissure, 381f, 382
  dysfunctions, 384–386
  olfactory apparatus, 377
  olfactory bulb, 379–382, 380f, 381f
  olfactory cortices, 383–384
    entorhinal cortex, 384
    piriform cortex, 384
  olfactory nerve, 379
  olfactory receptors, 377–379
  pathways, 382–383, 383f
Olfactory tract, 381f, 382
Olfactory tubercle, 382
Oligodendrocytes, 22, 22f–23f
  interfascicular, 22, 22f
  myelin from, 26–28
  perivascular, 22
Olivocerebellar tract, 87, 88f
Olivocochlear bundle, 358, 358f
Olivopontocerebellar atrophy, 88
Ondine's curse, 431, 457
One-and-a-half syndrome, 369
Open-angle glaucoma, 324
Ophthalmic artery, 328, 328f
Oppenheim's sign, 442
Opsoclonus, 341
Optic chiasma, 111, 112, 331–332, 331f, 333f, 334f. *See also* Visual system
Optic disc pallor, 333
Optic nerve, 267, 269–270, 270f, 322f, 329–331, 329f. *See also* Visual system
Optic nerve drusen, 330–331
Optic neuritis, 326, 330
Optic tract, 332–333, 332f–335f. *See also* Visual system
Optokinetic test, 371–372
Orbital apex (Jacod) syndrome, 273, 274
Orbitofrontal cortex, 119, 410
Organic visceral sensations, 421
Organ of Corti, 353, 353f, 354
  damage to, 354
  and nerve deafness, 359
Orientation columns, 337
Oscillopsia, 342
Otitis media, 351
  acute, 351
  acute suppurative, 351
  bacterial, 351
  chronic, 352
  chronic suppurative, 352
  serous, 351–352
  viral, 351
Otosclerosis, 351
Otoxic effects, of medications, 360
Oval window fistulas, 352
Oxytocin neurons, 114

P

P_0, myelin protein, 26
P_1, myelin protein, 26
P_2, myelin protein, 26
PACAP. *See* Pituitary adenylate cyclase–activating peptide (PACAP)
Pachygyria, 9
Pacinian corpuscles, 416, 417
Pain sensations, 429f
Pain transmission, 422, 426
Palatal myoclonus, 477
Paleocerebellar lesions, 96
Paleocerebellum, 84
Paleopallium, 141
Pallidotegmental tract, 471
Palmomental reflex, 134

Pancoast tumor, 203
Papez circuit of emotion, 121, 392f, 393, 404, 405f
Papilledema (choked disc), 270, 330
Parabigeminal nucleus, 69, 70
Parabrachial nuclear complex, 310
Parachute reflex, 444
Paradoxical sleep, 77. *See also* Rapid eye movements (REM)
Parahippocampal gyrus, 136, 404, 404f
Paralysis, 456
  sleep, 78
Paralysis agitans, 473
Paralytic polio, 446
Paramedian nuclei, 62, 74, 76f
Paramedian pontine reticular formation (PPRF), 65, 131, 339, 367, 369
Paranodal bulbs, 27
Parastriate cortex, 144
Parasympathetic nervous (craniosacral) system, 197, 198f, 205–208, 206f, 207f
  cranial part, 208
  sacral part, 208
Paraterminal gyrus, 138
Paraventricular nucleus (PVN), 114, 405
Paravertebral ganglia, 198f, 200–204, 201t, 202f–203f
Parietal apraxia, 157
Parietal lobe, 134–136, 134f–135f
  angular gyrus, 135
  hematoma, 135
  paracentral lobule, 135–136
  postcentral gyrus, 135
  precuneus, 135
Parinaud syndrome, 125, 192, 339, 344, 344f
Parkinsonian depression, 467
Parkinson's disease, 68, 245, 473, 473f
  acetylcholine and, 310
  anosmia and, 379
  dopamine in, depletion of, 312
  GABAergic neurons and, 308
  voice disorder in, 299
Pars compacta, 467
Pars reticulata, 467
Partial seizures, 162
Partial sensory seizures, 162
Parvalbumin, 86, 470
Past-pointing, 371
Patellar clonus, 442
Patients with decorticate rigidity, 451
PAX genes, 12
PAX-1 genes, 12
PAX-3 genes, 12
PAX-5 genes, 12
PAX-6 genes, 12
PAX-7 genes, 12
PAX-8 genes, 12
Pedunculopontine nucleus, 69, 75, 76–77, 310
Pellagra, 452t, 455
Periaqueductal gray matter, 120
Periglomerular olfactory neurons, 37
Perihypoglossal nuclei, 61
Perikaryon (soma), 24–25
Perineurium, 224
Peripheral gustatory receptors, 407–409
  salty taste, 407–408
  sour taste, 408
  sweet taste, 407
  tastants, classification of, 407
  taste buds, 408–409
  umami taste, 408
Peripheral myelin protein (PMP22), 27
Peripheral nerve fibers, 415
Peripheral nervous system (PNS), 21, 22, 197
  ganglia of, 201t
  MAG, 26, 27
  myelin, 27
  P_1 and P_2 proteins, 26
  PMP22, 27
  regeneration, 31, 32
  retrograde (indirect Wallerian) degeneration in, 28
  synapses in, 37–38
  tumors of, 23
Peripheral scotoma, 324
Peristriate cortex, 144
Perivascular oligodendrocytes, 22
Periventricular nucleus, 112
Pernicious anemia, 40
Pertussis toxin, 311
Pervasive developmental disorder not otherwise specified (PDD-NOS), 166
PFC. *See* Prefrontal cortex (PFC)
Phagocytosis
  of cellular debris, microglia and, 23
Phalen's maneuver, 243
Phantogeusia, 411. *See also* Gustatory dysfunctions
Phantom limb pain, 429
Phenylketonuria (PKU), 28, 33
Pheromones, 377
Phonemic tics, 479
Phosphatidylinositol-4,5-bisphosphate (PIP2), 379
Phosphoinositides, 27
Phospholipase A_2, 309
Phospholipase C (PLC), 379
Photoreceptors, 325, 326f, 416
  cones, 325–327
  rods, 327
Phrenic nerve, 227
  accessory, 228
  palsy, 227
Physiological scotoma, 326
Physiologic cup, 326, 329
Physiologic myoclonus, 477
Physostigmine, 207
Phytanic acid, 36
Pia mater (pia-glial layer), 11, 15, 22, 101, 183–184
  spinal cord, 46, 46f, 48, 51
Pick's disease, 137, 159, 160
Picrotoxin, 308
Pill rolling, 473
Pilocarpine, 207, 324
PIN. *See* Posterior interosseous nerve (PIN)
Pineal gland, 123, 124–125, 124f
Pinealoma, 125
Piriformis syndrome, 260
Pituitary adenylate cyclase–activating peptide (PACAP), 214
Pituitary gland, 112, 122–123, 122f
  dysfunctions, 123
  tumors, 123
PKU. *See* Phenylketonuria (PKU)
Placing reflex or reaction, 444
Plantar reflex, 264
Plexuses, autonomic, 210–213, 211f–212f. *See also specific* types
PLP. *See* Proteolipid protein (PLP)
Pluripotent stem cells
  differentiation into neural tissue, 3–4
PMP22. *See* Peripheral myelin protein (PMP22)
PNS. *See* Peripheral nervous system (PNS)
Polar cortex, 143
Polio infections, 446
Poliomyelitis, 225, 272, 442t, 446
Poliovirus, 446
Polyribosomes, 24
Pons, 62–66, 63f–64f
  caudal, 64–65, 64f
  embryonic developmental, 15, 15f–16f
  midpons, 65–66, 65f
  rostral, 66, 66f
Pontine arteries
  occlusion of, 63
Pontine flexure, 10, 10f
Pontine lesions, 448, 457
Pontine reticulospinal tract, 74
Pontocerebellar tract, 89, 90f, 94f
Pontocerebellum, 84
Pontomedullary junction, 62, 74
Port-wine discoloration, 282f
Positive myoclonus, 477
Positive swinging flashlight test, 323
Postcentral gyrus, 135
Posterior callosal vein, 174
Posterior cerebral artery, 171–172
  infarction of, 338
  occlusion of, 150–151, 333f, 337, 338
  plays, 102
Posterior commissure, 123, 124f, 126, 151
Posterior cord
  terminal branches of, 230f, 244–246
Posterior femoral cutaneous nerve, 257–258
  injury to, 258
Posterior funiculus
  ascending tracts in, 54, 54f
Posterior inferior cerebellar artery (PICA), 84, 456
  occlusion of, 291
Posterior interosseous nerve (PIN), 246
Posterior knee fibers of von Willebrand, 331
Posterior spinal arteries, 49f–50f, 50
Posterior thalamic zone (PTZ), 105
Posteroventral pallidotomy, 474
Postganglionic (postsynaptic) neurons, of sympathetic nervous system, 197, 202f
Postsympathectomy, 217
Postsynaptic axons, 420
Postsynaptic fibers, 410
Postsynaptic inhibitory potentials (IPSPs), 308
Postsynaptic receptors, 200
Posttraumatic apallic syndrome, 402
Posture and gait abnormalities (ataxia), 96
Pott's (Dupuytren's) fracture, 263
PPRF. *See* Paramedian pontine reticular formation (PPRF)
Prader–Willi syndrome–induced obesity, 118
Prazosin, 311
Precentral gyrus, 131
Preeclampsia, 310
  magnesium sulfate in, 310
Prefrontal cortex (PFC), 132–134, 404, 404f
  damage to, 132–133
Preganglionic (presynaptic) neurons, of sympathetic nervous system, 197, 202f
Pregeniculate nucleus, 103
Prehippocampal rudiment, 396
Premotor cortex, 146, 146t, 437
Preoptic nucleus, 112

Prerubral nucleus, 126
Presbycusis, 359
Presbyopia, 322, 325
Presynaptic receptors, 200
Pretectal syndrome, 344
Prevertebral ganglia, 198f, 202f, 204
Primary annulospiral fibers, 417
Primary aphakia, 322
Primary auditory cortex, 144
Primary motor cortex, 145, 145t, 437, 438
Primary sensory cortex, 143–144, 143f
Primary somesthetic cortex, 143
Primary visual cortex, 143–144
Principal mammillary fasciculus, 121
Principal (chief) trigeminal sensory nucleus, 65, 65f
Procaine, 39
Progressive nyctalopia, 327
Projection fibers, 153
Pronator syndrome, 242
Pronator teres, 239–240, 240f
Proprioceptors, 416
Prosencephalon, 10, 10f, 11t, 17
Prostaglandin D2, 118
Prostatic plexus, 213
Protein kinase C (PKC), 379
Proteolipid protein (PLP), 26, 27
Protopagnosia, 158
Protopathic sensations, 415
Protoplasmic astrocytes, 22
Pseudobulbar palsy, 442t
Pseudocoma, 450, 450f
Pseudomonas osteomyleitis of temporal bone. *See* External otitis
Pseudotumor cerebri, 330
Pseudounipolar neurons, 32, 36, 36f
Psychomotor activity
  dopamine role in, 133
Ptosis, 273f, 274
PTZ. *See* Posterior thalamic zone (PTZ)
Pudendal nerve, 256f, 258, 258f
  injury to, 258
Pulmonary plexus, 211
Pulvinar, 101f–102f, 104f, 106f, 107–108, 108f
Pupillary light reflex, 209–210
Pupillary-skin (ciliospinal) reflex, 210
Pure agraphia (aphasic agraphia), 156
Pure alexia, 156
Purkinje cell layer, cerebellar cortex, 84, 85–86, 85f, 86f
Purkinje cells, 16, 22
Purkinje neurons, 36, 87, 95, 470
  functions, 95
Putamen, 464, 475
Pyramidal cells, 141
Pyramidal decussation, 59
Pyramidal neurons, 36
Pyramidal tract, 438
Pyridoxal phosphate, 307

Q

Quadrangular space syndrome, 244
Quadranopsia, 328

R

Rabbit syndrome, 474
Rachischisis, 4, 6
  anencephaly (exencephaly), 6
  cranioschisis, 6
Radial nerve, 244–246, 245f
  injury to, 245–246
Radicular arteries, 49, 49f–50f, 50
Radicular veins, 51
Ramsay Hunt syndrome, 286. *See also* Herpes zoster oticus
Ranibizumab, 326
Raphe nuclei, 73–74, 74f, 120
  dorsal nucleus of raphe, 73
  lesions of, 73
  paramedian nuclei, 74
  raphe magnus, 73
  raphe obscurus, 73
  raphe pallidus, 73
  superior central nucleus, 74
Rapid eye movements (REMs), 73, 118
  characterization, 77
  locus ceruleus and, 77
Rathke's pouch, 17, 122
Raymond–Cestan syndrome, 447t
Raymond's syndrome, 449
Raynaud's disease, 217, 219
Raynaud syndrome, 232
Rebound phenomenon of Holmes, 97
Recklinghausen's disease, 227
Rectal reflex, 210
Recurrent laryngeal nerve, 298–299
Red nucleus, 71, 93, 467–468
Reduplicative paramnesia, 158–159
Referred pain, 226, 248, 249, 426
Reflexes. *See also specific* types
  autonomic, 208–210
  lesion and, 209
  overview, 208–209
  spinal, 264
    deep reflexes, 265–266, 265f
    superficial reflexes, 264–265
  superficial, 209
  visceral, 209–210
Reflex vasovagal syncope, 298
Refractive disorders, 324, 325f
  anisometropia, 325
  astigmatism, 325
  hyperopia, 325
  myopia, 324–325
  presbyopia, 325
Refsum's disease, 32, 35, 452t
Regeneration, 31–32
  in CNS, 31–32
  defined, 31
  factors, 31
  of nerve terminals, 409
  in PNS, 31, 32
  signs of, 31
Reissner's membrane, 353
Relative afferent pupillary defect (RAPD), 323
REM. *See* Rapid eye movements (REM)
Remak bundle, 225
Renal plexus, 212
Renshaw cells, 52
Reproductive cycle
  disorders in, 114
Reserpine, 311, 314
Resum's disease, 27
Reticular formation, 62, 65, 66, 73–78, 468–470
  ascending reticular activating system, 74–75
  damage to, 75
  lateral reticular zone, 75–77, 76f–77f
  medial reticular zone, 74
  overview, 73
  raphe nuclei, 73–74, 74f
  sleep and associated disorders, 77–78
Reticular neurons, 73
Reticular nuclear group, 102, 110
Reticulocerebellar circuit, 93, 93f
Reticulocerebellar tract, 89, 90f, 91f
Reticulospinal tracts, 55
Reticulotegmental nucleus, 66, 75, 76f
Retina, 120, 325–329
Retinal detachment, 325
Retinin, 327
Retinitis pigmentosa, 171, 327
Retinoic acid, 12
Retrobulbar neuritis, 326
Retrocochlear deafness, 360
Retrograde (indirect Wallerian) degeneration, 28
Rett syndrome, 166
Rhegmatogenous retinal detachment, 325
Rhombencephalon, 10, 10f, 11t, 12, 81
Rhombic lip, 81
  of alar plate, cerebellum from, 15–16, 16f
Rhomboid fossa, 57, 58
Rhombomeres, 12
Ribosomes, 25
Riche–Cannieu anastomosis, 237
Right infranuclear facial palsy, 290f
Rigidity, 473
Riley–Day syndrome, 217, 218
Riluzole, 309
Ring scotoma, 327
Rinne test, 361f, 362
Risperidone, 390
Roemheld syndrome, 226
Romberg's sign, 421, 456
Romberg test, 372
Roof plate, 11f, 12, 14, 15
  derivatives of, 16–17
Rooting reflex, 444
Roots, terminal branches of, 232–233
Rostral medulla, 62, 62f
Rostral midbrain, 70–71, 70f
  magnocellular part, 71
  oculomotor nuclear complex, 71
  parvocellular part, 71
  red nucleus, 71
  superior colliculus, 70
Rostral pons, 66, 66f
Rostral spinocerebellar tract, 432
Round window fistulas, 352
Rubral tremor, 473
Rubrocerebellar circuits, 93, 94f
Rubro-olivary tract, 468
Rubrospinal tract, 55, 440f
Ruffini endings, 417, 418f

S

Saccadic eye movement, 340–341
Sacral fibers, 438
Sacral part, parasympathetic nervous system, 208
Sacral plexus, 256–257, 256f–257f
  terminal branches of, 257–264, 257f–259f, 261f–263f
Sacral sparing, 425, 429, 430f
Sacral spinal nerves, 223, 256
Sacral spinal segments, 53, 53f
Saddle block (Caudal anesthesia), 47
Salicylates, effect of, on hearing, 360
Salty taste, 407–408
Saphenous nerve, 254
Satellite cells, 5

Saturday night palsy, 245
Saxitoxin, 39
Schizophrenia, 389–390. *See also* Hippocampal gyrus
- age of onset of, 389
- development of, role of hormones in, 390
- dopamine hypothesis and, 390
- glutamate in, role of, 309
- incidence of, 389
- manifestations of, 389
- ontogenic hypothesis of, 389
- stress-diathesis model of, 389
- treatment of, 390

Schlemm canal, 321
Schmidt–Lanterman incisures, 27
Schmidt's syndrome, 447t, 448
Schober index, 430
Schwann cells, 5, 22–23, 27, 28, 31, 39, 224, 225
- myelin from, 26–28

Schwann glial cells, 28
Schwannomas, 23
Sciatica, 260, 455
Sciatic nerve, 256f, 258–259, 259f
- injury to, 259–260

Scintillating scotoma (teichopsia), 326
SCN. *See* Suprachiasmatic nucleus (SCN)
Scotoma (focal blindness), 331
Secondary aphakia, 322
Secondary sensory cortex, 143f, 144, 425
Secondary somesthetic area, 144
Secondary tics, 479
Secondary visual cortex, 144
Second-order neurons, 418
Seizures, 77, 161–163
- absence, 163
- atonic, 163
- atypical absence, 163
- clonic, 163
- frontal lobe, 131–32
- generalized, 162
- nonconvulsive generalized, 163
- partial, 162
- partial sensory, 162
- tonic, 163
- tonic–clonic, 162

Selective serotonin reuptake inhibitors (SSRIs), 314
Selegeline, 385
Selegiline, 78
Sella turcica, enlargement of, 122
Semantic dementia, 394
Semicircular canals, 363–364, 363f
Semon's law, 299
Sensorineuronal hearing loss, 359–360
Sensory cortex, 143–144, 143f
- primary, 143–144, 143f
- secondary, 143f, 144

Sensory decussation level, caudal medulla, 59–60, 59f–60f
Sensory deficits, 444
Sensory dysfunctions, 428t
Sensory fibers, 415–416
Sensory loss, 453
Sensory receptors, 416–419, 416f, 418f, 419f. *See also* Extrapyramidal system
- capsulated receptors, 416
- encapsulated receptors, 416
- Golgi tendon organs, 417
- Merkel's disks (tactile menisci), 418
- Muscle (neuromuscular) spindles, 416
- nonencapsulated receptors, 418
- Ruffini endings, 417
- tactile corpuscles of Meissner, 417, 418f

Septal area, 119, 122, 396–399, 397f, 404, 405f
Septum pellucidum, 130, 140
Serotonin, 37, 68, 73, 78, 313–315
- function of, changes in, 313
- receptors, 313
  - 5-HT_1, 313–314
  - 5-HT_2, 314
  - 5-HT_3, 314
  - 5-HT_4, 314
  - 5-HT_5, 314
  - 5-HT_6, 314
  - 5-HT_7, 314
- role of, 314–315
- storage of, 313
- synthesis of, 313
- uptake of, 314
- vesicular transport of, 313

Serotoninergic neurons, 32, 314–315
Serotonin reuptake inhibitors, 78
Serous otitis media, 351
Sheehan's syndrome, 123
Shh gene. *See* Sonic hedgehog (Shh) gene
Shiff–Sherrington reflex, 264
Shy–Drager syndrome, 217, 220
Sialic acid (*N*-acetylneuraminic acid), 27
Sickle-cell anemia, 171
Sigmoid sinuses, 179
Simple partial seizures, 162
Simple tics, 479
Simpson's test, 40–41
Simultanagnosia, 158
Sinus arrhythmia, 215
Sjögren syndrome, 276
Sleep
- and associated disorders, 77–78
- cataplexy, 78
- hypnagogic hallucination, 78
- mechanism, substances involved in, 78
- multiple sleep latency test (MSLT), 78
- narcolepsy, 78
- NREM sleep, 77–78
- paralysis, 78
- REM sleep, 77–78

Sleep–wake cycle, 118
- disorders in, 114
- stages and types, 118

Sleepwalking (*somnambulism*), 77
Slow-spiking pacemakers, 467
Slow-wave sleep, 77
Sluder's neuralgia, 277–278
Smooth pursuit eye movement, 341
Snellen eye chart, 269, 331
Sniff test, 227
Snout reflex, 134
Sodium chloride, 407
Solitary nuclear complex, 61
Solitary tract, 61
Solubility in saliva, 407
Soma (perikaryon), 24–25
Somato-somatic synapses, 38
Somatostatin, 85, 213
Somatovisceral reflexes, 209–210
Somites
- formation of, 6

Sonic hedgehog (Shh) gene, 12
Sour taste, 408
Spasticity, 441, 441f
Special somatic afferents (SSA), 15, 347
- fibers, 15

Special visceral efferent fibers (SVEs), 15, 61
Special visceral efferent neurons (SVE), 445f, 446
Spectrin, 25, 27
Speech audiometry, 361
Speech reception threshold (SRT), 361
Sphenoparietal sinus, 179, 180f
Sphingomyelin, 27
Spike-timing-dependent plasticity (STDP), 394–395
Spina bifida, 3, 7–9, 8f, 452t, 455
- cystica (spina bifida manifesta/spina bifida aperta), 8
  - Arnold–Chiari syndrome in. *See* Arnold–Chiari syndrome
  - diastematomyelia in, 9–10
  - meningocele in, 8
  - meningomyelocele in, 9
- occulta, 7–8, 8f
  - dermoid sinuses in, 8
  - lipoma-covered defect in, 8

Spina bifida cystica, 455
Spina bifida occulta, 455
Spinal anesthesia, 48
Spinal cord, 5, 10, 45–55, 424f, 438
- arachnoid mater, 46f, 47–48
- ascending pathways, 54f, 54t
  - in lateral funiculus, 54
  - in posterior funiculus, 54
  - in ventral funiculus, 54–55
- blood supply, 48–51, 49f, 50f
- cauda equina, 45–46, 45f–46f
- conus medullaris, 45
- descending pathways, 54t, 55f
  - in lateral funiculus, 55
  - in ventral funiculus, 55
- dura mater (pachymeninx), 46–47, 46f
- embryonic developmental, 13–14, 13f
- endosteal layer, 46, 46f
- epidural space, 46, 46f
- internal organization
  - gray matter, 45, 51–52, 52f
  - white matter, 45, 52–53
- lesions
  - autonomic disturbances and, 214–215
- neural tube transformation into, 14
- overview, 45–48
- pathways, 54–55
- pia mater, 46, 46f, 48
- segments, 45, 53, 53f
- structure, 45–48, 45f–46f
- subdural space, 48
- transection of, 453
- venous drainage, 51

Spinal dysraphism (defective fusion), 7
Spinal epidural abscesses, 47
Spinal epidural hematoma, 47
Spinal lemniscus, 68
Spinal lesions, 447, 452
Spinal nerves, 45, 223–266
- brachial plexus, 229–230, 230f, 230t
  - lateral cord, terminal branches of, 234
  - medial cord, terminal branches of, 230f, 234–244, 235f–237f, 239f–243f
  - posterior cord, terminal branches of, 230f, 244–246
  - roots, terminal branches of, 232–233
  - superior trunk, terminal branches of, 233–234
  - trunk injuries, 230–232, 231f

cervical, 226–227
cervical plexus, 226–227, 228f, 229f
terminal branches, 227–229
components, 223–226, 225f
distribution, 223–226, 224f
dorsal rami, 225–226, 225f
endoneurium, 224
in entrapment neuropathies, 226
epineurium, 224
formation, 13–14, 223–226, 223f
injury, 226
lumbar, 249
lumbar plexus
terminal branches of, 249–256, 250f–251f, 253f–255f
overview, 223
pathological conditions, 225–226
perineurium, 224
recurrent meningeal branches, 224
sacral, 256
sacral plexus, 256–257, 256f–257f
terminal branches of, 257–264, 257f–259f, 261f–263f
spinal reflexes, 264
deep reflexes, 265–266, 265f
superficial reflexes, 264–265
thoracic, 247–249
vasa nervorum, 225
ventral rami, 225, 225f
Spinal reflexes, 264
deep reflexes, 265–266, 265f
superficial reflexes, 264–265
Spinal shock, 454
Spinal trigeminal nucleus, 65
Spinocerebellar tracts, 88, 89f, 90f, 432
Spinocerebellum, 84
Spinomedullary junction, 59, 59f
Spino-olivary tract, 54–55, 433
Spinoreticular tract, 427
Spinoreticulothalamic tract, 74, 109, 427
Spinospinal tract (fasciculus proprius), 55, 433
Spiral ganglia, 354
Spongioform encephalopathy, 477
SSPE. *See* Subacute sclerosing panencephalitis (SSPE)
Stapedius muscle palsy, 288
Stapedius reflex, 358
Star cells. *See* Astrocytes
Status epilepticus, 162–163
STDP. *See* Spike-timing-dependent plasticity (STDP)
Stellate cells, 86
Stellate ganglion syndrome, 217, 219
Stellate (star) neurons, 36
Stenosis
of basilar artery, 97
Sternocleidomastoid muscle (SCM), 228
Sternocleidomastoid muscle muscle, paralysis of, 300
Stiff-man syndrome, 307
norepinephrine in, role of, 312
Stomodeum, 17
Streptococcus pneumoniae, 350
Stria medullaris thalami, 123, 124f, 125
Striatal efferent fibers, 472
Striatal neurons contain γ-aminobutyric acid (GABA), 465
Striati of Brockhaus, 464
Striatonigral fibers, 470
Stridor, 9
Striopallidal fibers, 471
Striosomes, 465
Struthers' ligament, 241
Stump pain, 430
Sturge–Weber syndrome, 282, 283f
Subacute combined system degeneration, 452t, 455
Subacute sclerosing panencephalitis (SSPE), 159
Subarachnoid hemorrhage, 182–184
Subarachnoid space, 57
Subcallosal gyrus, 138
Subclavian artery, 49, 50
occlusion of, 97
Subcortical dementia, 161
Subcortical pathologic myoclonus, 477
Subcortical sensations, 432–433. *See also* Extrapyramidal system
Subcortical sensory aphasia, 156
Subcostal nerve, 247
Subdural abscess, 48
Subdural hematoma, 48, 176–177, 176f
Subdural space, 176
spinal cord, 48
Subfalcial herniation, 138–139
Subicular complex, 390–391
Subiculum, 141
Subluxation, 45
Submucosal plexus, 213
Subscapular nerve, 246
Substance P (SP), 68, 73, 316
Substantia nigra, 25, 68, 106, 466–467, 466f
lesion of, 68
Subthalamic nucleus, 126, 468, 469, 469f
Subthalamus, 101, 125f–126f, 126
embryonic developmental, 16–17
Succinic semialdehyde dehydrogenase, 307
deficiency of, 308
Suck reflex, 134
Sulci limitans, 57
Sulfahydryl groups, 411
Sumatriptan, 314
Sunflower cataract, 322
Superficial abdominal reflex, 442
Superficial fibular (peroneal) nerve, 260–261
injury to, 261
Superficial middle cerebral vein, 172
Superficial reflexes, 209
Superior alternating hemiplegia, 447t
Superior anastomotic vein of Trolard, 172
Superior central (median) nucleus, 66, 74, 76f
Superior cerebellar artery, 83
Superior cerebellar artery syndrome, 452t, 457
Superior cerebellar peduncles, 66, 69, 81
Superior cerebral veins, 172
Superior cervical ganglion, 201
Superior colliculus, 67, 70, 70f
Superior frontal gyrus, 131
Superior gluteal nerve, 256f, 257, 257f
injury to, 257
Superior hypogastric (presacral nerve) plexus, 202f, 213
Superior laryngeal nerve, lesion of, 298
Superior longitudinal fasciculus, 152
Superior medullary velum syndrome (bilateral trochlear nerve palsy), 66, 276
Superior mesenteric plexus, 202f, 212, 212f
Superior occipitofrontal fasciculus, 153
Superior orbital fissure syndrome, 274
Superior petrosal sinus, 180f, 181
Superior rectus muscle, ipsilateral palsy of, 270f
Superior sagittal sinus, 178
obstruction of, 179, 179f
Superior salivatory nucleus (SSN), 15, 64, 277
Superior trunk
terminal branches of, 233–234
Superior vestibular nucleus, 64
Supplementary motor area (Brodmann area), 134
Supplementary motor cortex, 145, 437
Suprachiasmatic nucleus (SCN), 112, 113–115
bilateral lesions of, 124
Supraclavicular nerves, 228
Suprameatal triangle, 350
Supranuclear facial palsy, 289f
Supranuclear gaze palsy, 467
Supraoptic nucleus, 114
Suprarenal plexus, 212
Suprascapular nerve, 230f, 233–234
SVEs. *See* Special visceral efferent fibers (SVEs)
Sweating, disorder of (Hyperhidrosis), 216–217
Sweet taste transduction, 407
Sweet-tasting stimuli, 409
Swinging light test, 269, 331
Sydenham's chorea (St. Vitus' dance), 475
Symmetrical axodendritic synapses, 38
Sympathectomy, 217
Sympathetic apraxia, 157
Sympathetic fibers, 203
distribution pattern, 204–205, 205f
ipsilateral disruption of, 201
Sympathetic nervous (thoracolumbar) system, 197, 200
activation of, 200
paravertebral ganglia, 198f, 200–204, 201t, 202f–203f
prevertebral ganglia, 198f, 202f, 204
Sympathetic presynaptic fibers, 223
Sympathetic (adrenergic) receptors, 200
Synapse(s), 37
asymmetric, 38
autonomic neurons and, 197–200, 198f–199f
axo-axonic, 38
axodendritic, 38
axosomatic, 37
chemical, 38
cholinergic, 38
classification, 38
in CNS, 37–38
dendrosomatic, 38
electrical, 38
formation, 3
Gray's type I, 38
Gray's type II, 38
α motor axon terminals, 38
multisynapses, 37
neuromuscular junction, 38–39, 38f
in PNS, 37–38
postsynaptic membrane, 37, 39
presynaptic membrane, 37, 39
somato-somatic, 38
synaptic cleft, 37
synaptic glomeruli, 37
transmission in, 38
vesicles, 37
Synaptic disorders
black widow spider (*Latrodectus mactans*) toxicity, 39
botulism, 39
bungarotoxin, 39
tetany, 40
Synaptic gaps
in ANS, 197
Synaptogenesis, 13
Synaptolemma, 37
SynCAM 1, cell adhesion molecule, 13

Synkinesis, 290
Syringobulbia, 7, 428t, 429
Syringomyelia, 7, 9, 428, 428t, 429, 429f
Systemic lupus, 31
Systemic lupus erythematosus, 40
Systemic sclerosis, 276

T

Tabes dorsalis, 420, 424f, 428t
Tactile agnosia, 158
Tactile aphasia, 156
Tactile corpuscles of Meissner, 417, 418f
TAG-1. *See* Transiently exposed axonal glycoprotein (TAG-1)
Tangier disease, 33
Tapia syndrome, 303
Tardive dyskinesia, 478
Tarsal tunnel syndrome, 263
Tastants
  classification of, 407. *See also* Peripheral gustatory receptors
  salty taste, 407
  sour taste, 408
  sweet taste, 407
  umami taste, 408
Taste, 407
  buds, 408–409, 410f. *See also* Peripheral gustatory receptors
  defined, 407
  determining eating habits, 407
  disorder, 410. *See also* Gustatory dysfunctions
  quality, 408, 409
  receptors, 407
Tau protein, 384
Tay–Sachs disease, 32, 34, 325
TCAs. *See* Tricyclic antidepressants (TCAs)
Tectum, 67
Tegmental nuclei, 120
Tegmentum, 67
Tela choroidea, 15, 101
Telencephalon, 10, 10f, 11t, 129–193
  basal (ganglia) nuclei, 167
  brain barriers, 184–188, 184f
    BBB *vs*. BCB, 186t
    circumventricular organs, 188, 188f
    functional and clinical significance, 186–188
  central lobe (insular cortex), 138, 138f
  cerebral cortex (gray matter), 140–147, 141f–142f
    association cortex, 146t, 147
    motor cortex, 143f, 144–146, 146t
    sensory cortex, 143–144, 143f
  cerebral dysfunctions
    agnosia, 158–159
    aphasia, 153–157, 154f
    apraxia, 157
    autism, 166–167
    cerebral dominance, 165–166
    coma, 163–165
    dementia, 159–161
    seizures, 161–163
  cerebral hemispheres, 129–131, 129f–130f
    blood supply, 167–172, 168f–172f
    venous drainage, 172–174, 172f–175f
  cerebral white matter, 149–153, 150f
    association fibers, 152–153, 152f–153f
    commissural fibers, 150–151, 150f–151f
    projection fibers, 153
  corpus callosum, 136f, 138f–139f, 139–140
  cortical afferents, 147
  cortical efferents, 147–149, 148f–149f
  embryonic developmental, 17–19, 18f
  frontal lobe, 131–132
    prefrontal cortex, 132–134
  limbic lobe, 135f–138f, 138–139
  meninges
    anteroinferior group of dural sinuses, 179–182, 180f–182f
    arachnoid mater, 182–183
    dural sinuses, 172f–175f, 177f, 178
    dura mater, 175–178, 175f–178f
    pia mater, 183–184
    posterosuperior group of dural sinuses, 172f–175f, 177f, 178–179, 180f
  occipital lobe, 131f, 135f, 137–138, 137f
  overview, 129
  parietal lobe, 134–136, 134f–135f
  temporal lobe, 136–137, 136f
    functional and clinical consideration, 136–137
  ventricular system and CSF, 188–193, 189f–191f
    hydrocephalus, 191–193, 191f–192f
Telencephalon impar, 17
Temperature regulation
  hypothalamic areas and, 118
Temperature sensations, 429
Temporal lobe, 136–137, 136f
  functional and clinical consideration, 136–137
  hematoma, 136
Temporal lobotomy, 396
Temporary threshold shift, 359
Tenascin (cytotactin), 12, 21
Tendon reflexes, 474
Tenth intercostal nerve, 247
Tentorium cerebelli, 81, 177, 177f
Teratoma, 8
Terminal boutons, 25
Tetanus, 217, 218–219
  autonomic dysfunction in, 214, 218
Tetany, 40
Tetracycline, 411
Tetrodotoxin, 39
TGF-β. *See* Transforming growth factor-β (TGF-β)
TGFβ_1. *See* Transforming growth factor beta1 (TGFβ_1)
Thalamic apoplexy, 110–111
Thalamic hand, 111
Thalamogeniculate arteries, 102
Thalamoperforating arteries, 102
Thalamostriate fibers, 470
Thalamus, 101–102, 101f–102f, 130, 410
  ARAS, 102
  blood supply, 102
  embryonic developmental, 16–17
  functional and clinical consideration, 110–111, 111t
  hematomas, 111
  intralaminar nuclear group, 102f, 105f–107f, 109–110, 109f
  lateral nuclear group
    lateral dorsal, 102f, 105f, 106f, 107, 107f
    lateral posterior, 102f, 105f, 106f, 107, 107f
    pulvinar, 101f–102f, 104f, 106f, 107–108, 108f
  medial nuclear group
    dorsomedial nucleus, 101f–104f, 106f–109f, 108–109, 110f
  midline nuclear group, 110
  nuclear group, 102–106, 102f
    anterior nucleus, 101f–103f, 103
    lateral geniculate body, 101f, 102f, 103–104, 104f
    medial geniculate body, 101f, 102f, 104, 104f
    ventral anterior nucleus, 102f–104f, 105–106, 106f
    ventral lateral nucleus, 102f–104f, 106, 106f
    ventral nuclear group, 103–106
    ventral posterior nucleus, 102f, 104–105, 104f, 106f
  reticular nuclear group, 110
Thermal sensations, 407, 429
Thermoanesthesia, 429
Third ventricle, 57, 101, 101f
Thoracic disk herniation, 248–249
Thoracic outlet syndrome (TOS), 231, 232
Thoracic spinal nerve roots, 248
Thoracic spinal nerves, 223, 247–249
Thoracic spinal segments, 53, 53f
Thrombin, 21
Thrombosis
  of cerebral veins, 174
  of inferior petrosal sinus, 181–182
TIA. *See* Transient ischemic attack (TIA)
Tibial nerve, 256f, 259f, 262–263, 262f
  injury to, 263–264
  palsy, 263
Tic disorders, 479
Tic douloureux. *See* Trigeminal neuralgia
Tics, 479
Timed up and go test (TUG), 372
Timolol, 324
Tinel's sign, 32, 243
Tingling sensations, 431
Tinnitus, 292, 360
  high-pitched, 360
  objective, 360
  pulsatile, 360
  subjective, 360
  unilateral, 360
T lymphocytes, 24, 30
Tocopherol (vitamin E), 474
Tolosa–Hunt syndrome, 273
Tonic–clonic seizure, 162
Tonic necl reflex, 444
Tonic pupil. *See* Adie pupil
Tonic seizures, 163
Torticollis, 276, 300
TOS. *See* Thoracic outlet syndrome (TOS)
Tourette's syndrome, 479
Tourettism, 479
Toxin-extruding antiporters (TEXANs), 313
Transcortical aphasia, 155
Transcortical motor aphasia, 155
Transcortical sensory aphasia, 155
Transcutaneous electrical nerve stimulation (TENS), 422
Transduction, 407
Transection, of spinal cord, 453
Transforming growth factor-β (TGF-β), 4
Transforming growth factor beta1 (TGFβ_1), 31
Transient ischemic attack (TIA), 97
Transiently exposed axonal glycoprotein (TAG-1), 12
Transient tic disorder, 479
Transneuronal (transynaptic) degeneration, 28
Transtentorial/central herniation, 178, 178f

Transverse (colli) cervical nerve, 228
Transverse sinuses, 179, 180f
Transynaptic (transneuronal) degeneration, 28
Trapezius muscle, paralysis of, 300
Trapezoid body, 64
Tremor, 473, 478
Trepidation sign, 442
Trichromats, 327
Tricyclic antidepressants (TCAs), 78, 314
  in glaucoma, 324
Trigeminal gangliopathy, 276
Trigeminal lemnisci, 68
Trigeminal lemniscus, 432
Trigeminal nerve, 5, 276–282, 277f
  anterior trunk of mandibular nerve, 280
  auriculotemporal nerve, 279–280, 279f
  corneal reflex, 277
  dorsal trigeminal lemniscus, 280
  frontal nerve, 276
  inferior alveolar nerve, 279, 279f
  injury to, 280
  lacrimal nerve, 276
  lacrimal reflex, 277
  lesion of, 280
  lingual nerve, 279
  mandibular branch, 278–279, 279f
  maxillary division, 278, 278f
  mesencephalic nucleus, 281
  motor trigeminal nucleus, 281
  mylohyoid nerve, 280
  nasal reflex, 278
  nasociliary branch, 276, 277f
  ophthalmic division, 276, 277f
  posterior trunk of mandibular nerve, 279
  principal sensory nucleus, 280
  spinal trigeminal nucleus, 280
  ventral trigeminal lemniscus, 280
  zygomatic nerve, 278
Trigeminal neuralgia, 281–282
Trigeminal neurinomas, 276
Trigeminal nucleus, 15
Trigeminocerebellar tract, 89–90, 90f, 432
Trochlear nerve, 274–276, 275f
  lesions of, 69
  palsy, 274, 275f, 276
Trochlear nucleus, 68
True claw hand, 244
Truncal vagotomy, 297
Trunk injuries, 230–232, 231f
Trypanosoma cruzi, 220
Trypsin, 21
Tryptophan hydroxylase, 313
Tubocurarine, 207
Tubulin, 27
Tumor of pontocerebellar angle. *See* Acoustic neuroma
Tunnel vision. *See* Bitemporal heteronymous hemianopsia
Tympanic cavity, drainage of inflammatory exudate from, 348–349
Tympanometry, 352
Tympanosclerosis, and hearing loss, 359
Tyrosine, 311
Tyrosine hydroxylase, 311

U

Uhthoff's sign, 323
Ulnar claw hand (griffe cubitale), 239, 239f
Ulnar nerve, 235–238, 237f
  injuries, 238–239, 239f
Umami taste, 408
UMN. *See* Upper motor neurons (UMN)
Uncal herniation, 136–137, 178
Uncinate fasciculus, 92, 152
Uncinate fits, 137, 268
Unilateral nasal hemianopsia, 332
Unilateral visual agnosia, 338
Unipolar neurons, 36
Upgoing toe reflex. *See* Babinski sign
Upper motor neurons (UMN), 437, 437f
  premotor cortex, 437
  primary motor cortex, 437
  supplementary motor cortex, 437
Upper motor neurons (UMN) palsies, 439–445, 441f. *See also* Lower motor neurons (LMN) palsies
  Babinski sign, 441
  cerebral palsy, 443
  Chaddock response, 442
  clonus, 442
  crossed extension reflex, 444
  dorsolateral syndrome, 443
  Gordon's finger sign, 442
  Gordon's leg sign, 441
  grasp reflex, 444
  Hoffman's sign, 442
  hyperreflexia, 442
  incurvation (Galant's) reflex, 444
  inverse myotatic reflex, 442
  Landau reflex, 444
  mass reflex of Riddoch, 444
  Moro (startle) reflex, 443–444
  motor dysfunctions, 442t
  Oppenheim's sign, 442
  parachute reflex, 444
  placing reflex or reaction, 444
  rooting reflex, 444
  spasticity, 441
  superficial abdominal reflex, 442
  tonic necl reflex, 444
  ventromedian syndrome, 442
Upper motor neuron (UMN) spastic palsy, 131
Urbach–Wiethe disease, 403
Ureteric plexus, 212
Urokinase, 21
Uvular/palatal reflex, 298

V

VA. *See* Vertebral artery (VA)
Vaccine-associated viruses, 446
Vagal trunks, 297
Vagus nerves, 5, 294–299, 296f–298f, 410f
  auricular branch, 295
  branches to carotid body, 295–296
  branches to celiac and superior mesenteric plexuses, 297–298
  cardiac branches, 296
  dorsal motor nucleus, 298
  esophageal branches, 297
  external laryngeal nerve, 296
  inferior laryngeal nerve, 296
  internal laryngeal nerve, 296
  meningeal branch, 295
  nucleus ambiguus, 298
  pharyngeal branch, 295
  pulmonary branches, 297
  solitary nucleus, 298
  spinal trigeminal nucleus, 298
  superior laryngeal branch, 296
Valleix's pressure points, 431
Vanilylmandelic acid (VMA), 311
VA nucleus. *See* Ventral anterior (VA) nucleus
Vasa nervorum, 225
Vascular dementia, 160
Vasoactive intestinal peptide (VIP), 205, 388
Vasoactive intestinal polypeptide (VIP), 213–214
Vasopressin, 123
Vasovagal syncope, 298
Vegas neuropathy, 238
Venous drainage
  caudal medulla, 59
  cerebellum, 83–84
  cerebral hemispheres, 172–174, 172f–175f
  spinal cord, 51
Venous lacunae, 178
Ventral anterior (VA) nucleus, 102f–104f, 105–106, 106f
Ventral cochlear nucleus, 64
Ventral funiculus, 53
Ventral lateral (VL) nucleus, 102f–104f, 106, 106f, 468
Ventral nuclear group, 103
Ventral pallidum, 466
Ventral posterior inferior (VPI) nucleus, 104, 105
Ventral posterior nucleus (VPN), 102f, 104–105, 104f, 106f
Ventral posterolateral (VPL) nucleus, 54, 60, 102, 104, 418, 420, 426
Ventral posteromedial (VPM) nucleus, 61, 102, 102f, 104–105, 104f, 105f, 106f, 432
Ventral roots, 223
Ventral spinocerebellar tract, 89, 91f, 432
Ventral spinothalamic tract (paleospinothalamic/anterior system), 54, 54f, 427f, 428
Ventral striatum, 465
Ventral tegmental (VT) nucleus, 69
Ventral tegmental (VT) radiation, 315
Ventral tegmental (VT) system, 467
Ventral trigeminal lemniscus, 432
Ventral trigeminal tract, 105
Ventricular system, 188–193, 189f–191f
  embryonic developmental, 17–19, 18f
  hydrocephalus, 191–193, 191f–192f
Ventromedial nucleus, 122
Ventromedian syndrome, 442, 442t
Verbal paraphasia, 155
Verger–Dejerine syndrome (Dejerine cortical sensory syndrome), 135–136
Vermian lesion, cerebellar, 341
Vernet's syndrome, 294
Vertebral artery (VA), 48–49, 49f–50f
  occlusion of, 97
Vertebral plexus, 51
Vertical gaze palsy, 341
Vertigo, 97, 291, 370
Vesical plexus, 202f, 213
Vestibular cortex, 144
Vestibular nerve, 290, 364, 365f. *See also* Vestibular system
Vestibular neuritis, 369
Vestibular nuclei, 364–366, 366f
  inferior, 365
  lateral, 366
  medial, 365–366
  superior, 365
Vestibular system, 363
  central vestibular pathways
    secondary vestibulocerebellar fibers, 366

vestibular nuclei, 364–366
vestibulo-ocular fibers, 367–368
vestibulospinal tracts, 366f, 367, 367f
dysfunctions, 368–371
acoustic neuroma, 370
ataxia, 371
combined vestibular and auditory deficits, 370
internuclear ophthalmoplegia, 368–369, 368f, 369f
labrynthitis, 369
Meniére disease, 369–370
nystagmus, 371
one-and-a-half syndrome, 369
past-pointing, 371
vertigo, 370–371
vestibular neuritis, 369
semicircular canals, 363–364, 363f
testing for
Berg balance scale, 373
caloric test, 372
dynamic gait index, 372–373
optokinetic test, 371–372
Romberg test, 372
timed up and go test (TUG), 372
vestibular receptors, 363, 363f
vestibule, 362f, 364
saccule, 364
utricle, 364
vestibular ganglion, 364
vestibular nerve, 364, 365f
Vestibulocerebellar fibers, 88, 89f, 90f
Vestibulocerebellum, 84
Vestibulocochlear nerve, 267, 290–292, 291f, 292f, 354
Vestibulo-ocular eye movement, 341
Vestibulospinal tracts, 55, 366f, 367, 367f
Vibratory sensation, 421
Villarreal's syndrome, 294
Vinblastine, 25
VIP. *See* Vasoactive intestinal peptide (VIP); Vasoactive intestinal polypeptide (VIP)
Visceral pain, 214, 426–427
Visceral reflexes, 209
Viscerosomatic reflexes, 209
Viscero-visceral reflexes, 209
Visual acuity, assessment of, 269, 331
Visual agnosia, 158
Visual deficit, 321, 323, 330, 336–338. *See also* Visual system
Visual disorders, 97–98
Visual field testing, 269, 331
Visual system, 321
eyeball, 321, 322f
anterior chamber of eye, 324
choroid layer, 321
ciliary body, 321–322
cornea, 321
iris and pupil, 323
lens, 322
retina, 325–329
sclera, 321
tunica fibrosa, 321
tunica nervosa, 325
tunica vasculosa, 321–323
gaze centers and, 343f, 344
lateral geniculate nucleus, 334
and ocular reflexes, 342–344
optic chiasma, 331–332, 331f, 333f, 334f
optic nerve, 322f, 329–331, 329f
optic radiation, 332f–335f, 334–336
optic tract, 332–333, 332f–335f
peripheral visual apparatus, 321
and refractive disorders, 324–325
visual cortex, 336–339
primary, 336–338
secondary, 338
tertiary, 338–339
visual impulses and, 321
Vitamin A, role of, in vision, 327
Vitreous body, eye, 322f, 324
VL nucleus. *See* Ventral lateral (VL) nucleus
VNO. *See* Vomeronasal organ (VNO)
Vocal tics, 479
Voice disorders, 299
Volkmann's ischemic contracture, 241
Voluntary facial palsy, 287
Vomeronasal organ (VNO), 377, 379
Vomiting reflex, 209
Von Recklinghausen's disease, 360
Von Recklinghausen's neurofibromatosis, 23
VPI. *See* Ventral posterior inferior (VPI) nucleus
VPL. *See* Ventral posterolateral (VPL) nucleus
VPLc nucleus. *See* Ventral posterolateral (VPL) nucleus
VPM. *See* Ventral posteromedial (VPM) nucleus
VPN. *See* Ventral posterior nucleus (VPN)

W

Waardenburg's syndrome, 12
Wallenberg's syndrome, 294, 456, 457
Wallerian (antergrade) degeneration, 28
Wartenberg's sign, 239
Weather for patients with Parkinsonism, 474
Weber's syndrome, 272, 447t, 449, 449f
Weber test, 361, 361f
Wernicke–Korsakoff's syndrome, 395–396
Wernicke's aphasia, 155
Wernicke's encephalopathy, 395
Wernicke's zone, 136
Westphal subtype, 476
White commissure, 52–53
White matter, 45, 52–53
of cerebral hemispheres, 149–153, 150f
association fibers, 152–153, 152f–153f
commissural fibers, 150–151, 150f–151f
projection fibers, 153
Willis, 115
Wilson's disease, 321, 477, 478
Wright's maneuver, 232
Writer's cramp, 477

X

Xerostomia, 411

Y

Yohimbine, 312

Z

Zinc, 409
Zona incerta, 126